Handbook of coordination catalysis in organic chemistry

Penny A Chaloner, MA, PhD (Cambridge)

School of Chemistry and Molecular Sciences,
University of Sussex

Butterworths
London Boston Durban Singapore Sydney Toronto Wellington

First published 1986

© Butterworth & Co. (Publishers) Ltd, 1986

British Library Cataloguing in Publication Data

Chaloner, P.A.
 Handbook of coordination catalysis in organic
 chemistry.
 1. Chemistry, Organic—Synthesis   2. Catalysis
 I. Title
 547′.2   QD262
 ISBN 0–408–10776–6

Library of Congress Cataloging in Publication Data

Chaloner, Penny A.
 Handbook of coordination catalysis in organic
chemistry.

 Bibliography: p.
 Includes index.
 1. Coordination compounds.   2. Catalysis.
3. Chemistry, Organic.   I. Title.
QD474.C47   1986   547.1′2242   85-30878
ISBN 0–408–10776–6

Printed and bound in Great Britain by Robert Hartnoll (1985) Ltd, Bodmin, Cornwall

For my mother

Preface

I suppose that all first-time authors begin their work filled with confidence that all will proceed smoothly and in high yield, and that they will meet the publisher's target date. Some five years later I am forced to confess that while the yield may be higher than expected the reaction has been much slower!

The purpose of this book is to show the increasing importance of homogeneous catalysis by metal complexes to organic chemists. In particular, I have aimed to highlight the special selectivities which can be obtained using coordination complexes as catalysts, since I believe that it is this selectivity which will encourage future developments in the field. With few exceptions I have confined my attention to homogeneous catalysts and closely related polymer supported complexes. Although it is not possible to be comprehensive (and I apologise for omitting any of readers' favourite reactions) I believe that most of the major areas have been discussed.

I owe a great deal to colleagues who have provided sympathy, advice and good cheer throughout the writing of the book. I am particularly indebted to Dr John Brown in whose laboratory I worked on homogeneous hydrogenation and whose infectious enthusiasm first led me to an interest in catalytic processes. Last, but by no means least, I must thank the secretarial staff at Oxford, Harvard, Rutgers and Sussex Universities for their impeccable typing and endless patience.

P.A.C.

Contents

1 Introduction   1

2 Hydrogenation and related reactions   9

3 Reactions of carbon monoxide   217

4 Other additions to carbon-carbon multiple bonds   307

5 Isomerisation reactions   403

6 Oxidation   451

7 Reactions of the carbonyl group   681

8 Formation of carbon-carbon bonds   731

9 Alkene metathesis   921

Glossary of terms and abbreviations   967

Index   985

1   Introduction

In recent years it will have become apparent, even to the most casual reader of the literature, that few organic syntheses do not involve at least one step in which a metal complex is involved. The advantages of using such species as catalysts are manifold. First and foremost is that catalysis improves reaction rates, and this may be translated into an ability to carry out the desired reaction under milder conditions than would otherwise be possible. For example, acetonitrile, MeCN, is hydrolysed only slowly in concentrated base at $100^{\circ}C$. In the metal bound species, $[(NH_3)_5Co(MeCN)]^{3+}$, hydrolysis is instantaneous at room temperature, representing an increase in rate by a factor of 10^6.

The second reason to prefer catalytic to stoichiometric reactions is economic. A catalyst is used in small quantities, typically in the region of 1 mole %. To compete with this a stoichiometric use of the reagent must be at least 99% efficient, a phenomenon which is relatively rare. Many of the catalysts which will be discussed in subsequent chapters are complexes of expensive metals and their cost is an important consideration. Additionally, some of the ligands used are constructed only by long and tedious syntheses. In theory, the catalyst may be recovered at the end of the reaction and reused but this is not always possible in practice. In this respect homogeneous catalysts are less successful than their heterogeneous analogues.

Most importantly, however, a coordination catalyst may be used to improve or alter the selectivity of a reaction. It is instructive to consider the types of selectivity which may be desired. Earliest came chemoselectivity, the selectivity between different functional groups, either in the same or different molecules. Figure 1 shows two classic examples, both reductions. It is frequently of value to reduce an alkyne to a <u>cis</u>-alkene but not to continue reduction to give the alkane. In the other example, an $\alpha\beta$-unsaturated aldehyde might be reduced to a saturated aldehyde, an allyl alcohol or a

1

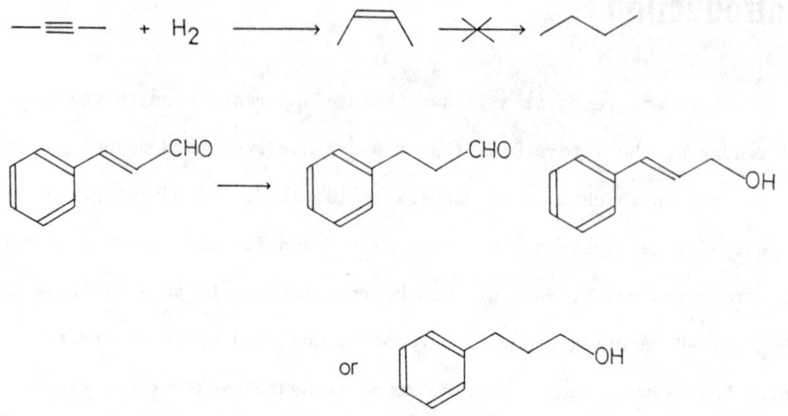

Figure 1 Examples of chemoselectivity in reduction

or a saturated alcohol. To take an example from the field of coordination catalysis, Bu₃SnH converts 1 to 2 with better than 99% selectivity in the presence of $(Ph_3P)_4Pd$[1,2]. By contrast, 3 gives 4 in excellent yield in a rhodium catalysed hydrosilylation reaction[3]. In 5, where the double bond is not conjugated with the carbonyl group, transfer hydrogenation gives up to 90% of the unsaturated alcohol, 6[4].

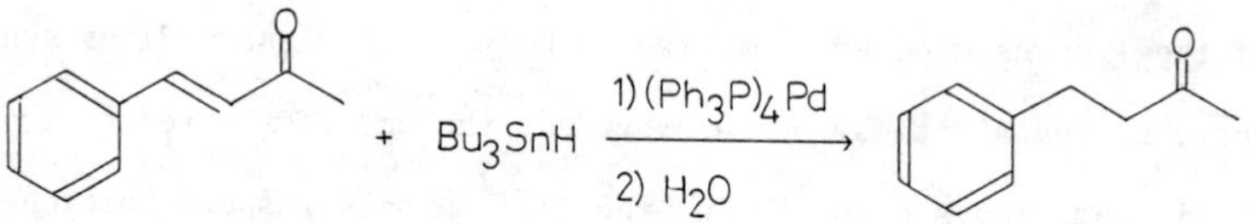

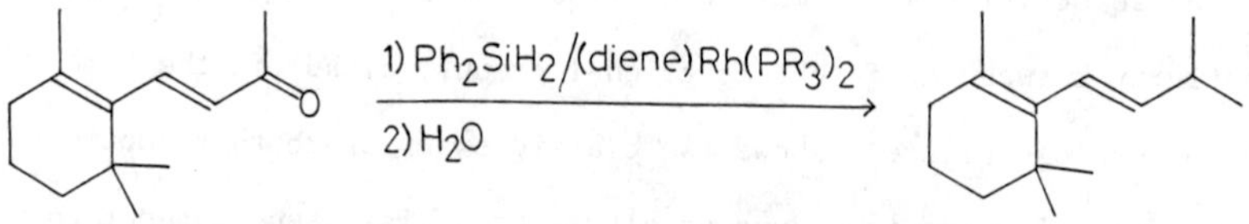

Regioselectivity involves reaction at one of two or more similar sites in a molecule. The well known Markownikov addition of HBr to alkenes is regiospecific. Many additions to double bonds catalysed by metal complexes are also regioselective or regiospecific. For example, rhodium catalysed hydroformylation of styrene gives mainly 2-phenylpropanal. Similarly, hydrosilylation of $\underline{7}$ gives mainly $\underline{8}$ in the presence of $HCo(CO)_4$[5] and radical addition of RSO_2Cl to $\underline{9}$ catalysed by $(Ph_3P)_3RuCl_2$ gives only $\underline{10}$[6]. In the reaction of $\underline{11}$ in the presence of $(Ph_3P)_4Pd$, the catalyst changes both the regiochemistry of the reaction and its outcome. In the absence of the catalyst Michael addition yields $\underline{12}$, but $(Ph_3P)_4Pd$ reacts with an isomer of $\underline{11}$, $\underline{13}$, to give the π-allyl complex, $\underline{14}$. This allyl complex is attacked regioselectively at the less hindered site to give $\underline{15}$[7].

A reaction is said to be diastereoselective when the product contains more of one possible diastereoisomer than the other. Two familiar examples are shown in Figure 2. A rather more complex example in the area of hydrogenations is the reduction of $\underline{16}$ to $\underline{17}$ and $\underline{18}$[8]. In the presence of $[(cod)Ir(py)(PCy_3)]^+$ as catalyst, the reaction is unselective, but with

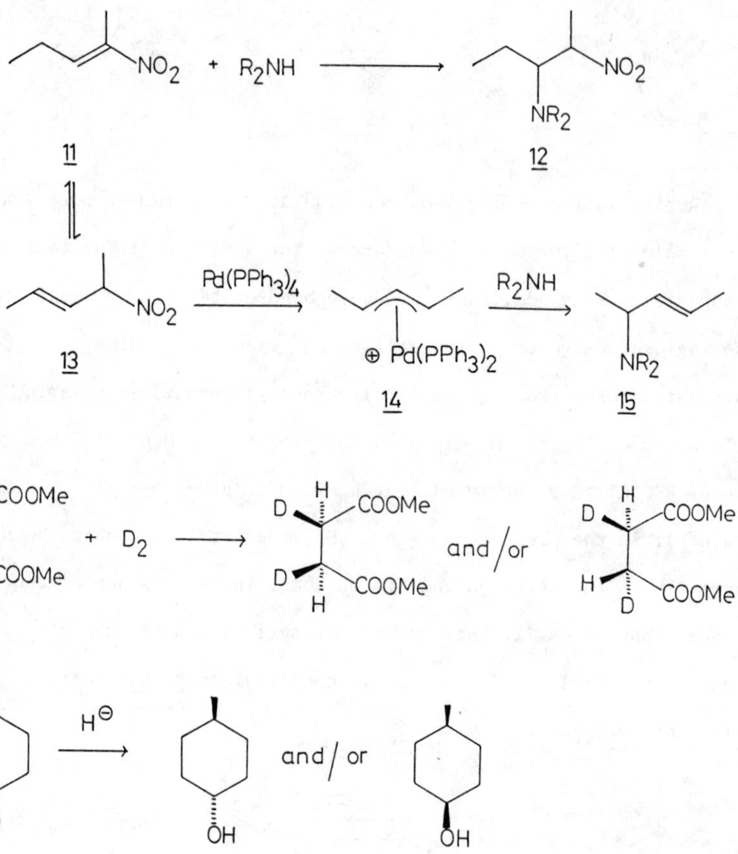

4

11

12

13

14

15

+ R₂NH → ... Pd(PPh₃)₄ ... ⊕ Pd(PPh₃)₂ ... R₂NH

+ D₂ → and/or

Hᐟ⊖ and/or

Figure 2   Examples of potentially diastereoselective reactions

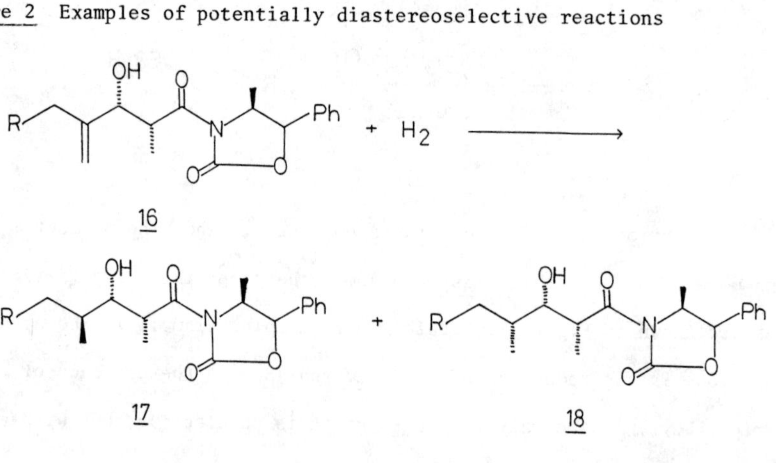

16

+ H₂ →

17   +   18

[(nbd)Rh(dppb)]$^+$ the ratio of 17:18 (R = Me) is 25:75 at 15 psi H$_2$ and 93:7

at 640 psi H$_2$. At low pressures isomerisation competes with reduction and

H$_2$ addition is rate-controlling. At elevated pressure complexation of the

substrate becomes rate-controlling. Various diastereoselectivities may be

achieved in the palladium catalysed substitution of allyl acetates (Figure 3)[9].

With nucleophiles such as NaCH(COOMe)$_2$ and lithium and tin enolates, path a,

giving retention, is followed. However, with Grignard reagents, PhZnCl and

CH$_2$=CHAlMe$_2$, path b predominates. Heteroatom nucleophiles such as R$_2$NH and

RCOO$^-$ give mixtures of products.

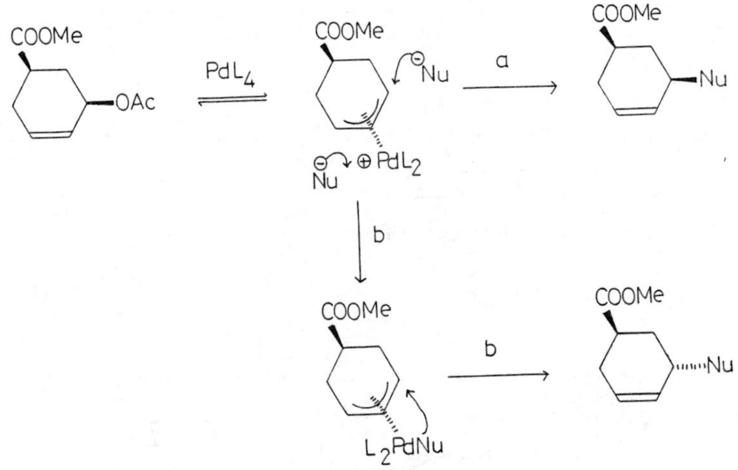

Figure 3 Palladium catalysed substitution of allyl acetates

Good enantioselectivity using coordination catalysts has only been

achieved in the last decade. Hydrogenation of 19 to 20 in the presence of

rhodium catalysts of chiral phosphines has been the most widely studied

reaction and optical yields up to 99% may be achieved. For example, the diene

rhodium complex of 21 gives 88% enantiomer excess in the reduction of 19

(R = Ph, R' = Me, R'' = H)[10]. Other recent examples of enantioselective

reactions are shown in Figure 4[11, 12].

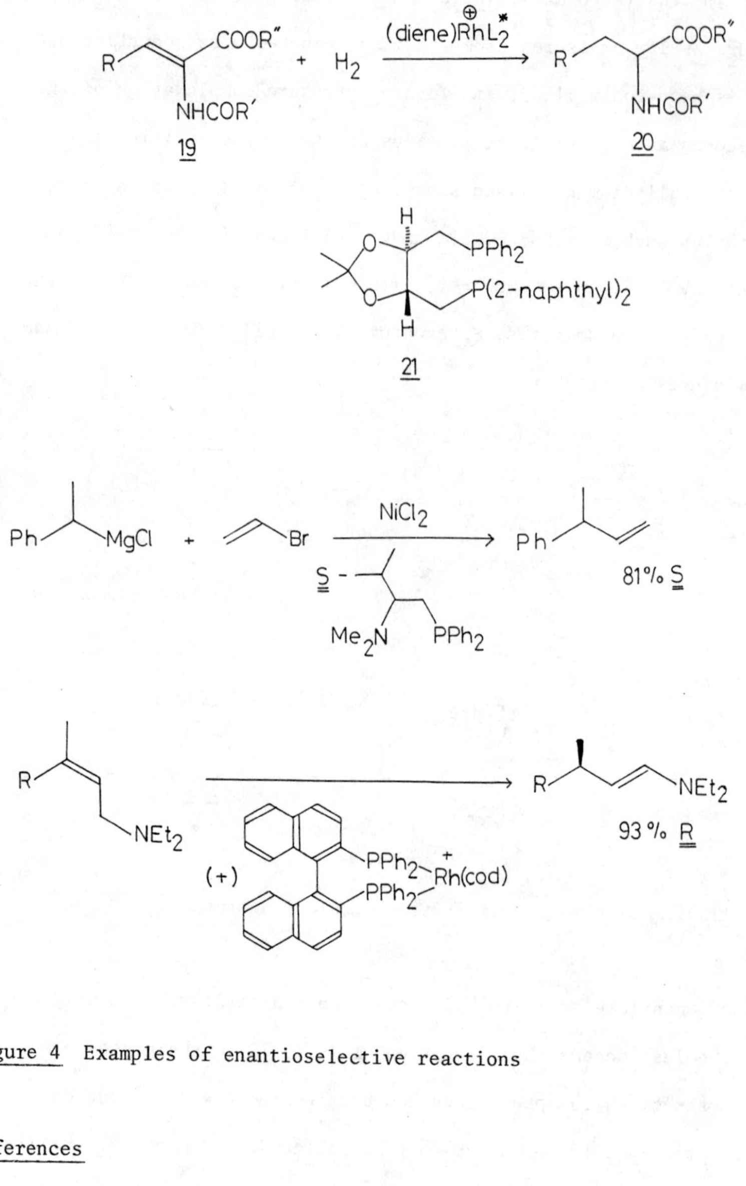

Figure 4 Examples of enantioselective reactions

References

1. E.Keinan and P.A.Gleize, Tetrahedron Lett., 23, 477 (1982).

2. P.Four and F.Guibe, Tetrahedron Lett., 23, 1825 (1982).

3. T.Kogure and I.Ojima, J. Organomet. Chem., 234, 249 (1982).

4. M.Visintin, R.Spogliarich, J.Kaspar and M. Graziani, J. Mol. Catal., 24 277 (1984).

5. G.K-I.Magomedov, G.V.Druzhkova, O.V.Shkol'nik, V.S.Nikitin, T.A.Shestkova and N.M.Bizyukova, J. Gen. Chem. USSR, 53, 1641 (1983).

6. N.Kamigata, H.Sawada, N.Suzuki and M.Kobayashi, Phosphorus and Sulphur, 19, 199 (1984).

7. R.Tamura, K.Hayashi, Y.Kai and D.Oda, Tetrahedron Lett., 25, 4437 (1984).

8. D.A.Evans and M.M.Morrissey, J. Amer. Chem. Soc., 106, 3866 (1984).

9. E.Keinan and Z.Roth, J. Org. Chem., 48, 1769 (1983).

10. T.Yamagashi, M.Yatagai, H.Hatakeyama and M.Hida, Bull. Chem. Soc. Jpn., 57, 1897 (1984).

11. T.Hayashi, M.Konishi, M.Fukushima, K.Kanehira, T.Hioki and M.Kumada, J.Org. Chem., 48, 2195 (1983).

12. K.Tani, T.Yamagata, S.Akutagawa, H.Kumobayashi, T.Taketomi, H.Takaya, A.Miyashita, R.Noyori and S.Otsuka, J. Amer. Chem. Soc., 106, 5208 (1984).

2 Hydrogenation and related reactions

2.1 Activation of hydrogen

  2.1.1 Oxidative addition of hydrogen

  2.1.2 Heterolytic cleavage of hydrogen

  2.1.3 Homolytic cleavage of hydrogen

2.2 Hydrogenation of simple alkenes

  2.2.1 Rhodium complexes

  2.2.2 Iridium complexes

  2.2.3 Cobalt complexes

  2.2.4 Ruthenium complexes

  2.2.5 Iron and Osmium complexes

  2.2.6 Nickel, Palladium and Platinum complexes

  2.2.7 Other metals

2.3 Reduction of functionalised alkenes

  2.3.1 Styrenes

  2.3.2 $\alpha\beta$-Unsaturated aldehydes

  2.3.3 $\alpha\beta$-Unsaturated ketones

  2.3.4 Other unsaturated carbonyl compounds

  2.3.5 Unsaturated nitriles and nitro compounds

  2.3.6 Alcohols, ethers and amines

2.4 Reduction of dienes

  2.4.1 Allenes

  2.4.2 Non-conjugated dienes

  2.4.3 Conjugated acyclic dienes

  2.4.4 Unsaturated fats

  2.4.5 Cyclic dienes

2.5   Hydrogenation of alkynes

   2.5.1   Reduction to alkanes

   2.5.2   Reduction of alkynes in the presence of alkenes

   2.5.3   Reduction of alkynes to alkenes

2.6   Hydrogenation of arenes

   2.6.1   Cobalt complexes

   2.6.2   Rhodium complexes

   2.6.3   Ruthenium complexes

   2.6.4   Ziegler catalysts

   2.6.5   Other catalysts

2.7   Hydrogenation of carbonyl groups

   2.7.1   Hydrogenation of aldehydes

   2.7.2   Hydrogenation of unactivated ketones

   2.7.3   Activated ketones

   2.7.4 $\alpha\beta$-Unsaturated carbonyl compounds

2.8   Nitrogen containing compounds

   2.8.1   Carbon-nitrogen double bonds

   2.8.2   Carbon-nitrogen triple bonds

   2.8.3   Nitrogen-nitrogen multiple bonds

   2.8.4   Nitro and nitroso groups

2.9   Transfer hydrogenation

   2.9.1   The hydrogen donor

   2.9.2   Reduction of ketones

   2.9.3   Reduction of alkenes

   2.9.4   Reduction of enones

   2.9.5   Other substrates

2.10 Hydrogenolysis

2.11 Polymer supported catalysts

  2.11.1 The supports

  2.11.2 Hydrogenation of alkenes

  2.11.3 Hydrogenation of alkynes

  2.11.4 Hydrogenation of dienes

  2.11.5 Hydrogenation of arenes

  2.11.6 Carbonyl groups and enones

  2.11.7 Other substrates

2.12 Asymmetric hydrogenation

  2.12.1 Reduction of carbon-carbon double bonds

    2.12.1.1 Rhodium phosphine complexes

    2.12.1.2 Other metals

  2.12.2 Asymmetric hydrogenation of ketones

  2.12.3 Asymmetric hydrogenation of carbon-nitrogen double bonds

  2.12.4 Asymmetric hydrogenation using non-phosphine ligands

  2.12.5 Asymmetric transfer hydrogenation

  2.12.6 Asymmetric polymer bound catalysts

2.13 Dehydrogenation

Homogeneous hydrogenation is one of the best understood and widely reviewed of all reactions catalysed by transition metal complexes.[1-7] In a chapter of this length it is difficult to do justice to the literature; the most extensive discussions will be of recent work and of areas which have not been extensively reviewed.

2.1 Activation of hydrogen

In order to be added to an organic substrate hydrogen must first be activated, and generally both it and the substrate must be brought into the coordination sphere of the metal catalyst. There are three common routes for activation and most of the catalysts discussed will fall into one of these categories. In the first, hydrogen is activated by oxidative addition to a metal (equation (1)).

$$L_nM + H_2 \longrightarrow L_mMH_2 \tag{1}$$

In the second case hydrogen is activated by heterolysis (equation (2)) and the final group homolyse the hydrogen molecule (equation (3)). Many examples of each route exist; in each case the best understood will be discussed.

$$L_nM + H_2 \longrightarrow L_mMH + H^+ \tag{2}$$

$$2L_nM + H_2 \longrightarrow 2L_mMH \tag{3}$$

2.1.1 Oxidative addition of hydrogen

Possibly the best known of all catalysts for homogeneous hydrogenation is that developed by Wilkinson, $(Ph_3P)_3RhCl$. Elucidation of the reaction mechanism was long and arduous;[8-13] it was necessary to distinguish between a hydride ($\underline{1} \rightarrow \underline{2} \rightarrow \underline{5} \rightarrow \underline{6}$) and an unsaturate ($\underline{1} \rightarrow \underline{2} \rightarrow \underline{3} \rightarrow \underline{6}$) route (Figure 1). A rôle for the unsaturate route was excluded by detailed kinetic experiments by Dutch[14] and French[15] groups. Dimers were very minor participants but solvation is important. The structures of some intermediates have been established by spectroscopic methods, including $\underline{7}$ and $\underline{8}$ (L = PPh_3, $P(\underline{p}\text{-tolyl})_3$)[8] and the C_2F_4 analogue of $\underline{8}$ studied by X-ray crystallography.[16] The reversibility of H_2 addition to $\underline{2}$ is shown by the interconversion of ortho- and para-hydrogen in the presence of the catalyst, loss of hydrogen being faster than alkene capture.[17] Calculations confirm that the expected picture for oxidative addition of hydrogen with sideways

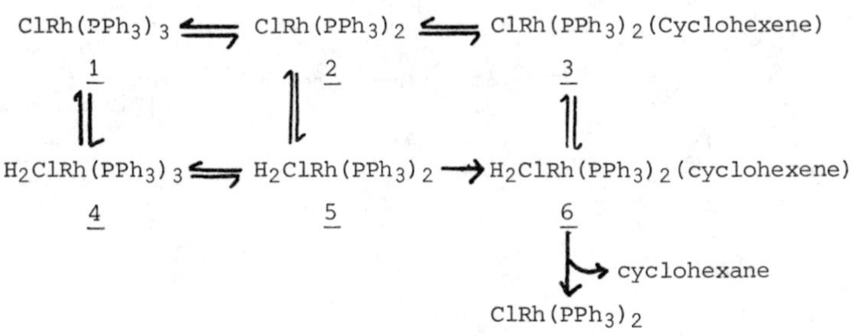
ClRh(PPh3)3
ClRh(PPh3)2
ClRh(PPh3)2(Cyclohexene)
1
2
3
H2ClRh(PPh3)3
H2ClRh(PPh3)2
H2ClRh(PPh3)2(cyclohexene)
4
5
6
cyclohexane
ClRh(PPh3)2

Figure 1 Mechanism of hydrogenation of cyclohexene in the presence of Wilkinson's catalyst.

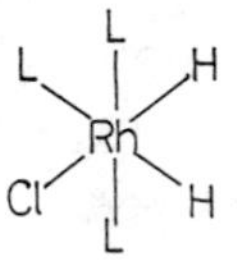
L
L
H
Rh
Cl
H
L

7

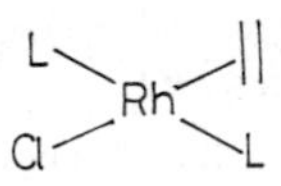
L
Rh
Cl
L

8

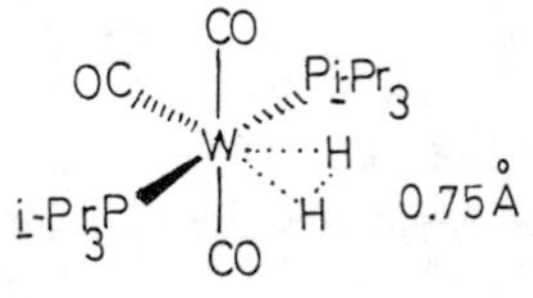
CO
OC
Pi-Pr3
W
H
i-Pr3P
H
0.75 Å
CO

9

approach of H_2 is approximately correct.[18] The molecular hydrogen complex, 9, may be used as a model for the transition state. The transition state for insertion of the alkene into an Rh-H bond in $H_2RhCl(PH_3)_2(C_2H_4)$ has been modelled and should be favourable[19] with only a modest energy barrier, though the precise trajectory may be complex.[20]

Clues as to the geometry of the intermediates of Figure 1 come from recent work of Brown.[21] In the complex 10 P_B dissociates rapidly at 25° to give 11 where H_A and H_B are equivalent. P_A exchanges more slowly with free phosphine. The intermediate, 12, with cis-phosphines is also important, suggesting that the true catalytic species may be 13 rather than 14 as previously assumed. Finally a note of caution must be sounded. ESCA measurements on both commercial and synthetic[22] Wilkinson's catalyst showed that these contained Rh(I) and Rh(III) in the ratio 3:2. An authentic complex was prepared starting with a Rh(I) species; this gave

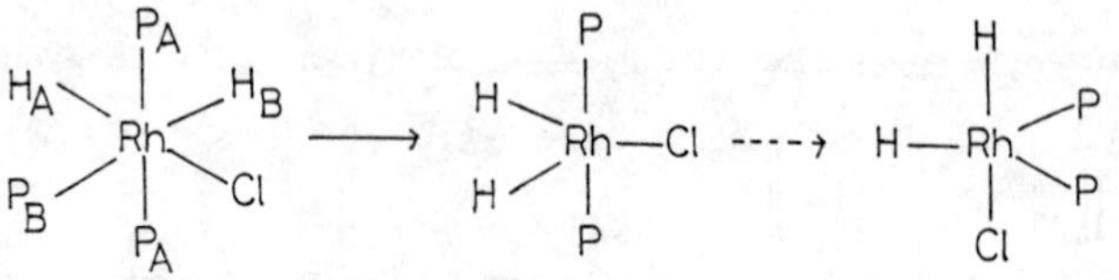

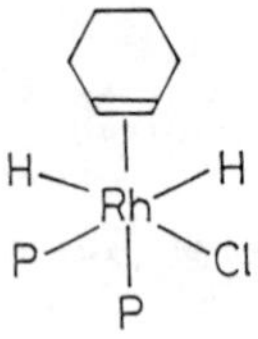

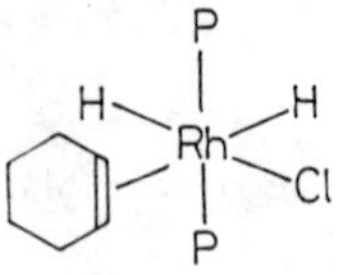

better results in catalysis but was oxidised in air to the mixture as

before.[23]

Wilkinson's work provoked the development of many similar catalysts

which activate H_2 by oxidative addition.[24]   For example, complexes

$[L_2Rh(diene)]^+A^-$ yield, on hydrogenation, 15 (L = PR_3, S = coordinating

solvent).[25]   Ortho $\rightleftharpoons$ para-hydrogen conversion is slower with these complexes

than for Wilkinson's catalyst, but still significant, indicating that the

dihydride, 15 is in equilibrium with a minor amount of a solvate, such as

16.[17]   When a substrate, α-benzamidocinnamic acid, is present no

equilibration occurs until it is consumed, indicating that H_2 addition to

the substrate is irreversible.   This mechanism, however, applies only to

monodentate phosphines; chelating biphosphines are unable to span the

mutually trans-positions.   When $[(diphos)\overset{+}{Rh}(diene)]A^-$ is hydrogenated in

methanol two moles of hydrogen are absorbed and the product is the cis-

biphosphine Rh(I) solvate, 17.[26-28]   The desolvated dimer, 18, may be

isolated and its structure was determined by X-ray crystallography.   Both

17 and its dppb analogue must be in equilibrium with small amounts of

<u>15</u>   <u>16</u>

<u>17</u>   <u>18</u>

dihydride since <u>ortho</u>- and <u>para</u>-hydrogen are interconverted, though again

this is suppressed by the presence of substrate, indicating that addition of

H_2 to a substrate complex is irreversible.   Studies of a wide range of

mono and bidentate phosphines indicate that the equilibrium between

dihydride and solvate is a finely balanced one.[29]

Further details of the pathway between the diene complex and <u>15</u>

have been elucidated by studying related iridium species.[30]   On treatment

with $H_2(0^\circ C, CH_2Cl_2)$ $[(cod)IrL_2]^+PF_6^-$ (L = R_3P, amine) yields <u>19</u> in a

reversible reaction, this being the first known diene metal dihydride

complex.   Analogous complexes were formed when L_2 = diphos[31] or

cyclooctadiene.[32]   On warming to room temperature H_2 may be lost to

renegerate the starting material or transferred to the diene to give an

intermediate which is irreversibly deactivated to $[Ir_2(\mu-H)_3H_2L_4](PF_6)_2$.

Under hydrogenation conditions, in the presence of excess cod, the major

intermediate is the stereoisomer, <u>20</u>.   Stereoelectronics prevents rapid

transfer of H_2 to cod and <u>20</u> is stable at $40^\circ C$.   Thus the reaction in the

early stages, where <u>19</u> is the major complex, is about 40 times as fast as in

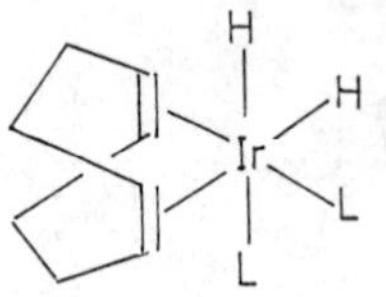

19 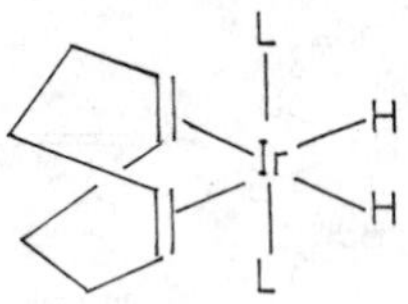   20

the subsequent period.[33]   A study of the effect of L on the rate of H_2

addition shows curious effects.   "Oxidative" addition occurs more rapidly

with decreasing electron density at the metal[34] and more oxidising addends

such as O_2 and MeI are not reactive.   Thus although these reactions fall

within the definition of oxidative addition, they are somewhat less

oxidative than usual.   The difference in reactivity between rhodium and

iridium complexes was highlighted in a study of 21.   For M = Rh, diene = nbd,

the product of hydrogenation is the cis-dihydride trans-diphosphine, 22.

However, 21 (M = Ir, diene = cod) gives, on reduction at -78°, the cis

diphosphine, cis-dihydride, 23.   On warming this is converted into 24 and

25.[26]

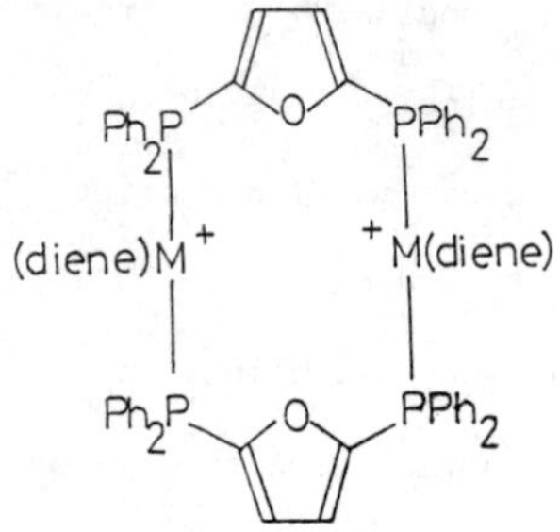

21 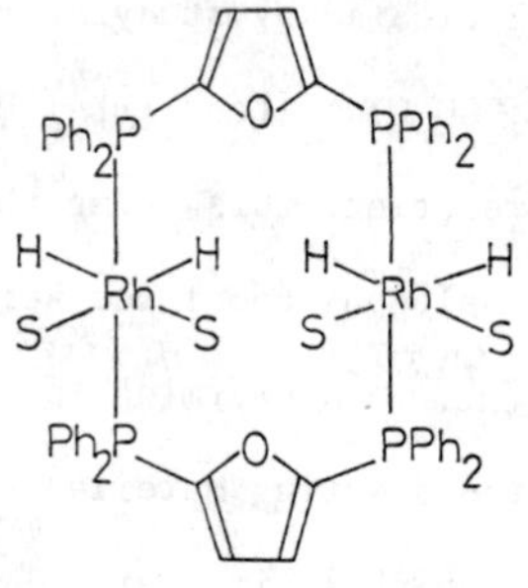   22

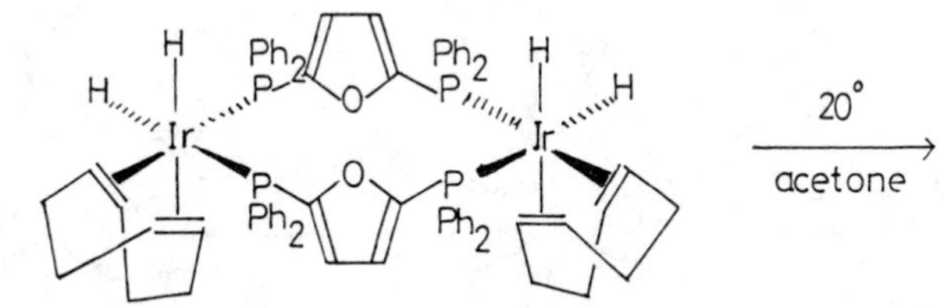

23

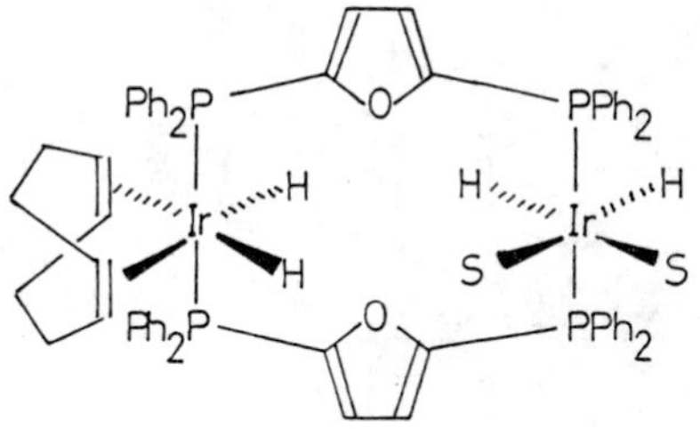

24 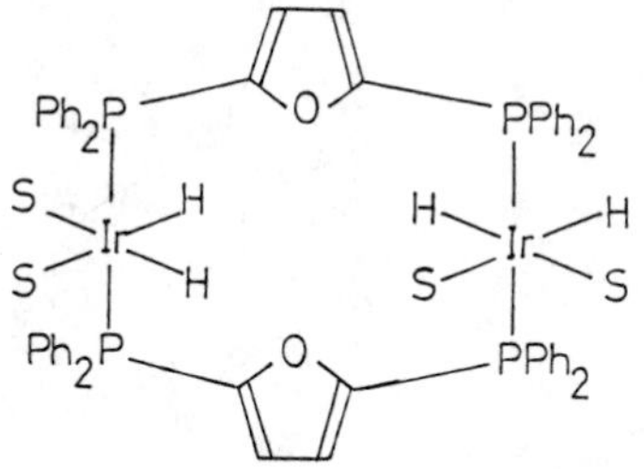   25

More recent work has focussed on the regiochemistry of H_2 addition.
For example 26 yields 27 as the kinetic and 28 as the thermodynamic
product.[35]   However, for 29, 30 is the sole product   under all conditions.
The predominance of an electronic over a steric effect was demonstrated by
the reaction of 31 which gives only 32.   The authors concluded that H_2
should be regarded as a 2e donor like ethylene.

Related work involves addition of H_2 to $IrXCOL_2$.   This is
reversible as shown by ortho ⇌ para-hydrogen equilibration.[36]   When L_2 =
diphos, X = Cl,Br,I,the sequence 33 ⇌ 34 ⇌ 35 is followed with 34 as the
kinetic and 35 as the thermodynamic product.[37]   When X = CN, 34 is more
stable than 35 and on heating isomerisation to 36 occurs.   It is clear
that subtle electronic factors are involved.   A number of other catalysts
reacting by this route are shown in Figure 2.

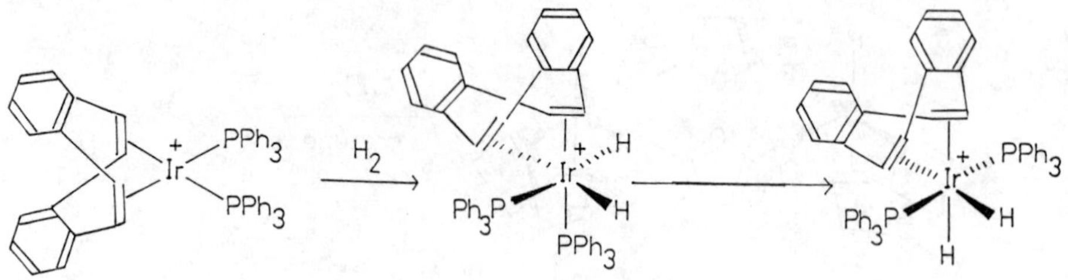

26   27   28

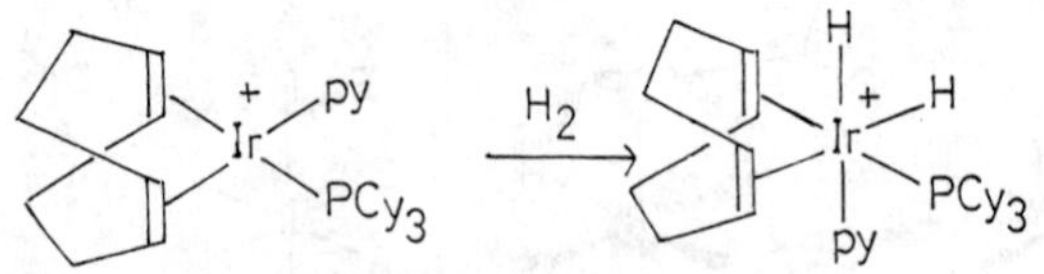

29   30

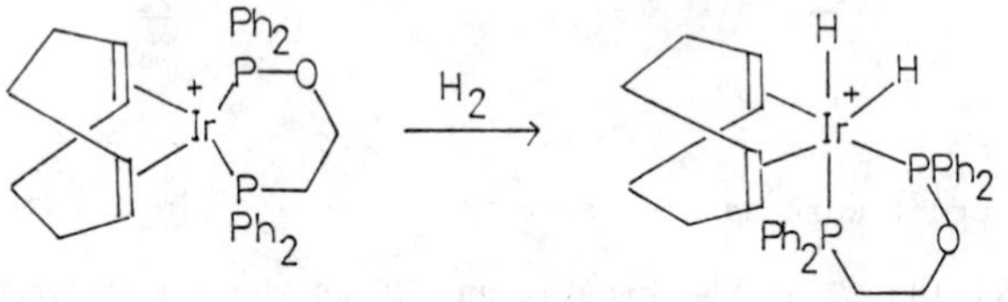

31   32

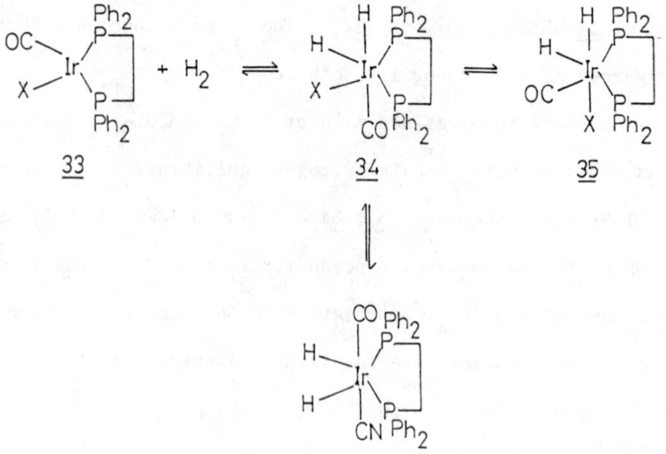

33   34   35

36

$(\underline{\eta}^3\text{-allyl})\text{Mn(CO)}_4\,|\,\text{PPh}_3$ [38] $\text{H}_2\text{Ru(PPh}_3)_4$ [40]

Arene Cr(CO)_3 [39] $\text{H}_4\text{Ru}_4\text{(CO)}_{12}$ [41]

$\text{Mo(CO)}_3(\text{PCy}_3)_2$ [18]

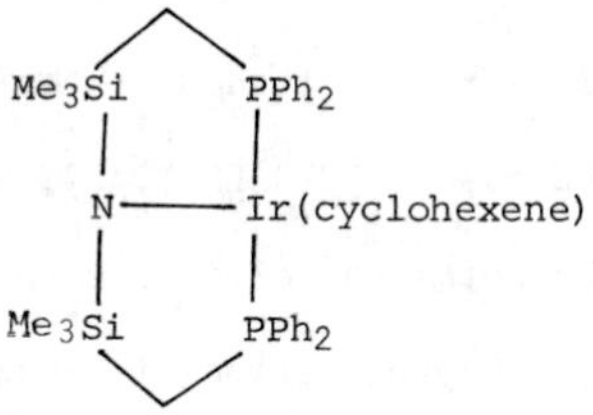

Figure 2   Catalysts which activate hydrogen by oxidative addition

2.1.2   Heterolytic cleavage of hydrogen

Although the term heterolytic cleavage correctly describes a mechanism of hydrogen activation it is often used more generally to describe any reaction in which a monohydride is formed by reaction of H_2 with a metal complex.   Two routes give rise to this observation.   In the first a concerted 4-centre reaction occurs (equation (4)).   The same stoichiometry arises from addition elimination (equation (5)).

$$M + H_2 + B \rightarrow \underset{\underset{\overset{|}{H}\cdots\overset{|}{H}}{}}{M\cdots B} \rightarrow MH + BH \tag{4}$$

$$M^{n+} + H_2 \rightleftharpoons MH_2^{(n+2)+} \xrightarrow{\ B\ } M\text{-}H^{(n+1)+} + B\overset{+}{H} \tag{5}$$

The latter does not represent true heterolytic activation but often cannot be distinguished from it by kinetic studies.   Whilst the orbital interactions in both cases have been thoroughly discussed,[43] as a first approximation we may exclude from the dihydride/base route metals where oxidative addition is unlikely.   It seems that a continuum of mechanisms may be the best representation.

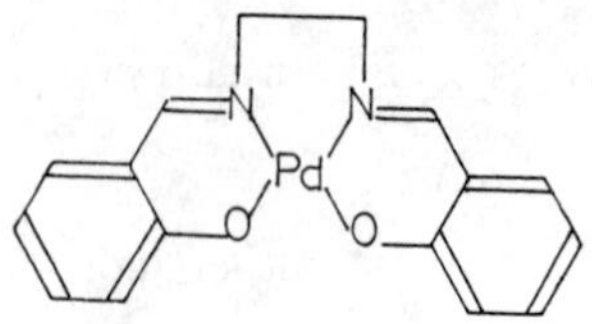

37

The best established of the systems which activate H_2 by a strictly heterolytic process is Pd(SALEN), 37, used as a model for hydrogenase. In particular the metal has electrons to interact with H_2 and heteroatoms to stabilise H^+. Hydrogenation of 1-hexene in the presence of 37 is inhibited by acid and enhanced by base. The mechanism proposed is shown in Figure 3[44] with heterolytic splitting of H_2 synchronised with scission of a Pd-O bond. A vacant coordination site is generated for incoming substrate. After insertion of the alkene into the Pd-H bond intramolecular protonation releases the product and regenerates 37. The heterolytic mechanism is guaranteed by the relative instability of Pd(IV) and the steric barriers to cis-oxidative addition. Other palladium complexes employing a similar mechanism include those of dimethylglyoxime and $(Ph_3P)_2Pd(OAc)$.[45]

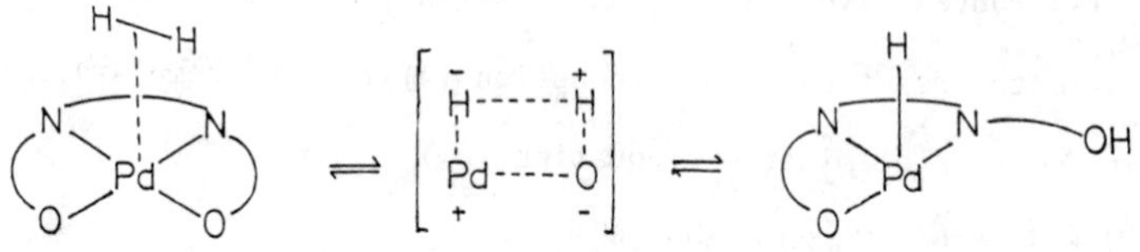

Figure 3 Mechanism of hydrogen activation by Pd(SALEN)

If Pd(SALEN) is the "purest" of the complexes activating hydrogen by a heterolytic mechanism, $(Ph_3P)_3RuCl_2$ is the most widely used. Dihydride formation would be unlikely for Ru(III) since Ru(V) is not favoured but might be practical for Ru(II); the data available are inadequate to distinguish the pathways. Using RuL_2X_3 (X = Cl,Br; L = PPh_3, $AsPh_3$) the route followed is given by equations (6) and (7). A base is essential to reaction[46] and $HRuXL_3$ and $(HRuXL_2)_2$ were isolated.

$$Ru(III) + H_2 + B \rightleftharpoons HRu(III) + BH^+ \qquad\qquad (6)$$

$$HRu(III) + Ru(III) \rightarrow 2Ru(II) + H^+ \qquad\qquad (7)$$

Ruthenium(II) complexes are more common catalysts. An induction period is noted for $(Ph_3P)_3RuCl_2$ during which $HRuCl(PPh_3)_3$ is formed. The mechanism proposed is phosphine dissociation, alkene coordination and insertion, followed by reaction with H_2 to yield product and regenerate catalyst.[47] Later work suggested that phosphine dissociation was negligble and alkene displaced phosphine with phosphine return accompanying insertion.[48]

In assessing the evidence for a heterolytic mechanism Brothers[43] cites examples where a base is essential and analogues where it is not.[49] The need for a basic cocatalyst is a significant part of the evidence for a heterolytic mechanism. Additionally the final reaction of alkyl-$RuCl(PPh_3)_n$ is supposed by some workers to involve dihydride formation and reductive elimination. The formation of Ru(IV) trihydrides by oxidative addition to $[HRu(dppp)_2]^+$ is known, providing a model for this route.[50] H_2/D_2 scrambling is catalysed by $HRuCl(PPh_3)_3$ but the initial HRu bond does not participate; equations (8) and (9) were invoked to account for this.[51] The literature in this area is somewhat speculative; kinetic data

$$HRuCl(PPh_3)_3 + D_2 = D_2Ru(PPh_3)_3 + HCl \qquad\qquad (8)$$

$$D_2Ru(PPh_3)_3 + H_2 = H_2D_2Ru(PPh_3)_3 = HDRu(PPh_3)_3 + HD \qquad (9)$$

in particular may usually be interpreted in several ways.[52]

A well established system involving oxidative addition/ deprotonation is provided by $[L_2Rh(diene)]^+A^-$. As we saw earlier $H_2RhL_2S_2$ (L ; phosphine, arsine; S = MeOH, Me$_2$CO) is the initial product but may be deprotonated to $HRhL_nS_m$,[25] the ease of deprotonation being related to the nature of L. These monohydrides are also hydrogenation catalysts but isomerisation is a major side reaction.

2.1.3   Homolytic cleavage of hydrogen

Homolytic activation of H_2 is described by equations (10) or (11). The best studied systems are cobalt complexes. $[Co(CN)_5]^{3-}$ reacts with H_2

$$2M^n + H_2 \rightarrow 2HM^{n+1} \tag{10}$$

$$M_2^n + H_2 \rightarrow 2HM^{n+1} \tag{11}$$

according to (10) to give $[HCo(CN)_5]^{3-}$; Co(II) is oxidised to Co(III). The system is complex since both pH and CN:Co ratio have a bearing on the precise nature of the species present.[53]   Two general theories are reasonable.   One involves a preformed dimer, $[Co_2(CN)_{10}]^{6-}$ which reacts to give $[(CN)_5CoH_2Co(CN)_5]^{6-}$, the latter then dissociating to give the product.[53,54]   The other invokes a sequential reaction (equations (12) – (14))[55] or a termolecular process.[56]   A recent suggestion[57] that a

$$[Co(CN)_5]^{3-} + H_2 \rightleftharpoons [H_2Co(CN)_5]^{3-} \tag{12}$$

$$[H_2Co(CN)_5]^{3-} + [Co(CN)_5]^{3-} \rightleftharpoons [(CN)_5Co\text{-}\text{-}H\text{-}\text{-}H\text{-}\text{-}Co(CN)_5]^{6-} \tag{13}$$

$$[(CN)_5Co\text{-}\text{-}\text{-}H\text{-}\text{-}\text{-}H\text{-}\text{-}\text{-}Co(CN)_5]^{6-} \rightleftharpoons 2HCo(CN)_5 \tag{14}$$

heterolytic route might be involved has been convincingly refuted.[58]

Dicobalt octacarbonyl $Co_2(CO)_8$ has been used more extensively in hydroformylation than hydrogenation but activation of H_2 occurs as in equation (15).   The presence of the hydridocarbonyl was demonstrated by

$$Co_2(CO)_8 + H_2 = 2HCo(CO)_4 \tag{15}$$

Orchin.[59]   Although the mechanism is not fully understood there is general agreement on a 4-centre transition state, 38.

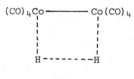

38

2.2   Hydrogenation of simple alkenes

Many hundreds of transition metal complexes catalyse the reduction of simple alkenes.   In considering selectivity we must discuss discrimination between alkenes with differing degrees of substitution and stereoselectivity.   The major competing reaction is isomerisation, both cis ⇄ trans and positional.

2.2.1   Rhodium complexes

Wilkinson's catalyst, $(Ph_3P)_3RhCl$, is the most widely discussed catalyst.[60]   With this in mind it is appropriate to summarise its particular features and give only illustrative and recent examples of its use.

1)   The catalyst is selective for the least substituted double bond in a molecule, since the substrate must be coordinated to a site adjacent to bulky triphenylphosphine ligands (Figure 4).[61]

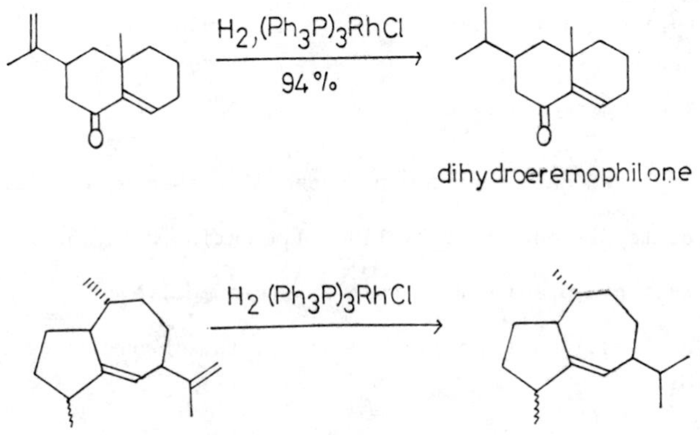

dihydroeremophilone

Figure 4   Selective reduction of the least hindered double bond in the presence of Wilkinson's catalyst·

2)   The relative rates of reduction of simple alkenes depend
     almost entirely on steric factors[6] (Table 1).   1-alkenes
     are reduced faster than internal alkenes and <u>cis</u>- faster
     than <u>trans</u>-compounds.   The rates for cycloalkenes
     reflect ring strain considerations.

<u>Table 1.</u> [62,63]   <u>Relative rates of alkene reduction in the presence
     of Wilkinson's catalyst</u>

Substrate	Relative rate
1-hexene	1.0
1-dodecene	1.18
2-methyl-1-pentene	0.91
<u>cis</u>-2-pentene	1.18
<u>trans</u>-3-hexene	0.066
<u>cis</u>-4-methyl-2-pentene	0.53
<u>trans</u>-4-methyl-2-pentene	0.11
3-ethyl-2-pentene	0.02
2,3-dimethyl-2-butene	2.3×10^{-3}
cyclopentene	1.47
cyclohexene	0.933
cycloheptene	0.75
1-methylcyclohexene	0.02

3)   The short-lived nature of the proposed alkylrhodium hydride
     intermediate discourages $\underline{\beta}$-hydride elimination with the
     result that hydrogenation is mainly site-specific.[64]
     Useful deuterations and tritiations have thus been achieved
     (Figure 5).

4)   Isomerisation is usually limited and with simple alkenes the
     isomerised product is often also reduced (Figure 6).

5)   Arene, carbonyl, hydroxy, halo, azo, ether, ester and carboxyl
     groups are not normally reduced.   Aldehydes and allyl
     alcohols may be decarbonylated (Figure 7).

Ref.

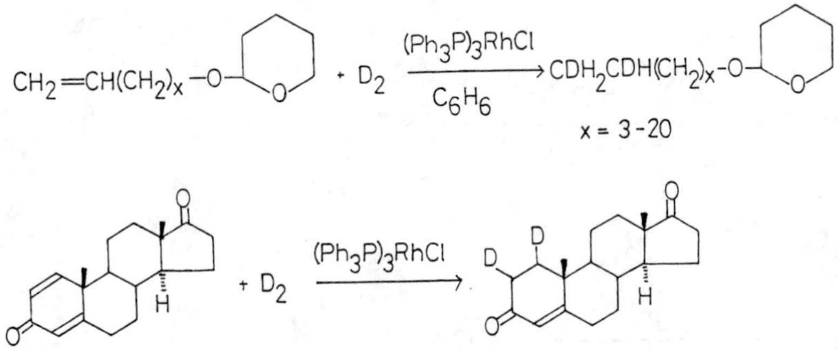

65

66

Figure 5 Catalytic deuteration in the presence of Wilkinson's catalyst.

Ref.

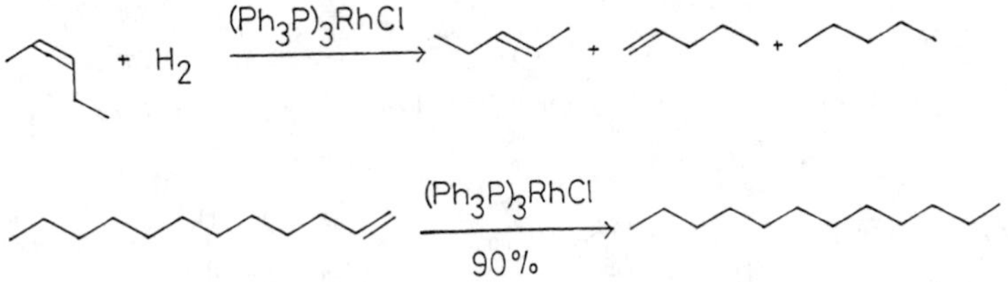

67

68

Figure 6 Reduction and isomerisation of alkenes in the presence of Wilkinson's catalyst.

6) Addition of hydrogen is normally rigorously cis. This has been demonstrated by deuterium labelling experiments such as 39 to 40[72] and 41 to 42.[73]

7) In the absence of groups able to coordinate to the metal, addition of hydrogen is to the less hindered face, with good selectivity (Figure 8).

66

96 % selectivity

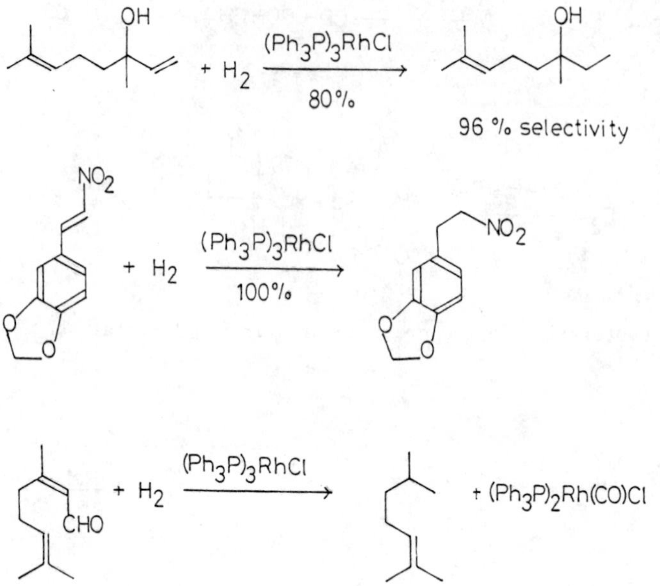

69

70,71

Figure 7 Group selectivity of Wilkinson's catalyst.

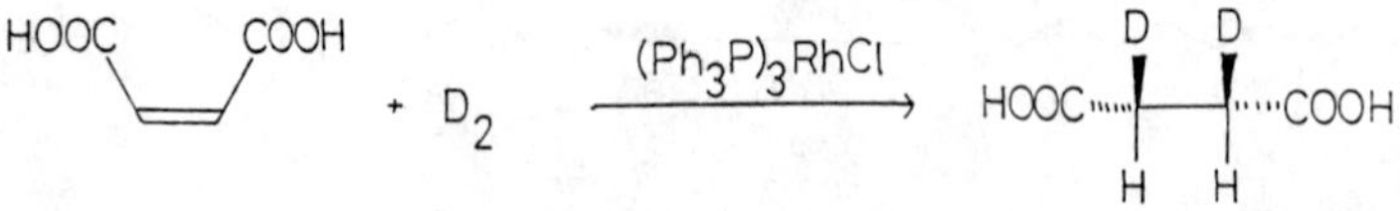

39   40

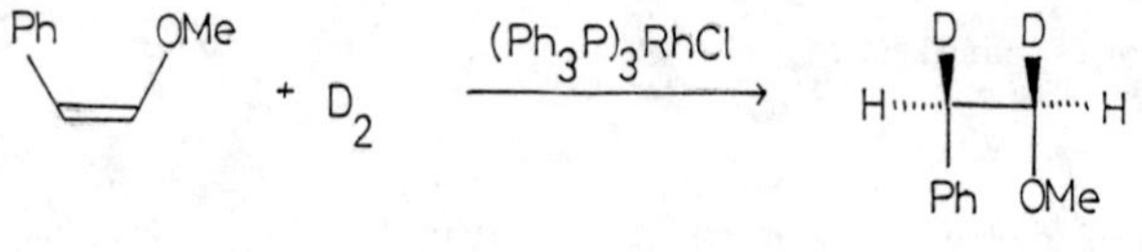

41   42

Ref.

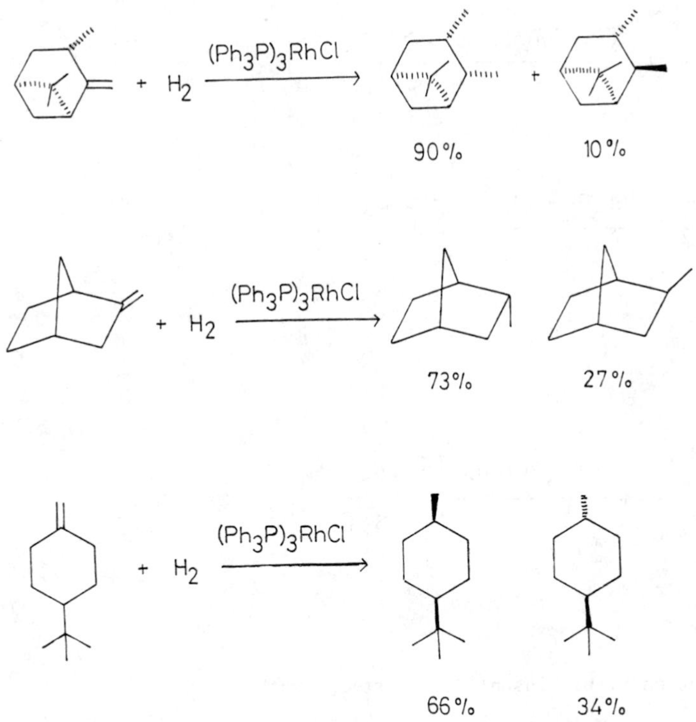

90%   10%   74

73%   27%   75

66%   34%   76

Figure 8   Stereoselectivity of reduction in the presence of Wilkinson's
catalyst.

8)   Some reports state that O_2 or peroxides enhance the reaction
rate, presumably by removal of Ph_3P as oxide[77] but such
additives are generally undesirable and for reproduceable
results O_2 should be excluded and substrates freed from
peroxide immediately before use.[78]   Sulphur containing
additives are less deleterious than for heterogeneous catalysts
and catalyst analogues and substrates are known containing
sulphur.[79]   Lewis acids increase the rate,[80] in particular
$AlBr_3$ which gives $(Ph_3P)_3RhBr$, known to be a better
catalyst.[81]

9)   Examples of use in synthesis are given in Figure 9.

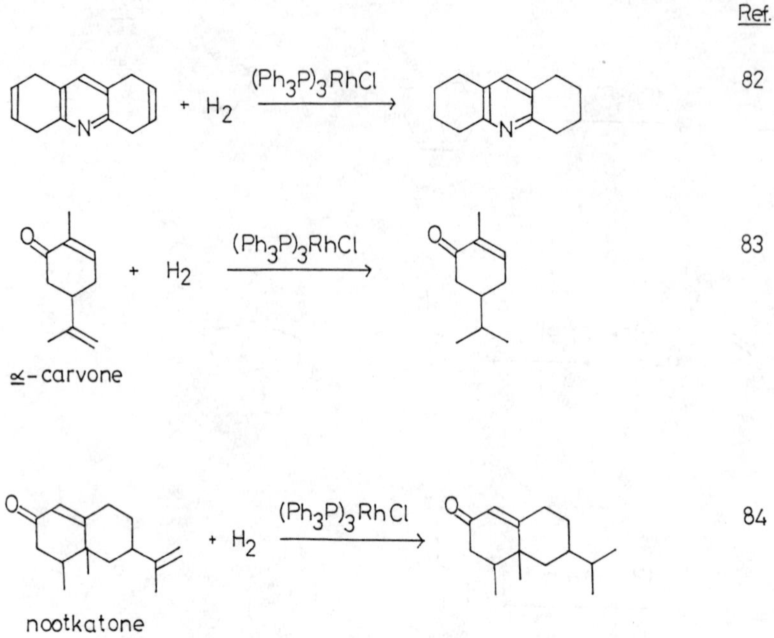

82

83

≼-carvone

84

nootkatone

Figure 9   Hydrogenation by Wilkinson's catalyst in synthesis

Many analogues of Wilkinson's catalyst have been prepared.   The most wide ranging study of changes of substitution in the phenyl ring is due to Halpern.[85]   Activity is enhanced in $((\underline{p}\text{-}XC_6H_4)_3P)_3RhCl$ when X is an electron donor whereas complexes, X = Cl, F, $COCH_3$, give lower rates.[86] These act by increasing the rate of insertion into the Rh-H bond by stabilising the alkyl product.   Bulky ortho-substituents, such as α-naphthyl, are deleterious.   Results with ortho-OR groups which may coordinate to the metal are complex.   43 gives a poor catalyst but (44)$_3$RhCl is very successful.[87]   Complexes of alkyl phosphines give poor rates; the increased basicity inhibits alkene coordination and hydrogen lability.[86]   Phosphite complexes are generally inactive.   Despite the presence of sulphur, 45 is only 2.5 times less active than $(Ph_3P)_3RhCl$,[88] and the sulphonic acids, 46, have been used in successful water soluble

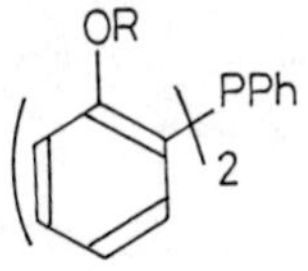

43

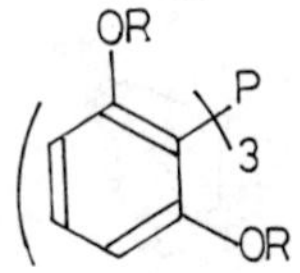

44

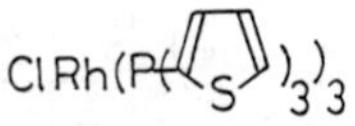

45

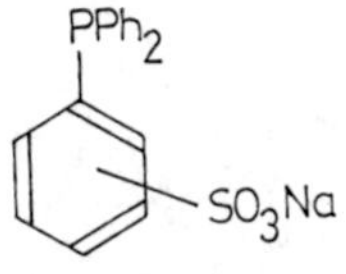

46

catalysts.[89]   Arsines and stibines give catalysts which are usually much less active than phosphine analogues.[67,90,91]   The presence of amino groups in the phosphine is not deleterious and may give useful altered selectivity.[67]

Replacement of the chloride in Wilkinson's catalyst with iodide or bromide gives more active species, and isomerisation is reduced.[92]   The carboxylates $(Ph_3P)_3Rh(OCOR)$, irrespective of R, give reduction at a similar rate to Wilkinson's catalyst and again little isomerisation.[93] $(Ph_3P)_3Rh(NO)$ is also a good catalyst, somewhat surprisingly for an odd electron species.[94,95]

Closely analogous to Wilkinson's catalyst are the cationic complexes 47.[25,96]   The dihydride, 15, and the monohydride, 48, are both hydrogenation catalysts but 48 is also a powerful isomerisation catalyst. Rates are higher than with Wilkinson's catalyst[85] and are not strongly dependent on the phosphine or other ligand.   No dissociation is required and steric requirements are lower.   When the phosphine is basic (e.g. $PhPMe_2$) deprotonation to 48, and consequently isomerisation, is suppressed. The range of suitable ligands is large, including amines and nitriles as well as phosphines.   49[97] and 50[98] have been particularly successful.

Complexes of chelating biphosphines are still more reactive with a rate increase of about 20-fold for diphos.   This is attributed to the favourability of intermediates with cis-phosphines in agreement with Brown's data.[21]   An increase of the chain length of the chelate using dppb gives a rate increase of 10^4-fold, attributed to the ease of attaining favourable geometries in the more flexible complex.[85]

The catalysts discussed so far have been derived from well defined compounds but many reports refer to catalysts formed in situ by addition of phosphines to $(Rh(diene)Cl)_2$ or analogues.   Ligands, activities, selectivities and experimental conditions reported vary widely and the effect of individual variables is difficult to quantify. In situ preparations give the best rates for P:Rh = 2 corresponding to the Osborn catalysts or the dissociated form of Wilkinson's catalyst.[86,99,100] A report that $(Rh(nbd)Cl)_2/R_3P$ gives isomerisation and hydrogenation in differing proportions depending on [Cl$^-$] and catalyst aging shows that these are complex systems.[101]   Some examples are given in Figure 10.

$HRh(PPh_3)_4$ has long been known and is a catalyst for reduction of 1-hexene.[108,109] $HRh(PF_3)(PPh_3)_3$ is a good catalyst for reduction but causes extensive isomerisation,[110] whilst $HRh(DBP)_4$ (DBP = dibenzophosphole,

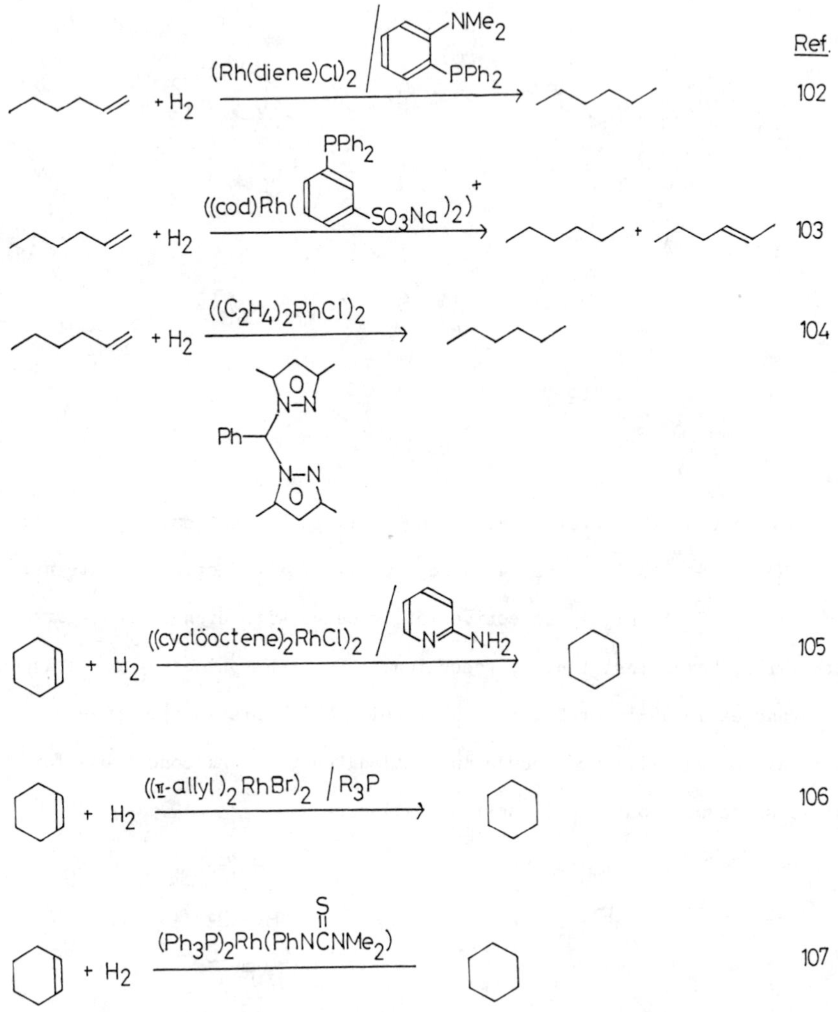

Figure 10 Hydrogenation in the presence of rhodium phosphine complexes
prepared in situ

51) was shown by a kinetic study to react via HRh(DBP)$_2$ as the active
intermediate.[111]   HRh(ttp), 52, is formed by reduction of the corresponding
chloride and its catalytic activity was cited as evidence for an HRh(PPh$_3$)$_3$
active species in HRh(PPh$_3$)$_4$ reductions.[112]   However, dissociation of
phosphine cannot be excluded and Et$_2$AlCl may also be involved.[113]   HRhL$_n$ is

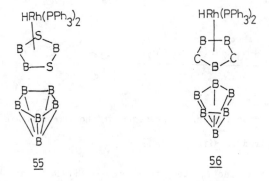

also thought to be the active species in catalysis by rhodium allyl complexes such as 53 and 54.[114,115] 53 has good activity for reduction of internal alkenes[116] and 54 is highly selective for monoenes with dienes, conjugated or otherwise, being resistent to reduction.[117] A phosphine may be replaced by a borane as in 55[118] or 56.[119] 55 requires high pressures, causes isomerisation, and will not reduce internal alkenes. The conditions for 56 are less severe, but isomerisation still occurs.

Many rhodium(I) carbonyl complexes are hydrogenation catalysts or may be readily activated. $HRh(CO)(PPh_3)_3$ is better known as a hydro-formylation catalyst but is also active for reduction and isomerisation of alkenes,[120] being more selective for 1-alkenes than $(Ph_3P)_3RhCl$.[121] The mechanism proposed involves dissociation of Ph_3P [122] and the formation of alkyl $Rh(CO)(PPh_3)_2$ [123], but the details would bear reinvestigation in the light of recent studies of hydroformylation by this complex. $(Ph_3P)_2RhCl(CO)$ catalyses reduction of ethylene and propene[124] but the reaction is very slow.[22] This may be attributed to the reluctance of the complex either to add H_2 [72] or to dissociate PPh_3. Some other carbon monoxide containing catalysts are shown in Figure 11.

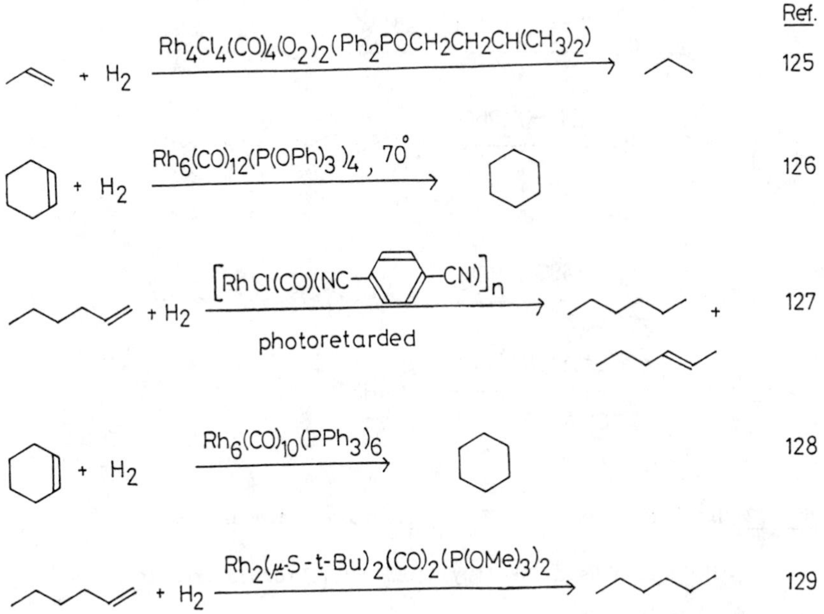

Ref.

125

126

127

photoretarded

128

129

Figure 11 Hydrogenation in the presence of rhodium carbonyl complexes.

A rather complex reaction scheme is needed to explain the action of $((C_5H_5)_2RhCl)_2$ and related complexes as catalysts.[130] C_5Me_5 is a more successful ligand.[131] $[(C_5Me_5Rh)_2(OH)_3]Cl$ is activated by air and monomeric $C_5Me_5Rh(OH)_2$ is the active species.[132] Even the presence of sulphur ligands in $[C_5Me_5Rh(SMe_2)_3]PF_6$ does not entirely deactivate the complex; cyclohexene is reduced quantitatively at 50°, 50 bar.[133]

Most of the catalysts and precursors discussed above were Rh(I) complexes and the active species in almost all rhodium catalysed hydrogenations is Rh(I). However, stable Rh(III) complexes are used and are reduced in situ to Rh(I). $RhCl_6{}^{3-}$ activates H_2 by a heterolytic route to give $[HRhCl_5]^{3-}$ [134] but this is not a generally useful reductant. With other ligands extensive use has been made of Rh(III) catalysts, especially L_3RhCl_3 (Figure 12).

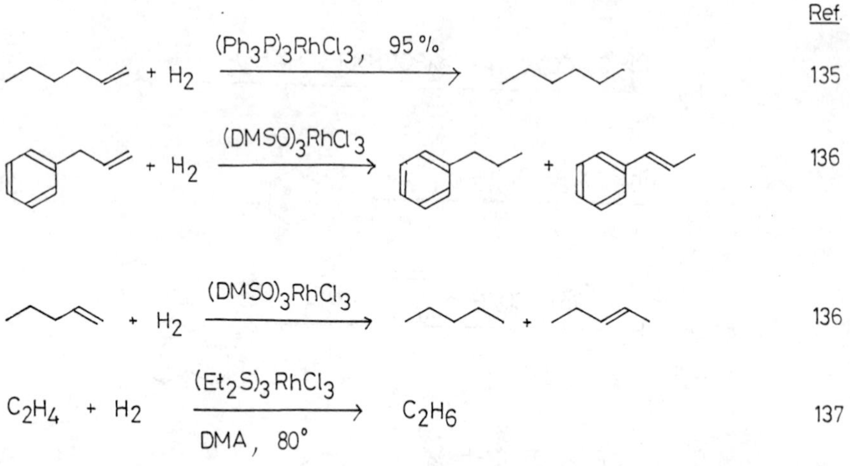

Ref.

135

136

136

137

Figure 12 Hydrogenation using L_3RhCl_3 catalyst precursors

McQuillin used [$RhCl(py)_2(DMF)BH_4$], formed from in situ reduction of $(py)_3RhCl_3$ with $NaBH_4$/DMF, as a catalyst for hydrogenation. The principal is similar to that above except that an external reductant rather than H_2 is used. The mechanism of hydrogenation is complex with hydrogen from both molecular H_2 and BH_4^- being transferred.[138] 1-alkenes and cycloalkenes are readily reduced with lower but significant rates for internal alkenes. Addition of H_2 is rigorously cis.

Catalysis by Rh(II) complexes has received less attention. Protonation of $Rh_2(OAc)_4$ in HBF_4 yields a green, air-stable, diamagnetic ion $Rh_2{}^{4+}$.[139] On addition of Ph_3P the solutions become red and active for alkene reduction at 1 atm. H_2, 25°C.   The variation in substrate reactivity

suggests that the active species is $H_2Rh(PPh_3)_2(solvent)_2$.   Hui used

$Rh_2(OAc)_4$ to catalyse reduction of 1-alkenes in DMF.[140]   In this case

Rh(II) may be the true catalyst, since it is recovered unchanged.   The use

of $(R_3P)_2RhCl_2/Et_3Al$ has been reported; an Rh(I) catalyst is proposed.[141]

2.2.2   Iridium complexes

It has frequently been tempting to regard iridum complexes as

paler reflections of their rhodium analogues and there are many similarities.

However, although $(Ph_3P)_3IrCl$[142] adds hydrogen readily to give

$H_2IrCl(PPh_3)_3$,[143] it is ineffective as a catalyst because PPh_3 dissociation

is poor.   Analogues of the Osborn catalysts (e.g. 57) are more active,

provided that the solvent used for reduction is non-cooordinating.   The

mechanism of the reaction was discussed in detail in section 2.1.   A number

of features are noteable.   Tri- and tetrasubstituted alkenes are reduced at

practicable rates;[144] 1-methylcyclohexene is reduced 75 times faster than

with Wilkinson's catalyst.[30]   With L = Ar_3P the best catalysts have

electron withdrawing groups in the para-position and hydrogenation and

isomerisation are enhanced to about the same extent.[145]   With 58 the nature

of R influences the rate, this being dependent on oxidative addition of

H_2.[34]   Good selectivity for carbon-carbon double bonds is observed; 59 is

reduced without carbonyl reduction or hydrogenolysis.[146]   A disadvantage is

that the catalyst is slowly deactivated to yield $[Ir_2(\underline{\mu}-H)_3H_2L_4](PF_6)_2$[31]

and may not be readily reactivated.

Various in situ catalysts are also known, formed from

$((cyclooctene)_2IrCl)_2/R_3P$ with P : Ir $\leqslant$ 2.[147,148] $(Ph_3P)_2IrCl$ catalyses

reduction of 1-hexene faster than the rhodium analogue but isomerisation is

also accelerated.   Fewer examples are reported than in rhodium chemistry

(Figure 13).

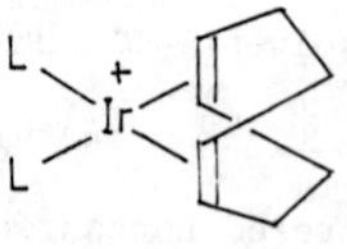

Figure 13 Hydrogenation using in situ iridium complexes

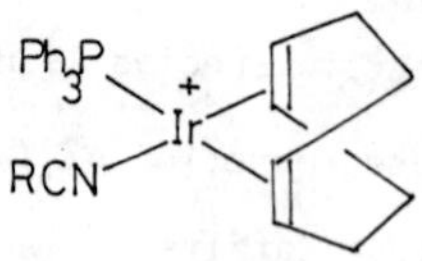

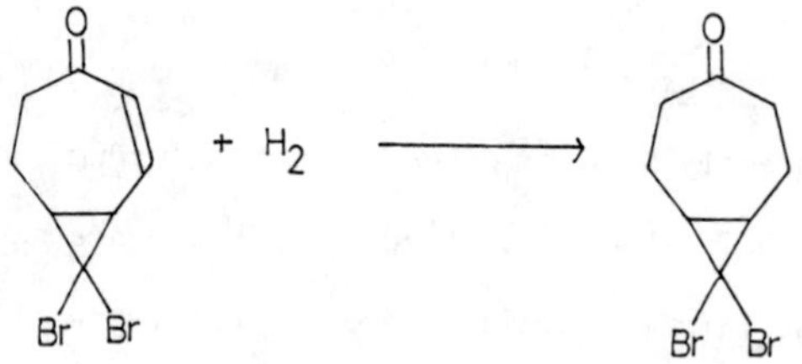

An early and popular catalyst was Vaska's compound, 61, which activates H_2 reversibly[36] at 25°C, 1 atm.[124,151,152] The mechanism for hydrogenation, however, follows an unsaturate pathway with alkene addition rate determining.[153] Reversible, stepwise transfer of hydrogen to substrate is indicated by extensive isomerisation.[154-156] Alkenes are reduced at 1 atm./40-60°C with a rate depending on steric factors (Table 2).[157] The variation of rate with phosphine depends on both steric and electronic effects.[158-161] Analogous cis-complexes such as 63 show substantially similar behaviour.[162,163]

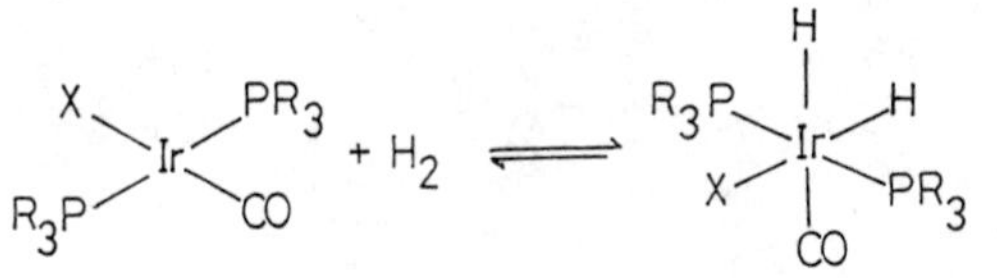

61   62   63

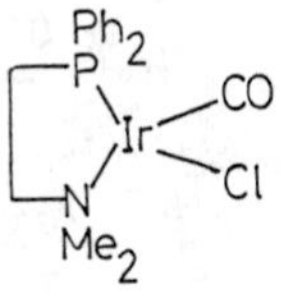

Table 2.   Rates of hydrogenation in the presence of Vaska's compound[a]

Substrate	Rate(mol l^{-1} min^{-1} x 10^{-3})
1-heptene	8.93
cis-2-heptene	0.97
trans-2-heptene	0.55
trans-3-heptene	0.72
cycloheptene	4.50

(a)   Catalyst 2 x 10^{-3} M, substrate 0.8M, toluene,
1 atm., 80°C

The related complexes $HIr(CO)(PPh_3)_3$ [120] and $HIrX_2L_2$ (X = halogen,
L = R_3P, R_3As) are also active for alkene reduction.[153,165-167]   Synthetic
applications are not extensive but the mechanism has been shown to involve
64 to 65.[153,167,168]   Reports of other alkene reductions using iridium
complexes are sparse (Figure 14).

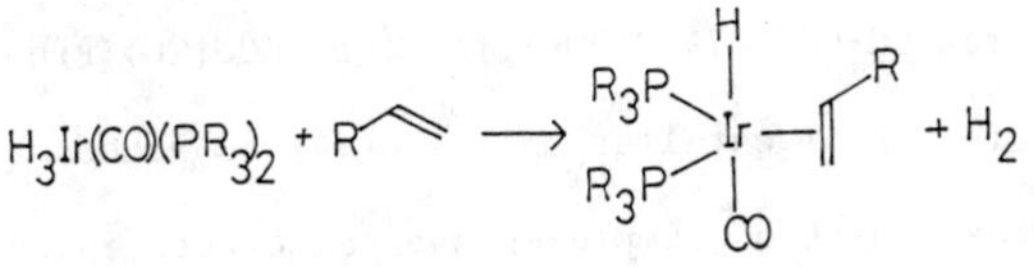

64   65

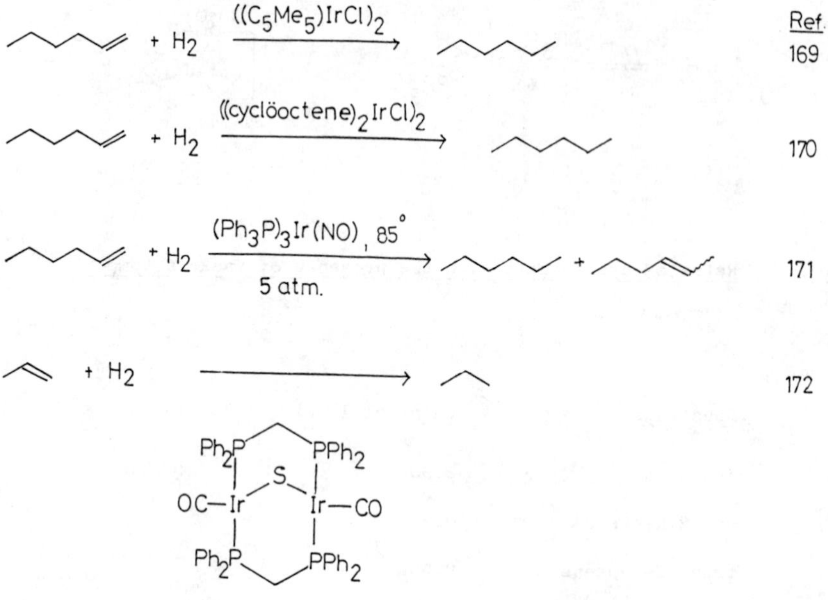

$$\text{alkene} + H_2 \xrightarrow{((C_5Me_5)IrCl)_2} \text{alkane} \qquad \text{Ref.} \quad 169$$

$$\text{alkene} + H_2 \xrightarrow{((\text{cyclöoctene})_2 IrCl)_2} \text{alkane} \qquad 170$$

$$\text{alkene} + H_2 \xrightarrow[\text{5 atm.}]{(Ph_3P)_3Ir(NO), 85°} \text{alkane} + \text{alkene} \qquad 171$$

$$\text{alkene} + H_2 \longrightarrow \text{alkane} \qquad 172$$

Figure 14   Reduction of alkenes in the presence of iridium complexes.

2.2.3   Cobalt complexes

Complexes of cobalt, the third member of the rhodium triad, are used as reduction catalysts, but are mostly selective for conjugated systems. $Co_2(CO)_8$ and derivatives are best known as hydroformylation catalysts but hydrogenation is also possible under specific conditions.   Increased alkyl substitution promotes hydrogenation over hydroformylation[173] but under a pure H_2 atmosphere decomposition to metal occurs.

Phosphine substituted analogues are more successful.   Cyclohexene is cleanly reduced in the presence of $HCo(CO)_3PBu_3$,[174] $HCo(CO)_2(PBu_3)_2$[175] and $[Co(CO)_2(PBu_3)_3]_3$.[176]   Except for the trimer, selection for 1-alkenes is good as it is also for $CH_3Co(CO)_2(P(OMe)_3)_2$.[177] $HCo(CO)(PPh_3)_3$, unlike its rhodium analogue, is a poor catalyst unless Et_3Al is added.

Cobalt hydride phosphine complexes have also been investigated (Table 3).   Kinetic studies show that many species are present under

Table 3.   Hydrogenation by cobalt hydride phosphine complexes

Catalyst	Substrates	Conditions	Ref.
$H_3Co(PPh_3)_3$	C_2H_4, cyclohexene	1 atm., 25°	178,179
$HCo(N_2)(PPh_3)_3$	1-butene, cyclohexene	1 atm., 25° a	180
$HCo(P(OAr)_3)_3$	1-butene	30p.s.i., 25°	181
$H_2((PhO)_3P)_3Co-P(OPh)_2$	1-butene	30p.s.i., 25°	181

(a)   Extensive isomerisation also occurs

catalytic conditions with the rate-determining step being $H_3Co(PPh_3)_2$(alkene)
to $H_2Co(PPh_3)_2$ alkyl.   The cobalt analogue of Wilkinson's catalyst,
$(Ph_3P)_3CoCl$ is very much less active because of its reluctance to undergo
oxidative addition.   1-Alkenes are reduced but under forcing conditions.[182]
Cobalt based Ziegler catalysts are well-known; they are stable and reusable
but relatively unreactive.[183,184]

2.2.4   Ruthenium complexes

Activation of H_2 by Ru(II) and Ru(III) phosphine complexes was
discussed in section 2.1. $(Ph_3P)_4RuCl_2$ and $(Ph_3P)_3RuCl_2$ both yield
$HRuCl(PPh_3)_3$ with hydrogen at 25°C.[185]   This is both reactive and selective
for 1-alkenes, the rate of reaction for cis- and trans-2-hexene being less
than 3×10^{-4} of that for 1-hexene,[47] with no measureable isomerisation.[186]
A range of arsine, stibine and sulphur ligands may replace PPh_3,   some of
them giving improved results for cyclooalkenes.[187]   Early studies of the
mechanism showed a simple picture (equations (16)-(18))[46]

$$HRuCl(PPh_3)_3 \rightleftharpoons HRuCl(PPh_3)_2 + PPh_3 \qquad (16)$$

$$HRuCl(PPh_3)_2 + alkene \rightleftharpoons Alkyl\ RuCl(PPh_3)_2 \qquad (17)$$

$$Alkyl\ RuCl(PPh_3)_2 + H_2 \rightarrow HRuCl(PPh_3)_2 + alkane \qquad (18)$$

There is considerable uncertainty as to the number of phosphines present, however, since the equilibrium constant for (16) is too low to measure.[48,188] An alternative, involving triphosphines, is given by equations (19)-(22). Stoichiometric reduction can occur using $HRuCl(PPh_3)_3$ in the absence of H_2,

$$HRuCl(PPh_3)_3 + alkene \rightleftharpoons HRuCl(PPh_3)_2alkene + PPh_3 \qquad (19)$$

$$HRuCl(PPh_3)_2alkene \rightleftharpoons Alkyl\ RuCl(PPh_3)_2 \qquad (20+$$

$$Alkyl\ RuCl(PPh_3)_2 + PPh_3 \rightleftharpoons Alkyl\ RuCl(PPh_3)_3 \qquad (21)$$

$$Alkyl\ RuCl(PPh_3)_3 + H_2 \rightarrow HRuCl(PPh_3)_3 + alkane \qquad (22)$$

the additional hydrogen coming from <u>ortho</u>-metallation to give $[(Ph_3P)_2ClRu(\underline{o}\text{-}C_6H_4PPh_2)]^+$. The <u>ortho</u>-metallated complex reacts with H_2 to regenerate $HRuCl(PPh_3)_3$ but this is not the preferred pathway under ordinary catalytic conditions.

Related ruthenium complexes abound. $HRu(OCOR)(PPh_3)_3$ also gives high selectivity for 1-alkenes, the nature of R not affecting the rate greatly.[189,190] Some other examples are shown in Figure 15. The analogues give rise to similar active species. For example, $Ru(OAc)_2(PPh_3)_2$ gives a catalyst on protonation[195] and photolysis of $HRuCl(CO)(PPh_3)_3$ yields $HRuCl(PPh_3)_3$.[196]

$H_2Ru(PPh_3)_4$[197] and $H_4Ru(PPh_3)_3$[198] are also catalysts for alkene reduction. The paths of operation proposed for them are related as shown in Figure 16.[199] Their relative importance depends on the strength of Ru-alkene bonding. The catalyst is selective for 1-alkenes and isomerisation occurs by an alternative <u>orthometallation</u> pathway. Related species include $[Ru(PPh_3)_3]^{2+}$[200] and <u>cis</u>-$H_2Ru(PF_3)_2(PPh_3)_2$ which gives less isomerisation.[40] Other examples are shown in Figure 17.

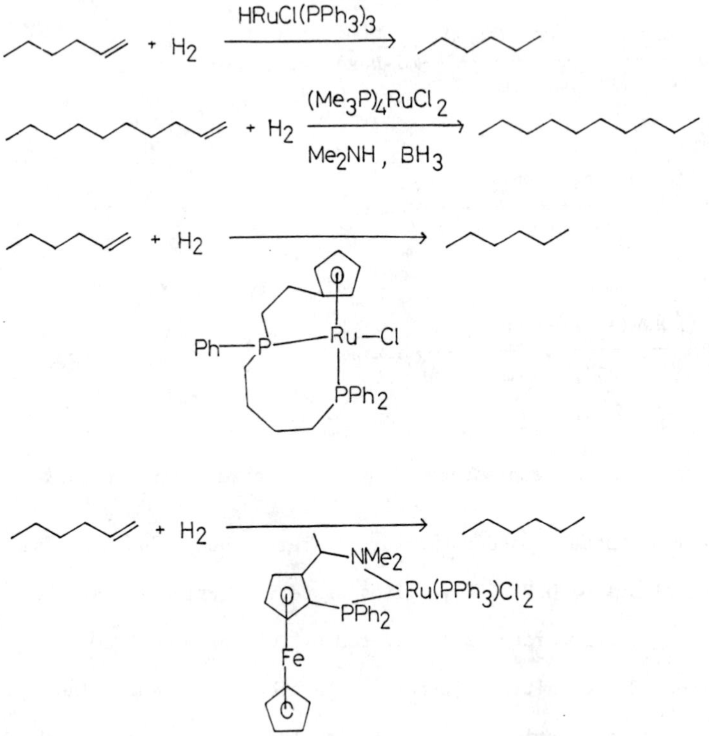

Ref.

191

192

193

194

Figure 15 Hydrogenation in the presence of ruthenium complexes

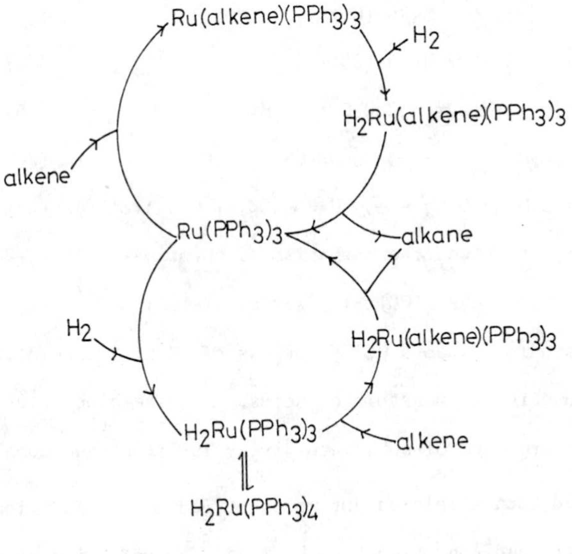

Figure 16 Mechanism of hydrogenation in the presence of $H_2Ru(PPh_3)_4$

Ref.

$$1\text{-hexene} + H_2 \xrightarrow[\text{45}^{\circ}, \text{ 1 atm}]{\text{cis-}H_2\text{ Ru (CO)}_2\text{(PPh}_3\text{)}_2} \text{2-hexenes, hexane}$$ 201

$$1\text{-hexene} + H_2 \xrightarrow[\text{20}^{\circ}, \text{ 1 atm}]{\text{HRu(BH}_4\text{) (PPh}_3\text{)}_3} \text{hexane}$$ 202

$$1\text{-hexene} + H_2 \xrightarrow[\text{25}^{\circ}, \text{ 1 atm}]{\text{HRu(BH}_4\text{) (PMePh}_2\text{)}_3} \text{hexane}$$ 203

<u>Figure 17</u> Hydrogenation in the presence of halide free ruthenium complexes

Recent reports have used ruthenium cluster carbonyls as catalyst precursors. Photolysis of $H_4Ru_4(CO)_{12}$ gives a species active for ethylene reduction.[204] The reaction rate is unaffected by CCl_4 or O_2 but CO depresses the rate and the initial cluster may then be recovered. The initial step is photodissociation of CO [equation (23)], well-known in the substitution of such complexes.[205] A combination of kinetic and deuterium

$$H_4Ru_4(CO)_{12} \xrightarrow[\text{heptane}]{h\nu} H_4Ru_4(CO)_{11} \qquad (23)$$

$$H_4Ru_4(CO)_{11} + H_2 \rightleftharpoons H_6Ru_4(CO)_{11} \qquad (24)$$

$$H_4Ru_4(CO)_{11} + C_2H_4 \rightleftharpoons H_3Ru_4(CO)_{11}C_2H_5 \qquad (25)$$

$$H_3Ru_4(CO)_{11}C_2H_5 + H_2 \longrightarrow H_4Ru_4(CO)_{11} + C_2H_6 \qquad (26)$$

$$H_3Ru_4(CO)_{11}C_2H_5 + C_2H_4 \longrightarrow H_3Ru_4(CO)_{11}(C_2H_5)(C_2H_4) \qquad (27)$$

labelling studies[41,206] showed the mechanism of equations (23) - (27). A related anionic system, $[H_3Ru_4(CO)_{12}]^-$, was also studied.[207]

In early studies aqueous HCl solutions of $[RuCl_4]^{2-}$ were reported to be active for reduction of unsaturated acids.[208] The blue solutions turned yellow in the presence of substrate giving Ru(II) alkene complexes. Simple alkenes formed such complexes but were not reduced. Additions are stereospecifically <u>cis</u> and conditions can be found for reduction of simple alkenes.[209]   The mechanism of action of Ru(arene)Cl$_2$ catalysts is probably similar, involving arene stabilised ruthenium hydrides.[210,211]

2.2.5 Iron and Osmium Complexes

These other members of the ruthenium triad do not offer the same scope in generation of hydrogenation catalysts.   Fe(CO)$_5$ may be used for reduction with HO$^-$/H$_2$O/MeOH but the products are homologated alcohols rather than alkanes.[212]   The reaction is well studied but falls outside the scope of this chapter.[213]   Various phosphine substituted versions of H$_2$Fe$_2$(CO)$_9$ NSiMe$_3$, after photolysis, catalyse the slow reduction of alkenes by molecular H$_2$.[214]   The cluster Fe$_4$S$_4$ is a model for the active site of bacterial hydrogenase and treatment of Fe$_4$S$_4$Cl$_4$(NBu$_4$)$_2$ with PhLi gives a catalyst for hydrogenation of 1-alkenes with good selectivity and little isomerisation.[215]

Osmium complexes have been similarly neglected.   HOsCl(CO)(PPh$_3$)$_3$ is active for reduction of acetylene to ethylene and ethane at 60°C/1 atm[165] and the corresponding bromide is active in hydrogenation of 1-alkenes and cycloalkenes, despite considerable competitive isomerisation.[216] HOsCl(PPh$_3$)$_3$[217], HOsCl$_2$(PPh$_3$)$_3$, H$_3$Os(PPh$_3$)$_3$ and H$_4$Os(PEtPh$_2$)$_3$[218]   are also active for 1-hexene reduction but isomerisation is also significant in all cases.

2.2.6 Nickel, Palladium and Platinum complexes

There has been relatively little work on the use of nickel complexes as hydrogenation catalysts.   [Ni(CN)$_4$]$^{2-}$/BH$_4$$^-$ hydrogenates monoenes and dienes but the complex must be in excess.   (Ph$_3$P)$_2$NiX$_2$ complexes catalyse reduction of 1-octene at 25°/1 atm.H$_2$ but much more slowly than palladium or platinum analogues.[219]   Addition of LiBH$_4$ enhances the rate.[220]   Nickel clusters have been reported to catalyse reduction of 1-alkenes[221] and alkynes but have not been further studied.[222]   There are numerous studies of nickel based Ziegler catalysts but they have not become widely popular.[223,224]

For palladium, early studies used $PdCl_2$ and $[PdCl_4]^{2-}$[225], though they are not very suitable for reduction of simple alkenes. $PdCl_2(DMSO)_2$ catalyses reduction and isomerisation of 1-alkenes but since internal alkenes are also reduced, alkanes are ultimately obtained.[226] Complexes such as 66 are rather ineffective alone but on addition of $SnCl_2$ a better catalyst is obtained.[227] 1-Octene is reduced (and isomerised) at 1 atm. $H_2/20^\circ C$.[219] An alternative method of activation is treatment with $AgBF_4$ in a coordinating solvent to give 66 (X = acetone). 1-Alkenes are then reduced under very mild conditions.[228] Allyl complexes such as 67 may be used after activation with $NaBH_4$.[229]

66   67

The hydrogenase models such as Pd(SALEN), 37, catalyse clean reduction of 1-alkenes to alkanes.[44] Palladium clusters have also been employed (Table 4); their mode of action is unknown but kinetic studies suggest that H_2 is activated by a heterolytic route.

$[PtCl_4]^{2-}$ solutions in DMF or DMA may be used for reduction of dicyclopentadiene (1 atm.,$20^\circ C$); they were not poisoned by thiophene indicating that the system is homogeneous.[232] With other substrates, however, metal is deposited. Zeise's dimer, $[PtCl_2(C_2H_4)]_2$ suffers from the same problem but conditions were found for ethylene reduction.[233] An important advance was the discovery that methanol solutions of H_2PtCl_6 gave a catalyst for ethylene reduction on treatment with $SnCl_2$.[234] Maximum activity is obtained for Sn:Pt > 5, the tin stabilising the system against reduction to metal. Variations in reaction parameters have been extensive[235-241] but all give qualitatively similar results. Both cis &

Table 4.   Palladium clusters as hydrogenation catalysts

Catalyst	Substrate	Comments	Ref.
$(Ph_3P)_2Pd_2$	cyclopentene	$(PhP)_2Pd_5$ is the true catalyst	230
$(Ph_3P)_2(Ph_2P)_2Pd_3Cl$	1-alkenes	–	231
$((Ph_3P)Pd(OAc)_2)_2$	alkenes	Dienes, alkynes, C=N, RNO_2 also reduced. $(PhP)_2Pd_5$ probably the true catalyst	45

trans and positional isomerisation precede reduction and in some cases are the sole reaction.   Proposed active species include $[Pt(SnCl_3)_5]^{3-}$ and $[HPtCl_2(SnCl_3)]^{2-}$.

$(Ph_3P)_2PtCl_2$ reacts with H_2 to yield $HPtCl(PPh_3)_2$ under severe conditions.[242]   Alkenes react reversibly with the hydride but catalytic reduction is disfavoured. $(Ph_3P)_2Pt(OCOCF_3)_2$ in TFA is reported to reduce 68, 69 and 70 in respectively 35%, 45% and 84% yield, but the mechanism is different, involving addition of TFA followed by hydrogenolysis of the trifluoroacetate.[243]   71 is reduced to 72 in the presence of $[PtCl_2(H_2O) P(C_6H_5)_3]$.

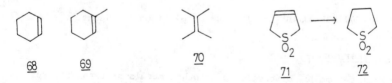

68   69   70   71   72

Promotion of the activity of H_2PtCl_6 with $SnCl_2$ led to the use of the same technique with $(R_3P)_2PtX_2$.   This yields initially $(R_3P)_2PtX(SnCl_3)$ where $X = Cl$,[244] $SnCl_3$,[234] H[242] or Ar.[245]   Hydrogen addition is enhanced and $HPt(SnCl_3)(PPh_3)_2$ is isolated from $(Ph_3P)_2PtCl_2/SnCl_2$ under H_2.   The results are qualitatively similar to those for H_2PtCl_6, with isomerisation as a major problem.[246] Analogous systems (which also isomerise) derive from $(Ph_3P)Pt(CO)Cl_2$[247] and $(Ph_3P)_3Pt$.[248] Some more exotic examples are shown in Figure 18.

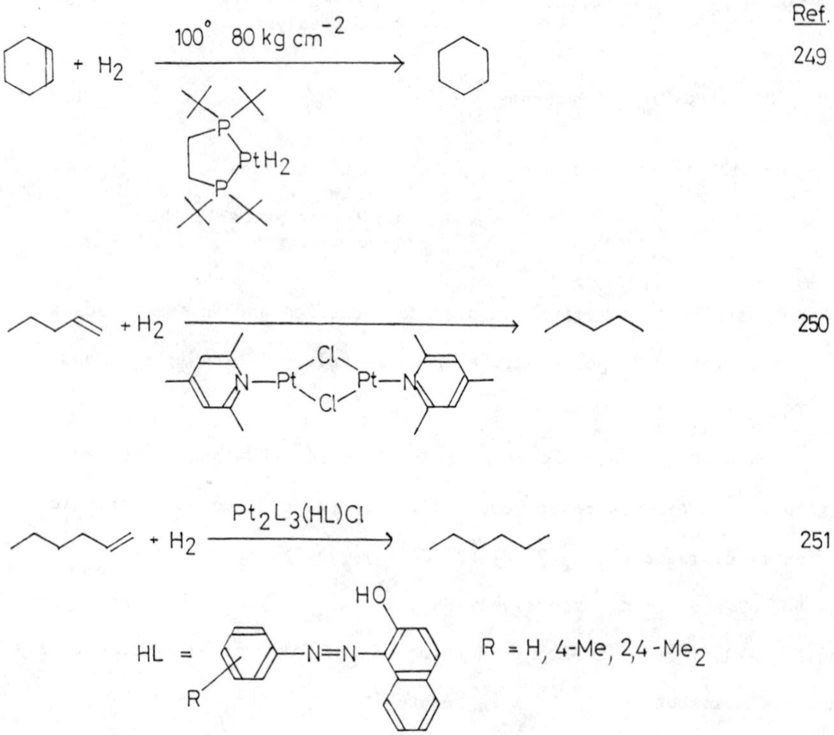

Ref. 249

Ref. 250

Ref. 251

Figure 18 Hydrogenation in the presence of platinum complexes.

2.2.7 Other Metals

Whilst examples of complexes of other metals as catalysts for alkene hydrogenation have been reported, most have not been systematically studied or widely used.

Titanium complexes have a long history as hydrogenation catalysts but much of the work relates to the Ziegler systems such as Cp_2TiCl_2/Et_3Al,

$Cp_2TiCl_2/h\underline{\nu}$, $Ti(O\underline{i}-Pr)_4/BuLi$, $Cp_2TiCl_2/BuLi/PBu_3$, $(Cp_2TiX)_2/LiAlH_4$, and $Ti(OR)_4/R_3Al$.[252-256] The reaction mechanism is not well understood and a full discussion will be deferred to Chapter 8. $Cp_2ZrCl_2/\underline{i}-Bu_2AlH$ is less efficient.[257]

Chromium, molybdenum and tungsten complexes have provided catalysts active for the reduction of conjugated and non-conjugated dienes but they are usually inactive for simple alkenes. In the next triad $Mn_2(CO)_{10}$ catalyses hydrogenation of 1-alkenes ($150^{\circ}C$, 200 atm H_2) but 2-alkenes react more slowly.[258] $(\pi-Allyl)Mn(CO)_4/PR_3$ gives an active, but short lived, catalyst for reduction of 1-alkenes under ambient conditions.[38] Despite the many known multihydride rhenium complexes they have not been used in catalysis. $ReCl_5/SnCl_2$ in polar solvents is reported for hydrogenation of 1-alkenes.[259] Patents briefly describe the use of $[Re_2(OCOR)_{4-n}]^{n+}$[260] and the recent preparation of $[HRe(MeCN)_4(PPh_3)_2]^{2+}$ may foreshadow some developments in this area.[261]

2.3 Reduction of Functionalised Alkenes

A study of the reduction of simple alkenes is essential to the understanding of the mode of action of catalysts but the transformations are rarely synthetically interesting, nor do they show the scope and limitations of the catalyst.

2.3.1 Styrenes

Most catalysts active for the reduction of 1-alkenes catalyse reduction of styrene (Table 5). With Wilkinson's catalyst styrene is reduced more rapidly than 1-alkenes, 2-6 times more rapidly than 1-hexene, for example.[262] There are few data on rates for substituted styrenes; both para-F and para-OMe increase the rate.[62,262] With $HRh(CO)(PPh_3)_3$, $(Ph_3P)_2Rh(CO)Cl$, $[L_2Rh(diene)]^+$ or $(Ph_3P)_2Ir(CO)Cl$, styrene reacts a little more slowly than 1-alkenes but the conflicts in the available data do not allow firm conclusions to be drawn.[97,120,121,157]

Table 5.   Catalysts for the Reduction of Styrene

Catalyst	Ref.
$(Ph_3P)_3RhCl$	62,63,91,94,99,262,263
$HRh(CO)(PPh_3)_3$	120,121
$(Ph_3P)_2Rh(CO)Cl$	157,160
$[L_2Rh(diene)]^+$	97
$((\underline{o}\text{-tolyl})_3P)_2RhCl_2/Et_3Al$	141
$(quinone)Rh(NO)(PPh_3)_2$	63,94,265
$HRh(DBP)_4$	111
$HIr(CO)(PPh_3)_3$	63,94,120,121
$(Ph_3P)_2Ir(CO)Cl$	157,159
$(Ph_3P)_3IrCl/H_2O_2$	266
$(Ph_3P)_3IrCl$	63,264
$(Ph_3P)_3Ir(NO)$	63
$Co_2(CO)_8$	267
$[Co(CN)_5]^{3-}$	55,268-270
$(Ph_3P)_3CoN_2$	271
$CH_3Co(CO)_2(P(OMe)_3)_2$	177
$H_2Ru(PPh_3)_4$	199
$HRu(NO)(PR_3)_3$	272
$H_2PtCl_6/SnCl_2(1:12)$	273
$\underline{cis}\text{-}PtCl_2(SR_2)(PPh_3)/SnCl_2$	247
$(2,4,6\text{-trimethylpyridyl})_2Pt_2(\underline{\mu}\text{-Cl})_2$	250
$[(diphos)Pd(acetone)_2]ClO_4$	228
$(diphos)PtCl(DMF)$	274
$MCl_3/Et_3Al(M=Fe,Co,Ni)$	275
$M(acac)_n R_n AlX_{3-n}(M=Fe,Cr,Ni)$	276
$Na(MeOCH_2CH_2O)_2AlH_2/Co(acac)_2$	184
Cp_2ZrH_2	257
$Cp_2Ti(CO)_2$	277,278

Styrene as substrate allows us to use some catalysts which are inactive for simpler alkenes. $[Co(CN)_5]^{3-}$ is useful for all types of conjugated double bond; in addition to styrene $\underline{\alpha}$-methyl and $\underline{\alpha}$-methoxy styrenes are reduced,though the $\underline{\beta}$-substituted compounds are inert.[269] $Co_2(CO)_8$ can be used to reduce simple alkenes but hydroformylation is the main reaction.   With styrene or 1-vinylnaphthalene appreciable amounts of ethylarene are obtained under hydroformylation conditions.[279]   In the reduction of α-methylstyrene the yield of iso-propylbenzene depends on solvent polarity with the most polar solvent, MeOH favouring hydrogenation by up to 93%.[280] $Cp_2Ti(CO)_2$ catalyses slow reduction of styrene, but not 1-alkenes.[277,278] $H_2PtCl_6/SnCl_2$ catalyses hydrogenation of 1-alkenes but is impractical due to the extensive isomerisation.   Styrene, unable to isomerise, gives satisfactory results.[273]

2.3.2 αβ-Unsaturated aldehydes

It is generally easier to hydrogenate carbon-carbon than carbon-oxygen double bonds, so the transformation 73 to 74 might be expected to be achieved in the presence of diverse catalysts.   Unfortunately this is not the case, with decarbonylation and overreduction to alcohol being major problems.

$$R\diagup\!\!\diagdown\!\!\diagup CHO \longrightarrow R\diagup\!\!\diagdown\!\!\diagup CHO$$

73   74

Using Wilkinson's catalyst the major side reaction is decarbonylation,[70,71,281]   this being essentially the only reaction for 75, 76 and 77.[83,282]   Propenal is reduced to propanal with only 5% propanol by judicious choice of conditions.   Bromo and iodo analogues given even more decarbonylation.   The use of absolute ethanol as solvent is useful in some cases and cinnamaldehyde is reduced to 78 in up to 60% yield with 40% ethyl benzene.[283]   Increased H_2 pressure enhances hydrogenation at the expense of decarbonylation but also promotes aldehyde reduction; at 50 atm.H_2, 75 suffers less CO loss but 18% 1-butanol is produced.   The complex $[(bipy)Rh(solvent)_2]^+$ is reported to give saturated aldehydes with moderate selectivity.[284]   The use of water gas and $Rh_6(CO)_{16}$ is also reported to be successful; cinnamaldehyde gives 78 in quantitative yield.[285]

75   76   77   78   79

The iridium complexes, $(Cy_3P)_2Ir(CO)Cl$ [286] and $H_3Ir(PPh_3)_3$ [287] have also been used but are not particularly selective.

$[Co(CN)_5]^{3-}$ is known for its specificity towards conjugated double bonds but is not especially efficient for unsaturated aldehydes. 75 is reduced quantitatively at $25°C/1$ atm.,[55,288,289] but 79 requires forcing conditions[290] and cinnamaldehyde gives 80.[268]

$Co_2(CO)_8$ might be expected to supress decarbonylation and reduction is largely successful though conditions are severe.   Some examples using this or substituted analogues are shown in Figure 19.

Although $(Ph_3P)_3RuCl_2$ does not catalyse reduction of propenal[282] $K+[(Ph_3P)_2Ph_2PC_6H_4-o-RuH_2]^-$ gives up to 75% selectivity for propanal at $100°C/690$ pKa.[295] $HOs(CO)(PPh_3)_3Br$ gives poor reduction of 75 with only 20% butanal formed, the other products being butanol and 81.[216]

Two palladium catalysts appear promising but have not been widely exploited.   An unknown active species is formed from $PdCl_2/NaBH_4$ and gives good selectivity in reduction of 75.   The analogue from $Ni(OAc)_2$ is equally active but less selective.[296] $(DMSO)_2PdCl_2$ catalyses reduction of acrolein to propanal without further hydrogenation whereas the rhodium analogue is rapidly poisoned by decarbonylation.[297] Cp_2MoH_2 catalyses reduction of 75 with good selectivity.[298]

80   81

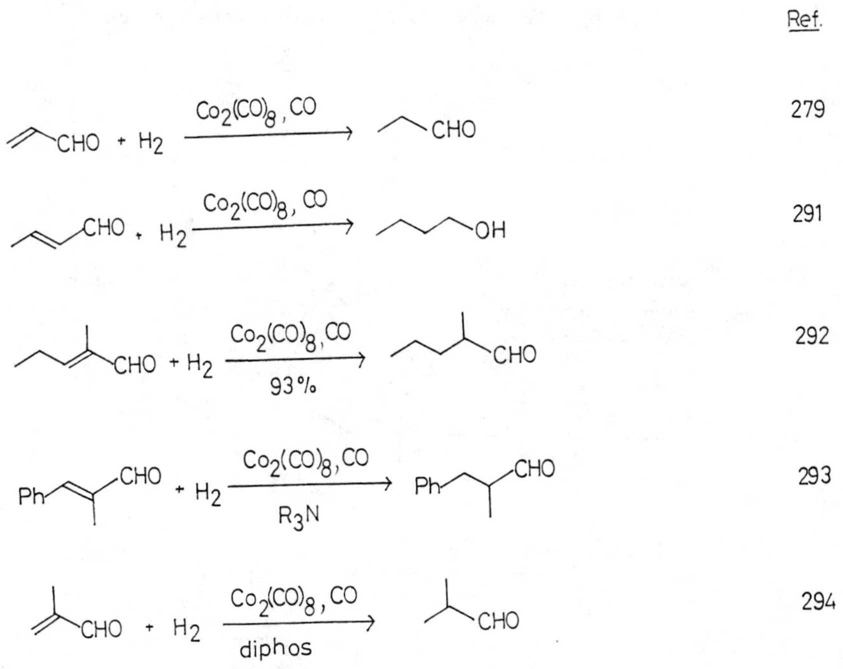

Ref.

$Co_2(CO)_8$, CO
CHO + H_2 $\longrightarrow$ CHO 279

$Co_2(CO)_8$, CO
CHO + H_2 $\longrightarrow$ OH 291

$Co_2(CO)_8$, CO
CHO + H_2 $\xrightarrow{\quad\quad}$ CHO 292
93%

$Co_2(CO)_8$, CO
Ph CHO + H_2 $\xrightarrow{\quad\quad}$ Ph CHO 293
R_3N

$Co_2(CO)_8$, CO
CHO + H_2 $\xrightarrow{\quad\quad}$ CHO 294
diphos

__Figure 19__   Hydrogenation in the presence of cobalt carbonyl complexes

2.3.3   αβ-Unsaturated Ketones

Since αβ-unsaturated ketones offer no ready pathway for decarbonylation the number of catalysts suitable for their reduction is enhanced.
Indeed, because of their greater affinity for metals they are reduced more
rapidly than simple alkenes.   Methyl vinyl ketone is reduced to 2-butanone
at 25°C/1 atm. in the presence of $(Cy_3P)_2Ir(CO)Cl$,[299] $[HCo(CN)_5]^{3-}$ [259],
$[L_2Rh(diene)]^{+}$ [25] and $HRh_2(N\text{-phenylanthranilate})_2Cl$.[300] $Co_2(CO)_8$,[279,292]
$Co_2(CO)_8/R_3P$ [294] and $H_3Co[P(O\text{-}\underline{i}\text{-}Pr)_3]_3$ [301]   require more severe conditions.
Mesityl oxide, __82__, is another popular substrate, though steric constraints
mean that its hydrogenation is slower.   Some examples are shown in Figure
20 and $H_2Rh_2(N\text{-phenylanthranilate})_2Cl$ and $NiCl_2/NaBH_4/DMF$ [302] are also active.
$(Ph_3P)_3RhCl$ is a poor catalyst deactivating at 70% conversion in toluene.

Adding H_2O_2 gives a more active, longer-lived, heterogeneous species.[305]

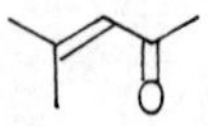

82

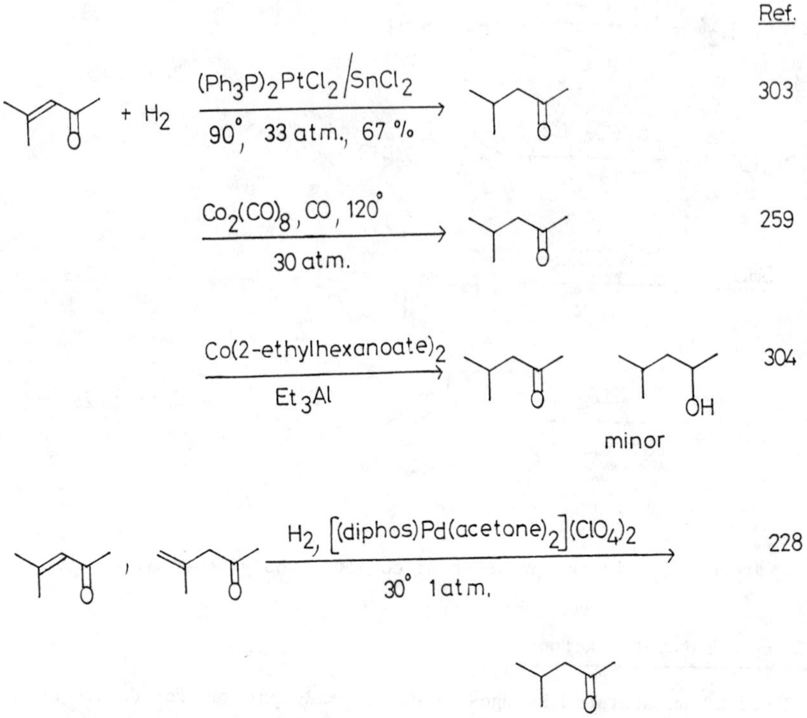

Ref.

303

259

minor

304

228

Figure 20 Catalytic hydrogenation of mesityl oxide

In some substrates activation by a neighbouring carbonyl group causes preferential reduction of a specific carbon carbon double bond (e.g. 83 to 84[306]).   As previously noted $[Co(CN)_5]^{3-}$ and related catalysts are useful only for double bonds activated by conjugation (Figure 21). $Rh_6(CO)_{16}$ is also only useful for activated double bonds.[285]   Other observed selectivities are steric in origin with the least substituted double bond being reduced as for 85[307,] and 87, santonin.[308]   Similarly in the presence of $(Ph_3P)_3RhCl$ 89 is reduced but 90 is not.[283]   By combining

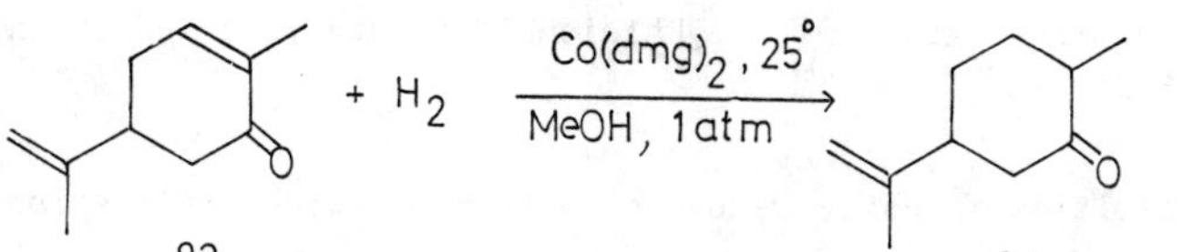

83

84

cis : trans = 3:7

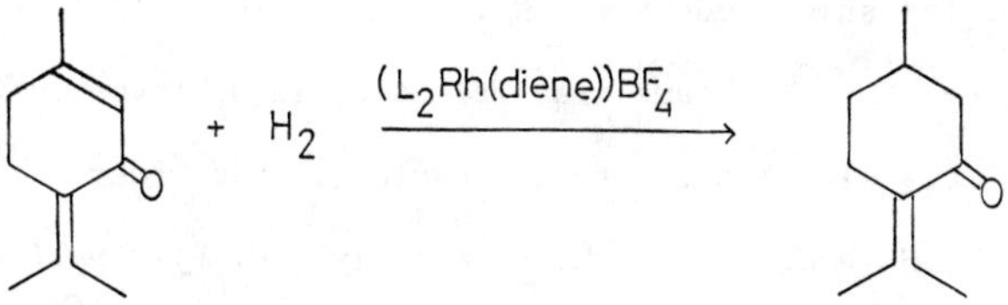

85

86

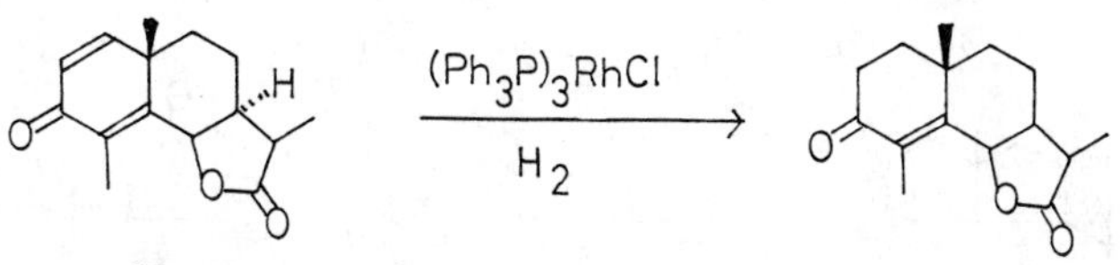

87

88

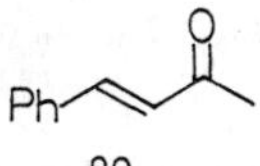

89

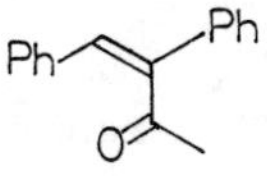

90

Ref.

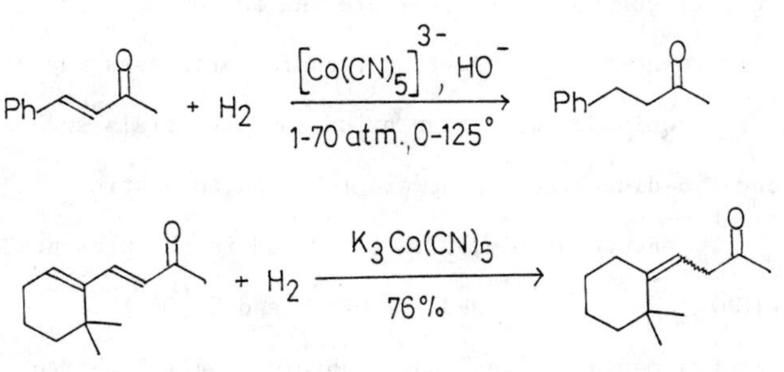

288, 290

289

<u>Figure 21</u> Hydrogenation of enones catalysed by cobalt complexes

reduction with the water gas shift $[L_2Rh(diene)]^+$ catalyses reduction of 89[309] with H_2O/CO.

Selectivities in enone reduction have found wide application in steroid chemistry. For example, androsta-1,4-diene-3,17-dione, 91, is reduced to 92 with good selectivity in the presence of $(Ph_3P)_3RhCl$.[66,310] The 4,6-diene gives the same product which is reduced more slowly to androstane-3,17-dione.[311] The same transformation is achieved in the presence of $(R_3P)_3RuCl_2$ with reactivity increasing in the order $R = \underline{p}-MeOC_6H_4 > \underline{p}-MeC_6H_4 > C_6H_5$.[312] More recently $[(Cy_3P)Ir(py)(cod)]^+PF_6^-$ has been used; 92 is obtained rapidly and the 5-α-androstane on exhaustive reduction.[311] The importance of the carbonyl group is emphasised by the

91 92

resistance to reduction of steroidal 4-enes lacking this feature. Trisub- stituted 4,5-double bonds are also difficult to reduce using $(Ph_3P)_3RhCl$ and $RhCl_3(py)_3/NaBH_4/DMF$ is a better catalyst.[138] Some further examples are shown in Figure 22.

The final group of enones to consider are the quinones. The mechanism of reduction is unknown since electron transfer processes may be expected to be important. Quinones with high oxidation potentials such as $\underline{\beta}$-naphthoquinones and 2,6-dinaphthoquinone are reported to destroy Wilkinson's catalyst.[313] $\underline{p}$-Benzoquinone, 93, is reduced in the presence of $(Ph_3P)_3RhCl$,[313] $[Co(CN)_5]^{3-}$[55,269,314], $PdCl_2/DMF$[232] and $Cu(OAc)_2$.[315] The products are usually hydroquinone, 94, and quinhydrone, a 1:1 charge transfer complex between 93 and 94. The substituted quinone, 95, is reduced at the less hindered double bond and naphthoquinones 97 and 99 (juglone) are also reduced.[312]

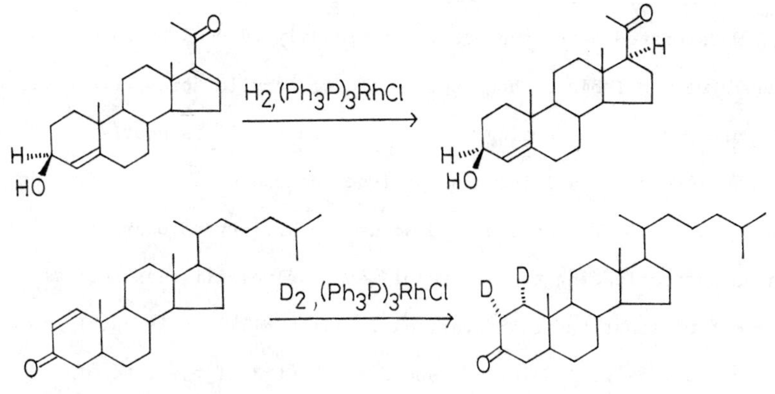

83

66,
311

Figure 22   Hydrogenation of steroids

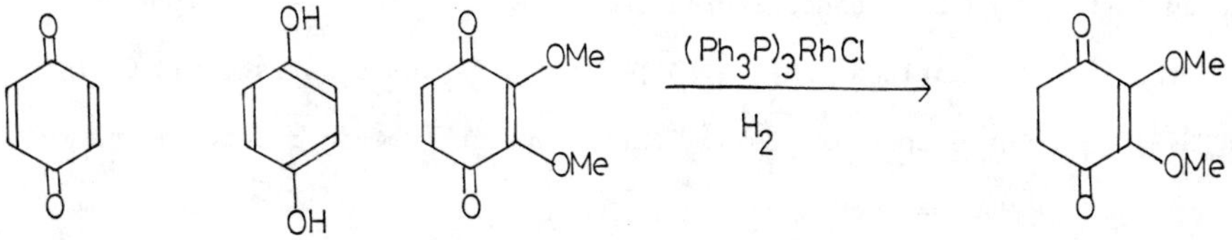

93   94   95   96

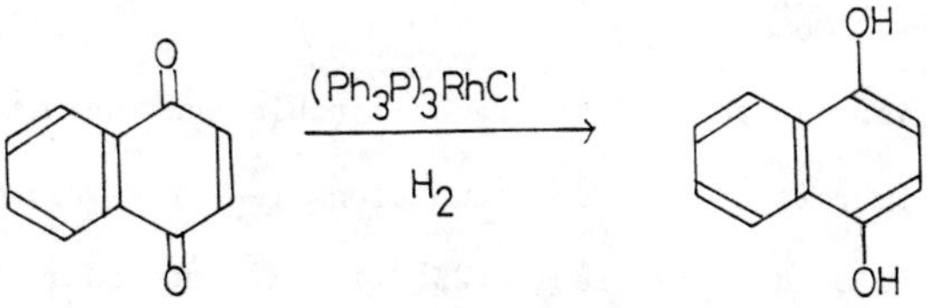

97   98

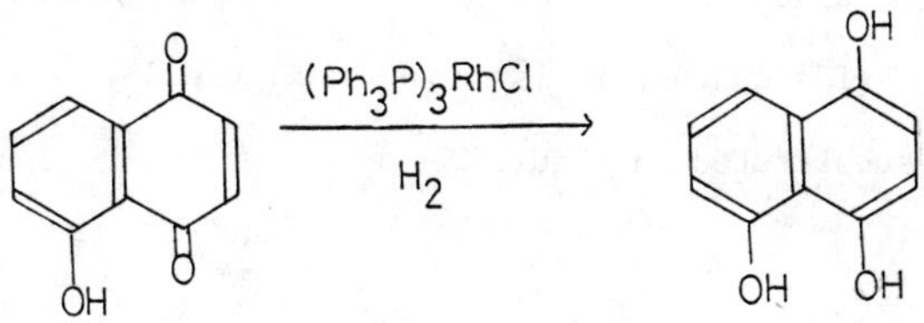

99   100

2.3.4 Other unsaturated carbonyl compounds

αβ-Unsaturated acids and esters are readily reduced to the saturated analogues. Table 6 shows some data for acrylic acid derivatives. A number of points are worth noting. Acrylic acid itself is poorly reduced in the presence of Wilkinson's catalyst because of the stability of the complex formed.[321] For unsaturated acids the ability to operate in water is an advantage leading to the use of water soluble ligands such as 46. The effect of substitution is variable. With Wilkinson's catalyst α-methylmethacrylate is successfully reduced at about half the rate of cyclohexene.[62] A similar result is observed for $(Ph_3P)_2Rh(CO)Cl$ and ethyl acrylate.[157] In the presence of $[Co(CN)_5]^{3-}$ α-methylmethacrylate is reduced faster than the unsubstituted compound.[288] This is explained by the mechanism of equations (28) to (31) where S is the substrate and Co is $[Co(CN)_5]^{3-}$. The σ-complex, CoSH, (101) plays no part in reduction and

$$2Co + H_2 \rightleftharpoons 2CoH \qquad (28)$$

$$CoH + S \longrightarrow Co + HS^{\cdot} \qquad (29)$$

$$HS^{\cdot} + CoH \longrightarrow Co + H_2S \qquad (30)$$

$$HS^{\cdot} + Co \rightleftharpoons CoSH \qquad (31)$$

with α-unsubstituted acrylic acids it is rather stable and inhibits reduction.[322] This is reflected in the conditions required for hydrogenation. Methacrylic acid is reduced at $25^{\circ}C/1$ atm. but under these conditions acrylic acid gives largely the hydrodimer α-methylglutaric acid. A clean reduction is obtained under 30 atm. H_2.[55,268,290] Reduction of α-substituted acrylic acids and esters is shown in Figure 23, β-substituted in Figure 24 and polysubstituted in Figure 25.

101

Table 6. Catalysts for reduction of acrylic acid derivatives

Acrylic acid	Acrylate esters	Acrylamide
$[RuCl_4]^{2-}$ [208]	$(Ph_3P)_3RhCl$ [63,77]	$[RuCl_4]^{2-}$ [316]
$[Co(CN)_5]^{3-}$ [55,62,268]	$(Ph_3P)_2M(CO)Cl$ [M=Rh,Ir] [157,317]	$(Ph_3P)_3RhCl$ [62]
$[RhCl_4]^{3-}$ [316]	$H_3Ir(PPh_3)_3$ [318]	$Rh_6(CO)_{16}$ [285]
$Rh(I)(Et_2S)_n$ [316]	$(Ph_3P)_3IrCl/H_2O_2$ [319]	$H_3Co(P(O-\underline{i}-Pr)_3)_3$ [301]
$H_3Ir(PPh_3)_3$ [318]	$(R_3P)_2Ir(CO)Cl$ [288,317]	
	$HRuCl(PPh_3)_3$ [191]	
	$(benzene)RuCl_2$ [210]	
	$NiCl_2/NaBH_4/DMF$ [320]	
	Cp_2MoH_2 [299]	
	$H_2Fe_2(CO)_9NSiMe_3/R_3P$ [211]	
	$Na(MeOCH_2CH_2O)_2AlH_2/Co(acac)_2$ [184]	

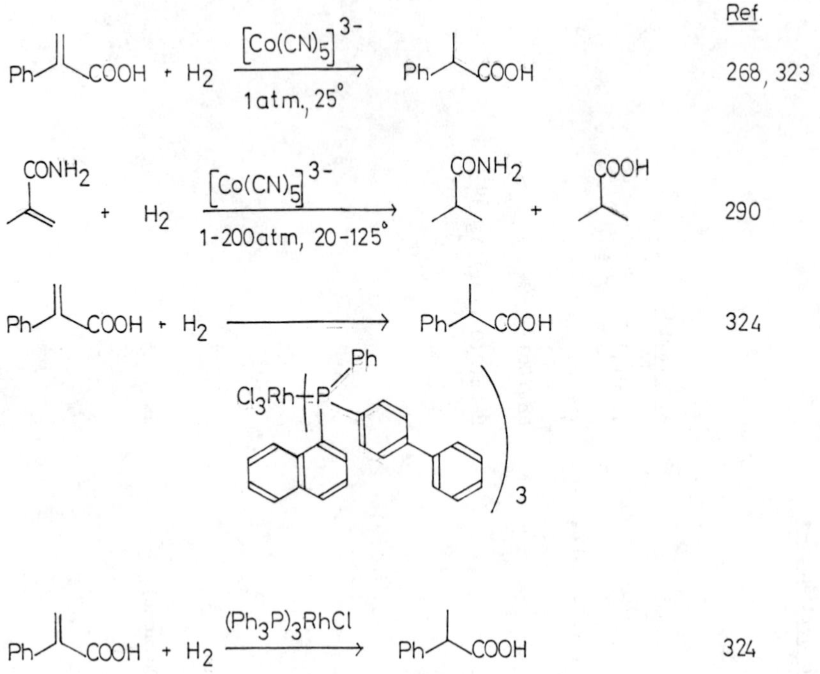

Ref.

$$Ph\text{—}C(=CH_2)\text{—COOH} + H_2 \xrightarrow[\text{1atm., 25°}]{[Co(CN)_5]^{3-}} Ph\text{—}CH(CH_3)\text{—COOH}$$ 268, 323

$$CH_2=C(CONH_2)\text{—} + H_2 \xrightarrow[\text{1-200atm, 20-125°}]{[Co(CN)_5]^{3-}} (CH_3)_2CH\text{—CONH}_2 + (CH_3)_2CH\text{—COOH}$$ 290

$$Ph\text{—}C(=CH_2)\text{—COOH} + H_2 \longrightarrow Ph\text{—}CH(CH_3)\text{—COOH}$$ 324

$$Cl_3Rh\left(P\begin{array}{c}Ph\\ \text{—naphthyl}\\ \text{—biphenyl}\end{array}\right)_3$$

$$Ph\text{—}C(=CH_2)\text{—COOH} + H_2 \xrightarrow{(Ph_3P)_3RhCl} Ph\text{—}CH(CH_3)\text{—COOH}$$ 324

Figure 23 Hydrogenation of α-substituted acrylic acids and esters.

       The relationship between reaction rate and substitution is not easily rationalised. For example, cinnamic acid is better reduced by an in situ iridium phosphine complex whereas a cationic one is better for α-methyl cinnamic. On the other hand ethyl cinnamate forms such a strong complex with the catalyst derived from $[(Ph_2MeP)_2Ir(cod)]^+PF_6^-$ that it is not reduced.[36] 102 is not reduced in the presence of $(Ph_3P)_3RhCl$ for steric reasons.

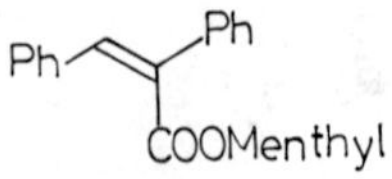

$$Ph\text{—}CH=C(Ph)\text{—COOMenthyl}$$

102

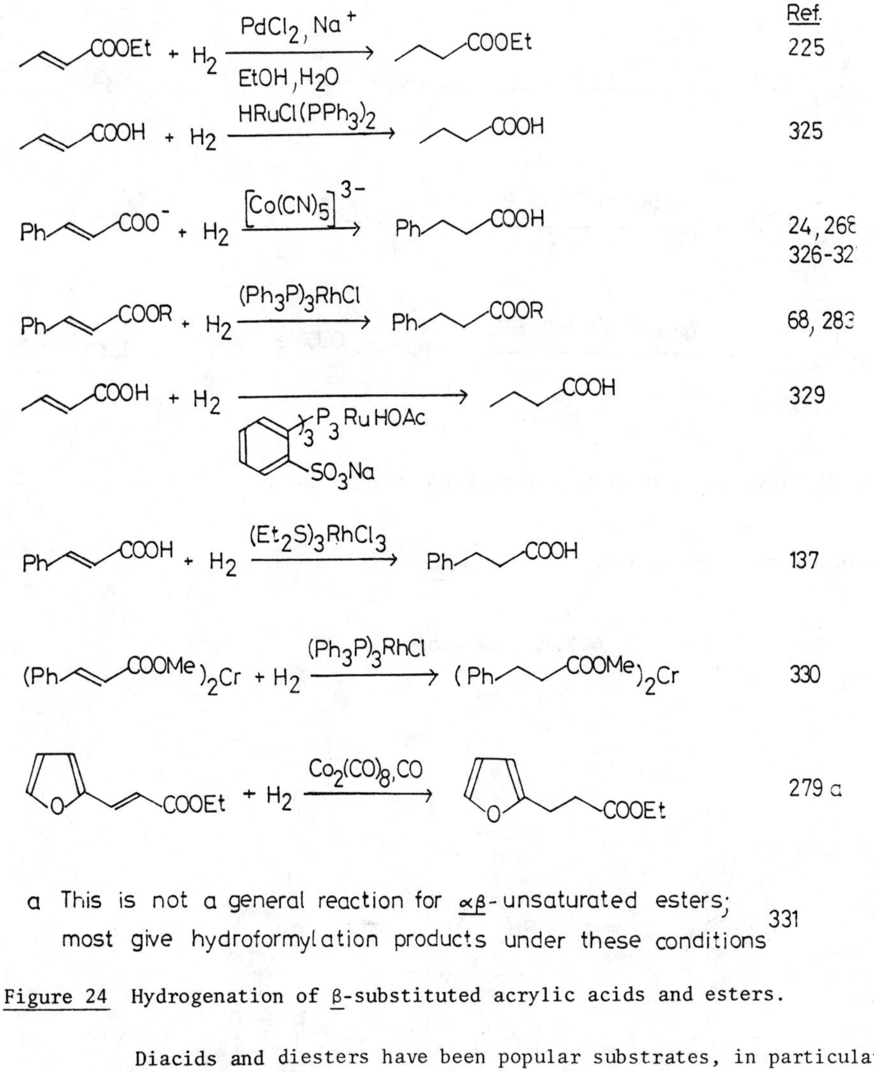

Ref.

$CH_3CH=CH{-}COOEt$ + H_2 $\xrightarrow[\text{EtOH}, H_2O]{\text{PdCl}_2,\ Na^+}$ $CH_3CH_2CH_2{-}COOEt$ — 225

$CH_2=CH{-}COOH$ + H_2 $\xrightarrow{\text{HRuCl(PPh}_3)_2}$ $COOH$ — 325

$Ph{-}CH=CH{-}COO^-$ + H_2 $\xrightarrow{[\text{Co(CN)}_5]^{3-}}$ $Ph{-}CH_2CH_2{-}COOH$ — 24, 268 326-32

$Ph{-}CH=CH{-}COOR$ + H_2 $\xrightarrow{(\text{Ph}_3P)_3RhCl}$ $Ph{-}CH_2CH_2{-}COOR$ — 68, 283

$CH_3CH=CH{-}COOH$ + H_2 $\xrightarrow{\left(\text{C}_6H_4\text{SO}_3Na\right)_3 P_3\ Ru\ HOAc}$ $COOH$ — 329

$Ph{-}CH=CH{-}COOH$ + H_2 $\xrightarrow{(\text{Et}_2S)_3RhCl_3}$ $Ph{-}CH_2CH_2{-}COOH$ — 137

$(Ph{-}CH=CH{-}COOMe)_2Cr$ + H_2 $\xrightarrow{(\text{Ph}_3P)_3RhCl}$ $(Ph{-}CH_2CH_2{-}COOMe)_2Cr$ — 330

furanyl-$CH=CH{-}COOEt$ + H_2 $\xrightarrow{\text{Co}_2(\text{CO})_8,\ CO}$ furanyl-$CH_2CH_2{-}COOEt$ — 279 a

a This is not a general reaction for $\alpha\beta$-unsaturated esters; most give hydroformylation products under these conditions [331]

Figure 24 Hydrogenation of β-substituted acrylic acids and esters.

Diacids and diesters have been popular substrates, in particular maleic and fumaric acids, the deuteration of which to give respectively meso- and dl-dideuterosuccinic has proved rigorously cis-addition for $(Ph_3P)_3RhCl$[68,72], $(py)_2(DMF)RhCl_2(BH_4)$[138], $HRh_2(N\text{-phenylanthranilate})_2Cl$[333], $[RuCl_4]^{2-}$[208], $(Et_2S)_3RhCl_3$[137] and $[Co(CN)_5]^{3-}$[288]. Rather more curious results are obtained with $[RuCl_4]^{2-}$ in DMA for fumaric acid. In addition to dl-d_2,d_3-dideuterosuccinic acid non-symmetric 103 is obtained. This is interpreted in terms of the scheme of Figure 26 where insertion into

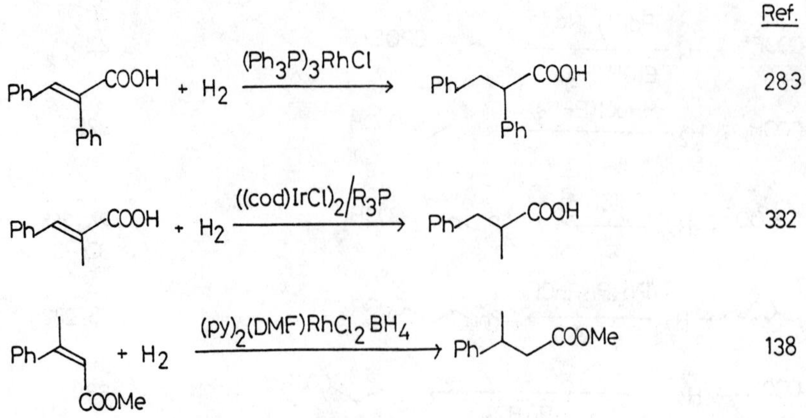

Ref.

Figure 25 Hydrogenation of polysubstituted acrylic acids and esters.

the Ru-H bond is reversible.[334]

$$HOOC-CD_2-CH_2-COOH$$

103

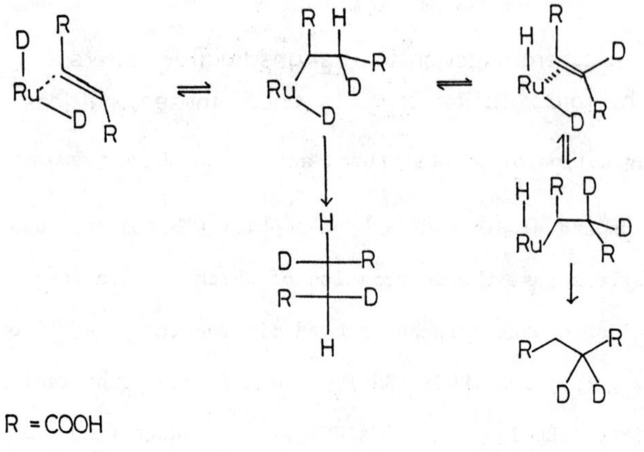

R = COOH

Figure 26 Mechanism of hydrogenation of fumaric acid in the presence of
$[RuCl_4]^{2-}$ /DMA

As expected with Wilkinson's catalyst, diethyl maleate is reduced 20 times faster than the fumarate.[262] For (Et$_2$S)$_3$RhCl$_3$, however, we observe the order cinnamic > fumaric > maleic acid, which is the inverse of the rhodium complex stability order.[137] This is also the case for (Ph$_3$P)$_2$Rh(CO)Cl[157] and (46)$_3$ RhCl.[335] Other catalysts (H$_4$Ru$_4$(CO)$_{11}$(P(OEt)$_3$)[336], H$_3$Ir(CO)(PPh$_3$)$_2$[337], [RuCl$_4$]$^{2-}$[334]) cause isomerisation of maleate at a rate faster than or comparable ((R$_3$P)$_2$Ir(CO)X[158], [Co(CN)$_5$]$^{3-}$[208]) with reduction rendering accurate comparisons difficult. Further examples of diacid and diester reductions are given in Figure 27.

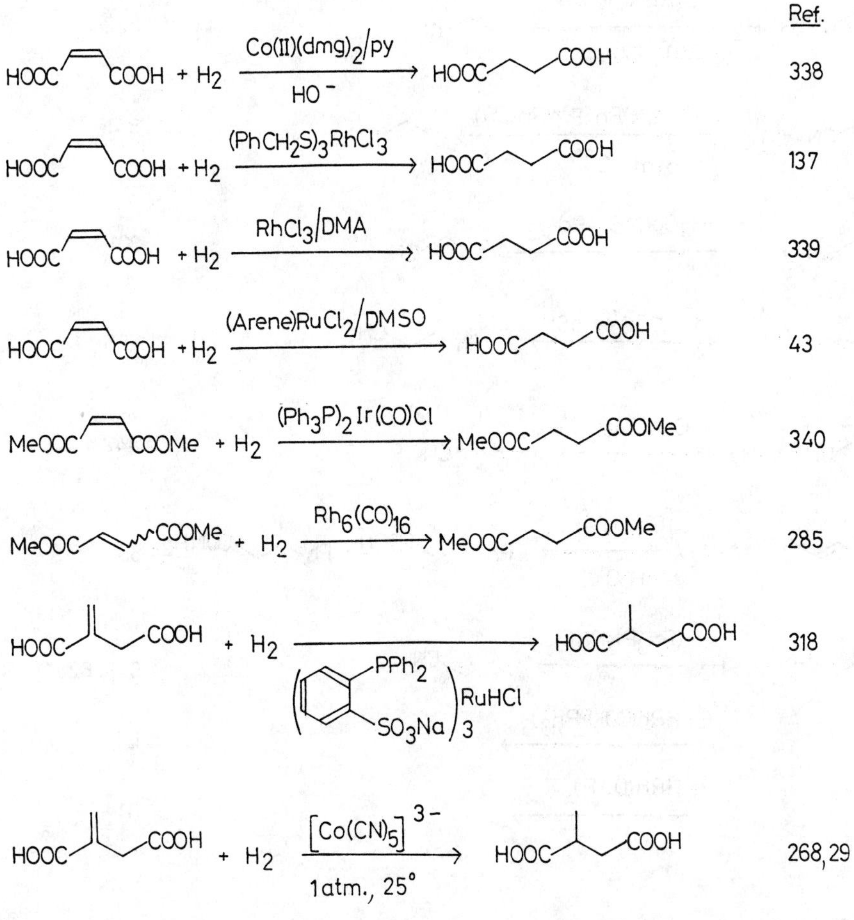

Figure 27 Hydrogenation of unsaturated diacids and diesters

2.3.5 Unsaturated nitriles and nitro compounds

These two groups have much in common with the unsaturated carbonyls.   Whilst with heterogeneous catalysts partial overreduction of -NO$_2$ and -CN is common most homogeneous catalysts give satisfactory and selective reduction of the carbon carbon double bond.   The available data are insufficient for generalisations but examples are shown in Figures 28 and 29.

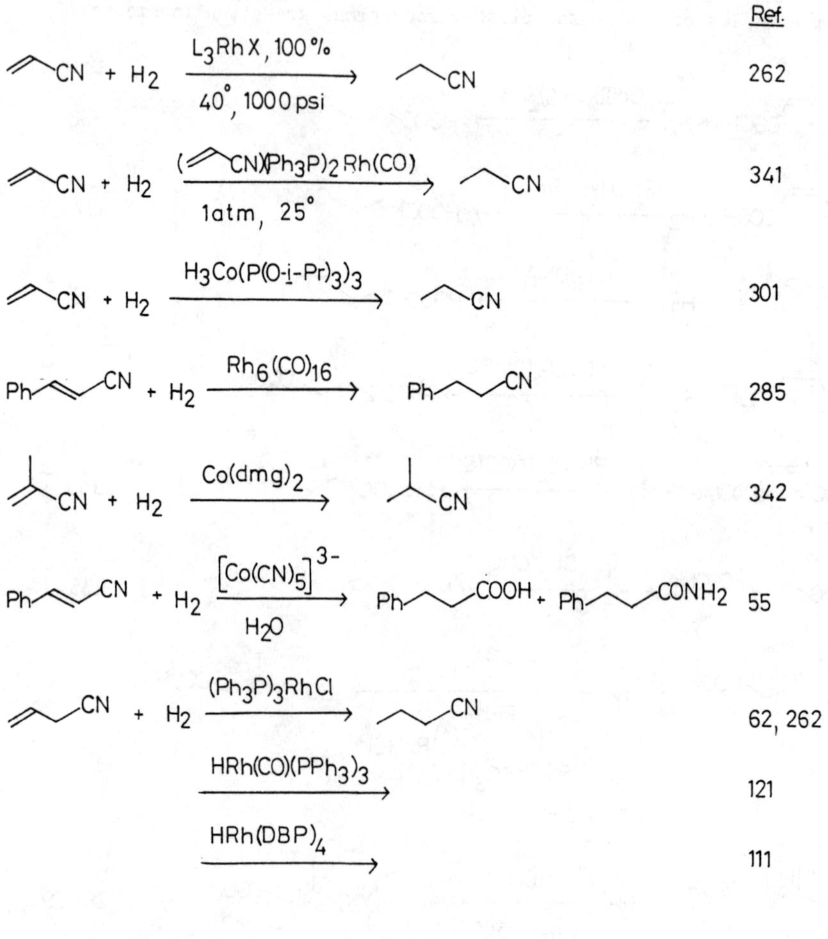

Figure 28 Hydrogenation of unsaturated nitriles

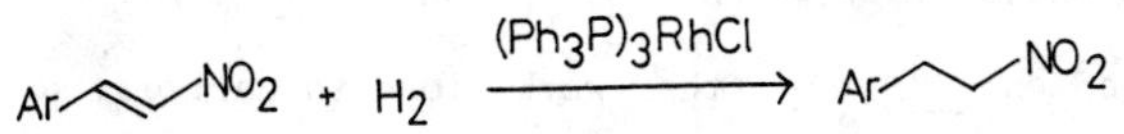

83, 283
312

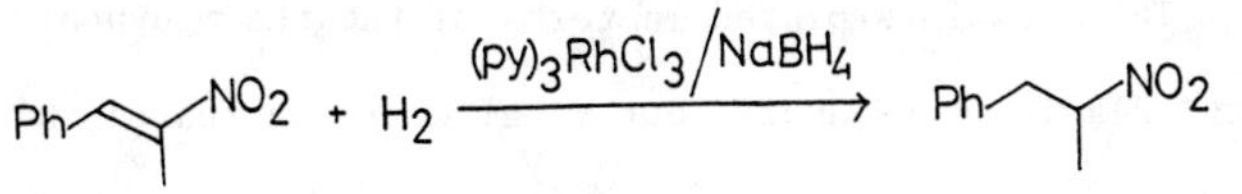

343

Figure 29   Hydrogenation of unsaturated nitro compounds

2.3.6 Alcohols, ethers and amines

Hydrogenation of unsaturated alcohols proceeds as for unfun-
ctionalised alkenes, with two classes of exceptions.   Allyl alcohol is
isomerised to propanal by some catalysts and subsequently deactivates them
by decarbonylation.[62,262]   The reduction of geraniol, 104, and its trans-
isomer, nerol, suffer from the same problem with the outcome depending on
the conditions.[82]   Decarbonylation is a problem with most rhodium
catalysts[95,111,121,281] and $(Ph_3P)_3RuCl_2$.[344] $Pd(OAc)_2$/phen[345],
$(DMSO)_2PdCl_2$[297] and $[(EDTA)RhCl]^-Na^+$[346] give some problems whilst
$[(PhArP)Pd(OH)(OH_2)_n]$ (Ar=$C_6H_4SO_3Na$) is reported to give a good result.[347]
Curiously 105 is reduced satisfactorily in the presence of $[Co(CN)_5]^{3-}$[55],
$(Ph_3P)_3RhCl$[89] and $(Ph_3P)_3RuCl_2$.

In another important range of substrates an OH group coordinates

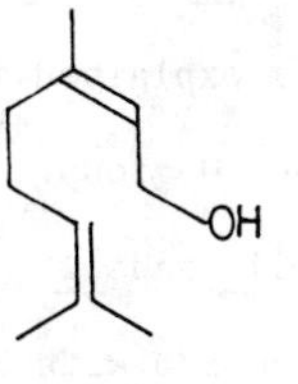

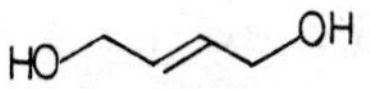

105

104

to the metal and directs the diastereoselectivity of reduction.   For

example, 106 is hydrogenated exclusively to the cis- product, 107.   The

transition state is envisaged as 108.[348]   Other work aimed to derive a set

of rules for cyclohexenols.   109 gives mainly trans-110 in the presence of

Wilkinson's catalyst.   This was interpreted in terms of the thermodynamic

stabilities of the metal alkyls or products, but metal coordination of the

OH may also be involved.[349]   In the reduction of 111 to 112 and 113 a

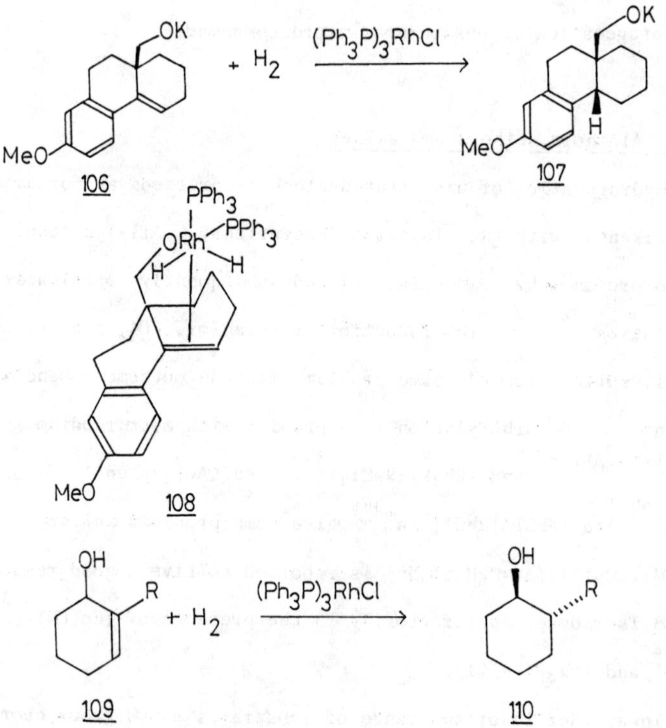

typical heterogeneous catalyst gives mainly 113 whilst $[(Cy_3P)(py)Ir(cod)]PF_6$

gives almost exclusively 112.   This is explained in terms of inter-

mediate,114,with iridium coordinated to the OH group.[350]   A similar

argument is invoked for the reductions of 115 and 118.[351]   Exocyclic

methylene groups, as in 121, may also be reduced with high stereoselection

where coordination is possible.[352]

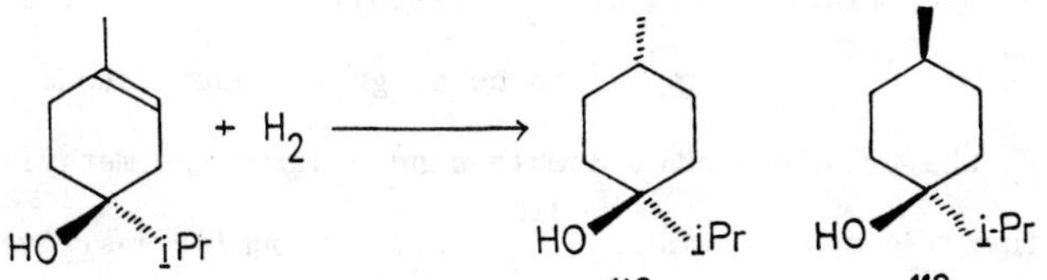

	112	113
Pd/C	20 %	80 %
(Cy$_3$P)Ir(cod)(py)PF$_6$	99.9 %	0.1 %

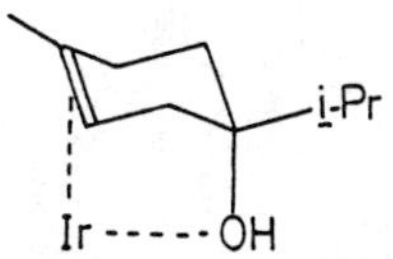

111

114

115 + H$_2$ $\xrightarrow{\text{(Cy}_3\text{P)(py)}\overset{+}{\text{Ir}}\text{(cod)}}$ 116 + 117

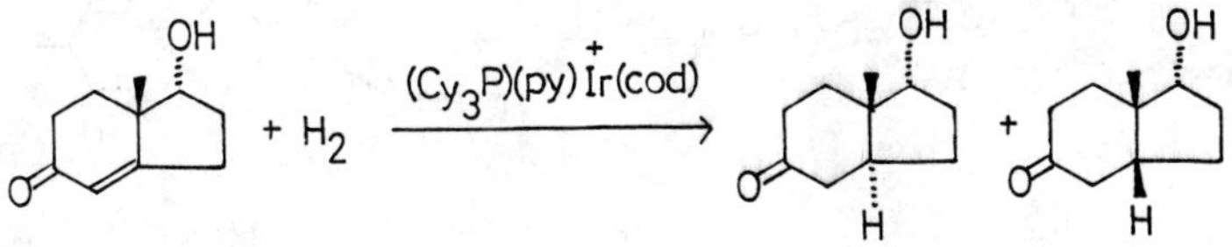

96 : 4

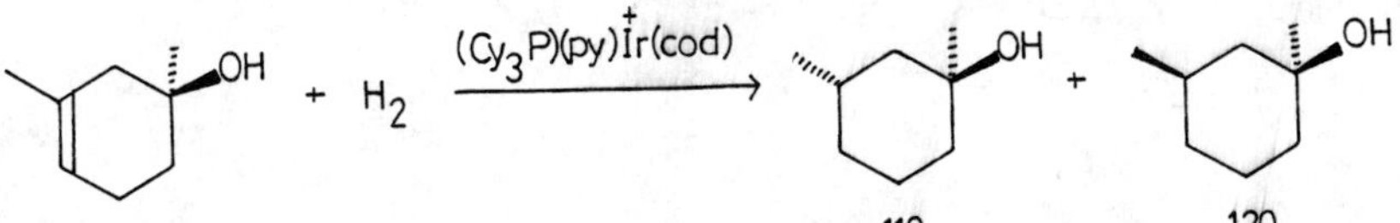

118 + H$_2$ $\xrightarrow{\text{(Cy}_3\text{P)(py)}\overset{+}{\text{Ir}}\text{(cod)}}$ 119 + 120

100 : 1

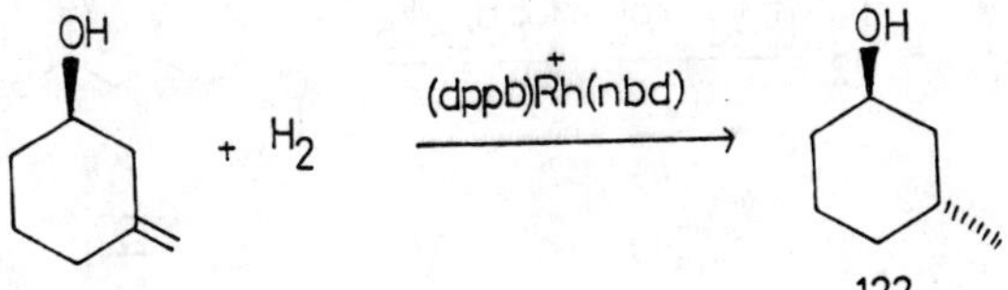

121 + H$_2$ $\xrightarrow{\text{(dppb)}\overset{+}{\text{Rh}}\text{(nbd)}}$ 122 > 98% selective

Although many vinyl ethers are successfully reduced in the presence of (Ph₃P)₃RhCl [83,262] they appear to be slightly deactivated.   In the reduction of 123 the trisubstituted double bond bearing the methyl rather than the methoxyl group is reduced. [312]   For compounds sensitive to hydrogenolysis (e.g. 125 [68]) or hydrolysis (e.g. 127 [353] and 129 [354]) homogeneous catalysts provide a milder route than heterogeneous ones where cleavage is common (Figure 30).   Vinyl and allyl esters are easily reduced as are allyl ethers (Figure 31).

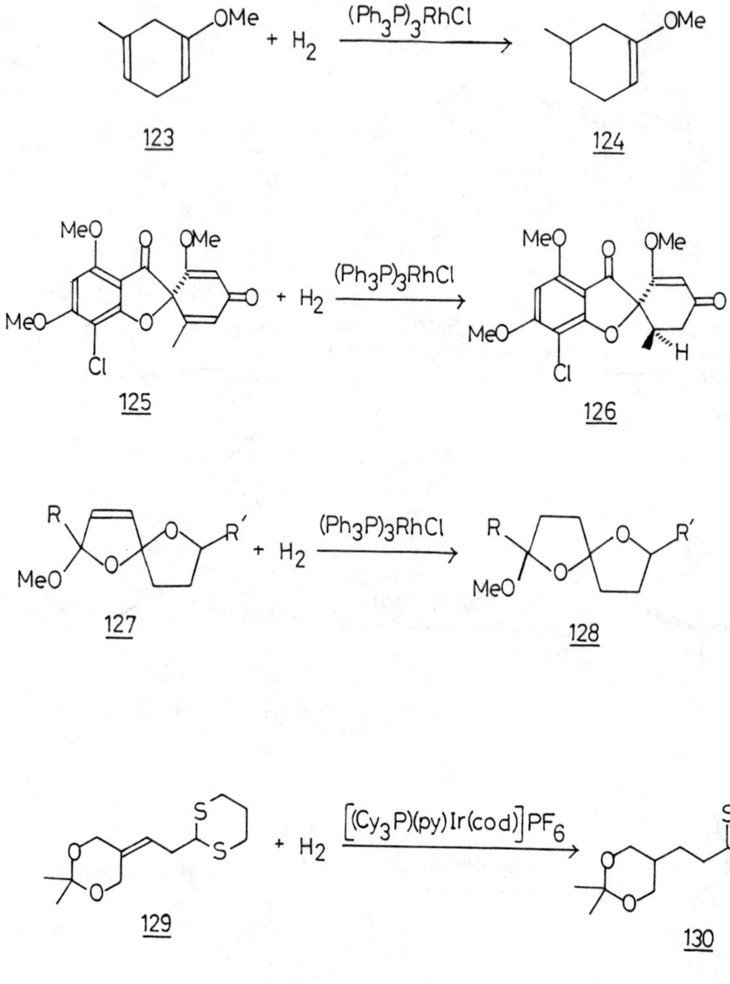

Figure 30 Hydrogenation of compounds sensitive to hydrolysis or hydrogenolysis

Ref.

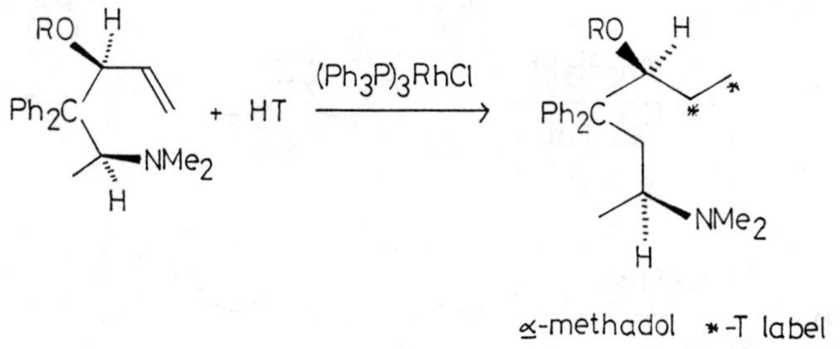

355

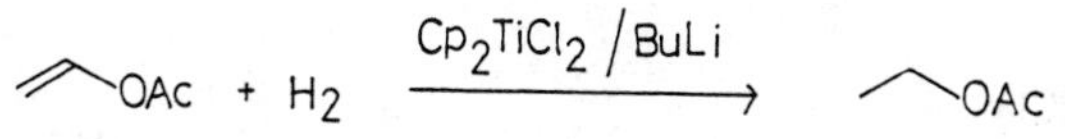

α-methadol $*$-T label

356

<u>Figure 31</u> Reduction of vinyl esters, allyl esters and ethers.

2.4 Reduction of Dienes

2.4.1 Allenes

There exist few reports of homogeneous hydrogenation of allenes but there is no reason to regard it as a difficult process. The usual product is a <u>cis</u>-alkene and the less substituted double bond reacts (Figure 32).

2.4.2 Non-conjugated dienes

In substrates where two or more double bonds are not conjugated we must distinguish catalysts which are capable of conjugating the bonds and those which are not. In the latter case the reactions resemble those of monoenes but may provide useful examples of selectivity.

Reduction of linear non-conjugated dienes in the presence of $(Ph_3P)_3RhCl$ proceeds sequentially, one double bond at a time.[262] $PhRh(C_8H_{12})(PPh_3)$, $Rh_2(OAc)_4/PPh_3$[360] and $HRh(CO)(PPh_3)_3$[120] are also active. In substrates with more than one type of double bond the least substituted is

Ref.

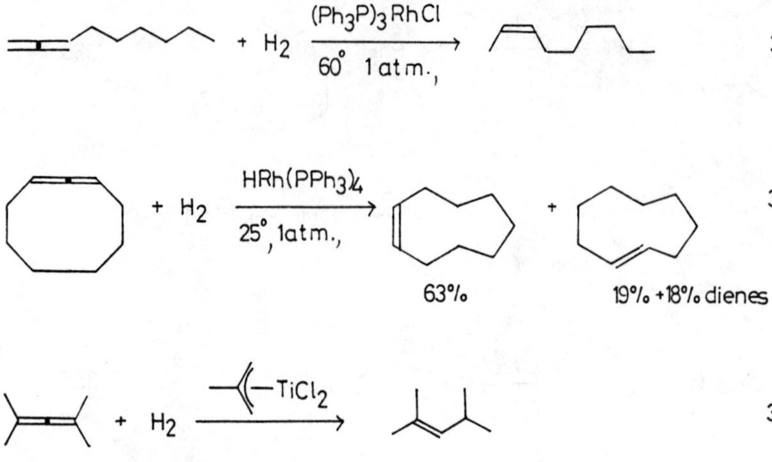

357

63%   19% +18% dienes

358

359

Figure 32   Homogeneous hydrogenation of allenes.

reduced first as noted earlier for eremopholine[61] (Figure 4), α-santonin, 87, and androsta-1,4-diene-3,17-dione.[361]

   $(Ph_3P)_3RuCl_2$ is fairly selective for the reduction of 1-alkenes but this and related catalysts give rather mixed results for linear non-conjugated dienes.[139,189]   Hexa-1,4-diene is reduced slowly to give 30% 2-hexenes.[47]   However, 131[49] and 133[362] are reduced with excellent selectivity. $(Ph_3P)_2Ru(OAc)_2/H^+$ gives up to 76% selectivity for 1-hexene in reduction of hexa-1,5-diene.[195]

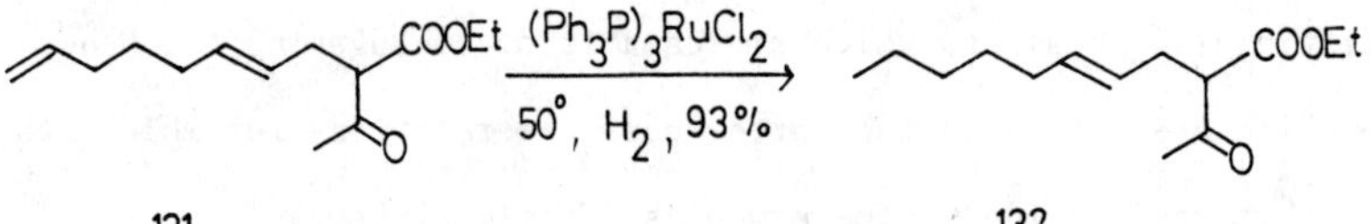

131   132

   The cobalt analogue of Wilkinson's catalyst, $(Ph_3P)_3CoCl$, catalyses reduction of hexa-1,5-diene, mainly to hexane[363] whilst $HCo(CO)_3PBu_3$ gives mainly hexenes.   135 is reduced using $HIrCl_2(PPh_3)_3/H_2O_2$ to give 136 with up to 87% selectivity and without decarbonylation.[365]

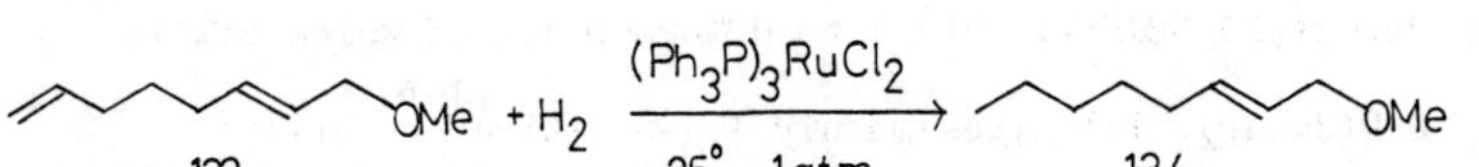

2.4.3 Conjugated acyclic dienes

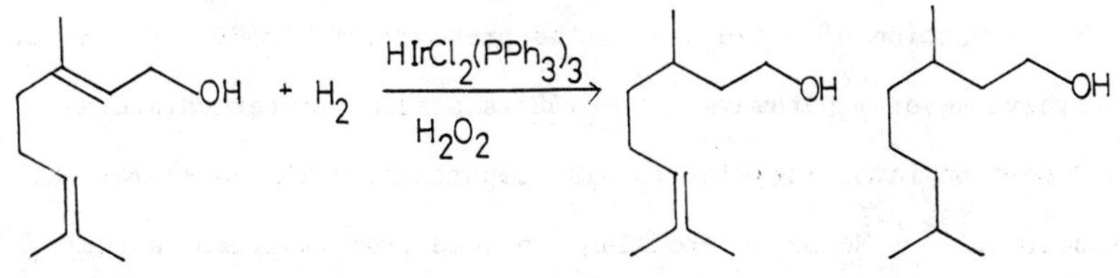

Most studies of hydrogenation of conjugated dienes have as their aim selective reduction of one or other double bond or 1,4-reduction. Rhodium complexes are not particularly useful. Conjugated dienes react more slowly than alkenes in the presence of Wilkinson's catalyst, butadiene being reduced at a reasonable rate only at 60°C.[72,262] Both the rate and the proportion of 1,4-addition increases with solvent polarity. $[L_2Rh(diene)]BF_4$ complexes give up to 99% selective reduction to monoenes, the regiochemistry depending on L (Table 7).[25] $((Ph_3P)Rh(CO)_2)_2/PEt_3$ also gives fair selectivity to monoenes with careful control of conditions;

Table 7.   Reduction of Dienes in the presence of $[L_2Rh(diene)]^+$

2,3-dimethylbutadiene → 2,3-dimethyl-1-butene + 2,3-dimethyl-2-butene

	%	%
L = diphos	43	57
$Ph_2AsCH_2CH_2AsPh_2$	80	20
$Ph_2PCH_2CH_2AsPh_2$	17	83

butadiene is reduced (50-100°C,15 atm.) to 90% monoenes of which 83% is 1-butene.[366]   HRh(dmgh)$_2$/PPh$_3$ gives mainly 1,4-addition[367] whilst (Ph$_3$P)$_2$RhCl(DMSO)/NaBH$_4$ gives up to 65% 1-butene from butadiene.[368]

The reduction of butadiene in the presence of HIr(CO)(PPh$_3$)$_3$ was found to involve several pathways.[153]   138 is a stable intermediate and ortho-C-H insertion involving PPh$_3$ is also important.   Monoenes are the primary products.[369]   Monoenes are also produced from isoprene in the presence of [((p-RC$_6$H$_4$)$_3$P)$_2$Ir(cod)]ClO$_4$, the rate of reaction increasing with phosphine basicity.[370]

138

Arene chromium tricarbonyl complexes are selective catalysts for reduction of 1,3-dienes by 1,4-hydrogen addition.   That addition is rigorously 1,4 was demonstrated by deuterium labelling studies.[371]   Early work employed 139 and the rate depends inversely on the stability of the arene complex.[372]   Phenanthrene, anthracene and naphthalene yield complexes from which the arene is more easily displaced and which are more active.[373]   Still milder conditions may be employed using (MeCN)$_3$Cr(CO)$_3$ or Cr(CO)$_6$/hν(1 atm,25°C).[371]   All the catalysts are relatively specific for trans,trans-dienes which can attain the cisoid conformation, 140, at chromium.[374]   Dienes where such coordination is hindered are reduced

139

140

slowly or, in the case of 141, isomerise via an ene-reaction before reduction to 143. Very hindered dienes such as 144 reduce specifically but slowly.[375] The mechanism involves the dihydride of 140 formed by oxidative addition. With $[CpCr(CO)_3]_2$, which appears to react in the same way, however, a monohydride intermediate $HCr(Cp)(CO)_3$ is proposed.[376]

Another group of catalysts specific for dienes are derived from $[Co(CN)_5]^{3-}$. The regioselectivity of reduction is dependent on the conditions. Butadiene yields mainly 1-butene for CN:CO > 5 and trans-2-butene for CN:Co < 5, in alcoholic or aqueous solution.[55] Early speculations suggested the mechanism followed equations (32) to (34);[377]

$$2[Co(CN)_5]^{3-} + H_2 \rightarrow 2[HCo(CN)_5]^{3-} \qquad (32)$$

$$[HCo(CN)_5]^{3-} + C_4H_6 \rightarrow [Co(CN)_5(C_4H_7)]^{3-} \qquad (33)$$

$$[Co(CN)_5(C_4H_7)]^{3-} + [HCo(CN)_5]^{3-} \rightarrow 2[Co(CN)_5]^{3-} + C_4H_8 \qquad (34)$$

$[Co(CN)_5(C_4H_7)]^{3-}$ is isolable.[378] Excess CN⁻ gives mainly σ-butenyl which yields 1-butene, whilst at lower [CN⁻] the syn-π-butenyl predominates and gives trans-2-butene. Some other examples are shown in Figure 33. A number of modified versions of this catalyst are known; one or more cyanide ions are replaced by diamines such as en, bipy or phen.[381] The differences are not fundamental, and mainly relate to changes in proportions of σ- and π-butenyls.

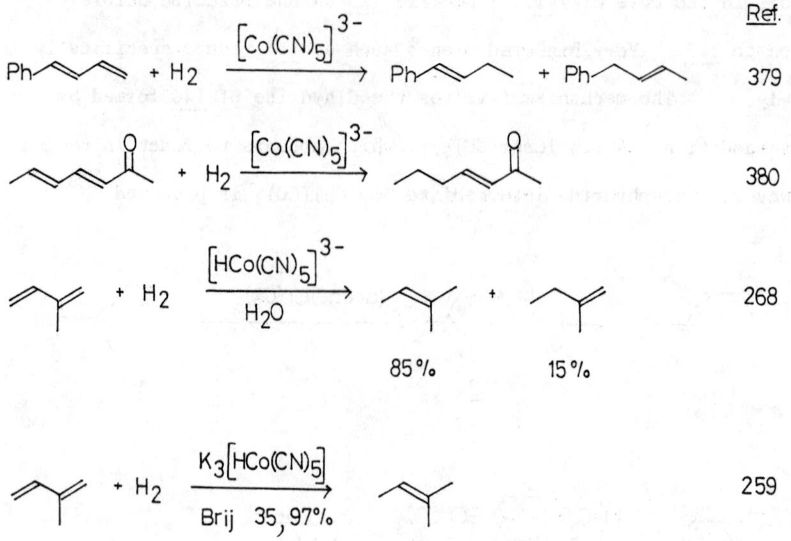

Ref.

$Ph\diagdown\diagup\diagdown\diagup$ + H$_2$ $\xrightarrow{\left[Co(CN)_5\right]^{3-}}$ $Ph\diagdown\diagup\diagdown$ + $Ph\diagdown\diagup\diagup$ 379

(structure) + H$_2$ $\xrightarrow{\left[Co(CN)_5\right]^{3-}}$ (structure) 380

(structure) + H$_2$ $\xrightarrow[H_2O]{\left[HCo(CN)_5\right]^{3-}}$ (structure) + (structure) 268

85% 15%

(structure) + H$_2$ $\xrightarrow[Brij\ 35,\ 97\%]{K_3\left[HCo(CN)_5\right]}$ (structure) 259

Figure 33 · Cobalt pentacyanide catalysed hydrogenation of dienes.

Another cobalt complex, (Ph$_3$P)$_3$CoCl, gives unselective reduction
of conjugated dienes.[363] However, in the presence of Lewis acids,
especially BF$_3$Et$_2$O, reduction occurs 1,2 at the more substituted double
bond (e.g. 146 to 147 and 148 to 149).[382] Other reports are more
scattered (Figure 34).

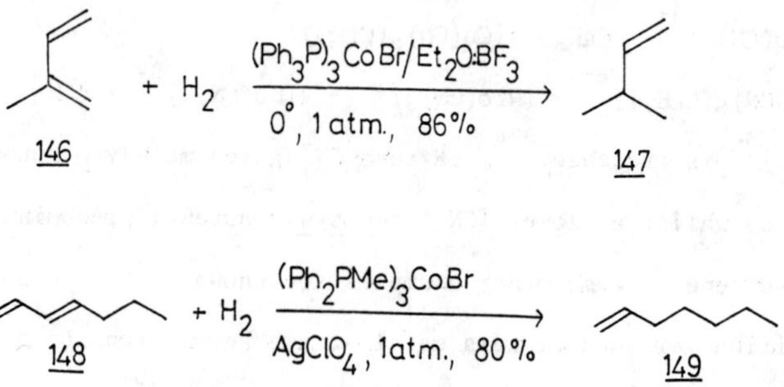

(structure) + H$_2$ $\xrightarrow[0°,\ 1\ atm.,\ 86\%]{(Ph_3P)_3CoBr/Et_2O:BF_3}$ (structure)

146 147

(structure) + H$_2$ $\xrightarrow[AgClO_4,\ 1atm.,\ 80\%]{(Ph_2PMe)_3CoBr}$ (structure)

148 149

Ref.

Butadiene + H$_2$ $\xrightarrow[\text{81\%}]{\text{Co(I) bipy, 1 atm., 25}^{\circ}}$ cis-2-butene   383

Butadiene + H$_2$ $\xrightarrow[\text{66}^{\circ},\ 15\ \text{atm}]{\text{(Co(CO)}_2\ \text{PBu}_3)_2}$ butenes   384

<u>Figure 34</u> Diene reduction in the presence of cobalt complexes

      Since (Ph$_3$P)$_3$RuCl$_2$ is selective for 1-alkenes one might expect it to be rather selective in promoting diene reduction. However, isomerisation competes and the usual product is a mixture of 2-alkenes.[47,385] Other ruthenium and iron catalysts are similarly unselective.[201,209,214]

      Numerous other catalysts have been reported to be active for diene hydrogenation but they have not been studied in detail or lack general application. The Ziegler catalysts including Cp$_2$TiCl$_2$/BuLi, CoCl$_2$/Et$_3$Al, Co(acac)$_2$/PBu$_3$/R$_3$Al, NiCl$_2$/NaBH$_4$,Cp$_2$TiR$_2$ and Cp$_2$V are all active, though most are unselective and require severe conditions.[254,386] Some other examples are shown in Figure 35.

2.4.4 Unsaturated fats

      There is a strong commercial motive to achieve selective reduction of unsaturated fats. Soybean and linseed oils consist of mixtures of mono-,di- and triene carboxylate esters. Linolenates (9,12,15-octadecatrienoates) give an unpleasant flavour and it is desirable to reduce them to linoleates (<u>cis</u>-9, <u>cis</u>-12-octadecadienoates). Control of the proportions of saturated and unsaturated fats is also desirable on nutritional grounds. The classic model for these fats has been methyl sorbate, <u>150</u> (R = Me). This may be reduced to the 2-, 3- and 4-enoates under appropriate conditions (Table 8).

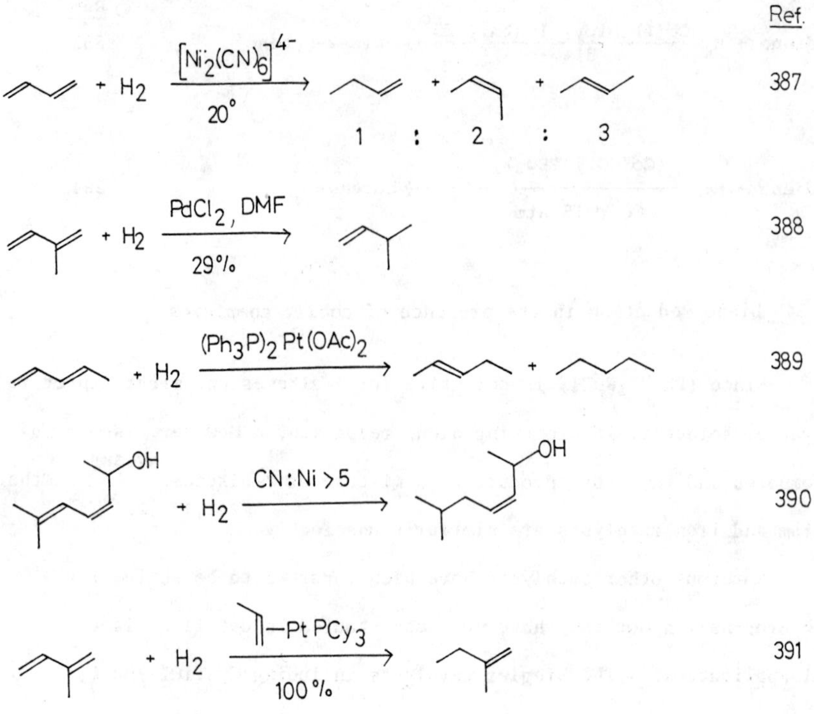

Ref.

387

388

389

390

391

Figure 35 Examples of diene hydrogenation.

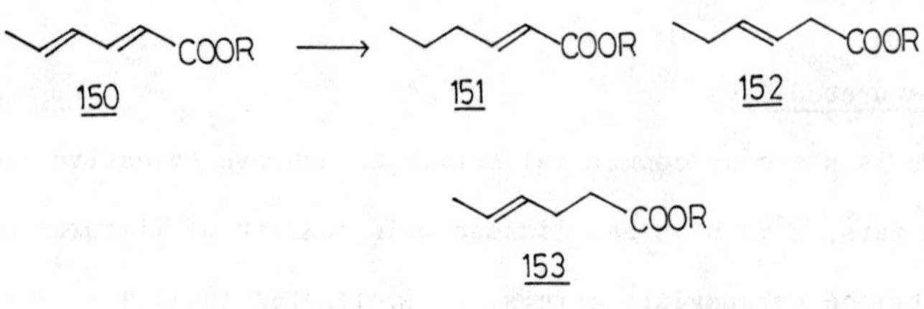

Reduction of more complex fats is not a solved problem.   Some

catalysts ($[Co(CN)_5]^{3-}$ and $(arene)Cr(CO)_3$) are selective for conjugated

dienes and some are capable of causing isomerisation.   With $[Co(CN)_5]^{3-}$,

9, 11, 13-octadecatrienoic acid was reduced to a diene, but this is a poor

isomerisation catalyst and largely insoluble in lipids.[400] $Co_2(CO)_8$ is

active for reduction of trans, trans-9, 11-dienoate and the cis, trans-

Table 8. Regioselectivity of Sorbate reduction

Catalyst	R	Conditions	%151	%152	%153	Ref.
$[Co(CN)_5]^{3-}$	H	1atm, 25°, H_2O	82	17	1	392
$[Co(CN)_5]^{3-}$	H	50atm, 70°	-	-	60	55
$K_3[HCo(CN)_5]$	Na	1atm, 25°, H_2O [a]	-	75	-	393
(Arene)Cr(CO)$_3$	Me	50atm, 150°	-	>90	-	374,394,395
$\boxed{P}$–C$_6$H$_4$Cr(CO)$_3$	Me	50atm, 70°	-	>90	-	396
Cr(CO)$_6$	Me	50atm, 165°	-	>90	-	397
(Arene)Mo(CO)$_3$	Me	40atm, 100–180°	-	major	-	372
(C$_2$H$_4$)Fe(CO)$_4$	Me	50atm, 175°	b	b	b	398
(Ph$_3$P)$_3$RuCl$_2$	Me	1atm, 25°	-	major [c]		399
NiCl$_2$/NaBH$_4$	Me	1atm, 25° DMF	80 [d]	5	-	400,402

(a)   Phase transfer conditions; without a phase transfer agent 90%
      151 is produced

(b)   All isomers produced in unspecified proportions; isomerisation
      occurs in the presence of the catalyst

(c)   Mainly trans

(d)   15% $CH_3(CH_2)_4COOMe$ is also produced

10, 12-compound to monoenes but skipped dienes are less reactive.[401]

(Arene)Cr(CO)$_3$ complexes isomerise non conjugated to conjugated systems,

though this is slower than reduction.[374]   Linoleate and linolenate give

a mixture of dienes and monoenes with a high cis-component.

(py)$_3$RhCl$_3$/NaBH$_4$ gives up to 95% monoenate from linoleate,

mostly the undesired trans-isomer; conjugation probably precedes reduction.[402] Cationic complexes, [L$_2$Rh(diene)]$^+$, can give useful results with careful control of conditions. At low pH (dihydride mechanism) a good yield of cis-monoene is obtained but with Et$_3$N(monohydride mechanism) trans-monoene and conjugated dienes are obtained.[403]

(Ph$_3$P)$_2$PtCl$_2$ catalyses reduction of soybean esters to monoenes.[404] Addition of SnCl$_2$ enhances reactivity but gives more conjugated product.[242] By using palladium, changing phosphine or additives, variations in selectivity are observed but some features are general. Conjugation precedes reduction, trans-alkene products predominate and the catalysts deteriorate in use.[242,405,406] (Ph$_3$P)$_2$NiX$_2$ catalyses isomerisation and selective hydrogenation of linoleate to monoenoate, whereas NiCl$_2$/NaBH$_4$/DMF reacts slowly to give 85% monoene with the desired cis-isomers predominating.[400,402]

2.4.5. Cyclic dienes

This class of dienes has provided few but popular substrates with opportunities for comparisons between catalysts. The transformations cyclopentadiene to cyclopentene and 1,3-cyclohexadiene to cyclohexene are particularly easy (Tables 9 and 10). 1,4-Cyclohexadienes are produced by Birch reduction of arenes. The less substituted double bond is reduced with good selectivity in the presence of rhodium (154 to 155 and 156)[83] and platinum (157 to 158)[236,239] catalysts. In other cases isomerisation

154   (Ph$_3$P)$_3$RhCl, 25°, 1atm   155   156

157   PtCl$_2$/SnCl$_2$, ROH   158

Table 9.   Hydrogenation of cyclopentadiene to cyclopentene

Catalyst	Conditions	Ref.
$[Co(CN)_5]^{3-}$	1atm, $25^{\circ}C$	55
$Li_3^+[HCo(CN)_5]^{3-}$	1atm, $25^{\circ}C$ CN:Co=4	407
$CpMoH_2$	160atm, $180^{\circ}C$ [a]	298
$PdCl_2$/o-phen	1atm, $25^{\circ}C$ [b]	391
$MX_2/NaBH_4$	1atm, $25^{\circ}C$ [b]	408

(a)   55% yield;   rest is oligomers

(b)   99% selective

Table 10.   Hydrogenation of 1,3-cyclohexadiene to cyclohexene

Catalyst	Conditions	Ref.
$[Co(CN)_5]^{3-}$	1 atm, $25^{\circ}C$	268,288,289
$Co(CO)_3(PPh_3)_2$	20-30 atm, $110-180^{\circ}C$ [a]	364
$(Arene)Cr(CO)_3$	40 atm, $100^{\circ}C$	374
$(CpCr(CO)_3)_2$	50 atm, $70^{\circ}C$	376
$(R_3P)_2Rh(CO)X$	1 atm, $25^{\circ}C$	262
$(Ph_3P)_2RhNO(quinone)$	1 atm, $60^{\circ}C$	265
$(Ph_3P)_2Ir(CO)Cl$	hν, no solvent, 1 atm, $25^{\circ}C$	369,409
$L(NO)Ir$ ortho$(C_6Br_4O_2)$	3 atm, $30^{\circ}C$	410
$NiCl_2/NaBH_4/DMF$	1 atm, $25^{\circ}C$	411
$Ni(acac)_2PPh_3/Et_3Al_2Cl_3$	(b)	412
$(Ph_3P)_3CoCl$	1 atm, $25^{\circ}C$	413

(a)   pCO = 5 atm

(b)   74% yield, 6% cyclohexane

precedes reduction; 1,4-cyclohexadiene yields 159 on deuteration using (arene)Cr(CO)$_3$ indicating 1,4-addition to a 1,3-diene.[374] When (Ph$_3$P)$_3$RhCl is used in reduction of isotetralin, 160, isomerisation must also occur since 161 and 162 are formed.[308] Other complexes including (Ph$_3$P)$_2$Ir(CO)Cl[369], [(p-RC$_6$H$_4$)$_3$P)$_2$Ir(cod)]ClO$_4$[370] and (acac)$_2$Ni/Et$_3$Al$_2$Cl$_3$/PPh$_3$[412] also catalyse reduction.

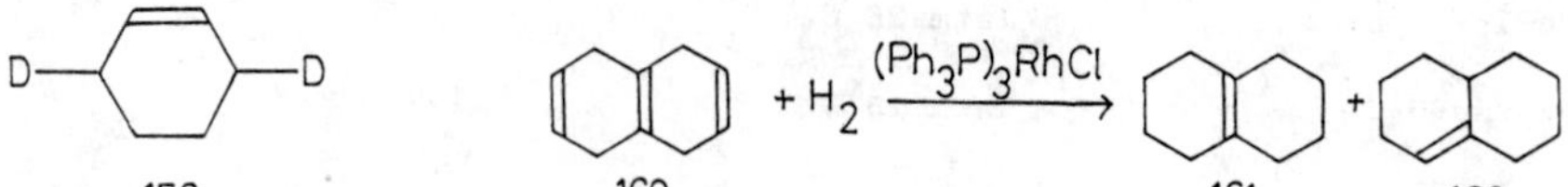

Cycloheptatriene is reduced in the presence of (arene)Cr(CO)$_3$ to cycloheptadiene and cycloheptene;[374] D-labelling implies 1,4-reduction followed by isomerisation occurs. Cp$_2$MoH$_2$ yields a mixture of diene and alkene at 140°C/160 atm. and alkene and alkane at 180°C.[298] Few selective reductions are on record; $\underline{n}^4$(cod)Ru$\underline{n}^6$(C$_8$H$_{10}$) gives cycloheptene specifically under mild conditions[414] and 163 gives cycloheptadiene selectively.[410]

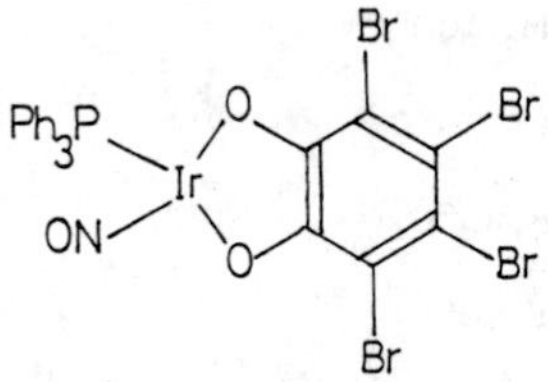

163

Both 1,3- and 1,5-cyclooctadiene have been popular substrates; again some catalysts are capable of isomerising the 1,5-diene. (Figure 36).

The commercial aim of this work is the selective hydrogenation of cyclododecatriene (from butadiene trimerisation) to cyclododecene. On ozonolysis this yields 1,10-decanedioic acid, an intermediate in the

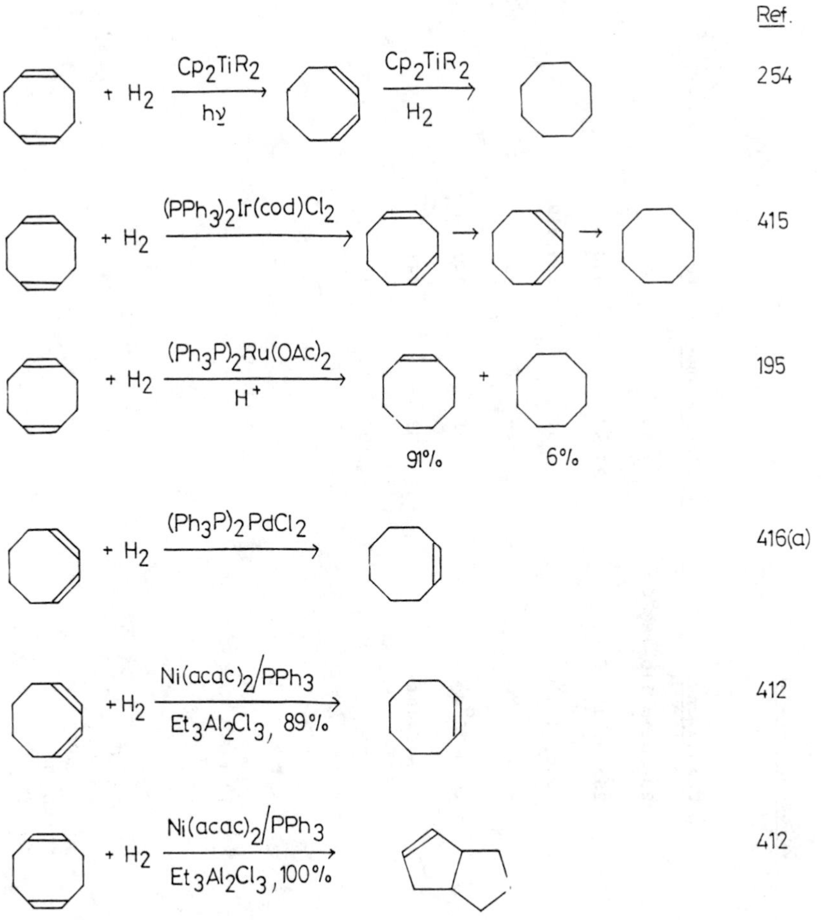

Ref.

254

415

195

91%   6%

416(a)

412

412

(a) 1,5-cod is reduced via the 1,3-isomer

Figure 36   Hydrogenation of cyclooctadienes

production of high quality polyamides.   All the common catalysts have been
tried (Table 11).

Bicyclic dienes such as norbornadiene have also been reduced.
The usual products of norbornadiene reduction are norbornene, norbornane
and nortricyclane, 164.   [(Ph₃P)₂Rh(nbd)]⁺ gives mainly norbornene by

Table 11.   Hydrogenation of cyclododecatriene

Catalyst	Conditions	% cyclododecene	Ref.
$(R_3P)_2Co(CO)_3Co(CO)_4$	20-30atm, 110-180°C	>90 [a]	417
$((Bu_3P)Co(CO)_3)_2$	25atm, 120°C	95-97	418
$(Ph_3P)_2Ru(CO)_2Cl_2/PPh_3$	1atm, 160°C	97	385,419
$(Ph_3P)_3RuCl_2$	10atm,25°C,Et$_3$N	87	49
$(py)_3RhCl_3/NaBH_4$	1atm,75°C	91 [b]	420
$[HPt(SnCl_3)_4]^{3-}/[Pt(SnCl_3)_5]^{3-}$	no solvent	87	421
$(Ph_3P)_2NiI_2$	-	>90	422

(a) cis:trans = 1:2

(b) 7-8% dienes

endo-H$_2$ addition.[25]   164 is the major product using [Co(CN)$_5$]$^{3-}$[269],

(Ph$_3$P)$_2$Rh(CO)$_2$Cl$_2$[385]   and (arene)Cr(CO)$_3$[372,375,423]   though small amounts of

norbornene are formed.   Complete reduction to norbornane occurs in the

presence of RuCl$_3$/PPh$_3$.[424]

164

2.5 Hydrogenation of alkynes

Carbon-carbon triple bonds are readily reduced by numerous

homogeneous and heterogeneous catalysts.   They are among the easiest

functional groups to reduce and interesting problems arise in achieving

selective reduction.

2.5.1 Reduction to alkanes

An enormous range of catalysts is available, including most of

those active for alkenes.   The alkene is usually an intermediate (Table 12).

Applications have been few;   most practical procedures employ hetero-

geneous catalysts.   One interesting use involved deuteration of 165 in a

synthesis of d^6-methylstearate.[426]

$$CH_3(CH_2)_3-C{\equiv}C-(CH_2)_2CH_2Cl$$

165

2.5.2 Reduction of Alkynes in the presence of alkenes

The ease of reduction of alkynes stems from their ability to bond

strongly to metals.   Thus in alkyne/alkene mixtures, alkyne is reduced

more readily.   1-Hexyne is reduced to hexene/hexane in a mixture with

1-octene in the presence of HRuCl(PPh$_3$)$_3$, (Ph$_3$P)$_4$Pd, (Ph$_2$PCH$_2$CH$_2$SPh)$_2$IrBr,

Table 12. Catalysts for reduction of alkynes to alkanes

Catalyst	Conditions	Substrates	Ref.
$(Ph_3P)_3MCl$ [M=Rh,Ir]	1atm,25°C	alkynes, Ph-C≡C-Ph	22,62,72,94,113 262,264,282
$(Ph_3P)_2Rh(CO)Cl$	1atm,70°C	1-alkynes	157,159,160
$(Ph_3P)_3M(NO)$ [M=Rh,Ir]	1atm,25°C	1-alkynes	94,95
$HM(PPh_3)_3(CO)$ [M=Rh,Ir]	1atm,25°C	1-alkynes	94,120
$(dppp)_2RhCl/Et_2AlCl$	1atm,25°C	1-alkynes	113
$(PPh_3)_2Ir(CO)X$	1atm,40-60°C	1-alkynes	124,157,159,160
$HRuCl(PPh_3)_3$	1atm,25°C	1-alkynes	47
$HPtCl(PPh_3)_2/SnCl_2$	-	$(CH_3)_2C(OH)C≡C-C(OH)(CH_3)_2$	424
$Cp_2Ti(CO)_2$	50atm,50°C	PhC≡CH,PhC≡CPh	425
$CpZrH_2$	50-100atm,80-120°C	PhC≡CH,alkynes	258
Ni(2-ethylhexanoate)$_2$/Et$_3$Al	-	3-hexyne	304

$((\eta^5\text{-Cp})\text{Fe}(\underline{\mu}_3\text{-CO}))_4$ and <u>166</u> (Y=Cl,Br; X=Cl,Br,PPh$_2$; R=Bu). [427-430]   Phenyl-

acetylene is reduced selectively in the presence of hexene and Co(acac)$_2$/

i-Bu$_3$Al. [431]

<u>166</u>

2.5.3 Reduction of alkynes to alkenes

The selective reduction of alkynes to alkenes requires that

alkynes are complexed more strongly and/or reduced more rapidly.   The

heterogeneous solution to this problem is the Lindlar catalyst.   Provided

that H$_2$ uptake is monitored, some homogeneous catalysts are also useful.

In the presence of (Ph$_3$P)$_3$RhCl 1-alkynes are reduced unselectively

but internal alkynes give <u>cis</u>-alkenes which are resistant to further

reduction.   But-2-yne-1,4-diol is reduced with better than 94%

selectivity. [432]   Alkynes bearing two electron withdrawing groups are not

reduced under mild conditions. [62]   Amongst the most successful catalysts

for this reduction is [L$_2$Rh(diene)]PF$_6$. [86]   (Figure 37).   Ethyl phenyl

propiolate gave a fully saturated product with the Lindlar catalyst.

(Ph$_3$P)Rh(OCOPh)(cod) is selective for reduction of 1-alkynes to 1-alkenes [433]

and Rh$_2$(OAc)$_4$/HBF$_4$ is similar. [139]   (py)$_3$RhCl$_3$/NaBH$_4$ gives <u>cis</u>-reduction of

<u>167</u> and <u>168</u> but diphenyl acetylene gives the <u>trans</u>-product, probably by

<u>ortho</u>-metallation of one of the rings. [138]   H[Rh$_2$(N-phenylanthranilate)$_2$Cl]

is an excellent catalyst for preparation of <u>cis</u>-D$_2$-ethylene from acetylene, [339]

but other alkynes are reduced rather slowly.

HOCH$_2$-C≡C-CH$_2$OH   MeOOC-C≡C-COOMe

<u>167</u>   <u>168</u>

Ref.

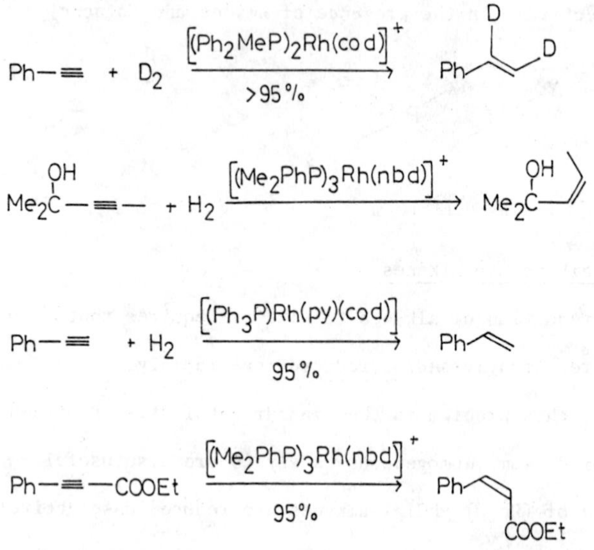

Ph—≡ + D₂ $\xrightarrow[>95\%]{[(Ph_2MeP)_2Rh(cod)]^+}$ [structure: Ph–CH=C(D)(CH_2D)]   433

Me₂C—≡— + H₂ $\xrightarrow[]{[(Me_2PhP)_3Rh(nbd)]^+}$ Me₂C(OH)–CH=CH₂   25

Ph—≡ + H₂ $\xrightarrow[95\%]{[(Ph_3P)Rh(py)(cod)]}$ Ph–CH=CH₂   97

Ph—≡—COOEt $\xrightarrow[95\%]{[(Me_2PhP)_3Rh(nbd)]^+}$ Ph–CH=CH–COOEt   433

Figure 37 Partial reduction of alkynes in the presence of cationic rhodium
complexes.

"A-frame" complexes such as 169[434] and 170[435] have been used for
selective reduction of alkynes to alkenes.   No alkene is reduced until the
alkyne is consumed, the reaction depending on the alkyne's ability to bridge
the metal atoms in a 1:1 complex.

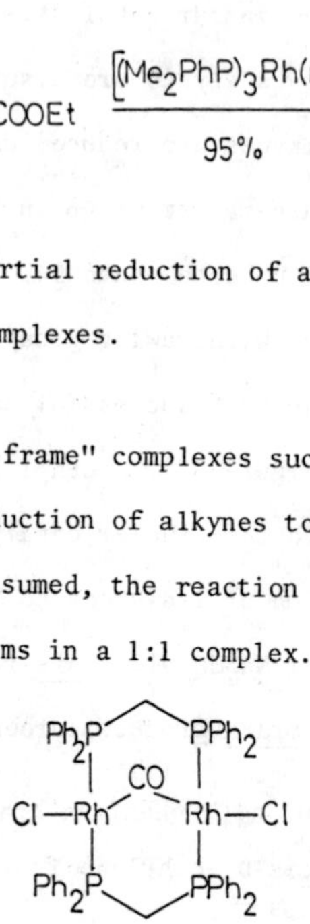

169

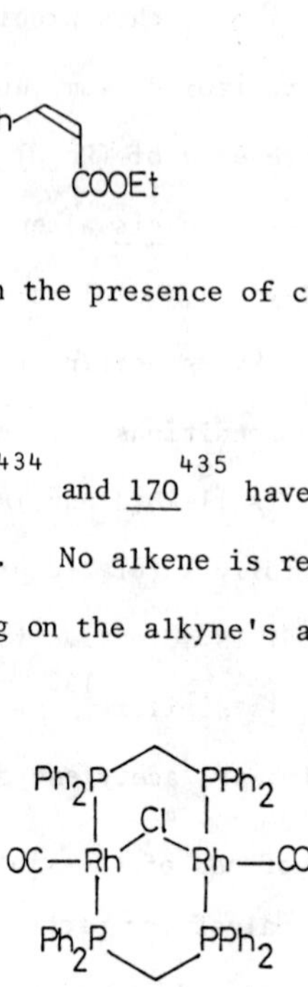

170

Analogous iridium "A-frame" complexes are less selective with
alkene and alkane produced simultaneously.[435]   Acetylene is reduced to
ethylene in the presence of HIr(CO)(PPh₃)₃ but the complex is less reactive
than the rhodium analogue and is slowly deactivated by an irreversible

reaction with acetylene.[120,165]   Other iridium complexes which give cis-

alkenes selectively include $H_3Ir(PPh_3)_3$, $(Ph_3P)_2Ir(CO)Cl/h\nu$[432]   and

$[Ir(\sigma\text{-carb})(CO)(PhCN)(PPh_3)]$ $(carb=7\text{-phenyl-}1,2\text{-}C_2B_{10}H_{12})$[436]   but they show

few advantages over rhodium analogues.

The stability of acetylene ruthenium complexes precludes selective

reduction to ethylene but 3-hexyne is reduced to cis-3-hexene in the

presence of $(Ph_3P)_3RuCl_2$.   Cis-addition of a ruthenium hydride to

complexed alkyne is thought to be the first step.[424] $Ru(OAc)_2/HBF_4/R_3P$

catalyses reduction of 1-alkynes to 1-alkenes.[139] $H_4Ru_4(CO)_{12}$ is fairly

selective for conversion of 1- and 2-pentyne to 1- and 2-pentene, activity

being enhanced by phosphine addition.[205]   Cis-2-pentene is the initial

product but cis-trans-isomerisation and double bond migration also occur.

Platinum and palladium complexes came later to this area and

despite some excellent selectivities they have not been widely used

(Figure 38).

$Cp_2Ti(CO)_2$ is a catalyst for selective reduction of 1-alkynes

to 1-alkenes.[425]   171 is an intermediate and conditions needed are

mild.[277,440]   A number of Ziegler catalysts are reported to give selective

171

reduction[441] but trimerisation to benzene derivatives is a major side

reaction.[442]   Other catalysts have received sparse attention (Figure 39).

All the catalysts discussed have been selective for the formation

of cis-alkene, though some also catalyse isomerisation.   However, for

trans-alkenes homogeneous catalysts are of little use and the most usual

route is an Na/NH_3 reduction.   One or two exceptions exist.   172 is

reduced to fumaric acid in the presence of $[Co(CN)_5]^{3-}$[445]   and

Ref.

1-heptyne + H$_2$ $\xrightarrow[>99\%]{\text{PdCl}_2 | \text{DMF}}$ 1-heptene   388

3-heptyne + H$_2$ $\xrightarrow[>95\%]{\text{Pd(OAc)}_2 | \text{PR}_3}$ cis-3-heptene   437

1-heptyne + H$_2$ $\xrightarrow[\text{NaBH}_4, \ 98-100\%]{\text{PdCl}_2 \text{(diamine)}}$ 1-heptene   390

Ph-C≡CH + H$_2$ $\xrightarrow{[\text{Pt(Sn(DMSO)Cl}_2)_5]\text{Cl}_2}$ PhCH=CH$_2$ 438

Ph-C≡CH + H$_2$ $\xrightarrow[98\%]{\text{Pt(C}_6\text{H}_{11}\text{NC)}_2 \text{(Mo(CO)}_3 (\underline{n}^5\text{C}_5\text{Me}_5)_2}$ PhCH=CH$_2$ + PhCH$_2$CH$_3$ 439
                                        66%   33%

<u>Figure 38</u> Selective reduction of alkynes in the presence of palladium and platinum complexes.

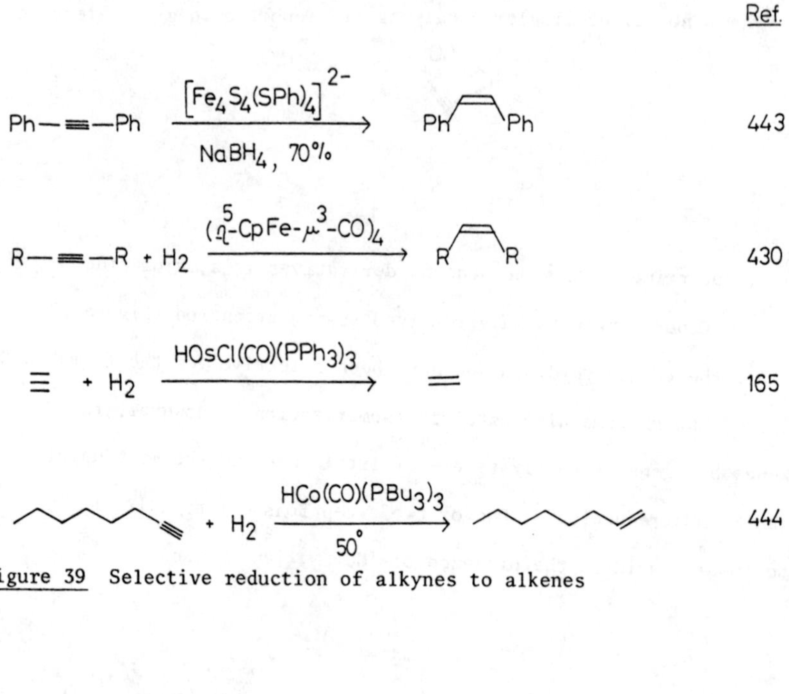

Ref.

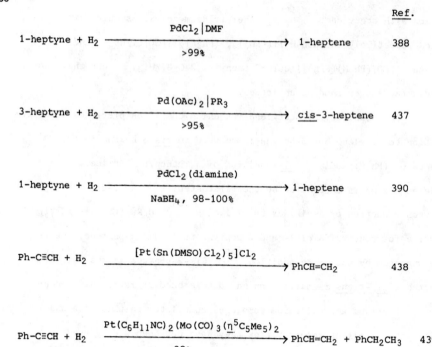

Ph—≡—Ph $\xrightarrow[\text{NaBH}_4, \ 70\%]{\left[\text{Fe}_4\text{S}_4\text{(SPh)}_4\right]^{2-}}$ Ph Ph   443

R—≡—R + H$_2$ $\xrightarrow{(\underline{n}^5\text{-CpFe-}\mu^3\text{-CO)}_4}$ R R   430

≡ + H$_2$ $\xrightarrow{\text{HOsCl(CO)(PPh}_3)_3}$ =   165

$\xrightarrow[50°]{\text{HCo(CO)(PBu}_3)_3}$ + H$_2$ 444

<u>Figure 39</u> Selective reduction of alkynes to alkenes

HOOC-C≡C-COOH

<u>172</u>

$H_2Rh(O_2COH)(P(\underline{i}\text{-Pr})_3)_2$ gives <u>trans</u>-stilbene from diphenyl acetylene, though this probably involves isomerisation.[446] A <u>trans</u>-vinyl compound is produced directly in the sequence <u>173</u> to <u>174</u> to <u>175</u>; although <u>trans</u>-alkene is the product, the catalyst is too short-lived to be useful.[447]

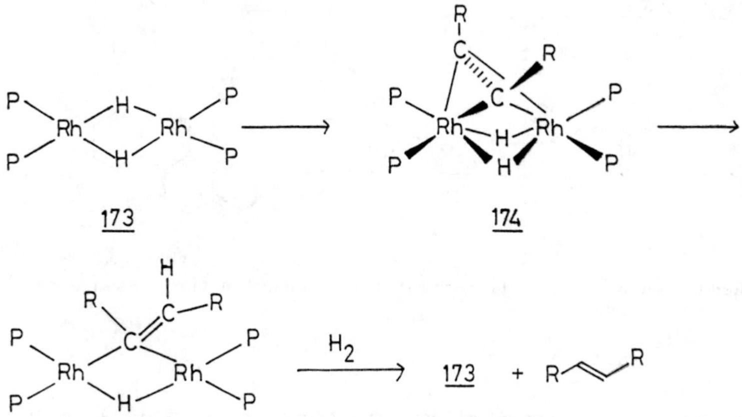

<u>173</u>   <u>174</u>

<u>175</u>

2.6 Hydrogenation of arenes

Reduction of arenes traditionally requires severe conditions and design of catalysts has been more concerned with activity than selectivity. A better understanding of metal arene chemistry has led to certain improvements.

2.6.1 Cobalt complexes

Benzene is poorly reduced under hydroformylation conditions in the presence of $Co_2(CO)_8$ but polycyclic aromatics give better results. Naphthalenes are reduced to tetralins[448] and anthracenes to 9,10-dihydro-compounds.[449] A radical mechanism operates and stereoselectivity is poor. Some other examples are shown in Figure 40.[448] $HCo(CO)_4$ is the active

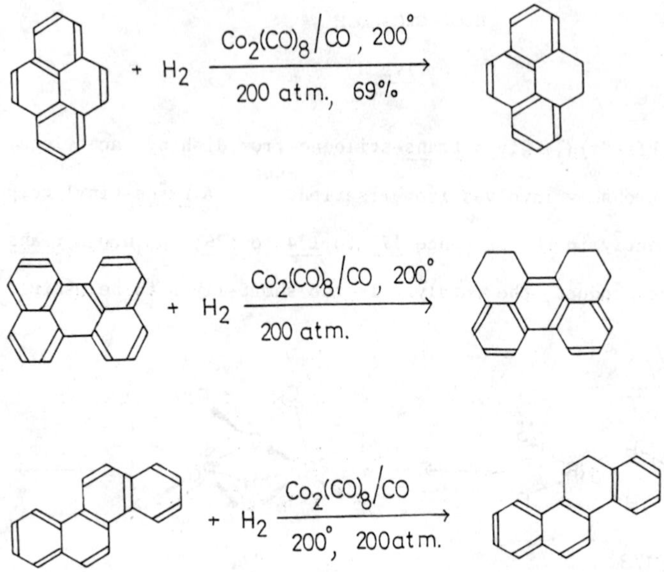

Figure 40 Reduction of polycyclic aromatic compounds in the presence of
$Co_2(CO)_8$

species and if this is preformed milder conditions may be used. Heterocyclic rings are reduced more easily, as with heterogeneous catalysts.
Pyridine yields N-formyl and N-methyl piperidine but carbonylation is
otherwise uncommon. 2-Phenylindole is reduced only in the heterocyclic
ring.[450] Thiophene derivatives are reduced (e.g. 174 to 175)[451] but
2-acetyl thiophene is also hydrogenolised to give 176 and 177.[291]
More modern work has been dominated by $(\eta^3-C_3H_5)Co(P(OMe)_3)_3$[452]
which catalyses reaction at 1 atm., $25^\circ C$. Hydrogenation of benzene yields
cyclohexane; cyclohexene and cyclohexadiene are not observed. The
reaction is slow but is speeded by using bulkier phosphine or phosphite
ligands.[453] Certain limits on substitution obtain. Four or more CH_3-
groups in sequence reduce the rate to zero, electron withdrawing groups
deactivate, and protic substituents (-OH,-COOH) destroy the catalyst.

174   175   176   177

In addition to operating under mild conditions this catalyst has
the advantage of essentially complete stereoselectivity.   Deuteration of
benzene yields all cis-$C_6H_6D_6$ without H/D exchange.[454]   Cis-dimethyl
cyclohexanes are produced from xylene, and 178 from anthracene.[453]   Cyclohexenes are not intermediates.   The mechanism has been well studied.   The
allyl complex does not undergo detectable phosphite exchange, nor is there
evidence for interaction with arenes.[455]   Hydrogen reacts in the absence
of arene to give propene and low temperature n.m.r. suggests a dihydride is
present.   The first two steps are given by equations (35) and (36);
$\sigma \rightleftharpoons \pi$-interconversion precedes oxidative addition.

$$(\eta^3\text{-allyl})Co(P(OR)_3)_3 \rightleftharpoons (\eta^1\text{-allyl})Co(P(OR)_3)_3 \qquad (35)$$

$$(\eta^1\text{-allyl})Co(P(OR)_3)_3 + H_2 \rightleftharpoons (\eta^1\text{-allyl})CoH_2(P(OR)_3)_3 \qquad (36)$$

Loss of phosphite makes room for the incoming arene and rates are reduced
in the presence of excess phosphite.[456]   The intermediate, 179, was
proposed, with hydrogen transfer via a series of allyl cobalt monohydrides
and dihydrides.   The analogous complex, 180, is also active and its structure
has been determined.[457]   With substituted arenes the rate correlates with
σ indicating an electrophilic rate controlling step.   The reaction is zero
order in arene, suggesting that complex formation is not limiting and a

178   179   180

$k_H/k_D = 1.3$ indicates that H_2 addition is significant. Russian workers believe that addition of H_2 to a catalyst arene complex is rate determining.[458]

Reduction of alkyl and alkenyl benzenes is somewhat more complex. Toluene gives d_6-,d_7-,d_8- and d_9-products, though no deuterium is incorporated into the starting material. Tert-butyl benzene is not deuterated in the side chain and the implication is that cobalt allyls such as 181, 182 and 183 are involved.[459]

181   182   183

2.6.2 Rhodium complexes

Early catalysts suffered less from poor activity than from short lifetimes. $[Rh_2(PPh_3)_2Cl_2O_5(C_6H_6)_2]$ and $[Rh(H_2O)(PPh_3)_3ClO]_n$ are active for benzene reduction but last only about 20 turnovers.[263] Allyl rhodium phosphites[460] and (acac) rhodium phosphites[461] are similarly short-lived. Other reported rhodium catalysts are a diverse group. Acetophenone is reduced to 1-phenylethanol and cyclohexylmethylketone in the presence of $((cod)RhCl)_2$ and phase transfer agents[462] whilst $[(diphos)Rh(anthracene)]^+$ is a catalyst for the unexpectedly selective reduction of anthracene to 184.[463] $[(\eta^5-C_5Me_5)RhCl_2]_2$ in the presence of base and 50 atm.H_2 gives a stereoselective reaction. Unprotected OH or COOH groups poison the catalyst but aryl ethers, esters and ketones give good results.[464] Aryl halides are hydrogenolised. Rhodium complexes of N-phenylanthranilic acid are effective[465] but the difficulty of obtaining well characterised complexes makes mechanistic study impossible.

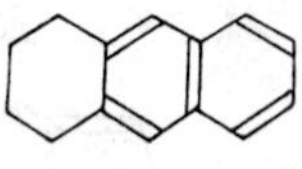

184

A fairly long-lived catalyst is provided by 185 which reacts reversibly with H_2 to give 186.[466] It is not, however, stereoselective, with up to 30% trans-products in xylene reduction. Also substantial H/D exchange in the unreacted arene is noted.

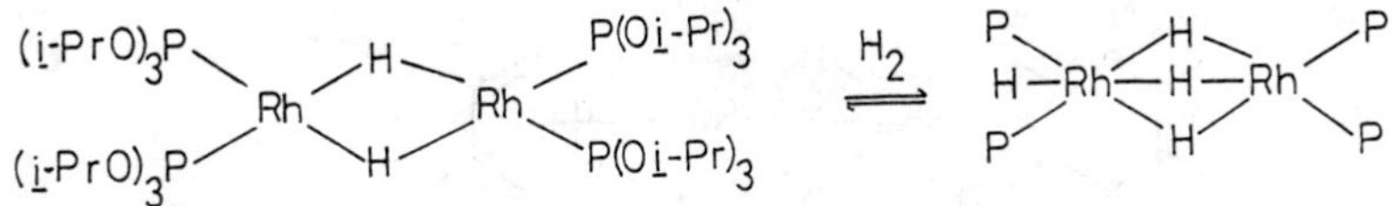

185

186

As usual heterocyclic rings are reduced more easily than benzene. (py)$_3$RhCl$_3$/NaBH$_4$/DMF catalyses the reduction of pyridine to piperidine and in quinoline only the heterocyclic ring is reduced.[467]

2.6.3 Ruthenium complexes

Arene complexation is a critical step in reduction and arene ruthenium complexes such as 187 are effective catalysts for benzene reduction.[468] Cyclohexenes are produced in non-trivial amounts and stereo-selection is limited, with o- and p-xylene giving respectively 10 and 18% trans-dimethylcyclohexanes. H/D exchange is extensive. The complex is in equilibrium with a bis-η^4-structure, 187b, which may react with H_2 in the first step of the cycle. Exchange to introduce the substrate is presumed to be rate limiting.

A number of other catalysts are known including the hexamethyl-benzene analogue of 187[469] and 188.[470] [Ru$_2$(η^6-C$_6$Me$_6$)$_2$H$_2$Cl$_2$] is a catalyst for reduction of substituted arenes; it is tolerant of aryl ethers, alcohols, esters and secondary amines but nitro groups are reduced and

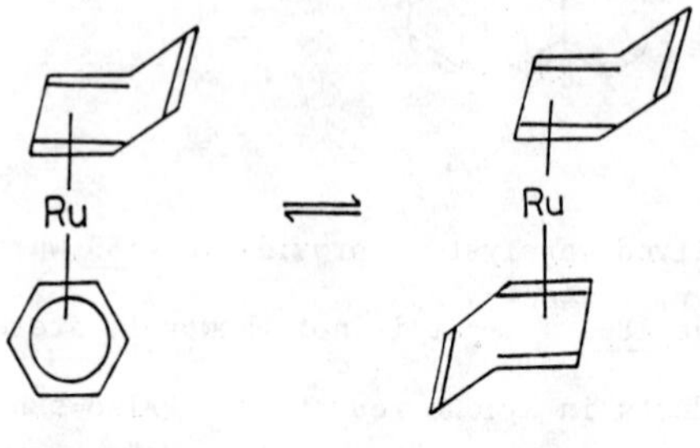

187a   187b

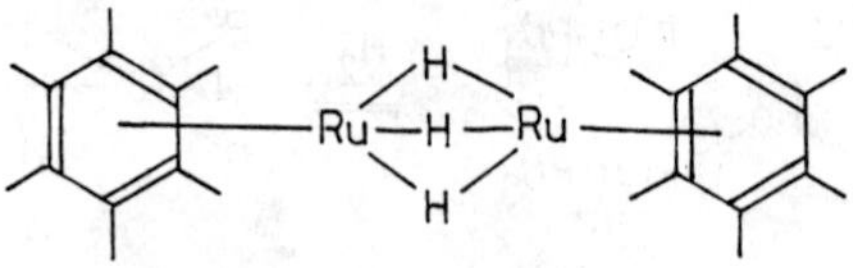

188

halides hydrogenolised.[471] Phosphine modified complexes are also useful; their distinctive feature is selective reduction of polycyclic compounds (Figure 41).

2.6.4 Ziegler catalysts

A wide range of catalytic reductions of arenes occur in the presence of Ziegler catalysts. However, the active species are poorly defined, conditions required are severe, and selectivity limited (Figure 42).

2.6.5 Other catalysts

η^6-Arene complexes of Ni and Co such as 189 are short lived

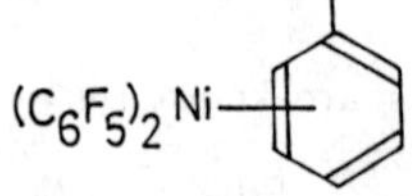

$(C_6F_5)_2 Ni$—

189

Ref.

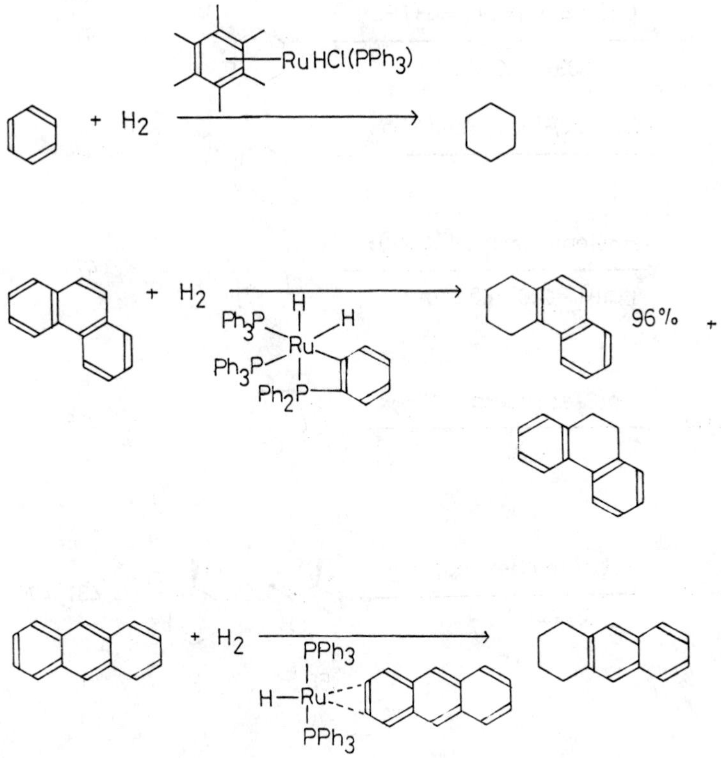

471

472

96% +

473

Figure 41 Reduction of arenes in the presence of ruthenium complexes.

catalysts for arene reduction.[478] Fe(CO)$_5$ and Mn$_2$(CO)$_8$(Bu$_3$P)$_2$ catalyse

the water gas shift and use H$_2$O/CO to reduce the heterocyclic ring of 190

with 100% selectivity.[479,480]

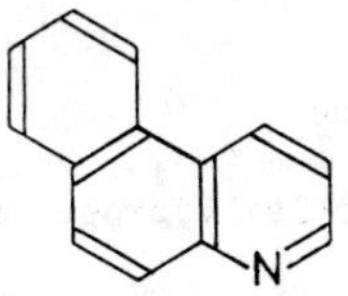

190

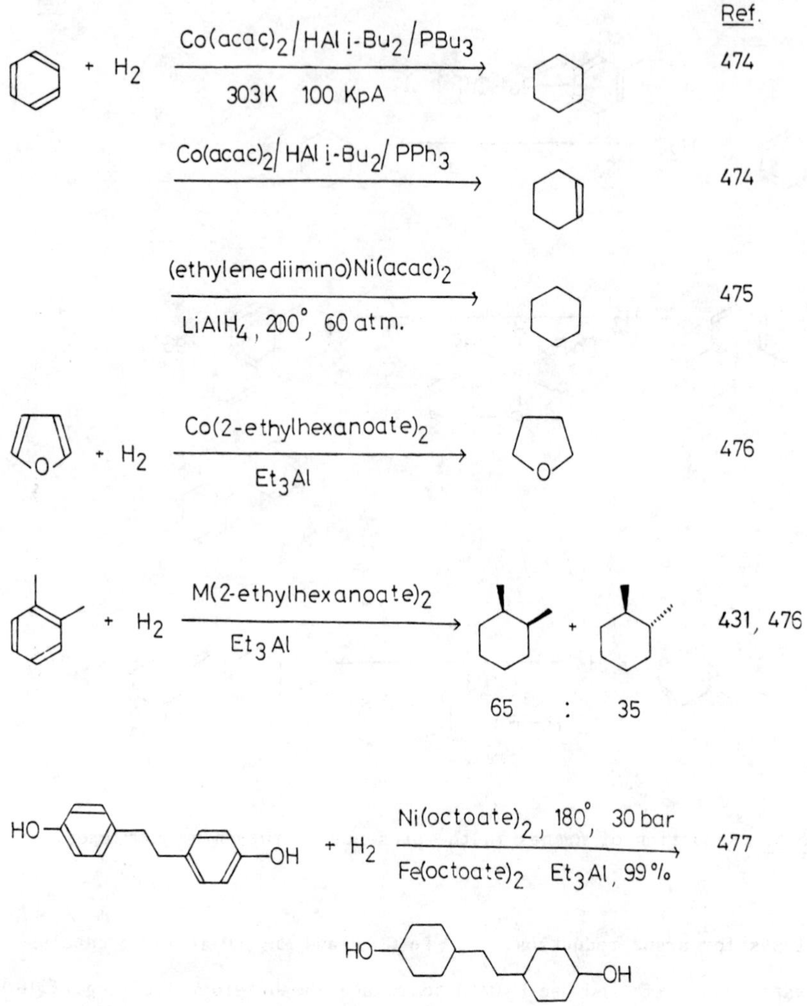

Figure 42   Reduction of arenes in the presence of Zeigler catalysts.

2.7 Hydrogenation of carbonyl groups

Catalytic addition of molecular hydrogen is rarely the method of choice for carbonyl reduction.   Alternatives are provided by hydrosilylation, hydride reduction, or transfer hydrogenation.   Addition of H_2 is relatively difficult since metal carbonyl dihydrides are few and the alcohol products poison many catalysts.[481]   A number of special cases are useful, however.

2.7.1 Hydrogenation of Aldehydes

Aldehydes are reduced to primary alcohols at 1 atm., 25°C in the presence of $[(R_3P)_2Rh(diene)]^+ClO_4^-$.[482]   Basic phosphines are more useful than PPh_3[25] and a trace of water is essential.   The lifetime of the catalyst is short with irreversible poisoning by decarbonylation.   Other useful rhodium catalysts are shown in Figure 43.

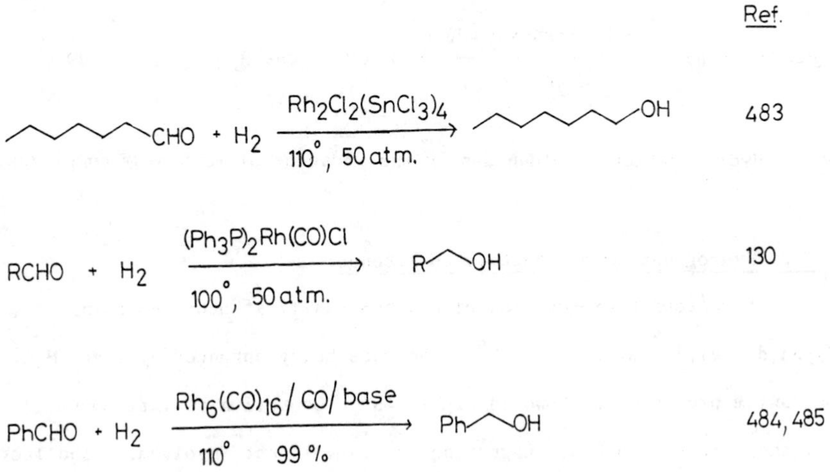

Ref.

Figure 43   Hydrogenation of aldehydes in the presence of rhodium complexes.

$H_3Ir(PPh_3)_3/HOAc$ is a very active catalyst with aldehyde reduction occurring at 1 atm., 50-100°C.[287,317]   Ketones require more forcing conditions (100 atm.) and the catalyst is long lived.

Neutral ruthenium catalysts have been successfully used for aliphatic and aromatic aldehydes.[49]   Some decarbonylation is unavoidable using $(Ph_3P)_3RuCl_2$ or $HRuCl(PPh_3)_3$ but the complex formed $(HRuCl(CO)(PPh_3)_3)$ is also active in reduction.[486]   Ketones and nitro groups are not reduced and selectivity, lifetime and oxygen stability are all high (Figure 44).

Ref.

$$RCHO + H_2 \xrightarrow{\text{HRuBr}(PEt_2Ph)_3} RCH_2OH$$

317

$$PhCHO + H_2 \xrightarrow[100^{\circ}C, 1000 \text{ p.s.i.}]{\text{HRuCl(CO)}(PPh_3)_3} PhCH_2OH$$

487

$$\underline{D}\text{-glucose} + H_2 \xrightarrow[\text{DMA, 3 atm., } 75^{\circ}C]{\text{HRuCl}(PPh_3)_3 \mid HCl} \underline{D}\text{-glucitol}$$

488

$$CH_3CH_2CHO + H_2 \xrightarrow[C_{10}H_8, \, Et_2O]{K^+[(Ph_3P)_2Ph_2\overline{PC_6H_4RuH_2}]^-} CH_3CH_2CH_2OH$$

489

Figure 44 Hydrogenation of aldehydes in the presence of ruthenium complexes.

2.7.2 Hydrogenation of unactivated ketones

Homogeneous hydrogenation of ketones was first achieved using $[(R_3P)_2Rh(diene)]^{+}$ [25] at 1 atm., $25^{\circ}C$, the rate being enhanced by trace H_2O. The mechanism proposed is shown in Figure 45. Deuterium is introduced only at the carbonyl carbon, suggesting the enol is not involved. The lack of a primary isotope effect indicates that H-transfer is not rate-controlling. [490]

"In situ " catalysts from $((diene)RhCl)_2/R_3P$ are most successful with basic unhindered phosphines. [491] Strongly basic conditions were favoured by Gargano using $(Ph_3P)RhCl(C_8H_{12})$ and $(Ph_3P)_2Rh_2H_2Cl_2(C_8H_{12})$. [492] Rhodium hydroxide complexes are thought to be the true catalysts and the enol form is reduced. Addition of R_3N to $(Ph_3P)_3RhCl$ gives another active catalyst [493] and there are other examples (Figure 46).

The iridium catalysts discussed in 2.7.1 are also active for ketone reduction, though more severe conditions are necessary. [497] The ruthenium

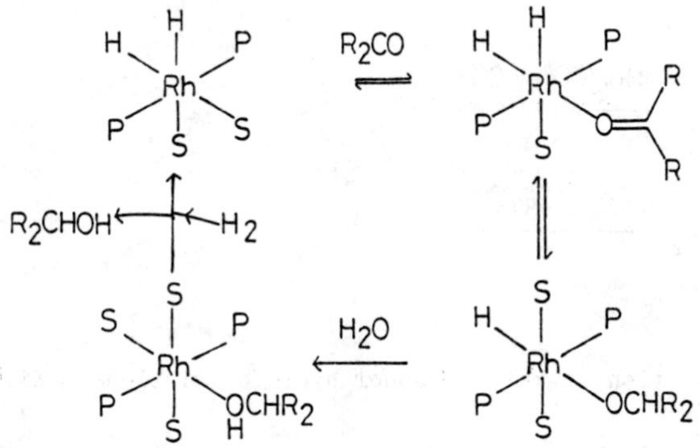

Figure 45   Mechanism of hydrogenation of ketones in the presence of
cationic rhodium complexes.

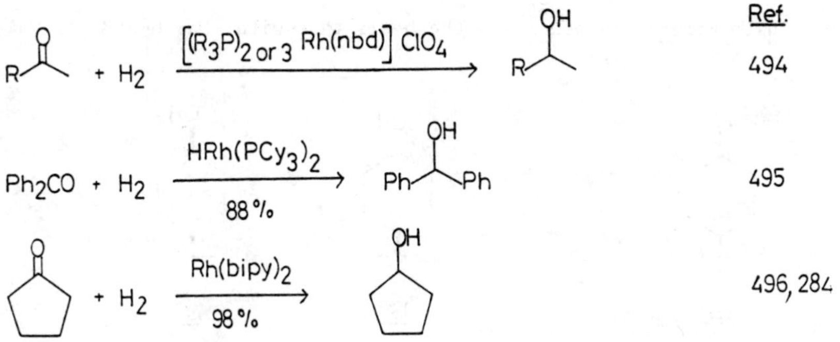

Figure 46   Hydrogenation of ketones in the presence of rhodium complexes.

species of 2.7.1 are also useful here.[487,490]   Other ruthenium complexes
used specifically for ketones include (46)$_2$RuCl$_2$(H$_2$O),[498]   H$_4$Ru$_4$(CO)$_{12}$ and
H$_4$Ru$_4$(CO)$_9$(PPh$_3$)$_3$.[499]

Hydroformylation conditions are severe but reduction may be
achieved[500]   (Figure 47).   Ziegler catalysts give poor results and are
incompatible with many other functional groups.[302,320,400,403]   Cyclohexanone, however, is cleanly reduced in the presence of
Ni(2-ethylhexanoate)$_2$/Et$_3$Al at 100-150°C, 70-100 atm.[304]

Ref.

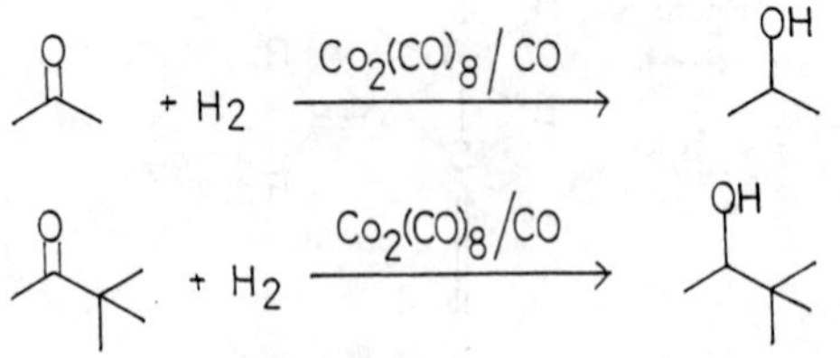

291

501

Figure 47 Hydrogenation of ketones under hydroformylation conditions.

Reduction of 4-tert-butylcyclohexanone, 190, yields two diastereomeric products. The trans-isomer, 191, is the more stable thermodynamically but 192 requires approach to the less hindered face of the carbonyl. Catalysts vary widely in their stereoselectivity (Table 13). These results were explained[502] on the basis that with PPh_3 $HRh(PR_3)_3$ was

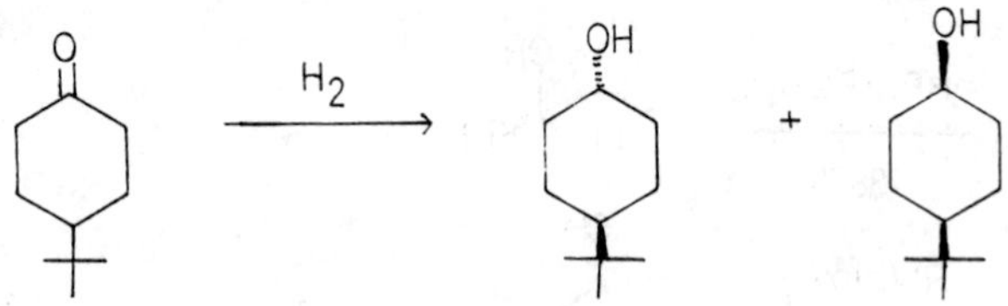

190 191 192

Table 13. Stereoselectivity in reduction of 4-tert-butylcyclohexanone

Catalyst	% trans	Solvent	Temp(°C)	Ref.
$[H_2Rh(PMe_2Ph)S_2]^+$	86	MeOH	25	25
$[(Ph_3P)_2Rh(cod)]^+/Et_3N$	63	PhH	20	97
$(Rh(nbd)Cl)_2/PPh_3/Et_3N$	27	PhH/MeOH	50	502
$(Rh(nbd)Cl)_2/PBu_3$	91	PhH/MeOH	50	502
$HRh(PPh_3)_4$	20	PhH/MeOH	50	502
$HRh(PCy_3)_2$	94	i-PrOH	20	495

the key intermediate. This coordinates the carbonyl edge-on from the less hindered face. More basic phosphines gave mainly $H_2Rh(PR_3)_2$ which coordinates the ketone as an n-donor.

2.7.3 Activated Ketones

Aryl diketones and α-dicarbonyl compounds are more difficult to reduce than aliphatic ketones using hydride reducing agents. The reverse is true for hydrogenation where additional unsaturated groups assist coordination. Acetophenone is reduced to 1-phenylethanol in the presence of $[(Me_2PhP)_2Rh(diene)]^+$ or $(py)_3RhCl_3/NaBH_4$ at 1 atm., $25°C.$[25,343] $(Rh(cod)Cl)_2/CTAB/bovine$ serum albumin is active but unselective; 21% 1-phenylethanol is obtained together with 28% cyclohexylmethylketone.[462] $Co_2(CO)_8/H_2/CO$ is unsuitable, as competitive hydrogenolysis to the alkyl-benzene occurs.[501] 193 is cleanly reduced to 194.

193   194

Keto esters are hydrogenated using a number of catalysts inactive for simple ketones. Wilkinson's catalyst is active giving 196 in 98% yield from 195.[503] (dppb)RhCl(S) is also active. $[Co(CN)_5]^{3-}$ is selective for conjugated carbonyls; benzaldehyde is reduced to benzyl alcohol[55] and benzil selectively to benzoin.[268] In contrast, $(py)_3RhCl_3/NaBH_4$ yields PhCHOHCHOHPh.[343] Pyruvic acid is reduced to lactic acid in the presence of $(46)_3HRu(OAc).$[318]

195        $+ H_2$ $(Ph_3P)_3RhCl, 20°$ 196
                        20 atm., THF

Hydrogenation of other carbonyl compounds is uncommon. Activated esters such as CF_3CO_2Me are cleanly reduced to alcohols in the presence of $K^+[(Ph_3P)_2(Ph_2P)Ru_2H_4].2C_6H_{14}O_3$ [489] but unactivated esters give poor conversions to complex mixtures. γ-Ketoacids such as 197 are reduced and cyclised in the presence of $(Ph_3P)_3RuCl_2$ [504] and diacids may also be reductively cyclised, the selectivity depending on the ring size (e.g. 199 and 201). [499] Diesters react in a similar way (e.g. 206 and 208). [505]

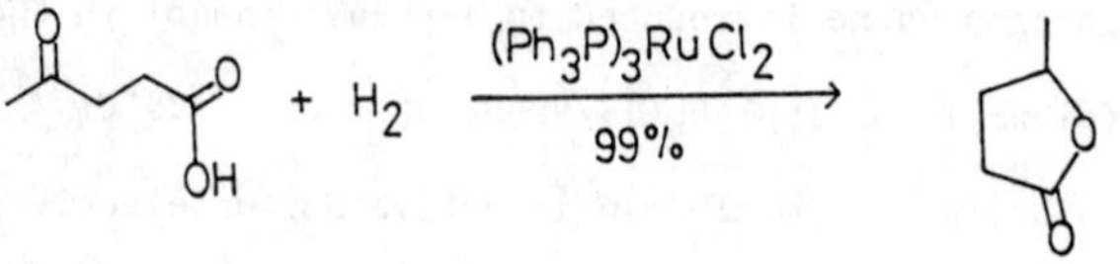

197   + H₂ $\xrightarrow[99\%]{(Ph_3P)_3RuCl_2}$ 198

199   + H₂ $\xrightarrow[100^\circ, 50\%]{H_4Ru_4(CO)_8(PBu_3)_4}$ 200

201   + H₂ $\xrightarrow[100^\circ]{H_4Ru_4(CO)_8(PBu_3)_4}$ 202   +   203   +

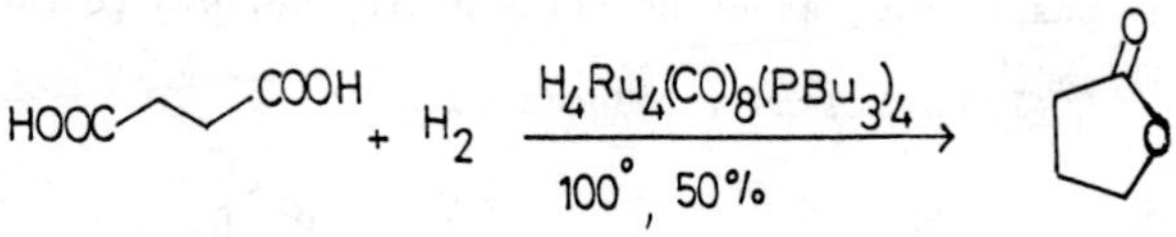

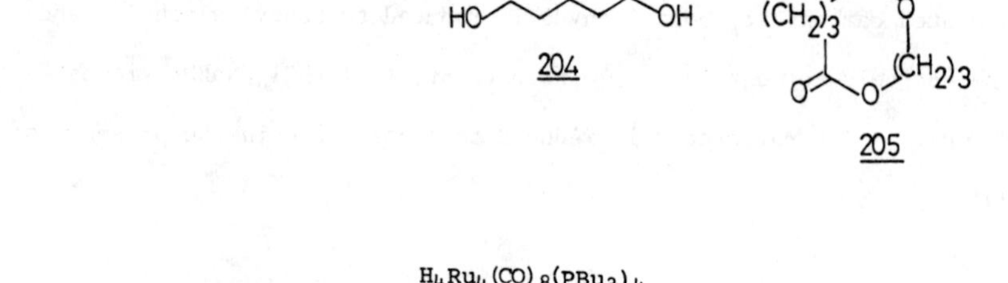

204   205

$MeOOC-COOMe$ + H₂ $\xrightarrow[180^\circ, 130\ atm., 100\%]{H_4Ru_4(CO)_8(PBu_3)_4}$ $HOCH_2COOMe$

206   207

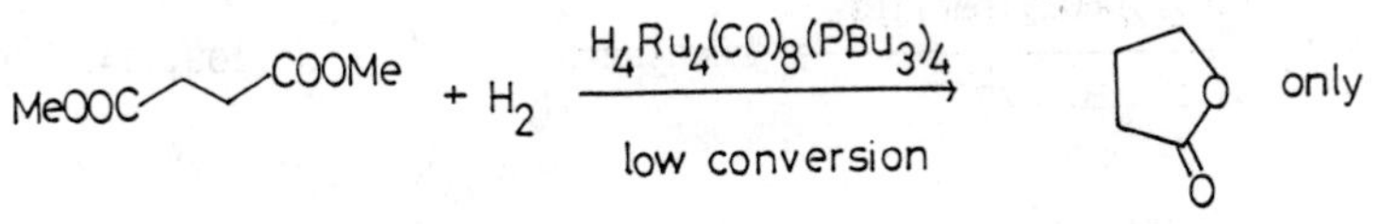

2.7.4 αβ-Unsaturated carbonyl compounds

Carbon-carbon double bond reduction in enones is easy and we saw numerous examples (2.3.3). Reduction to allyl alcohols is much more difficult and little success has been achieved to date. Some examples are shown in Figure 48 but this remains a reaction best achieved via hydrosilylation or hydride reduction.

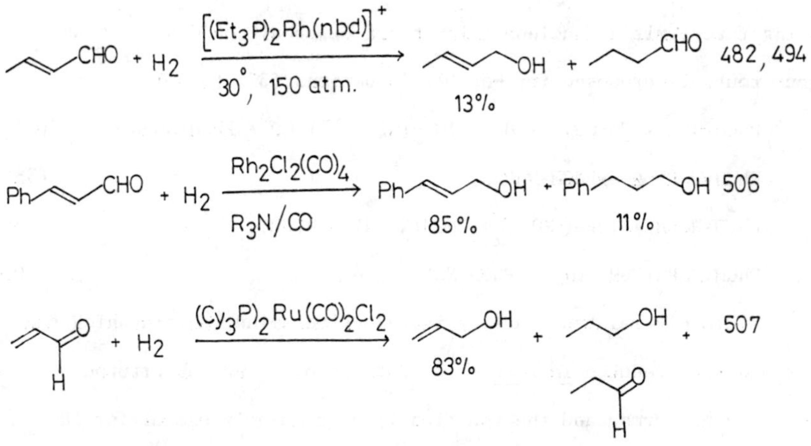

Ref.

Figure 48 Reduction of αβ-unsaturated aldehydes to allyl alcohols.

2.8 Nitrogen containing compounds

2.8.1 Carbon-nitrogen double bonds

The reduction of C=N has been sought as a synthetic route to substituted amines. The reduction is facile but not widely used.

Ref.

$$PhCH=NPh + H_2 \xrightarrow[\text{1 atm., 25}^{\circ}]{[(py)_2RhCl_2(DMF)]BF_4} PhCH_2NHPh$$

138, 343, 467

$$RN=CHR' + H_2 \xrightarrow[\text{DMF}]{(PhP)_2Pd_5} RHNCH_2R'$$

508

$$PhN=CHPh + H_2 \xrightarrow{Fe(CO)_5} PhNHCH_2Ph$$

509

Figure 49 Homogeneous hydrogenation of imines.

(Figure 49). Few data give any clue to the mechanism. Using $ArCH=NAr^1$ and $[Co(CN)_5]^{3-}$ the rate increases with electron withdrawing groups in Ar^1 suggesting rate limiting nucleophilic attack of cobalt on C=N.[510] An analogous route is proposed for $Fe(CO)_5$ (equations (37) to (40)).[509]

$$PhCH=NPh + Fe(CO)_5 + H_2 \rightarrow [HFe(CO)_4]^- + CO + PhCH=\overset{+}{N}HPh \quad (37)$$

$$PhCH=\overset{+}{N}HPh \longleftrightarrow Ph\overset{+}{C}H-NHPh \quad (38)$$

$$Ph\overset{+}{C}H-NHPh + [HFe(CO)_4]^- \rightarrow PhCH(NHPh)-FeH(CO)_4 \quad (39)$$

$$PhCH(NHPh)FeH(CO)_4 \rightarrow PhCH_2NHPh + Fe(CO)_4 \quad (40)$$

A closely related reaction is reductive amination, in which the imine is made and reduced in situ.[510] Both primary and substituted amines[511] may be formed and the reaction is particularly useful for the synthesis of amino acids from α-keto acids.[512] Fairly severe conditions of temperature and pressure are required (Figure 50).

Oximes are reduced to primary amines but there is little advantage over heterogeneous catalysts. Both reduction of C=N and hydro-genolysis of N-O occurs (e.g. 210 to 211).[510]

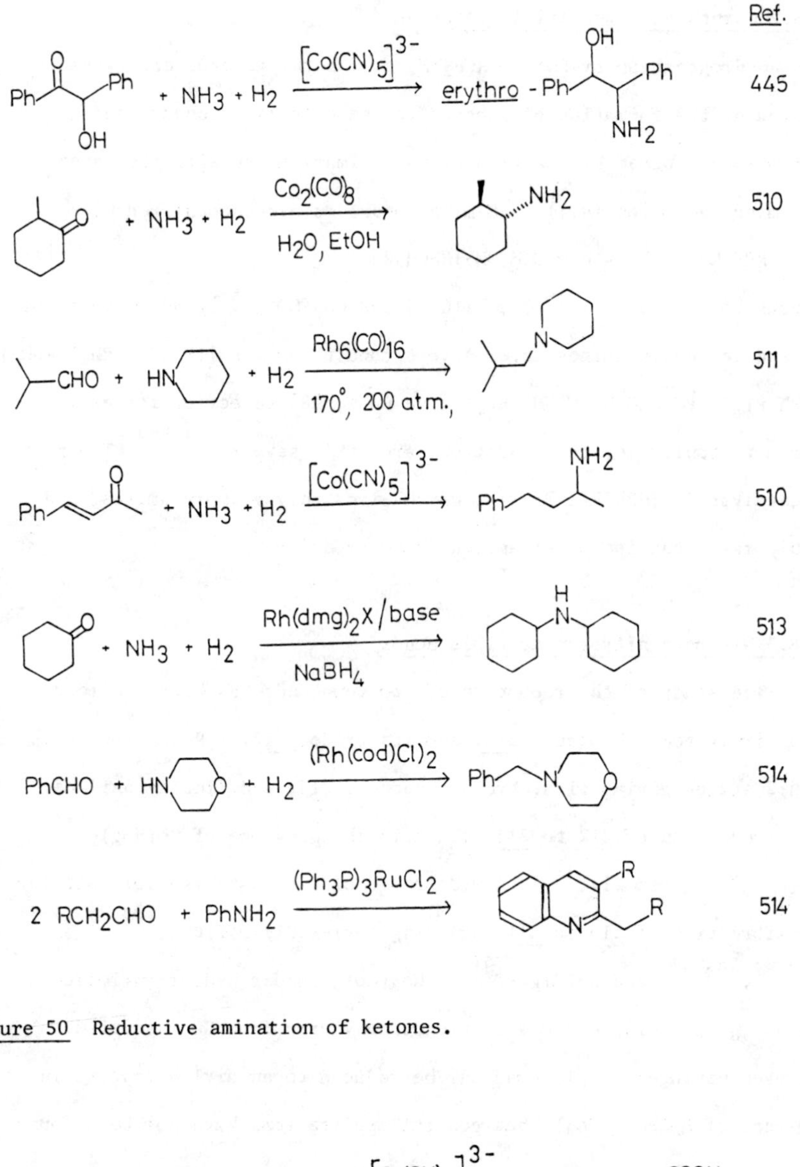

Ref.

Ph–C(O)–CH(OH)–Ph + NH$_3$ + H$_2$ $\xrightarrow{[Co(CN)_5]^{3-}}$ erythro - Ph–CH(OH)–CH(NH$_2$)–Ph   445

(2-methylcyclohexanone) + NH$_3$ + H$_2$ $\xrightarrow[H_2O, EtOH]{Co_2(CO)_8}$ (2-methylcyclohexylamine)   510

(CH$_3$)$_2$CH–CHO + HN(piperidine) + H$_2$ $\xrightarrow[170°, 200 atm.,]{Rh_6(CO)_{16}}$ (isobutylpiperidine)   511

Ph–CH=CH–C(O)–CH$_3$ + NH$_3$ + H$_2$ $\xrightarrow{[Co(CN)_5]^{3-}}$ Ph–CH$_2$–CH$_2$–CH(NH$_2$)–CH$_3$ 510

(cyclohexanone) + NH$_3$ + H$_2$ $\xrightarrow[NaBH_4]{Rh(dmg)_2X/base}$ dicyclohexylamine   513

PhCHO + HN(morpholine) + H$_2$ $\xrightarrow{(Rh(cod)Cl)_2}$ Ph–CH$_2$–N(morpholine)   514

2 RCH$_2$CHO + PhNH$_2$ $\xrightarrow{(Ph_3P)_3RuCl_2}$ (3-R-2-(CH$_2$R)quinoline)   514

Figure 50 Reductive amination of ketones.

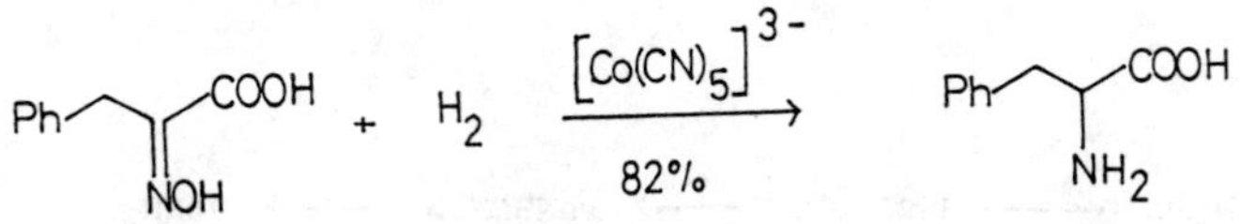

Ph–CH$_2$–C(=NOH)–COOH + H$_2$ $\xrightarrow[82\%]{[Co(CN)_5]^{3-}}$ Ph–CH$_2$–CH(NH$_2$)–COOH

210   211

2.8.2 Carbon nitrogen triple bonds

Hydrogenation of RCN to RCH_2NH_2 over heterogeneous catalysts is complicated by the formation of substantial amounts of secondary and tertiary amines, formed by reaction of the primary amine with the intermediate imine (equation (41)). Similar problems are encountered with

$$RCH_2NH_2 + RCH=NH \rightarrow RCH(NH_2)NHCH_2R \qquad (41)$$

homogeneous catalysts. $Fe(CO)_5$, $Ni(CO)_4$ and $Co_2(CO)_8$[515] give mixtures of primary and secondary amines under severe conditions. $K_2[(Ph_3P)_2(Ph_2P)RuH_2]$ and $K[(Ph_3P)_2Ph_2\overline{PC_6H_4Ru}H_2]C_{10}H_8.Et_2O$ give up to 98% selection for ethylamine from acetonitrile, but conditions are still severe.[295,489] The most useful catalyst is $[HRh(P(\underline{i}-Pr)_3)_3]$ which gives only primary amines, and reasonably fast reaction under ambient conditions.[516]

2.8.3 Nitrogen nitrogen multiple bonds

The study of the reduction of azo compounds has been confined almost entirely to azobenzene, 213, and its oxide, 212. Reduction takes place in 2 stages giving first hydrazobenzene, 214, and finally aniline. Selective reduction of 212 to 213 occurs in the presence of $Co(dmg)_2$[517] but $[Co(CN)_5]^{3-}$ gives 214.[518] A wide range of catalysts is available for the transformation of 213 to 214 including $(py)_2RhCl_2(DMF)BH_4$,[138,343,467] $[Co(CN)_5]^{3-}$,[55,518] and $Co(dmg)_2$.[517] $Co_2(CO)_8$, under hydroformylation conditions, gives aniline and a little diphenylurea from CO insertion. The nitrogen nitrogen triple bond may be reduced to an aryl hydrazine in the presence of $H_2RhCl(PPh_3)_2$ but general applications have not been found.[519]

$$\underset{\underset{+}{212}}{\overset{\overset{O^-}{|}}{Ph-N=N-Ph}} \longrightarrow \underset{213}{PhN=NPh} \longrightarrow \underset{214}{PhNHNHPh} \longrightarrow PhNH_2$$

2.8.4 Nitro and Nitroso groups

Reduction of the nitro group proceeds in a stepwise fashion and the intermediate may, on occasion, be isolated (Figure 51). Additionally

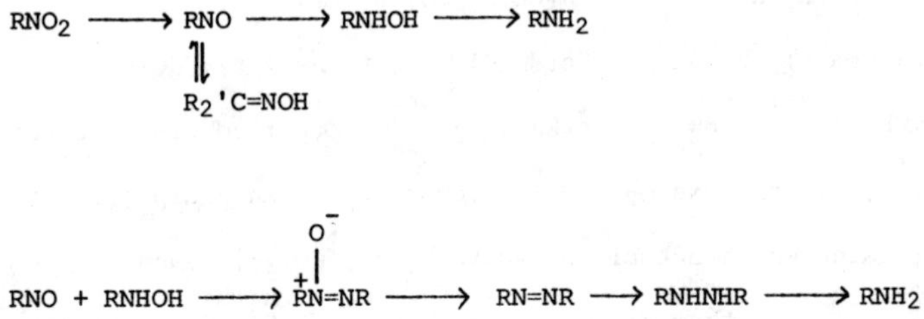

$$RNO_2 \longrightarrow RNO \longrightarrow RNHOH \longrightarrow RNH_2$$

$$R_2{'}C=NOH$$

$$RNO + RNHOH \longrightarrow \overset{O^-}{\underset{}{R\overset{+}{N}=NR}} \longrightarrow RN=NR \longrightarrow RNHNHR \longrightarrow RNH_2$$

Figure 51 Routes for reduction of nitro groups.

they may react together to give azobenzene so this is a complex reaction. Partial reduction to oximes is known for aliphatic nitro compounds, with the tautomerisation of the intermediate nitroso compound. The catalyst is usually Cu(I) and the conditions are quite severe.[520] (e.g. 215 to 216). PhNHOH may be obtained from PhNO$_2$ using catalysts capable of 1-electron transfer; 99% selectivity was observed using the rhodium indigosulphonic acid, 217.[521]

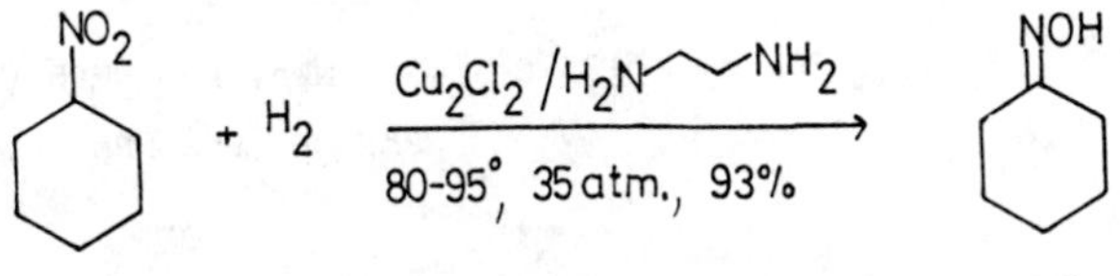

215 + H$_2$ $\xrightarrow[\text{80-95}^\circ, \ 35\,\text{atm.,} \ 93\%]{\text{Cu}_2\text{Cl}_2/\text{H}_2\text{N}\frown\text{NH}_2}$ 216

215

216

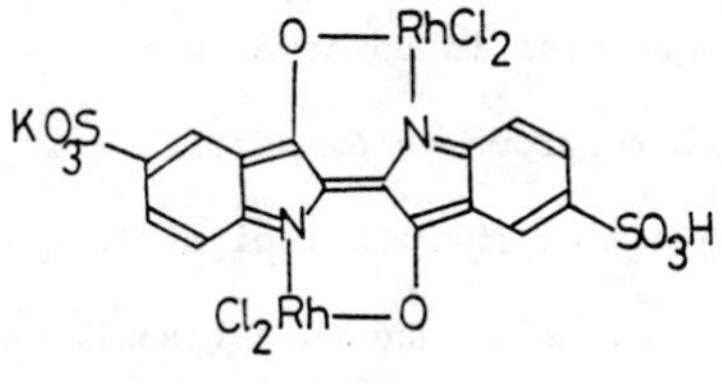

217

Dimerised products are isolated with many catalysts but complete reduction may usually be achieved using more severe conditions or longer reaction times (Figure 52).   Generally we wish to obtain amines in high yield.   Suitable catalysts include $(py)_2RhCl_2(DMF)BH_4$ [138,343,467], $[Co(CN)_5]^{3-}$ [55], $Ru_3(CO)_{12}$ [523], $HRh(CO)_4$ [524], $Co_2(CO)_8$ [525], $Ni(2-ethylhexanoate)_2/Et_3Al$ [431], $Co(dmgH)_2$ [526], $\underline{trans}-(py)_2PdCl_2$ [527], $(acac)Rh(P(OPh)_3)_2$ [461] and $(\underline{n}\ C_5Me_5RhCl_2)_2$ [464].   Other complexes catalyse the water gas shift and use H_2O/CO for reduction.   An example is provided by $Rh_6(CO)_{16}/amine$ which actually involves $[Rh_{12}(CO)_{30}]^{2-}$ and $[Rh_7(CO)_{16}]^{3-}$ [528].   Another employs $(Ph_3P)_2PtCl_2/SnCl_4/Et_3N$ but the mechanism is more complex;   no H_2 is detected and the water gas shift does not occur in the absence of substrate. [529-531]

	Conditions	Products	Ref.
$PhNO_2 + H_2 \xrightarrow{[Co(CN)_5]^{3-}}$	2 moles NaOAc	PhN=NPh	55
	1.2 moles NaOAc	PhNHNHPh	55
	1 atm., 25°C	PhN=NPh, PhNHNHPh	55, 268
	30 atm., 80°C	PhN=NPh, PhNHNHPh, PhNH₂	55
	70 atm., 25°C	PhN=NPh	55, 268
	70 atm., 125°C	PhN=NPh, PhNHNHPh, PhNHOH, PhN̟=NPh $\overset{\mid}{O^-}$	522

Figure 52   Reduction of nitrobenzene in the presence of $[Co(CN)_5]^{3-}$.

The reaction mechanism is unknown in most cases but may involve electron transfer, the nitro group being particularly susceptible to this type of reduction.   The use of rhodium indigosulphonic acids was mentioned; these give diradicals on reduction and then transfer electrons to yield $PhNO_2^{-\cdot}$, and ultimately $PhNH_2$ (Figure 53). [532]   Catalysts derived from

$$2PhNO_2^{-\cdot} + 2H^+ \longrightarrow PhNO_2 + PhNO + H_2O$$

$$PhNO + 2PhNO_2^{-\cdot} + 4H^+ \longrightarrow PhNHOH + 2PhNO_2 + H_2O$$

$$PhNHOH + 2PhNO_2^{-\cdot} + 2H^+ \longrightarrow PhNH_2 + 2PhNO_2 + H_2O$$

Figure 53 Electron transfer mechanism for nitrobenzene reduction.

Co(dmgH)$_2$[526,533] also involve paramagnetic intermediates as do those using chloranilic acid (218)/K$_2$IrCl$_6$/NaBH$_4$[534] and ($\underline{n}$-allyl PdCl)$_2$/NaBH$_4$/P(NR)$_3$[229].[3]

Aromatic nitro groups are more readily reduced than alkenes or carbonyls so conversion of 219 to 220 is achieved with good selectivity.[517]

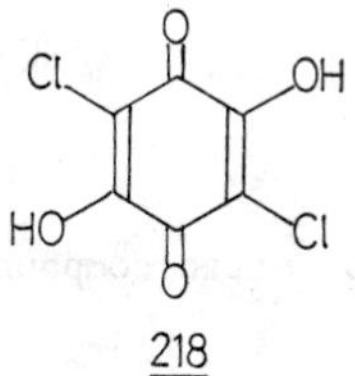

218

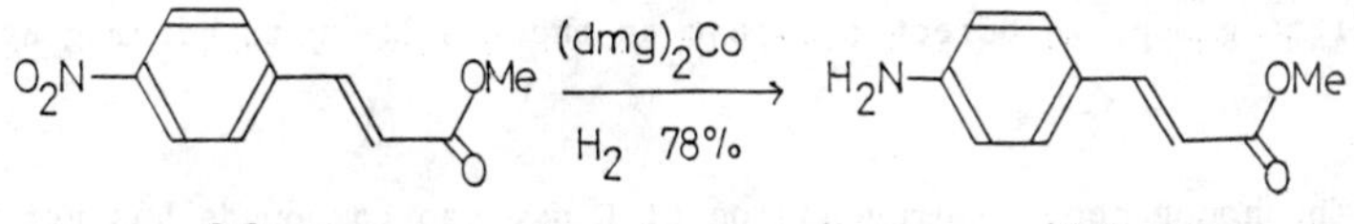

219   220

The position is reversed with aliphatic nitro compounds.[283] (Section 2.3.5)

Another long standing problem is selective reduction of one of two nitro groups on an aromatic ring. Selectivity depends on the fact that PhNO is reduced faster than PhNO$_2$. Some typical results are shown in Figure 54. It is notable that reduction of 221 occurs at the more

Ref.

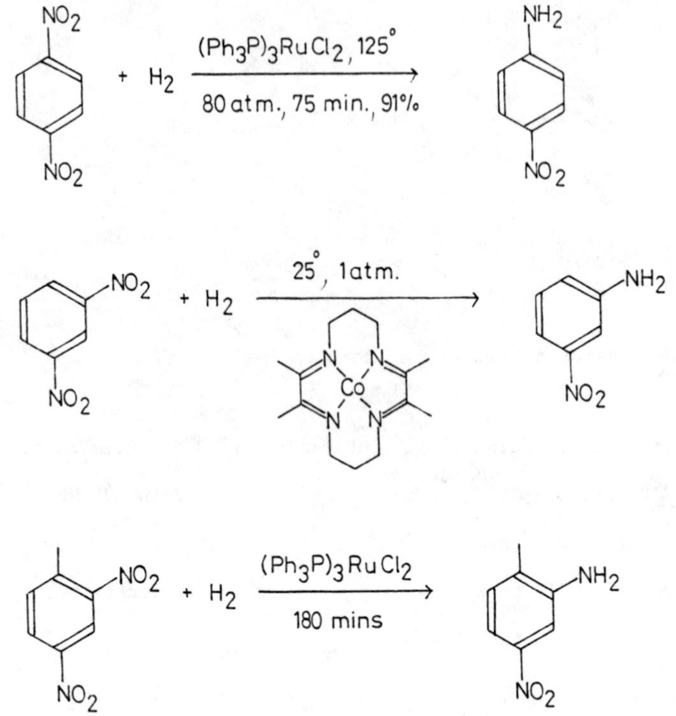
NO2
NO2
+ H2
(Ph3P)3RuCl2, 125°
80 atm., 75 min., 91%
NH2
NO2
NO2
NO2
+ H2
25°, 1 atm.
N
N
Co
N
N
NH2
NO2
NO2
NO2
+ H2
(Ph3P)3RuCl2
180 mins
NH2
NO2

535

536

535

221

Figure 54 Selective reduction of polynitro compounds.

hindered nitro group, in direct contrast to the results with heterogeneous catalysts.

The homogeneous hydrogenation of C-nitroso compounds has not been studied independently but there are isolated reports of the reduction of N-nitroso compounds to secondary amines (e.g. 222 to 223).[288,378,522]

$$\underset{\mathbf{222}}{Ph_2N\text{-}N{=}O} + H_2 \xrightarrow[\text{70 atm., }60^{\circ}]{[Co(CN)_5]^{3-}} \underset{\mathbf{223}}{Ph_2NH}$$

2.9 Transfer Hydrogenation

The use of an organic donor as a source of reducing power dates back to 1903 when Knoevenagel reported that dimethyl-1,4-dihydroterephthalate, 224, disproportionated in the presence of Pd black. With homogeneous catalysts, the reaction does not need pressure apparatus and some unusual selectivities are noted.[537]

224   225   226

2.9.1 The hydrogen donor

The donor may, in principle, be any organic compound whose oxidation potential is low enough for H-transfer under mild conditions. In practice, donors are usually hydroaromatics and alcohols. Both 1° and 2° alcohols are useful, the 2° being more reactive and the 1° more selective.[538-541] For example, with $(Ph_3P)_3RuCl_2$ as catalyst 1° alcohols reduce only $\alpha\beta$-unsaturated C=C whereas 2° alcohols also reduce carbonyls. Isopropanol is the most common reductant; it is inexpensive and the product acetone is volatile. Benzylic alcohols are better donors than aliphatic analogues, the reaction being driven by conjugation in the product[542] but other bulky groups reduce the rate.[543]

Hydroaromatic compounds used as donors include indoline, tetrahydroquinoline[544], pyrrolidine, dihydrofuran, dioxan[545] and dihydroanthracene.[546] Reactivity order seems to depend on the substrate. Formic acid[547] and formaldehyde[548] are also popular donors having the advantage that the oxidation product, CO_2, is volatile.

2.9.2 Reduction of ketones

Ketones are reduced by H-transfer from i-PrOH in the presence of several rhodium, ruthenium and iridium complexes (Figure 55).   Reaction times are long but conversion selective.   Other alcohols, both 1° and 2°, have been used, but the position of the equilibrium is less favourable.

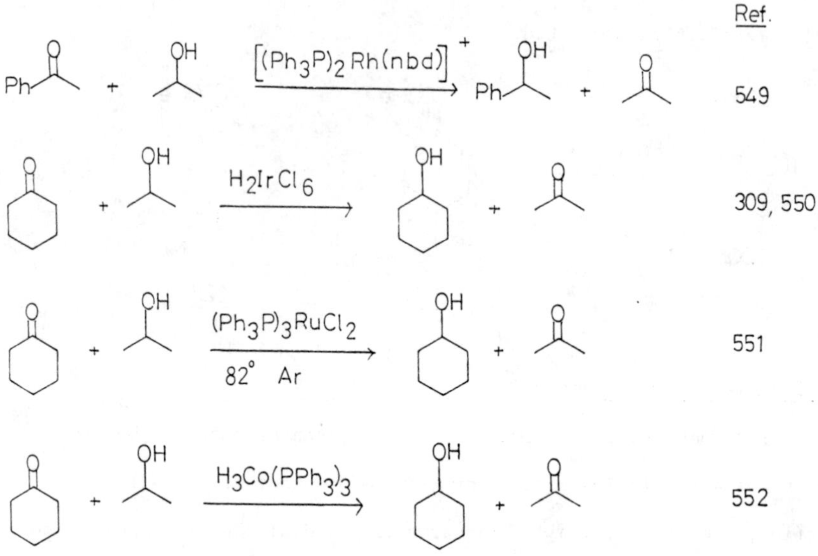

Ref.

Ph—C(O)—CH₃ + (CH₃)₂CHOH $\xrightarrow{[(Ph_3P)_2Rh(nbd)]^+}$ Ph—CH(OH)—CH₃ + (CH₃)₂C=O   549

cyclohexanone + (CH₃)₂CHOH $\xrightarrow{H_2IrCl_6}$ cyclohexanol + (CH₃)₂C=O   309, 550

cyclohexanone + (CH₃)₂CHOH $\xrightarrow[82°\ Ar]{(Ph_3P)_3RuCl_2}$ cyclohexanol + (CH₃)₂C=O   551

cyclohexanone + (CH₃)₂CHOH $\xrightarrow{H_3Co(PPh_3)_3}$ cyclohexanol + (CH₃)₂C=O   552

Figure 55 Transfer hydrogenation of ketones by iso-propanol.

For substituted cyclohexanones the thermodynamically disfavoured axial alcohol is usually the major product.   Typical selectivities for t-butylcyclohexanone are shown in Table 14.   Similar results are obtained for other substrates.   4-Methylcyclohexanone gives cis-alcohol with Rh(I)/PPh₃ in situ but the trans-isomer predominates with Ph₂PNR₂.[557] 227 yields the axial alcohol with (Ph₃P)₃RhCl but the equatorial isomer with (Ph₃P)₃RuCl₂.[353]   The 3-keto groups of steroids give axial alcohols but 4- and 17-keto compounds are unaffected.[558]   With few exceptions the catalyst approaches the carbonyl from the less hindered side, the degree of preference reflecting steric factors.

R—N (structure with methyl and =O)

<u>227</u>

Table 14.   Reduction of 4-<u>tert</u> butyleyclohexanone by <u>i</u>-PrOH

Catalyst	% trans	Ref.
$H_2IrCl_6/(MeO)_3P$	98	553
$[IrCl_6]^{2-}/H_2O/DMSO$	75	539,550
$IrCl_4/H_3PO_3$	97	550
$Ir(C_2H_4)_2(bipy)$	97	554
$[(4,7-Me_2phen)_2RhCl_2]Cl/KOH$	20	555
$[(diphos)Rh(nbd)]^+/KOH$	67	556
$HRh(PCy_3)_2$	94	495

Differences in rate between catalysts depend on the ligands.   In
the series $[P_2Rh(diene)]^+$, diphos > dppp ∿ PPh_3 > dppb ∿ DIOP > PMe_2Ph >
Ph_2PCH_2Ph > $PMePh_2$ and PCy_2Ph > P(<u>p</u>-tolyl)$_3$ ∿ P(<u>p</u>-MeOC$_6$H$_4$)$_3$ ∿ PPh_3 >
$PCyPh_2$ > PCy_3 ∿ PMe_2Ph > $PMePh_2$ > P(<u>o</u>-tolyl)$_3$ > P(<u>t</u>-Bu)$_3$ > P(<u>o</u>-MeOC$_6$H$_4$)$_3$.
This reflects two opposing trends;   steric hindrance slows reaction but
basicity enhances it.[559]

Hydroaromatics are also useful donors, dihydrofuran being superior
to alcohols.[547]   Some examples using these or formic acid are given in
Figure 56.

The reaction mechanism may follow one of two general routes
(Figure 57).[560]   In the first, A, the donor is coordinated and then the
acceptor whilst in B, this order is reversed.   Additional complications are
provided by <u>bis</u>-donor and acceptor complexes and the order of mixing may
affect the rate.[561]   Different catalysts have different contributions from

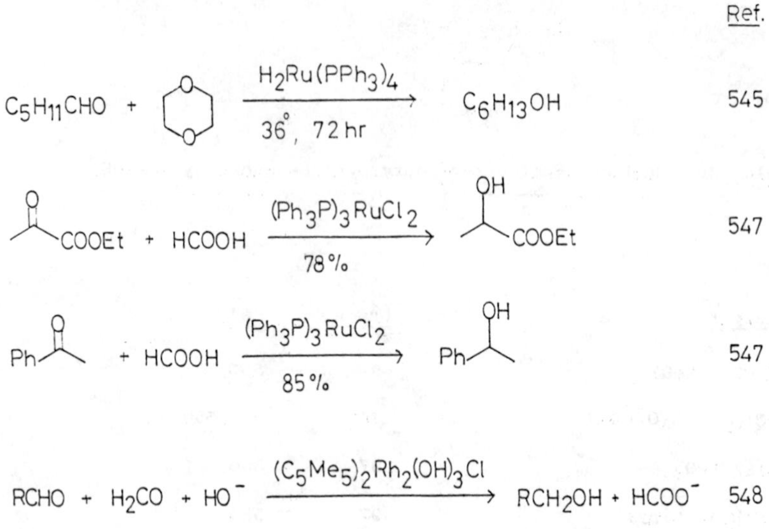

$$C_5H_{11}CHO \; + \; \text{(dioxane)} \quad \xrightarrow[36^\circ, \; 72\,hr]{H_2Ru(PPh_3)_4} \quad C_6H_{13}OH \qquad\qquad 545$$

$$\text{CH}_3\text{C(O)COOEt} \; + \; HCOOH \quad \xrightarrow[78\%]{(Ph_3P)_3RuCl_2} \quad \text{CH}_3\text{CH(OH)COOEt} \qquad 547$$

$$Ph\text{-C(O)CH}_3 \; + \; HCOOH \quad \xrightarrow[85\%]{(Ph_3P)_3RuCl_2} \quad Ph\text{-CH(OH)CH}_3 \qquad 547$$

$$RCHO \; + \; H_2CO \; + \; HO^- \quad \xrightarrow{(C_5Me_5)_2Rh_2(OH)_3Cl} \quad RCH_2OH \; + \; HCOO^- \quad 548$$

Figure 56   Transfer hydrogenation of aldehydes and ketones.

each route.   For example, reduction of cyclohexanone with 1-phenylethanol/
(Ph$_3$P)$_2$RuCl$_2$ gives 70% route B and the rest via A.[561]   Route A is predominent with H$_2$Ru(PPh$_3$)$_4$[545], HRh(CO)(PPh$_3$)$_3$[538], Ir(bipy)(cod)Cl[496,552]
and H$_3$Co(PPh$_3$)$_3$.[552]

2.9.3   Reduction of alkenes

This reaction has not been extensively studied since many
alternatives exist.   Extensive isomerisation is often a serious problem.
Figure 58 shows some examples.   The major difference from the ketone
reactions is that a significant kinetic isotope effect is observed for
(CH$_3$)$_2$CDOH, implying that H-transfer is rate limiting.[545]

2.9.4   Reduction of enones

Transfer hydrogenation of enones is usually selective for
C=C reduction.   Most of the catalysts discussed so far are active, and as
before i-PrOH and 1-phenylethanol are the most well-used donors   (Figure 59).

$$HRh(PPh_3)_4 \longrightarrow Rh(PPh_3)_3 = RhP_3$$

Route A

$$RhP_3 + D \rightleftharpoons RhP_3D$$

$$RhP_3D + A \rightleftharpoons RhP_3DA$$

$$RhP_3DA \xrightarrow{k_D} products$$

Route B

$$RhP_3 + A \rightleftharpoons RhP_3A$$

$$RhP_3A + D \rightleftharpoons RhP_3AD$$

$$RhP_3AD \xrightarrow{k_A} products$$

Formation of bis-complexes

$$RhP_3D + D \rightleftharpoons RhP_3D_2$$

$$RhP_3A + A \rightleftharpoons RhP_3A_2$$

D = Donor

A = Acceptor

Figure 57 Mechanism of transfer hydrogenation.

Other donors include sugar alcohols[565], aryl aldehydes[566], ethylene glycol[567] and tin formates[568]. $\alpha\beta$-Unsaturated aldehydes are reported to yield allyl alcohols with $HIrCl_2(DMSO)_3$[539].

Both the routes of Figure 57 also operate in these reactions

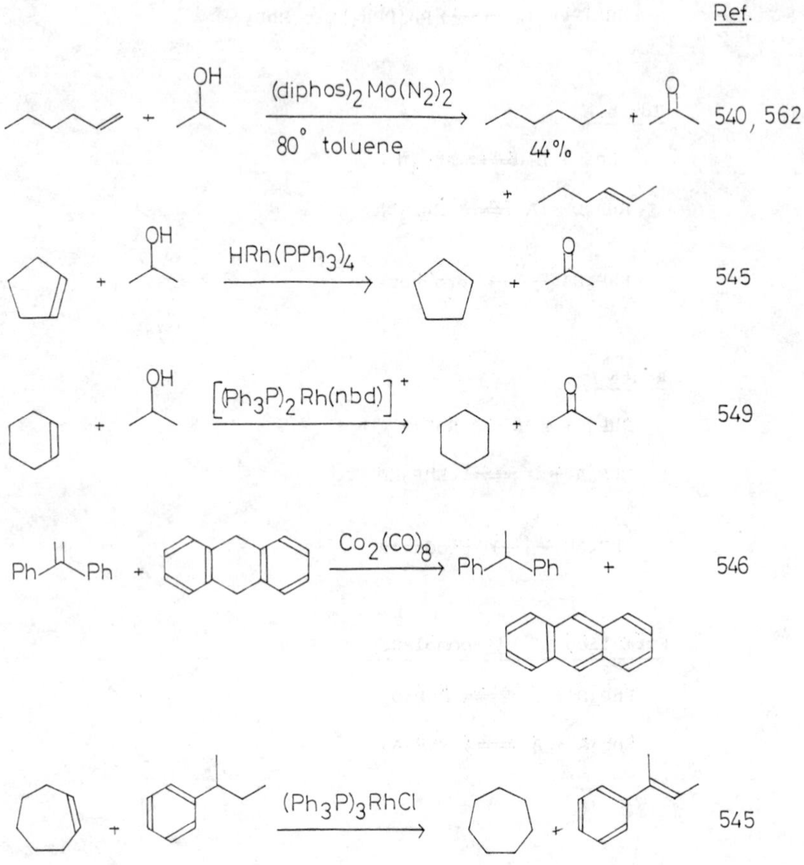

114

Figure 58 Transfer hydrogenation of alkenes.

as confirmed by kinetic and isotope studies.[569]   With (Ph$_3$P)$_3$RuCl$_2$ the

rate limiting step is transfer of the $\underline{\alpha}$-hydrogen of the alcohol to the

$\underline{\beta}$-carbon of the enone.[541]   With HRh(PPh$_3$)$_4$ the same orientation of

addition is observed but the rate controlling step is breaking of the

OH bond with route A dominating.[560]

$$\text{cyclohexenone} + Ph\text{-CH(OH)CH}_3 \xrightarrow{\text{HRh(PPh}_3)_4} \text{cyclohexanone} + Ph\text{-COCH}_3 \qquad 560$$

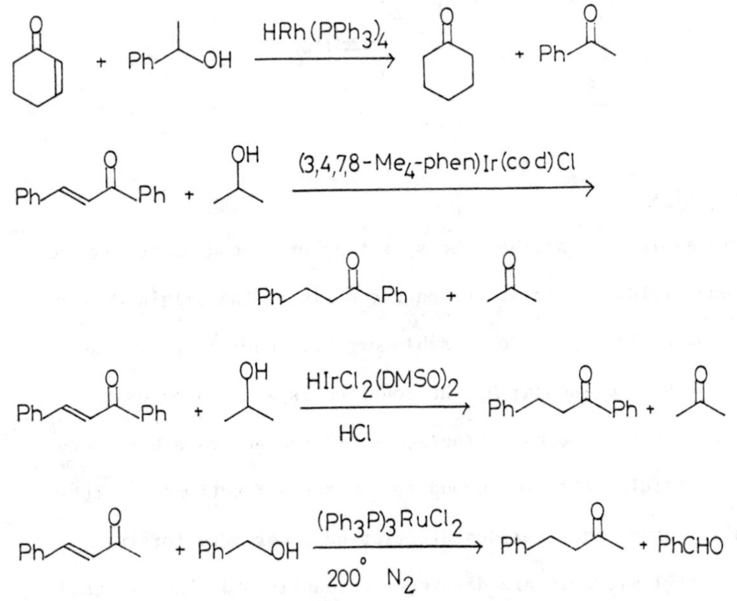

$$Ph\text{-CH=CH-CO-Ph} + (CH_3)_2CHOH \xrightarrow{(3,4,7,8\text{-Me}_4\text{-phen})Ir(cod)Cl} \qquad 563$$

$$Ph\text{-CH}_2\text{CH}_2\text{-CO-Ph} + (CH_3)_2CO$$

$$Ph\text{-CH=CH-CO-Ph} + (CH_3)_2CHOH \xrightarrow[\text{HCl}]{\text{HIrCl}_2(\text{DMSO})_2} Ph\text{-CH}_2\text{CH}_2\text{-CO-Ph} + (CH_3)_2CO \qquad 539$$

$$Ph\text{-CH}_2\text{-CH=CH-CO} + Ph\text{-CH}_2OH \xrightarrow[200^{\circ}\ N_2]{(Ph_3P)_3RuCl_2} Ph\text{-CH}_2\text{CH}_2\text{-CO} + PhCHO \qquad 564$$

Figure 59   Transfer hydrogenation of enones.

2.9.5   Other substrates

The stereochemistry of alkyne reduction is variable.   Cis-
alkenes are obtained with IrCl$_2$(DMSO)$_3$, 228 being isolated as an intermediate.[539]   (Ph$_3$P)$_2$Ru(CO)(OCOCF$_3$)$_2$ also gives cis-products[570]   but trans-
stilbene is obtained after in situ isomerisation by H$_3$Ir(PPh$_3$)$_2$.[571]

Imines are reduced to 2° amines in the presence of (Ph$_3$P)$_3$RhCl[572],
(Ph$_3$P)$_3$RuCl$_2$[573]   and HOsCl(CO)(PPh$_3$)$_3$[574] , the reactions being similar to
those of ketones.   In 229 electron donors in the para-position of Ar^2
increase the rate implying that H-transfer to C is not rate-controlling.
ArNO$_2$ is reduced with good selectivity to ArNH$_2$ by 2° alcohols in the
presence of (bipy)Ir(C$_2$H$_4$)$_2$Cl[575], Rh$_2$Cl$_2$(CO)$_4$[543]   and RuCl$_3$.[545]

Ph⟍＝⟋IrCl$_2$(DMSO)$_3$

Ph

__228__

Ar^1⟍
   ⟋＝NAr$_2$
R^1

__229__

2.10 Hydrogenolysis

Hydrogenolysis describes the splitting of a bond AB to give AH
and BH. We may divide the reactions on the basis of the origin of the
hydrogen, viz. molecular H_2, an oxidisable organic donor or a hydride
reducing agent. With molecular H_2 the reaction is well known using
heterogeneous catalysts. Benzyl ethers, esters and amines are reduced
making $PhCH_2^-$ a useful protecting group for O- and N-functions. Other
ethers, acetals, esters and epoxides are cleaved under more forcing
conditions. Benzyl alcohols are deoxygenated easily and other alcohols
with more difficulty. Hydrogenolysis of alkyl and aryl halides is
achieved over palladium and that of acyl halides is the Rosemund reduction.

The earliest homogeneous reactions involved $[Co(CN)_5]^{3-}$ as
catalyst for cleavage of carbon halogen bonds.[378] The reaction proceeds
via steps (41) to (43). Kinetic and stereochemical studies, rearranged

$$2[Co(CN)_5]^{3-} + H_2 \rightleftharpoons 2[HCo(CN)_5]^{3-} \qquad (41)$$

$$[Co(CN)_5]^{3-} + RX \rightarrow [Co(CN)_5X]^{3-} + R^\cdot \qquad (42)$$

$$R^\cdot + [HCo(CN)_5]^{3-} \rightarrow RH + [Co(CN)_5]^{3-} \qquad (43)$$

products[576] and trapping with acrylonitrile[269] confirm the intermediacy of
a radical. $[Co(CN)_5]^{3-}$ is also active for deoxygenation of allylic and
homoallylic alcohols; from __230__ and __231__ trans-2-butene is the main product
for $[CN^-]:[Co] < 5$.[577] When $[CN^-]:[Co] > 5$ 80% of the product from __230__
is 1-butene, __231__ is unreactive and rates are 10^2 lower. Benzyl esters
(e.g. __232__) are also cleaved, providing an excellent route for deprotection
of amino acids under mild, non-racemising conditions.[578] Some other cobalt

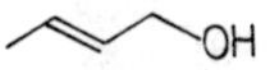

230   231   232

catalysed reactions are shown in Figure 60.

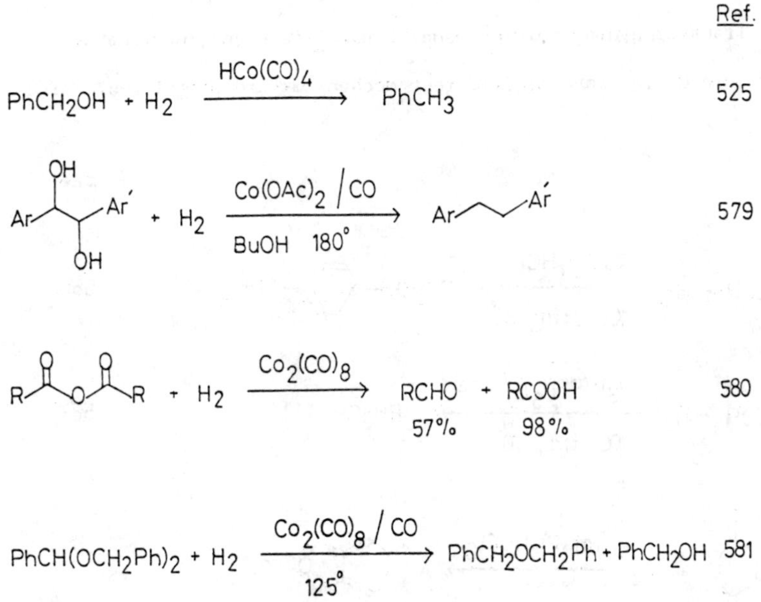

Ref.

$$PhCH_2OH + H_2 \xrightarrow{HCo(CO)_4} PhCH_3 \qquad 525$$

579

580

$$PhCH(OCH_2Ph)_2 + H_2 \xrightarrow[125°]{Co_2(CO)_8 / CO} PhCH_2OCH_2Ph + PhCH_2OH \quad 581$$

Figure 60 Hydrogenolysis in the presence of cobalt complexes.

The hydrogenolysis of activated (e.g. PhCOCH₂Cl)[538] and unactivated (e.g. cyclohexyliodide)[582] halides is catalysed by rhodium complexes in the presence of base. Aryl halides are reduced with base/ Pd(II) phthalocyanine.[583] Homogeneous analogues of the Rosemund reduction are known in the presence of (Ph₃P)₂PdCl₂[584] and Pd(I)phthalocyanine.

Epoxides substituted with vinyl and aryl groups are easily hydrogenolised. 233 is converted to 234, 230 and 231, the proportions depending on R. Hydrogenation of the double bond competes.[585] Styrene oxide is converted to 2-phenylethanol in up to 68% yield with

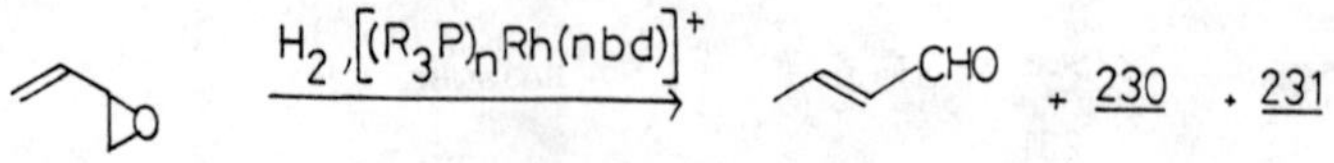

$$233 \quad \xrightarrow[]{H_2 ,\,[(R_3P)_nRh(nbd)]^+} \quad 234 \quad \text{—CHO} + \underline{230} . \underline{231}$$

$[(Et_3P)_2Rh(nbd)]^+$ as catalyst.[482]   A number of other hydrogenolyses using H_2 are known but have not been widely used (Figure 61).

Transfer hydrogenolyses usually use i-PrOH and indoline as donors (Figure 62).   Several different mechanisms are postulated.   The

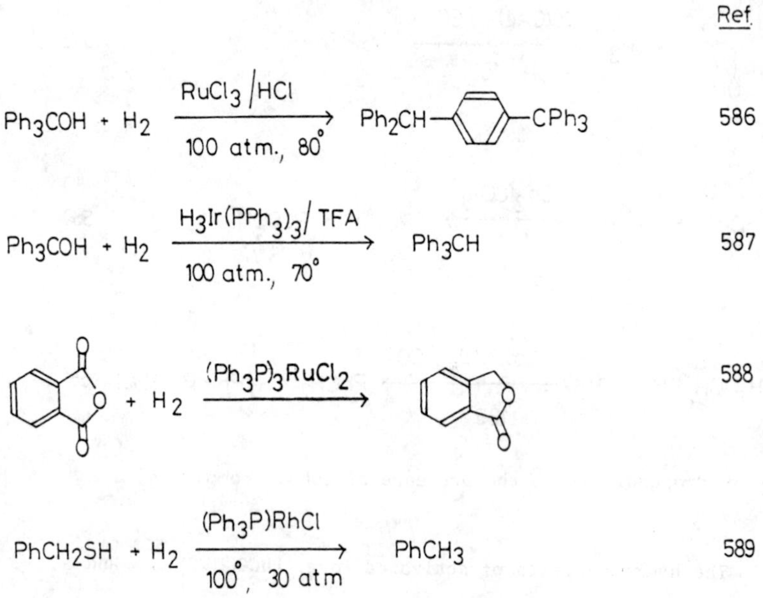

Ref.

$$Ph_3COH + H_2 \xrightarrow[\text{100 atm., 80°}]{RuCl_3/HCl} Ph_2CH\text{—}\bigcirc\text{—}CPh_3 \qquad 586$$

$$Ph_3COH + H_2 \xrightarrow[\text{100 atm., 70°}]{H_3Ir(PPh_3)_3/TFA} Ph_3CH \qquad 587$$

$$\bigcirc + H_2 \xrightarrow{(Ph_3P)_3RuCl_2} \bigcirc \qquad 588$$

$$PhCH_2SH + H_2 \xrightarrow[\text{100°, 30 atm}]{(Ph_3P)RhCl} PhCH_3 \qquad 589$$

Figure 61 Hydrogenolyses using molecular hydrogen.

palladium catalysed reaction proceeds via reduction of Pd(II) to Pd(O) by the donor, DH_2, oxidative addition of RX and hydrogenolysis of C-Pd. ((44) to (47)).   The ruthenium catalysed

$$PdCl_2 + DH_2 \rightarrow Pd(O) + D + 2HCl \qquad (44)$$

$$Pd(O) + ArX \rightarrow ArPdX \qquad (45)$$

Ref.

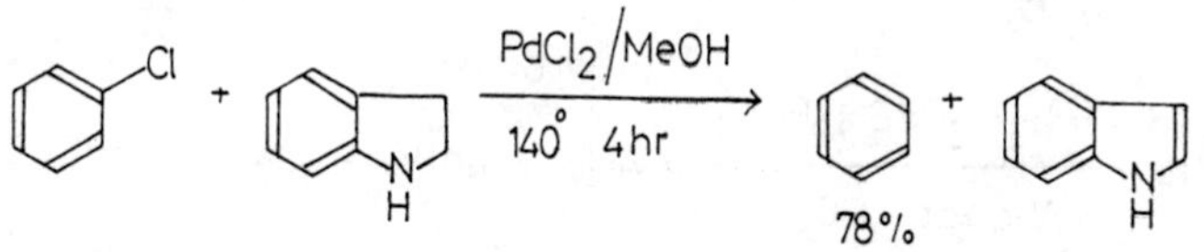

590

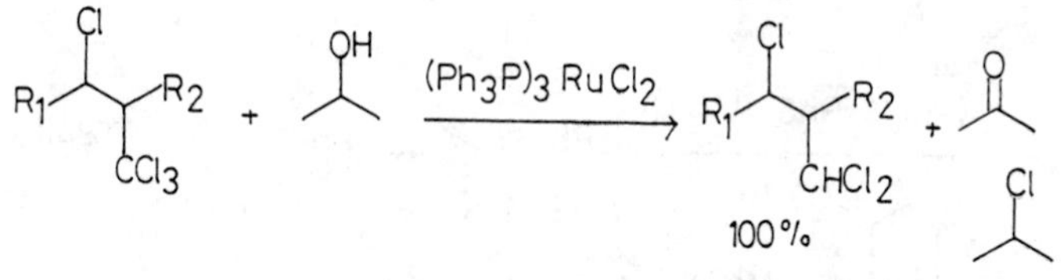

591

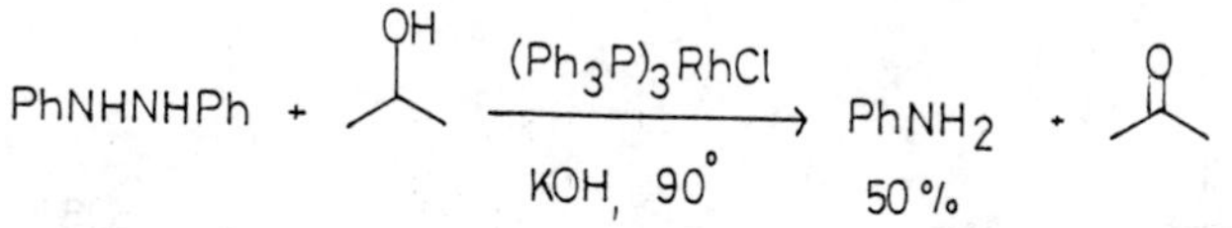

592

Figure 62   Transfer hydrogenolyses using indoline and iso-propanol donors.

$$ArPdX + DH_2 \rightarrow ArH + HPdX + D \tag{46}$$

$$HPdX \rightarrow Pd(0) + HX \tag{47}$$

reduction involves $HRuCl(PPh_3)_x$ (solvent) which adds RX and eliminates RH.
More esoteric donors have also been employed (Figure 63), the most
interesting being those involving NAD(H).

The final group of hydrogenolyses are truly hydride reductions
and are discussed only for completeness.   Various complex hydrides have
been employed (e.g. 235 to 236,[597]   237 to 238[598]   and 239 to 240[599]).   Most
of the reaction mechanisms are unknown, but with 239 a palladium allyl
complex is a likely intermediate;   the pathway has been determined in the
analogous case of 241.[600]

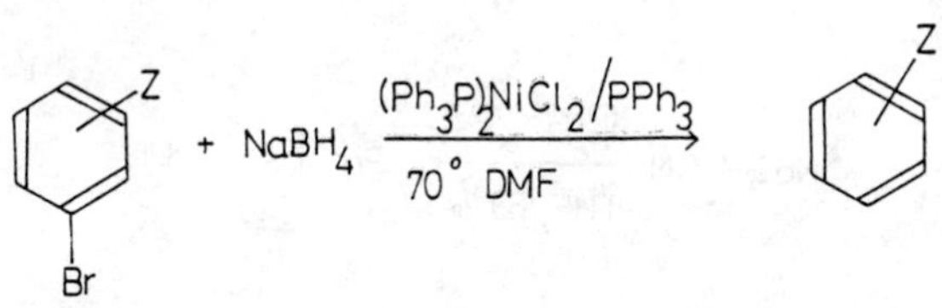

235   236

$$ArBr + HCOONa \xrightarrow[\text{Aliquat } 336]{(Ph_3P)_2PdCl_2/PPh_3} ArH + NaBr + CO_2 \qquad 593$$

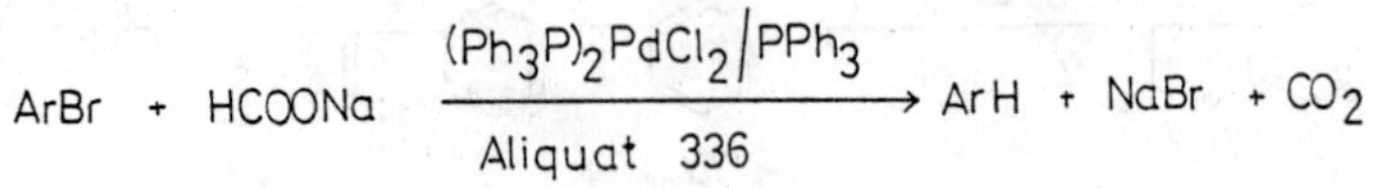

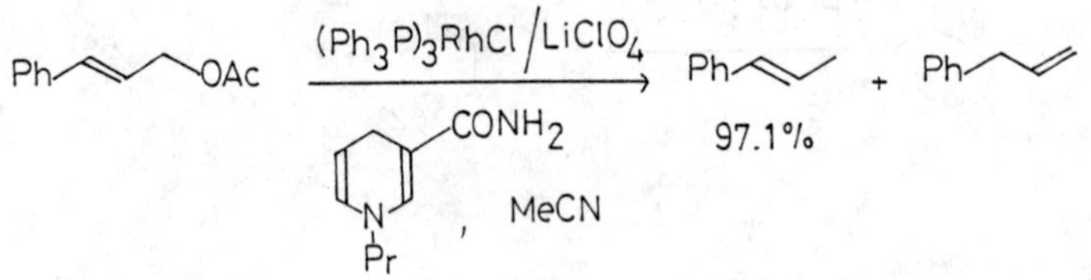

594

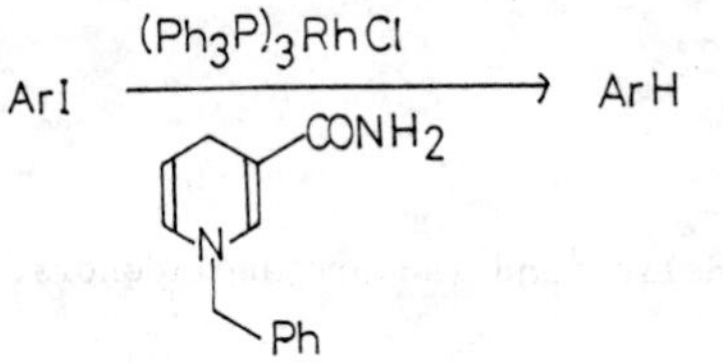

595

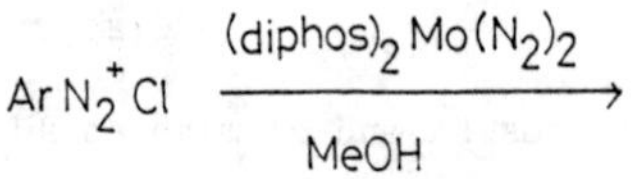

$$ArN_2^+Cl \xrightarrow[\text{MeOH}]{(diphos)_2 Mo(N_2)_2} ArH \qquad 596$$

Figure 63   Transfer hydrogenolyses using other donors.

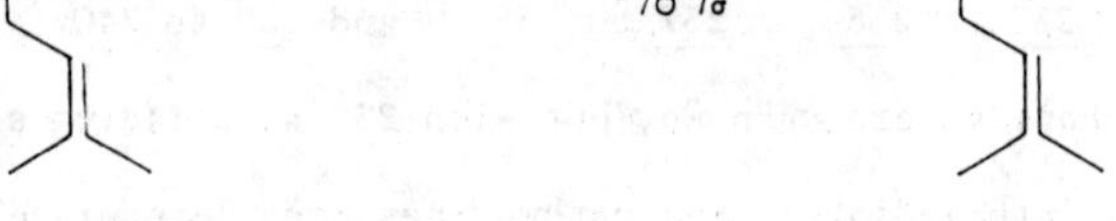

237   238

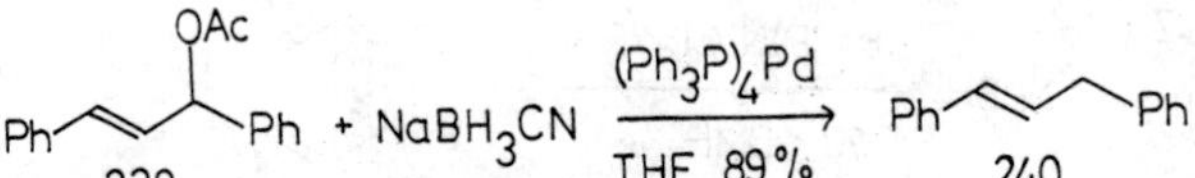

239   240

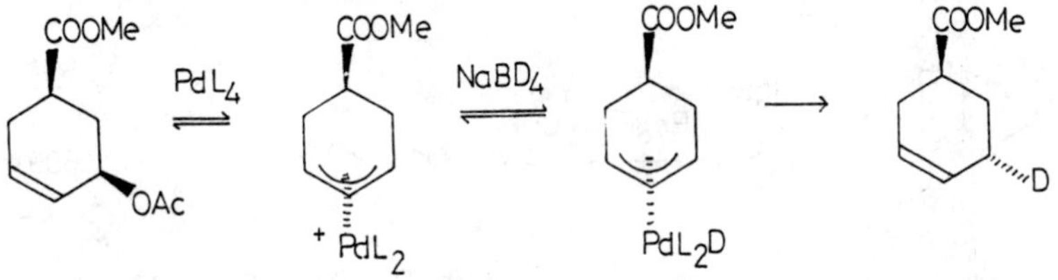

Another useful source of $H^{\ominus}$ is the SnH bond, particularly in conjunction with Pd(0) catalysis. Allyl acetates are reduced[600] and acyl halides converted to aldehydes (e.g. 245 to 246)[601] The reaction of allyl ethers and esters (247 and 248) constitute removal of an allyl protecting group.[602] Grignard reagents bearing a β-hydrogen may also act as hydride donors (Figure 64).

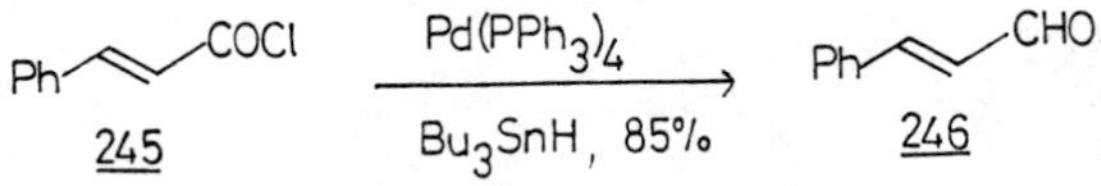

$$Ph\diagdown\diagup COCl \xrightarrow[\text{Bu}_3\text{SnH , 85\%}]{\text{Pd(PPh}_3)_4} Ph\diagdown\diagup CHO$$

245   246

$$ArOCH_2CH=CH_2 + Bu_3SnH \xrightarrow{(Ph_3P)_4Pd} ArOH + CH_3CH=CH_2 + Bu_3SnX$$

247

$$RNHCO_2CH_2CH=CH_2 + Bu_3SnH \xrightarrow[\text{HX}]{(Ph_3P)_4Pd} RNH_2 + CO_2 + CH_3CH=CH_2 + Bu_3SnX$$

248

2.11 Polymer supported catalysts

This area lies at the frontier of homogeneous and heterogeneous catalysis and the published data are a microcosm of those for homogeneous catalysis. Supported catalysts are somewhat less active than homogeneous analogues but are less sensitive to O_2 and eliminate solubility problems.

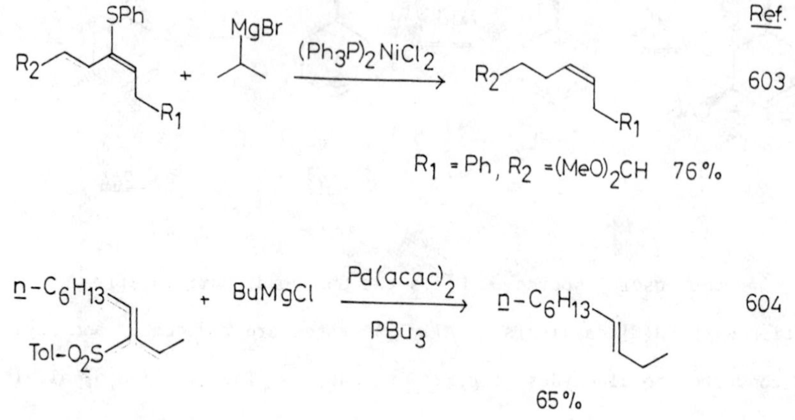

$R_1 = Ph$, $R_2 = (MeO)_2CH$ 76%

$$\underset{\text{Tol-O}_2\text{S}}{n\text{-}C_6H_{13}} + \text{BuMgCl} \xrightarrow[\text{PBu}_3]{\text{Pd(acac)}_2} n\text{-}C_6H_{13}$$

604

65%

Figure 64 Hydrogenolyses by Grignard reagents.

Catalyst recovery is easy, though re-use is sometimes limited by leaching of metal from the support.

2.11.1 The supports

The most common support is polystyrene cross linked with divinyl benzene. It is chemically fairly inert, available in a number of forms and easily modified. Without cross linking the catalysts are normally soluble and removed by membrane filtration. With low cross linking we obtain a resin which swells in polar solvents whilst 20-60% cross linking gives macroreticular or macroporous resins with a high surface area.[605]

Phosphination of the polymer to yield 249 involves bromination, lithiation and treatment with Ph_2PCl, whilst 250 comes from treatment of chloromethylated polymer with LiPPh$_2$. A bidendate ligand may be introduced in the same way (251 to 252). The distribution of ligands varies with polymer type, being fairly uniform in those with low cross-linking but concentrated near the surface of macroreticular beads. Ligands other than phosphines have also been attached. Anthranilic acid

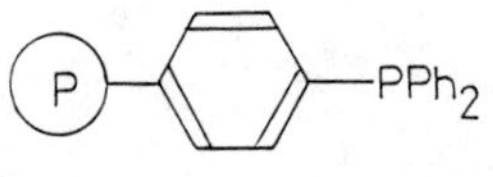

249

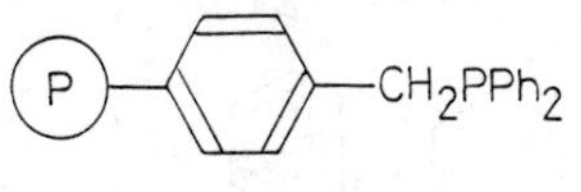

250

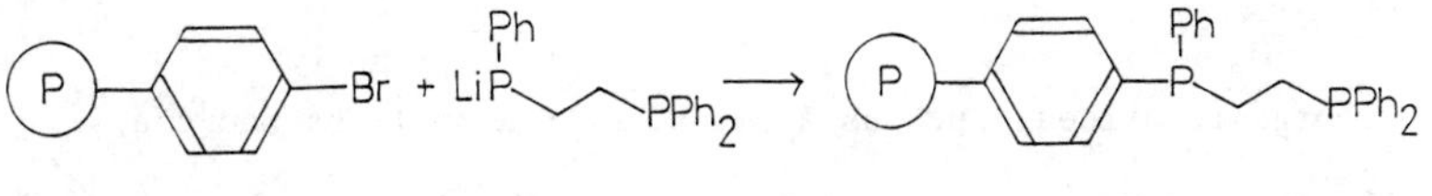

251 + LiP(Ph)—PPh₂ → 252

derivatives such as 253 give hydrogenation catalysts when loaded with
rhodium or palladium.[606]   Acac, cyclopentadiene and sulphur groups have
been used for rhodium coordination.[607,608]

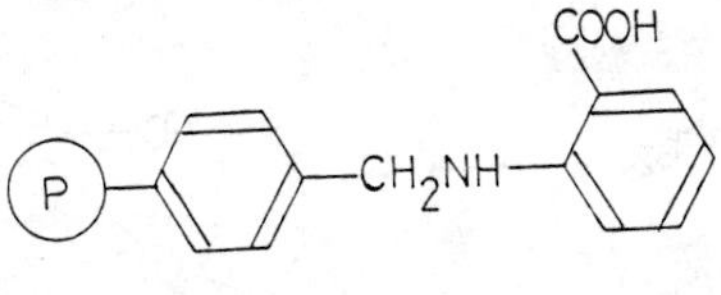

253

As well as polystyrene, polyvinylpyridine and polyacrylic acid
have been used to support ruthenium complexes.[609]   Polyacrylic acid with
more or less cross-linking is used for rhodium[610] and $PdCl_2(DMSO)_2/NaBH_4$
is stabilised on N-polyvinylpyrollidone.   Polyphenylene, isophthalimide,
polysiloxanes, polyvinylchloride, polysulphoratohexadiyne, polyoximes
and polyamides have also been useful.

An alternative route to support material is the polymerisation
of functionalised monomers.   Thus 254, 255 and 256 in varying proportions
are copolymerised to give phosphines with a range of ligand loadings and
physical properties.[611,612]

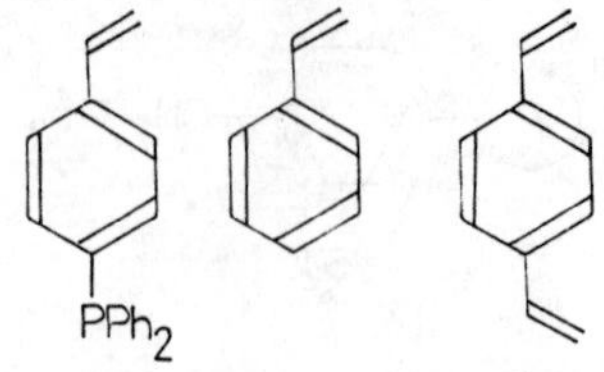

254   255   256

Inorganic oxides, such as alumina, silica, zeolites and clays offer an alternative range of supports which are thermally stable and mechanically robust. The active complex is usually on the surface and accessible to substrate. Whilst some metal complexes may be bound directly to the oxide[613] phosphines may be grafted on to the surface (Figure 65).[614] Different properties are obtained by varying the distance

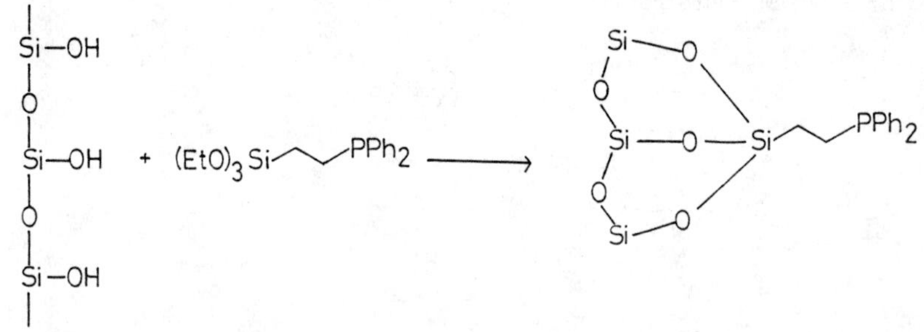

Figure 65 Grafting of a phosphine ligand onto silica.

between silicon and phosphorus or introducing chelating phosphines as in 257.[615] Amine and amide modified silicas are useful[616] and glass, alumina and ion-exchange resins may also be modified. Most active catalysts are prepared by exchange with metal phosphine complexes but a few are formed by grafting preformed complexes on to the polymer.

2.11.2 Hydrogenation of alkenes

If the substrate must diffuse into the polymer to react, then its size and the catalyst framework will control reaction rate. In

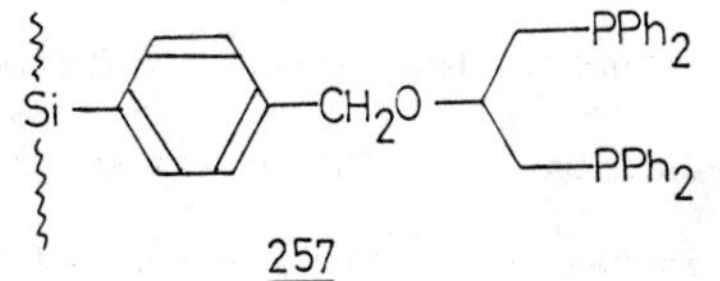

257

particular, larger substrates may be expected to react more slowly. Cyclohexene is reduced faster than Δ^2-cholestene with polymer bound $(Ph_3P)_3RhCl$ whereas the rate difference with homogeneous catalysts was negligible.[617] Heterogenised versions of Wilkinson's catalyst are the most popular, with activity depending both on the support and the method of preparation (Figure 66). A large excess of phosphine is deleterious, as is O_2, though less so than in the homogeneous case. Kinetic studies suggest that the mechanism is similar to that in free solution[621] but this is difficult to define with certainty since the polymer may act as a mono, bi or tridendate ligand.

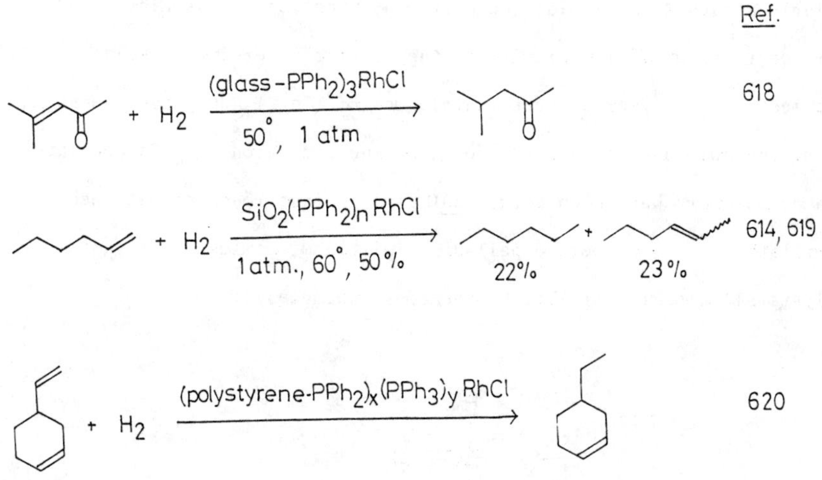

Ref.

618

614, 619

620

Figure 66   Alkene hydrogenation in the presence of polymer supported Wilkinson's catalyst.

A few polymer bound catalysts are more active than solution analogues. $[Rh(C_2H_4)_2Cl]_2$ on phosphinated silica is 50 times more active than $(Ph_3P)_3RhCl$ for alkene reduction.[622] The origin of the effect is unclear with some authors attributing it to the presence of metal crystallites. Polymer supported analogues of most other rhodium catalysts are known (e.g. for 258 to 259)[623]; most are longer lived than homogeneous species.

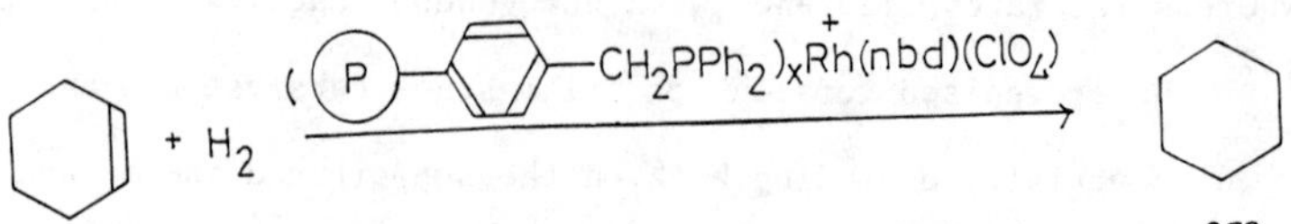

258   259

Iridium based catalysts are similar to, but less active than rhodium analogues. Vaska's complex, attached to 1% cross-linked poly-styrene, gives rates which depend on P:Ir ratio.[624] The same catalyst can employ HCOOH as a hydrogen source.[625] Iridium pyridine catalysts similarly attached are rapidly deactivated; the metal dissociates and dimerises in solution rather than on the polymer, where it is well dispersed.[626] Polymer supported analogues of $(Ph_3P)_3RuCl_2$ are almost as good as the homogeneous catalyst for 1-hexene reduction.[627] Carboxylate and acac polymers have been used; 260 gives a fast reaction but much isomerisation.[609] Supported palladium, platinum, osmium and titanium catalysts all closely parallel homogeneous analogues.

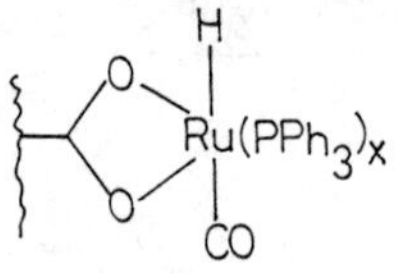

260

2.11.3 Hydrogenation of alkynes

A number of polymer supported catalysts are active for selective or unselective reduction of alkynes but most reports are incidental to the catalyst's other function and a range of potentially suitable species have not been tested.[606]   Polystyrene supported $PdCl_2$ gives mainly cis-2-heptene from 2-heptyne[628] whilst $Rh_2(CO)_4Cl_2/Al_2O_3$ gives both alkenes and alkanes from 1-alkynes.[629]

2.11.4 Hydrogenation of dienes

With few exceptions polymer supported rhodium catalysts yield alkanes in good yield[610] ; there is one report of a successful partial reduction of cyclododecatriene.[620]   However, using palladium or platinum complexes the reduction can be stopped at the monoene stage.[628]   The most commonly supported species are $MCl_2(PhCN)_2$ and reduction of soybean esters is fastest in alcohols, but most selective in CH_2Cl_2.[630]   For methyl sorbate to methyl-2-hexenoate one of the most selective catalysts is 253/Pd(0); it is reusable at least 3 times.[631]   Cyclic dienes, whether conjugated or not, are readily transformed to cycloalkenes with good selectivity.[610,628]

Other complexes have been studied but reports are scattered. $[(Polymer-PPh_2)_xIr(CO)Cl(PPh_3)_{2-x}]$[624] and $[(polymer-PPh_2)_2Ru(CO)_2Cl_2]$[632] are used to reduce 1,5-cod to cyclooctene. Polystyrene supported $Cr(CO)_3$ is up to 99.8% selective for 3-hexenoate in reduction of methyl sorbate, a significant improvement on the homogeneous catalyst.[633]

2.11.5 Hydrogenation of arenes

The reaction is known but studies have been difficult. Severe conditions are often needed and metal is produced.[623]   Metal crystallites on polymers are active catalysts but fall outside the scope of this work.

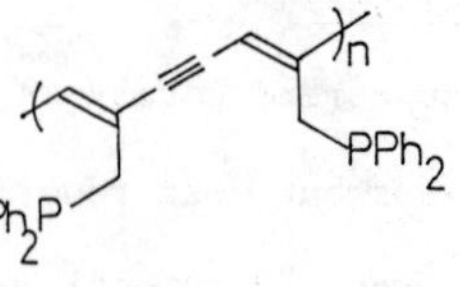

261

For example 261 loaded with (Rh(nbd)Cl)$_2$ gives quantitative reduction of benzene under ambient conditions, though the homogeneous analogue is inactive.[634] The only catalyst, for which there is <u>specific</u> evidence for the non-involvement of metal, is Ru(0) on phosphinated polystyrene; benzene but not naphthalene is reduced whereas it is known that the metal reduces naphthalene readily.[635]

2.11.6 Carbonyl groups and enones

Activities and selectivities of the useful supported catalysts again differ little from homogeneous analogues. Ketones are reduced in the presence of [(Ⓟ-PPh$_2$)$_2$Rh(diene)]$^+$ and 262[607], though these may involve Rh(II) intermediates. In reduction of enones with Rh$_6$(CO)$_{16}$/amines a polymeric amine enhances the rate since dimerisation to larger inactive clusters is prevented.[636] αβ-Unsaturated aldehydes are reduced to allyl alcohols in the presence of (RhCl(CO)$_2$)$_2$ on cross linked polystyrene modified with pyrrolidine.[637]

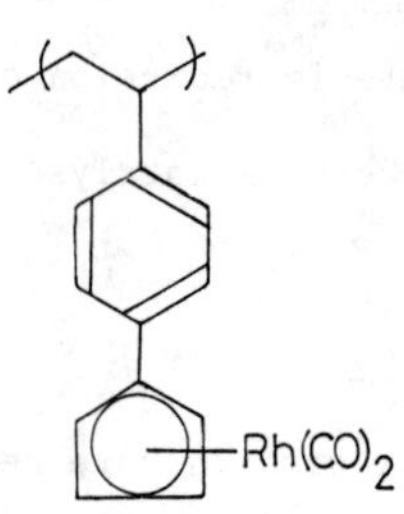

262

2.11.7 Other substrates

Various supported catalysts are active for nitrogen containing functional groups.   Few general conclusions may be drawn and examples are given in Figure 67.

Ref.

$$PhNO_2 + H_2 \xrightarrow[80^\circ,\ 1000\ \text{p.s.i.},\ 97\%]{253 | Pd} PhNH_2$$

606

$$PhCN + H_2 \xrightarrow[700\ \text{p.s.i.},\ 60\%]{253 | Pd} PhCH_2NHCH(NH_2)CHPh +$$

606

$$PhCH_2N=CHPh$$

33%

66%

$$ArNO_2 + H_2 \xrightarrow[PdCl_2,\ 1\ \text{atm.},\ 30^\circ]{SiO_2-CH_2CH_2NCO(CH_2)_3} PhNH_2$$

616

Figure 67   Hydrogenation of nitrogen containing substrates in the presence of polymer supported catalysts.

2.12   Asymmetric Hydrogenation

The goal of synthesising chiral molecules in an optically active form has attracted chemists for many years, and asymmetric hydrogenation is the most successful and best understood of catalytic enantioselective reactions.[638,639]

2.12.1 Reduction of carbon carbon double bonds

2.12.1.1 Rhodium phosphine complexes

In principle a range of tertiary asymmetric centres may be

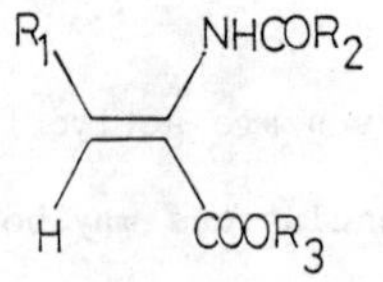

263

generated in one step by enantioselective hydrogenation, but in practice
the range of substrates is more restricted.   The most notable are
dehydroamino acids, 263, and there now exist well over 100 phosphine
complexes capable of catalysing their reduction in good optical yield.

Monophosphines have provided few good examples.   They may be
divided into two groups: those possessing chirality at phosphorus or at
carbon.   Fair optical yields are obtained when a resolved phosphine has
an additional functional group able to bind to metal.[640]   Thus 263
(R$_1$=Ph, R$_2$=Me, R$_3$=H) is reduced in the presence of 264, 265[641], 266[642]
and 267[643] in respectively 88%, 58%, 80% and 13% optical yields.   Various
m- and p-substituted analogues gave < 20% enantiomer excess (e.e.).

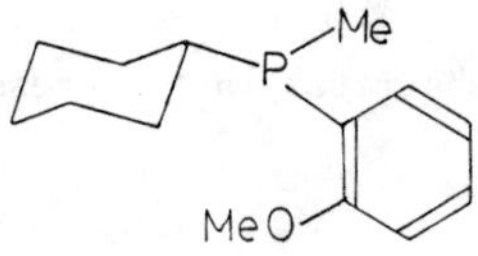

264 CAMP

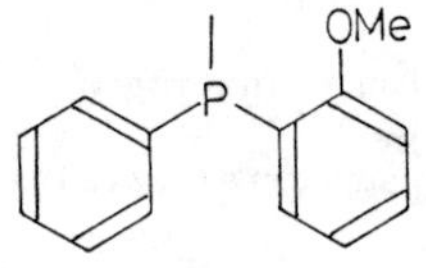

265 PAMP

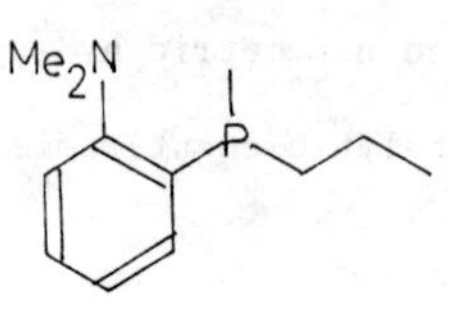

266

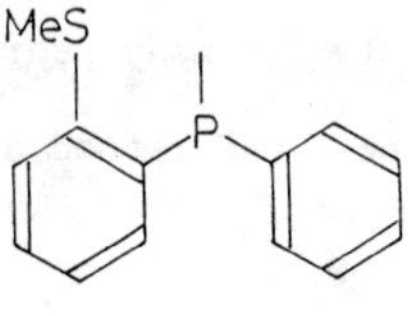

267

Monophosphines chiral at carbon are more numerous.   Some are
derived from natural products such as neomenthyldiphenyl phosphine,
(NMDPP)[644]   and 268 from a sugar.[645,646]   Table 15 shows a range of the

Table 15.   Asymmetric hydrogenation using chiral monophosphine complexes

Ligand	Enantiomer excess (a)					Ref.
	Atropic acid	α-Acetamido cinnamic acid	Itaconic acid	α-Methyl Styrene	α-Methyl cinnamic acid	
Sec-butyl diphenylphosphine	0.2R			3.1S		647
NMDPP	28S		6R	4.5S	60R	647
268a	16.7R	9R	14R			645,646
268b		67R	49S			645,646

(a)   The enantiomer excesses are the highest reported to date for each ligand/substrate combination

268 a X = CH$_2$

268 b X = CH$_2$O

available results.   This has been applied to the synthesis of chiral

dihydrogeranic   acid, used to make vitamin E (269 to 270).[648]   Again the

introduction of a functional group capable of metal binding is beneficial.

269

Rh—NMDPP
————————→
MeOH, Et$_3$N, H$_2$

270

79% R

The ferrocenyl derived aminophosphine, PPFA, (271) has been shown to

chelate rhodium through nitrogen and phosphorus in [PPFA Rh(nbd)]$^+$PF$_6$.[649]

This catalyses the reduction of 263 in up to 84% optical yield.   Related

but less structurally rigid aminophosphines 272-274 have also been studied

but give disappointing results.[650,651]   Monophosphines chiral at both

carbon and phosphorus give no substantial improvement.[648]   The mechanism

of hydrogenation using chiral monophosphines is not as well studied as that

for diphosphines.   The first important intermediate is [L$_2$RhH$_2$(solvent)$_2$]$^+$,

as for achiral analogues.[25]   However, the equilibrium between dihydride and

solvate is a sensitive one depending on the phosphine.[29]   With PAMP, the

OMe group occupies a coordination site and the solvate predominates.

The number of phosphines forming 5-membered chelate rings is

small but this group is much the most successful for the hydrogenation of

dehydroamino acids.   Most of the examples (Figure 68) are chiral by virtue

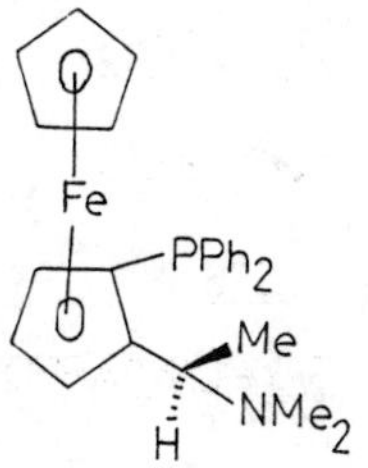

271

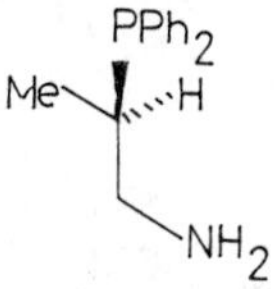

272

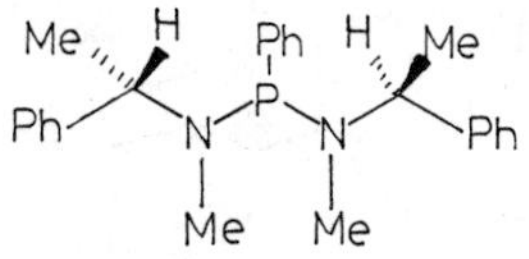

273

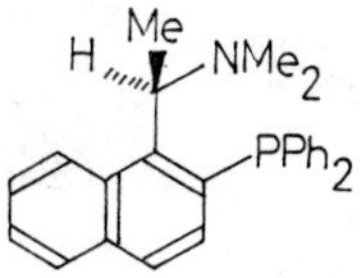

274

of asymmetry in the interphosphine chain but DIPAMP and analogues are chiral at phosphorus.[652]   Early experiments used "in situ" catalysts but later a well-defined cationic complex was preferred.   The range of substrates giving good optical yields is small.   In addition to the carbon carbon double bond it is essential that there is at least one polar group able to coordinate to metal, and the best results are obtained with substrates capable of tridentate binding.   Results for dehydroamino acids are shown in Table 16 and αβ-unsaturated mono- and diacids in Table 17.

        Catalysis by 5-ring chelating phosphine rhodium complexes has proved rather amenable to mechanistic study.   On hydrogenation in methanol 2 moles H_2 are consumed to yield a solvate.   A Z-dehydroamino acid displaces solvent to give an air-sensitive complex which has been fully characterised spectroscopically.[28,169]   Coordination is via the carbon carbon double bond and the amide carbonyl.   With E-substrate the position is more complex;   in the presence of base isomerisation gives the Z-enamide complex, with further substrate isomerised more slowly.   Isomerisation under hydrogenation conditions has been studied using deuterium labelling.

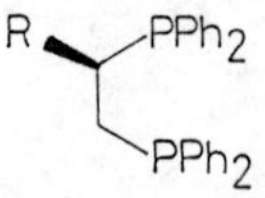

134

275

(a) R = Me   R-PROPHOS[653]
(b) R = Ph   R-PHEPHOS[654]
(c) R = Cy   R-CYCPHOS[655]

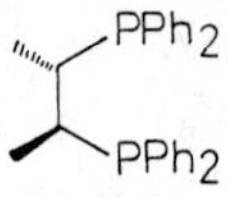

276

S,S - CHIRAPHOS[656]

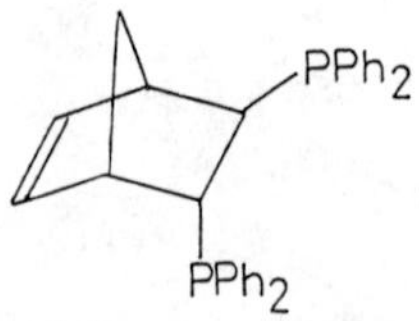

277

R,R - NORPHOS[657]

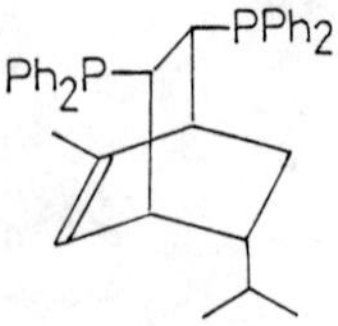

278

PHELLANPHOS[658]

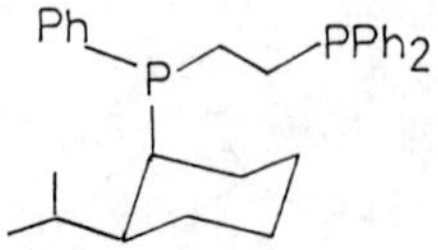

279[654]

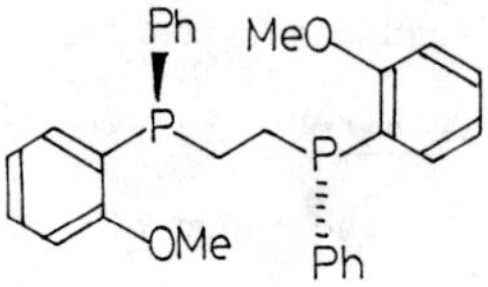

280

R,R - DIPAMP[659]

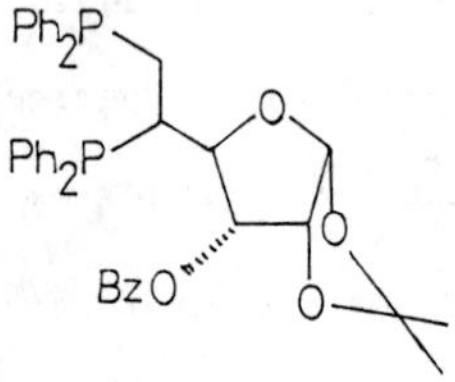

281

GLUCPHOS[660]

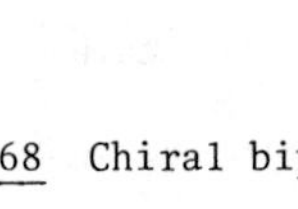

282[661]

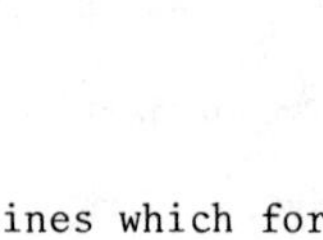

283[662]

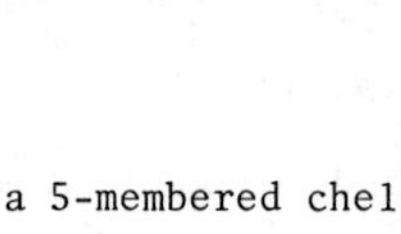

284[663]

Figure 68   Chiral biphosphines which form a 5-membered chelate ring.

Table 16. Asymmetric hydrogenation of dehydroamino acids in the presence of 5-ring chelate complexes

Ligand	Optical yield [a]					Ref.
	Acetamido-acrylic acid	Acetamido-cinnamic acid	Methyl acetamido-cinnamate	Benzamido-cinnamic acid	Methyl benzamido-cinnamate	
DIPAMP	94	94	96	93		659,664
CHIRAPHOS	91	89		99	83 [b]	656
PROPHOS	90	91	94	93	88 [c]	653
CYCPHOS	96	91		94	90 [c]	655,665
PHEPHOS	85	82		84	76	654
NORPHOS	90	94.6	88.4	89	84.6	657,663,666
PHELLANPHOS	95	94.5	66	85	58 [c]	658,667
279		62		100	100	654
282					100	661
284	91	87	86	92	[d]	662,663

(a) Best reported optical yield

(b) Benzyl ester

(c) Ethyl ester

(d) Not reduced

Table 17. Asymmetric hydrogenation of unsaturated acids in the presence of 5-ring chelate complexes

Ligand	Itaconic acid	Atropic acid	Optical yield [a]				Ref.
			α-Acetoxy ethyl- acrylate	Dimethyl itaconate	β-Methyl itaconate	α-Methyl itaconate	
DIPAMP	77		94	88	55	88	668
PROPHOS			81				653
NORPHOS	60						666
PHELLANPHOS		12					667
284	18,29	0		50			662,663

(a) Best reported to date

It is solvent dependent and in C_6H_6 little isomerisation occurs and reduction is optically efficient.[670]   Crystal structure studies are also significant.[671]   Chiral rhodium biphosphine diene complexes crystallise well and all the structures with C_2 symmetry have a special feature recognised by Knowles.[672]   If the P-Ph groups are viewed from the remote side of the coordination plane then the axial pair appear edge-on and the equatorial pair face-on.   This is well enough defined to permit prediction

285

of the sense of asymmetric hydrogenation.   When face oriented rings occupy the upper left and lower right quadrants, 285 (E=edge, F=face), the predominant product will be S.   Structures of diene complexes of DIPAMP, CHIRAPHOS and NORPHOS, as well as larger chelates, conform to this pattern. CYCPHOS[673] shows some deviations with neither polymorph conforming exactly.

A further development was the crystallisation of rhodium enamide complexes of diphos and CHIRAPHOS[674] confirming earlier spectroscopic studies regarding the nature of enamide binding.   However, the CHIRAPHOS complex has the opposite configuration to that of the amino acid ester produced on its hydrogenation.   The solid state structure was shown to persist in solution by analysis of the CD spectrum.[675]

Kinetic studies show that the hydrogenation of Z-methyl acetamidocinnamate by (diphos)Rh$^+$ is some 60 times faster than that with the CHIRAPHOS analogue.[674]   Brown's work on DIPAMP complexes demonstrated that two diastereomers of similar structure, 286 and 287 are formed on enamide complexation in the ratio 10:1 at room temperature.[28]   When these or the analogous diphos complexes are exposed to H_2 at -80°C a hydrido

alkyl, 288 is formed.[676]   It is stable at -80°C but decomposes at -50°C.

288 is derived entirely from the minor diastereomer, 287, and the major

observed species is not significant in the catalytic cycle.

Most asymmetric hydrogenations occur conveniently at 25°C/1 atm.

It is generally the case that optical yields are lower at increased

pressures.[677,678]   It should be possible to derive mechanistic information

from these data; increased pressure may lead to a new mechanism or a

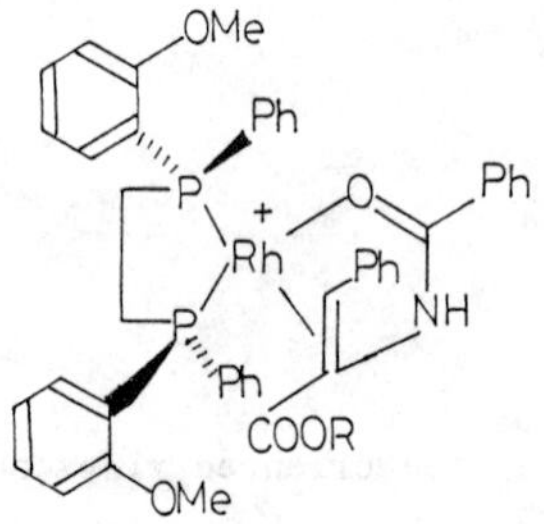

286

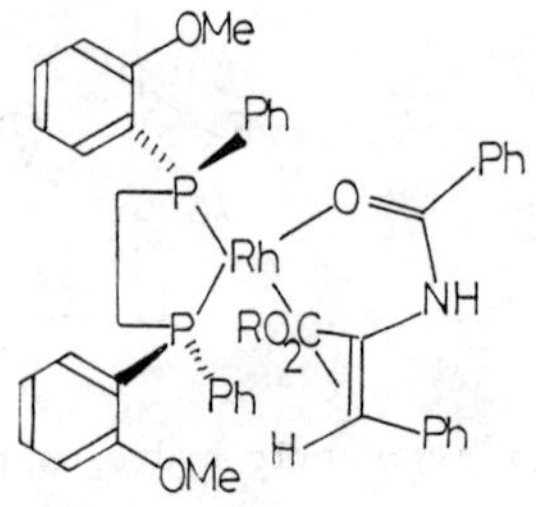

287

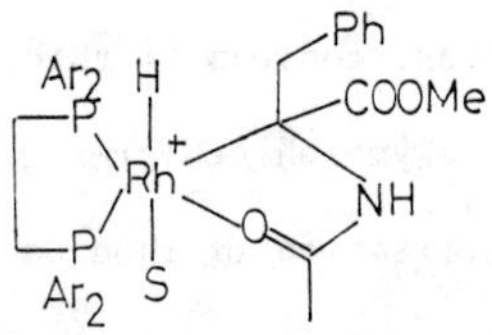

288

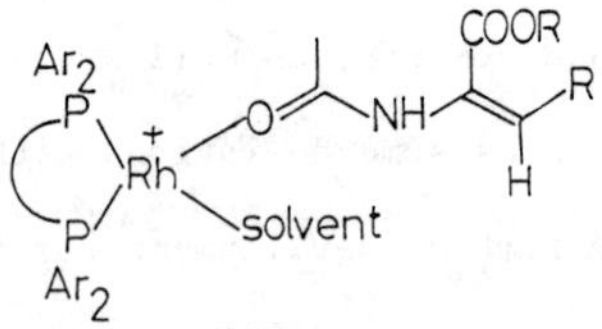

289

change in the rate determining step.   However, a further complication is

provided by the observation that 286 and 287 interconvert via 289 at a

rate much faster than that provided by dissociative exchange of

substrate.[679]   The combination of crystallographic and n.m.r. experiments

gives a clear indication of the intermediates on the reaction pathway, but

does not define the rate-controlling step.   Hydrogen addition in the

presence of substrate is irreversible as defined by the lack of

ortho $\rightleftharpoons$ para H_2 conversion.[17]   Isotope effect studies strongly suggest

that H$_2$ addition is rate determining and is followed by rapid H-transfer
to give <u>288</u>.[680]

Stereochemical considerations in 6-membered rings are complex.
In the absence of substituents a chair conformation is favoured and with
one substituent this is also the case, since it may occupy an equatorial
site.   This approximates to σ-symmetry and suggests that there will be
little discrimination in binding.   However, with two substituents, as in
SKEWPHOS, <u>290</u>, one would have to be axial in a chair conformation so that
the "skew" conformation, in which both are equatorial is favoured.   This
is confirmed crystallographically for [(SKEWPHOS)Rh(nbd)]$^+$ and the
dissymmetric conformation would be expected to give good results in
reduction.   This is born out in practice for SKEWPHOS and CHAIRPHOS,
<u>291</u> (Figure 69).[681,682]   Few other 6 ring phosphines are known;   <u>292</u> gives
poor results.[683]

<u>290</u> S,S-SKEWPHOS   <u>291</u> R-CHAIRPHOS   <u>292</u>

By far the largest number of chiral complexes used in asymmetric
hydrogenation are those containing a seven-membered chelate ring.   These
are generally more versatile than the 5-ring chelates in the range of
substrates reduced.   A selection is shown in Figure 70 and the results in
Table 18.   Whilst 5-ring chelate complexes gave poor results for bidentate
substrates with 7-ring chelates simple enamides and αβ-unsaturated acids
may also be reduced in good optical yield.

The many available phosphines may be grouped conveniently.
DIOP has been extensively modified with the intention of optimising the

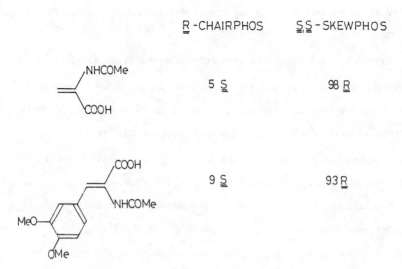

R-CHAIRPHOS   S,S-SKEWPHOS

5 S   98 R

9 S   93 R

Figure 69   Asymmetric hydrogenation using SKEWPHOS and CHAIRPHOS complexes.

yield for particular reactions.   The effects of substituting the phenyl ring were studied.[695]   A m-Me group increases rate and e.e. whilst p-substituents have little effect.   o-Substituents cause a more profound change giving lower rates and e.e.'s.[696]   Moreover, with PAMPOP reversal of chirality occurred and studies of the enamide complexes suggested 308 as a stable intermediate.[692]   The effects are illustrated in Table 19 which shows how a series of modified ligands were used to reduce 309, a precursor of L-tryptophan.

308

309

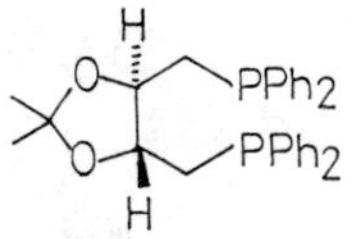

293 R,R-DIOP

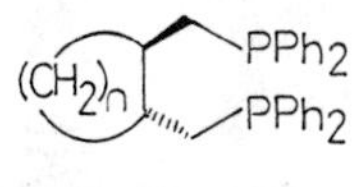

294 n=1
295 n=2
296 n=3
297 n=4

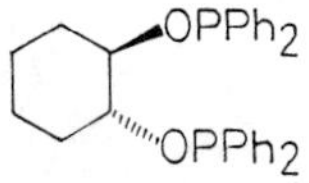

298

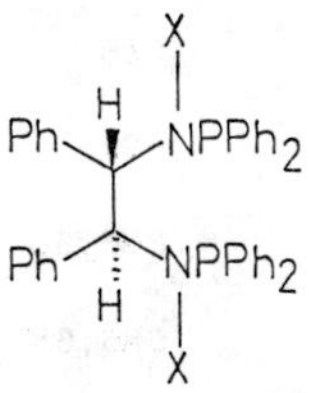

299 X = H PNNP
300 X = Me PNNPMe

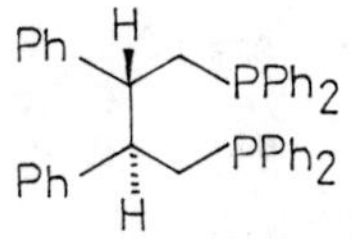

301 S,S-PCCP

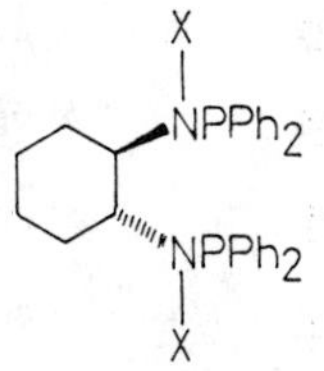

302 X = H
303 X = Me

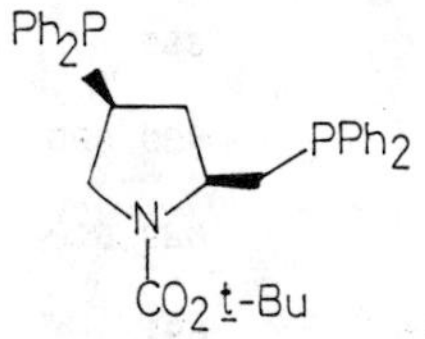

304 BPPM

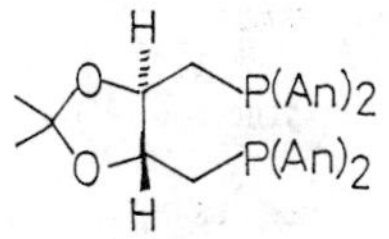

305 PAMPOP
An = o-Anisyl

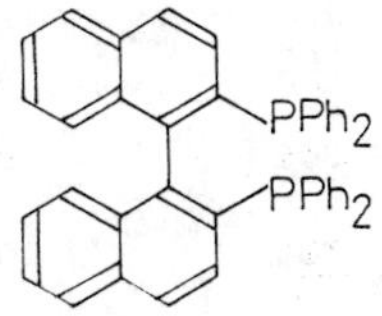

306 BINAP

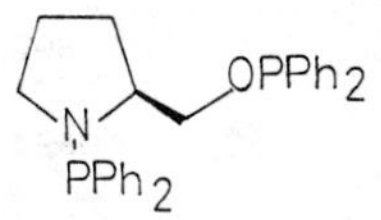

307 S-PROLOPHOS

Figure 70   Chiral biphosphines which form 7-ring chelate complexes.

Table 18.   Reduction of dehydroamino acids with 7-ring chelate complexes

Ligand	Optical yield			Ref.
	Acetamido-acrylic acid	Acetamido-cinnamic acid	Methyl acetamido-cinnamate	
R,R-DIOP	73R	82R	69R	684,685
294	23R	15R		686
295	72R	87R	44R	687
296	72R	63R	38R	686,687
297	40R	35R	1S	686,687
298	79S	68.5S		688
299	92S	94S,71S		689
300	86.3R	68R		689
301	61R	54R		689
302		70R		689,690
303	94S	89.5S		689,690
304	100R	91R [a]	15R [a]	691
305	8S	31S	71S	692
306		84R		693
307	80S	78S		694

(a)   Et$_3$N added

Table 19.   R,R-DIOP analogues for reduction of 309

Arylrings	Enantiomer excess	Relative rate
$(3\text{-}MeC_6H_4)_2$	84.4R	2.0
$(4\text{-}MeC_6H_4)_2$	72R	0.8
$(C_6H_5)_2$	73R	1.0
$(3,5\text{-}Me_2C_6H_3)_2$	86R	3.0
$(3\text{-}OMeC_6H_4)_2$	72R	0.7
$(2\text{-}OMeC_6H_4)_2$	27S	0.002
$(3\text{-}ClC_6H_4)_2$	54R	0.008
$(3,4\text{-}Me_2C_6H_3)_2$	78R	2.0

Carbocyclic analogues of DIOP, 294-297, have also been prepared. Like DIOP the R,R-isomers give R amino acids. The optical yield decreases as the size of the ring increases.[697] Phosphinites and aminophosphines provide a similar group with a CH_2 group replaced by O, NH or NMe. We can make some generalisation about their action. 310 gives poorer but 298 better optical yields than the carbon analogues, the S,S ligand giving S-acids in both cases.[698] 302 and 303 both give good ee's for dehydroamino acids (70-90%) but with opposite stereochemical senses. R-product is obtained from S,S-303 and S-product from S,S-302.[689,690] Analogous observations are made with 299 and 300.

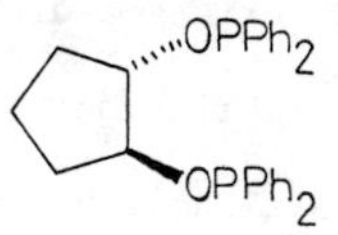

310

Sugars have provided a rich source of phosphinites, most of which can be prepared rather simply.[699] Examples include 311 (from glucose), 312 from galactose, and 313 from xylose. Complexes of these and related ligands are effective for dehydroamino acid reduction, and to a lesser extent for $\alpha\beta$-unsaturated acids. Conditions tend to be severe and the mechanism has not been studied.

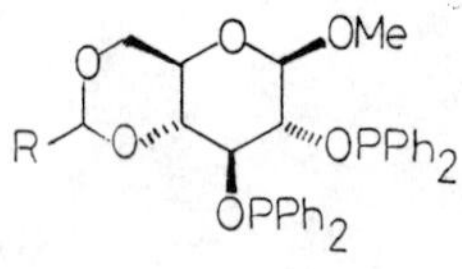

311

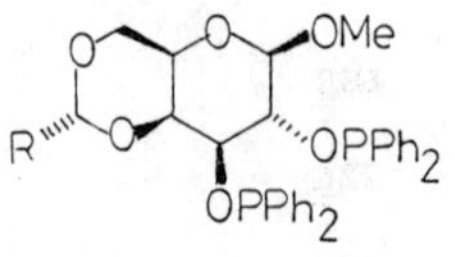

312

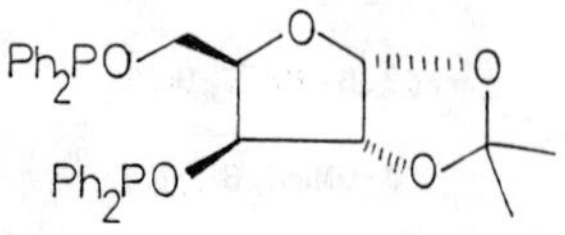

313

The synthesis of 2S, 4S-pyrrolidine biphosphines complexes, 314, has been very extensively studied. They have proved remarkably versatile in the range of substrates reduced. (Figure 71). A wide variety of R-groups in 314 has been used but the optical yield is not very sensitive to the nature of R. This inspired the synthesis of the lipophilic compound, 314 (R=CO$_2$C$_{27}$H$_{45}$),[700] in which the long alkyl chain confers solubility in non-polar solvents.

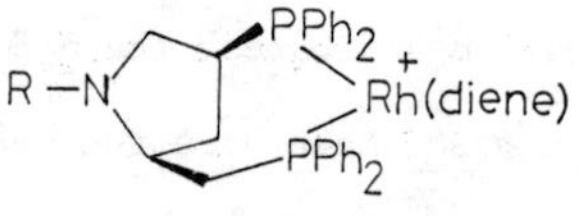

314

It is apparent that there exists a disconcerting variation in optical yield with relatively minor variations in ligand structure. A number of mechanistic studies are available but the origin of the stereo-selectivity is not clear. Just as for the 5-ring chelates the first intermediate is a solvate[29] and the second an enamide complex, bound

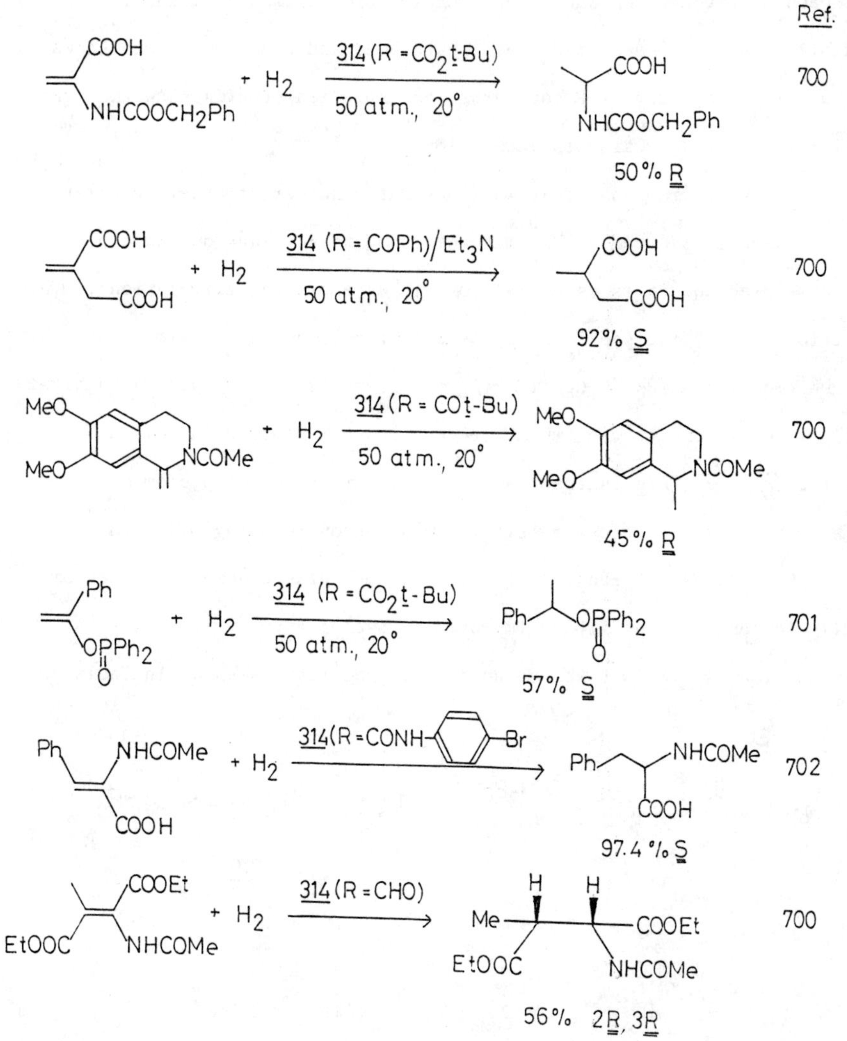

Ref.

COOH
$+$ H$_2$ → 314 (R = CO$_2$t·Bu) / 50 atm., 20° → COOH 700
NHCOOCH$_2$Ph   NHCOOCH$_2$Ph
50% R

COOH
$+$ H$_2$ → 314 (R = COPh)/Et$_3$N / 50 atm., 20° → COOH 700
COOH   COOH
92% S

MeO   314 (R = COt-Bu) MeO
$+$ H$_2$ → 50 atm., 20° → NCOMe 700
MeO NCOMe   MeO
45% R

Ph   314 (R = CO$_2$t-Bu)
$+$ H$_2$ → 50 atm., 20° → Ph OPPh$_2$ 701
OPPh$_2$ O
O   57% S

Ph NHCOMe $+$ H$_2$ → 314(R = CONH⟨⟩Br → Ph NHCOMe 702
COOH   COOH
97.4% S

COOEt   314 (R = CHO)   H H
$+$ H$_2$ → → Me COOEt 700
EtOOC NHCOMe   EtOOC NHCOMe
56% 2R, 3R

Figure 71 Asymmetric hydrogenation using chiral pyrrollidinobiphosphines.

through carbon carbon double bond and amide carbonyl.[703] However, [31]P
n.m.r. gives broadened spectra at 25°C showing that dynamic exchange is
much more significant. There is more structural variation, and Z-enamide
complexes with 2:1 stoichiometry, tridentate binding and σ-bonding have
all been spectroscopically characterised.[692] No hydridoalkyl intermediates

have been observed. An additional complication arises from the relative
flexibility of the 7-membered ring; both boat and chair forms are known
in the solid state for DIOP complexes, and solution studies show that the
barrier to interconversion is rather low.[704]

Moving from enamides, we find that a number of other substrates
are successfully reduced. The mechanisms of these reactions have been
little studied and it is essential to rely on empirical correlations. The
reduction of dehydrodipeptides may be expected to be rather similar to that
for dehydroamino acids. Considering the reduction of 315 with an achiral
catalyst we may expect the two diastereomers to be formed in unequal
amounts.[705,706] This is the case; the S,S-diastereomer predominates from
S-substrate. An analogous result is obtained on reducing 317 with
$(Rh(cod)Cl)_2/Ph_2P(CH_2)_nPPh_2$.[707] However, such diastereomer excesses are
usually low and may readily be improved by use of asymmetric catalysts.
The reduction of 318 is taken as an example and results shown in Table 20.

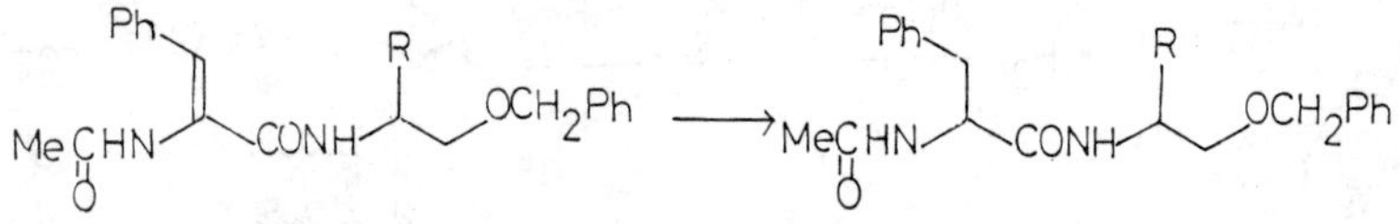

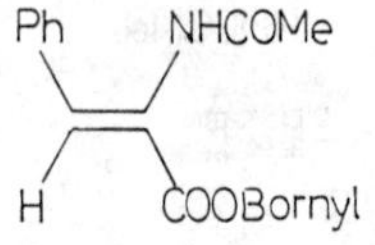

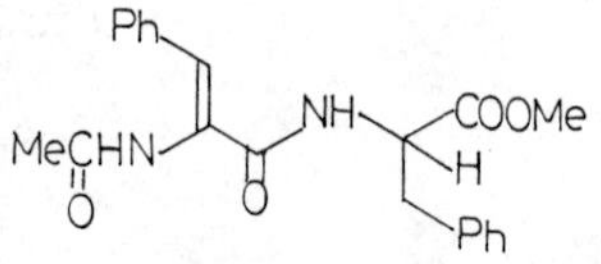

In most cases the enantioselectivity is dominated by the chirality of the
ligand and whilst it is superior to that for simple dehydroamino acids,
the direction is the same. The one exception is provided by DIOXOP, 319,
a large ring chelating biphosphine; the existing chiral centre is

Table 20.   Asymmetric hydrogenation of 318

Ligand	R - 318 → R,S + R,R		S - 318 → S,S + S,R		Ref.
S,S-DIOP	90	10	95	5	707
R,R-DIPAMP			95	5	707
BPPM	2	98	1	99	705,706
314 (R=p-BrC₆H₄)	1	99			705,706

important and high optical yields are obtained only when this is S.[708]
Bis-dehydrodipeptides have been studied by Kagan's group[709] and a complex
algorithm devised to explain the results, which are not very impressive.
This reaction has been used in a synthesis of enkephalin analogues

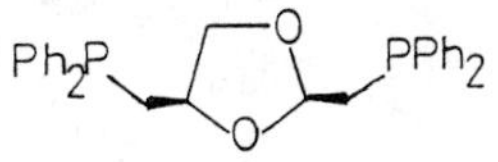

319

(Figure 72).

        Unsaturated acids have been well investigated and a number of
complexes studied in solution.[710]   Many structures are possible with mono
and diacids; carboxylates bind better than free acids and diastereomer
ratios in solution do not correlate with observed optical yields. Itaconic
acid has been a particularly popular substrate (Table 21).   Itaconic acid
has the potential and does, form tridentate complexes.   Other diacids, in
particular mesaconic and citraconic, have been less successful. Asymmetric
reduction of atropic acid, 327, is very desirable since anti-inflammatory

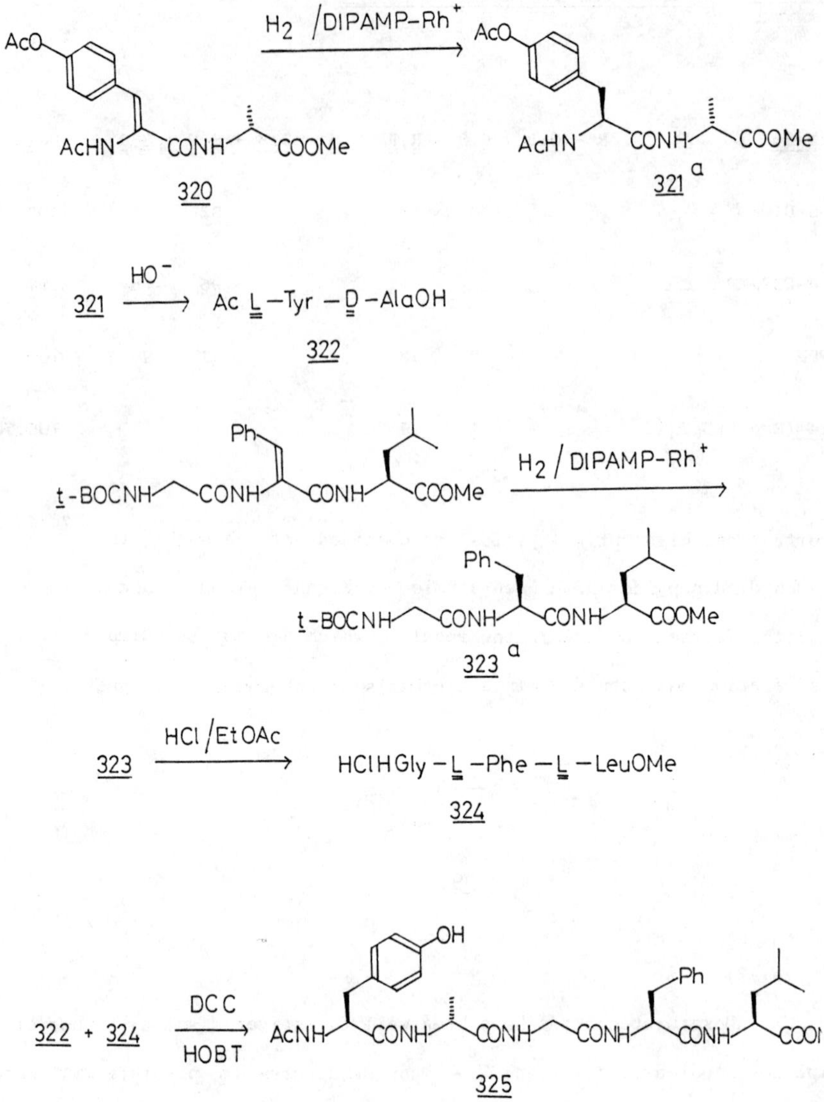
H2 /DIPAMP-Rh+
AcO
AcHN
CONH
COOMe
320
321 a
321
HO−
Ac L−Tyr −D−AlaOH
322
Ph
t-BOCNH
CONH
CONH
COOMe
H2 / DIPAMP-Rh+
323 a
HCl /EtOAc
323
HClHGly −L −Phe −L −LeuOMe
324
OH
322 + 324
DCC
HOBT
AcNH
325

a One recrystallisation gives 100% diastereomeric purity.

Figure 72 Asymmetric synthesis of an enkephalin analogue.

Table 21. Asymmetric hydrogenation of itaconic acid

Ligand	Conditions	Et$_3$N	Optical yield	Ref.
299	1 atm, 25°C	-	71.4R	688,689
299	1 atm, 25°C	✓	85R	688,689
300	5 atm, 25°C	-	5.8S	688,689
301	1 atm, 25°C	-	24S	688,689
326	1 atm, 25°C	✓	82R	688,689
304	50 atm, 25°C	✓	94.5S	711
307	1 atm, 20°C	-	20R	712

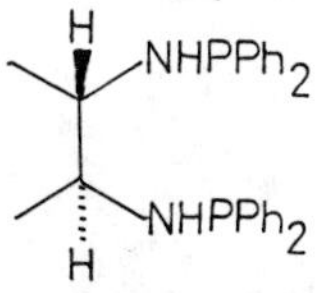

326 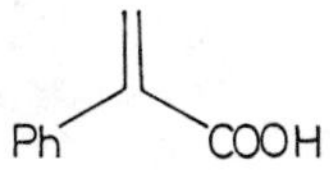   327

analogues are active only in one enantiomeric form.   However, most

attempts have met with limited success.   The best reported (63S) is for

reduction of the carboxylate anion using "Rh(DlOP)Cl".[695]   Simple

αβ-unsaturated acids such as 328 and 329 were also studied.   Modest

(30-40%) optical yields were obtained for E or Z-328 with little variation

between phosphines.[648]   Not only is 329 reduced in the presence of

Rh(DlOP)Cl in 88% e.e. but the catalyst is not poisoned by the thiophene.[713]

Relatively few larger ring chelates have been studied in detail

but they are quite versatile.   A good example is provided by the ferrocenyl

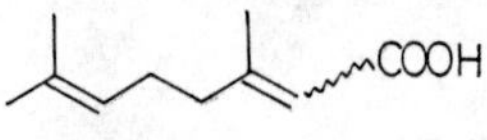

328

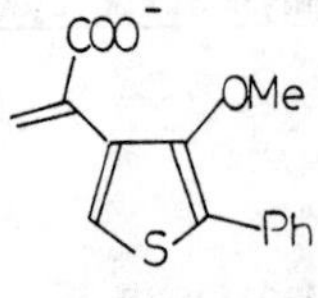

329

714
phosphines 330 and 331.   Their adaptability is related to the -OH or -NMe$_2$ functional groups which can interact with an appropriate centre of a bound substrate.   Figure 73 illustrates some uses.   With BPPFA it is found that dehydroamino acids are reduced more efficiently than the esters and that Et$_3$N reduces the optical yield.   H-bonding between carboxyl and tertiary amine is thought to control stereoselection.

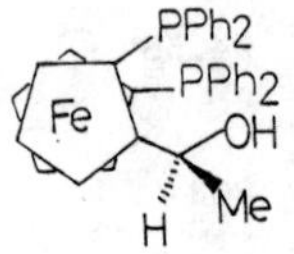

330   S,R-BPPFOH 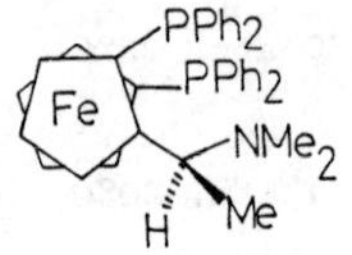   331   S,R-BPPFA

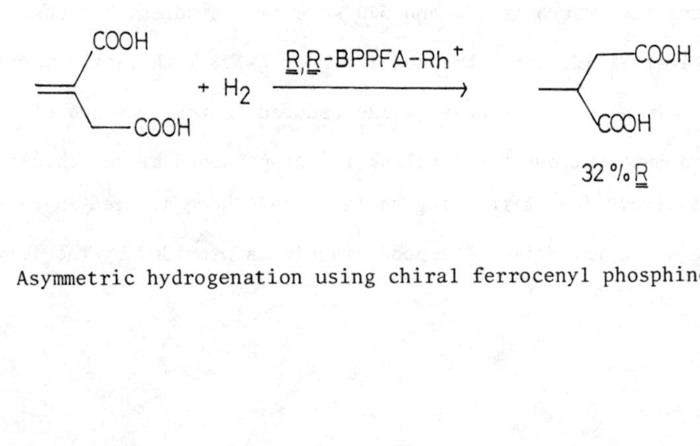

Figure 73 Asymmetric hydrogenation using chiral ferrocenyl phosphines.

Monosaccharides have been a prolific source of medium ring chelating biphosphines. The most useful is DIOXOP, <u>319</u>, derived from D-glucose,[715] which reduces enamides and unsaturated acids in good optical yield in the presence of an amine. Its mechanism of action is rather different and it was the first example of a stable dihydride from a chelating phosphine complex. Coordination of the acetal oxygen is significant in several steps of the catalytic cycle.[716] Examples of its use are shown in Figure 74.[645,717] Other sugar derived biphosphines include <u>332</u> and <u>333</u>; it is noteable that <u>333</u>, which is constrained to act as a <u>trans</u>-chelating ligand gives poor results.[21]

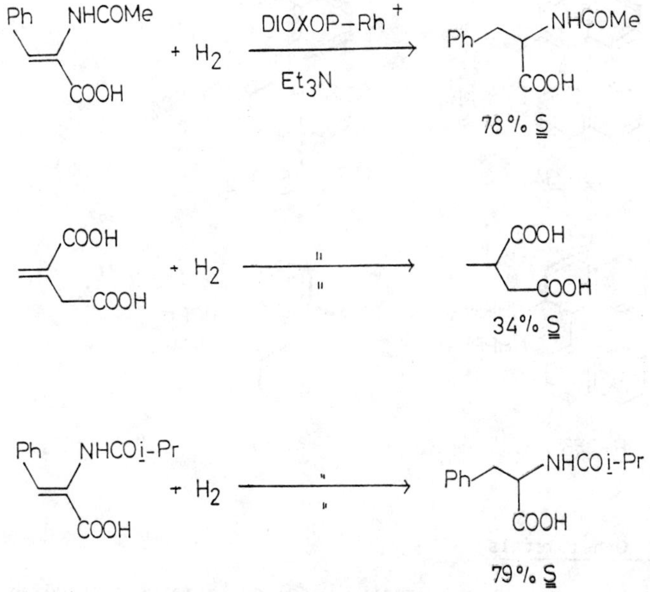

Figure 74 Asymmetric hydrogenation using DIOXOP complexes.

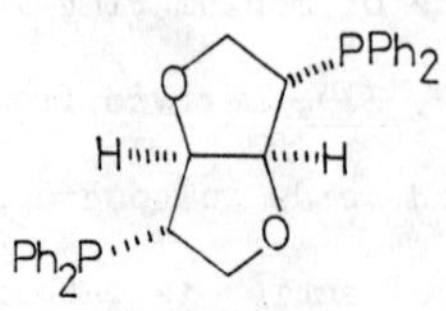

332

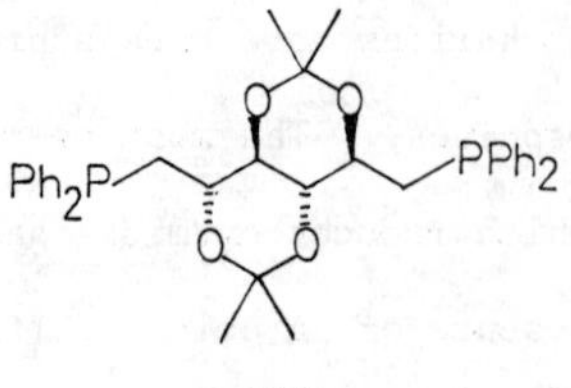

333

Other large ring chelates have been less interesting.   334-336
are comparable, the direction of optical induction being analogous, $\underline{S}$ from
334[718] and 335[719] and $\underline{R}$ from 336.[720]   For comparable substrates the
optical efficiency is in the order 336 > 334 > 335.   Curiously, however,
the $\underline{S}$-biphenyl, 337, gives $\underline{R}$-product (up to 81% e.e) in reduction of
acetamidoacrylic acid.[721]

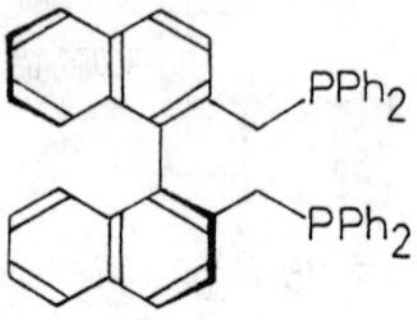

$\underline{S}$-334

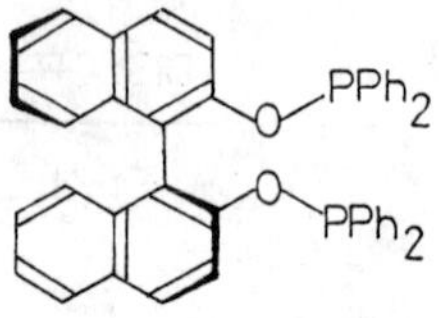

$\underline{S}$-335

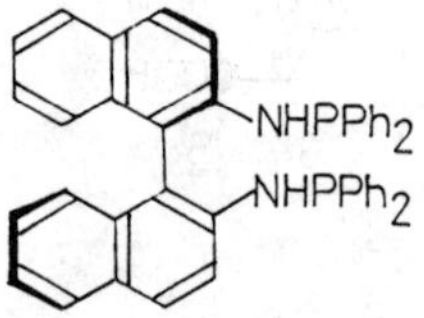

$\underline{R}$-336

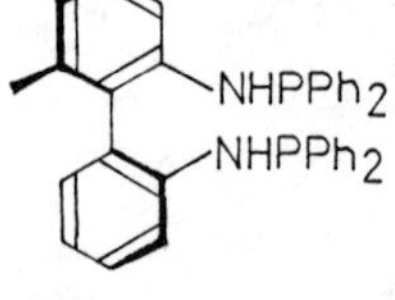

$\underline{S}$-337

2.12.1.2   Other metals

Rhodium is a somewhat expensive metal so there is considerable
interest in finding alternatives.   A catalyst solution containing a
cyanocobalt(II) chiral amine complex was inefficient both optically and
chemically for atropic acid reduction.[722] $Co_2(CO)_6(NMDPP)_2$ has been used
for reduction of 338 but optical yields are low (16%$\underline{S}$).[723]   Other NMDPP
cobalt complexes are formed in Ziegler systems such as $Co(acac)_2/Et_3Al/NMDPP$.

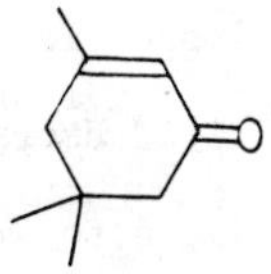

338

Optical yields are modest but non-polar substrates are reduced quite
effectively, with up to 25% e.e. for α-ethylstyrene.[724]

Ruthenium(II) complexes have found most use in reduction of
unsaturated acids; the higher charge on the metal may assist coordination.
The work done has been rather sporadic and general conclusions are
difficult to draw (Table 22). The mechanism is also difficult to
elucidate since many species are present in solution. $Ru_2Cl_4(DIOP)_3$
yields both $HRuCl(DIOP)_2$ and $(RuCl_2(DIOP))_2$ on reduction. With cluster

Table 22. Asymmetric hydrogenation by ruthenium (+)-DIOP complexes

Catalyst precursor	Enantiomer excess				Ref.
	Itaconic acid	Citraconic acid	Mesaconic acid	Atropic acid	
$Ru_2Cl_4(DIOP)_3$	23R	0		40R	725
$HRuCl(DIOP)_2$	38R			40R	726
$RuCl_2(DIOP)(DIOP*)$ [a]	38R				726
$H_4Ru_4(CO)_8(DIOP)_2$	3.2R	1.1R	8R		727

(a)   DIOP* is monodentate

catalysts optical yields are higher for monoacids rising to 58% for
E-α-methyl cinnamic acid.[728]   Optical yield decreases with increased

pressure and esters, which bind less well, also give lower e.e's.   Some

non-polar alkenes, including 339 and 340, have also been reduced;   the

optical yields are low and extensive isomerisation influences the outcome

of reduction.[729]

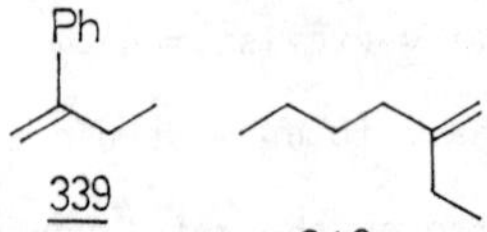

339

340

Iridium complexes have not been much used in asymmetric hydrogenation.   Ir₄(CO)₁₀(DIOP) and Ir(DIOP)(cod)Cl are known but have not been

used in catalysis.   [(NMDPP)Ir(cod)(PhCN)]⁺ gives low yields in enamide

reduction.

Palladium complexes have been investigated by Soviet workers.

The catalyst is prepared by treating (η^3-allyl PdCl)₂ with phosphine

followed by NaBH₄.   Ligands include 341 and 342 but optical yields both

for enamides and itaconates are low.[730]

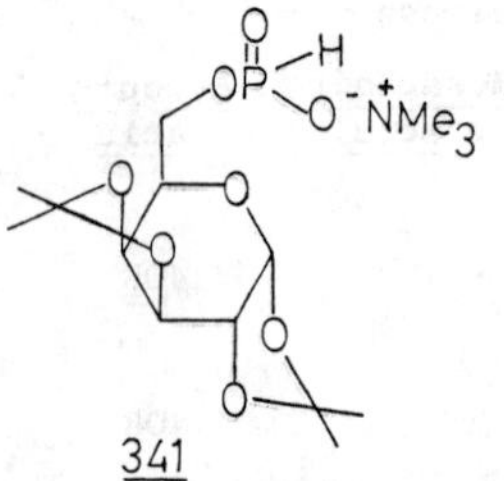

341

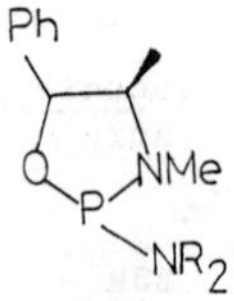

342

2.12.2 Asymmetric hydrogenation of ketones

The first enantioselective reductions employed [H₂RhL₂S₂]⁺

where L is a chiral phosphine such as PMePhBz or DIOP.[731,732]   Some

results for acetophenone are shown in Table 23.   The influence of phosphine

structure is considerable but reliable correlations are elusive.   Chiral

Table 23.   Asymmetric hydrogenation of acetophenone

Catalyst	Conditions	Enantiomer excess	Ref.
$[(PhMe\underline{i}-PrP)_2Rh(nbd)]^+$	55°C, 60psi	4.2	733
$[(PhB\underline{z}MeP)_2Rh(nbd)]^+$	50°C, 1atm	8.6	731
$[(PhEtMeP)_2Rh(nbd)]^+$	50°C, 1atm	0.24	734
$[(DIOP)Rh(nbd)]^+$	50°C, 1atm	2.8	735
$[(BPPFOH)Rh(cod)]^+$	0°C, 50atm	43	736
$[(CAMP)_2Rh(nbd)]^+$		56	733
$[(\underline{343})Rh(cod)]^+$	20°C, 12atm	16.7	737
$[(\underline{310})Rh(nbd)]^+$	25°C, 200atm	22	738

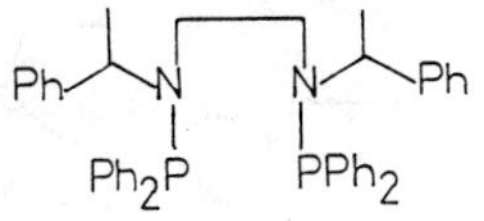

$$Ph\diagdown_N\diagup^N\diagdown Ph$$
$$Ph_2P \qquad PPh_2$$

$$\underline{SS}\text{-}\underline{343}$$

bidentate phosphines are less suitable because of low reaction rates.   An

<u>in situ</u> catalyst from DIOP gives an enantiomer excess up to 51% for aceto-

phenone and addition of Et$_3$N increases this to 81%, adding credence to the

theory that the rate controlling step is attack of a   rhodium monohydride

on the carbonyl.[739,740]

Reaction conditions are important, but again not readily

correlated with results. Arylalkyl ketones give better results than

dialkyl ketones and substrates with the potential for additional coordination are also successful.   Pyruvates are readily reduced to

lactates (Table 24).   The dicarbonyl compound, 344, is a popular substrate

since the product, 345, is pantolactone, a precursor of D-(+)-pantothenate.

BPPM and analogues have been best studied with 87% as the best optical

yield.[743]

Table 24.   Asymmetric hydrogenation of pyruvates, $CH_3COCOOR$

Ligand	R	Enantiomer excess	Ref.
BPPFA (331)	H	83%	741
BPPFOH (330)	H	83%	736
BPPM (304)	i-Bu	65-75%	736
DIOP (293)	n-Pr	42%	742

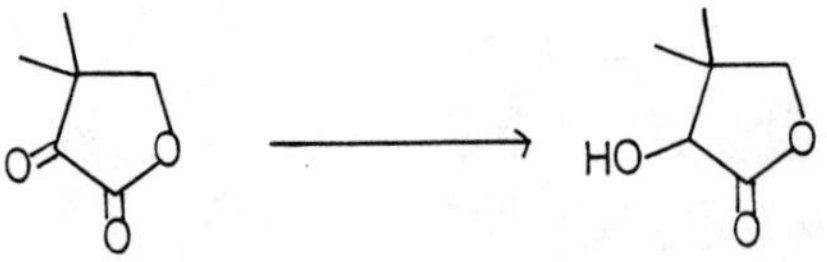

344   345

Adjacent amino groups are also an advantage and this permits a

useful synthesis of β-blockers such as 348 and 349 in good optical

yield.[744,745]   Modest diastereomer excesses are reported in reduction of

350.[706]

Several prochiral ketones have been reduced in the presence of

$H_4Ru_4(CO)_8(DIOP)_2$.[746]   Optical yields are low, the best being for

PhCOCOOH.[747]   Both ketones and anhydrides[748] are reduced in the presence

of $Ru_2Cl_4(DIOP)_3$.   Acetophenone gives poor chemical yields but fair (37%)

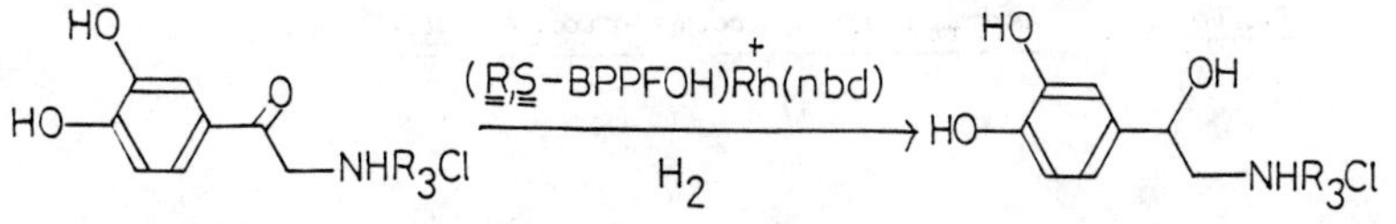

346 → 347

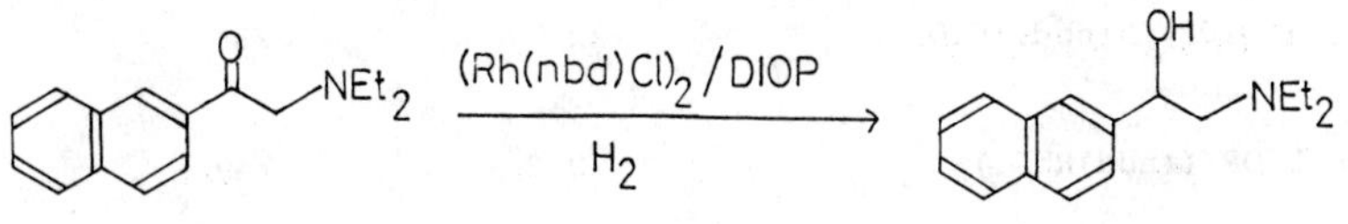

348 → 349

optical excess. 351 is reduced to 352 in fair chemical and low optical yield.

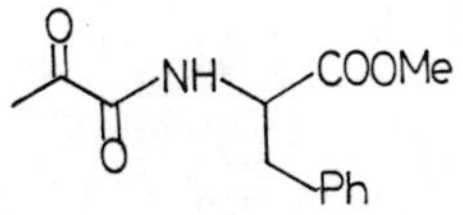

350

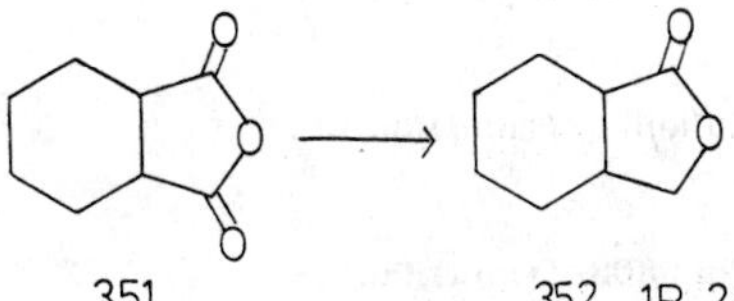

351 ⟶ 352 1R, 2S

2.12.3 Asymmetric hydrogenation of carbon nitrogen double bonds

This area has received relatively little attention and the data are too few to justify generalisations. Some data for 353 are shown in Table 25. Very modest yields are obtained for reduction of ketoximes using $H_4Ru_4(CO)_8(DIOP)_2$.

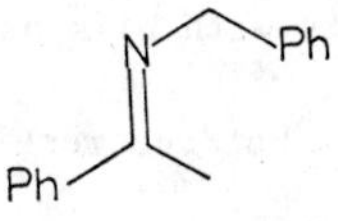

353

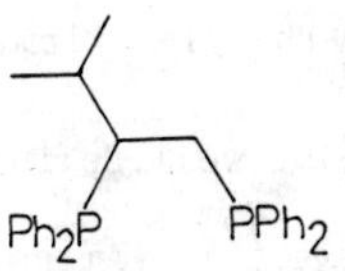

354 R-VALPHOS

Table 25.   Asymmetric hydrogenation of 353

Catalyst	Optical yield	Ref.
[(DIOP)Rh(nbd)]ClO$_4$	22	731
DIOP/((nbd)RhCl)$_2$	3.3	749
BPPM/((nbd)RhCl)$_2$	1.5	749
VALPHOS/((nbd)RhCl)$_2$	66	750
[(BPPFOH)Rh(nbd)]$^+$	48	744
PhMeBzP/((nbd)RhCl)$_2$	1.5	749
PHEPHOS/((nbd)RhCl)$_2$	64	750
DIPAMP/((nbd)RhCl)$_2$	4	750

2.12.4 Asymmetric hydrogenation using non-phosphine ligands

Asymmetric hydrogenation has been achieved in the presence of (py)$_3$RhCl$_3$/NaBH$_4$/chiral amide.[751]   Spectral and conductivity measurements supported the formulation [(py)$_2$(amide)RhCl(BH$_4$)]$^+$X$^-$ and typical results are shown in Figure 75.   More recent work has involved chiral amines, amidines and Schiff's bases.   α-Phenylethylamine interacts with Wilkinson's catalyst and PdCl$_2$ to give well defined complexes but the results in hydrogenation are unimpressive.[753]   Amino acids have been used as auxiliaries; for example (Ile)$_2$Co/Et$_3$Al catalyses reduction of α-ethylstyrene in 25% optical yield.[754]

Schiff's bases have been more successful (Figure 76).   The reaction has not been studied systematically but appears to have promise.

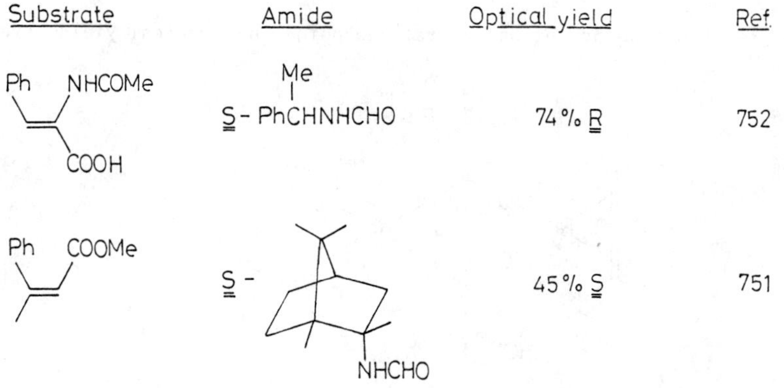

Substrate	Amide	Optical yield	Ref.

Figure 75 Asymmetric hydrogenation using rhodium amide complexes.

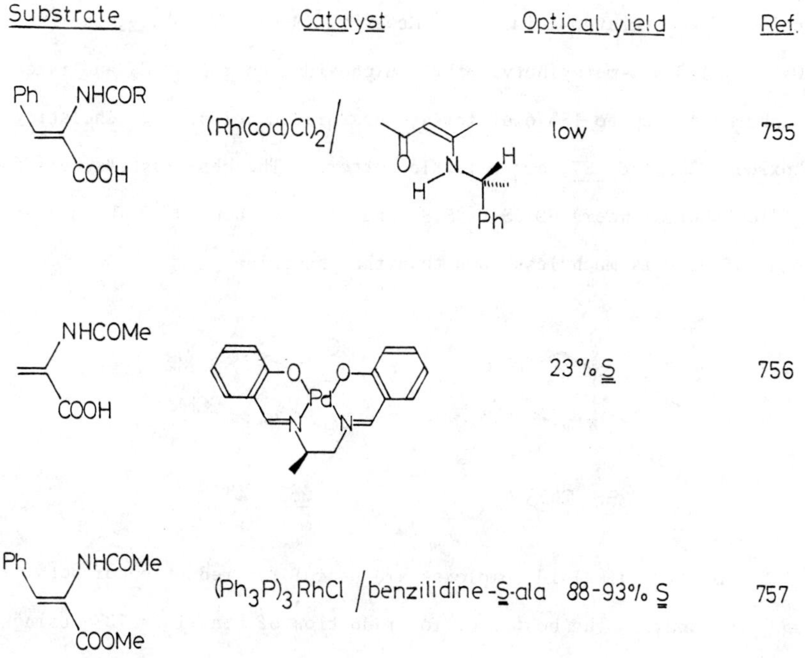

Substrate	Catalyst	Optical yield	Ref.

Figure 76 Asymmetric hydrogenation using Schiff's base ligands.

Chiral rhodium carboxylate complexes such as 355 are very active in reduction, in the presence of an achiral phosphine, but optical yields are low.[758]

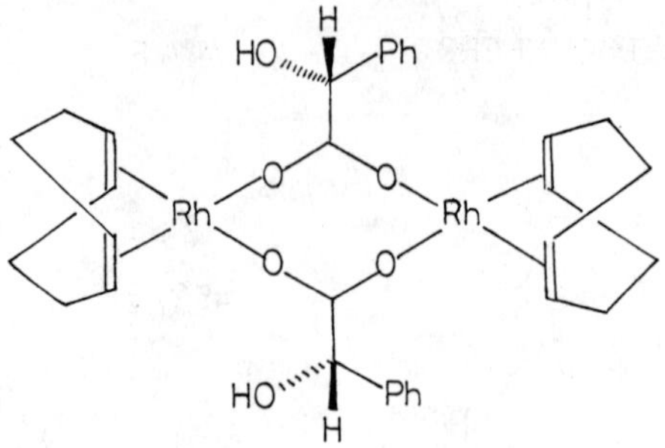

355

Monodentate sulphoxides with chirality at C or S have been coordinated to rhodium but are generally ineffective for asymmetric reduction of unsaturated acids.[759]   However, $[RuCl_2(MBMSO)_2]_3$, where MBMSO is S-bonded 2-methylbutylmethyl sulphoxide, chiral at C_2 and racemic at sulphur, gives up to 15% e.e. for $\alpha\beta$-unsaturated acids.[760]   Chelating sulphoxides, 356 and 357, are a little better.   The best results were for $RuCl_2(DIOS)(DDIOS)$ where DDIOS is S,S- and DIOS S,O-bonded.   The arsenic analogue of DIOP is much less good than the phosphine.[761]

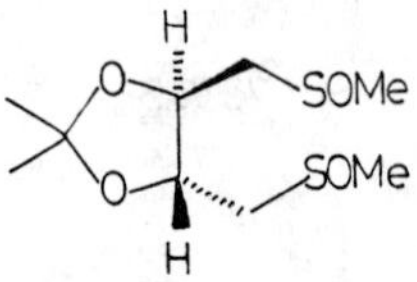

356 DIOS

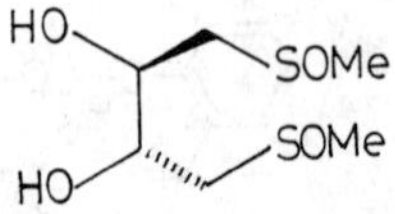

357 DDIOS

Co(dmg)$_2$/alkaloid complexes are useful for reduction of both C=C and C=O bonds.   The best e.e. for reduction of benzil is 78%S using quinine as the chiral auxiliary.[722]   The related complex, 358, gives 78% e.e. in the same reaction and also reduces keto esters with up to 50% optical yield.[762]   Other chiral amines and alkaloids have been studied as auxiliaries but none approaches quinine in effectiveness.   Considering

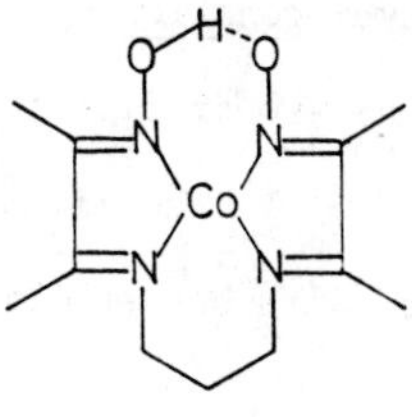

$\underline{358}$

other substrates, keto esters are moderately successful but alkenes are
not.[722]   Some evidence is available for the mechanism of equations (48)
to (53).[763] (Q = quinine)

$$2Co(II)L + H_2 \rightleftharpoons 2HCo(III)L \tag{48}$$

$$2HCo(III)L \rightleftharpoons 2H^+ + 2Co(I)L \tag{49}$$

$$2H^+ + 2Q \rightleftharpoons 2HQ^+ \tag{50}$$

$$Co(I)L + HQ^+ + R_1R_2CO \rightarrow LCo(III)C(OH)R_1R_2 \tag{51}$$

$$LCo(III)C(OH)R_1R_2 + HQ^+ \rightarrow R_1R_2CHOH + Q + Co(III)L \tag{52}$$

$$Co(III)L + Co(I)L \rightarrow 2Co(II)L \tag{53}$$

The stereochemistry of the product is determined in (51) as Co(I) attacks
the prochiral carbon atom and a proton is transferred from quinine.   The
transition state is predicted to be $\underline{359}$.[764]

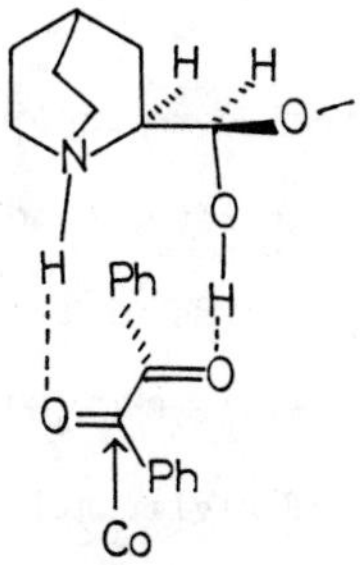

$\underline{359}$

Soluble Ziegler catalysts consisting of
$(\underline{i}\text{-BuO})_3Al/Ti(O\text{-menthoxide})_4$ catalyse non-chiral reduction of 3,4-dimethyl-
1-pentene.   However, (menthylcyclopentadiene)$_2$TiCl$_2$/LiAlH$_2$(OCH$_2$CH$_2$OCH$_3$)$_2$
gives 28%$\underline{S}$ e.e. in reduction of $\underline{\alpha}$-ethylstyrene.[765]

2.12.5 Asymmetric transfer hydrogenation

Two types of reaction, enantiomer discrimation and enantioface discrimination may be distinguished. In the first case (equation (54)) a racemic alcohol is dehydrogenated with one enantiomer reacting faster than the other (see 2.13).

$$R_1CHOHR_2 + R_3CH=CHR_4 \rightarrow R_1COR_2 + R_3CH_2CH_2R_4 +$$
$$(\underline{R},\underline{S}) \qquad\qquad R_1CHOHR_2 \left(\left| \underline{R}-\underline{S} \right| > 0 \right) \qquad\qquad (54)$$

In enantioface discrimination we focus on the prochiral acceptor (equation (55)); hydrogenation gives new chiral centres.

$$DH_2 + R_1R_2C=CHR_3 \xrightarrow{\text{catalyst}} D + R_1R_2CHCH_2R_3 \qquad (55)$$

The donor, D, may be chiral, achiral or racemic. Rhodium and iridium complexes have been little used.[556] Ketones are reduced by i-PrOH in the presence of (Ir(cod)Cl)$_2$/NMDPP in good chemical but poor optical yield.[766] [L$_2$Ir(cod)]PF$_6$(L$_2$=PROPHOS) gave 30%$\underline{S}$ in acetophenone reduction whilst L$_2$ = $\underline{360}$ gave 21%$\underline{R}$.[767]

$\underline{360}$

Ruthenium catalysts have been more important. Using a chiral catalyst and an achiral donor, $\alpha\beta$-unsaturated mono- and diacids are reduced (Figure 77). Monocarboxylic acids give better results than diacids, and the donor is important, since PhCH$_2$OH and p-MeOC$_6$H$_4$CH$_2$OH give differing results in reduction of tiglic acid. Esters are reduced with fair conversion but low e.e. Studies of the mechanism are hampered by the multiplicity of species in solution. Isotope labelling studies imply that the rate controlling step is H-abstraction from CH$_2$ of PhCH$_2$OH and it is suggested that the chirality determining step is preferential coordination of one prochiral face of the alkene.[770]

Ref.

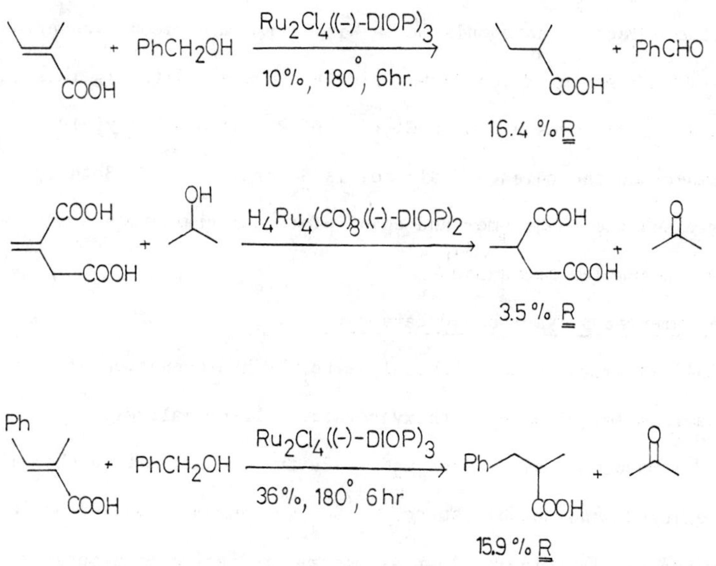

768

769

768

Figure 77 Asymmetric transfer hydrogenation of unsaturated acids.

Where a chiral donor is the only source of chirality the first reports involved a sugar donor and an enone acceptor. Rather high inductions were observed (up to 34%S for 363).[771] With tiglic acid as substrate and 362 as donor 8.9%R e.e. is achieved.[772]

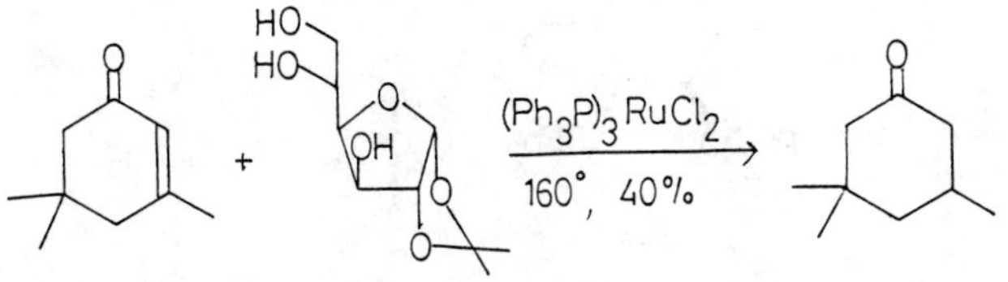

361 362 363

When 362 is used as donor and $Ru_2Cl_4((-)-DlOP)_3$ as catalyst a cooperative effect is observed in tiglic acid reduction (22.5%R e.e.). With (+)-DlOP the effect is antagonistic. With a racemic alcohol enantiomer and enantioface discrimination occurs simultaneously. With tiglic acid and α-phenethanol in the presence of $RuCl_2((-)-DlOP)_3$ an optical yield of 26.4%R is obtained and the unreacted alcohol is S-enriched.[725] Both in this and other cases the enantiomer and enantioface discrimination are unequal, arguing a complex mechanism.

2.12.6 Asymmetric polymer bound catalysts

In 1940 Nakamura obtained chiral amines by hydrogenation of oximes over platinum modified by menthoxyacetate. Later palladium on silk was used for reduction of C=C and C=N. Further metals with a chiral modifier have enjoyed considerable success.[773] Palladium on poly-S-valine and poly-S-leucine catalyse inefficient asymmetric reduction of α-methyl and α-acetamido cinnamic acid.[774] Similarly ruthenium on poly-L-methylethylene imine has been used for mesityl oxide and methyl acetoacetate.[775] There is one example of achiral catalyst on a cellulose backbone. 364 is closely related to 298 and gives up to 77% optical yield in reduction of α-ethylstyrene.[688,738] However, reaction is very slow.

364

Another, larger, group of catalysts are polymers modified with groups analogous to the phosphines used in homogeneous hydrogenation. Synthetic routes are shown in Figure 78. Polystyrene is chloromethylated, oxidised and reacted with a diol to build the dioxolan ring of DlOP.[776]

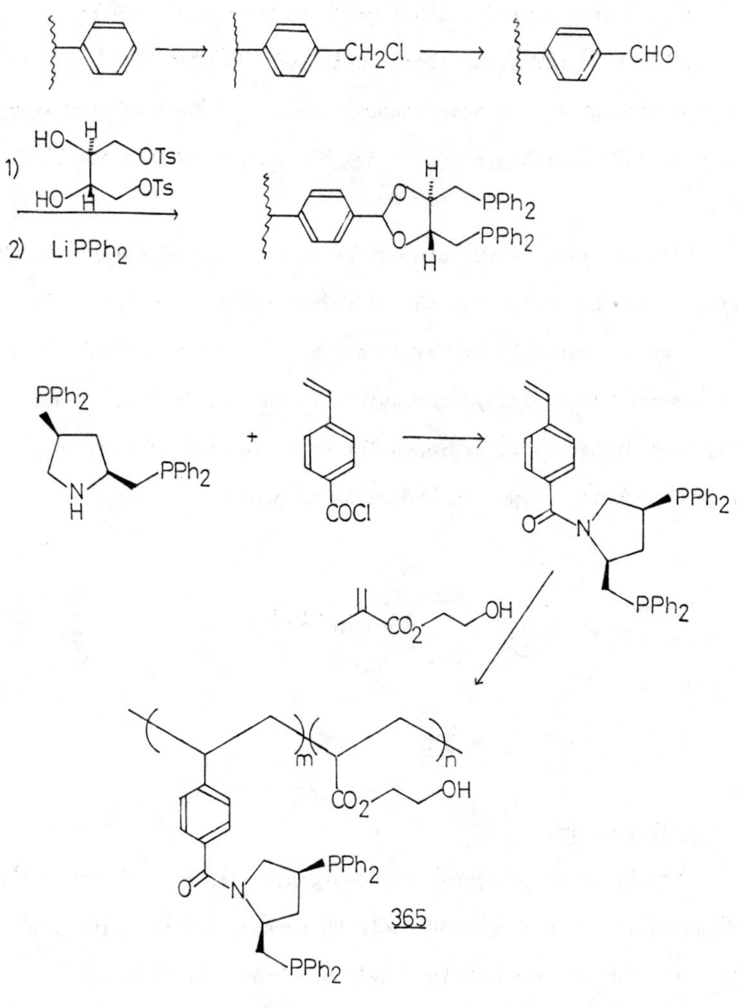

Figure 78 Preparation of chiral polymers.

Alternatively an optically active monomer is copolymerised with a
methacrylate.[777]   In both cases the polymer is subsequently reacted with
a metal complex bearing a replaceable ligand to give the active catalyst.

"DlOP" has been bonded on to several types of functionalised
polystyrene supports, the results being comparable with the homogeneous
case, except that the reaction is diffusion limited and slower.[778]

Silica is also a good support[779] but graphite gave poor results.[780]
Other catalysts have modifying groups analogous to BPPM, again giving
results very similar to the homogeneous cases.[777]   For example, itaconate
dianion is reduced in 82% e.e. (c.f. 89% for 314, R=Ph) and 344 in
76% e.e..[781]

Non-chelating chiral phosphines have also been polymer bound;
all derive their chirality from one or more P-menthyl groups.   The
problems of mono-versus bi- or tridentate binding are encountered here.
Few examples are particularly successful.   366 is relatively inactive
requiring high pressures of hydrogen but gives results comparable with
menthyl and neomenthyl diphenyl phosphine complexes.[782]

366

2.13 Dehydrogenation

Catalytic dehydrogenation of organic compounds is more difficult
than hydrogenation, since thermodynamic factors favour the saturated
species.   In some stoichiometric dehydrogenations the reaction is driven
by the favourable formation of a metal complex of the unsaturated product
(e.g. 368[783] and 370[784]).   Cyclohexene has been dehydrogenated to benzene
in the presence of [HRh ($\underline{i}$-PrO)$_3$P)$_2$]$_2$ with concomitant formation of 371,[785]
and the stoichiometric dehydrogenation of pentane is achieved using
(Ar$_3$P)$_2$ReH$_7$ to give 372.[786]   Cyclopentane is converted into the cyclopentadienyl complex, 373, in the presence of a hydrogen acceptor.[144]

A number of catalytic dehydrogenations rely for their effectiveness on the continual removal of H$_2$.   Many data refer to dehydrogenation

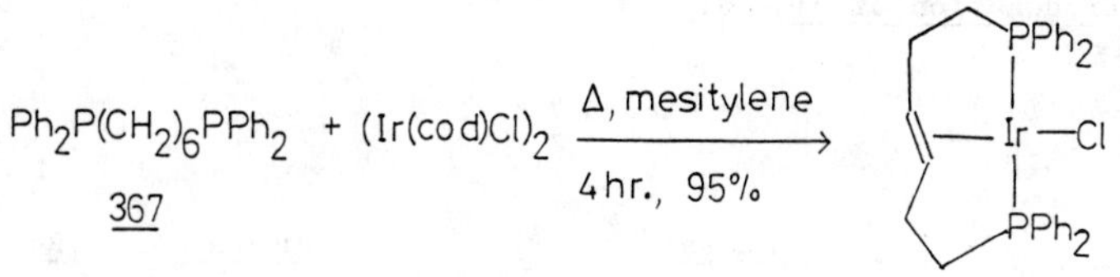

$$Ph_2P(CH_2)_6PPh_2 \ + \ (Ir(cod)Cl)_2 \xrightarrow[\text{4 hr., 95\%}]{\Delta,\ \text{mesitylene}}$$

367

368

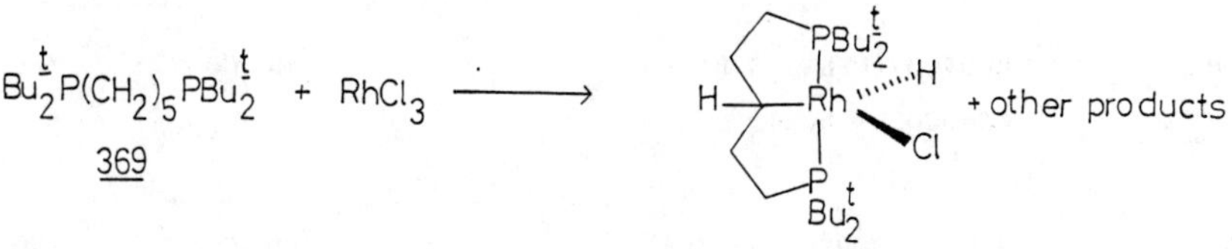

$$Bu^t_2P(CH_2)_5PBu^t_2 \ + \ RhCl_3 \longrightarrow \qquad + \ \text{other products}$$

369

370

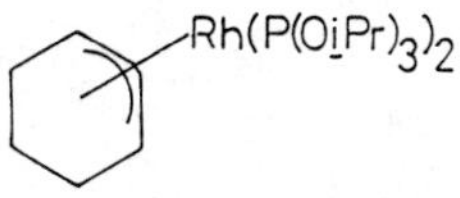

371

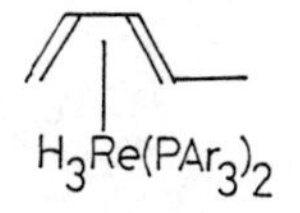

372

$$H_2\overset{+}{Ir}S_2L_2 \ + \ \bigcirc \ + \ 3 \diagup\!\!\!\diagdown\!\!\!\diagup \xrightarrow[80°]{CH_2Cl_2} \ \bigcirc\!-\overset{+}{Ir}HL_2 \ + \ 3 \diagup\!\!\!\diagdown\!\!\!\diagup$$

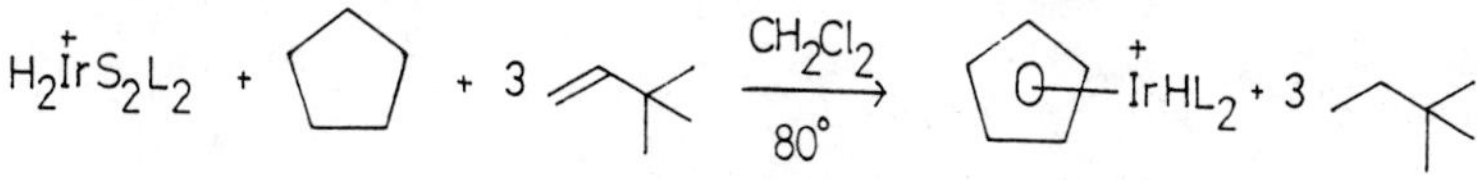

373

of alcohols to aldehydes and ketones (Table 26). The mechanisms proposed are similar for Rh and Ru complexes (Figure 79). Conversion of alcohols to esters is achieved in the presence of $H_2Ru(PPh_3)_4$, which does not require H^+, and involves a dihydride rather than a monohydride (e.g. 374 to 375).[791] Reactions are also known for dehydrogenation of alkanes, alkenes, and dihydroaromatic and heterocyclic compounds but these are not general.

A further group of reactions employ a cooxidant to regenerate the active catalyst from a reduced intermediate. The most widely used

Table 26. Dehydrogenation of alcohols

Substrate	Catalyst	Product	Ref.
i-PrOH	$RhCl_3$, $SnCl_2$, HCl	Me_2CO	787
i-PrOH	$HRu(CO)(PPh_3)_2X/HX$ ($X=OCOCF_3, NO_3$)	Me_2CO	788
i-PrOH	SiO_2-$C_6H_4PPh_2/Rh_2(OAc)_4$	Me_2CO	789
n-PrOH	$(Polystyrene$-$PPh_2)_mRu(OCOCF_3)_2(PPh_3)_n$	CH_3CH_2CHO	790

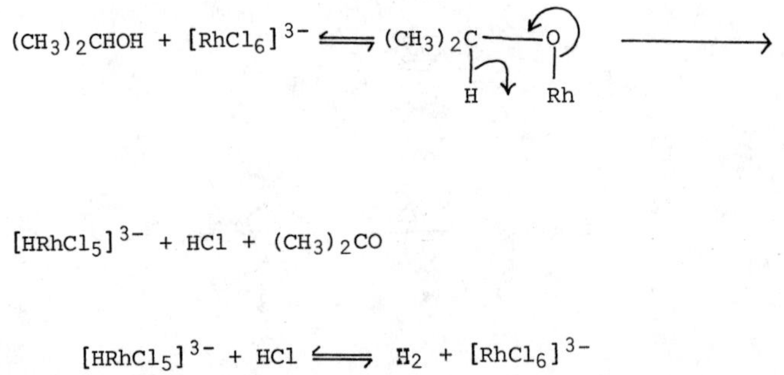

$$(CH_3)_2CHOH + [RhCl_6]^{3-} \rightleftharpoons (CH_3)_2C-O \quad \longrightarrow$$

$$[HRhCl_5]^{3-} + HCl + (CH_3)_2CO$$

$$[HRhCl_5]^{3-} + HCl \rightleftharpoons H_2 + [RhCl_6]^{3-}$$

Figure 79 Mechanism of catalytic dehydrogenation using rhodium complexes.

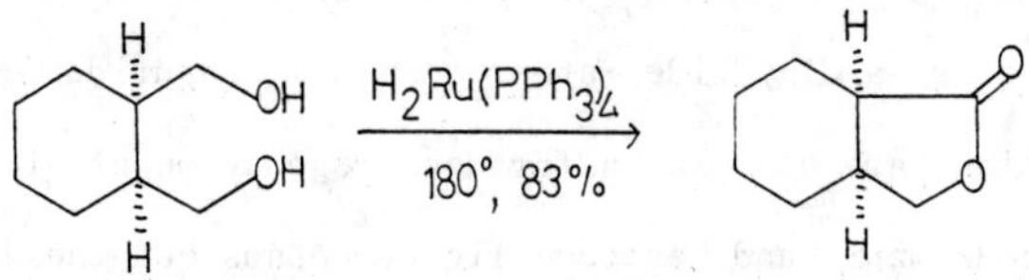

374 → 375

$H_2Ru(PPh_3)_4$, $180°$, 83%

system is Pd(II)/Pd(0) with Cu(II) and/or molecular oxygen as reoxidant

(Figure 80).[792] A recent development has been the use of cobalt

complexes to dehydrogenate easily synthesised indolines, 376, to indoles, 377,

in the presence of oxygen. The reaction is thought to proceed via

a cobalt dioxygen complex coordinated to the nitrogen of indoline.[793]

$$RO_2CCH_2CH=CHCH_2CO_2R + 2\,CuCl_2 \xrightarrow{PdCl_2} 2CuCl + 2\,HCl +$$

$$RO_2CCH=CH-CH=CHCO_2R$$

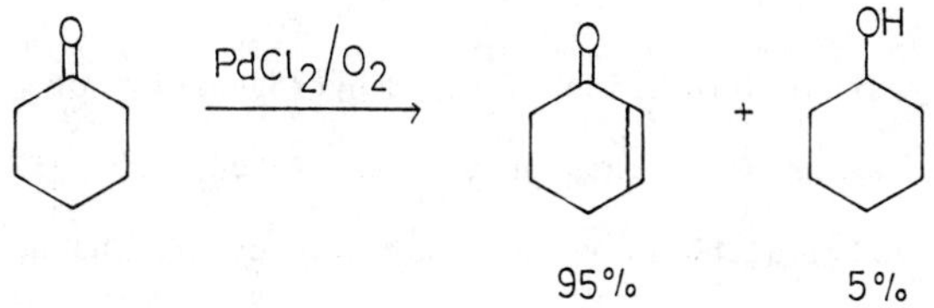

$$\xrightarrow{PdCl_2/O_2}$$

95%   5%

Figure 80   Coupled dehydrogenation/oxidation reactions.

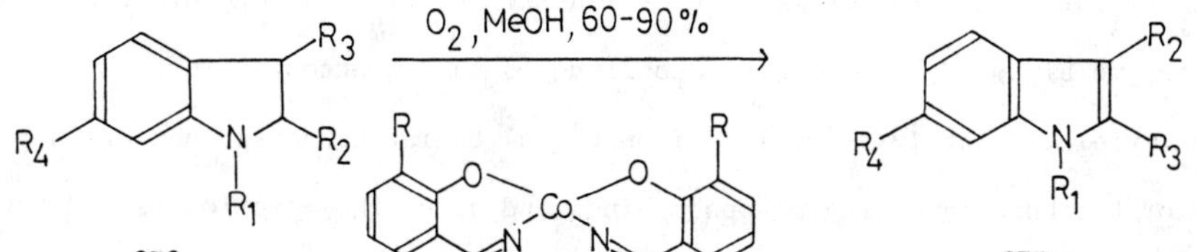

$$\xrightarrow{O_2,\,MeOH,\,60-90\,\%}$$

376   377

The final method by which catalytic dehydrogenation is achieved

to transfer hydrogenation. Disproportionation, for example of

1,4 cyclohexadiene to benzene and cyclohexene, in the presence of

(Ph₃P)₂Ir(CO)Cl, or 378 to 379 and 380, are special examples.[794] Other

acceptors include cyclooctadiene, quinones and imines.[795] 3,3-Dimethyl

butene has been used in a number of reactions which are stoichiometric or

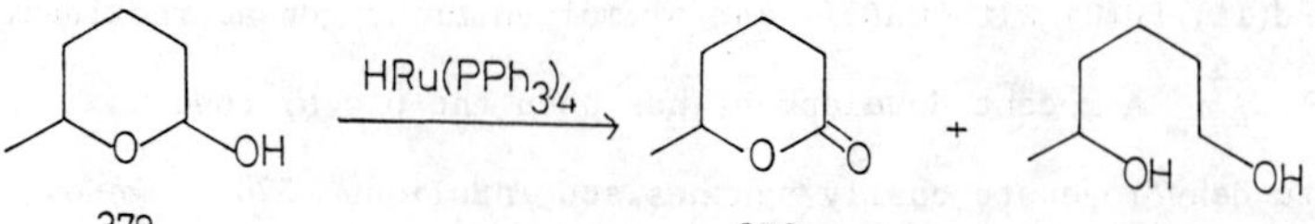

378   HRu(PPh₃)₄   379   +   380

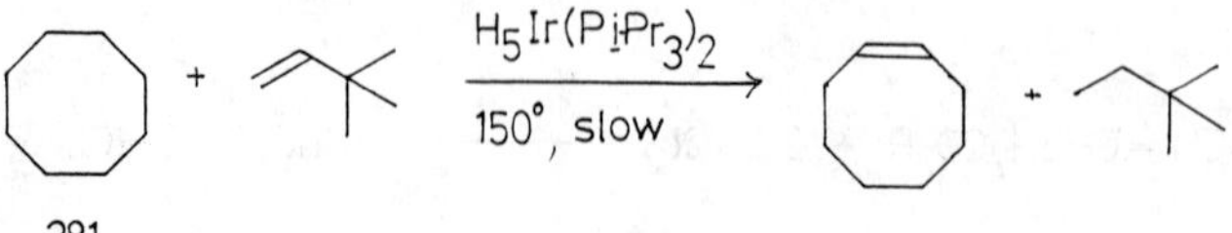

381   +   H₅Ir(PⁱPr₃)₂, 150°, slow   382   +

barely catalytic, but show promise for future development (e.g. 381 to
382).[796]

        The most widely studied reaction is the dehydrogenation of
secondary alcohols to ketones using ketones or αβ-unsaturated carbonyl
compounds as acceptors. Allyl alcohols react under mild conditions at
100°C in the presence of $(Ph_3P)_3RuCl_2$, saturated alcohols requiring
160-190°C for good conversion.[797] Dehydrogenation of a racemic 2°
alcohol (e.g. 383) may lead to a kinetic resolution. Both rhodium and
ruthenium catalysts are active, and H-transfer occurs with both donor and
acceptor metal bound, since bulky substituents on the acceptor influence
optical yield. Catalysts derived from chiral chelating phosphines are not
superior to those involving monophosphines and since the reaction is
inhibited by excess phosphine,dissociation must precede the rate limiting
step. The maximum value of k_R/k_S = 1.3 for $Ru_2Cl_4((-)-DIOP)_3$ ensuring
that this is of greater mechanistic than synthetic value.[798-800]

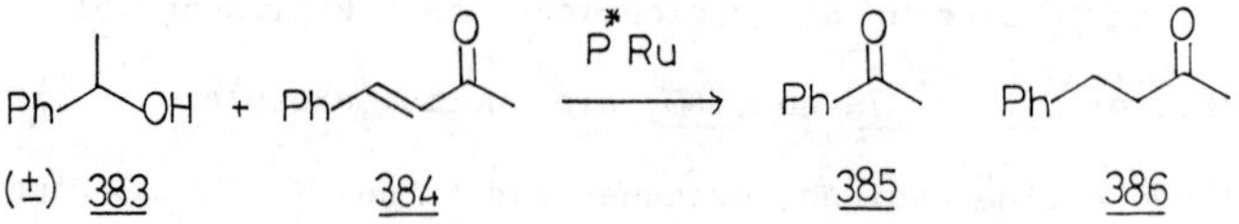

Ph—OH (±) 383   +   Ph— 384 $\overset{*}{P}$ Ru   Ph— 385   Ph— 386

When the substrate is a diol, further reactions may occur.

1,2-Diols such as 387 give diketones but 1,3- and 1,4-giols give lactones

(e.g. 389).[801]   Asymmetric versions of this reaction are known (392 to

393) but optical yields are modest.

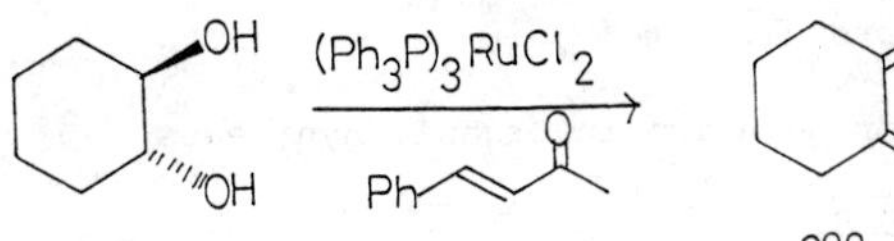

387   388

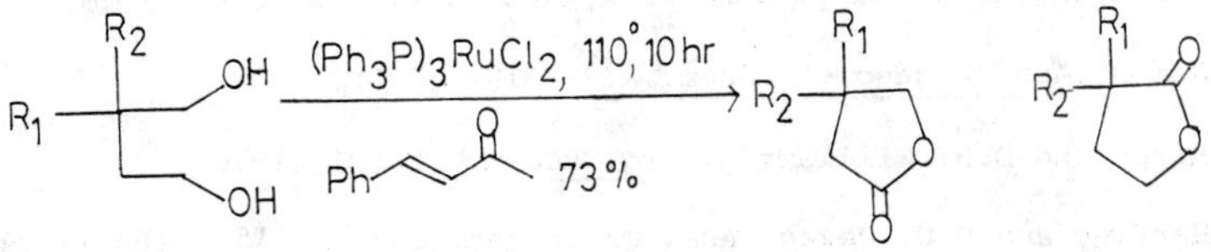

389   390   391

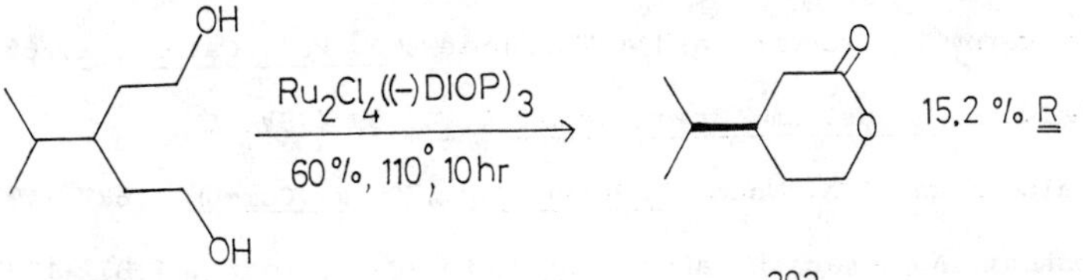

392   393

References

1   B.R. James, Homogeneous Hydrogenation, Wiley Interscience, New York, 1973.

2   H.M. Colquhoun, J. Holton, D.J. Thompson and M.V. Twigg, New Pathways for Organic Synthesis; Practical Applications of Transition Metals, Plenum, New York, 1984, Chapter 7.

3   P.N. Rylander, Catalytic Hydrogenation in Organic Syntheses, Academic Press, New York, 1979.

4   M. Freifelder, Catalytic Hydrogenation in Organic Synthesis, Procedures and Commentary, John Wiley and Sons, New York, 1978.

5   B.R. James, Adv. Organomet. Chem., 17, 319 (1979).

6   A.J. Birch and D.H. Williamson, Org. React., 24, 1 (1976).

7   F.R. Hartley and P.N. Vezey, Adv. Organomet. Chem., 15, 189 (1979).

8   C.A. Tolman, P.Z. Meakin, D.L. Lindner and J.P. Jesson, J. Amer. Chem. Soc., 96, 2762 (1974).

9   J. Halpern, T. Okamoto and A. Zakhariev, J. Mol. Catal., 2, 65 (1977).

10 J. Halpern, Trans. Am. Cryst. Assoc., 14, 59 (1978).

11 J. Halpern and C.S. Wong, J. Chem. Soc., Chem. Commun., 629 (1973).

12 Y. Ohtani, A. Yamagishi and M. Fujimoto, Chem. Lett., 1187 (1977).

13 P. Søgaard-Anderson and J. Ulstrup, Acta Chem. Scand., A37, 585 (1983).

14 C. Rousseau, M. Evrard and F. Petit, J. Mol. Catal., 3, 309 (1977/78).

15 M.H.J.M. de Croon, P.F.M.T. van Nisselrooij, H-J.A.M. Kuipers and J.W.E. Coenen, J. Mol. Catal., 4, 325 (1978).

16 P.B. Hitchcock, M. McPartlin and R. Mason, J. Chem. Soc., Chem. Commun., 1367 (1969).

17 J.M. Brown, L.R. Canning, A.J. Downs and A.M. Forster, J. Organomet. Chem., 255, 103 (1983).

18 J-Y. Saillard and R. Hoffmann, J. Amer. Chem. Soc., 106, 2006 (1984).

19 A. Dedieu, Inorg. Chem., 19, 375 (1980).

20   J. Arriau, J. Fernandez and E. Melendez, J. Chem. Res., 106 (1984).

21   J.M. Brown, private communication.

22   J.F. Young, J.A. Osborn, F.H. Jardine and G. Wilkinson, J. Chem. Soc.,
     Chem. Commun., 131 (1965).

23   M. Carvalho, L.F. Wieserman and D.M. Hercules, Appl. Spec., 36, 290
     (1982).

24   R.S. Drago, J.G. Miller, M.A. Hoselton, R.D. Farris and M.J. Desmond,
     J. Amer. Chem. Soc., 105, 444 (1983).

25   R.R. Schrock and J.A. Osborn, J. Chem. Soc., Chem. Commun., 567 (1970);
     J. Amer. Chem. Soc., 93, 2397, 3089 (1971), 98, 2134, 2143, 4450 (1976).

26   J. Halpern, D.P. Riley, A.S.C. Chan and J.J. Pluth, J. Amer. Chem.
     Soc., 99, 8055 (1977).

27   D.A. Slack and M.A. Baird, J. Organomet. Chem.,142, C69 (1977).

28   J.M. Brown and P.A. Chaloner, J. Amer. Chem. Soc., 102, 3040 (1980).

28   J.M. Brown, P.A. Chaloner, A.G. Kent, B.A. Murrer, P.N. Nicholson,
     D. Parker and P.J. Sidebottom, J. Organomet. Chem., 216, 263 (1981).

30   R.H. Crabtree, Acc. Chem. Res., 12, 331 (1979).

31   R.H. Crabtree, H. Felkin and G.E. Morris, J. Organomet. Chem., 141,
     205 (1977).

32   R.H. Crabtree, H. Felkin, T. Khan and G.E. Morris, J. Organomet. Chem.,
     144, C15 (1978), 168, 183 (1979).

33   R.H. Crabtree, D.F. Chodosh, J.M. Quirk, H. Felkin, T. Fillebeen-Khan
     and G.E. Morris, Fund. Res. Homog. Catal., 3, 475 (1979).

34   R.H. Crabtree and S.M. Morehouse, Inorg. Chem., 21, 4210 (1982).

35   R.H. Crabtree and R.J. Uriarte, Inorg. Chem., 22, 4152 (1983).

36   M.E. Tadros and L. Vaska, J. Colloid Interface Sci., 85, 389 (1982).

37   C.E. Johnson, B.J. Fisher and R. Eisenberg, J. Amer. Chem. Soc., 105,
     7772 (1983).

38   L.S. Stuhl and E.L. Muetterties, Inorg. Chem., 17, 2148 (1978).

174

39   M. Cais and D. Fraenkel, Ann. N.Y. Acad. Sci., 333, 23 (1980).

40   R.A. Head and J.F. Nixon, J. Chem. Soc., Dalton, 913 (1978).

41   Y. Doi, S. Tamura and K. Koshizuka, J. Mol. Catal., 19, 213 (1983).

42   M.D. Fryzuk and P.A. MacNeil, Organometallics, 2, 682 (1983).

43   P.J. Brothers, Prog. Inorg. Chem., 28, 1 (1981).

44   G. Henrici-Olivé and S. Olivé, Angew. Chem., Int. Ed., Engl., 13, 549 (1974).

45   A.S. Berenblyum, A.G. Knizhnik, S.L. Mund and I.I. Moiseev, J. Organomet. Chem., 234, 219 (1982).

46   B.R. James, A.D. Rattray and D.K.W. Wang, J. Chem. Soc., Chem. Commun., 792 (1976).

47   P.S. Hallman, B.R. McGarvey and G. Wilkinson, J. Chem. Soc., A, 3143 (1968).

48   P.R. Hoffman and K.G. Caulton, J. Amer. Chem. Soc., 97, 4221 (1975).

49   J. Tsuji and H. Suzuki, Chem. Lett., 1083, 1085 (1977).

50   T.V. Ashworth and E. Singleton, J. Chem. Soc., Chem. Commun., 705 (1976).

51   J. Halpern and B.R. James, Canad. J. Chem., 44, 671 (1966).

52   R.S. Patil, R.V. Chaudhari and D.N. Sen, J. Mol. Catal., 23, 51 (1984).

53   J. Bayston, N.K. King and M.E. Winfield, Adv. Catal., 9, 312 (1957).

54   B. de Vries, J. Catal., 1, 489 (1962).

55   J. Kwiakek, I.L. Mador and J.Y. Seyler, Adv. Chem., 37, 201 (1963).

56   J. Halpern and M. Pribanić, Inorg. Chem., 9, 2616 (1970).

57   M.B. Mooiman and J.M. Pratt, J. Chem. Soc., Chem. Commun., 33 (1981).

58   J. Halpern, Inorg. Chim. Acta Lett., 77, L105 (1983).

59   M. Orchin, L. Kirch and I. Goldfarb, J. Amer. Chem. Soc., 78, 5450 (1956).

60   F.H. Jardine, Prog. Inorg. Chem., 28, 63 (1981).

61   M. Brown and L.W. Piszkiewicz, J. Org. Chem., 32, 2013 (1967).

62   F.H. Jardine, J.A. Osborn and G. Wilkinson, J. Chem. Soc., A, 1574
     (1967).

63   W. Strohmeier and R. Endres, Z. Naturforsch. B, 25, 1068 (1970).

64   G.K. Koch and J.W. Dalenberg, J. Label. Cmpds. Radiopharm., 6, 395
     (1970).

65   R.O. Adlof, W.R. Miller and E.A. Emken, J. Label. Cmpds. Radiopharm.,
     15, 625 (1978).

66   B. Zeeh, G. Jones and C. Djerassi, Chem. Ber., 100, 3204 (1967).

67   A.S. Hussey and Y. Takeuchi, J. Org. Chem., 35, 643 (1970).

68   J-F. Biellmann and H. Liesenfelt, C.R. Acad. Sci. Paris, 263C, 251
     (1966).

69   R.E. Harmon, S.K. Gupta and D.J. Brown, Chem. Rev., 73, 21 (1973).

70   K. Ohno and J. Tsuji, J. Amer. Chem. Soc., 90, 99 (1968).

71   M.C. Baird, C.J. Nyman and G. Wilkinson, J. Chem. Soc., A, 348 (1968).

72   J.A. Osborn, F.H. Jardine, J.F. Young and G. Wilkinson, J. Chem. Soc.,
     A, 1711 (1966).

73   D.A. Slack and M.C. Baird, J. Amer. Chem. Soc., 98, 5539 (1976).

74   M. Barthélémy, A. Gianfermi and Y. Bessière, Bull. Soc. Chim. Fr.,
     1821 (1976).

75   C. Rousseau, M. Evrard and F. Petit, J. Mol. Catal., 5, 163 (1979).

76   R.L. Augustine and R.J. Pellet, J. Chem. Soc., Dalton, 832 (1979).

77   W. Strohmeier and E. Hitzel, J. Organomet. Chem., 87, 353 (1975).

78   M.J. Bennett and P.B. Donaldson, Inorg. Chem., 16, 1581, 1585 (1977);
     J. Amer. Chem. Soc., 93, 3307 (1971).

79   A.J. Birch and K.A.M. Walker, Tetrahedron Lett., 1935 (1967).

80   S.H. Strauss and D.F. Shriver, Inorg. Chem., 17, 3069 (1978).

81   R.A. Porter and D.F. Shriver, J. Organomet. Chem., 90, 41 (1975).

82   A.J. Birch and H.H. Mantsch, Aust. J. Chem., 22, 1103 (1969).

83   A.J. Birch and K.A.M. Walker, J. Chem. Soc., C, 1894 (1966)

84　H.J. Odom and A.R. Pinder, J. Chem. Soc., Perkin I, 2193 (1972).

85　J. Halpern, Phosphorus and Sulfur, 18, 307 (1983).

86　L. Horner, H. Büthe and H. Siegel, Tetrahedron Lett., 4023 (1968);

　　Ann., 751, 135 (1971).

87　L. Horner and G. Simons, Z. Naturforsch., B, 39, 497 (1984).

88　Ya. L. Gol'dfarb, E.I. Klabunovskii, A.A. Dudinov, L.N. Sukhobok,

　　V.A. Pavlov, B.D. Polkovnikov and V.P. Litvinov, Bull. Acad. Sci.

　　U.S.S.R., Div. Chem. Sci., 32, 1704 (1983).

89　Y. Dror and J. Manassen, J. Mol. Catal., 2, 219 (1977).

90　J.T. Mague and G. Wilkinson, J. Chem. Soc., A, 1736 (1966).

91　R. Stern, Y. Chevallier and L. Sajus, Compt. Rend. Acad. Sci., Paris,

　　264C, 1740 (1967).

92　G.C. Bond and R.A. Hillyard, Discuss. Farad. Soc., 46, 20 (1968).

93　R.W. Mitchell, J.D. Ruddick and G. Wilkinson, J. Chem. Soc., A, 3224

　　(1971).

94　W. Strohmeier and R. Endres, Z. Naturforsch., B, 27, 1415 (1972).

95　G. Dolcetti, Inorg. Nucl. Chem. Lett., 9, 705 (1973).

96　J.R. Shapley, R.R. Schrock and J.A. Osborn, J. Amer. Chem. Soc., 91,

　　2816 (1969).

97　R.H. Crabtree, A. Gautier, G. Giordano and T. Khan, J. Organomet. Chem.,

　　141, 113 (1977).

98　R. Usón, L.A. Oro, J. Artigas and R. Sariego, J. Organomet. Chem., 179,

　　65 (1979).

99　Y. Chevallier, R. Stern and L. Sajus, Tetrahedron Lett., 1197 (1969).

100　J-C. Poulin, T-P. Dang and H.B. Kagan, J. Organomet. Chem., 84, 87

　　(1975).

101　J.G. Santos, Transition Met. Chem., 9, 155 (1984).

102　L. Horner and G. Simons, Z. Naturforsch., B, 39, 504 (1984).

103   A.F. Borowski, D.J. Cole-Hamilton and G. Wilkinson, Nouv. J. Chim., 2, 137 (1978).

104   P.Y. Leung and L.K. Peterson, J. Organomet. Chem., 219, 409 (1981).

105   M. Zuber,   B. Banaś and F. Pruchnik, J. Mol. Catal., 10, 143 (1981).

106   H. Pasternak, T. Glowiak and F. Pruchnik, Inorg. Chim. Acta, 19, 11 (1976).

107   A.W. Gal and F.H.A. Bolder, J. Organomet. Chem., 142, 375 (1977).

108   M. Takesada, H. Yamazaki and N. Hagihara, Bull. Chem. Soc. Jpn., 41, 270 (1968).

109   G. Gregorio, G. Pregaglia and R. Ugo, Inorg. Chim. Acta, 3, 89 (1969).

110   J.F. Nixon and J.R. Swain, J. Organomet. Chem., 72, C15 (1974).

111   D.E. Budd, D.G. Holah, A.N. Hughes and B.C. Hui, Canad. J. Chem., 52, 775 (1974).

112   D.L. Dubois and D.W. Meek, Inorg. Chim. Acta, 19, L29 (1976).

113   J. Niewahner and D.W. Meek, Inorg. Chim. Acta Lett., 64, L123 (1982); A.C.S., Adv. Chem. Ser., 196, 257 (1982).

114   C.A. Reilly and H. Thyret, J. Amer. Chem. Soc., 89, 5144 (1967).

115   J. Powell and B.L. Shaw, J. Chem. Soc., A, 583 (1968).

116   M.D. Fryzuk, Canad. J. Chem., 61, 1347 (1983).

117   M. Bottrill and M. Green, J. Organomet. Chem., 111, C6 (1976).

118   D.A. Thompson and R.W. Rudolph, J. Chem. Soc., Chem. Commun., 770 (1976).

119   T.E. Paxson and M.F. Hawthorne, J. Amer. Chem. Soc., 96, 4674 (1974).

120   W. Strohmeier and S. Hohmann, Z. Naturforsch, B., 25, 1309 (1970).

121   C. O'Connor and G. Wilkinson, J. Chem. Soc., A, 2665 (1968).

122   M. Yagupsky and G. Wilkinson, J. Chem. Soc., A, 941 (1970).

123   G. Yagupsky, C.K. Brown and G. Wilkinson, J. Chem. Soc., Chem. Commun., 1244 (1969):, J. Chem. Soc., A, 1392 (1970).

124   L. Vaska and R.E. Rhodes, J. Amer. Chem. Soc., 87, 4970 (1965).

125   W.R. Cullen, B.R. James and G. Strukul, Canad. J. Chem., 56, 1965
       (1978).

126   W. Abboud, Y. Ben Taarit, R. Mutin and J.M. Basset, J. Organomet.
       Chem., 220, C15 (1981).

127   A. Efraty and I. Feinstein, Inorg Chem., 21, 3115 (1982).

128   W. Reimann, W. Abboud, J.M. Basset, R. Mutin, G.L. Rempel and
       A.K. Smith, J. Mol. Catal., 9, 349 (1980).

129   P. Kalck, R. Poilblanc, R-P. Martin, A. Rovera and A. Gaset,
       J. Organomet. Chem., 195, C9 (1980).

130   D. Evans, J.A. Osborn and G. Wilkinson, J. Chem. Soc., A, 3133 (1968).

131   M.R. Churchill and S.W-Y. Ni, J. Amer. Chem. Soc., 95, 2150 (1973).

132   J.E. Hamlin, K. Hirai, V.C. Gibson and P.M. Maitlis, J. Mol. Catal.,
       15, 337 (1982).

133   M.J.H. Russell, C. White, A. Yates and P.M. Maitlis, J. Chem. Soc.,
       Dalton, 849 (1978).

134   B.R. James and G.L. Rempel, Canad. J. Chem., 44, 233 (1966).

135   G. Wilkinson, U.S. Patent, 3, 857, 900 (1974); Chem. Abs., 82, 170055g
       (1975).

136   N.M. Nazarova, L.Kh. Freidlin, Yu. A. Kopyttsev and T.I. Varava,
       Bull. Acad. Sci. U.S.S.R., Div. Chem. Sci., 21, 1376 (1972).

137   B.R. James and F.T.T. Ng, J. Chem. Soc., Dalton, 355, 1321 (1972).

138   P. Abley, I. Jardine and F.J. McQuillin, J. Chem. Soc., C, 840 (1971).

139   P. Legzdins, R.W. Mitchell, G.L. Rempel, J.D. Ruddick and
       G. Wilkinson, J. Chem. Soc., A, 3322 (1970).

140   B.C.Y. Hui, W.K. Teo and G.L. Rempel, Inorg. Chem., 12, 757 (1973).

141   J.P. Howe, K. Lung and T.A. Nile, J. Organomet. Chem., 208, 401 (1981).

142   M.A. Bennett and D.L. Milner, J. Chem. Soc., Chem. Commun., 581
       (1967); J. Amer. Chem. Soc., 91, 6983 (1969).

143   J.P. Collman, M. Kubota, F.D. Vastine, J.Y. Sun and J.W. Kang,
       J. Amer. Chem. Soc., 90, 5340 (1968).

144 R.H. Crabtree, P.C. Demou, D. Eden, J.M. Michelcic, C.A. Parnell, J.M. Quirk and G.E. Morris, J. Amer. Chem. Soc., 104, 6994 (1982).

145 R. Uson, L.A. Oro and M.J. Fernandez, J. Organomet. Chem., 193, 127 (1980).

146 R.H. Crabtree in Homogeneous Catalysis by Metal Phosphine Complexes, ed. L. Pignolet, Plenum, New York, 1983.

147 H. van Gaal, H.G.A.M. Cuppers and A. van der Ent, J. Chem. Soc., Chem. Commun., 1694 (1970).

148 A. van der Ent and T.C. van Soest, J. Chem. Soc., Chem. Commun., 225 (1970).

149 B. Olgemöller, H. Bauer and W. Beck, J. Organomet. Chem., 213, C57 (1981).

150 W. de Aquino, R. Bonnaire and C. Potvin, J. Organomet. Chem., 154, 159 (1978).

151 L. Vaska, J. Amer. Chem. Soc., 88, 4100 (1966).

152 R.C. Taylor, J.F. Young and G. Wilkinson, Inorg. Chem., 5, 20 (1966).

153 M.G. Burnett, R.J. Morrison and C.J. Strugnell, J. Chem. Soc., Dalton, 632, 1663 (1974).

154 W. Strohmeier and R. Fleischmann, J. Organomet. Chem., 42, 163 (1972).

155 W. Strohmeier, R. Fleischmann and W. Rehder-Stirnweiss, J. Organomet. Chem., 47, C37 (1973).

156 W. Strohmeier and W. Diehl, Z. Naturforsch., B, 28, 207 (1973).

157 W. Strohmeier, W. Rehder-Stirnweiss and R. Fleischmann, Z. Naturforsch., B, 25, 1481 (1970).

158 W. Strohmeier and T. Onoda, Z. Naturforsch, B, 24, 461, 515, 1493 (1969).

159 W. Strohmeier and R. Fleischmann, Z. Naturforsch., B, 24, 1217 (1969); J. Organomet. Chem., 29, C39 (1971).

180

160 W. Strohmeier and W. Rehder-Stirnweiss, Z. Naturforsch., B, 24, 1219
(1969), 26, 61 (1971); J. Organomet. Chem., 18, P28 (1969).

161 W. Strohmeier, R. Fleischmann and T. Onoda, J. Organomet. Chem., 28,
281, 32, 137 (1971).

162 D.M. Roundhill, R.A. Bechtold and S.G.N. Roundhill, Inorg. Chem., 19,
284 (1980).

163 T.B. Rauchfuss, J.L. Clements, S.F. Agnew and D.M. Roundhill,
Inorg. Chem., 16, 775 (1977).

165 L. Vaska, Inorg. Nucl. Chem. Lett., 1, 89 (1965).

166 F. Glockling and M.D. Wilbey, J. Chem. Soc., A, 1675 (1970).

167 M.G. Burnett and R.J. Morrison, J. Chem. Soc., A, 2325 (1971).

168 M. Drouin and J.F. Harrod, Inorg. Chem., 22, 999 (1983).

169 M.R. Churchill and S.A. Julis, Inorg. Chem., 18, 1215 (1979).

170 C.Y. Chan and B.R. James, Inorg. Nucl. Chem. Lett., 9, 135 (1973).

171 J.P. Collman, N.W. Hoffman and D.E. Morris, J. Amer. Chem. Soc., 91,
5659 (1969).

172 C.P. Kubiak, C. Woodcock and R. Elsenberg, Inorg. Chem., 19, 2733
(1980).

173 M. Freund, L. Markó and J. Laki, Acta Chim. Acad. Sci. Hung., 31, 77
(1962).

174 I. Ogata and A. Misono, Discuss. Farad. Soc., 46, 72 (1968).

175 G.F. Ferrari, A. Andreeta, G.F. Pregaglia and R. Ugo, J. Organomet.
Chem., 43, 213 (1972).

176 G. Pregaglia, A. Andreeta, G. Ferrari and R. Ugo, J. Chem. Soc.,
Chem. Commun., 590 (1969).

177 T.S. Janik, M.F. Pyszczek and J.D. Attwood, J. Mol. Catal., 11, 33
(1981).

178 A. Misono, Y. Uchida, T. Saito and K.M. Song, J. Chem. Soc., Chem.
Commun., 419 (1967).

179   J.L. Hendrikse, J.H. Kaspersma and J.W.E. Coenen, Int. J. Chem. Kinet.,
      7, 557 (1975).

180   E. Balogh-Hergovich, G. Speier and L. Markó, J. Organomet. Chem., 66,
      303 (1974).

181   L.W. Gosser, Inorg. Chem., 14, 1453 (1975), 15, 1348 (1976).

182   R.C. White and J.G. Thatcher, Ger. Offen., 2, 245, 832 (1974);
      Chem. Abs., 81, 3324s (1974).

183   S. Tyrlik and M. Michalski, J. Organomet. Chem., 102, 93 (1975).

184   T. Ueno and H. Miya, Chem. Lett., 39 (1973).

185   T.A. Stephenson and G. Wilkinson, J. Inorg. Nucl. Chem., 28, 945
      (1966).

186   P.S. Hallman, D. Evans, J.A. Osborn and G. Wilkinson, J. Chem. Soc.,
      Chem. Commun., 305 (1967).

187   M.M. Taqui Khan, R. Mohiuddin, S. Vancheesan and B. Swamy, Ind. J.
      Chem., 20A, 564 (1981).

188   B.R. James, L.D. Markham and D.K.W. Wang, J. Chem. Soc., Chem.
      Commun., 439 (1974).

189   D. Rose, J.D. Gilbert, R.P. Richardson and G. Wilkinson, J. Chem. Soc.,
      A, 2610 (1969).

190   G. Sbrana, G. Braca and E. Giannetti, J. Chem. Soc., Dalton, 1847
      (1976).

191   W. Strohmeier and G. Buckow, J. Organomet. Chem., 110, C17 (1976).

192   H. Singer, E. Hademer, U. Oehmichen and P. Dixneuf, J. Organomet.
      Chem., 178, C13 (1979).

193   T. Kauffmann and J. Olbrich, Tetrahedron Lett., 1967 (1984).

194   G.E. Rodgers, W.R. Cullen and B.R. James, Canad. J. Chem., 61, 1314
      (1983).

195   A. Spencer, J. Organomet. Chem., 93, 389 (1975).

196   G.L. Geoffroy and M.G. Bradley, Inorg. Chem., 16, 744 (1977).

182

197   A. Yamamoto, S. Kitazume and S. Ikeda, J. Amer. Chem. Soc., 90, 1089 (1968).

198   W.H. Knoth, J. Amer. Chem. Soc., 90, 7172 (1968).

199   S. Komiya and A. Yamamoto, Bull. Chem. Soc. Jpn., 49, 2553 (1976); J. Mol. Catal., 5, 279 (1979).

200   R.W. Mitchell, A. Spencer and G. Wilkinson, J. Chem. Soc., Dalton, 846 (1973).

201   F. Porta, S. Cenini, S. Giordano and M. Pizzotti, J. Organomet. Chem., 150, 261 (1978).

202   D.G. Holah, A.N. Hughes, B.C. Hui and C.T. Kan, J. Catal., 48, 340 (1977).

203   R.H. Crabtree and A.J. Pearman, J. Organomet. Chem., 157, 335 (1978).

204   Y. Doi, S. Tamura and K. Koshizuka, Inorg. Chim. Acta Lett., 65, L63 (1982).

205   J.L. Graff and M.S. Wrighton, Inorg. Chim. Acta, 63, 63 (1982); J. Amer. Chem. Soc., 102, 2123 (1980).

206   Y. Doi, K. Koshizuka and T. Keii, Inorg. Chem., 21, 2732 (1982).

207   G. Süss-Fink and J. Reiner, J. Mol. Catal., 16, 231 (1982).

208   J. Halpern, J.F. Harrod and B.R. James, J. Amer. Chem. Soc., 83, 753 (1961), 88, 5150 (1966).

209   L.Kh. Freidlin, E.F. Litvin and K.G. Karimov, Bull. Acad. Sci. U.S.S.R., Div. Chem. Sci., 23, 785 (1974).

210   A.G. Hinze, Rec., 92, 542 (1973).

211   R. Iwata and I. Ogata, Tetrahedron, 29, 2753 (1973).

212   W. Reppe and H. Vetter, Ann., 582, 133 (1953).

213   H. Sternberg, R. Markby and I. Wender, J. Amer. Chem. Soc., 79, 6116 (1957), 78, 5704 (1956).

214   I. Fischler, R. Wagner and E.A. Koerner von Gustorf, J. Organomet. Chem., 112, 155 (1976).

215   H. Inoue and M. Sato, J. Chem. Soc., Chem. Commun., 983 (1983).

216   R.A. Sánchez-Delgado, A. Audriollo and N. Valencia, J. Chem. Soc.,
       Chem. Commun., 444 (1983).

217   A. Oudeman, F. van Rantwijk and J. van Bekkum, J. Coord. Chem., 4,
       1 (1974).

218   T. Mizuta, H. Samejima and T. Kwan, Bull. Chem. Soc. Jpn., 41, 727
       (1968); J. Chem. Soc. Jpn., 89, 1028 (1968).

219   P. Abley and F.J. McQuillin, Discuss. Farad. Soc., 46, 31 (1968).

220   V. Kadlec and H. Kadlecova, Czech., CS, 190, 174 (1981); Chem. Abs.,
       96, 122182p (1982).

221   J. Müller, B. Passon and S. Schmitt, J. Organomet. Chem., 195, C21
       (1980).

222   M.G. Thomas, E.L. Muetterties, R.O. Day and V.W. Day, J. Amer. Chem.
       Soc., 98, 4645 (1976).

223   M. Tzinmann, D. Cuzin and F. Coussemant, J. Mol. Catal., 4, 191
       (1978).

224   T. Thangaraj, S. Vancheesan, J. Rajaram and J.C. Kuriacose,
       Ind. J. Chem., 19A, 404 (1980).

225   E.B. Maxted and S.M. Ismail, J. Chem. Soc., 1750 (1964).

226   L.Kh. Freidlin, N.M. Nazarova and Y.A. Kopytsev, Bull. Acad. Sci.
       U.S.S.R., Div. Chem. Sci., 21, 196 (1972).

227   M. Sakakibara, Y. Takahashi, S. Sakai and Y. Ishii, Inorg. Nucl. Chem.
       Lett., 5, 427 (1969).

228   J.A. Davies, F.R. Hartley, S.G. Murray and G. Marshall, J. Mol. Catal.
       10, 171 (1981); J. Chem. Soc., Dalton, 2246 (1980).

229   A.T. Teleshev, N.V. Kuznetsova, I.D. Rozhdestvenskaya and
       É.E. Nifant'ev, J. Gen. Chem. U.S.S.R., 49, 652 (1979).

230   A.S. Berenblyum, A.G. Knizhnik, S.L. Mund and I.I. Moiseev, Bull.
       Acad. Sci. U.S.S.R., Div. Chem. Sci., 31, 1111 (1982).

231 A.S. Berenblyum, A.P. Aseeva, L.I. Lakhman and I.I. Moiseev, J. Organomet. Chem., 234, 237 (1982).

232 P.N. Rylander, N. Himelstein, D.R. Steele and J. Kreidl, Englehard Ind. Tech. Bull., 3(2), 61 (1962); Chem. Abs., 57, 15864d (1962).

233 K.E. Hayes, Nature, 210, 412 (1966).

234 R.D. Cramer, E.L. Jenner, R.V. Lindsey and U.G. Stolberg, J. Amer. Chem. Soc., 85, 1691 (1963).

235 C.-H. Cheng, L. Kuritzkes and R. Eisenberg, J. Organomet. Chem., 190, C21 (1980).

236 H. van Bekkum, J. van Gogh and G. van Minnen-Pathuis, J. Catal., 7, 292 (1967).

237 L.P. van't Hof and B.G. Linsen, J. Catal., 7, 295 (1967).

238 R.D. Cramer, R.V. Lindsey, C.T. Prewitt and U.G. Stolberg, J. Amer. Chem. Soc., 87, 658 (1965); 88, 3534 (1966).

239 H. van Bekkum, F. van Rantwijk, G. van Minnen-Pathuis, J.D. Remijnse and A. van Veen, Rec., 88, 911 (1969).

240 J.F. Young, R.D. Gillard and G. Wilkinson, J. Chem. Soc., 5176 (1964).

241 A.P. Khrushch, L.A. Tokina and A.E. Shilov, Kinet. Catal., 7, 793 (1966).

242 J.C. Bailar and H. Itatani, J. Amer. Chem. Soc., 89, 1592 (1967).

243 M.I. Kalinkin, Z.N. Parnes, D.Kh. Shaapini, N.P. Shevlyakova and D.N. Kursanov, Dokl. Chem., 229, 492 (1976).

244 M.C. Baird, J. Inorg. Nucl. Chem., 29, 367 (1967).

245 R.V. Lindsey, G.W. Parshall and U.G. Stolberg, J. Amer. Chem. Soc., 87, 658 (1965).

246 R.W. Adams, G.E. Batley and J.C. Bailar, J. Amer. Chem. Soc., 90, 6051 (1968); Inorg. Nucl. Chem. Lett., 4, 455 (1968).

247 G.K. Anderson, H.C. Clark and J.A. Davies, Inorg. Chem., 22, 427, 434, 439 (1983).

248   K. Kushi, H. Kanai, K. Tamara and S. Yoshida, Chem. Lett., 539 (1972).

249   T. Yoshida, T. Yamagata, T.H. Tulip, J.A. Ibers and S. Otsuka,

      J. Amer. Chem. Soc., 100, 2063 (1978).

250   R. Rumin, J. Organomet. Chem., 247, 351 (1983).

251   A.S. Todozhokova, P.S. Chekrii and M.L. Khidekel', Bull. Acad. Sci.

      U.S.S.R., Div. Chem. Sci., 24, 1579 (1975).

252   D.S. Breslow and N.R. Newburg, J. Amer. Chem. Soc., 81, 81 (1959).

253   F.A. Shmidt, N.M. Ryntina, V.V. Saraev, V.A. Gruznykh and V.A. Makarov,

      React. Kinet. Catal. Lett., 5, 101 (1976).

254   E. Samuel, J. Organomet. Chem., 198, C65 (1980).

255   G.P. Pez and S.C. Kwan, J. Amer. Chem. Soc., 98, 8079 (1976).

256   E.V. Evdokimova, B.M. Bulychev and G.L. Soloveichik, Kinet. Catal.,

      22, 144 (1981).

257   P.C. Wailes, H. Weigold and A.P. Bell, J. Organomet. Chem., 43, C32

      (1972).

258   T.A. Weil, S. Metlin and I. Wender, J. Organomet. Chem., 49, 227

      (1973).

259   A.P. Krushch and A.E. Shilov, Kinet. Catal., 10, 389 (1969).

260   Johnson Matthey and Co., Fr. Patent, 2,055,060 (1971); Chem. Abs.,

      76, 50629b (1972).

261   R.H. Crabtree, G.G. Hlatky, C.P. Parnell, B.E. Segmüller and

      R.J. Uriarte, Inorg. Chem., 23, 354 (1984).

262   J.P. Candlin and A.R. Oldham, Discuss. Farad. Soc., 46, 60 (1968).

263   R.L. Augustine and J.F. van Peppen, J. Chem. Soc., Chem. Commun.,

      495, 497, 571 (1970).

264   W. Strohmeier and R. Endres, Z. Naturforsch., B, 26, 730 (1971).

265   S. Ceninini, R. Ugo and F. Porta, Gazz. Chim. Ital., 111, 293 (1981).

266   F. van Rantwijk, Th.G. Spek and H. van Bekkum, Rec., 91, 1057 (1972).

186

267 T. Takegami, C. Yokokawa, Y. Watanabe, H. Masada and Y. Okuda,

     Bull. Chem. Soc. Jpn., 37, 1190 (1964).

268 J. Kwiatek, I.L. Mador and J.K. Seyler, J. Amer. Chem. Soc., 84, 304

     (1962).

269 J. Kwiatek and J.K. Seyler, A.C.S. Adv. Chem., 70, 207 (1968).

270 J. Halpern and L.Y. Wong, J. Amer. Chem. Soc., 90, 6665 (1968).

271 F.K. Shmidt, V.V. Sarayev, L.O. Nindakova, V.A. Gruznykh,

     S.M. Krasnopolskaya and Y.S. Levkovskii, React. Kinet. Catal. Lett.,

     9, 113 (1978); Kinet. Catal., 9, 249 (1978).

272 S.T. Wilson and J.A. Osborn, J. Amer. Chem. Soc., 93, 3068 (1971).

273 W. Strohmeier and M. Pföhler, Z. Naturforsch., B, 31, 390, 941 (1977).

274 J.A. Davies, F.R. Hartley and S.G. Murray, Inorg. Chem., 19, 2299

     (1980).

275 Y. Takegami, T. Ueno and T. Fujii, Bull. Chem. Soc. Jpn., 42, 1663

     (1969).

276 I.V. Kalechits and F.K. Shmidt, Kinet. Catal., 7, 541 (1966).

277 K. Sonogashira and N. Hagihara, Bull. Chem. Soc. Jpn., 39, 1178 (1966).

278 G. Fachinetti and C. Floriani, J. Chem. Soc., Chem. Commun., 66 (1974).

279 H. Adkins and G. Krsek, J. Amer. Chem. Soc., 71, 3051 (1949).

280 D.M. Rudkovskii and N.S. Imyanitov, J. Appl. Chem. U.S.S.R., 35, 2608

     (1962).

281 J.G. Wadkar and R.V. Chaudhari, J. Mol. Catal., 22, 103 (1983).

282 F. Jardine and G. Wilkinson, J. Chem. Soc., C, 270 (1967).

283 R.E. Harmon, J.L. Parsons, D.W. Cook, S.K. Gupta and J. Schoolenberg,

     J. Org. Chem., 34, 3684 (1969).

284 G. Mestroni, R. Spogliarich, A. Camus, F. Martinelli and

     G. Zassinovich, J. Organomet. Chem., 157, 345 (1978).

285 T. Kitamura, N. Sakamoto and T. Joh, Chem. Lett., 379 (1973).

286 W. Strohmeier and M. Lukacs, J. Organomet. Chem., 133, C47 (1977).

287 W. Strohmeier and H. Steigerwald, J. Organomet. Chem., 129, C43 (1977).

288 J. Kwiatek, Catal. Rev., 1, 37 (1967).

289 D.L. Reger, M.M. Habib and D.J. Fauth, J. Org. Chem., 45, 3860 (1980).

290 J. Kwiatek and I.L. Mador, U.S. Patent, 3,185,727 (1965); Chem. Abs.,
     63, 2900g (1965).

291 I. Wender, R. Levine and M. Orchin, J. Amer. Chem. Soc., 72, 4375
     (1950); U.S. Patent 2,614,107 (1952); Chem. Abs., 47, 5422d (1953).

292 E. Ucciani, R. Laï and L. Tanguy, Compt. Rend. Acad. Sci., Paris,
     281C, 877 (1975).

293 K. Kogami and J. Kumanotani, Bull. Chem. Soc. Jpn., 46, 3562 (1973).

294 K. Murata and A. Matsuda, Bull. Chem. Soc. Jpn., 54, 1899 (1981).

295 G.P. Pez, R.A. Grey and J. Corsi, J. Amer. Chem. Soc., 103, 7528
     (1981).

296 J.W. Russell, D.M. Duncan and S.C. Hansen, J. Org. Chem., 42, 551
     (1977).

297 L.Kh. Freidlin, Yu.A. Kopyttsev and N.M. Nazarova, Bull. Acad. Sci.
     U.S.S.R., Div. Chem. Sci., 22, 687 (1973).

298 A. Nakamura and S. Otsuka, Tetrahedron Lett., 4529 (1973).

299 W. Strohmeier, M. Michel and L. Weigelt, Z. Naturforsch., B, 35, 648
     (1980).

300 N.V. Borunova, L.Kh. Freidlin, M.L. Khidekel', S.S. Daniélova
     V.A. Avidov and P.S. Chekrii, Bull. Acad. Sci. U.S.S.R., Div. Chem.
     Sci., 432 (1968).

301 M.C. Rakowski and E.L. Muetterties, J. Amer. Chem. Soc., 99, 739 (1977).

302 W. Strohmeier and H. Steigerwald, Z. Naturforsch., B, 31, 1149 (1977).

303 H.A. Tayim and J.C. Bailar, J. Amer. Chem. Soc., 89, 4330 (1967).

304 S.J. Lapporte, Ann. N.Y. Acad. Sci., 158, 510 (1969).

305 W. Strohmeier and E. Hitzel, J. Organomet. Chem., 102, C37 (1975),
     91, 373 (1975).

188

306   M.N. Ricroch and A. Gaudemer, J. Organomet. Chem., 67, 119 (1974).

307   J. Solodar, J. Org. Chem., 43, 1787 (1978).

308   J.J. Sims, V.K. Honwad and L.H. Selman, Tetrahedron Lett., 87 (1969).

309   J. Kaspar, R. Spogliarich, A. Cernogoraz and M. Graziani,
       J. Organomet. Chem., 255, 371 (1983).

310   H.J. Brodie, K.J. Kripalani and G. Possanza, J. Amer. Chem. Soc., 91,
       1241 (1969).

311   W. Voelter and C. Djerassi, Chem. Ber., 101, 58, 1154 (1968).

312   S. Nishimura, O. Yomuto, K. Tsuneda and H. Mori, Bull. Chem. Soc.
       Jpn., 48, 2603 (1975).

313   A.J. Birch and K.A.M. Walker, Tetrahedron Lett., 3457 (1967);
       Aust. J. Chem., 24, 513 (1971).

314   A.A. Vlček and J. Hanzlík, Inorg. Chem., 6, 2053 (1967), 8, 669 (1969).

315   M. Calvin, Trans Farad. Soc., 34, 1181 (1938); J. Amer. Chem. Soc.,
       61, 2230 (1939).

316   B.C. Hui and B.R. James, Canad, J. Chem., 52, 3760 (1974); B.R. James
       and F.T.T. Ng, Canad J. Chem., 53, 797 (1975).

317   W. Strohmeier and E. Hitzel, J. Organomet. Chem., 110, C22 (1976).
       W. Strohmeier and J.-P. Stasch, Z. Naturforsch., B, 34, 755 (1979).

318   R.S. Coffey, J. Chem. Soc., Chem. Commun., 923 (1967).

319   J. Joó and Z. Tóth, J. Mol. Catal., 8, 369 (1980).

320   W. Strohmeier and H. Steigerwald. Z. Naturforsch., B, 30, 468 (1975).

321   M.A. Bennett and P.A. Longstaff, Chem. Ind., 846 (1965).

322   L.M. Jackman, J.A. Hamilton and J.M. Lawler, J. Amer. Chem. Soc.,
       90, 1914 (1968).

323   K. Ohkubo, T. Kawabe, K. Yamashita and S. Sakaki, J. Mol. Catal.,
       24, 83 (1984).

324   R.E. Harmon, J.L. Parsons and S.K. Gupta, J. Chem. Soc., Chem.
       Commun., 1365 (1969).

325 Z. Tóth, F. Joó and M.T. Beck, Inorg. Chim. Acta, 42, 153 (1980).

326 K. Ohkubo, K. Tsuchihashi, H. Ikebe and H. Sakamoto, Bull. Chem. Soc. Jpn., 48, 1114 (1975).

327 R. Lykvist and R. Larsson, J. Mol. Catal., 19, 1 (1983).

328 A. Bergmann, R. Karlsson and R. Larsson, J. Catal., 38, 418 (1975).

329 F. Joó, Z. Tóth and M.T. Beck, Inorg. Chim. Acta, 25, L61 (1977).

330 A.N. Nesmeyanov, L.P. Yur'eva, N.N. Zaitseva, V.K. Latov, S.D. Soinov, L.M. Samoilenko, K.K. Babievskii and A.G. Makarvoskaya, Bull. Acad. Sci. U.S.S.R., Div. Chem. Sci., 25, 903 (1976).

331 J. Falbe, N. Huppes and F. Korte, Chem. Ber., 97, 863 (1964).

332 J. Solodar, J. Org. Chem., 37, 1840 (1972).

333 O.N. Efimov, M.L. Khidekel', V.A. Avilov, P.S. Chekrii, O.N. Eremenko and A.G. Ovcharenko, J. Gen. Chem. U.S.S.R., 38, 2581 (1968).

334 B. Hui and B.R. James, J. Chem. Soc., Chem. Commun., 198 (1969).

335 F. Joó, L. Somsák and M.T. Beck, J. Mol. Catal., 24, 71 (1984).

336 P.M. Lausorot, G.A. Vaglio and M. Valle, Gazz. Chim. Ital., 109, 127 (1979).

337 M.G. Burnett and C.J. Strugnell, J. Chem. Res., S, 250 (1977).

338 P. Shanthalakshmy, S. Vancheesan, J. Rajaram and J.C. Kuriacose, Ind. J. Chem., 19A, 901 (1980).

339 B.R. James and G.L. Rempel, Discuss. Farad. Soc., 46, 48 (1968).

340 M.G. Burnett, R.J. Morrison and C.J. Strugnell, J. Chem. Soc., Dalton, 701 (1973).

341 T. Kwon, J.C. Woo and C.S. Chin, Polyhedron, 2, 1225 (1983).

342 S. Takeuchi, Y. Ohgo and J. Yoshimura, Bull. Chem. Soc. Jpn., 47, 463 (1974).

343 C.J. Love and F.J. McQuillin, J. Chem. Soc., Perkin I, 2509 (1973).

344 S.R. Patil, S.M. Sen and R.V. Chaudhari, J. Mol. Catal., 19, 233 (1983).

190

345 A.S. Berenblyum, S.L. Mund, T.P. Goranskaya and I.I. Moiseev, Bull. Acad. Sci. U.S.S.R., Div. Chem. Sci., 30, 2041 (1981).

346 V.F. Kuznetsov, O.A. Karpeiskaya, A.A. Belyi and M.E. Vol'pin, Bull. Acad. Sci. U.S.S.R., Div. Chem. Sci., 30, 900 (1981).

347 A.S. Berenblyum, T.V. Turkova and I.I. Moiseev, Bull. Acad. Sci. U.S.S.R., Div. Chem. Sci., 30, 361 (1981).

348 H.W. Thompson and E. McPherson, J. Amer. Chem. Soc., 96, 6232 (1974).

349 Y. Senda, T. Iwasaki and S. Mitsui, Tetrahedron, 28, 4059 (1972).

350 R.H. Crabtree and M.W. Davis, Organometallics, 2, 681 (1983).

351 G. Stork and D.E. Kahne, J. Amer. Chem. Soc., 105, 1072 (1983).

352 J.M. Brown and S.A. Hall, Tetrahedron Lett., 25, 1393 (1984).

353 V.Z. Sharf, É.A. Mistryukov, L.Kh. Freidlin, I.S. Portyakova and V.N. Krutii, Bull. Acad. Sci. U.S.S.R., Div. Chem. Sci., 28, 1322 (1979).

354 S.L. Schreiber and T.J. Sommer, Tetrahedron Lett., 24, 4781 (1983).

355 H.H. Seltzman, S.D. Wyrik and C.G. Pitt, J. Label. Cmpds. Radiopharm., 18, 1365 (1981).

356 R.H. Grubbs, C. Gibbons, L.C. Kroll, W.D. Bonds and C.H. Brubaker, J. Amer. Chem. Soc., 95, 2373 (1973).

357 M.M. Bhagwat and D. Devaprabhakara, Tetrahedron Lett., 1391 (1972).

358 S. Siegel and G. Perot, J. Chem. Soc., Chem. Commun., 114 (1978).

359 H.A. Martin and R.O. De Jongh, Rec, 90, 713 (1971).

360 M. Green, T.A. Kuc and S.H. Taylor, J. Chem. Soc., Chem. Commun., 1553 (1970).

361 S. Nishimura, K. Tsuneda, H. Mori and M. Sawai, Japan. Kokai, 73, 10, 059 (1973); Chem. Abs., 78, 136539a (1973).

362 S. Takahashi, H. Yamazaki and N. Hagihara, Bull. Chem. Soc. Jpn., 41, 254 (1968).

363   Texaco Development Corp., Fr. Demande, 2,203,791 (1974).

      Chem. Abs., 81, 104722a (1974).

364   A. Misono and I. Ogata, Bull. Chem. Soc. Jpn., 40, 2718 (1967);

      Discuss. Farad. Soc., 46, 72 (1968).

365   Y. Takagi, S. Takahashi, J. Nakayama and K. Tanaka, J. Mol. Catal., 10,

      3 (1981).

366   G.F. Pregaglia, G.F. Ferrari, A. Andreetta, G. Capparella, F. Genorii

      and R. Ugo, J. Organomet. Chem., 70, 89 (1974).

367   S.N. Zelenin and M.L. Khidekel', Russ. Chem. Rev., 39, 103 (1970).

368   L.Kh. Freidlin, Yu.A. Kopyttsev, E.F. Litvin and N.M. Nazarova,

      J. Org. Chem. U.S.S.R., 10, 434 (1974).

369   J.E. Lyons, J. Catal., 30, 490 (1973).

370   R. Uson, L.A. Oro and M.J. Fernandez, J. Organomet. Chem., 193, 127

      (1980).

371   M.A. Schroeder and M.S. Wrighton, J. Organomet. Chem., 74, C29 (1974);

      J. Amer. Chem. Soc., 95, 5764 (1973).

372   E.N. Frankel and F.L. Little, J. Am. Oil Chem. Soc., 46, 256 (1969),

      47, 11 (1970).

373   M. Cais, M. Kaftory, D.H. Kohn and D. Tartarsky, J. Organomet. Chem.,

      184, 103 (1979).

374   E.N. Frankel, E. Selke and C.A. Glass, J. Org. Chem., 34, 3930, 3936

      (1969).

375   Y. Eden, D. Fraenkel, M. Cais and E.A. Halevi, Is. J. Chem., 15, 223

      (1976/77).

376   A. Miyake and H. Kondo, Angew. Chem., Int. Ed. Engl., 7, 631 (1968).

377   T. Funabiki, M. Matsumoto and K. Tarama, Bull. Chem. Soc. Jpn., 45,

      2723 (1972).

378   J. Kwiatek and J.K. Seyler, J. Organomet. Chem., 3, 421 (1965).

379   T. Funabiki, M. Mohri and K. Tarama, J. Chem. Soc., Dalton, 1813 (1973).

380 L.I. Gvinter, L.Kh. Freidlin, N.V. Borunova and G.Ya. Starodubskaya, Bull. Acad. Sci. U.S.S.R., Div. Chem. Sci., 21, 2248 (1972).

381 D.L. Reger and A. Gabrielli, J. Mol. Catal., 12, 173 (1981).

382 K. Kawakami, T. Mizoroki and A. Ozaki, Chem. Lett., 846 (1976); J. Mol. Catal., 5, 175 (1979).

383 H. Kanai, N. Yamamoto, K. Kishi, K. Mizuno and K. Tarama, J. Catal., 73, 228 (1982).

384 G.F. Pregaglia, A. Andreetta, G.F. Ferrari, G. Montrasi and R. Ugo, J. Organomet. Chem., 33, 73 (1971).

385 D.R. Fahey, J. Org. Chem., 38, 80, 3343 (1973).

386 Y. Tajima and E. Kunioka, J. Org. Chem., 33, 1689 (1968); J. Catal., 11, 83 (1968).

387 M.G. Burnett, J. Chem. Soc., Chem. Commun., 507 (1965).

388 A. Sisak, I. Jablonkai and F. Ungváry, Acta Chim. Acad. Sci. Hung., 103, 33 (1980).

389 A.S. Berenblyum, T.P. Goranskaya, S.L. Mund and I.I. Moiseev, Bull. Acad. Sci. U.S.S.R., Div. Chem. Sci., 28, 1280 (1979).

390 L.Kh. Freidlin, L.I. Gvinter, L.N. Suvorova and S.S. Daniélova, Bull. Acad. Sci. U.S.S.R., Div. Chem. Sci., 22, 2204 (1973).

391 R. Bertani, G. Carturan and A. Scrivanti, Angew. Chem., Int. Ed. Engl., 22, 246 (1983).

392 W.K. Rohwedder, A.F. Mabrouk and E. Selke, J. Phys. Chem., 69, 1711 (1965).

393 D.L. Reger and M.M. Habib, J. Mol. Catal., 7, 365 (1980).

394 P. Le Maux, J.Y. Saillard, D. Grandjean and G. Jaouen, J. Org. Chem., 45, 4524 (1980).

395 P. Le Maux, G. Jaouen and J.Y. Saillard, J. Organomet. Chem., 212, 193 (1981).

396   R.A. Awl, E.N. Frankel, J.P. Friedrich and E.H. Pryde, J. Amer. Oil

       Chem. Soc., 55, 577 (1978).

397   E.N. Frankel, R.A. Awl and J.P. Friedrich, J. Amer. Oil Chem. Soc.,

       56, 965 (1979).

398   N. Maoz and M. Cais, Is. J. Chem., 6, 32P (1968).

399   L.Kh. Freidlin, E.F. Litvin and K.G. Karimov, J. Gen. Chem. U.S.S.R.,

       44, 2489 (1974).

400   A.G. Hinze and D.J. Frost, J. Catal., 24, 541 (1972).

401   E. Ucciani, A. Pelloquin and G. Cecchi, J. Mol. Catal., 3, 363

       (1977/78).

402   P. Abley and F.J. McQuillin, J. Catal., 24, 536 (1972); J. Chem. Soc.,

       Chem. Commun., 477 (1968).

403   C. Andersson and R. Larsson, J. Amer. Oil Chem. Soc., 58, 54 (1981).

404   J.C. Bailar and H. Itatani, J. Amer. Oil Chem. Soc., 43, 337 (1966).

405   D.H. Goldsworthy, F.R. Hartley and S.G. Murray, J. Mol. Catal., 19,

       257, 269 (1983).

406   E.N. Frankel, H. Itatani and J.C. Bailar, J. Amer. Oil Chem. Soc., 49,

       132 (1972).

407   G. Pregaglia, D. Morelli, F. Conti and G. Gregorio, Discuss. Farad.

       Soc., 46, 110 (1968).

408   L.P. Shuikina, A.I. El'natanova, L.S. Koraleva, O.P. Parenago and

       V.M. Frolov, Kinet. Catal., 22, 149 (1981).

409   W. Strohmeier and L. Weigelt, J. Organomet. Chem., 82, 417 (1974).

410   B. Giovannitti, M. Ghedeni, G. Dolcetti and G. Denti, J. Organomet.

       Chem., 157, 457 (1978).

411   W. Strohmeier and H. Steigerwald, Z. Naturforsch., B, 30, 643 (1975),

       32, 111 (1977).

412   M. Sakai, F. Harada, Y. Sakakibara and N. Uchino, Bull. Chem. Soc.

       Jpn., 55, 343 (1982).

413   R.C. White and J.G. Thatcher, Brit. Pat., 1,407,987 (1975);

       Chem. Abs., 84, 16721f (1976).

414   M. Airoldi, G. Deganello, G. Dia and G. Gennaro, J. Organomet. Chem.,

       187, 391 (1980).

415   M. Gargano, P. Giannoccaro and M. Rossi, J. Organomet. Chem., 84,

       389 (1975).

416   G. Strukul and G. Carturan, Inorg. Chim. Acta, 35, 99 (1979).

417   K.E. Atkin, U.S. Patent 3,308,177 (1967); Chem. Abs., 67, 21504r

       (1967).

418   T.E. Zhesko, D.V. Mushenko, N.S. Barinov, A.G. Nikitina, E.G. Novikova,

       A.P. Khvorov and S.V. Shapkin, Kinet. Catal., 17, 1052 (1976).

419   B.R. James, L.D. Markham, B.C. Hui and G.L. Rempel, J. Chem. Soc.,

       Dalton, 2247 (1973).

420   A.D. Shebaldova, V.I. Bystrenina, V.N. Kravtsova and M.L. Khidekel',

       Bull. Acad. Sci. U.S.S.R., Div. Chem. Sci., 24, 1986 (1975).

421   G.W. Parshall, J. Amer. Chem. Soc., 94, 8716 (1972).

422   T.A. Zhesko, Yu.N. Kukushkin, A.G. Nikitina, V.P. Kotel'nikov and

       N.S. Barinov, J. Gen. Chem. U.S.S.R., 49, 1980 (1979).

423   M.J. Mirbach, T.N. Phu and A. Saus, J. Organomet. Chem., 236, 309

       (1982).

424   I. Jardine, R.W. Howsam and F.J. McQuillin, J. Chem. Soc., C, 260

       (1969).

425   K. Sonogashira and N. Hagihara, Bull. Chem. Soc. Jpn., 39, 1178

       (1976).

426   R.O. Adlof and E.A. Emken, J. Label. Cmpds. Radiopharm., 18, 419

       (1981).

427   I.C.I., Fr. Patent 1,538,700 (1968); Chem. Abs., 71, 83317y (1969).

428   M.C.K. Willott, K.K. Joshi, A.R. Oldham and R.W. Dunning, Brit. Pat.

       1,154,937 (1969); Chem. Abs., 71, 49213b (1969).

429   K.A. Taylor, Adv. Chem., 70, 195 (1968).

430   C.U. Pittman, R.C. Ryan, J. McGee and J.P. O'Connor, J. Organomet.

       Chem., 178, C43 (1979).

431   W.R. Kroll, J. Catal., 15, 281 (1969).

432   W. Strohmeier and K. Grünter, J. Organomet. Chem., 90, C45 (1975).

433   R.H. Crabtree, J. Chem. Soc., Chem. Commun., 647 (1975).

434   M. Cowie and T.J. Southern, Inorg. Chem., 21, 246 (1982),

       J. Organomet. Chem., 193, C46 (1980).

435   A. Sanger, Canad. J. Chem., 60, 1363 (1982).

436   F. Morandini, B. Longato and S. Bresadola, J. Organomet. Chem., 239,

       377 (1982).

437   A. Sisak, F. Ungváry and G. Kiss, J. Mol. Catal., 18, 223 (1983).

438   N.V. Borunova, P.G. Antonov, Yu.N. Kukushkin, Ya.G. Mukhtarov,

       A.N. Shan'ko and L. Kh. Freidlin, Bull. Acad. Sci. U.S.S.R., Div.

       Chem. Sci., 25, 948 (1976).

439   A. Fusi, R. Ugo, R. Psaro, P. Braunstein and J. Dehand, J. Mol.

       Catal., 16, 217 (1982).

440   G. Fachinetti, C. Floriani, F. Marchetti and M. Mellini, J. Chem.

       Soc., Dalton, 1398 (1978).

441   F.K. Shmidt, V.G. Lipovich, S.M. Krasnopol'skaya and I.V. Kalechits,

       Kinet. Catal., 11 , 486 (1970).

442   D.V. Sokol'skii, G.N. Sharifkanova and N.F. Noskova, Dokl. Chem.,

       194, 694 (1970).

443   T. Itoh, T. Nagano and M. Hirobe, Tetrahedron Lett., 21, 1343 (1980).

444   G. Pregaglia, R. Castelli and A. Andreetta, Fr. Patent, 1,514,495

       (1968); Chem. Abs., 70, 79633k (1969).

445   M. Murakami, K. Suzuki and J. Kang, J. Chem. Soc. Jpn., 83, 1226

       (1962).

196

446 T. Yoshida, W.J. Youngs, T. Sakaeda, T. Ueda, S. Otsuka and J.A. Ibers, J. Amer. Chem. Soc., 105, 6273 (1983).

447 R.R. Burch, E.L. Muetterties, R.G. Teller and J.M. Williams, J. Amer. Chem. Soc., 104, 4257 (1982).

448 S. Friedman, S. Metlin, A. Svedi and I. Wender, J. Org. Chem., 24, 1287 (1959).

449 P.D. Taylor and M. Orchin, J. Org. Chem., 37, 3913 (1982).

450 J.T. Shaw and F.T. Tyson, J. Amer. Chem. Soc., 78, 2538 (1956).

451 H. Greenfield, S. Metlin, M. Orchin and I. Wender, J. Org. Chem., 23, 1054 (1958).

452 E.L. Muetterties and F.J. Hirsekorn, J. Amer. Chem. Soc., 95, 5419 (1973), 96, 4063, 7920 (1974).

453 L.S. Stuhl, M.R. Du Bois, F.J. Hirsekorn, J.R. Bleeke, A.E. Stevens and E.L. Muetterties, J. Amer. Chem. Soc., 100, 2405 (1978).

454 E.L. Muetterties, M.C. Rakowski, F.J. Hirsekorn, W.D. Larson, V.J. Basus and F.A.L. Anet, J. Amer. Chem. Soc., 97, 1266 (1975).

455 F.J. Hirsekorn, M.C. Rakowski and E.L. Muetterties, J. Amer. Chem. Soc., 97, 237 (1975).

456 V.A. Zadov, A.V. Finkel'shtein and A.A. Vaisburd, Kinet. Catal., 23, 704 (1982).

457 M.R. Thompson, V.W. Day, K.D. Tau and E.L. Muetterties, Inorg. Chem., 20, 1237 (1981).

458 V.E. Zadov, A.V. Finkel'shtein, A.A. Vaisburd and O.U. Vangolov, Kinet. Catal., 23, 532 (1982); Dokl. P. Chem., 263, 242 (1982).

459 J.R. Bleeke and E.L. Muetterties, J. Amer. Chem. Soc., 103, 556 (1981).

460 V.W. Day, M.F. Fredrich, G.S. Reddy, A.J. Sivak, W.R. Pretzer and E.L. Muetterties, J. Amer. Chem. Soc., 99, 8091 (1977).

461 D. Pieta, A.M. Trzeciak and J.J. Ziółkowski, J. Mol. Catal., 18, 193 (1983).

462   K.R. Januszkiewicz and H. Alper, Organometallics, 2, 1055 (1983).

463   C.R. Landis and J. Halpern, Organometallics, 2, 840 (1983).

464   M.J. Russell, C. White and P.M. Maitlis, J. Chem. Soc., Chem. Commun.,
      427 (1977).

465   O.N. Efimov, O.N. Eremenko, A.G. Ovcharenko, M.L. Khidekel' and
      P.S. Chekrii, Bull. Acad. Sci. U.S.S.R., Div. Chem. Sci., 778 (1969).

466   E.L. Muetterties and J.R. Bleeke, Acc. Chem. Res., 12, 324 (1979).

467   I. Jardine and F. McQuillin, J. Chem. Soc., Chem. Commun., 626
      (1970).

468   M.Y. Darensbourg and E.L. Muetterties, J. Amer. Chem. Soc., 100, 7425
      (1978).

469   J.W. Johnson and E.L. Muetterties, J. Amer. Chem. Soc., 99, 7395
      (1977).

470   M. Bennett, Chemtech, 10, 444 (1980).

471   M.A. Bennett, T.N. Huang and T.W. Turney, J. Chem. Soc., Chem. Commun.,
      582 (1978).

472   R.A. Grey, G.P. Pez and A. Wallo, J. Amer. Chem. Soc., 102, 5948 (1980).

473   R. Wilczynski, W.A. Fordyce and J. Halpern, J. Amer. Chem. Soc., 105,
      2066 (1983).

474   F.K. Shmidt, Yu.S. Levkovskii, N.M. Ryutina and T.I. Bakunina,
      Kinet. Catal., 23, 299 (1982).

475   P. Patnaik and S. Sarkar, J. Ind. Chem. Soc., 56, 266 (1979).

476   G. Bressan and R. Broggi, Chim. Ind. (Milan), 50, 1194 (1968).

477   D. Durand, G. Hillion, C. Lassau and L. Sajus, Ger. Offen., 2, 421,
      934 (1974); Chem. Abs., 82, 97160f (1975).

478   K.J. Klabunde, B.B. Anderson, M. Bader and L.J. Radonovich, J. Amer.
      Chem. Soc., 100, 1313 (1978).

479   R.H. Fish, A.D. Thormodsen and G.A. Cremer, J. Amer. Chem. Soc., 104,
      5234 (1982).

480   T.J. Lynch, M. Banah, H.D. Kaesz and C.R. Porter, J. Org. Chem., 49, 1266 (1984).

481   B. Heil, L. Markó and S. Törös in "Homogeneous Catalysis by Metal Phosphine Complexes," ed. L. Pignolet, Plenum, New York, 1983.

482   H. Fujitsu, E. Matsumura, K. Takeshita and I. Mochida, J. Org. Chem., 46, 2287, 5353 (1981).

483   J.A. Osborn, G. Wilkinson and J.F. Young, J. Chem. Soc., Chem. Commun., 17 (1965).

484   K. Kaneda, M. Yasumura, T. Imanaka and S. Teranishi, J. Chem. Soc., Chem. Commun., 935 (1982).

485   B.R. Cho and R.M. Laine, J. Mol. Catal., 15, 383 (1982).

486   W. Strohmeier and L. Weigelt, J. Organomet. Chem., 145, 189 (1978).

487   R.A. Sanchez-Delgado and O.L. De Ochoa, J. Mol. Catal., 6, 303 (1979).

488   W.M. Kruse and L.W. Wright, Carbohydrate Res., 64, 293 (1978).

489   R.A. Grey, G.P. Pez and A. Wallo, J. Amer. Chem. Soc., 103, 7536 (1981).

490   R.A. Sanchez-Delgado, A. Andriollo, O.L. De Ochoa, T. Suarez and N. Valencia, J. Organomet. Chem., 209, 77 (1981), 202, 427 (1980).

491   S. Vastag, B. Heil and L. Markó, J. Mol. Catal., 5, 189 (1979).

492   M. Gargano, P. Giannoccaro and M. Rossi, J. Organomet. Chem., 129, 239 (1977).

493   B. Heil, S. Törös, J. Bakos and L. Markó, J. Organomet. Chem., 175, 229 (1979).

494   H. Fujitsu, E. Matsumura, K. Takeshita and I. Mochida, J. Chem. Soc., Perkin I, 2650 (1981).

495   K. Tani, K. Suwa, E. Tanigawa, T. Yoshida, T. Okano and S. Otsuka, Chem. Lett., 261 (1982).

496   G. Mestroni, R. Spogliarich, A. Camus, F. Martinelli and G. Zassinovich, J. Organomet. Chem., 140, 63 (1977); Inorg. Nucl.

Chem. Lett., 12, 865 (1976).

497   W. Strohmeier, H. Steigerwald and M. Lukács, J. Organomet. Chem., 144, 135 (1978).

498   F. Joó and M.T. Beck, React. Kinet. Catal. Lett., 2, 257 (1975).

499   M. Bianchi, G. Menchi, F. Francalanci, F. Piacenti, U. Matteoli, P. Frediani and C. Botteghi, J. Organomet. Chem., 188, 109 (1980).

500   L. Markó, B. Heil and S. Vastag, Adv. Chem. Ser., 132, 27 (1974).

501   I. Wender, S. Metlin and M. Orchin, J. Amer. Chem. Soc., 73, 5704 (1951).

502   S. Törös, L. Kollár, B. Heil and L. Markó, J. Organomet. Chem., 255, 377 (1983).

503   I. Ojima, T. Kogure and K. Achiwa, J. Chem. Soc., Chem. Commun., 428 (1977).

504   K. Osakada, T. Ikariya and S. Yoshikawa, J. Organomet. Chem., 231, 79 (1982).

505   U. Matteoli, M. Bianchi, G. Menchi, P. Frediani and F. Piacenti, J. Mol. Catal., 22, 353 (1984).

506   T. Mizoroki, K. Seki, S-i. Meguro and A. Ozaki, Bull. Chem. Soc. Jpn., 50, 2148 (1977).

507   W. Strohmeier and K. Holke, J. Organomet. Chem., 193, C63 (1980).

508   A.S. Berenblyum, S.L. Mund, L.G. Danileva and I.I. Moiseev, Bull. Acad. Sci. U.S.S.R., Div. Chem. Sci., 29, 1244 (1980).

509   M.A. Radhi and L. Markó, J. Organomet. Chem., 262, 359 (1984).

510   T. Yamaguchi, K. Toujima and T. Tsumura, J. Catal., 26, 274 (1972). M. Murakami, K. Suzuki, M. Fujishige and J.W. Kang, J. Chem. Soc. Jpn., 85, 223, 235 (1964).

511   L. Markó and J. Bakos, J. Organomet. Chem., 81, 411 (1974).

512   Asahi Chem. Ind. Co., Jap. Pat., 21,058 (1963); Chem. Abs., 60, 3093a (1964).

513   M.V. Klyuev, B.G. Rogachev and M.L. Khidekel', Bull Acad. Sci.
       U.S.S.R., Div. Chem. Sci., 27, 2344 (1978).

514   Y. Watanabe, Y Tsuji, Y. Ohsugi and J. Shida, Bull Chem. Soc. Jpn.,
       56, 2452 (1983).

515   D.R. Levering, U.S. Patent, 3,152,184 (1964); Chem. Abs., 62, 427f
       (1965).

516   T. Yoshida, T. Okano and S. Otsuka, J. Chem. Soc., Chem. Commun.,
       870 (1979).

517   Y. Ohgo, S. Takeuchi and J. Yoshimura, Bull. Chem. Soc. Jpn., 44,
       283 (1971).

518   A. Kasahara and T. Hongu, J. Chem. Soc. Jpn., 86, 1343 (1965).

519   L. Toniolo, G. de Luca, C. Panattoni and G. Deganello, Gazz. Chim.
       Ital., 104, 961 (1974).

520   J.F. Knifton, J. Catal., 33, 289 (1974); J. Org. Chem., 38, 3296
       (1973).

521   S.D. Kushch, E.N. Izakovich, M.L. Khidekel' and V.V. Strelets,
       Bull. Acad. Sci. U.S.S.R., Div. Chem. Sci., 30, 1201 (1981).

522   J. Kwiatek and I.L. Mador, U.S. Patent, 3,205,217 (1965); Chem. Abs.,
       63, 17978g (1965).

523   F.L'Eplattenier, P. Matthys and F. Calderazzo, Inorg. Chem., 9, 342
       (1970).

524   M. Takesada and H. Wakamatsu, Bull. Chem. Soc. Jpn., 43, 2192 (1970).

525   I. Wender, H. Greenfield, S. Metlin and M. Orchin, J. Amer. Chem. Soc.,
       74, 4079 (1952).

526   S. Tyrlik, M. Kwiecinski, A. Rockenbauer and M. Cujor, J. Coord.
       Chem., 11, 205 (1982).

527   S. Bhattacharya, P. Khandual and C.R. Saha, Chem. Ind., 600 (1982).

528   H. Alper and S. Amaratunga, Tetrahedron Lett., 2603 (1980).

529   E. Alessio, F. Vinzi and G. Mestroni, J. Mol. Catal., 22, 327 (1984).

530   R.C. Ryan, G.M. Wilemon, M.P. Dalsanto and C.U. Pittman, J. Mol. Catal.,
      5, 319 (1979).

531   Y. Watanabe, Y. Tsuji, T. Ohsumi and R. Takeuchi, Tetrahedron Lett.,
      24, 4121 (1983).

532   E.G. Chepaikin, M.L. Khidekel', V.V. Ivanova, A.I. Zakhariev and
      D.M. Shopov, J. Mol. Catal., 10, 115 (1981).

533   A. Rockenbauer, M. Györ, M. Kwienciński and S. Tyrlik, Inorg. Chim.
      Acta, 58, 237 (1982).

534   S.I. Kondrat'ev, A.V. Bulatov and M.L. Khidekel', Bull. Acad. Sci.
      U.S.S.R., Div. Chem. Sci., 31, 419 (1982).

535   J.F. Knifton, Tetrahedron Lett., 2163 (1975); J. Org. Chem., 40, 519
      (1975).

536   J. Bremer, V. Dexheimer and K. Madeja, J. Prakt. Chem., 323, 857
      (1981).

537   G. Brieger and T.J. Nestrick, Chem. Rev., 74, 567 (1974).

538   R. Grigg, T.R.B. Mitchell and S. Sutthivaiyakit, Tetrahedron, 37,
      4313 (1981).

539   J. Trocha-Grimshaw and H.B. Henbest, J. Chem. Soc., Chem. Commun.,
      544 (1967), J. Chem. Soc., Perkin I, 601 (1974).

540   T. Tatsumi, M. Shibagaki and H. Tominaga, J. Mol. Catal., 13, 331
      (1981).

541   Y. Sasson and J. Blum, J. Org. Chem., 40, 1887 (1975).

542   V.Z. Sharf, L.Kh. Freidlin and V.N. Krutii, Bull. Acad. Sci. U.S.S.R.,
      Div. Chem. Sci., 22, 2207 (1973).

543   K.F. Liou and C.H. Cheng, J. Org. Chem., 47, 3018 (1982).

544   T. Nishiguchi, H. Imai, Y. Hirose and K. Fukuzumi, J. Catal., 41,
      249 (1976).

545 H. Imai, T. Nishiguchi and K. Fukuzumi, J. Org. Chem., 39, 1622 (1974),

41, 665, 2688 (1976), 42, 431 (1977); Chem. Lett., 807 (1975).

546 Y. Matsui, T.E. Nalesnik and M. Orchin, J. Mol. Catal., 19, 303

(1983).

547 Y. Watanabe, T. Ohta and Y. Tsuji, Chem. Lett., 1585 (1980);

Bull. Chem. Soc. Jpn., 55, 2441 (1982).

548 J. Cook and P.M. Maitlis, J. Chem. Soc., Chem. Commun., 924 (1981).

549 R. Uson, L.A. Oro, R. Sariego and M.A. Esteruelas, J. Organomet.

Chem., 214, 399 (1981).

550 Y.M.Y. Haddad, H.B. Henbest, J. Husbands, T.R.B. Mitchell and

J. Trocha-Grimshaw, J. Chem. Soc., Perkin I, 596 (1974).

551 L.Kh. Freidlin, V.Z. Sharf, V.N. Krutii and T.V. Lysyak, Bull. Acad.

Sci. U.S.S.R., Div. Chem. Sci., 22, 1807 (1973).

552 E. Małunowicz, S. Tyrlik and Z. Lasocki, J. Organomet. Chem., 72

269 (1974).

553 P.A. Browne and D.N. Kirk, J. Chem. Soc., C, 1653 (1969).

554 F. Vinzi, G. Zassinovich and G. Mestroni, J. Mol. Catal., 18, 359

(1983).

555 G. Zassinovich, G. Mestroni and A. Camus, J. Organomet. Chem., 168,

C37 (1979).

556 R. Spogliarich, G. Zassinovich, G. Mestroni and M. Graziani,

J. Organomet. Chem., 179, C45 (1979), 198, 81 (1980).

557 P. Svoboda and J. Hetflejš, Coll. Czech. Chem. Commun., 42, 2177

(1977).

558 J.C. Orr, M. Mersereau and A. Sanford, J. Chem. Soc., Chem. Commun.,

162 (1970).

559 R. Spogliarich, A. Tencich, J. Kaspar and M. Graziani, J. Organomet.

Chem., 240, 453 (1982).

560 D. Beaupère, L. Nadjo, R. Uzan and P. Bauer, J. Mol. Catal., 14, 129
     (1982), 1C, 73, 20, 165, 195 (1983); J. Organomet. Chem., 238, C12
     (1982).

561 S.M. Pillai, S. Vancheesan, J. Rajaram and J.C. Kuriacose, J. Mol.
     Catal., 20, 169 (1983).

562 T. Tatsumi, M. Shibagaki and H. Tominaga, J. Mol. Catal., 24, 19
     (1984).

563 A. Camus, G. Mestroni and G. Zassinovich, J. Organomet. Chem., 184,
     C10 (1980).

564 Y. Sasson and J. Blum, Tetrahedron Lett., 2167 (1971).

565 G. Descotes and D. Sinou, Tetrahedron Lett., 4083 (1976).

566 J. Blum, Y. Sasson and S. Iflah, Tetrahedron Lett., 1015 (1972).

567 Y. Sasson, J. Blum and E. Dunkelblum, Tetrahedron Lett., 3199 (1973).

568 J.D. Wuest and B. Zacharie, J. Org. Chem., 49, 166 (1984).

569 J. Azran, O. Buchman, M. Orchin and J. Blum, J. Org. Chem., 49, 1327
     (1984).

570 A. Dobson, D.S. Moore and S.D. Robinson, J. Organomet. Chem., 177,
     C8 (1979).

571 R. Zanella, F. Canziani, R. Ros and M. Graziani, J. Organomet. Chem.,
     67, 449 (1974).

572 V.Z. Sharf, L.Kh. Freidlin, I.S. Portyakova and V.N. Krutii,
     Bull. Acad. Sci. U.S.S.R., Div. Chem. Sci., 28, 1324 (1979).

573 I.S. Shekoyan, G.V. Varnakova, V.N. Krutii, K.I. Karpeiskaya and
     V.Z. Sharf, Bull. Acad. Sci. U.S.S.R., Div. Chem. Sci., 24, 2700
     (1975).

574 R. Grigg, T.R. Mitchell and N. Tongpenyai, Synthesis, 442 (1981).

575 G. Mestroni, G. Zassinovich, C. del Bianco and A. Camus, J. Mol.
     Catal., 18, 33 (1983).

576 P.B. Chock and J. Halpern, J. Amer. Chem. Soc., 91, 582 (1969).

577   T. Funabiki, Y. Yamazaki and T. Tarama, J. Chem. Soc., Chem. Commun.,
       63 (1978).

578   G. Losse and H.U. Stiehl, Z. Chem., 21, 188 (1981).

579   D. Forster and G.F. Schaefer, U.S. Patent 4,087,463 (1978);
       Chem. Abs., 89, 108624h (1978).

580   N. Yamagami, H. Wakamatsu and J. Furukawa, Ger. Offen., 2,016,061
       (1970); Chem. Abs., 74, 12609w (1971).

581   B.I. Fleming and H.I. Bolker, Canad. J. Chem., 54, 685 (1976).

582   P. Kvintovics, B. Heil, J. Palágyi and L. Markó, J. Organomet. Chem.,
       148, 311 (1978).

583   H. Eckert, G. Fabry, Y. Kiesel, G. Raudaschl and C. Seidel, Angew.
       Chem., Int. Ed. Engl., 22, 881 (1983).

584   A. Schoenberg and R.F. Heck, J. Amer. Chem. Soc., 96, 7761 (1974).

585   H. Fujitsu, E. Matsumura, S. Shirahama, K. Takeshita and I. Mochida,
       J. Chem. Soc., Perkin I, 855 (1982).

586   M.I. Kalinkin, Z.N. Parnes, G. Zh. Dyskina and D.N. Kursanov,
       Bull. Acad. Sci. U.S.S.R., Div. Chem. Sci., 25, 1999 (1976).

587   S.M. Markosyan, M.I. Kalinkin, Z.N. Parnes and D.N. Kursanov,
       Dokl. Chem., 255, 533 (1981).

588   J.E. Lyons, J. Chem. Soc., Chem. Commun., 412 (1975).

589   A.S. Berenblyum, L.K. Ronzhin, M.V. Ermolaev, I.V. Kalechits and
       M.L. Khidekel', Bull. Acad. Sci. U.S.S.R., 22, 2596 (1973).

590   H. Imai, T. Nishiguchi, M. Tanaka and K. Fukuzumi, Chem. Lett., 855,
       (1976); J. Org. Chem., 42, 2309 (1977).

591   Y. Sasson and G.L. Rempel, Synthesis, 448 (1975).

592   B.M. Savchenko, V.Z. Sharf, V.N. Krutii and L.Kh. Freidlin,
       Bull. Acad. Sci. U.S.S.R., Div. Chem. Sci., 28, 2448 (1979).

593   R. Bar, Y. Sasson and J. Blum, J. Mol. Catal., 16, 175 (1982).

594   K. Nakamura, A. Ohno and S. Oka, Tetrahedron Lett., 24, 3335 (1983).

595   S. Yasui, K. Nakamura and A. Ohno, Chem. Lett., 377 (1984).

596   V.S. Lenenko, A.P. Borisov, U.D. Makhaev, E.I. Mysov, V.B. Shur and
       M.E. Vol'pin, Bull. Acad. Sci. U.S.S.R., Div. Chem. Sci., 32, 858
       (1983).

597   R.J.P. Corriu and B. Meunier, J. Organomet. Chem., 65, 187 (1974).

598   N. Shimizu, K. Watanabe and Y. Tsuno, Chem. Lett., 1877 (1983).

599   R.O. Hutchins, K.Learn and R.P. Fulton, Tetrahedron Lett., 21, 27
       (1980).

600   E. Keinan and N. Greenspoon, Tetrahedron Lett., 23, 241 (1982);
       J. Org. Chem., 48, 3545 (1983).

601   F. Guibe, P. Four and H. Riviere, J. Chem. Soc., Chem. Commun., 432
       (1980).

602   P. Four and F. Guibe, Tetrahedron Lett., 23, 1825 (1982).

603   B.M. Trost and P.L. Ornstein, Tetrahedron Lett., 22, 3463 (1981).

604   J.-L. Fabre and M. Julia, Tetrahedron Lett., 24, 4311 (1983).

605   W. Heitz, Adv. Polym. Sci., 23, 1 (1977).

606   N.L. Holy, J. Org. Chem., 43, 4686 (1978); J. Chem. Soc., Chem.
       Commun., 1074 (1978); Fund. Res. Homog. Catal., 3, 691 (1979).

607   A. Sekiya and J.K. Stille, J. Amer. Chem. Soc., 103, 5096 (1981).

608   H.-S. Tsung and C.H. Brubaker, J. Organomet. Chem., 216, 129 (1981).

609   G. Sbrana, G. Braca, G. Valentini, G. Pazienza and A. Altomare,
       J. Mol. Catal., 3, 111 (1977/78).

610   Y. Nakamura and H. Hirai, Chem. Lett., 165 (1976), 645, 809 (1974).

611   J. Manassen, Is. J. Chem., 8, 5p (1970).

612   A.J. Naaktgeboren, R.J.M. Nolte and W. Drenth, Rec., 97, 112 (1978);
       J. Amer. Chem. Soc., 102, 3350 (1980).

613   M.D. Ward and J. Schwartz, J. Amer. Chem. Soc., 103, 5253 (1981).

614   K.G. Allum, R.D. Hancock, I.V. Howell, T.E. Lester, S. McKenzie,
       R.C. Pitkethly and P.J. Robinson, J. Catal., 43, 322, 331 (1976);

J. Organomet. Chem., 87, 203 (1975), 107, 393 (1976).

615 M. Bartholin, Ch. Graillat, A. Guyot, G. Coudurier, J. Bandiera and C. Naccache, J. Mol. Catal., 3, 17 (1977/78).

616 Y-J. Li and Y.Y. Jiang, J. Mol. Catal., 19, 277 (1983).

617 S.L. Regen and D.P. Lee, Is. J. Chem., 17, 284 (1978).

618 R.H. Grubbs and L.C. Kroll, J. Amer. Chem. Soc., 93, 3062 (1971).

619 M. Czaková and M. Čapka, J. Mol. Catal., 11, 313 (1981);
Z.M. Michalska, M. Čapka and J. Stoch, J. Mol. Catal., 11, 323 (1981).

620 C.U. Pittman, L.R. Smith and R.M. Hanes, J. Amer. Chem. Soc., 97, 1742, 1749 (1975).

621 M.J.M. de Croon and J.W.E. Coenen, J. Mol. Catal., 11, 301 (1981).

622 J. Conan, M. Bartholin and A. Guyot, J. Mol. Catal., 1, 375 (1975/76).

623 G. Strukul, P. D'Olimpio, M. Bonivento, F. Pinna and M. Graziani, J. Mol. Catal., 1, 309 (1978/76), 2, 179 (1977).

624 C.U. Pittman, S.E. Jacobson and H. Hiramoto, J. Amer. Chem. Soc., 97, 4774 (1975).

625 J. Azran, O. Buchman, G. Höhne, H. Schwartz and J. Blum, J. Mol. Catal., 18, 105 (1983).

626 M. Bartholin, C. Graillat and A. Guyot, J. Mol. Catal., 10, 361, 377 (1981).

627 C.P. Nicolaides and N.J. Coville, J. Organomet. Chem., 222, 285 (1981).

628 K. Kaneda, M. Terasawa, T. Imanaka and S. Teranishi, Chem. Lett., 1005 (1975); Fund. Res. Homog. Catal., 3, 671 (1979).

629 P.M. Lausorot, G.A. Vaglio and M. Valle, J. Organomet. Chem., 204, 249 (1981).

630 H.S. Bruner and J.C. Bailar, Inorg. Chem., 12, 1465 (1973); J. Amer. Oil. Chem. Soc., 49, 533 (1972).

631 E.N. Frankel, J.P. Friedrich, T.R. Bessler, W.F. Kwolek and N.L. Holy, J. Amer. Oil. Chem. Soc., 57, 349 (1980).

632   C.U. Pittman and G. Wilemon, Ann. N.Y. Acad. Sci., 333, 67 (1980).

633   C.U. Pittman, B.T. Kim and W.M. Douglas, J. Org. Chem., 40, 590 (1975).

634   T. Okano, K. Tsukiyama, H. Konishi and J. Kiji, Chem. Lett., 603
      (1982).

635   P. Pertici, G. Vitulli, C. Carlini and F. Ciardelli, J. Mol. Catal.,
      11, 353 (1980).

636   T. Kitamura, T. Joh and N. Haghihara, Chem. Lett., 379 (1973),
      203 (1975).

637   T. Mizoroki, K. Seki, S. Meguro and A. Ozaki, Bull. Chem. Soc. Jpn.,
      50, 2148 (1977).

638   B. Bosnich and M.D. Fryzuk, Top. Stereochem., 12, 119 (1981).

639   J.M. Brown and P.A. Chaloner in Homogeneous Catalysis with Metal
      Phosphine Complexes, ed. L. Pignolet, Plenum, New York, 1983.

640   T.V. Ashworth, E. Singleton, R.B. English and M.M. de V. Steyn,
      S. Afr. J. Chem., 36, 97 (1983).

641   W.S. Knowles, M.J. Sabacky and B.D. Vineyard, Adv. Chem. Ser., 132,
      274 (1974); Ann. N.Y. Acad. Sci., 214, 119 (1973).

642   L. Homei and B. Schlotthauer, Phosphorus and Sulphur, 4, 155 (1978).

643   L. Horner and G. Simons, Z. Naturforsch., B, 39, 512 (1984).

644   J.D. Morrison, R.E. Burnett, A.M. Aguiar, C.J. Morrow and C. Phillips,
      J. Amer. Chem. Soc., 93, 1301 (1971).

645   D. Lafont, D. Sinou and G. Descotes, J. Organomet. Chem., 169, 87
      (1979).

646   M. Yamashita, K. Hiramatsu, M. Yamada, N. Suzuki and S. Inokawa,
      Bull. Chem. Soc. Jpn., 55, 2917 (1982).

647   A.M. Aguiar, C.J. Morrow, J.D. Morrison, R.E. Burnett, W.F. Masler
      and N.S. Bhacca, J. Org. Chem., 41, 1545 (1976).

648   D. Valentine, K.K. Johnson, W. Priester, R.C. Sun, K. Toth and
      G. Saucy, J. Org. Chem., 45, 3698 (1980). D. Valentine, R.C. Sun and

208

K. Toth, J. Org. Chem., 45, 3703 (1980).

649 W.R. Cullen, F.W.B. Einstein, C.-H. Huang, A.C. Willis and E.-S. Yey,

J. Amer. Chem. Soc., 102, 988 (1980).

650 G. Pracejus and H. Pracejus, Tetrahedron Lett., 3497 (1977).

651 K. Yamamoto, A. Tomita and J. Tsuji, Chem. Lett., 3 (1978).

652 W.S. Knowles, M.J. Sabacky and B.D. Vineyard, Ger. Offen., 2,456,937

(1974); Chem. Abs., 83, 164367q (1975); G.L. Bachman and B.D. Vineyard,

Ger. Offen., 2,638,071d (1975); Chem. Abs., 87, 5591z (1977).

653 M.D. Fryzuk and B. Bosnich, J. Amer. Chem. Soc., 100, 5491 (1978).

654 R.B. King, J. Bakos, C.D. Hoff and L. Markó, J. Org. Chem., 44, 1729

(1979).

655 D.P. Riley and R.E. Shumate, J. Org. Chem., 45, 5187 (1980).

656 M.D. Fryzuk and B. Bosnich, J. Amer. Chem. Soc., 99, 6262 (1977).

657 H. Brunner and W. Pieronczyk, Angew. Chem., Int. Ed. Engl., 18, 620

(1979).

658 M. Lauer, O. Samuel and H.B. Kagan, J. Organomet. Chem., 177, 309

(1979).

659 B.D. Vineyard, W.S. Knowles, M.J. Sabacky, G.L. Bachman and

D.J. Weinkauff, J. Amer. Chem. Soc., 99, 5946 (1977).

660 T.H. Johnson and G. Rangarajan, J. Org. Chem., 45, 62 (1980).

661 D.L. Allen, V.C. Gibson, M.L.H. Green, J.F. Skinner, J. Bashkin and

P.D. Grebenik, J. Chem. Soc., Chem. Commun., 895 (1983).

662 D. Lafont, D. Sinou and G. Descotes, J. Chem. Res., S, 117 (1982).

663 J.P. Amma and J.K. Stille, J. Org. Chem., 47, 468 (1982).

664 W.S. Knowles, M.J. Sabacky, B.D. Vineyard and D.J. Weinkauff,

J. Amer. Chem. Soc., 97, 2567 (1975).

665 D.P. Riley, J. Organomet. Chem., 234, 85 (1982).

666 H. Brunner and M. Pröbster, Inorg. Chem. Acta, 61, 129 (1982).

667 O. Samuel, R. Couffignal, M. Lauer, S.Y. Zhang and H.B. Kagan,

Nouv. J. Chim., 5, 15 (1981).

668 W.C. Christopfel and B.D. Vineyard, J. Amer. Chem. Soc., 101, 4406
     (1979).

669 J.M. Brown and P.A. Chaloner, J. Chem. Soc., Chem. Commun., 613 (1979).

670 K.E. Koenig and W.S. Knowles, J. Amer. Chem. Soc., 100, 7561 (1978).

671 J. Halpern, Pure Appl. Chem., 55, 99 (1983).

672 W.S. Knowles, Acc. Chem. Res., 16, 106 (1983).

673 J.D. Oliver and D.P. Riley, Organometallics, 2, 1032 (1983).

674 A.S.C. Chan, J.J. Pluth and J. Halpern, Inorg. Chim. Acta, 37, L477
     (1979); J. Amer. Chem. Soc., 102, 5952 (1980).

675 P.C. Chua, N.K. Roberts, B. Bosnich, S.J. Okrasinski and J. Halpern,
     J. Chem. Soc., Chem. Commun., 1278 (1981).

676 J.M. Brown and P.A. Chaloner, J. Chem. Soc., Chem. Commun., 344 (1980).

677 I. Ojima, T. Kogure and N. Yoda, J. Org. Chem., 45, 4728 (1980).

678 D. Sinou, Tetrahedron Lett., 22, 2987 (1981).

679 J.M. Brown, P.A. Chaloner and G.A. Morris, J. Chem. Soc., Chem. Commun.,
     664 (1983).

680 J.M. Brown and D. Parker, Organometallics, 1, 950 (1982).

681 H.B. Kagan, J.C. Fiaud, C. Hoornaert, D. Meyer and J.C. Poulin,
     Bull. Soc. Chim. Belg., 88, 923 (1979).

682 P.A. MacNeil, N.K. Roberts and B. Bosnich, J. Amer. Chem. Soc., 103,
     2273 (1981).

683 K. Kellner, A. Tzschach, N. Nagy-Magos and L. Markó, J. Organomet. Chem.,
     193, 307 (1980).

684 H.B. Kagan and T.-P. Dang, J. Amer. Chem. Soc., 94, 6429 (1972).

685 R. Glaser, S. Geresh and J. Blumenfeld, J. Organomet. Chem., 112, 355
     (1976).

686 P. Aviron-Violet, Y. Colleuille and J. Varagnat, J. Mol. Catal., 5,
     41 (1979).

210

687   R. Glaser, M. Twaik, S. Geresh and J. Blumenfeld, Tetrahedron Lett.,
       4635 (1977).

688   M. Tanaka and I. Ogata, J. Chem. Soc., Chem. Commun., 735 (1975).

689   M. Fiorini and G.M. Giongo, J. Mol. Catal., 5, 303 (1979), 7, 411
       (1980).

690   K. Onuma, T. Ito and A. Nakamura, Bull. Chem. Soc. Jpn., 53, 2012,
       2016 (1980); Tetrahedron Lett., 3163 (1979).

691   K. Achiwa, J. Amer. Chem. Soc., 98, 8265 (1976).

692   J.M. Brown and B.A. Murrer, Tetrahedron Lett., 21, 581 (1980);
       J. Chem. Soc., Perkin II, 489 (1982).

693   A. Miyashita, A. Yasuda, H. Takaya, K. Toriumi, T. Ito, T. Souchi
       and R. Noyori, J. Amer. Chem. Soc., 102, 7932 (1980); Tetrahedron, 40,
       1245 (1984).

694   M. Petit, A. Mortreux, F. Petit, G. Buono and G. Peiffer, Neur. J.
       Chim., 7, 593 (1983).

695   H.B. Kagan, Pure Appl. Chem., 43, 401 (1975).

696   U. Hengartner, D. Valentine, K.K. Johnson, M.E. Larscheid, F. Pigott,
       F. Schiedl, J.W. Scott, R.C. Sun, J.M. Townsend and T.H. Williams,
       J. Org. Chem., 44, 3741 (1979).

697   R. Glaser and S. Geresh, Tetrahedron, 35, 2381 (1979).

698   T. Hayashi, M. Tanaka, Y. Ikeda and I. Ogata, Bull. Chem. Soc. Jpn.,
       52, 2605 (1979).

699   R. Jackson and D.J. Thompson, J. Organomet. Chem., 159, C29 (1978).

700   K. Achiwa, Chem. Lett., 777 (1977); Tetrahedron Lett., 3735 (1977),
       1475,2583 (1978); Hetercycles, 8, 247 (1977).

701   T. Hayashi, K. Kanehira and M. Kumada, Tetrahedron Lett., 22, 4417
       (1981).

702   I. Ojima and N. Yoda, Tetrahedron Lett., 21, 1051 (1980).

703   J.M. Brown and P.A. Chaloner, J. Chem. Soc., Perkin II, 711 (1982).

704 P.A. Chaloner, J. Organomet. Chem., 266, 191 (1984).

705 I. Ojima, Pure Appl. Chem., 1, 99 (1984).

706 I. Ojima, N. Yoda, M. Yatabe, T. Tanaka and T. Kogure, Tetrahedron,
     40, 1255 (1984).

707 R. Glaser, S. Geresh, J. Blumenfeld, B. Vainas and M. Twaik,
     Is. J. Chem., 15, 17 (1976).

708 D. Sinou, D. Lafont, G. Descotes and A.G. Kent, J. Organomet. Chem.,
     217, 119 (1981).

709 J.-C. Poulin and H.B. Kagan, J. Chem. Soc., Chem. Commun., 1261 (1982);
     Chimia, 247 (1982).

710 J.M. Brown and D. Parker, J. Org. Chem., 47, 2722 (1982).

711 K. Achiwa, Heterocycles, 15, 1023 (1981).

712 E. Cesarotti, A. Chiesa, G. Ciani and A. Sironi, J. Organomet. Chem.,
     251, 79 (1983).

713 A.-P. Stoll and R. Süess, Helv. Chim. Acta, 57, 2487 (1974).

714 T. Hayashi and M. Kumada, Acc.Chem. Res., 15, 395 (1982).

715 G. Descotes, D. Lafont and D. Sinou, J. Organomet. Chem., 150, C14
     (1978).

716 G. Descotes, D. Lafont, D. Sinou, J.M. Brown, P.A. Chaloner and
     D. Parker, Nouv. J. Chim., 5, 167 (1981).

717 D. Lafont, D. Sinou, G. Descotes, R. Glaser and S. Geresh, J. Mol.
     Catal., 10, 305 (1981).

718 K. Tamao, H. Yamamoto, H. Matsumoto, M. Miyake, T. Hayashi and
     M. Kumada, Tetrahedron Lett., 1389 (1977).

719 R.H. Grubbs and R.A. de Vries, Tetrahedron Lett., 1879 (1977).

720 S. Miyano, M. Nawa and H. Hashimoto, Chem. Lett., 729 (1980).

721 A. Uehara, T. Kubota and R. Tsuchiya, Chem. Lett., 441 (1983).

722 Y. Ohgo, Y. Natori, S. Takeuchi and Y. Yoshimura, Chem. Lett., 709,
     1327 (1974).

212

723　P. le Maux and G. Simonneaux, J. Organomet. Chem., 252, C60 (1983).

724　L.O. Nindakova, F.K. Shmidt, E.I. Klabunovskii, V.N. Sheveleva and
　　　V.A. Pavlov, Bull. Acad. Sci. U.S.S.R., Div. Chem. Sci., 30, 2177
　　　(1981).

725　K. Ohkubo, I. Terada, K. Sugahara and K. Yoshinaga, J. Mol. Catal.,
　　　7, 421 (1980).

726　B.R. James, R.S. McMillan, R.H. Morris and D.K.W. Wang, Adv. Chem.
　　　Ser., 167, 122 (1978).

727　M. Bianchi, F. Piacenti, P. Frediani, U. Matteoli, C. Botteghi,
　　　S. Gladiali, and E. Benedetti, J. Organomet. Chem., 141, 107 (1977).

728　C. Botteghi, S. Gladiali, M. Bianchi, U. Matteoli. P. Frediani,
　　　P.G. Vergamini and E. Benedetti, J. Organomet. Chem., 140, 221 (1977).

729　M. Bianchi, U. Matteoli, P. Frediani, G. Menchi, F. Piacenti,
　　　C. Botteghi and M. Marchetti, J. Organomet. Chem., 252, 317 (1983).

730　E.E. Nifant'ev, T.S. Kukhareva, M.Yu. Antipin, Yu.T. Struchkov and
　　　E.I. Klabunovsky, Tetrahedron, 39, 797 (1983).

731　A. Levi, G. Modena and G. Scorrano, J. Chem. Soc., Chem. Commun.,
　　　6 (1975).

732　T. Hayashi, T. Mise and M. Kumada, Tetrahedron Lett., 4351 (1976).

733　J. Solodar, Chemtech., 5, 421 (1975).

734　M. Tanaka, Y. Watanabe, T. Mitsudo, H. Iwane and Y. Takegami,
　　　Chem. Lett., 239 (1973).

735　S. Törös, B. Heil and L. Markó, J. Organomet. Chem., 159, 401 (1978).

736　I. Ojima, T. Kogure and K. Achiwa, J. Chem. Soc., Chem. Commun.,
　　　428 (1977).

737　M. Fiorini, F. Marcati and G.M. Giongo, J. Mol. Catal., 3, 385
　　　(1977/78).

738　T. Hayashi, M. Tanaka and I. Ogata, Tetrahedron Lett., 295 (1977).

739　L. Markó, Pure Appl. Chem., 51, 2211 (1979).

740 S. Tőrős, B. Heil, L. Kollár and L. Markó, J. Organomet. Chem., 197
85 (1980).

741 M. Kumada, T. Hayashi, T. Mise, A. Katsumura, N. Nagashima and
M. Fukushima, Kenkyu Hokoku, 37, 69 (1980); Chem. Abs., 96, 85708s
(1982).

742 I. Ojima and T. Kogure, J. Organomet. Chem., 195, 239 (1980).

743 I. Ojima, Fund. Res. Homog. Catal., 2, 181 (1978).

744 T. Hayashi, A. Katsumara, M. Konishi and M. Kumada, Tetrahedron Lett.,
425 (1979).

745 S. Tőrős, L. Kollár, B. Heil and L. Markó, J. Organomet. Chem., 232,
C17 (1982).

746 M. Bianchi, U. Matteoli, G. Menchi, P. Frediani, S. Pratesi,
F. Piacenti and C. Botteghi, J. Organomet. Chem., 198, 73 (1980).

747 K. Ohkubo, M. Setoguchi and K. Yoshinaga, Inorg. Nucl. Chem. Lett.,
15, 235 (1979).

748 K. Osakada, M. Obana, T. Ikariya, M. Saburi and S. Yoshikawa,
Tetrahedron Lett., 22, 4297 (1981).

749 S. Vastag, B. Heil, S. Tőrős and L. Markó, Transition Met. Chem.,
2, 58 (1977).

750 S. Vastag, J. Bakos, S. Tőrős, N.E. Takach, R.B. King, B. Heil and
L. Markó, J. Mol. Catal., 22, 283 (1984).

751 P. Abley and F. McQuillin, J. Chem. Soc., C, 844 (1971).

752 V.A. Pavlov, E.I. Klabunovskii, G.S. Barysheva, L.N. Kaigorodova
and Y.S. Airapetov, Bull. Acad. Sci. U.S.S.R., Div. Chem. Sci., 24
2262 (1975).

753 L.F. Godunova, E.S. Levitina, E.I. Karpeiskaya and E.I. Klabunovskii,
Bull. Acad. Sci. U.S.S.R., Div. Chem. Sci., 30, 595 (1981).

754 L.O. Nindakova, F.K. Shmidt, E.I. Klabunovskii, V.N. Shevaleva and
V.A. Pavlov, Bull. Acad. Sci. U.S.S.R., Div. Chem. Sci., 30, 2177 (1981).

214

755   H. Brunner and A.F.M. Mokhlesur Rahman, Z. Naturforsch, B, 38,
1332 (1983).

756   A.T. Baryshnikov, V.M. Belikov, V.K. Latof and M.B. Saporovskaya,
Bull. Acad. Sci. U.S.S.R., Div. Chem. Sci., 30, 1206 (1981).

757   L.M. Koroleva, V.K. Latov, M.B. Saporovskaya and V.M. Belikov,
Bull. Acad. Sci. U.S.S.R., Div. Chem. Sci., 28, 2214 (1978).

758   Z. Nagy-Magos, S. Vastag, B. Heil and L. Markó, J. Organomet. Chem.,
171, 97 (1979).

759   B.R. James and R.S. McMillan, Canad. J. Chem., 55, 2353, 3927 (1977).

760   Y. Ohgo, S. Takeuchi and J. Yoshimura, Bull. Chem. Soc., Jpn., 44,
583 (1971).

761   A.D. Calhoun, W.J. Kobos, T.A. Nile and C.A. Smith, J. Organomet.
Chem., 170, 175 (1979).

762   R.W. Waldron and J.H. Weber, Inorg. Chem., 16, 1220 (1977).

763   M.M. Kucharska and S. Tyrlik, React. Kinet. Catal. Lett., 15, 145
(1980).

764   Y. Ohgo, S. Takeuchi, Y. Natori and J. Yoshimura, Bull. Chem. Soc.
Jpn., 54, 2124 (1981).

765   E. Cesarotti, R. Ugo and R. Vitiello, J. Mol. Catal., 12, 63 (1981).

766   R. Spoliarich, G. Zassinovich, J. Kaspar and M. Graziani, J. Mol.
Catal., 16, 359 (1982).

767   G. Zassinovich, C. del Bianco and G. Mestroni, J. Organomet. Chem.,
222, 323 (1981).

768   K. Yoshinaga, T. Kito and K. Ohkubo, Bull. Chem. Soc. Jpn., 56, 1786
(1983).

769   M. Bianchi, U. Matteoli. G. Menchi, P. Frediani, F. Piaceni and
C. Botteghi, J. Organomet. Chem., 195, 337 (1980).

770   K. Yoshinaga, T. Kito and K. Ohkubo, J. Chem. Soc., Perkin II, 469
(1984).

771 G. Descotes and D. Sinou, Tetrahedron Lett., 4083 (1976).

772 K. Ohkubo, K. Sugahara, I. Terada and K. Yoshinaga, Inorg. Nucl. Chem. Lett., 14, 297 (1978).

773 M.J. Fish and D.F. Ollis, Catal. Rev. Sci. Eng., 18, 259 (1978).

774 R.L. Beamer and W.D. Brown, J. Pharm. Sci., 60, 583 (1971).

775 H. Hirai and T. Furuta, J. Polym. Sci. B, Polym. Lett., 9, 459, 729 (1971).

776 J.C. Poulin, W. Dumont, T.P. Dang and H.B. Kagan, Compt. Rend. Acad. Sci. Paris, 277C, 41 (1973).

777 K. Achiwa, Chem. Lett., 905 (1978).

778 G.L. Baker, S.J. Fritschel and J.K. Stille, J. Org. Chem., 46, 2960 (1981).

779 I. Kolb, M. Cerný and J. Hetflejš, React. Kinet. Catal. Lett., 7, 199 (1977).

780 H.B. Kagan, T. Yamagishi, J.C. Motte and R. Setton, Is. J. Chem., 17, 274 (1978).

781 Fuji Chemicals, Jpn. Kokai Tokkyo Toho, 80, 129, 277 (1980); Chem. Abs., 94, 174853m (1981).

782 H.W. Krause, React. Kinet. Catal. Lett., 10, 243 (1979).

783 P.W. Clark, J. Organomet. Chem., 110, C13 (1976).

784 C. Crocker, R.J. Errington, R. Markham, C.J. Moulton and B.L. Shaw, J. Chem. Soc., Dalton, 387 (1982).

785 R.R. Burch, E.L. Muetterties, and V.W. Day, Organometallics, 1, 188 (1982).

786 D. Baudry, M. Ephritikhine, H. Felkin and J. Zakrzewski, J. Chem. Soc., Chem. Commun., 1235 (1982).

787 H. Moriyama, T. Aoki, S. Shinoda and Y. Saito, J. Chem. Soc., Perkin II, 369 (1982).

788 P.B. Critchlow and S.D. Robinson, Inorg. Chem., 17, 1896, 1902 (1978).

789   S. Shinoda, Y. Tokushige, T. Kojima and Y. Saito, J. Mol. Catal., 17, 77, 81 (1982).

790   W.K. Rybak and J.J. Ziołkowski, J. Mol. Catal., 11, 365 (1981).

791   S.I. Murahashi, K-i. Ito, T. Naota and Y. Maeda, Tetrahedron Lett., 22, 5327 (1981).

792   R.J. Theissen, J. Org. Chem., 36, 752 (1971).

793   A.A. Inada, Y. Nakamura and Y. Morita, Chem. Lett., 1287 (1980).

794   L. Cottier, G. Descotes and J. Sabadie, J. Mol. Catal., 7, 337 (1980).

795   S.-I. Murahashi, K. Kondo and T. Hakata, Tetrahedron Lett., 23, 229 (1982).

796   H. Felkin, T. Fillebeen-Khan, Y. Gault, R. Holmes-Smith and J. Zakrzewski, Tetrahedron Lett., 25, 1279 (1984).

797   S.M. Pillai, S. Vancheesan, J. Rajaram and J.C. Kuriakose, J. Mol. Catal., 16, 349 (1982).

798   K. Okhubo, K. Hirata, K. Yoshinaga and M. Okada, Chem. Lett., 183, 577 (1976).

799   K. Ohkubo, T. Ohgushi and K. Yoshinaga, Chem. Lett., 775 (1970); J. Coord. Chem., 6, 185 (1977), 8, 195 (1979).

800   K. Ohkubo, I. Terada and K. Yoshinaga, Chem. Lett., 1467 (1977); Inorg. Nucl. Chem. Lett., 13, 443 (1977).

801   Y. Ishii, K. Osakada, T. Ikariya, M. Saburi and S. Yoshikawa, Tetrahedron Lett., 24, 2677 (1983), Chem. Lett., 1179 (1982).

3 Reactions of carbon monoxide

3.1 Hydroformylation

  3.1.1 Cobalt Catalysts

    3.1.1.1 Side Reactions

    3.1.1.2 Functionalised Alkenes

  3.1.2 Rhodium Catalysts

    3.1.2.1 Chemoselectivity

    3.1.2.2 Regioselectivity

    3.1.2.3 Variations of Substrate

  3.1.3 Other Metals

  3.1.4 Polymer Supported Hydroformylation Catalysts

  3.1.5 Asymmetric Hydroformylation

      Non-chelating Phosphines

      Chelating Biphosphines

3.2 Hydrocarboxylation and Related Reactions

  3.2.1 Conversion of Alkenes to Esters, Lactones and Acids

  3.2.2 Carbonylation of Alkenes in the Presence of Other Nucleophiles

  3.2.3 Reactions of Alkynes

  3.2.4 Reactions of Dienes

  3.2.5 Functionalised Alkenes

  3.2.6 Asymmetric Hydrocarboxylation

3.3 Carbonylation of Alcohols

  3.3.1 Cobalt Catalysed Reactions

  3.3.2 Rhodium Catalysts

  3.3.3 Iridium Catalysts

  3.3.4 Other Metals

3.4 Carbonylation of Halides

  3.4.1 Carbonylation of Aryl Halides

3.4.2 Vinyl Halides

3.4.3 Benzyl and Allyl Halides

3.4.4 Other Halides

3.5 Carbonylation of Organometals

3.6 Carbonylation of Nitrogen Containing Functional Groups

3.6.1 Amine Carbonylation

3.6.2 Carbonylation of Aromatic Nitro Compounds

3.6.3 Other Nitrogen Containing Substrates

3.7 Miscellaneous Carbonylations

3.8 Fischer Tropsch Reactions: The Hydrogenation of Carbon Monoxide

3.9 Decarbonylation

3.9.1 Decarbonylation of Acyl Halides

3.9.2 Decarbonylation of Aldehydes

3.9.3 Decarbonylation of Other Substrates

The use of carbon monoxide as a chemical feedstock has become more important with decreased availability and increased cost of petroleum feedstock. Hydroformylation and hydrocarboxylation of alkenes leads to oxygen functionalised products and is used industrially on a large scale. Carbonylation of halides and of nitrogen compounds is important in the academic laboratory while carbonylation of methanol is a commercial source of acetic acid. The Fischer Tropsch reaction involves reduction of carbon monoxide to alkanes and alcohols and has recently attracted new attention, though the most useful catalysts are heterogeneous.

The fundamental reaction in transition metal catalysed carbon monoxide chemistry is insertion of CO into a metal carbon bond, viz. 2 to 3. The genesis of 1 and the fate of 3 depend on the substituents both at metal and carbon, and other species present. Both steps shown are reversible, and decarbonylation of appropriate substrates is possible.

$$L_nM-C \xrightarrow{\quad CO \quad} L_nM-\overset{\overset{\displaystyle O}{\displaystyle \|}}{C}-C \longrightarrow L_nM-\overset{\overset{\displaystyle O}{\displaystyle \|}}{C}-C$$

$$\underline{1} \qquad\qquad\qquad \underline{2} \qquad\qquad\qquad \underline{3}$$

3.1 Hydroformylation

Hydroformylation is the addition of hydrogen and carbon monoxide

to a carbon–carbon multiple bond, e.g. $\underline{4}$ to $\underline{5}$. It was first discovered

in 1938[1] and is the most intensively studied carbonylation reaction.

The combination of CO and H_2 is commonly known as syngas and is readily

available from coal as well as from petroleum feedstock.

This reaction provides a particularly apt illustration of the

types of selectivity discussed in Chapter 1. In most cases the object of

the reaction is the production of aldehydes. However, many hydro-

$$\text{(alkene)} + CO + H_2 \xrightarrow{\quad catalyst \quad} \overset{\displaystyle}{\underset{H \qquad CHO}{\text{(product)}}}$$

$$\underline{4} \qquad\qquad\qquad\qquad\qquad \underline{5}$$

formylation catalysts also catalyse isomerisation reactions or hydro-

genation of the carbon–carbon double bond, or of the carbonyl of the

product leading respectively to hydrocarbons and saturated alcohols.

Additionally the product may undergo self condensation to yield aldol

products. A highly chemoselective catalyst is essential.[2-7]

When a non–symmetric alkene such as styrene, $\underline{6}$, is hydroformylated

two regioisomeric aldehydes may be formed, $\underline{7}$ and $\underline{8}$. The n–aldehyde is

generally industrially more useful and extensive effort has been devoted

to increasing the n/iso ratio in the products of the reaction with

terminal alkenes. The problem of diastereoselectivity does arise with more

$$Ph\diagdown\hspace{-0.3em}\diagup \quad + \; CO \; + \; H_2 \quad \xrightarrow{\text{catalyst}} \; Ph\diagdown\diagup\diagdown^{CHO} \; + \; Ph\diagdown\diagup\big|_{CHO}$$

6   7   8

highly substituted alkenes but since these are rather unreactive it has
been more usual to work with deuterated alkenes such as 9. Finally when
the substrate is a prochiral alkene and the catalyst bears chiral ligands
enantioselective hydroformylation is possible as in the production of 8.

$$\diagup\hspace{-0.3em}\diagdown \quad + \; CO \; + \; H_2 \quad \xrightarrow{\text{catalyst}} \quad D\cdots\big|\cdots D \; + \; \cdots\big|\cdots D$$

9   10   11

3.1.1 Cobalt Catalysts

Cobalt carbonyl complexes were the first used hydroformylation
catalysts and represent a significant proportion of industrial production.
It was recognised that $HCo(CO)_4$ is the species responsible for catalysis
in the presence of $Co_2(CO)_8$ and a mechanism was proposed by Heck and
Breslow.[8] While some modifications have been made since, their conclusions
are generally accepted and are shown in Figure 1. For 1-alkenes an
alternative insertion possibility exists giving rise to $RCH(CH_3)Co(CO)_3$;
this will yield $RCH(CH_3)CHO$ on carbonylation.

Another catalyst involves modification of $Co_2(CO)_8$ with
phosphine ligands. Information about both these and the unmodified
catalysts has been provided by high pressure IR studies of the reacting
mixtures. For the cobalt/1-octene system at 250 atm. and $150°C$ the
acyl complex $C_8H_{17}COCo(CO)_4$ was observed in a steady state, suggesting
that its hydrogenolysis was the rate-determining step.[9] However, in the
presence of Bu_3P the complex $HCo(CO)_3PBu_3$ predominates throughout the
reaction and it was concluded that with slower reacting systems such as

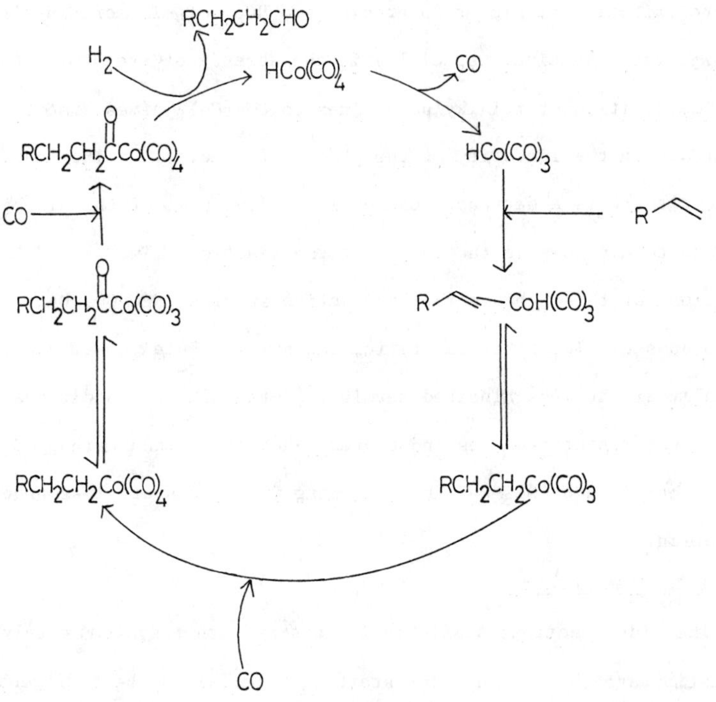

RCH₂CH₂CHO is shown as RCH_2CH_2CHO

$$RCH_2CH_2CHO$$

$H_2 \longrightarrow HCo(CO)_4 \longleftarrow CO$

$RCH_2CH_2\overset{O}{\overset{\|}{C}}Co(CO)_4 \qquad HCo(CO)_3$

$CO \longrightarrow$

$RCH_2CH_2\overset{O}{\overset{\|}{C}}Co(CO)_3 \qquad R \diagup\!\!\!\diagdown - CoH(CO)_3$

$RCH_2CH_2Co(CO)_4 \qquad RCH_2CH_2Co(CO)_3$

CO

$R \diagup\!\!\!\diagdown$

Figure 1 Mechanism of cobalt catalysed hydroformylation.

these interaction of metal and alkene might be rate determining.[10,11]

        The aim of industrial hydroformylation is to produce linear aldehydes in high selectivity. When $Co_2(CO)_8$ is used as catalyst a mixture of linear and branched products (branched predominating) is obtained. α-Methyl aldehydes may arise in two ways. Firstly insertion of the terminal alkene may give a branched metal alkyl. Alternatively isomerisation of the starting material (with or without release from cobalt[12]) to give an internal double bond leads inevitably to a branched alkyl. A third but less likely possibility involves isomerisation of the acyl cobalt complexes. The most influential reaction parameter is the partial pressure of carbon monoxide, increasing pressure giving rise to a higher proportion of linear products. This has been interpreted in terms of $HCo(CO)_3$ and $HCo(CO)_4$ being competitive as hydroformylation catalysts,

$HCo(CO)_4$ predominating at higher CO pressures. $HCo(CO)_4$ is more specific and $HCo(CO)_3$ more reactive, especially in the absence of free CO.[7]

The addition of trialkylphosphines to $Co_2(CO)_8$ gives rise to a marked increase in the linearity of the product formed.[13,14] The reaction is slower and there is a decreased chemoselectivity, some of the aldehydes being reduced to alcohols in the course of the reaction. Two interpretations of this behaviour of the active species, $HCo(CO)_3PR_3$[10], have been proposed. The first is steric, the bulkier catalyst attacking preferentially at the less hindered terminus of the alkene. Additionally phosphines are stronger σ-donors and poorer π-acceptors than carbon monoxide so that in $HCo(CO)_3PR_3$ the remaining CO ligands are less prone to dissociation.

3.1.1.1 Side Reactions

The side reactions available in this and other systems catalysing hydroformylation are three fold. The starting material may be hydrogenated or isomerised and the product may also be hydrogenated. While undesirable in a practical sense, studies of these reactions, most particularly isomerisation, have shed light on the mechanism involved.

Interception of the alkylcobalt carbonyl before insertion and hydrogen addition gives rise to alkanes. For simple alkenes and unmodified cobalt catalysts this is relatively minor representing only 0.5 – 3% of the total reaction. However, some ligand modified cobalt catalysts give higher proportions of reduction, up to 15%.[15-17] Conjugated systems are more prone to hydrogenation, ethyl benzene being obtained in >50% yield from styrene[18-20] and even more alkane from α-methylstyrene.[21] $\alpha\beta$-Unsaturated aldehydes or ketones are reduced at 125°C, under milder conditions than are necessary for hydroformylation.[22] The reaction is thought to proceed via 1,4-addition to produce an intermediate π-oxapropenyl complex, 12, analogous to the isolable allyl complex, 13. For both steric and electronic reasons $\alpha\beta$-unsaturated esters do not form such complexes and may be hydroformylated

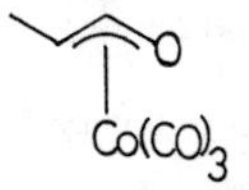

12 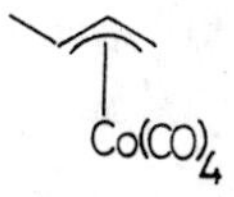   13

with good chemoselectivity.[23,24]

Almost all crude hydroformylation products contain significant
amounts of alcohols resulting from reduction of the product aldehydes.[25-27]
When the reaction temperature is $185^{\circ}C$ the alcohol is the main product.[28]
The reaction is thought to proceed as shown in Figure 2 and the rate
expression is given by $d[RCH_2OH]/dt = K[RCHO][Co][H_2][CO]^{-2}$ explaining the
dependence of product composition on feed gas composition. Whilst the
alcohols may themselves be valuable the formation of acetals presents
problems in separation and it is usually preferable to adjust the reaction
conditions to produce either alcohol or carbonyl exclusively.

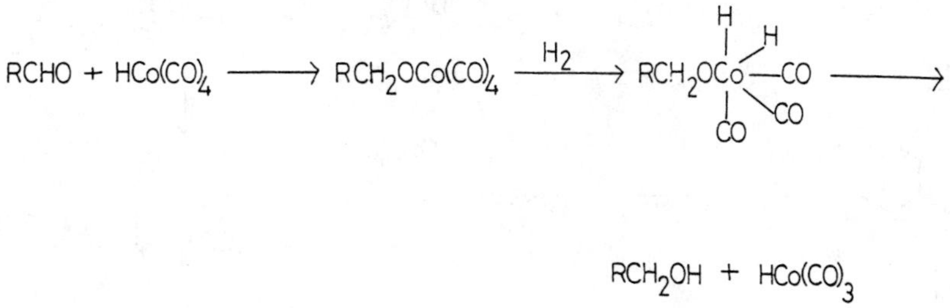

Figure 2   Mechanism of cobalt catalysed reduction of carbonyl compounds.

Catalytic hydroformylation of 2-pentene yields predominently
hexanol, but little 1-pentene is detected.[29] Isomerisation is accelerated
when CO pressure is low and temperatures high. Early work suggested that
the observations were due to the addition and elimination of cobalt
carbonyl hydride but the introduction of carbonyl at a site not originally
part of the double bond, and the absence of free isomerised alkene makes

this unlikely. Piacenti studied the hydroformylation of $1-^{14}$C-propene and

ω-deuterated 1-alkenes.[12,30] Even for 1-hexene the formyl group could be

attached to all the carbon atoms of the starting material. Deuterium was

found only on carbons 2, 3 and ω of the aldehydes. Additionally if chiral

S-(+)- 3-methyl-1-hexene is hydroformylated 3% of R - 3-ethylhexanol is

obtained having 70% optical purity.[31] S - 4-methyl-trans-2-hexene and

S - 2,2,5-trimethyl-trans-3-heptene are also reduced with excellent optical

purity.[32] This implies that the π-complexes do not dissociate

significantly under catalytic conditions , since this would result in loss

of stereochemistry at the chiral centre. Also if 3-deutero-3-methyl-1-

hexene, 14, is hydroformylated migration of deuterium occurs concurrently

with isomerisation (Figure 3).[33] The cobalt moves down the hydrocarbon

chain without ever being completely eliminated from it.

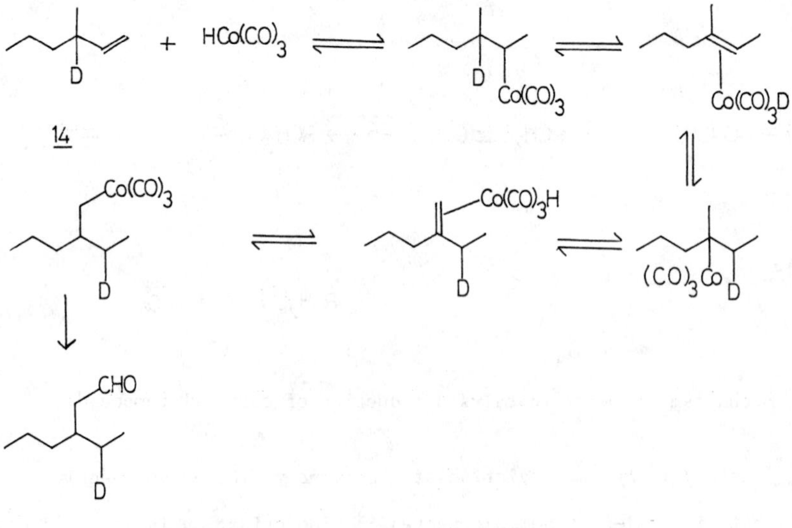

Figure 3 Cobalt catalysed hydroformylation and isomerisation.

Formation of 3-pentanone from ethylene and CO was noted by Roelen

when hydroformylation was first discovered.[7] Bertrand considered that

ketones were derived from coupling of alkyl (15) and acylcobalt (16)

$$RCH_2COCo(CO)_4 + RCH_2Co(CO)_4 \longrightarrow Co_2(CO)_8 + R\text{-}CO\text{-}R$$

15   16

complexes.[34] The reaction is very sensitive to steric hindrance;

3-pentanone has been obtained in greater than 50% yield[35] and 4-heptanone

is obtained from propene in the presence of cobalt complexes modified by

chelating biphosphines.[36] The reaction is favoured by low total pressure

and high alkene concentration.

3.1.1.2 Functionalised Alkenes

Few cobalt catalysed hydroformylations of functionalised alkenes

have found commercial applications, and few provide surprises. As noted

before $\alpha\beta$-unsaturated aldehydes and ketones are hydrogenated rather than

carbonylated but they may hydroformylated indirectly via their acetals.[37]

$\alpha\beta$-Unsaturated esters may be carbonylated with reasonable chemoselectivity

and variable regioselectivity. For example, methyl acrylate gives

preference for β-reaction by a factor of 18.3:1 in the presence of $Co_2(CO)_8$

and diphos.[24,38] Methyl crotonate, however, gives remote hydroformylation

to yield 17 in 73% yield.[23,24]

Unsaturated alcohols, particularly allyl alcohol, are

unsatisfactory substrates since isomerisation to the aldehyde is the major

pathway. Some other functionalised substrates are shown in Figure 4.

$$OHC\text{-}CH_2CH_2\text{-}COOMe$$

17

3.1.2 Rhodium Catalysts

Rhodium carbonyl catalysis of hydroformylation is similar to

that based on cobalt. The catalyst precursor is either $Rh_4(CO)_{12}$[42] or

$Rh_6(CO)_{16}$[43] and the active species is $HRh(CO)_3$. $(Rh(CO)_2Cl)_2$ is

converted to $Rh_4(CO)_{12}$ under mild carbonylation conditions.[44] The

proposed mechanism is shown in Figure 5. For 1-heptene hydrogenation of

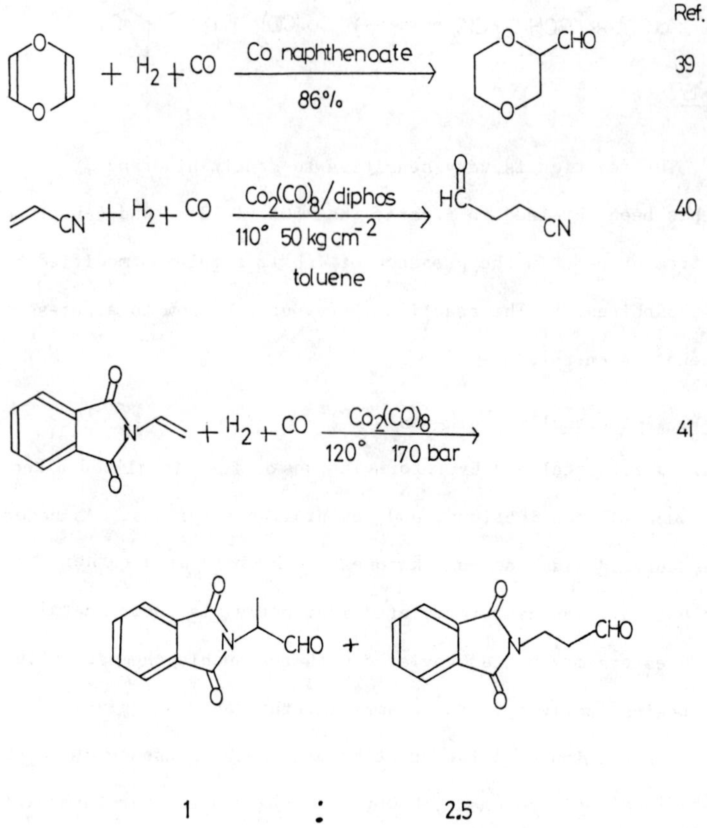

Figure 4 Cobalt catalysed hydroformylation of functionalised alkenes.

the acyl is thought to be rate determining and the metal acyl is observed in the IR spectrum of the reacting mixture.[45]

Most modern rhodium catalysed hydroformylations involve catalysts modified by various ligands so that the active species is $HRh(CO)_2L_n$. Wilkinson focussed on the mechanism for reactions in which $HRhCo(PPh_3)_3$ is the catalyst precursor.[46-48] He proposed associative and dissociative pathways shown in Figure 6 and subsequent discussion of mechanism long used some version of this scheme.

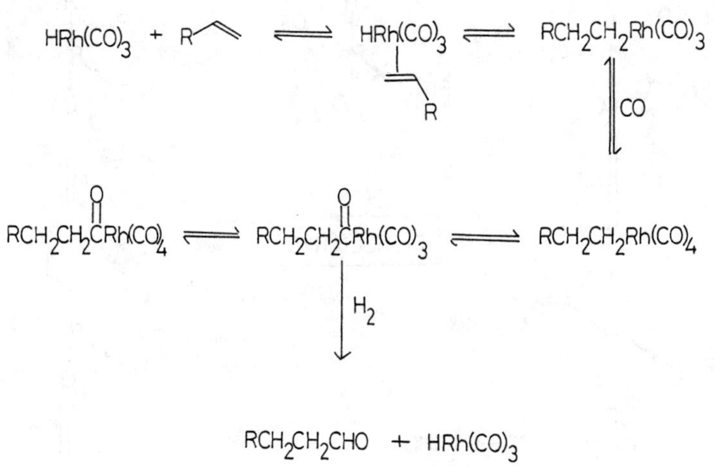

Figure 5 Mechanism of rhodium carbonyl catalysed hydroformylation.

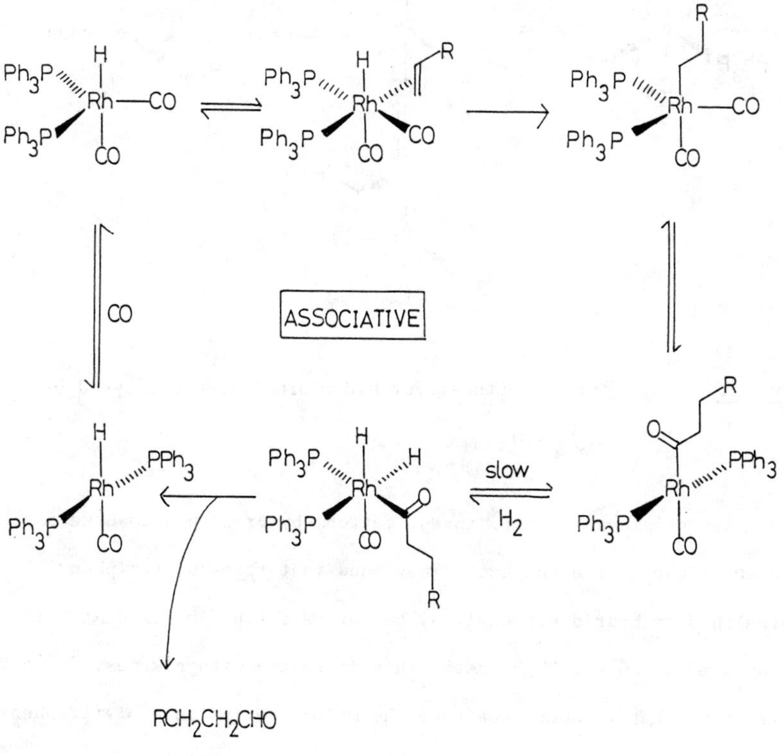

Figure 6a Associative pathway for hydroformylation catalysed by HRh(PPh₃)₃(CO)

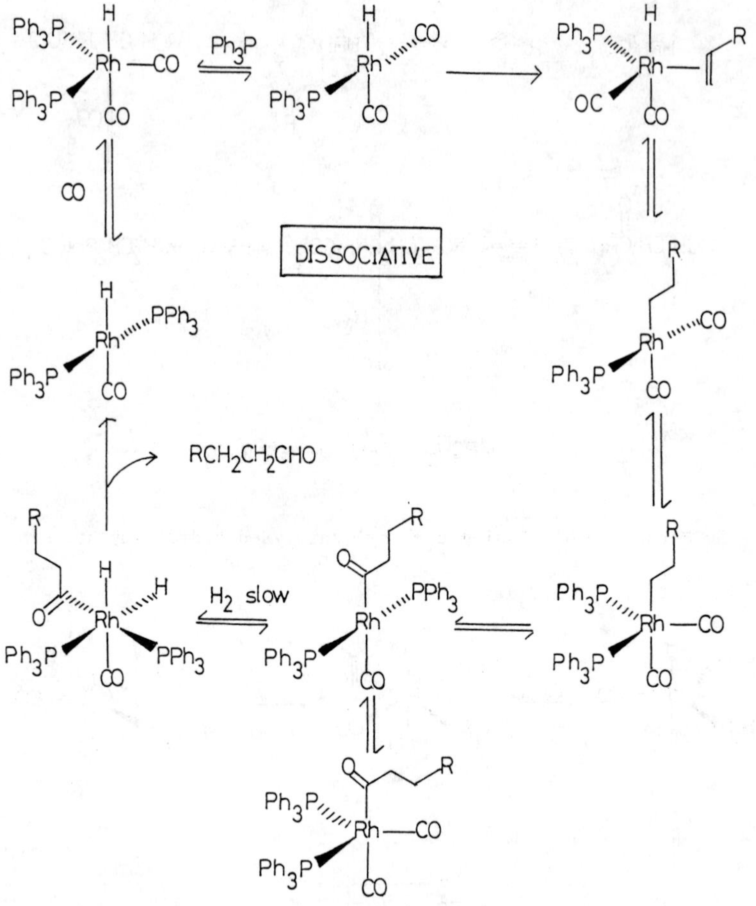

DISSOCIATIVE

RCH$_2$CH$_2$CHO

H$_2$ slow

Figure 6a cntd. Dissociative pathway for hydroformylation catalysed by
HRh(PPh$_3$)$_3$(CO).

Recent elegant n.m.r. investigations by Brown and coworkers throw

a different light on the subject. They show that trans-bis(triphenyl-

phosphine)rhodium hydridocarbonyl, A, cannot be a true intermediate since

it is captured by CO and Ph$_3$P faster than it reacts with alkenes.[49] In the

presence of CO, but without substrate the major species is a bis(triphenyl-

phosphine) bis(CO)rhodium hydride, which exists as a rapidly

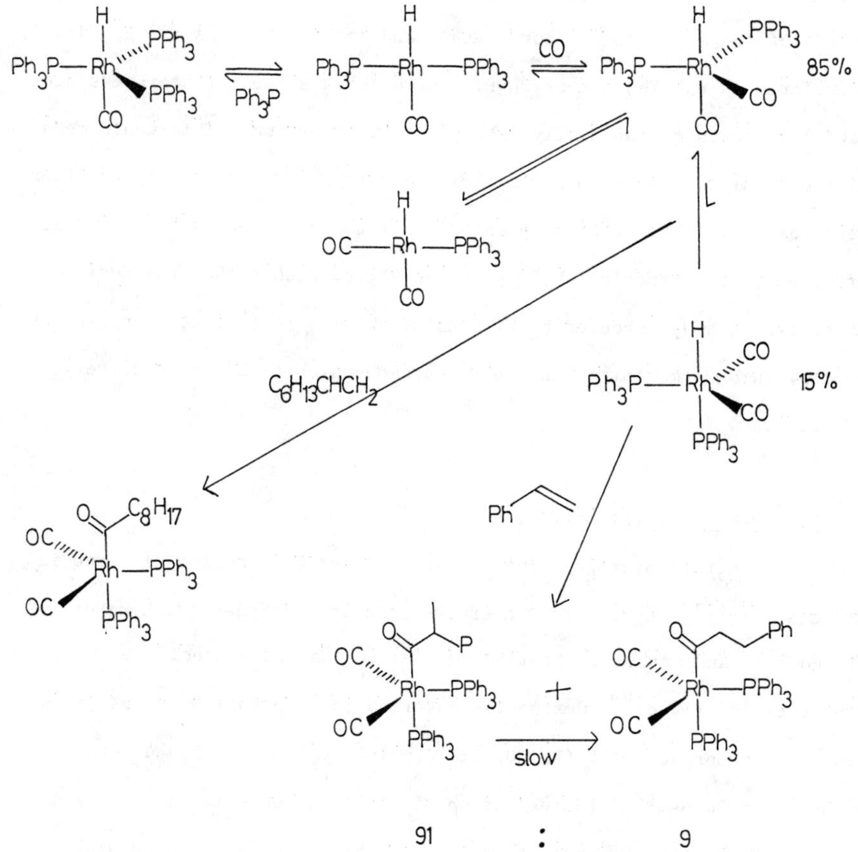

Figure 6b Brown's mechanism for rhodium catalysed hydroformylation.

interconverting mixture of stereoisomers (Figure 6b). Alkene and carbon
monoxide give acyl complexes, and those derived from octene and styrene
were observed in detail. At 178K ^{13}C labelling and ^{13}C and ^{31}P n.m.r.
establish the structure of the octene derivative conclusively. On warming
several dynamic processes become important, notably exchange of Ph$_3$P and
transfer of the alkyl group between carbon monoxides, probably accompanied
by dissociative exchange with free CO. With styrene two acyl complexes
are formed in the ratio 91:9, this being very similar to the ratio of
2-phenylpropanal to 3-phenylpropanal formed in hydroformylation of styrene

with 1:1 H_2-CO under ambient conditions. On examining the proton n.m.r.
spectrum of the major isomer both methyl and methine signals are broadened,
particularly in the absence of Ph_3P. The authors suggest that reversible
ligand loss leads to rapid alkyl⇌acyl interconversion. This would mean
that the catalytic cycle involves their formation and trapping by H_2 after
ligand loss. It would then be possible that the observed isomer ratio is
controlled by the kinetic lability of the acyls, rather than the dual
mechanistic pathway favoured by Wilkinson. Whilst further work is clearly
in order, this work represents a new approach which promises to be very
productive.

3.1.2.1 Chemoselectivity

Substrate hydrogenation is rarely a serious problem with rhodium
catalysts.[46,50,51] Hydrogenation of the product aldehydes is, however,
well known[52] and may be controlled to give the useful alcohols 19 and 20
in the ratio of 1:2.8.[53] Hexene is converted to n-hexanol in 97.9% yield
at $140°$ in the presence of $(Rh(CO)_2Cl)_2$, polyethylene glycol and
N-methylpyrrolidone.[54] Isomerisation of higher alkenes is also catalysed
by many rhodium complexes but in most cases this does not compete with
hydroformylation.[55]

3.1.2.2 Regioselectivity

As before the desired product is a linear aldehyde and modified catalysts have been quite successful. Without modifiers the catalyst is active but unselective; in one example 1-octene gave only 31% nonanal. However, with a high phosphine (or other ligand) concentration and low CO pressure much higher isomer ratios were obtained, with minimal loss to hydrogenation. It is thought that these conditions favour species with more than one phosphine ligand, which present a more sterically hindered environment for the incoming alkene and hence favour an insertion to a linear metal alkyl.   Additionally rates are improved. Some examples are given in Figure 7.

Apart from the steric effect of the large phosphine ligands it is clear that their electronic character is also significant. When styrene derivatives, $\underline{21}$, are hydroformylated in the presence of $(Rh(CO)_2Cl)_2$, selectivity for $\underline{23}$ increases with σ_p. In the presence of Ph_3P more $\underline{23}$ is produced but its proportion now decreases with σ_p.   Also complexes of the type $RhClCO(Ph_{3-x}P(NR_1R_2)_x)_2$ have been studied by ESCA, IR and ^{31}P n.m.r. spectroscopy. In the free aminophosphines donation is from nitrogen to phosphorus, the reverse being true in the rhodium complexes.   The more electron withdrawing the groups R_1 and R_2 the higher the proportion of linear aldehydes, the relationship being linear with the electron distribution between ligand and metal. They conclude that the regio-

$\underline{21}$ $\underline{22}$ $\underline{23}$

selective process in the reaction scheme is governed by the insertion of alkene into Rh-H in which the partial charge on rhodium is very important. The species are shown in Figure 8, the regioselection being in accord with stabilisation of the developing charge on carbon.

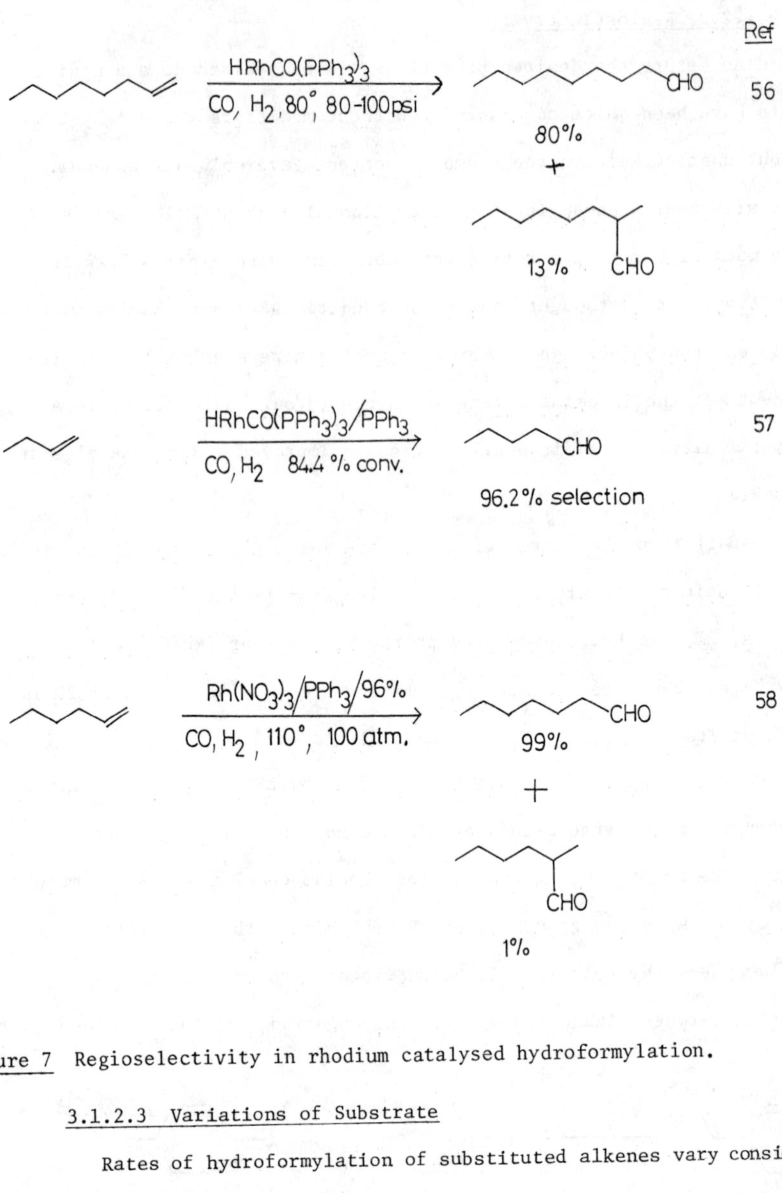

Ref

80%

+

13%   CHO

56

96.2% selection

57

99%

+

CHO

1%

58

Figure 7 Regioselectivity in rhodium catalysed hydroformylation.

3.1.2.3 Variations of Substrate

Rates of hydroformylation of substituted alkenes vary considerably. Aryl substituted double bonds react the most rapidly and hindered double bonds rather slowly.[61,62] This observation has been used in several patented processes for the hydroformylation of limonene, 24, to yield 25, which is active for the destruction of Candida albicans on wood.[63] Other examples are shown in Figure 9.

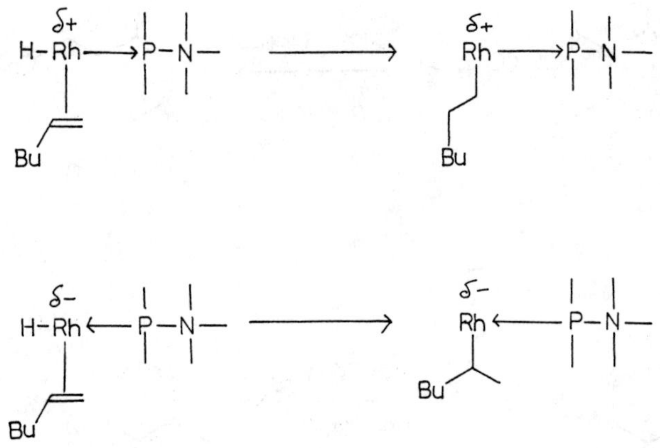

Figure 8   Insertion of alkenes into Rh-H bond.

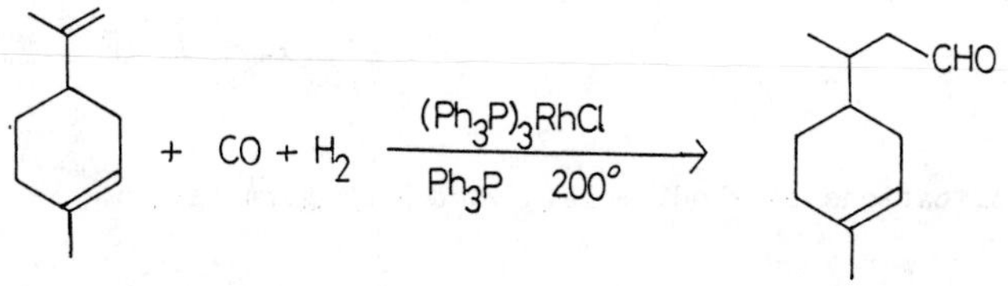

24   25

Besides obtaining monoaldehydes, dialdehydes also result from reactions with non-conjugated double bonds. Few of the reactions are very specific (e.g. 26 to 27, 28 and 29[66]) and they have not found many applications. Conjugated dienes might appear to be promising hydroformylation substrates because the expected dialdehydes could be converted to useful products. However, reaction is slow, yielding mainly monoaldehydes and monoalcohols with unmodified rhodium catalysts.[67] A catalyst modified by tributylphosphine is effective for the partial bis-hydroformylation of butadiene,[68] but the only practical process is a two-stage reaction as shown in Figure 10.[69] Selective mono-hydroformylation has been achieved in the presence of catalysts modified by Ph_3P[67] or Et_2PPEt_2[70]; in the former case a rhodium allyl intermediate is thought to be important.

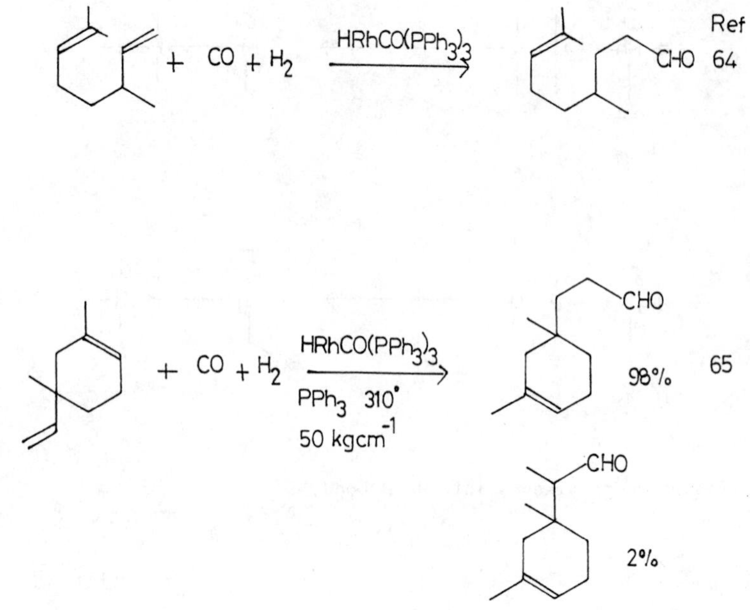

+ CO + H$_2$ $\xrightarrow{\text{HRhCO(PPh}_3)_3}$ CHO

Ref 64

+ CO + H$_2$ $\xrightarrow[\substack{PPh_3 \ 310° \\ 50 \ kgcm^{-1}}]{\text{HRhCO(PPh}_3)_3}$

CHO 98% 65

CHO 2%

Figure 9 Applications of rhodium catalysed hydroformylation.

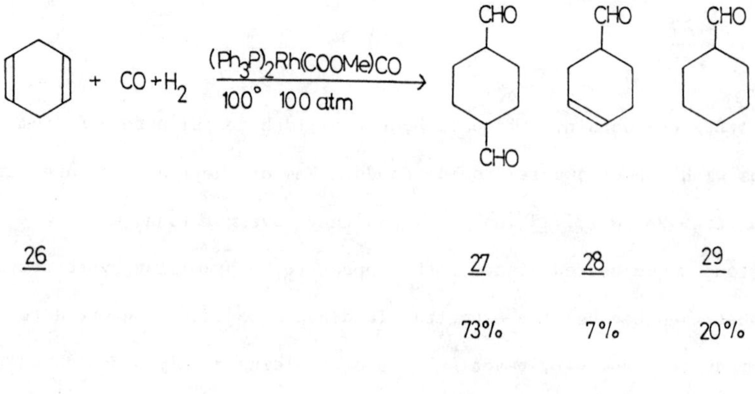

+ CO + H$_2$ $\xrightarrow[100° \ 100 \ atm]{(Ph_3P)_2Rh(COOMe)CO}$

CHO / CHO   CHO   CHO

26

27   28   29

73%   7%   20%

αβ-Unsaturated aldehydes and ketones are hydrogenated, just as with cobalt catalysts, though once again the problem may be circumvented by reaction of the acetals (e.g. 30 to 31 and 32[71]). αβ-Unsaturated esters react successfully, the regioselectivity depending on the conditions employed. With ethyl acrylate, 33, both α- and β- isomers are obtained[72]

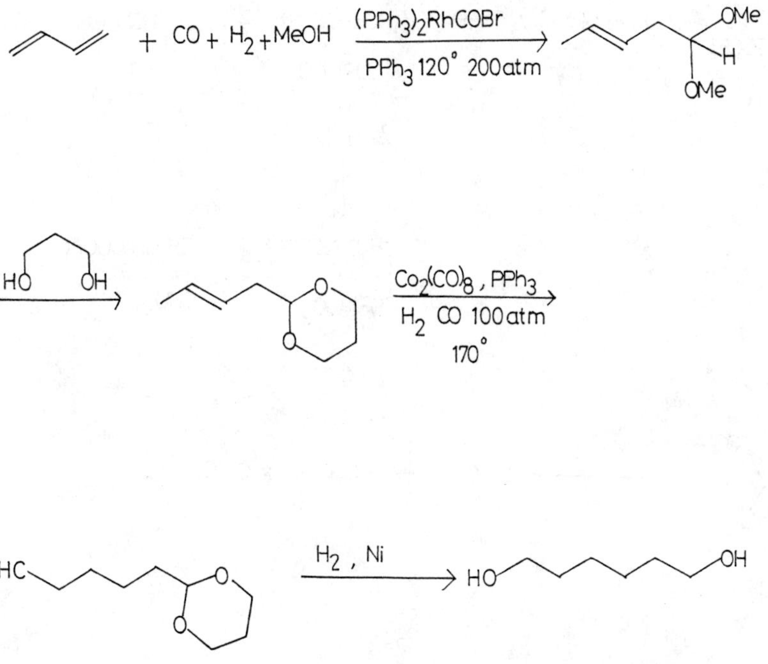

Figure 10 Two stage process for the hydroformylation of butadiene.

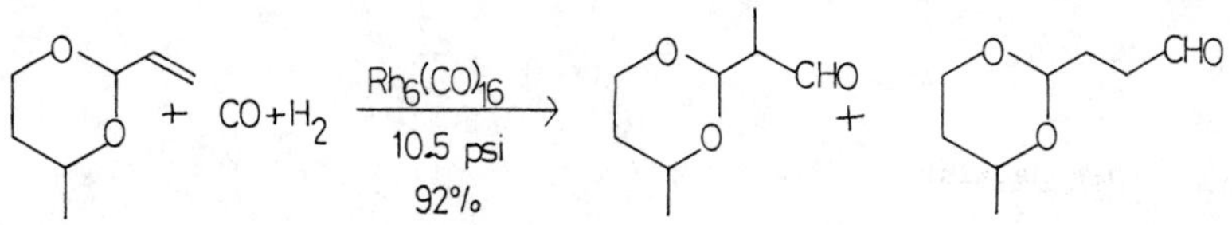

30   31   32

14%   86%

(35 and 37). High temperatures promote isomerisation and give more 37 while high CO pressure inhibits the rearrangement of 34. High H_2 pressure accelerates reduction before isomerisation. For methyl methacrylate, 38, the porportion of β-isomer, 40, also increases with temperature suggesting again that the branched alkyl, 41, is the initial insertion product.[73] The increased proportion of 40 with increasing phosphine concentration for $R_3P = (PhO)_3P$ is in line with previously observed steric effects.[56].

$$\text{33} \quad \underset{\text{COOEt}}{\diagup\!\!\!\diagup} + CO + HRh(CO)_3 \longrightarrow \underset{\underset{CORh(CO)_3}{|}}{CH_3-CH-COOEt} \xrightarrow{H_2} \underset{\underset{CHO}{|}}{CH_3CHCOOEt}$$

33   34   35

$$\underset{\underset{CORh(CO)_3}{|}}{CH_2CH_2COOEt} \xrightarrow{H_2} \underset{\underset{CHO}{|}}{CH_2CH_2COOEt}$$

36   37

$$\underset{\text{COOMe}}{\diagup\!\!=\!\!\diagdown} \longrightarrow OHC-\!\!\!\!\overset{|}{\underset{|}{}}\!\!\!\!-COOMe + OHC\diagdown\!\!\!\diagup^{COOMe}$$

38   39   40

$$\overset{COOMe}{\underset{|}{-\!\!\!\!\overset{|}{}\!\!\!\!-RhL_n}} \longrightarrow L_nRh\diagdown\!\!\!\diagup^{COOMe}$$

41   42

3.1.3 Other Metals

While only cobalt and rhodium catalysed reactions are of commercial significance other catalysts are known. Ruthenium species were thoroughly studied by Wilkinson; hexene is converted to aldehydes with high chemoselectivity in the presence of a variety of precursors[74,75] but regioselectivity is modest, n/iso ratios of 2 to 2.5 being typical. On the basis of a detailed kinetic analysis it was concluded that $Ru(CO)_3(PPh_3)_2$ is the active species and the mechanism is shown in Figure 11.

A number of recent papers report platinum catalysed reactions, particularly in the presence of stannous chloride. Complexes of both monophosphines $((R_3P)_2PtCl_2|SnCl_2)$ and chelating biphosphines are useful.

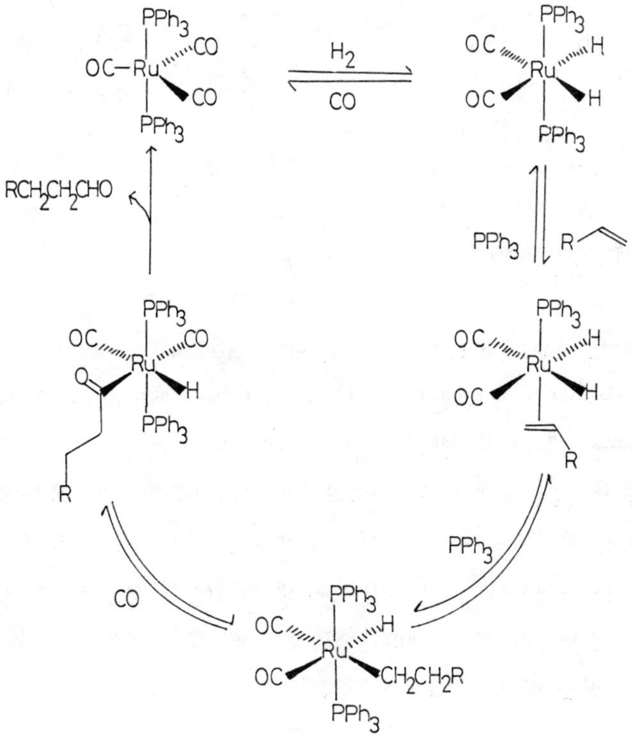

Figure 11 Ruthenium catalysed hydroformylation.

Chemoselectivity is not invariably high and careful attention must be paid to reaction conditions to obtain good yields of aldehyde.[76] Competitive hydrogenation is a particular problem, especially with substituted alkenes.[77] Some excellent regioselectivities have been reported; in the presence of $HPt(SnCl_3)(CO)(Ph_3P)_2$, 1-pentene is converted into >95% hexanol.[78] With chelating biphosphines yield and selectivity is sharply maximised with $Ph_2P(CH_2)_nPPh_2$ for n = 4.[79,80] The shorter the alkene, the lower the observed linearity, which suggests a steric origin for the effect. Mechanistic studies of this stystem are at an early stage but 43 and 44 may be isolated under suitable conditions and would certainly seem

$$\underline{43}$$

$$\underline{44}$$

to be likely intermediates.[81]

3.1.4 Polymer Supported Hydroformylation Catalysts

Both rhodium and cobalt catalysts have been bonded to organic and inorganic supports. Typical examples are given in Figure 12. Selectivity for n-aldehydes is generally rather lower than for homogeneous systems, though not outside the normal range of variation. This seems to be due, in part, to the multiplicity of catalytically active sites. The catalysts are less sensitive to oxygen and may be reused several times without leaching of the metal or loss of activity.

3.1.5 Asymmetric Hydroformylation

Optically active aldehydes may be obtained by the hydro-formylation of prochiral alkenes.[85] When a trisubstituted alkene such as 45 is used as substrate two new chiral centres may be generated; the reaction is generally stereospecifically cis so only one diastereomer should be formed.[86] Few really successful results have been reported. One of the major problems seems to be that the optical yield deteriorates as the reaction proceeds; the product aldehydes are very susceptible to racemisation and the frequently severe reaction conditions almost certainly accelerate this. The modification of cobalt catalysts with chiral ligands has been particularly unsuccessful; early work involved (+)-N-(1-phenethyl)salicyladimine as ligand giving 2.8% maximum optical yield for conversion of α-methylstyrene to 3-phenylbutanal.[87] Neither

Ref

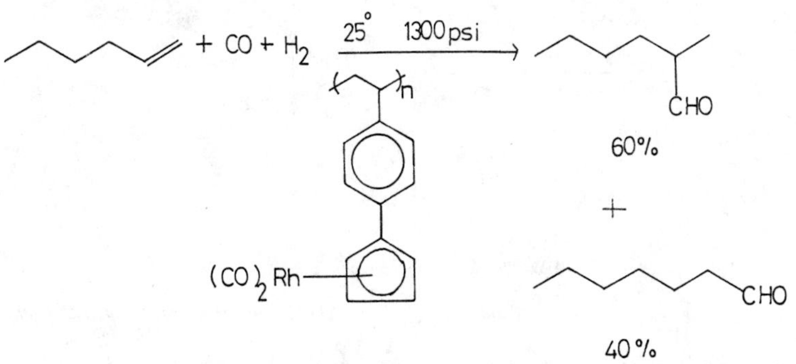

82

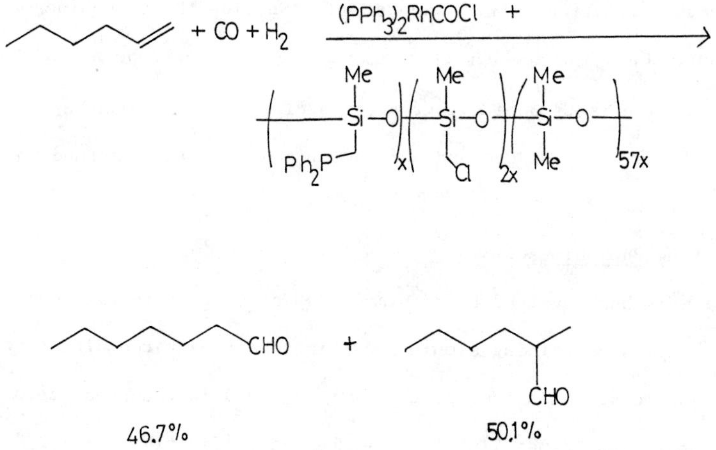

83

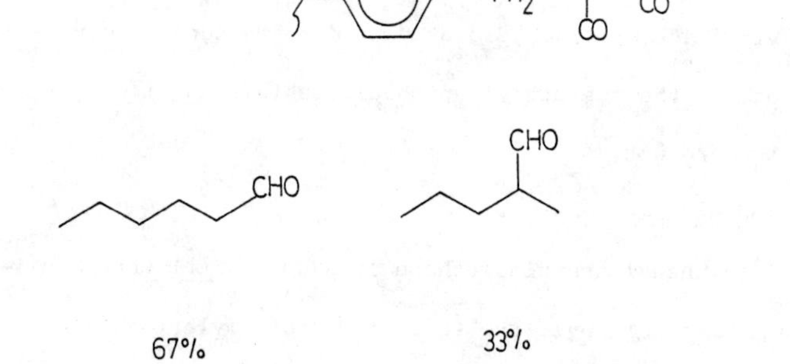

84

<u>Figure 12</u> Polymer bound hydroformylation catalysts.

$$R_1 - \overset{R_2}{\underset{R_3}{\overset{|}{\underset{|}{C}}}} \overset{}{\underset{H}{}} \longrightarrow R_1 \overset{R_2}{\underset{R_3}{\overset{}{\cdots}}} \overset{H}{\underset{H}{\cdots CHO}}$$

<u>45</u>   <u>46</u>

DIOP[88] nor MePhnPrP[89] were any more useful.

The milder conditions under which rhodium derived catalysts may be used have ensured that these species have been most prominent in asymmetric hydroformylation. It is convenient to divide the phosphines into two groups, chelating and non-chelating though it should be noted that this distinction is to some extent artificial. There is mounting evidence to indicate that bridged and non-chelating species are formed in the latter case also.[90]

Non-chelating Phosphines

Most work has involved hydroformylation of styrene in the presence of catalysts containing simple phosphines bearing three different alkyl groups. Up to 42% enantiomer excess was achieved in the presence of n-PrMePhP.[91] A number of generalisations may be made. Firstly S-hydtropaldehyde is produced from catalysts containing R-phosphines. Secondly the enantiomer excess decreases with increasing temperature and markedly with conversion; this may imply that the product is racemising under the reaction conditions and/or that the catalyst deteriorates with time. Results with functionalised substrates have been sparse but it is noteworthy that vinyl acetate is hydroformylated in reasonable optical yield; the presence of the acetate group should inhibit <u>in situ</u> racemisation of the product.[92]

Chelating Biphosphines

When <u>bis</u>(diphenylphosphino)ethane is added to <u>tris</u>(triphenylphosphine)rhodium chlorocarbonyl the rate of hydroformylation is drastically reduced. Thus comparatively few experiments have been attempted using phosphines which form 5-membered chelate complexes. E-2-butene, 2,3-dimethyl-1-butene and α-methyl styrene have all been reacted in the presence of $(Rh(nbd)Cl)_2$ and S,S-CHIRAPHOS to give products with optical yields between 15% and 25%.[93] DIPAMP was used with limited success in the catalytic conversion of 47 to 48.[94]

7-ring chelating biphosphines, in particular DIOP, have been more extensively tested. Some examples using DIOP-based catalysts are given in Figure 13.[85] Fine tuning of the ligand in the last example can improve the optical yield to 51% when a dibenzophosphole group replaces

47   48   49

PPh₂.[95]

The precise origins of the reaction's stereoselectivity remain somewhat obscure despite a useful empirical model proposed by Pino and coworkers.[93] The fact that C_{2v} alkenes such as cis-2-butene, dihydrofuran and bicyclo[2.2.2]octene all give optically active products on hydroformylation indicates that some chiral discrinination must occur after the olefin binding step, since this necessarily produces only one complex. The fact that cis- and trans-2-butene give different optical yields, despite having the same sec-alkyl rhodium complex, 50, as an intermediate suggests that the optical induction takes place in the formation of the alkyl rather than subsequently. It seems that optical selectivity is largely dependent on steric bulk and that polar factors play a comparatively minor role.

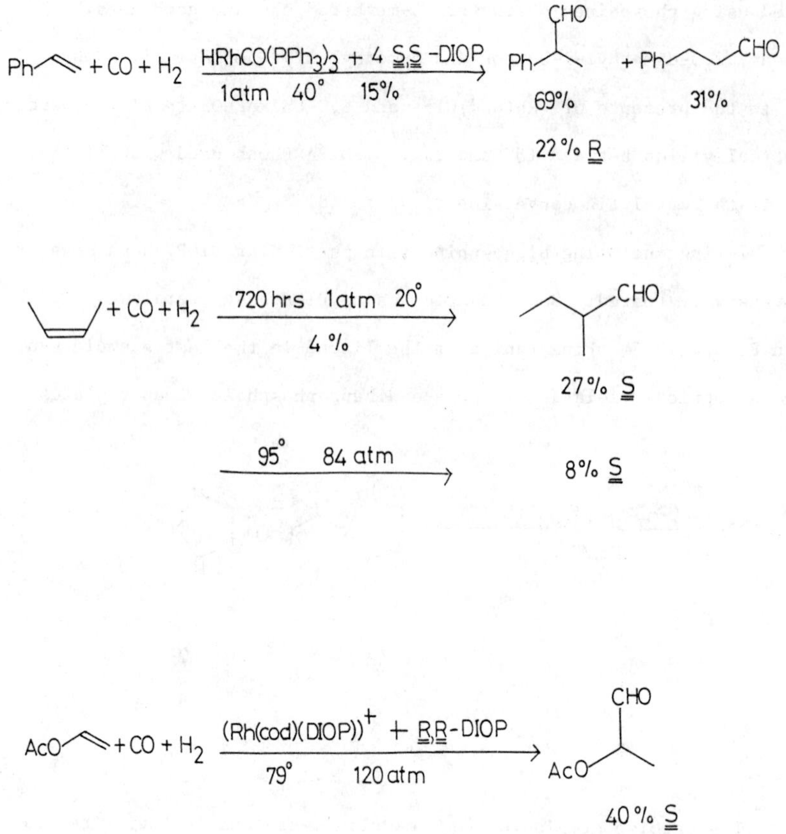

Ph⌒⤙ + CO + H₂ $\xrightarrow[\text{1atm} \quad 40° \quad 15\%]{\text{HRhCO(PPh}_3)_3 + 4 \underline{SS}\text{-DIOP}}$ Ph⌒CHO (69%) + Ph⌒⌒CHO (31%)

22% R

⌲⌴ + CO + H₂ $\xrightarrow[4\%]{\text{720 hrs} \quad \text{1atm} \quad 20°}$ ⌲⤙CHO

27% S

$\xrightarrow{95° \quad 84 \text{ atm}}$ 8% S

AcO⌒⤙ + CO + H₂ $\xrightarrow[79° \quad 120 \text{atm}]{(\text{Rh(cod)(DIOP)})^+ + \underline{RR}\text{-DIOP}}$ AcO⌒⤙CHO

40% S

Figure 13   Rhodium (DIOP) catalysed asymmetric hydroformylations.

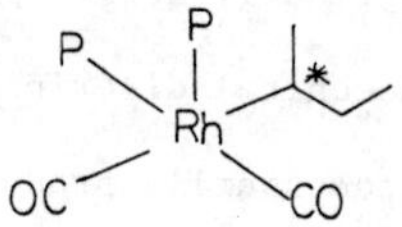

50

A number of platinum complexes such as 51 and 52 were early reported to be good hydroformylation catalysts in the presence of stannous chloride promoter.[96][97] The reaction is thought to proceed in the same manner as its achiral analogue. Despite early reports to the contrary[96] it has been shown that asymmetric induction takes place (as for rhodium),

$$\underline{51} \qquad \xrightarrow{\text{SnCl}_2} \qquad \underline{52}$$

prior to metal alkyl formation, since the isomeric butenes give products
of different optical yields.[98] In the absence of stannous chloride it has
recently been noted that the selectivity to aldehydes is much reduced
(60-80%) but that the enantiomer excess is dramatically increased, to 95%
$\underline{S}$ for styrene in the presence of PtCl_2 and (-)DBP-DIOP.[99]   Whether this
unprecedented result may be generalised remains to be seen but it is
certainly a most promising development.

3.2 Hydrocarboxylation and Related Reactions

The hydrocarboxylation reaction ($\underline{53}$ to $\underline{54}$ and $\underline{55}$) is closely

$$R \diagup\!\!\!\diagdown \;+\; CO + H_2O \;\xrightarrow{\text{catalyst}}\; R\underset{\underline{54}}{\overset{\text{COOH}}{\diagup\!\!\diagup}} \;+\; R\diagup\!\!\!\diagdown\!\!\!\diagup\text{COOH}$$

$$\underline{53} \qquad\qquad\qquad\qquad \underline{54} \qquad\qquad \underline{55}$$

related to hydroformylation and is the prototype of a variety of reactions
in which a metal acyl, $\underline{56}$, is reacted with a nucleophile.

$$R\diagup\!\!\!\diagdown\overset{\text{ML}_n}{\underset{O}{\diagdown}} \;+\; Nu \;\longrightarrow\; R\diagup\!\!\!\diagdown\overset{Nu}{\underset{O}{\diagdown}} \;+\; ML_n$$

$$\underline{56} \qquad\qquad\qquad\qquad \underline{57}$$

3.2.1 Conversion of Alkenes to Esters, Lactones and Acids

The hydroesterification of 1-alkenes, like hydroformylation, is most profitably directed toward the formation of linear esters. The contents of the isomeric mixture are influenced by catalyst, solvent and additives. Many catalysts are known; $Ni(CO)_4$ was one of the earliest used, but although yields can be excellent[100], unless the alkene is particularly activated[101] the conditions required are rather severe[102] ($\sim$200°C, 100–300 atm). This leads to poor selectivity in acid and ester formation, and extensive isomerisation of unreacted starting material.[103] Cobalt catalysts are active under slightly milder conditions and linear esters may account for up to 75% of the product.[104] With internal alkenes the catalysts rapidly isomerise the double bond to the 1-position so that the linear ester is again the major product.[105,106]

Recently, however, most of the catalysts reported have been palladium complexes $(R_3P)_2PdCl_2$ or $(R_2PPR_2)PdCl_2$ since they exhibit high selectivity under relatively mild conditions. A careful analysis of the factors controlling regioselectivity was performed by Fenton.[107] Increasing CO pressure reduced the n/iso ratio which was a maximum at a temperature of 125°. A small amount of water improved yield. Fine tuning of the catalyst yields variable results as Figure 14 shows. Much more active catalysts are obtained on treatment with stannous chloride which is also said to improve selection to linear esters.[111–114]

The mechanism of the reaction has been studied by several groups. The scheme of Figure 15 was suggested by Japanese workers[115] for cyclohexene. Support for the involvement of $HPdCl(Ph_3P)_2$ as the active species comes from the observation of an increased rate in the presence of HCl.[116] $Pd(CO)(PPh_3)_3$ may be isolated in 80% yield from reaction of cyclohexene with $PdCl_2(PPh_3)_2|H_2|CO|EtOH$ or from reaction of $Pd(PPh_3)_4$ and CO at room temperature, but it is inactive as a catalyst in the absence of HCl.[117] The most encouraging support for the scheme comes from isolation

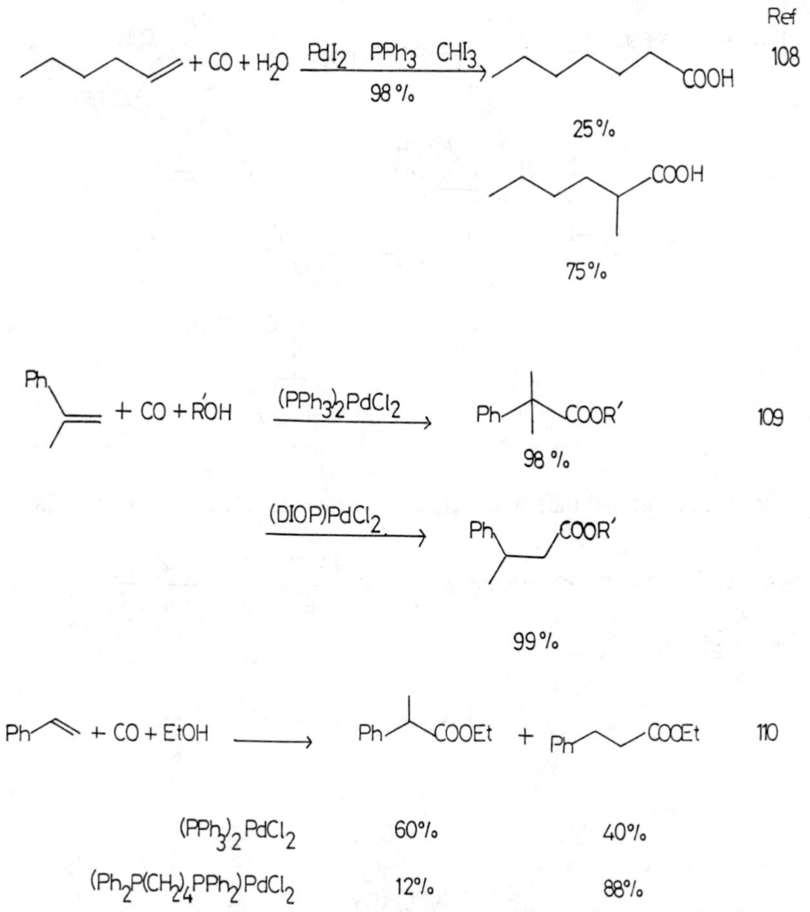

Ref

$$\text{(alkene)} + CO + H_2O \xrightarrow[\text{98\%}]{PdI_2 \ \ PPh_3 \ \ CHI_3} \text{(products)}$$

108

25%

75%

$$\text{(alkene)} + CO + R\dot{O}H \xrightarrow{(PPh_3)_2PdCl_2} Ph\text{—}COOR'$$

98 %

109

$$\xrightarrow{(DIOP)PdCl_2} Ph\text{—}COOR'$$

99%

$$Ph\text{—} + CO + EtOH \longrightarrow Ph\text{—}COOEt + Ph\text{—}COOEt$$

110

$(PPh_3)_2PdCl_2$	60%	40%
$(Ph_2P(CH_2)_4PPh_2)PdCl_2$	12%	88%

Figure 14   Palladium catalysed hydrocarboxylation and hydroesterification.

of trans-Pd($COPr_n$)Cl(PPh_3)$_2$ from propene hydroformylation.[118]

Platinum complexes do not catalyse the hydrocarboxylation of

non-terminal alkenes but good yields of linear acids and esters may be

obtained under comparatively mild conditions as in the reactions of 58 and

60.[119,111]

A rather unusual reaction involves catalysis by copper[120] or

silver[121] cations in concentrated acid.   Carbonium ions are the inter-

mediates and only tertiary carboxylic acids are formed, in rather variable

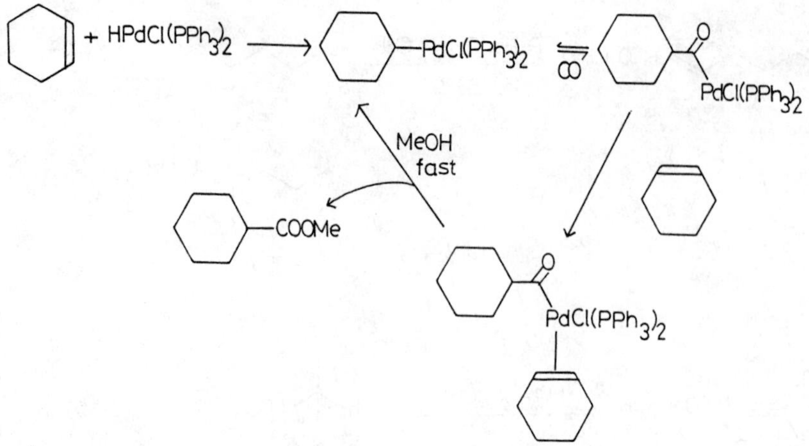

Figure 15 Mechanism of palladium catalysed hydroesterification of cyclohexene.

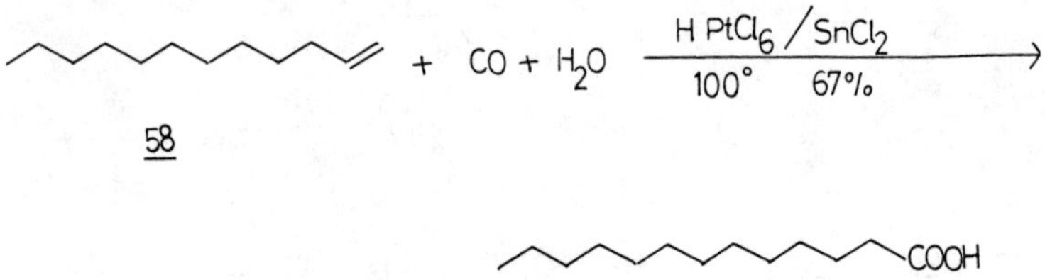

selectivity. A typical scheme is shown in Figure 16.[122,123]

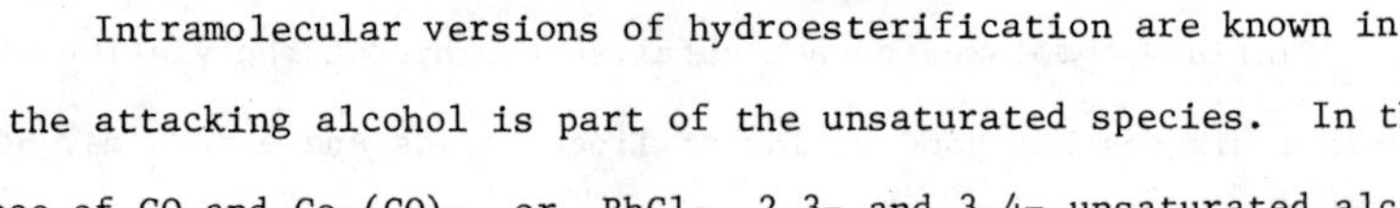

Intramolecular versions of hydroesterification are known in which the attacking alcohol is part of the unsaturated species. In the presence of CO and $Co_2(CO)_8$ or $RhCl_3$, 2,3- and 3,4- unsaturated alcohols yield the corresponding γ- or δ-lactones. The usual conditions are 100-300° and 70-300 atm. CO. The reaction of 62 is typical[124-126] and we shall see analogues of this reaction with halides and alkynes which have received wider synthetic use.

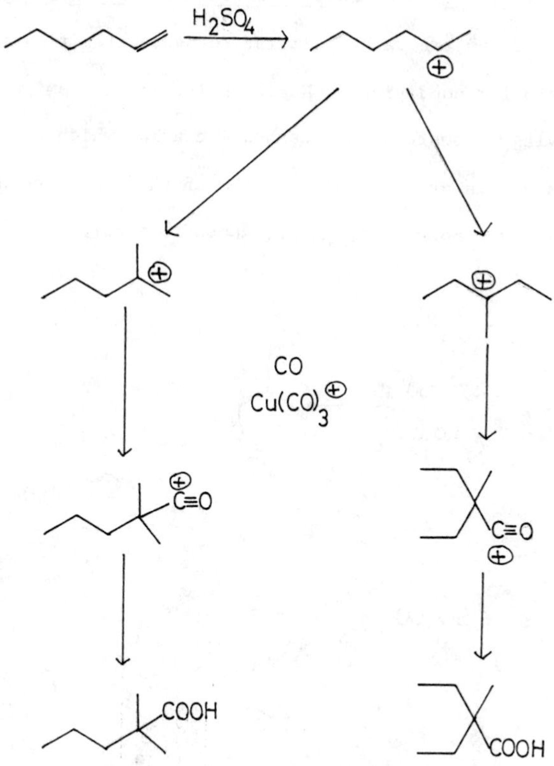

45 %   27%

Figure 16 Carboxylation of alkenes catalysed by copper and silver cations.

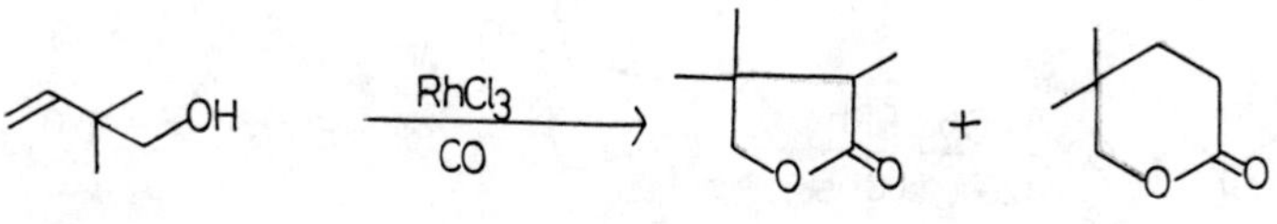

62   63 51 %   64 14 %

3.2.2 Carbonylation of Alkenes in the Presence of Other Nucleophiles

These reactions are rather similar to hydroesterifications and take place under similar conditions. Nucleophiles include amines, yielding amides, thiols giving thioesters, and carboxylic acids which are converted to anhydrides. Catalysts are variable and few of the reactions have found wide synthetic utility. Some examples are shown in Figure 17.

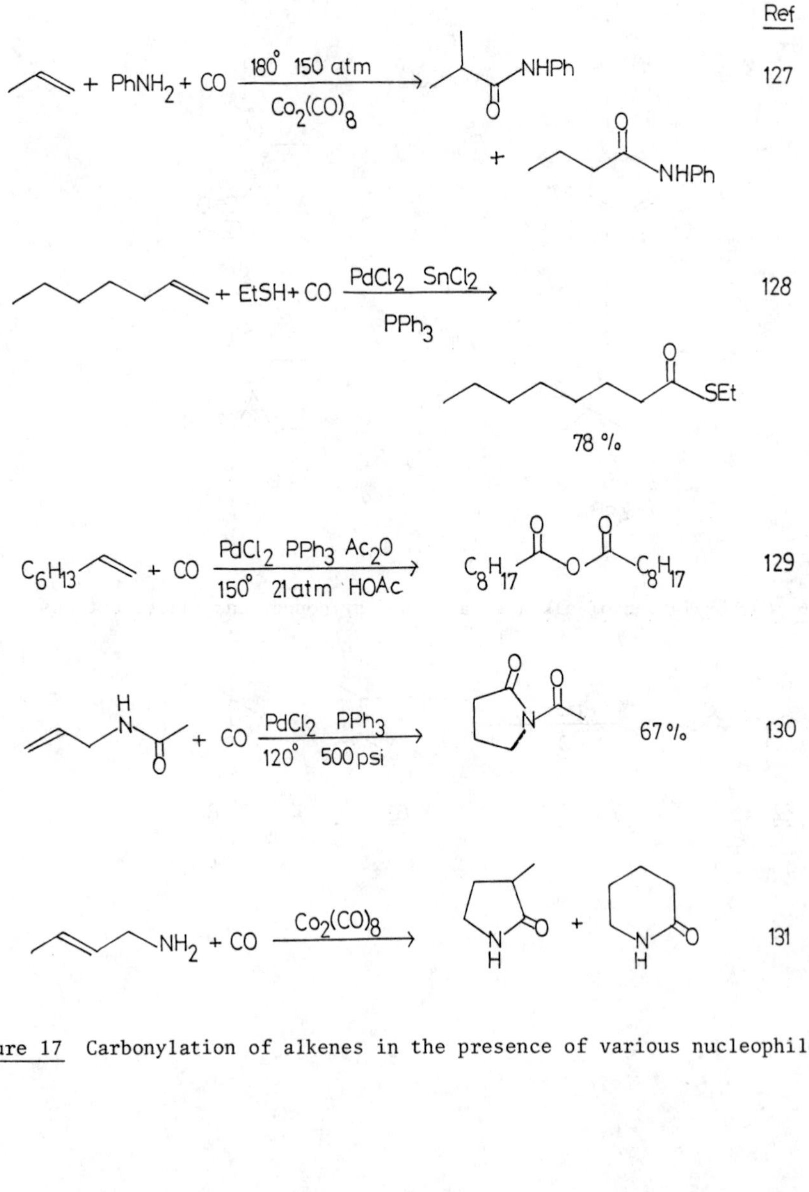

Figure 17 Carbonylation of alkenes in the presence of various nucleophiles.

3.2.3 Reactions of Alkynes

The hydrocarboxylation of alkynes yields unsaturated acids - acrylic acid, its derivatives and homologues. In contrast to the reaction of alkenes there is usually no isomerisation during reaction. The product yield and distribution depends on the catalyst metal, the ligands and the solvent. With nickel catalysts the addition takes place in a cis- manner to give predominently the Markovnikoff product. For example propyne gives methyl methacrylate in the presence of $Ni(CO)_4|H^+$.[132] With palladium catalysts addition is still cis but the anti- Markovnikoff product now predominates[133] (e.g. 65 to 66).[134] Competing reactions include dicarbonylation[135] and oligomerisation[136] but these do not seriously diminish yield.

$$R-C\equiv C-H \; + \; CO + R'OH \quad \xrightarrow[SnCl_2 \quad 65\%]{(Me_2Ph)_2PdCl_2} \quad$$

R   81% selectivity

COOR'

65   66

The intramolecular analogue of this reaction has received more attention since it may give rise to a series of biologically significant α-methylenelactones. A typical reaction is the conversion of 67 to 68.[137,138] Subsequently a mixture of $PdCl_2$, Bu_3P and $SnCl_2$ in MeCN was found to be a better catalyst. Careful deuterium labelling studies elucidated the mechanism as shown in Figure 18 and the reaction has been shown to work well with a wide variety of substrates.[139,140]

$$\text{67} \xrightarrow[\substack{PdCl_2 \\ H_2N \overset{S}{\underset{}{\parallel}} NH_2}]{CO \; (50 \, psi) \quad 50°} \text{68}$$

94%

67   68

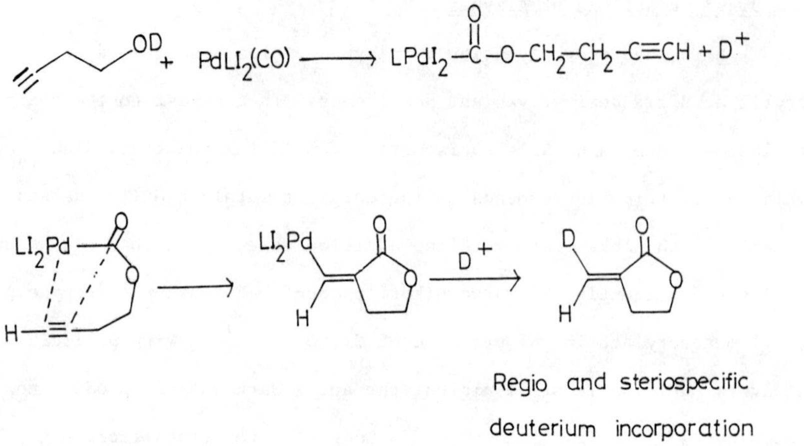

Regio and steriospecific

deuterium incorporation

Figure 18 Mechanism of the carbonylation-cyclisation reaction of acetylenic

alcohols.

3.2.4 Reactions of Dienes

The hydrocarboxylation of non-conjugated dienes and polyenes
does not differ greatly from the reaction of isolated double bonds. Mono-
and polyesters are obtained in various proportions depending on the
precise conditions[141] and there is little isomerisation. For example in
the reaction of 1,5,9-cyclododecatriene, 69, monoester, 70, may be
obtained almost exclusively at 50° in 2:1 EtOH: 69 and 10% HCl. At 100°
71 predominates to the extent of about 60%.[142]

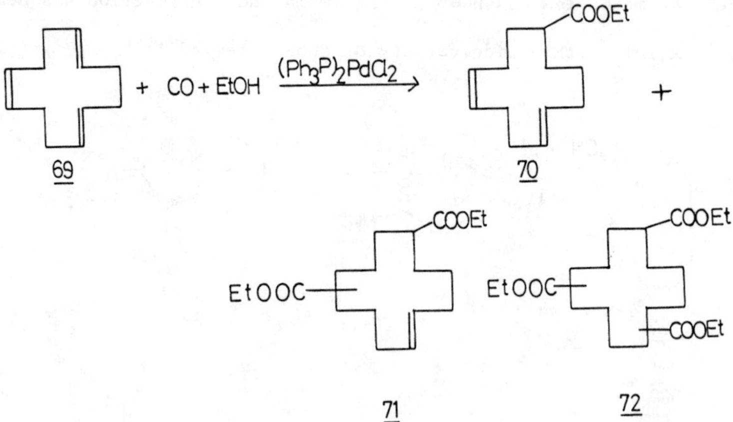

For conjugated dienes simple carbonylation and dimerisation carbonylation take place in alcohol depending on the catalyst. When $PdCl_2$ is used, with or without Ph_3P, methyl-3-pentenoate, 74, is the exclusive product.[143] This is also obtained in good yield in the presence of $Co_2(CO)_8$|pyridine.[144] However, with halide-free palladium catalysts 3,8-nonadienoates, 75, are obtained. The formation of these products may be justified by the scheme of Figure 19. In the presence of chloride ion the bis-allyl complex, 77, (an intermediate in many palladium catalysed telomerisations) may not be formed and a pentenoate is formed from 76.[145] A combination of two carboxylation reactions was used in the synthesis of Royal Jelly acid, 80.[146]

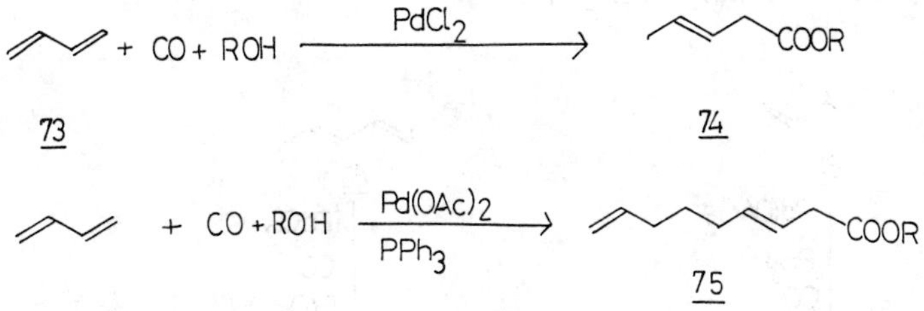

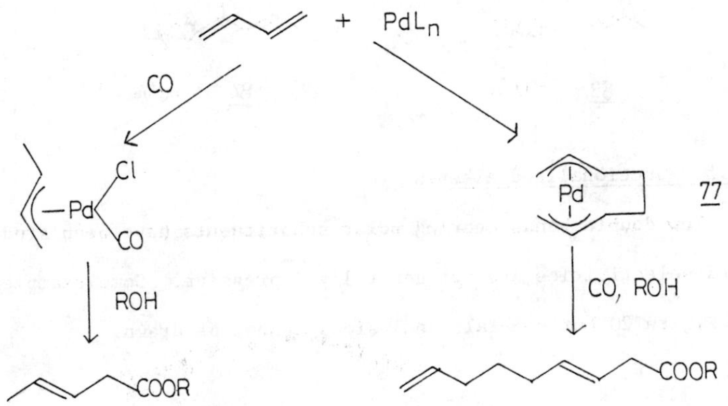

Figure 19 Carbonylation and dimerisation of butadiene catalysed by palladium complexes.

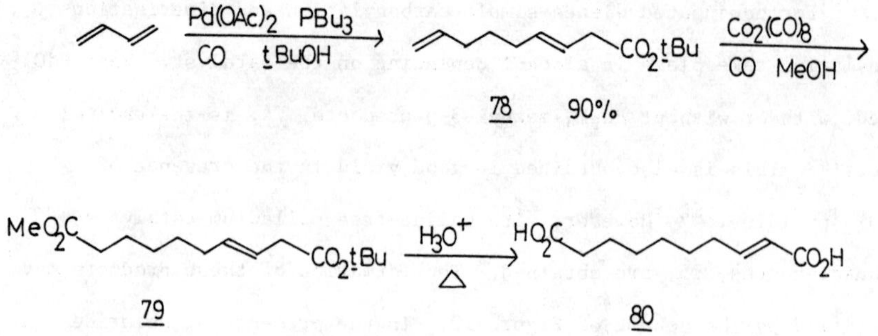

Substituted butadienes such as 81 and 82 do not dimerise, irrespective of the catalyst used, but are carboxylated slowly. Isoprene gives selective reaction at the less substituted end,[143] while piperylene gives mainly the branched ester 84.[147]

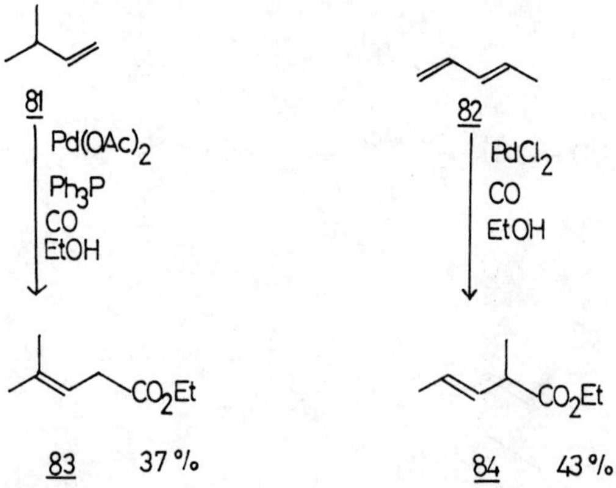

3.2.5 Functionalised Alkenes

Few double bonds bearing polar substituents have been studied in detail and selectivities are not generally impressive. Some examples are shown in Figure 20 but general conclusions cannot be drawn.

3.2.6 Asymmetric Hydrocarboxylation

Much less work has been undertaken in this area than for related hydroformylations. α-Methyl styrene has been the most popular substrate

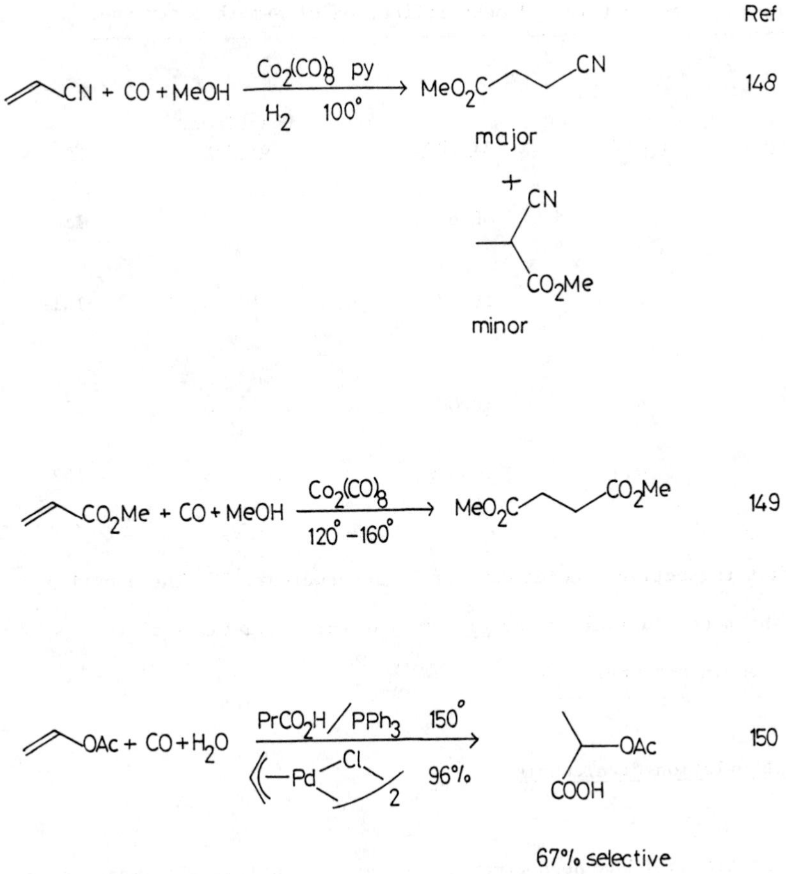

Ref

148

major

minor

149

150

67% selective

Figure 20 Hydroesterification of functionalised alkenes.

and some typical results for 85 to 86 are shown in Table 1. Other

substrates include both aryl and alkyl substituted alkenes, the latter

being universally unsuccessful.[150,153] It is likely that chelating

biphosphines do not remain chelated during the reaction since the

enantiomer excess is maximised at a ratio of P:Pd of 0.5; kinetic studies

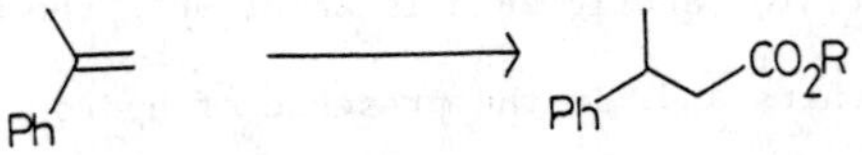

85   86

Table 1. Asymmetric hydroesterification of α-methyl styrene.

Phosphine	P : Pd	Alcohol	Enantiomer excess		Ref.
DIOP	2.	EtOH	10	S	150
	2	iPrOH	40	S	151
DIOP DBP	4	iPrOH	43	S	151
DIOP	0.4	t-BuOH	58.6	S	152

suggest that the active species may not be mononuclear.[154] The addition
has been shown conclusively to be cis[152] but further details of the
mechanism remain obscure.

3.3 Carbonylation of Alcohols

Acetic acid has been produced in large quantities for many years
and currently some 3 million tons a year is manufactured. Carbonylation
of methanol represents almost a half of this. Many catalysts are known
but rhodium, iridium and cobalt are the most significant from an industrial
point of view.

3.3.1 Cobalt Catalysed Reactions

Typical conditions for the reaction are 230°C, 500-700 atm. and
10^{-1} M metal. Selectivity for acetic acid is about 90%, the main by-
products being esters, ethers and, in the presence of hydrogen, (produced
in situ by the water gas shift reaction) methane, acetaldehyde and ethanol.
The cobalt is usually added in the form of CoI_2 and is converted in situ

to HCo(CO)$_4$. This also yields HI which is essential to the reaction
scheme proposed in Figure 21.[155] Reaction rate is increased but
selectivity is reduced when synthesis gas is used; acetaldehyde is then
the primary product and this is reduced to ethanol.[156] In the most
favourable cases (CoI$_2$, Ph$_2$P(CH$_2$)$_6$PPh$_2$) selectivity may be up to 89%.[157]
In the presence of an amine or alcohol, instead of water, as nucleophile
amides and esters may be obtained, as in the conversion of 87 to 88 and 89.
The reaction is a one-pot synthesis and yields ranging from 20% to 80% are
obtained.[158]

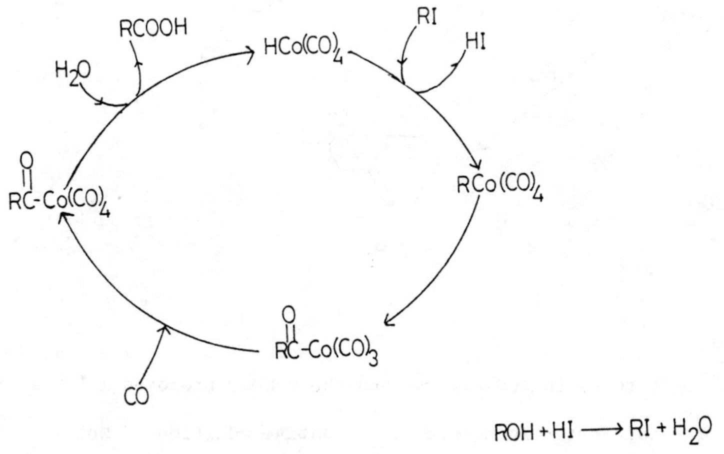

$$ROH + HI \longrightarrow RI + H_2O$$

Figure 21   Cobalt catalysed carbonylation of methanol.

3.3.2 Rhodium Catalysts

Catalyst systems based on rhodium may be used under much milder
conditions than their cobalt analogues. These are typically 180°C, 30-40
atm., 10^{-3} M metal and methyl iodide promoter. Selectivities for acetic
acid are better than 99% so that this process is economically efficient
despite the high cost of rhodium. Kinetic studies of the reaction show
that it is comparatively simple and the iodide promoter may be added in
several forms without noted differences (HI, I$_2$, MeI). The active species

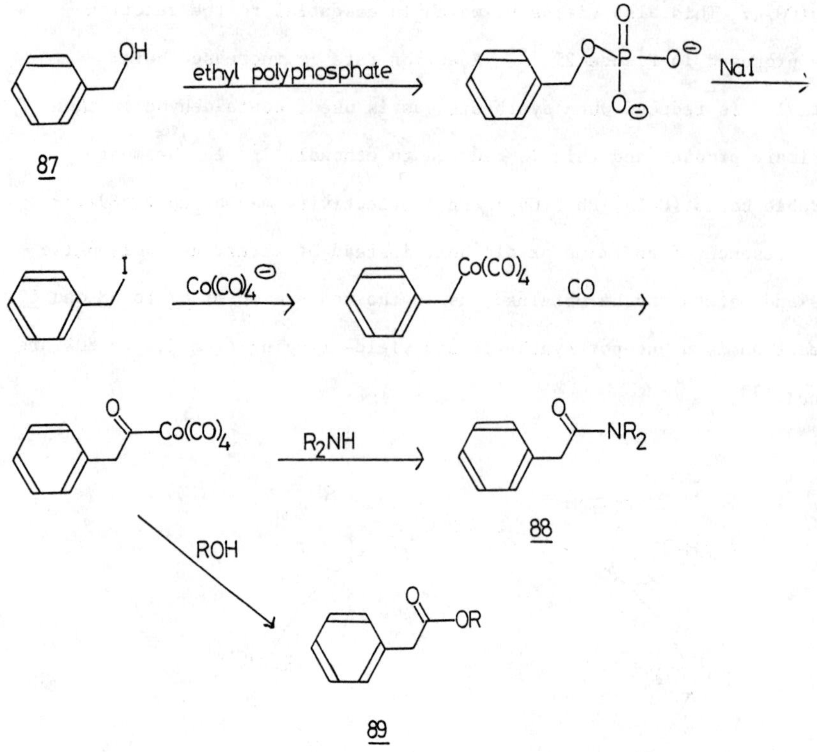

87

88

ROH

89

is thought to be $[Rh(CO)_2I_2]^{\ominus}$ and the scheme proposed is shown in Figure
22. The rate-determining step is oxidative addition of MeI to
$[Rh(CO)_2I_2]^{\ominus}$.[159] The reaction has been studied by IR at room
temperature,[160] showing that an acyl complex is the first detectable
intermediate. X-ray crystallography has determined the structure of the
Me$_3$PhN$^{\oplus}$ salt as the dimer 91.[161] 91 reacts rapidly with carbon monoxide
at room temperature to give a new species designated as B although its
stereochemistry is uncertain. Other rhodium catalysts give essentially
analogous reactions with kinetics which conform to the proposed scheme.
Examples of suitable conditions are given in Figure 23.

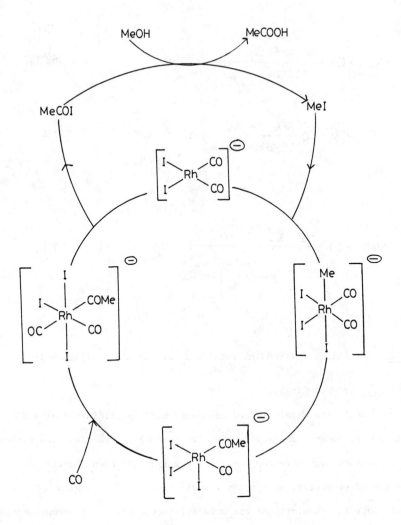

<u>Figure 22</u> Mechanism of rhodium catalysed carbonylation of methanol.

<u>91</u>

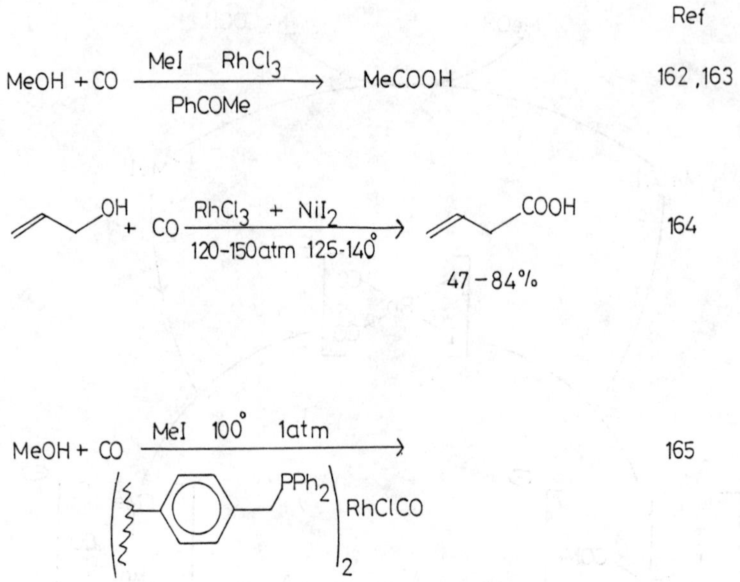
Ref
MeOH + CO
MeI RhCl3
PhCOMe
MeCOOH
162,163
OH
+ CO
RhCl3 + NiI2
120-150atm 125-140°
COOH
47 – 84%
164
MeOH + CO
MeI 100° 1atm
PPh2
RhClCO
2
165

Figure 23 Conditions for rhodium catalysed carbonylation of alcohols.

3.3.3 Iridium Catalysts

While the catalysts and promoters used in iridium catalysed reactions are superficially very similar to those for rhodium, the system is clearly very much more complex. The reaction has been closely investigated by several groups, by kinetic studies,[166] *in situ* spectroscopy and chemistry of observed intermediates.[167] The conclusions are shown in Figure 24, with the same organic reactions as before. There are two relatively independent catalytic cycles with the concentration of $I^{\ominus}$ determining which is followed. Thus the ionic cycle is important under conditions where iodide ion can exist i.e. high water levels or ionic additives. Although it gives a nucleophilic iridium species capable of rapid reaction with methyl iodide it also inhibits the formation of an acyl *via* conversion of $[CH_3Ir(CO)_2I_3]^{\ominus}$ to $[CH_3COIr(CO)_2I_3]^{\ominus}$. In few cases are the relative stereochemistries of the species well defined; in an analogous system $CH_3\overset{O}{\overset{\|}{C}}Ir(CO)I_2(Ph_3P)_2$ was isolated as a complex stereoisomeric mixture.[168]

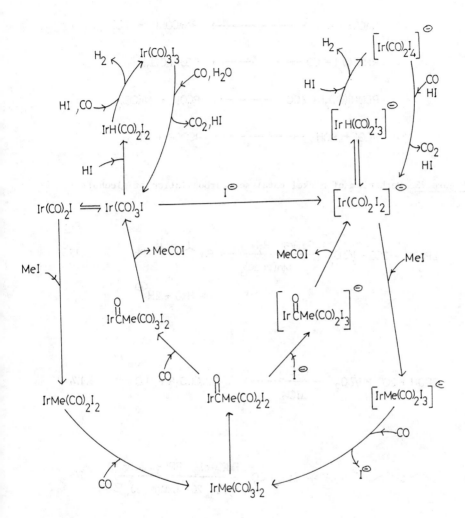

Figure 24   Mechanism of iridium catalysed carbonylation of methanol.

3.3.4   Other Metals

A variety of nickel complexes, as well as the metal itself in the presence of iodine, have been used as catalysts. Heck proposed a mechanism shown in Figure 25.[169] While conditions are generally rather severe the metal is at least inexpensive and a number of diols have been usefully

$$\text{ROH} + \text{HI} \longrightarrow \text{RI} + \text{H}_2\text{O}$$

$$\text{Ni(CO)}_4 + \text{RI} \longrightarrow \text{RNi(CO)}_2\text{I} + 2\text{CO}$$

$$\text{RNi(CO)}_2\text{I} + \text{CO} \longrightarrow \text{RCONi(CO)}_2\text{I}$$

$$\text{RCONi(CO)}_2\text{I} + 2\text{CO} \longrightarrow \text{RCOI} + \text{Ni(CO)}_4$$

$$\text{RCOI} + \text{ROH} \longrightarrow \text{RCOOR} + \text{HI}$$

Figure 25 Mechanism of nickel catalysed carbonylation of alcohols.

Ref

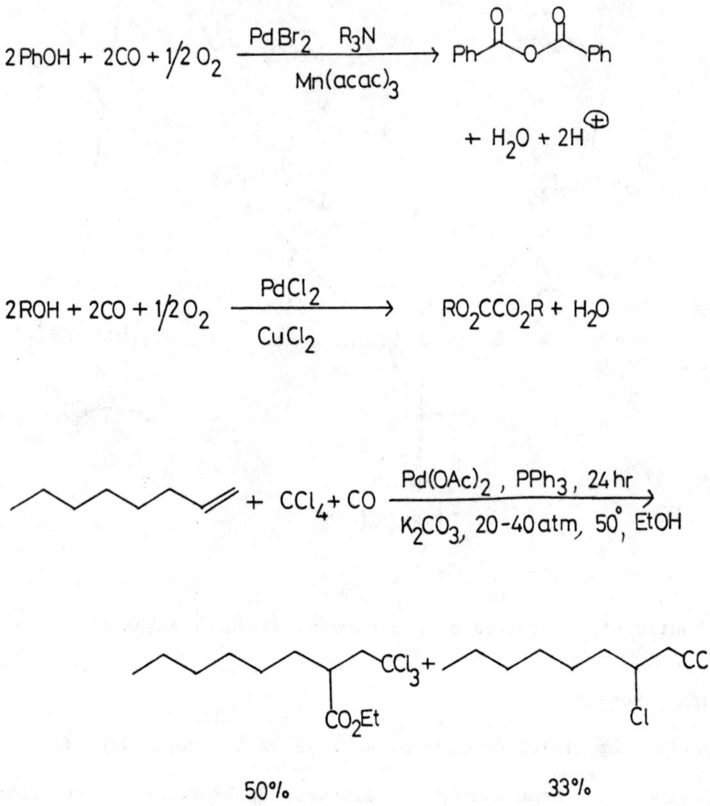

$$2\,\text{PhOH} + 2\text{CO} + \tfrac{1}{2}\text{O}_2 \xrightarrow[\text{Mn(acac)}_3]{\text{PdBr}_2 \quad \text{R}_3\text{N}} \text{Ph} \overset{\text{O}}{\underset{}{\text{C}}}\text{O}\overset{\text{O}}{\underset{}{\text{C}}}\text{Ph}$$

$$+ \text{H}_2\text{O} + 2\text{H}^{\oplus}$$

173

$$2\,\text{ROH} + 2\text{CO} + \tfrac{1}{2}\text{O}_2 \xrightarrow[\text{CuCl}_2]{\text{PdCl}_2} \text{RO}_2\text{CCO}_2\text{R} + \text{H}_2\text{O}$$

174

$$+ \text{CCl}_4 + \text{CO} \xrightarrow[\text{K}_2\text{CO}_3,\ 20\text{-}40\,\text{atm},\ 50^\circ,\ \text{EtOH}]{\text{Pd(OAc)}_2,\ \text{PPh}_3,\ 24\,\text{hr}}$$

175

50%   33%

Figure 26 Palladium catalysed oxidative carbonylations.

carbonylated to diacids in high yield.[171]

Palladium catalysts may be used to carbonylate allyl alcohols, leading to retention of the double bond in the products,[172] but their main application is in oxidative carbonylations such as those in Figure 26.

As previously noted the cobalt catalysed reaction of MeOH with CO may give acid, aldehyde or alcohol. The additional presence of ruthenium, however, significantly improves the selectivity for alcohol[176-179] though the nature of the active species is unknown. Ruthenium alone gives rather varied results with varied substrates; some examples are shown in Figure 27.[180]

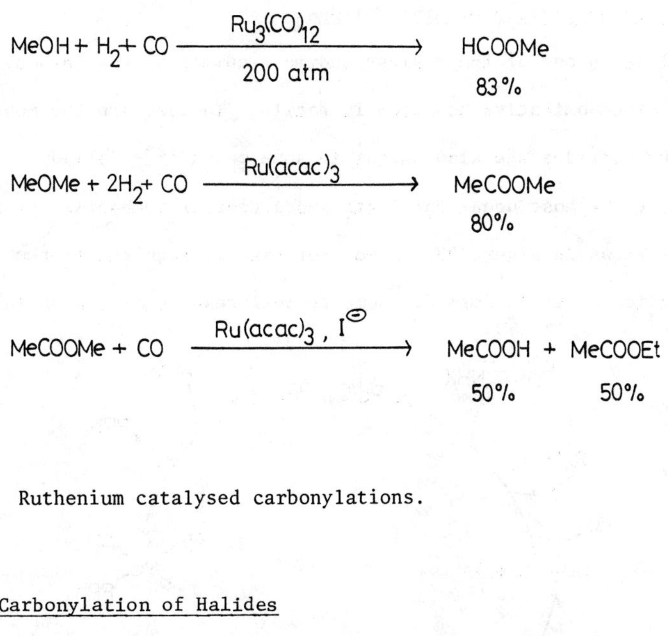

$$MeOH + H_2 + CO \xrightarrow[\text{200 atm}]{Ru_3(CO)_{12}} HCOOMe$$
$$83\%$$

$$MeOMe + 2H_2 + CO \xrightarrow{Ru(acac)_3} MeCOOMe$$
$$80\%$$

$$MeCOOMe + CO \xrightarrow{Ru(acac)_3,\ I^{\ominus}} MeCOOH + MeCOOEt$$
$$50\% \qquad 50\%$$

Figure 27 Ruthenium catalysed carbonylations.

3.4 Carbonylation of Halides

Carbonylation of saturated, unsaturated and aryl halides has been discussed in several reviews.[181,182] Various products may be obtained as shown in Figure 28. The important process in most cases is the oxidative addition of the halide to a metal complex. This is followed by

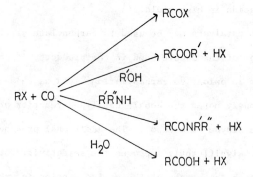

RCOX

RCOOR' + HX

R'OH

RX + CO

R'R"NH

RCONR'R" + HX

H₂O

RCOOH + HX

Figure 28 Pathways in halide carbonylation.

CO insertion and either reductive elimination or nucleophilic attack.

3.4.1 Carbonylation of Aryl Halides

This is one of the easiest and most common of the carbonylation reactions, since oxidative addition is facile. Iodides are the most reactive but bromides are also useful in some cases.[183] Palladium complexes are the most usual catalysts and a general mechanism, proposed by Heck, is shown in Figure 29. A mole of base is required to remove hydrogen halide as it is formed. Some typical results are shown in Figure 30.

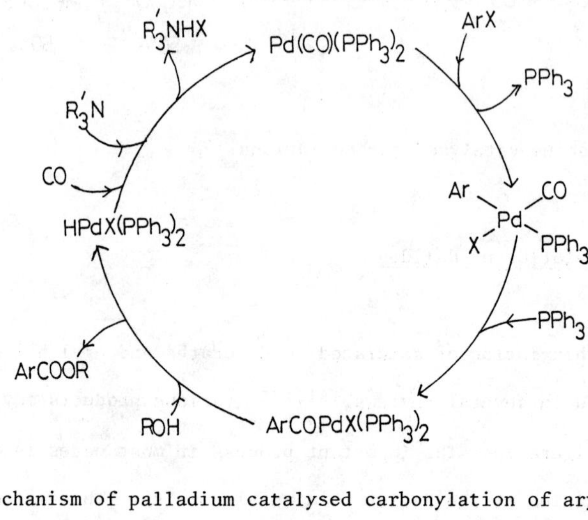

$R_3'NHX$

$Pd(CO)(PPh_3)_2$

ArX

PPh₃

$R_3'N$

CO

Ar   CO

HPdX(PPh₃)₂

X   Pd   PPh₃

PPh₃

ArCOOR

ROH

ArCOPdX(PPh₃)₂

Figure 29 Mechanism of palladium catalysed carbonylation of aryl halides.

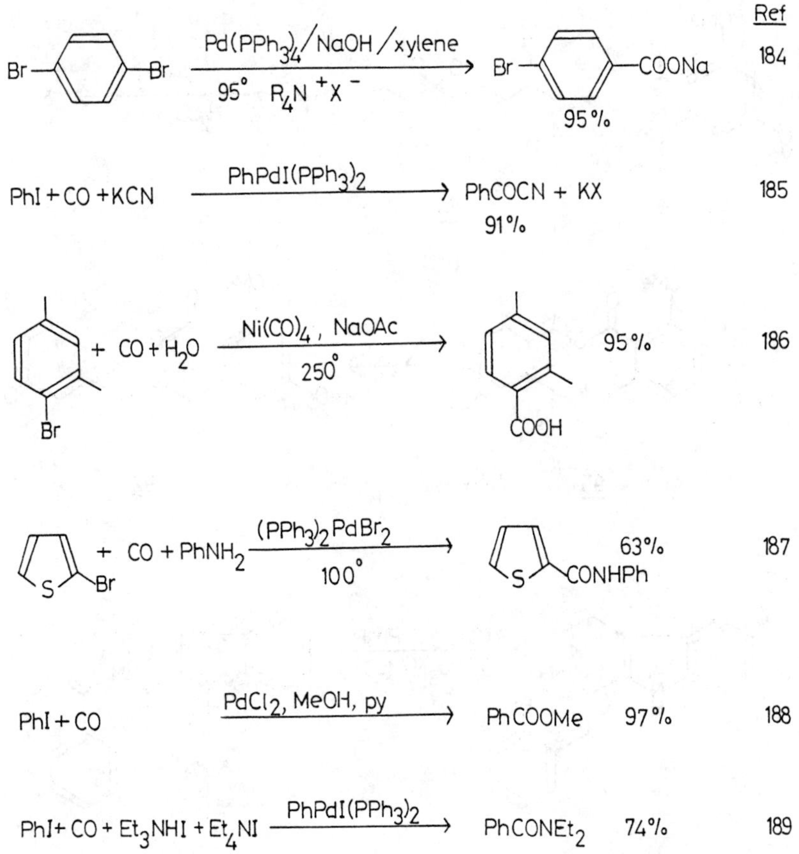

Ref

Br—⟨benzene⟩—Br $\xrightarrow[\text{95}^\circ \ R_4N^+X^-]{\text{Pd(PPh}_3)_4/\text{NaOH}/\text{xylene}}$ Br—⟨benzene⟩—COONa   184

95%

PhI + CO + KCN $\xrightarrow{\text{PhPdI(PPh}_3)_2}$ PhCOCN + KX   185

91%

⟨structure with Br⟩ + CO + H$_2$O $\xrightarrow{\text{Ni(CO)}_4, \text{ NaOAc}}$ ⟨structure with COOH⟩   95%   186

250°

⟨thiophene⟩ + CO + PhNH$_2$ $\xrightarrow[\text{100}^\circ]{\text{(PPh}_3)_2\text{PdBr}_2}$ ⟨thiophene-CONHPh⟩   63%   187

PhI + CO $\xrightarrow{\text{PdCl}_2, \text{MeOH, py}}$ PhCOOMe   97%   188

PhI + CO + Et$_3$NHI + Et$_4$NI $\xrightarrow{\text{PhPdI(PPh}_3)_2}$ PhCONEt$_2$ 74%   189

Figure 30   Examples of palladium catalysed carbonylation of aryl halides.

The efficiency of this reaction has led to its use in several
important synthetic sequences. In the synthesis of the orsellenic acid
type macrolide zearalenone, 92, the critical ester linkage is made by
reaction of 93 and 94 in the presence of CO and palladium chloride.[190]
Cyclisations may give both lactones and lactams directly. For example
cyclisation of 96 gives the yohimbane skeleton.[191] Reaction of 98 under
the same conditions gives a precursor to sendaverin 100[192] while 101
gives 102, a precursor of the important diazepam drugs.[193]

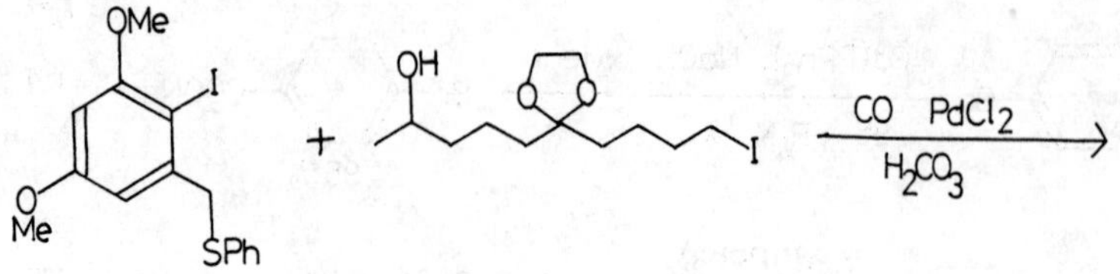
OMe
I
O
Me
SPh
+
OH
O
O
I
CO PdCl2
H2CO3

93 94

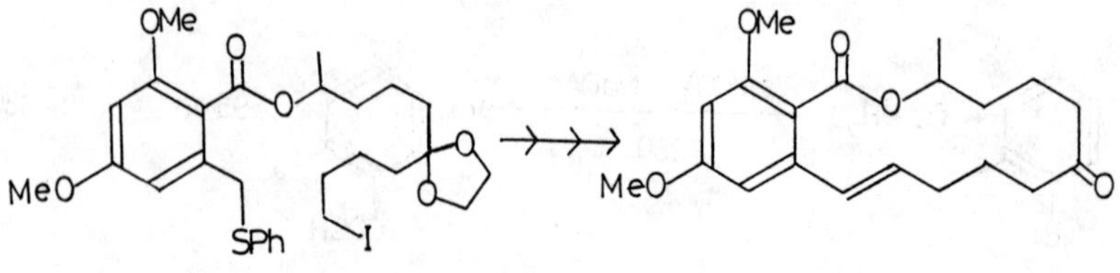
OMe
O
O
O
O
MeO
SPh
I
OMe
O
O
MeO
O

95 70% 92

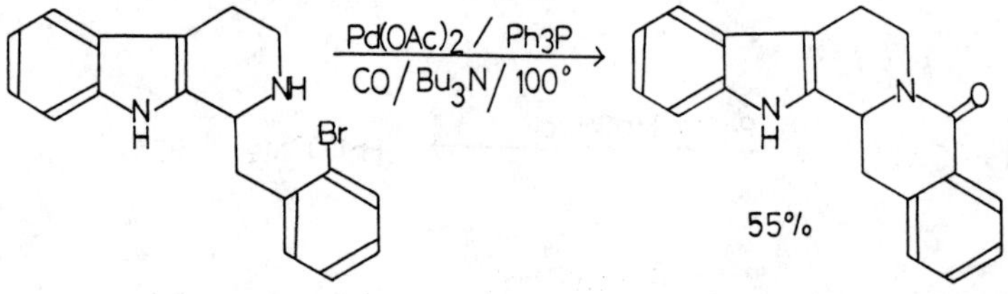
NH
N
H
Br
Pd(OAc)2 / Ph3P
CO/Bu3N/100°
N
H
N
O
55%

96 97

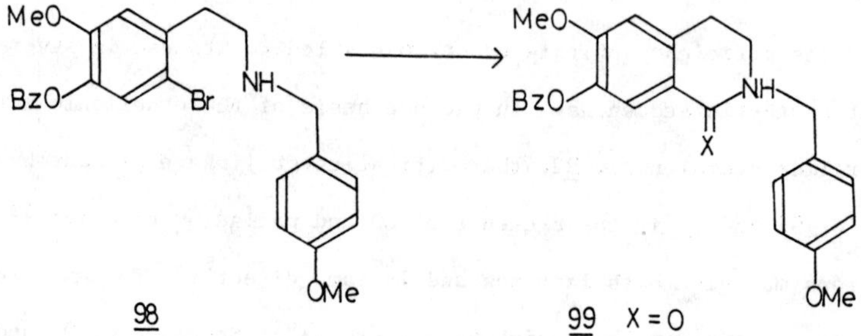
MeO
BzO
Br
NH
OMe
MeO
BzO
NH
X
OMe

98

99 X = O

100 X = H_2

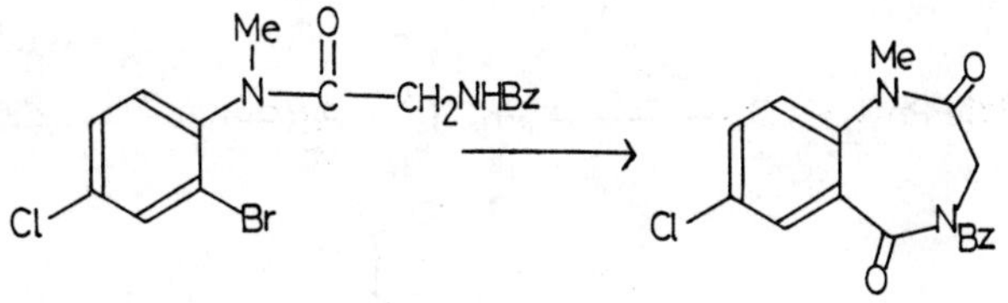

101   102

Lactones such as 104 are equally accessible although yields are
only satisfactory in the presence of a large excess of Ph₃P and when 5-
and 6- membered rings are being formed.[194] It should be noted that these
reactions occur under very much milder conditions than the cyclisations of
ω-amino and ω-hydroxyalkenes; temperatures are moderate and excess CO
pressure is not necessary. An interesting variation on the theme of
oxygen nucleophiles is the use of strained cyclic ethers. The rate of
reaction depends on the degree of ring strain with epoxides being
particularly successful. Figure 31 shows a typical sequence; aryl, vinyl
and benzyl halides are all reactive.[195]

Other useful transformations are achieved when carbon

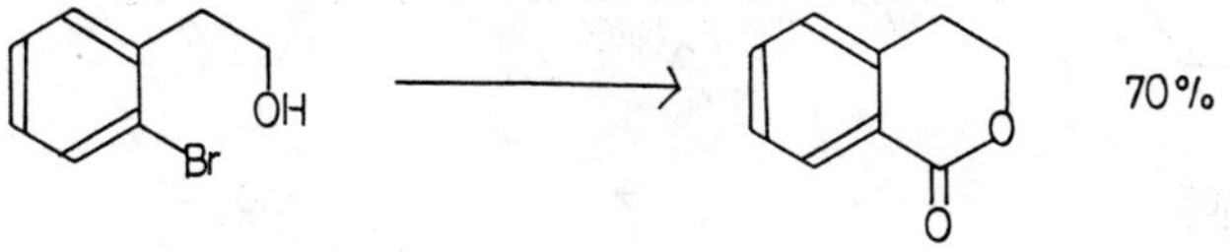

103   104

nucleophiles intercept the intermediate metal acyls. Methyl ketones are
synthesised with tetramethyltin and either palladium[196] or nickel[197]
catalysts, in excellent yield (e.g. 105 to 106). In the presence of base,
even terminal acetylenes will act as nucleophiles. Again yields are
excellent and aryl and vinyl bromides are reactive (e.g. 107 to 108).[198]

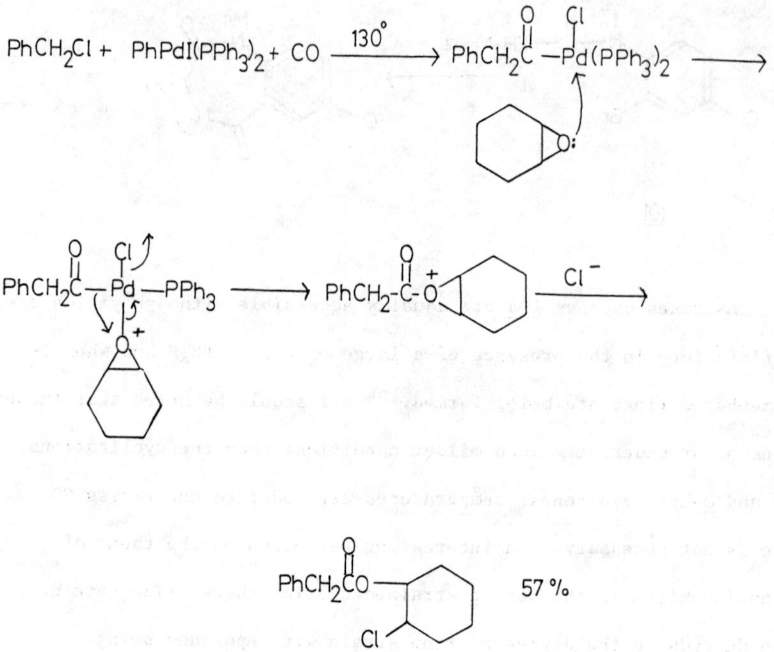

266

PhCH₂Cl + PhPdI(PPh₃)₂ + CO $\xrightarrow{130°}$ PhCH₂C̈—Pd(PPh₃)₂ $\longrightarrow$

PhCH₂C̈—Pd—PPh₃ $\longrightarrow$ PhCH₂-C̈-O⁺ $\quad$ Cl⁻ $\longrightarrow$

PhCH₂CO—⟨cyclohexane ring⟩ $\quad$ 57%
Cl

Figure 31 Carbonylation of halides in the presence of epoxides.

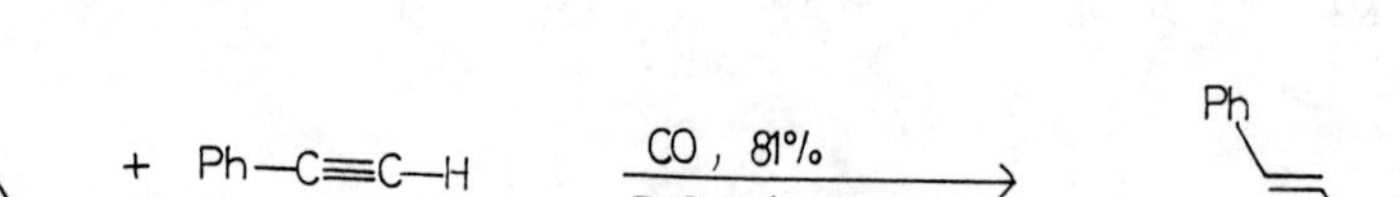

MeO—⟨benzene ring⟩—I + Me₄Sn $\xrightarrow[\text{(Ph₃P)₂Ni(CO)₂}]{\text{CO} \quad 94\%}$ MeO—⟨benzene ring⟩—C(O)CH₃

105 $\qquad\qquad\qquad\qquad\qquad$ 106

Ph—CH=CH—Br + Ph—C≡C—H $\xrightarrow[\text{PdCl₂/}]{\text{CO , 81\%}}$ Ph—CH=CH—C(O)—C≡C—Ph

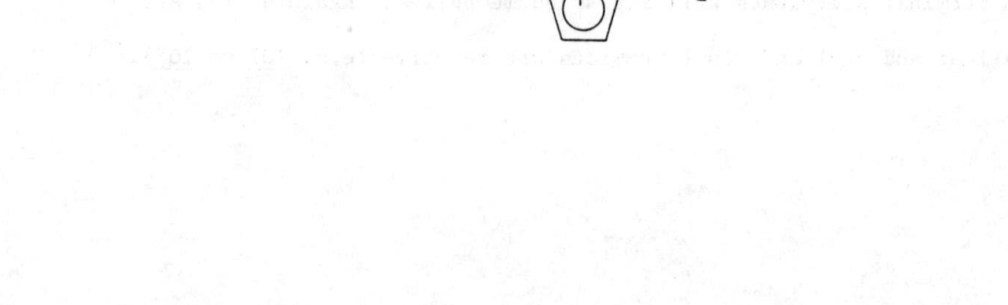

107 $\qquad\qquad\qquad\qquad\qquad\qquad\qquad\qquad$ 108

3.4.2 Vinyl Halides

Vinyl halides show very similar reactions to the aryl compounds but have found fewer applications.[199] Where it is known, stereochemical retention is excellent. Examples are given in Figure 32. The most interesting application has been the synthesis of butenolides accomplished by Stille. 110 is formed by $LiAlH_4$ reductions of 109 in the presence of I_2. Conditions for carbonylation are mild (25°, 1-3 atm. CO) and yields excellent.[202]

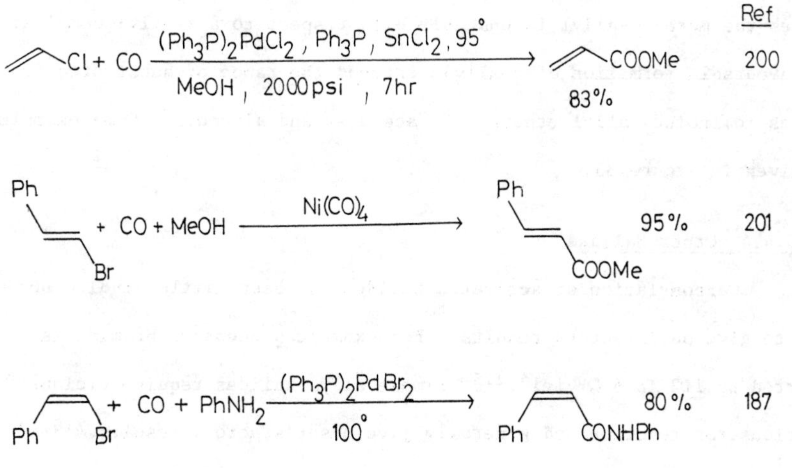

Figure 32 Carbonylation of vinyl halides.

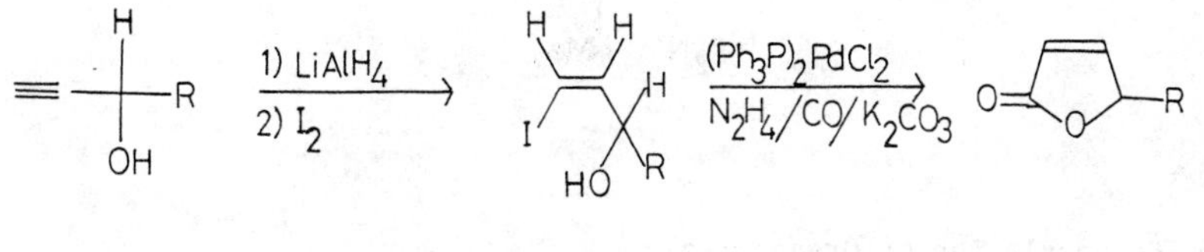

3.4.3 Benzyl and Allyl Halides

While these groups of compounds react with CO under similar conditions the reaction mechanisms are frequently different. Benzyl halide reactions proceed as described for vinyl and aryl derivatives except that oxidative addition is a little slower. Using cobalt catalysts partly alleviates this problem since $Co(CO)_4^{\ominus}$ may be produced in the presence of base and displaces halide (or sulphonate) in an S_N2 reaction.[203] In this way numerous excellent conversions have been reported.[204-208]

While an oxidative addition pathway is available to allyl halides the metal $\underline{\sigma}$-allyl is unstable with respect to a $\underline{\pi}$-allyl complex. The favourable formation of π-allyls extends the range of substrates from halides to include allyl ethers,[209] acetates and alcohols. Some examples are given in Figure 33.

3.4.4 Other Halides

Carbonylation of activated halides has been little studied but seems to give satisfactory results. For example $\underline{\alpha}$-phenacyl bromide is converted to 112 in 6 % yield.[215] Simple alkyl halides require vicious conditions for reaction and generally give unsatisfactory results.[216,217]

3.5 Carbonylation of Organometals

The reaction of carbon dioxide with organolithium compounds or Grignard reagents is well known, but with less electropositive metals the carbon-metal bond is not sufficiently polar to react directly with CO or CO_2. A case in point is the reaction of organomercury compounds, which

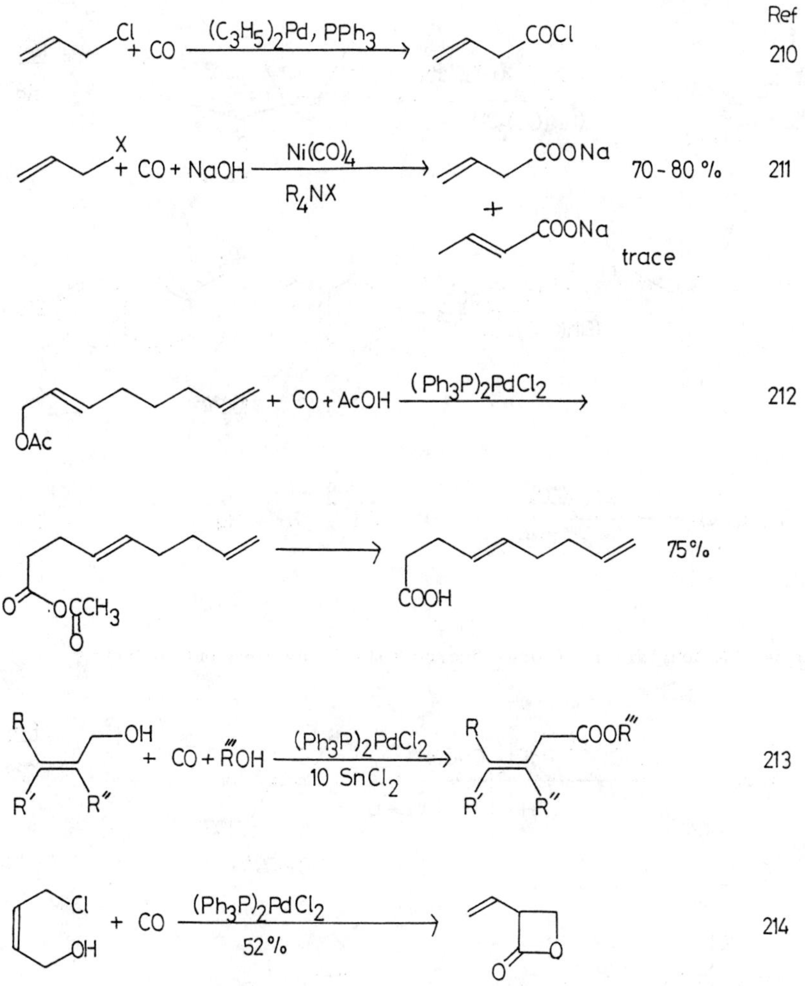

Figure 33 Carbonylation of allyl halides.

have the advantage that they are easily obtained and insensitive to air and

moisture. Two general reaction types are known. In the first a hydro-

carbon solvent is used and ketones are obtained generally in yields greater

than 90% (Figure 34).

When the solvent is an alcohol the products are esters which are

again obtained in excellent yield (Figure 35). The mechanisms of the

Ref 218, 218, 219

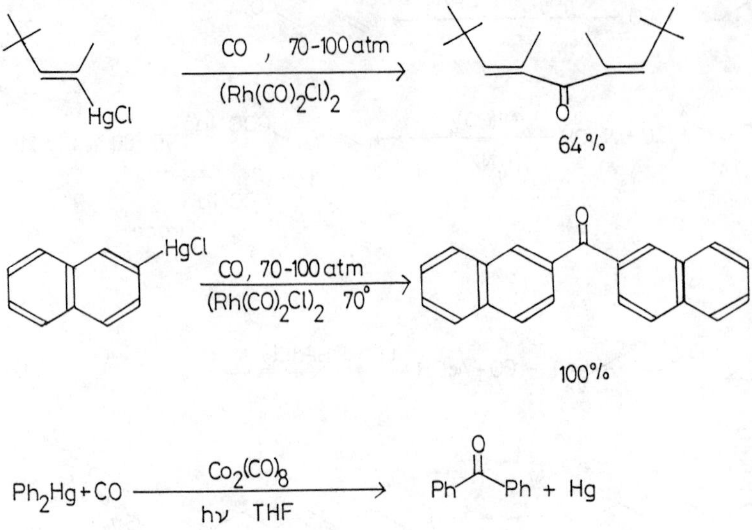

Figure 34 Carbonylation of organomercurials in hydrocarbon solvents.

Ref 220, 221, 222

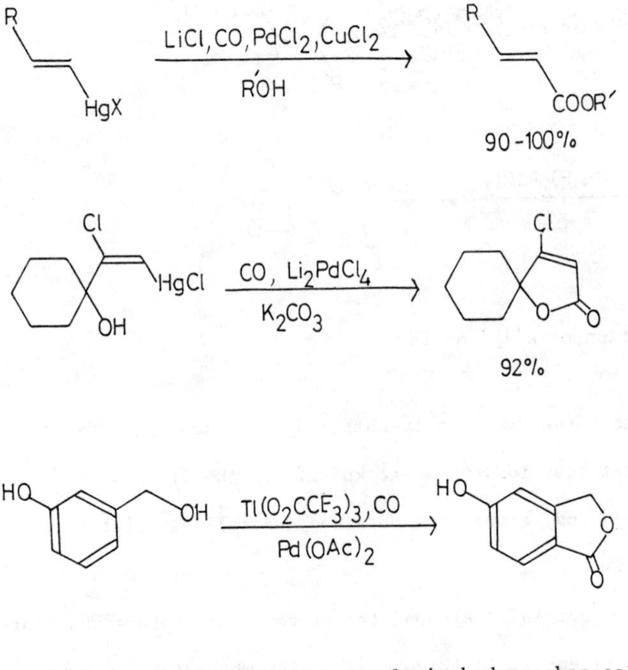

Figure 35 Carbonylation of organometals in hydrocarbon solvents.

rhodium and palladium catalysed reactions are different; RHgX adds oxidatively to Rh(I) complexes and 113 has been isolated as an intermediate.[223] Palladium, however, is thought to undergo exchange with mercury or thallium, carbonylation, and displacement of Pd(0) from an intermediate such as 114. A cooxidant is thus essential for catalytic reaction. Analogously a compound as feebly metallic as a vinyl borane may be carbonylated stereospecifically in yields up to 95%[224] (e.g. 115 to 116).

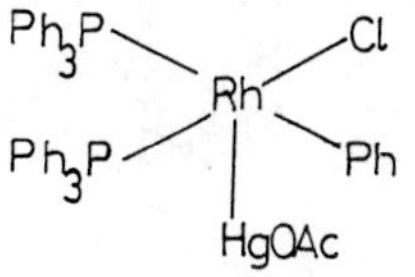

113

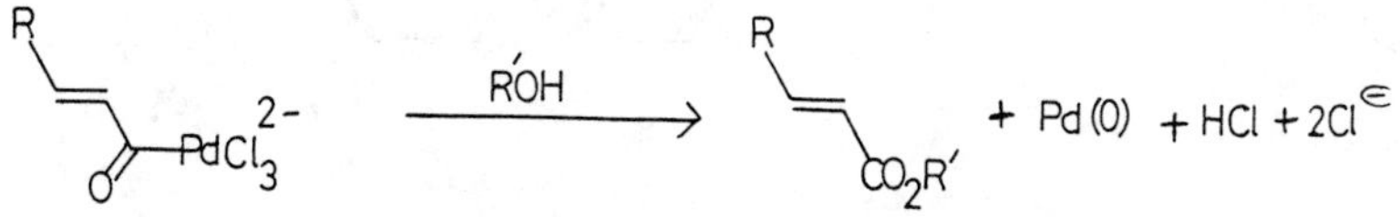

114

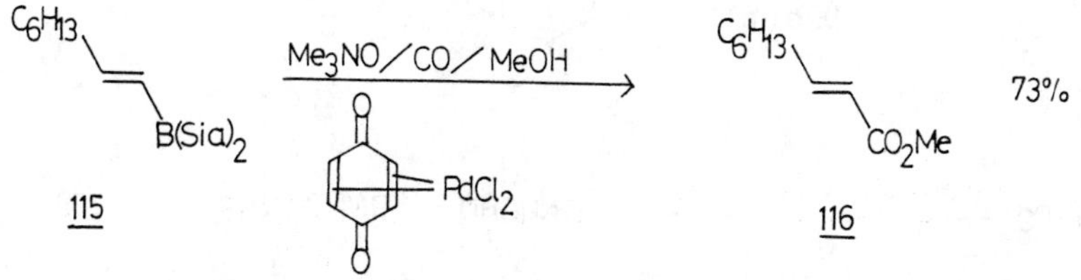

115   116   73%

3.6 Carbonylation of Nitrogen Containing Functional Groups

A number of nitrogen containing organic molecules react with carbon monoxide in the presence of metal catalysts. The reactions are a rather diverse group, which have not been extensively commercially exploited.

3.6.1 Amine Carbonylation

Reaction of a primary or secondary amine with carbon monoxide may yield a formamide or a urea depending on conditions.[225] As a general rule aliphatic amines give N-formyl derivatives, and aromatic amines, ureas, but many exceptions are known. Some examples are given in Figure 36.

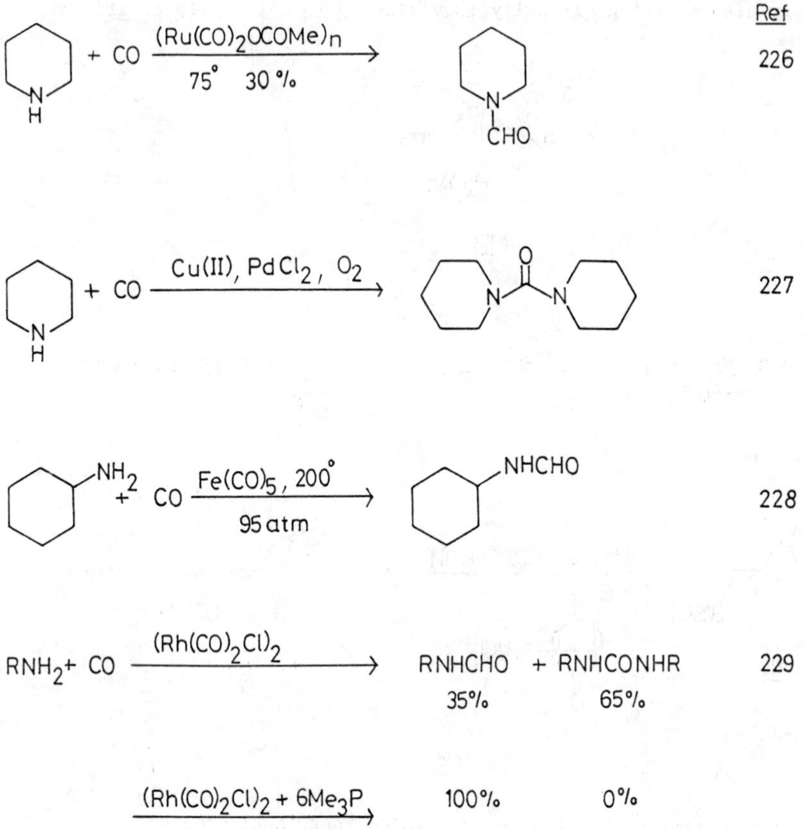

Ref

$+ CO \xrightarrow[75°\quad 30\%]{(Ru(CO)_2OCOMe)_n}$ 226

$+ CO \xrightarrow{Cu(II),\ PdCl_2,\ O_2}$ 227

$+ CO \xrightarrow[95\,atm]{Fe(CO)_5,\ 200°}$ 228

$RNH_2 + CO \xrightarrow{(Rh(CO)_2Cl)_2}$ RNHCHO + RNHCONHR   229

35%   65%

$\xrightarrow{(Rh(CO)_2Cl)_2 + 6Me_3P}$ 100%   0%

Figure 36 Carbonylation of amines.

3.6.2 Carbonylation of Aromatic Nitro Compounds

This process yields aromatic isocyanates, which are used in the commercial manufacture of polyurethanes. Interest in it has been mainly industrial and conditions of reaction tend to be severe. Both palladium and rhodium complexes are active catalysts as the reactions of 117[230] and 118[231] show. Little definitive mechanistic information is available but speculation has centered on nitrenes[232] and their metal complexes (such as 119) as intermediates.[233]

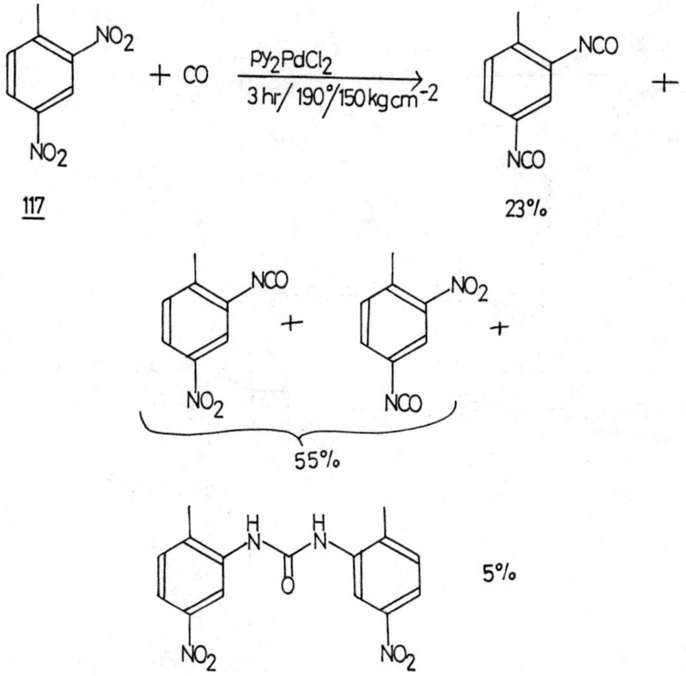

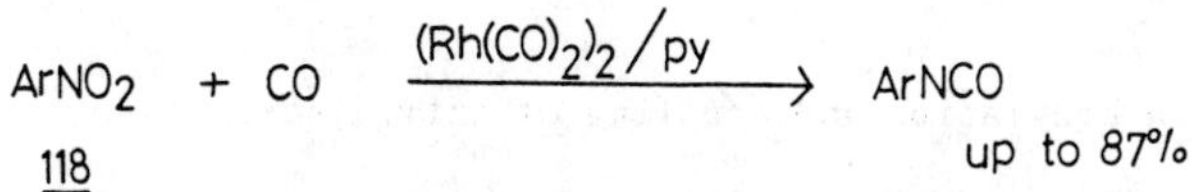

3.6.3 Other Nitrogen Containing Substrates

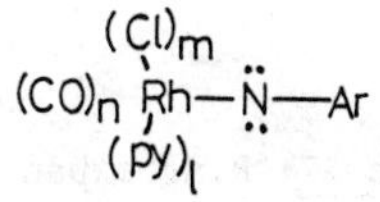

Other substrates have received little investigation. The best understood is probably the ring expansion and coupling reactions of aziridines as exemplified in Figure 37.[234] Aryl diazonium compounds, 120, may be converted to anhydrides,[235] but more interestingly to aryl ketones, 121. The reaction succeeds with 3- and 4-nitrodiazonium salts in contrast to Friedel-Crafts acylation reactions.[236]

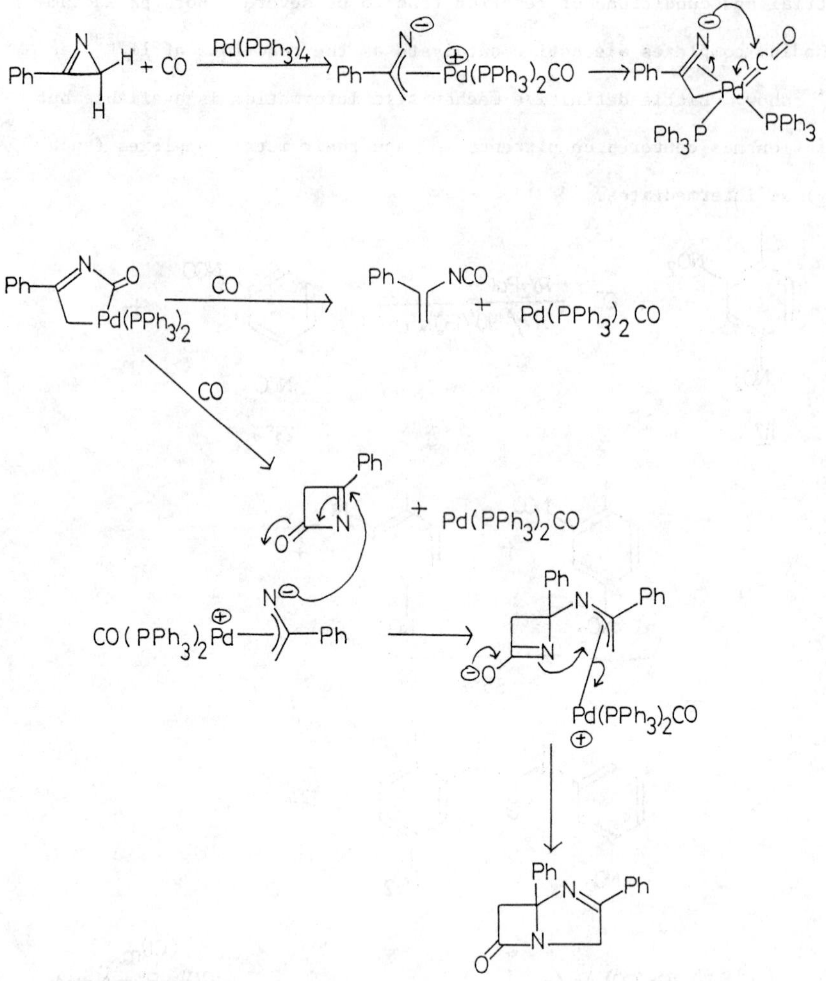

Figure 37 Ring expansion, carbonylation and coupling of aziridines.

$$ArN_2^+BF_4^- + CO + R_4Sn \xrightarrow{Pd(OAc)_2} ArCOR \quad 70\text{-}90\%$$

<u>120</u>   <u>121</u>

3.7. Miscellaneous Carbonylations

Small strained rings, both carbo- and heterocyclic, may be opened in the presence of transition metals. This reaction may be combined with carbonylation to give ketones from cyclopropanes. (e.g. reactions of 122[237,238] and 123[239]). Epoxides and oxetanes may undergo carbonylation or carboxylation as Figure 38 shows.

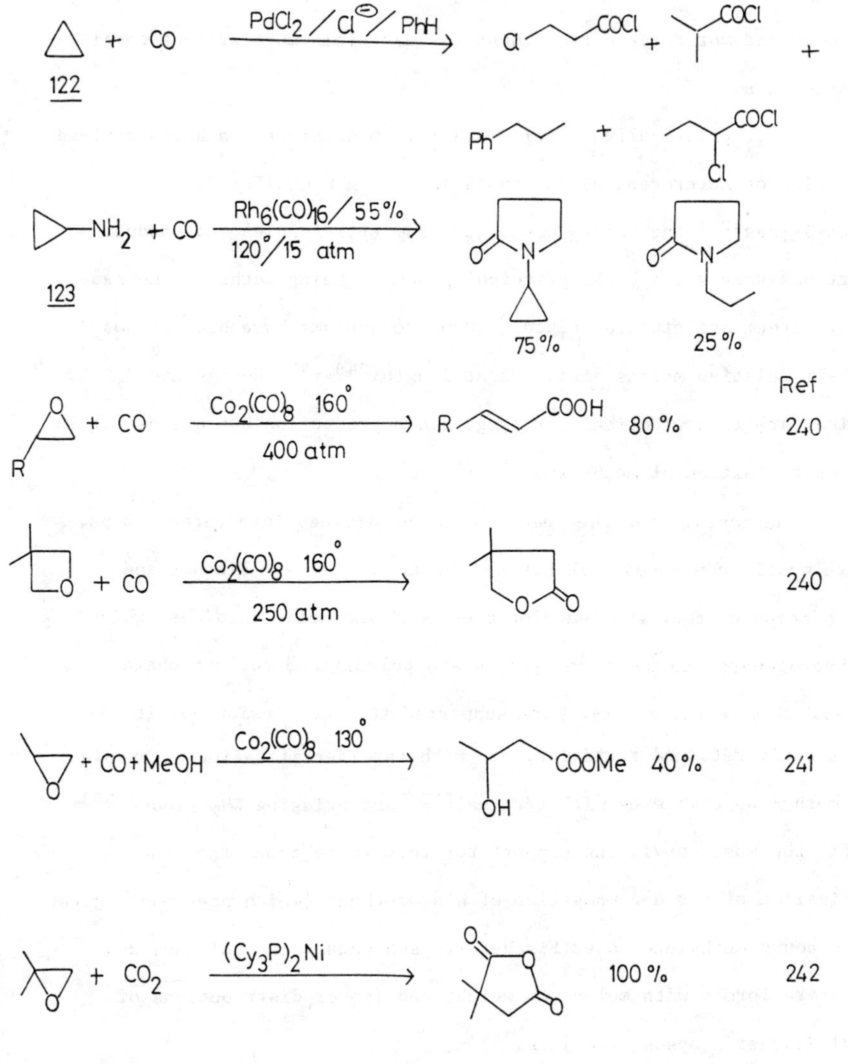

Figure 38 Carbonylation and carboxylation of epoxides and oxetanes.

3.8 Fischer Tropsch Reactions: The Hydrogenation of Carbon Monoxide

Hydrocarbons and alcohols may be produced catalytically from synthesis gas at or above atmospheric pressure at a few hundred degrees Celsius. The first synthesis of methane in this way was in 1913[243] and although the process enjoyed some vogue in the 1930's the only commercial plants currently in operation are in South Africa.[244] In the current era of energy shortage, however, any process which can produce chemical feedstocks and motor fuels from cheap raw materials must attract considerable attention.

Most practically useful Fischer Tropsch processes have involved a diversity of heterogeneous catalysts including $K_2CO_3|Fe$[245], ZnO, $Cr_2O_3|base$[246] and $Ca[Ir_3Rh_9(CO)_{30}]$.[247] Efficiencies and product proportions vary widely, the principal products being methanol, methane, dimethyl ether and ethylene glycol. Many mechanisms have been proposed and their relative merits discussed at length.[248,249] Homogeneous reactions are less practical for large scale production but are useful in terms of definition of mechanism.[250,251]

Mechanisms for the reaction may be divided into three groups, and are mostly concerned with heterogeneous processes. Fischer and Tropsch proposed that the reaction proceeded via metal carbides, which were hydrogenated to methylene groups and polymerised to long chain species. Some early kinetic work supported this conclusion but it has only recently returned to favour,[251] with the identification of stable metal methylene complexes with terminal[252] and bridging CH_2 groups.[253] Probably the most convincing support for this route comes from the investigation of the decomposition of diazomethane (which presumably gives surface bound methylene) over Fischer Tropsch catalysts. Alkanes and alkenes are formed with molecular weight and isomer distributions of typical Fischer Tropsch reactions.[254]

A second theory explained the formation of oxygenated products by making a hydroxycarbene intermediate, 125, responsible for carbon-carbon bond formation.[255,256] Termination takes place by reductive elimination from 127 or 128. This mechanism has few specific organometallic precedents and is probably incorrect in its explanation of chain growth, but carbenes may well be reasonable intermediates. The route of Pichler and Schultz[257] also involves these species, this time produced via CO insertion and subsequent reduction (129 to 134). Their scheme accounts in detail for all the observed products and these ideas have been incorporated into several subsequent models.[258,259]

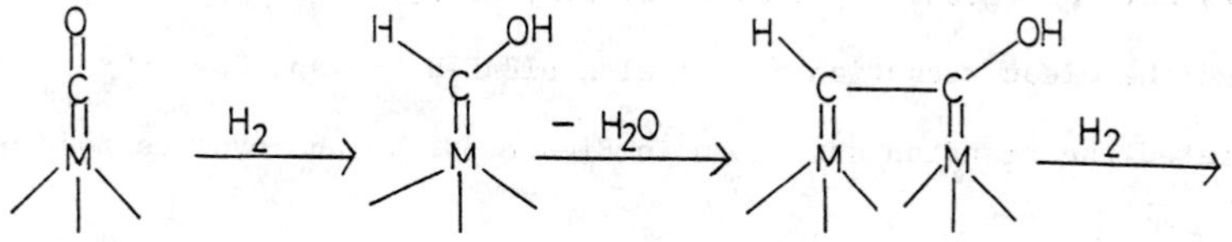

124

125

126

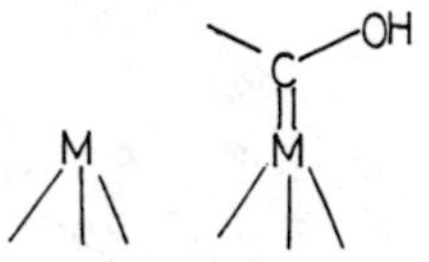

127

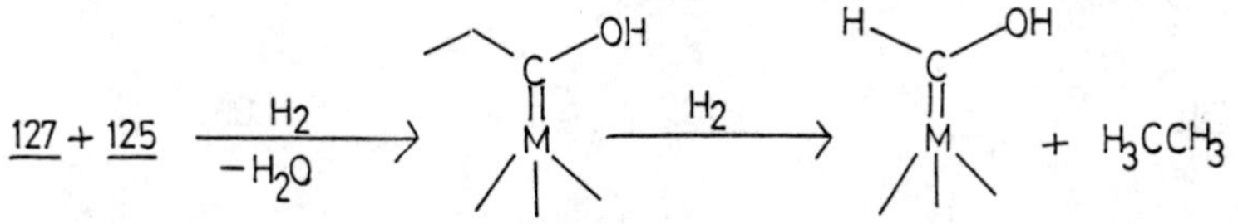

127 + 125

128

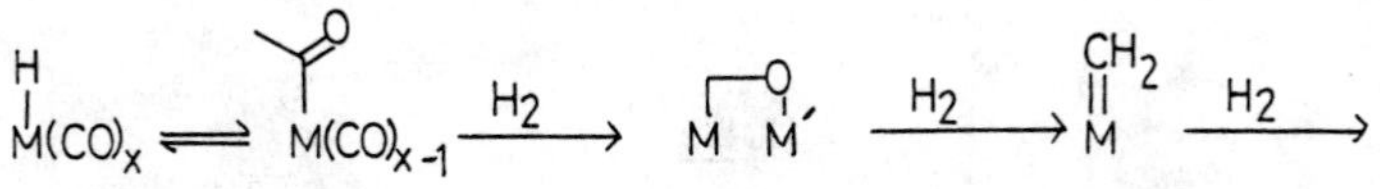

129

130

131

132

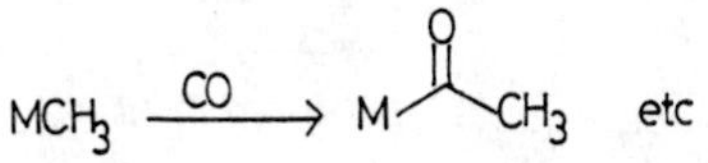

133

134

The final route involves the stepwise hydrogenation of a metal coordinated carbon monoxide. The formation of the first hydrogen carbon bond by insertion into a metal hydride is crucial, and without precedent. Insertion of CO into a metal alkyl is well known and metal formyl complexes are well documented,[260-262] if rather prone to decarbonylation.[263] An alternative suggested formulation of this intermediate is 135.[262] This type of bonding has been demonstrated in the acyl complexes $Cp_2ZrCOCH_3(CH_3)$ and $Cp_2TiCOCH_3(Cl)$[264] and the species is implicated in the first well documented reduction of CO by molecular hydrogen (136 to 140). Formyl complexes may, on reduction, lead to hydroxy methyl derivatives. Casey described the clean reduction of 141 with DIBAL-H[263] and Sweet[265] demonstrated the reaction set shown in Figure 39, which involves most of

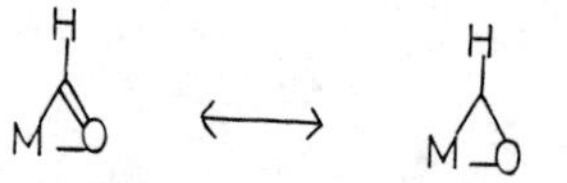

135a   135b

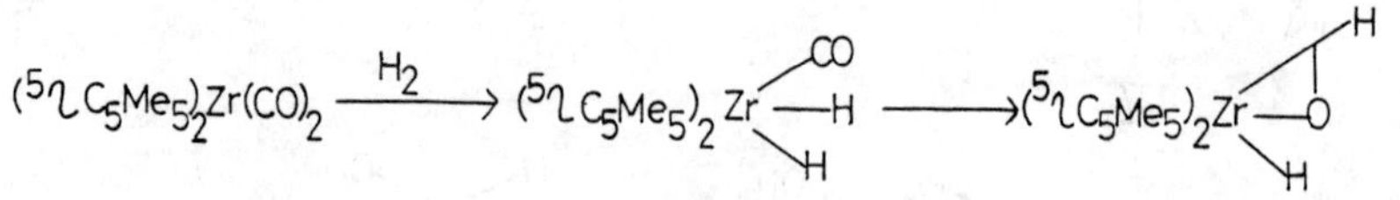

136   137   138

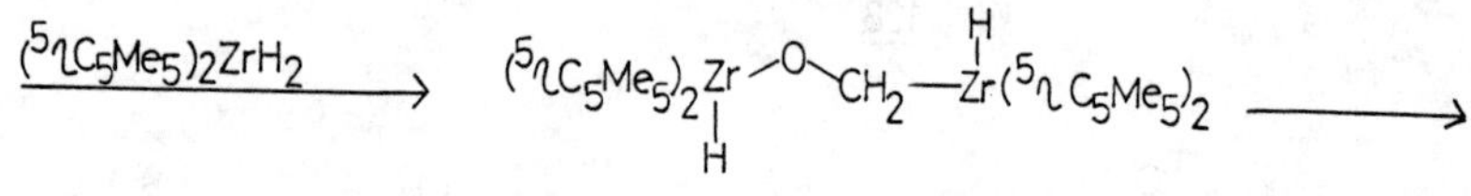

139

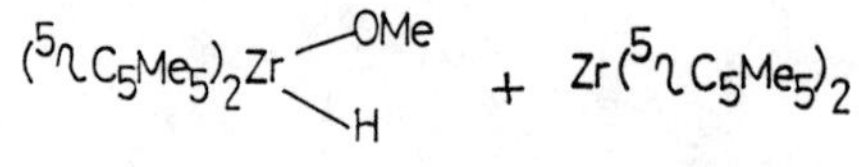

140

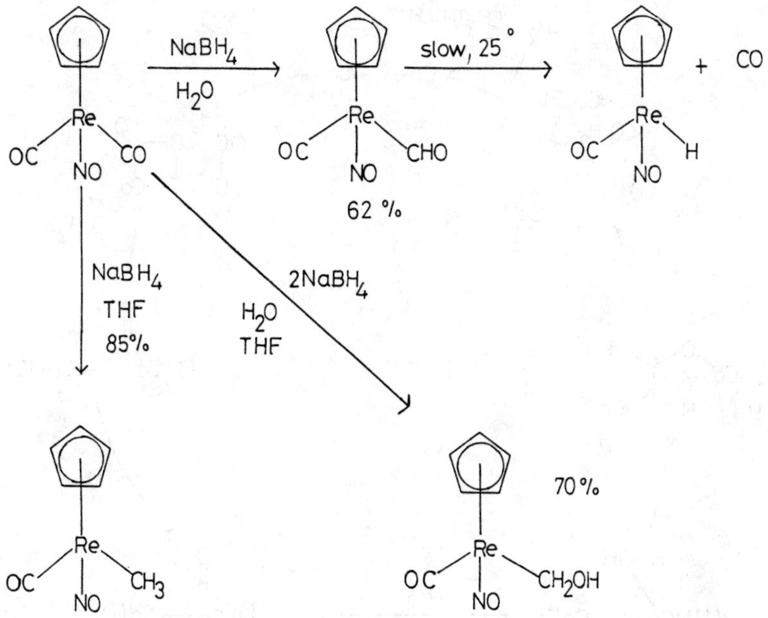

Figure 39   Reductions of coordinated carbon monoxide as a model for the

Fischer Tropsch reaction.

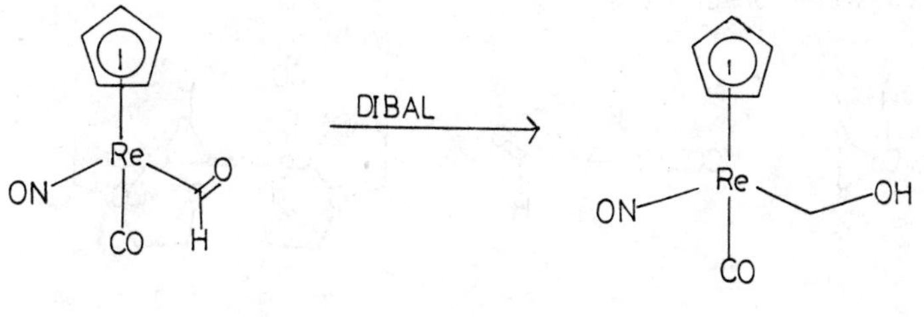

141   142

the proposed Fischer Tropsch intermediate.

There is little evidence for the true mechanism of the few homogeneous reactions known. Multimetallic catalysts are so common, however, that intermediates with more than one metal are likely to be important. An illustrative proposal is due to Wilkinson and is shown in Figure 40.[266]

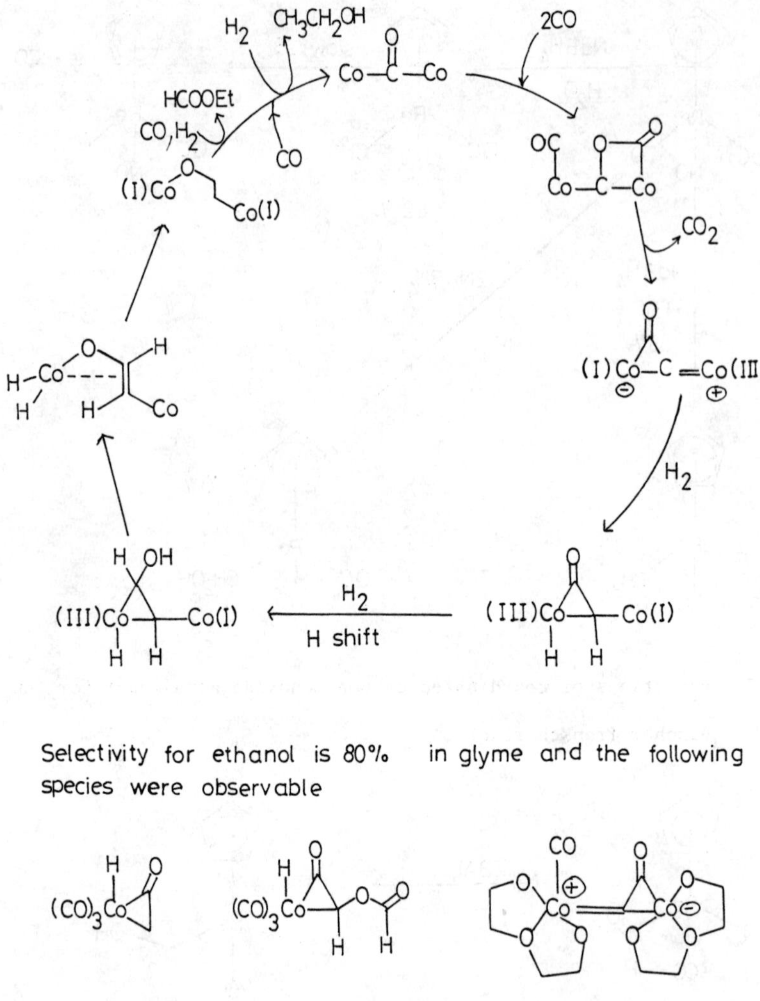

H_2
CH_3CH_2OH
2CO
HCOOEt
CO, H_2
CO
(I)Co
Co(I)
CO_2
(I) Co—C =Co(III)
H_2
(III)Co—Co(I)
H shift
Selectivity for ethanol is 80% in glyme and the following species were observable
$(CO)_3Co$

Figure 40 Proposed mechanism of a homogeneous Fischer Tropsch reaction.

3.9 Decarbonylation

Removal of CO from carbonyl compounds may be considered as the formal reverse of its addition. Only for aldehydes and acyl halides is it at all general;[267] the first required step is oxidative addition to a metal complex (143 to 144) which is difficult for ketones or esters, such reactions as are known for these substrates generally proceeding by other

$$RCOQ + ML_n \longrightarrow R-\overset{\overset{\displaystyle O}{\|}}{C}-M\overset{Q}{\underset{L_n}{\diagdown}} \longrightarrow R-\overset{\overset{\displaystyle Q}{|}}{\underset{\underset{\displaystyle CO}{|}}{M}}-L_n \longrightarrow RQ + ML_nCO$$

<u>143</u>   <u>144</u>   <u>145</u>   <u>146</u>   <u>147</u>

routes. The R group then migrates from CO to metal (as in <u>145</u>) and the cycle is completed by the elimination of RQ, <u>146</u>.

3.9.1 Decarbonylation of Acyl Halides

Both catalytic and stoichiometric reactions are known using Rh(I) complexes. In the stoichiometric reaction several intermediates are isolable and give useful information as to the reaction mechanism. A rhodium acyl such as <u>148</u>[268] is initially formed and several such have been isolated as crystalline solids.[269, 270] X-ray crystallography and [31]P n.m.r.[268] shows that they are rigid trigonal bipyramidal structures. Reaction with an acyl halide labelled with [36]Cl gives scrambling between complex and starting material, showing that addition is rapidly reversible. Optically active halides give chiral acyls such as <u>149</u>, but on heating racemisation in the acyl $\rightleftharpoons$ alkyl rearrangement step gives achiral product. Metal alkyls are less stable and few have been directly observed. The ethyl derivative <u>151</u> may be prepared by addition of

<u>148</u>   <u>149</u>

$$(Ph_3P)_2RhHCl_2 + C_2H_4 \longrightarrow (Ph_3P)_2RhEtCl_2 \overset{CO}{\longrightarrow} Et\overset{\overset{\displaystyle O}{\|}}{C}RhCl_2(Ph_3P)_2$$

<u>150</u>   <u>151</u>   <u>152</u>

ethylene to a rhodium hydride; carbonylation gives the propionyl derivative 152, stable at low temperature.[271] With $IrClCO(Ph_3P)_2$ aroyl chlorides are not decarbonylated but $ArCIrCl_2CO(Ph_3P)_2$ is observed as a transient in the stoichiometric production of $ArIrCl_2CO(Ph_3P)_2$.[272] Thus for stoichiometric decarbonylation by $(Ph_3P)_3RhCl$ we may write the scheme of Figure 41 with confidence. In order for reaction to occur catalytically an analogous oxidative addition must occur to 156, with loss of CO. In this sequence, however, intermediates have not been studied.

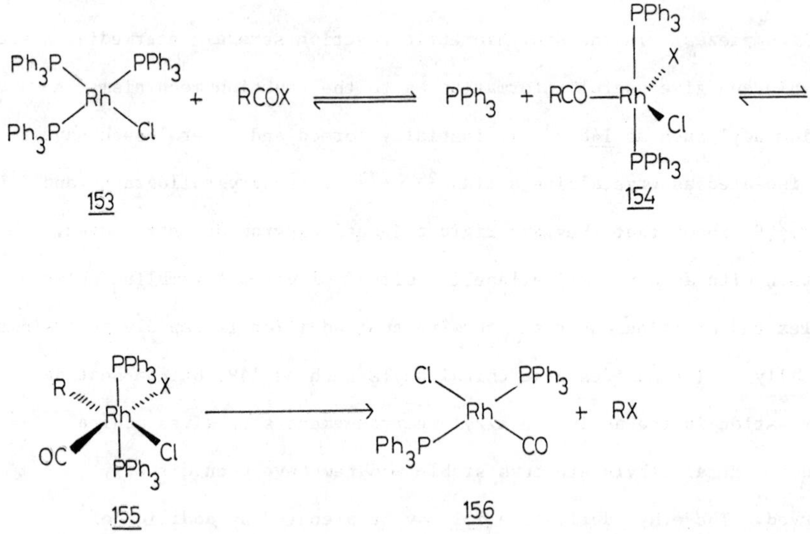

Figure 41 Mechanism of stoichiometric decarbonylation of halides by Wilkinson's catalyst.

A wide range of acyl halides are reactive. Iodides are decarbonylated faster than bromides or chlorides, but even fluorides react slowly[273] and there is one example of an acyl phosphide being decarbonylated stoichiometrically to the corresponding phosphine.[274] Halides are obtained cleanly and in high yield where there is no hydrogen β to the acyl group (Figure 42). However, when a β-hydrogen is available

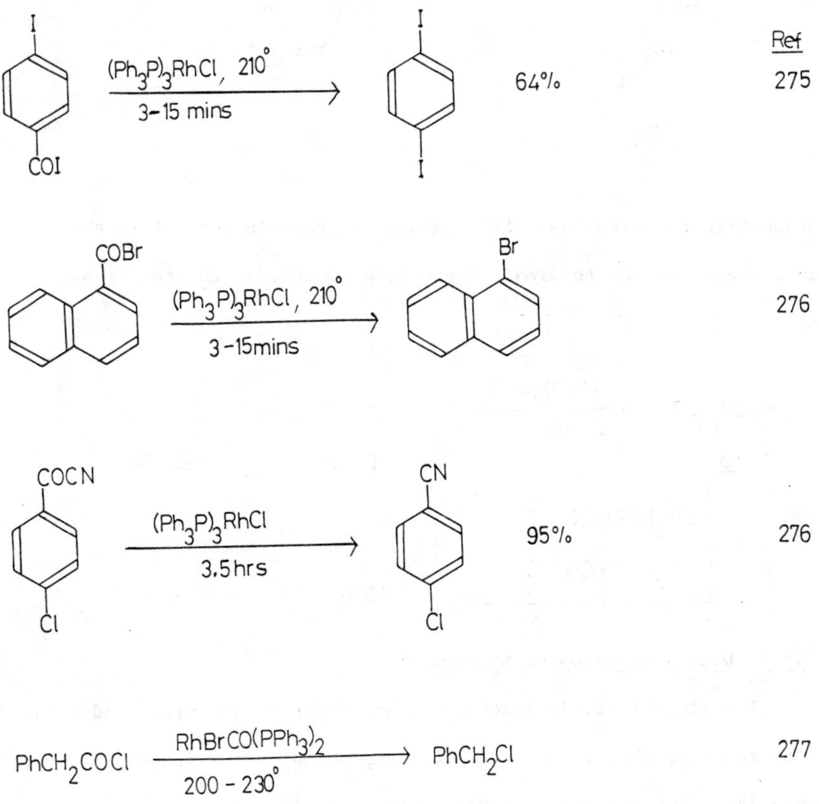
Ref
COI
(Ph3P)3RhCl, 210°
3-15 mins
64%
275
COBr
Br
(Ph3P)3RhCl, 210°
3-15mins
276
COCN
CN
(Ph3P)3RhCl
3.5hrs
95%
276
Cl
PhCH2COCl
RhBrCO(PPh3)2
200 - 230°
PhCH2Cl
277

Figure 42 Examples of the catalytic decarbonylation of acyl halides.

alkenes rather than halides are obtained, with rather variable selectivity When $CH_3(CH_2)_8COCl$ is treated with $RhCl_3$ or $PdCl_2$ a mixture of nonenes is obtained.[278] The reaction has been studied in detail by Stille and his coworkers.[268,270,279] Erythro- and threo- 2,3-diphenylbutanoyl chloride gave E- and Z - 157 respectively, consistent with retention of stereochemistry in acyl $\rightleftharpoons$ alkyl rearrangement and cis- β - H elimination. The alternative path of rate determining cis-elimination, without an alkyl intermediate, is also consistent with this observation and with the high deuterium isotope effect observed (K_H/K_D = 7 for $C_5D_5CD_2CH_2COCl$). Selectivity toward a particular alkene is potentially high; in the

157a   157b

stoichiometric reaction Saytzeff elimination predominates but in the
catalytic reaction the terminal alkene is produced, as the reactions of
158 show.

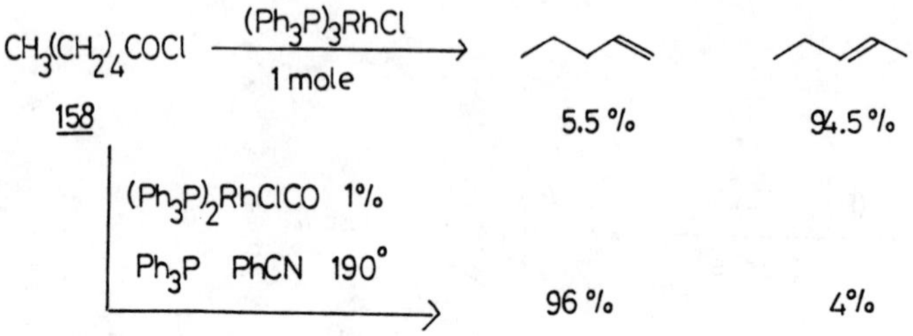

3.9.2 Decarbonylation of Aldehydes

The stoichiometric reaction occurs under rather mild conditions
in the presence of $(Ph_3P)_3RhCl$, the rate depending on the degree of steric
hindrance.[269]   The reaction has been most useful for the removal of
formyl groups from steroids and sugars (e.g. the reactions of 159[280] and
161[281]). It is thought to proceed in a manner analogous to that for acyl
halides, but the metal acyl hydride is not normally sufficiently stable to
be observed. An exception is provided by 163, which is stabilised by
chelation[282] and an iridium analogue, 167, may be isolated in high yield.
On heating decarbonylation occurs and 165 is regenerated.[283] An analogous
species, 169, is invoked to explain the cyclisation of 168 in the presence
of a rhodium catalyst, and the insertion of ethylene into the acyl–Rh or
alkyl–Rh bond.[284] The reaction products are generally saturated
hydrocarbons, alkenes being only a minor component. This may reflect the
greater ease with which reductive elimination occurs as the reaction of
172 shows.[285] Stereochemical control has been tested both with αβ–

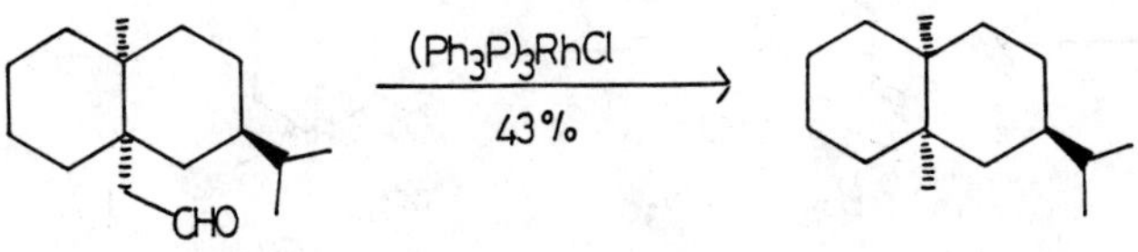

159   160

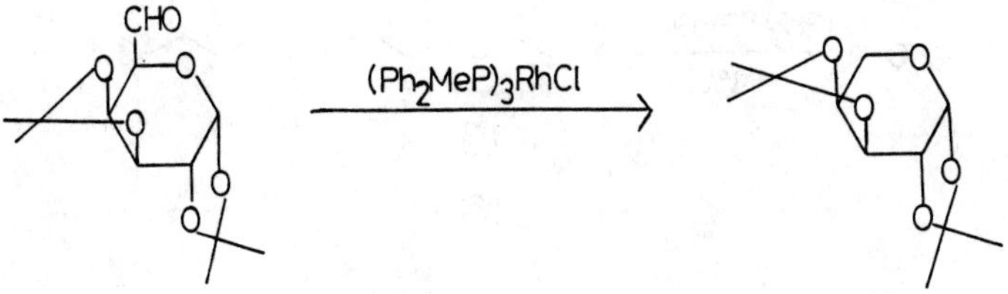

161   162

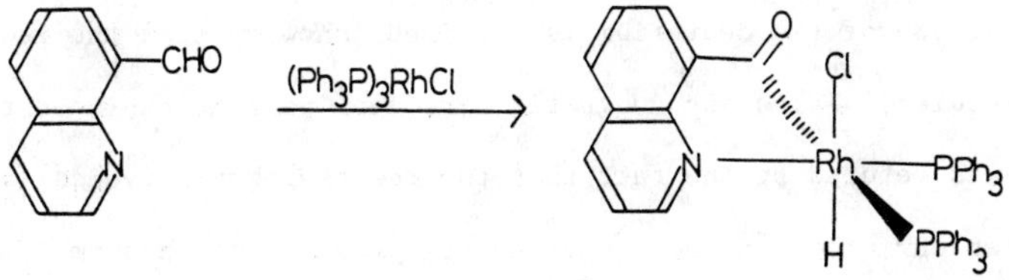

163

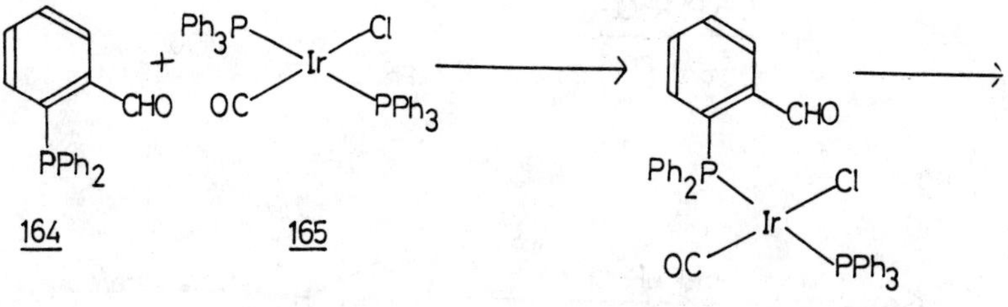

164   165   166

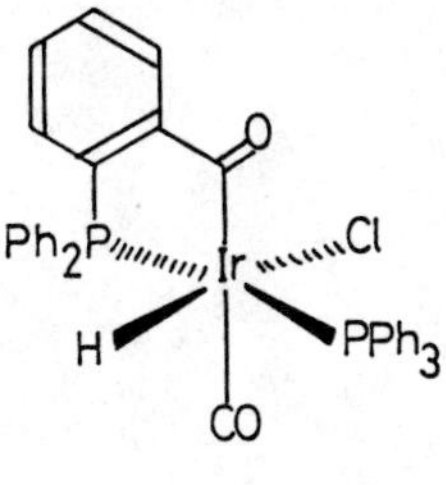

167

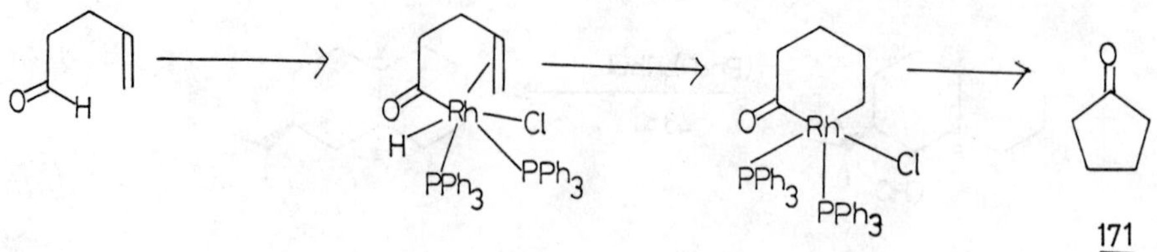

168   169   170   171

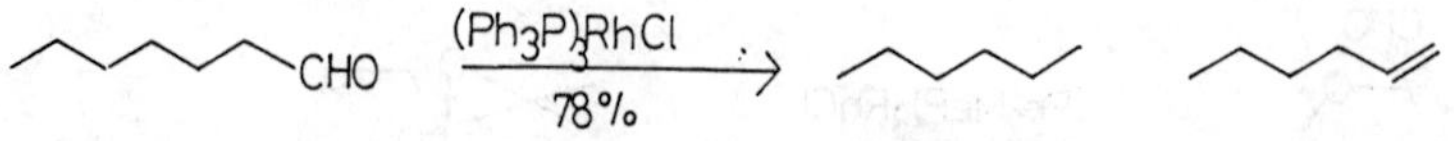

172   173   174
                                 86%   14%

unsaturated aldehydes, such as 175, and chiral aldehydes, such as 177; in
both cases the stereochemistry is largely retained.[286] With deuterated
aldehyde 179, 93% of the deuterium is retained, showing that the reaction
is intramolecular. Walborsky suggested a radical pair mechanism but the
contention was refuted by the fact that the rearrangements typical of
radical decarbonylations do not occur in the presence of rhodium
complexes[287] (e.g. the reaction of 181).

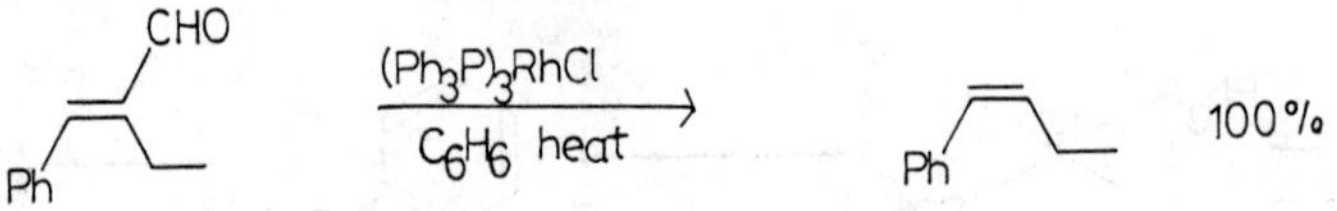

175   176   100%

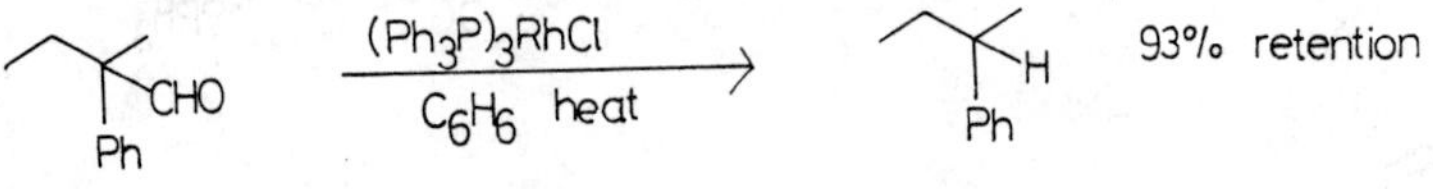

177   178   93% retention

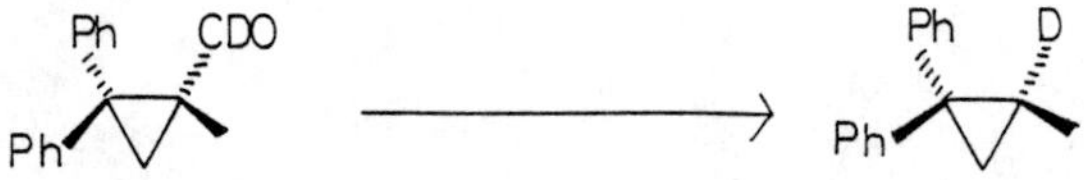

179   180

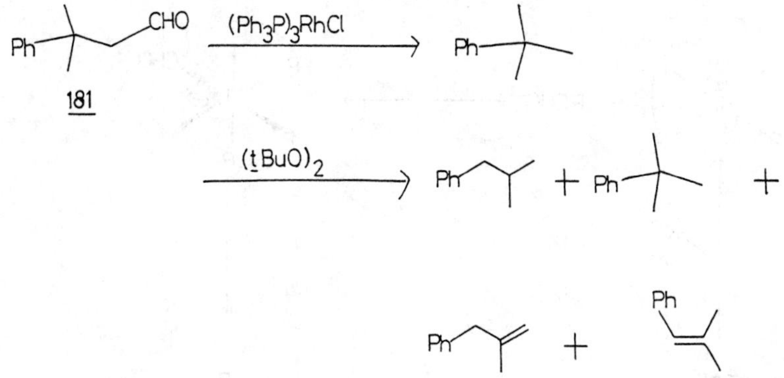

181

The difficulty in achieving catalytic decarbonylation is that
the product of the stoichiometric reaction, trans-$(Ph_3P)_2RhCl(CO)$ does not
lose CO under mild conditions, so the catalyst is not regenerated. For
high boiling aldehydes the reaction may be made catalytic above 180° but
it is thought that carbon monoxide loss is directly from the metal
acyl,[276,285,288] and the reaction is confined to aryl aldehydes. Rhodium
cations, however, bind CO much more weakly than neutral species and
$[Rh(diphos)_2]^{(+)}$ does not react significantly with CO at 1 atm. Thus,
cations might be expected to follow the cycle of Figure 43 where CO loss is
not the rate determining step. The reaction is successful for
$[Rh(diphos)_2]^{(+)}$, $[Rh(Ph_2P(CH_2)_3PPh_2)_2]^{(+)}$ and still better for
$[Rh(Ph_2PCH_2CH_2AsPh_2)_2]^{(+)}$, the conditions are mild ($<$100°C) and the catalyst
is stable for more than 100,000 turnovers.[289] Like its stoichiometric
counterpart it is sensitive to steric effects and also gives complete
retention of stereochemistry in the decarbonylation of R - 182.[290]

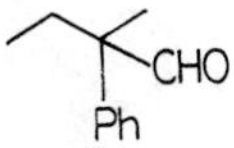

182

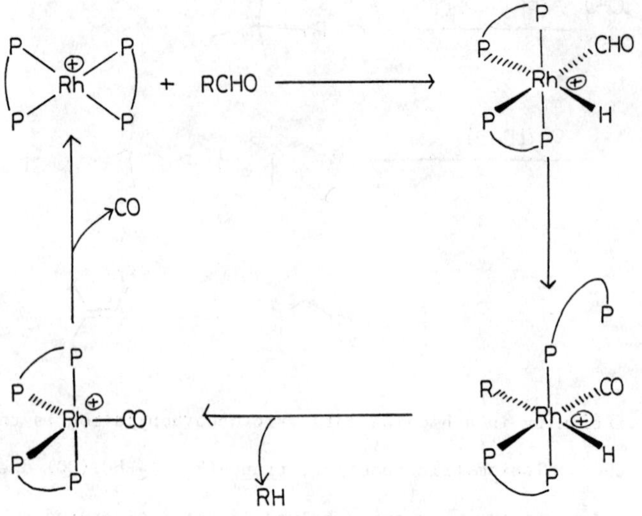

Figure 43 Proposed cycle for aldehyde decarbonylation in the presence of
cationic rhodium complexes.

3.9.3 Decarbonylation of Other Substrates

The decarbonylation of allyl alcohols is readily achieved since
they may be converted into aldehydes in the presence of $(Ph_3P)_3RhCl$.
Hydrocarbons are the major products[291]   (e.g. 183 to 184 and 185).

As previously noted, decarbonylation of other carbonyl compounds
is difficult, principally because they do not readily undergo oxidative
addition to metals. Some examples are shown in Figure 44. The mechanisms
are poorly defined but it is noteworthy that all the substrates have an
additional binding site close to the carbonyl which is to be lost.

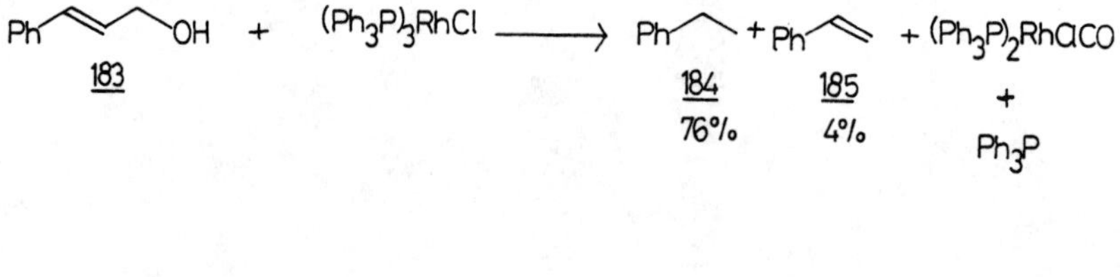

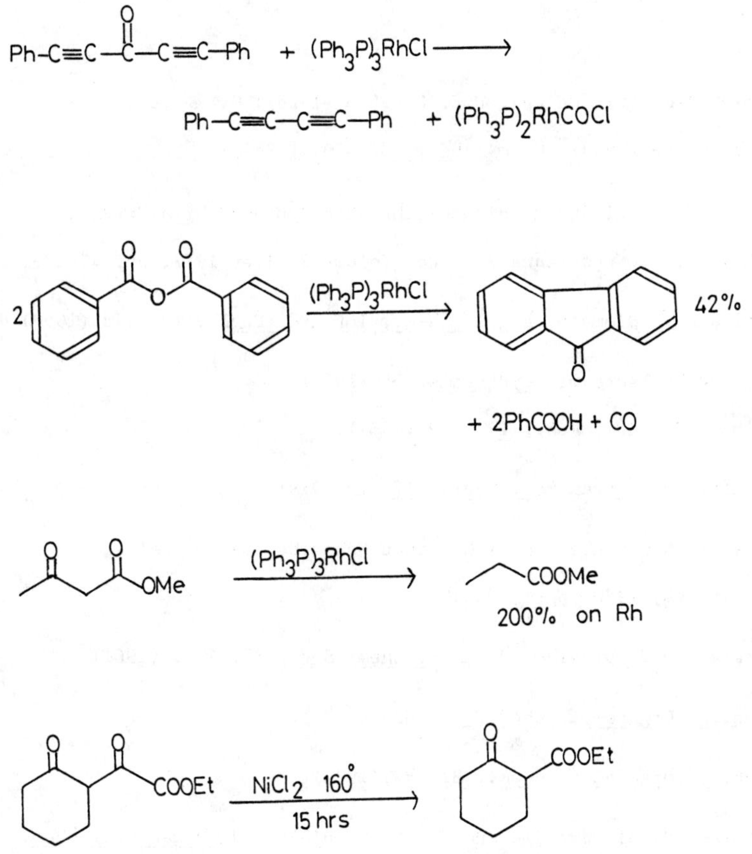

Ph–C≡C–$\overset{\displaystyle O}{\overset{\|}{C}}$–C≡C–Ph + (Ph₃P)₃RhCl ⟶

Ph–C≡C–C≡C–Ph   + (Ph₃P)₂RhCOCl

2 [benzoic anhydride structure] $\xrightarrow{(Ph_3P)_3RhCl}$ [fluorenone structure] 42%   293

+ 2PhCOOH + CO

[methyl acetoacetate structure] $\xrightarrow{(Ph_3P)_3RhCl}$ [structure]COOMe   294

200% on Rh

[2-(ethoxycarbonyl)cyclohexanone structure]COOEt $\xrightarrow[15\ hrs]{NiCl_2 \quad 160°}$ [cyclohexanone structure]COOEt   295

Figure 44   Decarbonylation of miscellaneous carbonyl compounds,

References

1   Ruhrchemie, A.G. (O. Roelen), D.E. 849.548 (1938); O. Roelen, U.S. Patent 2,327,066 (1943); Chem. Abs., 38, 550₁ (1944).

2   P. Pino, F. Piacenti and M. Bianchi, Organic Syntheses via Metal Carbonyls, Ed. I. Wender and P. Pino, Volume 2 Wiley 1977, pp. 43-233.

3   H. Siegel and W. Himmele, Angew. Chem., Int. Ed. Engl., 19, 178 (1980).

4   M. Ryang and S. Tsutsumi, Synthesis, 55 (1971).

5   M. Orchin, Acc. Chem. Res., 14, 259 (1981).

6   R.L. Pruett, Adv. Organomet. Chem., 17, 1 (1979).

7   B. Cornils in New Synthesis with Carbon Monoxide, Ed. J. Falbe, Springer Verlag, 1980, pp. 1-225.

8   R.F. Heck and D.S. Breslow, J. Amer. Chem. Soc., 83, 4023 (1961).

9   R. Whyman, J. Organomet. Chem., 66, C23 (1974).

10   R. Whyman, J. Organomet. Chem., 81, 97 (1974).

11   M. van Boven, N. Alemdaroglu and J.M.L. Penninger, J. Organomet. Chem. 84, 65 (1975).

12   M. Bianchi, F. Piacenti, P. Frediani and U. Matteoli, J. Organomet. Chem., 137, 361 (1977).

13   W. Rupilius, J.J. McCoy and M. Orchin, Ind. Eng. Chem., Prod. Res. Develop. 10(2), 142 (1971).

14   E.R. Tucci, Ind. Eng. Chem., Prod. Res. Develop., 7(2), 125, 7(3), 227 (1968).

15   J. Falbe, J. Organomet. Chem., 94, 213 (1975).

16   C.K. Brown and G. Wilkinson, J. Chem. Soc. A, 2753 (1970).

17 G.F. Pregaglia, A. Andreetta, G.F. Ferrari and R. Ugo, J. Organomet. Chem., 30, 387 (1971).

18 C. Botteghi, G. Consiglio and P. Pino, Chimia, 26, 141 (1972).

19 C. Botteghi, M. Branca, M. Marchetti and A. Saba, J. Organomet. Chem., 161, 197 (1978).

20 R. Laï and E. Ucciani, J. Mol. Catal., 4, 401 (1978).

21 D.M. Rudkovskii and N.S. Imyanitov, J. Appl. Chem. U.S.S.R., 35, 2608 (1962).

22 R. W. Goetz and M. Orchin, J. Amer. Chem. Soc., 85, 2782 (1963).

23 R. Laï and E. Ucciani, Adv. Chem. Ser., 132, 1 (1974).

24 K. Murata and A. Matsuda, Bull. Chem. Soc. Jpn., 53, 214 (1980).

25 L. Marko, Proc. Chem. Soc., London, 67 (1962).

26 C.L. Aldridge and H.B. Jonassen, J. Amer. Chem. Soc., 85, 886 (1963).

27 R.W. Goetz and M. Orchin, J. Org. Chem., 27, 3698 (1962).

28 I. Wender, R. Levine and M. Orchin, J. Amer. Chem. Soc., 72, 4375 (1950).

29 I.J. Goldfarb and M. Orchin, Adv. Catal., 14, 1 (1957).

30 F. Piacenti, M. Bianchi, P. Frediani, U. Matteoli and A. Lo Moro, J. Chem. Soc., Chem. Commun., 789 (1976).

31 F. Piacenti, S. Pucci, M. Bianchi, R. Lazzaroni and P. Pino, J. Amer. Chem. Soc., 90, 6847 (1968).

32 F. Piacenti, M. Bianchi and P. Frediani, Adv. Chem. Ser., 132, 283 (1973).

33 C.P. Casey and C.R. Cyr, J. Amer. Chem. Soc., 95, 2240 (1973).

34   J.A. Bertrand, C.A. Aldridge, S. Husebye and H.B. Jonassen, J. Org. Chem., 29, 790 (1964).

35   E. Naragon, A. Millendorf and J. Vergilio, U.S. Patent, 2,699,453 (1955); Chem. Abs., 50, 1893h (1956).

36   K. Murata and A. Matsuda, Chem. Lett.; 11 (1980); Bull. Chem. Soc. Jpn., 54, 249 (1981).

37   J. Habeshaw and R.W. Rae, British Patent, 702, 201 (1954); Chem. Abs., 49, 5514b (1955).

38   Agency of Industrial Sciences and Technology, Jpn. Kokai Tokkyo Koho, 80, 134, 643 (1980); Chem. Abs., 94, 37138t (1981).

39   T.C. Snapp and A.E. Blood, U.S. Patent, 3,888,880 (1975); Chem. Abs., 83 79254f (1975).

40   Agency of Industrial Sciences and Technology, Jpn. Kokai Tokkyo Koho, 81 78,637 (1981); Chem. Abs., 95, 176531z (1981).

41   S. Sato, Nippon Kagaku Zasshi, 90, 404 (1969); Chem. Abs., 71, 21828g (1969).

42   P. Chini, S. Martinengo and G. Garlaschelli, J. Chem. Soc., Chem. Commun., 709 (1972).

43   R.M. Laine, J. Amer. Chem. Soc., 100, 6451 (1978).

44   R.B. King, A.D. King and M.Z. Iqbal, J. Amer. Chem. Soc., 101, 4893 (1979).

45   G. Csontos, B. Heil and L. Marko, Ann. N.Y. Acad. Sci., 239, 47 (1974). B. Heil, L. Marko and G. Bor, Chem. Ber., 104, 3418 (1971).

46   D. Evans, J.A. Osborn and G. Wilkinson, J. Chem. Soc. A, 3133 (1968). C.K. Brown and G. Wilkinson, J. Chem. Soc. A, 2753 (1970).

47   G. Yagupsky, C.K. Brown and G. Wilkinson, J. Chem. Soc. A, 1392 (1970).

48 G. Wilkinson, Bull. Soc. Chim. Fr., 5055 (1968).

49. J.M. Brown, A.G. Kent and P.J. Sidebottom, J. Chem. Soc., Chem.
    Commun., in press.

50 R. Uson, L.A. Oro, C. Claver, M.A. Garralda and J.M. Moreto, J. Mol.
    Catal., 4, 231 (1978).

51 T. Kitamura, N. Sakamoto and T. Joh, Chem. Lett., 379 (1973).

52 B. Heil and L. Marko, Chem. Ber., 99, 1086 (1966).

53 K. Kaneda, M. Yasumura, M. Hiraki, T. Imanaka and S. Teranishi, Chem.
    Lett., 1763 (1981).

54 F.J. McQuillan, British Patent, 1,555,331 (1979); Chem. Abs., 93,
    12948y (1980).

55 S. Franks and F.R. Hartley, J. Mol. Catal., 12, 121 (1981).

56 R.L. Pruett and J.A. Smith, J. Org. Chem., 34, 327 (1969).

57 T.F. Strevels and N. Harris, Eur. Pat. Appl. 16, 285 (1980); Chem.
    Abs., 94, 83610p (1980).

58 M. Yamaguchi and T. Onoda, Japan. 76 27,649 (1976); Chem. Abs., 86,
    89173e (1977).

59 T. Hayashi, M. Tanaka and I. Ogata, J. Mol. Catal., 13, 323 (1981).

60 J. Grimblot, J.P. Bonnelle, A. Mortreux and F. Petit, Inorg. Chim. Acta,
    34, 29 (1979). J. Grimblot, J.P. Bonnelle, C. Vaccher, A. Mottreux,
    F. Petit and G. Peiffer, J. Mol. Catal., 9, 357 (1980). C. Vaccher,
    A. Mortreux and F. Petit, J. Mol. Catal., 12, 329 (1981).

61 B. Heil and L. Marko, Chem. Ber., 102, 2238 (1969). B. Heil, L.
    Marko and G. Bor, Chem. Ber., 104, 3418 (1971).

62 I. Wender, S. Metlin, S. Ergun, H.W. Sternberg and H. Greenfield, J.

Amer. Chem. Soc., 78, 5401 (1956).

63   J. Hagen, R. Lehmann and K. Bansemir, Ger. Offen., 2,914,090 (1980); Chem. Abs., 94, 25719v (1981).

64   E.H. Pommer, B. Zeeh and F. Linhart, Ger. Offen., 2,455,082 (1974); Chem. Abs., 85, 142972x (1976).

65   M. Takeda, H. Iwane and T. Hashimoto, Jpn. Kokai Tokkyo Koho 80 28,969 (1980); Chem. Abs., 93, 114041q (1980).

66   A. Spencer, J. Organomet. Chem., 124, 85 (1977).

67   B. Fell, W. Boll and J. Hagen, Chem- Ztg., 99, 452,485 (1975).

68   B. Fell and W. Rupilius, Tetrahedron Lett., 2721 (1969).

69   R. Kummer, Ger. Offen., 2,451,473 (1976); Chem. Abs., 85, 62676p (1976).

70   H. Bahrmann and B. Fell, J. Mol. Catal., 8, 329 (1980).

71   C.C. Cumbo, Eur. Pat. Appl., 3,753 (1979); Chem. Abs., 92, 75859v (1980).

72   Y. Takegami, Y. Watanabe and H. Masada, Bull. Chem. Soc. Jpn., 40, 1459 (1967).

73   C.U. Pittmann, W.D. Honnick and J.J. Yang, J. Org. Chem., 45, 684 (1980).

74   R.A. Sanchez-Delgado, J.S. Bradley and G. Wilkinson, J. Chem. Soc. Dalton, 399 (1976).

75   D. Evans, J.A. Osborn and G. Wilkinson, J. Chem. Soc. A, 3133 (1968).

76   H.C. Clark and J.A. Davies, J. Organomet. Chem., 213, 503 (1981).

77   S.C. Tang and L. Kim, J. Mol. Catal., 14, 231 (1982).

78   C-Y. Hsu and M. Ochin, J. Amer. Chem. Soc., 97, 3553 (1975).

79 Y. Kawabata, T. Hayashi and I. Ogata, J. Chem. Soc., Chem. Commun.,
    462 (1979).

80 T. Hayashi, Y. Kawabata, T. Isoyama and I. Ogata, Bull. Chem. Soc.
    Jpn., 54, 3438 (1981).

81 R. Bardi, A.M. Piazzesi, G. Cavinato, P. Cavoli and L. Toniolo, J.
    Organomet. Chem., 224, 407 (1982).

82 A. Sekiya and J.K. Stille, J. Amer. Chem. Soc., 103, 5096 (1981).
    B.H. Chang, R.H. Grubbs and C.H. Brubaker, J. Organomet. Chem., 172,
    81 (1979).

83 M.O. Farrell, C.H. van Dyke, L.J. Boucher and S.J. Metlin, J.
    Organomet. Chem., 172, 367 (1979).

84 G.O. Evans, C.U. Pittman, R. McMillan, R.T. Beach and R. Jones, J.
    Organomet. Chem., 67, 295 (1974).

85 G. Consiglio and P. Pino, in press.

86 A. Stefani, G. Consiglio, C. Botteghi and P. Pino, J. Amer. Chem.
    Soc., 99, 1058 (1977).

87 P. Pino, C. Saloman, C. Botteghi and G. Consiglio, Chimia, 26, 655
    (1972).

88 F. Piacenti, G. Menchi, P. Frediani, U. Matteoli and C. Botteghi,
    Chim. Ind. (Milan), 60, 808 (1978). Chem. Abs., 90, 86616y (1979).

89 P. Pino, G. Consiglio, C. Botteghi and C. Saloman, Adv. Chem. Ser.,
    132, 295 (1974).

90 J.D. Uhruh and J.R. Christenson, J. Mol. Catal., 14, 19 (1982).
    O.R. Hughes and D.A. Young, J. Amer. Chem. Soc., 103, 6636 (1981).

91 M. Tanaka, Y. Watanabe, T. Mitsudo and Y. Takegami, Bull. Chem. Soc.,
    Jpn., 47, 1698 (1974).

92 H.B. Tinker and A.J. Solodar, U.S. Patent, 4,268,688 (1981); Chem.
    Abs., 95, 114798h (1981).

93   G. Consiglio, F. Morandini and P. Pino, private communication.

94   Y. Becker, A. Eisenstadt and J.K. Stille, J. Org. Chem., 45, 2145 (1980).

95   C.F. Hobbs and W.S. Knowles, J. Org. Chem., 46, 4422 (1981).

96   G. Consiglio and P. Pino, Helv. Chim. Acta, 59, 642 (1976).

97   P.S. Pregosin and S.N. Sze, Helv. Chim. Acta, 61, 1848 (1978).

98   Y. Kawabata, T.M. Suzuki and I. Ogata, Chem. Lett., 361 (1978).

99   C.U. Pittman, Y. Kawabata and L.I. Flowers, J. Chem. Soc., Chem. Commun., 473 (1982).

100 J. Falbe, Carbon Monoxide in Organic Synthesis, Chapter 2. Springer-Verlag, New York, 1970.

101 C.W. Bird, R.C. Cookson, J. Hudec and R.O. Williams, J. Chem. Soc., 410 (1963).

102 C.W. Bird, Chem. Rev., 62, 283 (1962).

103 B. Fell and J.M.C. Tetteroo, Angew. Chem., Int. Ed. Engl., 4, 790 (1965).

104 R.L. Shubkin, U.S. Patent 3,644,443 (1972); Chem. Abs., 76, 99122v (1972).

105 A. Matsuda and H. Uchida, Bull. Chem. Soc. Jpn., 38, 710 (1965).

106 P. Pino and R. Ercoli, Chim. Ind. (Milan), 36, 536 (1954); Chem. Abs., 50, 195f (1956).

107 D.M. Fenton, J. Org. Chem., 38, 3192 (1973).

108 M. Yamaguchi and K. Tano, Japan., 74 43,930; Chem. Abs., 83, 27589z (1975).

109 G. Consiglio and M. Marchetti, Chimia, 30, 26 (1976).

110 Y. Sugi, K. Bando and S. Shin, Chem. Ind., 397 (1975).

111 J.F. Knifton, J. Org. Chem., 41, 793, 2885 (1976).

112 J.F. Knifton, Ger. Offen., 2,552,218 (1976); Chem. Abs., 85, 32458b (1976), 86, 105965x (1977).

113   F. Knifton, Ger. Offen., 2,303,118 (1973); Chem. Abs., 79, 146008m
      (1973).

114   J.J. Mrowca, U.S. Patent, 4,257,973 (1981); Chem. Abs., 95, 97089h
      (1981).

115   H. Yoshida, N. Sugita, K. Kudo and Y. Takezaki, Bull. Chem. Soc.
      Jpn., 49, 2245 (1976).

116   G. Cavinato and L. Toniolo, J. Mol. Catal., 6, 111 (1979).

117   F. Morandini, G. Consiglio and F. Wenzinger, Helv. Chim. Acta, 62,
      59 (1979).

118   R. Bardi, A. del Pra, A.M. Piazzesi and L. Toniolo, Inorg. Chim.
      Acta, 35, L345 (1979).

119   L.J. Kehoe and R.A. Schell, J. Org. Chem., 35, 2846 (1970).

120   Y. Soma and H. Sano, Japan. Kokai, 74 61,092 (1974); Chem. Abs., 83,
      137525h (1975).

121   Y. Souma and H. Sano, Bull. Chem. Soc. Jpn., 47, 1717 (1974).

122   Y. Souma, H. Sano and J. Iyoda, J. Org. Chem., 38, 2016 (1973).

123   Y. Souma and H. Sano, J. Org. Chem., 38, 3633 (1973).

124   J. Falbe, Angew. Chem., Int. Ed. Engl., 5, 435 (1966).

125   A. Matsuda, Bull. Chem. Soc. Jpn., 41, 1876 (1968).

126   J. Falbe, H-J. Schulze-Steinen and F. Korte, Chem. Ber., 98, 886
      (1965).

127   P. Pino, F. Piacenti, M. Bianchi and R. Lazzaroni, Chim. Ind.
      (Milan), 50, 106 (1968); Chem. Abs., 68, 86796c (1968).

128   J.F. Knifton, U.S. Patent 3,880,898 (1975); Chem. Abs., 83, 113698r
      (1975).

129   D.M. Fenton, U.S. Patent 3,641,071 (1972); Chem. Abs., 76, 99132y
      (1972).

130   V.P. Kurkov, U.S. Patent, 4,113,735 (1978); Chem. Abs., 90, 54818z
      (1979).

131   J. Falbe and F. Korte, Chem. Ber., 98, 1928 (1965).

132   S. Kunichika, Y. Sakakibara and T. Nakamura, Bull. Chem. Soc. Jpn., 41, 390 (1968).

133   J. Tsuji, M. Takahashi and T. Takahashi, Tetrahedron Lett., 21, 849 (1980).

134   J.F. Knifton, J. Mol. Catal., 2, 293 (1977).

135   I.A. Orlova, N.F. Alekseeva, A.D. Troitskaya and O.N. Temkin, J. Gen. Chem. U.S.S.R., 49, 1398 (1979).

136   M. Foà and L. Cassar, Gazz. Chim. Ital., 102, 85 (1972).

137   J.R. Norton, K.E. Shenton and J. Schwartz, Tetrahedron Lett., 51 (1975).

138   T.F. Murray, V. Varma and J.R. Norton, J. Chem. Soc., Chem. Commun., 907 (1976).

139   T.F. Murray and J.R. Norton, J. Amer. Chem. Soc., 101, 4107 (1979).

140   T.F. Murray, E.G. Samsel, V. Varma and J.R. Norton, J. Amer. Chem. Soc., 103, 7520 (1981).

141   J. Tsuji, S. Hosaka, J. Kiji and T. Susuki, Bull. Chem. Soc. Jpn., 39, 141 (1966).

142   K. Bittler, N. v. Kutepov, D. Neubauer and H. Reis, Angew. Chem. Int. Ed. Engl., 7, 329 (1968).

143   J. Tsuji and H. Yasuda, Bull. Chem. Soc. Jpn., 50, 553 (1977).

144   A. Matsuda, Bull. Chem. Soc. Jpn., 46, 524 (1973).

145   J. Tsuji, Adv. Organomet. Chem., 17, 141 (1979).

146   J. Tsuji and H. Yasuda, J. Organomet. Chem., 131, 133 (1977).

147   C. Bordenca and W.E. Marsico, Tetrahedron Lett., 1541(1967).

148   S. Sato, Y. Ono, S. Tatsumi and H. Wakamatsu, Nippon Kagaku Zasshi, 92, 178 (1971); Chem. Abs., 76, 33755x (1972).

149   A. Matsuda, Bull. Chem. Soc. Jpn., 42, 571 (1969).

150   C. Botteghi, G. Consiglio and P. Pino, Chimia, 27, 477 (1973).

151   T. Hayashi, M. Tanaka and I. Ogata, Tetrahedron Lett., 3925 (1978).

152   G. Consiglio and P. Pino, Chimia, 30, 193 (1976).

153   G. Consiglio, Helv. Chim. Acta, 59, 124 (1976).

154   G. Consiglio, J. Organomet. Chem., 132, C26 (1977).

155   H. HohenSchutz, N. von Kutepov and W. Himmele, Hydrocarbon Process,
       45 (11), 141 (1966).

156   T. Mizoroki and M. Nakayama, Bull. Chem. Soc. Jpn., 41, 1628 (1968).

157   Y. Sugi, K-i. Bando and Y. Takami, Chem. Lett., 63 (1981).

158   T. Imamoto, T. Kusumoto and M. Yokoyama, Bull. Chem. Soc. Jpn., 55,
       643 (1982).

159   D.E. Morris and H.B. Tinker, Chem. Technol., 2, 554 (1972).

160   D. Forster, J. Amer. Chem. Soc., 98, 846 (1976), Adv. Organomet.
       Chem., 17, 255 (1975).

161   G.W. Adamson, J.J. Daly and D. Forster, J. Organomet. Chem., 71, C17
       (1974).

162   T. Matsumoto, K. Mori, T. Mizoroki and A. Ozaki, Bull. Chem. Soc.
       Jpn., 50, 2337 (1977).

163   F.E. Paulik and J.F. Roth, J. Chem. Soc., Chem. Commun., 1578 (1968).

164   H. Fernholz and D. Freudenberger, Ger. Offen., 2,2175344 (1973);
       Chem. Abs., 80, 26763c (1974).

165   M.S. Jarrell and B.C. Gates, J. Catal., 40, 255 (1975).

166   D. Brodzki, B. Denise and G. Pannetier, J. Mol. Catal., 2, 149
       (1977).

167   D. Forster, J. Chem. Soc., Dalton, 1639 (1979).

168   T. Matsumoto, T. Mizoroki and A. Ozaki, J. Catal., 51, 96 (1978).

169   R.F. Heck, J. Amer. Chem. Soc., 85, 2013 (1963).

170   H.J. Hagemeyer, U.S. Patent 2,739,169 (1956); Chem. Abs., 50,
       16835d (1956).

171   W. Reppe, Ann., 582, 1 (1953).

172   J. Tsuji, M. Morikawa and J. Kiji, Tetrahedron Lett., 1061 (1963).

173   J.E. Hallgren and G.M. Lucas, J. Organomet. Chem., 212, 135 (1981).

174   D.M. Fenton and P.J. Steinwand, J. Org. Chem., 39, 701 (1974).

175   J. Tsuji, K. Sato and H. Nagashima, Tetrahedron Lett., 23, 893
      (1982).

176   R.A. Fiato, U.S. Patent 4,233,466 (1980); Chem. Abs., 94, 83601m
      (1981).

177   G.N. Butter, U.S. Patent 3,285,948 (1966); Chem. Abs., 66, 65072g
      (1967).

178   W.R. Pretzer, T.P. Kobylinski and J.E. Bozik, U.S. Patent, 4,133,966
      (1979); Chem. Abs., 90, 120998m (1979).

179   M. Hidai, M. Orisako, M. Ue, Y. Uchida, K. Yasufuku and H. Yamazaki,
      Chem. Lett., 143 (1981).

180   G. Braca, G. Sbrana, G. Valentini, G. Andrich and G. Gregorio,
      Fundam. Res. Homogeneous Catal., 3, 221 (1979).

181   Ya. T. Eidus, A.L. Lapidus, K.V. Puzitskii and B.K. Nefedov, Russ.
      Chem. Rev., 42, 199 (1973).

182   T.A. Weil, L. Cassar and M. Foà in Organic Synthesis via Metal
      Carbonyls, ed. I. Wender and P. Pino, Volume 2, Wiley, New York,
      1977; pp. 517-543.

183   A. Schoenberg and R.F. Heck, J. Amer. Chem. Soc., 96, 7761 (1974).

184   L. Cassar, M. Foà and A. Gardano, J. Organomet. Chem., 121, C55
      (1976).

185   M. Tanaka, Bull. Chem. Soc. Jpn., 54, 637 (1981).

186   M. Nakayama and T. Mizoroki, Bull. Chem. Soc. Jpn., 42, 1124 (1969).

187   A. Schoenberg and R.F. Heck, J. Org. Chem., 39, 3327 (1974).

188   M. Toyoguchi and H. Yoshida, Japan. Kokai, 75 05,347 (1975); Chem.
      Abs., 82, 139745h (1975).

189   T. Kobayashi and M. Tanaka, J. Organomet. Chem., 231, C12 (1982).

190   T. Takahashi, T. Nagashima and J. Tsuji, Chem. Lett., 369 (1980).

191   G.D. Pandey and K.P. Tiwari, Synth. Commun., 10, 523 (1980).

192   M. Mori, K. Chiba and Y. Ban, Heterocycles, 6, 1841 (1977).

193   M. Mori, M. Ishikura, T. Ikeda and Y. Ban, Heterocycles, 16, 1491
      (1981).

194   M. Mori, K. Chiba, N. Inotsume and Y. Ban, Heterocycles, 12, 921
      (1979).

195   M. Tanaka, M. Koyanagi and T. Kobayashi, Tetrahedron Lett., 22, 3875
      (1981).

196   M. Tanaka, Tetrahedron Lett., 2601 (1979).

197   M. Tanaka, Synthesis, 47 (1981).

198   T. Kobayashi and M. Tanaka, J. Chem. Soc., Chem. Commun., 333 (1981).

199   J-J. Brunet, C. Sidot and P. Caubere, Tetrahedron Lett., 22, 1013
      (1981).

200   J.F. Knifton, U.S. Patent 3,991,101 (1976); Chem. Abs., 86, 55907g
      (1979).

201   E.J. Corey and L.S. Hegedus, J. Amer. Chem. Soc., 91, 1233 (1969).

202   A. Cowell and J.K. Stille, Tetrahedron Lett., 133 (1979).

203   R.F. Heck and D.S. Breslow, J. Amer. Chem. Soc., 85, 2779 (1973).

204   M. El-Chahawi and H. Richtzenhain, Ger. Offen. 2,259,072 (1974);
      Chem. Abs., 81, 77679h (1974).

205   M. El-Chahawi, H. Richtzenhain and W. Vogt, Ger. Offen., 2,410,782
      (1975); Chem. Abs., 84, 4684h (1976).

206   H. Alper and H. des Abbayes, J. Organomet. Chem., 134, C11 (1977).

207   L. Cassar and M. Foa, J. Organomet. Chem., 134, C15 (1977).

208   H. des Abbayes and A. Buloup, Tetrahedron Lett., 21, 4343 (1980).

209   S. Imamura and J. Tsuji, Tetrahedron, 25, 4187 (1969).

210   H.C. Volger, K. Vrieze, J.W.F.M. Lemmers, A.P. Praat and P.W.N.M.
      van Leeuwen, Inorg. Chim. Acta, 4, 435 (1970).

211 M. Foà and L. Cassar, Gazz. Chim. Ital., 109, 619 (1979).

212 S. Hattori, H. Fukutani, M. Tokizawa and H. Okada, Japan. 72
     47,005 (1972); Chem. Abs., 78, 97141t (1973).

213 J.F. Knifton, J. Organomet. Chem., 188, 223 (1980).

214 A. Cowell and J.K. Stille, J. Amer. Chem. Soc., 102, 4193 (1980).

215 J.K. Stille and P.K. Wong, J. Org. Chem. 40, 532 (1975).

216 H. Erpenbach, K. Gehrmann, W. Lork and P. Prinz, Ger. Offen., DE
     3,016,900 (1981); Chem. Abs., 96, 19680a (1982).

217 F. Piacenti, M. Bianchi, P. Frediani and U. Matteoli, J. Organomet.
     Chem., 87, C54 (1975).

218 R.C. Larock and S.S. Hershberger, J. Org. Chem., 45, 3840 (1980).

219 D. Seyferth and R.J. Spohn, J. Amer. Chem. Soc., 90, 540 (1968),
     91, 3037 (1969).

220 R.C. Larock, J. Org. Chem., 40, 3237 (1975).

221 R.C. Larock and B. Reifling, Tetrahedron Lett., 4661 (1976).

222 R.C. Larock and C.A. Fellows, J. Amer. Chem. Soc., 104, 1900 (1982).

223 W.C. Baird and J.H. Surridge, J. Org. Chem., 40, 1364 (1975).

224 N. Miyaura and A. Suzuki, Chem. Lett., 879 (1981).

225 C.W. Bird, Chem. Rev., 62, 283 (1962).

226 G.L. Rempel, W.K. Teo, B.R. James and D.V. Plackett, Adv. Chem.
     Ser., 132, 166 (1974).

227 V.A. Golodov, Yu. L. Sheludyakov and D.V. Sokol'skii, Fundam. Res.
     Homogeneous Catal., 3, 239 (1979).
     Yu. L. Sheludyakov, V.A. Golodov and D.V. Sokol'skii, Dokl. P.
     Chem., 249, 998 (1979).

228 B.D. Dombek and R.J. Angelici, J. Catal., 48, 433 (1977).

229 D. Durand and C. Lassau, Tetrahedron Lett., 2329 (1969).

230 T. Yamahara, T. Deguchi, S. Takamatsu, M. Usui, H. Yoshihara and
     K. Hirose, Japan. Kokai, 74 18,848 (1974); Chem. Abs., 80, 133007f

(1974).

231   S.S. Novikov, V.I. Manov-Yuvenskii, A.V. Smetanin and B.K.
      Nefedov, Dokl. Chem., 251, 135 (1980).

232   B.N. Nefedov and V.I. Manov-Yuvenskii, Bull. Acad. Sci. U.S.S.R.,
      Div. Chem. Sci., 28, 540 (1979).

233   V.I. Manov-Yuvenskii and B.K. Nefedov. Bull. Acad. Sci. U.S.S.R.,
      Div. Chem. Sci., 30, 816 (1981).

234   H. Alper, C.P. Petera and F.R. Ahmed, J. Amer. Chem. Soc., 103,
      1289 (1981).
      H. Alper and C.P. Mahatantila, Organometallics, 1, 70 (1982).
      T. Sakakibara and H. Alper, J. Chem. Soc., Chem. Commun., 458 (1979).

235   K. Kikukawa, K. Kono, K. Nagira, F. Wada and T. Matsuda, J. Org.
      Chem., 46, 4413 (1981).

236   K. Kikukawa, K. Kono, F. Wada and T. Matsuda, Chem. Lett., 35 (1982).

237   J. Tsuji, M. Morikawa and J. Kiji, Tetrahedron Lett., 817 (1965).

238   M. Hidai, M. Orisaku and Y. Uchida, Chem. Lett., 753 (1980).

239   A.F.M. Iqbal, Tetrahedron Lett., 3381 (1971).

240   R.F. Heck, Organic Syntheses via Metal Carbonyls, ed. I. Wender and
      P. Pino, Volume 1, p.373, Wiley, New York (1968).

241   J.L. Eisenmann, R.L. Yamartino and J.F. Howard, J. Org. Chem., 26,
      2102 (1961).

242   R.J. DePasquale, J. Chem. Soc., Chem. Commun., 157 (1973).

243   BASF Ger. Patent 293,787 (1913); Chem. Abs., 11, 2582 (1917).

244   J.G. Kronseder, Hydrocarbon Process, 55 (7), 56F (1976).

245   H.H. Storch, Adv. Catal., 1, 113 (1948).

246   G. Natta, Giorn. Chimici ind. e applicata, 12, 13 (1930); Chem.
      Abs., 24, 2717 (1930).

247   E.S. Brown, U.S. Patent 3,974,259 (1976); Chem. Abs., 85, 162790q
      (1976).

248   C.K. Rofer-De Poorter, Chem.. Rev., 81, 447 (1981).

249   C.D. Frohning, React. and Struc., 11, 309 (1980).

250   C. Masters, Adv. Organomet. Chem., 17, 61 (1979).

251   W.A. Herrmann, Angew. Chem. Int. Ed. Engl., 21, 117 (1982).

252   R.R. Schrock, Acc. Chem. Res., 12, 98 (1979).

      W.A. Herrmann, B. Reiter and H. Biersack, J. Organomet. Chem., 97,
      245 (1975).

253   W.A. Herrmann, Angew. Chem., Int. Ed. Engl., 17, 800 (1978).

      J.C. Hayes, G.D.N. Pearson and N.J. Cooper, J. Amer. Chem. Soc.,
      103, 4648 (1981).

254   R.C. Brady and R. Pettit, J. Amer. Chem. Soc., 102, 6181 (1980).

255   H.H. Storch, N. Golumbic and R.B. Anderson, "The Fischer Tropsch
      and Related Syntheses", Wiley, New York, 1951.

256   Yu. T. Eidus, Russ. Chem. Rev., 36, 338 (1967).

257   H. Pichler and H. Schulz, Chem. Ing.-Tech., 42, 1162 (1970).

258   I. Wender, Catal. Rev. Sci. Eng., 14, 97 (1976).

259   G. Henrici-Olive and S. Olive, Angew. Chem., Int. Ed. Engl., 15,
      136 (1976).

260   J.P. Collmann and S.R. Winter, J. Amer. Chem. Soc., 95, 4089
      (1973).

261   C.P. Casey and S.M. Neumann, J. Amer. Chem. Soc., 98, 5395 (1976).

262   J.E. Bercaw, Lect. Joint Conf. Chem. Inst. Can. Am. Soc., 1977,
      Paper INOR 085.

263   C.P. Casey, S.M. Neumann, M.A. Andrews and D.R. McAlister, Pure
      Appl. Chem., 52, 625 (1980).

      J.A. Gladysz, Fundam. Res. Homogeneous Catal. 2, 173 (1978).

      J.A. Gladysz, Aldrichimica Acta, 12, 13 (1979).

264   G. Fachinetti, C. Floriani, F. Marchetti and S. Merlino, J. Chem.
      Soc., Chem. Commun., 522 (1976).

265 J.R. Sweet and W.A.G. Graham, J. Amer. Chem. Soc., 104, 2811 (1982).

266 R.J. Daroda, J.R. Blackborrow and G. Wilkinson, J. Chem. Soc., Chem. Commun., 1098 (1980).

267 J. Tsuji in "Organic Syntheses via Metal Carbonyls" ed. I. Wender and P. Pino, Volume 2, Wiley 1977.

268 K.S.Y. Lau, Y. Becker, F. Huang, N. Baenziger and J.K. Stille, J. Amer. Chem. Soc., 99, 5664 (1977).

269 J. Tsuji and K. Ohno, Tetrahedron. Lett., 4713 (1966). J. Amer. Chem. Soc., 90, 99 (1968).

270 J.K. Stille and R.W. Fries, J. Amer. Chem. Soc., 96, 1514 (1974).

271 M.C. Baird, J.T. Mague, J.A. Osborn and G. Wilkinson, J. Chem. Soc., A, 1347 (1967).

272 J. Blum, S. Kraus and Y. Pickholtz, J. Organomet. Chem., 33, 227 (1971).

273 G.A. Olah and P. Kreienbühl, J. Org. Chem., 32, 1614 (1967).

274 E. Lindner and A. Thasitis, Chem. Ber., 107, 2418 (1974).

275 J. Blum, H. Rosenman and E.D. Bergmann, J. Org. Chem., 33, 1928 (1968).

276 J. Blum, E. Oppenheimer and E.D. Bergmann, J. Amer. Chem. Soc., 89, 2338 (1967).

277 W. Strohmeier and P. Pföhler, J. Organomet. Chem., 108, 393 (1976).

278 J. Tsuji and K. Ohno, J. Amer. Chem. Soc., 90, 94 (1968).

279 J.K. Stille, F. Huang and M.T. Regan. J. Amer. Chem. Soc., 96, 1518 (1974).

280 D.J. Dawson and R.E. Ireland, Tetrahedron Lett., 1899 (1968).

281 D.J. Ward, W.A. Szarek and J.K.N. Jones, Chem. Ind., 162 (1970).

282 J.W. Suggs, J. Amer. Chem. Soc., 100, 640 (1978).

283 T.B. Rauchfuss, Fundam. Res. Homogeneous Catal., 3, 1021 (1979).

284 C.F. Lochow and R.G. Milter, J. Amer. Chem. Soc., 98, 1281 (1976).

285   K. Ohno and J. Tsuji, J. Amer. Chem. Soc., 90, 99 (1968).

286   H.M. Walborsky and L.E. Allen, Tetrahedron Lett., 823 (1970);
      J. Amer. Chem. Soc., 93, 5465 (1971).

287   M.D. Johnson, M.L. Tobe and L.-Y. Wong, J. Chem. Soc., Chem. Commun.,
      298 (1967), J. Chem. Soc. A, 923 (1968), 929, 2516 (1969).

288   D.A. Clement and J.F. Nixon, J. Chem. Soc., Dalton, 195 (1973).

289   D.H. Doughty and L.H. Pignolet, J. Amer. Chem. Soc., 100, 7083
      (1978).

290   D.H. Doughty, M.F. McGuiggan, H. Wang and L.H. Pignolet, Fundam.
      Res. Homogeneous Catal., 3, 909 (1979).

291   A. Emery, A.C. Oehschlager and A.M. Unrau, Tetrahedron Lett.,
      4401 (1970).

292   E. Müller and A. Segnitz, Ann., 1583 (1973).

293   J. Blum and Z. Lipshes, J. Org. Chem., 34, 3076 (1969).

294   K. Kanada, H. Azuma, M. Wayaku and S. Teranishi, Chem. Lett., 215
      (1974).

295   S.I. Zav'yalov and L.B. Kochanova, Bull. Acad. Sci. U.S.S.R.,
      Div. Chem. Sci., 27, 2161 (1978).

4 Other additions to carbon-carbon multiple bonds

4.1 Hydrosilylation

    4.1.1. Hydrosilylation of Alkenes.

    4.1.2. Hydrosilylation of Dienes.

    4.1.3. Hydrosilylation of Alkynes.

    4.1.4. Asymmetric Hydrosilylation of Alkenes.

    4.1.5. Polymer Supported Catalysts.

    4.1.6. Mechanisms of Hydrosilylation.

    4.1.7. Conclusions.

4.2 Cyclopropanation.

    4.2.1. The Simmons Smith Reaction.

    4.2.2. Decomposition of Diazo Compounds by Transition Metal Complexes.

    4.2.3. Asymmetric Cyclopropanation.

4.3 Hydrocyanation

    4.3.1. Hydrocyanation of Simple Alkenes.

    4.3.2. Strained Alkenes.

    4.3.3. Functionalised Alkenes.

    4.3.4. Reactions of Dienes and Synthesis of Adiponitrile.

    4.3.5. Hydrocyanation of Alkynes.

    4.3.6. Mechanism of Alkene Hydrocyanation.

4.4 Hydrohalogenation and Related Reactions.

4.5 Hydrometallation and Carbometallation.

    4.5.1. Hydrometallation of Alkenes.

    4.5.2. Hydrometallation of Alkynes.

    4.5.3. Carbometallation.

4.6 Addition of Heteroatomic Nucleophiles to Alkenes and Related Reactions.

    4.6.1. Addition of Amines, Alcohols and Water to Alkenes.

    4.6.2. Addition of Amines and Alcohols to Metal Allyls.

4.6.3. Addition to Alkynes.

4.6.4. Nucleophilic Substitutions.

4.1 Hydrosilylation

Hydrosilylation is the term used to describe the addition of a silicon hydride across a carbon-carbon or a carbon-heteroatom multiple bond. Addition to ketones and imines is used mainly as an alternative method of reduction and is discussed in the appropriate chapter. Recent years have seen great advances in the use of organosilicon intermediates in organic synthesis[1,2] and hydrosilylation provides a route to a wide variety of functionalised silanes. Polyorganosiloxanes may also be made in this way and hydrosilylation is extensively used for the crosslinking of polymers. Reviews of the literature in this area have been provided by Speier[3] and Lukevics[4].

4.1.1. Hydrosilylation of Alkenes

Complexes of many transition metals provide useful hydrosilylation catalysts. Speier's (H_2PtCl_6) and Wilkinson's (($PPh_3)_3RhCl$) catalysts have been studied in most detail but there are reports of vanadium, manganese, nickel, palladium, ruthenium, cobalt and iron based systems. There is a single example of catalysis by an osmium complex.[5] Iridium complexes are catalytically inactive but have been used as models for the corresponding rhodium species.[6]

For hydrosilylation of alkenes in the presence of chloroplatinic acid (H_2PtCl_6) the reactivity of the substrate increases with increased electron density in the double bond but is decreased by the presence of bulky groups.[7,8] For 1-alkenes two products may be obtained (Figure 1); 2 is the product of Markownikov addition and 3 that of anti-Markownikov reaction (in early work this is sometimes said to be in accord with Farmer's rule). Terminal adducts are obtained almost exclusively in reactions

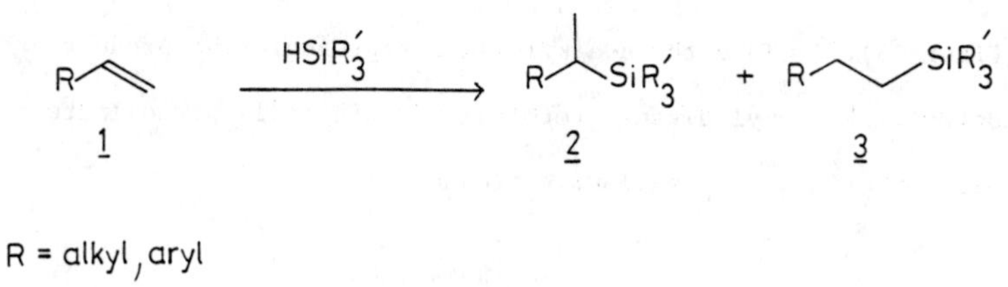

R = alkyl, aryl

Figure 1 Hydrosilylation of 1-alkenes.

catalysted by H_2PtCl_6 and other platinum complexes; examples are given in

Figure 2. It appears that the reaction is essentially irreversible with no

equilibrium between the product silane and its precursor; no racemisation

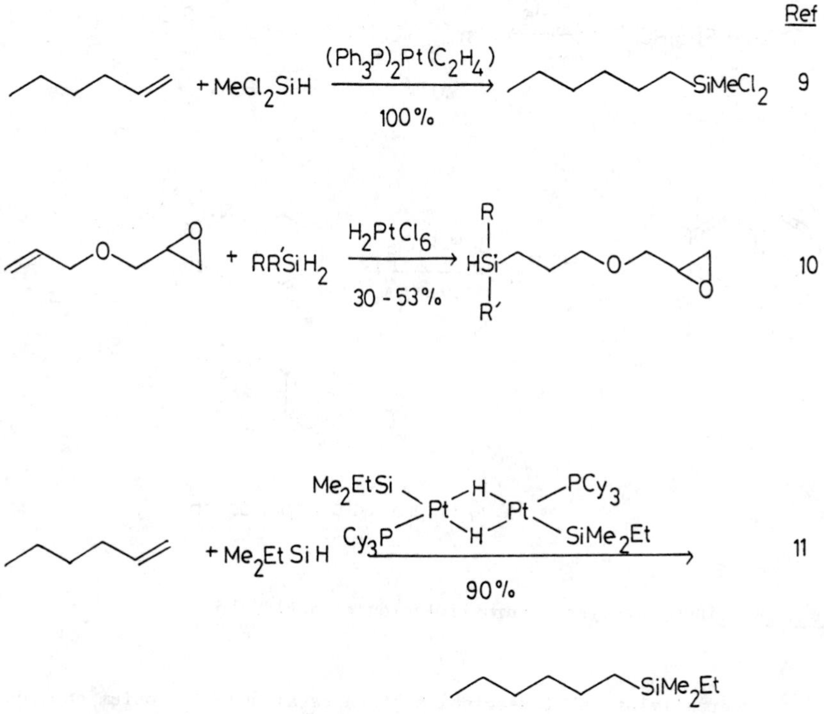

Ref

Figure 2 Hydrosilylation catalysed by platinum complexes.

of optically active 4 occurred after 24 hours at 100° in the presence of

H_2PtCl_6 (Figure 3).[12] Thus the n-alkylsilane represents the product of

kinetic control. For aryl alkenes terminal adducts still predominate but

are no longer invariably the exclusive products (Figure 4).

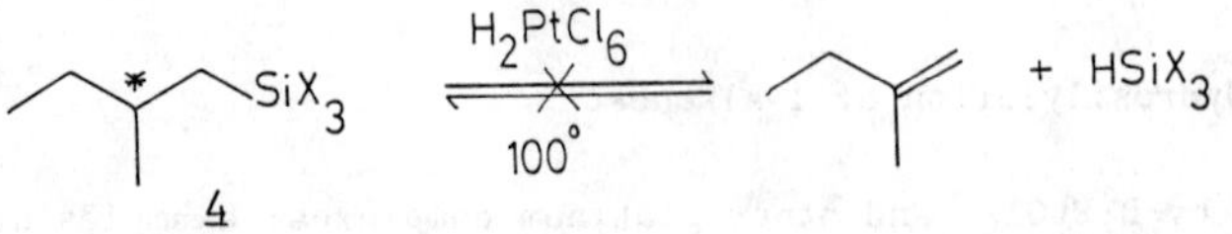

Figure 3 Attempted racemisation of a chiral silane by H_2PtCl_6.

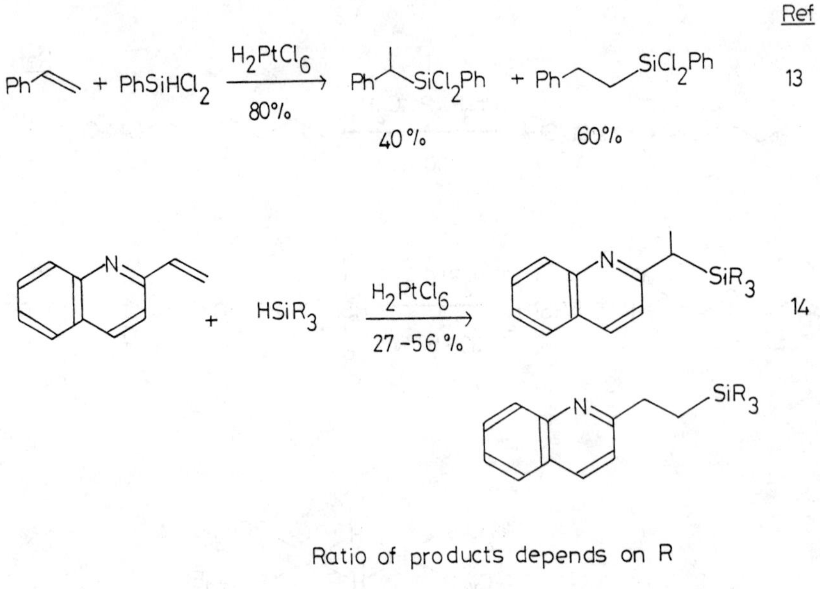

Ref

13

14

Ratio of products depends on R

Figure 4 Platinum catalysed hydrosilylation of aryl alkenes.

Hydrosilylation of terminal alkenes catalysed by rhodium complexes

also tends to give terminal adducts (Figure 5) as do reactions catalysed by

vanadium,[17] cobalt[18] and palladium[19] complexes.

More diverse results are obtained by the use of nickel complexes

as catalysts. Terminal silanes are obtained from the reaction of

Ref

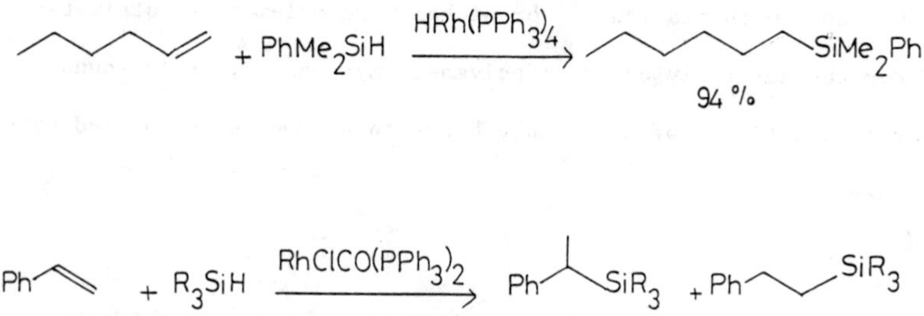

SiMe$_2$Ph 15

94 %

16

Figure 5   Rhodium catalysed hydrosilylation of terminal alkenes.

dichloromethylsilane with 1-octene catalysed by most chelating biphosphine nickel dichloride complexes.[20,21] (hydrogenolysis of the Si-Cl bond also occurs). However, using 1,2-bis(dimethylphosphino)-1,2-dicarba-closo-decaborane nickel dichloride as catalyst gives up to 40% of the branched adduct (with no hydrogenolysis).[22] For addition of trichlorosilane to styrene mixtures of α- and β- adducts are obtained, depending on the precise catalyst. Nickel chloride in the presence of PPh$_3$ and cuprous chloride gives a 75% yield of silanes of which 95% is the α-adduct.[23] With a dicyclopentadienyl nickel catalyst the addition of dichloromethylsilane to both styrene and 1-octene gives exclusively the branched product, the result of Markownikov addition.[24]

Examples of the hydrosilylation of di- and tri-substituted alkenes are less numerous and their reactions tend to be less specific. Terminal disubstituted alkenes give terminal adducts[25] though more slowly than monosubstituted analogues.[26] The hydrosilylation of internal alkenes with both platinum[27] and rhodium[28] catalysts is also slow. The reaction of 1-heptene with trichlorosilane at 85° is 54% complete in two hours, whereas cis- and trans-2-heptene react to the extent of only 10% and 1% under the same conditions[28] (85°, 2 hour, $((C_2H_4)_2RhCl)_2$). For molecules containing more than one type of double bond an estimate of relative reactivities may be obtained. 4-vinylcyclohexene reacts only at the exocyclic double bond with platinum[29] or nickel[30] complexes as catalysts (Figure 6). For the naturally occuring polyene, myrcene, 5, it is found that the order of reactivity of the double bonds in a reaction catalysed by H_2PtCl_6 is a>b>c.[31]

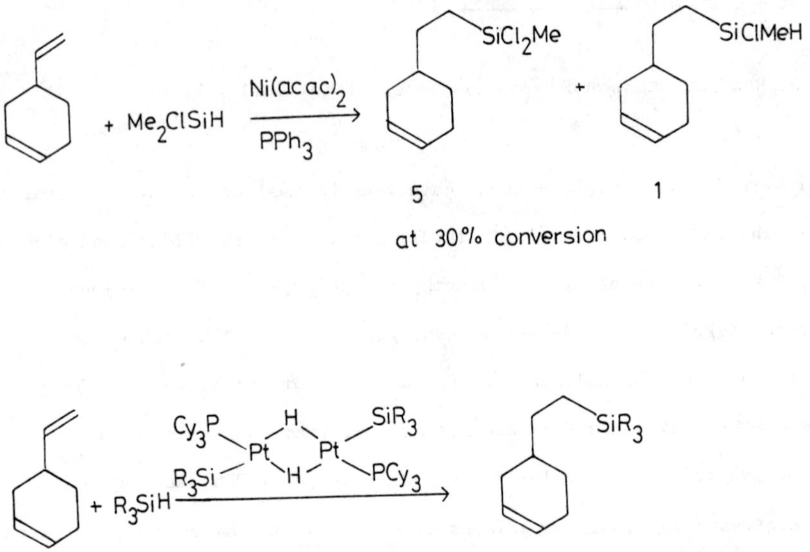

at 30% conversion

Figure 6 Hydrosilylation of 4-vinylcyclohexene.

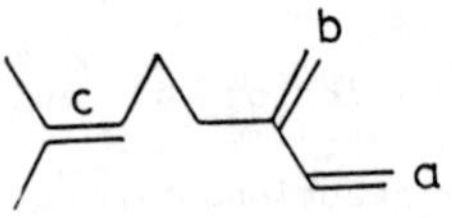

5

A problem associated with selective hydrosilylation of alkenes is the possibility of competitive double-bond isomerisation. When 1-hexene is hydrosilylated in the presence of Wilkinson's catalyst, $(PPh_3)_3RhCl$, isomerisation to cis-2-hexene, trans-2-hexene and 3-hexenes occurs during the reaction, the relative rates depending on catalyst concentration.[32]

Conversely many internal alkenes give only terminal adducts, where such a

path is available to them by a series of isomerisations (Figure 7).

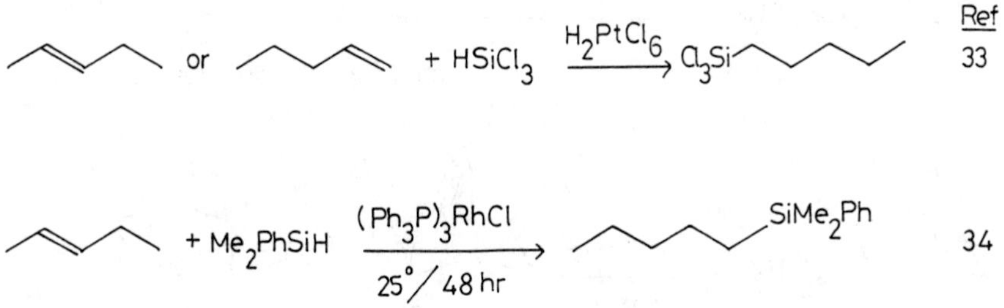

Figure 7   Isomerisation and hydrosilylation of disubstituted alkenes.

Cycloalkenes present a different problem. A few, such as

cycloocta-1,5-diene[20] and norbornene,[25] give comparatively pure products

(Figure 8). Cyclohexene derivatives are rather unreactive and extensively

isomerised products are often obtained. While 1-ethylcyclohexene reacts

with trichlorosilane in the presence of H_2PtCl_6 to give a 95% yield of the

terminal adduct, 1-propylcyclohexene gives only 2% of the corresponding

silane, together with recovered starting material and five other products

in varying amounts[35] (Figure 9). The only hydrosilylation of a cyclohexene

reported to proceed in a specific manner at a reasonable rate involves the

catalyst   formed by the reaction of $VO(acac)_2$ with a trialkylaluminum in

the presence of phosphine. The anti-Markownikov adduct 7 is obtained from

6 in 80% yield.[36]

The hydrosilylation of alkenes bearing electron donating or

withdrawing groups also, generally, proceeds in an anti-Markownikov manner.

Figure 10 shows examples, all of which produce potentially useful

organosilicon monomers. Multiply substituted alkenes tend to give mixtures

of products and only extensive testing of catalysts and conditions will

give the desired result. Unsaturated esters and ketones may additionally

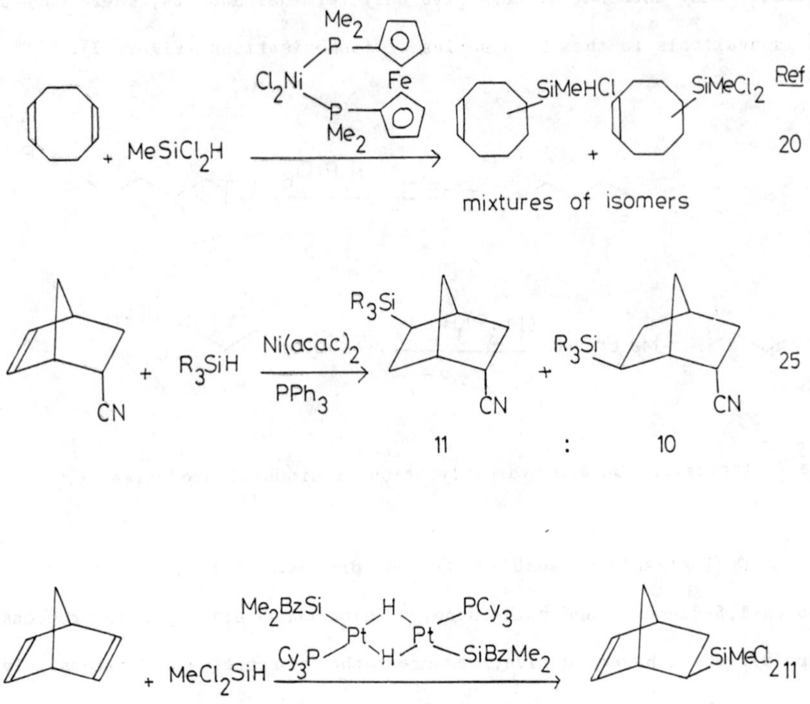

mixtures of isomers

11 : 10

Figure 8   Hydrosilylation of cycloalkenes.

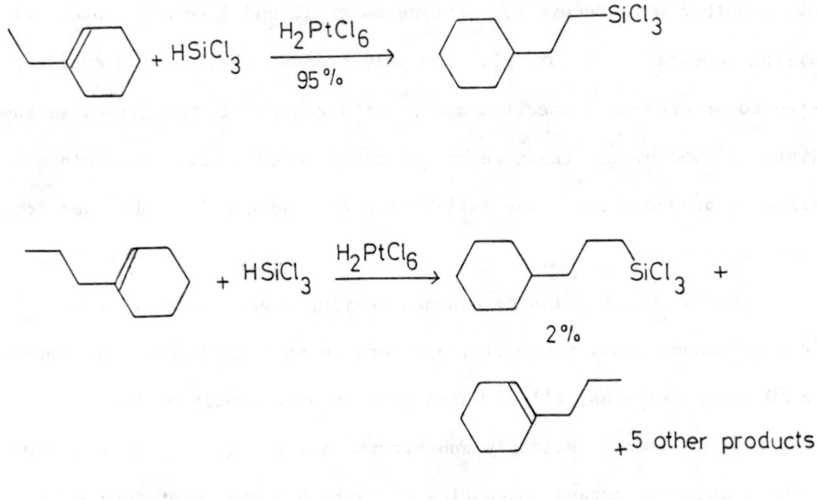

95 %

2 %

+ 5 other products

Figure 9   Hydrosilylation of substituted cyclohexenes.

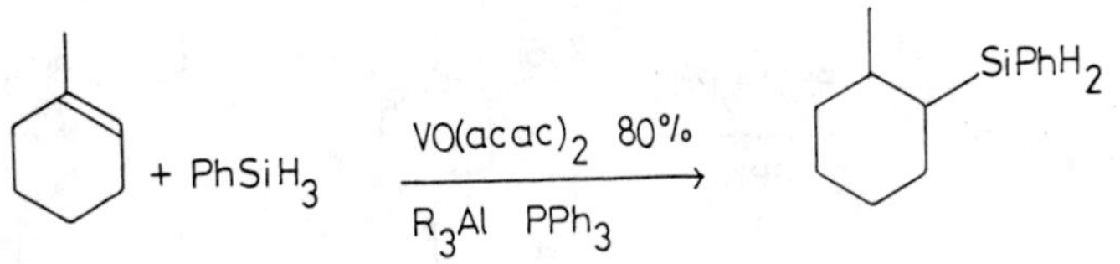

Reaction of 6:

A methylcyclohexene $+ PhSiH_3$ reacting via $VO(acac)_2$, R_3Al, PPh_3 (80%) to give a methylcyclohexane bearing $SiPhH_2$ (7).

6   7

		Ref
$\diagup\!\!\diagdown\!OAc + HSiCl_3 \xrightarrow[7hr, 120°]{Ni(acac)_2/PPh_3} Cl_3Si\diagdown\diagup\!OAc$		37
$\diagup\!O\diagdown\!\!\vartriangle\!\!O + HSiEt_3 \xrightarrow{H_2PtCl_6} Et_3Si\diagdown\diagup\!O\diagdown\!\!\vartriangle\!\!O$		38
$\diagup\!\!\diagdown\!CN + HSiCl_3 \xrightarrow[80\%]{Ni(acac)_2/PPh_3} Cl_3Si\!-\!\overset{CN}{\underset{}{\diagup\!\!\diagdown}}$		39
$\diagup\!\!\diagdown\!COOEt + Me_2PhSiH \xrightarrow[76\%]{(Ph_3P)_3RhCl} Me_2PhSi\diagdown\diagup\!\diagdown\!COOEt$		40

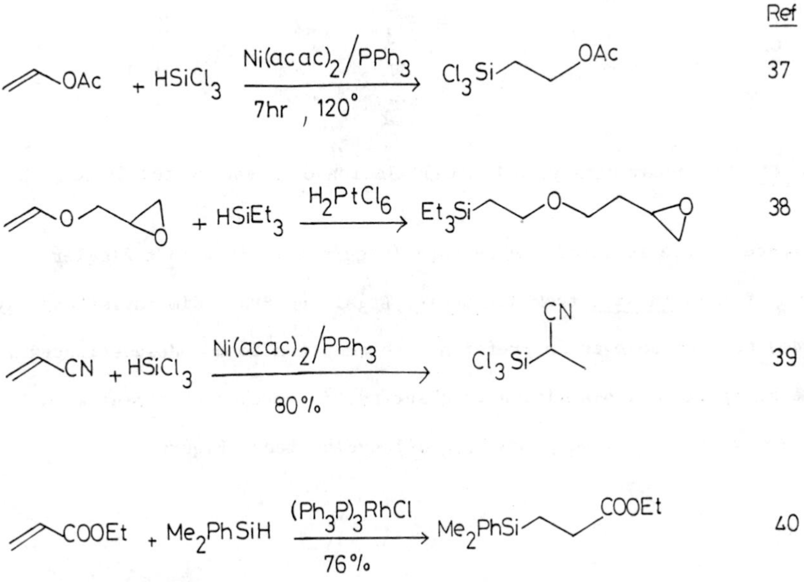

Figure 10 Hydrosilyation of alkenes bearing polar groups.

give rise to addition to carbon–oxygen as well as carbon–carbon double bonds.

4.1.2. Hydrosilylation of Dienes

While H_2PtCl_6 is not particularly active for the hydrosilylation of conjugated dienes very rapid reactions occur at room temperature catalysed by palladium and nickel complexes. 8 reacts with hydrosilanes in the presence of $(PhCN)_2PdCl_2|PPh_3$ to give 1,4-addition to the diene, in

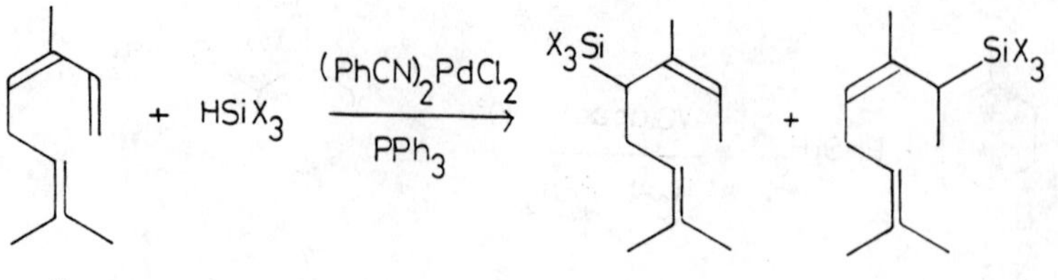

8 9 10

X_3	9	10
Cl_3	42	58
$MeCl_2$	100	0
Me_2Cl	100	0

Figure 11 Palladium catalysed hydrosilylation of a conjugated diene.

preference to the isolated double bond (Figure 11).[41] With a Ziegler catalyst formed in situ from $Ni(acac)_2$, Et_3Al and PPh_3, dimethylsilane may be added to a diene even in preference to addition to a monosubstituted double bond; again 1,4-addition is observed.[42] Further treatment with H_2PtCl_6 gave the cyclised product, a silacyclooctene (Figure 12).

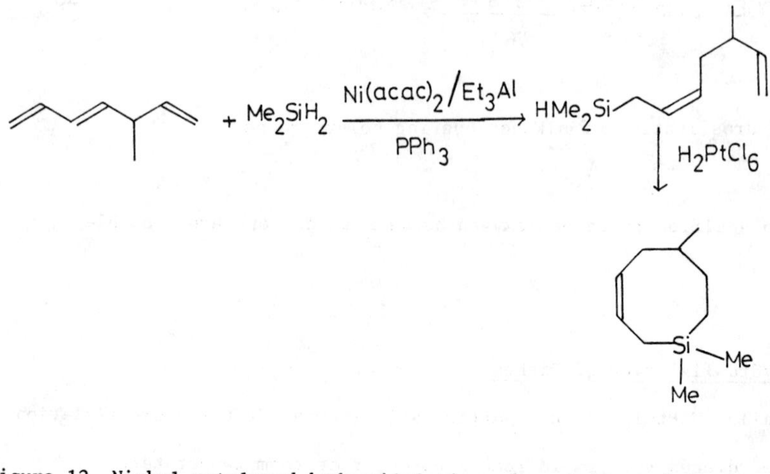

Figure 12 Nickel catalysed hydrosilylation of a conjugated diene.

In each of these cases 1,4-addition to the diene to yield a cis-alkene such as **11** was observed. However, two other products, **12** and **13** are possible even for the reaction of butadiene. Selectivity with platinum,[43] palladium[44] and rhodium[45] complexes as catalysts is generally poor; platinum catalysed reactions also tend to give substantial amounts of bis-silylated products. Reactions in the presence of nickel complexes are more successful; 1,4-addition generally predominates and in some cases reasonable stereoselectivity is achieved (Figure 13).

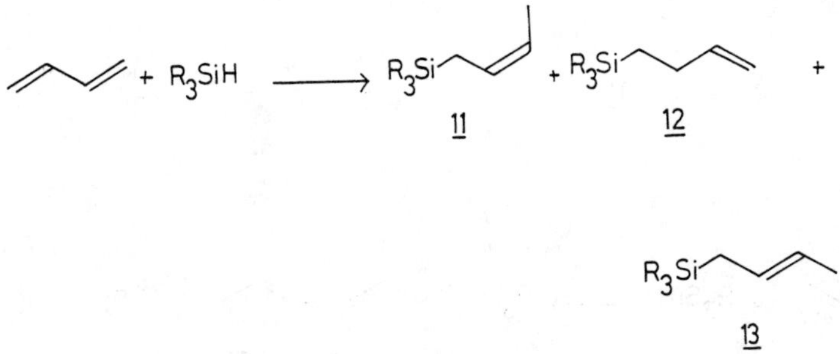

$\diagdown\diagup + R_3SiH \longrightarrow R_3Si\diagdown\diagup$ **11** $+ R_3Si\diagdown\diagup$ **12** $+$

$R_3Si\diagdown\diagup$ **13**

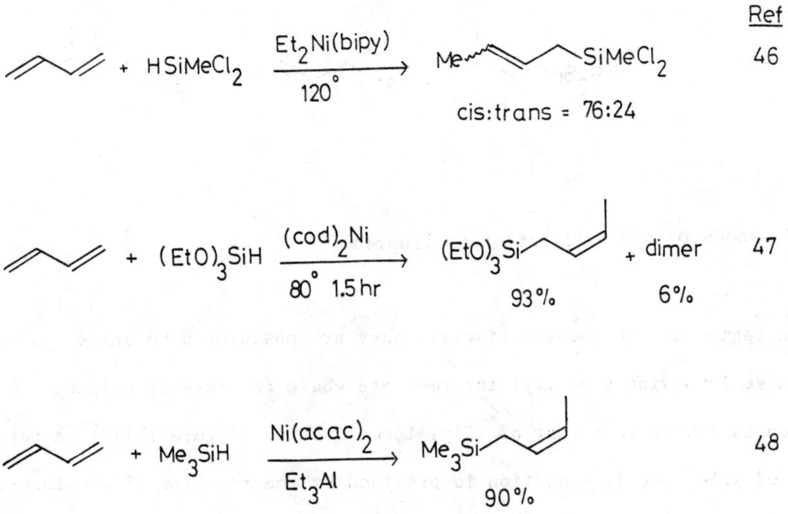

Ref

$\diagdown\diagup + HSiMeCl_2 \xrightarrow[120°]{Et_2Ni(bipy)} Me\diagdown\diagup SiMeCl_2$ 46

cis:trans = 76:24

$\diagdown\diagup + (EtO)_3SiH \xrightarrow[80° \ 1.5hr]{(cod)_2Ni} (EtO)_3Si\diagdown\diagup + dimer$ 47

93%   6%

$\diagdown\diagup + Me_3SiH \xrightarrow[Et_3Al]{Ni(acac)_2} Me_3Si\diagdown\diagup$ 48

90%

Figure 13 Hydrosilylation of butadiene catalysed by nickel complexes.

318

When considering addition to substituted dienes the further complication of regiochemistry arises and selective reactions are exceptional. Most studies have used isoprene, 14, or trans-piperylene, 15, as the substrate. H_2PtCl_6 is an unsatisfactory catalyst for hydrosilylation of isoprene; a complex mixture of products is obtained, the proportions of which depend on the substituents on the silane (Figure 14).[49] However, in the presence of $(PPh_3)_2PdCl_2$[50] or $(PPh_3)_4Pt$[51] silanes of type 18 may be

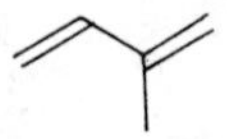

14    15

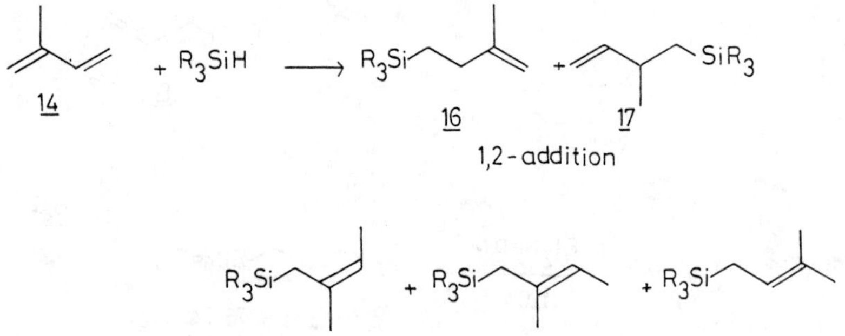

14 + R₃SiH ⟶ R₃Si~~⟍ 16 + ⟍~~SiR₃ 17

1,2-addition

R₃Si 18 + R₃Si 19 + R₃Si 20

Figure 14 Modes of hydrosilylation of isoprene.

produced regio- and stereoselectively. They are postulated to arise via a mechanism involving a π-allyl intermediate where the rate of silane insertion is faster than that of allyl isomerisation (Figure 15).[52] A rare example of selective 1,2-addition is provided by the reaction of dichloromethylsilane with trans-piperylene catalysed by $(PPh_3)_4Pt$; the silane is added to the less substituted double bond to give 86% 21.[53] This selectivity is extended to a number of substituted dienes which give mixed

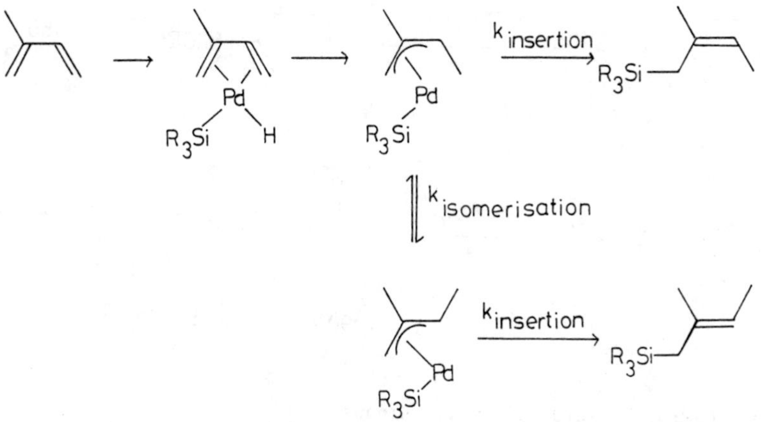

Figure 15   Hydrosilylation of isoprene catalysed by palladium complexes.

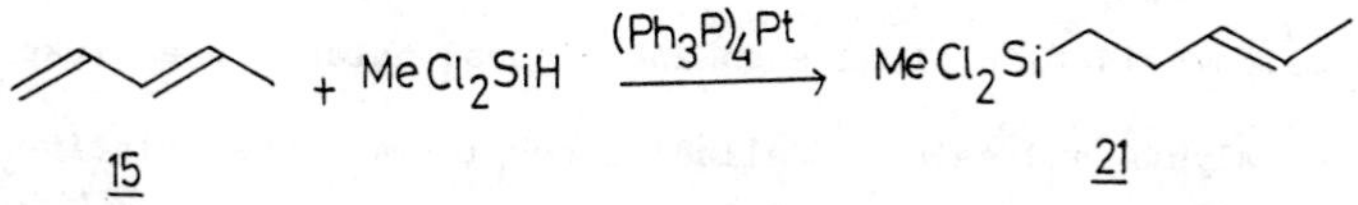

products under conditions of H_2PtCl_6 catalysis. 1,4-addition to both cis-

and trans-piperylene is also catalysed by a mixture of $Ni(acac)_2$ and

Et_3Al.[54] Very selective addition is obtained with a rather unusual

catalyst; when 15, trimethylsilane and $Cr(CO)_6$ are photolysed, 90% of pure

cis-22 is obtained together with 10% of the regioisomer 23.[55] Little work

has dealt with more highly substituted acyclic dienes but hydrosilylation of

cyclic dienes gives mainly 1,4-addition (Figure 16).

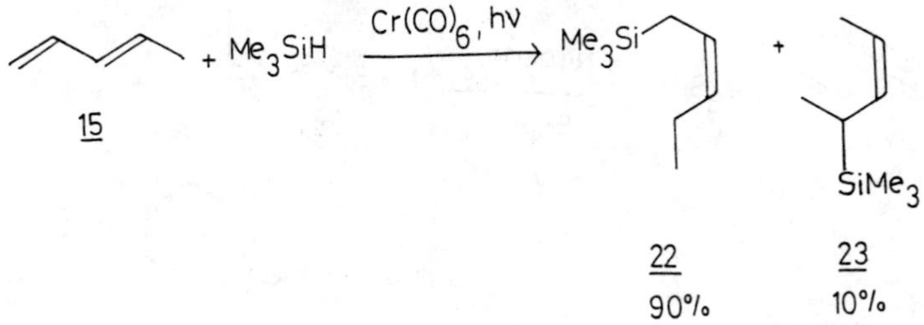

Ref

56

57

90%   5%

Figure 16   Hydrosilylation of cyclic dienes.

A number of catalysts used for hydrosilylation are also capable
of catalysing oligomerisation of alkenes and dienes, notably the nickel-
based Ziegler catalysts and several palladium complexes. The relative
abundance of products depends not only on the ratio of butadiene to silane
(substituted dienes show little tendency to telomerise under these
conditions[58]) but also on the nature of the ligands at the metal centre.
For example in the reaction of butadiene with trimethylsilane in the
presence of Ni(acac)$_2$ and Et$_3$Al 10% of an octadiene, 25, is obtained (with
a ratio of diene to silane of 2.5:1) but this yield is increased to 50% on
addition of PPh$_3$.[48] Under the same conditions the octadiene dimer is the
sole product in the reaction with phenylsilane and insertion takes place
into only one of the silicon-hydrogen bonds.[59]

24

25

An alternative type of octadiene dimer, <u>27</u>, is obtained using Ni(cod)$_2$ as catalyst,[49] the geometry probably arising from a <u>bis-π-allyl</u> nickel intermediate such as <u>28</u>. Analogous species are obtained in reactions

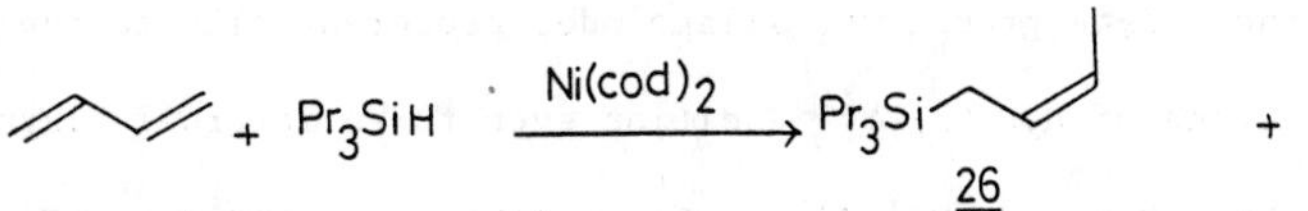

<u>26</u>

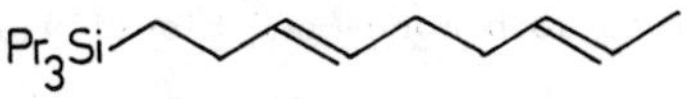

<u>27</u>

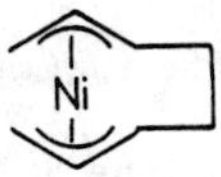

<u>28</u>

catalysed by palladium complexes giving in favourable cases complete selectivity for the octadiene product (Figure 17).

<u>Ref</u>

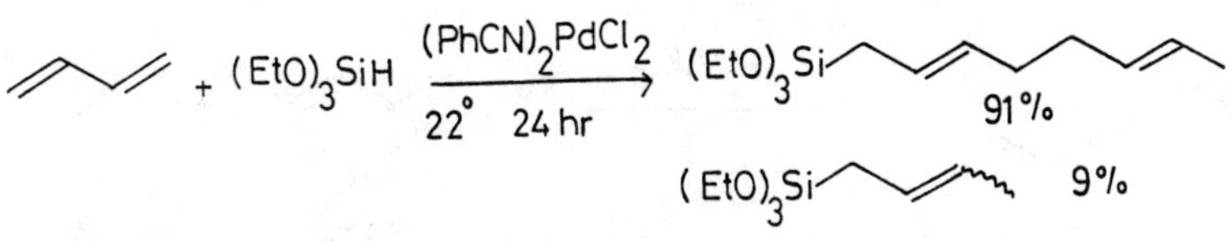

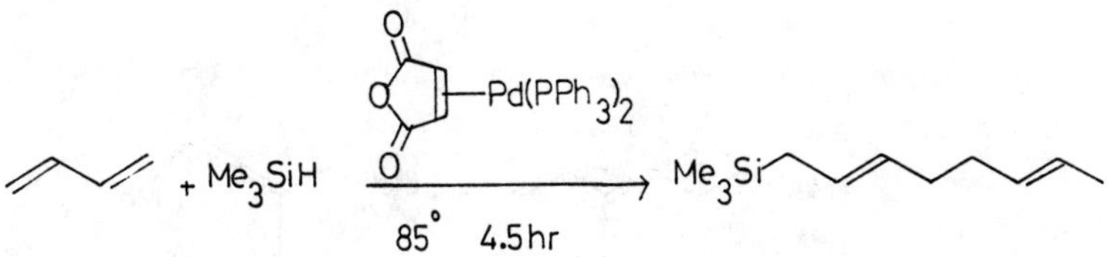

However, Cl$_3$SiH and PhMe$_2$SiH give only the 1:1 adduct under these conditions.

<u>Figure 17</u> Palladium catalysed hydrosilylation/dimerisation of butadiene.

4.1.3. Hydrosilylation of Alkynes

Hydrosilylation of alkynes generally takes place under milder conditions than those required for alkenes. For 29 which contains both terminal alkene and alkyne groups the silane adds preferentially to the alkyne in the presence of H_2PtCl_6.[61] Addition even to an internal alkyne is preferred to that to a terminal alkene but a mixture of regioisomeric products is obtained (Figure 18).[62] The hydrosilylation of acetylene may give rise to both mono- and bis-silylated products. The vinylsilane is generally the more useful of the two[63,64] but it is only recently that any systematic investigation has been directed toward practical laboratory syntheses. Watanabe and his coworkers[65] tested a wide range of both catalysts and silanes. With trichlorosilane 90% of the vinylsilane was

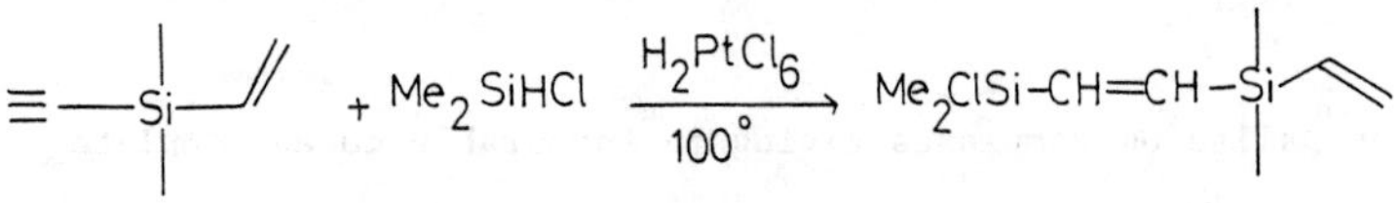

29

30

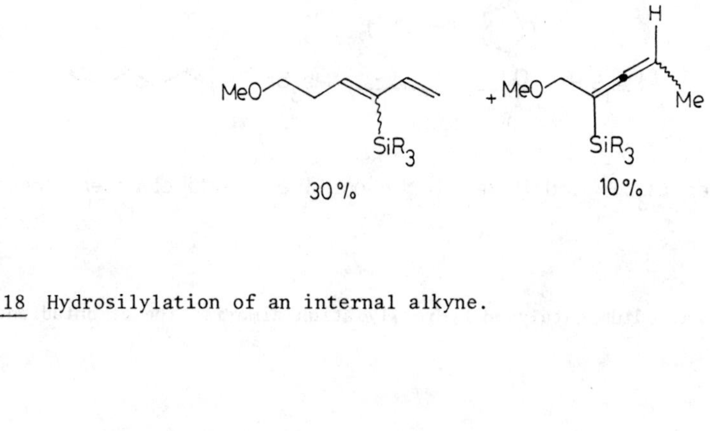

60 %

30 %

10 %

Figure 18 Hydrosilylation of an internal alkyne.

produced in reactions catalysed by $RuCl_2(PPh_3)_3$ or H_2PtCl_6 though the latter

is much the more active catalyst. For triethoxysilane $PtCl_2(PPh_3)_2$ was the

optimal catalyst giving a ratio of 97:3 of mono:bis adducts. Methyl-

dichlorosilane gave generally poor selectivity; the best result was in a

reaction catalysed by $RhH(PPh_3)_4$ which gave 54% vinylsilane and 17% bis-

silylalkane. The same catalyst was effective for addition of methyl-

diethoxysilane to give only monosilylated product in 89% yield. Finally

with n-hexyldichlorosilane the best catalyst was $Pt(PPh_3)_4$ giving 87% mono-

and 3% bis-adduct. In all these later cases H_2PtCl_6 catalysed bis-

silylation exclusively in a reaction exothermic at room temperature.

Russian workers have also prepared mono- and bis-silane adducts with

acetylene in the presence of rhodium complexes; their main aim was to

achieve the transformation under mild conditions.[66] Despite the importance

of vinylsilanes in organic synthesis, hydrosilylation is rarely the method

of choice for their preparation. The most common route involves silylation

of alkenyl metals (e.g. 32) the reaction being stereospecific.[2,67,68]

Wurtz-like coupling of vinyl chlorides such as 34 is also successful[69] and

a wide variety of vinylsilanes have been prepared from the corresponding

silylated alkynes[70] (e.g. 37 → 38). While hydrosilylation reactions may be

adjusted to give useful yields and selectivities by extensive optimisation,

there are few general guidelines of predictive value.

The characteristic product of addition of a silane to a terminal

alkyne in the presence of H_2PtCl_6 is a trans-alkene, the product of cis-

addition[71] (Figure 19). The proportion of the α-isomer depends on the

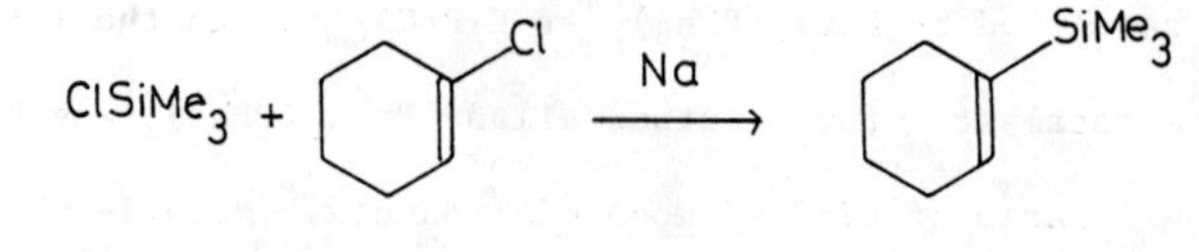

Figure 19 Hydrosilylation of a terminal alkyne.

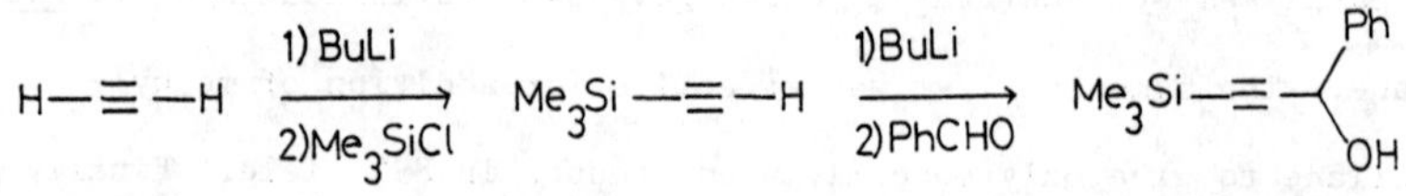

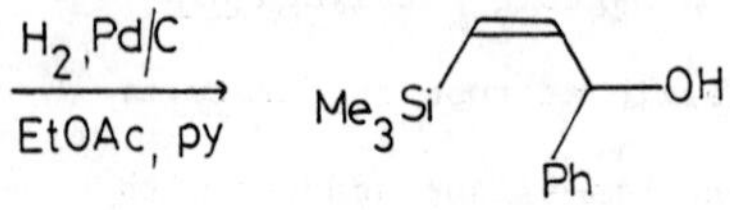

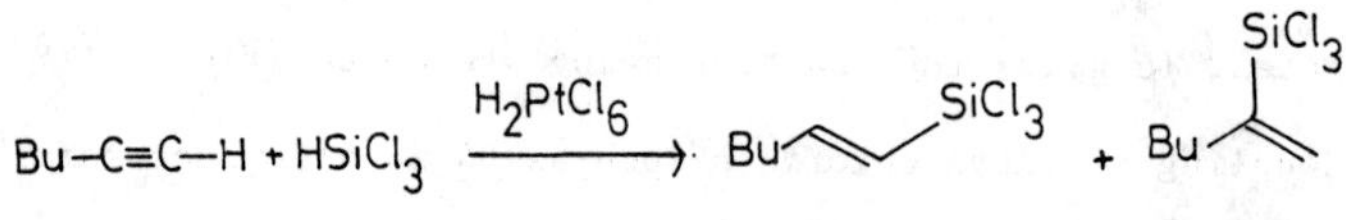

silane, substrate and solvent. For the addition of triethylsilane to RC≡CH the proportion of the α-adduct decreased for electron donating R-groups; α:β = 10:90 for R= t-Bu and 55:45 for R= COMe.[72] The presence of a donor solvent such as THF increases the proportion of β-adduct.[73,74] Chvalovský et al.[75] have studied the effect of variation of the silane but were not able to draw firm guidelines. The binuclear platinum catalysts such as 39, developed by Stone and co-workers do, however, give rather selective reactions. For example the addition of trichlorosilane to 1-heptyne proceeded in 93% overall yield to give β- and α-adducts in the ratio of 96:4.[76]

$$39$$

Data on the hydrosilylation of alkynes by rhodium complexes are more sparse. $(PPh_3)_3RhCl$ catalyses the addition of trialkylsilanes to give almost exclusively the β-adduct but as a mixture of cis- and trans-isomers. It is suggested that the trans-product is formed initially but is isomerised under the reaction conditions probably by $RhHCl(SiEt_3)(PPh_3)_2$.[77] No hydrosilylation is observed in the presence of this catalyst, however, if the alkene is rigorously purified and oxidants are excluded. Their rôle appears to be the removal of a mole of triphenylphosphine as its oxide to give a coordinatively unsaturated complex; when the ratio of phosphine to rhodium is less than two the catalyst is active in the absence of oxygen.[78] When complexes of other phosphines are used the cis:trans ratio in the product may be significantly altered. If a catalyst is formed in situ from R_3P and $(RhCl(cyclooctene)_2)_2$ a good yield of vinylsilane is obtained with P:Rh= 0.5 to 1, falling off rapidly in the presence of excess phosphine. For R= PhO the trans:cis ratio for the addition of triethylsilane to hexyne is 63:30 but for R= o-C_6H_4-OMe trans:cis = 9:82.[79,80] It is suggested that the reaction proceeds as in Figure 20 and that the degree of isomerisation depends on the ability of the donor phosphine to stabilise the dipolar intermediate, 40.

There are few examples of the hydrosilylation of simple disubstituted alkynes. Diphenylacetylene[81] and but-2-yne[82] add silanes in the cis-manner in the presence of H_2PtCl_6. A similar reaction may be achieved under milder conditions using the binuclear catalysts of Stone.[83]

It is necessary to mention the hydrosilylation of alkynes with an oxygen or nitrogen atom β to the triple bond, if only because of the

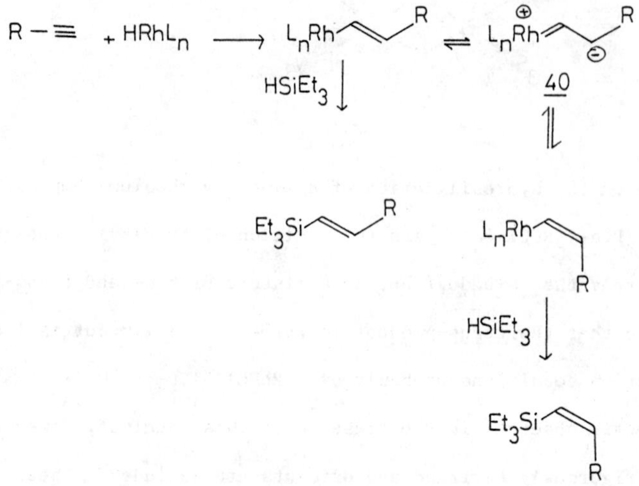

$$R-\!\!\equiv\; +\; \mathrm{HRhL}_n \longrightarrow \mathrm{L}_n\mathrm{Rh}\!\!\diagup\!\!\diagdown\!\!\diagup R \;\rightleftharpoons\; \mathrm{L}_n\overset{\oplus}{\mathrm{Rh}}\!\!\diagup\!\!\diagdown\!\!\diagup\overset{\ominus}{}\!\! R$$

HSiEt$_3$ ↓   40

$$\mathrm{Et_3Si}\!\!\diagup\!\!\diagdown\!\!\diagup R \qquad\qquad \mathrm{L_nRh}\!\!\diagup\!\!\diagdown\!\! R$$

HSiEt$_3$ ↓

$$\mathrm{Et_3Si}\!\!\diagup\!\!\diagdown\!\! R$$

Figure 20 Rhodium catalysed hydrosilylation and isomerisation.

numerous reports in this field. Unfortunately some of these do not bear
close examination; frequently only the major product is isolated and
stereochemistry is ignored. Examples are given in Figure 21.

4.1.4. Asymmetric Hydrosilylation of Alkenes

        The addition of silanes to prochiral alkenes catalysed by
transition metal complexes bearing chiral ligands has not received a great
deal of attention and achievement has been correspondingly modest. This may
well reflect the fact that, as yet, relatively few direct uses for chiral
silanes have been developed. Asymmetric hydrosilylation of geminally
disubstituted alkenes has been achieved using platinum, rhodium, nickel and
palladium complexes of chiral phosphines. With cis-ethylene-(R)-benzyl-
methylphenylphosphine platinum dichloride, 41, as catalyst dichloromethyl-

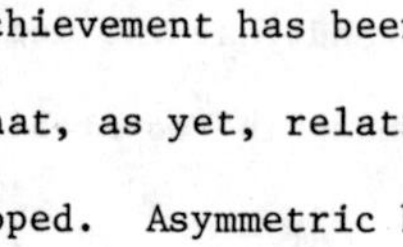

$$\mathrm{PhMeBzP}\diagdown\!\!\underset{\mathrm{Pt}}{}\!\!\diagup\!\!\overset{\mathrm{Cl}}{\diagdown}\,\mathrm{Cl}$$

41

Ref

84

20-40%

8 - 10%

85

68%

32%   trace

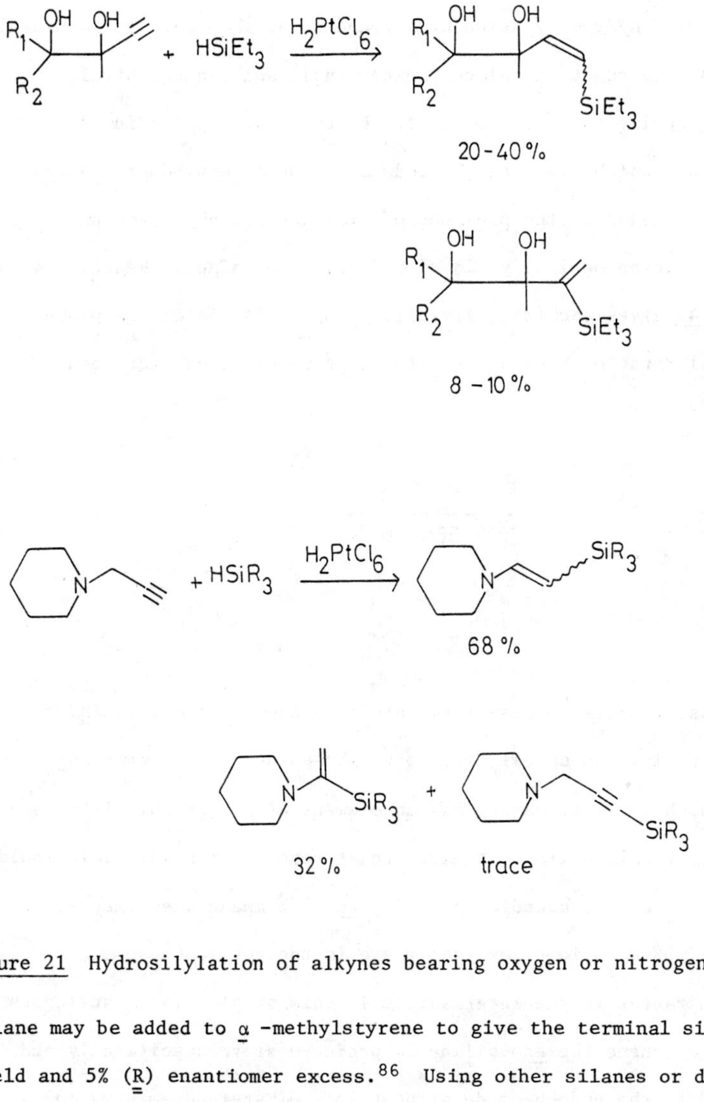

Figure 21 Hydrosilylation of alkynes bearing oxygen or nitrogen functions.

silane may be added to α -methylstyrene to give the terminal silane in 43%

yield and 5% (R) enantiomer excess.[86] Using other silanes or different

platinum complexes gave even less successful results. The same reaction

catalysed by trans - bis(phosphine)nickel dichloride complexes gives a lower

(31%) yield but a better enantiomer excess (21% R). However, unfortunately

some hydrogenolysis of Si-Cl bonds also occurs.[87] The rhodium catalyst,

(DIOP)RhCl(solvent), so successful for asymmetric hydrogenation, is also

inefficient here giving only a 10% (S) optical yield in the reaction of

$\underline{\alpha}$- methylstyrene with trimethylsilane.[88]

Asymmetric hydrosilylation of styrene cannot be achieved in nickel or platinum catalysed reactions since the terminal, and hence achiral, silane always predominates in the product. However, with palladium (II) phosphine complexes which give the $\underline{\alpha}$ - adduct as the sole product, when trichlorosilane is added in the presence of bis(menthyldiphenylphosphine) palladium dichloride an optical yield of 34% ($\underline{S}$) is obtained. Reaction with cyclopentadiene, 42, gives a chiral allylsilane, 43, in 81% yield; the product has a substantial rotation but α_D for the pure material has not been established.[89,90]

$$\text{42} \qquad + \quad HSiCl_3 \quad \xrightarrow[120°,\ 58hr\ ,\ 81\%]{(MDPP)_2\ PdCl_2} \qquad \text{43}$$

The most impressive result to date is achieved with a palladium catalyst incorporating the chiral ferrocenyl phosphine, $\underline{R}$ - $\underline{S}$ PFPA, 44, which contains both a chiral centre and an element of planar chirality. Catalysed by 44, trichlorosilane is added to styrene to give 45. This could be converted into the corresponding alcohol, 46 (52% enantiomer excess) or the bromide, 47 (10% enantiomer excess; there is thought to be some carbonium ion character in the intermediate in this step). In an analogous reaction with norbornene the exo-silane is produced stereospecifically and may be converted to the endo-bromide without loss of stereochemistry (53% enantiomer excess).[91]

44

4.1.5. Polymer Supported Catalysts

Polymer supported hydrosilylation catalysts have a wide application in industrial processes. They may be easily removed from the reaction medium and often reused with little loss of activity. Rhodium, platinum and palladium complexes bonded to either organic or inorganic supports have proved valuable.

Catalysts formed by treatment of a variety of functionalised styrene-divinylbenzene copolymers or polymethylacrylates with H_2PtCl_6 successfully promote the addition of triethoxysilane to 1-hexene giving 90-99% yields in 2 hr at 80°. Three successive reuses of the catalyst reduced the activity only to 98%. In this case the terminal silane was obtained as the sole product but with the less reactive triethylsilane some branched isomer was detected together with isomerisation of the starting material.[92] H_2PtCl_6 bonded to an anion exchange resin also gives a very stable catalyst which showed no loss of activity for the regiospecific addition of dimethylphenylsilane to 1-heptene after 180 hr use.[93] Unlike their soluble analogues rhodium catalysts bonded to functionalised polystyrene are air-stable, losing no activity after 3 days in air.[94] Additionally regioselective addition is observed, giving a yield of 83% terminal silane for the addition of triethylsilane to 1-hexene.[95]

Platinum, palladium and rhodium have also been bonded to inorganic silica or alumina supports. Supported ammonium and phosphonium platinum (II) and (IV) salts give similar regio- and stereoselectivity to homogeneous platinum catalysts in the hydrosilylation of alkenes and alkynes but the supported species are more stable and reusable.[96] Other platinum on silica catalysts give more β -adduct in the addition of trichlorosilane to styrene than H_2PtCl_6 and this selectivity may be increased by donor additives (PPh$_3$ or pyridine).[97] The rhodium catalysts show similar activity and selectivity to their soluble counterparts and a moderate lifetime; when the catalyst is reused for the third time the reaction rate is slowed by 50%. It has been found to be advantageous for the rhodium to be bound very close to the catalyst surface; when the support is silica-$(CH_2)_n$-PPh$_2$ the catalyst where n=1 gives reaction about ten times as fast as those where n=2-6. This may be accounted for on the basis that when n is large dimeric structures may be formed where the rhodium is coordinatively saturated.[98] The only reported palladium catalyst supported on silica is of high activity but may not be reused as the metal is slowly abstracted into solution.[97]

While it is not possible to discuss in detail the industrial processes involving hydrosilylation the main classes of important reactions will be illustrated. The preparation of organosiloxanes and polyorgano-siloxanes has been achieved using platinum, palladium and rhodium catalysts. For example, the hydrosilylation polymerisation of dihydroorganosiloxanes catalysed by (PPh$_3$)$_3$RhCl gives highly stereoregular polycarboorgano-siloxanes with the reduced formation of $\underline{\alpha}$ -addition byproducts.[99] The addition of 1,3,5,7-tetramethyl-1,3,5,7-tetravinylcyclotetrasiloxane, <u>49</u>, to 1,3,5,7-tetramethyl-1,3,5,7-tetrahydrocyclotetrasiloxane, <u>50</u>, is catalysed at 120° by (PPh$_3$)$_4$Pt giving a transparent, hard, glass-like organopolysiloxane.[100]

Hydrosilylation is extensively used for the cross-linking of polymers and, in particular, for the vulcanisation of silicone rubber. For example [P(OPh)$_3$]$_4$Pt and a tin (II) salt cross-link a siloxane polymer to a silicone rubber with lower viscosity than that produced using H_2PtCl_6 as

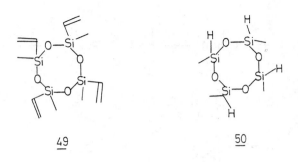

49   50

catalyst.[101] The copolymerisation of vinylmesitylene and vinyldimethyl-

chlorosilane gave a vinylsilane terminated polymer of molecular weight

approximately 6100. This was hydrosilylated with an Si-H terminated

dimethylsiloxane of molecular weight 9300 in the presence of H_2PtCl_6 to

give a block polymer of good tensile strength and heat resistance.[102]

4.1.6. Mechanisms of Hydrosilylation

It is not possible to discuss here the mechanisms of hydro-

silylation by the whole range of catalysts which have provided examples in

the previous sections. Many still lie within the realms of speculation

while others lack generality. Only the most widely used catalysts will be

considered and the emphasis will be on trends which relate to their

selectivity rather than details of intermediates or kinetics.

The mechanism proposed by Chalk and Harrod is generally accepted

for hydrosilylation catalysed by H_2PtCl_6[103] (Figure 22). However, it is

necessary to note that there is direct evidence for few of these species

and some catalysts are thought not to give strictly homogeneous solutions.

The mechanism accounts for alkene isomerisation via the reversible formation

of a metal-alkyl and for the exchange of deuterium into various positions

of the adduct from an Si-D bond.[104] The rate of hydrosilylation depends on

the stability of the platinum styrene complex for a series of substituted

styrenes, suggesting the pre-rate determining step formation of such a

complex.[105] The relative reactivity of the silanes decreases with an

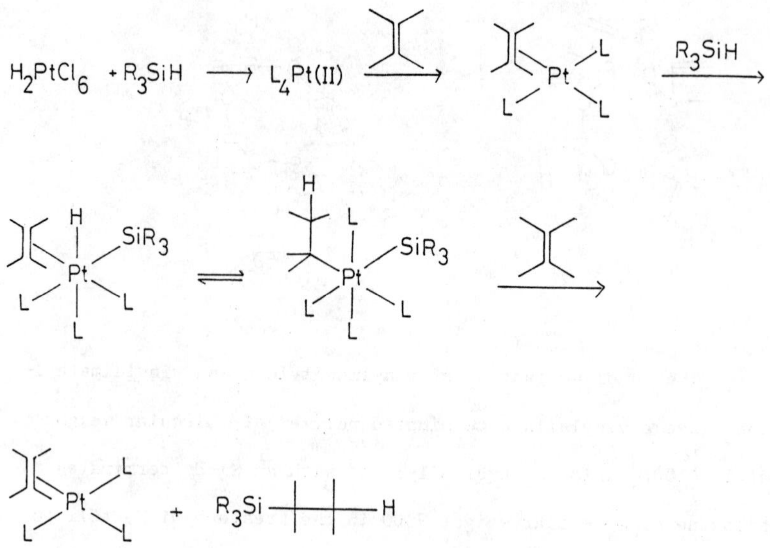

$$H_2PtCl_6 + R_3SiH \longrightarrow L_4Pt(II) \longrightarrow \quad \xrightarrow{R_3SiH}$$

Figure 22 Mechanism of hydrosilylation catalysed by H_2PtCl_6.

increase of electron density on the hydrogen atom and the following series have been noted: $HSiCl_3$ > $HSiCl_2Et$ > $HSiClEt_2$ > $HSiEt_3$ > and $HSiPr(OEt)_2$ > $HSiPr_2(OEt)$ > $HSiPr_3$.[106] The changes do not correlate directly with the Taft parameter so steric effects must also be involved.

Very similar mechanisms are proposed for reactions catalysed by rhodium and nickel complexes.[20] For catalysis by $(R_3P)_3RhCl$ studies have led to the isolation of many adducts of the type $(R_3P)_2Rh\ H(SiR_3')Cl$[107,108] and it has been found that the reactivity of the silane in hydrosilylation is inversely related to the stability of these adducts. However, the effect of oxygen has been previously noted[79,109] and a number of complexes (e.g. $((C_2H_4)_2RhCl)_2$[28]) which do not contain phosphine ligands also give good catalysts, so it is clear that a single mechanism is not general.

There has been little specific investigation into the mechanisms of hydrosilylation of alkynes but there has been considerable speculation about the pathways followed in the reactions of conjugated dienes. Much of this hinges on the relative stabilities of π- allyl intermediates and there are clear contrasts between nickel and palladium complex catalysed reactions. In the addition of a silane to butadiene palladium catalysts give trans- alkenes while those based on nickel give cis- alkenes.[48] (Figure 23). This dichotomy of behaviour is preserved in telomerisation reactions yielding octadienylsilanes[44,60] It seems likely that the stereo-isomers arise from different allyl isomers (Figure 24). For nickel catalysts the rate of hydrosilylation is greater than that of allyl isomerisation, the converse obtaining for palladium.

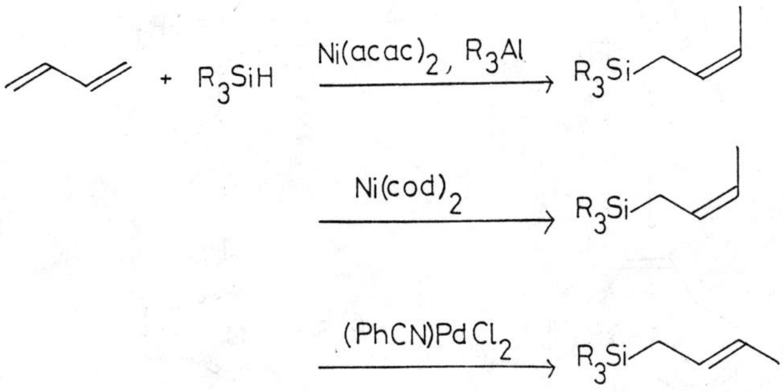

Figure 23 Stereochemistry of hydrosilylation of butadiene.

4.1.7. Conclusions

Recent advances in organosilicon chemistry as a tool for organic synthesis has led to a demand for a variety of functionalised organosilanes. Hydrosilylation, even with the wealth of catalysts currently available, can accomplish only a few transformations in high yield and selectivity. The production of industrially important organosilicon monomers has to date attracted most attention but the application of these methods in organic synthesis will surely gather pace in the future.

334

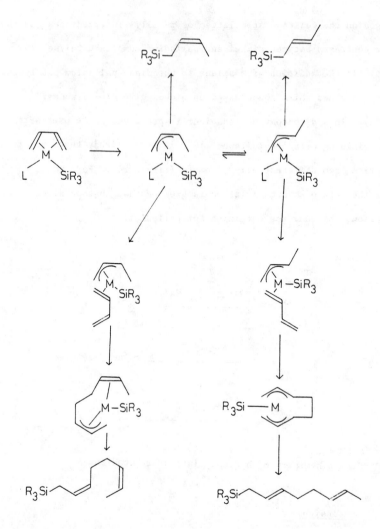

Figure 24   Mechanism of hydrosilylation/dimerisation of butadiene.

4.2.   Cyclopropanation

4.2.1.   The Simmons-Smith Reaction

A wide variety of natural products contain cyclopropane rings and their synthesis is most frequently accomplished by the insertion of a carbene, or carbene equivalent, into a carbon-carbon double bond.   The earliest general method used to accomplish this was reported in 1958[110] by

Simmons and Smith and it has been used widely since that time.[111] It

involves the reaction of an organozinc reagent prepared from methylene

iodide and a zinc copper couple with a double bond (Figure 25). It is

normally stereospecific with respect to the double bond and may be used for

large-scale preparations. The precise nature of the organozinc compound has

been the subject of some dispute but it is clear that the copper functions

only to activate the zinc surface to reaction.[112]

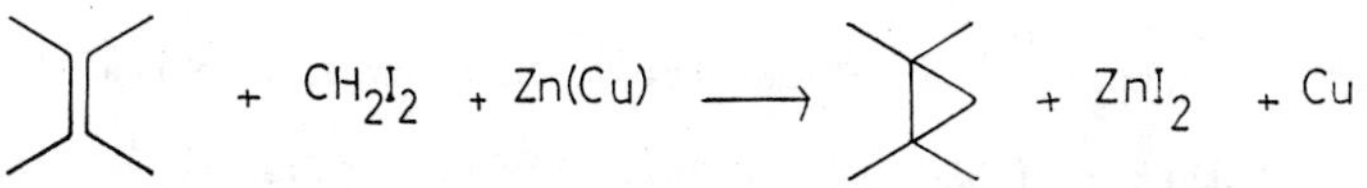

Figure 25   Simmons Smith reaction.

A number of modifications of the Simmons-Smith reaction have been

developed. One of the products of the reaction is zinc iodide and its Lewis

acid character has proved deleterious in the isolation of sensitive products.

This problem may be avoided by using glyme, (in which the salt is insoluble)

in place of the more usual ether, as the reaction solvent. Under these

conditions 52 may be isolated in 70% yield.[113] The use of silver in place

of copper has been reported to give a more active reagent.[114] Under these

conditions cyclohexene is converted to bicyclo[4.1.0]heptane in 95% yield in

2 hr at 0°C, the conventional Simmons-Smith reaction taking 24 hr at 25°C to

achieve the same result. Addition of pyridine to remove the zinc salts

replaced hydrolysis. The carbenoid derived from diethylzinc and methylene

iodide is also highly reactive and unlike the Simmons-Smith reagent does not

cause polymerisation of vinyl ethers. Thus propenyl iso-butyl ether 53 may

be cyclopropanated to 54 in 80% yield by this method.[115]

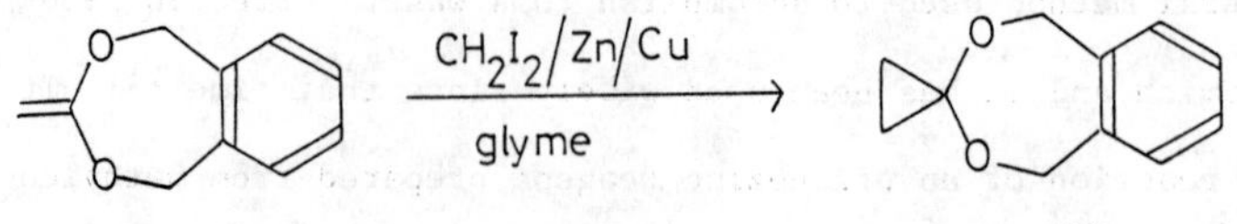

$CH_2I_2/Zn/Cu$
glyme

51 52

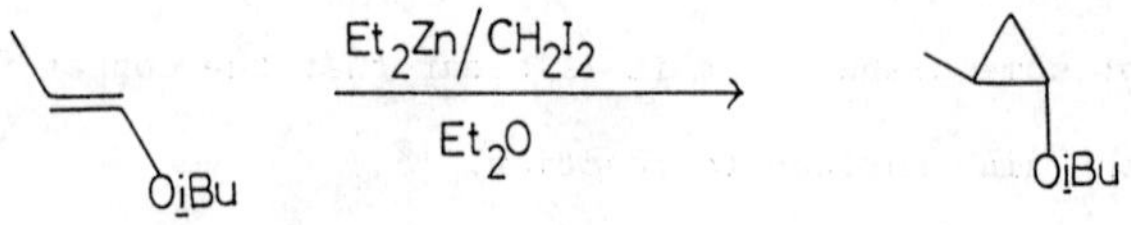

Et_2Zn/CH_2I_2
Et_2O
OiBu
OiBu

53 54

The syn/anti stereoselectivity achieved in simple reactions of this type is rather variable (Figure 26). However, diastereoselectivity in the presence of a nearby oxygen functionality is excellent, attack occuring from the same side as the oxygen, presumably via pre-coordination of the zinc. This was used to prove the relative configuration of the bicyclic sesquiterpene thujopsene, 56, a constituent of wood oil.[118] Cyclopropanation of the cis-allylic alcohol 55 gave material identical with that from natural sources.

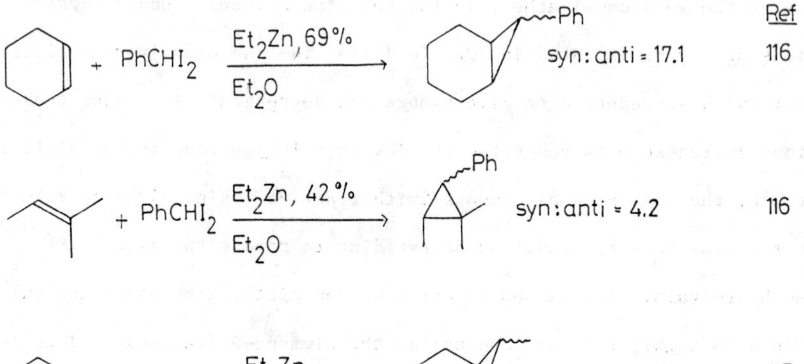

Ref
+ $PhCHI_2$
Et_2Zn, 69%
Et_2O
Ph
syn:anti = 17.1
116
+ $PhCHI_2$
Et_2Zn, 42%
Et_2O
Ph
syn:anti = 4.2
116
+ CH_3CHI_2
Et_2Zn
syn:anti = 1.5
117

Figure 26 Syn/anti stereoselectivity of the Simmons Smith reaction.

55   56

4.2.2. Decomposition of Diazo Compounds by Transition Metal Complexes

Strictly, the Simmons-Smith reaction is neither catalytic nor homogeneous, but it is still the most common synthesis of cyclopropanes. Catalytic cyclopropanation is achieved by the decomposition of diazo compounds catalysed by transition metals and their salts. Although there is little direct evidence for their existence the formation of transient metal carbenoids as intermediates is now universally accepted. The reaction may be homogeneous or heterogeneous and intra- or intermolecular (Figure 27). The most common catalysts are copper salts of which a wide range has proved useful. As might be expected for reactions involving carbenes the major competing pathways are dimerisation (thus diazoacetates yield maleate and fumarate derivatives) and insertion into allylic carbon-hydrogen bonds. This latter is a particularly serious problem when the carbene precursor is a diazomalonate, but may be partly overcome by the addition of peroxides, which also enhance the reaction rate.[122]

The diverse nature of the copper containing catalytic systems has led to conflicting interpretations of mechanistic investigations. Much of the debate has generated more heat than light. Both copper (I) and copper (II) salts are able to function as catalysts or their precursors and the oxidation state of the active species is in some doubt. Kochi[123] argues that copper (II) salts with poorly coordinating anions such as triflate and tetrafluoroborate are reduced to copper (I) during the reaction. However, Wulfman reinterprets these data and cites his own

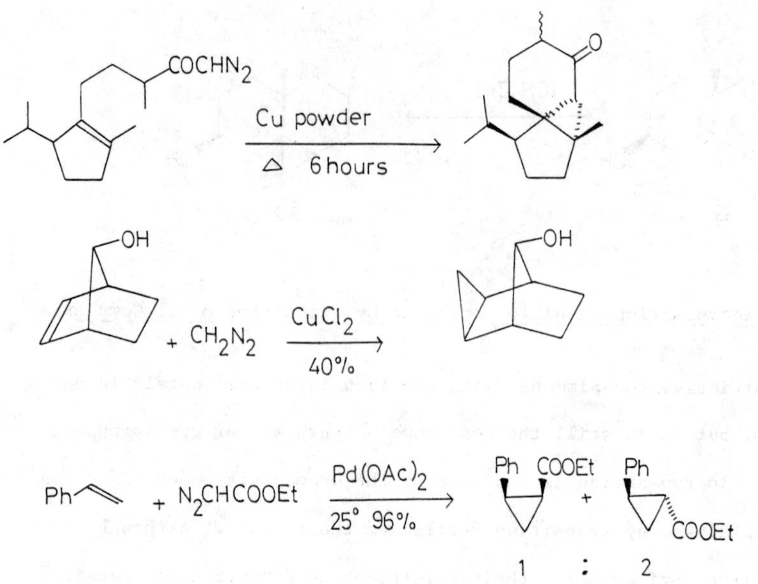

Figure 27   Catalytic decomposition of diazo compounds by transition metals
and their salts.

extensive work in this field to support the theory of a copper (II)

catalysed reaction.[124]   Subsequently it has been postulated that catalysts

such as copper (I) triflate, with poorly coordinating counterions, are able

to bind alkene and carbene simultaneously; they are thus fairly selective

for unhindered alkenes.   On the other hand, for catalysts with a strongly

coordinating anion only the carbene is metal bound and it is subsequently

transferred to the double bond with a selectivity dependent on the

electronic and steric character of both.

This distinction between catalysts having a single vacant

coordination site and those having multiple accessible sites is well

illustrated by a comparison of the decomposition of diazo compounds by

rhodium and palladium catalysts.   Rhodium carboxylates have only one

potential site for the coordination of an extra ligand so that the carbene

but not the alkene becomes bound to the metal.[125]   This results in a rather active catalyst which does not discriminate efficiently between alkene

substrates. It is useful for internal alkenes such as *trans*-4-octene, which

is rather unreactive under other conditions[126] and trisubstituted alkenes

such as 57 and 59.[127] Rhodium acetate also catalyses the successful

synthesis of cyclopropenes from alkynes and a rather selective ring

expansion of benzene (Figure 28).

In contrast palladium (II) carboxylates have several binding

sites and it is possible that both carbene and alkene are coordinated

simultaneously. Cyclopropanation of strained (e.g. norbornene[130]) or

activated (e.g. styrene) alkenes was early reported, but hindered substrates

were less reactive. The reactivities of alkenes followed approximately the

order of their binding constants to silver ion, confirming that pre-

coordination is important.[125] More recently mild conditions have been found

in which unactivated mono- and disubstituted alkenes give good yields of

cyclopropanes (Figure 29). In the latter cases the less substituted double

bond reacts selectively; the Simmons-Smith reaction gives mixed products

in these cases.

One interesting example of the contrast between rhodium and

palladium catalysts arises from their behaviour in the cyclisation of the

diazocarbonyl compound 61.[133] (Figure 30). Palladium (II) is able to form

Ref

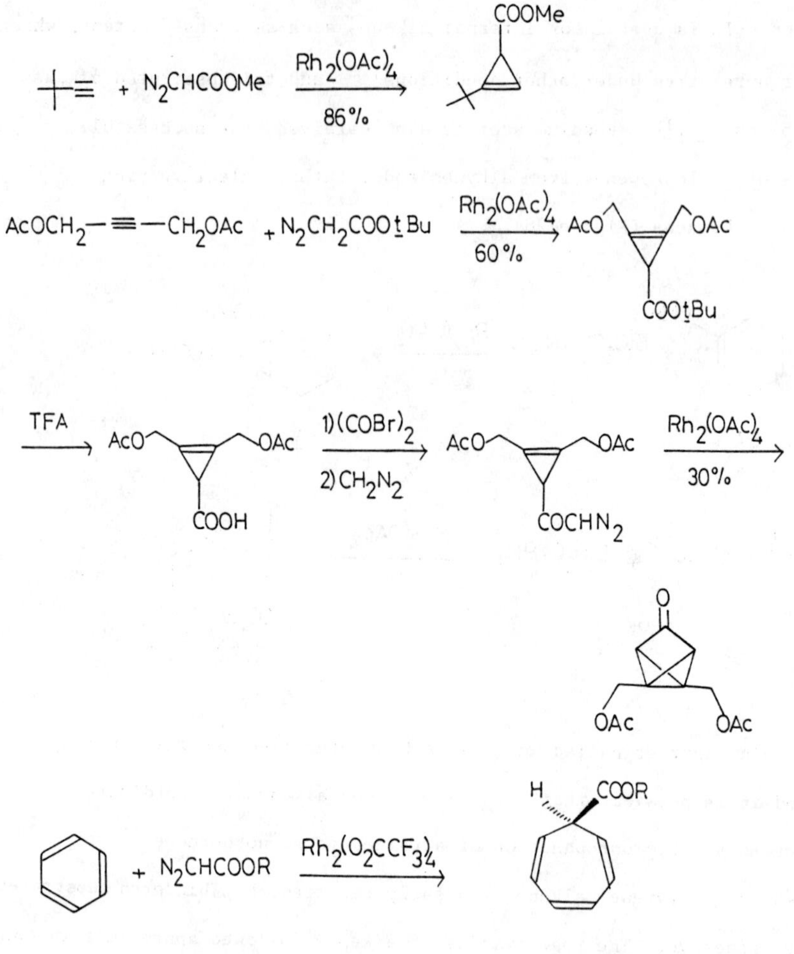

128

129

126

Figure 28 Rhodium acetate catalysed carbene additions.

a complex with both alkene and carbene, thus holding them in the correct orientation for insertion. Rhodium (II) formed only a carbene complex which then reacts by the favoured 5-exo-trig pathway to give ring closure.[134]

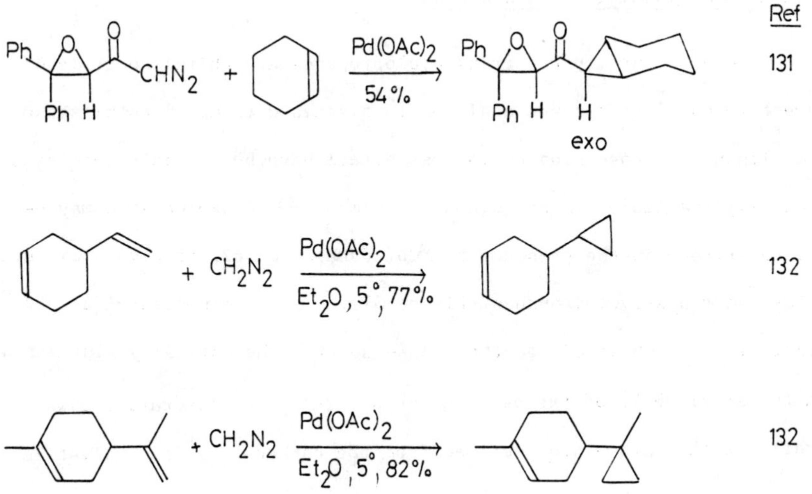

Ref

131

132

132

Figure 29 Palladium acetate catalysed carbene additions.

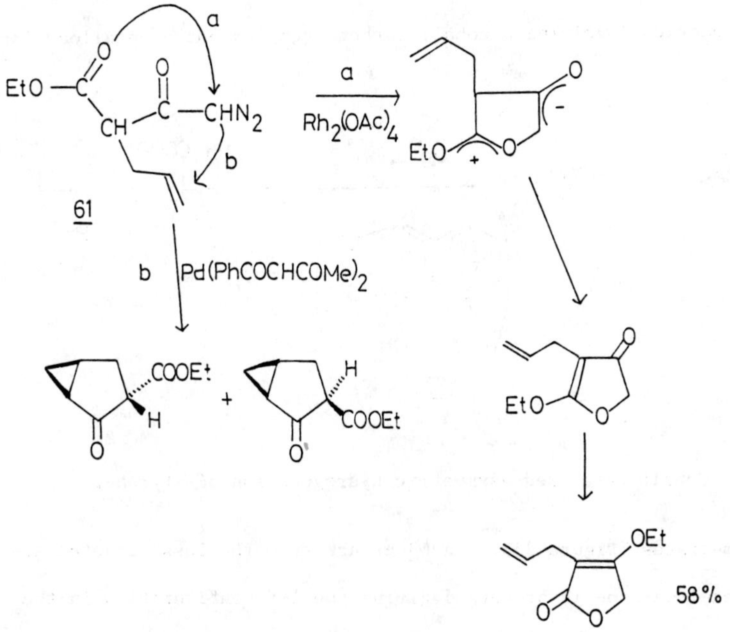

Figure 30 Alternative modes of reaction of a diazocarbonyl compound in

the presence of rhodium or palladium catalysts.

4.2.3. Asymmetric Cyclopropanation

Many naturally occurring cyclopropanes are chiral so it is of interest to catalyse carbene addition to alkenes with metal complexes of chiral ligands. Those studied in most detail have been cobalt complexes of the readily available camphorquinone dioximes.[135] Diazoacetates may be added to styrene in the presence of $Co(\alpha cqDH)_2H_2O$, 62, to give a mixture of cis- and trans-cyclopropanes (Figure 31). When R = neopentyl an 88% optical yield is obtained for the trans-isomer. The optical yield increases with the steric bulk of the ester group and the S- configuration always prevails at the carbon atom derived from the carbene. This catalyst is also reactive enough to catalyse cyclopropanation of electron deficient alkenes, $\alpha\beta$- unsaturated esters and nitriles, and its reaction with conjugated dienes is regioselective for the terminal double bond   (Figure 32). A convincing rationalisation is presented for the enantioselectivity of the reaction involving a cobalt carbene complex and a metallocyclobutane

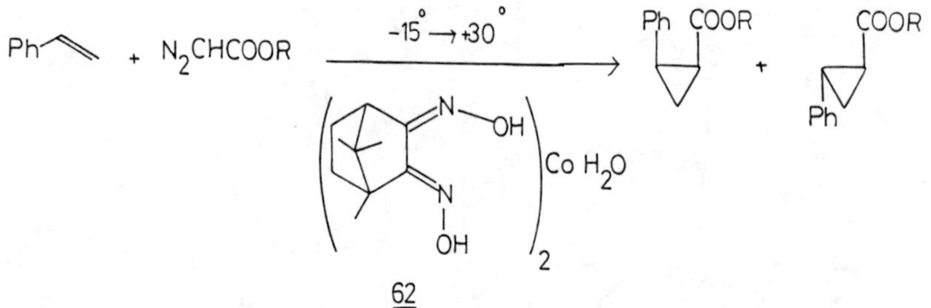

Figure 31 Cobalt catalysed asymmetric hydrogenation of styrene.

as intermediates (Figure 33). In 63 attack from the less hindered si- face of the metal-carbene is preferred giving the 1-S configuration in the product.

The pyrethrin family of insecticides have attracted considerable attention because of their high activity and low mammalian toxicity. Simple

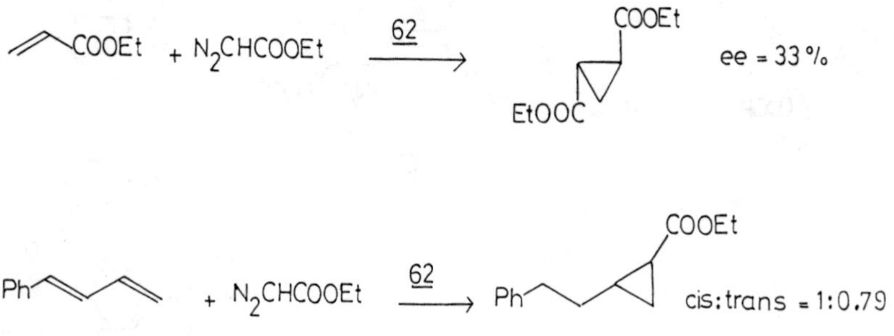

Figure 32   Cobalt catalysed cyclopropanation of conjugated substrates.

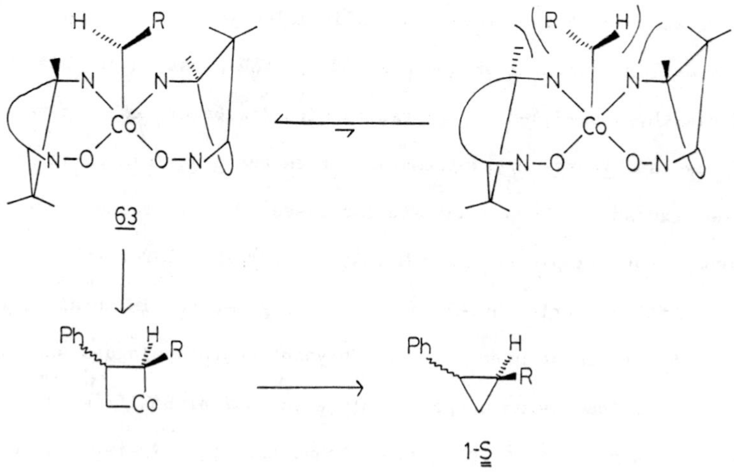

Figure 33   Proposed transition state for cyclopropanation by cobalt camphorquinone oxime catalysts.

esters of (+)-trans - chrysanthemic acid 64 show activity as high as that of the more complex naturally occurring esters.[136] Unnatural halogenated pyrethroids such as 65 have been found to have improved photostability.[137] The "natural" stereoisomer in all cases shows the highest biological activity so there has been considerable interest in their asymmetric synthesis.   The most successful catalysts to date have been copper complexes of the chiral Schiff's bases formed from aromatic aldehydes and amino alcohols. For example 66 is formed from salicylaldehyde and an amino alcohol derived from a natural amino acid.[138] In the presence of this complex 2,5-dimethylhexa-2,4-diene, 67, reacts with diazoacetates to give mixtures of cis- and trans-chrysanthemates. Both the trans/cis ratio and the enantiomer excess in the product are increased by an increased bulk in the ester group; for example for R_1 = Me, R_2 = 5-tert-butyl-2-octyl-oxyphenyl and R = 2,3,4-trimethyl-3-pentyl, the trans-ester is obtained in 92% yield and 88% (+) enantiomer excess. Chrysanthemate analogues such as 65 have been synthesised using copper complexes of chiral Schiff's bases derived from aminosugars.[139] For 65 it is found that the cis-1-R isomer is the most active and the optimal catalyst is found to be the copper complex of the Schiff's base formed between pyridine-2-carbaldehyde 70 and methyl-4,6-O-benzylidene-2-amino-2-deoxy-α-D-allopyranoside 71. The cis-isomer is obtained in 80% yield and 50% enantiomer excess (R).

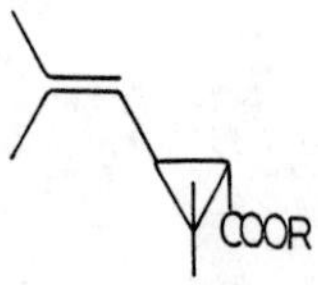

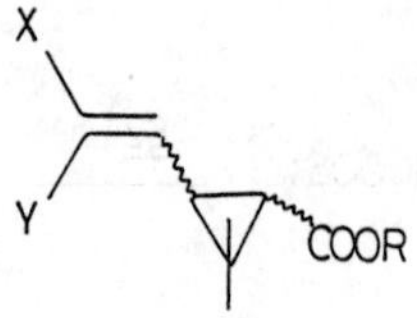

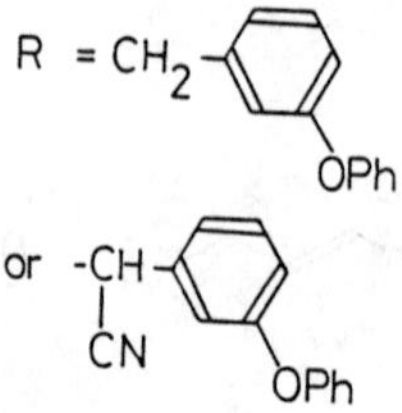

These last reactions hold great promise for future developments in cyclopropanation chemistry. The metal used is cheap and readily available and the ligands are derived from natural chiral sources rather than needing a lengthy synthesis or tedious resolution.

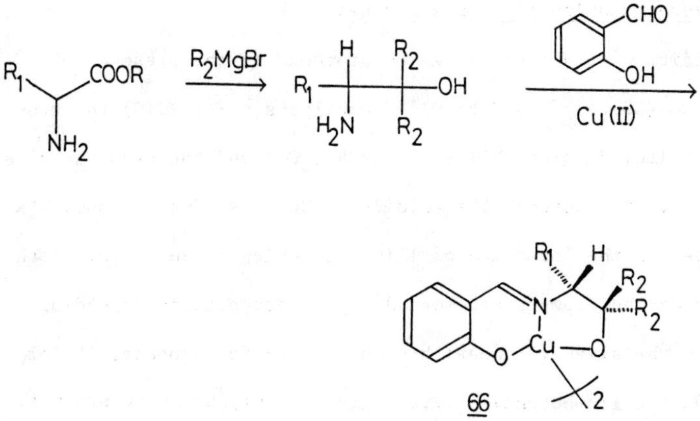

4.3 Hydrocyanation

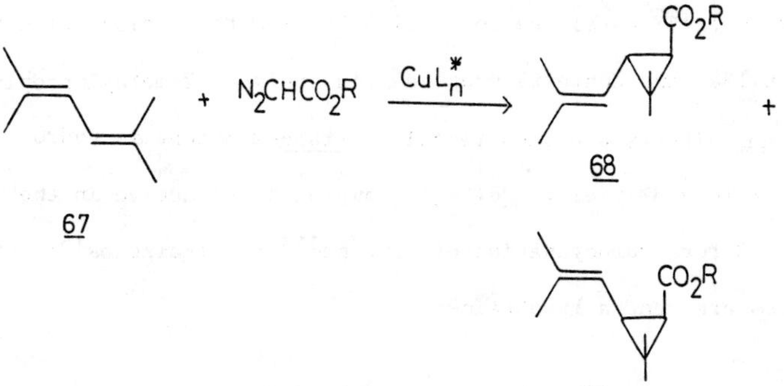

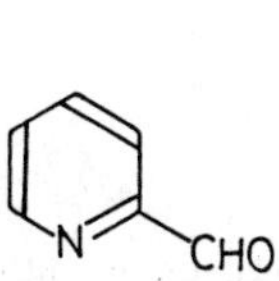

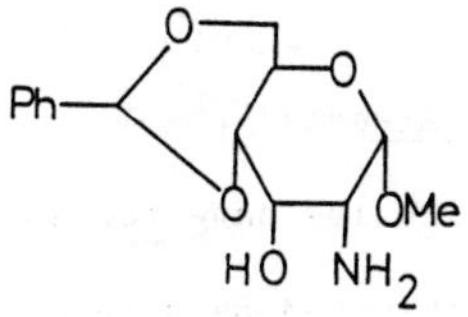

The addition of HCN to activated alkenes, particularly $\alpha\beta$-unsaturated carbonyl compounds, is promoted by a variety of catalysts, mainly bases,[140] but including some transition metal complexes. Unactivated alkenes and alkynes react in the presence of metal catalysts; most reports appear in the patent literature and few mechanistic studies exist.[141,142]

4.3.1. Hydrocyanation of Simple Alkenes

Addition of HCN to 1-alkenes proceeds in the presence of Ni(0)[142-5], $Co_2(CO)_8$[146-8] and Pd(0)[149] catalysts. For Ni(0) the true catalyst is believed to be $[HNiL_3]^+$ (L=PR_3, CO) and the presence of a Lewis acid, able to increase the acidity of HCN, is advantageous. Excess ligand suppresses the formation of $Ni(CN)_2L_2$ which is inactive. Both linear and branched isomers are formed, their proportions depending critically on the alkene; linear: branched = 1.5 for propene, 19 for n-hexene and 99 for iso-butene. Styrene gives mainly branched product, presumably due to electronic effects.[150]

$Co_2(CO)_8$[146] catalyses conversion of C_2H_4 to propionitrile in good yield at 130° and converts propene exclusively to 2-methyl propionitrile. Whilst internal alkenes are less reactive, trans-2-butene is hydrocyanated to 72 in 9.3% yield. Pd(PR_3)$_4$ complexes are active in the presence of R_3B for hydrocyanation of styrene[150] and 1-alkenes[151] but mixed products are generally obtained.

72

4.3.2. Strained Alkenes

Various studies have focussed on the addition of HCN to disubstituted alkenes activated by strain, though few of the products find useful applications. Norbornene, 73, is hydrocyanated to give the exo-, 74 or endo-product, 75. In the presence of Pd(P(OR)$_3$)$_4$ | P(OR)$_3$, 74 is produced selectively,[151,152] but $Co_2(CO)_8$ or $Co_2(CO)_6(PR_3)_2$ as catalysts yield mixtures.[153] Norbornadiene gives dicyanides with similar catalysts and few remarkable selectivites. With Pd(0) there is modest selection in the first step (exo: endo = 86:14) but only exo- reaction in the second.[152] Other reactions of strained alkenes are shown in Figure 34.

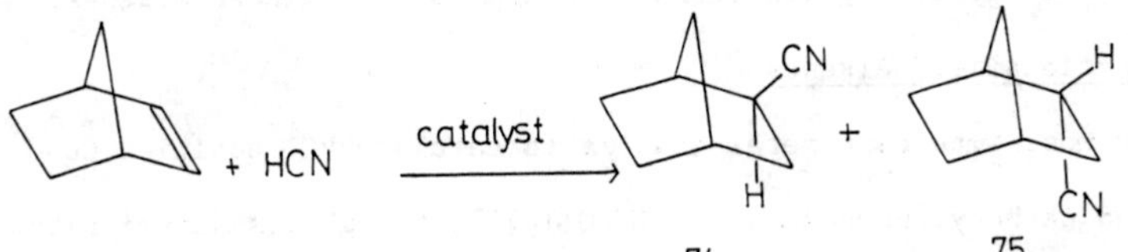

73 74 75

Ref

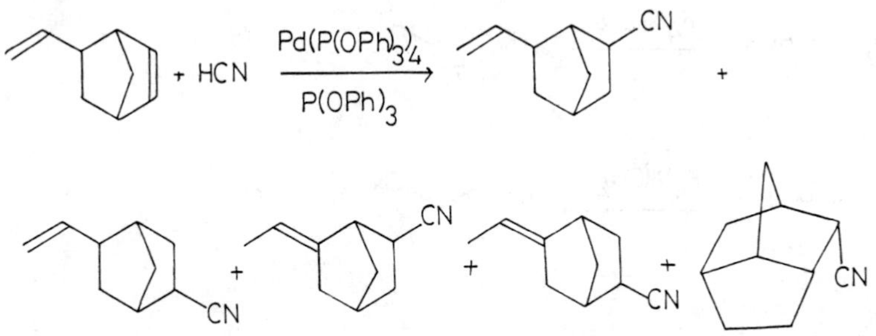

152

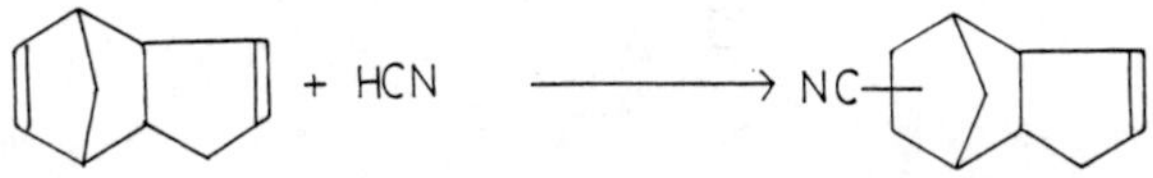

$Co_2(CO)_8$ 24% 146-8

$Pd(P(OPh)_3)_4$, $P(OPh)_3$ 35% 152

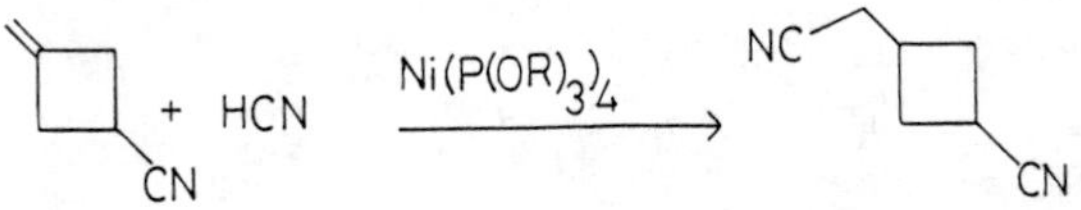

150

Figure 34 Hydrocyanation of strained alkenes.

The products of addition to norbornene or norbornadiene are chiral and asymmetric addition is achieved in the presence of Pd((+)-DIOP)$_2$|(+)-DIOP.[154-156] From norbornene, 1S,2S,4R-74 is obtained in 31% enantiomer excess. The same catalyst gives terminal nitriles from linear alkenes.

4.3.3. Functionalised Alkenes

The most important metal complex to catalyse HCN addition to

$\alpha\beta$-unsaturated carbonyl compounds is $[Ni(CN)_4]^{4-}$, though its superiority

to basic catalysts is unclear.[143,157] Conventional addition occurs in most

cases (reactions of 76, 78 and 80). There are a few other reports but no

systematic studies (e.g. 82 to 83 and 84).[158]

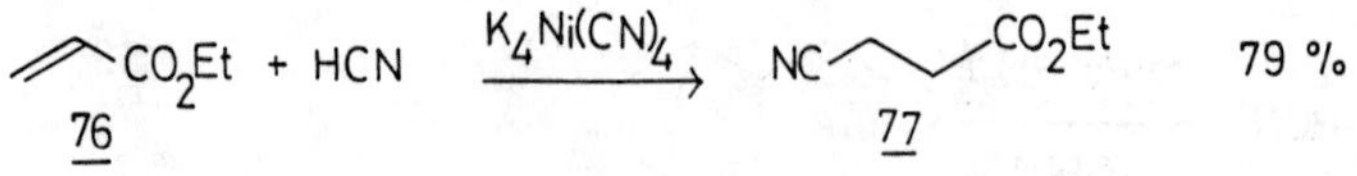

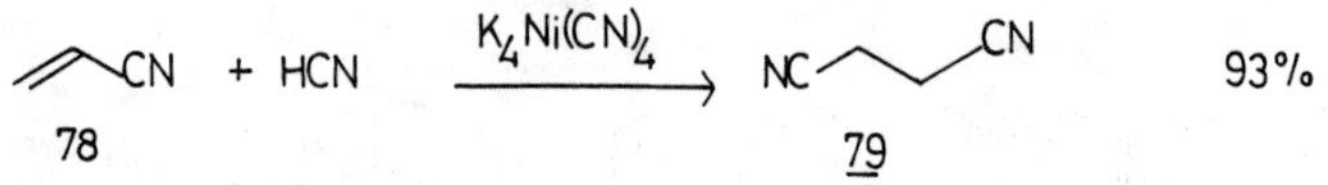

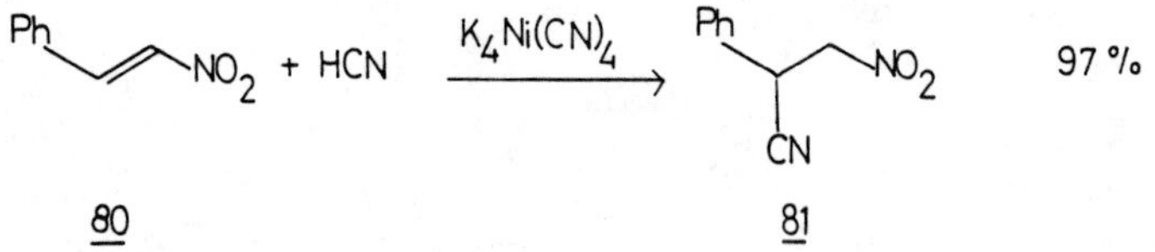

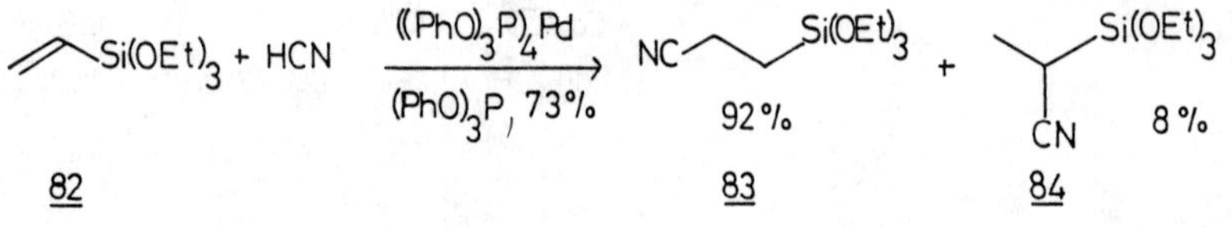

4.3.4. Reactions of Dienes and the Synthesis of Adiponitrile

Motivation for the study of hydrocyanation is its potential use

in conversion of butadiene to adiponitrile, 85. Addition of 1 mole HCN

to yield 3-pentenenitrile, 86, was one of the first reported hydrocyana-

tions.[148] The catalyst was $Ni(CO)_4$ and added phosphines or arsines

enhanced yields. Controlled monoaddition is facile and 86 is always the

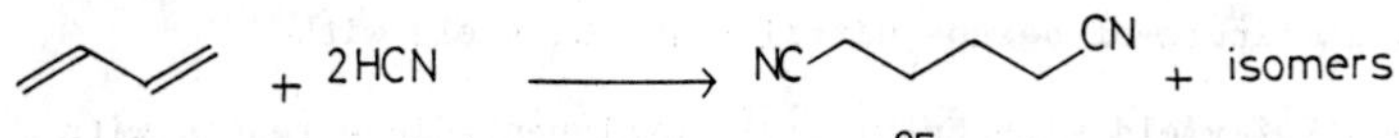

major product. (Figure 35). Subsequent conversion of 86 requires
isomerisation followed by hydrocyanation of the more reactive terminal
alkene. A wide variety of catalysts is to be found in the patent liter-
ature including $((RO)_3P)_3W(CO)_3|Ph_3B$[170], $Ni(P(OR)_3)_4$[151,153,169,171-177],
$(Ph_3P)_3Ni|ZnCl_2$[178], $Pd(P(OPh)_3)_4|Ph_3B$[149], $HCo(P(OPh)_3)_4$[141],
$RuCl_2(PPh_3)_3$[179] and $(Ph_3P)_3RhCl$.[141] The best selectivites for adiponitrile are around 80%.[171]

Other conjugated dienes have received less attention but

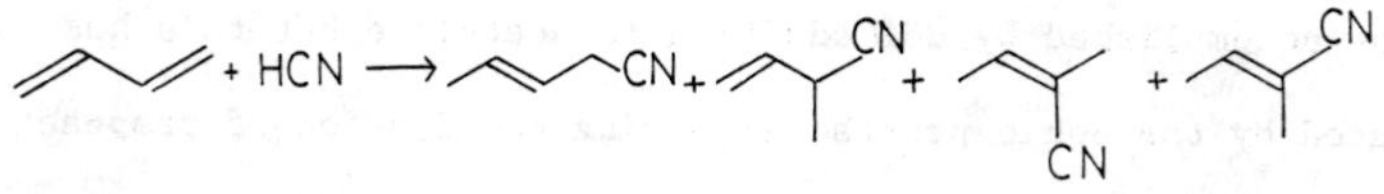

Catalyst	86	87	88	89	Ref
CuCl,  thiophene, 17 hr 120°	91.6 %	8.4%	–	–	160, 161
$Ni(P(OR)_3)_4$, 100°, 8 hr	54 %	42%	3%	0.5%	150, 162
CuCl $(NC)_2C=C(CN)_2$ 60° 100 hr	88.3%	–	–	–	163, 164
CuBr, LiBr, $NC(CH_2)_4CN$	92%	–	–	–	165, 166
$Ni(P(-O-C_6H_4-)_3)_3$ $(NCCH_2CH=CHCH_3)_3$ 	66%	33%	–	–	167-9

Figure 35 Hydrocyanation of butadiene.

isoprene gives a mixture of hexene nitriles in 50% yield with $Co_2(CO)_8$[146] and 90% yield with Cu_2Cl_2.[180] Cyclopentadiene reacts with good selectivity to 3-cyanocyclopentene, 90, but subsequent reaction to dicyanocyclopentanes is less satisfactory.[181,182]

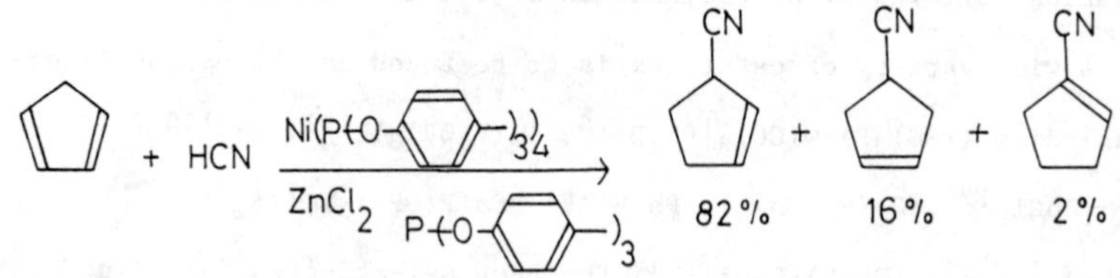

4.3.5. Hydrocyanation of Alkynes

Until the end of the 1960's the industrial synthesis of acrylonitrile was accomplished by HCN addition to acetylene but this has since been replaced by the Sohio process involving cooxidation of propene and ammonia.[183] Addition takes place in the presence of Cu(I) salts, HCl and NH_4Cl, the major biproducts being acetylene dimer and acetaldehyde.[184-188] The reaction kinetics have been thoroughly studied.[189-191] 1-Alkynes yield linear $\alpha\beta$-unsaturated nitriles with Cu(I) catalysts[189] and mixtures of linear and branched products with $Ni(P(OR)_3)_4$.[192,193] Conversions are excellent, a yield of 93% 93 being achieved from diphenylacetylene with $((PhO)_3P)_4Ni|(PhO)_3P|ZnCl_2$. The mechanism is thought to involve attack of CN^- inter- or intramolecularly (from the metal) on a coordinated alkyne. Hydrocyanation|hydrogenation occurs with $[Ni(CN)_4]^{2-}|$ $NaBH_4$[194] and $[Co(CN)_5]^{3-}$[195] (Figure 36).

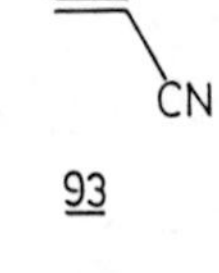

93

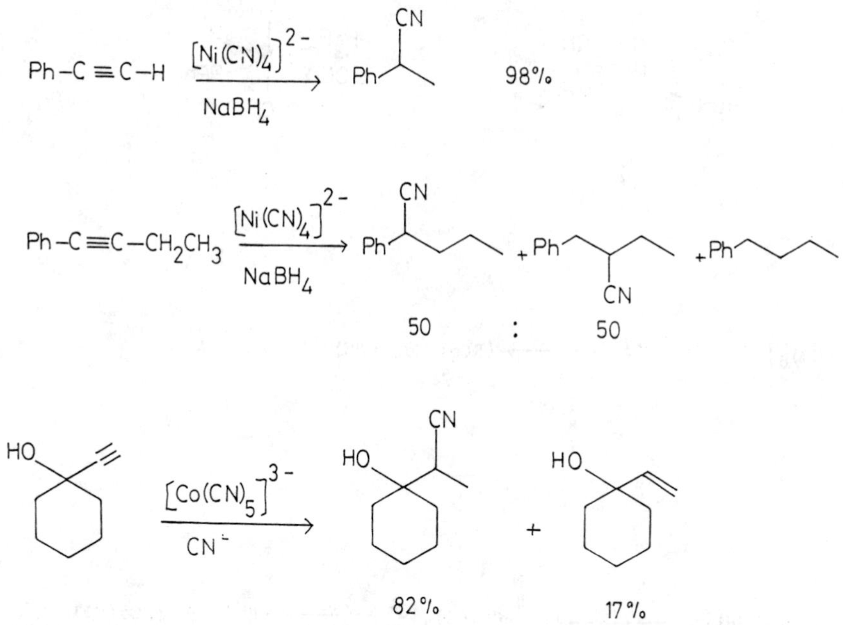

Ph−C≡C−H $\xrightarrow[\text{NaBH}_4]{[\text{Ni(CN)}_4]^{2-}}$ Ph— (with CN group) 98%

Ph−C≡C−CH$_2$CH$_3$ $\xrightarrow[\text{NaBH}_4]{[\text{Ni(CN)}_4]^{2-}}$ Ph— (with CN) + Ph— (with CN) + Ph—

50 : 50

HO— (cyclohexane with ethynyl) $\xrightarrow[\text{CN}^-]{[\text{Co(CN)}_5]^{3-}}$ HO— (with CN) + HO— (with vinyl)

82% 17%

Figure 36 Hydrocyanation/hydrogenation of alkynes.

4.3.6. Mechanism of Alkene Hydrocyanation.

The mechanism is not established but some related observations
may be cited as relevant. HCN adds oxidatively to Wilkinson's catalyst
or Vaska's compound to give respectively 94 and 95.[196,197]
Pt(PPh$_3$)$_3$ or 4 yields HPt(CN)(PPh$_3$)$_2$[198] and Ni(P(OEt)$_3$)$_4$, HNi(P(OEt)$_3$)$_3$CN
and HNi(P(OEt)$_3$)$_4$CN.[199] A metal hydride intermediate is also indicated
by the isomerisation of 1-pentene by Ni(Ph$_2$P(CH$_2$)$_4$PPh$_2$)$_2$ in the presence
of HCN [200,201] which is attributed to a detectable 5-coordinate complex,
HNi(CN) PH$_2$P(CH$_2$)$_4$PPh$_2$)$_2$, with one mono- and one bidentate phosphine
ligand. The stoichiometric reaction of [RCo(CN)$_5$]$^{3-}$ is known to
proceed via the pathway of Figure 37[196,202,203] and there is no direct
precedent for elimination of RCN from R−M−CN.[204] However, addition of
HCN is strictly cis with a Pd(DIOP)$_2$ catalyst[205] and largely cis with

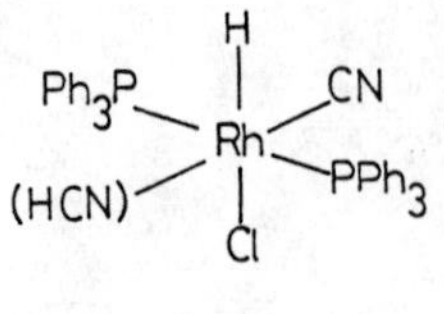

94

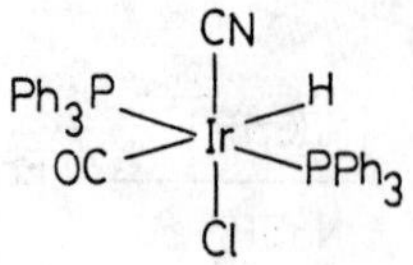

95

$$[RCo(CN_5)]^{3-} \ + \ H+ \ \longrightarrow \ [RCo(CN_4(CNH)]^{2-}$$

$$[RCo(CN_4(CNH)]^{2-} \longrightarrow [R\overset{N-H}{\underset{\parallel}{C}}-Co(CN_4]^{2-} \longrightarrow RCN \ + \ [HCo(CN_4]^{2-}$$

Figure 37 Stoichiometric reaction of an alkyl cobalt cyanide complex.

$Ni(P(OPh)_3)_4$.[206] This is best rationalised by cis-addition of a metal
hydride followed by reductive elimination of RCN with retention
(Figure 38).

4.4 Hydrohalogenation and Related Reactions

Hydrohalogenation of alkenes and alkynes does not normally require catalysis and the effect of metal complexes is little studied.
Addition of HCl to cyclohexene is accelerated by $SnCl_4$ acting as an
electrophilic catalyst[207] and it is likely that TaF_5, used for HF addition
to tetrachloroethylene acts similarly[208]. Addition of one mole HCl to
chloroprene, 96, gives the 1,4-product, 97, in >99% yield in the presence
of Cu_2Cl_2.[209,210]

The commercial importance of vinyl chloride has led to a study
of HCl addition to acetylene. Mercuric[211] and cuprous chlorides[212] are

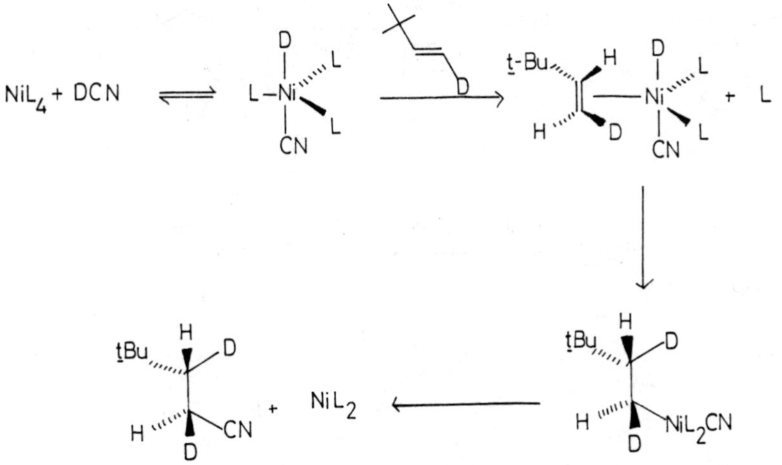

Figure 38 Proposed mechanism of alkene hydrocyanation.

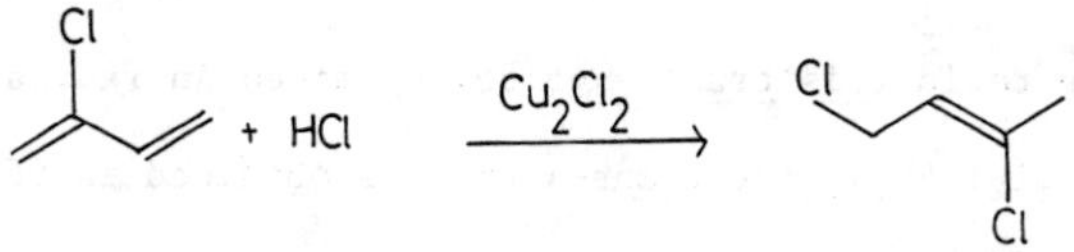

96   97

the most popular catalysts, and despite their low solubility, the catalytic

reaction is genuinely homogeneous.[213] Conversions are excellent and the

catalyst is long lived.[214] The reaction is not invariably regioselective

and anti-Markovnikov products are isolated via the intermediacy of an

alkynyl copper species.[215] Hydrofluorination of acetylene is catalysed

by Ph_2Hg in quantitative yield by an unknown mechanism.[216]

Addition of halogenated hydrocarbons to double bonds is initiated

by radical promoters, usually unselectively. Palladium and ruthenium

complexes catalyse selective reaction under mild conditions.

$Pd(OAc)_2 | PPh_3 | K_2CO_3$ gives the conversion 98 to 99 at room temperature

whereas $(PhCO_2)_2$, Cu_2Cl_2 and $RuCl_2(PPh_3)_3$ have no activity below 40°.[217]

If the substrate is an allyl alcohol further reaction occurs to give a

ketone (100 to 101).[218] With the same catalyst cyclohexene to 102 proceeds with the diastereomer ratio (cis:trans = 45:55) obtained in radical

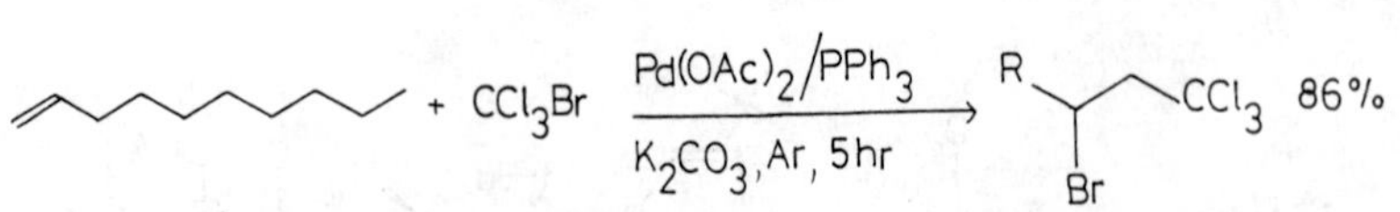

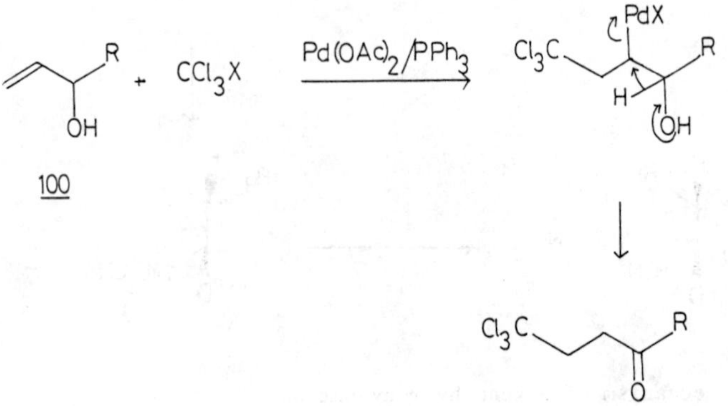

reactions but with $RuCl_2(PPh_3)_3$ the trans-isomer is obtained in 96% yield.

Asymmetric addition is achieved in the presence of ((-)-DIOP)RhCl, styrene

being converted to 103 in > 32% (S) optical yield.[219] $RuCl_2(PPh_3)_3$ also

catalyses addition of chlorinated acid halides (cycloheptene to 104); under

radical conditions considerable polymerisation occurs.[220]

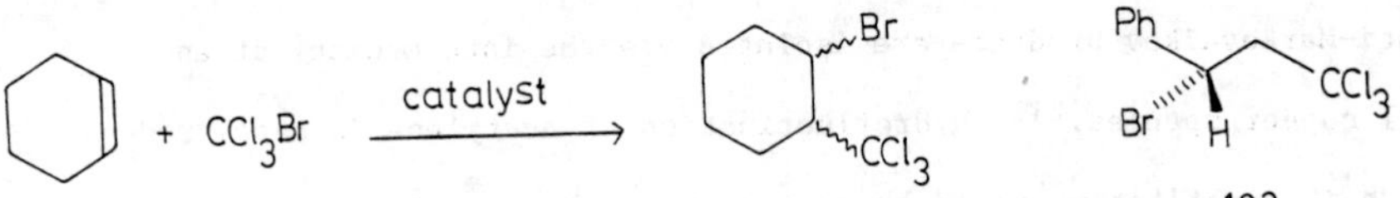

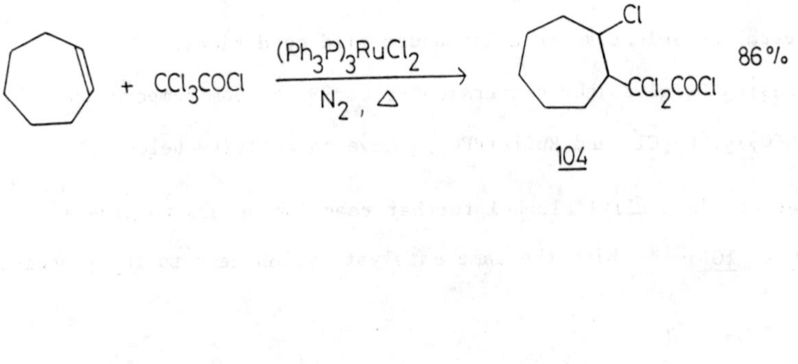

4.5 Hydrometallation and Carbometallation

Addition of metal hydrides or metal alkyls to carbon-carbon multiple bonds leads to useful synthetic intermediates. Whilst many uncatalysed reactions are known both regio- and stereochemical control may be enhanced in the presence of metal complexes.

4.5.1. Hydrometallation of Alkenes

Titanium tetrachloride catalyses addition of $LiAlH_4$[221] or R_3AlH[222] to carbon carbon double bonds. The process is solvent dependent with THF or diglyme, but not ether, suitable for 105 to hexane.[223] Substitution decreases the reaction rate with $RCH=CH_2 > R_2C=CH_2 > RCH=CHR$.[224-227] The rate for internal alkenes is not too slow to be useful, however; trans-2-hexene reacts with $HAl(NR_2)_2$ in the presence of Cp_2TiCl_2 to give an organoaluminium compound which is protonated (or deuterated) to n-hexane in quantitative yield.[228] The reaction is sensitive to steric hindrance with $(CH_3)_3CCH=CH_2$ giving low yields.[229]

$$C_4H_9 \diagdown \quad \xrightarrow[\text{2) } H_2O]{\text{1) } LiAlH_4/TiCl_4} \quad C_6H_{14}$$

105

Many aluminium hydrides are useful; $LiAlH_4$, $NaAlH_4$, $LiAlMe_3H$, $NaAlMe_3H$, $NaAl(OCH_2CH_2OCH_3)_2H_2$ all give similar results.[229] $HAl(NiPr_2)_2$ has been particularly useful in Ashby's hands[225,228] and is advocated because of its easy preparation. Addition of DIBAH in the presence of $Ti(OBu)_4$ gives addition, as does i-Bu_3Al, where the coproduct is isobutene. Catalysts include $TiCl_4$, $Ti(OR)_4$, Cp_2TiCl_2, $TiCl_3$[225], $ZrCl_4$[221], $ZrCl_4|$ $4ROH$[230] and Cp_2ZrCl_2.[231,232] Changing the atmosphere from nitrogen to argon increases the yield, possibly by supressing N_2 fixation by titanium.

The reaction mechanism is thought to involve hydrometallation by a titanium hydride, then titanium-aluminium exchange (Figure 39). Authentic hydrometallation is only proven by deuterium incorporation on

$$LiAlH_4 + Cp_2TiCl_2 \longrightarrow Cp_2Ti(H)Cl + LiAlH_3Cl$$

$$RCH = CH_2 + Cp_2Ti(H)Cl \longrightarrow \begin{matrix} H \text{----} TiCp_2Cl \\ RHC \text{====} CH_2 \end{matrix} \longrightarrow RCH_2CH_2TiCp_2Cl$$

$$RCH_2CH_2TiCp_2Cl + LiAlH_4 \longrightarrow RCH_2CH_2AlH_3Li + Cp_2Ti(H)Cl$$

Figure 39 Mechanism of titanium catalysed hydroalumination.

deutrolysis of the product. Metal addition is normally to C-1 as predicted on both electronic and steric grounds, an exception being provided by styrene which gives 106 and 107 in the ratio 9:1 after deuterolysis. Organoalanes react with many electrophiles including H^+, halogens[221], NCS[226] and CH_3COCl[222], and may be oxidised to alcohols[233] and acetates[234]. Homo coupling ($Cu(OAc)_2$) yields alkanes[235] Cu_2Cl_2 catalyses coupling with 108[224] (See also Chapter 8).

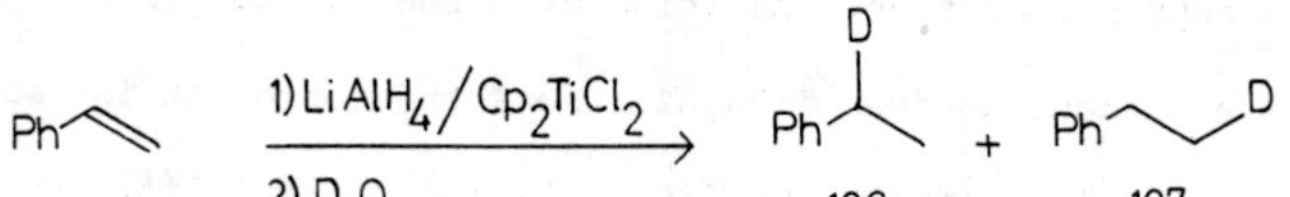

106 107

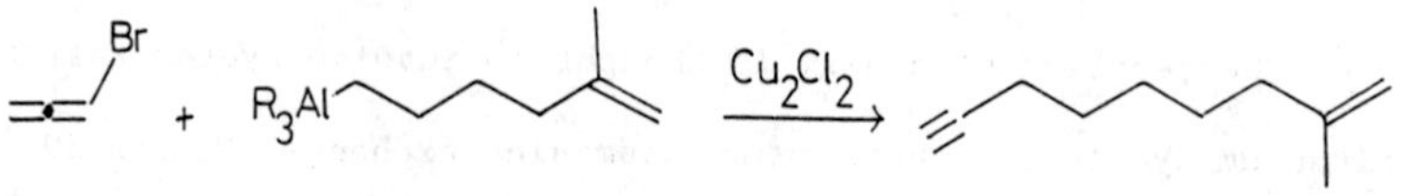

108 109 110

Few other metal hydrides are well studied. MgH$_2$ adds in
the presence of Cp$_2$TiCl$_2$[236] or NiCl$_2$|Cp$_2$ZrCl$_2$[237] and deuterium incorpora-
tion shows that there is a true alkyl magnesium present. More commonly
the source of the hydride is a Grignard reagent which reacts with alkene
elimination (e.g. 111 to 112)[238] and the reaction is driven by removal
of the volatile product. Isomerisation competes significantly with
addition.[239] The product Grignards react conventionally and a synthesis of
spirolactones is achieved by a carboxylation of the product. (113 to
116)[240]

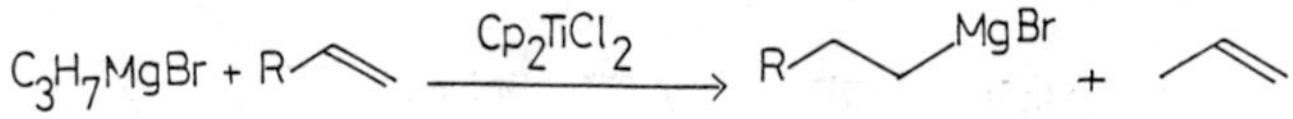

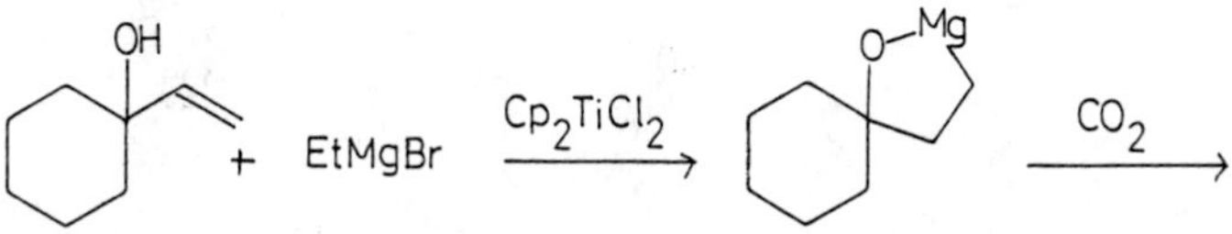

4.5.2. Hydrometallation of Alkynes

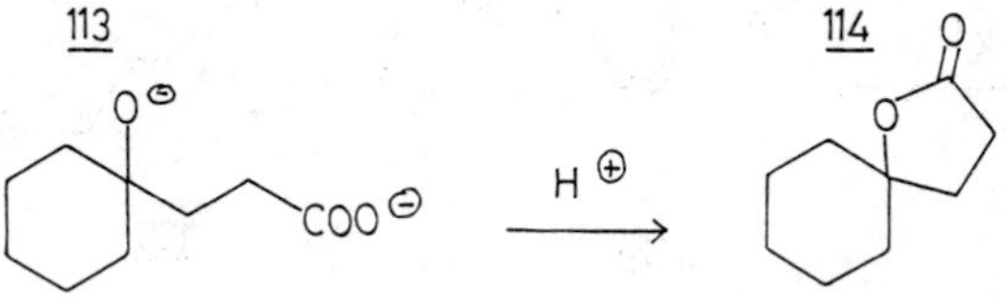

Aluminium and magnesium hydrides are added to alkynes in the
same way as to alkenes. Treatment of 4-octyne with LiAlH$_4$|Cp$_2$TiCl$_2$
followed by D$_2$O gives 100% cis-reduction and 100% D incorporation.[229]
However, 117 gives 118 and 119 and similar lack of regioselection is
observed in the reaction of 2-hexyne with HAl(NiPr$_2$)$_2$|Cp$_2$TiCl$_2$[241] without
catalysis. LiH, RAlH$_2$, AlH$_3$[242] and Cp$_2$ZrH$_2$[243] all add trans to alkynes.
R$_2$AlH adds cis but only at temperatures which cause isomerisation and
coupling.

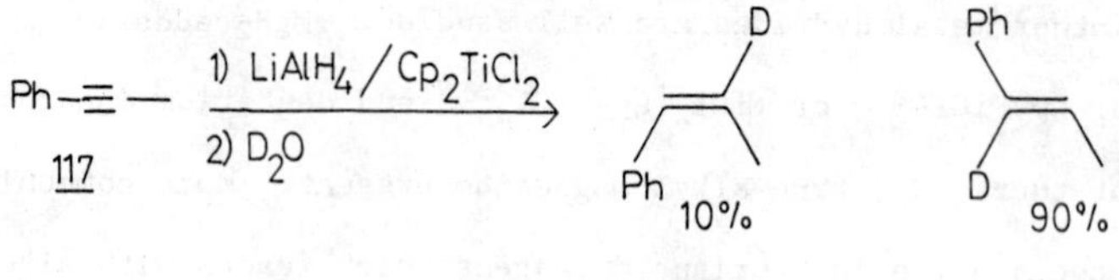

$$\text{Ph}-\!\!\equiv\!\!- \quad \xrightarrow[\text{2) D}_2\text{O}]{\text{1) LiAlH}_4 / \text{Cp}_2\text{TiCl}_2}$$

117

118   10%

119   90%

Isobutyl magnesium bromide gives predominently cis-hydrometalla-tion (120 to 121) with 3-hexyne[244] but if the substrate bears a trimethyl-silyl group (123) isomerisation of the product Grignard occurs under the reaction conditions.[245]

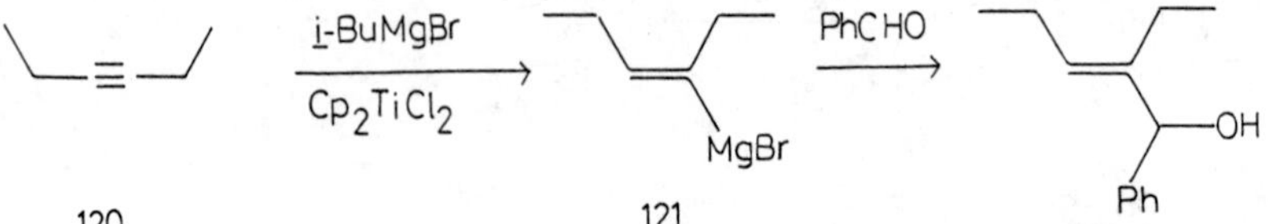

120

$$\xrightarrow{\text{i-BuMgBr}} \quad \text{Cp}_2\text{TiCl}_2$$

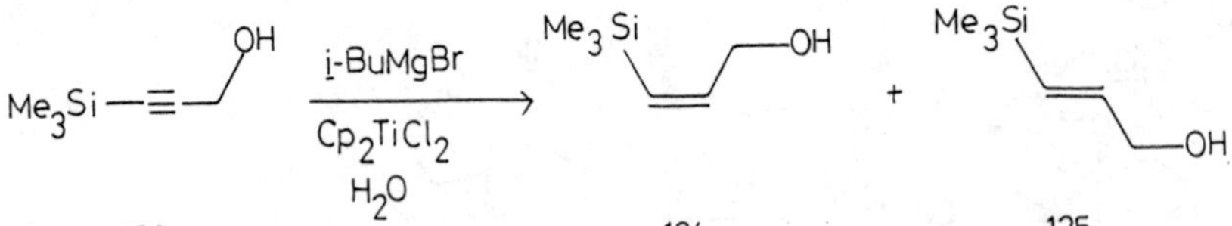

121

$$\xrightarrow{\text{PhCHO}}$$

122

$$\text{Me}_3\text{Si}-\!\!\equiv\!\!-$$

123

$$\xrightarrow[\text{H}_2\text{O}]{\text{i-BuMgBr} \atop \text{Cp}_2\text{TiCl}_2}$$

124

+

125

4.5.3. Carbometallation

Carbometallation of alkynes is used for the synthesis of stereo-specifically substituted alkenes. The reaction is exemplified by 126 to 127 and stereospecificity is high. We shall be concerned only with cases where the vinyl metal is well characterised since only these are amenable to further elaboration.

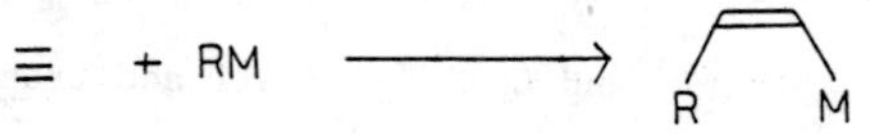

$$\equiv \quad + \quad \text{RM} \quad \longrightarrow \quad$$

126

R   M

127

Grignard reagents add to triple bonds in low yield under severe conditions. For example 128 is converted to 129 after 6 days at 100°.[246],[247] In the presence of Cu(I) salts the rate is improved (Figure 40). Other catalysts including Ti(III)[250], Rh(III)[251] and Fe(III) are less effective.

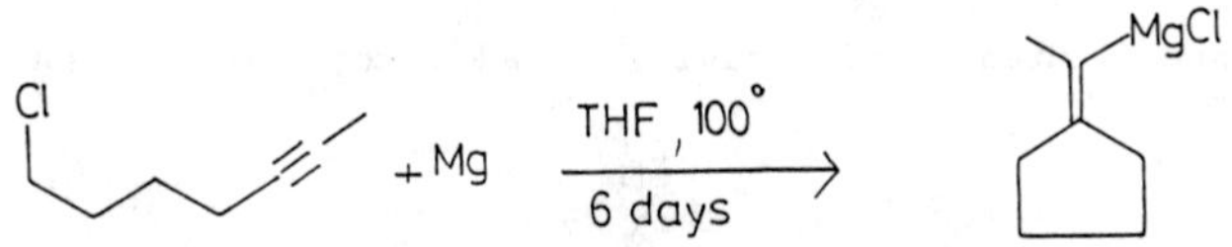

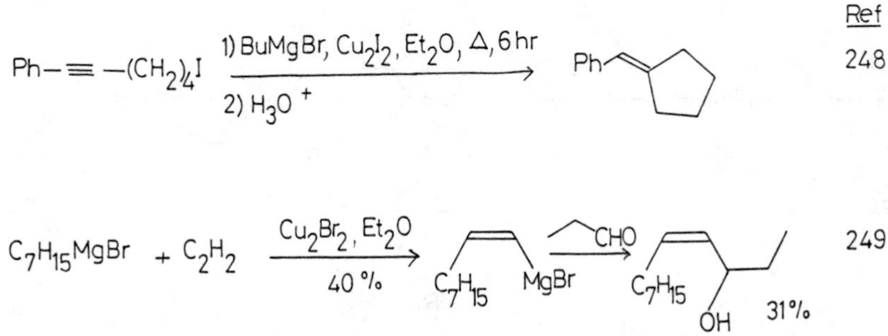

Ref

248

249

Figure 40 Cu(I) catalysed addition of Grignard reagents to alkynes.

For disubstituted alkynes the most widely used catalyst is (Ph$_3$P)$_2$NiCl$_2$; this catalyses cis-addition of MeMgBr to diphenyl acetylene in good yield, though reducing Grignards give less satisfactory results.[252] The intermediacy of the vinyl Grignard was confirmed by deuterium labelling.[253],[254] With arylalkylacetylenes yields are modest but regio- and stereoselection are high.[255] (130 to 131). Eneynes give specific reaction at the triple bond to give gem-disubstituted dienes (134a, b) with high regio- but low stereo selection.[256],[257]

Alkynes bearing other functional groups give more complex reactions. Propargylic alcohols (135) give stereo specifically anti-addition, the reaction occurring faster in the presence of Cu_2I_2.[258,259] Again deuterium labelling confirms the intermediacy of 136. Regio selectivity depends on whether the alcohol is primary, secondary or tertiary, the more substituted species giving attack remote from the OH group

Ph —≡— R + RMgX $\xrightarrow{(Ph_3P)_2NiCl_2}$ [structure 131] $\xrightarrow{H_2O}$ [structure 132]

130   131   132

R —≡—⟋ + PhMgBr $\xrightarrow{(Ph_3P)_2NiCl_2}$ [structure 134a] + [structure 134b] + Ph–Ph

133   134 a   134 b

—≡—⟋OH + RMgBr $\xrightarrow{Cu_2I_2}$ [structure 136] $\xrightarrow{D_2O}$ [structure 137]

135   136   137

and also undergoing elimination to yield allenes (e.g. 140).[259,260] The presence of a β-amino group also usually gives anti-addition, the reaction regiochemistry depending on R_1 and R_2 (141 to 142 and 143).[258,261,262] In an unusual exception the amino group of 144 is postulated to provide a binding site for magnesium and to direct the addition in a regio- and stereoselective manner.[263]

The presence of a trimethylsilyl group on the triple bond deactivates it towards carbometallation and the size of the group promotes

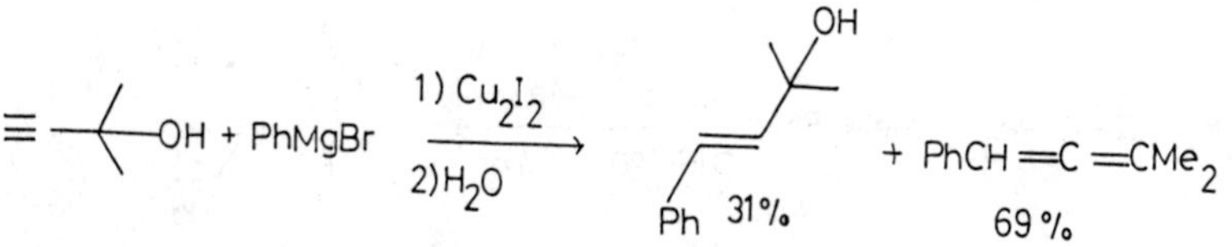

138   139   140

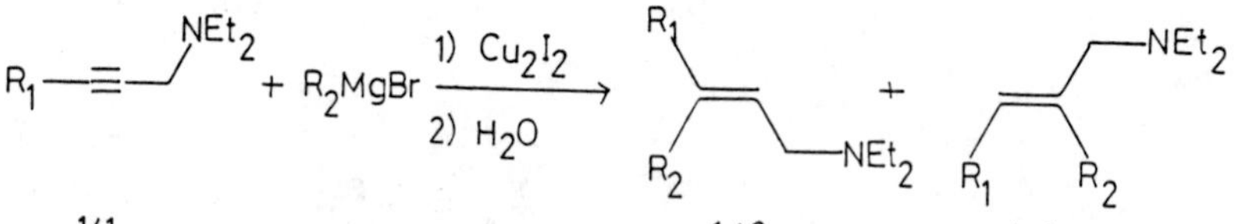

141   142   143

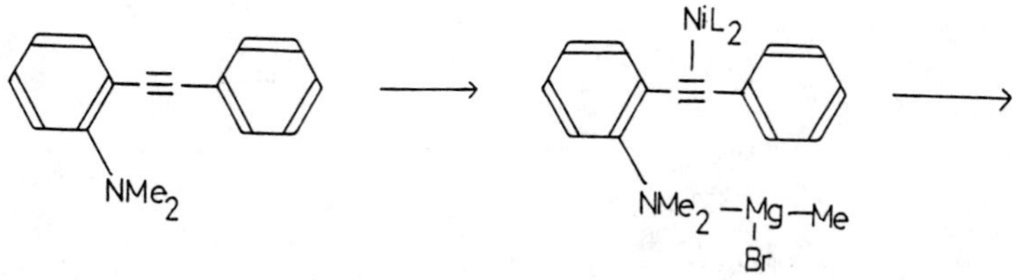

144   145

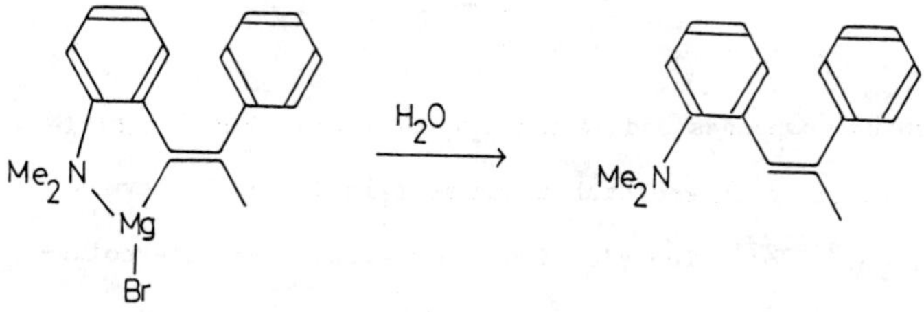

146   147

product isomerisation so that stereoselection is not maintained. The reaction is, however, regiospecific with metallation only at the carbon bearing silicon (148 to 149).[264,265] With silylated eneynes the deactivating effect of Si is sufficient to direct carbometallation to the usually less reactive double bond; 150 is converted to 151 in modest yield.[266,267]

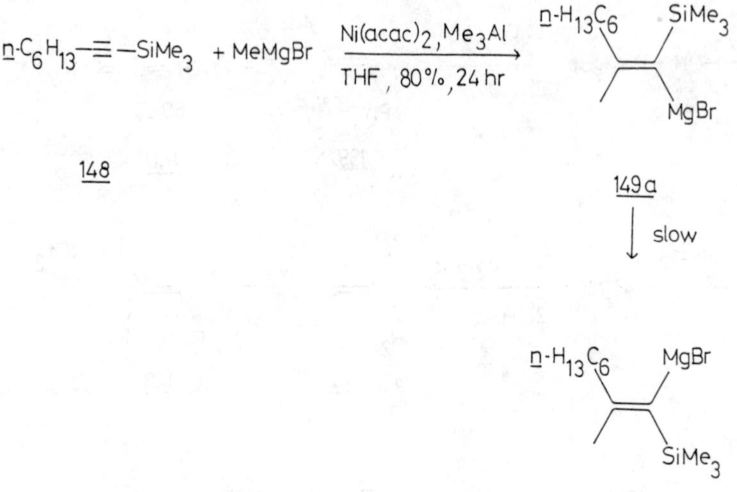

$\underline{n}\text{-}C_6H_{13}\text{—}\equiv\text{—}SiMe_3$ + MeMgBr $\xrightarrow[\text{THF}, 80\%, 24\,hr]{\text{Ni(acac)}_2, \text{Me}_3\text{Al}}$

148

149a

| slow

149b

$Me_3Si\text{—}\equiv\text{—}/\!/$ $\xrightarrow[\text{NiCl}_2]{\text{PhMgBr}}$ $Me_3Si\text{—}\equiv\text{—}\diagup\!\!\diagdown Ph$

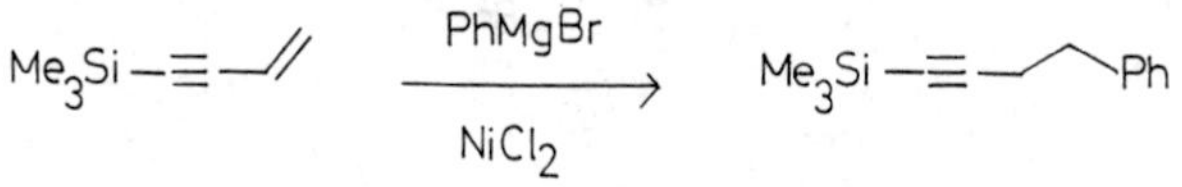

150   151

Organoaluminium compounds add mainly <u>syn</u> to alkynes but reactivities are low, regioselection unpredictable and metallation of 1-alkynes competes with addition.[268-271] The distribution of regio- and stereoisomers is altered by using nickel catalysts, but the major reaction in these cases is alkyne oligomerisation.[272] The most successful catalyst is Cp_2ZrCl_2, widely exploited by Negishi's group in a variety of syntheses.[273] Addition is highly stereo- (for <u>syn</u>-addition) and regioselective (for <u>gem</u>-disubstitution) as exemplified by the reaction of <u>152</u>.[274,275] The presence of vinylaluminium compounds is confirmed by deuteration and iodination.[276-8] The reaction is slow when Cp_2ZrCl_2 is used catalytically, taking typically 24 hrs so that many of the practical applications employ stoichiometric quantities. Many functional

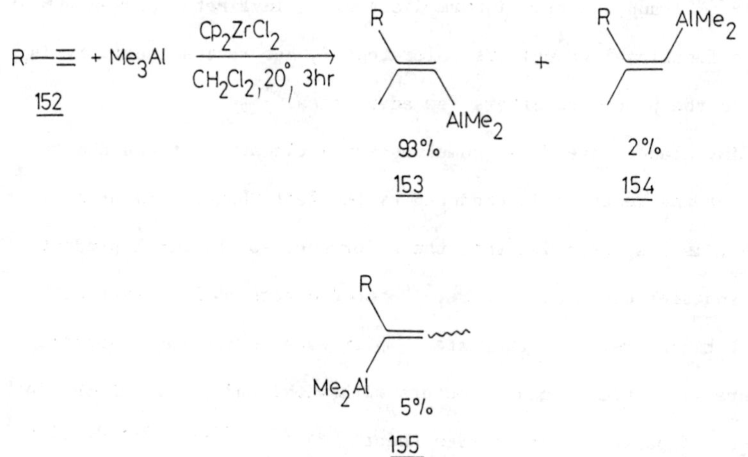

$$R-\!\!\equiv\ +\ Me_3Al \xrightarrow[\text{CH}_2\text{Cl}_2,\ 20^\circ,\ 3\text{hr}]{\text{Cp}_2\text{ZrCl}_2}$$

152

153 93%

154 2%

155 5%

groups are tolerated; 156 (n=1,2; Z=OH, OSiMe$_2$t-Bu, SPh, I) gives products with excellent regio- and stereoselection.[279] Disubstituted
alkenes are reactive, but larger amounts of zirconium are now essential;

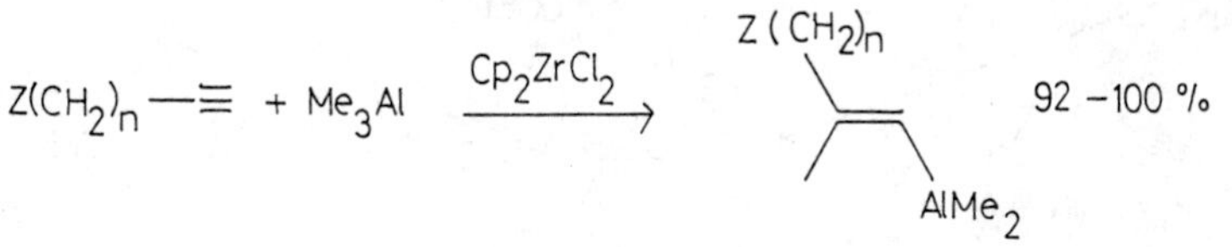

$$Z(CH_2)_n -\!\!\equiv\ +\ Me_3Al \xrightarrow{\text{Cp}_2\text{ZrCl}_2}$$

156

157 92 – 100 %

again stereoselection is high. A change to other trialkylaluminium
compounds gives less predictable results. n-Pr$_3$Al and 1-octyne give
linear and branched addition products in the ratio 1:3 together with up
to 50% hydrometallation.[280] With i-Bu$_3$Al the only product is that of
hydrometallation. Using n-Pr$_2$AlCl suppresses the undesired reaction but
regioselectivity with these complex alanes remains capricious.

The other metals of the zirconium triad have been less studied.
Hafnium complexes behave similarly but reaction rates are faster.
Cp$_2$TiCl$_2$ is a catalyst but again is more often used stoichiometric-
ally.[281,282] Stereospecificity is comparable to that achieved with

364

Cp$_2$ZrCl$_2$[282-286] and the true intermediacy of a vinyl metal proven. Most of the same functional groups are tolerated[283], but regioselectivity is variable and the procedure offers few advantages.

Vinylalanes have found numerous synthetic applications and the usefulness of the reaction is enhanced by the fact that the minor regio-isomer, 155, is less reactive than the major one, so the final product is regio- and stereochemically pure.[274] Good electrophiles react with the alane but conversion to the -ate complex extends the range appreciably (Figure 41). Coupling also occurs with vinyl and allyl halides in the presence of palladium salts (See Chapter 8).[274,275,277,278,287,289]

Ref

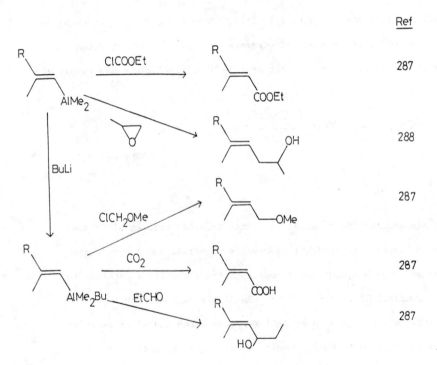

287

288

287

287

287

Figure 41 Reactions of vinylalanes.

4.6 Addition of Heteroatomic Nucleophiles to Alkenes and Related Reactions

4.6.1. Addition of Amines, Alcohols and Water to Alkenes

Nucleophilic attack of nitrogen, oxygen and sulphur on metal alkene complexes is well studied.[290,291] The usually nucleophilic alkene becomes an electron acceptor, undergoing "umpolung" reactions. Addition may take place by two routes (Figure 42), either direct attack on the Π-system wihich gives a trans-product (route a) or by attack at the metal and cis-transfer to the alkene (route b). Theoretical studies suggest that while electrostatic attraction is important it is not the only

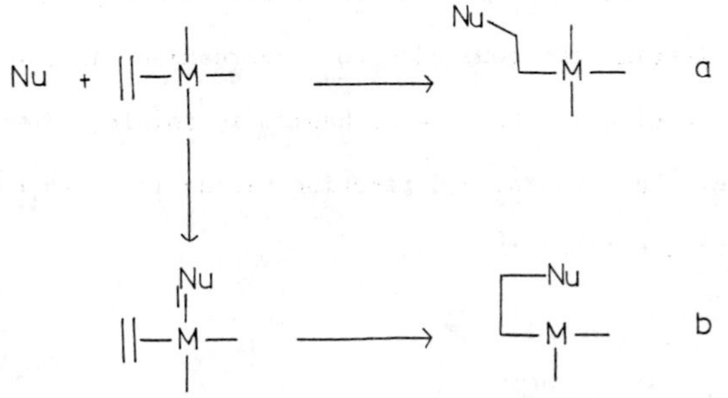

Figure 42 Modes of nucleophilic attack on metal alkene complexes.

factor and charge transfer from alkene to metal is difficult to estimate.[292] Activation arises from slippage of the metal along the alkene towards an η^1-arrangement and regioselectivity is in accord with the slippage hypothesis. For example attack of MeO⁻ on $CpFe(CO)_2(CH_2=$ *CH-OR) occurs only at C*[293] and a crystal structural study of 158 shows extremely non-symmetric bonding in the ground state which results in enhanced electrophilicity.[294]

Addition of water to alkenes is a facile, acid catalysed process not requiring metal catalysis. However, some metal salts are industrially important including $Ti(SO_4)_2$ [295], $Mo(VI)$ oxalate[296],

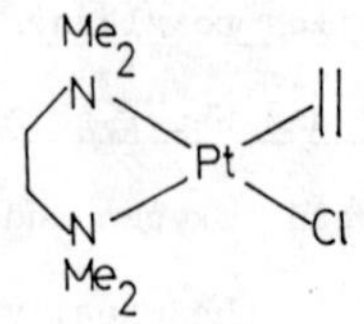

158

Cu^{2+} [297-9] and $H_6P_2Mo_{20}O_{68}.48H_2O$.[300] Palladium complexes have attracted
some attention because the reaction resembles the Wacker oxidation; in
these cases[301] and in a platinum analogue[302] the water is delivered to
the alkene from the metal. Ethanol and acetic acid may also be added to
alkenes, for example myrcene, 159, which gives ethers in low yields and
poor selectivity.[303] Acetic acid (OD) adds to norbornadiene in the
presence of $(Ph_3P)_3Pt$ to give 164 but the mechanism is complex, involving
163 as an intermediate. The uncatalysed reaction yields 165 with considerable scrambling of the label.[304]

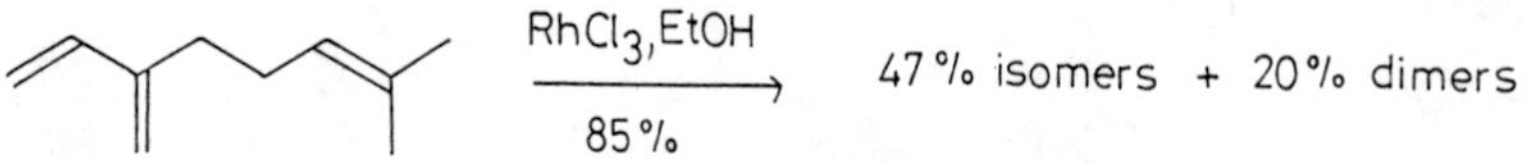

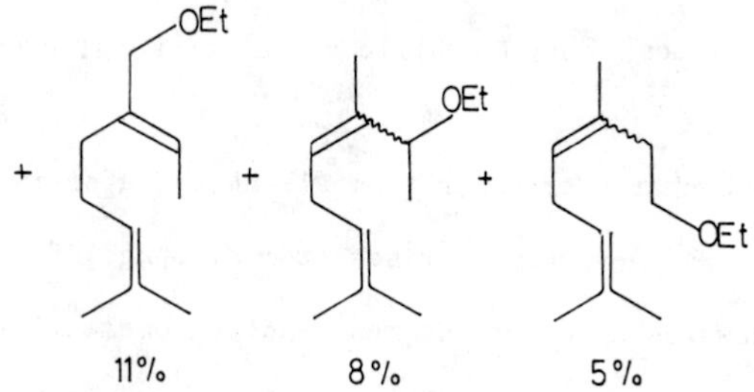

Addition of alcohols to vinyl ethers to yield acetals or
exchanged vinyl ethers is catalysed by palladium (II) complexes.[305]
Acetal formation is supressed by buffer. For substituted vinyl ethers,
166, stereospecific inversion to 167 occurs indicating that the alcohol

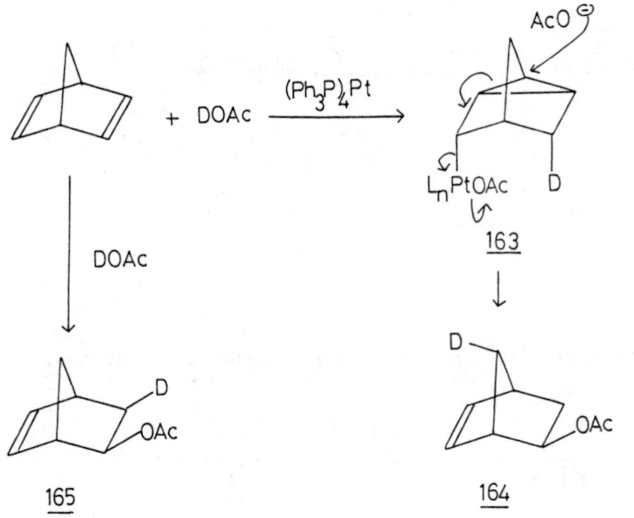

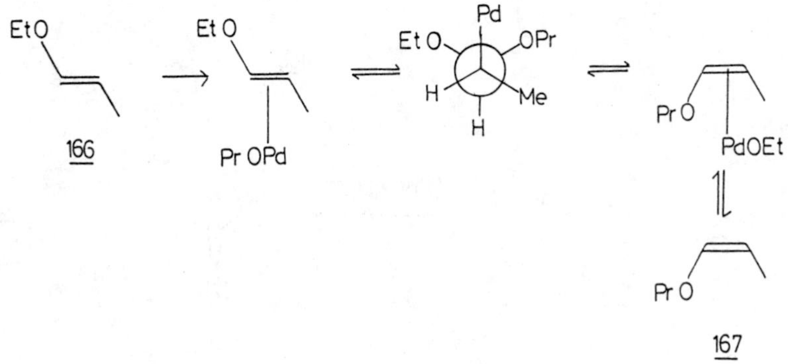

is added _via_ palladium followed by stereospecific elimination of the other alkoxyl group. Other additions are known, but their mechanisms have not been studied (Figure 43).

Trans-attack of amines on alkene-Pd complexes is well documented to β-palladated amines, _169_.[308,309] These may be treated with Pb(OAc)$_4$ to give the acetoxy compound with retention of stereo-chemistry[309], or Br$_2$|R$_2$NH to yield a _cis_-diamine with inversion.[310] With a chiral amine modest enantiomer excesses were achieved.[311] Regiochemistry

$$\text{COOMe} + \text{MeOH} \xrightarrow[\text{NaOH} \quad 100\%]{\text{Ni(acac)}_2, \text{Et}_3\text{Al}, \text{Bu}_3\text{P}} \text{MeO} \diagdown \text{COOMe}$$ 306

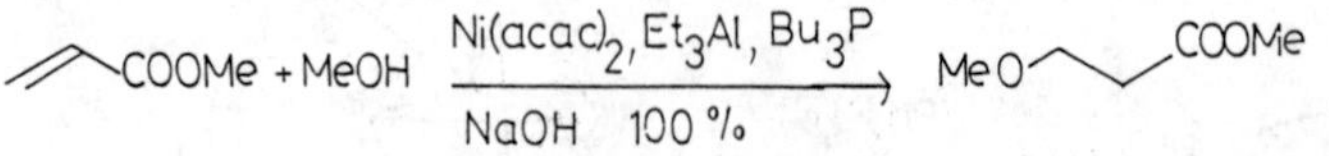

$$\diagup\diagdown + \text{EtOH} \xrightarrow{\text{RuCl}_3, \text{SnCl}_2} \diagup\diagdown\text{OEt}$$ 307

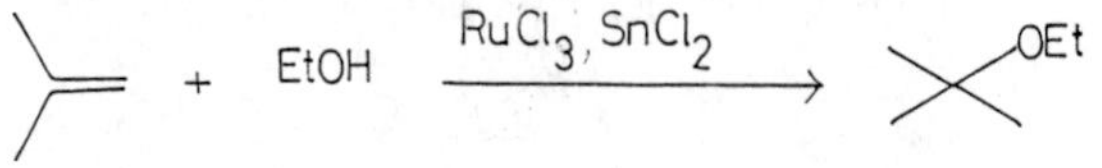

Figure 43   Catalysed addition of oxygen nucleophiles to alkenes.

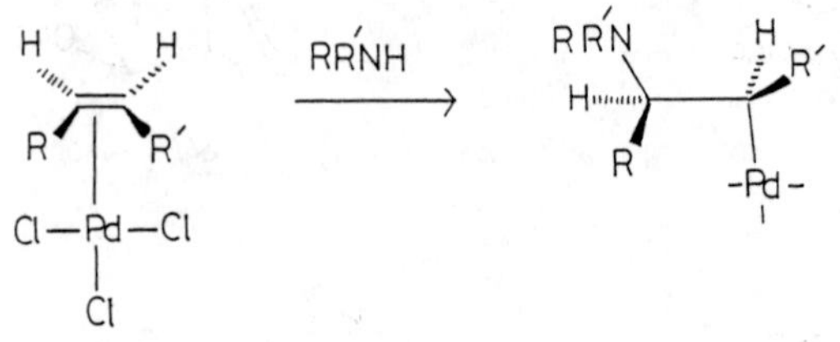

168

169

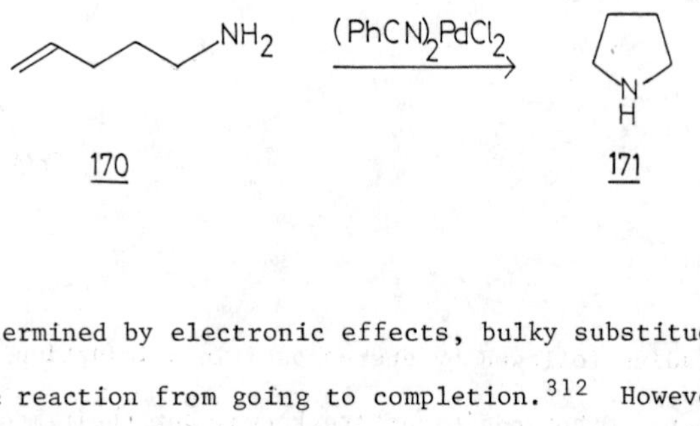

170

171

is largely determined by electronic effects, bulky substituents merely
preventing the reaction from going to completion.[312] However, both
these reactions and their intramolecular analogues[313] (170 to 171) are
stoichiometric and attempts to make them catalytic are hampered by
deactivation of Pd(II) by exess amine.[314] A more useful route involves
elimination of PdH to regenerate alkene and reoxidation of the Pd(0) by
benzoquinones. Both inter- and intramolecular reactions have been
reported (Figure 44).[315,316] Genuinely catalytic additions are also known
in simple systems; ethylene and dimethylamine gives tertiary amines in

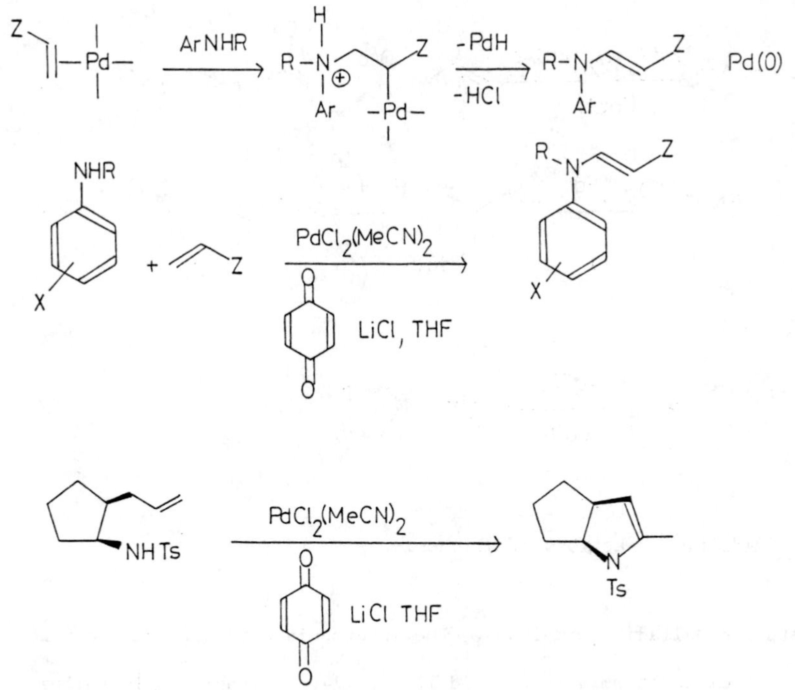

Figure 44   Palladium catalysed amination of alkenes.

92% yield in the presence of $RuCl_3$[317], 13% with $((cod)RhCl)_2$ and 50% with $RhCl_3$.[318]   αβ-Unsaturated nitriles react with ammonia[319,320] and arylamines[321] in the presence of Cu(II) to give addition in low yield.

Allenes react with both oxygen and nitrogen nucleophiles though the reactions are complicated by oligomerisation, and any selectivity is unusual, except in intramolecular reactions.(Figure 45)

4.6.2. Addition of Amines and Alcohols to Metal Allyls

Many additions of nucleophiles to unsaturated systems proceed via metal allyls, a reaction which is generally well understood. 1,2- and 1,4-addition to 1,3-dienes is accomplished in this way and the most popular

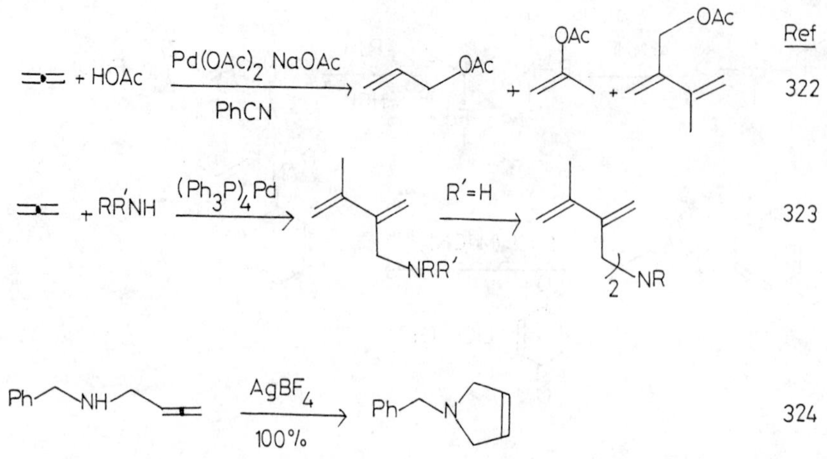

Figure 45 Addition reactions of allenes.

catalysts are palladium complexes. The intermediate is 172 which may be
attacked at either terminus to yield 173 or 174. Suitable nucleophiles
include HOAc[325], ROH[326], RSH[327], oximes[328], NH_3[329] and primary and
secondary amines.[330] The major competing reaction is formation of butadiene telomers and oligomers,[331-3] and for ammonia the formation of
secondary amines, 175 and 176.[329] Suppression of oligomerisation may be
achieved by use of trialkylphosphine ligands[332], or other metals as

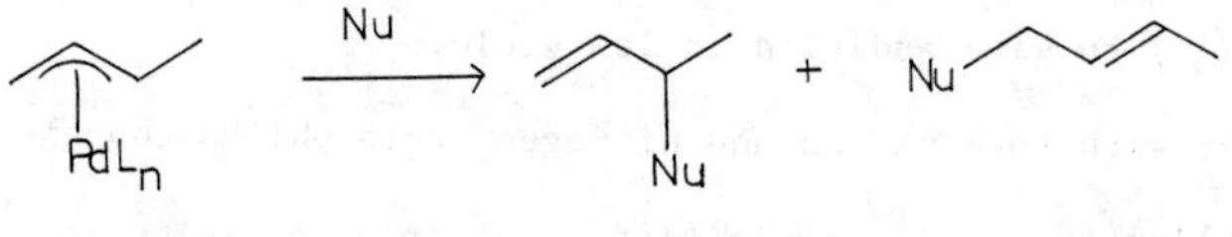

172   173   174

catalysts including $RhCl_3$[330], $RuCl_3$[325], and $NiCl_2PhP(Oi-Pr)_2$|$NaBH_4$|
ROH[334,335] and $Pt(PPh_3)_4$|PPh_3.[336] (e.g. production of 177). Substituted
dienes are less likely to be oligomerised and isoprene reacts with
alcohols in the presence of $RhCl_3$ to give the 1,2-addition product, 178, in

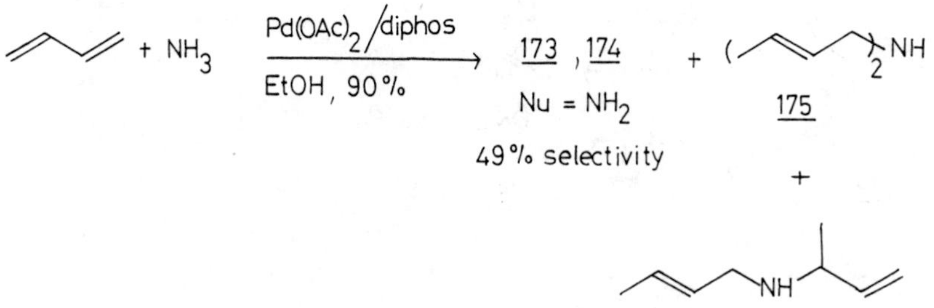

good yield.[337] Several studies make it clear that the branched isomer,
173, is the kinetic product of the reaction but isomerises slowly to
174 in the presence of the catalyst[334,335,338-40]. If the reaction is
performed in the presence of chiral phosphines[341] or phosphites[342],
chiral amines are obtained in low optical excess.

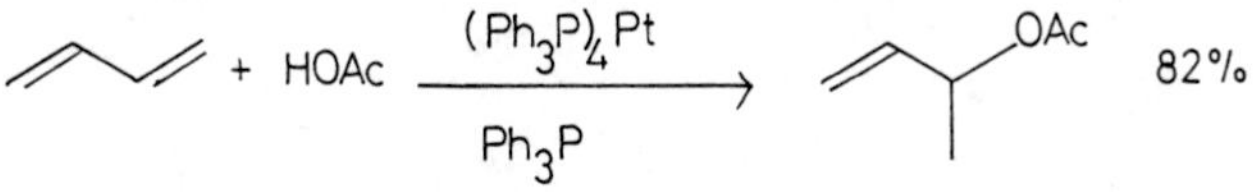

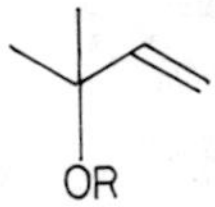

The other important method of generating metal allyl complexes
is from labile allyl derivatives, 179. Suitable leaving groups include
Cl[343], OAc[344], OCOOR[345], OPh[346], OBz[347], OH[34], OPO(OR)$_2$[348] and SO$_2$Ar.[349]
As before, kinetic attack on allyls generated in this way yields branched
products which isomerise to the thermodynamically more stable species
(e.g. 181 to 182 and 183).[350] Reports of stereoselectivity are varied.
Although most heteroatomic nucleophiles and stable carbanions attack the
allyl at the face remote from palladium, acetate may also add cis.[345]

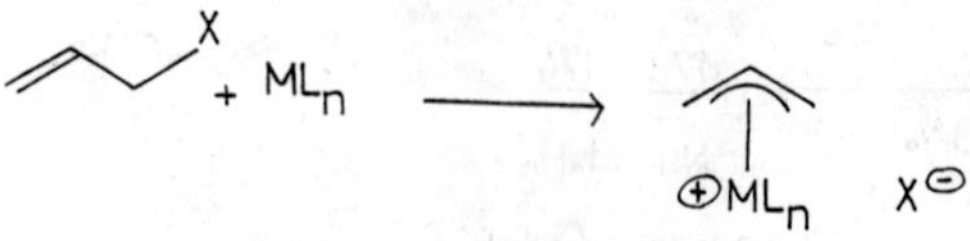

179   180

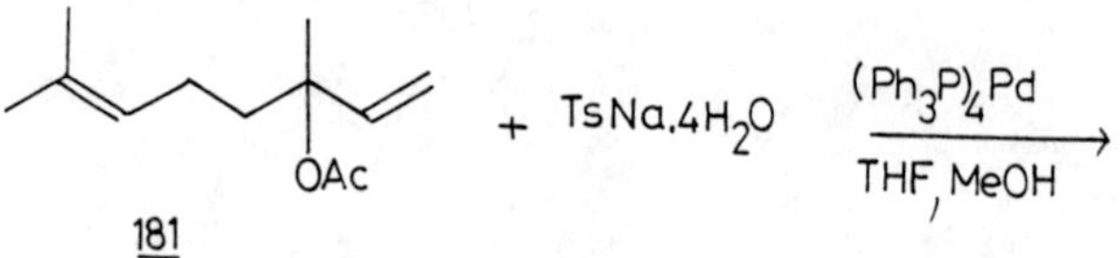

181   + TsNa.4H$_2$O $\xrightarrow[\text{THF, MeOH}]{(Ph_3P)_4Pd}$

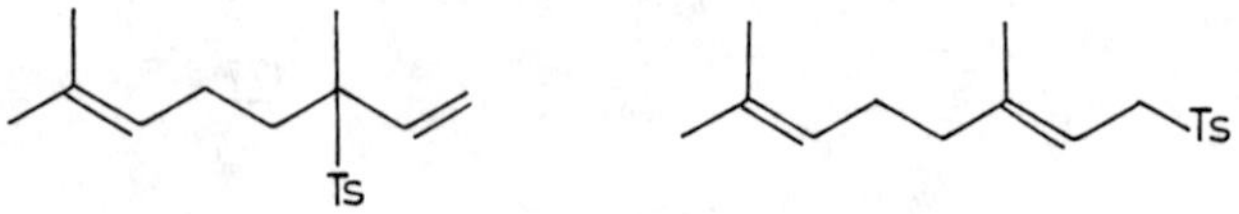

182   183

0° 3hr   78%   9%

25° 12hr   —   84%

Whilst 184 reacted to give mainly retention, implying *trans*-attack, 185 gave mixtures of products. It is clear that the results were complicated by starting material isomerisation. Good stereospecificity is known however; 186 is converted to 187 without loss of chirality.[351] Some applications of the reaction in synthesis are shown in Figure 46.

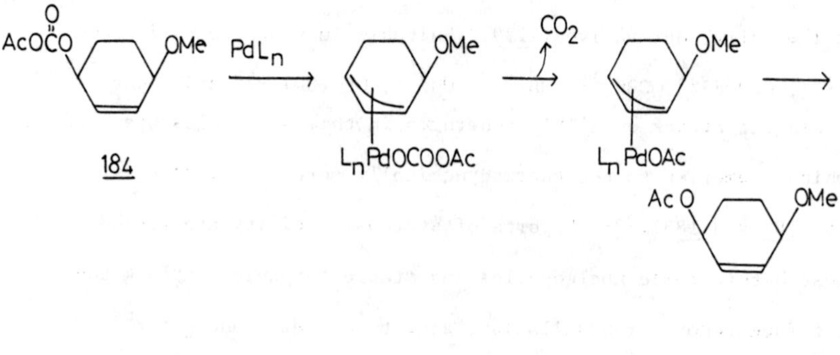

184   L_nPdOCOOAc L_nPdOAc

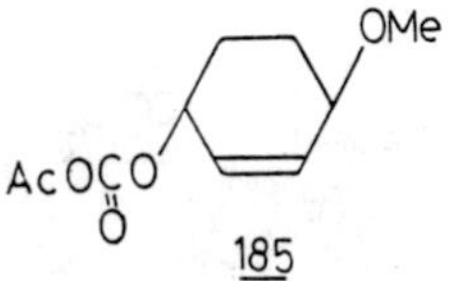

4.6.3. Addition to Alkynes

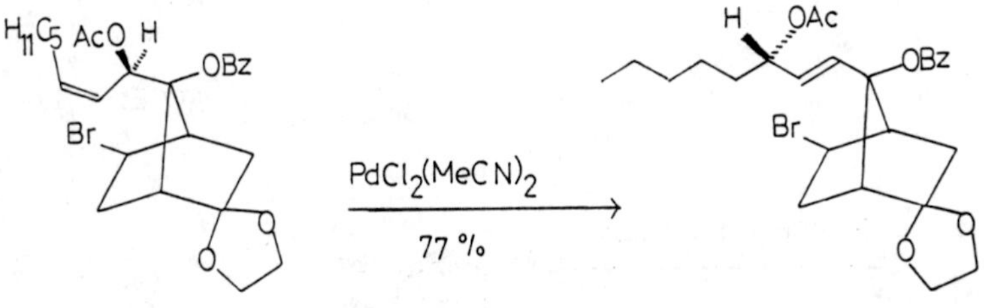

In this area hydration has received most attention. The commonly used catalyst is Hg^{2+} [356] and the reaction proceeds by attack of water on the coordinated alkyne. The predominence of intra- or extrasphere attack depends on exact reaction conditions [357] but an analogue with an acid nucleophile gives the cis-product, 189, indicating oxygen delivery from mercury. [358] Copper salts are also catalysts [359] and

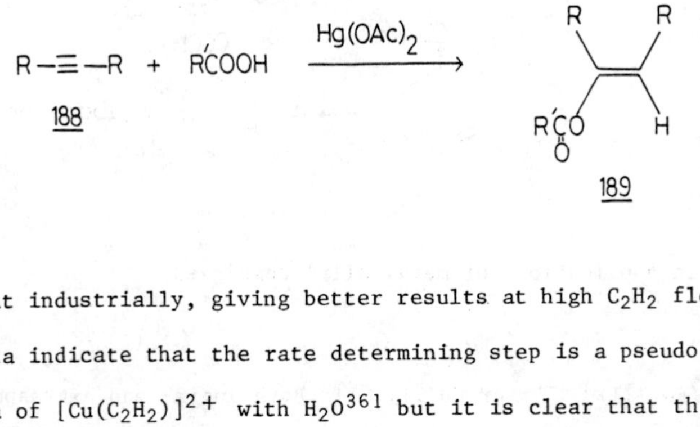

important industrially, giving better results at high C_2H_2 flow rates. [360] Some data indicate that the rate determining step is a pseudo first-order reaction of $[Cu(C_2H_2)]^{2+}$ with H_2O [361] but it is clear that this does not obtain under all conditions and more complex equilibria [362] (including some involving Cu-alkyne σ-complexes which give anti-Markovnikov products [363]) are also important.

Palladium complexes also catalyse alkyne hydration particularly

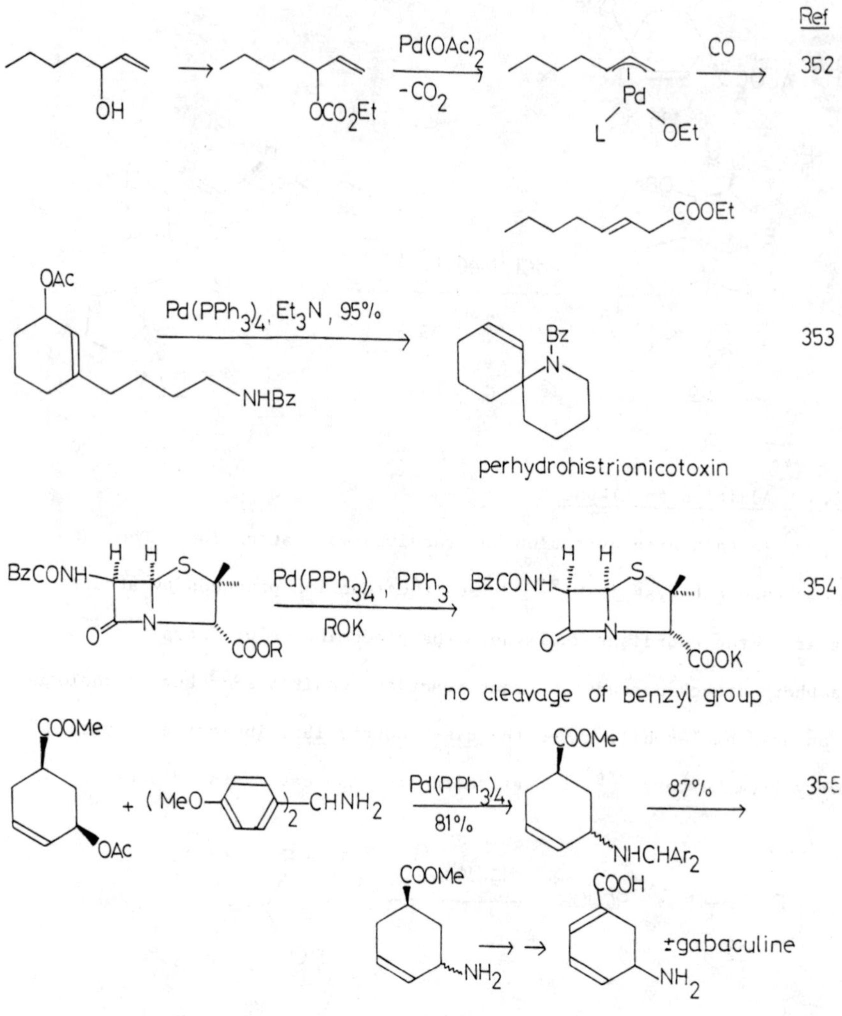

352

353

perhydrohistrionicotoxin

354

no cleavage of benzyl group

355

±gabaculine

Figure 46 Synthetic applications of metal allyl complexes.

when activated by Fe(III)[364-66] or Cu(II).[367] Both intra- and extrasphere

mechanisms operate. An intramolecular attack of -NH$_2$ on a palladium

alkyne complex leads to a synthesis of substituted pyrroles.[368] (190 to

191)

190   191

4.6.4. Nucleophilic Substitutions

I include catalysed nucleophilic substitutions here, not for any mechanistic similarity with the previous sections, but because the products are frequently related. Ullmann coupling of heteroatomic nucleophiles with aryl halides is catalysed by soluble copper salts. Chlorides[369] and bromides[370] are more reactive than iodides[371] since halide exchange occurs readily and Cu_2I_2 is a poor catalyst. Suitable nucleophiles include ammonia [372,373], primary amines[374] and alcohols.[375] Intra- and intermolecular couplings have found many synthetic applications (Figure 47). Mechanistic studies show that a solvent in which the copper salt is soluble is desirable[378] and that the reaction rate correlates with σ with a ρ of 1.4 indicating that the rate determining step is anion attack.[379]

Other metal complexes (in particular of Ni and Pd) are active for aryl and vinyl halide substitution, but there have been few systematic studies. As with the Ullmann coupling the effect of substituents on the catalysed rate is not so great as on the uncatalysed[380] and a possible route involves oxidative addition to M(O). However, $-NO_2$ groups inhibit the reaction suggesting that electron transfer may be critical and that a Ni(I) complexed radical anion is an intermediate. R_2NH[380], CN^- [387], RS^- [382] and RO^- [380] all react with aryl halides and some interesting selectivities may be obtained (e.g. the reactions of 192 and 194).[381,383] With vinyl halides the reaction proceeds with predominent retention of stereochemistry about the double bond (196 to 197)[384,385] the major side reaction being vinyl dimersation.[386]

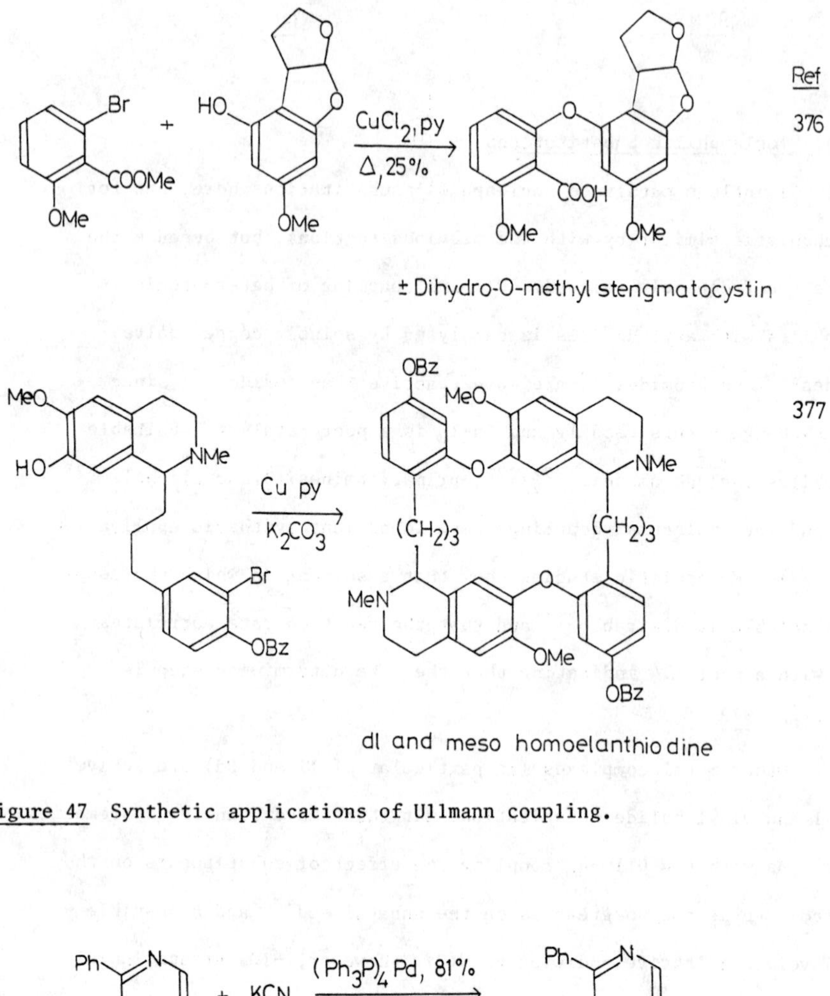

Ref 376

± Dihydro-O-methyl stengmatocystin

377

dl and meso homoelanthiodine

Figure 47 Synthetic applications of Ullmann coupling.

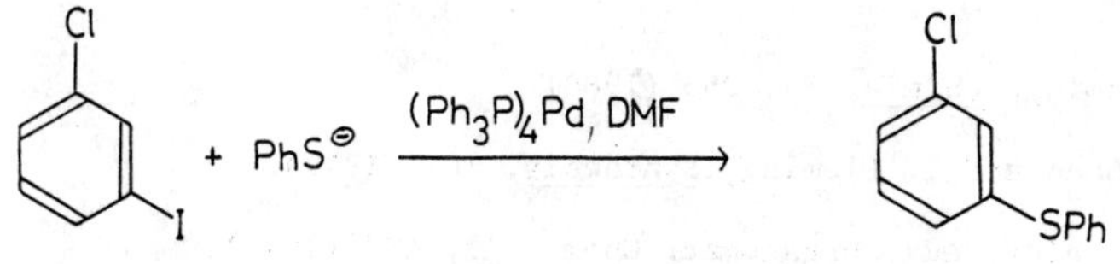

194 + PhS⊖ →(Ph₃P)₄Pd, DMF→ 195

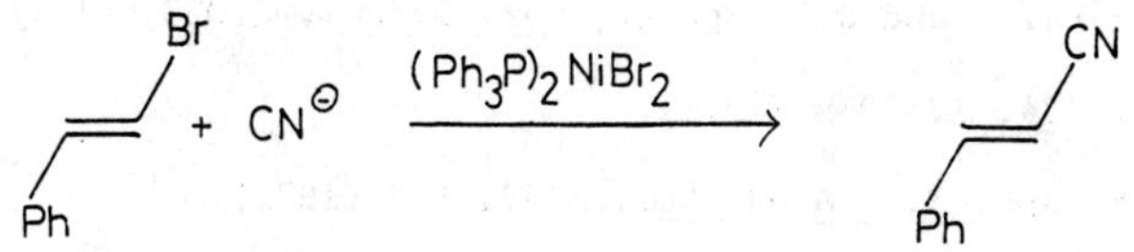

196 + CN⊖ →(Ph₃P)₂NiBr₂→ 197

References

1   I. Fleming, Chimia, 34, 265 (1980).

2   T.H. Chan and I. Fleming, Synthesis, 761 (1979).

3   J.L. Speier, Adv. Organomet. Chem., 17, 407 (1979).

4   E. Lukevics, Z.V. Belyakova, M.G. Pomerantseva and M.G. Voronkov,
    Organomet. Chem. Libr., 5, 1 (1977).

5   A.C. Chuan-Ching and J.L. Speier, Ger. Offen., 2,033,661 (1971);
    Chem. Abs., 74, 125833h (1971).

6   A.J. Chalk, Ann. N.Y. Acad. Sci., 172, 533 (1971).

7   M. Čapka, P. Svoboda, V. Bažant and V. Chvalovsky, Chem. Prum.,
    21, 324 (1971); Chem. Abs., 75, 129074e (1971).

8   V.A. Ponomarenko, 2nd Dresden Symposium on Organo- and Non-Silicate
    Silicon Chemistry, Mitt. Chem. Ges. DDR, Dresden, 52 (1964).

9   K. Yamamoto, T. Hayashi and M. Kumada, J. Organomet. Chem., 28,
    C37 (1971).

10 S.I. Sadykh-zade and M. Sh. Sultanova, J. Gen. Chem. U.S.S.R., 44,
    1750 (1974).

11 M. Green, J.A.K. Howard, J. Proud, J.L. Spencer, F.G.A. Stone and
    C.A. Tsipis, J. Chem. Soc., Chem. Commun., 671 (1976).

12. L. Spialter and D.H. O'Brien, J. Org. Chem., 32, 222 (1967).

13. A.I. Nogaideli, L.I. Nakaidze and V.S. Tskhovrebashvili,
    J. Gen. Chem. U.S.S.R., 45, 1283 (1975).

14. I.D. Segal, Tezisy Dokl. Resp. Konf. Molodykh Uch.-Khim., 2nd, 1,
    9 (1977); Chem. Abs., 89, 109710v (1978).
    J. Lukevics, T. Lapina, E. Liepins and I. Segals, Khim. Geterotsikl.
    Soedin., 962 (1977); Chem. Abs., 87, 168124p (1977).

15. H. Kono, N. Wakao, I. Ojima and Y. Nagai, Chem. Lett., 189 (1975).

16. J. Rejhon and J. Hetflejš, Collect. Czech. Chem. Commun., 40, 3680
    (1975).

17. V.P. Yur'ev, I.M. Salimgareeva, V.V. Kaverin and G.A. Tolstikov, J. Gen. Chem. U.S.S.R., *47*, 329 (1977).

18. N.J. Archer, R.N. Haszeldine and R.V. Parish, J. Chem. Soc., Dalton, 695 (1979).

19. M. Hara, K. Ohno and J. Tsuji, J. Chem. Soc., Chem. Commun., 247 (1971).

20. M. Kumada, Y. Kiso and M. Umeno, J. Chem. Soc., Chem. Commun., 611 (1970).

21. Y. Kiso, M. Kumada, K. Maeda, K. Sumitami and K. Tamao, J. Organomet. Chem., *50*, 311 (1973).

22. M. Kumada, K. Sumitani, Y. Kiso and K. Tamao, J. Organomet. Chem., *50*, 319 (1973).

23. E.W. Bennett and P.J. Orenski, J. Organomet. Chem., *28*, 137 (1971).

24. V.O. Reikhsfel'd, T.P. Khvatova, M.I. Astrakhanov and G.I. Saenko, J. Gen. Chem. U.S.S.R., *47*, 661 (1977).

25. V.P. Yuryev, I.M. Salimgareyeva, V.V. Kaverin, K.M. Khalilov and A.A. Panasenko, J. Organomet. Chem., *171*, 167 (1979).

26. V.N. Tarasenko, G.V. Odabstiyan and A.I. Lakhtikov, Tr. Mosk. Khim.-Tekhnol. Inst., *70*, 138 (1972); Chem. Abs., *78*, 135228m (1973).

27. R.A. Benkesev and W.C. Muench, J. Amer. Chem. Soc., *95*, 285 (1973).

28. P. Svoboda, M. Čapka and J. Hetflejš, Collect. Czech. Chem. Commun., *37*, 3059 (1972).

29. M. Green, J.L. Spencer, F.G.A. Stone and C.A. Tsipis, J. Chem. Soc., Dalton Trans., 1519 (1977).

30. V.V. Kaverin, I.M. Salimgareeva, N.V. Kovaleva and V.P. Yur'ev, J. Gen. Chem. U.S.S.R., *49*, 1575 (1979).

31. L.D. Nasiak and H.W. Post, J. Organomet. Chem., *23*, 91 (1970).

32. R.N. Haszeldine, R.V. Parish and R.J. Taylor, J. Chem. Soc., Dalton Trans., 2311 (1974).

33. R.J. Fessenden and W.D. Kray, J. Org. Chem., 38, 87 (1973).

34. A.J. Chalk, J. Organomet. Chem., 21, 207 (1970).

35. R.A. Benkeser, S. Dunny, G.S. Li, P.G. Nerlekar and S.D. Work,
    J. Amer. Chem. Soc., 90, 1871 (1968).

36. V.P. Yur'ev, I.M. Salimgareeva and V.V. Kaverin, J. Gen. Chem.
    U.S.S.R., 47, 541 (1977).

37. V.P. Yur'ev, I.M. Salimgareeva and V.V. Kaverin, Bull. Acad. Sci.
    U.S.S.R., Div. Chem. Sci., 25, 1588 (1976).

38. R.A. Sultanov, I.A. Khudayarov and S.I. Sadykh-zade, J. Gen. Chem.
    U.S.S.R., 39, 373 (1969).

39. V.V. Kaverin, I.M. Salimgareeva and V.P. Yur'ev, J. Gen. Chem.
    U.S.S.R., 48, 103 (1978).

40. Y. Nagai, I. Ojima and M. Tajima, Japan. Kokai 76 98, 255 (1976);
    Chem. Abs., 86, 106768x (1977).

41. I. Ojima and M. Kumagai, J. Organomet. Chem., 157, 359 (1978).

42. V.P. Yur'ev, I.M. Salimgareeva, O. Zh. Zhebarov and L.M. Khalilov,
    J. Gen. Chem. U.S.S.R., 47, 1414 (1977).

43. Z.V. Belyakova, M.G. Pomerantseva, K.K. Popkov, L.A. Efremova and
    S.A. Golubstov, J. Gen. Chem. U.S.S.R., 42, 879 (1972).

44. J. Langova and J. Hetflejs, Collect. Czech. Chem. Commun., 40, 420
    (1975).

45. J. Rejhon and J. Hetflejs, Collect. Czech. Chem. Commun., 40. 3190
    (1975).

46. Y. Kiso, K. Tamao and M. Kumada, J. Organomet. Chem., 76, 95 (1974).

47. M. Capka and J. Hetflejs, Collect. Czech. Chem. Commun., 40, 3020
    (1975).

48. V.P. Yur'ev, I.M. Salimgareeva, O. Zh. Zheharov and G.A. Tolstikov,
    J. Gen. Chem. U.S.S.R., 46, 368 (1976).

49. R.A. Benkeser, F.M. Merritt and R.T. Roche, _J. Organomet. Chem._, _156_, 235 (1978).

50. V.S. Vaisarova, J. Schraml and J. Hetflejs, _Collect. Czech. Chem. Commun._, _43_, 265 (1978).

51. J. Tsuji, M. Hara and K. Ohno, _Tetrahedron_, _30_, 2143 (1974).

52. I. Ojima, _J. Organomet. Chem._, _134_, Cl (1977).

53. W. Fink, _Helv. Chim. Acta_, _54_, 1304 (1971).

54. A.J. Cornish, M.F. Lappert and T.A. Nile, _J. Organomet. Chem._, _132_, 133 (1977).

55. M.S. Wrighton and M.A. Schroeder, _J. Amer. Chem. Soc._, _96_, 6235 (1974).

56. A.J. Cornish, M.F. Lappert, G.A. Filatrovs and T.A. Nile, _J. Organomet. Chem._, _172_, 153 (1979).

57. V. Vaisarova and J. Hetflejs, _Collect. Czech. Chem. Commun._, _41_, 1906 (1976).

58. M.L. Lappert, T.A. Nile and S. Takahashi, _J. Organomet. Chem._, _72_, 425 (1974).

59. V.P. Yur'ev, I.M. Salimgareeva, O. Zh. Zhebarov and G.A. Tolstikov, _Bull. Acad. Sci. U.S.S.R._, _Div. Chem. Sci._, _24_, 1772 (1975).

60. S. Takahashi, T. Shibano and N. Hagihara, _J. Chem. Soc._, _Chem. Commun._ 161 (1969).

61. T.K. Gar, A.A. Buyakov, A.V. Kisin and V.F. Mironov, _J. Gen. Chem. U.S.S.R._, _41_, 1596 (1971).

62. A.I. Kakhniashvili, T. Sh. Gvaliya and D. Sh. Ioramashvili, _Soobshch Akad. Nauk Gruz. S.S.R._, _68_, 609 (1972); _Chem. Abs._, _78_, 97755w (1973).

63. I. Fleming, _Chem. Soc. Rev._, _10_, 83 (1981).

64. P. Magnus, _Aldrichimica Acta_, _13_, 43 (1980).

65. H. Watanabe, M. Asami and Y. Nagai, _J. Organomet. Chem._ _195_, 363 (1980).

66. M.G. Voronkov, V.B. Pukhnarevich, I.I. Tsykhanskaya and Yu. S. Varshavskii, Dokl. Chem., 254, 449 (1980).

67. H. Neumann and D. Seebach, Tetrahedron Lett., 4839 (1976).

68. H. Neumann and D. Seebach, Chem. Ber., 111, 2785 (1978).

69. V. Mironov, N.G. Maksimova, V.V. Nepomnina, Bull. Acad. Sci. U.S.S.R., Div. Chem. Sci., 313 (1967).

70. G. Stork and E. Colvin, J. Amer. Chem. Soc., 93, 2080 (1971).

71. V.B. Pukhnarevich, L. Kopylova, B.A. Trofimov and M.G. Voronkov, J. Gen. Chem. U.S.S.R., 45, 2600 (1975).

72. V.B. Pukhnarevich, S.P. Sushchinskaya, V.A. Pestunovich and M.G. Voronkov, J. Gen. Chem. U.S.S.R., 43, 1274 (1973).

73. V.O. Reikhsfel'd, V.B. Pukhnarevich, S.P. Sushchinskaya and A.M. Evdokimov, J. Gen. Chem. U.S.S.R., 42, 159 (1972).

74. M.F. Shostakovskii, S.P. Sushchinskaya, V.B. Pukhnarevich, A.I. Borisova, A. Kh. Filippova and V.G. Sakharovskii, Bull. Acad. Sci. U.S.S.R., Div. Chem. Sci., 2443 (1969).

75. V. Chvalovsky, J. Pola, V.B. Pukhnarevich, L.I. Kopylova, E.O. Tsetlina, V.A. Pestunovich, B.A. Trofimov and M.G. Voronkov, Collect, Czech. Chem. Commun., 41, 391 (1976).

76. C.A. Tsipis, J. Organomet. Chem., 187, 427 (1980).

77. H.M. Dickers, R.N. Haszeldine, A.P. Mather and R.V. Parish, J. Organomet. Chem., 161, 91 (1978).

78. H.M. Dickers, R.N. Haszeldine, L.S. Malkin, A.P. Mather and R.V. Parish, J. Chem. Soc., Dalton Trans., 308 (1980).

79. K.A. Brady and T.A. Nile, J. Organomet. Chem., 206, 299 (1981).

80. J.P. Howe, K. Lung and T.A. Nile, J. Organomet. Chem., 208, 401 (1981).

81. A.I. Nogaideli, D. Ya. Zhinkin, L.I. Nakaidze, A.S. Shapatin and V.S. Tskhovrebashivili, J. Gen. Chem. U.S.S.R., 46, 1044 (1976).

82. J.W. Ryan and J.L. Speier, J. Org. Chem. 31, 2698 (1966).

83. M. Green, J.L. Spencer, F.G.A. Stone and C.A. Tspsis, J. Chem. Soc.,
    Dalton Trans., 1525 (1977).

84. K.I. Cherkezishvili, I.M. Gverdtsiteli and R.I. Kubashvili,
    J. Gen. Chem. U.S.S.R., 44, 1009 (1974).

85. E. Lukevits, A.E. Pestunovich, V.A. Pestunovich, E.E. Liepin'sh and
    M.G. Voronkov, J. Gen. Chem. U.S.S.R., 41, 1591 (1971).

86. K. Yamamoto, T. Hayashi and M. Kumada, J. Amer. Chem. Soc., 93,
    5301 (1971).

87. K. Yamamoto, T. Hayashi, M. Zembayashi and M. Kumada, J. Organomet.
    Chem., 118, 161 (1976).

88. K. Yamamoto, Y. Uramoto and M. Kumada, J. Organomet. Chem., 31,
    C9 (1971).
    K. Yamamoto, T. Hayashi, Y. Uramoto, R. Ito and M. Kumada, J.
    Organomet. Chem., 118, 331 (1976).

89. Y. Kiso, K. Yamamoto, K. Tamao and M. Kumada, J. Amer. Chem. Soc.,
    94, 4373 (1972).

90. K. Yamamoto, Y. Kiso, R. Ito, K. Tamao and M. Kumada, J. Organomet.
    Chem., 210, 9 (1981).

91. T. Hayashi, K. Tamao, Y. Katsuro, I. Nakae and M. Kumada,
    Tetrahedron Lett., 21, 1871 (1980).

92. M. Čapka, P. Svoboda, M. Krause and J. Hetflejs, Chem. Ind.
    (London), 650 (1972).

93. V.O. Reikhsfel'd, V.N. Vinogradov and N.A. Filippov, J. Gen. Chem.
    U.S.S.R., 43, 2207 (1973).

94. M. Čapka, P. Svoboda, M. Černý and J. Hetflejs, Tetrahedron Lett.,
    4787 (1971).

95. V. Bazant, M. Čapka, I. Dietzmann, H. Fuhrmann, J. Hetflejs and H.
    Pracejus, Czech. Patent, 182,315 (1980); Chem. Abs., 94, 30911f (1981).

96. V.S. Brovko, N.K. Skvortsov and V.O. Reikhsfel'd, J. Gen. Chem. U.S.S.R., 51, 335 (1981).

97. M. Čapka and J. Hetflejš, Collect. Czech. Chem. Commun., 39, 154 (1974).

98. Z.M. Michalska, M. Čapka and J. Stoch, J. Mol. Catal., 11, 323 (1981).

99. J. Soucek, A.A. Kuzma, L.M. Chananasvili and V.M. Myasina, Czech. Patent 179,188 (1979); Chem. Abs., 92, 59426w (1980).

100. J. Tsuji and M. Hara, Japan. 73 42,959 (1973); Chem. Abs., 81, 121448h (1974).

101. M. Hatanaka and S. Nagashima, Jpn. Kokai Tokkyo Koho, 78, 146,993 (1978); Chem. Abs., 90, 153308q (1979).

102. C. Prud' homme, P. Chaumont and J. Herz, Eur. Pat. Appl., 8,997 (1980); Chem. Abs., 93, 73424w (1980).

103. A.J. Chalk and J.F. Harrod, J. Amer. Chem. Soc., 87, 16 (1965).

104. L.H. Sommer, J.E. Lyons and H. Fujimoto, J. Amer. Chem. Soc., 91, 7051 (1969).

105. V.O. Reikhsfel'd, M.I. Astrakhanov and E.G. Kagan, J. Gen. Chem. U.S.S.R., 40, 670 (1970).

106. M. Čapka, P. Svoboda, V. Bažant and V. Chvalovsky, Collect. Czech. Chem. Commun., 36, 2785 (1971).

107. F. de Charentenay, J.A. Osborn and G. Wilkinson, J. Chem. Soc., A, 787 (1968).

108. R.N. Haszeldine, R.V. Parish and D.J. Parry, J. Chem. Soc., A, 683 (1969).

109. G. Kuncova and V. Chvalovsky, Collect. Czech. Chem. Commun., 45 2240 (1980).

110. H.E. Simmons and R.D. Smith, J. Amer. Chem. Soc., 80, 5323 (1958).

111. H.E. Simmons, T.L. Cairns, S. Vladuchick and C.M. Hoiness, Org. React., 20, 1 (1973).

112. E.P. Blanchard and H.E. Simmons, J. Amer. Chem. Soc., 86, 1337 (1964).

113. H.H. Wasserman, G.M. Clark and P.C. Turley, Topics in Current Chemistry, 47, 73 (1974).

114. J.M. Denis, C. Girard and J.M. Conia, Synthesis, 549 (1972).

115. J. Furukawa, N. Kawabata and J. Nishimura, Tetrahedron, 24, 53 (1968).

116. J. Nishimura, J. Furukawa, N. Kawabata and H. Koyama, Bull. Chem. Soc. Jpn., 44, 1127 (1971).

117. J. Furukawa, N. Kawabata and J. Nishimura, Tetrahedron Lett., 3495 (1968).

118. W.G. Dauben and A.C. Ashcroft, J. Amer. Chem. Soc., 85, 3673 (1963).

119. J.F. Ruppert, M.A. Avery and J.D. White, J. Chem. Soc., Chem. Commun. 978 (1976).

120. R.E. Pincock and J.I. Wells, J. Org. Chem., 29, 965 (1964).

121. R. Paulissen, A.J. Hubert and Ph. Teyssie, Tetrahedron Lett., 1465 (1972).

122. B.W. Peace and D.S. Wulfman, J. Chem. Soc., Chem. Commun., 1179 (1971).

123. R.G. Salomon and J.K. Kochi, J. Amer. Chem. Soc., 95, 3300 (1973).

124. D.S. Wulfman and D. Poling, Reactive Intermediates, 1, 321; Edited R.A. Abramovic, Plenum Press, 1980.

125. A.J. Anciaux, A.J. Hubert, A.F. Noels, N. Petiniot and P. Teyssie, J. Org. Chem., 45, 695 (1980).

126. N. Petiniot, A.F. Noels, A.J. Anciaux, A.J. Hubert and P. Teyssie, Fundamental Research in Homogeneous Catalysis, 3, 421 (1979).

127. M.P. Doyle, D. Van Leusen and T.H. Tamblyn, Synthesis, 787 (1981).

128. N. Petiniot, A.J. Anciaux, A.F. Noels, A.J. Hubert and Ph. Teyssie, Tetrahedron Lett., 1239 (1978).

129. H. Irngartinger, A. Goldman, R. Schappert, P. Garner and P. Dowd, J. Chem. Soc., Chem. Commun., 455 (1981).

130. J. Kottwitz and H. Vorbrüggen, Synthesis, 636 (1975).

131. F.L.M. Smeets, L. Thijs and B. Zwanenburg, Tet., 36, 3269 (1980).

132. M. Suda, Synthesis, 714 (1981).

133. S. Bien, A. Gillon and S. Kohen, J. Chem. Soc., Perkin I, 489 (1976).

134. J.E. Baldwin, J. Chem. Soc., Chem. Commun., 734 (1976) and following papers.

135. A. Nakamura, A. Konishi, Y. Tatsuno and S. Otsuka, J. Amer. Chem. Soc., 100, 3443 (1978).
A. Nakamura, A. Konishi, R. Tsujitani, M-a. Kudo and S. Otsuka, J. Amer. Chem. Soc., 100, 3449 (1978).
A. Nakamura, Pure Appl. Chem., 50, 37 (1978).

136. M. Elliott, A.W. Farnham, N.F. Janes, P.H. Needham and B.C. Pearson, Nature, 213, 493 (1967).
M. Elliott, N.F. Janes and B.C. Pearson, J. Chem. Soc., C, 2551 (1971).

137. M. Elliott, A.W. Farnham, N.F. Janes, P.H. Needham, D.A. Pulman and J.H. Stevenson, Nature, 246, 169 (1973).

138. T. Aratani, Y. Yoneyoshi and T. Nagase, Tetrahedron Lett., 2599 (1977).

139. D. Holland, D.A. Laidler and D.J. Milner, J. Mol. Catal., 11, 119 (1981).
J. Crosby, D. Holland, D.A. Laidler and D.J. Milner, Eur. Pat. Appl. 22,608 (1981); Chem. Abs., 95, 114886k (1981).

140. W. Nagata and M. Yoshioka, Org. React., 25, 255 (1977).

141. E. S. Brown in Organic Syntheses via Metal Carbonyls ed. P. Wender, Volume II p. 655 (1977).

142. B. R. James, Comprehensive Organometallic Chemistry, 8, 353 (1982).

143. B. W. Taylor and H. E. Swift, J. Catal., 26, 254 (1972).

144. H. E. Shook, U. S. Patent 4,082,811 (1978); Chem. Abs., 89, 109935x (1978).

145. C. M. King, W. C. Seidel and C. A. Tolman, U. S. Patent 3,925,445 (1975); Chem. Abs., 84, 88921u (1976).

146. P. Arthur, D. C. England, B. C. Pratt and G. M. Whitman, J. Amer. Chem. Soc., 76, 5364 (1954).

147. P. Arthur and B. C. Pratt, U. S. Patent 2,666,748 (1954); Chem. Abs., 49, 1774c (1955).

148. P. Arthur and B. C. Pratt, U. S. Patent 2,666,780 (1954); Chem. Abs., 49, 1776a (1955).

149. W. C. Drinkard and R. V. Lindsey, Ger. Offen. 1,806,098 (1969); Chem. Abs., 71, 49343u (1969).

150. E. I. du Pont de Nemours & Co., Brit. Pat. 1,104,140 (1968); Chem. Abs., 68, 77795z (1968).

151. E. S. Brown and E. A. Rick, J. Chem. Soc., Chem. Commun., 112 (1969).

152. E. S. Brown and E. A. Rick, Amer. Chem. Soc., Div. Petrol. Chem., Prepr., 14, 1329 (1969).

153. C. M. King and M. T. Musser, U. S. Patent 3,864,380 (1975); Chem. Abs., 82, 155421e (1975).

154. W. R. Jackson and P. S. Elmes, J. Amer. Chem. Soc., 101, 6128 (1979).

155. K. Brown and P. A. Chaloner, J. Organomet. Chem., 217, C25 (1981).

156. P. S. Elmes and W. R. Jackson, Ann. N. Y. Acad. Sci., 333, 225 (1980).

157. G. R. Coraor and W. Z. Heldt, U. S. Patent 2,904,581 (1959); Chem. Abs., 54, 4393f (1960).

158. E. S. Brown, E. A. Rick and F. D. Mendicino, J. Organomet. Chem.,
     38, 37 (1972).

159. G. W. Parshall, J. Mol. Catal., 4, 256 (1978).

160. D. Y. Waddan and G. R. Crooks, Ger. Offen. 2,450,863 (1975);
     Chem. Abs., 83, 58183s (1975).

161. D. Y. Waddan, Ger. Offen. 2,344,767 (1974); Chem Abs., 80, 145481u
     (1974).

162. W. C. Drinkard and R. V. Lindsay, Fr. Patent 1,544,656 (1968);
     Chem. Abs., 71, 123622c (1969).

163. D. Y. Waddan, Ger. Offen. 2,336,852 (1974); Chem. Abs., 80, 120342m
     (1974).

164. D. D. Coffman, L. F. Salisbury and N. D. Scott, U. S. Patent
     2,509,859 (1950); Chem. Abs., 44, 8361c (1950).

165. R. J. Benzie and D. Y. Waddan, Ger. Offen. 2,805,760 (1978); Chem.
     Abs., 89, 214924r (1978).

166. R. J. Benzie and D. Y. Waddan, Ger. Offen. 2,812,151 (1978);
     Chem. Abs., 90, 5941a (1979).

167. W. C. Seidel and C. A. Tolman, U. S. Patent 3,850,973 (1974);
     Chem. Abs., 82, 97704m (1975).

168. P. Albanese, L. Benzoni, G. Carnisio and A. Crivelli, Ital. Pat.
     869,900 (1970); Chem. Abs., 78, 135701k (1973).

169. E. I. du Pont de Nemours and Co., Brit. Pat. 1,178,950 (1970);
     Chem. Abs., 72, 89831d (1970).

170. W. C. Drinkard and B. W. Taylor Ger. Offen., 1,807,089 (1969);
     Chem. Abs., 71, 101350k (1969).

171. R. G. Downing and R. A. Fouty, Fr. Demande 2,010,935 (1970); Chem.
     Abs., 73, 66068s (1970).

172. E. I. du Pont de Nemours and Co., Brit. Pat. 1,112,539 (1968);
     Chem. Abs., 69, 26810p (1968).

173. W. C. Drinkard, Ger. Offen. 1,806,096 (1969); Chem. Abs., 71, 30093r (1969).

174. C. M. King, W. C. Seidel and C. A. Tolman, Ger. Offen. 2,237,703 (1973); Chem. Abs., 78, 135700j (1973).

175. W. C. Drinkard and R. J. Kassal Fr. Patent 1,529,134 (1968); Chem. Abs., 71, 30092q (1969).

176. Y. I. Mok, U. S. Patent 3,846,474 (1974); Chem. Abs., 82, 124818k (1975).

177. H. E. Shook, Ger. Offen. 2,421,081 (1974); Chem. Abs., 82, 155842t (1975).

178. W. C. Drinkard, Ger. Offen. 2,055,747 (1971); Chem. Abs., 75, 64569y (1971).

179. W. C. Drinkard and B. W. Taylor, Ger. Offen. 1,807,088 (1969); Chem. Abs., 71, 38383f (1969).

180. W. A. Schulze and J. E. Mahan U. S. Patent 2,422,859 (1947); Chem. Abs., 42, 205d (1948).

181. Gulf Research and Development Co., Jpn. Kokai Tokkyo Koho 79 81,243 (1979); Chem. Abs., 91, 157344s (1979).

182. H. E. Swift and C. Y. Wu, U. S. Patent 4,215,068 (1980); Chem. Abs., 93, 238927c (1980).

183. G. W. Parshall, Homogeneous Catalysis, Chapters 4 and 8, Wiley, 1980.

184. G. M. Sitnikov, Plast. Massy, 68 (1977); Chem. Abs., 86, 156013t (1977).

185. F. Zaripova, E. Bobrovitskaya, Mater. Resp. Nauchno-Tekh. Konf. Molodykh Uch. Pererab. Nefti Neftekhim., 2nd 107 (1974); Chem. Abs., 85, 78387f (1976).

186. L. V. Reshetnikova, L. N. Reshetova, N. A. Rastrogina, A. A. Khorkin, O. N. Temkin and R. M. Flid, Khim. Prom-st. (Moscow) 262 (1975); Chem. Abs., 84, 10419s (1976).

187. L. N. Reshetova, A. A. Khorkin, L. V. Reshetnikova, B. S. Vorob'ev, A. I. Nefedov, E. P. Grigoryan, R. M. Flid and O. N. Temkin, U.S.S.R. Patent 475,357 (1975); Chem. Abs., 83, 98148r (1975).

188. L. V. Reshetnikova, V. S. Gerasimova, L. N. Reshetova, O. N. Temkin and R. M. Flid, Khim. Prom. (Moscow), 490 (1974); Chem. Abs., 81, 169880v (1974).

189. Sh. O. Badanyan, M. G. Voskanyan and Zh. A. Chobanyan, Russ. Chem. Rev., 50, 1075 (1981) and references therein.

190. L. M. Mel'nikova, L. N. Reshetova, O. N. Temkin and R. M. Flid, Khim. Atsetilina Tekhnol. Karbida Kal'tsiya 361 (1972); Chem. Abs., 80, 36680n (1974).

191. R. M. Flid, O. N. Temkin, E. A. Alfanas'eva, L. N. Reshetova, L. V. Mel'nikova, A. A. Khorkin, T. G. Sukhova and N. A. Rastrogina, Khim. Prom., 43, 486 (1967); Chem. Abs., 68, 29228v (1968).

192. W. C. Drinkard and R. V. Lindsey, Fr. Patent 1,533,557 (1968); Chem. Abs., 71, 101353p (1969).

193. W. R. Jackson and C. G. Lovel, J. Chem. Soc., Chem. Commun., 1231 (1982).

194. T. Funabiki and Y. Yamazaki, J. Chem. Soc., Chem. Commun., 1110 (1979).

195. T. Funabiki. Y. Yamazaki and K. Tarama, J. Chem. Soc., Chem. Commun., 63 (1978).

196. H. Singer and G. Wilkinson, J. Chem. Soc., A, 2516 (1968).

197. L. Benzoni, C. Zanzottera, M. Camia, V. M. Tacchi and M. De Innocentiis, Chim. Ind. (Milano), 50, 1227 (1968); Chem. Abs., 70, 43537t (1969).

198. F. Cariati, R. Ugo and F. Bonati, Inorg. Chem., 5, 1128 (1966).

199. W. C. Drinkard, D. R. Eaton, J. P. Jesson and R. V. Lindsey, Inorg. Chem., 9, 392 (1970).

   R. A. Schunn, Inorg. Chem., 9, 394 (1970).

200. J. D. Druliner, A. D. English, J. P. Jesson, P. Meakin and C. A. Tolman, J. Amer. Chem. Soc., 98, 2156 (1976).

201. B. Corain and G. Puosi, J. Catal., 30, 403 (1973).

202. M. D. Johnson, M. L. Tobe and L. Y. Wong, J. Chem. Soc., A 923, 929 (1968).

203. M. D. Johnson, M. L. Tobe and L. Y. Wong, J. Chem. Soc., Chem. Commun., 298 (1967).

204. J. Chatt, R. S. Coffee, A Gough and D. T. Thompson, J. Chem. Soc. A, 190 (1968).

205. W. R. Jackson and C. G. Lovel, Tetrahedron Lett., 23, 1621 (1982).

206. J. E. Bäckvall and O. S. Andell, J. Chem. Soc., Chem. Commun., 1098 (1981).

207. G. B. Sergeev, V. V. Smirnov, T. N. Rostovshchikova, V. A. Polyakov and O. S. Korinfskaya, Kinet. Catal., 20, 1212 (1979).

208. A. E. Feiring, J. Fluorine Chem., 14, 7 (1979).

209. F. Hagedorn, K. Wedemeyer and R. Mayer-Mader, Ger. Offen. 2,318,115 (1974); Chem. Abs., 82, 30940s (1975).

210. D. K. Burchett, N. Yamazaki and H. Yoshiike, Ger. Offen. 2,413,739 (1974); Chem. Abs., 82, 3761c (1975).

211. N. F. Alekseeva, O. N. Temkin and R. M. Flid, Kinet. Catal., 11, 1319 (1970).

212. L. A. Gasparyan, N. G. Karapetyan, A. S. Tarkhanyan, R. M. Mnatsakanyan, T. K. Manukyan, S. S. Kazazyan and M. G. Ierusalimskaya, Arm. Khim. Zh., 22, 434 (1969); Chem. Abs., 72, 2948a (1970).

213. R. Vestin, B. Wahlund and T. Lindblom, Acta Chem. Scand., 21, 2335, 2351 (1967).

214. G. K. Oparina, R. N. Gurskii, I. L. Vaisiman, L. V. Terent'eva, R. V. Istratova, N. F. Alekseeva and R. M. Flid, Khim. Alsetilena, 459 (1968); Chem. Abs., 71, 29975y (1969).

215. G. H. Shestakov, A. V. Massal'skaya, S. M. Airyan and O. N. Temkin, Kinet. Catal., 18, 337 (1977).

216. F. E. Kung, Fr. Patent 1,542,025 (1968); Chem. Abs., 71, 101277s (1969): U. S. Patent 3,646,230 (1972); Chem. Abs., 76, 112658w (1972).

217. J. Tsuji, K. Sato and H. Nagashima, Chem. Lett., 1169 (1981).

218. H. Nagashima, K. Sato and J. Tsuji, Chem. Lett., 1605 (1981).

219. S. Murai, R. Sugise and N. Sonoda, Angew. Chem., Int. Ed. Engl., 20, 475 (1981).

220. T. Nakano, Y. Shimada, R. Sako, M. Kayama, H. Matsumoto and Y. Nagai, Chem. Lett., 1255 (1982).

221. M. Sato and F. Sato, Jpn. Kokai Tokkyo Koho 78, 103,426 (1978); Chem. Abs., 90, 5881f (1979).

222. F. Sato, H. Kodama, Y. Tomuro and M. Sato, Chem. Lett., 623 (1979).

223. J. J. Eisch, Comp. Organomet. Chem., 1, 557 (1982).

224. F. Sato, H. Kodama and M. Sato, Chem. Lett., 789 (1978).

225. E. C. Ashby and S. A. Noding, J. Org. Chem., 44, 4364 (1979).

226. F. Sato, S. Sato and M. Sato, J. Organomer. Chem., 131, C26 (1977).

227. F. Sato, Y. Mori and M. Sato, Chem. Lett., 833 (1978).

228. E. C. Ashby and S. A. Noding, Tetrahedron Lett., 4579 (1977).

229. E. C. Ashby and S. A. Noding, J. Org. Chem., 45, 1035 (1980).

230. G. A. Tolstikov, V. M. Dzhemilev, O. S. Vostrikova and A. G. Tolstikov, Bull. Acad. Sci. U.S.S.R., Div. Chem. Sci., 31, 596 (1982).

231. E-i. Negishi and T. Yoshida, Tetrahedron Lett., 21, 1501 (1981).

232. F. Sato, Y. Tomuro, H. Ishikawa and M. Sato, Chem. Lett., 99 (1980).

233. J. R. Zietz, G. C. Robinson and K. L. Lindsay, Comp. Organomet. Chem., 7, 366 (1982).

234. F. Sato, Y. Mori and M. Sato, Tetrahedron Lett., 1405 (1979).

235. K. Isagawa, M. Ohige, K. Tatsumi and Y. Otsuji Chem. Lett., 1155 (1978).

236. E. C. Ashby and T. Smith, J. Chem. Soc., Chem. Commun., 30 (1978).

237. L. Yu Gubaidullin, R. M. Sultanov and U. M. Dzhemilev, Bull. Acad. Sci. U.S.S.R., Div. Chem. Sci., 31, 638 (1982).

238. F. Sato. H. Ishikawa and M. Sato, Tetrahedron Lett., 21, 365 (1980).

239. E. Colomer and R. Corriu, J. Organomet. Chem., 82, 367 (1974).

240. J. J. Eisch and J. E. Galle, J. Organomet. Chem., 160 C8 (1978).

241. E. C. Ashby and S. R. Noding, J. Organomet. Chem., 177, 117 (1979).

242. J. J. Eisch and M. W. Foxton, J. Organomet. Chem., 12, P33 (1968).

243. P. C. Wailes, H. Weigold and A. P. Bell, J. Organomet. Chem., 27, 373 (1971).

244. F. Sato, H. Ishikawa and M. Sato, Tetrahedron Lett., 22, 85 (1981).

245. F. Sato, H. Watanabe, Y. Tanaka and M. Sato, J. Chem. Soc., Chem. Commun., 1126 (1982).

246. H. G. Richey and A. M. Rothman, Tetrahedron Lett., 1457 (1968).

247. E. A. Hill, J. Organomet. Chem., 91, 123 (1975).

248. J. K. Crandall, P. Battioni, J. T. Wehlacz and R. Bindra, J. Amer. Chem. Soc., 97, 7171 (1975).

249. A. Alexakis, G. Cahiez and J. F. Normant, J. Organomet. Chem., 177, 293 (1979).

250. J. R. C. Light and H. H. Zeiss, J. Organomet. Chem., 21, 517 (1970).

251. M. Michman and M. Balog, J. Organomet. Chem., 31, 395 (1971).

252. J-G. Duboudin and B. Jousseaume, J. Organomet. Chem., 44, C1 (1972).

253. J-G. Duboudin and B. Jousseaume, C. R. Acad. Sci. Paris, 276, 1421 (1973).

254. J-G. Duboudin and B. Jousseaume, J. Organomet. Chem., 96, C47 (1975).

255. J. G. Duboudin and B. Jousseaume, J. Organomet. Chem., 162, 209 (1978).

256. L. M. Zubritskii, L. N. Cherkasov, T. N. Fomina and Kh. V. Bal'yan, J. Org. Chem., U.S.S.R., 11, 204 (1975).

257. L. M. Zubritskii, T. N. Fomina and Kh. V. Bal'yan, J. Org. Chem., U.S.S.R., 17, 63 (1981), 18, 1209 (1982).

258. R. Mornet and L. Gouin, Bull. Soc. Chim. Fr., 737 (1977).

259. B. Jousseaume and J-G. Duboudin, J. Organomet. Chem., 91, C1 (1975).

260. J. G. Duboudin and B. Jousseaume, J. Organomet. Chem., 168, 1 (1979).

261. H. G. Richey, W. F. Erickson and A. S. Heyn, Tetrahedron Lett., 2183 (1971).

262. B. Jousseaume, Ph.D. Thesis, Bordeaux, 1977.

263. R. W. M. Ten Hoedt, G. Van Koten and J. G. Noltes, J. Organomet. Chem., 170, 131 (1979).

264. J. F. Normant and A. Alexakis, Synthesis, 841 (1981).

265. B. B. Snider, R. S. E. Conn and M. Karras, Tetrahedron Lett., 1679 (1979).

266. L. M. Zubritskii, L. N. Cherkasov, T. M. Fomina and Kh. V. Bal'yan, J. Gen. Chem., U.S.S.R., 46, 440 (1976).

267. L. N. Cherkasov, Izv. Vyssh. Uchebn. Zaved., Khim. Khim. Tekhnol., 20, 363 (1977); Chem. Abs., 87, 135534u (1977).

268. J. J. Eisch and W. C. Kaska, J. Organomet. Chem., 2, 184 (1964).

269. T. Mole and J. R. Surtees, Aust. J. Chem., 17, 1229 (1964).

270. R. Rienäcker and D. Schwengers, Ann. 737, 182 (1970).

271. L. Lardicci, A. M. Caporusso and G. Giacomelli, J. Organomet. Chem., 70, 333 (1974).

272. A. M. Caporusso G. Giacomelli and L. Lardicci, J. Org. Chem., 42, 914 (1977), 44, 231 (1979); J. Chem. Soc., Perkin I, 1900 (1981).

273. E-i. Negishi, Pure Appl. Chem., 53, 2333 (1981).

274. E-i. Negishi, N. Okukado, A. O. King, D. E. Van Horn and B. I. Spiegel, J. Amer. Chem. Soc., 100, 2254 (1978).

275. E-i. Negishi, D. E. Van Horn, A. O. King and N. Okukado, Synthesis, 501 (1979).

276. D. E. Van Horn and E-i. Negishi, J. Amer. Chem. Soc., 100, 2252 (1978).

277. E-i. Negishi, L. F. Valente and M. Kobayashi, J. Amer. Chem. Soc., 102, 3298 (1980).

278. T. Yoshida, Chem. Lett., 293 (1982).

279. C. L. Rand, D. E. Van Horn, M. W. Moore and E-i. Negishi, J. Org. Chem., 46, 4093 (1981).

280. E-i. Negishi and T. Yoshida, Tetrahedron Lett., 21, 1501 (1980).

281. D. E. Van Horn, L. F. Valente, M. J. Idacavage and E-i. Negishi, J. Organomet. Chem., 156, C20 (1978).

282. B. B. Snider and M. Karras, J. Organomet. Chem., 179, C37 (1979).

283. D. C. Brown, S. A. Nichols, A. B. Gilpin and D. W. Thompson, J. Org. Chem., 44, 3457 (1979).

284. L. C. Smedley, H. A. Tweedy, R. A. Coleman and D. W. Thompson, J. Org. Chem., 42, 4147 (1979).

285. R. A. Coleman, C. M. O'Doherty, H. E. Tweedy, T. V. Harris and D. W. Thompson, J. Organomet. Chem., 107, C15 (1976).

286. H. E. Tweedy, R. A. Coleman and D. W. Thompson, J. Organomet. Chem., 129, 69 (1977).

287. N. Okukado and E-i. Negishi, Tetrahedron Lett., 2357 (1978).

288. D. B. Malpass, S. C. Watson and G. S. Yeargin, J. Org. Chem., 42, 2712 (1977).

289. H. Matsushita and E-i. Negishi, J. Amer. Chem. Soc., 103, 2882 (1981).

396

290. B. Åkermark, J-E. Bäckvall and K. Zetterberg, Acta Chem. Scand., B36, 577 (1982).

291. S. G. Davies, M. L. H. Green and D. M. P. Mingos, Tetrahedron, 34, 3047 (1978).

292. O. Eisenstein and R. Hoffmann, J. Amer. Chem. Soc., 103, 4308 (1981).

293. P. Lennon, M. Madhavarao, A. Rosan and M. Rosenblum, J. Organomet. Chem., 108, 93 (1976).

294. L. Maresca and G. Natile, J. Chem. Soc., Chem. Commun., 40 (1983).

295. O. Matsumoto, Y. Tachibana, S. Oshima, O. Idai and Y. Matsui, Japan. Kokai 77,113,704 (1977); Chem. Abs., 88, 50270a (1978).

296. S. N. Massie, U. S. Patent 3,705,912 (1972); Chem. Abs., 79, 4994f (1973).

297. M. L. Spector, J. H. Craddock and F. E. Caropreso, U. S. Patent 3,328,469 (1967); Chem. Abs., 67, 116554w (1967) .

298. F. Matsuda and T. Kato, Jpn. Kokai Tokkyo Koho 79 27,505 (1979); Chem. Abs., 91, 19890r (1979).

299. T. Izumi, K. Yamazaki, R. Ohkami and M. Toda, Japan. 74 36,204 (1974); Chem. Abs., 82, 139310h (1975).

300. F. G. Mesich, Ger. Offen. 1,954,896 (1970); Chem. Abs., 73, 44916g (1970).

301. P. M. Henry, J. Org. Chem., 38, 2766 (1973).

302. T. Yoshida, T. Matsuda, T. Okano, T. Kitani and S. Otsuka, J. Amer. Chem. Soc., 101, 2027 (1979).

303. R. J. H. Duprey, W. D. Fordham, J. F. Janes, D. V. Banthorpe and M. R. Young, Chem. Ind., 847 (1973).

304. E. F. Magoon and L. H. Slaugh, J. Organomet. Chem., 55, 409 (1973).

305. J. E. McKeon, P. Fitton and A. A. Griswold, Tetrahedron, 28, 227,233 (1972).

306. V. M. Dzhemilev, R. N. Kakhretdinov and G. A. Tolstikov, Bull. Acad. Sci. U.S.S.R., Div. Chem. Sci., 26, 327 (1977).

307. J. E. Hamlin and P. M. Maitlis, J. Mol. Catal., 11, 129 (1981).

308. B. Åkermark, J. E. Bäckvall, K. Siirala-Hansen´, K. Sjöberg and K. Zetterberg, Tetrahedron Lett., 1363 (1974).

309. J-E, Bäckvall, Tetrahedron Lett., 2225 (1975).

310. J-E. Bäckvall, Tetrahedron Lett., 163 (1978).

311. J-E. Bäckvall, E. E. Björkman, S. E. Byström and A. Solladié-Cavallo, Tetrahedron Lett., 23, 943 (1982).

312. M. Green, J. K. K. Sarhan and I. M. Al-Najjar, J. Chem. Soc., Dalton, 1565 (1981).

313. B. Pugin and L. M. Venanzi, J. Organomet. Chem., 214, 125 (1981).

314. J-E. Bäckvall and S. E. Byström, J. Org. Chem., 47, 1126 (1982).

315. J. J. Bozell and L. S. Hegedus, J. Org. Chem., 46, 2561 (1981).

316. L. S. Hegedus and J. M. McKearin, J. Amer. Chem. Soc., 104, 2444 (1982).

317. D. M. Gardner and R. T. Clark, Eur. Pat. Appl. EP 39,061 (1981); Chem. Abs., 96, 85039z (1982).

318. D. R. Coulson, U. S. Patent 3,758,586 (1973); Chem. Abs., 79, 125808g (1973).

319. D. R. Coulson, Tetrahedron Lett., 429 (1971).

320. E. W. Kluger, T-K.Su and T-J. Thompson, U.S. Patent 4,260,556 (1981); Chem. Abs., 95, 42409g (1981).

321. S. Appa Rao, A. Kumar, H. Ila and H. Junjappa, Synthesis, 623 (1981).

322. J. E. Lloyd, Brit. Pat. 1,146,707 (1969); Chem. Abs., 70, 105924c (1969).

323. D. R. Coulson, J. Org. Chem., 38, 1483 (1973).

324. A. Claesson, C. Sahlberg and K. Luthman, Acta Chem. Scand., B33, 309 (1979).

398

325. J. Thivolle-Cazat and I. Tkatchenko, J. Chem. Soc., Chem. Commun.,
     1128 (1982).

326. K. C. Dewhirst, J. Org. Chem., 32, 1297 (1967).

327. U. M. Dzhemilev, R. V. Kunakov and R. L. Gaisin, Bull. Acad. Sci.
     U.S.S.R., Div. Chem. Sci., 30, 2213 (1981)

328. R. Baker and M. S. Nobbs, Tetrahedron Lett., 3759 (1977).

329. C. F. Hobbs and D. E. McMakins, U.S. Patent 4,120,901 (1978); Chem.
     Abs., 90, 54459h (1979).

330. R. Baker and D. E. Halliday, Tetrahedron Lett., 2773 (1972).

331. Y. Tamaru, R. Suzuki, M. Kagotani and Z. Yoshida, Tetrahedron Lett.,
     21, 3787, 3791 (1980).

332. D. Commereuc and Y. Chauvin, Bull. Soc. Chim. Fr., 652 (1974).

333. W. E. Walker, R. M. Manyik, K. E. Atkins and M. L. Farmer,
     Tetrahedron Lett., 3817 (1970).

334. R. Baker, A. Onions, R. J. Popplestone and T. N. Smith, J. Chem.
     Soc., Perkin II, 1133 (1975).

335. R. Baker, A. H. Cook, E. E. Halliday and T. N. Smith, J. Chem. Soc.,
     Perkin II, 1511 (1974).

336. E. I. Leupold and H. J. Arpe, Ger. Offen., 2,252,514 (1974);
     Chem. Abs., 81, 25101u (1974).

337. H. Kawazura, T. Takagaki and Y. Ishii, Bull. Chem. Soc. Jpn.,
     48, 1949 (1975).

338. R. Baker, D. E. Halliday and T. N. Smith, J. Chem. Soc., Chem.
     Commun., 1583 (1971).

339. J. Kiji, Y. Yamamoto, E. Sasakawa and J. Furukawa, J. Chem. Soc.,
     Chem. Commun., 770 (1973); J. Organomet. Chem., 77, 125 (1974).

340. K. Takahashi, A. Miyake and G. Hata, Bull. Chem. Soc. Jpn., 45,
     1183 (1972).

341. C. F. Hobbs and D. E. McMackins, U. S. Patent 4,204,997 (1980); Chem. Abs., 93, 185742e (1980).

342. U. M. Dzhemilev, R. N. Fakhretodinov, A. G. Telin, G. A. Tolstikov A. A. Panasenko and E. V. Vasil'eva, Bull. Acad. Sci. U.S.S.R., Div. Chem. Sci., 29, 1943 (1980).

343. D. G. Brady, J. Chem. Soc., Chem. Commun., 434 (1970).

344. T. Yamamoto, J. Ishizu and A. Yamamoto, J. Amer. Chem. Soc., 103, 6863 (1981).

345. J-E. Bäckvall, R. E. Nordborg and J. Vågberg, Tetrahedron Lett., 24, 411 (1983).

346. G. Hata, K. Takahashi and A. Miyake, J. Chem. Soc., Chem. Commun., 1392 (1970).

347. J. Tsuji and T. Mitsuyasu, Japan. 74 20,162 (1974); Chem. Abs., 82, 30949b (1975).

348. Y. Tanigawa, K. Nishimura, A. Kawasaki and S-i. Murahashi, Tetrahedron Lett., 23, 5549 (1982).

349. K. Ogura N. Shibuna and H. Iida, Tetrahedron Lett., 22, 1519 (1981).

350. K. Inomata, T. Yamamoto and H. Kotake, Chem. Lett., 1357 (1981).

351. P. A. Grieco, P. A. Tuthill and H. L. Sham, J. Org. Chem., 46, 5005 (1981).

352. J. Tsuji, K. Sato and H. Okumoto, Tetrahedron Lett., 5189 (1982).

353. S. A. Godleski, J. D. Meinhart, D. J. Miller and S. Van Wallendael, Tetrahedron Lett., 22, 2247 (1981).

354. P. D. Jeffrey and S. W. McCombie, J. Org. Chem., 47, 587 (1982).

355. B. M. Trost and E. Keinan, J. Org. Chem., 44, 3451 (1979).

356. Fieser and Fieser, Reagents for Organic Synthesis, Volume 1, page 650, Wiley Interscience (1967).

400

357. D. V. Sokol'Skii, Ya. A. Dorfman, S. S. Segizbaeva, I. A. Kazantseva
     and B. Yu. Nogerbekov, Zh. Prikl. Khim. (Leningrad), 43, 502 (1970);
     Chem. Abs., 73, 39102s (1970).

358. M. Camps and J-P. Montheard, C. R. Acad. Sci. Paris, C283, 215
     (1976).

359. D. V. Sokol'skii, B. Yu. Nogerbekov, Ya. A. Dorfman, Khim.
     Atsetilena Tekhnol. Karbida Kal'tsiya, 371 (1972); Chem. Abs., 79,
     125455g (1973).

360. I. K. Khristich, Issled. Termogr. Katal., 53 (1966); Chem. Abs.,
     68, 86797d (1968).

361. S. S. Agrawal, G. S. Natarajan and K. A. Venkatchalan Ind. Chem.
     Eng., 16, 7 (1974).

362. D. V. Sokol'skii, Ya. A. Dorfman, N. A. Karazhanova and B. Yu.
     Nogerbekov, J. Org. Chem. U.S.S.R., 7,1641 (1971).

363. G. K. Shestakov, A. V. Massal'skaya, S. M. Airyan and O. N. Temkin,
     Kinet. Catal., 18, 337 (1977).

364. D. V. Sokol'skii, Ya. A. Dorfman, S. S. Segizbaeva and I. A.
     Kazantseva, Kinet. Catal., 10, 541 (1969), Zh. Fiz. Khim., 44,
     98 (1970); Chem. Abs., 73, 13797x (1970).

365. D. V. Sokol'skii and S. S. Segizbaeva, Khim. Atsetilena, 468 (1968);
     Chem. Abs., 71, 30003m (1969).

366. D. V. Sokol'skii, S. S. Segizbaeva and Ya. A. Dorfman, Khim.
     Atsetilena Tekhnol. Karbida Kal'tsiya, 384 (1972); Chem. Abs., 79,
     125487b (1973).

367. S. M. Brailovskii, O. N. Temkin, U. S. Shestakova and A. F. Kuperman,
     Kinet. Catal., 22, 1149 (1981).

368. K. Utimoto, H. Miwa and H. Nozaki, Tetrahedron Lett., 22, 4277
     (1981).

369. M. Koshi, S. Kawakami and H. Horikiri, Japan. Kokai 75,100,028 (1975); Chem. Abs., 84, 16923y (1976).

370. B. D. Howarth and R. J. Kobylecki, Swiss Pat. 597,124 (1978); Chem. Abs., 89, 23948s (1978).

371. M. Sato, S. Ebine and S. Akabori, Synthesis, 472 (1981).

372. H. Azuma, J. Ota and H. Watanabe, Japan. Kokai 76 59,824 (1976); Chem. Abs., 85, 94934u (1976).

373. T. Kamiyama, S. Enomoto and M. Inoue, Yuki Gosei Kagaku Kyokaishi, 36, 784 (1978); Chem. Abs., 90, 71845y (1979).

374. M. Matsuoka, Y. Makino, K. Yoshida and T. Kitao, Chem. Lett., 219 (1979).

375. P. A. T. Hoye, Brit. U K Pat. Appl. 2,025,403 (1980); Chem. Abs., 93, 94977u (1980).

376. M. J. Rance and J. C. Roberts, Tetrahedron Lett., 277 (1969).

377. T. Kametani, S. Takano and K. Haga, Chem. Pharm. Bull., 16, 633 (1968).

378. A. L. Williams, R. E. Kinney and R. F. Bridger, J. Org. Chem., 32, 2501 (1967).

379. V. V. Litvak and S. M. Shein, J. Org. Chem. U.S.S.R., 10, 555 (1974).

380. R. Cramer and D. R. Coulson, J. Org. Chem., 40, 2267 (1975).

381. Y. Akita, M. Shimazaki and A. Ohta, Synthesis, 974 (1981).

382. H. J. Cristau, B. Chabaud, A. Chêne and H. Christol, Synthesis, 892 (1981).

383. T. Migata, T. Shimizu, Y. Asami, J-i. Shiobara, Y. Kato and M. Kosugi, Bull. Chem. Soc. Jpn., 53, 1385 (1980).

384. G. Axelrad, S. Laosooksathit and R. Engel, Synth. Comm., 11, 405 (1981).

385. Y. Sakakibara, N. Yadani, I. Ibuki, M. Sakai and N. Uchino, Chem. Lett., 1565 (1982).

386. K. Tagaki, N. Hayama and S. Inokawa, Chem. Lett., 1435 (1978).

5 Isomerisation reactions

5.1 The Fundamental Reactions

5.2 The Isomerisation of Alkenes

   5.2.1 Double Bond Migration

   5.2.2 Skeletal Isomerisation of Alkenes

   5.2.3 Applications of Alkene Isomerisation

5.3 Isomerisation of Alkenes Substituted by Polar Groups

   5.3.1 Allyl Alcohols and Allyl Ethers

   5.3.2 Allyl Amines and Amides

   5.3.3 Unsaturated Carbonyl Compounds

5.4 Isomerisation of Strained Ring Compounds

   5.4.1 Strained Four-membered Rings

   5.4.2 Strained Three-membered Rings

5.5 Miscellaneous Rearrangements

   5.5.1 Epoxide Rearrangements

   5.5.2 Azirine and Aziridine Rearrangements

   5.5.3 Allylic Transposition

Thus far isomerisation reactions have been regarded only as a side reaction of the catalytic process under review. This is not always undesirable; both hydrosilylation and hydroformylation of internal alkenes may lead to good yields of the terminal adducts. Isomerisation is, however, useful in its own right.

5.1 The Fundamental Reactions

The most important process in isomerisation reactions is the interconversion of π- and σ-bonded organometallic complexes. The bond between a metal and an alkene is described by the well-known Chatt-Dewar-Duncanson

403

model[1] (Figure 1). Electron density is donated from the filled π-orbital of the alkene into an empty metal orbital, while back-bonding occurs from filled metal orbitals to the π*-orbital of the bound alkene. Another type of π-bonded ligand is the allyl group as in 1. This is a three electron donor and may be formally considered to have a σ- and a π-bond to the metal.

Figure 1 Chatt-Dewar-Duncanson model for metal alkene binding.

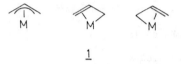

1

Transition metal complexes in which the metal is σ-bonded to carbon were long considered to be impossible to prepare because of their inherent instability. However, a large number are now known and calculations[2] have shown that the M-C bond energy is comparable with that of the C-C bond in alkanes[3]. The reason for the instability of metal alkyls is kinetic in origin; many low energy pathways are available for decomposition. An early approach to the preparation of metal alkyls was the addition of a metal hydride to a coordinated double bond (2 ⇌ 3). For most transition metals, particularly those of Group VIII, this equilibrium lies far to the left, and the reverse process, β-hydride elimination, is a major route for alkyl decomposition. The relative stabilities of σ- and π-bonded complexes are dependent both on the electronic nature of the organic species and on the other ligands bonded to the metal. For example, σ-donor ligands tend to stabilise π-complexes, but σ-complexes are destabilised by the trans-effect of other σ-donors[3].

$$\underline{2} \qquad \underline{3}$$

When considering allyl ligands the position is more complex; σ-$\underline{4}$
and π-allyl $\underline{5}$ are both known as stable compounds. Any non-symmetrical allyl
complex such as $\underline{6}$ has differing bonds to each terminus, depending on the
nature of R^1 and R^2. Several stereochemical forms are possible, and may be
interconverted; the terminal groups are described as being syn and anti with
respect to the hydrogen atom or substituent at C_2.

$$\underline{4} \qquad \underline{5}$$

$$\underline{6a} \qquad\qquad \underline{6b} \qquad\qquad \underline{6c} \qquad\qquad \underline{6d}$$

5.2 The Isomerisation of Alkenes

5.2.1 Double Bond Migration

Alkene isomerisation may be achieved in the presence of a wide
variety of reagents. Strong acids and bases are efficient but incompatible
with many sensitive functional groups. Transition metal complexes act as
catalysts both for cis⇌trans interconversion and for migration of double
bonds along a carbon chain. Theoretically the product consists of a mixture
of alkenes in proportions related to their thermodaynamic stability, but

kinetic factors do not always permit the attainment of equilibrium and proportions of products may be quite different in the early stages of reaction.

Four mechanisms were proposed by Ugo[4] for the migration of double bonds (Figure 2). Of these only 2-I, the metal hydride, and 2-III, the π-allyl, mechanisms are at all common. It is instructive at this stage to consider the implications of the two mechanisms in terms of product distribution and isotope labelling studies. For this a simple substrate is most useful and the isomerisation of 1-pentene has been studied in the presence of a range of catalysts.

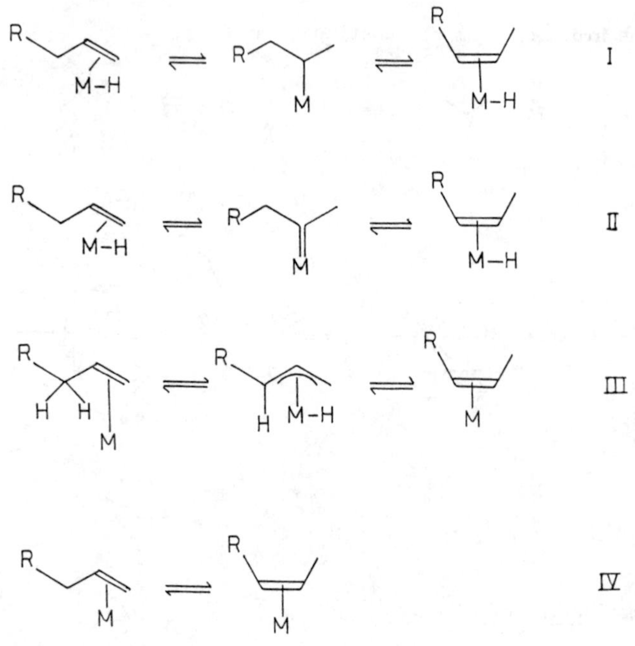

Figure 2 Proposed mechanisms for alkene isomerisation.

1-pentene is isomerised to a mixture of cis- and trans-2-pentene; the reaction of the internal alkenes is much slower. In simple cases where the mechanism is thought to involve metal hydride and σ-alkyl intermediates it is found that the ratio of cis- to trans-2-pentene in the early stages of the reaction is higher than the equilibrium value eventually attained. This has been explained[5] in terms of the greater stability of cis- over trans-alkene complexes. Conversely, for isomerisations proceeding via a metal-allyl intermediate, high proportions of trans-isomers are found throughout, reflecting the greater stability of the syn-allyl 6a over the anti-isomer 6d[6]. Unfortunately this simple rationalisation does not apply in all cases and varying the kinetic model for the reaction can give quite different results. Some examples are given in Figure 3.

For 3-I, the active catalyst is thought to be $HCo(PF_3)_2(PPh_3)$ and the reaction to procede via a metal hydride. 3-II appears to involve the same mechanism but little cis-2-heptene is obtained. For 3-III the intermediacy of an allyl hydride has been conclusively demonstrated from labelling studies and the ratio between the products reflects the expected stability of the syn- and anti-intermediates. Kinetic analysis of the isomerisation of 1-butene in 3-IV is indicative of a π-allyl mechanism, but the major product early in the reaction is cis-2-butene derived from the less stable syn-allyl. 3-V shows a most impressive selectivity for the production of cis-alkene; this falls off only when all the 1-butene has been consumed. Both the nature of the catalyst and the mechanism are obscure, but esr signals due to Ni(I) are observed for the catalytic solution. Thus the prediction of product stereochemistry directly from a knowledge of reaction mechanism is unpromising. Details of the kinetics of individual steps are essential since the relative rates of several processes determine the composition of the product (for example, Figure 4).

In distinguishing between π-allyl and σ-alkyl mechanisms deuterium labelling studies have been valuable. Consider the isomerisation of a terminal alkene dideuterated at C_3 (Figure 5). In the π-allyl mechanism the metal abstracts deuterium from C_3 and returns it to C_1, which is thus specifically

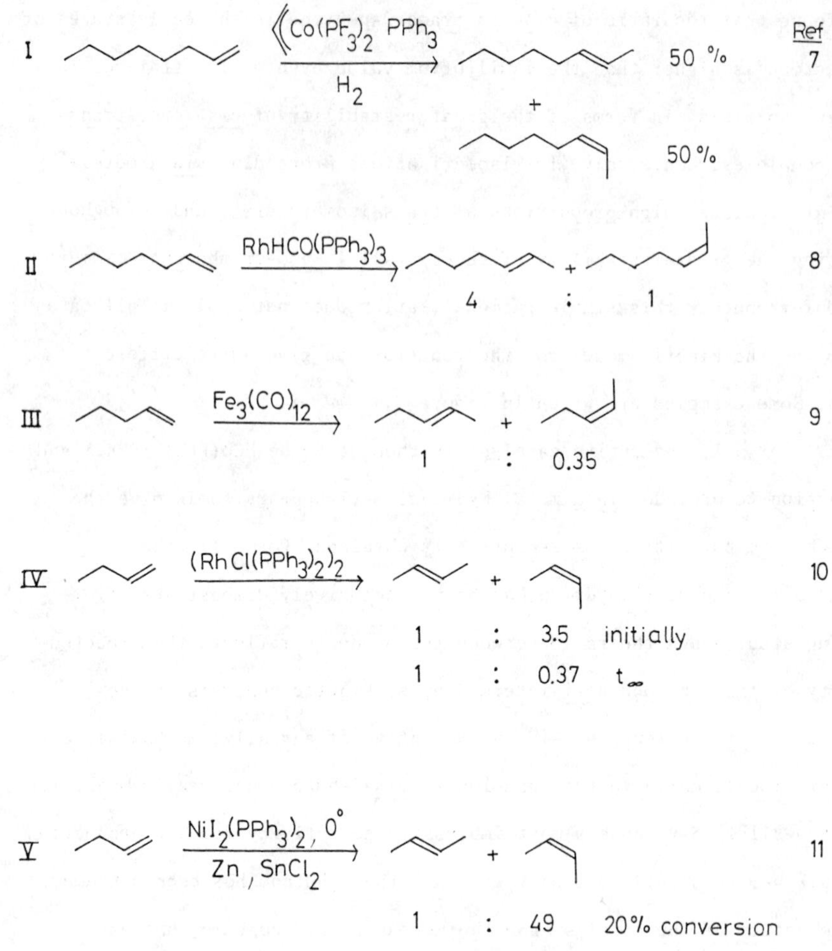

$$
\begin{array}{lll}
\text{I} & \xrightarrow[\text{H}_2]{\langle\!\langle Co(PF_3)_2\ PPh_3} & \quad 50\ \% \qquad \underline{Ref}\ 7 \\
& & + \\
& & \quad 50\ \%
\end{array}
$$

I $\xrightarrow[\text{H}_2]{\langle\!\langle Co(PF_3)_2\ PPh_3}$ 50 % $\underline{Ref}$ 7

+

50 %

II $\xrightarrow{RhHCO(PPh_3)_3}$ 4 : 1   8

III $\xrightarrow{Fe_3(CO)_{12}}$ 1 : 0.35   9

IV $\xrightarrow{(RhCl(PPh_3)_2)_2}$ 10

1 : 3.5   initially

1 : 0.37 t_{∞}

V $\xrightarrow[Zn,\ SnCl_2]{NiI_2(PPh_3)_2,\ 0^{\circ}}$ 11

1 : 49   20% conversion

<u>Figure 3</u>   Isomerisation of terminal alkenes.

deuterated. A further cycle may remove protium from C_1 and return it to C_3, thus deuterating the vinyl group in recovered starting material. At no stage should deuterium be incorporated from the solvent and there should be no transfer of deuterium from labelled to unlabelled material in a cross-over experiment. These criteria are satisfied for the isomerisation of the deuterated alkene, <u>7</u>, in the presence of $Fe_3(CO)_{12}$[12]. <u>8</u> is essentially the

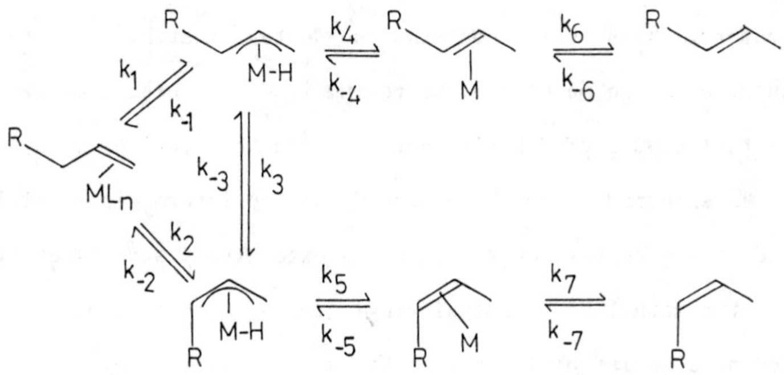

<u>Figure 4</u> The kinetics of alkene isomerisation.

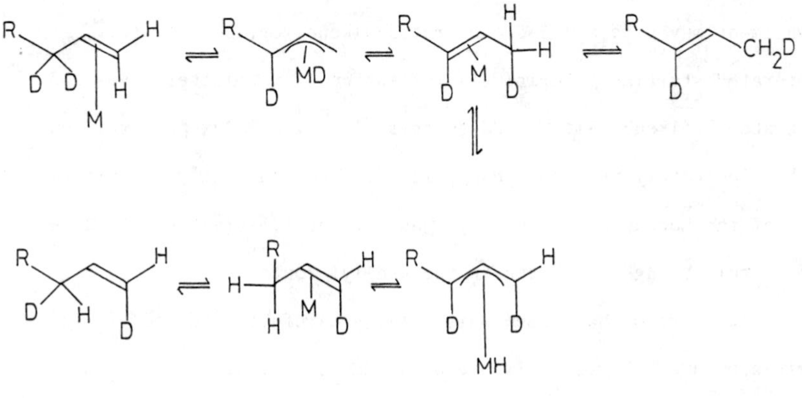

<u>Figure 5</u> Isomerisation of a C_3 dideuterated 1-alkene by a <u>π</u>-allyl mechanism.

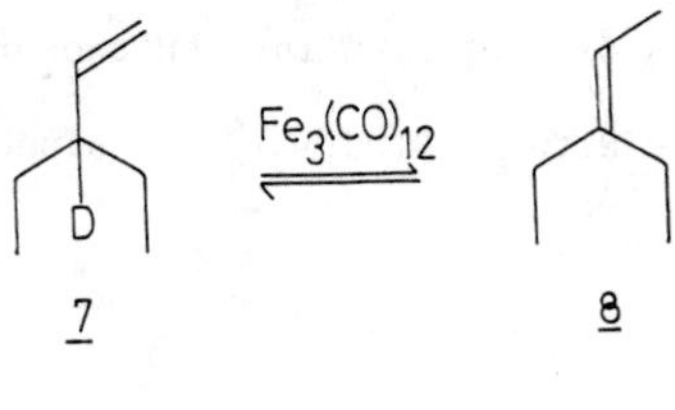

7 8

sole product of the reaction, accounting for 97% of the equilibrium mixture. Mass spectrometry shows that 8 contains one deuterium atom, consistent with an intramolecular exchange mechanism. Deuterium is distributed equally between the three methyl groups of 8 but is not found in the vinyl or methylene positions - as expected since these were C_2 of the intermediate allyl. If 7 is recovered at low conversion the label is extensively scrambled with deuterium in the methyl and terminal vinyl groups. All these data may be accommodated by a series of 1,3-shifts via a π-allyl hydride.

Isomerisation via a metal alkyl intermediate gives a more complex picture for a labelled substrate (Figure 6). The initial stages of the reaction involve metal complexation and insertion to give the alkyl, 9. This now collapses to give a metal deuteride. Release of the isomerised alkene (containing only one deuterium) and coordination of further alkene gives 10. Rearrangement may give a 1,3-dideuterated alkene, or, degenerately, trideuterated starting material. Isomerisation of the latter gives 2,3-dideuterated 2-alkene, 11. To add to this already complex pattern, many metal hydride catalysts exchange hydride with the solvent. The critical feature of the mechanism is that 1,2- as well as 1,3-transfer of deuterium is possible, this being prohibited in the π-allyl route.

Two systems have been particularly carefully studied. 12 is isomerised by $HCo(CO)_4$ to a mixture of products, all of which may be accounted for in terms of 1,2-addition of a metal hydride[13] (Figure 7). In the isomerisation of 3,3-D_2-1-butene in the presence of $(Ph_3P)_3NiX$ extra deuterium was found to be quickly incorporated into the starting material and both cis- and trans-2-butene were deuterium deficient at this stage[14]. Tolman has studied the incorporation of deuterium into butenes from the solvent in the isomerisation of 1-butene with $Ni(P(OEt)_3)_4$ in sulphuric acid, but it is much slower than isomerisation[15].

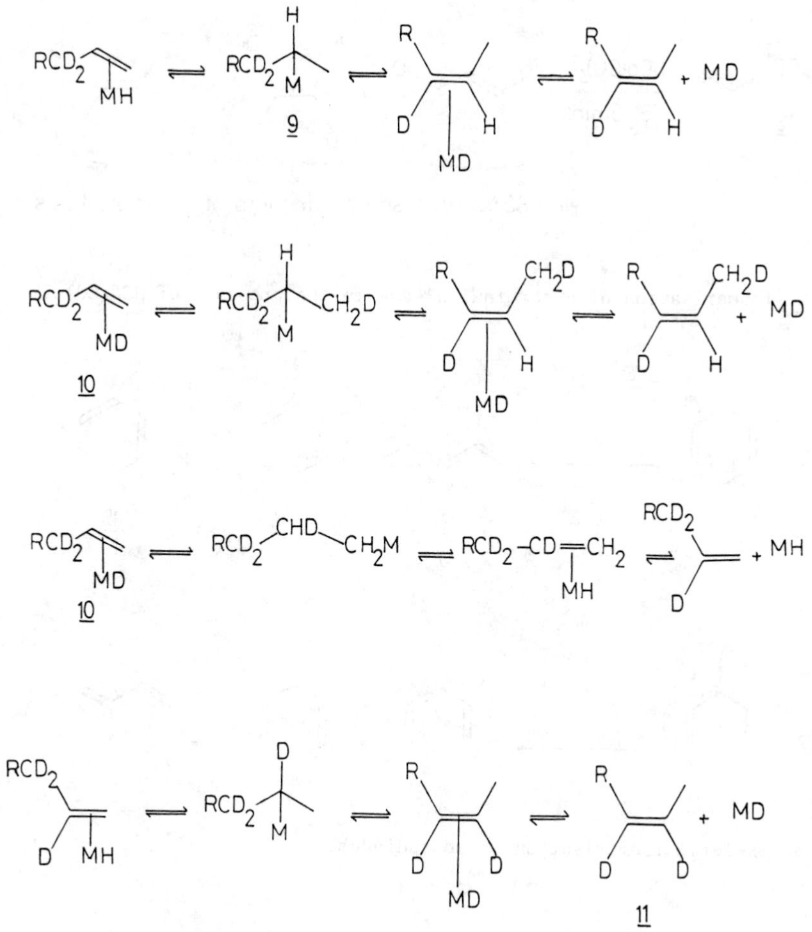

Figure 6   Isomerisation of a C_3 dideuterated 1-alkene by a metal hydride

complex.

5.2.2 Skeletal Isomerisation of Alkenes

Catalysed carbon skeletal isomerisation is a much rarer process than double bond migration. All the known catalysts are nickel hydride complexes formed either by protonation of complexes of the type $Ni(PR_3)_3$ or by the reaction of $(R_3P)_2NiCl_2$ with diisobutyl aluminium chloride[16]. Two types of skeletal isomerisation have been studied[17] (Figure 8).

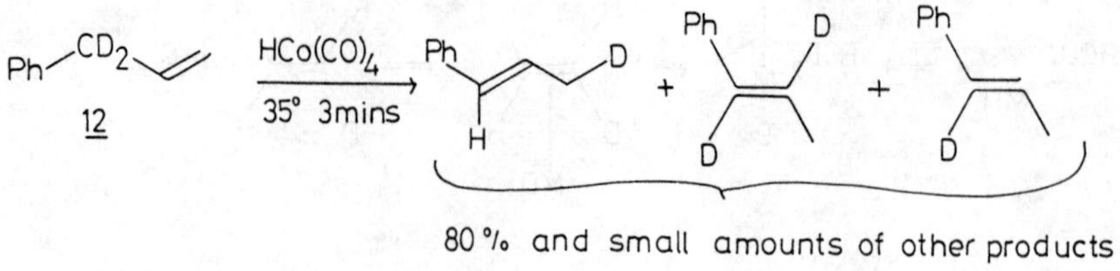

$$\text{Ph}-\text{CD}_2\diagup \quad \xrightarrow[\text{35° 3mins}]{\text{HCo(CO)}_4} \quad \ldots$$

12

80 % and small amounts of other products

Figure 7   Isomerisation of a terminal alkene in the presence of HCo(CO)_4.

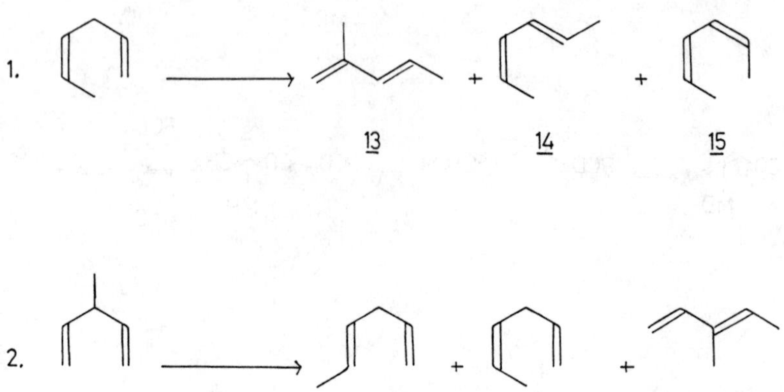

1.   **13**   **14**   **15**

2.

Figure 8   Skeletal isomerisation of pentadienes.

The products **14** and **15** of the first type of process require only double bond isomerisation and may be accounted for as described previously, though in this case the process appears to be irreversible. In the production of **13** a series of deuterium labelling studies established the fate of each carbon atom; the mechanism of Figure 9 was proposed. This scheme also accounts for the rate enhancement noted when the reaction is carried out in the presence of ethylene and many of the steps have close analogies in oligomerisation reactions[18].

Analogous studies of the second type of process suggest a different mechanism involving the reversible fragmentation of a metal alkyl into an

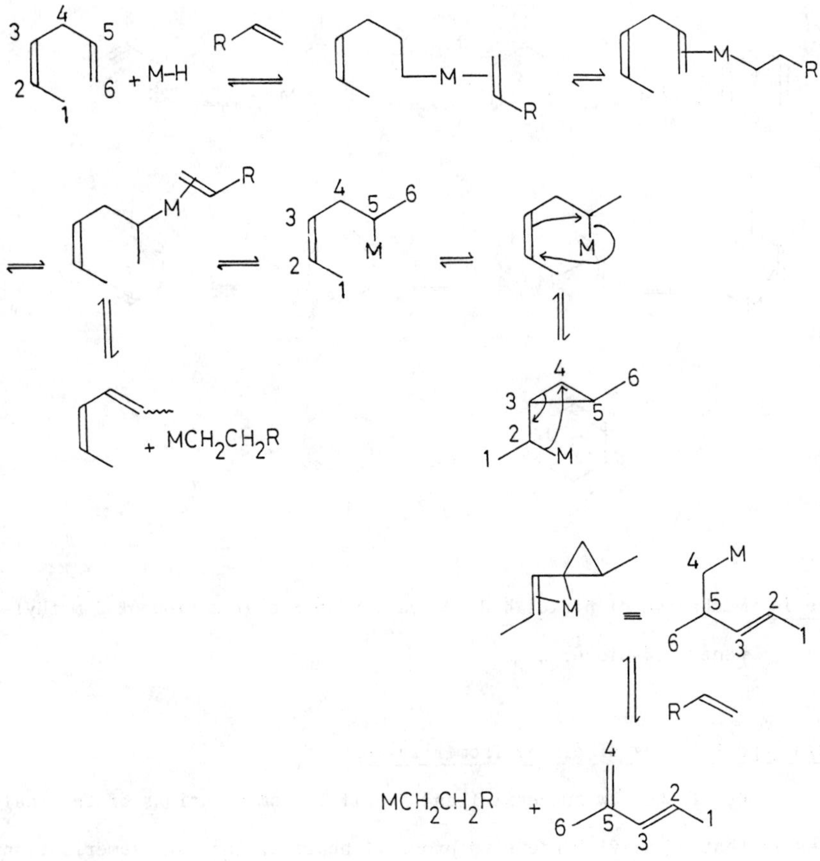

<u>Figure 9</u> Mechanism of nickel hydride catalysed isomerisation of <u>cis</u>-
hexa-1,4-diene.

allyl metal alkene complex. Evidence for this unprecedented reaction came
from the incorporation of added propene into the products. The suggested
mechanism (Figure 10) bears an even closer relation to that of alkene
oligomerisation, since the critical fragmentation is the reverse of the
normal chain lengthening step[19].

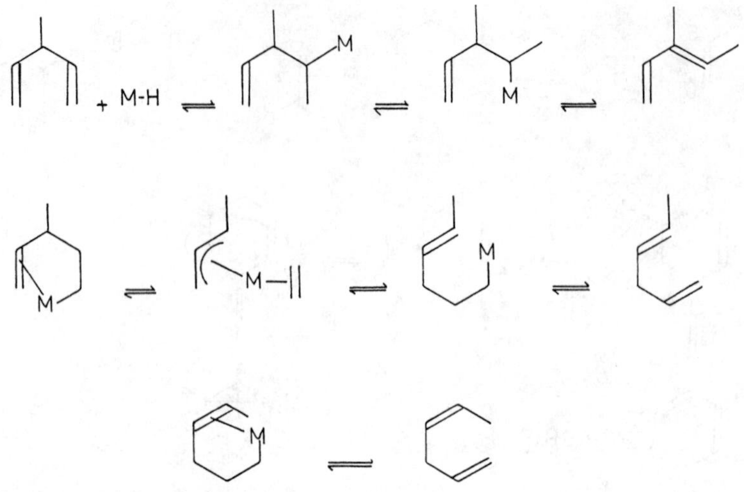

Figure 10 Mechanism of nickel hydride catalysed isomerisation of 3-methyl-
          penta-1,4-diene.

5.2.3 Applications of Alkene Isomerisation

One of the few successful and specific isomerisations of terminal
alkenes is that of allyl benzene to propenyl benzene. Further isomerisation
is impossible and the product is stabilised by conjugation between the double
bond and the aromatic ring. Also the trans:cis ratio at equilibrium for the
product is high, about 95:5, and this equilibrium is readily attained in the
presence of a variety of metal complexes. The dimeric nickel complex,
$Ni_2(CN)_2(Ph_2P(CH_2)_4PPh_2)_3$ is a very active catalyst, giving 100% conversion
in 20 minutes at room temperature[20]. Ruthenium[21], cobalt[22] and palladium[23]
complexes have also been useful. Anethole, 17, is used as a flavour and a
fragrance and is prepared industrially by the prolonged isomerisation of
estragole, 16, with strong base at 200 °C[24]. Under these conditions 56%
conversion is achieved after 12 hours with a trans:cis ratio of 82:18 in the
product. 17 with the same isomer ratio is produced in 93% yield after 8 hours

at 100 °C in the presence of Fe(CO)$_5$, providing a significant advantage over the present procedure[25].

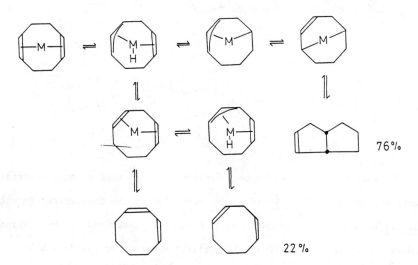

Transformation of non-conjugated cyclic dienes to their conjugated isomers may also be routinely achieved. The conversion of cycloocta-1,5-diene to the 1,3-isomer was the first to be studied. When catalysed by RhCl$_3$[26], H$_2$Ir$_2$Cl$_2$(1,5-cod)(PPh$_3$)$_2$[27] or Cp$_2$Ti(μ-η_1,η_5-Cp)TiCp[28], the reaction proceeds stepwise via the 1,4-isomer, but no 1,4-cod may be detected with Fe(CO)$_5$ as catalyst. A more unusual isomerisation of this substrate in the presence of a catalyst derived from (Bu$_3$P)$_2$NiBr$_2$ and ZnCl$_2$ in anisole gives rather selective formation of [3.3.0] bicyclooctene[30] (Figure 11).

Figure 11   Isomerisation of cycloocta-1,5-diene catalysed by a nickel complex.

416

The isomerisation of the products of Birch reductions catalysed by $(Ph_3P)_3RhCl$ is a useful way of generating conjugated cyclohexadienes, and enables anisole to be used as a synthon for cyclohexenone[31] (Figure 12).

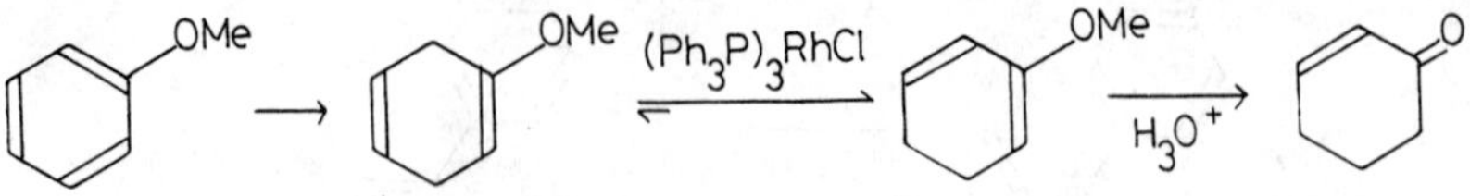

Figure 12 Conversion of anisole to cyclohexenone.

5-Ethylidene bicyclo[2.2.1]hept-2-ene, 19, is the best third monomer for ethylene propylene diene rubbers, and may be prepared by isomerisation of vinyl bicycloheptene, 18, itself derived from the Diels Alder reaction of cyclopentadiene and butadiene. In the presence of $Co(N_2)(PPh_3)_3$ both stereoisomers, 19a and 19b, are obtained[32], whereas with $Fe(CO)_5$/aniline only 19b is formed[33].

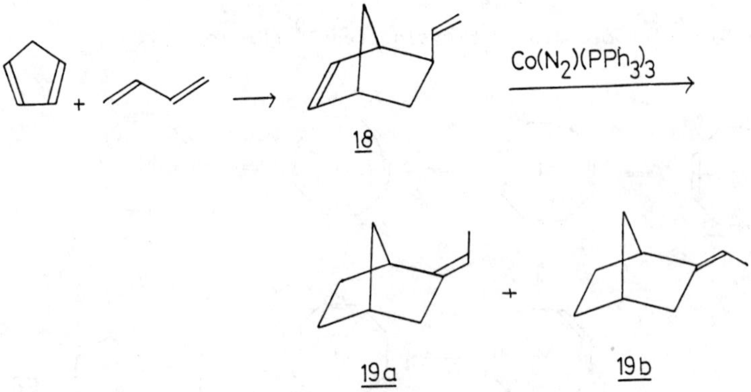

A nickel catalysed process has been used as part of an industrial process for pseudoionone. 21 and 22 are generated by the Cope rearrangement of 20. 21 is the primary product and is partially isomerised to the desired 22 under the reaction conditions. Isomerisation may be achieved with 97% selectivity (i.e. 22:21 = 97:3) and 95% recovery in the presence of $Ni(acac)_2$ at 165 °C[34].

Enantioselective isomerisation of alkenes is rare but when 4-methyl hexene is treated with the Ziegler polymerisation catalysts derived from tetrakismenthoxy titanate[35] or (DIOP)NiCl$_2$[36] and triisobutyl aluminium both recovered starting material and isomerised products are found to exhibit very small rotations.

5.3 Isomerisation of Alkenes Substituted by Polar Groups

We have seen that the regiochemistry of double bond migrations is dependent on the relative stability of the various isomers. Thus, substituents capable of conjugation with a carbon carbon double bond may direct the course of isomerisation. For example, products containing αβ-unsaturated carbonyl compounds, vinyl ethers or enamides will all be strongly favoured.

5.3.1 Allyl Alcohols and Allyl Ethers

The isomerisation of an allyl alcohol yields, after tautomerisation, an aldehyde or ketone. The furfuryl alcohol, 23, is converted to 2,3-dihydro-2-furaldehyde, 24, in the presence of RhCl$_3$ at 100 °C, but recovery is low due to polymerisation of these sensitive materials[37]. A more successful catalyst

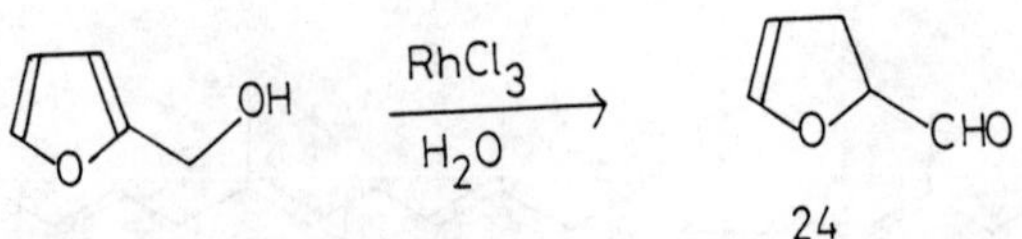
OH
RhCl3
H2O
CHO

23 24

was $(Rh(CO)_2Cl)_2$ in a two phase system with sodium hydroxide solution in the presence of a phase transfer catalyst[38]. The active species is thought to be $Rh(CO)_2OH$ and the reaction proceeds via a π-allyl mechanism (Figure 13) to give complete conversion in 6-10 hours at room temperature.

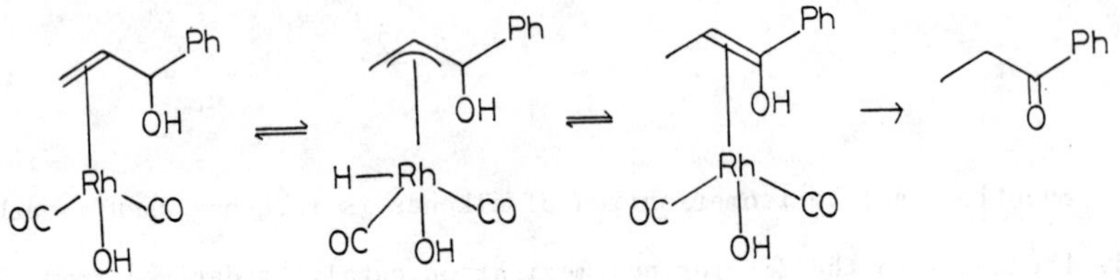
Ph
OH
Rh
OC
CO
OH
H—Rh
OC
CO
OH

Figure 13 Allyl alcohol isomerisation catalysed by $Rh(CO)_2OH$

Certain iridium complexes have proved to be more active; for example, 0.1 mole % of $(Ph_2MeP)_2Ir(cod)\ PF_6$ causes quantitative isomerisation of 25 to 26 in 30 minutes at 20°[39]. More highly substituted allyl alcohols were found to be less reactive. Similar transformations are catalysed by nickel and osmium complexes but the catalysts are less active and shorter lived (Figure 14).

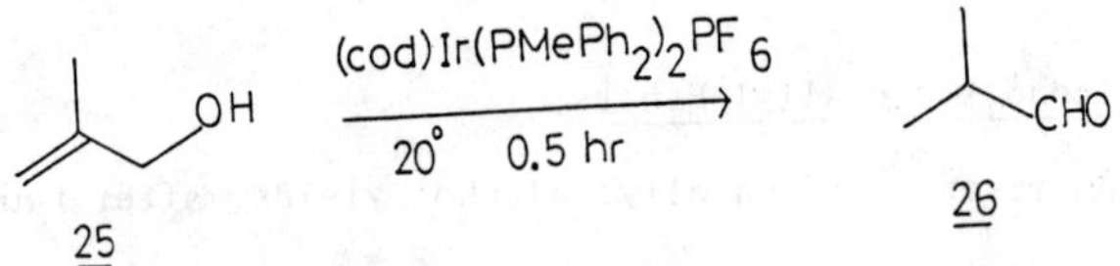
OH
(cod)Ir(PMePh2)2PF6
20° 0.5 hr
CHO

25 26

Examples of enantioselective isomerisation are few and disappointing.

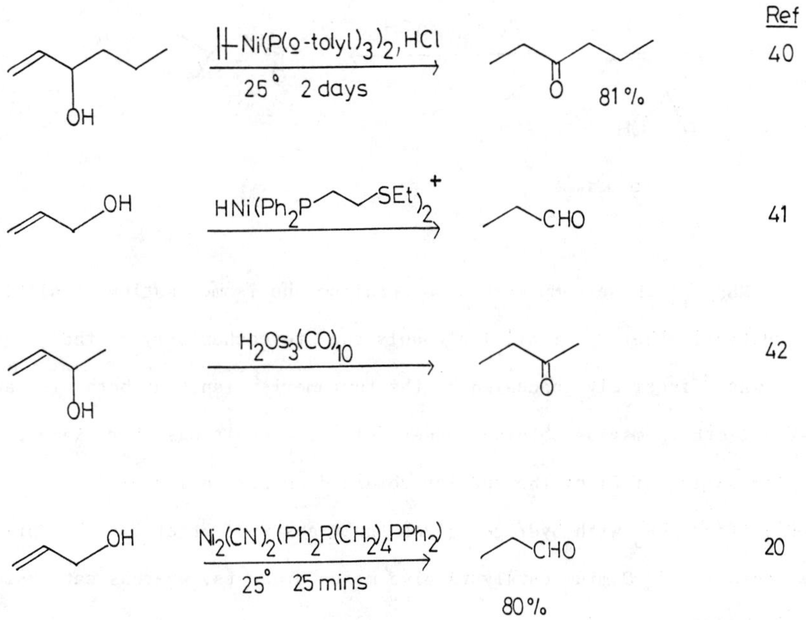
Ref
Ni(P(o-tolyl)3)2, HCl
25° 2 days
OH
81%
40
OH
HNi(Ph2P SEt)2 +
CHO
41
H2Os3(CO)10
OH
42
OH
Ni2(CN)2(Ph2P(CH2)4PPh2)
25° 25 mins
CHO
80%
20

Figure 14 Isomerisation af allyl alcohols by nickel and osmium complexes.

Using $HRh(CO)(PPh_3)_3$ and (-)-DIOP as catalyst, 27 is isomerised to 28 in 91% chemical but only 4% optical yield[43]. Even in the isomerisation of the optically active alcohol, 29, chirality transfer is only about 40% efficient suggesting that the transition state is not sufficiently rigid for good stereoselectivity[44]. This clearly remains an area for exploitation with new chiral catalysts.

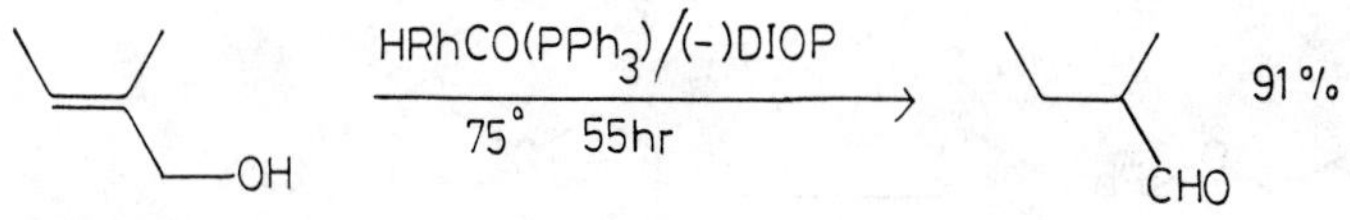
HRhCO(PPh3)/(-)DIOP
75° 55hr
OH
CHO
91%

27 28

29   30

Many of these complexes also catalyse the isomerisation of allyl ethers to vinyl ethers. For allyl alcohols the stereochemistry of the reaction was effectively concealed by the tautomerisation, but both cis- and trans-vinyl ethers may be obtained under suitable conditions. For example, the isomerisation of 31 by the species obtained on activation of $(Ph_2MeP)_2Ir(cod)$ PF_6 with hydrogen gives the trans-enol ether via the more stable syn-allyl[45]. Osmium catalysts give mixed products, whereas catalysis by nickel hydrides yields only the cis-product in the isomerisation of phenyl allyl ether, 33[40]. This reaction was used for the conversion of the arylmethylene chroman-4-one, 35, to the homoisoflavone skeleton, 36; the starting material is unchanged by acid and decomposes in base[46].

31   32

33   34

35 → 36

Since the products of isomerisation are enol ethers they may be hydrolysed to give carbonyl compounds; thus alcohols may be protected as their allyl ethers. Deprotection of 37 is achieved by $(Ph_3P)_3RhCl$ catalysed isomerisation, then gentle hydrolysis[47]. However, in more complex alcohols some material is lost by reduction to the propyl derivative. A cleaner reaction in the presence of hydrogen activated $(Ph_2MeP)_2Ir(cod) PF_6$ has been used for the isomerisation of allyl ethers of sugars such as 38, it being noteworthy that the benzyl protecting groups are unaffected[48]. 39 is obtained homogeneous in 90% yield and may be hydrolysed in the presence of $HgCl_2$.

5.3.2 Allyl Amines and Amides

The conversions of allyl nitrogen compounds are similar to their oxygen analogues and many of the same catalysts are useful. Isomerisation of allyl amines yields enamines, which may subsequently be hydrolysed to carbonyl compounds. Thus, neryl or geranyl cyclohexylamine, 41, may be converted to citronellal enamine, 42, with a cobalt catalyst. 42 is then hydrolysed to give 85% citronellal, 43[49]. The isomerisation of a prochiral

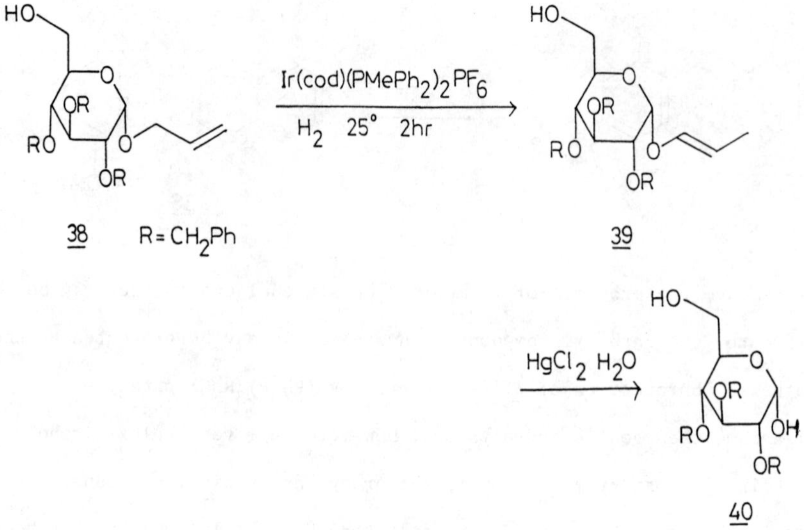

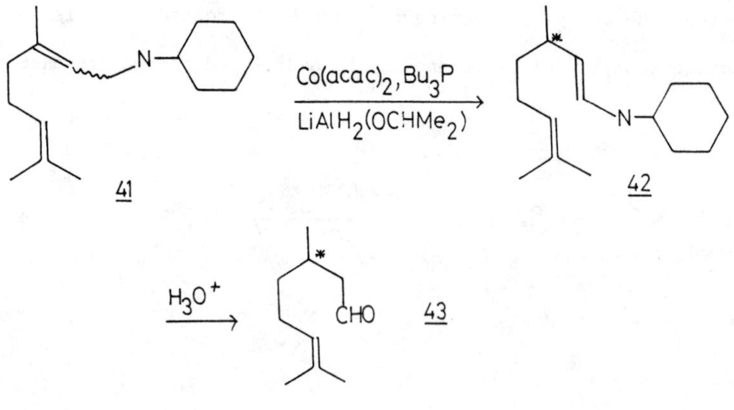

38   R = CH₂Ph

39

40

41

42

43

allyl amine, 44, may give a chiral enamine. When N,N-diethyl nerylamine is

isomerised in the presence of a cobalt(II) nephthenoate/Et₃Al/(+)-DIOP

catalyst, citronellal trans-enamine is obtained with the 3R configuration in

32% enantiomer excess; the chemical yield, however, is low (23%)[50].

    There are few general methods for the preparation of enamides, so

isomerisation of allyl amides is an attractive route. Stille has reported

such processes with HRu(PPh₃)₃Cl or HRh(PPh₃)₄ as catalysts. The cis-isomer,

$$R_1R_2C=CCH_2NR_2 \quad \rightleftharpoons \quad R_1R_2\overset{*}{C}H-CH=CHNR_2$$

44   45

48, predominates in the products and the reaction may be regarded as being under conditions of kinetic control, since 47 and 48 do not interconvert in the presence of these catalysts[51].

46   47   48

5.3.3 Unsaturated Carbonyl Compounds

Although it might be expected that a carbon-carbon double bond could be induced to migrate into conjugation with a carbonyl group, few reactions of this type catalysed by metal complexes are known. αβ-Unsaturated carbonyl compounds are transformed to more substituted isomers by rhodium and iridium catalysts (Figure 15), the isomerisation proceeding by successive migrations around the ring. A nickel hydride catalyst converts γδ-aldehydes and esters to their βγ-isomers but does not give conjugated products[40]. However, βγ-unsaturated nitriles, 49, yield the αβ-unsaturated isomers, 50, in the presence of another nickel hydride complex[41].

49   50

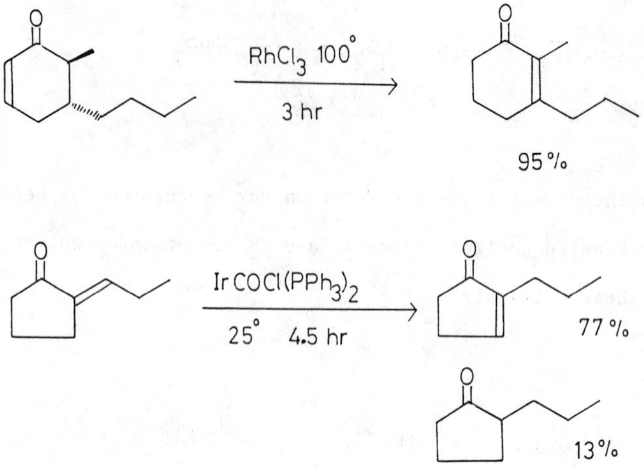

Figure 15   Isomerisation of αβ-unsaturated carbonyl compounds.

5.4   Isomerisation of Strained Ring Compounds

The isomerisation of saturated hydrocarbons is of great importance to the petrochemical industry, but these reactions are normally achieved in the presence of heterogeneous catalysts, and few proceed via well-defined intermediates. Molecular rearrangements of a number of highly strained hydrocarbons are catalysed by transition metal complexes, however, and these have been used to prepare previously inaccessible carbon skeletons.

5.4.1   Strained Four-membered Rings

Cyclobutene may be opened thermally to butadiene in a conrotatory mode, a process allowed by the Woodward Hoffmann rules. However, if the ring is fused, as for example in the Dewar benzene, 51, the allowed mode of opening would produce a highly strained trans-double bond in the 6-membered ring, and thus 51 has a half-life of several hours at room temperature. The rearrangement of 51 to the benzene derivative is accomplished rapidly in the presence of various metal catalysts. Two alternative rôles for the metal have been proposed. The first suggests that the rearrangement proceeds through the

51   52

allowed mode and that the metal stabilises the strained product, which is subsequently isomerised[54]. Alternatively the interaction of the orbitals of the cyclobutene with those of a transition metal might lower the barrier to the disallowed, but geometrically feasible, disrotatory opening[55]. The controversy is discussed by Bishop in an extensive review[56]. Many catalysed rearrangements originally proposed to proceed via orbital symmetry disallowed routes have, however, been shown to involve discrete metal bound intermediates.

    51 is transformed into 52 in the presence of $(Rh(C_2H_4)_2Cl)_2$; kinetic studies show that the reaction proceeds via a complex set of equilibria, the reactive species involving two substrate molecules bound to the metal[57]. The same conversion is catalysed by $(PhCN)_2PdCl_2$[58]. Analogously, 53 is rearranged to 55 in 10 seconds in the presence of 1 mole of Ag^+, the therm l reaction being complete only after 4-5 hours at 180 °C. The intermediacy of 54 is demonstrated by a trapping reaction with maleic anhydride.

53   54
                                          55

    Cubane, 56, undergoes two distinct types of rearrangement in the

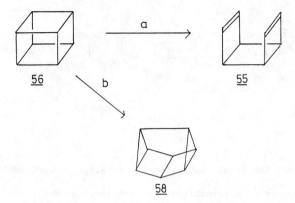

56   a   55

b

58

Figure 16   Pathways for the rearrangement of cubane.

presence of appropriate catalysts (Figure 16). The product is either

tricyclooctadiene, 57, or the pentacyclic hydrocarbon cuneane, 58. Formally

process a is a reversion of a disallowed π_{2s} + π_{2s} cycloaddition, whereas b

is a σ_{2a} + σ_{2a} reaction, also disallowed. In general rhodium catalysed

reactions follow pathway a, while route b is adopted in the presence of silver

ion. Two mechanisms may account economically for all the data so far

published.

The rhodium catalysed reaction is thought to proceed via the

insertion of Rh(I) into one of the carbon-carbon bonds of cubane to give the

Rh(III) intermediate, 59 (Figure 17). Subsequent extrusion of Rh(I) yields the

bicyclooctadiene. Support for this mechanism comes from the relative

insensitivity of the reaction rate to substituent effects and the isolation

of a rhodium acyl derivative, 60, from the reaction of $Rh_2(CO)_4Cl_2$;

subsequent treatment with PPh_3 gives the ketone, 61[60].

Ag[+] coordinates to alkenes so it might be expected to have some

affinity for a σ-bond such as that in a strained cyclobutane, which has high

p-character. Opening of this precoordinated complex leads to an

argentocarbonium ion, 62(Figure 18). Paquette has demonstrated reversible

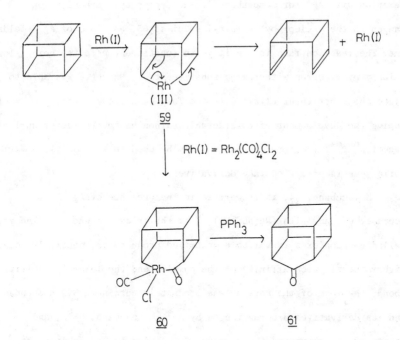

Figure 17 Cubane isomerisation catalysed by rhodium complexes.

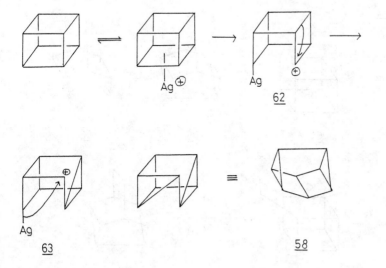

Figure 18 Cubane isomerisation catalysed by Ag$^+$.

complexation with Ag^+ for homocubane[61]. The cyclobutyl carbonium ion rearranges to the cyclopropyl carbinyl species, <u>63</u>, and loss of Ag^+ yields cuneane. The reaction rate is sensitive to substituent effects, being slowed appreciably by electron withdrawing substituents[62]. Paquette was able to correlate the substituent effect with the Taft constant $\underline{\sigma}^*$, giving $\underline{\rho}^* = -1.5$, indicating the development of considerable carbonium ion character during the reaction[63,64]. These two mechanisms may be used to rationalise a variety of rearrangements of cyclobutane derivatives.

Homocubane, <u>64</u>, is isomerised in the presence of Ag^+[65] to norsnoutane, <u>65</u>, and with $(Rh(nbd)Cl)_2$[66] to the tricyclodiene, <u>66</u>. The route for palladium complexes is variable and mixed products are generally obtained. The higher the electron affinity of the complex and the lower its ability to back bond, the more of the norsnoutane product is obtained. <u>Bis</u>-homocubane, <u>67</u>, and its derivatives are rearranged by Ag^+ to snoutane, <u>68</u>[67], and basketene, <u>69</u>, is converted to snoutene, <u>70</u>[68]. Kinetic and substituent analysis has shown that bond switching is triggered by electrophilic attack of Ag^+ at C_2-C_5 in <u>71</u> with direct formation of the cyclopropyl carbinyl cation, <u>72</u>[63].

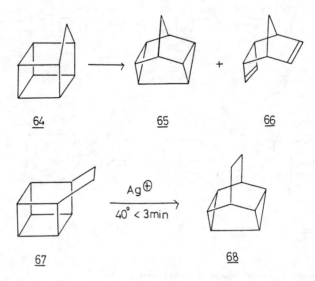

<u>64</u>   <u>65</u>   <u>66</u>

$Ag^{\oplus}$
$40° < 3min$

<u>67</u>   <u>68</u>

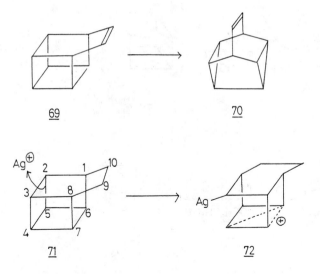

69   70

71   72

Rh(I) complexes give diene products such as 74 and 75[64], as does a catalyst formed _in situ_ from Ni(cod)$_2$ and a phosphine[69]. Palladium complexes again give mixtures, the composition of which depends on the other ligands[70]. The isomerisation of the bis-homocubane, 77, has led to a convenient synthesis of gram quantities of semibullvalene, 79 (Figure 19)[71].

73   74   75

Quadricyclane, 80, may be produced by photolysis of norbornadiene, 81, and has been suggested as a potential store for solar energy. Consequently, the disallowed $\pi_{2s} + \pi_{2s}$ reversion to the diene has been extensively studied. Early work[72] involved Rh(I) catalysts and the capture of 83 on insertion of CO provided strong evidence for a mechanism similar to

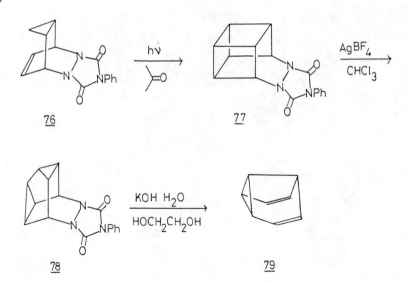

76   77

78   79

Figure 19   Synthesis of semibullvalene involving silver ion catalysed
isomerisation of a bis-homocubane.

that proposed for the cubane derivatives[73]. The reaction catalysed by
(Rh(nbd)(μ-OAc))$_2$ has been studied in detail, and a complex mechansim
accounts for the formation of two dimers as well as norbornadiene[24].

80   81

82   83

Nickel(0) complexes are also active catalysts for the isomerisation of quadricyclane, Ni(cod)$_2$ giving 86% conversion in 48 hours at -16 $^\circ$C for 84. In the presence of a suitable alkene, cycloadducts such as 86 are formed[75]. A complex of Ni(II) with cyclopropene is also extremely active. The cyclopropenyl ligand is formally a 4e donor, but because of the stability of the cyclopropenium cation the metal has the electronic character of Ni(0) while retaining the steric attributes of Ni(II)[76]. Catalysts also include cobalt complexes of porphyrins, phthalocyanines and glyoximes. Manassen found that only square planar species were active catalysts[77]. Despite lengthy dispute neither the mechanism nor the nature of the catalytically active species are known[78,79].

5.4.2 Strained Three-membered Rings

The bonds of a cyclopropane ring have higher p-character than those of a cyclobutane, and thus might be expected to coordinate more readily to metals.The isomerisation of quadricyclane is initiated by metal bonding to a cyclopropane-cyclobutane bond.

Cyclopropanes may be isomerised if conjugated with an aryl ring or a carbon-carbon double bond. For example, 87 is isomerised by 0.03 mole% of $Rh_2(CO)_4Cl_2$ to a mixture of products, but 88 is quite inert[80]. The reaction is thought to proceed via precoordination of Rh(I) to the cyclopropane which is favourable only when there is conjugation with a further unsaturated group. The cyclopropane then opens predominently to 90 (Figure 20) where rhodium is coordinated to the aryl ring. Hydride shift and loss of Rh(I) completes the rearrangement. Vinylcyclopropanes react similarly in the presence of $Rh_2(CO)_4Cl_2$; examples are given in Figure 21[81]. Here the intermediates should be allyl cations and the intermediacy of metal allyl complexes has been demonstrated in the stoichiometric rearrangements of rhodium complexes of divinyl cyclopropanes[82,83]. Nickel hydrides catalyse similar reactions but more varied products are obtained, since the initially produced species may be further isomerised[84].

Metal catalysed rearrangements of cyclopropenes are also known. Some are thought to proceed via a metal carbene such as 94 (Figure 22). The subsequent fate of the carbene depends on the precise nature of the substituents, and examples are known involving insertion into a C-H bond[85], capture by a nucleophile[86] or rearrangement[87] have been observed.

Electrophilic addition of Ag^+ to one of the double bonds of a biscyclopropenyl compound, 95, initiates a complex reorganisation to Dewar

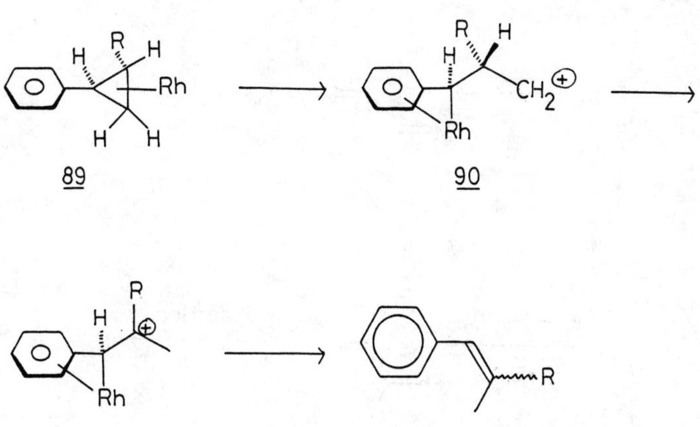

89   90

91   92

Figure 20   Mechanism of the Rh(I) catalysed opening of an aryl cyclopropane.

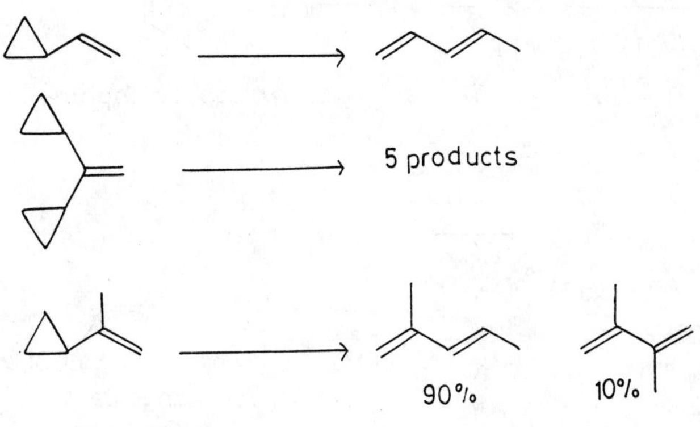

5 products

90%   10%

Figure 21   Rh(I) catalysed rearrangement of vinyl cyclopropanes.

benzene or benzene derivatives[88], the regioselectivity being dependent on the substituents (Figure 23).

The strain in fused three-membered rings allows more facile opening in the presence of transition metal catalysts. The reactions are varied and are most easily classified in mechanistic terms. Ag⁺ reactions are

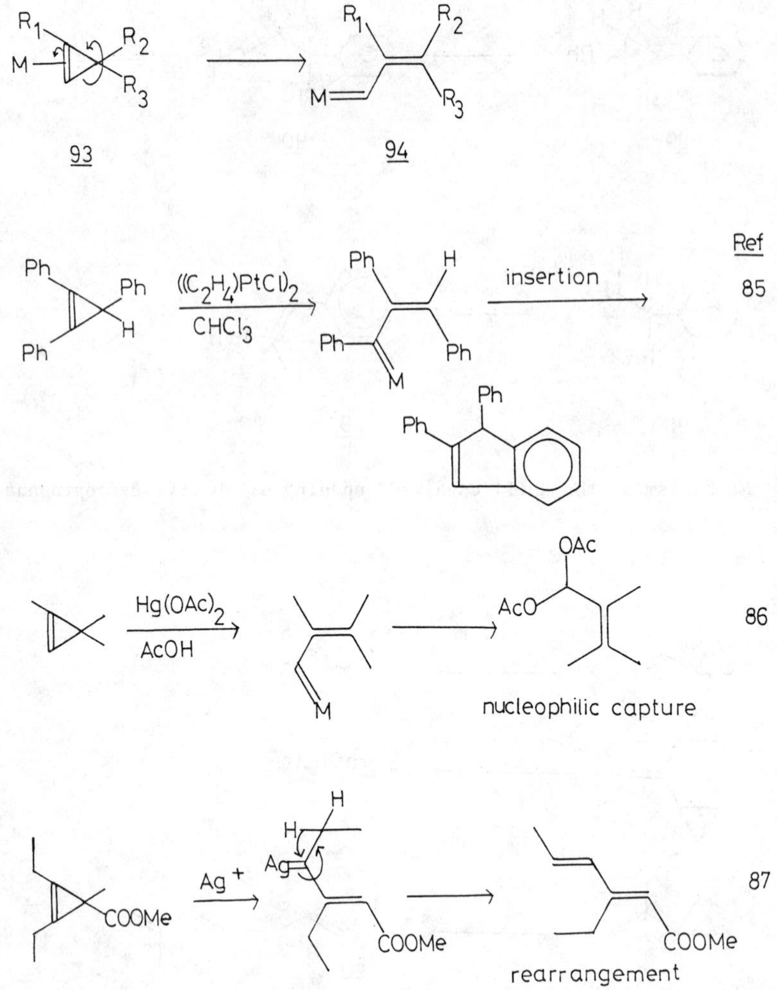

Figure 22 Cyclopropene rearrangement _via_ metal carbene intermediates.

of an electrophilic type, the first step being the formation of an

argentocarbonium ion (Figure 24)[89]. With Rh(I) complexes the reaction proceeds

through a discrete Rh(III) intermediate and evidence for a rhodium acyl is

obtained from IR spectra in the presence of CO[90]. Yet a further mechanism

involving a metal carbene complex has been postulated for the rearrangement

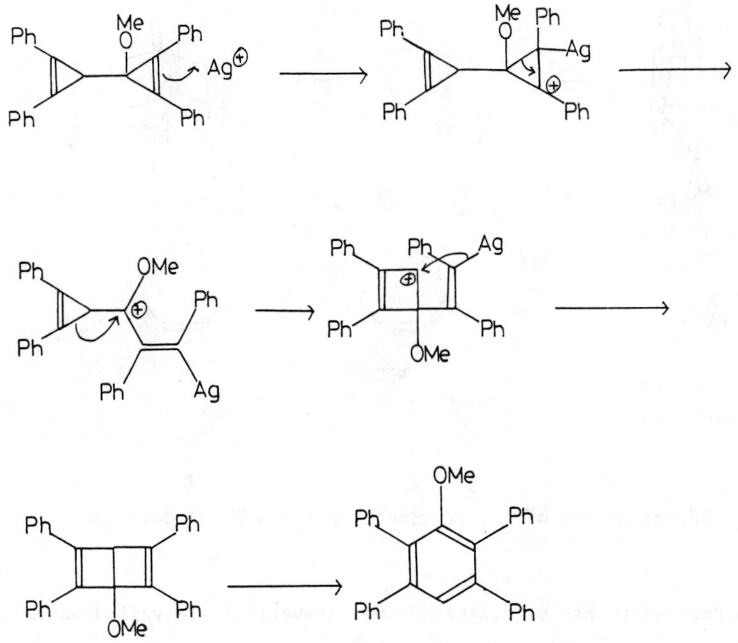

Figure 23 Ag⁺ catalysed rearrangement of a bis-cyclopropenyl compound.

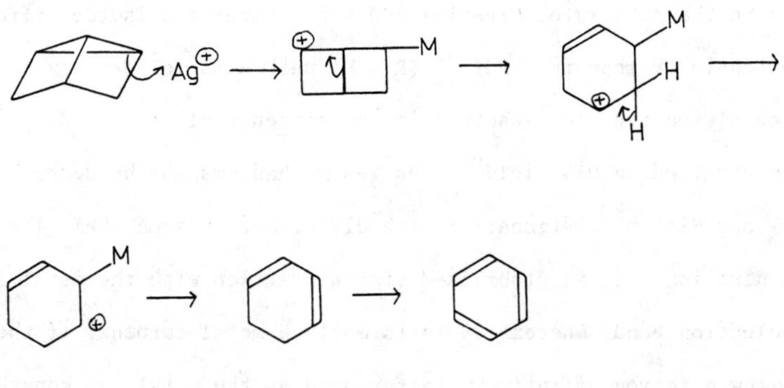

Figure 24 Ag⁺ catalysed rearrangement of a fused cyclopropane.

of the tricyclooctane, 96 by $(Ir(CO)_3Cl)_2$ (Figure 25)[91].

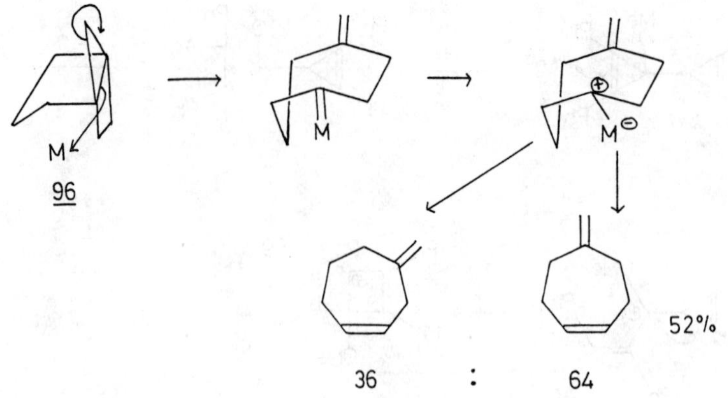

96

36 : 64   52%

Figure 25   Rearrangement of a tricyclooctane via a metal carbene.

Rearrangements of bicyclobutane derivatives are varied and have been the subject of extensive controversy. Taking the substituted bicycloheptane, 97, as a prototype Paquette[92] distinguishes four separate reaction pathways, all of which are known under suitable conditions (Figure 26). The product depends both on the catalyst used and the substituents on the substrate. Kinetics and substituent and isotope effects have been extensively reported. For 97 (R = H) path α is followed exclusively in the Ag^+ catalysed reaction, whereas in the presence of $(Ir(CO)_3Cl)_2$ the β-product is obtained in 91% yield[93]. The two mechanisms may be described as in Figure 27 and Bishop[56] rationalises the difference between them. The metallocarbonium ion, 98, is stabilised via interaction with the metal in a 3-centre 2-electron bond, whereas 99 is formally a metal carbene. If the metal has a low electron affinity 99 is favoured as the metal can donate electron density directly to the positive centre. In contrast, catalysts with high electron affinities and poor ability to back bond favour 98.

Introduction of alkyl substituents into the 2- and 6-positions of bicycloheptane has little effect on the Ag^+ catalysed rearrangement to

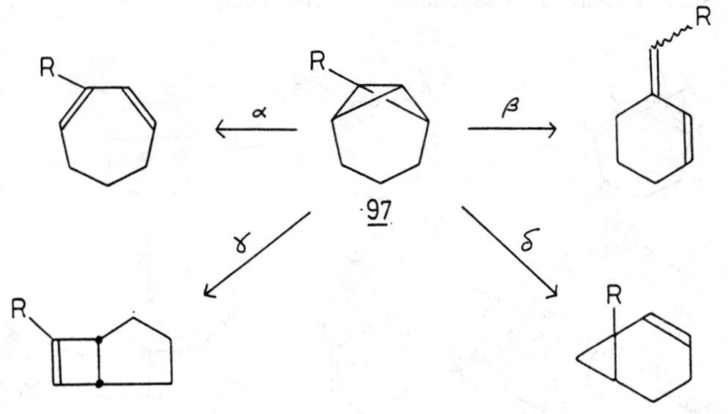

Figure 26   The routes available for bicycloheptane isomerisation.

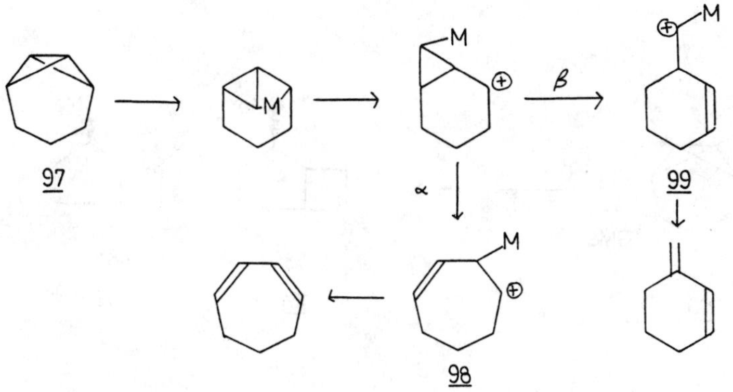

Figure 27   Mechanism of isomerisation of bicycloheptane.

heptadienes. The reaction is slightly accelerated since the intermediate

carbonium ion is stabilised. Substitution at positions 1 or 7 leads to a more

complex mixture of products with the γ-route now predominating[92,95]

(Figure 28). With a further substituent in the 5-position, as in 100, the

rearrangement is quite stereospecific to 101 and 102[95]. Cleavage of bonds

a-b or a-d in 103 is ruled out by the nature of the product, and cleavage of

a-c followed by a hydride shift would not give a stereospecific reaction.

Cleavage of b-c is preferred over cleavage of c-d because of the stabilising effect of the substituent on the carbonium ion formed.

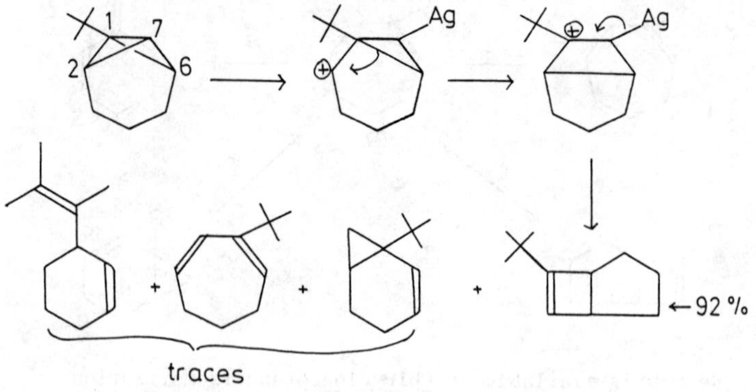

Figure 28 Isomerisation of 1-tert-butyl bicycloheptane.

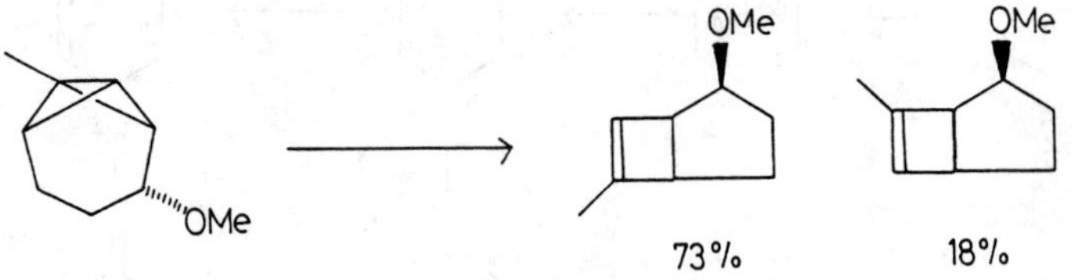

100 101 102

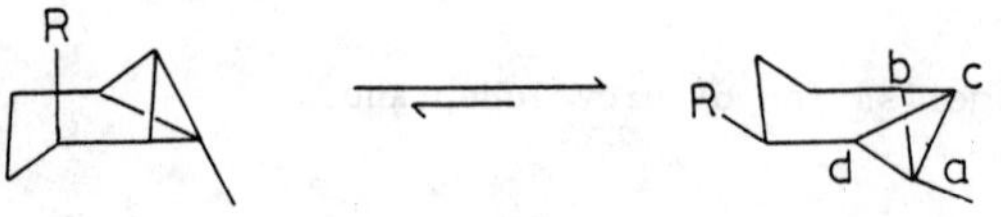

103a 103b

The rearrangement of other bicyclobutanes follows a similar pattern. In the presence of Ag^+, the sole products are dienes, which are often formed stereospecifically[96] (e.g. 104 to 105). With Rh(I) catalysts a metal carbene may be involved and in favourable cases is trapped in an intramolecular reaction (Figure 29)[97]. The pattern of the rearrangement may usually be predicted from an understanding of the realtive energies of the carbonium ions and carbenes involved.

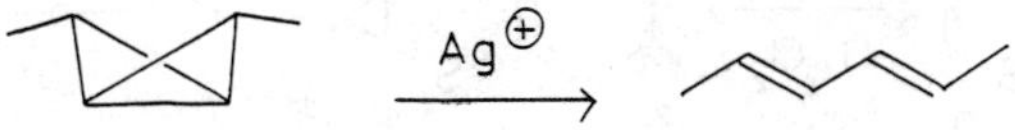

$$104 \qquad \xrightarrow{Ag^{\oplus}} \qquad 105$$

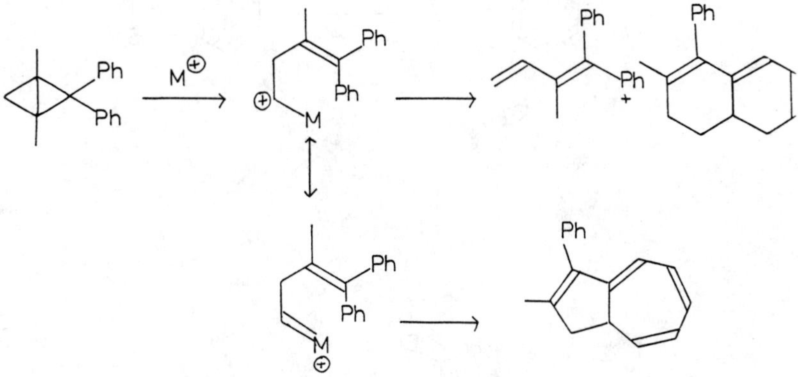

Figure 29 Rh(I) catalysed isomerisation of a substituted bicyclobutane.

While of considerable theoretical interest, few bicyclobutane

rearrangements have found practical application in synthetic organic

chemistry. One example is provided by the rearrangement of 106 to the

substituted thiepin, 107[98]; these rather fragile compounds cannot be

prepared by other routes. Successive carbene addition and carbenoid

rearrangement provide a simple route to a variety of bicyclobutanes; these are

subsequently rearranged to give ring expanded products (Figure 30)[99].

5.5 Miscellaneous Rearrangements

5.5.1 Epoxide Rearrangements

The transformation of epoxides to aldehydes and ketones has been

accomplished in the presence of acid[100] and base[101], but for sensitive

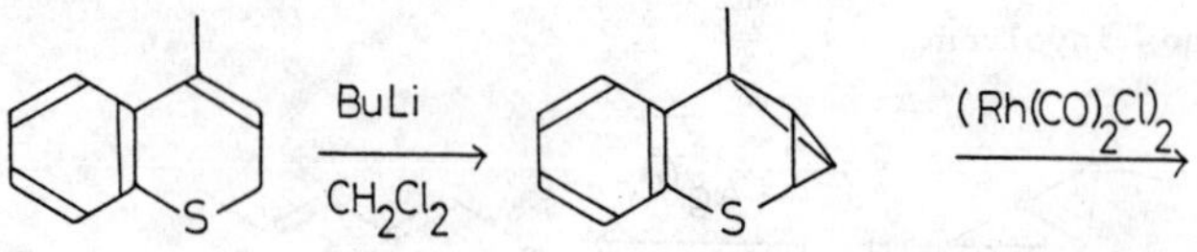

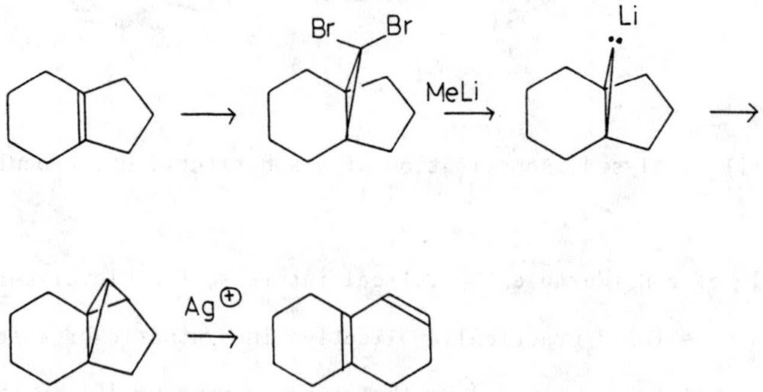

Figure 30 Carbene addition and bicyclobutane rearrangement providing a route for ring expansion.

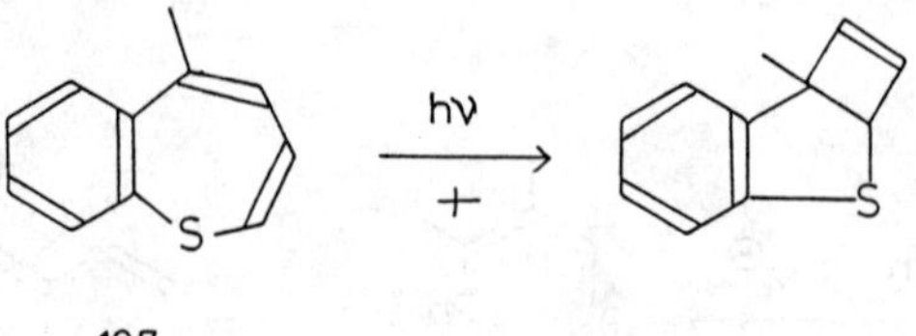

compounds the neutral conditions offered by transition metal catalysts are more attractive. Diene epoxides have been studied and may be rearranged to allyl alcohols or ketones depending on the conditions (Figure 31).

5.5.2 Azirine and Aziridine rearrangements

3-Membered rings containing nitrogen are also isomerised in the

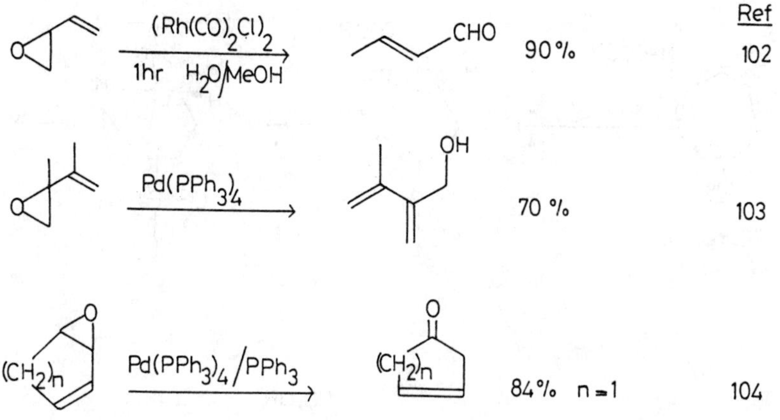

		Ref
	90%	102
	70%	103
	84% n=1	104

Figure 31 Catalysed rearrangements of epoxides.

presence of a transition metal catalyst. $(PhCN)_2PdCl_2$ catalyses the rearrangement of 108 to 109, providing a new route to the skeleton of the tropane alkaloids such as cocaine, atropine and scopalamine. The reaction is thought to proceed via an addition elimination mechanism. The final loss of $PdCl_2$ may occur via a 1,2-palladium shift followed by elimination or loss of HCl and then loss of PdCl with a hydride shift (Figure 32)[105].

The treatment of azirines with rhodium[106] or palladium[107] catalysts may give ring expansion to indole derivatives. The reaction probably proceeds through a metal nitrene; whilst this is analogous to the intermediate proposed for the thermal reaction, the metal catalysed process takes place under milder conditions being complete for 110 (R = H) in less than 3 minutes at 30 °C.

5.5.3 Allylic transposition

The facile formation of palladium allyl complexes leads to an easy route for allylic transposition. For example, the cyclohexenyl acetate, 111, is readily epimerised in the presence of $Pd(PPh_3)_4$, both isomers losing

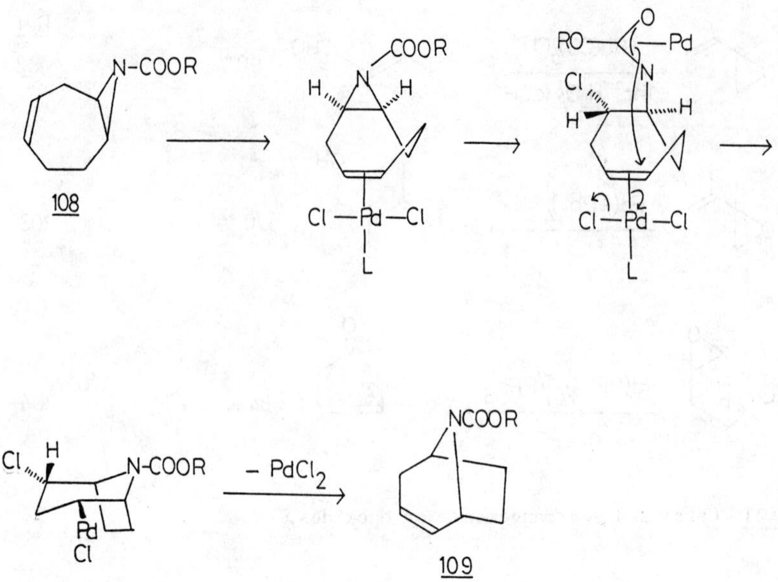

Figure 32 Pd(II) catalysed rearrangement of an aziridine.

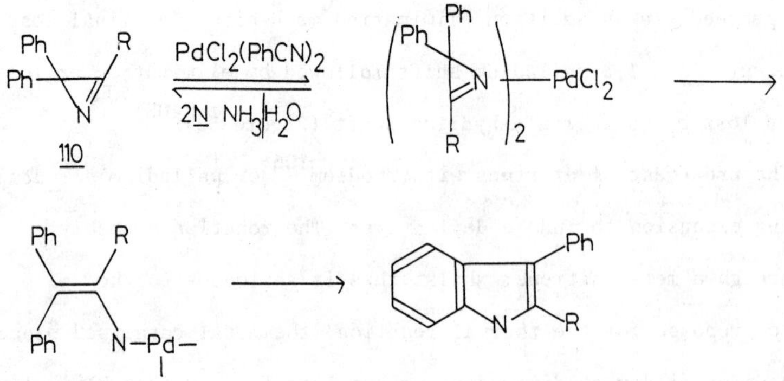

acetate slowly to give a cyclohexadiene[108]. Onder the same conditions for

allyl thiophosphate esters, 112, the allyl group may be rearranged from

oxygen to sulphur[109]. An allyl palladium complex is claimed as an

intermediate, but the transposition is formally the result of a pericyclic

reaction and a number of analogues are known. Diallyl ethers are rearranged

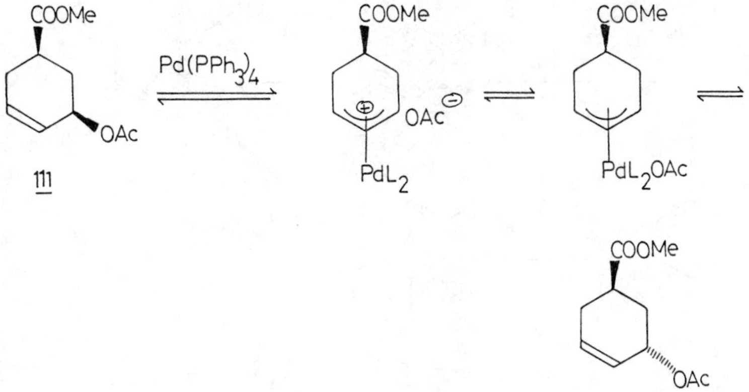

COOMe

Pd(PPh$_3$)$_4$

111

COOMe

OAc

PdL$_2$

COOMe

PdL$_2$OAc

COOMe

OAc

MeO

S

P

MeO

O

Pd(PPh$_3$)$_4$

MeO

O

P

MeO

S

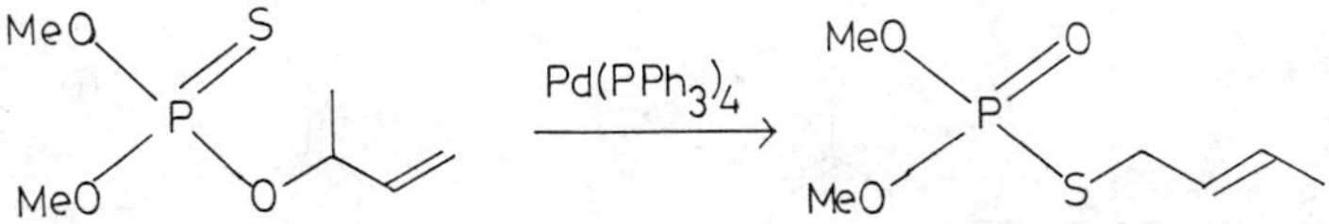

112

113

in the presence of (Ph$_3$P)$_3$RuCl$_2$ to $\gamma\delta$-unsaturated ketones. The reaction is regioselective for unsymmetrical ethers giving rise only to the less substituted carbonyl compound[110]. This is thought to arise because the transformation of allyl to vinyl ether by Ru(II) is faster in the unsubstituted case. Even prolonged exposure to the catalyst causes no rearrangement of the product (Figure 33).

Some rearrangements of alkynes may be initiated by siver ion; the mechanism involves coordination of Ag$^+$ to the triple bond, making it more susceptible to attack by a remote double bond[111]. The transformation of Figure 34 may be achieved thermally only at 200 °C. An analogous ester rearrangement promoted by Cu$_2$Cl$_2$ or AgBF$_4$ has been used in the synthesis of allenic acetates, 115, useful in Diels Alder reactions[112].

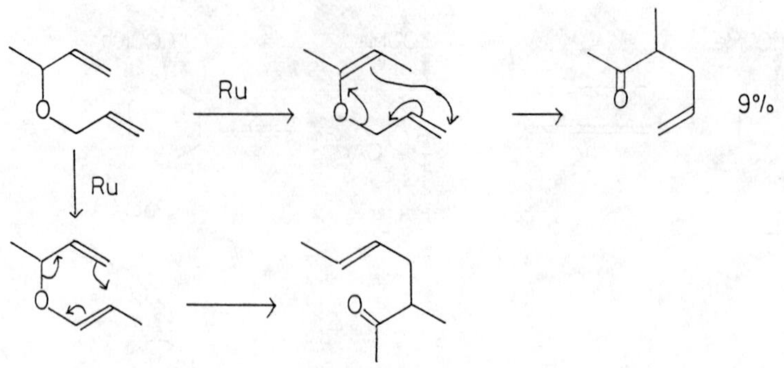

Figure 33 Ru(II) catalysed isomerisation of diallyl ethers.

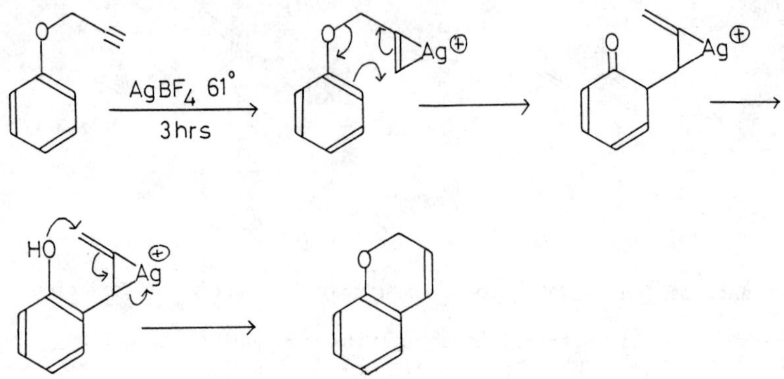

Figure 34 Ag^+ catalysed rearrangement of a propargyl ether.

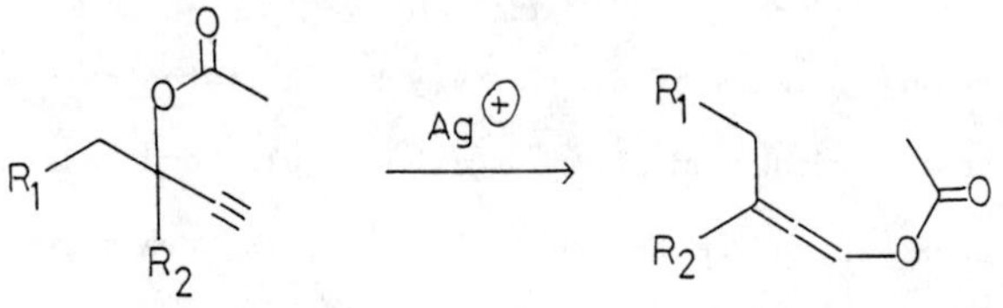

114

115

References

1   M.J.S.Dewar, Bull. Soc. Chim. Fr., C71 (1951); M.J.S.Dewar and G.P.Ford,
    J.Amer. Chem. Soc., 101, 783 (1979) and intervening papers.

2   A.K.Rappé and W.A.Godard, J. Amer. Chem. Soc., 99, 3966 (1977).

3   K.Tatsumi and M.Tsutsui, Fundamental Research in Homogeneous Catalysis,
    3, 55 (1979).

4   F.Conti, L.Raimondi, G.F.Pregaglia and R.Ugo, J. Organomet. Chem.,
    70, 107 (1974).

5   J.F.Harrod and A.J.Chalk, J. Amer. Chem. Soc., 86, 1776 (1964).

6   J.W.Faller, M.T.Tulley and K.J.Laffey, J. Organomet. Chem., 37, 193
    (1972); J.Lukas, J.E.Ramakers-Blom, T.G.Hewitt and J.J. de Boer, J.
    Organomet. Chem., 46, 167 (1972).

7   M.A.Cairns and J.F.Nixon, J.Organomet. Chem., 87, 109 (1975).

8   W.Strohmeier and W.Rehder-Sturweiss, J. Organomet. Chem., 22, C27 (1970).

9   D.Bingham, B.Hudson, D.E.Webster and P.B.Wells, J. Chem. Soc., Dalton,
    1521 (1974).

10 F.d'Amico, J. von Jouanne and H.Kelm, J. Mol. Catal., 6, 327 (1979)

11 H.Kanai, J. Chem. Soc., Chem. Commun., 203 (1972)

12 C.P.Casey and C.R.Cyr, J. Amer. Chem. Soc., 95, 2248 (1973).

13 W.T.Hendrix and J.L.von Rosenberg, J. Amer. Chem. Soc., 98, 4850 (1976).

14 M.J.d'Aniello and E.K.Barfield, J. Amer. Chem. Soc., 100, 1474 (1978).

15 C.A.Tolman, J. Amer. Chem. Soc., 94, 2994 (1972).

16 L.W.Gosser and G.W.Parshall, Tetrahedron Lett., 2555 (1971).

17 R.G.Miller, P.A.Pinke, R.D.Stauffer, H.J.Golden and D.J.Baker, J. Amer.
    Chem. Soc., 96, 4211 (1974).

18 P.A.Pinke and R.G.Miller, J. Amer. Chem. Soc., 96, 4221 (1974).

19 H.J.Golden, D.J.Baker and R.G.Miller, J. Amer. Chem. Soc., 96, 4235 (1974).

20 B.Corrain, Gazz. Chim. Ital., 102, 687 (1972).

21 J.E.Lyons, Ger. Offen., 2,147,323 (1972); Chem. Abs., 77, 5130s (1972);
    E.O.Sherman and M.Olsen, J. Organomet. Chem., 172, C13 (1979).

22 A.J.Hubert and H.Reimlinger, Synthesis, 405 (1970).

23 B.I.Cruikshank and N.R.Davies, Aust. J. Chem., 26, 2635 (1973); N.R.Davies
    and A.D.Di Michiel, Aust. J. Chem., 26, 1529 (1973).

24 A.P.Wagner, Manuf. Chemist, 23, 56 (1952).

25 R.J. De Pasquale, Synth. Commun., 10, 225 (1980).

26 J.K.Nicholson and B.L.Shaw, Tetrahedron Lett., 3533 (1965).

27 M.Gargano, P.Giannoccaro and M.Rossi, J. Organomet. Chem., 84, 389 (1975).

28 G.P.Pez and S.C.Kwan, J. Amer. Chem. Soc., 98, 8079 (1976).

29 J.Arnet and R.Pettit, J. Amer. Chem. Soc., 83, 2954 (1961).

30 T.Nishiguchi, H.Imai and K.Fukuzumi, J.Catal., 39, 375 (1975).

31 A.J.Birch and G.S.R. Subba Rao, Tetrahedron Lett., 3797 (1968).

32 J.Kovács, W.Pritzkow, G.Speier and L.Markó, Acta. Chim. Acad. Sci. Hung.,
    88, 177 (1976).

33 Yu.G.Osokin, M.Ya.Grinberg, V.Sh.Feldblyum, O.A.Yasinskii,
    V.V.Plachtinskii, E.R.Kofanov, V.A.Ustinov and G.S.Mironov, React.Kinet.
    Catal. Lett., 9, 189 (1978).

34 T.Onishi, Y.Fugita and T.Nishida, Chem. Lett., 765 (1979).

35 C.Carlini, E.Chiellini and R. Solaro, J. Polm. Sci., Polym. Chem. Ed.,
    18, 2129 (1980).

36 G.Giacomelli, L.Lardicci, R.Menicagli and L.Bertero, J. Chem. Soc.,
    Chem. Commun., 633 (1979).

37 R.E.Rinehart, U.S.Patent, 3,433,808 (1969); Chem. Abs., 70, 106368m
    (1969).

38 H.Alper and K.Hachem, J. Org. Chem., 45, 2269 (1980)

39 D.Baudry, M.Ephitikhine and H.Felkin, Nouv. J. Chim., 2, 355 (1978).

40 C.F.Lochow and R.G.Miller, J. Org. Chem., 41, 3020 (1976).

41 P.Rigo, M.Bressan and M.Basato, Inorg. Chem., 18, 860 (1979).

42 A.J.Deeming and S.Hasso, J.Organomet. Chem., 114, 313 (1976).

43 C.Botteghi and G.Giacomelli, Gazz. Chim. Ital., 106, 1131 (1976).

44 W.Smadja, G.Ville and C.Georgoulis, J. Chem. Soc., Chem. Commun.,
    594 (1980).

45 D.Baudry, M.Ephitikhine and H.Felkin, J. Chem. Soc., Chem. Commun.,
    694 (1978).

46 J.Ardrieux, D.H.R.Barton and H.Patin, J. Chem. Soc., Perkin I, 359 (1977).

47 E.J.Corey and J.W.Suggs, J. Org. Chem., 38, 3224 (1973).

48 J.J.Oltvoort, C.A.A. van Boeckel, J.H. de Koenig and J.H. van Boom,
    Synthesis, 305 (1981).

49 T.Taketomi, H.Kumobayashi, S.Akutagawa, T.Yamanaka and T.Yoshida, Jpn.
    Kokai Tokkyo Koho, 79,05,906 (1979); Chem. Abs., 91, 20803w (1979)

50 H.Komobayashi, S.Akutagawa and S.Otsuka, J. Amer. Chem. Soc., 100, 3949
    (1978).

51 J.K.Stille and Y.Becker, J. Org. Chem., 45, 2139 (1980).

52 P.A.Grieco, M.Nishizawa, N.Marinovic and W.J.Ehmann, J. Amer. Chem. Soc.,
    98, 7102 (1976).

53 N.Katsin and R.Ikan, Synth. Commun., 7, 185 (1977).

54 R.Pettit, H.Sugahara, J.Wristers and W.Merk, Discuss. Faraday Soc.,
    47, 71 (1969).

55 F.D.Mango, Coord. Chem. Rev., 15, 109 (1974).

56 K.C.Bishop, Chem. Rev., 76, 461 (1976).

57 H.C.Volger and M.M.P.Gaasbeeck, Rec., 86, 1290 (1968).

58 H.Dietl and P.M.Maitlis, J. Chem. Soc., Chem. Commun., 759 (1967).

59 W.Merk and R.Pettit, J. Amer. Chem. Soc., 89, 4788 (1967).

60 L.Cassar, P.E.Eaton and J.Halpern, J. Amer. Chem. Soc., 92, 3515 (1970).

61 L.A.Paquette, J.S.Ward, R.A.Boggs and W.B.Farnham, J. Amer. Chem. Soc.,
    97, 1101 (1975).

62 L.Cassar, P.E.Eaton and J.Halpern, J. Amer. Chem. Soc., 92, 6366 (1970).

63 L.A.Paquette, R.S.Beckley and W.B.Farnham, J. Amer. Chem. Soc., 97,
    1089 (1975).

64 L.A.Paquette, R.A.Boggs, W.R.Farnham and R.S.Beckley, J. Amer. Chem. Soc., 97, 1112 (1975).

65 L.A.Paquette and J.C.Stowell, J. Amer. Chem. Soc., 92, 2584 (1970).

66 L.A.Paquette, R.A.Boggs and J.S.Ward, J. Amer. Chem. Soc., 97, 1118 (1975).

67 L.A.Paquette and J.C.Stowell, J. Amer. Chem. Soc., 93, 2459 (1971).

68 L.A.Paquette, Acc. Chem. Res., 4, 280 (1971).

69 R.Noyori, M.Yamakawa and H.Takeya, J. Amer. Chem. Soc., 98, 1471 (1976).

70 W.G.Dauben and A.J.Kielbania, J. Amer. Chem. Soc., 93, 7345 (1971).

71 L.A.Paquette, J. Amer. Chem. Soc., 92, 5765 (1970).

72 H.Hogeveen and H.C.Volger, J. Amer. Chem. Soc., 89, 2486 (1967).

73 L.Cassar and J.Halpern, J. Chem. Soc., Chem. Commun., 1082 (1970).

74 M.Chen and H.M.Feder, Inorg. Chem., 18, 1864 (1979).

75 R.Noyori, I.Umeda, H.Kawauchi and H.Takaya, J. Amer. Chem. Soc., 97, 812 (1975).

76 R.B.King and S.Ikai, Inorg. Chem., 18, 949 (1979).

77 J.Manassen, J. Catal., 18, 38 (1970).

78 M.J.Chen and H.M.Feder, J. Catal., 55, 105 (1978).

79 H.D.Wilson and R.G.Rinker, J. Catal., 42, 268 (1976).

80 P-W.Chum and J.A.Roth, J. Catal., 39, 198 (1975).

81 H.W.Voight and J.A.Roth, J. Catal., 33, 91 (1974).

82 V.Aris, J.M.Brown, J.A.Coneely, B.T.Golding and D.H.Williamson, J. Chem. Soc., Perkin II, 4 (1975).

83 J.M.Brown, B.T.Golding and J.J.Stofko, J. Chem. Soc., Perkin II, 436 (1978).

84 P.A.Pinke, R.D.Stauffer and R.G.Miller, J. Amer. Chem. Soc., 96, 4229 (1974).

85 J.A.Walker and M.Orchin, J. Chem. Soc., Chem. Commun., 1239 (1968).

86 T.Shirafuji, Y.Yamamoto and N.Nozaki, Tetrahedron Lett., 4713 (1971)

87 J.H.Leftin and G.Gil-Av, Tetrahedron Lett., 3367 (1972).

88 R.Weiss and S.Andrae, Angew. Chem., Int. Ed. Engl., 12, 150, 152 (1973).

89 J.Hogeveen and J.Thio, Tetrahedron Lett., 3463 (1973).

90 T.J.Katz and S.Cerefice, J. Amer. Chem. Soc., 91, 2405 (1969); H.C. Volger, H.Hogeveen and M.M.P.Gaasbeek, J. Amer. Chem. Soc., 91, 218 (1969).

91 P.G.Gassman and E.A.Armour, Tetrahedron Lett., 1431 (1971)

92 L.A.Paquette and G.Zon, J. Amer. Chem. Soc., 96, 203 (1974).

93 P.G.Gassman and T.J.Atkins, J. Amer. Chem. Soc., 94, 7748 (1972).

94 L.A.Paquette and G.Zon, J. Amer. Chem. Soc., 96, 224 (1974).

95 G.Zon and L.A.Paquette, J.Amer. Chem. Soc., 96, 215 (1974).

96 M.Sakai, H.Yamaguchi, H.H.Westberg and S.Masamune, J. Amer. Chem. Soc., 93, 1043 (1971).

97 P.G.Gassman, Angew. Chem.,Int. Ed. Engl., 11, 323 (1972).

98 I.Murata and T.Tatsuoka, Tetrahedron Lett., 2697 (1975).

99 L.A.Paquette, E. Chamot and A.R.Browne, J. Amer. Chem. Soc., 102, 637 (1980).

100 D.H.R.Barton, A.F.Gosden, G.Mellows and D.Widdowson, J. Chem. Soc., Chem. Commun., 1067 (1968).

101 A.C.Cope and B.D.Tiffany, J. Amer. Chem. Soc., 73, 4158 (1951).

102 D.C.Lini, K.C.Ramey and W.B.Wise, U.S.Patent, 3,465,043 (1969); Chem. Abs., 72, 12106p (1970).

103 M.Suzuki, Y.Oda and R.Noyori, J. Amer. Chem. Soc., 101, 1623 (1979).

104 R.Noi, M.Suzuki, S.Kurozimi and Y.Hashimoto, Jpn. Kokai Tokkyo Koho, 79,154,725 (1980); Chem Abs., 92, 214982v (1980).

105 G.R.Wiger and M.F.Rettig, J. Amer. Chem. Soc., 98, 4168 (1976).

106 H.Alper and J.E.Prickett, J. Chem. Soc., Chem. Commun., 483 (1976).

107 K.Isomura, K.Uto and H.Taniguchi, J. Chem. Soc., Chem. Commun., 664 (1977).

108 B.M.Trost, T.R.Verhoeven and J.M.Fortunak, Tetrahedron Lett., 2301 (1979).

109 Y.Yamada, K.Mukai, H.Yoshioka, V.Tamura and Z-i.Toshida, Tetrahedron Lett., 5015 (1979).

110 J.M.Reuter and R.G.Salomon, J. Org. Chem., 42, 3360 (1977).

111 U.Koch-Pomeranz, H-J.Hansen and H.Schmid, Helv. Chim. Acta, 56, 2981
     (1973).

112 R.C.Cookson, M.C.Cramp and P.J.Parsons, J. Chem. Soc., Chem. Commun.,
     197 (1980).

6 Oxidation

6.1   Mechanisms of catalytic oxidation.

6.2   Oxidation of alkenes to yield epoxides.

   6.2.1   Epoxidations using hydrogen peroxide.

   6.2.2   Epoxidations using molecular oxygen.

   6.2.3   Epoxidations using alkyl hydroperoxides.

      6.2.3.1   Variation of the catalyst.

      6.2.3.2   Variation of the hydroperoxide.

      6.2.3.3   By-products and side reactions.

      6.2.3.4   Variation of alkene structure.

      6.2.3.5   Stereoselectivity.

   6.2.4   Epoxidations using iodosobenzene.

   6.2.5   Epoxidations using sodium hypochlorite.

   6.2.6   Asymmetric epoxidation.

6.3   Oxidation of alkenes to carbonyl compounds.

   6.3.1   Palladium catalysts.

   6.3.2   Rhodium catalysts.

   6.3.3   Iridium catalysts.

   6.3.4.   Other metals.

6.4   Allylic oxidation of alkenes.

   6.4.1   Manganese catalysts.

   6.4.2   Cobalt catalysts.

   6.4.3   Rhodium catalysts.

   6.4.4   Iridium catalysts.

   6.4.5   Palladium catalysts.

   6.4.6   Copper catalysts.

   6.4.7   Other metals.

452

6.5   Other alkene oxidations.

    6.5.1   Conversion to cis-diols using osmium catalysts.

    6.5.2   Oxidation in the presence of palladium complexes.

    6.5.3   Cobalt catalysts.

    6.5.4   Manganese catalysts.

    6.5.5   Ruthenium catalysts.

    6.5.6   Copper catalysts.

    6.5.7   Other metals as catalysts.

6.6   Oxidation of alkynes.

6.7   Oxidation of arenes.

6.8   Oxidation of alkyl arenes.

    6.8.1   Manganese catalysts

    6.8.2   Cobalt catalysts.

    6.8.3   Cobalt/manganese catalysts.

    6.8.4   Peroxodisulphate oxidations.

    6.8.5   Copper catalysts.

    6.8.6   Other metals.

6.9   Oxidation of saturated hydrocarbons.

6.10   Oxidation of alcohols.

    6.10.1   Oxidation of primary alcohols.

        6.10.1.1   Oxidation by peroxodisulphate.

        6.10.1.2   Ruthenium catalysts.

        6.10.1.3   Other metals.

    6.10.2   Oxidation of secondary alcohols.

        6.10.2.1   Ruthenium catalysts.

        6.10.2.2   Oxidation by peroxodisulphate.

        6.10.2.3   Palladium catalysts.

        6.10.2.4   Copper catalysts.

        6.10.2.5   Cobalt catalysts.

        6.10.2.7   Other metals.

6.10.3   Oxidation of diols.

6.10.4   Oxidation of α-ketols.

6.11   Oxidation of aldehydes.

6.12   Oxidation of ketones.

6.13   Oxidation of carboxylic acids and their derivatives.

    6.13.1   Monocarboxylic acids.

    6.13.1   Dicarboxylic acids.

    6.13.3   Hydroxy acids.

    6.13.4   Esters and amides.

    6.13.5   Amino acids.

    6.13.6   Ascorbic acid.

6.14   Oxidation of ethers.

6.15   Oxidation of phenols and related compounds.

    6.15.1   Oxidation of monohydric phenols.

        6.15.1.1   Cobalt catalysts.

        6.15.1.2   Copper catalysts.

        6.15.1.3   Other metals.

    6.15.2   Oxidation of catechols.

        6.15.2.1   Copper catalysts

        6.15.2.2   Iron catalysts.

        6.15.2.3   Other metals.

    6.15.3   Oxidation of anilines.

6.16   Oxidation of nitrogen containing compounds.

    6.16.1   Oxidation of amines.

        6.16.1.1   Primary amines.

        6.16.1.2   Secondary amines.

        6.16.1.3   Tertiary amines.

    6.16.2   Oxidation of isocyanides.

    6.16.3   Oxidation of nitroso compounds.

    6.16.4   Oxidation of azobenzenes.

6.16.5   Oxidation of hydrazines, hydrazides and hydrazones.

6.16.6   Oxidative cleavage.

6.17   Oxidation of phosphorus containing compounds.

6.18   Oxidation of sulphur containing compounds.

6.18.1   Oxidation of thiols to disulphides.

6.18.2   Oxidation of sulphides and sulphoxides.

6.1   <u>Mechanisms of Catalytic Oxidation</u>

The most obvious method for transition metal catalysed oxidation is to activate both molecular oxygen and substrate by coordination, and then to transfer an oxygen atom to the substrate within the coordination sphere.   This would leave metal oxide which must subsequently be reduced to obtain a catalytic cycle.   It is instructive to consider the range of metal complexes of molecular oxygen.

Complexes fall into two classes, peroxo and superoxo,which are formally likened to $[O_2]^{2-}$ and $[O_2]^{-\bullet}$.   In the superoxo complexes only one oxygen is metal bound, $[O_2]^{-\bullet}$ acts as a two electron donor and the MÔO angle is about 120°.[1,2]   Some examples are given in Figure 1.

Peroxo complexes, of type <u>4</u>, are often referred to as side or π-bonded oxygen complexes.   They are usually diamagnetic and represent a formal two electron transfer from metal to O_2.   Both oxygens are metal bound and O_2 acts as a four electron donor, but the M-O bonds are often of unequal length.   Some examples are shown in Figure 2.

Transition metal μ-peroxo complexes of type <u>10</u> are also diamagnetic, with O_2 donating two electrons to each of two metals.   The bonding is complex, involving a four centre molecular orbital.[11]   Some examples are shown in Figure 3.

Dioxygen activated as a superoxo complex may act either as a base (equation (1)) or as a radical, able to abstract H$^\bullet$ (equation (2)).

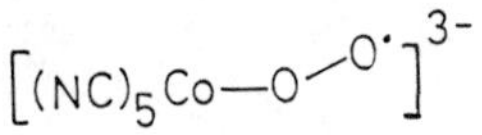

$$\left[(NC)_5 Co-O-O-O' \right]^{3-}$$

$\underline{1}^3$

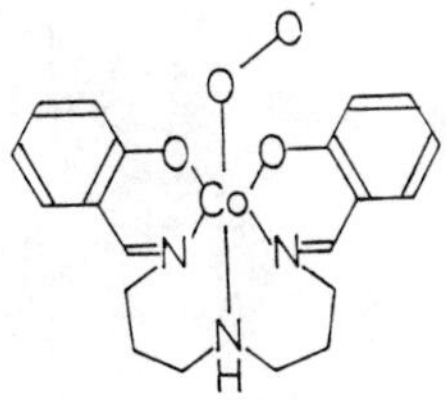

$\underline{2}^4$

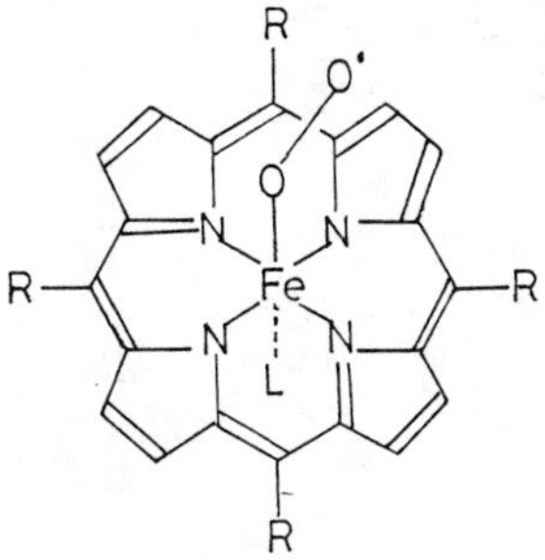

$\underline{3}^5$

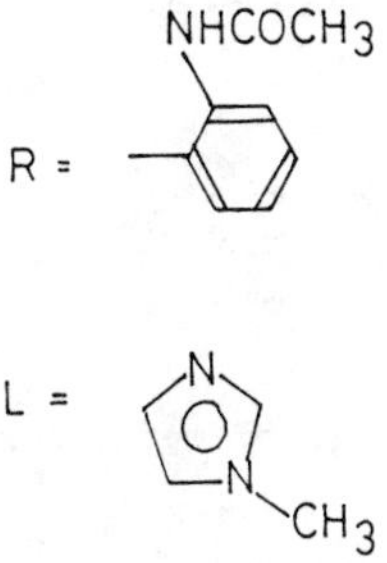

R =

L =

Figure 1 Examples of metal superoxo complexes.

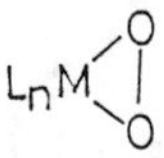

$\underline{4}$

Subsequently, radical reactions of the substrate predominate in both cases

$$MO_2^{-\cdot} + XH \longrightarrow HO_2^{\cdot} + M + X^- \qquad (1)$$

$$MO_2^{-\cdot} + RH \longrightarrow [MOOH]^- + R^\cdot \qquad (2)$$

and detailed mecanisms will be discussed later[16]. Stoichiometric oxidations
by peroxo complexes have proved useful models for catalytic reactions. Those
shown in Figure 4 transfer both oxygen atoms to the substrate within the

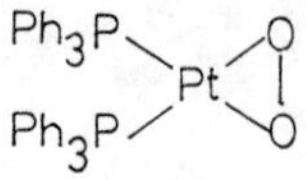

Ph$_3$P—Pt(O$_2$) with Ph$_3$P

5 6

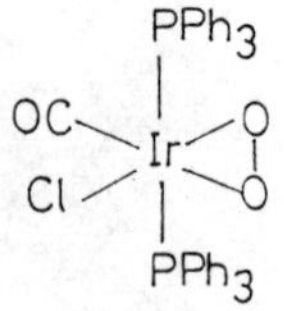

OC—Ir(O$_2$)(Cl)(PPh$_3$)$_2$

6 7

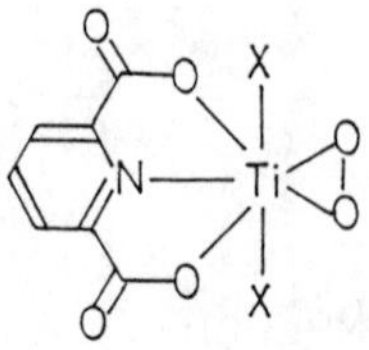

Ti complex with pyridine dicarboxylate, X, X and O$_2$

7 8

M complex with L–L and O$_2$, O

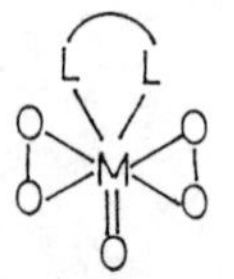

8 M = Cr, L L = bipy 9

9 M = Mo, L L = C$_2$O$_4^{2-}$ 10

Figure 2 Examples of metal peroxo complexes.

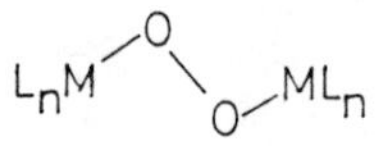

L_nM—O—O—ML_n

10

metal coordination sphere[17]. These examples all involved Group VIII metals, but Group VI peroxo complexes also interact with appropriate substrates. For example, **14** oxidises cyclohexene to its epoxide[18,19]. The transfer of oxygen from μ-peroxo complexes is relatively uncommon, an example being provided by the reaction **15** to **16**.[20] Generally they are not involved in catalytic reactions, and since their formation tends to be irreversible, it is an important route for deactivation of other catalysts.

The use of molecular oxygen as oxidant is highly desirable economically, but direct activation in a catalytic cycle is often difficult. A relay mechanism, shown schematically in Figure 5, is a useful alternative.

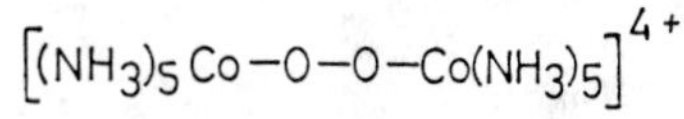

$\left[(NH_3)_5 Co-O-O-Co(NH_3)_5\right]^{4+}$

$\underline{11}$ [12]

$= Fe$

$Fe(III)-O{\sim}O-Fe(III)$

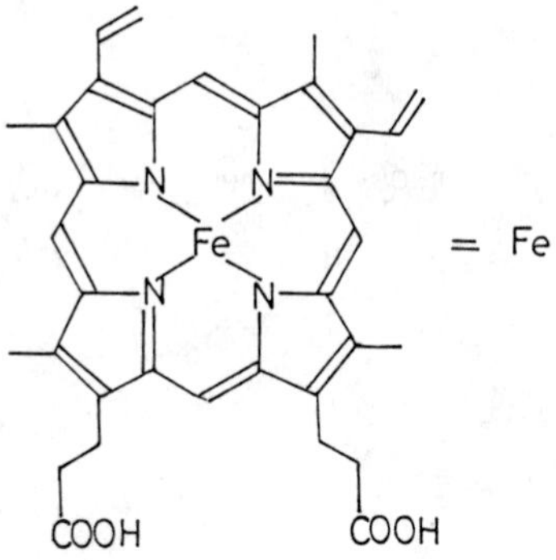

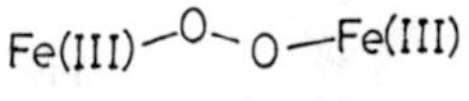

$\underline{12}$ [13,14]

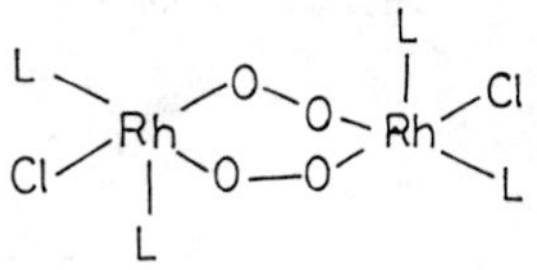

$L = PPh_3$

$\underline{13}$ [15]

Figure 3   Examples of μ-peroxo complexes.

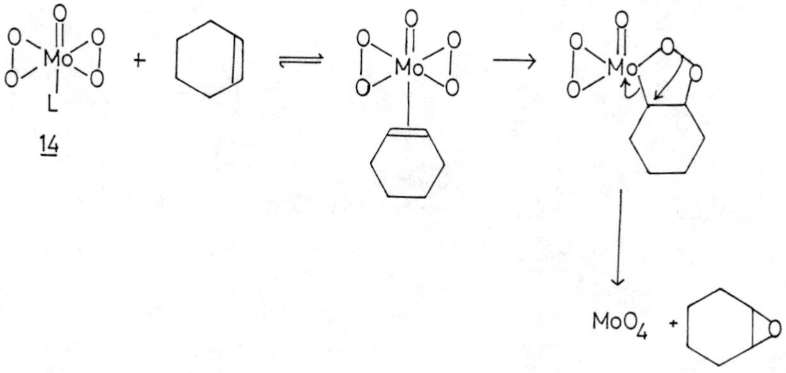

$\underline{14}$

$MoO_4 +$

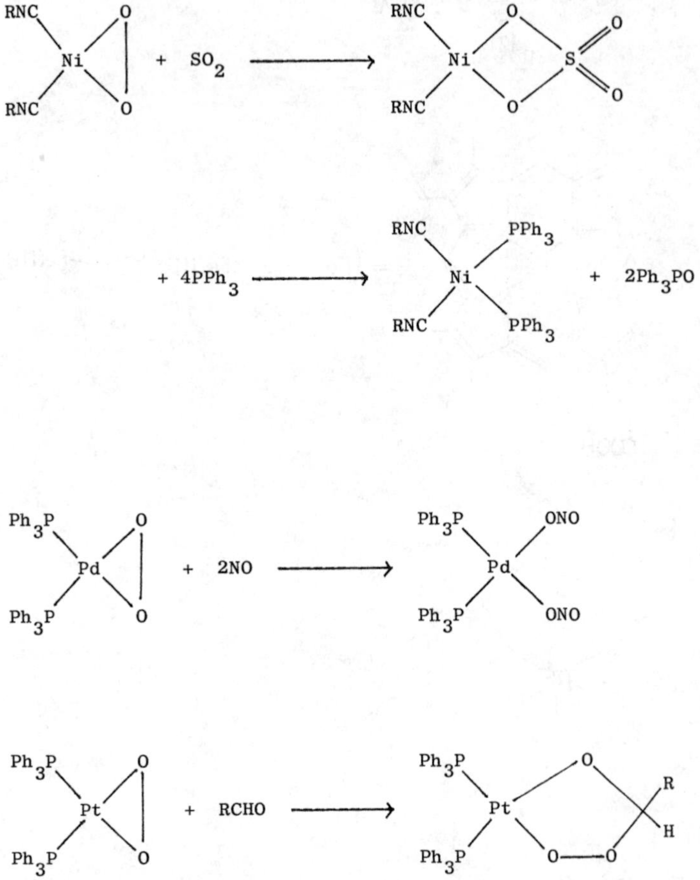

Figure 4 Stoichiometric oxidations by metal peroxo complexes.

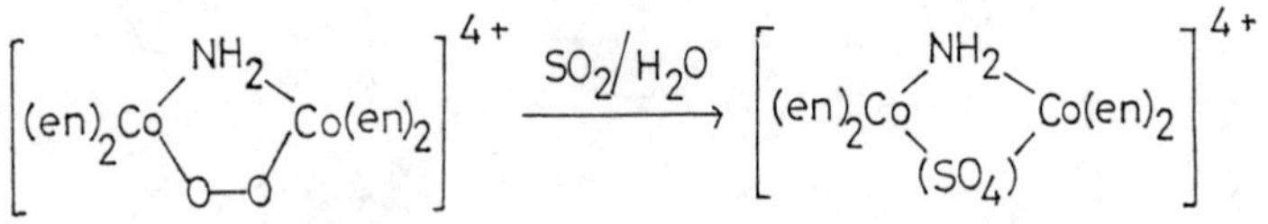

Substrate + Metal A(oxidised) $\longrightarrow$ Substrate(oxidised) + Metal A(reduced)

Metal A(reduced) + Metal B(oxidised) $\longrightarrow$ Metal A(oxidised) + Metal B(reduced)

Metal B(reduced) + O_2 $\longrightarrow$ Metal B(oxidised)

Figure 5 Relay scheme for oxidation using molecular oxygen.

The substrate is oxidised by metal A in a high oxidation state, the metal being reduced.   Reduced A is oxidised by metal B in a high oxidation state.   Reduced B is then oxidised by molecular oxygen or some other oxidant.   The best known example of this process is the Wäcker oxidation of ethylene to acetaldehyde (equations (3) - (5)).   Metal A is palladium and metal B copper.

$$C_2H_4 + PdCl_2 + H_2O \longrightarrow CH_3CHO + 2HCl + Pd(0) \qquad (3)$$

$$Pd(0) + 2CuCl_2 \longrightarrow PdCl_2 + 2CuCl \qquad (4)$$

$$2CuCl + 2HCl + \tfrac{1}{2}O_2 \longrightarrow 2CuCl_2 + H_2O \qquad (5)$$

An entirely different group of reactions is provided by metal catalysed autoxidations.   In some cases these could more properly be described as initiations, since the metal is not involved in every catalytic cycle.   Metal oxidants may react with organic compounds to give radicals in two different ways.   In the first, (equations (6),(7)) electron transfer from RH to metal gives an organic radical cation which decomposes to R$^\cdot$ and H$^+$.   The ease of electron transfer depends on the

$$M(OAc)_3 + RH \longrightarrow [M(OAc)_3]^- + RH^{+\cdot} \qquad (6)$$

$$RH^{+\cdot} \longrightarrow R^\cdot + H^+ \qquad (7)$$

ionisation potential of the organic molecule.   Alternatively, the first step is electrophilic substitution, followed by homolysis of the carbon-metal bond to give, again, R$^\cdot$ (equations (8),(9)).   The ease of this

$$M(OAc)_3 + RH \longrightarrow RM(OAc)_2 + HOAc \qquad (8)$$

$$RM(OAc)_2 \longrightarrow R^\bullet + M(OAc)_2 \qquad (9)$$

reaction also roughly parallels ionisation potential and distinctions

between the two pathways are frequently difficult. The usual fate of

the radical is reaction with O_2 to give a hydroperoxide. This may, in

turn, be decomposed in a metal catalysed pathway to give alcohols or

carbonyl compounds (Figure 6).

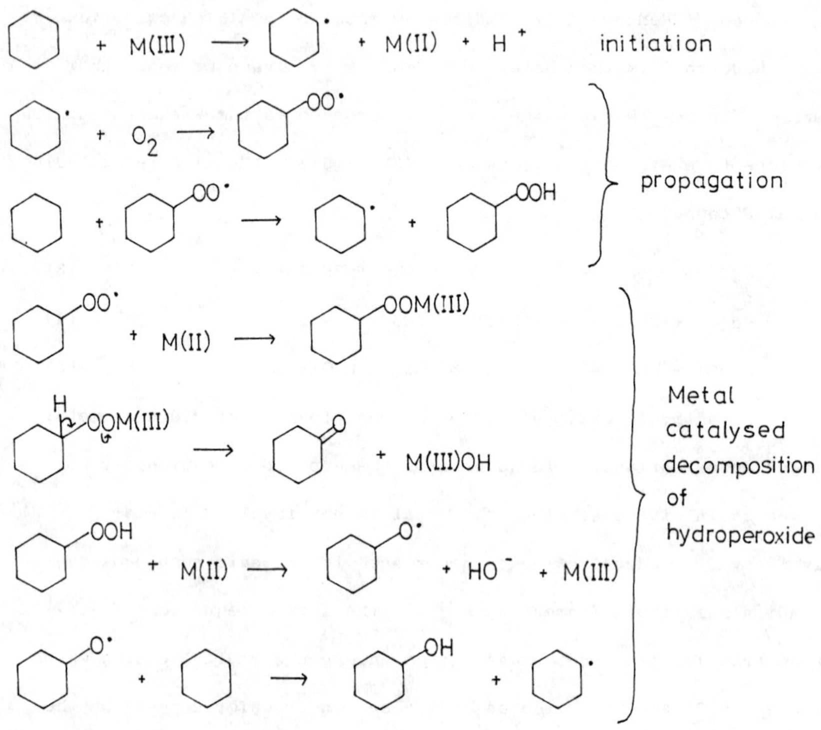

Figure 6 Metal catalysed autoxidation.

The final general mechanism for catalytic oxidation does not

employ molecular oxygen as the source of oxidising power, but instead uses

H_2O_2, RO_2H or RCO_3H. Whilst more expensive than O_2, their cost is not

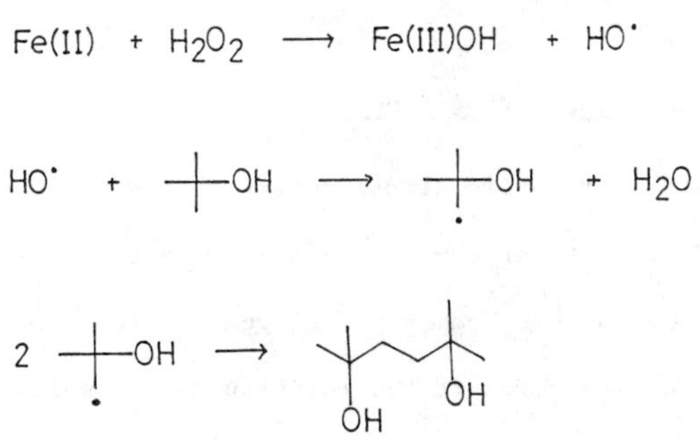

461

$$Fe(II) + H_2O_2 \longrightarrow Fe(III)OH + HO^\cdot$$

$$HO^\cdot + \;\begin{array}{c}\rule{0pt}{0pt}\end{array}\!\!-OH \longrightarrow \begin{array}{c}\rule{0pt}{0pt}\end{array}\!\!-OH + H_2O$$

$$2\;\begin{array}{c}\rule{0pt}{0pt}\end{array}\!\!-OH \longrightarrow$$

Figure 7   Homolytic decomposition of H_2O_2 in synthesis.

prohibitive, and the major problems derive from the thermal and shock
sensitivity of these species.   Both H_2O_2 and RO_2H are readily decomposed
to give radicals by certain metal complexes.   This has been used
synthetically (Figure 7) but does not differ greatly from the radical
reactions discussed earlier.[21]   Such radical processes predominate when
the catalyst is a complex of cobalt, manganese, iron or copper.   Heterolysis
of peroxides is accomplished in the presence of molybdenum, tungsten,
vanadium and titanium catalysts.   All these metals have high Lewis acidity
and low oxidation potential.   The true catalyst is usually the metal in
a high oxidation state, an induction period being noted while this is
formed.   The mechanism of alkene epoxidation by H_2O_2 in the presence of
MoO_3 involves formation of the inorganic peracid $HOMo(O)_2OOH$, which then
reacts in much the same way as an organic peracid.   When alkyl hydro-
-peroxides are the oxidants the active species is L_nMOOR.

6.2   Oxidation of alkenes to yield epoxides

The conversion of alkenes to epoxides is important synthetically,
since it allows the simultaneous functionalisation of two adjacent carbon
atoms.   The usual oxidising agents are electrophilic oxygen compounds

including H_2O_2, ROOH and RCO_3H.

6.2.1 Epoxidations using hydrogen peroxide

On the basis of relative electrophilicity one would not anticipate large differences between H_2O_2 and ROOH in epoxidation. In practice, however, H_2O_2 is less commonly employed, since it may be used only in polar solvents, which retard the reaction and in which many substrates are insoluble. Additionally, the epoxide is often opened to the glycol under aqueous conditions.

Many acidic metal oxides were early found to be able to catalyse conversion of alkenes to glycols by H_2O_2, the reaction proceeding via the epoxide.[22-27] When neutral or basic conditions are used, however, epoxidation may be achieved.[28-33]

The metal oxides used as catalysts form stable inorganic peracids which are the active epoxidising agents. It is usually assumed that the mechanism is that of 17, entirely analogous to organic peracids.[34] Intramolecular reactions can be visualised for unsaturated alcohols and acids, which may become metal coordinated and are epoxidised particularly easily.

17

The scope of the reaction with simple alkenes is limited and yields are generally poor. Reaction of 2-methylpropene with anhydrous H_2O_2 in benzene/pentanol in the presence of $MoO_2(acac)_2$ gives a mixture of epoxide and hydroperoxide. Cyclohexene may be epoxidised using MoO_3[35]

or Mo(CO)$_6$[36] as catalyst, provided that water, which inhibits the reaction, is removed as it is formed. Fe(acac)$_3$ has been used in a reaction generating mainly <u>trans</u>-stilbene oxide from <u>cis</u>- or <u>trans</u>-stilbene. The starting alkene, rather than the epoxide, is isomerised under the reaction conditions.[37] Allyl bromides and chlorides are epoxidised in good yield using H$_2$MoO$_4$, the active species being either [HMoO$_5$]$^-$ or [HMoO$_6$]$^-$.[38] Larger ring cycloalkenes give rather better results using H$_2$WO$_4$, H$_2$MoO$_4$ or V$_2$O$_5$, but the reaction is rather slow.[33] More recently, simple alkenes have been epoxidised by a dilute H$_2$O$_2$ solution using <u>18</u> as catalyst. In this case the mechanism involves attack of HOO$^-$ on an alkene activated by metal coordination.[39]

18

19

The reactions of unsaturated acids, alcohols and esters are more successful, involving intramolecular oxygen transfer as in <u>19</u>. Some examples are shown in Figure 8. In molecules such as <u>20</u>, where there is more than one type of double bond, epoxidation by H$_2$O$_2$ in the presence of VO(acac)$_2$[45] or (NH$_4$)$_6$Mo$_7$O$_{24}$[46] gave only <u>21</u>.

20

21

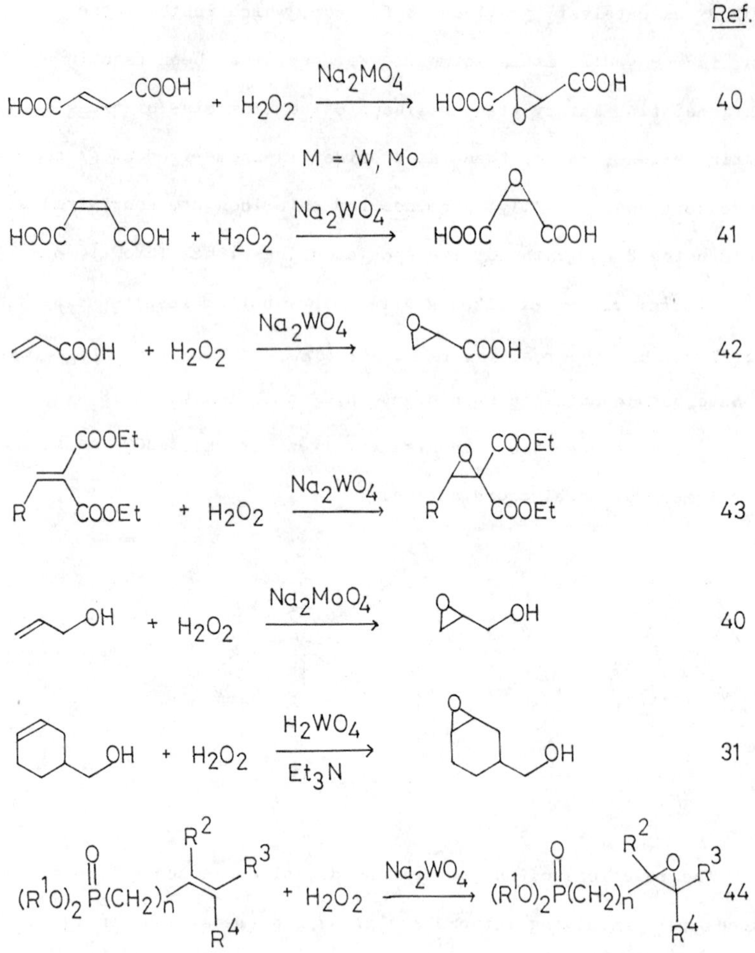

HOOC⌒COOH + H_2O_2 $\xrightarrow{Na_2MO_4}$ HOOC△COOH ⟶ 40

M = W, Mo

HOOC⌣COOH + H_2O_2 $\xrightarrow{Na_2WO_4}$ HOOC△COOH 41

⌒COOH + H_2O_2 $\xrightarrow{Na_2WO_4}$ △COOH 42

R⌒($COOEt$)($COOEt$) + H_2O_2 $\xrightarrow{Na_2WO_4}$ R△($COOEt$)($COOEt$) 43

⌒OH + H_2O_2 $\xrightarrow{Na_2MoO_4}$ △OH 40

(cyclohexene)CH$_2$OH + H_2O_2 $\xrightarrow[Et_3N]{H_2WO_4}$ (epoxide)OH 31

$(R^1O)_2\overset{O}{\underset{\|}{P}}(CH_2)_n\text{–}CR^2R^3R^4$ + H_2O_2 $\xrightarrow{Na_2WO_4}$ $(R^1O)_2\overset{O}{\underset{\|}{P}}(CH_2)_n\text{–}\triangle R^2R^3R^4$ 44

Figure 8 Epoxidation of unsaturated acids, alcohols and esters by H_2O_2.

6.2.2 Epoxidations using molecular oxygen

Most interactions between alkenes and molecular oxygen are allylic autoxidations, (see 6.4) dominated by radical pathways. However, there is considerable interest in using O_2 to generate epoxides by non-radical routes. For example, ethylene oxide is prepared industrially by vapour phase oxidation of ethylene over a supported silver catalyst.[47-48]

One group of epoxidations is thought to proceed via the
formation of an allyl hydroperoxide which subsequently epoxidises the
double bond.   For example, cyclohexene is converted to 22 with moderate
selectivity in the presence of CpV(CO)₄.   The probable pathway is shown
in Figure 9,[49-52] the stereoselectivity to the syn-epoxy alcohol being

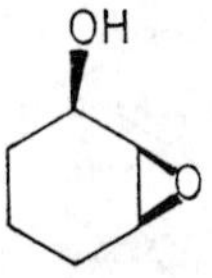

22

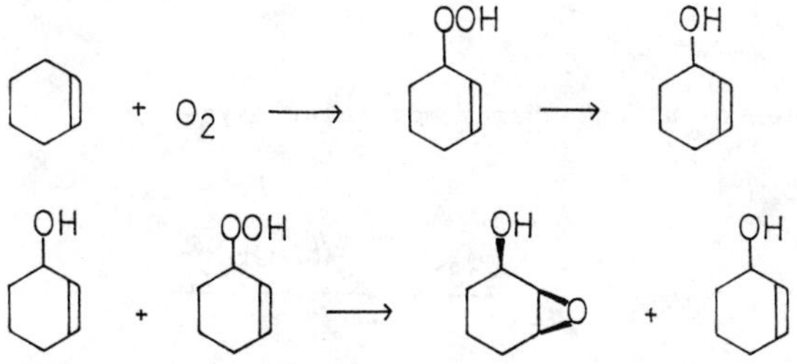

Figure 9   Reactions of cyclohexene with O_2 in the presence of vanadium
           complexes.

similar to that observed in t-BuOOH epoxidations in the presence of
vanadium complexes.   The selectivity towards epoxidation of the allyl
alcohol is high using vanadium catalysts, but less marked with other
species (Figure 10).   Other radical initiated epoxidations are shown in
Figure 11.   A final and rather interesting example uses as catalyst
[Fe₃O(piv)₆(MeOH)₃].   The reaction of 23 is regioselective (similar to
that using peracids and distinct from that in the presence of vanadium
catalysts) and the radical nature of the reaction was proven by inhibitor
studies.[63]

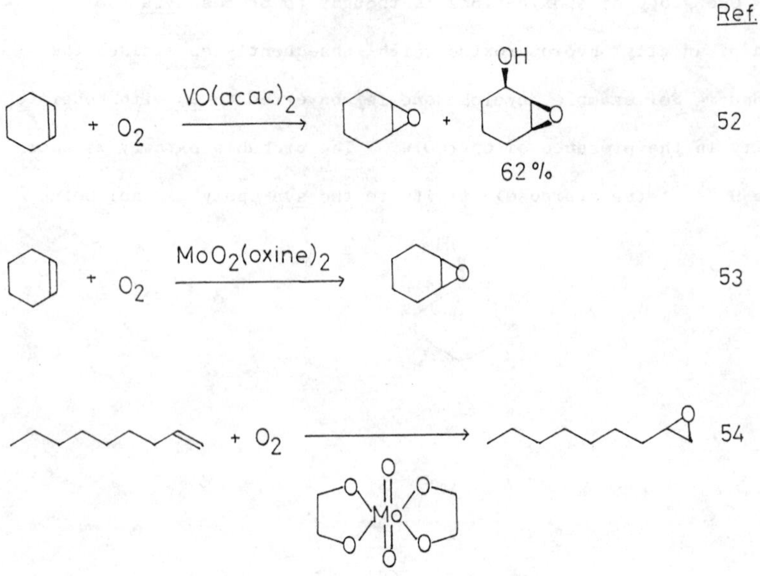

52

53

54

Figure 10   Epoxidation of alkenes using molecular oxygen.

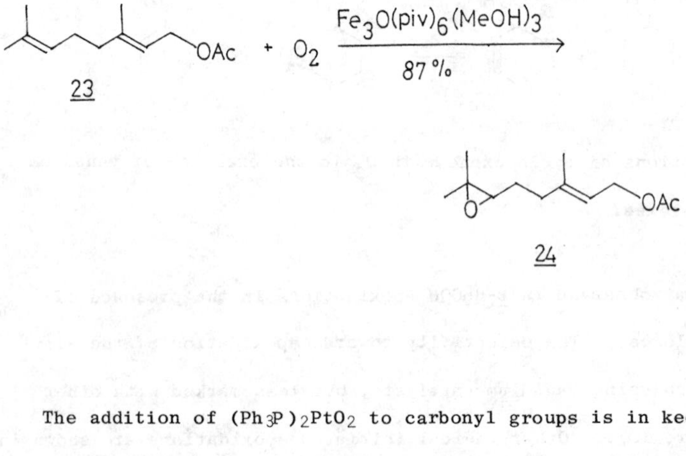

The addition of $(Ph_3P)_2PtO_2$ to carbonyl groups is in keeping with the nucleophilic character of the coordinated oxygen molecule, and an analogue of the base catalysed epoxidation of enones by H_2O_2 might be expected.   Cyclic peroxy adducts such as 25 are readily formed with alkenes bearing electron withdrawing groups and thermal decomposition may give ketones via 26, or epoxides via 27.[64,65]

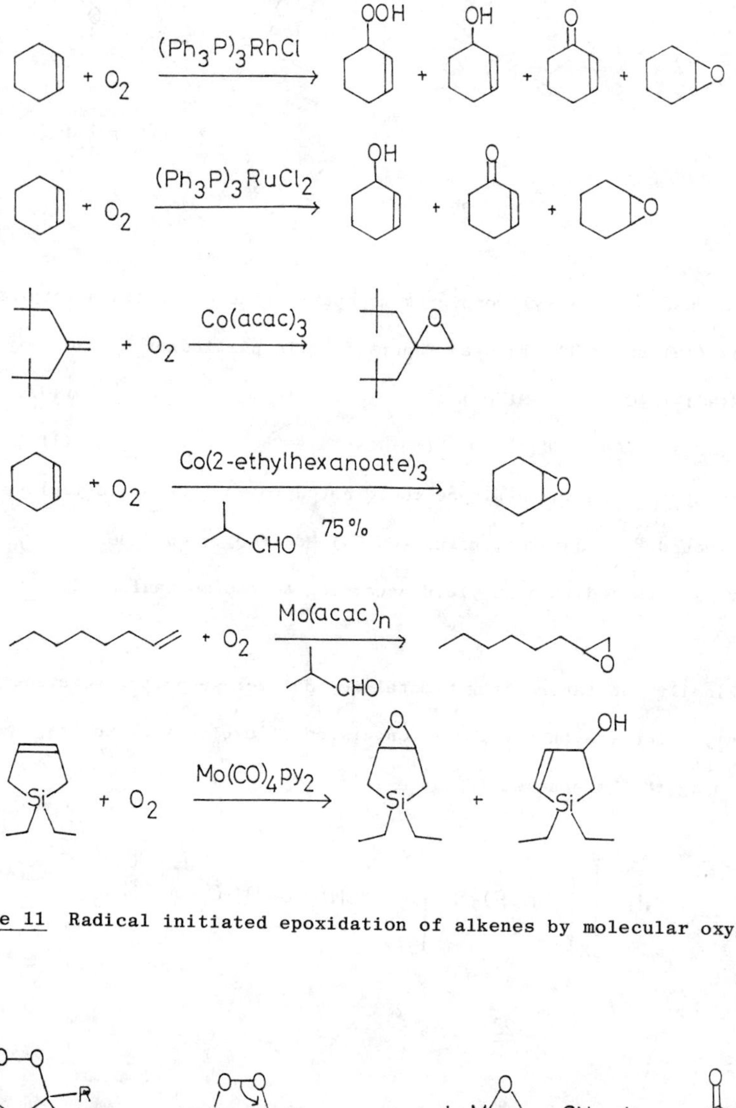

Ref.

(Ph₃P)₃RhCl → 55

$(Ph_3P)_3RhCl$ 55

$(Ph_3P)_3RuCl_2$ 56 -58

$Co(acac)_3$ 59

$Co(2-ethylhexanoate)_3$ 75%   60

$Mo(acac)_n$ 61

$Mo(CO)_4py_2$ 62

Figure 11 Radical initiated epoxidation of alkenes by molecular oxygen.

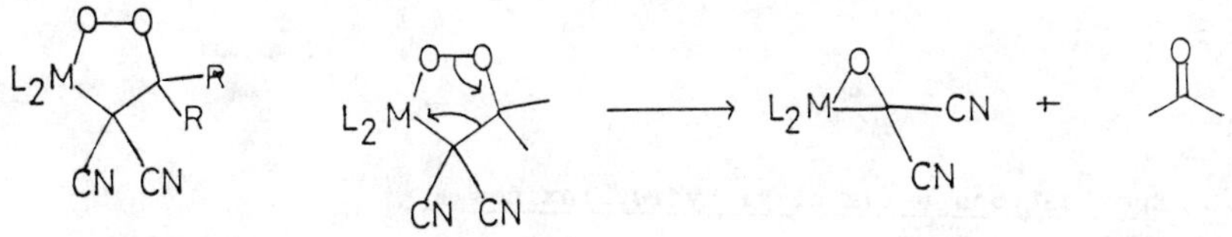

25

26

$$\underline{27}$$

The use of nitrosyl complexes as epoxidation catalysts according to equations (10) and (11) has been reported. In particular, L_nCoNO_2 may

$$M(NO) + \tfrac{1}{2}O_2 \longrightarrow M(NO_2) \tag{10}$$

$$M(NO_2) + RCH = CH_2 \longrightarrow M(NO) + RCH \overset{O}{-} CH_2 \tag{11}$$

transfer an oxygen atom to an alkene coordinated to palladium, to yield a carbonyl compound.[66] However, using $(CH_3CN)_2PdCl(NO_2)$, epoxides or ketones may be obtained in high yield according to the mechanism of Figure 12.[67]

Finally, an interesting bimetallic, polymer supported species, $\underline{28}$, is a very selective catalyst for conversion of cyclohexene to its oxide according to the scheme of Figure 13.[68]

$$\underline{28}$$

6.2.3 Epoxidations using alkyl hydroperoxides

Reaction of t-butyl hydroperoxide with alkenes occurs in the presence of a variety of complexes of molybdenum, vanadium and titanium.[69,70] The selectivity of the reaction depends on avoiding

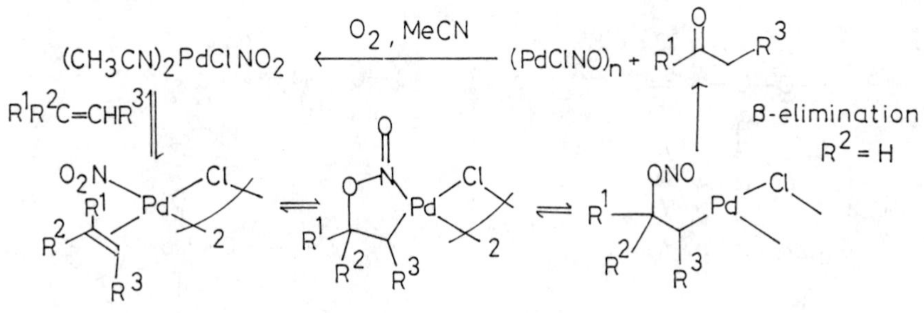

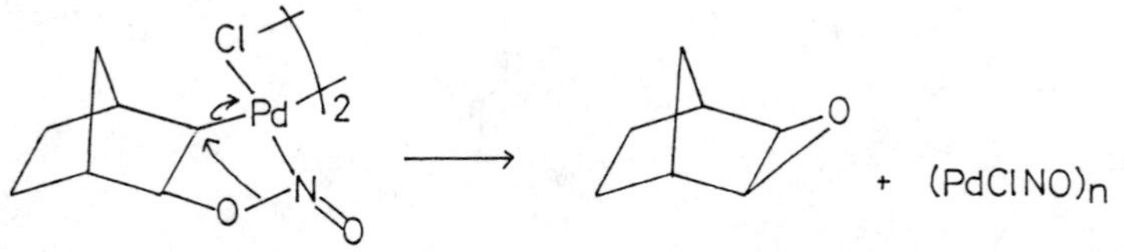

Figure 12   Oxygen atom transfer from a metal nitro complex to an alkene.

homolytic reactions of the hydroperoxide (equations (12) and (13)). Such

reactions are the characteristic ones in the presence of Co(III) and

$$M^{n+} + ROOH \longrightarrow M^{(n-1)+} + RO_2 \cdot + H^+ \qquad (12)$$

$$M^{(n-2)+} + ROOH \longrightarrow M^{n+} + RO \cdot + HO^- \qquad (13)$$

Mn(III), and iron, rhodium, nickel, platinum and copper species also give

poor yields of epoxides.   However, the reaction in the presence of

vanadium and molybdenum complexes involves formation of 29, followed by

heterolysis of the O-O bond.

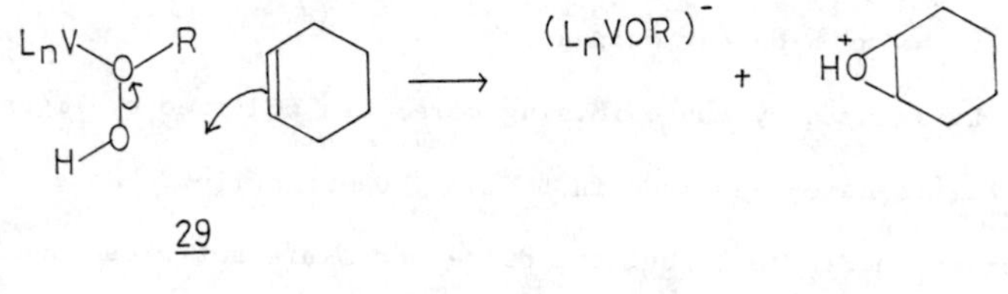

29

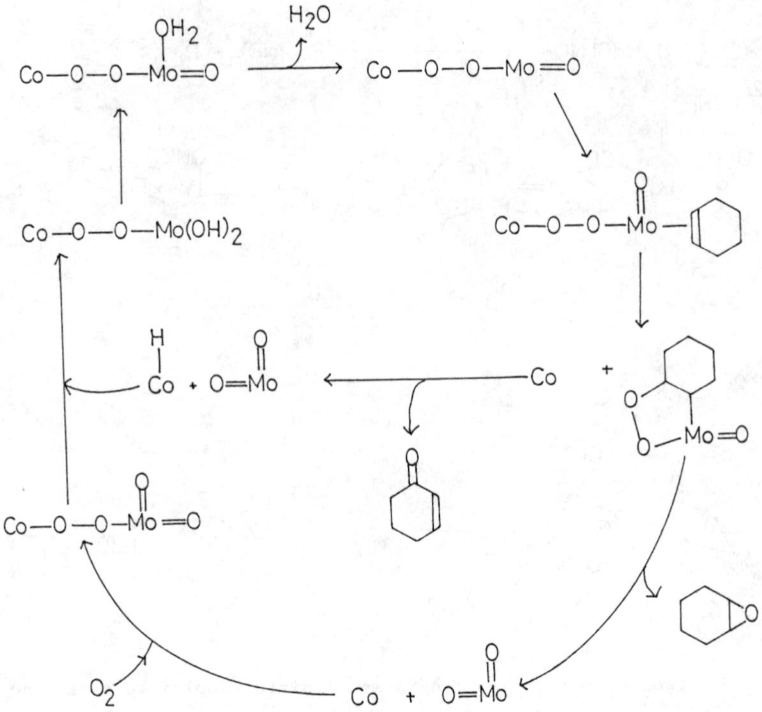

Co—O—O—Mo=O (with OH₂ above Mo)   → H₂O → Co—O—O—Mo=O

Co—O—O—Mo(OH)₂

Co—O—O—Mo— (cyclohexene)

H
Co + O=Mo   ←   —Co   +

(cyclohexenone)

Co—O—O—Mo=O

O₂   Co + O=Mo

(epoxide structure)

Figure 13 Mechanism of epoxidation of cyclohexene using a bimetallic, polymer supported catalyst.

6.2.3.1 Variation of the catalyst

Sheldon made a very thorough study of epoxidation of cyclohexene with t-BuOOH in the presence of a wide variety of catalysts. Some of the results are shown in Table 1.[71] In order to be both active and selective for epoxides the catalyst should be a fairly strong Lewis acid and a poor oxidant. These requirements are best met by Mo(VI) and W(VI); V(V) gives lower selectivity being a better oxidant.

Both Lewis acidity and oxidising power are modified by ligands and extensive trials have been made in the molybdenum series. Some examples are shown in Figure 14, but the differences are not great and may

Table 1

Catalysts for epoxidation of cyclohexene by t-BuOOH in benzene at 90°

Catalyst	Time (hrs)	Conversion (%)	Epoxide yield (%)
$Mo(CO)_6$	2	98	94
$MoO_2(acac)_2$	1	98	94
$W(CO)_6$	18	95	89
$Ti(OBu)_4$	20	80	66
$UO(acac)_2$	18	55	35
$VO(acac)_2$	2.5	96	13
$Nb(OBu)_4$	20	33	30

also be related to induction periods, solubility and rate of ligand degradation.   Displacement of one or more ligands is common and $Mo_2O_6(EDTA)$, where the ligands are very strongly bound, is inactive.[77]

Whilst other metals were generally less suitable for reactions of simple alkenes under Sheldon's conditions some excellent selectivities have been achieved by careful control of reaction parameters.   Most of the examples derive from the patent literature (Figure 15) so the effect of systematic variations of ligands and conditions is difficult to assess.

6.2.3.2 Variation of the hydroperoxide

t-Butyl hydroperoxide is commercially available and is the most widely used of the alkyl hydroperoxides.[83]   However, many others are accessible.   Alkylaromatic hydroperoxides are most reactive when the ring bears an electron withdrawing group on the aryl ring, since this enhances

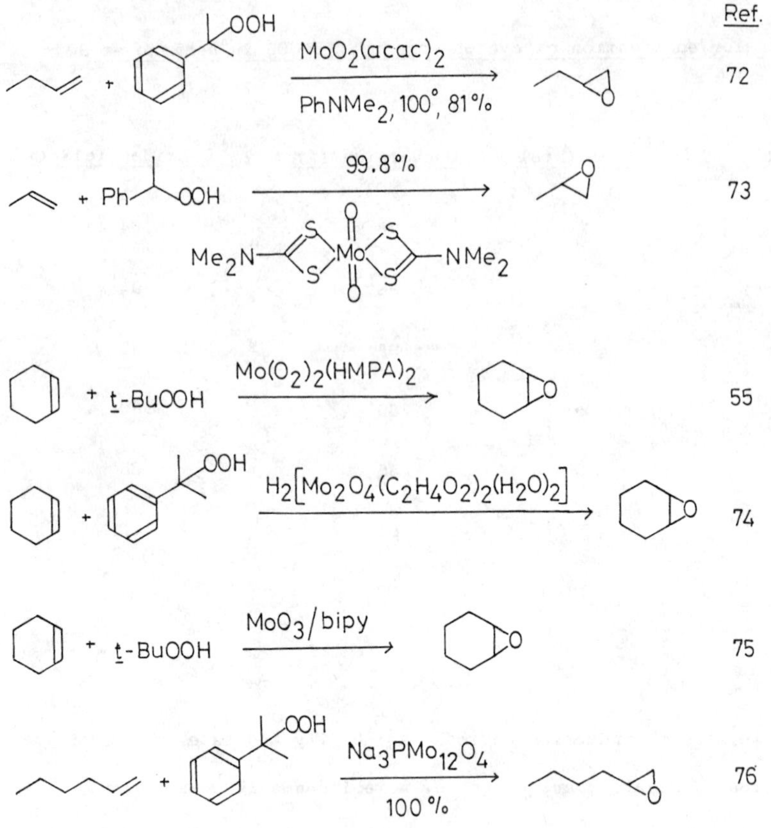

Figure 14 Molybdenum complexes as epoxidation catalysts.

the electrophilicity of the oxygen and stabilises the hydroperoxide against homolytic cleavage.[84,85]   The relative reactivities of ethylbenzene, t-butyl and cumene hydroperoxides have been reported to be 2.4:1.3:1.[86]

Some recent work has used peresters and perethers as alternatives to hydroperoxides (Figure 16).[87,88]

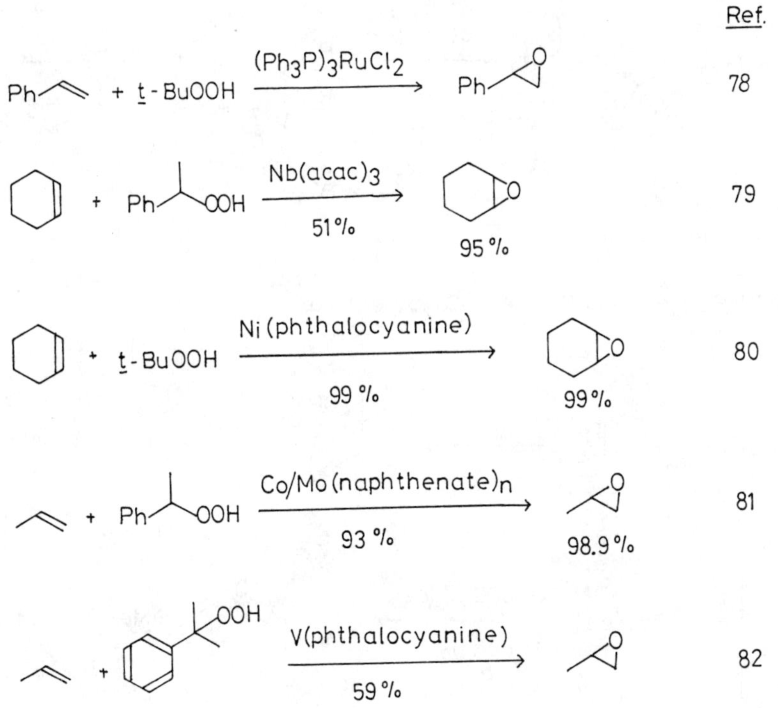

Ref.

Ph⌃⌃ + t-BuOOH → (Ph₃P)₃RuCl₂ → Ph⌃△O ... 78

$Ph\diagup\!\!\!\diagup$ + t-BuOOH $\xrightarrow{(Ph_3P)_3RuCl_2}$ Ph-epoxide   78

cyclohexene + Ph$\diagup$OOH $\xrightarrow[51\%]{Nb(acac)_3}$ cyclohexene oxide   79

95%

cyclohexene + t-BuOOH $\xrightarrow[99\%]{Ni(phthalocyanine)}$ cyclohexene oxide   80

99%

$\diagup\!\!\diagup$ + Ph$\diagup$OOH $\xrightarrow[93\%]{Co/Mo(naphthenate)_n}$ epoxide   81

98.9%

$\diagup\!\!\diagup$ + (cumene)OOH $\xrightarrow[59\%]{V(phthalocyanine)}$ epoxide   82

Figure 15 Catalytic epoxidation of simple alkenes.

6.2.3.3 By-products and side reactions

In order to achieve good selectivity in epoxidation it is necessary to supress homolytic cleavage of the hydroperoxide[89,90] to give the radicals $RO_2^{\cdot}$ and $RO^{\cdot}$. These may react with solvent or alkene to give the products of allylic oxidation. For example, cholesteryl acetate reacts with t-BuOOH to give both β-epoxide and allylic oxidation in the presence of Fe(acac)₃, whereas Mo(CO)₆ gives good selection for the epoxide.[91] The hydroperoxides formed may themselves epoxidise further alkene, giving complex mixtures.[92] The presence of radical scavengers usually supresses allylic oxidation.[93] Cleavage of the epoxide to give diols may provide a further complication.

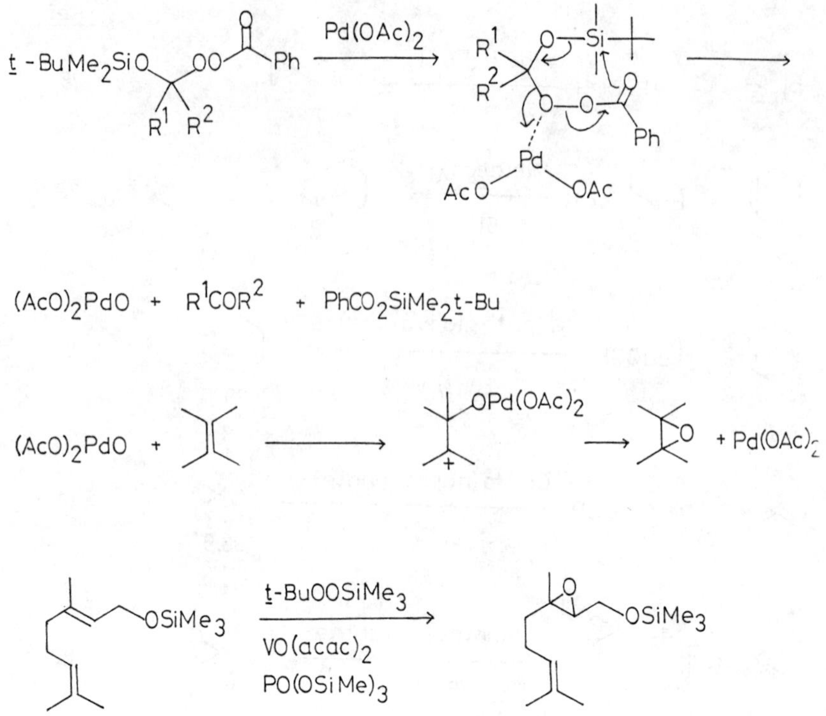

Figure 16 Catalytic epoxidation using peresters and perethers.

6.2.3.4 Variation of alkene structure

Molybdenum catalysts

Extensive kinetic studies have established the route of Figure 17 for alkene epoxidation in the presence of metal catalysts. Such a scheme, which exists in several forms, explains the autoretardation by excess alcohol, the extent of this depending on the equilibrium constants for the various steps. Molybdenum complexes are not particularly badly affected by such autoretardation.[84,89] Exchange rates for ROH, ROOH and cyclohexene in the coordination sphere of $H_2[Mo_2O_4(oxalate)_2(H_2O)_2]$ have been measured.[94]

(1) Activation of the catalyst by oxidation and/or dissolution of a precursor.

(2) $ROOH + M^{n+} \rightleftharpoons M^{n+}(ROOH)$

(3) Rate controlling step

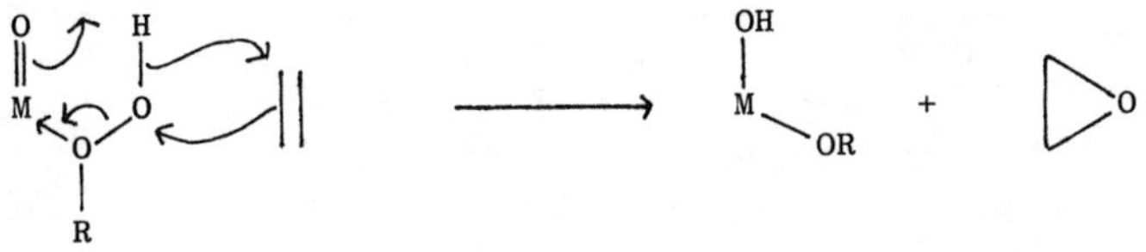

$$\begin{array}{c} O \\ \| \\ M-O \\ | \\ O \\ | \\ R \end{array} \cdots H \cdots O \quad \longrightarrow \quad \begin{array}{c} OH \\ | \\ M \\ \diagdown OR \end{array} \quad + \quad \triangle O$$

or variant

(4) $$\begin{array}{c} OH \\ | \\ M \\ \diagdown OR \end{array} \rightleftharpoons \begin{array}{c} O \\ \| \\ M \end{array} \quad + \quad ROH$$

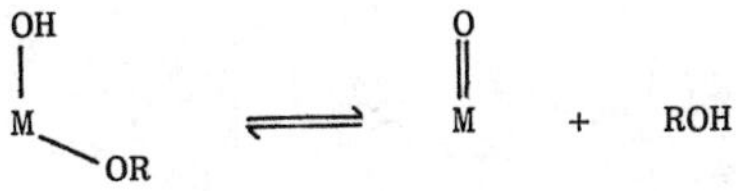

(5) $$\begin{array}{c} O \\ \| \\ M \end{array} \quad + \quad ROOH \rightleftharpoons \begin{array}{c} O \quad H \\ \| \quad | \\ M \quad O \\ \diagdown O \diagup \\ | \\ R \end{array}$$

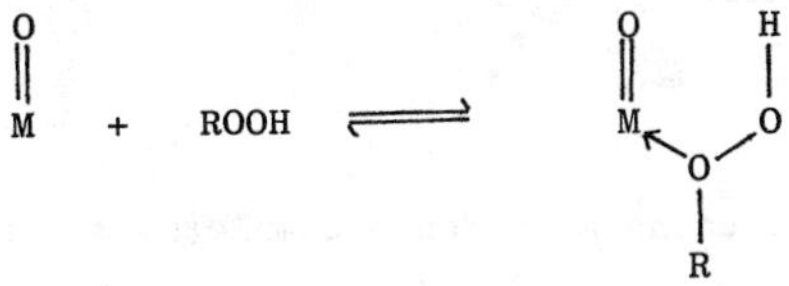

Figure 17 Mechanism of metal catalysed epoxidation of alkenes.

        Precoordination of the alkene to the metal is also possible and rates increase with the nucleophilicity of the alkene. This may be attributed either to easier coordination or to the electrophilic nature of the rate controlling step.[90,95] Molybdenum complexes are usually the catalysts of choice for simple unfunctionalised alkenes (Figures 18 and 19). A disubstituted double bond reacts selectively in the presence of a mono-substituted one (30 to 31).[113]

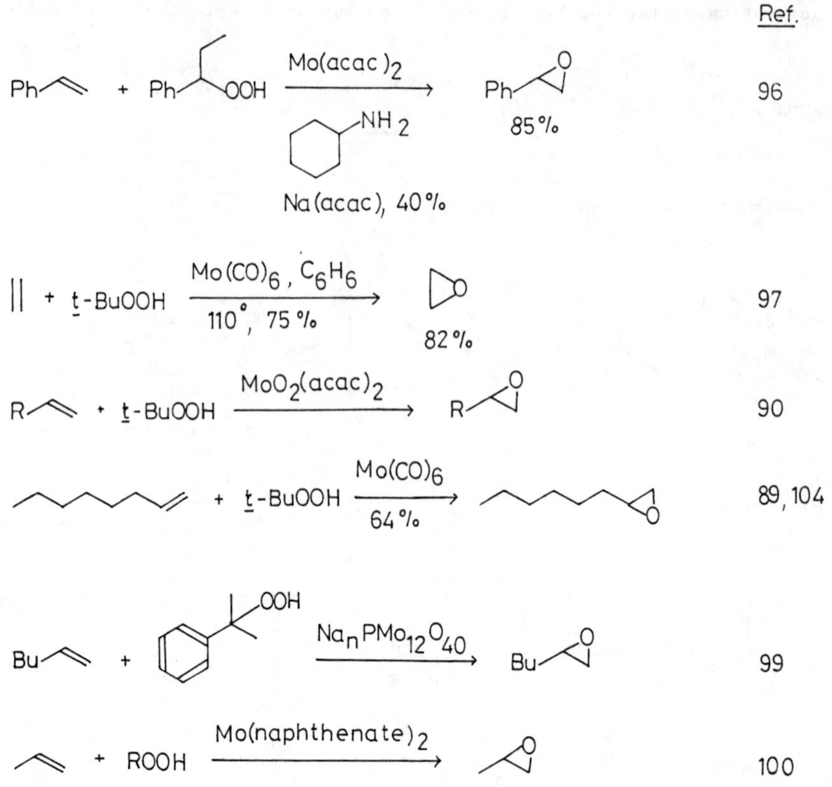

Ref.

Ph⟍⟋ + Ph⟍CH(CH₂CH₃)OOH →[Mo(acac)₂] [cyclohexyl-NH₂] [Na(acac), 40%] Ph⟍▵O 85%   96

|| + t-BuOOH →[Mo(CO)₆, C₆H₆] [110°, 75%] ▷O 82%   97

R⟍⟋ + t-BuOOH →[MoO₂(acac)₂] R⟍▵O   90

⟍⟍⟍⟍⟍⟋ + t-BuOOH →[Mo(CO)₆] [64%] ⟍⟍⟍⟍⟍▵O   89, 104

Bu⟍⟋ + [(CH₃)₂C(OOH)Ph] →[Na_n PMo₁₂O₄₀] Bu⟍▵O   99

⟍⟋ + ROOH →[Mo(naphthenate)₂] ▵O   100

Figure 18 Epoxidation of 1-alkenes using molybdenum complexes as catalysts.

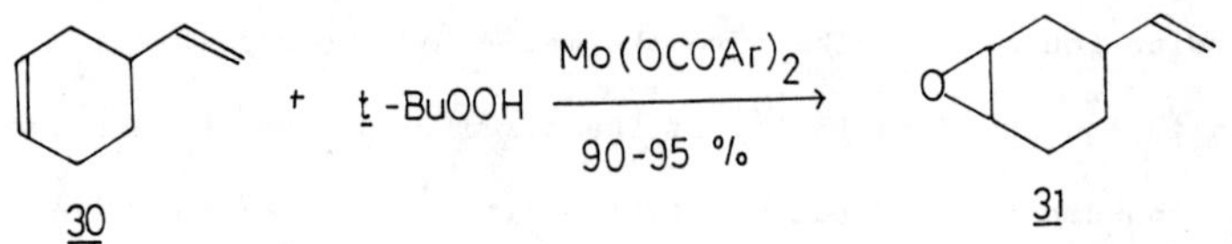

[cyclohexenyl-vinyl] 30 + t-BuOOH →[Mo(OCOAr)₂] [90-95 %] [epoxide] 31

$$H_2(Mo_2O_4(oxalate)_2(H_2O)_2)$$ 101

$$MoO_2(oxine)_2$$ 102

$$MoO_5(Bu_3PO)_2$$ 103

$$H_2MoO(diphos)Cl_5$$ 104

$$Mo_2O_4(OH)_2(OPh)_2$$ 105

$$(TPP)MoO(OMe)$$ 106

$$MoCl_2(NO)_2L_2$$ 107

$$Mo(CO)_6$$ 108

$$MoO_2(acac)_2$$ 109

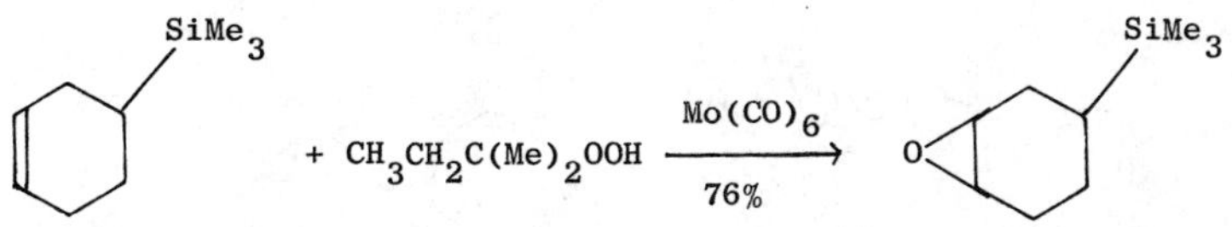

110

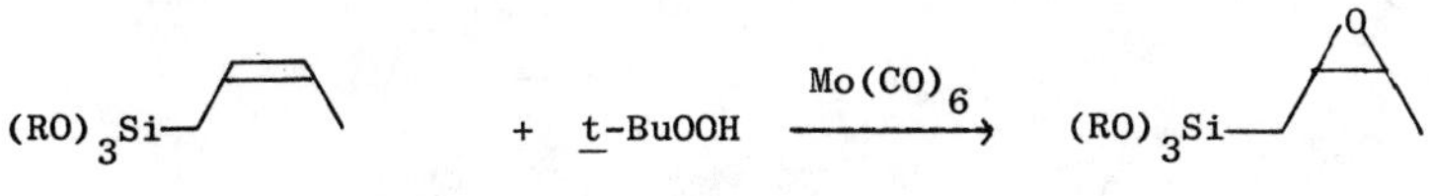

111

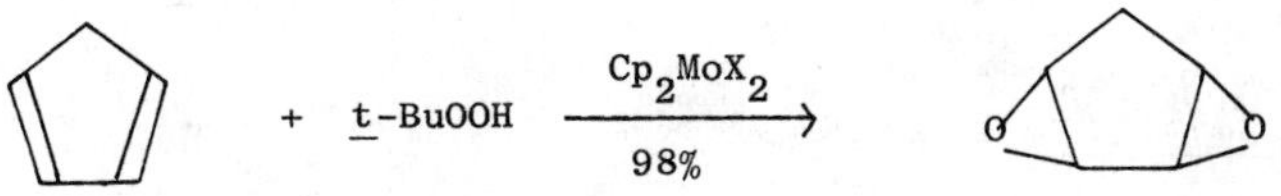

112

Figure 19   Epoxidation of disubstituted alkenes using molybdenum complexes as catalysts.

In the presence of electron withdrawing substituents in the allylic position epoxidation is retarded. For example, allyl chlorides are epoxidised 10 times more slowly than alkenes.[90,114,115] Cyano groups retard the reaction when in the allylic position (reactions of 32 and 34) or inhibit it in acrylonitrile.[84,116] Allyl alcohols give poor reactions in the presence of molybdenum catalysts since they do not complex as well as with the vanadium species. Molybdenum complexes are the catalysts of choice for particularly electron poor double bonds (Figure 20) but enol ethers and esters are also smoothly epoxidised (36 to 37).[119]

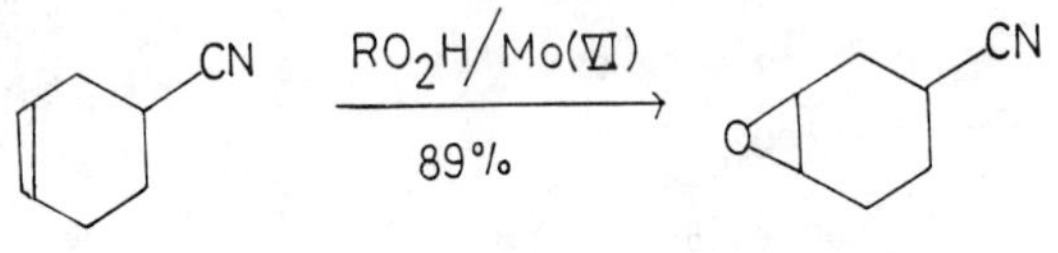
CN
RO2H/Mo(VI)
89%
CN

32 33

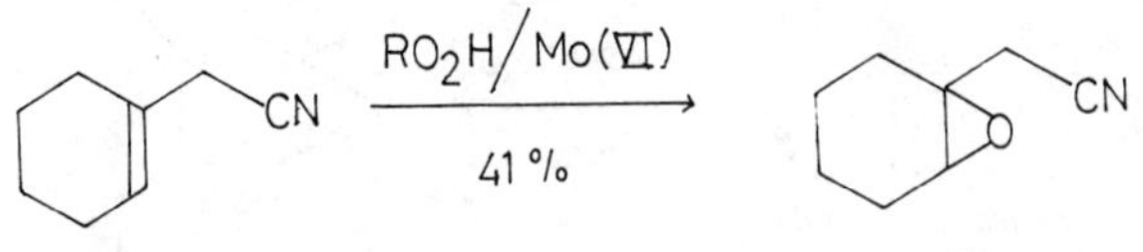
CN
RO2H/Mo(VI)
41 %
CN

34 35

Ref.

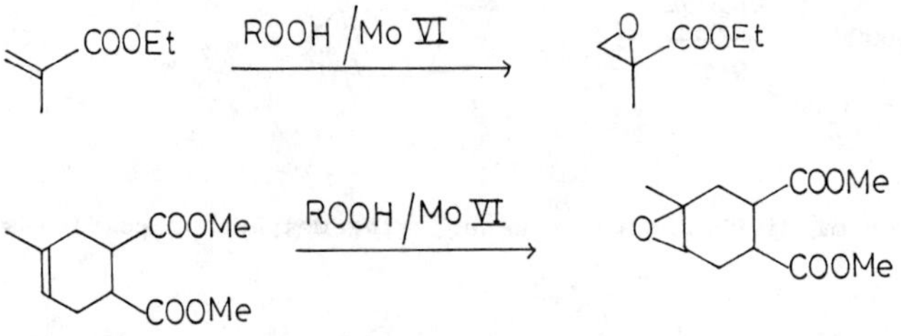
COOEt
ROOH/Mo VI
COOEt
COOMe
ROOH/Mo VI
COOMe
COOMe
COOMe

117

118

Figure 20 Molybdenum somplexes as catalysts for epoxidation of electron poor double bonds.

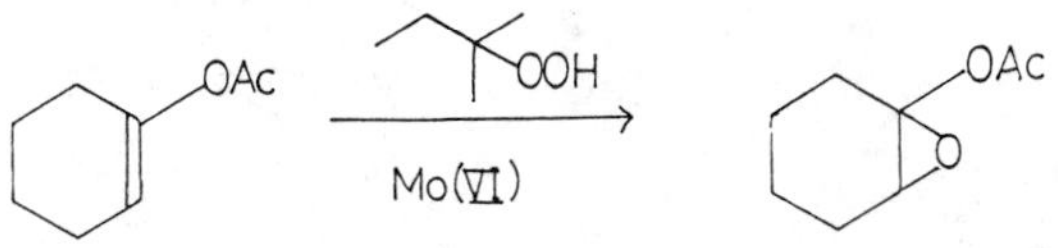

36   37

Vanadium complexes

Whilst there are numerous examples of the use of vanadium complexes to catalyse the epoxidation of simple alkenes (Figure 21), they are rarely the catalysts of choice since molybdenum analogues provide higher rates and are less subject to autoretardation. However, with allyl alcohols both rates and yields are substantially superior with vanadium. The transition state, 38, involves complexation of the alcohol to vanadium in the correct orientation for oxygen delivery.[124] Thus, in molecules containing more than one type of double bond (39 and 41), that which has an allylic OH is epoxidised with excellent selectivity.[125,126] Some further examples are shown in Figure 22.

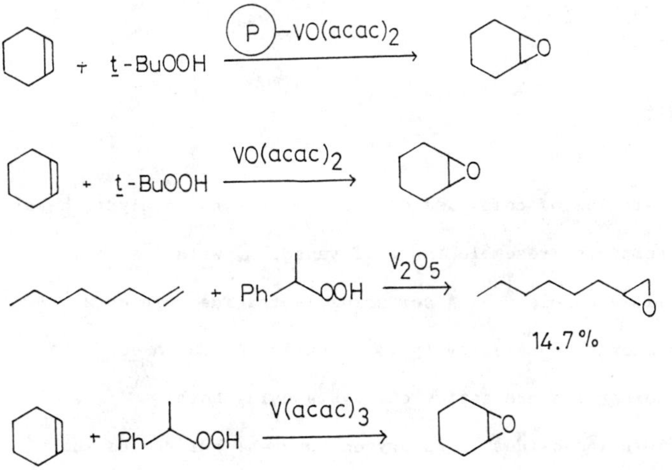

Ref

120,121

122

123

14.7%

124

Figure 21 Vanadium complexes as catalysts for epoxidation of simple alkenes.

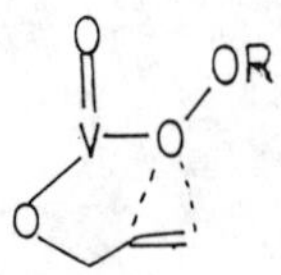
OR
V
O

or

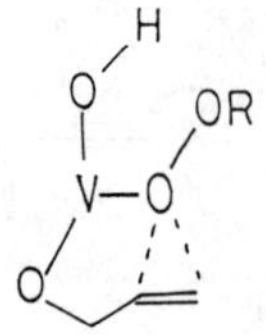
H
O
OR
V
O

38a 38b

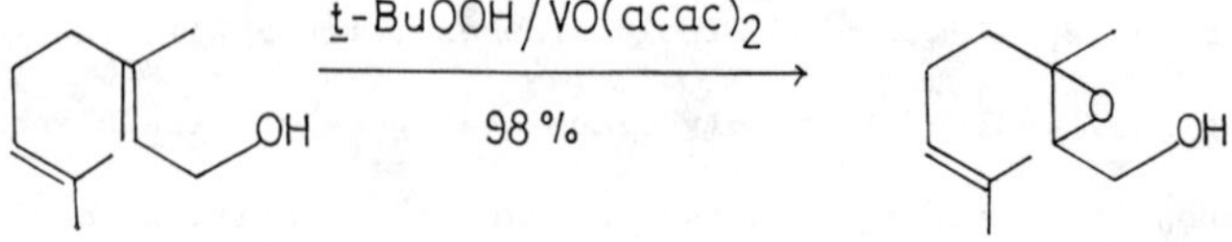
t-BuOOH/VO(acac)2
98%
OH
OH

39 40

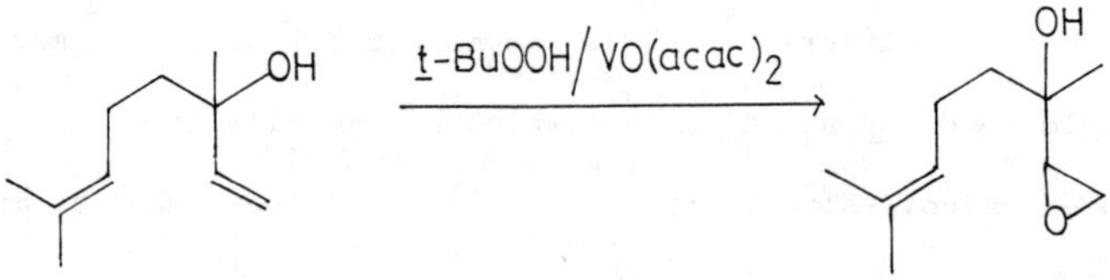
t-BuOOH/VO(acac)2
OH
OH

41 42

Other metals

Systematic studies of complexes of other metals as catalysts have been few. Titanium complexes resemble those of vanadium, with a strong preference for allylic alcohols.[130] A series of lanthanide complexes (Y, Sm and Lu) also show some selectivity in oxidation of 39 to 40.[131] Manganese and iron porphyrins are active catalysts using both hydroperoxides[132] and amine oxides[133] as oxygen sources, but yields and selectivities are low and the mechanism of action is quite different.

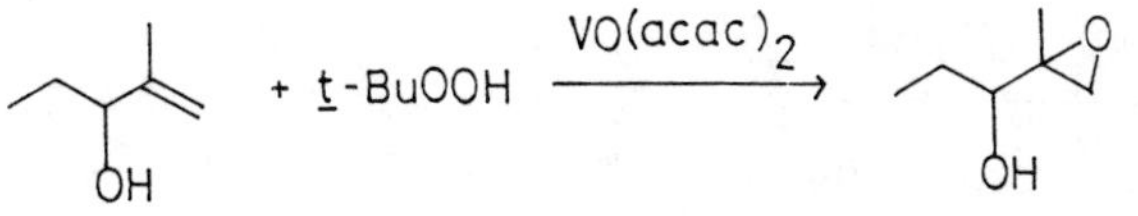

127

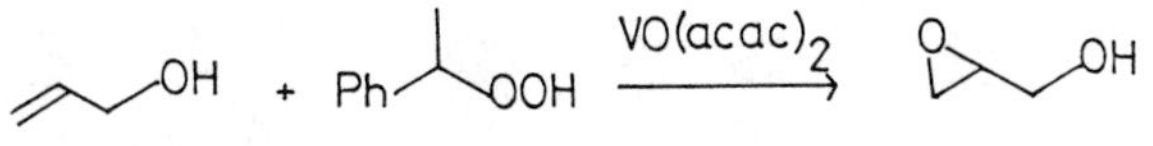

128

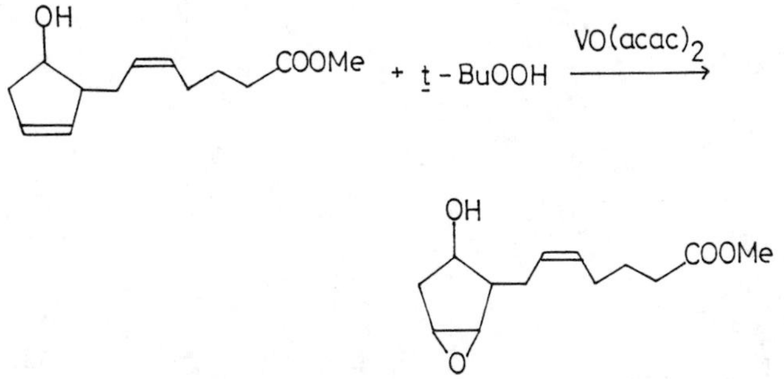

129

Figure 22 Vanadium complexes as catalysts for epoxidation of allyl alcohols.

6.2.3.5 Stereoselectivity

The stereochemistry of epoxidation of alkenes not bearing polar
functional groups is determined by steric factors, with the epoxidising
agent attacking from the less hindered side of the double bond. Some examples
are shown in Figure 23; stereoselectivities are only slightly better than
those obtained using peracids.

In contrast, alkenes with polar functional groups show different
selectivities when epoxidised with metal/ROOH than those noted with
peracids. Some examples using molybdenum complexes are shown in Figure 24;
in all cases attack is from the face of the molecule bearing the functional
group, arguing that this is coordinated to the metal. Curiously, the last

482

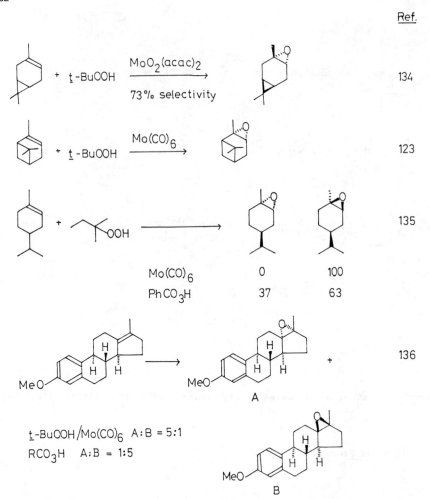

Figure 23 Stereoselective epoxidation of simple alkenes.

example, which does not employ ROOH, gives selectivity dependent only on steric factors.

More interesting, however, are the reactions of allyl alcohols, particularly those catalysed by vanadium complexes. A further example of the regioselectivity of this reaction is the epoxidation of 43, a crucial step in a juvenile hormone synthesis.[126] The syn-directive effect of the

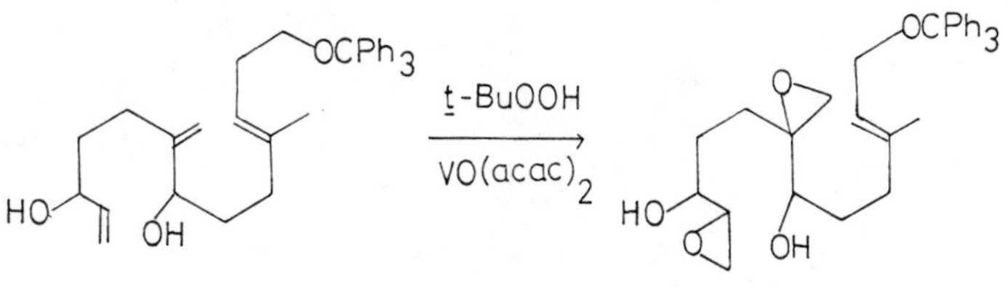

$$\underset{\underline{43}}{} \xrightarrow[\text{VO(acac)}_2]{t\text{-BuOOH}} \underset{\underline{44}}{}$$

OH-group was known from early work using peracids, but is, in this case,

relatively weak.[140] With VO(acac)$_2$/t-BuOOH, however, the effect is strong

and reactions are essentially stereospecific. Figure 25 shows the

selectivities with cyclohexenols, the effect being most pronounced for

the homoallylic alcohol.[125] Later work suggests that selectivity decreases

as alkene conversion increases, particularly with molybdenum complexes,

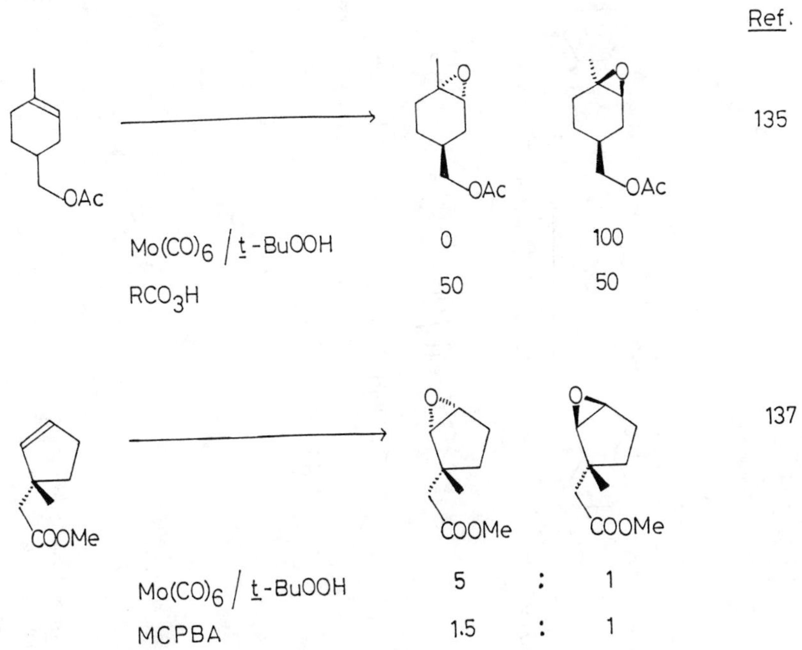

Ref.

135

Mo(CO)$_6$ / t-BuOOH   0   100

RCO$_3$H   50   50

137

Mo(CO)$_6$ / t-BuOOH   5   :   1

MCPBA   1.5   :   1

Figure 24 Epoxidation of alkenes bearing polar functional groups in the
presence of molybdenum complexes.

Figure 24 cntd.

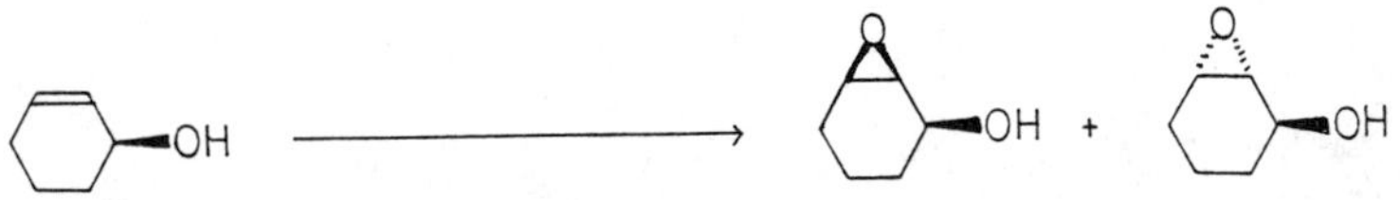

t-BuOOH / Mo(CO)$_6$	98	2
t-BuOOH / VO(acac)$_2$	98	2
PhCO$_3$H	92	8

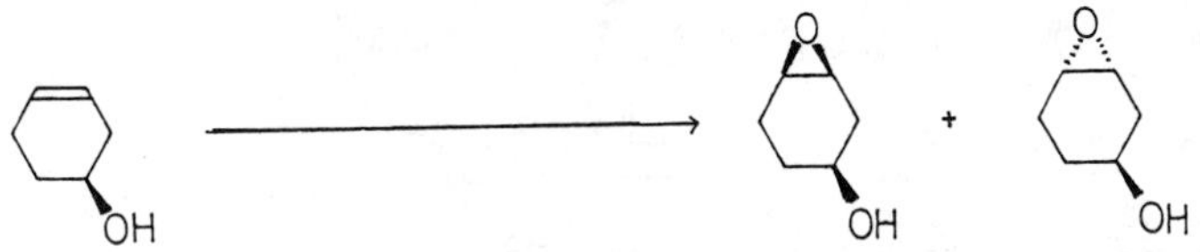

t-BuOOH / Mo(CO)$_6$	98	2
t-BuOOH / VO(acac)$_2$	98	2
PhCO$_3$H	60	40

Figure 25   Stereoselective epoxidation of cyclohexenols.

due to blocking of coordination by the product alcohol.[141]   The syn-
directing effect results from preferential transfer of oxygen to one face
of a double bond within a metal/substrate/hydroperoxide complex[142]; the
complex, 45, which has been defined by X-ray crystallography, is a useful
model.[143]   Careful studies of cyclic allylic alcohols reinforce this point.
With 5- and 6-membered rings both t-BuOOH/VO(acac)$_2$ and MCPBA give mainly
syn-epoxide. With medium rings t-BuOOH/VO(acac)$_2$ still gives syn-products
whereas MCPBA gives anti-compounds[144].   The different geometries were
rationalised on the basis of different sized rings in the transition
states.   With 46, the reaction is more complex, resulting in transannular
reactions as well as   epoxidation.   It appears that the correct orientation
for epoxide formation is not available without strain to the vanadium

45

46		47	48	49	50
$MoO_2(acac)_2/\underline{t}$-BuOOH		52		48	4
$VO(acac)_2/\underline{t}$-BuOOH		19		21	60
MCPBA		46		54	0.5

complex, and dehydrogenation occurs. With 2,4-dienols, such as 51, vanadium catalysed epoxidation leads to the formation of 53 by trans-annular rearrangement of the initially formed syn-epoxide.[145]

Stereoselectivity is not restricted to cyclic systems and recently epoxidation has been used to establish several contiguous centres of defined relative stereochemistry. In the epoxidation of allyl alcohols, erythro-epoxyalcohols predominate to an extent which depends on conditions, catalyst and substituents. The results of several studies are shown in Table 2. Recently it has been shown that the presence of a silyl

Table 2

Stereoselectivity in epoxidation of allyl alcohols

Alcohol	Conditions	% threo	% erythro	Ref.
	t-BuOOH/VO(acac)$_2$	20	80	146
	t-BuOOH/Mo(CO)$_6$	44	56	146
	MCPBA	60	40	146
	t-BuOOH/VO(acac)$_2$	5	95	146
	t-BuOOH/Mo(CO)$_6$	16	84	146
	MCPBA	45	55	146
	t-BuOOH/VO(acac)$_2$	71	29	146
	t-BuOOH/Mo(CO)$_6$	84	16	142
	MCPBA	95	5	142
54	t-BuOOH/VO(acac)$_2$	34	66	147
54	t-BuOOH/Mo(CO)$_6$	61	39	147
54	MCPBA	62	38	147

substituent enhances stereoselectivity and may readily be removed with
retention of configuration. Thus, 55 is epoxidised with 75% selectivity
for the erythro-epoxyalcohol by t-BuOOH/VO(acac)₂ but the silylated
analogue, 56, gives 57 in 99% selectivity.[148]

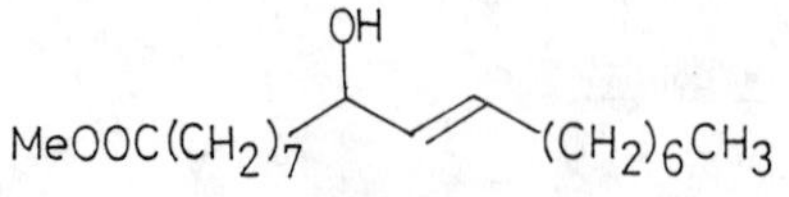

54

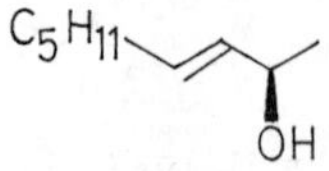

55

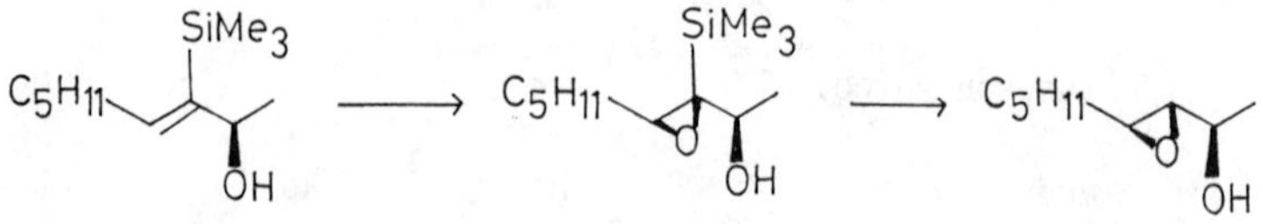

56 → 57 → 58

Whilst modest selectivities are obtained even using peracids with allyl alcohols, homoallylic alcohols are stereoselectively epoxidised only in metal catalysed reactions. Table 3 shows the results with some simple systems. A transition state, 63, which minimises steric interactions between substituents has been proposed. Much use has been made of such epoxidations in complex syntheses (Figure 26).

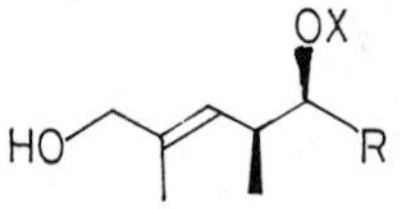

59

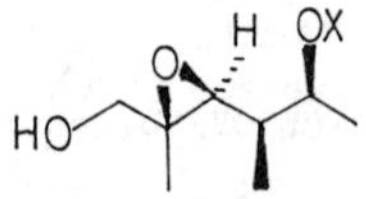

60

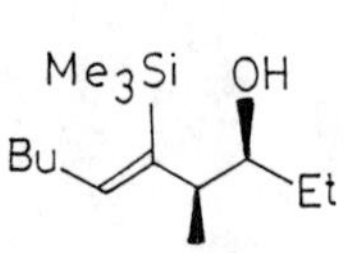

61

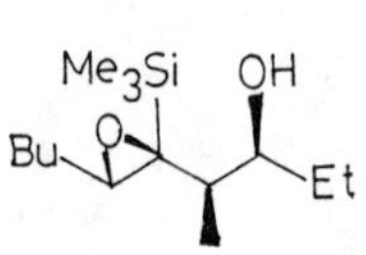

62

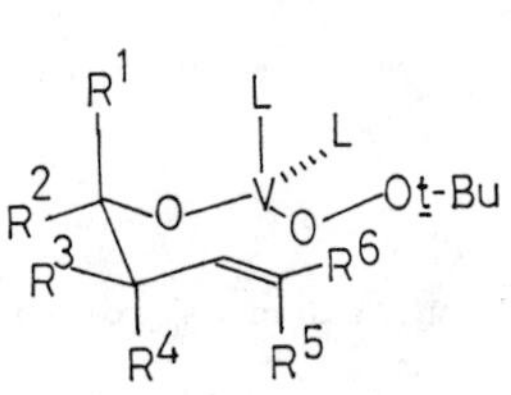

63

Table 3

Stereoselective epoxidation of homoallylic alcohols

Alcohol	Conditions	Major product	Diastereomer Ratio	Ref.
	t-BuOOH/VO(acac)$_2$		>400	149
	t-BuOOH/VO(acac)$_2$		>40	149
	t-BuOOH/VO(acac)$_2$		4.6	149
59, X = SiMe$_2$t-Bu, R=(CH$_2$)$_4$OSiMe$_2$t-Bu	t-BuOOH/VO(acac)$_2$	60	1.5	130
	MCPBA		<0.04	130
	t-BuOOH/Ti(OiPr)$_4$		10	130
61	t-BuOOH/VO(acac)$_2$	62	sole product reported	150

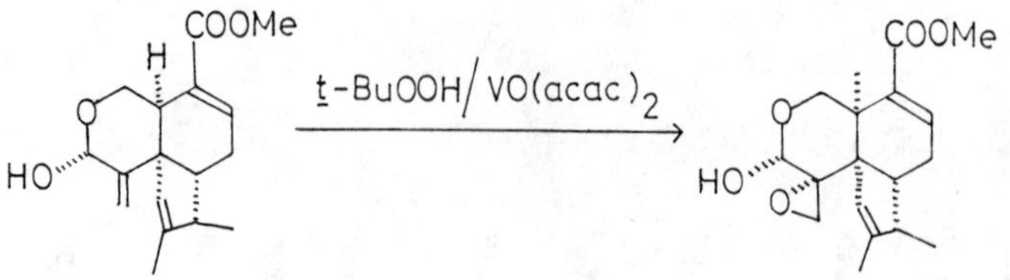

151

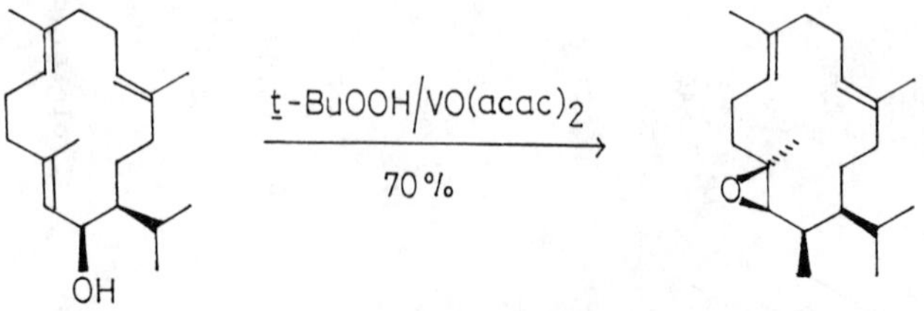

152

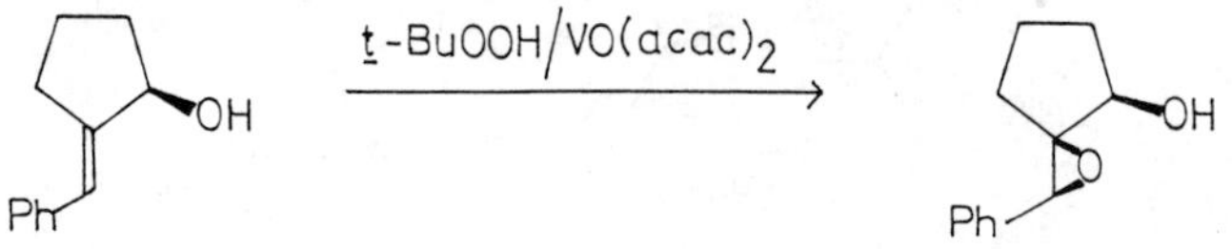

153

Figure 26 Uses of stereoselective epoxidations in synthesis.

6.2.4 Epoxidations using Iodosobenzene.

Oxidation using iodosobenzene, PhIO, is frequently proposed to proceed _via_ the mechanism of equations (14) and (15) and reactions are confined to metals which can form stable M=O species.[154] Among these are

$$M + PhIO \longrightarrow M=O + PhI \qquad (14)$$

$$M=O + substrate \longrightarrow M + substrate(O) \qquad (15)$$

tetraphenylporphyrin (TPP) complexes of iron, manganese and chromium which are models for the active site of Cytochrome P450.[155]

PhIO in the presence of Fe(TPP)Cl rapidly epoxidises _cis_-stilbene

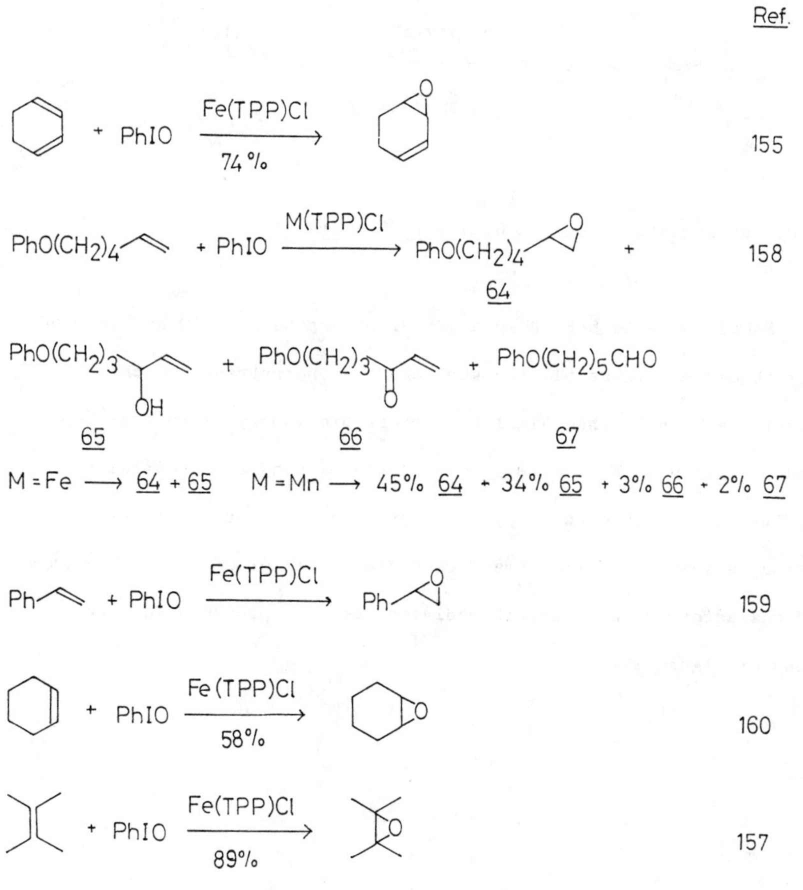

Ref.

Figure 27 Catalytic epoxidations using iodosobenzene.

to cis-stilbene oxide. By contrast, trans-stilbene is unreactive.[156] The
relative reactivity of alkenes depends both on steric and electronic
factors with rates in the order p-MeOC$_6$H$_4$CH=CH$_2$ > PhCH=CH$_2$ >
p-O$_2$NC$_6$H$_4$CH=CH$_2$.[157] There is also a substantial preference for
trisubstituted double bonds, confirming that the mechanism is electrophilic.
Some examples are shown in Figure 27. Products originating from radicals
are thought to be formed from the Fe(IV)-O· form of the catalyst. Cu^{2+}
has also been used as a catalyst [161] and the reaction of PhI=NTs with
alkenes to give aziridines, **68**, is entirely analogous.[162]

Ph⌕ + PhI=NTs → Mn(TPP)Cl → Ph⌕N—Ts

80%

68

6.2.5 Epoxidations using sodium hypochlorite.

NaOCl seems to behave in a manner analogous to PhIO and is used to generate oxo complexes of iron and manganese porphyrins[163]. The subsequent reaction is then similar to those previously discussed. Some examples are given in Figure 28. The presence of base is essential to selectivity; in the absence of pyridine substantial amounts of the products of allylic oxidation are formed from an intermediate with little oxenoid character. Metal oxo-intermediates are also proposed for the reactions of Figure 29.

Ref.

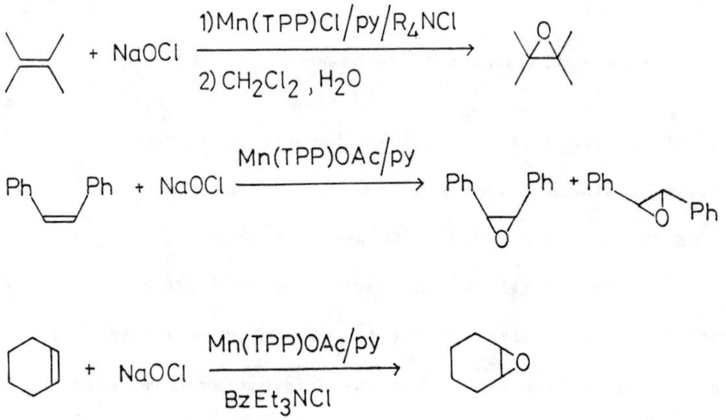

1) Mn(TPP)Cl/py/R_4NCl
+ NaOCl →
2) CH_2Cl_2, H_2O

164

Mn(TPP)OAc/py
Ph⌕Ph + NaOCl → Ph⌕Ph + Ph⌕Ph

165

Mn(TPP)OAc/py
+ NaOCl →
BzEt$_3$NCl

166

Figure 28 Catalytic epoxidations using sodium hypochlorite.

Ph⌒Ph $\xrightarrow[\text{CH}_2\text{Cl}_2/\text{H}_2\text{O}]{\text{RuCl}_3/\text{NaIO}_4}$ Ph△Ph   167

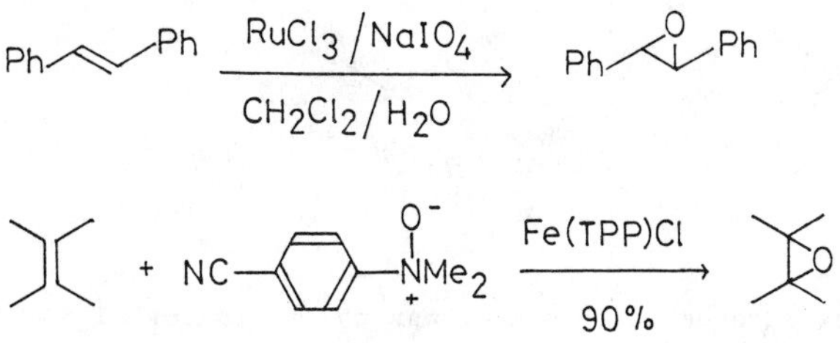

（＋NC─⟨⟩─NMe₂）$\xrightarrow[\text{90\%}]{\text{Fe(TPP)Cl}}$ ⟨epoxide⟩   168

Figure 29 Other catalytic epoxidations using metal oxo intermediates.

6.2.6 Asymmetric epoxidation

The demand for chiral compounds of good optical purity has also touched the field of epoxidation. Chiral peracids were studied some years ago but optical yields were generally poor.[169] The first studies with transition metal complexes involved allyl alcohols and **69** with cumene hydroperoxide as oxidant. Enantiomer excesses ranged from 13 - 33%.[170,171] Another group used the ligand, **70**, forming an *in situ* catalyst with $\text{MoO}_2(\text{acac})_2$ but optical yields were less than 50% and the ligand was rapidly degraded.[172] Hydroxamic acids such as **71** bind well to V and Mo and are not easily degraded but optical yields in epoxidation of **72** do not exceed 50%.[173]

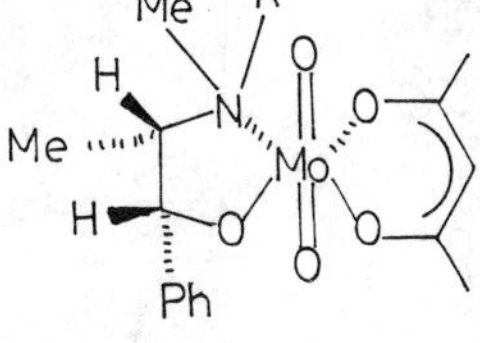

69 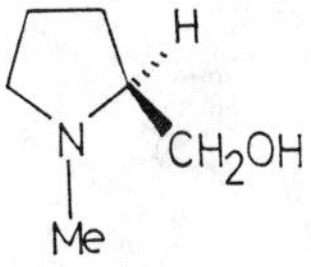   **70**

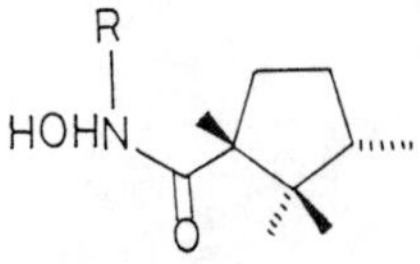

71 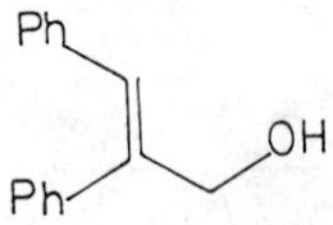   72

The most important advance in this area was the introduction of Ti(O-i-Pr)₄ as catalyst and dialkyl tartrates as chiral modifiers, and this method is now widely used for enantioselective epoxidation of allyl alcohols. Kinetic studies give the rate law of equation (16) but the

$$\text{Rate} = \frac{k[\underline{t}\text{-BuOOH}]\ [\text{Ti(OR)}_2(\text{tartrate})]\ [\text{allyl alcohol}]}{[\underline{t}\text{-BuOH}]} \qquad (16)$$

mechanism is complex. The active species, 73 (R = i-Pr, X = NCH₂Ph) and 74 (R = i-Pr, R' = Et) have been isolated and studied by X-ray crystallography.[174] Loss of OR and dissociation of the carbonyl groups exposes mer-positions on each titanium for coordination of alcohol, and a ternary hydroperoxide complex, 75, is thought to be formed. A very convincing rationalisation of the optical induction in a wide range of allyl alcohols may thus be obtained.[175] The direction of optical induction is determined by the chirality of the tartrate modifier so that both epoxide enantiomers may be readily obtained. Examples are shown in Figure 30.

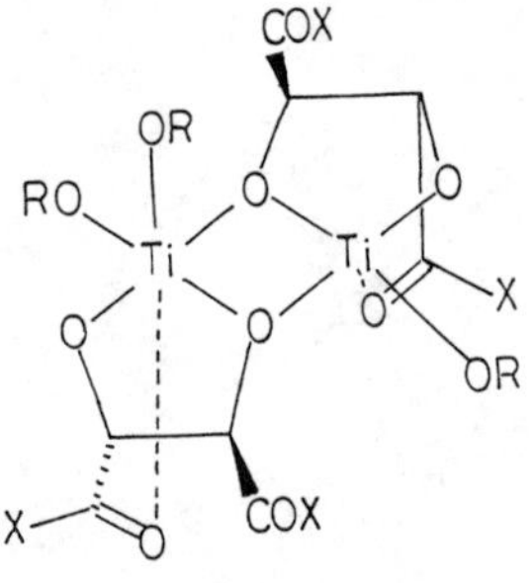

73 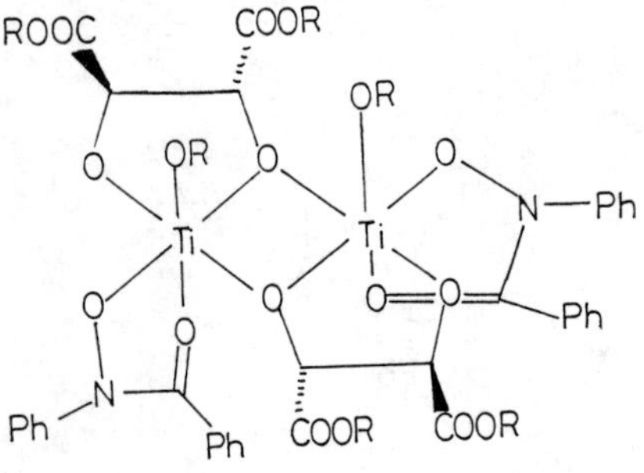   74

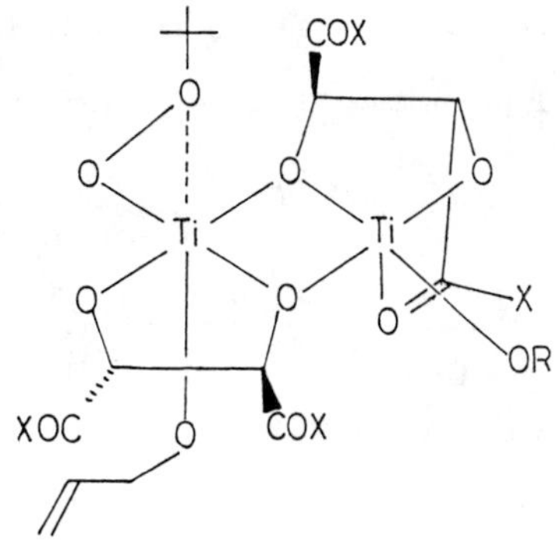
COX
O
O
O
O
Ti
Ti
O
O
O
X
OR
XOC
O
COX

75

Ref.

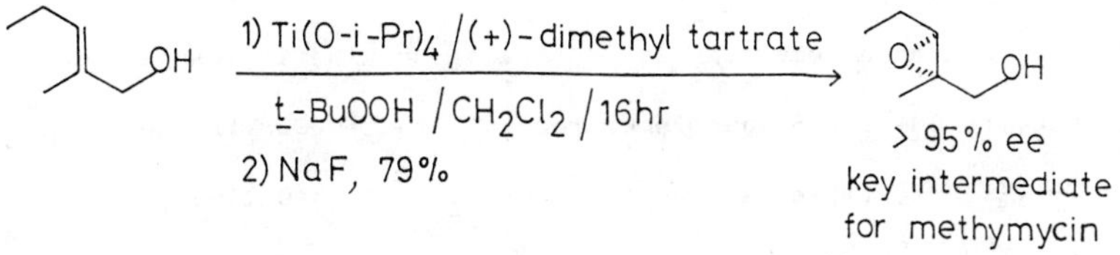
OH
1) Ti(O-i-Pr)4 /(+)-dimethyl tartrate
t-BuOOH / CH2Cl2 / 16hr
2) NaF, 79%
O
OH
> 95% ee
key intermediate
for methymycin

176

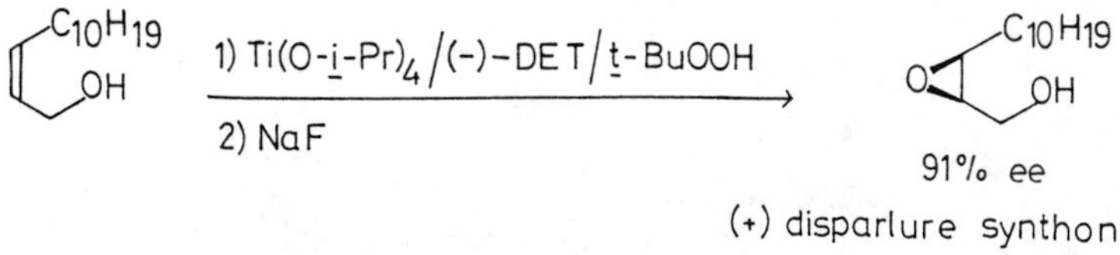
C10H19
OH
1) Ti(O-i-Pr)4 /(-)-DET / t-BuOOH
2) NaF
C10H19
O
OH
91% ee
(+) disparlure synthon

176

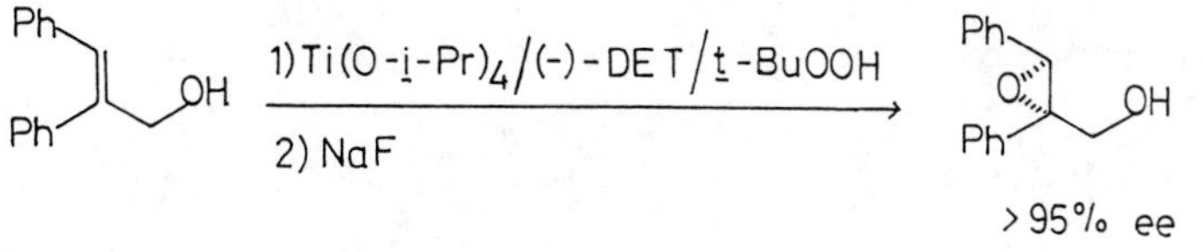
Ph
Ph
OH
1) Ti(O-i-Pr)4 /(-)-DET / t-BuOOH
2) NaF
Ph
O
OH
Ph
>95% ee

177

Figure 30 Asymmetric epoxidation using Ti(O-i-Pr)$_4$/dialkyl tartrate catalysts.

This reaction has been widely used as the chirality generating step in syntheses of natural products. 76 gives 77 in 88% optical purity;

this is a key step in the synthesis of 2,3-dihydro-2-isopropyl-2,5-dimethyl furan, a sex specific compound in the female of the beetle, Hylecoetus Dermestoides L.[178] A high but unspecified optical yield is obtained in oxidation of 78, a precursor of the ant venom alkaloid, 80.[179] With 81 enantiomer and enantioface discrimination occur to give mainly the 2S,3R- form of 82 in 91-95% enantiomer excess.[180] Analogous kinetic resolution occurs with racemic 83. The S-enantiomer reacts with t-BuOOH in the presence of Ti(O-i-Pr)$_4$/(+)-diisopropyltartrate some 15-140 times faster than the R-enantiomer. S-83 gives an erythro:threo ratio of 98:2 whilst that for R-83 is 38:62. High e.e. is obtained in the unreacted allyl alcohol (>96% for 84).[181]

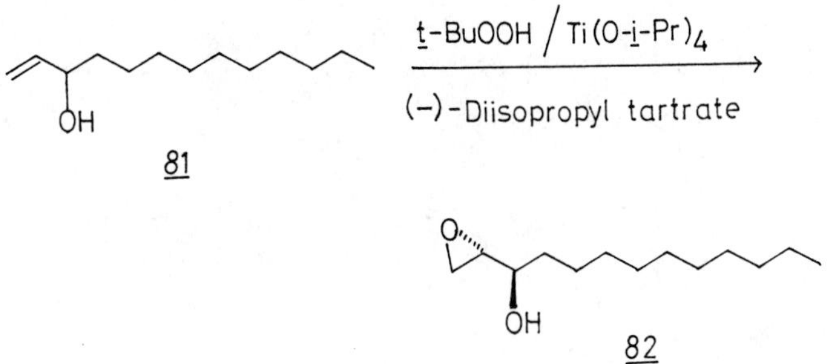

81

82

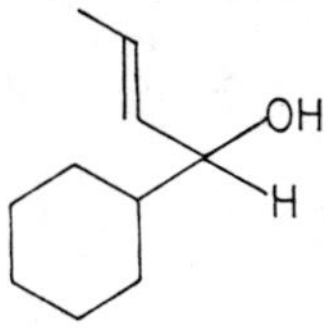

83

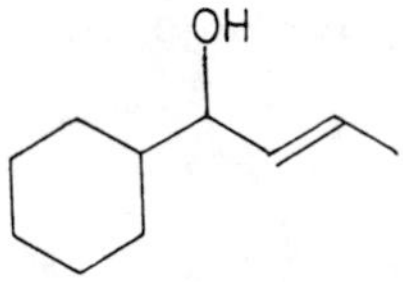

84

Sugars have been synthesised by similar reactions; 85 is epoxidised to give 86 and 87 in the ratio 99:1, and appropriate homologation leads to the aldohexoses [182]. In the steroid field, 88 is epoxidised in 98% ee according to the Sharpless rules, and this provides an asymmetric synthesis of 24R-24,25-dihydroxyvitamin D3 and 24S-24,25-dihydroxycholesterol [183].

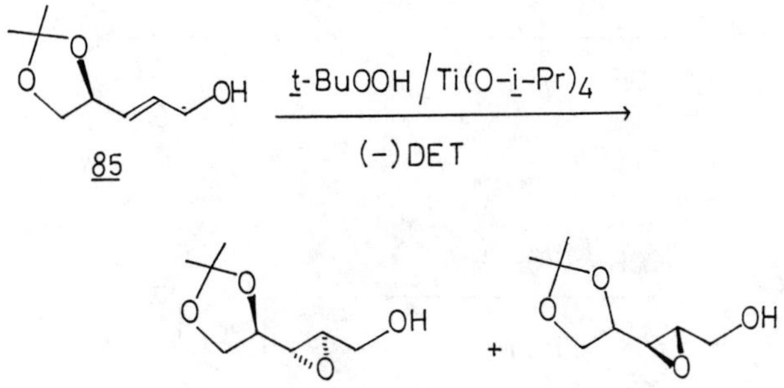

85

86   +   87

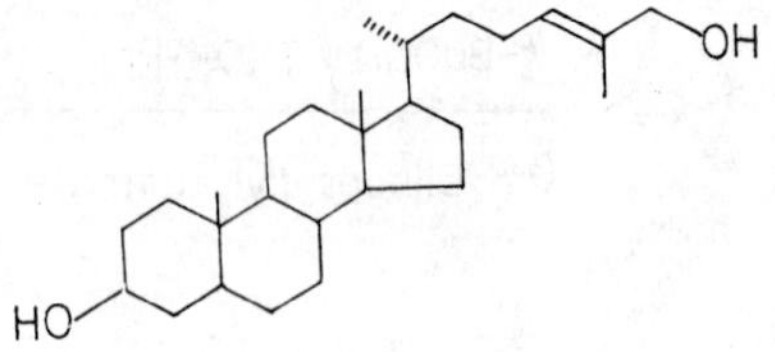

88

The reaction is limited in molecules containing an ester group.
Thus, 89 (X = COOMe) gives unsatisfactory results and the ester must be
masked as a 4,5-diphenyloxazolyl group. 90 is then epoxidised in good chemical
and optical yield.[184]

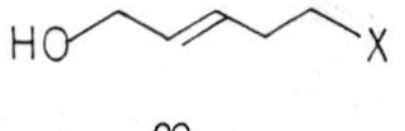

89

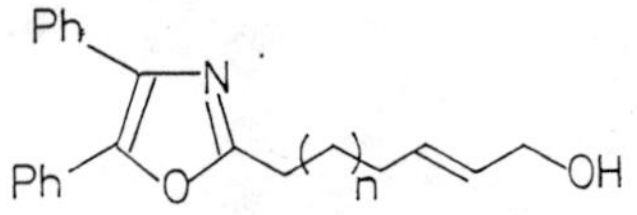

90

Allylic eneynes, such as 91, react only at the double bond.[185]
Homoallylic alcohols, 93, are much less satisfactory giving optical yields
only around 50%.[186]

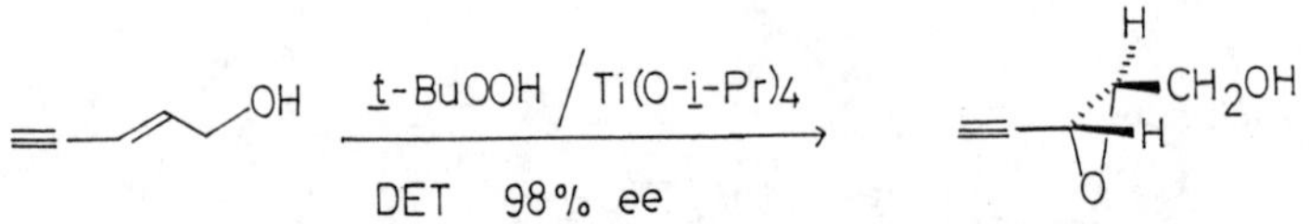

91

92

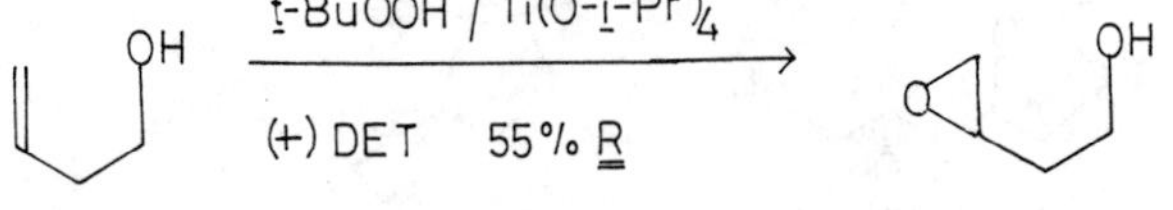

93

94

Polymer supported tartrates were not particularly successful in respect of stability or optical yields,[187] but a series of tartramides, 95, have proved useful.[188]

Few other optically acive catalysts have been systematically studied (Figure 31).

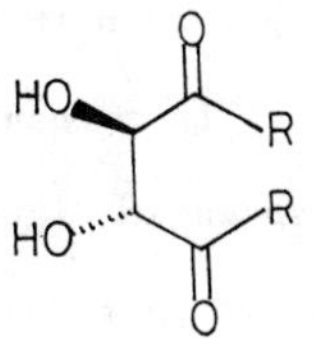

95 a R = $NHCH_2Ph$
b R = $N(CH_2)_4$
c R = NHPr

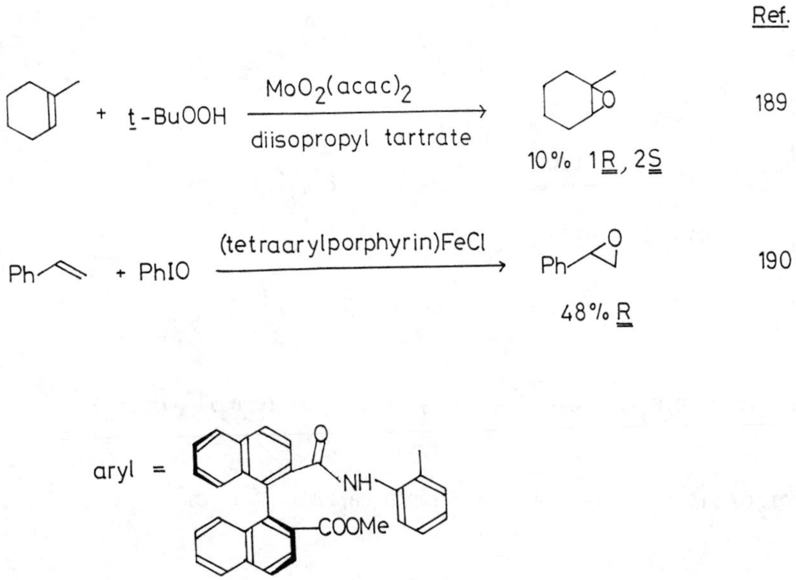

Ref.
MoO2(acac)2
+ t-BuOOH
diisopropyl tartrate
10% 1R, 2S
189
(tetraarylporphyrin)FeCl
Ph
+ PhIO
Ph
48% R
190
aryl =
NH
COOMe

Figure 31 Asymmetric epoxidations using various catalysts.

6.3 Oxidation of Alkenes to carbonyl compounds.

6.3.1 Palladium catalysts.

The Wäcker process, the oxidation of ethylene to acetaldehyde, was developed more than 25 years ago and has been thoroughly studied.[191-193] (See also 6.1) Although certain details of the mechanism are still debated the generally accepted route is shown in Figure 32.[194-196] A summary of the relevant experimental data is pertinent.

$$[PdCl_4]^{2-} + C_2H_4 \; \rightleftharpoons \; \left[\begin{array}{c} Cl \\ Pd \\ Cl \quad Cl \end{array} \right]^{-} + Cl^{-} \qquad (17)$$

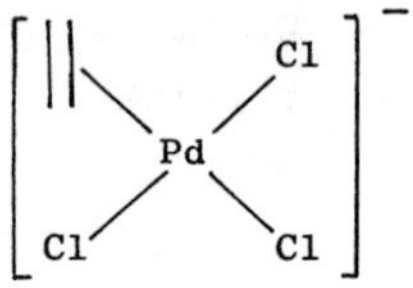

$$\left[\begin{array}{c} Cl \\ Pd \\ Cl \quad Cl \end{array} \right]^{-} \; \xrightarrow{H_2O} \; [PdCl_2(H_2O)(C_2H_4)] + Cl^{-} \qquad (18)$$

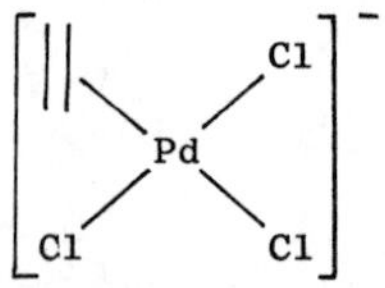

$$[PdCl_2(H_2O)(C_2H_4)] \xrightarrow{H_2O} [PdCl_2(OH)(C_2H_4)]^{-} + H_3O^{+} \qquad (19)$$

$$[PdCl_2(OH)(C_2H_4)]^{-} \longrightarrow [HOCH_2CH_2PdCl] + Cl^{-} \qquad (20)$$

$$[HOCH_2CH_2PdCl] \longrightarrow CH_3CHO + Pd(0) + HCl \qquad (21)$$

Figure 32   Mechansim of the Wacker oxidation.

1) The reaction is inhibited by Cl$^-$ in accord with equations (17) and (18).[197]

2) The reaction is inhibited by H$^+$ in accord with equation (19).

3) The solvent kinetic isotope effect (k_H/k_D = 4.05) is consistent with equation (19).[198]

4) The reaction is first order in C_2H_4 and Pd(II).[197]

5) No deuterium is incorporated from D_2O, indicating that H-transfer is intramolecular.

6) There is a competitive isotope effect with $CH_2=CD_2$.

7) There is no kinetic isotope effect with C_2D_4 implying that decomposition of 96 is not rate determining. Conversion to 97 is slow then transfer of hydrogen fast.[199]

8) Deuterium labelling studies imply that external rather than coordinated water attacks the coordinated alkene.[200]

Alternative versions of the reaction, using other catalyst systems for relay oxidation, are shown in Figure 33.

The reaction of higher alkenes under Wäcker conditions proceeds more slowly and gives rise to many side products. Some examples in aqueous solution are known, but most alkenes are insoluble in such systems.[203]

$$C_2H_4 \; + \; O_2 \; \xrightarrow[\text{PPh}_3]{\text{Pd(OAc)}_2} \; CH_3CHO$$

201

$$C_2H_4 \; + \; O_2 \; \xrightarrow[\text{HPA-n}]{\text{Pd(II)}} \; CH_3CHO$$

202

$$C_2H_4 \; + \; O_2 \; \xrightarrow[\text{benzoquinone}]{\text{Pd(II)}} \; CH_3CHO$$

202

Figure 33 Oxidations analogous to the Wacker oxidation.

Selective oxidation of 1-alkenes to methyl ketones may, however, be achieved in DMF.[204] The reaction distinguishes between terminal and internal alkenes and is tolerant of aldehyde, acid, ester, alcohol, ether, acetal and halide groups. Some examples are shown in Figure 34. Steric hindrance considerably affects the rate of oxidation. Thus, reaction of 99 can be effected in two stages with 100 isolable in good yield.[204]Similarly, one of the terminal double bonds of 102 is oxidised in 3 hours, while the other takes 3 days.

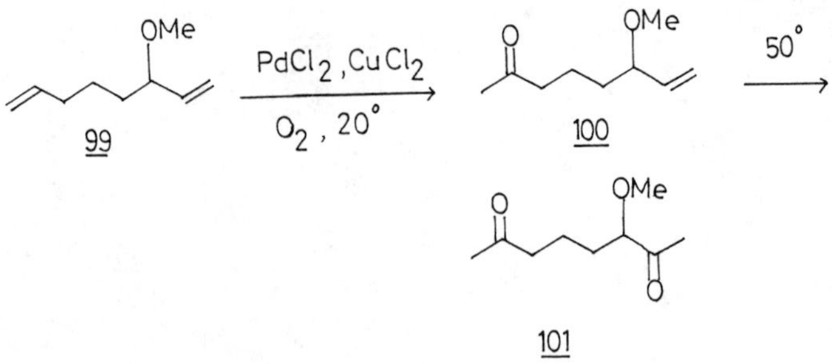

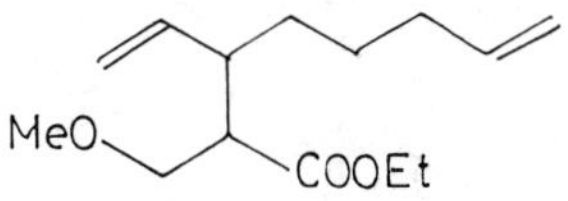

102

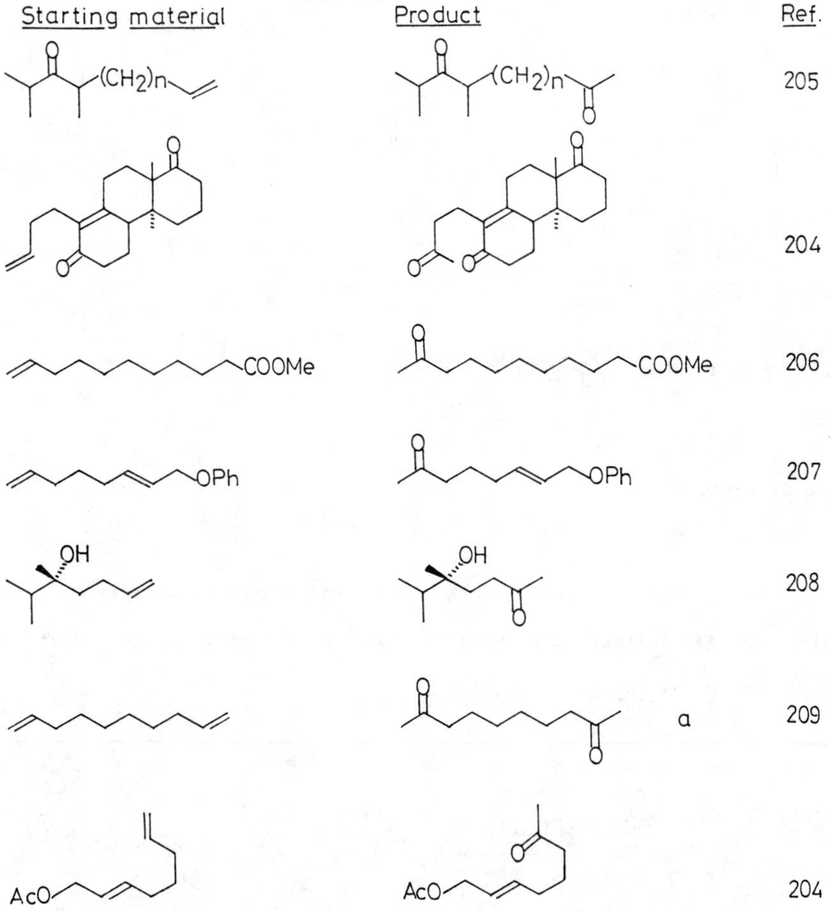

Starting material	Product	Ref.

Figure 34 Catalytic oxidation of 1-alkenes to methyl ketones in the presence

of $PdCl_2/CuCl_2/DMF/O_2$.

Starting material	Product	Ref.
		210
		211
		212
		213

a conditions are $H_2O/C_6H_6/R_4NX$

Figure 34 cntd.

The effect of adjacent polar functional groups is usually deleterious with neither allyl alcohols nor acetates giving good yields (e.g. reactions of 103 and 106).[214]

103    $\xrightarrow[O_2]{PdCl_2, CuCl_2}$ 104   29%   +   105   8%

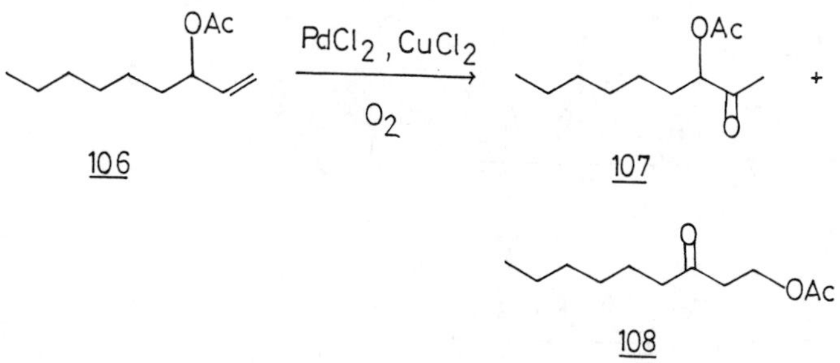

Other palladium based oxidising systems for 1-alkenes are many,
but they have not been so widely used. Most differ in the relay agent or
the reoxidant. For example, Figure 35 shows reactions where quinones are
used as oxidants and Figure 36 shows other palladium complexes employed.
Palladium peroxo complexes have also been useful. t-BuOOH converts
$(RCO_2)_2Pd$ to $(RCO_2PdOO-\underline{t}-Bu)_4$, and on reaction with alkenes, ketones are
formed in good yield. H_2O_2 may also be used as an oxidant and the reaction
mechanism is given in Figure 37.[220,221] Similar intermediates are proposed
for stoichiometric oxidation with $(Ph_3P)_2PdO_2$.[222]

Ref.

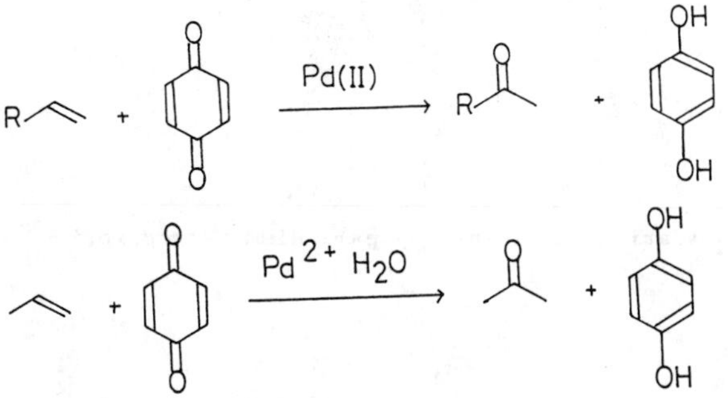

215

216

Figure 35   Catalytic oxidation of 1-alkenes by quinones in the presence of
palladium complexes.

$$\text{(ClPd NO)}_n / h\nu \qquad 217$$

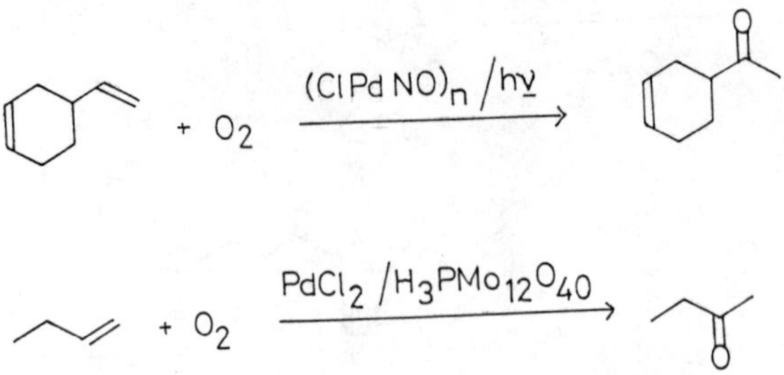

$$\text{PdCl}_2 / H_3 PMo_{12}O_{40} \qquad 78, 218, 219$$

Figure 36 Catalytic oxidation of 1-alkenes in the presence of palladium complexes.

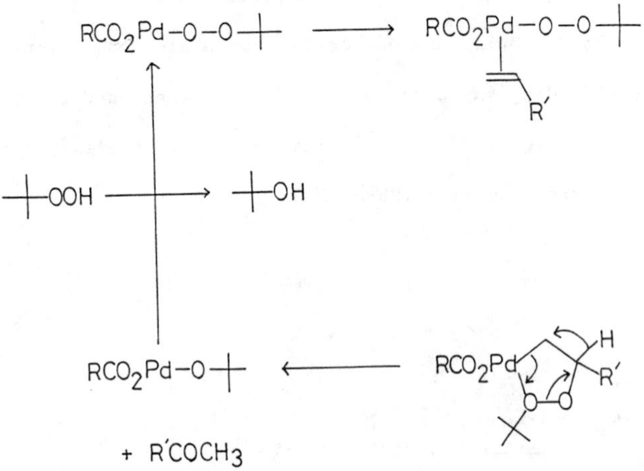

Figure 37 Mechanism of oxidation of alkenes by palladium hydroperoxide complexes.

In the presence of alcohols, acetals are obtained directly (Figure 38; note the unusual selectivity of the reaction with styrene). An intramolecular version of the reaction was used in a synthesis of 110.[226] In analogous manner, 111 is converted to 112.[227]

$$Ph\diagup\!\!\!\diagdown \ + \ O_2 \ \xrightarrow[\text{THF}/\text{MeOH}]{\text{PdCl}_2/\text{CuCl}_2} \ Ph\diagdown\!\!\diagdown\diagup\substack{\text{OMe}\\\text{OMe}}$$

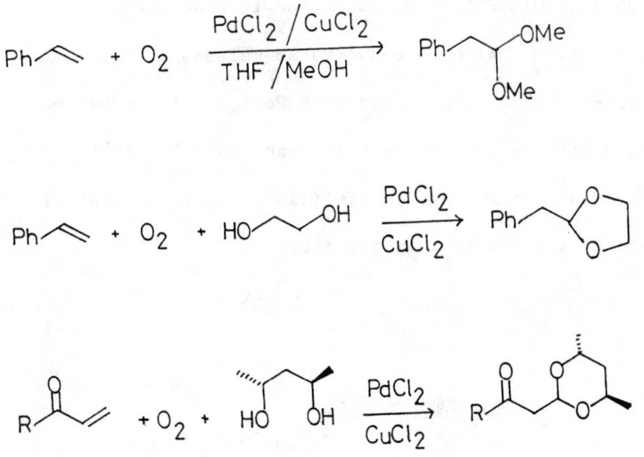

223

$$Ph\diagup\!\!\!\diagdown \ + \ O_2 \ + \ HO\diagdown\!\!\diagdown\diagup OH \ \xrightarrow[\text{CuCl}_2]{\text{PdCl}_2} \ Ph\diagdown\!\!\diagdown$$

224

$$R\overset{O}{\diagdown\!\!\diagdown} \ + O_2 + \ HO \quad OH \ \xrightarrow[\text{CuCl}_2]{\text{PdCl}_2} \ R\overset{O}{\diagdown\!\!\diagdown\diagup}$$

225

Figure 38 Oxidation of alkenes in the presence of alcohols.

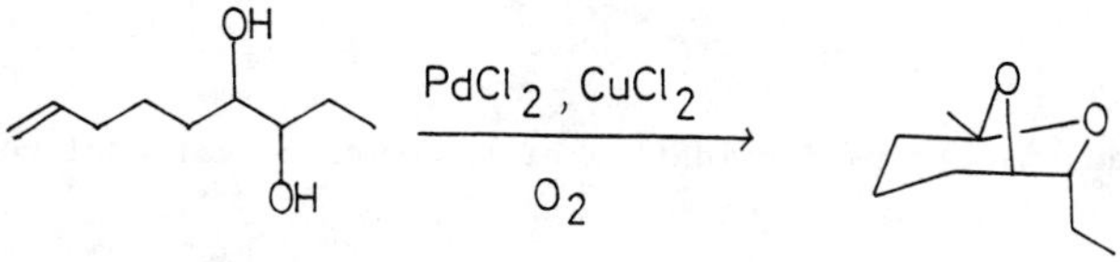

$$\text{109} \quad \xrightarrow[O_2]{\text{PdCl}_2, \text{CuCl}_2} \quad \text{110}$$

109   110

$$\diagup\!\!\!\diagdown\!\!\diagdown\diagup NH_2 \quad \xrightarrow[O_2]{\text{PdCl}_2, \text{CuCl}_2} \quad$$

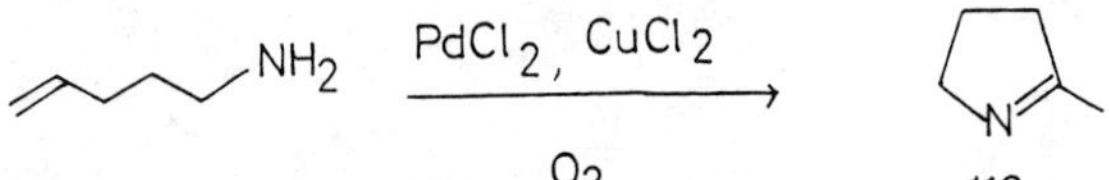

111   112

Compared with 1-alkenes, internal alkenes react slowly and with low selectivity. Cycloalkenes are the best substrates and cyclohexene is oxidised to cyclohexanone in the presence of $PdSO_4/H_3PMo_6W_6O_{40}$.[228] Some success has been achieved in oxidising $\alpha\beta$- and $\beta\gamma$-unsaturated carbonyl compounds. Under the usual conditions of $PdCl_2/CuCl_2$/DMF the reaction is slow, but with t-BuOOH/Na_2PdCl_4/HOAc/H_2O_2 oxidation is smooth and regioselective (Figure 39).[229] With βγ-unsaturated ketones or esters π-allyl palladium complexes, 113, are formed with $PdCl_2/CuCl_2/O_2$/DMF but in aqueous dioxan, oxidation to 114 is the main reaction.[230] Oxidation of allyl ethers and esters is also regioselective (115 to 116) and that of homoallyl esters, 117, only a little less selective.[231]

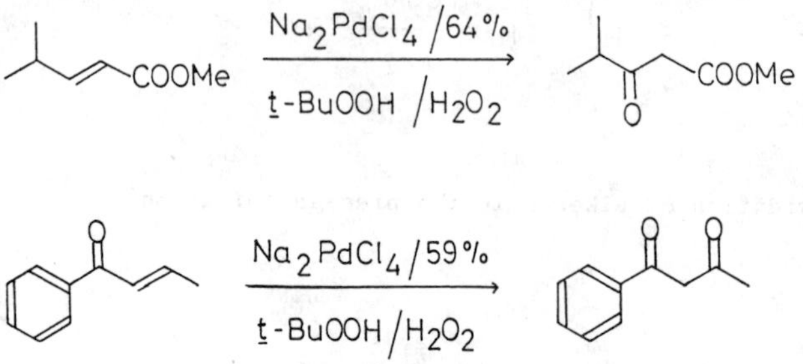

Figure 39 Palladium catalysed oxidation of unsaturated carbonyl compounds.

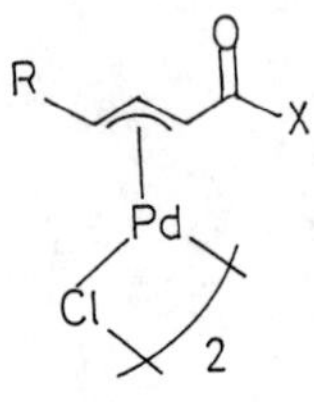

113

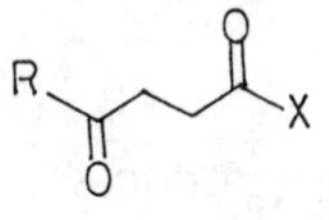

114

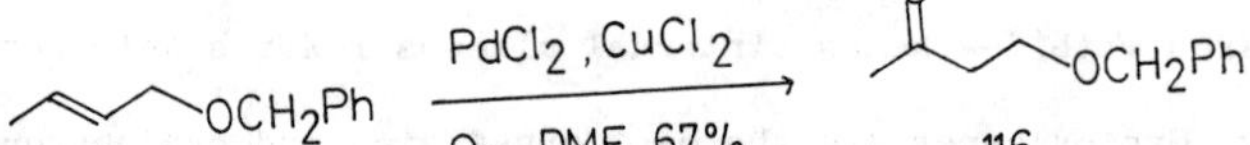

115 116

117

91 : 9
118   119

Many such reactions have found application in natural product synthesis; Figure 40 shows the syntheses of cupranone[232] and nortestosterone,[233] both of which use a palladium catalysed oxidation as a key step.

6.3.2 Rhodium catalysts

Rh(I) forms a range of RhO_2 complexes and oxygen may be transferred to alkenes. As with the palladium peroxo complexes, the mechanism probably involves peroxometallation, 120 to 122.[95,124] Evidence for this mechanism has been sought and several experiments are pertinent. With $[(Ph_3As)_4RhO_2^{18}]ClO_4$, an alkene, $RCH=CH_2$, is converted into $RCO^{18}CH_3$. $(Ph_3P)_3RhCl$ promotes cooxidation of alkenes and Ph_3P and $RhCl_3/Cu(ClO_4)_2$, $Rh(ClO_4)_3$ and $RhCl_3/FeCl_3$ are also active.[234-236] The O^{18} labelling studies strongly disfavour a mechanism involving coordinated hydroperoxide. Although a Wacker mechanism was proposed for $RhCl_3$,[235] the reaction is inhibited by H_2O and excessive oxygen pressures, suggesting that this is not the usual route. The regioselectivity is determined by Markownikoff peroxometallation to give 123 from $RCH=CH_2$, and 124 is isolable.[235]

An alternative route for the decomposition of the peroxometallocycle is represented by 125. The two pathways were distinguished by a D-labelling study. 50% D is lost from $RCD=CH_2$, and about 50% incorporated into $RCOCH_3$

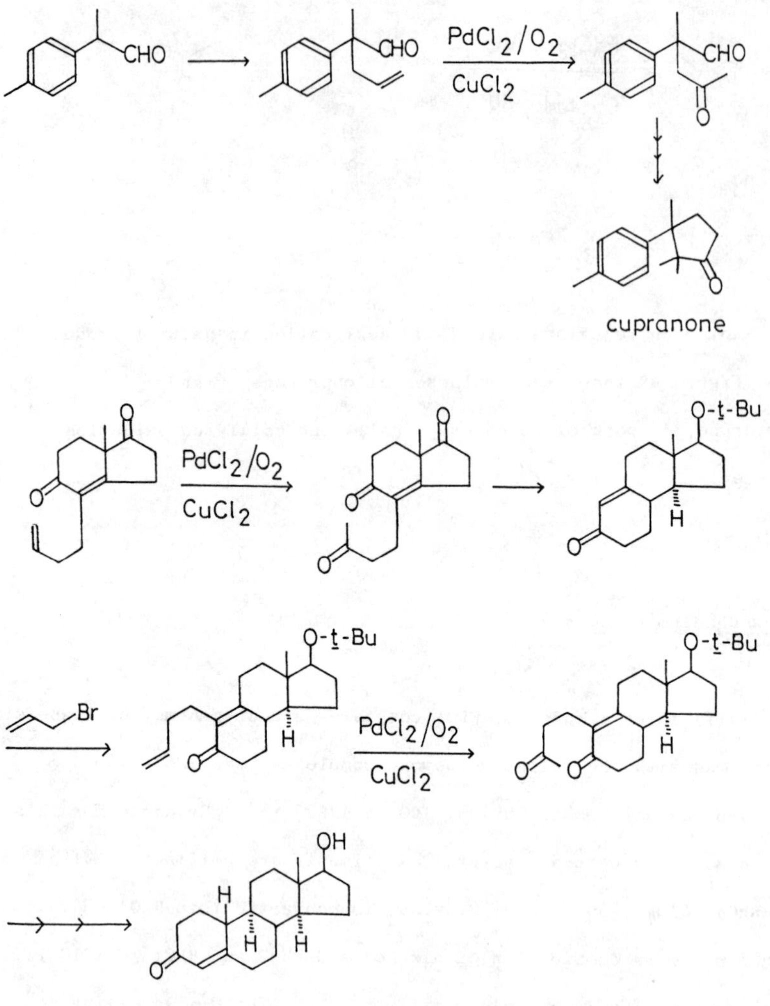

Figure 40 Palladium catalysed oxidation of alkenes in natural product synthesis.

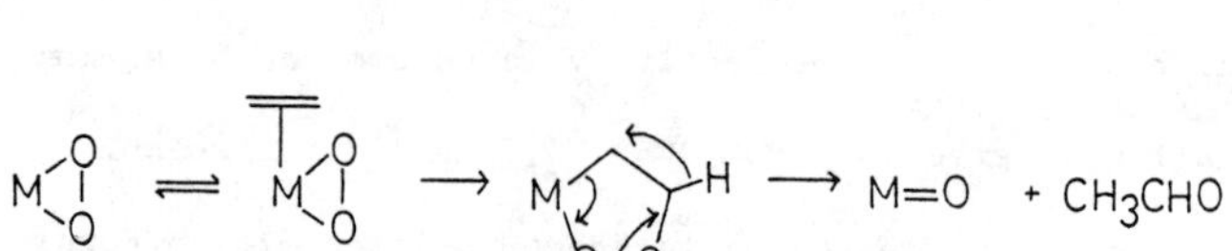

123

124

125

from EtOD, suggesting 125 is correct, since the original route, 122, does
not predict D-exchange.[237]

The reaction scheme is incomplete at this point since both
oxygens are transferred to substrate, and a Wacker pathway is proposed for
conversion of XRh(III)O to Rh(I) (Figure 41).

$XRh(III)=O \xrightarrow{HY} XRh(III)OH \atop Y \xrightarrow{RCH=CH_2}$

$XRh(RCH=CH_2) \longleftarrow$ RCOCH_3

Figure 41 Wacker route for regeneration of Rh(I) from XRh=O.

Selectivity for methyl ketones from 1-alkenes is usually good; the examples of Figure 42 use O_2 as an oxidant. Cu(II) is often also present, but its function is not clear.

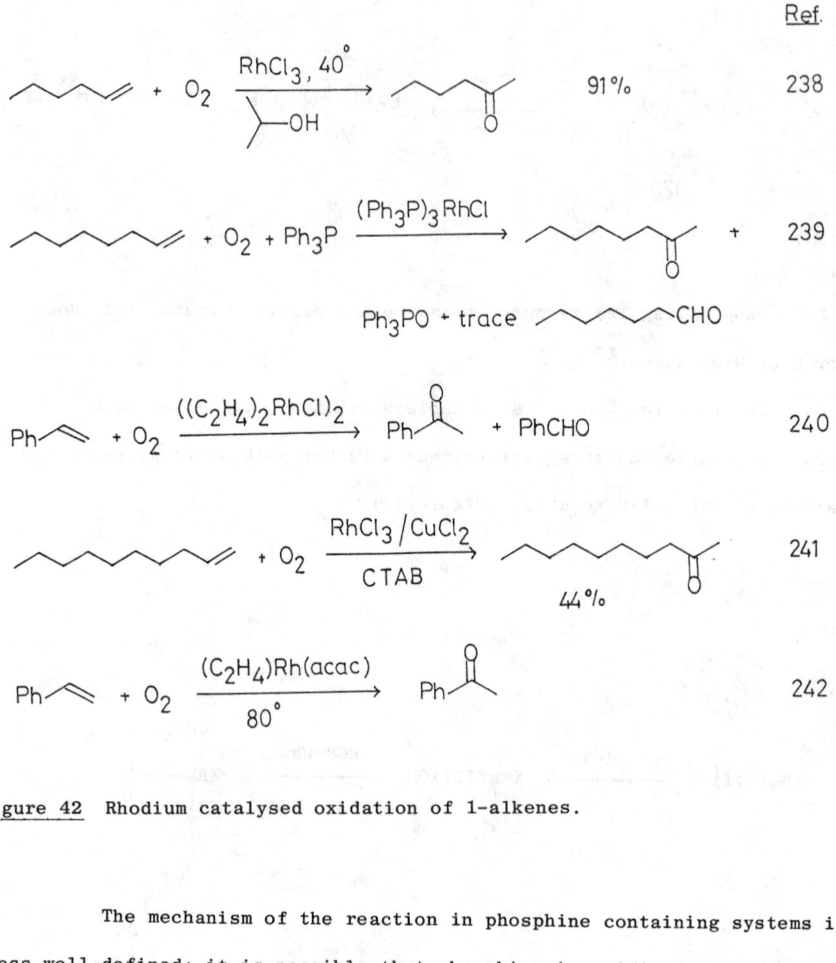

Ref.

$\text{1-alkene} + O_2 \xrightarrow[\text{(CH}_3)_2\text{CHOH}]{\text{RhCl}_3, 40°} \text{methyl ketone} \quad 91\% \quad 238$

$\text{1-alkene} + O_2 + Ph_3P \xrightarrow{(Ph_3P)_3RhCl} \text{ketone} + \quad 239$

$Ph_3PO + \text{trace} \quad \text{CHO}$

$Ph\text{-alkene} + O_2 \xrightarrow{((C_2H_4)_2RhCl)_2} Ph\text{-C(O)CH}_3 + PhCHO \quad 240$

$\text{1-alkene} + O_2 \xrightarrow[\text{CTAB}]{\text{RhCl}_3/\text{CuCl}_2} \text{ketone} \quad 44\% \quad 241$

$Ph\text{-alkene} + O_2 \xrightarrow[80°]{(C_2H_4)Rh(acac)} Ph\text{-C(O)CH}_3 \quad 242$

Figure 42   Rhodium catalysed oxidation of 1-alkenes.

The mechanism of the reaction in phosphine containing systems is less well defined; it is possible that phosphine is oxidised in a separate cycle (see 6.17) but two other mechanisms have been proposed (Figure 43)[239,243] The major competing reactions are alkene cleavage to give aldehydes[244] and alkene isomerisation.[241] Minor amounts of alcohols are formed by Rh-H reduction of the ketones.[245] In a few cases, the use of a rhodium catalyst

Mechanism A

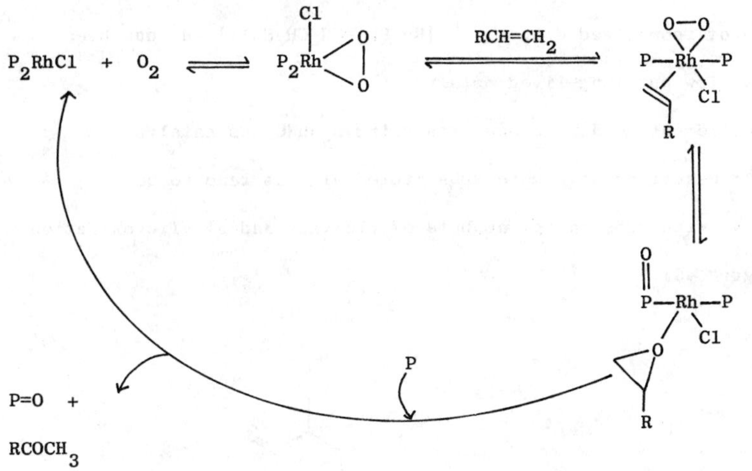

Mechanism B

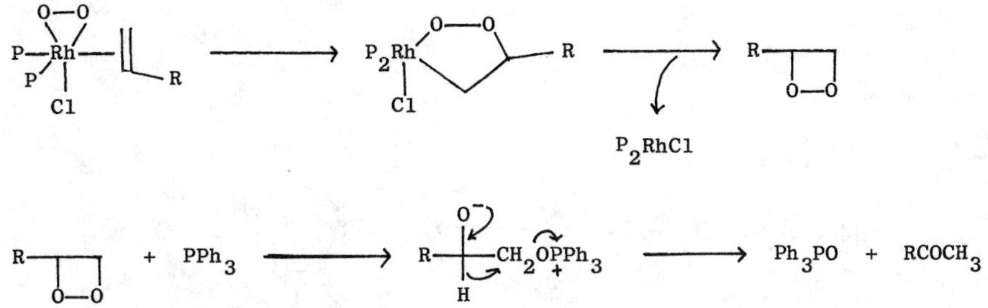

P = PPh$_3$

Figure 43 Mechanism of rhodium catalysed cooxidation of alkenes and
phosphines.

is preferred to a palladium species, despite the extra cost. For example, **126** is converted to **127** by $PdCl_2/CuCl_2/O_2$ but the yield is low. Using $RhCl_3/FeCl_3/O_2/DMF$, **127** is obtained in 80% yield, contaminated only by minor amounts of isomerised dienes.[246] $[Rh(Ph_2PCH_2CH_2SR)_2]^+/H^+$ has been reported as a slow but long-lived catalyst.[247]

A different mechanism operates using L_nRhNO_2 as catalyst (Figure 44).[248] Reactions with more substituted alkenes tend to be unsatisfactory, with substantial amounts of cleavage and allylic oxidation products (Figure 45).

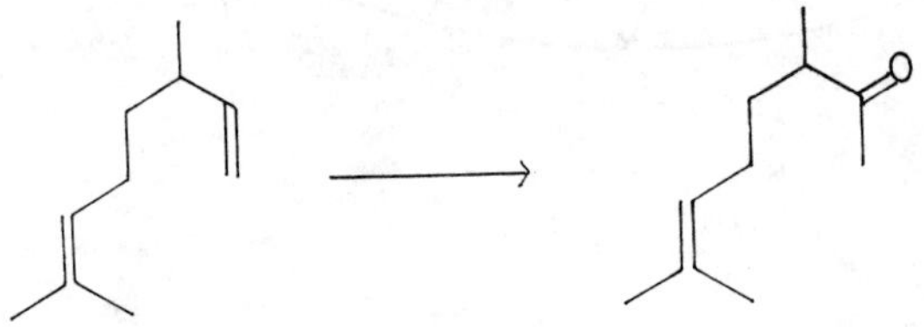

126   **127**

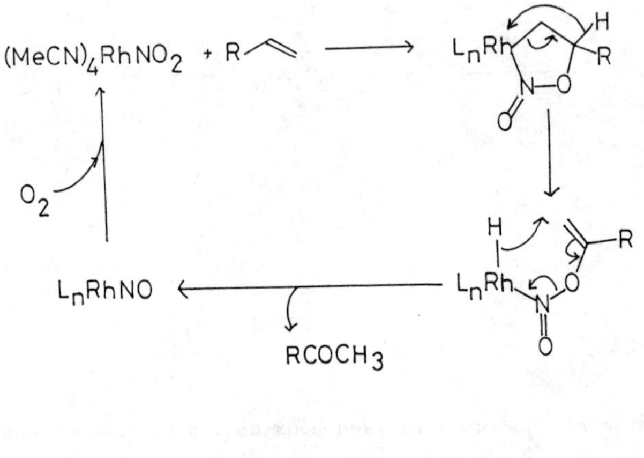

$(MeCN)_4RhNO_2 + R$ ⟶ L_nRh R

O_2

L_nRhNO ⟵ L_nRh

$RCOCH_3$

Figure 44 Mechanism of alkene oxidation in the presence of $(MeCN)_4RhNO_2$

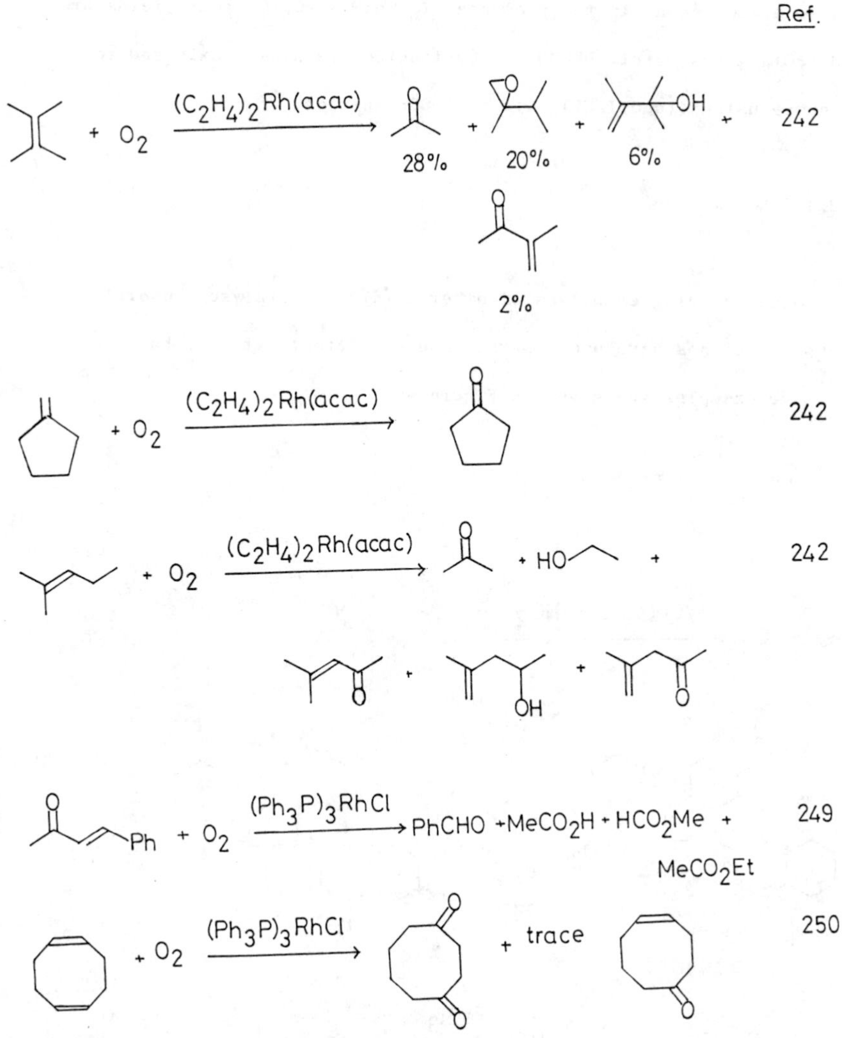

Ref.

242

28%   20%   6%

2%

242

242

+ HO⟶ +

242

+ O₂ →(Ph₃P)₃RhCl→ PhCHO + MeCO₂H + HCO₂Me +   249

MeCO₂Et

+ O₂ →(Ph₃P)₃RhCl→   + trace   250

Figure 45 Rhodium catalysed oxidation of substituted alkenes.

6.3.3 Iridium catalysts

Although formation of dioxygen adducts is facile for iridium

complexes, their use as oxidation catalysts has been limited. Styrene is oxidised to acetophenone in the presence of $(Ph_3P)_2IrCl(CO)$ but yields are poor and $IrCl_3$ gives mainly PhCHO.[251] Cyclooctene is slowly oxisised to cyclooctanone using $((cod)IrCl)_2$ <u>via</u> an IrOOH species.[252]

6.3.4 Other metals

Studies using complexes of other metals to catalyse conversion of alkenes to ketones have been sparse, and consistent patterns have not emerged. Some examples are given in Figure 46.

Ref.

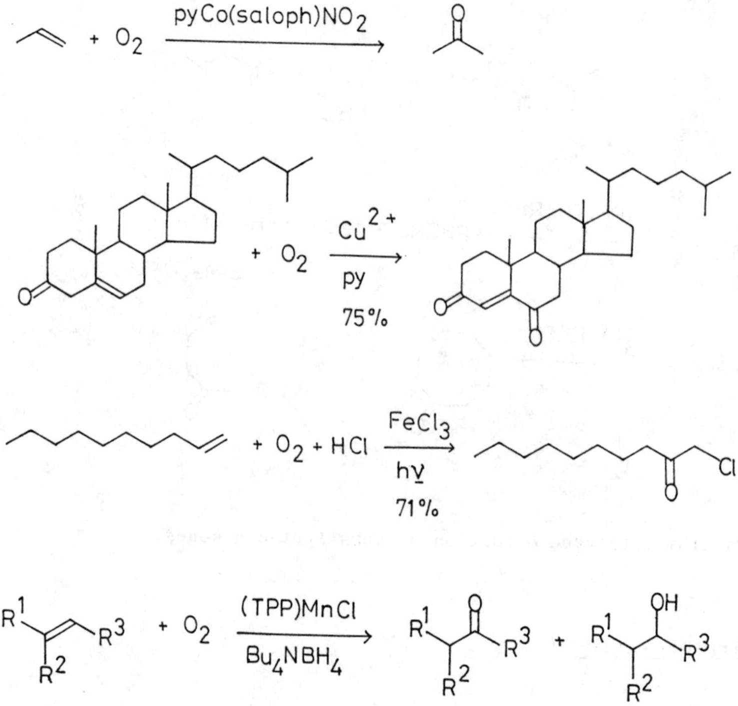

66

253

254

255

<u>Figure 46</u> Catalytic oxidation of alkenes to ketones.

6.4 Allylic oxidation of alkenes.

6.4.1 Manganese catalysts

Allylic oxidation of alkenes is usually initiated by H-abstraction by a radical, shown for Mn(III)/Br$^-$ in equations (22) - (24).[256] In the

$$Mn(III) + Br^- \longrightarrow Mn(II) + Br^{\cdot} \qquad (22)$$

$$Br^{\cdot} + RCH_2CH=CH_2 \longrightarrow HBr + R\overset{\cdot}{C}HCH=CH_2 \qquad (23)$$

$$R\overset{\cdot}{C}HCH=CH_2 + Mn(OAc)_3 \longrightarrow RCH(OAc)CH=CH_2 +$$

$$RCH=CHCH_2OAc + Mn(OAc)_2 \qquad (24)$$

absence of AcO$^-$ or other anion, using (TPP)MnCl/NaBH$_4$, epoxides may be formed and reduced in situ to alcohols.[257] Without NaBH$_4$, however, cyclohexene is converted mainly to **128** and **129**.[258]

80% + 18% + 2%

128   **129**

6.4.2 Cobalt catalysts

Radical pathways also play a major rôle in cobalt catalysed allylic oxidation. Thus, hydroperoxide, **131**, is isolated as the major product of autoxidation of **130** in the presence of Co(OAc)$_3$.[259] When **132** is oxidised in the presence of cobalt stearate, and the hydroperoxides reduced in situ some 20 compounds were obtained of which **133** and **134** were the major components.[260]

The initially formed hydroperoxides may also be decomposed in a cobalt catalysed reaction, the kinetics of which have been studied in detail (equations (25) - (31)).[261]

$$Co^{2+} + ROOH \longrightarrow Co^{3+} + RO^{\cdot} + HO^{-} \tag{25}$$

$$Co^{3+} + RH \longrightarrow Co^{2+} + R^{\cdot} + H^{+} \tag{26}$$

$$R^{\cdot} + O_2 \longrightarrow RO_2^{\cdot} \tag{27}$$

$$RO_2^{\cdot} + RH \longrightarrow ROOH + R^{\cdot} \tag{28}$$

$$ROOH \longrightarrow RO^{\cdot} + HO^{\cdot} \tag{29}$$

$$HO^{\cdot} + RH \longrightarrow H_2O + R^{\cdot} \tag{30}$$

$$RO^{\cdot} + RH \longrightarrow ROH + R^{\cdot} \tag{31}$$

The most popular substrate for mechanistic study is cyclohexene and wider applications have not been sought. 128 and 129 are the main products, and among the catalysts used have been Co(naphthenate)$_2$,[262] Co(acac)$_2$,[263] Co(salen),[264] Na[Co(acac)$_2$(NO$_2$)$_2$],[265] cobalt porphyrins[266] and phthalocyanines.[267]

6.4.3 Rhodium catalysts

Since rhodium species form diamagnetic oxygen complexes, it was of interest to determine whether an ene pathway (135 to 136) could compete with radical pathways. The reaction of carvomenthone, 137, under photochemical ene conditions gives 139, whereas the rhodium catalysed

135   136

137   138   139

140   141   142

reaction gives 141 and 142, as well as the alcohols, suggesting a radical intermediate.[268]

However, other reactions are less simple. Cyclooctene gives cyclooctenone and cyclooctanone in the presence of Rh_2(cyclooctene)$_4Cl_2$ and a radical inhibitor.[269,270] The mechanism suggested is shown in Figure 47.A more complex scheme is needed to explain the cooxidation of

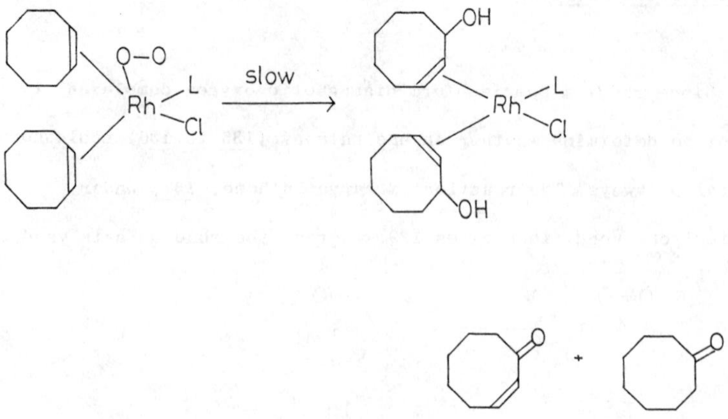

Figure 47 Rhodium catalysed oxidation of cyclooctene.

of cyclooctene (to cyclooctenol) and Ph_3P in the presence of Rh(I).
Hydroquinone has no effect and D-labelling studies show that attack is at
the vinyl hydrogen. This eliminates both previous mechanistic propositions,
that a π-allyl complex is attacked by coordinated O_2 or OOH, or that
oxygen is inserted into an allylic CH. The route now suggested involves
formation of <u>143</u>, followed by elimination of Ph_3PO to give <u>144</u>, and
β-hydride shift of the allyl hydrogen to rhodium.[271]

Many reactions in the presence of $(Ph_3P)_3RhCl$ are radical in
character and pronounced induction periods are observed when alkene
carefully freed from trace peroxide is used.[272] However, the routes for
specific reactions are still open to speculation and O_2 may also be
transferred from $Rh(I)O_2$ or $Rh(II)O_2^-$ within the coordination sphere. Some
examples are shown in Figure 48. In the presence of
octaethylporphyrin Rh(III)Cl, cyclohexanol is formed from cyclohexene with
the anti-Markownikoff alcohol predominating in substituted cases[275]. t-BuOOH
may also be used as an oxidant, giving enones and allyl acetates by a

non-radical route[276].

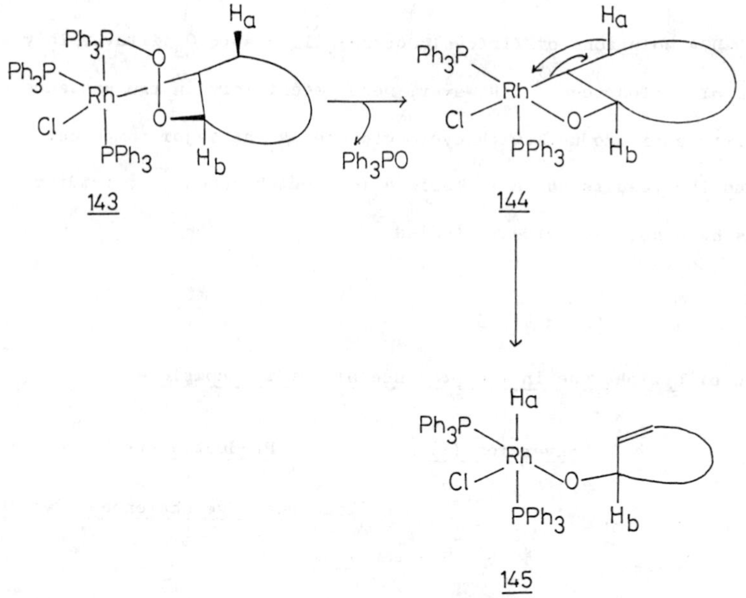

Ref.

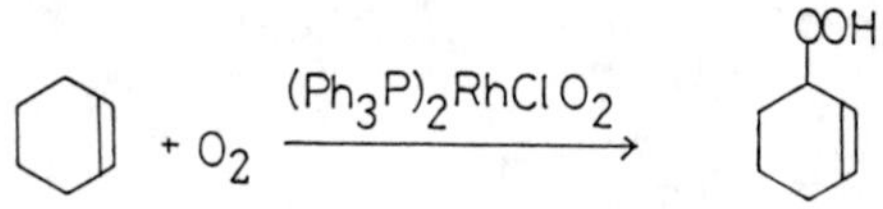

272

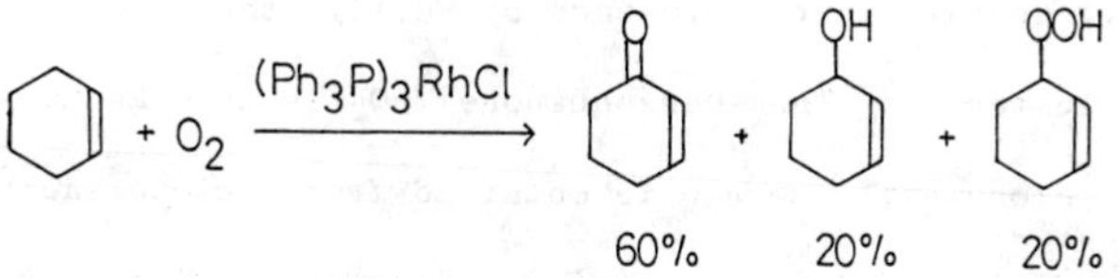

273

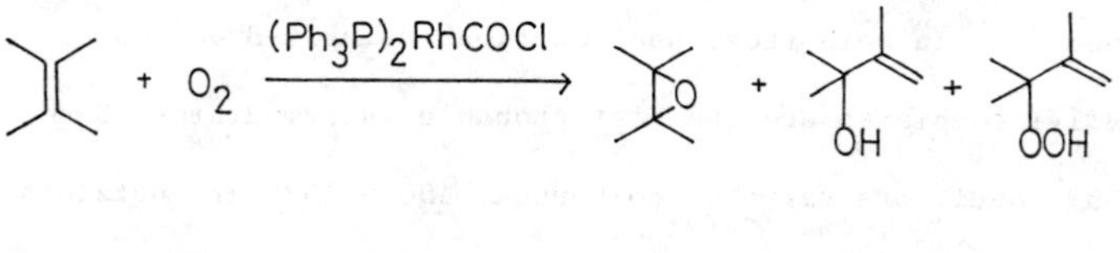

274

Figure 48 Rhodium catalysed allylic oxidation of alkenes.

6.4.4 Iridium catalysts

DMA solutions of (Ir(cyclooctene)$_2$Cl$_2$ absorb O$_2$ irreversibly with expulsion of cyclooctene[277]. However, more recent work in the presence of H$^+$ gave a mixture of products with cyclooctenone as the major component[278]. Other than the results shown in Table 4 for cyclohexenone[279], iridium catalysts have not been widely studied.

Table 4

Oxidation of cyclohexene in the presence of iridium complexes

Catalyst	Conversion (%)	Product distribution		
		Cyclohexene oxide	Cyclohexenone	Cyclohexenol
(Ph$_3$P)$_2$Ir(CO)(O$_2$)Cl	72	0.64	1.01	traces
(Ph$_3$P)$_2$Ir(CO)Cl	25	2.82	15.19	11.05

6.4.5 Palladium catalysts

Palladium catalysts are best known for oxidising alkenes to ketones or vinyl derivatives, but efficient allylic oxidation is known. Thus, propene gives allyl acetate in the presence of Pd(II), the true catalyst being a Pd(0) cluster.[280] When benzoquinone/MnO$_2$ is used as an oxidant a 95% yield of cyclohexenyl acetate is obtained from cyclohexene.[281] The regiochemistry of oxidation of 146 was studied; using benzoquinone as oxidant the ratio of 147:148 is 2:3 but with duroquinone a 24:1 selectivity to 148 is obtained.[282] In both these and reactions using t-BuOOH as oxidant,[283] π-allyl complexes are the most probable intermediates. Under slightly different conditions carbonyl compounds, 150 - 152, are obtained, again via π-allyl complexes.[284]

6.4.6 Copper catalysts

Most of the reactions in this group are radical chain in character. Cyclohexene is oxidised by $S_2O_8^{2-}$ in the presence of a carboxylate and Cu(II) according to equations (32) - (36), the function of the copper being to convert the radical to a cation and subsequently an allyl carboxylate. In the absence of copper, H-abstraction from solvent occurs to give cyclohexyl carboxylates.[285]

$$S_2O_8^{2-} + Cu(I) \longrightarrow SO_4^{-\cdot} + SO_4^{2-} + Cu(II) \qquad (32)$$

$$SO_4^{-\cdot} + RCOO^- \longrightarrow RCOO^\cdot + SO_4^{2-} \qquad (33)$$

$$RCOO^\cdot + C_6H_{10} \longrightarrow RCOOH + \qquad (34)$$

$$RCOO^\cdot + C_6H_{10} \longrightarrow \qquad (35)$$

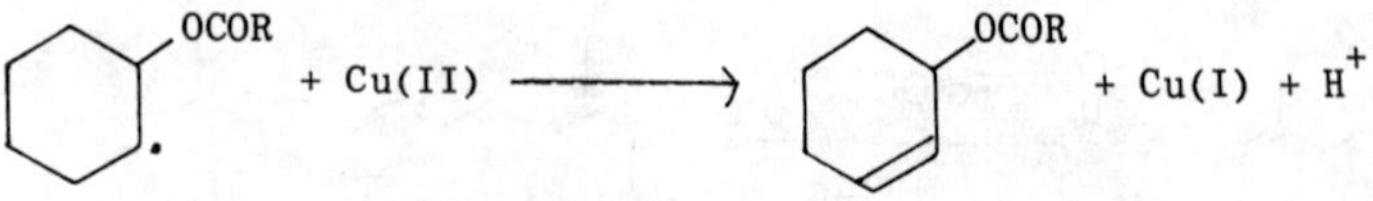

$$\text{(cyclohexyl-OCOR radical)} + \text{Cu(II)} \longrightarrow \text{(cyclohexenyl-OCOR)} + \text{Cu(I)} + \text{H}^+ \quad (36)$$

Other reports are rather sparse; 153 catalyses conversion of cyclohexene to cyclohexenyl hydroperoxide as the only primary product[286] and a copper complex of chiral 154 gives asymmetric allylic oxidation of cyclohexene.[287]

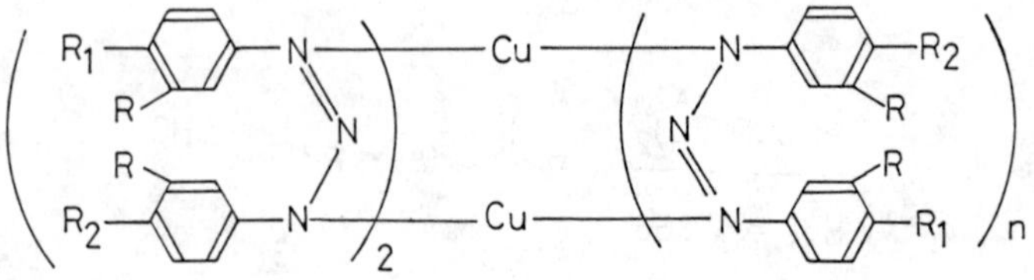

153

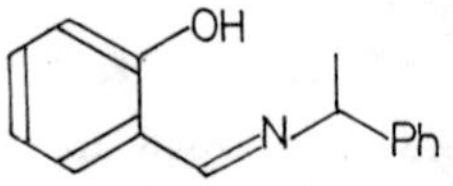

154

6.4.7 Other metals

Few systematic studies have involved complexes of other metals and most of the examples (Figure 49) involve radical autoxidation.

6.5 Other alkene oxidations

6.5.1 Conversion to cis- diols using osmium catalysts.

Stoichiometric conversion of alkenes to diols using OsO_4 is usually satisfactory, but on a large scale the cost and the toxicity of

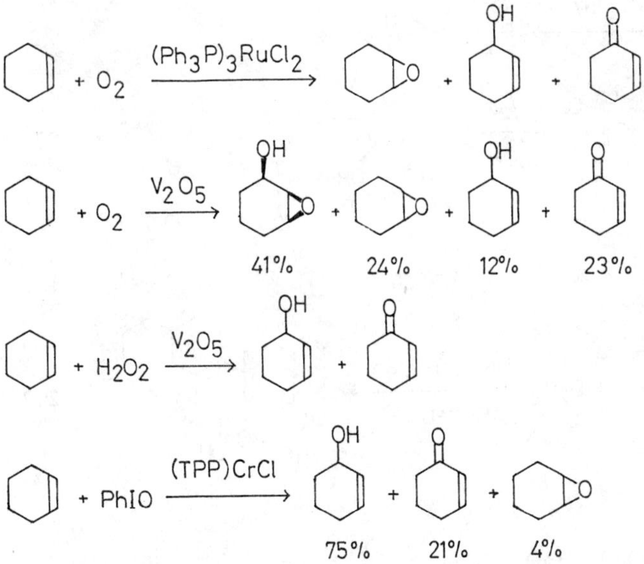

Ref.

57

288

24

289,29

Figure 49 Catalytic allylic oxidation of alkenes.

osmium suggest the use of a catalytic reaction.[291] This is achieved using OsO_4 with an oxidant which hydrolyses the intermediate Os(VI) ester oxidatively to regenerate Os(VIII). A number of such oxidants exist and hydroxylation is stereospecifically cis.

Oxidation by O_2 is attractive but is little used. Air oxidises Os(VI) to Os(VIII) in aqueous solution at a rate dependent on pH.[292] Unfortunately cleavage of the cis- diols also often occurs. Ethylene glycol is obtained from C_2H_4 at pH 8.5 - 9.5 at 50°C but 1-alkenes give cleavage products.[292]

A catalytic amount of OsO_4 in the presence of H_2O_2 oxidises alkenes

Ref.

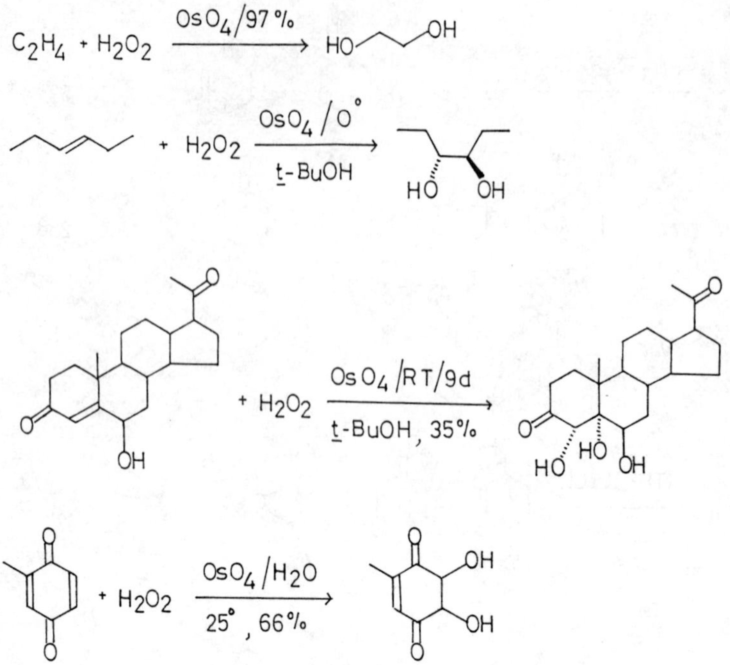

$C_2H_4 + H_2O_2 \xrightarrow{OsO_4/97\%}$ HO⁀OH

22

$+ H_2O_2 \xrightarrow[\underline{t}\text{-BuOH}]{OsO_4/0°}$

22

$+ H_2O_2 \xrightarrow[\underline{t}\text{-BuOH, 35\%}]{OsO_4/RT/9d}$

293

$+ H_2O_2 \xrightarrow[25°, 66\%]{OsO_4/H_2O}$

294

Figure 50   Cis-hydroxylation of alkenes using OsO_4/H_2O_2.

to cis-diols as the major products. There are many reported examples

(Figure 50). The reaction proceeds via formation of peroxyosmic acid, H_2OsO_6,

which reacts rapidly with alkenes to give 155. Hydrolysis gives OsO_4 and the

cis-diol[295]. The major disadvantage is that overoxidation with cleavage to

carbonyls is common and yields are usually low.

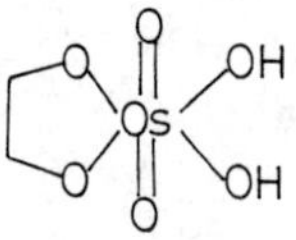

155

KClO$_3$ in the presence of OsO$_4$ gives <u>cis</u>-diols from alkenes. Free HOCl may be formed during the reaction, since chlorohydrins are common side products. Ester formation is thought to be rate determining[296] and chlorate is reduced to chloride.A wide range of ClO$_3^-$ salts are useful (Figure 51). Where there is more than one double bond, that which is more electron rich is preferentially attacked. Thus, conversion of <u>156</u> to <u>157</u> in 80% yield is a key step in the synthesis of the antibacterial fulvene, fulvoplumierin,[301] and in <u>158</u> the enone is relatively unreactive.[302] Attack is predominently from the less hindered side of the double bond; thus, <u>160</u> is obtained as an intermediate for prostaglandin synthesis.[303]

<u>156</u>   <u>157</u>

<u>158</u>

<u>159</u>   <u>160</u>

Ref.

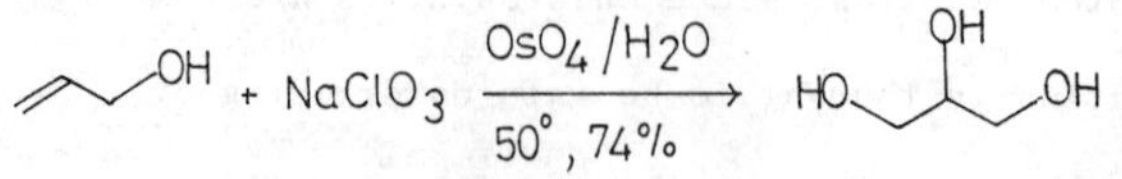

297

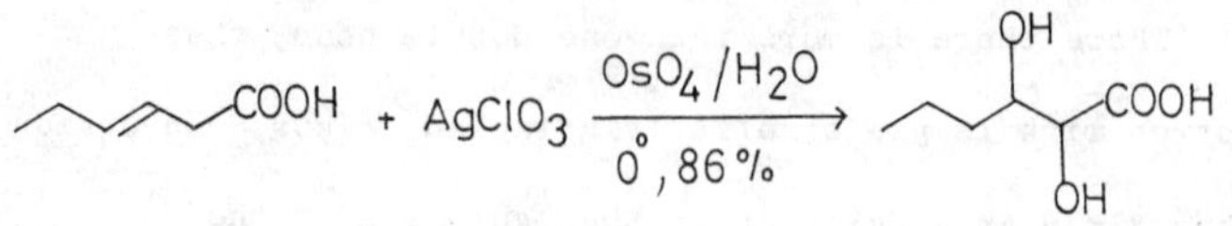

298

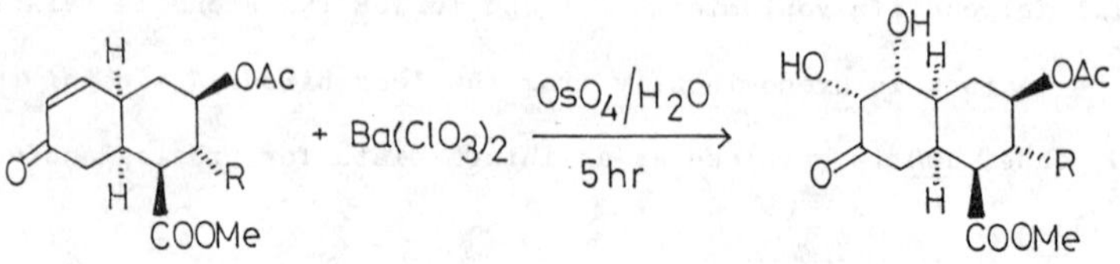

299

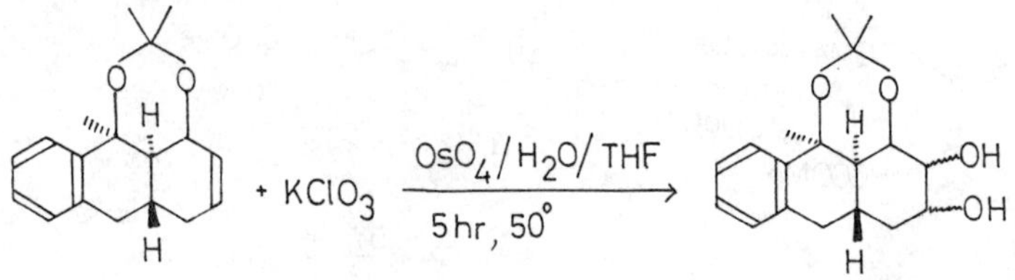

300

Figure 51   Cis-hydroxylation of alkenes using OsO_4/ClO_3^-.

The possible formation of HOCl in perchlorate oxidations suggested
that NaOCl might be a suitable oxidant. The reagent seems promising and
allyl alcohol is converted to glycerol in 98% yield.[304]

Since vicinal diols are readily cleaved by $NaIO_4$, the use of this
as oxidant gives carbonyl derivatives as the exclusive products (Figure 52).

The use of t-BuOOH as oxidant in the presence of Et_4NOH gives
yields of diols in the region of 70 - 80%, and overoxidation is minor
(e.g. 161 to 162 and 163[309] and 164 to 165[310]). Hindered alkenes give poor
results.[311]

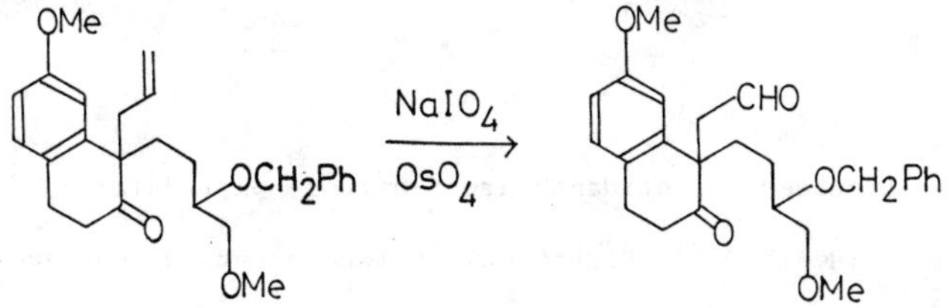

305

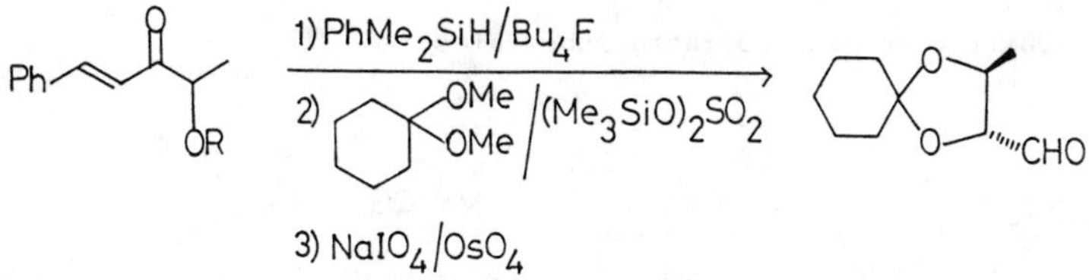

306

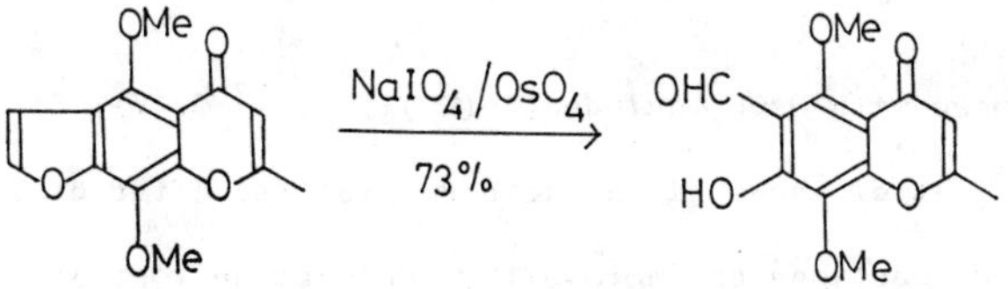

307

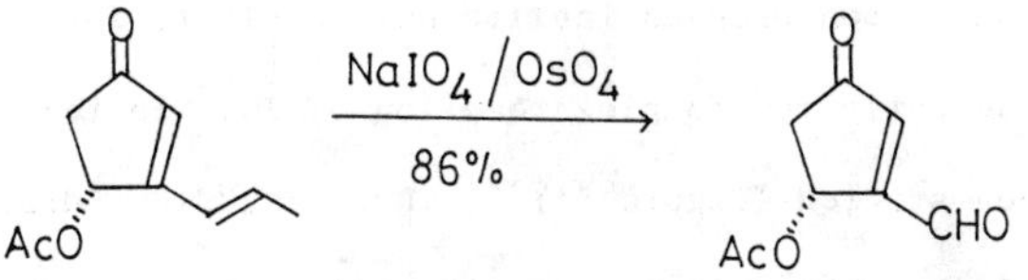

308

Figure 52 Hydroxylation and cleavage of alkenes using OsO_4/IO_4^-.

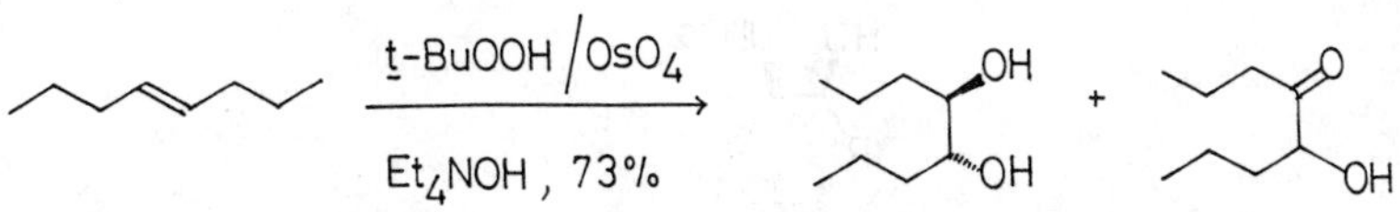

161

10 : 1

162 163

CH₃(CH₂)₇ ... (CH₂)₇CH₂OH →[t-BuOOH/OsO₄ / Et₄NOH, 51%] CH₃(CH₂)₇ ... (CH₂)₇CH₂OH

$$CH_3(CH_2)_7 \quad (CH_2)_7CH_2OH \xrightarrow[Et_4NOH,\ 51\%]{t\text{-}BuOOH/OsO_4} \quad CH_3(CH_2)_7 \quad (CH_2)_7CH_2OH$$

164   165

Among the most successful oxidants are N-oxides, particularly
N-methylmorpholine N-oxide (NMO)[312](Figure 53). Attack occurs predominently
from the less hindered side of the molecule (166 to 167[315]). Tri-, and
particularly tetrasubstituted alkenes,are less efficiently oxidised and
$Me_3N^+\text{-}O^-$ is superior to NMO for these substrates.[316]

MeOOC ... MeOOC →[NMO / OsO₄] MeOOC ... MeOOC —OH, —OH

166   167

Other oxidants used with OsO_4 include $[Fe(CN)_6]^{3-}$,[317] $Cr(VI)$,[318]
$Ce(IV)$[319] and HNO_3, the last of these giving acid derivatives after diol
cleavage.[320] Some related reactions are noteworthy. Chloramine T gives
168 from alkenes in the presence of OsO_4, with $O_3Os=NTs$ as the effective
reagent.[321] CrO_3 and V_2O_5 have been used as inefficient catalysts for
hydroxylation by H_2O_2.[22] Recently the complex reaction of IO_4^- in the
presence of Rh(III) has been studied (Figure 54).[322] $RuCl_3/NaOCl$ oxidises
cyclohexene to the cis-diol, but this cleaves to give adipic acid in good
yield.[323]

HO NHTs

R ... R

168

Ref.

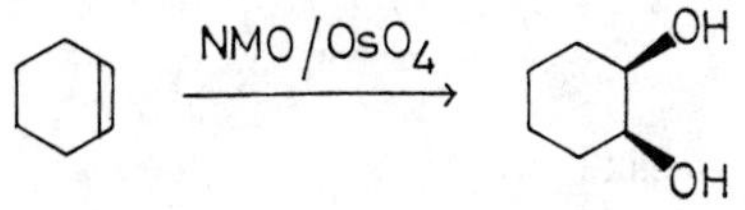

313

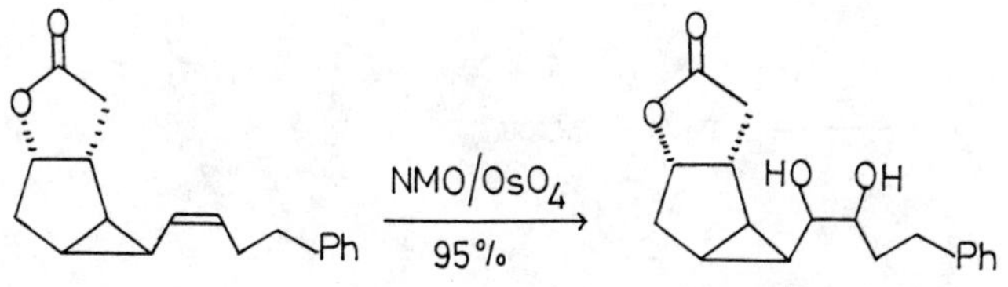

312

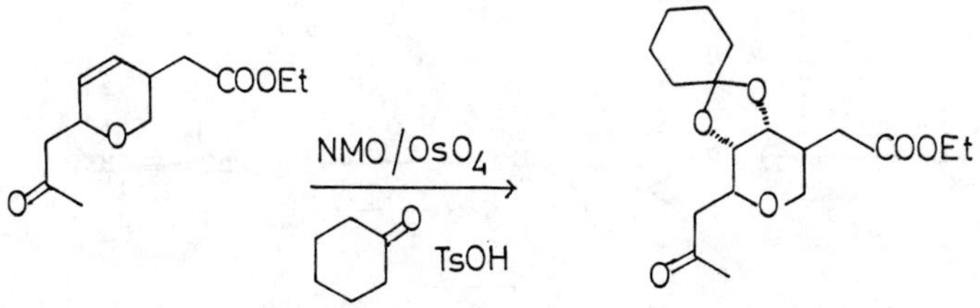

314

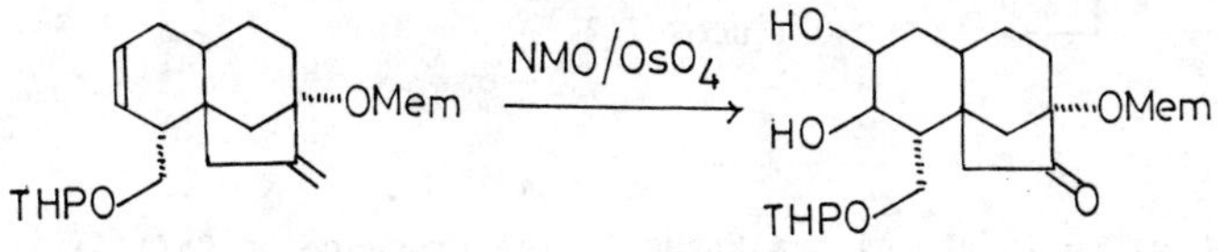

315

Figure 53 **Hydroxylation of alkenes using OsO_4/NMO.**

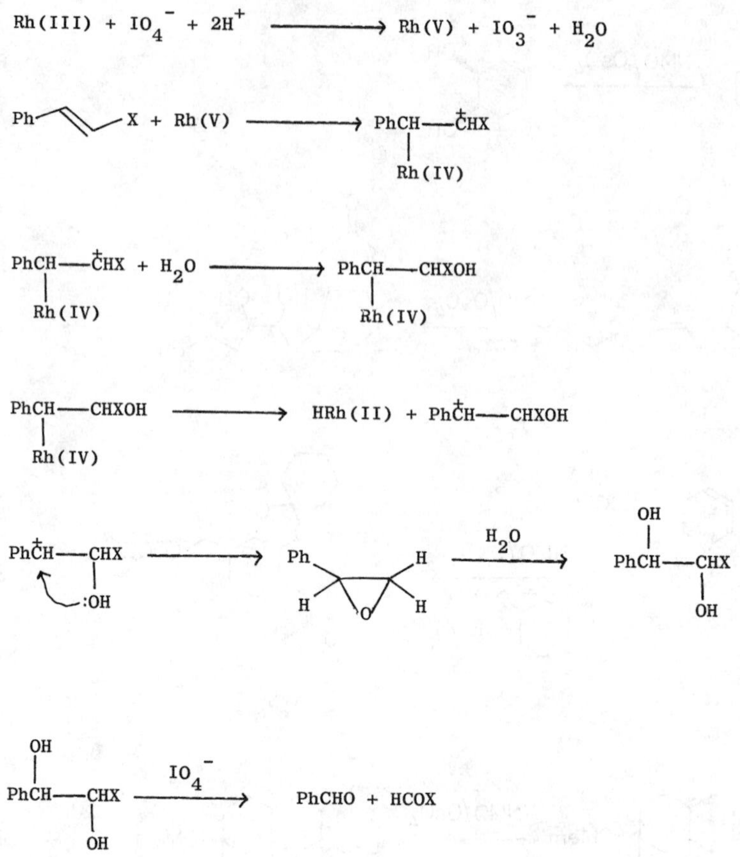

$$Rh(III) + IO_4^- + 2H^+ \longrightarrow Rh(V) + IO_3^- + H_2O$$

$$Ph\diagdown{=}\diagup{-}X + Rh(V) \longrightarrow \underset{\underset{Rh(IV)}{|}}{PhCH}{-\!-}\overset{+}{C}HX$$

$$\underset{\underset{Rh(IV)}{|}}{PhCH}{-\!-}\overset{+}{C}HX + H_2O \longrightarrow \underset{\underset{Rh(IV)}{|}}{PhCH}{-\!-}CHXOH$$

$$\underset{\underset{Rh(IV)}{|}}{PhCH}{-\!-}CHXOH \longrightarrow HRh(II) + Ph\overset{+}{C}H{-\!-}CHXOH$$

$$Ph\overset{+}{C}H{-\!-}\underset{\underset{:OH}{\curvearrowleft}}{CHX} \longrightarrow \underset{H}{\overset{Ph}{\diagdown}}\triangle\underset{H}{\overset{H}{\diagup}} \xrightarrow{H_2O} \underset{\underset{OH}{|}}{PhCH}{-\!-}\overset{\overset{OH}{|}}{CHX}$$

$$\underset{\underset{OH}{|}}{PhCH}{-\!-}\overset{\overset{OH}{|}}{CHX} \xrightarrow{IO_4^-} PhCHO + HCOX$$

Figure 54 Reaction of periodate with alkenes in the presence of Rh(III).

6.5.2 Oxidation in the presence of palladium complexes.

Numerous palladium catalysed oxidations are related to the Wäcker reaction (see 6.3.1). Vinyl acetate is formed from ethylene in the presence of $PdCl_2/CuCl_2/O_2$ in non-aqueous solvents.[324] Both rate and selectivity are increased by NaOAc. The reaction follows the pathway of equations (37) – (39).[325,326] Some acetaldehyde is usually formed as a by-product. Both

$$[Pd(OAc)_3]^- + C_2H_4 \longrightarrow [Pd(OAc)_3(C_2H_4)]^- \qquad (37)$$

$$[Pd(OAc)_3(C_2H_4)]^- \longrightarrow (AcOCH_2CH_2PdOAc) + AcO^- \qquad (38)$$

$$AcOCH_2CH_2PdOAc \dashrightarrow AcOCH=CH_2 + Pd(0) + HOAc \qquad (39)$$

$CuCl_2$ and heteropolyacids may be used as relays for reoxidation.[76] Oxidation of higher alkenes has been studied but there are many products and yields are low, due to alkene isomerisation and allyl complex formation. With 1-alkenes the terminal enol acetate, 169, is the major product,[327] formed by acetoxypalladation and elimination of HPdOAc.[328] With more substituted alkenes, such as 173, a single product, 174, is obtained, whilst internal alkenes give allyl acetates.[329] Analogous reactions give vinyl ethers and acetals.[330,331] Intramolecular reactions, such as those of 175 and 177 also give vinyl ethers.[332] An enantioselective version of this reaction (179 to 180) indicates that the palladium allyl remains intact throughout, and a pathway involving β-hydride elimination and HCl loss is excluded. Reoxidation is postulated to occur via a bimetallic complex, 182.

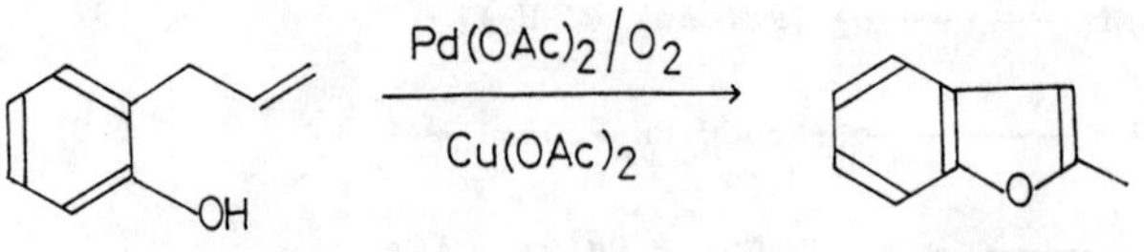
Pd(OAc)2/O2
Cu(OAc)2
OH

175 176

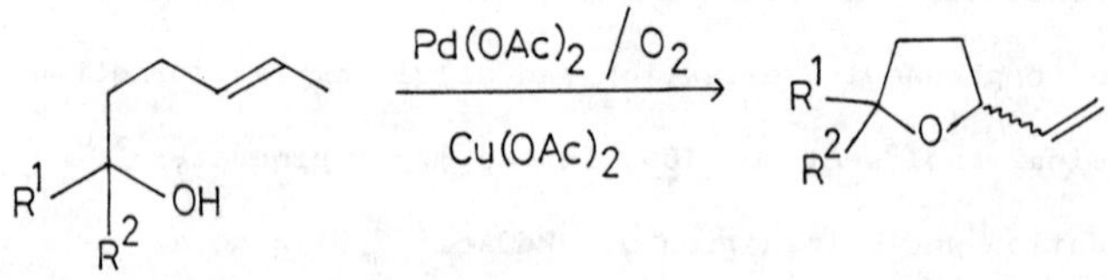
Pd(OAc)2 /O2
Cu(OAc)2
R1
R2
OH

177 178

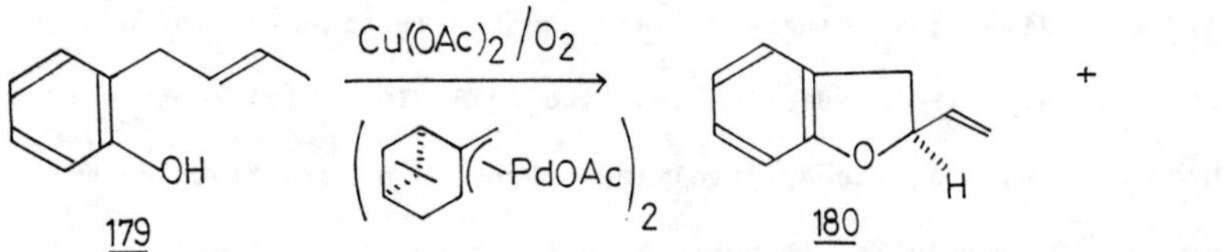
Cu(OAc)2 /O2
PdOAc
2
OH
H
+

179 180 18 % ee 181

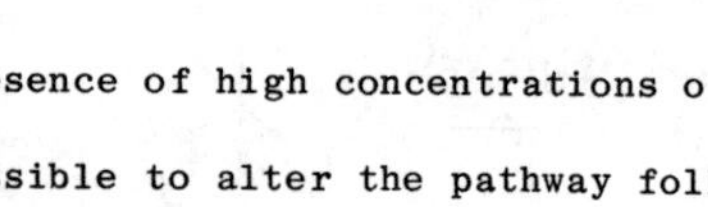
OOH
OAc
Pd
Cu
OAc

182

In the presence of high concentrations of nucleophiles and Cu(II) or NO_3^-, it is possible to alter the pathway followed by the intermediate, 183, to give glycol derivatives.(equation (40)). The rôle of Cu(II) or NO_3^- in promoting nucleophilic displacement is unclear. NO_3^- is less

$$AcOCH_2CH_2PdX \xrightarrow{\text{Nu}} AcOCH_2CH_2Nu \qquad (40)$$

$$\underline{183}$$

strongly Pd-coordinated than Cl^- or AcO^- and ionisation gives a more electrophilic metal as the leaving group. However, this would also enhance β-hydride elimination. In the presence of Cu(II) the route of equations (41) - (45) may be followed. The range of efficient reactions is

$$AcOCH_2CH_2PdX + CuX \longrightarrow AcOCH_2CH_2{}^{\cdot} + PdX^+ + CuX + X^- \qquad (41)$$

$$AcOCH_2CH_2{}^{\cdot} + PdX^+ \longrightarrow AcOCH_2CH_2{}^+ + PdX \qquad (42)$$

$$AcOCH_2CH_2{}^{\cdot} + CuX_2 \longrightarrow AcOCH_2CH_2X + CuX \qquad (43)$$

$$AcOCH_2CH_2{}^+ + X^- \longrightarrow AcOCH_2CH_2X \qquad (44)$$

$$PdX + CuX_2 \longrightarrow PdX_2 + CuX \qquad (45)$$

considerable (Table 5). The process in the presence of $NO_3{}^-$ has been carefully studied. It is postlated that palladium nitrites, formed early in the reaction are the true catalysts [339] (equations (46) - (52)). If the $NO_3{}^-$

$$Pd(II) + C_2H_4 \xrightarrow{H_2O} CH_3CHO + Pd(0) \qquad (46)$$

$$Pd(0) + 2HNO_2 + 2H^+ \longrightarrow Pd(II) + 2NO + 2H_2O \qquad (47)$$

$$2NO + 2HNO_3 \rightleftharpoons 2NO_2 + 2HNO_2 \qquad (48)$$

$$2NO_2 + H_2O \rightleftharpoons HNO_2 + HNO_3 \qquad (49)$$

$$Pd(II) + NO_2{}^- \longrightarrow [Pd(NO_2)]^+ \qquad (50)$$

$$[Pd(NO_2)]^+ + C_2H_4 \xrightarrow{HOAc} [Pd(NO)]^+ + HOCH_2CH_2OAc \qquad (51)$$

$$[Pd(NO)]^+ + NO_2 \longrightarrow [Pd(NO_2)]^+ + NO \qquad (52)$$

is labelled with ^{17}O, the label is found in both NO and the acetate carbonyl, suggesting the sequence $\underline{184}$ to $\underline{187}$ or $\underline{188}$ to $\underline{191}$.[340] Similar reactions occur with higher alkenes but are less useful (e.g. to give $\underline{192}$[334]).

Table 5

Conversion of ethylene to glycol derivatives in the presence of Pd(II)

Conditions	Products	Ref.
$PdCl_2$, LiCl, $CuCl_2$, $Cu(OAc)_2$, 65°	$AcOCH_2CH_2OAc$ (90%)	333
Pd(II), HNO_3	$O_2NOCH_2CH_2ONO_2$	334
Li_2PdCl_6, LiCl, $CuCl_2$	$ClCH_2CH_2OAc$ + $AcOCH_2CH_2OAc$	335
Pd(II), HOAc, NO_3^-	$HOCH_2CH_2OAc$	336
$PdCl_2$, HOAc, $Fe(NO_3)_2$	$HOCH_2CH_2OAc$	337
$PdCl_2$, $CuCl_2$	$CH_3CH_2OCH_2CH_2Cl$	338

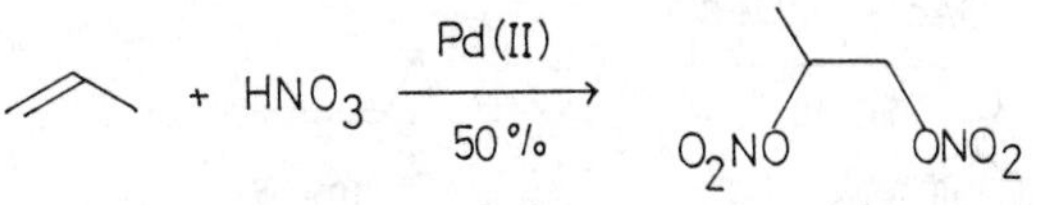

$$\text{/\!\!=\!\!\textbackslash} \quad + \quad HNO_3 \quad \xrightarrow[50\%]{Pd(II)} \quad O_2NO \diagdown\diagup ONO_2$$

192

Analogous oxidations of dienes are known, typically 193 to 194.[341] Acetate addition gives a 1,4-product with good selectivity. The reaction proceeds via a π-allyl complex, 195, which is preferentially attacked at the less hindered site. With other nucleophiles this preference is less marked; 196 and 197 are produced in comparable amounts in ROH[342] and 198 is the sole product with ethylene glycol.[343] Stereoselectivity has been studied using cyclic dienes. With Pd(OAc)$_2$/LiOAc/benzoquinone, cyclohexadiene gives the trans-product, 199 with >90% selectivity, but with Li$_2$PdCl$_4$, or on addition of LiCl, the cis-product, 200, predominates. This is explained by initial trans-acetoxypalladation to give 201. If acetate is coordinated

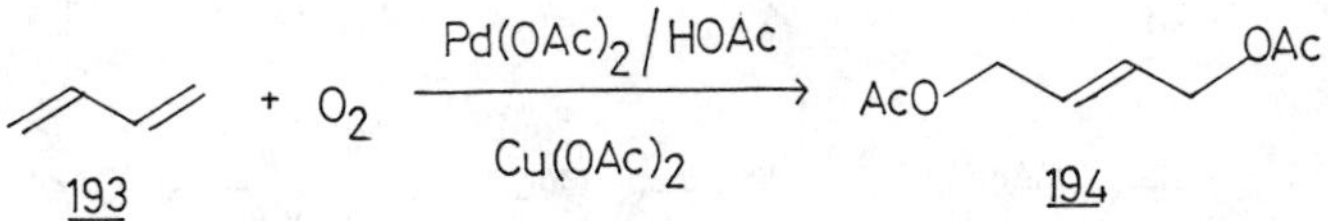

$$\text{193} \quad + \quad O_2 \quad \xrightarrow[Cu(OAc)_2]{Pd(OAc)_2 / HOAc} \quad AcO \diagdown\diagup\diagdown\diagup OAc \quad \text{194}$$

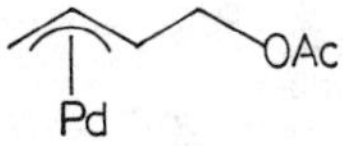

195

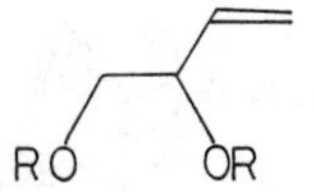

196

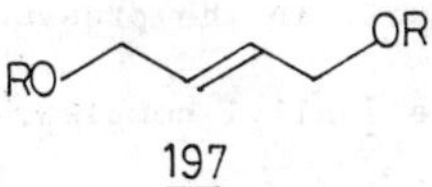

197

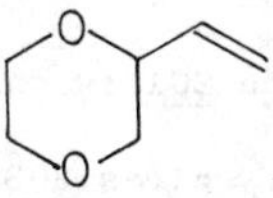

198

to palladium _cis_-migration gives the _trans_-product. In the presence of Cl⁻ this path is blocked and 201 is converted to the π-allyl complex, 202. Exchange of AcO⁻ for Cl⁻ gives 203 which is attacked by external acetate to give 200.

6.5.3 Cobalt catalysts

Co(III) is able to transfer one electron to an alkene to give a radical cation (equation (53)).[345] In aqueous media this gives a mixture

$$RCH=CH_2 + Co(III) \longrightarrow R\overset{+}{C}H-CH_2^{\cdot} + Co(II) \qquad (53)$$

of aldehydes, ketones, acids and dienes. With C_2H_4, ethylene dibromide and ethylene glycol diacetate are obtained with 40 - 60% selectivity at 30 - 60% conversion in the presence of $Co(OAc)_2/HOAc/NaBr/O_2$.[346] With the more powerful oxidant $Co(OCOCF_3)_3$ ethylene is cleanly converted to ethylene glycol di(trifluoroacetate) (equations (54) - (57)).[347]

$$C_2H_4 + [Co(OCOCF_3)_2]^+ \longrightarrow {}^+CH_2\text{-}CH_2^{\cdot} + Co(OCOCF_3)_2 \quad (54)$$

$${}^+CH_2\text{--}CH_2^{\cdot} + CF_3COOH \longrightarrow CF_3COOCH_2\overset{\cdot}{C}H_2 + H^+ \qquad (55)$$

$$CF_3COOCH_2\overset{\circ}{C}H_2 + [Co(OCOCF_3)_2]^+ \longrightarrow CF_3COOCH_2CH_2^+ +$$
$$Co(OCOCF_3)_2 \qquad (56)$$

$$CF_3COOCH_2CH_2^+ + CF_3COOH \longrightarrow CF_3COOCH_2CH_2OOCCF_3 + \overset{..}{H}{}^+ \quad (57)$$

Quite a different mechanism operates in oxidations catalysed by (TPP)Co complexes. HCoL is formed in situ and adds to the alkene (equation (58)). Homolysis (equation (59)) gives the radical which is autoxidised to the hydroperoxide. Cobalt catalysed decomposition of the hydroperoxide gives alcohol and ketone, which is reduced in situ by BH_4^- to the alcohol.[348]

$$HCoL + PhCH=CH_2 \rightleftharpoons PhCH(CH_3)CoL \qquad (58)$$

$$PhCH(CH_3)CoL \longrightarrow PhCHCH_3 + Co(II)L \qquad (59)$$

6.5.4 Manganese catalysts

Oxidation of alkenes by $Mn(OAc)_3$ in HOAc gives γ-lactones, but the process is not catalytic.[349] However, conversion of ethylene to its glycol is achieved catalytically in the presence of $Mn(OAc)_3$/KI (equations (60) – (63)).[350] Hydroperoxide intermediates may also be involved since direct reoxidation of $Mn(OAc)_2$ is known to be difficult.

$$Mn(III) + I^- \longrightarrow Mn(II) + I^{\cdot} \tag{60}$$

$$I^{\cdot} + CH_2=CH_2 \rightleftharpoons ICH_2CH_2^{\cdot} \tag{61}$$

$$ICH_2CH_2^{\cdot} + Mn(OAc)_3 \longrightarrow ICH_2CH_2OAc + Mn(OAc)_2 \tag{62}$$

$$ICH_2CH_2OAc + AcO^- \longrightarrow AcOCH_2CH_2OAc + I^- \tag{63}$$

In a different kind of reaction $Mn(II)$ catalyses the oxidation of equation (64).[351] The mechanism involves activation of the double bond by

$$3HO_2CCH=CHCO_2H + 4BrO_3^- \xrightarrow[H^+]{Mn(II)} 6HCOOH + 6CO_2 + 4Br^- \tag{64}$$

coordination, then oxidation by BrO_3^- to an $Mn(III)$ complex. Attack of water and diol cleavage completes the sequence. The course of the $S_2O_8^{2-}$ oxidation of 1-decene, is substantially altered in the presence of $Mn(OAc)_2$ (equation (65)). A radical chain mechanism is postulated but the details are unknown.[352]

$$C_8H_{17}CH=CH_2 + S_2O_8^{2-} \xrightarrow{HOAc}$$

$$\underset{\substack{| \quad |\\ OAc \quad OAc}}{C_8H_{17}CH{-}{-}CH} + C_8H_{17}CH=CHOAc + C_8H_{17}CH(Me)OAc \tag{65}$$

no catalyst	10%	6%	21%
$Mn(OAc)_2$	43%	4%	8%

6.5.5 Ruthenium catalysts

RuCl$_3$ catalyses oxidative cleavage of carbon carbon double bonds in the presence of various oxidants but the mechanism is incompletely understood (Figure 55).

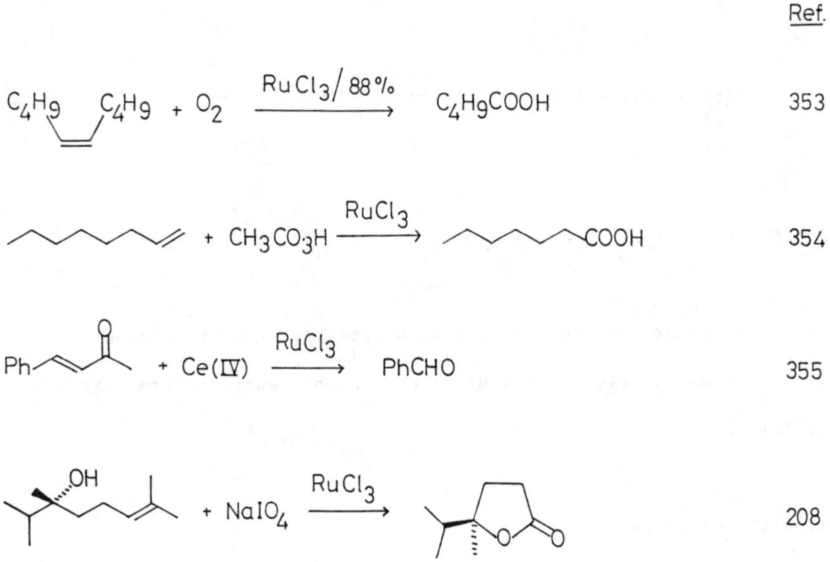

<u>Ref.</u>

C$_4$H$_9$ C$_4$H$_9$ + O$_2$ $\xrightarrow{\text{RuCl}_3/\,88\%}$ C$_4$H$_9$COOH   353

+ CH$_3$CO$_3$H $\xrightarrow{\text{RuCl}_3}$ ~~~~~COOH   354

Ph + Ce(IV) $\xrightarrow{\text{RuCl}_3}$ PhCHO   355

+ NaIO$_4$ $\xrightarrow{\text{RuCl}_3}$ 208

Figure 55   Ruthenium catalysed oxidative cleavage of alkenes.

6.5.6 Copper catalysts

CuCl$_2$ catalyses the conversion of propene to its glycol diacetate by a complex pathway (equations (66) - (72)). [Cu(OAc)$_{4-n}$Cl$_n$]$^{2-}$ complexes are detected and I$_2$ is enhanced in electrophilicity by complexation with copper.[356]

$$I_2 + HOAc \rightleftharpoons I_2.HOAc \tag{66}$$

$$I_2 + CuCl_2 \rightleftharpoons CuCl_2.I_2 \tag{67}$$

$$I_2.HOAc + CH_3CH=CH_2 \longrightarrow I_2.(CH_3CH=CH_2) + HOAc \tag{68}$$

$$CuCl_2.I_2 + CH_3CH=CH_2 \longrightarrow (CuCl_2.\ I_2.(CH_3CH=CH_2)) \tag{69}$$

$$I_2 + CH_3CH=CH_2 \longrightarrow CH_3CHICH_2I \tag{70}$$

$$(CuCl_2.I_2.(CH_3CH=CH_2)) \xrightarrow{HOAc} CH_3CH(OAc)CH_2I + $$
$$HCuCl_2 + \tfrac{1}{2}I_2 \tag{71}$$

$$CH_3CH(OAc)CH_2I + AcO^- \longrightarrow CH_3CH(OAc)CH_2OAc + I^- \tag{72}$$

6.5.7 Other metals as catalysts.

A range of other metal complexes have been used in alkene oxidation. The mechanisms of this diverse group of reactions are largely unknown (Figure 56).

6.6 Oxidation of Alkynes.

Oxidation reactions of alkynes have been little studied. OsO_4 in the presence of a tertiary amine, L, gives 205 which is readily hydrolysed to a diketone.[291] The use of an oxidant in stoichiometric quantities renders the reaction catalytic in osmium. Analogous reactions are found in the presence of $(Ph_3P)_2RuCl_2$ and RuO_2 (Figure 57). With Ru-based catalysts cleavage occurs more readily when alkyl rather than aryl groups are attatched to the triple bond. $Mo(CO)_6/\underline{t}$-BuOOH gives extensive cleavage even with Ph-C≡C-Ph, which gives benzil in only 35% yield.[363]

1-Alkynes undergo oxidative coupling to diynes in the presence of copper salts (e.g. 206 to 207).[364] ESR measurements show that Cu^{2+}

is present throughout and R-C≡C-Cu has been proposed as an intermediate. However, there is also an autocatalytic process not involving Cu^{2+}.[365]

$$C_3H_7\diagdown\diagup C_3H_7 + (NH_4)_2S_2O_8 \xrightarrow{Fe(OAc)_3} C_3H_7\cdots\overset{OAc}{\underset{H}{|}}\cdots\overset{H}{\underset{OAc}{|}}\cdots C_3H_7 \qquad 357$$

$$C_2H_4 + O_2 \xrightarrow[AcOH, NaOAc]{RhCl_3, CuCl_2} AcO\diagup\diagdown \qquad\qquad 358$$

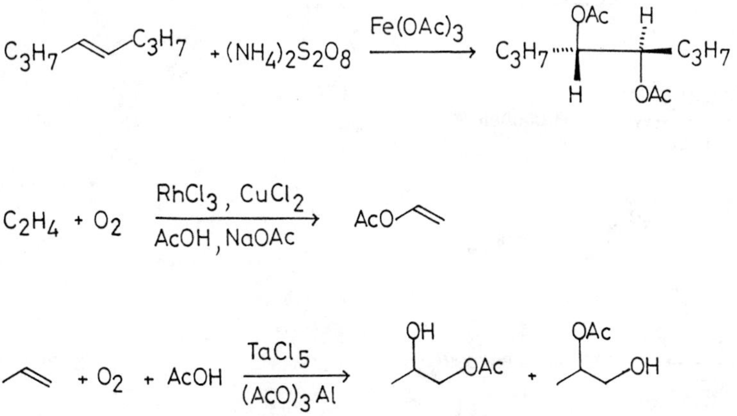

$$\diagup\diagdown + O_2 + AcOH \xrightarrow[(AcO)_3Al]{TaCl_5} \overset{OH}{|}\diagup\diagdown OAc + \overset{OAc}{|}\diagup\diagdown OH$$

Figure 56 Various metals as catalysts for alkene oxidation.

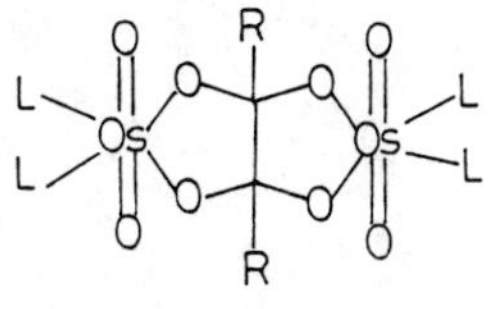

205

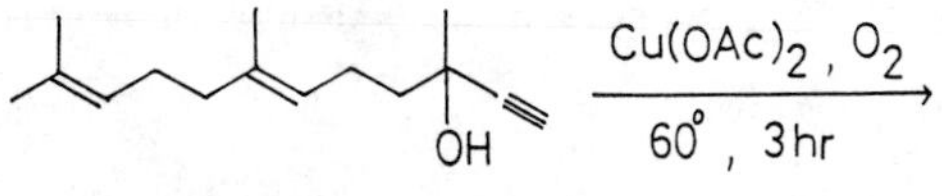

206

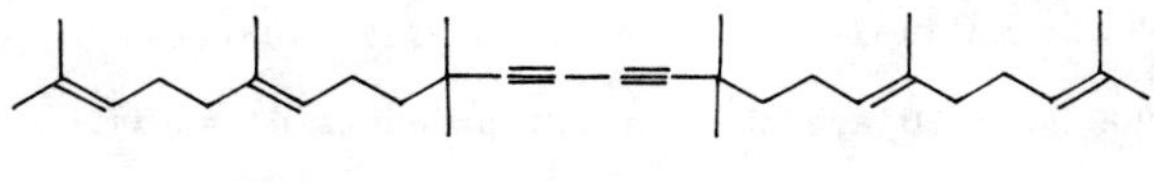

$$\xrightarrow[60°, 3hr]{Cu(OAc)_2, O_2}$$

207

$$EtC{\equiv}COMe \xrightarrow[\text{KClO}_3]{\text{OsO}_4} EtCOCOOMe$$

360

$$PhC{\equiv}CPh \xrightarrow[\text{NMO}]{\text{OsO}_4} PhCOCOPh$$

359

$$PhC{\equiv}CPh \xrightarrow[\text{PhIO}]{\text{(Ph}_3\text{P)}_3\text{RuCl}_2} PhCOCOPh$$

361

$$RC{\equiv}CNR'_2 \xrightarrow[\text{PhIO}]{\text{(Ph}_3\text{P)}_3\text{RuCl}_2} RCOCONR'_2$$

361

$$BuC{\equiv}CBu \xrightarrow[\text{NaOCl}]{\text{RuO}_2} BuCOCOBu + BuCOOH$$

70%   19%

362

Figure 57   Oxidation of alkynes to diketones in the presence of ruthenium
and osmium catalysts.

6.7 Oxidation of arenes.

Studies of selective oxidation of arenes are scattered, though
many metal complexes are active. Few mechanisms are known with certainty
but radical chain processes are common.

Palladium complexes catalyse the oxidative coupling of arenes
to biaryls in the presence of O_2, a reaction discussed in more detail
in Chapter 8. The initial step is electrophilic addition of palladium to
the arene. 209 and 210 are the initial products of addition of Pd(OAc)$_2$
to an electron rich arene and they lose HPdOAc to give meta-acetates.[366]
This acetoxylation is too slow to be used in synthesis. Addition of

$Cr_2O_7{}^{2-}$, $NO_3{}^-$ or $S_2O_8{}^{2-}$ increases the rate but decreases the selectivity[366,367]. Oxidative carbonylation (211 to 212)[368] and oxidative coupling to alkenes (213 to 214)[369] may also be achieved.

208   Pd(OAc)$_2$ 209   +   210

$$ArH + CO_2 \xrightarrow[\text{t-BuOOH}]{Pd(II)} ArCOOH$$

211   212

$$211 + CO \xrightarrow[\text{t-BuOOH/CH}_2\text{=CHCH}_2\text{Cl}]{Pd(OAc)_2/HOAc} 212$$

$$ArH + RCH=CHY \xrightarrow[\text{PhCO}_3\text{t-Bu}]{Pd(OCOR)_2} ArC(R)=CHY$$

213   214

R = Ph, Y = COMe, COOR

Radicals are involved in copper[370] and cobalt[371] mediated oxidation of anthracene to anthroquinone. Selectivity is high and the mechanism is shown in Figure 58. Intermediacy of hydroperoxides was postulated in an analogous rhodium catalysed reaction, which may also use t-BuOOH as oxidant.[372]

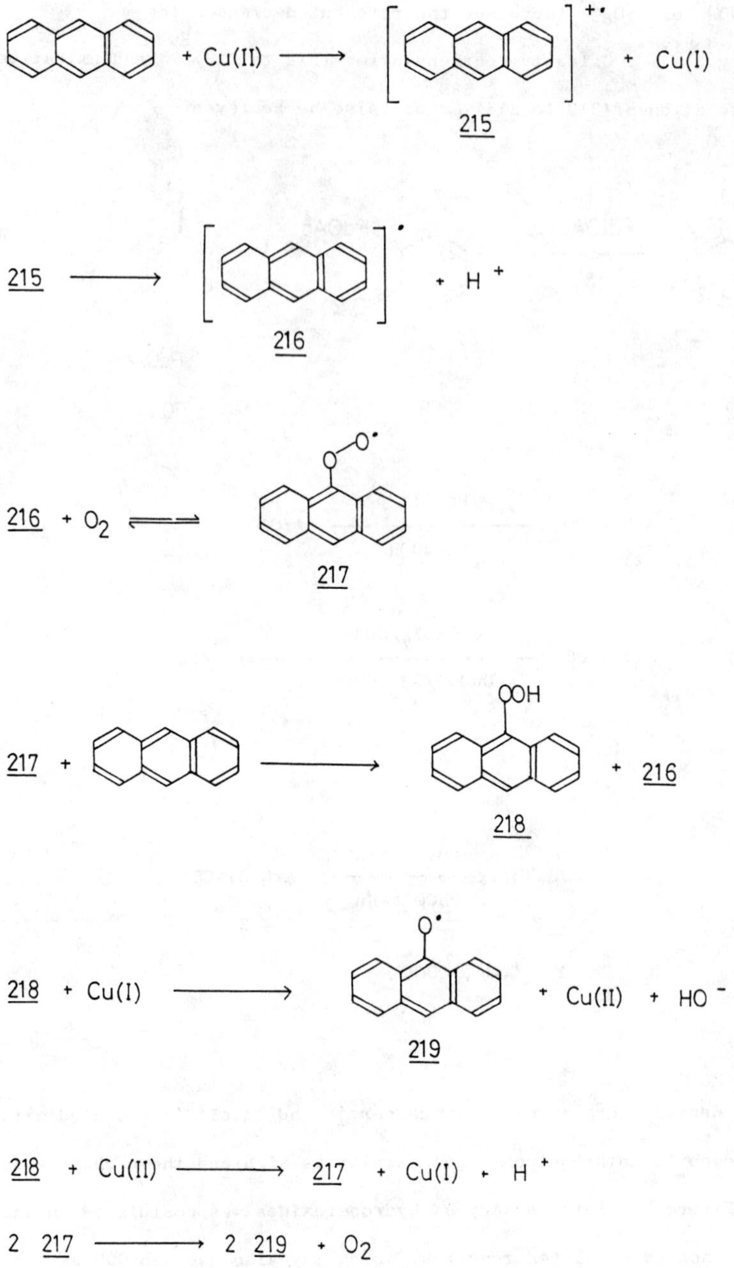

$$\text{anthracene} + Cu(II) \longrightarrow [\text{anthracene}]^{+\bullet} + Cu(I)$$

<u>215</u>

$$\underline{215} \longrightarrow [\text{anthracene}]^{\bullet} + H^+$$

<u>216</u>

$$\underline{216} + O_2 \rightleftharpoons \text{anthracene-OO}^{\bullet}$$

<u>217</u>

$$\underline{217} + \text{anthracene} \longrightarrow \text{anthracene-OOH} + \underline{216}$$

<u>218</u>

$$\underline{218} + Cu(I) \longrightarrow \text{anthracene-O}^{\bullet} + Cu(II) + HO^-$$

<u>219</u>

$$\underline{218} + Cu(II) \longrightarrow \underline{217} + Cu(I) + H^+$$

$$2\ \underline{217} \longrightarrow 2\ \underline{219} + O_2$$

Figure 58 Mechanism of oxidation of anthracene to anthroquinone.

Figure 58   cntd.

Electron transfer is the initiating step in a number of other cases. In the presence of Fe^{2+}, $S_2O_8{}^{2-}$ gives $SO_4{}^-$˙ which converts naphthalene to **223**. Acetate gives **224**, which in the presence of Cu^{2+}, gives acetoxynaphthalene.[373] Without Cu^{2+}, acetate radical is formed and further coupling occurs.[374] Some further examples are shown in Figure 59.

223   **224**

6.8 Oxidation of alkyl arenes.

Oxidation of alkyl arenes is an important industrial process. The mechanism usually involves electron transfer followed by autoxidation. The catalysts of choice are manganese and/or cobalt acetates with various promoters.

6.8.1 Manganese catalysts

Two pathways are known for oxidation of ArMe in the presence of Mn(III). In the first, electron transfer gives a radical cation which cleaves to $ArCH_2$˙ and H^+ in the rate determining step (equations (73) - (76)).

$$ArMe + Mn(III) \rightleftharpoons [ArMe]^{+˙} + Mn(II) \qquad (73)$$

$$[ArMe]^{+˙} \xrightarrow{\text{slow}} ArCH_2{}˙ + H^+ \qquad (74)$$

$$ArCH_2{}˙ + Mn(III) \longrightarrow ArCH_2{}^+ + Mn(II) \qquad (75)$$

$$ArCH_2{}^+ + HOAc \longrightarrow ArCH_2OAc + H^+ \qquad (76)$$

Ref.

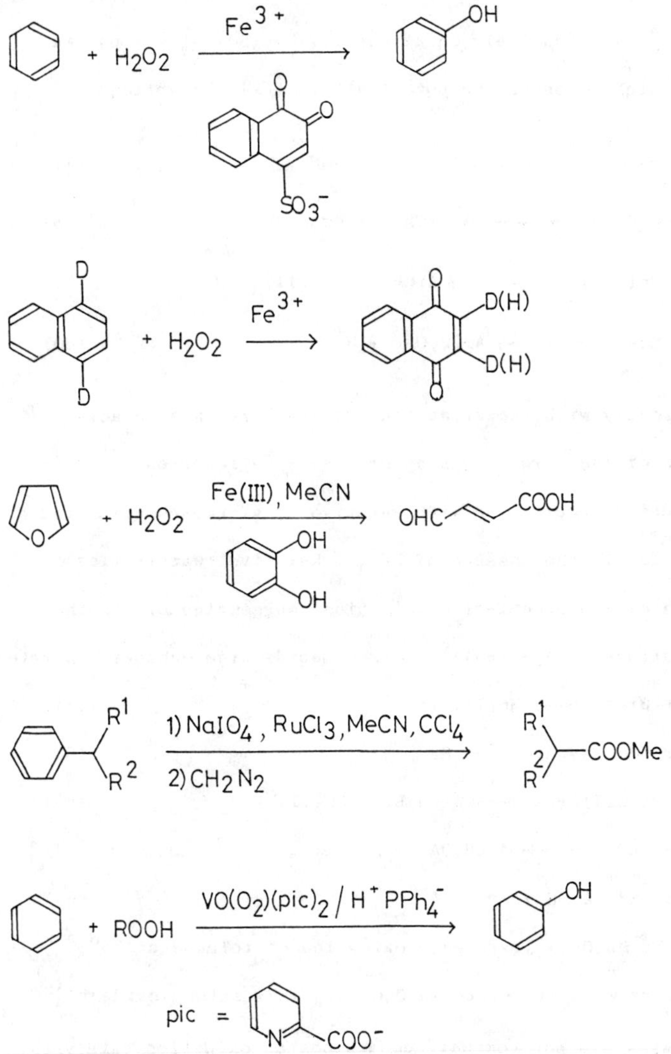

375

376

377

378

379

Figure 59 Catalytic oxidation of arenes.

The rate is dependent on the ionisation potential of the aromatic substrate and this route is only avaliable for hydrocarbons with ionisation potentials less than 8eV.[380,381] Alternatively,equations (77) - (80) give a pathway for substates with higher ionisation potential. In a side reaction

$$Mn(OAc)_3 \xrightarrow{\text{HOAc}} Mn(OAc)_2 + {}^{\cdot}CH_2COOH \tag{77}$$

$$ArMe + {}^{\cdot}CH_2COOH \longrightarrow ArCH_2{}^{\cdot} + HOAc \tag{78}$$

$$ArCH_2{}^{\cdot} + Mn(III) \longrightarrow ArCH_2{}^{+} + Mn(II) \tag{79}$$

$$ArCH_2{}^{+} + HOAc \longrightarrow ArCH_2OAc + H^{+} \tag{80}$$

${}^{\cdot}CH_2COOH$ reacts directly with the arene ring to give aryl acetic acids.[382]

Catalysis of these reactions by Br^- has been reported.[381][383] Here a radical reaction occurs even with reactive alkyl arenes which would give electron transfer in the absence of Br^-. Relative reactivities correspond to those of photochemical bromination, suggesting $Br^{\cdot}$ is the chain carrier (equations (81) - (85)). Strong acids also enhance the rate;

$$Mn(III) + Br^- \longrightarrow Mn(II) + Br^{\cdot} \tag{81}$$

$$ArMe + Br^{\cdot} \longrightarrow ArCH_2{}^{\cdot} + HBr \tag{82}$$

$$ArCH_2{}^{\cdot} + Mn(III)Br \longrightarrow ArCH_2Br + Mn(II) \tag{83}$$

$$ArCH_2Br + AcO^- \longrightarrow ArCH_2OAc + Br^- \tag{84}$$

$$AcO^- + HBr \rightleftharpoons AcOH + Br^- \tag{85}$$

H_2SO_4 accelerates the $Mn(OAc)_3$ catalysed oxidation of toluene at $25°$.[384] Activation also occurs with TFA since $Mn(OCOCF_3)_3$ is a strong oxidant.[382]

Mn(III) salts are not commonly used alone as oxidation catalysts. Some examples are given in Figure 60.

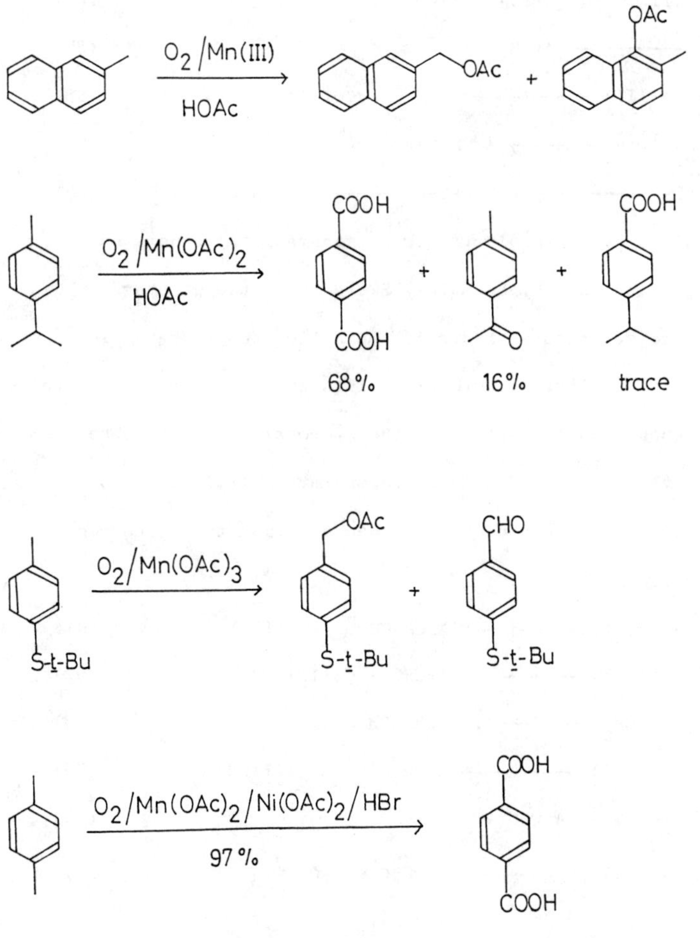

Ref.

385

386

381

387

Figure 60 Oxidation of alkyl arenes in the presence of Mn(III) salts.

6.8.2 Cobalt catalysts

Reactions using $Co(OAc)_3$ as catalyst proceed exclusively by electron transfer, and the reactivity of the arene is correlated with its ionisation potential.[388] Also, the relative rates of oxidation of alkyl benzenes are the reverse of those expected from a radical pathway.[380,386,389]

The mechanism has been thoroughly studied (equations (86) - (90)). Proton

$$ArMe + Co(III) \rightleftharpoons [ArMe]^{+\bullet} + Co(II) \qquad (86)$$

$$[ArMe]^{+\bullet} \longrightarrow ArCH_2^{\bullet} + H^+ \qquad (87)$$

$$ArCH_2^{\bullet} + Co(III) \longrightarrow ArCH_2^+ + Co(II) \qquad (88)$$

$$ArCH_2^+ + HOAc \longrightarrow ArCH_2OAc + H^+ \qquad (89)$$

$$ArCH_2^{\bullet} + O_2 \longrightarrow ArCH_2\text{-}O\text{-}O^{\bullet} \longrightarrow products \qquad (90)$$

loss may be controlled by stereoelectronic considerations rather than product stability; thus, 225 gives mainly 226 despite the greater stability of the more substituted radical. Kinetic studies have proved complex.[390,392] In particular, $Co(OAc)_3$ exists as a dimer and the rate determining step depends on the nature of the hydrocarbon. In subsequent steps $ArCH_2\text{-}O\text{-}O^{\bullet}$ may abstract hydrogen or regenerate Co(III) from Co(II) (equations (91) - (96)). The aldehyde is then oxidised to the corresponding benzoic acid.

$$ArCH_2OO^{\bullet} + Co(II) \longrightarrow ArCH_2COO^- + Co(III) \qquad (91)$$

$$ArCH_2OO^{\bullet} + Co(II) \longrightarrow ArCHO + Co(III)OH \qquad (92)$$

$$ArCH_2OO^{\bullet} + ArCH_3 \longrightarrow ArCH_2OOH + ArCH_2^{\bullet} \qquad (93)$$

$$ArCH_2OOH + Co(II) \longrightarrow ArCH_2O^{\bullet} + Co(III)OH \qquad (94)$$

$$ArCH_2O^{\bullet} \longrightarrow ArCHO \qquad (95)$$

$$ArCH_2\text{-}OOH + Co(II) \longrightarrow ArCHO + H_2O \qquad (96)$$

225     $\xrightarrow[\text{HOAc}]{\text{Co(OAc)}_2 / O_2}$ 226 (90%) + 227

The presence of halide ions considerably enhances the reaction rate.[393,394] Again Br· is the chain carrying species (equations (97) - (100)) and the reaction has the characteristics of a radical process.[395]

$$Co(III) + Br^- \longrightarrow Co(II) + Br· \qquad (97)$$

$$Br· + ArMe \longrightarrow ArCH_2· + HBr \qquad (98)$$

$$ArCH_2· + O_2 \longrightarrow ArCH_2O_2· \qquad (99)$$

$$ArCH_2O_2· + Co(II) \longrightarrow ArCHO + Co(III)OH \qquad (100)$$

Cl^- also enhances the rate, but selectivities suggest an electron transfer mechanism still operates, and rate acceleration is due to the formation of a Co(III) complex of different oxidation potential.[396] The rate is also improved in the presence of TFA and other strong acids due to the formation of $[Co(OCOCF_3)_2]^+$ and analogues.[397]

Moving to consider various substrates, the simplest is toluene, from which PhCHO or PhCOOH may be isolated under appropriate conditions. Some of the many available data are summarised in Table 6. The oxidation of xylenes gives a wider range of potential products, for example, 228 - 232, from para-xylene. Since aldehydes are oxidised more rapidly than methyl groups, 228 and 230 are uncommon. Some results are shown in Table 7. Terephthalic acid, 232, is an important product industrially, and a number of patents also report its preparation by oxidation of 228[408] and 229.[409,410] Tri- and tetramethyl benzenes are similarly oxidised to mixtures of carboxylic acids (Table 8). Other functional groups including halogen, OR and ester do not interfere with the oxidation (Figure 61). Methyl heterocycles, naphthalenes and biphenyls are also readily oxidised. (Figure 62).

Table 6

Oxidation of toluene in the presence of Co(III) catalysts.

Catalyst	Conditions	Products	Ref.
CoBr$_3$	ZrCl$_4$/HOAc	PhCOOH (>90%)	398
Co(octanoate)$_3$	Air, 375°F, 150 psi	PhCOOH (40 - 50%)	399
Co(OAc)$_3$	O$_2$, 60°	PhCHO	390
Co(OAc)$_2$	NaBr/HOAc	PhCHO then PhCOOH	400
Co(OAc)$_3$	HOAc, 87°	PhCOOH	391

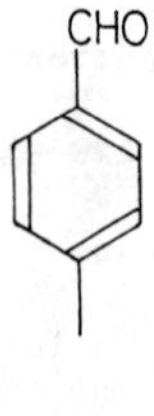

228 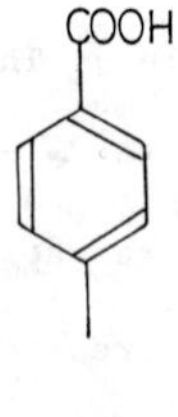   229 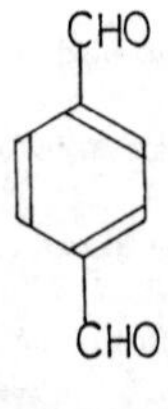   230 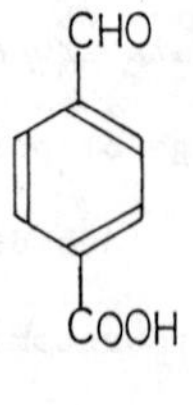   231 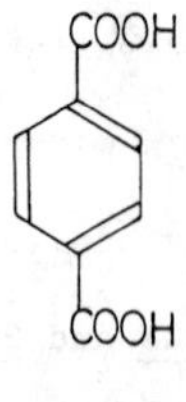   232

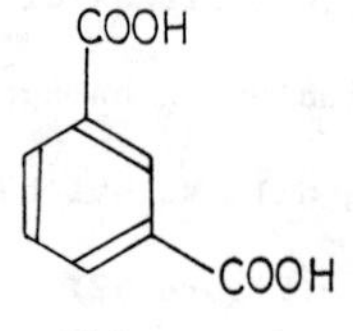

233 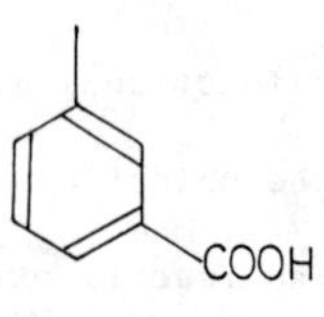   234 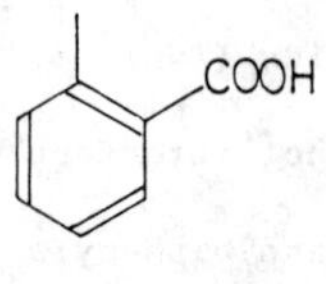   235

Table 7

Oxidation of xylenes catalysed by Co(III).

Xylene	Conditions	Products	Ref.
para	$Co(OAc)_2$, 300 psi, 260°F	$\underline{229}$ (10%) + $\underline{232}$ (90%)	401
para	$Co(OAc)_3$, Zr^{4+}, 1 atm., 100°	$\underline{232}$ (100%)	402
para	$Co(OAc)_2$, HCl, PhCl, NaBr, HOAc	$\underline{232}$ (94%)	394
para	$Co(OAc)_2$, $La(OAc)_3$, HBr, HOAc, 190°, 20 kg cm^{-2}	$\underline{232}$ (95%)	403
para	$Co(OAc)_2$, NH_4Br, $N(CH_2CH_2CN)_3$, 1 atm.	$\underline{232}$ (78.5%), $\underline{229}$ (2.9%), p-HOC_6H_4COOH (5.5%)	404
meta	$CoBr_2$, HOAc, 140°, 3 atm., air	$\underline{233}$ (90%)	405
meta	$Co(OAc)_2$, NH_4Br, $N(CH_2CH_2CN)_3$, 1 atm.	$\underline{233}$ (84%)	404
meta	$Co(naphthenate)_2$	$\underline{234}$ (79%)	406
ortho	$(\underline{236})_2Co$	$\underline{235}$ (86%)	407

Table 8

Oxidation of trimethyl benzenes in the presence of Co(III).

Substrate	Conditions	Product	Ref.
237	Co(OAc)$_2$, NH$_4$Br, N(CH$_2$CH$_2$CN)$_3$, 1 atm.	238 (85%)	404
237	CoBr$_2$, CH$_3$CHO	238 (99%)	411
237	CoBr$_2$, HOAc, 100°	239 (42%)	412
240	Co(OAc)$_2$, HBr, AcOH, 3 hr., 115°	241	413

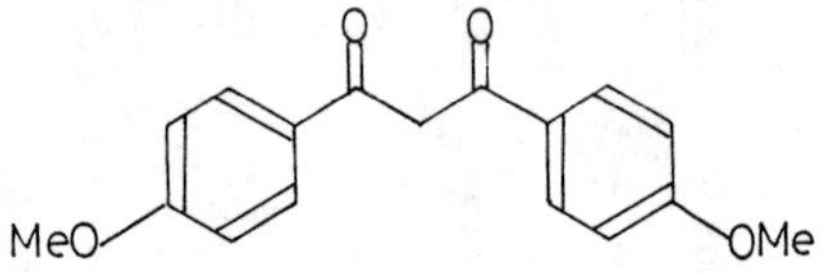

236

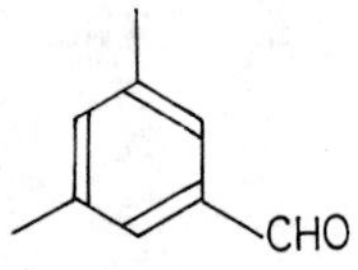

239

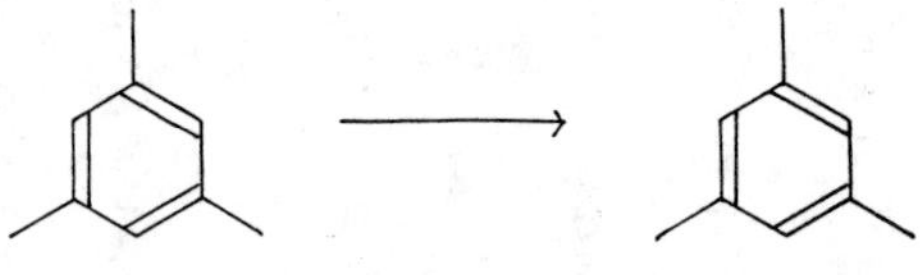

237

238

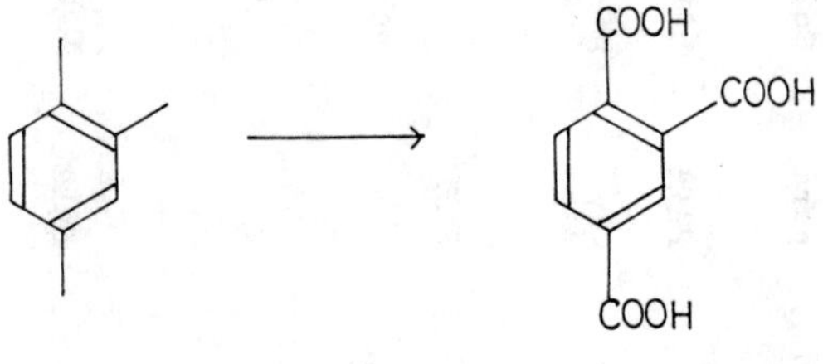

240

241

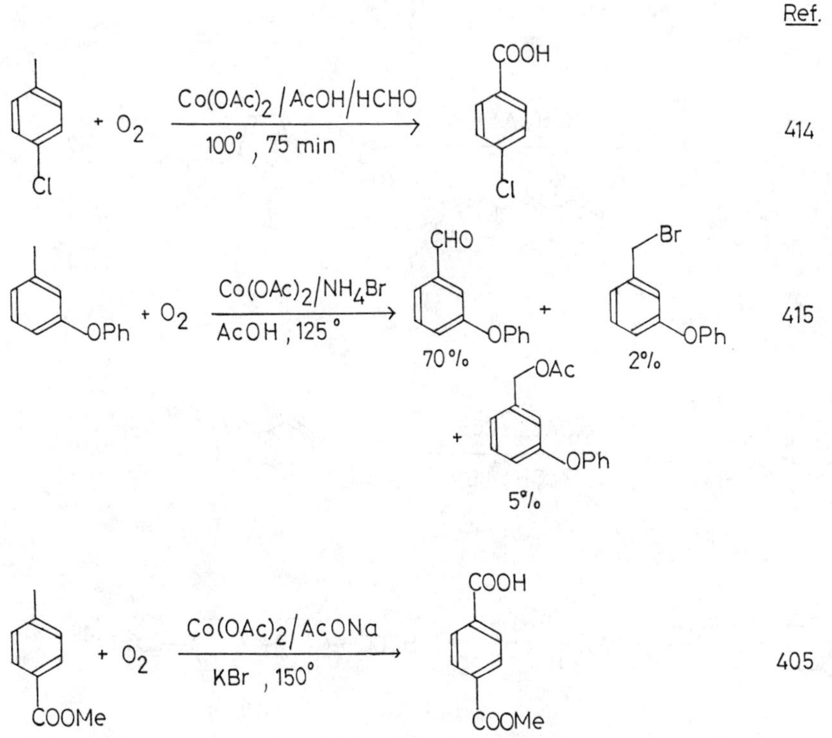

414

415

405

Figure 61   Cobalt catalysed oxidation of functionalised alkyl arenes.

Ref.

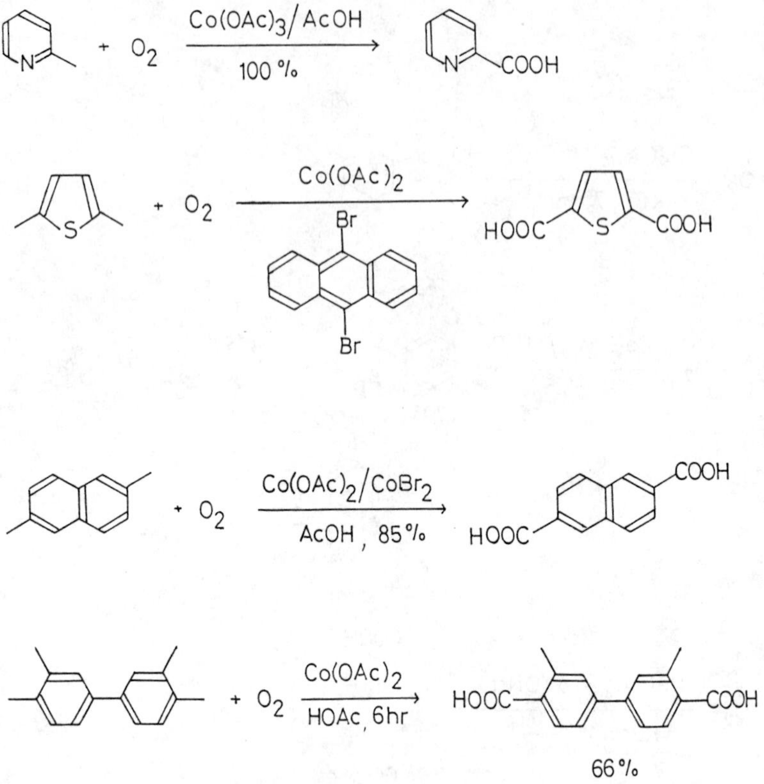

416

417

418

419

Figure 62 Cobalt catalysed oxidation of methyl heterocycles, naphthalenes, and biphenyls.

Oxidation of ethyl benzene gives acetophenone as the major product,[420] and both substituted and heterocyclic analogues are similarly reactive.[421] However, p-ethyl toluene gives products derived from oxidation of the methyl groups. The relative ease of oxidation of methyl groups, noted earlier, is quite general (Figure 63). Other alkyl groups are oxidised with lower efficiency; hydroperoxides may sometimes be isolated,[265] but alcohols or ketones are the usual products.[390,424-426]

Ref.

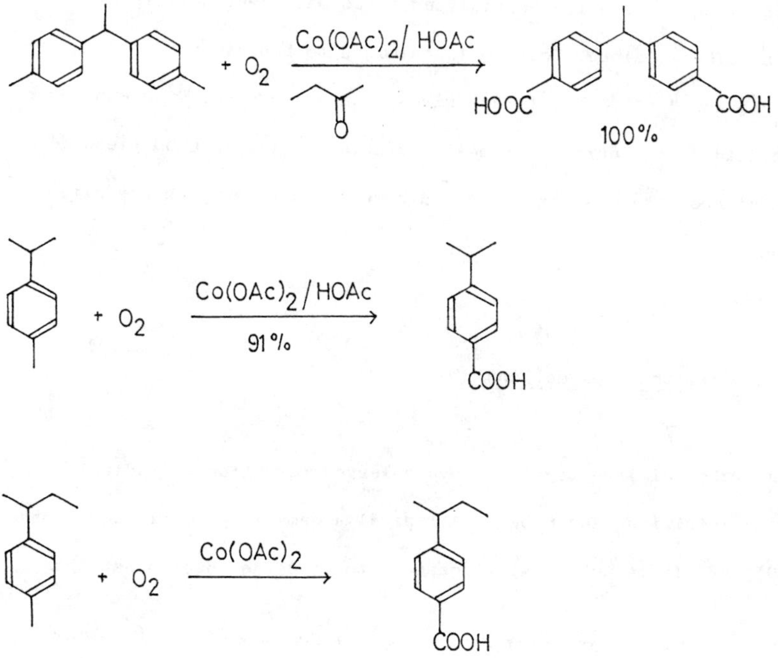

422

423

386

Figure 63 Cobalt catalysed oxidation of dialkyl benzenes.

6.8.3 Cobalt/manganese catalysts.

It is often found that better rates are obtained in alkyl benzene oxidation when both Mn(III) and Co(III) are present in the reaction mixture. Different selectivities may also be observed. The kinetics of the reaction are very complicated and it seems that the presence of Co(III) improves the generation of Br , whereas Mn(III) accelerates the decomposition of RO_2 .[427,428] In a recently studied peracid oxidation it was found that the synergising effect of Co(III) on $Mn(OAc)_2$/HBr lies in its ability

to promote oxidation of Mn(II) by peracid. The synergising effect of

Mn(III) on Co(OAc)Br is due to its ability to accelerate electron transfer

from Co(III). Cu^{2+} has an antagonistic effect on both processes since it

catalyses Br˙ recombination at the expense of H-abstraction.[429]

The range of reactions catalysed by $Co(OAc)_3$/$Mn(OAc)_3$ is similar

to that with $Co(OAc)_3$ alone. Toluene is oxidised to benzoic acid and

xylenes to diacids.[430,431] Mesitylene gives trimesitic acid,[432] pseudocumene

trimellitic acid[433] and durene pyromellitic acid,[434] all in good yields.

Methyl naphthalenes are also oxidised and some further examples are given

in Figure 64.

6.8.4 Peroxodisulphate oxidations

A number of investigators have observed formation of radical

cations using metal ions and $S_2O_8^{2-}$. A typical scheme is given in equations

(101) - (105).[439] It is not usually possible to stop the reaction at the

$$Cu^+ + S_2O_8^{2-} \longrightarrow Cu^{2+} + SO_4^{2-} + SO_4^{-}\cdot \qquad (101)$$

$$SO_4^{-}\cdot + ArMe \longrightarrow SO_4^{2-} + [ArMe]^{+}\cdot \qquad (102)$$

$$[ArMe]^{+}\cdot \longrightarrow ArCH_2\cdot + H^+ \qquad (103)$$

$$ArCH_2\cdot + Cu^{2+} + H_2O \longrightarrow ArCH_2OH + Cu^+ + H^+ \qquad (104)$$

$$ArCH_2OH \xrightarrow[Cu^{2+}]{S_2O_8^{2-}} ArCHO \qquad (105)$$

alcohol stage since this is of comparable reactivity to the starting

material. An exception is the oxidation of 4,5-dimethylimidazole, which

gives good selectivity for 4-hydroxymethyl-5-methylimidazole, an intermediate

in the industrial synthesis of cimetidine.[440] The high yields of aldehydes

Ref.

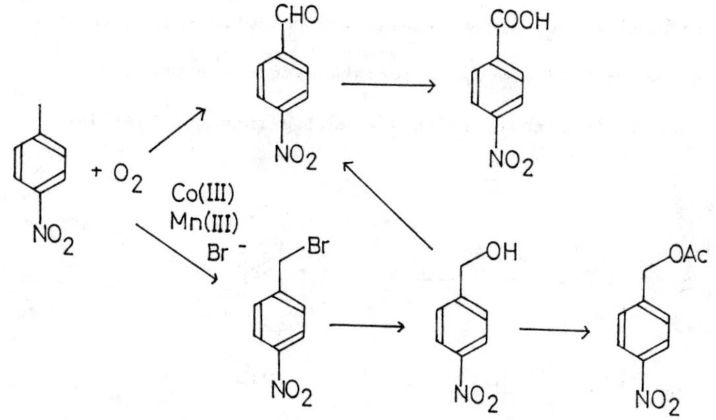

435

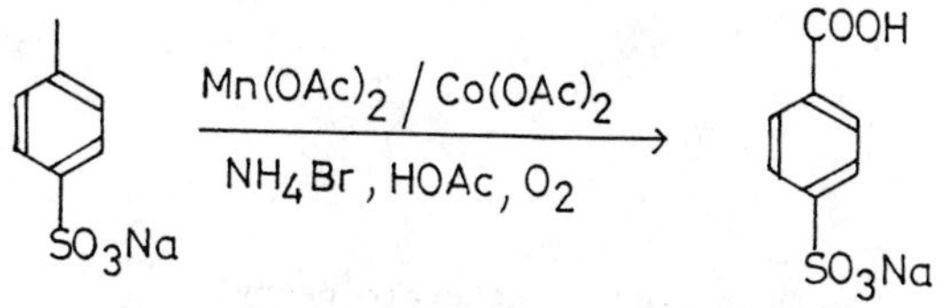

436

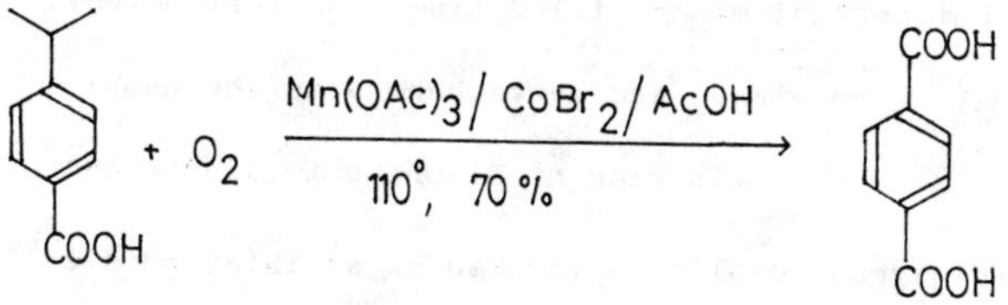

437

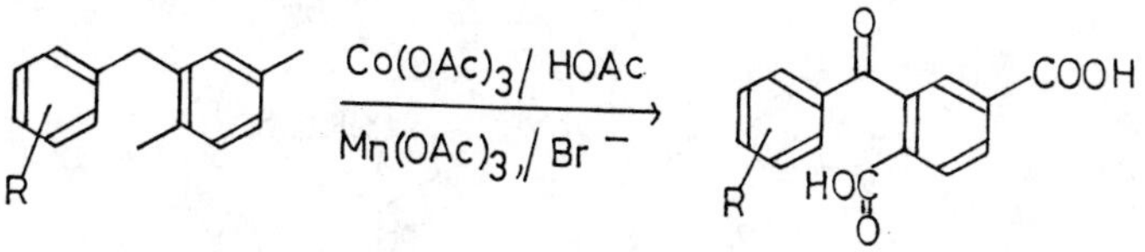

438

Figure 64 Catalytic oxidation of alkyl arenes in the presence of Co(III) and Mn(III).

usually obtained (e.g. 242 to 243) are consistent with an electron transfer

process, since the aldehydes have higher ionisation potentials than the

methyl arenes.[441] In acetic acid, benzylic acetate esters are obtained

in good yield, presumably since they are less reactive than the starting

materials.[442,443]

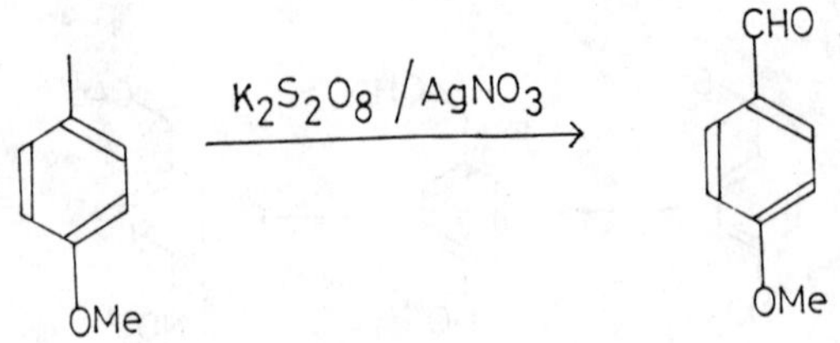

242      $K_2S_2O_8 / AgNO_3$ 243

6.8.5 Copper catalysts.

The Cu(II)/Cu(I) couple has been used to generate benzylic

radicals for autoxidation, but mechanisms are ill-defined. Hydroperoxides,

such as 245[444] and 247[445] may be obtained, and acetophenone is the usual

product from ethyl benzene.[446] In the oxidation of fluorene to fluorenone

a ternary substrate/copper/dioxygen complex is invoked as an intermediate.[447]

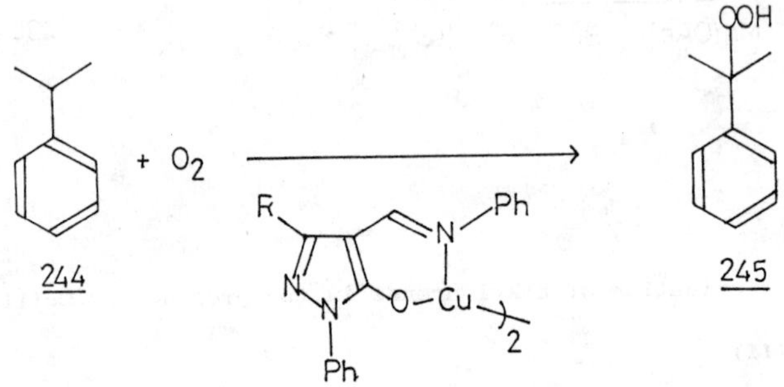

244   + O_2 245

6.8.6 Other metals.

Whilst various other metal complexes have been used to oxidise alkyl arenes and some good selectivities obtained, the reactions have not been widely used, nor the mechanisms well studied. Some examples are given in Figure 65.

6.9 Oxidation of saturated hydrocarbons.

Oxidation of saturated hydrocarbons occurs mainly by a radical mechanism, is unselective, and requires extreme conditions. Metal complexes as initiators provide some acceleration but selectivity usually remains low. For example, alkanes are hydroxylated in the presence of $Co(OAc)_2$/ $HOAc/O_2$. Methane is converted to methanol and methyl chloride using H_2PtCl_6/O_2 in a stoichiometric process and the use of $Na_8HPMo_6V_6O_{40}$ as a relay for reoxidation permits a catalytic reaction.[452]

Butane is oxidised at 170° to give a variety of products with 40% selectivity to acetic acid, in the presence of various metal ions. Using $Co(OAc)_2$/butanone allows the reaction to occur at 100°- 125° and selectivity to HOAc rises to 83%.[453]

$$ArMe + O_2 \xrightarrow[\text{HBr}]{V_2O_5} ArCOOH$$

448

$$PhCH_2Ph \xrightarrow[\text{Ru(II)}]{PhI(OAc)_2} PhCOPh$$

449

$$\underline{p}\text{-}CH_3C_6H_4CH_3 \xrightarrow[\text{Ru(II)}]{IO_4^-} \underline{p}\text{-}CH_3C_6H_4COOH$$

449

$$PhCH_2CH_3 + O_2 \xrightarrow{(Ph_3P)_3RhCl} PhCOCH_3$$

450

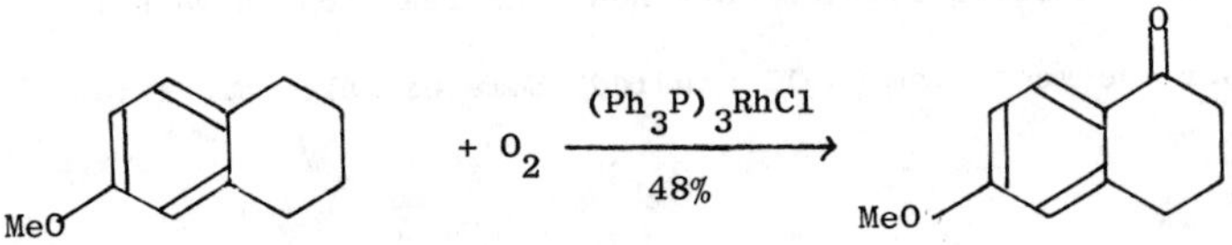

451

Figure 65 Other metal complexes as catalysts for alkyl arene oxidation.

A certain measure of selectivity is achieved in the presence of metal tetraphenylporphyrin complexes. Both ROOH and PhIO have been used as oxidants and heptane is conveted to 2-, 3- and 4-heptanols and heptanones. Yields do not depend strongly on the ligand using ROOH, but with PhIO the nature of the aryl ring is relevant. This is explained by the mechanisms of Figure 66.[454,455]

Tertiary CH bonds are curiously unreactive; iso-butane is less reactive than n-butane. This has been interpreted in terms of rate-limiting electron transfer (equation (106)); however, many anomalies exist.

$$M + ROOH \longrightarrow MOH + RO^{\cdot}$$

$$RO^{\cdot} + R'H \longrightarrow R'^{\cdot} + ROH$$

$$R'^{\cdot} + MOH \longrightarrow R'OH + M$$

$$M + PhIO \longrightarrow M=O + PhI$$

$$M=O + R'H \longrightarrow MOH + R'^{\cdot}$$

$$R'^{\cdot} + MOH \longrightarrow R'OH + M$$

M = tetraphenylporphyrin metal complex

<u>Figure 66</u>   Mechanisms of alkane oxidation in the presence of metal

tetraphenylporphyrin complexes.

$$RH + Co(III) \longrightarrow [RH]^{+\cdot} + Co(II) \qquad (106)$$

Rearrangements, such as that of <u>249</u>, are fairly common.[456] In the presence

of TFA some impressive selectivities have been obtained; n-heptane is

converted to 2-heptyl acetate in 81% ($Co(OAc)_3$/HOAc/TFA/N_2), 2-heptanone

in 83%($Co(OAc)_3$/HOAc/TFA/O_2), and 2-chloroheptane in 80% ($Co(OAc)_3$/

Cl_3CCOOH/HOAc/N_2) selectivities.[393,457]

Ph — 248

Ph — 249

Ph — 250

Ph —N3
21%
251

Ph —OH
1%
252

Ph —N3
10%
253

Ph —OH
1%
254

Cycloalkanes, in particular cyclohexane, have been popular substrates. With $Co(OAc)_3$/HOAc at 90° the rates are in the order $C_5 > C_6 > C_7- C_{12}$, which the authors interpret in terms of rate determining cobalt complexation.[458] In the absence of oxygen the products of Co(III) oxidation are **255** and **256** but with O_2, adipic acid is obtained in up to 83% selectivity.[458,459]

255

256

The oxidation of cyclohexane by ROOH and PhIO is also catalysed by metal tetraarylporphyrin complexes. Cyclohexanone and cyclohexanol are obtained in varying proportions (Table 9). In the presence of X^-, cyclohexylX is also obtained.[160,456,461-463] The radical nature of the process was proven by the reaction of **257**, which gave, after hydroxylation with (TPP)FeCl/PhIO, a mixture of **258** and **259**.[464] In addition, with norcarane, **260**, the intermediate, **261** is both captured and rearranged.[465] The best yielding reaction of this type is provided by 5,10,15,20-tetra(pentachlorophenyl)porphinato Fe(III)Cl/PhIO, which gives cyclohexanol in up to 73% yield.[466]

Table 9

Oxidation of cyclohexane in the presence of metal tetraphenylporphyrin complexes

Catalyst	Oxidant	% Cyclohexanol	% Cyclohexanone	Ref.
(TPP)FeCl	PhCMe$_2$OOH	40	20	106
(TPP)FeCl	t-BuOOH	20	12	460
(TPP)FeCl	PhIO	12	1	155,460
(TPP)MnCl	PhCMe$_2$OOH	1	15	106

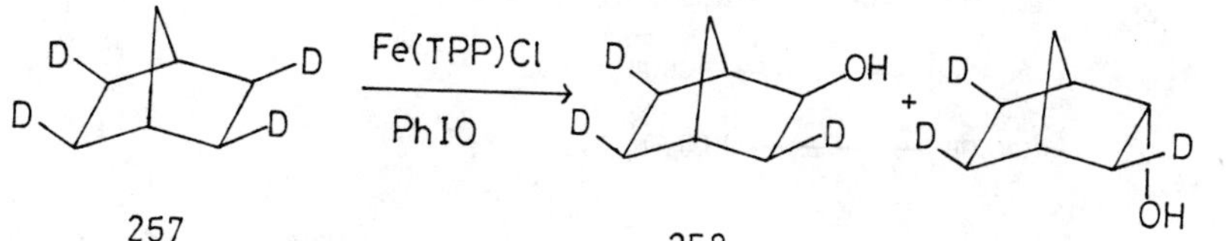

257 → Fe(TPP)Cl / PhIO → 258 + 259

260 + (TPP)MCl → 261 + (TPP)MCl

262 + 263 → 264

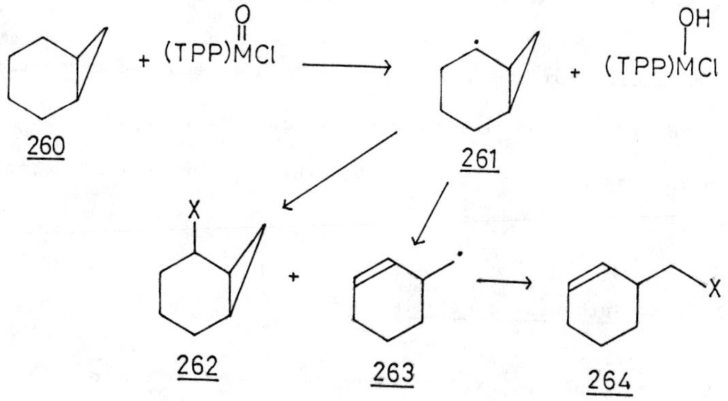

6.10 Oxidation of alcohols

6.10.1 Oxidation of primary alcohols

Primary alcohols may be oxidised in two steps to yield aldehydes and carboxylic acids. Representatives of most types of oxidant and most mechanisms are known.

6.10.1.1 Oxidation by peroxodisulphate

Silver ion in conjunction with $S_2O_8^{2-}$ initiates radical oxidation of primary alcohols. Radical generation may follow several routes (equations (107) - (110)). The carbon centred radical is the

$$S_2O_8{}^{2-} + Ag^+ \longrightarrow Ag^{2+} + SO_4{}^{-\bullet} + SO_4^{2-} \tag{107}$$

$$Ag^{2+} + RCH_2OH \longrightarrow R\overset{\bullet}{C}HOH + Ag^+ + H^+ \tag{108}$$

$$SO_4{}^{-\bullet} + RCH_2OH \longrightarrow R\overset{\bullet}{C}HOH + SO_4^{2-} + H^+ \tag{109}$$

$$Ag^{2+} + RCH_2OH \longrightarrow RCH_2O^{\bullet} + Ag^+ + H^+ \tag{110}$$

usual intermediate; reaction with further $S_2O_8{}^{2-}$ gives RCHO (equation (111)) and selectivity is good.[467][468] ESR studies suggest that O- and C-centred

$$R\overset{\bullet}{C}HOH + S_2O_8{}^{2-} \longrightarrow RCHO + HSO_4^- + SO_4{}^{-\bullet} \tag{111}$$

radicals are in equilibrium and H-abstraction in a 6-membered transition state (265 to 266) is invoked in conversion of quinoline in good selectivity to 267 and 268.[469] Cu^+ may also be used to initiate electron transfer.

6.10.1.2 Ruthenium catalysts

Ruthenium complexes oxidise alcohols by abstraction of α-hydrogens to give carbocations, $R\overset{+}{C}HOH$. The reduced ruthenium is then reoxidised by a variety of oxidants, one of the most popular being

$[Fe(CN)_6]^{3-}$.   A typical scheme is given by equations (112) - (115).[470,471]

Other kinetic schemes use Ru(VI) as the active catalyst.[472]   Some reactions

$$[Ru(III)OH]^{2+} + PhCH_2OH \rightleftharpoons [Ru(III)PhCH_2OH]^{3+} + HO^- \qquad (112)$$

$$[Ru(III)PhCH_2OH]^{3+} \xrightarrow{\text{slow}} Ph\overset{+}{C}HOH + [HRu(III)]^{2+} \qquad (113)$$

$$[HRu(III)]^{2+} + 2HO^- + 2[Fe(CN)_6]^{3-} \longrightarrow [Ru(III)OH]^{2+}$$
$$+ 2[Fe(CN)_6]^{4-} + H_2O \qquad (114)$$

$$Ph\overset{+}{C}HOH \longrightarrow PhCHO + H^+ \qquad (115)$$

are reported to be selective for aldehydes but the usual product is a

carboxylic acid.[473]   When $S_2O_8^{2-}$ is used as an oxidant, acids are the

exclusive products.[474,475]   Some other examples are shown in Figure 67.

6.10.1.3 Other metals

Osmium complexes may be used with an oxidant in the same way as

ruthenium, but the toxicity of the metal renders this less attractive.[479]

Copper complexes have been quite widely used and in the presence of

phen/O_2 are selective for oxidation of primary alcohols.[480]   Some further

examples are given in Figure 68.

$$\text{epoxide-CH}_2\text{OH} \xrightarrow{\text{RuCl}_3/\text{NaIO}_4} \text{HO-CO-CH}_2\text{-epoxide} \qquad 186$$

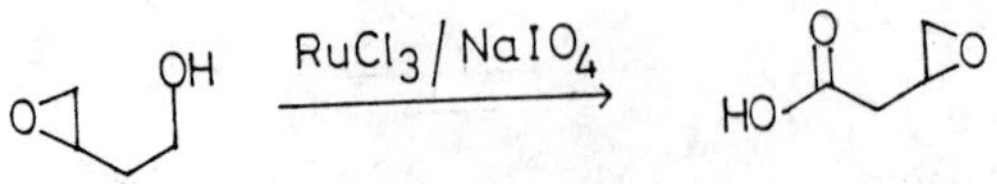

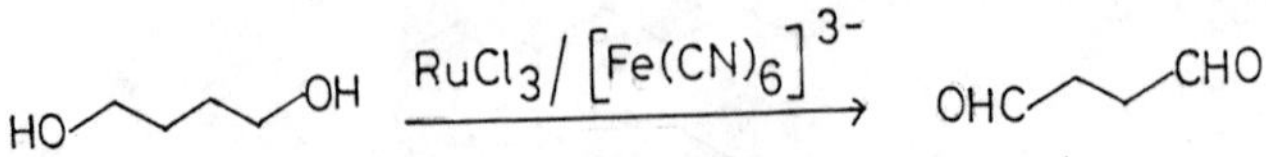

$$\text{HO-CH}_2\text{CH}_2\text{CH}_2\text{CH}_2\text{-OH} \xrightarrow{\text{RuCl}_3/\left[\text{Fe(CN)}_6\right]^{3-}} \text{OHC-CH}_2\text{CH}_2\text{-CHO} \qquad 476$$

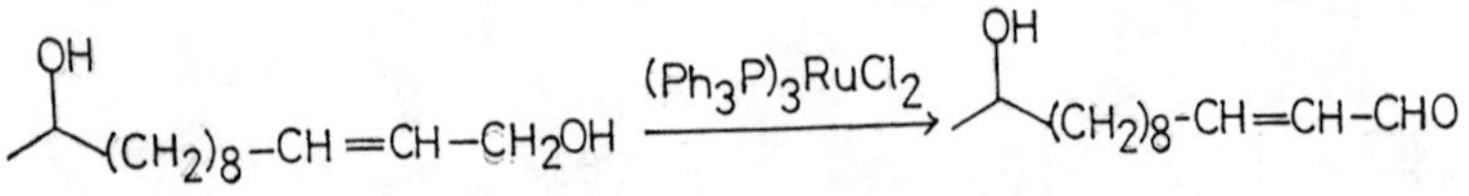

$$\overset{\text{OH}}{|}\text{-(CH}_2)_8\text{-CH=CH-CH}_2\text{OH} \xrightarrow{\text{(Ph}_3\text{P)}_3\text{RuCl}_2} \overset{\text{OH}}{|}\text{-(CH}_2)_8\text{-CH=CH-CHO} \qquad 477$$

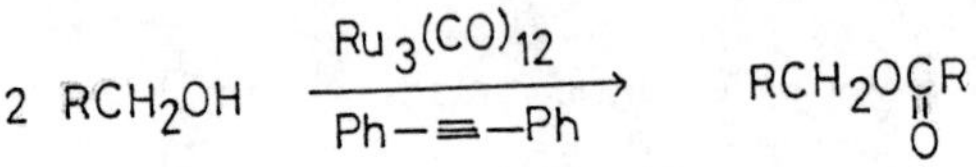

$$2\ \text{RCH}_2\text{OH} \xrightarrow[\text{Ph}-\equiv-\text{Ph}]{\text{Ru}_3\text{(CO)}_{12}} \text{RCH}_2\text{O}\overset{\text{O}}{\underset{\|}{\text{C}}}\text{R} \qquad 478$$

Figure 67   Oxidation of primary alcohols catalysed by ruthenium complexes.

6.10.2   Oxidation of secondary alcohols

6.10.2.1   Ruthenium catalysts

Oxidation of secondary alcohols to ketones in the presence of ruthenium complexes is similar to that of primary alcohols.   The ruthenium may be added as $RuCl_3$,[485] K_3RuCl_6,[486] $(Ph_3P)_3RuCl_2$,[487] or RuO_2,[488]   but the mechanism is unchanged.   Reoxidants include $[Fe(CN)_6]^{3-}$,[476] $S_2O_8^{2-}$,[474] O_2[489], t-BuOOH[490], PhIO[487] and $NaIO_4$.[491]   Osmium catalysed reactions are similar, but not widely used.[492]   The ketone is the sole product in high yield (Figure 69).

$$\text{(n-alkyl)}{-}OH \xrightarrow[96\%]{O_2/Co(OAc)_3} \text{(n-alkyl)}{-}COOH \qquad 481$$

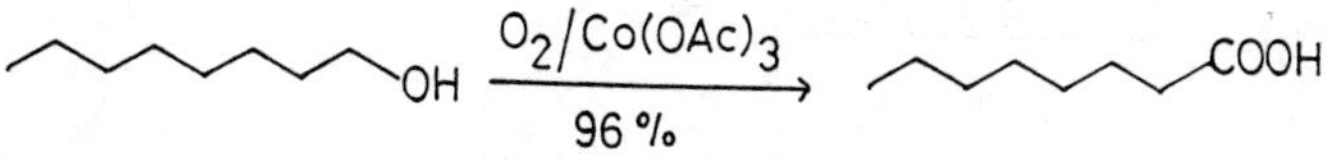

$$RCH_2OH + \underline{t}\text{-BuOOH} \xrightarrow{Zr(OAc)_2} RCHO \qquad 482$$

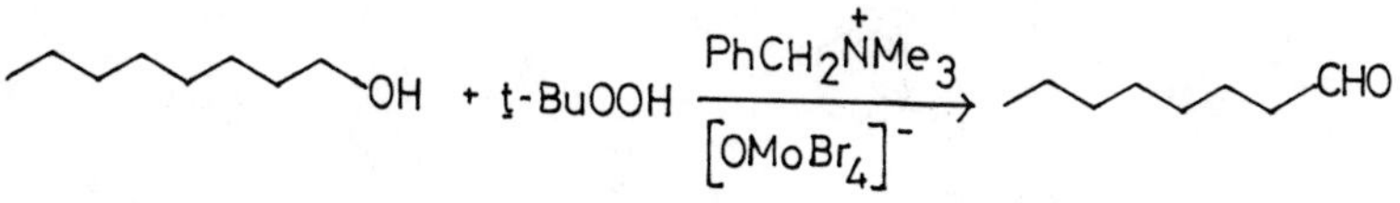

$$\text{(n-alkyl)}{-}OH + \underline{t}\text{-BuOOH} \xrightarrow[{[OMoBr_4]^-}]{PhCH_2\overset{+}{N}Me_3} \text{(n-alkyl)}{-}CHO \qquad 483$$

Steroid with COCH$_2$OH substituent, R^2, R^3, R^4, R^5, R^6 $\xrightarrow{air, Cu(OAc)_2}$ Steroid with COCHO substituent, R^2, R^3, R^4, R^5, R^6 484

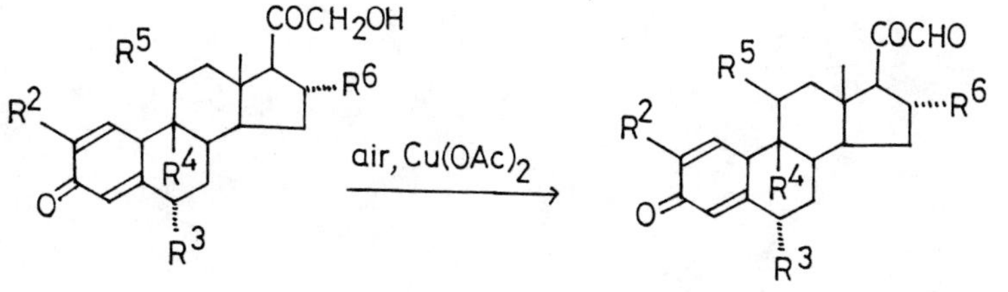

Figure 68 Catalytic oxidation of primary alcohols.

6.10.2.2 Oxidation by peroxodisulphate

Again these are similar to those of primary alcohols; initiation occurs by Ag$^+$ or Cu$^+$ catalysed decomposition of S$_2$O$_8^{2-}$ followed by $\underline{\alpha}$-H abstraction. Cyclopentanol is thus oxidised to cyclopentanone.[469] In other cases Cu^{2+} oxidises the intermediate radical[496] or rearrangement occurs.

Ref.

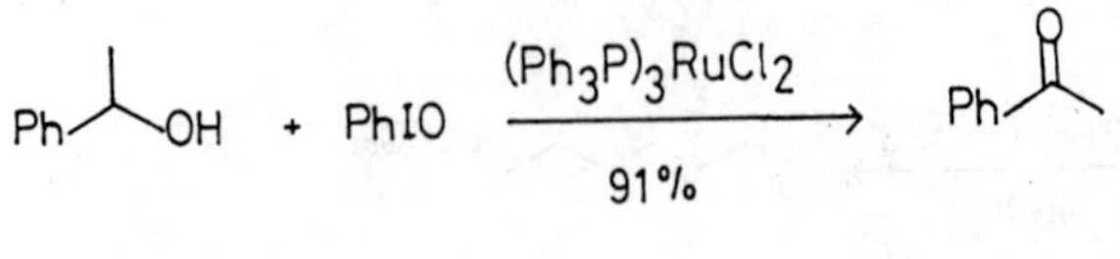

487

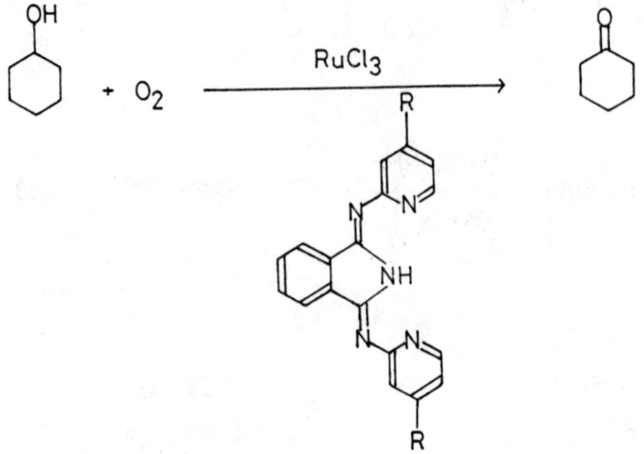

493

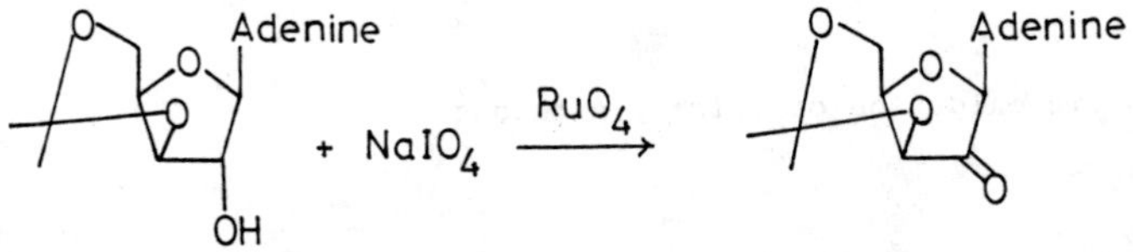

494

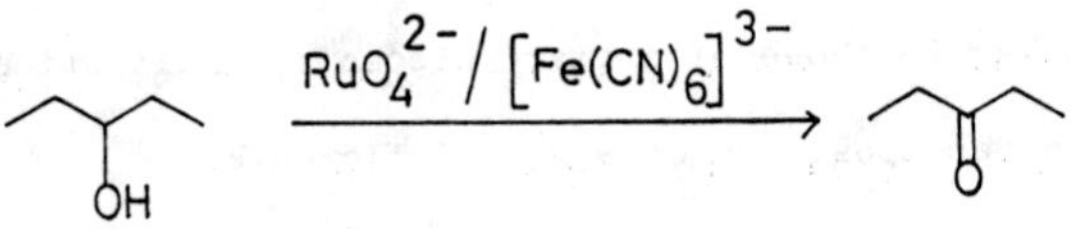

495

Figure 69 Oxidation of secondary alcohols in the presence of ruthenium complexes.

6.10.2.3 Palladium catalysts

Most palladium catalysed oxidations involve a relay mechanism. Pd(II) oxidises the alcohol to a ketone and Pd(0) is reoxidised by another oxidant. α-Hydrogen abstraction is the usual mechanism, but β-hydrogen transfer may also be important. Figure 70 shows a range of oxidants and substrates.

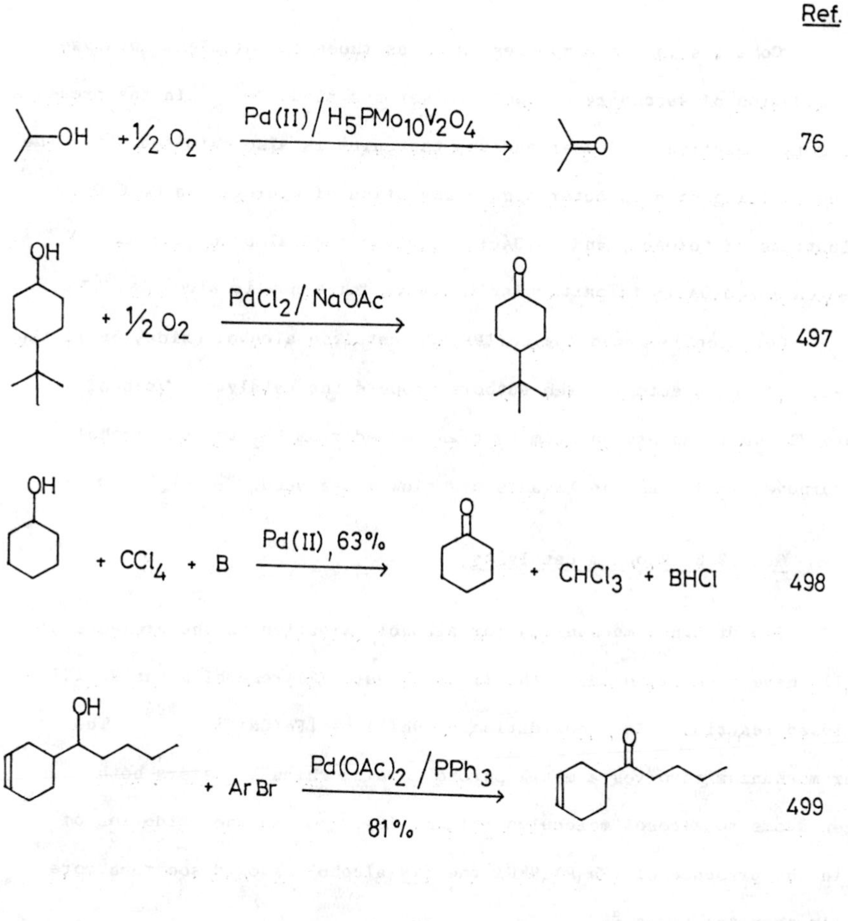

<u>Ref.</u>

76

497

498

499

<u>Figure 70</u> Oxidation of secondary alcohols in the presence of palladium complexes.

6.10.2.4 Copper catalysts

Copper catalysts operate in the presence of molecular oxygen, but a reoxidation, rather than a metal oxygen complex, is involved. With Cu_2Cl_2/phen secondary alcohols give poorer yields than primaries.[500] Copper and other metal $\underline{\beta}$-phthalocyanine complexes have also been used as catalysts.[501]

6.10.2.5 Cobalt catalysts

Cobalt superoxo complexes, such as those of Co(salen) catalyse slow oxidation of secondary alcohols to ketones at 60-70°. In the presence of Ph_3P the reaction is faster and the phosphine is also oxidised.[502] The rate determining step is outer sphere migration of hydrogen to $Co-O-O^{\cdot}$. Combinations of $Co(OAc)_2$ and $Co(OAc)_3$ catalyse radical oxidation at $60°$[503] and $Mn(OAc)_2/Co(OAc)_2$ is particularly active for benzylic alcohols.[504]

$Co(Saloph)NO_2$ and $(py)Co(TPP)NO_2$ catalyse alcohol oxidation in the presence of Lewis acids. The authors propose the catalytic cycle of Figure 71, where an oxygen atom is transferred from NO_2 to the alcohol. The turnover is low as the ligands are slowly degraded.[505]

6.10.2.6 Rhodium catalysts

Two distinct mechanisms for alcohol oxidation in the presence of Rh(III) have been proposed. The first (Figure 72) resembles the Ru(III) catalysed reaction, with reoxidation of Rh(I) by $[Fe(CN)_6]^{3-}$.[506] The other mechanism involves a metal peroxo complex which transfers both oxygen atoms to alcohol molecules (Figure 73).[507] In the oxidation of 269 in the presence of $(Ph_3P)_3RhCl$ the $\underline{cis}$-alcohol is oxidised much more rapidly than the $\underline{trans}$.[508]

LCoNO2
A
R1R2CHOH
OH
H
LCo—N—O—R1
O—A
R2
½O2
LCoNO
A
R1R2CO + H2O

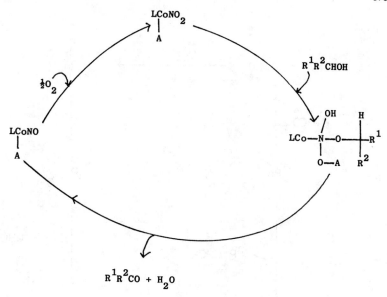

A = Lewis acid

Figure 71 Mechanism of alcohol oxidation in the presence of cobalt nitro complexes.

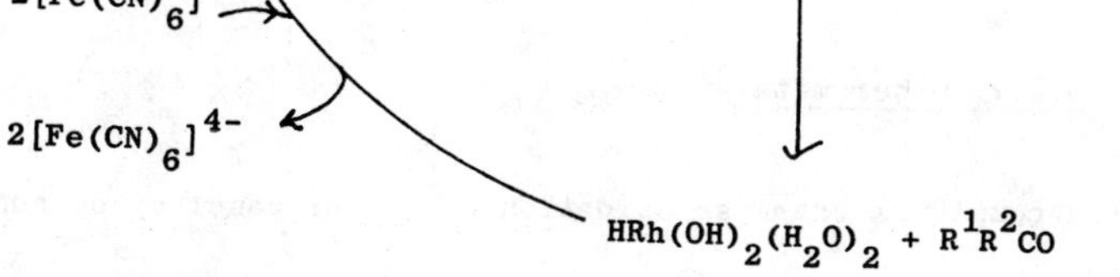

Rh(H2O)3(OH)3 + R1R2CHO-
(H2O)3(OH)2Rh—O—R2
R1
H
2[Fe(CN)6]3-
2[Fe(CN)6]4-
HRh(OH)2(H2O)2 + R1R2CO

Figure 72 Mechanism of alcohol oxidation in the presence of Rh(III) complexes.

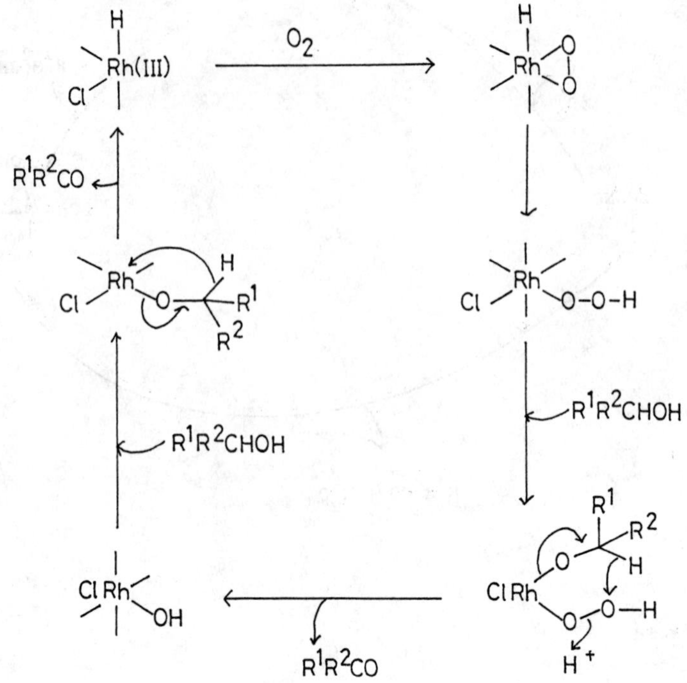

Figure 73 Mechansim of alcohol oxidation by rhodium peroxo complexes.

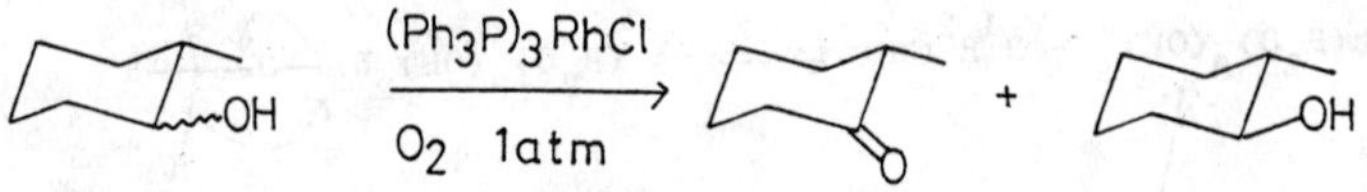

269   270   271

6.10.2.7 Other metals

Iron complexes catalyse oxidation of iso-propanol by peracetic acid,[509] and a selective remote hydroxylation of cyclohexanol proceeds via the mechanism of Figure 74.

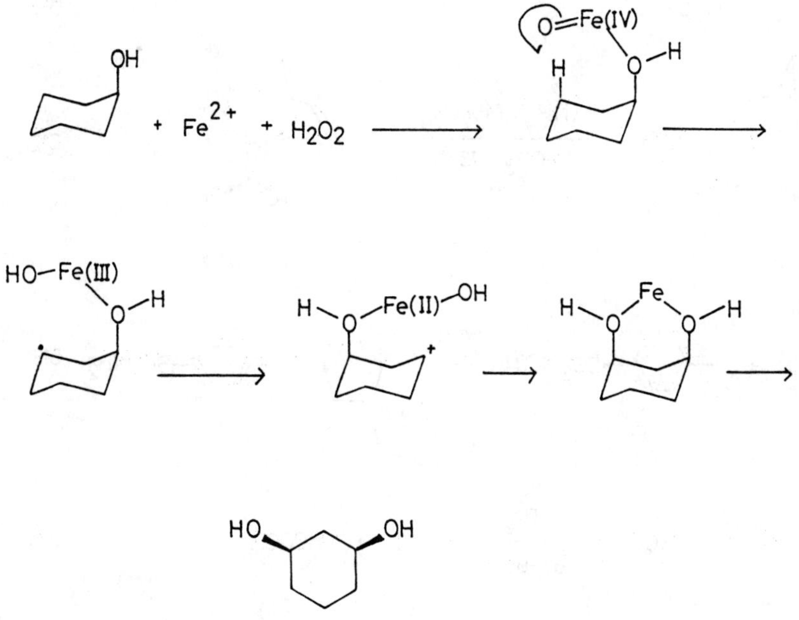

Figure 74 Mechanism of hydroxylation of cyclohexanol in the presence of iron complexes.

Ce(IV) may be used as an oxidant in the presence of Ir(III), or as a catalyst, using $NaBrO_3$ as stoichiometric oxidant. In the latter case, there is a strong selectivity for oxidation of secondary rather than primary alcohols (272 to 273).[511] Secondary alcohols are also oxidised selectively by $R_4N^+[OMoBr_4]^-/\underline{t}-BuOOH$ in a two phase system,[483] or $(NH_4)_6Mo_7O_{24}/H_2O_2$.[512] Some further examples are shown in Figure 75.

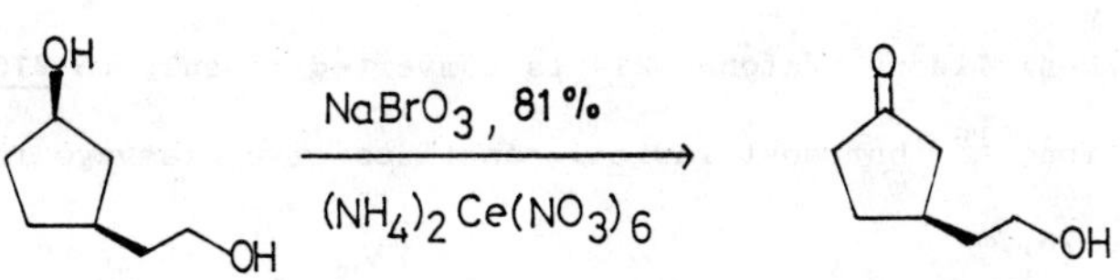

Ref.

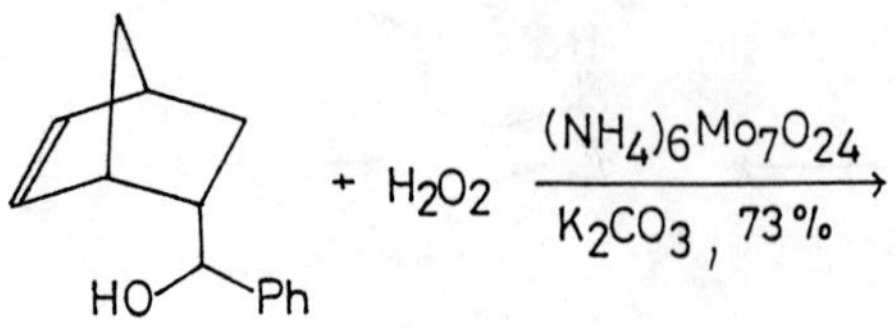

512

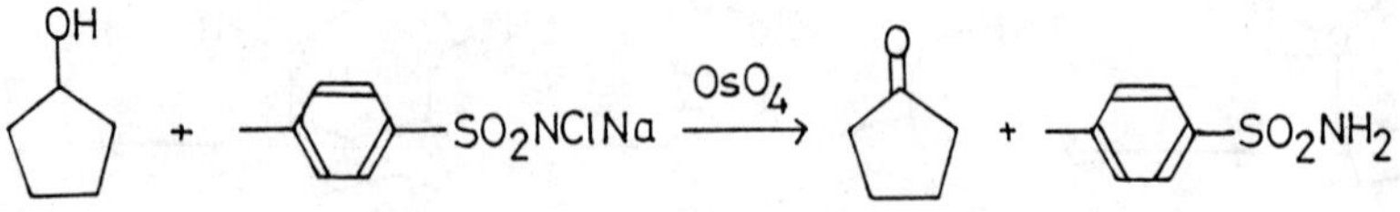

513

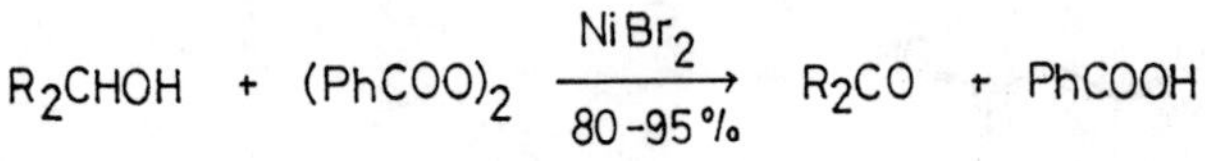

$$R_2CHOH + (PhCOO)_2 \xrightarrow[80-95\%]{NiBr_2} R_2CO + PhCOOH$$

514

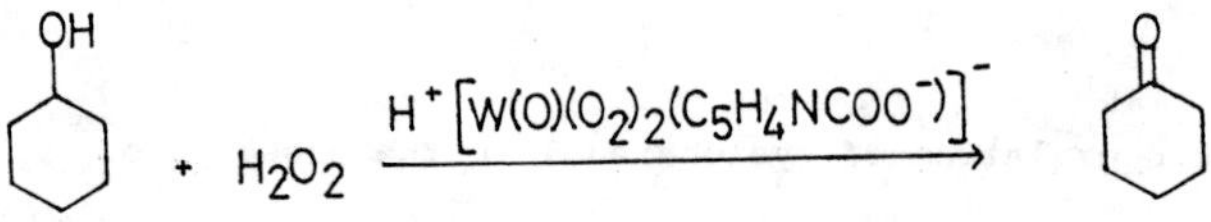

515

Figure 75 Oxidation of secondary alcohols in the presence of various metal complexes.

6.10.3 Oxidation of diols

Oxidation of vicinal diols may follow a different course from that of alcohols, because of the possibility of cleavage reactions. With $(Ph_3P)_3RuCl_2$/benzylidene acetone, 274 is converted cleanly to 275, without further reaction,[516] but most radical reactions give cleavage in good yield (Figure 76).

579

$CH_3(CH_2)_7CHOHCH_2OH \xrightarrow[100°]{O_2/Co(OAc)_2} CH_3(CH_2)_7COOH$ 90% 517

$CH_3(CH_2)_7CHOHCHOH(CH_2)_7COOH \xrightarrow[BuOAc]{O_2/Co(OAc)_2} HOOC(CH_2)_7COOH + CH_3(CH_2)_7COOH$ 83.2% 86.7% 518

$PhCHOHCH_2OH \xrightarrow{t\text{-}BuOOH/VO(acac)_2} PhCOOH$ 94% 519

$HOCH_2CHOHCH_2OH \xrightarrow{S_2O_8{}^{2-}/Ag^+} HCOOH + HOCH_2COOH$ 520

Ref.

Figure 76 Catalytic oxidative cleavage of diols.

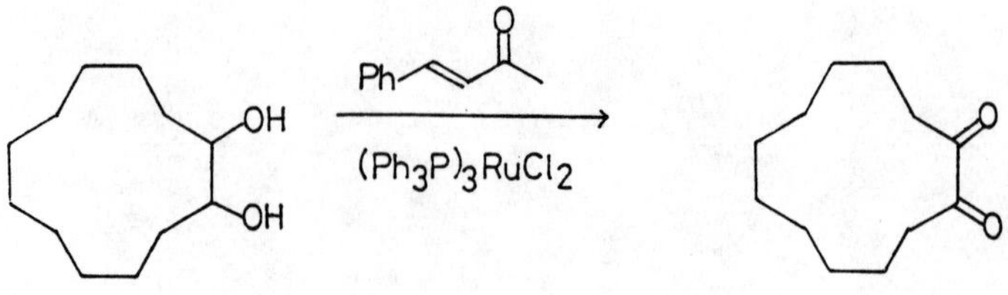

274   275

6.10.4 Oxidation of α-ketols

Whilst oxidation of vicinal diols usually gives cleavage, via diketones, the reaction of α-ketols has rarely been directed towards this end. Benzoin is oxidised to benzil in the presence of Co(II), Cu(II) or Ni(II).[521] Only in the case of Co(II) has the mechanism been studied (Figure 77).[522] An analogous ruthenium catalysed reaction has been used for conversion of 276 to 277.[523]

$$2Co(II) \rightleftharpoons [Co(II)]_2$$

$$[Co(II)]_2 + PhCHOHCOPh \longrightarrow [(Co(II))_2 PhCHOHCOPh]$$

$$[(Co(II))_2 \overset{OH}{\underset{|}{PhCHCOPh}}] \longrightarrow [(Co(II))_2 \overset{O}{\underset{|}{PhCHCOPh}}] \xrightarrow{O_2}$$

$$[(Co(III))_2 \overset{O_2^-}{\underset{|}{PhCHCOPh}}\overset{O^-}{\underset{|}{}}] \longrightarrow products$$

Figure 77 Mechanism of oxidation of benzoin in the presence of Co(II).

276   O_2 /RuCl$_3$ 277

6.11 Oxidation of aldehydes

The oxidation of aldehydes is of commercial importance because it provides a route for the conversion of linear aldehydes (from hydroformylation of alkenes) to carboxylic acids. Aldehydes are readily oxidised in air by a radical chain route (equations (116) - (118)) and it has long been known that metal ions accelerate the process.

$$PhCHO + R^{\bullet} \longrightarrow Ph\overset{\bullet}{C}O + RH \qquad (116)$$

$$Ph\overset{\bullet}{C}O + O_2 \longrightarrow PhCO_3^{\bullet} \qquad (117)$$

$$PhCO_3^{\bullet} + PhCHO \longrightarrow PhCO_3H + Ph\overset{\bullet}{C}O \qquad (118)$$

The reaction of PhCHO in the presence of Co(III) is typical; the first step is radical formation (equation (119)) and is rate determining.[524] A somewhat more complex kinetic scheme has subsequently

$$Co(III) + PhCHO \longrightarrow Co(II) + Ph\overset{\bullet}{C}O + H^{+} \qquad (119)$$

been proposed.[525] A similar reaction occurs in the presence of Fe(III), but with Ni(OAc)$_2$ the reoxidation step becomes rate-determining (equations (120), (121)).[524] In substituted cases, for example p-tolualdehyde, the

$$Ni(III) + PhCHO \xrightarrow{\text{fast}} Ni(II) + Ph\overset{\bullet}{C}O + H^{+} \qquad (120)$$

$$Ni(II) + PhCO_3H \xrightarrow{\text{slow}} Ni(III) + PhCO_2^{\bullet} + HO^{-} \qquad (121)$$

aldehyde is selectively oxidised by a range of catalysts.[526] Both palladium[527] and molybdenum[528] complexes have also been used to catalyse benzaldehyde oxidation by a radical pathway.

A number of metal oxygen complexes catalyse oxidation of PhCHO to $PhCO_2H$ and $PhCO_3H$. Reactivity is in the order $(Ph_3P)_2RhCl(CO)$ > $(Ph_3P)_2PdO_2$ > $(Ph_3P)_4Pd$ > $(Ph_3P)_3RhCl$ > $((Ph_3P)_2RhCl)_2$ > $(Ph_3P)_2IrO_2Cl(CO)$ ⤳ $(Ph_3P)_2PtO_2$.[529] A radical pathway is not involved and the proposed mechanism is shown in equations (122) - (125). Using ruthenium complexes an alternative

$$(Ph_3P)_2PdO_2 + PhCHO \rightleftharpoons (Ph_3P)_2PdO_2 \cdot PhCHO \qquad (122)$$

$$(Ph_3P)_2PdO_2 \cdot PhCHO \longrightarrow (Ph_3P)_2Pd \overset{O-O}{\underset{O}{\diagup \diagdown}} CHPh \qquad (123)$$

$$(Ph_3P)_2Pd \overset{O-O}{\underset{O}{\diagup \diagdown}} CHPh + O_2 \longrightarrow (Ph_3P)_2PdO_2 + \qquad (124)$$
$$PhCO_3H$$

$$PhCO_3H + PhCHO \longrightarrow 2PhCOOH \qquad (125)$$

route (equations (126) - (129) is proposed.[530]

$$PhCHO + H_2O \rightleftharpoons PhCH(OH)_2 \qquad (126)$$

$$PhCH(OH)_2 + [Ru(H_2O)_6]^{3+} \rightleftharpoons complex \qquad (127)$$

$$complex \longrightarrow PhCH(OH)=OH^+ + H_2O + [HRu(H_2O)_5]^{2+} \qquad (128)$$

$$[HRu(H_2O)_5]^{2+} \xrightarrow{IO_4^-} [Ru(H_2O)_6]^{3+} + IO_3^- \qquad (129)$$

Oxidation of acetaldehyde may yield acetic or peracetic acid in high selectivity according to catalyst and conditions. The proportions of the two products depend on the relative rates of oxidation and metal catalysed peracid decomposition. Thus, iron salts give mainly peracetic acid whilst manganese complexes give acetic acid. More generally oxygen consumption rates are in the order Fe > Co >> Mn > Ni > Cr,V and selectivity to peracid Fe,Co > none >> Ni, Cr, V, Cu >> Mn.[531] Acetaldehyde reacts with peracetic acid to give $CH_3CH(OH)OOCOCH_3$ and the overall mechanism is then given by equations (130) - (135).[532,533] In the presence of Cu(II) the "catalyst"

$$Co(III) + CH_3CHO \longrightarrow CH_3CO + Co(II) + H^+ \qquad (130)$$

$$CH_3CO + O_2 \longrightarrow CH_3CO_3 \qquad (131)$$

$$CH_3CO_3 + CH_3CHO \longrightarrow CH_3CO_3H + CH_3CO \qquad (132)$$

$$CH_3CO_3H + Co(II) \longrightarrow Co(III) + CH_3CO_2 + HO^- \qquad (133)$$

$$CH_3CH(OH)OOCOCH_3 + Co(II) \longrightarrow Co(III) + CH_3CO_2$$
$$+ CH_3CO_2^- \qquad (134)$$

$$CH_3CO_2 + CH_3CHO \longrightarrow CH_3CO_2H + CH_3CO \qquad (135)$$

competes successfully for the radicals and acetic anhydride is the major

product (equations (136), (137)).[534] Metal phthalocyanines catalyse

$$CH_3CO + Cu(II) \longrightarrow CH_3\overset{+}{CO} + Cu(I) \qquad (136)$$

$$CH_3\overset{+}{CO} + CH_3CO_2H \longrightarrow CH_3COOCOCH_3 + H^+ \qquad (137)$$

the oxidation <u>via</u> an M-O_2 complex which abstracts H to give the acetyl

radical.[535]

The oxidation of higher aldehydes is fairly similar to that of

acetaldehyde and good selectivities to either acid or peracid are reported

(Figure 78).

With a limited O_2 supply or a high catalyst:substrate ratio

other reactions intervene. For example, under Co(III) catalysis acyl radicals

add to alkenes (equations (138) - (140)).[540] In the presence of stoichiometric

$$RCHO + Co(III) \longrightarrow RCO + Co(II) + H^+ \qquad (138)$$

$$RCO + R'CH=CH_2 \longrightarrow R'CHCH_2COR \qquad (139)$$

$$R'CHCH_2COR + RCHO \longrightarrow RCO + R'CH_2CH_2COR \qquad (140)$$

Mn(III) or Cu^{2+}/py, abstraction of the $\underline{\alpha}$-hydrogen to give RCHCHO occurs.

Oxygenation gives a hydroperoxide which is subsequently cleaved.[541]

Ref.

$$CH_3CH_2CHO + O_2 \xrightarrow{Mn(OAc)_2} CH_3CH_2CO_2H$$ 536

$$\xrightarrow[\text{amine}]{Cu(NO_3)_2} CH_3CH_2CO_3H$$ 537

$$\xrightarrow{Co(TPP)X} CH_3CH_2CO_3H$$ 538

$$CH_3CH_2CH_2CHO + O_2 \xrightarrow{PdY_2/R_4P^+X^-} CH_3CH_2CH_2CO_2H + CH_3CH_2CH_2CO_3H$$ 539

$$C_5H_{11}CHO + O_2 \xrightarrow{(NH_4)_6Mo_7O_{24}} C_5H_{11}CO_2H$$ 512

$$PhCH_2CH_2CHO + O_2 \xrightarrow{(bipy)_2Cu_2(OH)_2} PhCH_2CH_2CO_2H$$ 500

Figure 78 Catalytic oxidation of linear aldehydes.

Ref.

$$(CH_3)_2CHCHO + O_2 \xrightarrow[97\%]{Cp_2Fe} (CH_3)_2CHCO_2H$$ 542

$$PhCH(CH_3)CHO + O_2 \xrightarrow{Mn(acac)_2} PhCH(CH_3)CO_2H$$ 543

$$CH_3CH(OAc)CHO + O_2 \xrightarrow[98\%]{Fe(OAc)_2} CH_3CH(OAc)CO_2H$$ 544

Figure 79 Catalytic oxidation of branched aldehydes.

Branched aldehydes are also readily oxidised to acids (Figure 79) and an interesting variant is provided by pivalaldehyde/Mn(III). The radical $(CH_3)_3CCO$ is easily decarbonylated to $(CH_3)_3C^{\cdot}$ and iso-butane and iso-butene are the major products.[545]

Some oxidations of unsaturated aldehydes follow an essentially analogous pathway, but yields and selectivities tend to be lower, since the reaction is complicated by radical induced polymerisation of both substrates and products. However, α-methylacrolein is oxidised to α-methylacrylic acid in the presence of $PdCl_2$, Co(naphthenate)$_2$ and $Mo_{12}PNbBi_3O_{45}$.[546] A different reaction occurs in the presence of Cu^{2+}/py. Formation of the dienolate, **278**, is rate controlling and this is oxidised by Cu(II) to **279**. Autoxidation gives the hydroperoxide, **280**, and **281** and **282** are the usual products.[253] Finally, in the presence of $NH_3/(RhCl(CO)_2)_2$, unsaturated aldehydes are cleanly oxidised to unsaturated nitriles.[529]

278	**279**	**280**

281	**282**

6.12 Oxidation of ketones

The ususal products of ketone autoxidation are aldehydes and carboxylic acids formed according to equation (141). The reaction may be

$$RCOR' + O_2 \xrightarrow{\text{catalyst}} RCHO + R'COOH \qquad (141)$$

exemplified by butanone (Figure 80). Among effective catalysts are Fe^{3+}/phen,[547] Cu^{2+}/py[548] and Cu^{2+}/phen.[549] An analogous reaction is provided by the conversion of acetophenone to benzoic acid in the presence of Mn(III) and/or Co(III).[550] A thorough mechanistic study of the oxidation of $PhCH_2COPh$ in the presence of Cu(II) reveals that the diketone is not an intermediate in the formation of PhCOOH.[551]

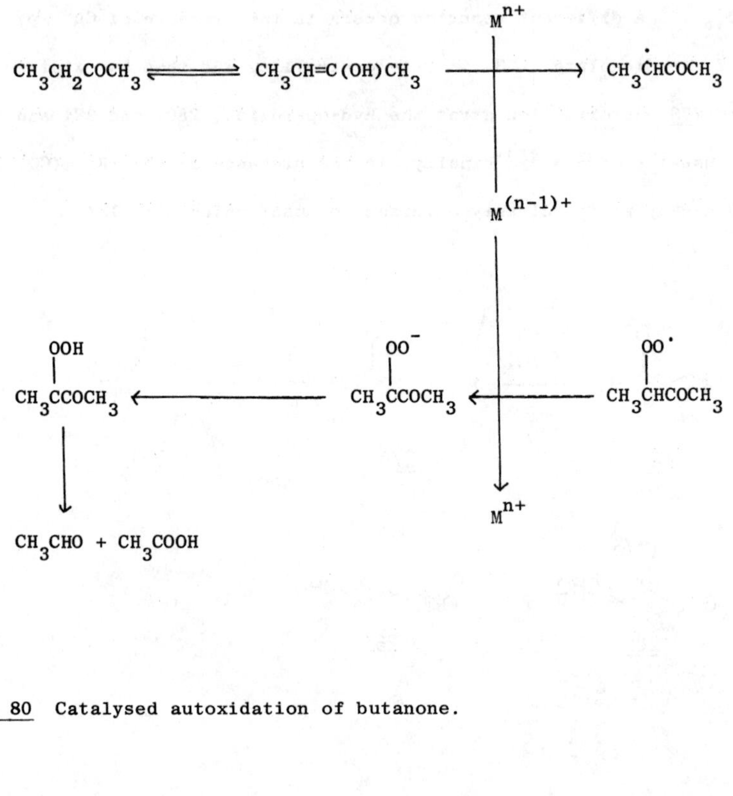

Figure 80 Catalysed autoxidation of butanone.

With $Cu^{2+}/S_2O_8^{2-}$ acetone is oxidised to CO_2 and acetic acid,[552] but if an alkene and arene are present coupling may occur to give 285 in good yield.[469] With other ketones, such as 286, 1,5 H-migration gives remote oxidation to 289.[553]

$$R\!=\!\! \diagup \;\;+\;\; \overset{\overset{\displaystyle O^{\bullet}}{|}}{\diagdown\!\!=} \;\;\longrightarrow\;\; \text{(ketone radical } \underline{283})$$

283

$$\underline{283} \;+\; \text{(4-methylquinolinium, } \underline{284}) \;\longrightarrow\; \text{(substituted product } \underline{285})$$

284   **285**

$$\underset{\underline{286}}{\text{(hexanone)}} \;\xrightarrow[\text{Fe}^{2+}]{\text{S}_2\text{O}_8{}^{2-}}\; \underset{\underline{287}}{\text{(radical cation)}^{+\bullet}} \;\longrightarrow\; \underset{\underline{288}}{(\overset{+}{\text{OH}}\text{ radical)}}$$

$$\underline{288} \;\downarrow$$

289

A number of apparently non-radical routes for ketone oxidation have also been reported. Thus, OsO_4 catalyses hydroxylation of acetone (equations (142) – (146)), the alcohol being oxidised by periodate,[554] and $Ru(III)/IO_4{}^-$ follows a similar route.[555] The oxidation of butanone

$$CH_3COCH_3 + HO^- \longrightarrow CH_3\text{-}\overset{\overset{\displaystyle O^-}{|}}{C}\text{=}CH_2 \tag{142}$$

$$CH_3\overset{\overset{\displaystyle O^-}{|}}{C}\text{=}CH_2 + OsO_4 \xrightarrow{\;\text{slow}\;} CH_3\overset{\overset{\displaystyle O\text{-}OsO_4{}^-}{|}}{C}\text{=}CH_2 \tag{143}$$

$$CH_3\overset{\overset{\displaystyle O\text{-}OsO_4{}^-}{|}}{C}\text{=}CH_2 \xrightarrow{\;HO^-\;} CH_3\overset{\overset{\displaystyle O}{\|}}{C}\text{-}CH_2OH + Os(VI) \tag{144}$$

$$Os(VI) + IO_4{}^- \longrightarrow Os(VIII) + IO_3{}^- \tag{145}$$

$$CH_3\overset{\displaystyle O}{\overset{\|}{C}}CH_2OH \xrightarrow{\ IO_4^-\ } CH_3\overset{\displaystyle O}{\overset{\|}{C}}CHO \xrightarrow{\ IO_4^-\ } CH_3COO^- + HCOO^- \qquad (146)$$

in the presence of Ru(III)/Ce(IV), however, proceeds by a different

pathway (equations (147) - (150)).[556]

$$CH_3CH_2\overset{\displaystyle O}{\overset{\|}{C}}CH_3 + H_3O^+ \rightleftharpoons CH_3CH_2\overset{\displaystyle OH}{\overset{|}{C}}-CH_3 \qquad (147)$$
$$\underset{+}{}$$

$$CH_3CH_2\overset{\displaystyle OH}{\overset{|}{\underset{+}{C}}}CH_3 + [RuCl_5(H_2O)]^{2-} \xrightarrow{\ slow\ } \qquad (148)$$

$$
\begin{array}{c}
CH_3CH_2 \qquad OH \\
\diagdown \quad \diagup \\
C \\
\diagup \quad \diagdown \\
CH_3 \qquad \overline{R}uCl_5
\end{array}
$$

290

$$290 \xrightarrow{\ fast\ } [HRuCl_5]^{3-} + CH_3CH_2\overset{\displaystyle O}{\overset{\|}{C}}CH_2OH + 2H^+ \qquad (149)$$

$$[HRuCl_5]^{3-} + 2Ce(IV) + H_2O \xrightarrow{\ fast\ } [RuCl_5(H_2O)]^{2-} +$$
$$2Ce(III) + H^+ \qquad (150)$$

The oxidation of cyclohexanone has been thoroughly studied
and some results are shown in Table 10. The mechanism in all cases involves
formation of 291, which is autoxidised to 292. In certain cobalt
catalysed reactions glutaric as well as adipic acid is formed; labelling
studies indicate that 298 and 299 are intermediates.[565]

Non- autoxidation mechanisms are also known. The reaction in
the presence of Ru(III)/IO_4^- proceeds via the enolate and cyclohexane-1,2-
dione.[555] A similar route is followed with Co(phen)O_2 complexes and the
1,2-dione may be isolated in this case.[566] Metal hydroperoxides, such as
$H^+[MoO(O_2)_2(C_5H_4NCOO)]$, give Baeyer Villegar oxidation to 293.[567] In the
presence of PdCl_2, dehydrogenation yields phenol, and with Pd(OAc)_2/HPA-2
cyclohexenone is obtained.[76] Few reactions of substituted cyclohexanones

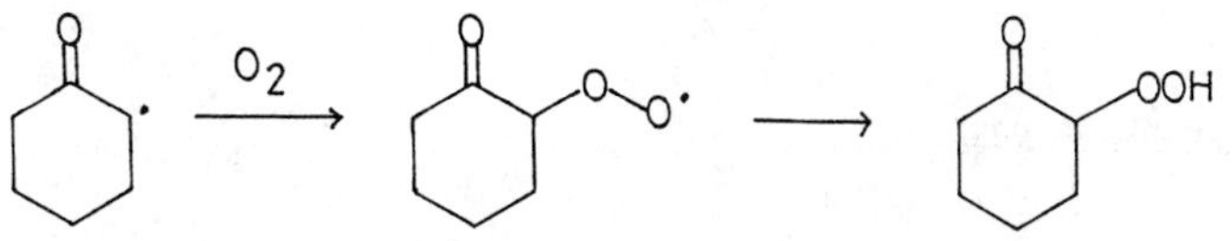

291   292   294

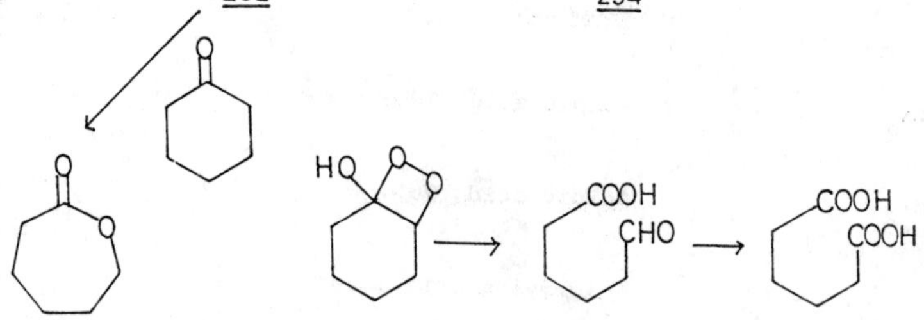

293   295   296   297

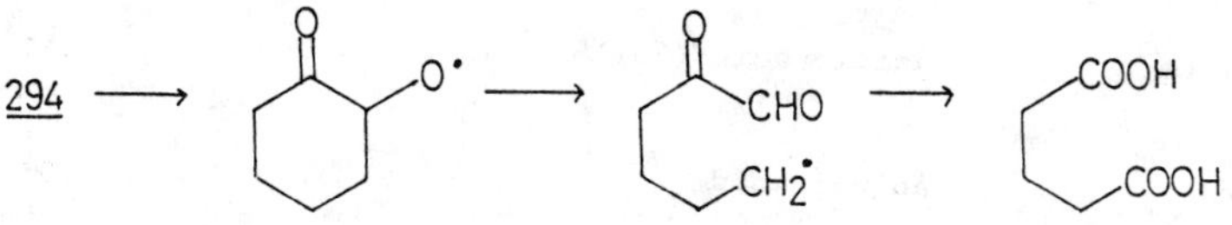

294 ⟶ 298 ⟶ 299 ⟶ 300

are known, but they seem to be regioselective. Thus, the enolate, 301, is autoxidised to 302 and rearranges to 303. Attack of methanol gives $CH_3CO(CH_2)_4COOMe$ in good yield.[568]

Reports of diketone oxidation are sparse. Pentane-2,4-dione

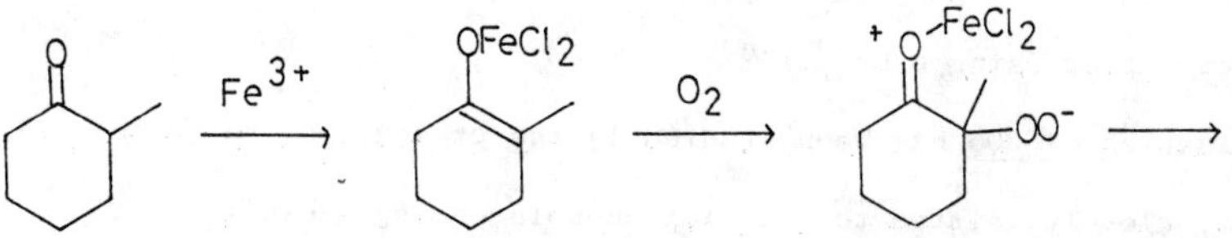

301   302

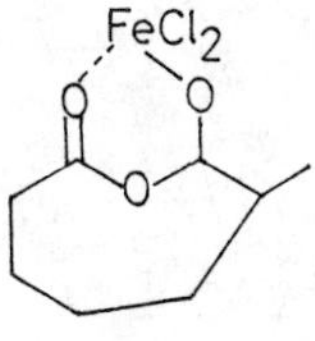

303

590

Table 10

Autoxidation of cyclohexanone

Catalyst	Products	Ref.
$Mn(OAc)_3$	Adipic acid (70%)	557
$Co(OAc)_2$	Adipic acid (70%)	558
Cu^{2+}	5-formylvaleric acid	559
Fe^{3+}	293 (70%)	560
$Co(OAc)_2/V(acac)_2$	Adipic acid (76%)	561
$Co(stearate)_3$	Adipic acid	562
$Re_2(CO)_{10}$	Adipic acid	563
$Rh_6(CO)_{16}$	Adipic acid	564

reacts via a radical route, coupling with $RCH=CH_2$ to yield 304, in the presence of $Mn(OAc)_3$ or $Co(OAc)_3$.[569]   Benzil is converted to dibenzoyl peroxide in two steps using L_2PtO_2.[570]

Oxidation of 305 has been studied in the presence of Mn(Salen), the mechanism, closely related to that for phenols, being shown in Figure 81.[571]

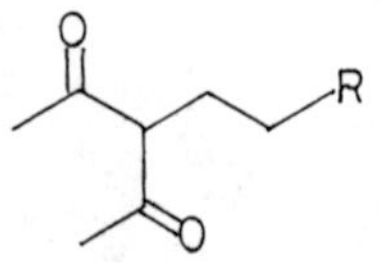

304

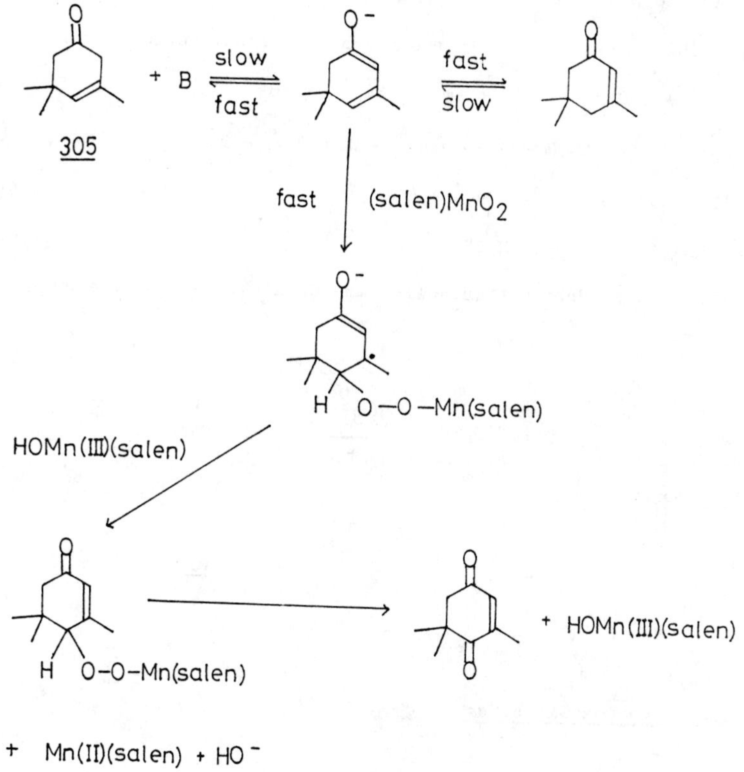

Figure 81 Mechanism of oxidation of 305 in the presence of Mn(salen) complexes.

6.13 Oxidation of Carboxylic acids and their derivatives

6.13.1 Monocarboxylic acids

Carboxylic acids are oxidatively decarboxylated in the presence of metal ions including Pb(IV), Mn(III),[382] Co(III) and Ce(IV)[572] according to equation (151). Such reactions are rarely catalytic or useful. With

$$\text{RCOOH} + M^{n+} \longrightarrow M^{(n-1)+} + R^{\cdot} + CO_2 + H^+ \tag{151}$$

acetic acid, $^{\cdot}CH_2COOH$ is formed, presumably because of the instability of $CH_3^{\cdot}$, the usual catalysts being Mn(III) or Ce(IV). Further steps involve H-abstraction or autoxidation.

In the presence of $S_2O_8^{2-}/Ag^+$ a catalytic reaction occurs

(equations (152) – (155)).[573]   In the absence of other reagents, the

$$Ag^+ + S_2O_8^{2-} \longrightarrow Ag^{2+} + SO_4^{2-} + SO_4^{-\circ} \tag{152}$$

$$Ag^{2+} + RCH_2COOH \longrightarrow Ag^+ + RCH_2COO^\circ + H^+ \tag{153}$$

$$RCH_2COO^\bullet \longrightarrow RCH_2^\bullet + CO_2 \tag{154}$$

$$2RCH_2^\bullet \longrightarrow RCH_2CH_2R \tag{155}$$

dimer is the main product but quinoline may be alkylated to give 306 in

good yield.[469]

306

6.13.2 Dicarboxylic acids

Dicarboxylic acids also undergo catalysed decarboxylation,
differences from the monoacids arising because of the subsequent reactions
of the radicals.   For example, $HOOC-CH_2CH_2CH_2^\bullet$, formed from glutaric acid
in the presence of $S_2O_8^{2-}/Ag^+/Cu^{2+}$ cyclises to 307 which is oxidised to
308.   Cyclisation becomes less successful as the ring size increases.[574]

307   308

Succinic and adipic acids are not decarboxylated by $S_2O_8^{2-}/Ag^+$
but form radicals, such as 309, which are subsequently hydroxylated and
further oxidised.   Both ketones and dihydroxy compounds are formed with

low selectivity.[575]   Malonic acid is cleaved under these conditions, and
oxalic acid is cleaved both by $S_2O_8^{2-}/Cu^{2+}$ and BrO_3^-/Mn^{2+}.[576]

6.13.3 Hydroxy acids

With an $\underline{\alpha}$-hydroxy acid as substrate one could envisage either
radical decarboxylation or alcohol oxidation to occur.   With $S_2O_8^{2-}/Ag^+$
decarboxylation predominates with ketones as the major products (equations
(156) - (160)).[597]   Thus lactic acid is converted to CH_3CHO.[578]   Radical

$$Ag^+ + S_2O_8^{2-} \longrightarrow Ag^{2+} + SO_4^{-\bullet} + SO_4^{2-} \tag{156}$$

$$SO_4^{-\bullet} + H_2O \longrightarrow HSO_4^- + HO^\bullet \tag{157}$$

$$HO^\bullet + R^1R^2CH(OH)COOH \longrightarrow R^1R^2CH(OH)COO^\bullet + H_2O \tag{158}$$

$$R^1R^2CH(OH)COO^\bullet \longrightarrow R^1R^2\dot{C}HOH + CO_2 \tag{159}$$

$$R^1R^2\dot{C}HOH + SO_4^{-\bullet} \longrightarrow R^1R^2CO + HSO_4^- + H^+ \tag{160}$$

decarboxylation is also involved in the Mn(II) catalysed Tℓ(III) oxidation
of lactic acid to acetic acid.[579]

Different products are formed in $[Fe(CN)_6]^{3-}$ oxidation catalysed
by OsO_4.   Glycollate is oxidised to oxalate $\underline{via}$ the aldehyde (equations
(161) - (163)).[580]   Oxalate is then slowly decarboxylated to formaldehyde.

$$HOCH_2COO^- + [OsO_4(OH)_2]^{2-} \longrightarrow [(HO)_3OsO_3-OCH_2COO^-]^{3-} \tag{161}$$

$$[(HO)_3OsO_3OCH_2COO^-]^{3-} \xrightarrow{HO^-} H\overset{O}{\overset{\|}{C}}COO^- + [OsO_3(OH)_4]^{4-} \tag{162}$$

$$[OsO_3(OH)_4]^{4-} + 2[Fe(CN)_6]^{3-} \longrightarrow [OsO_4(OH)_2]^{2-} +$$

$$2[Fe(CN)_6]^{3-} + H_2O \qquad (163)$$

With PhCH(OH)COOH decarboxylation is the major process, and PhCHO the

primary product.[581] Some further examples are shown in Figure 82.

Ref.

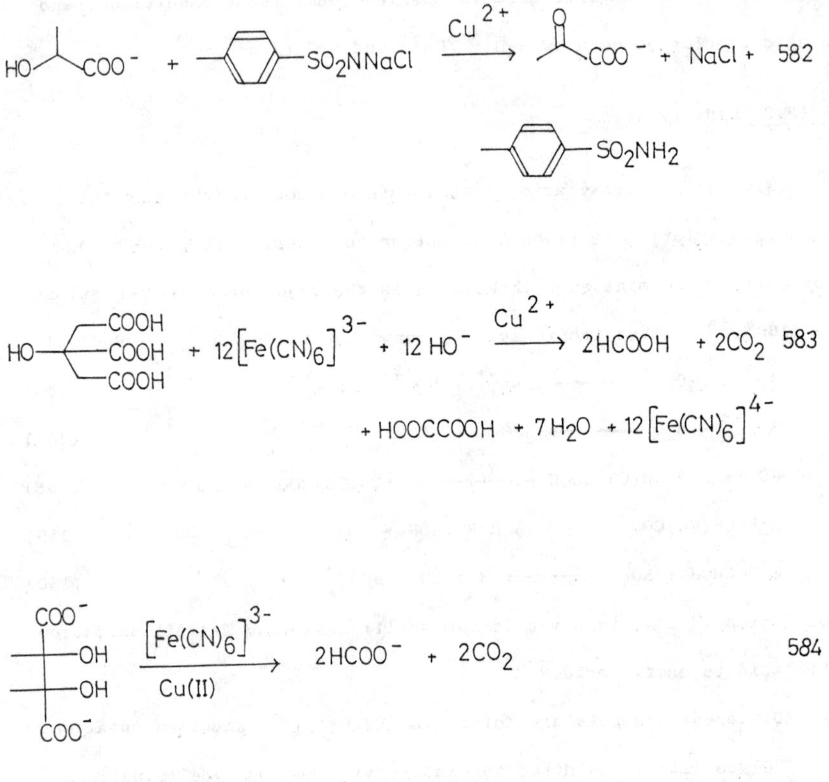

Figure 82 Catalytic oxidation of α-hydroxy acids

6.13.4 Esters and amides

Oxidation of esters is relatively easy but synthetic uses are limited. $Ce(IV)/Ag^+$ gives a radical reaction[585] and Ru(III) catalysed oxidation proceeds via hydride abstraction (equations (164) - (167)).[586]

$$R^1COOCH_2R^2 + Ru(III) \rightleftharpoons R^1-\overset{\overset{O}{\|}}{C}-O-\underset{\underset{Ru(III)}{\overset{|}{H}}}{\overset{\overset{H}{|}}{C}}-R^2 \qquad (164)$$

$$R^1-\overset{\overset{O}{\|}}{C}-\underset{\underset{Ru(III)}{|}}{O}-CH_2R^2 \longrightarrow Ru(I) + H^+ + R^1\overset{\overset{O}{\|}}{C}-O-\overset{+}{C}H-R^2 \qquad (165)$$

$$R^1\overset{\overset{O}{\|}}{C}-O-\overset{+}{C}H-R^2 \longrightarrow R^1COOH + R^2CHO \qquad (166)$$

$$Ru(I) + IO_4^- + 2H^+ \longrightarrow Ru(III) + IO_3^- + H_2O \qquad (167)$$

Some further examples are shown in Figure 83.

Ref.

$(CH_3)_2CHCOOMe \xrightarrow{O_2/Cu(OAc)_2} CH_2=C(CH_3)COOMe$ 587

$PhCH_2OAc \xrightarrow[NaBr]{O_2/Co(OAc)_2} PhCOOH + CH_3COOH$ 588

$CH_3CH_2OAc \xrightarrow{O_2/Co(OAc)_2} 2AcOH$ 589

Figure 83 Catalytic oxidation of esters.

Little work has been done in the area of amide oxidation. Decarboxylation of DMF gives Me_2NH in the presence of Fe^{3+} and Co^{2+} bipy complexes[590], and radical induced cleavage also occurs with $S_2O_8{}^{2-}/Ag^{+}$[591]. Cyclic compounds, such as 310 and 312, give dicarbonyl derivatives[592].

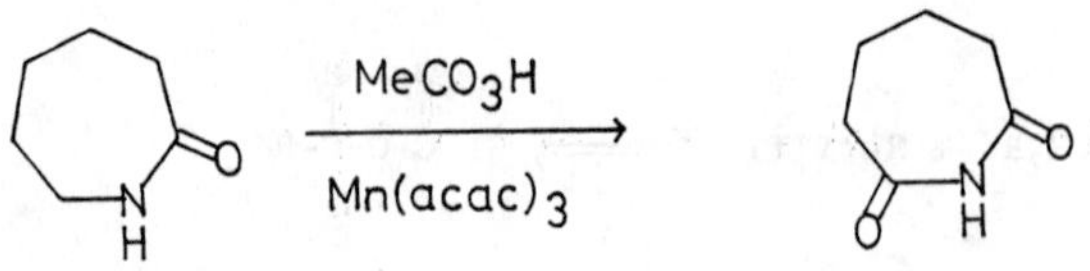

310   311

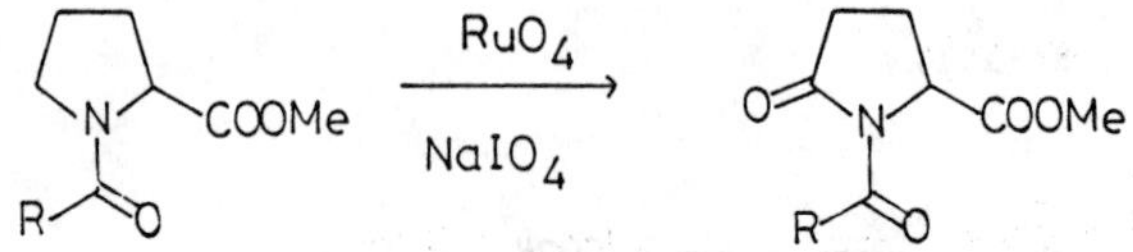

312   313

6.13.5 Amino acids

α-Amino acids are oxidised by O_2 to α-keto acids in the presence of pyridoxal and first row transition metal ions.[593] It is proposed that a metal-O_2-Schiff base complex is the reactive intermediate (Figure 84). In model systems Mn(II) is the most effective catalyst, but the enzymic system uses Cu(II).

Simpler systems, however, give aldehydes according to equation (168). Both radical[594] (equations (169) - (172)) and coordination mechanisms[595] (equations (173) - (176)) are known. The presence of Cu(II)

$$RCH(NH_2)COOH \xrightarrow{O_2} RCHO + CO_2 + NH_3 \qquad (168)$$

$$S_2O_8{}^{2-} + Ag^{+} \longrightarrow SO_4{}^{-\cdot} + SO_4{}^{2-} + Ag^{2+} \qquad (169)$$

$$RCH(NH_2)COOH + Ag^{2+} \longrightarrow RCH(NH_2)COO + Ag^+ + H^+ \quad (170)$$

$$RCH(NH_2)COO \longrightarrow RCHNH_2 + CO_2 \quad (171)$$

$$RCHNH_2 + S_2O_8{}^{2-} + H_2O \longrightarrow RCHO + NH_3 + HSO_4{}^-$$
$$+ SO_4{}^-{} \cdot \quad (172)$$

$$RCH(NH_2)COO^- + [OsO_4(OH)_2]^{2-} \longrightarrow$$
$$[OsO_4(OH)RCH(NH_2)COO^-]^{3-} + H_2O \quad (173)$$

$$[OsO_4(OH)RCH(NH_2)COO^-]^{3-} + HO^- \longrightarrow RCH=NH +$$
$$[OsO_4(OH)_2]^{4-} + CO_2 + H_2O \quad (174)$$

$$RCH=NH + H_2O \longrightarrow RCHO + NH_3 \quad (175)$$

$$[OsO_4(OH)_2]^{4-} + 2[Fe(CN)_6]^{3-} \longrightarrow [OsO_4(OH)_2]^{2-}$$
$$+ 2[Fe(CN)_6]^{4-} \quad (176)$$

is advantageous in the radical pathway, since it both accelerates decomposition of $S_2O_8{}^{2-}$ and radical collapse. Additional examples are provided by the reactions of 315[596] and 316,[597] but the process is of more mechanistic than synthetic interest.

$$H_2NCH_2COOH + Ce(IV) \xrightarrow{Ag^+} HCOOH + NH_3 + CO_2 + Ce(III)$$

315

$$CH_3CH(NH_2)COOH + H_2O_2 \xrightarrow{Fe(II)} CH_3CHO + NH_3 + CO_2$$

316

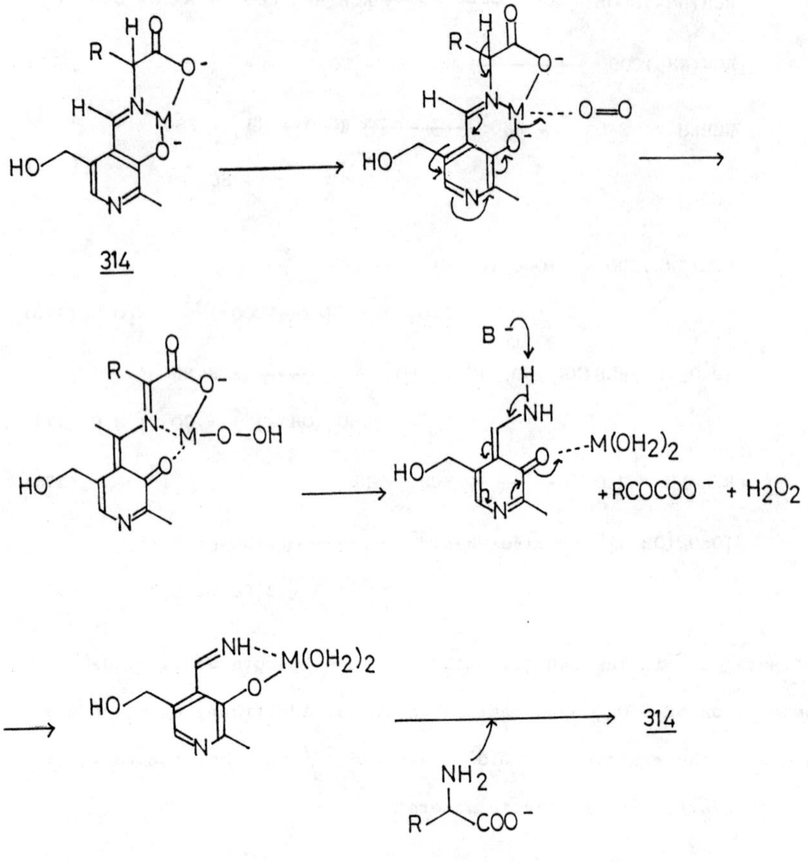

314

Figure 84 Mechanism of catalytic oxidation of amino acids.

6.13.6 Ascorbic acid

Catalytic oxidation of ascorbic acid, H_2A, 317, to dehydroascorbate, A, 318, has been studied by many groups as a model for copper containing oxygenase enzymes. For the reaction in the presence of Cu^{2+}, the mechanism of equations (177) - (183) has been proposed.[598] The effect of anions

317 $\xrightarrow{\text{O}_2}$ 318

$$2Cu^{2+} + HA^- \longrightarrow [Cu_2A]^{2+} + H^+ \qquad (177)$$

$$[Cu_2A]^{2+} + O_2 \rightleftharpoons [Cu_2A.O_2]^{2+} \qquad (178)$$

$$[Cu_2A.O_2]^{2+} \longrightarrow [Cu(A)O_2]^{-\cdot} + Cu^{3+} \qquad (179)$$

$$Cu^{3+} + H_2O \rightleftharpoons Cu^{2+} + HO^\cdot + H^+ \qquad (180)$$

$$HO^\cdot + H_2A \longrightarrow A^{-\cdot} + H_3O^+ \qquad (181)$$

$$A^{-\cdot} + [Cu_2A.O_2]^{2+} \longrightarrow [CuA.O_2]^{-\cdot} + Cu^{2+} + A \qquad (182)$$

$$[CuA(O_2)]^{-\cdot} \longrightarrow Cu^{2+} + A^{\cdot -} + O_2^{2-} \qquad (183)$$

and micelles as well as other ligands has been studied.[599] A similar reaction is found in the presence of Fe, Zn, Co, V, Ni and U salts.[600-602] Metal chelates of Cu(II) and Fe(III), such as those of nitriloacetic acid and EDTA, have a lower catalytic effect and a rate independent of pO_2; this is interpreted in terms of a 1-electron transfer between the metal and the ascorbate monoanion. $A^{-\cdot}$ is then oxidised in a second step and O_2 reoxidises the reduced metal.

Cu(II) is also active in promoting oxidation by $S_2O_8^{2-}$, via equations (184) - (188).

$$S_2O_8^{2-} + Cu^+ \longrightarrow Cu^{2+} + SO_4^- + SO_4^{2-} \qquad (184)$$

$$Cu^{2+} + H_2A \longrightarrow Cu^+ + HA^\cdot + H^+ \qquad (185)$$

$$H_2A + S_2O_8^{2-} \longrightarrow HA^\cdot + H^+ + SO_4^{-\cdot} + SO_4^{2-} \qquad (186)$$

$$HA^- + S_2O_8^{2-} \longrightarrow HA^\cdot + SO_4^{-\cdot} + SO_4^{2-} \qquad (187)$$

$$HA^\cdot + SO_4^{-\cdot} \longrightarrow A + H^+ + SO_4^{2-} \qquad (188)$$

Cobalt complexes, such as Co(salen), are oxidised to superoxo derivatives, and ascorbic acid acts as a π-donor to stabilise a ternary complex. Electron transfer to give $H_2A^{+\cdot}$ is rate determining.[604] Other reported catalysts include Zr^{4+}/flavin,[605] $[(NH_3)_5RuORu(NH_3)_4ORu(NH_3)_5]^{7+}$,[606] vanadyl and cobalt phthalocyanines.[607]

6.14 Oxidation of ethers

Oxidation of ethers yields esters, usually with good selectivity. Two mechanisms were discussed in the reaction of THF with H_2O_2 catalysed by $[Rh_2(C_5Me_5)(\underline{\mu}OH)_3]Cl$ (Figure 85). Mechanism B is preferred.[608] $RuCl_3/NaIO_4$ effects a similar transformation, and in an aprotic solvent the ester or lactone is obtained with good selectivity.[353] With a two phase system diacids are also obtained.[609] The reaction of 319 is a key step in the synthesis of picrotoxinin[619] and some further examples are given in Figure 86.

319   320

Mechanism A

$$RH + Y^{\cdot} \longrightarrow R^{\cdot} + YH$$

$$R^{\cdot} + O_2 \longrightarrow RO_2^{\cdot}$$

$$RO_2^{\cdot} + RH \longrightarrow RO_2H + R^{\cdot}$$

RH =

$Y^{\cdot}$ is a metal centred radical

Mechanism B

$$Rh(III)(O_2H)X + THF \longrightarrow [Rh(III)(THF)X]^+ + HO_2^-$$

$$[Rh(III)(THF)X]^+ + THF + HO_2^- \longrightarrow$$

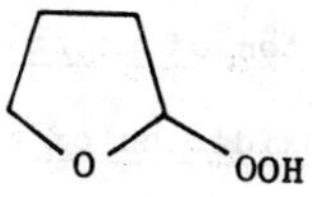

$$+ [Rh(I)(THF)] + HX$$

$$Rh(I)(THF)_n + \tfrac{1}{2}O_2 + H_2O \longrightarrow Rh(III)(O_2H) + THF$$

$$[Rh(III)_2(OH)_3]Cl \rightleftharpoons 2[Rh(III)(OH)X]$$

$$[Rh(III)(OH)X] + H_2O_2 \longrightarrow [Rh(III)(OH_2)(O_2H)X]$$

C_5Me_5 ligands are omitted for clarity

Figure 85   Mechanism for the oxidation of THF by H_2O_2 in the presence of $[Rh_2(C_5Me_5)_2(\underline{\mu}\text{-OH})_3]Cl$.

$$\text{(1,4-dioxane)} + O_2 \xrightarrow{RhCl_3} HCOOCH_2CH_2OOCH + HOCH_2CH_2OOCH \qquad 611$$

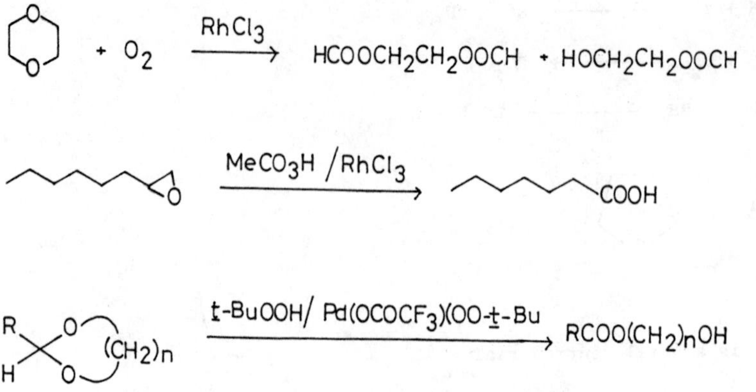

$$\text{(epoxide)} \xrightarrow{MeCO_3H /RhCl_3} \text{(chain)}-COOH \qquad 354$$

$$\underset{H}{\overset{R}{>}}\!\!C\!\!\underset{O}{\overset{O}{<}}(CH_2)_n \xrightarrow{\text{t-BuOOH/ Pd(OCOCF}_3\text{)(OO-\underline{t}-Bu}} RCOO(CH_2)_nOH \qquad 612$$

Figure 86 Catalytic oxidation of ethers.

6.15 Oxidation of phenols and related compounds

6.15.1 Oxidation of monohydric phenols

6.15.1.1 Cobalt catalysts

The superoxo derivatives of Co(II) Schiff's base complexes catalyse oxidation of phenols to p-quinones. The most important species are Co(salen), 321, Co(salpn), 322 and Co(salpr), 323. The reaction mechanism for the formation of quinones and dimers is shown in Figure 87.[613] The precise course followed depends both on the catalyst and the reaction conditions. Quinones are favoured by high [Co] concentration and low temperatures, whereas high temperatures and low [Co] favours dimers.[614] Some examples are given in Figure 88.

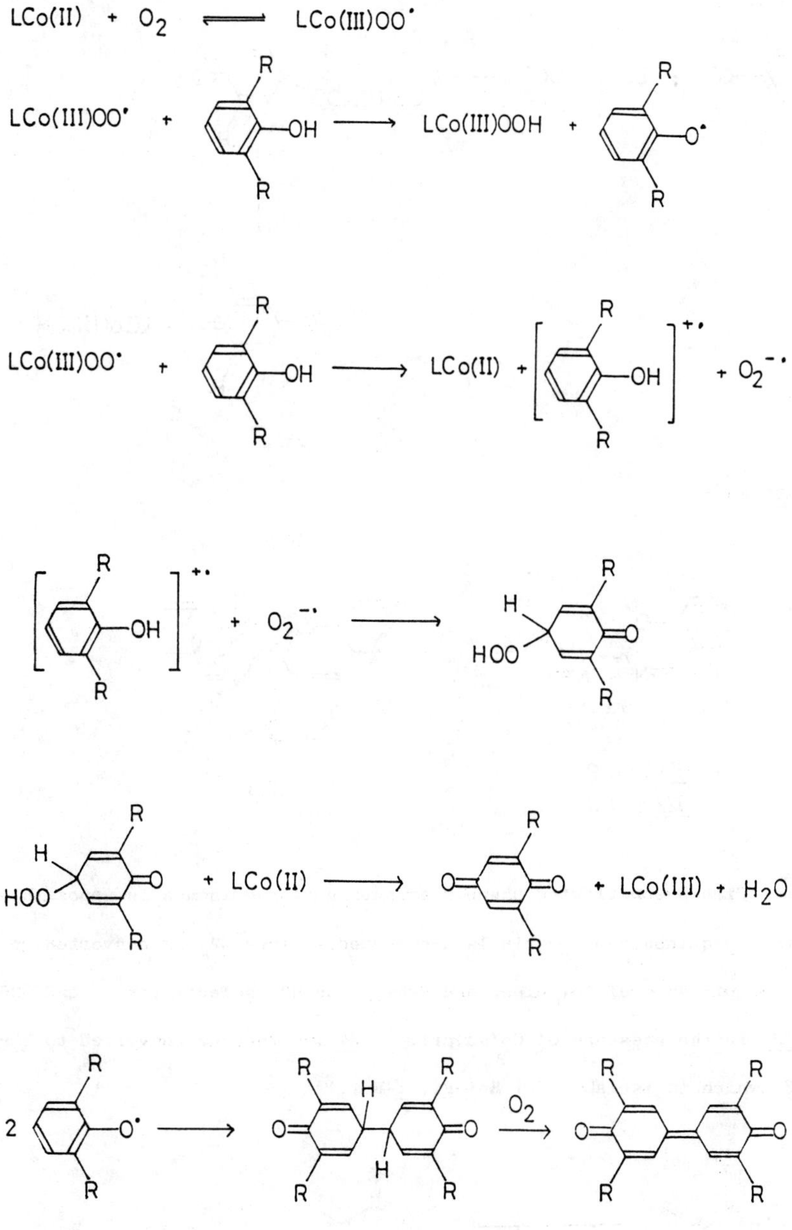

Figure 87 Mechanism of cobalt catalysed oxidation of phenols.

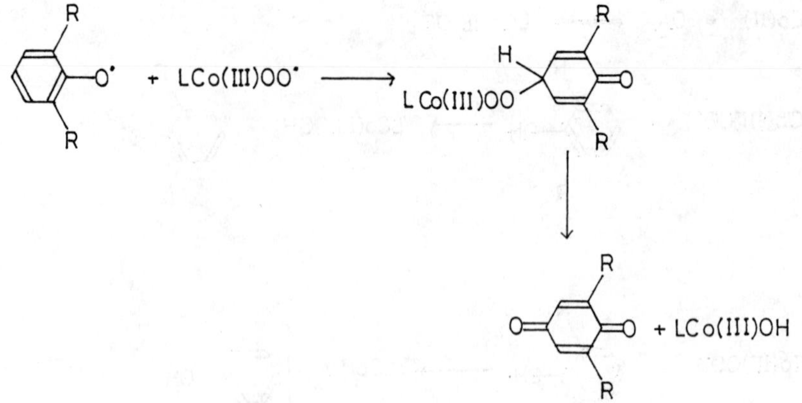

$$\text{[phenoxy radical]} + LCo(III)OO^{\bullet} \longrightarrow LCo(III)OO\text{[cyclohexadienone]}$$

$$\longrightarrow \text{[p-quinone]} + LCo(III)OH$$

Figure 87 cntd.

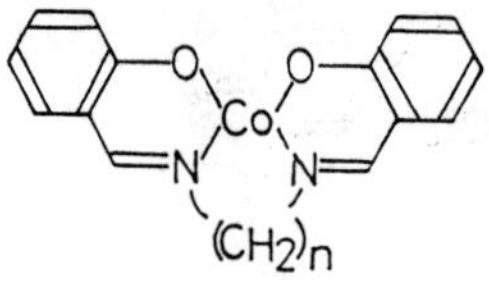

321 n = 2
322 n = 3

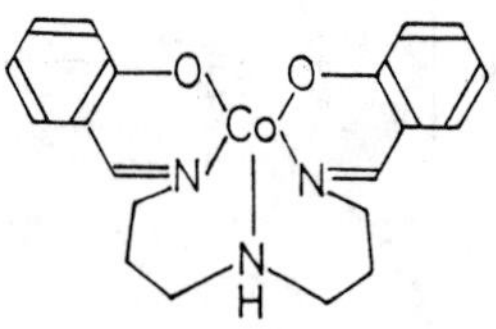

323

With p-substituted phenols oxidation to p-quinones is impossible
and either o-quinones or p-quinols are formed.   Thus 324 is converted to
325 in the presence of Co(salen) and PPh$_3$, with 99% selectivity[620] and 326
gives 327 in the presence of Co(salpr).   327 may then be converted to 329,
via 328 (which is isolable for R=i-Pr, t-Bu).[621]

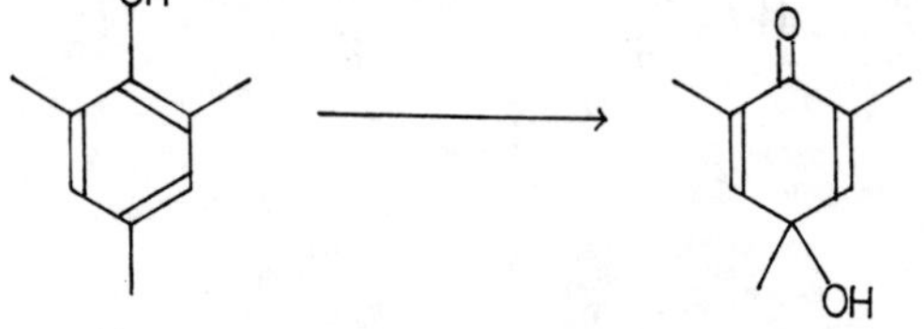

324   325

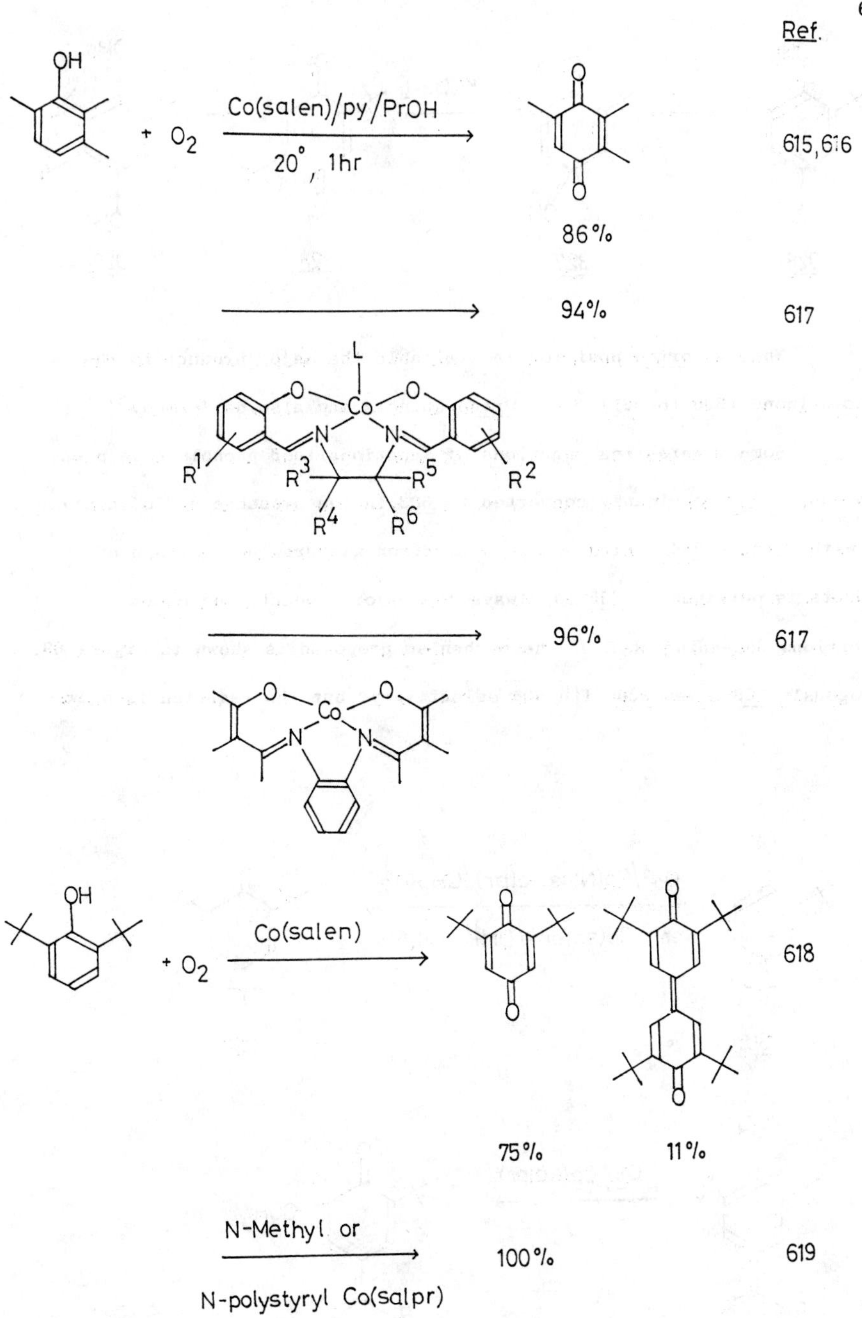

Figure 88   Cobalt catalysed oxidation of phenols.

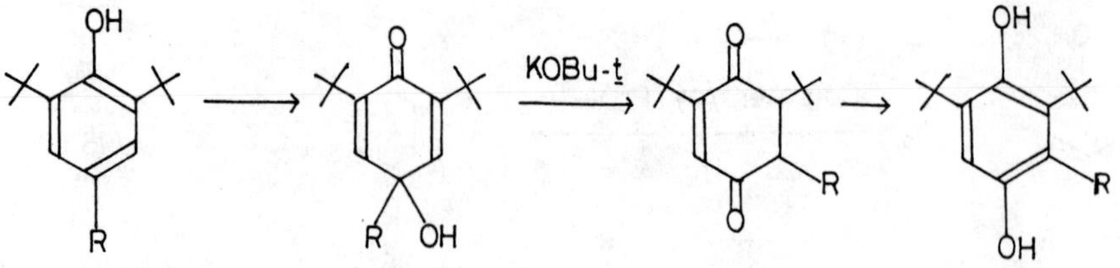

326 327 328 329

When an *ortho*-position is available the major product is the
ortho-quinone (330 to 331).[618]   Ortho-quinols may also be formed.[622]

Some interesting reactions of functionalised phenols have been
reported.   332 is cleanly converted to 333 in the presence of Co(salpr),
but with imines, 334, which are less electron withdrawing, a range of
products is obtained.   335 is always the major product, the exact
proportions depending on R.   The mechanism proposed is shown in Figure 89.[623]
Analogously 338 gives 339 with 60% selectivity, but the reaction is slow.

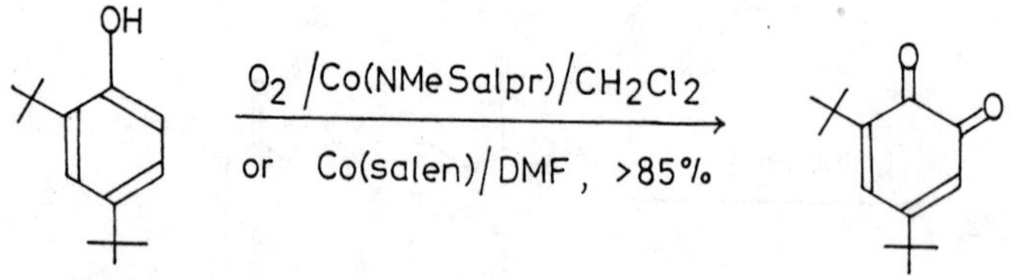

330 331

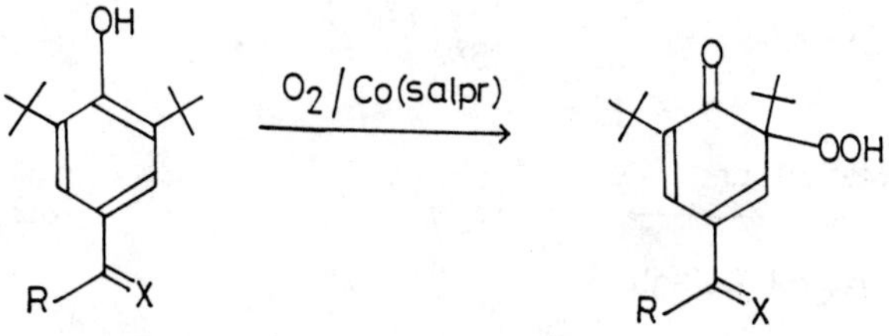

332 333

X = O, NOMe

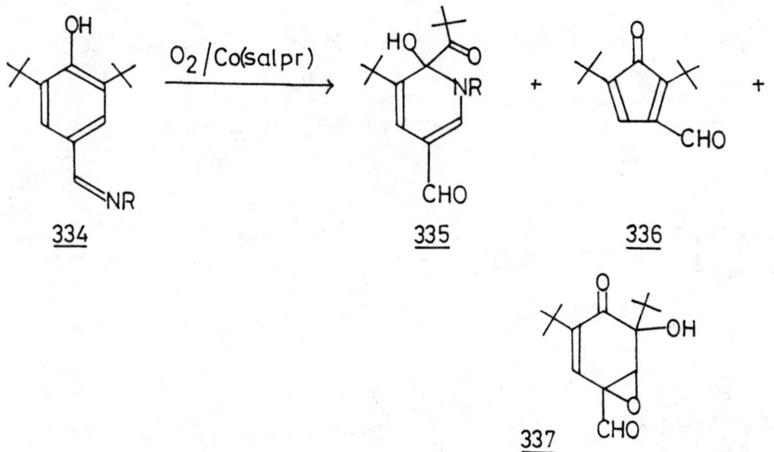

334   335   336

337

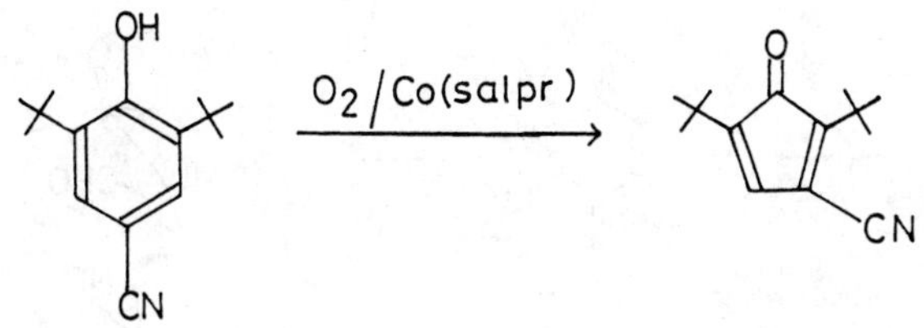

338   339

A number of other cobalt dioxygen complexes have been used to catalyse phenol oxidation. Some interesting selectivities have been achieved but rationalising the reactions is difficult and synthetic applications are few (Figure 90).

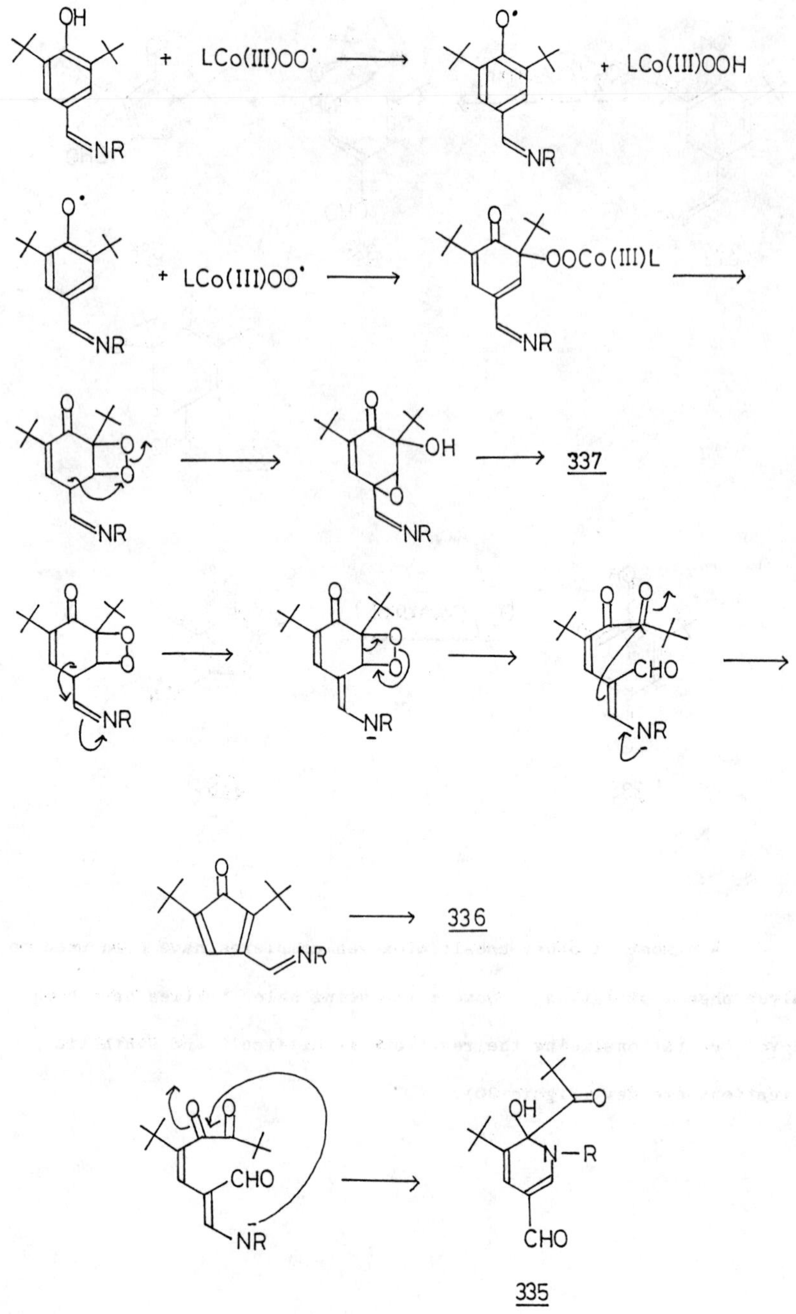

Figure 89 Mechanism of oxidation of imine functionalised phenols.

Ref.

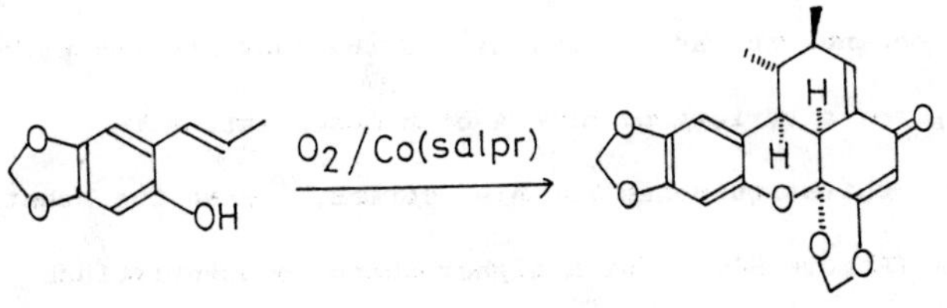

624

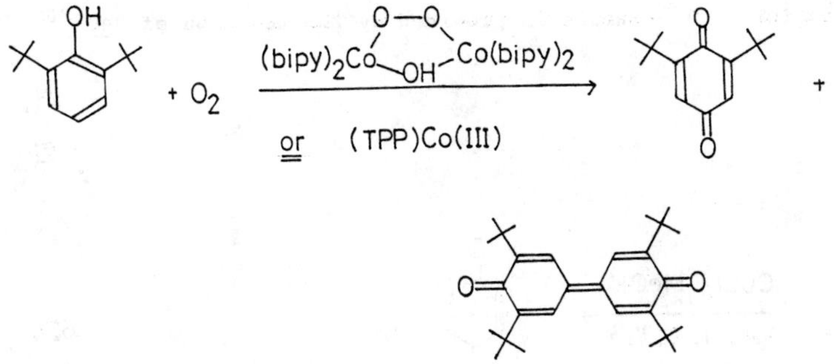

625
626
627

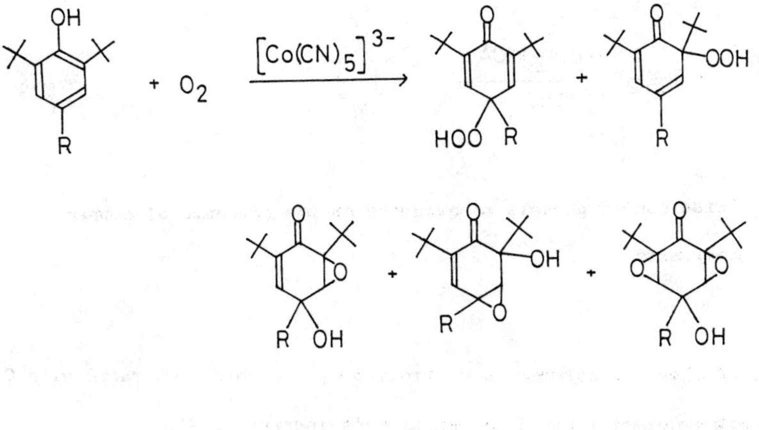

613

__Figure 90__ **Oxidation of phenols in the presence of cobalt dioxygen complexes.**

6.15.1.2. Copper catalysts

Oxidation of phenols in the presence of copper ions is initiated by electron transfer from phenoxide to Cu(II) to give Cu(I) and phenoxy radical. Cu(I) is reoxidised to Cu(II) by molecular oxygen. The fate of the phenoxy radical depends on the conditions and the substitution pattern. Oxidative coupling gives a variety of dimers or polymerisation to polyphenylene ethers, whilst quinones are also formed, when the substrate concentration is low (Figure 91). With higher phenol concentrations dimers and polymers are the major products. Amines are used to enhance phenolate formation. An example is provided by the reaction of 340[630-633]

Ref.

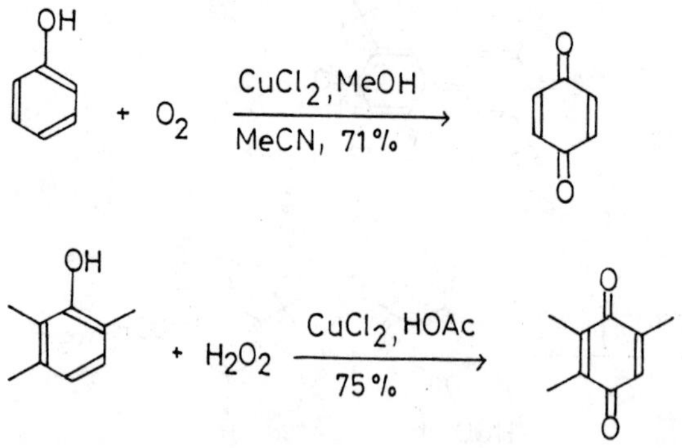

628

629

Figure 91 Oxidation of phenols to quinones in the presence of copper complexes.

The ratio of dimer to polymer is controlled by the amine: Cu ratio with C-O coupling and polymerisation favoured as this increases.[630]

Dimerisation rather than polymerisation is preferred for phenols bearing very bulky groups. Thus 330 gives mainly 341 with either Cu_2Cl_2[634] or Cu(II)/(Et$_2$NCH$_2$CH$_2$NEt$_2$).[635]

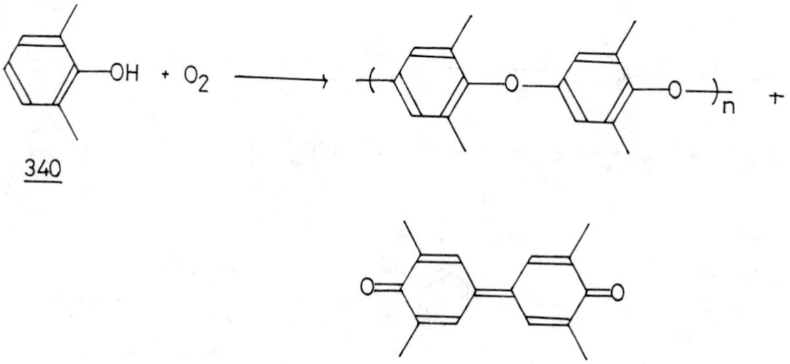

340

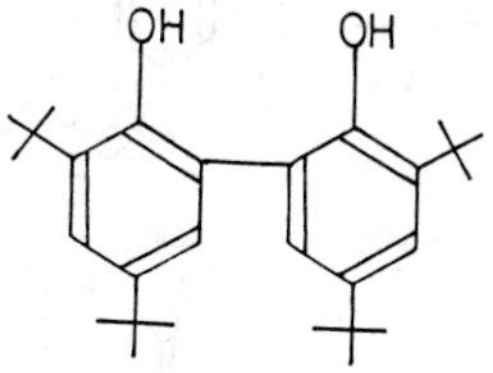

341

6.15.1.3. Other metals

Whilst complexes of other metals have not been extensively
studied, many of the reactions fall into the categories already discussed.
Thus reactions with several M(TPP)X complexes proceed via M-OO$^{\bullet}$ giving
mainly 343 from 342.[627]   Catalysis by Fe(II) phthalocyanines involves a
similar mechanism with 343 as the sole product.[618]

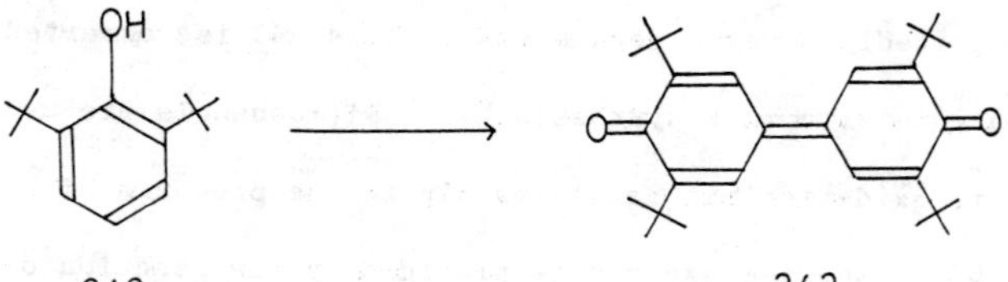

342   343

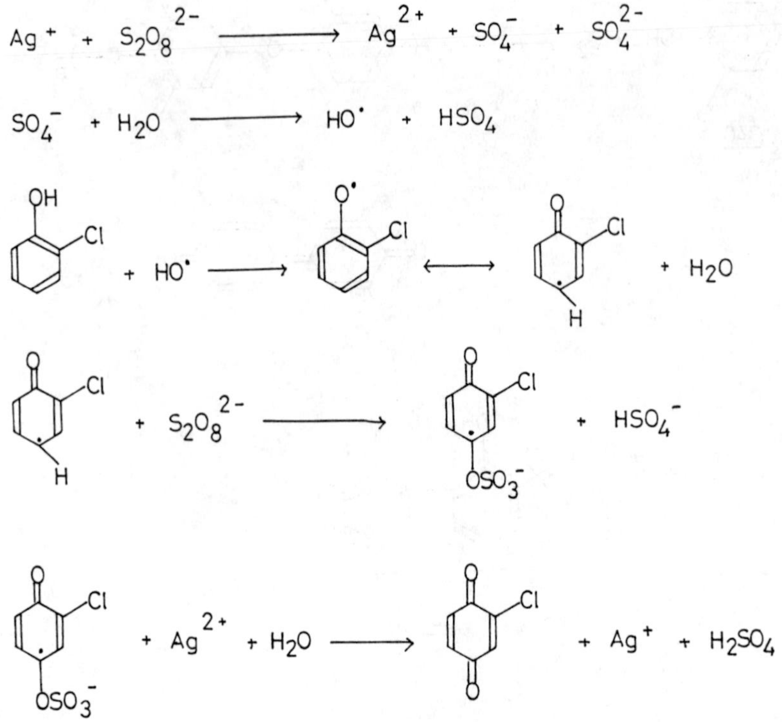

Figure 92   Radical initiated oxidation of phenols with $Ag^+/S_2O_8^{2-}$.

Radical initiated reactions are also common, with $Ag^+/S_2O_8^{2-}$ providing their source for a number of substituted phenols (Figure 92).[636]

Oxidation of phenols to quinones in the presence of Ru(III) occurs via reportedly diverse mechanisms.   Thus 344 is converted to 345, an intermediate in vitamin E synthesis.[637]   Nitrophenols are usually rather inert to oxidation but react readily in the presence of Ru(III)/IO_4^-.[637]   Another example is provided by the reaction of 346.[638]

344   345

346   347

6.15.2. Oxidation of catechols

6.15.2.1 Copper catalysts

The oxidation of catechols to o-quinones in the presence of
Cu^{2+} depends on $[O_2]$, $[Cu^{2+}]$ and [substrate] and this was quickly
interpreted in terms of a ternary complex.[639]   Since then, however, the
reaction has been studied in more detail.   In the stoichiometric
conversion of 348 to 349 only the complex, 350, is needed.[640,641]   The
mechanism involves a copper catecholate complex, 351,[642] and the anaerobic
oxidation depends on the Cu(I)/Cu(II) redox potential. Oxidation of
the quinone to 352 also occurs under anaerobic conditions[640], but the
reaction in the presence of O_2 is more vigorous and probably involves a
different mechanism. The prolonged study of this process reflects its
importance as a model for pyrocatechase [640,643]. ^{18}O labelling suggests that
the quinone is converted into products via 353.

To shed further light on the mechanism several substituted catechols have been studied. 354 gives both 355 and 356 with low selectivity.[644] 359 has been a useful substrate and 360 may be isolated as the first product when the catalyst is bis(1-phenyl-1,3,5-hexanetrionato) copper,[640] Cu(bipy)$_2$Cl,[645] cobalt porphyrins,[646] Cu$_2$(CH$_3$COCH$_2$COCH$_2$COCH$_3$)$_2$[647] or 1,4-(di-2'-pyridyl)aminophthalazine Cu(II).[648] However, using CuCl/MeOH/Py regioselective ring cleavage yields 361 which cyclises to 362 in fair yield.[641]

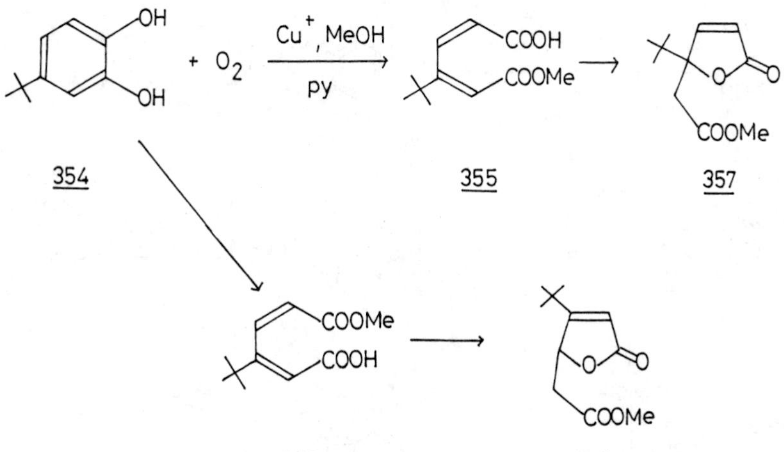

354   355   357

356   358

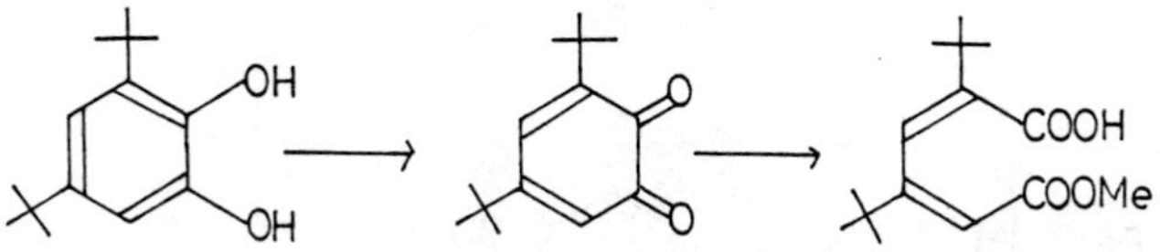

359   360   361

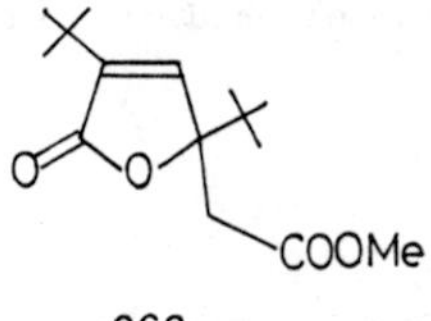

362

6.15.2.2. Iron catalysts

Fe(II) and Fe(III) complexes are also catalysts for catechol oxidation but the mechanism proposed is quite different from that with copper salts. 348 is oxidised to muconic acid using Fe(III)/H_2O_2 via the mechanism of Figure 93.[649] The reaction with molecular O_2 is analogous. The iron complex of 359, 363, forms a superoxo complex, 364.

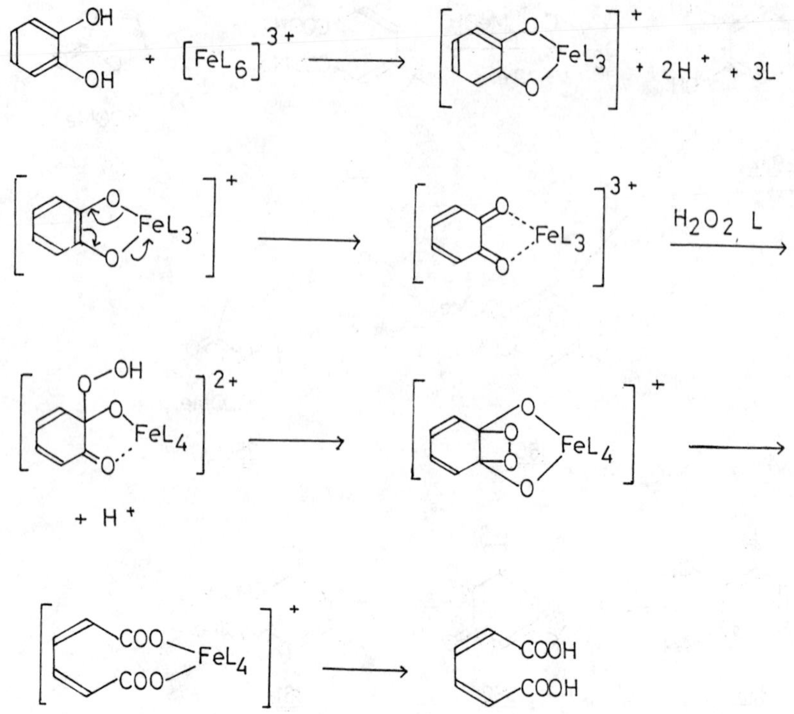

$$\text{catechol} + [FeL_6]^{3+} \longrightarrow [\text{catecholate-}FeL_3]^{+} + 2H^{+} + 3L$$

$$[\cdots FeL_3]^{+} \longrightarrow [\cdots FeL_3]^{3+} \xrightarrow{H_2O_2, L}$$

$$[\text{OOH}\cdots FeL_4]^{2+} + H^{+} \longrightarrow [\cdots FeL_4]^{+} \longrightarrow$$

$$[\text{COO}\cdots FeL_4 \cdots \text{COO}]^{+} \longrightarrow \text{COOH, COOH (muconic acid)}$$

Figure 93 Mechanism of oxidation of catechol in the presence of Fe(III).

The radical attacks the ring to give 365 which collapses to 366, analogous to an intermediate in the previous scheme.[650]   366 gives the lactone, 367, via the muconic acid as the major product (80-85%) with iron nitriloacetate salts.[651]   Although the o-quinone may be isolated from reaction in the presence of Fe(II)(bipy) it is not converted to the lactone.[652]

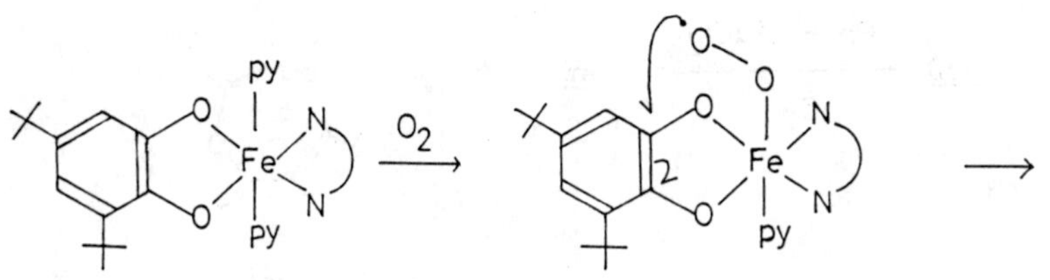

363

364

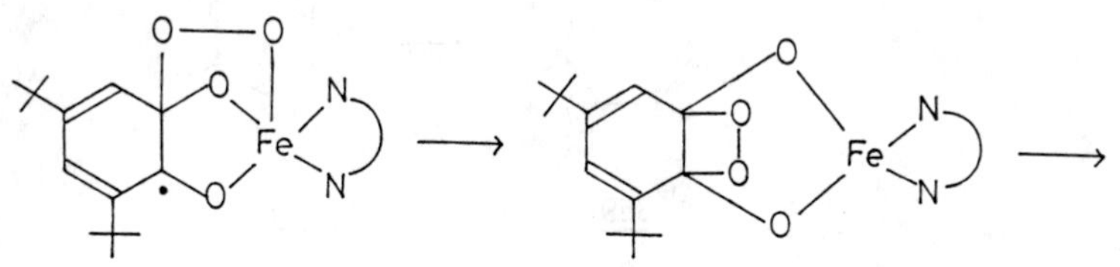

365

366

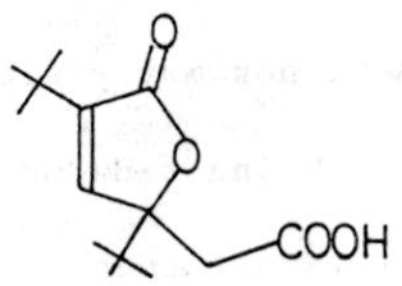

367

6.15.3.2 Other metals

Catechol oxidation by molecular oxygen proceeds smoothly in the presence of (Ph$_3$P)$_3$RuCl$_2$ to give 360, 368, 369 and 370 from 359.[653] The true catalytic species is unknown since the Ph$_3$P is all oxidised to Ph$_3$PO. t-BuOOH may also be used as oxidant; in this case the quinone is obtained in high yield.[654] Further examples are shown in Figure 94.

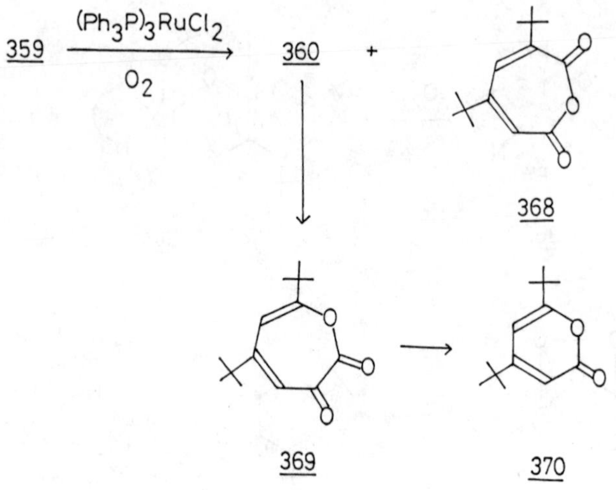

$$359 \xrightarrow[\text{O}_2]{\text{(Ph}_3\text{P)}_3\text{RuCl}_2} 360 \quad +$$

368

369   370

6.15.3. Oxidation of anilines

Oxidation of anilines may yield nitrobenzenes or coupled products depending on conditions.   Using t-BuOOH as oxidant and VO(acac)$_2$ or Mo(CO)$_6$ as catalyst, ArNO$_2$ is obtained.   ArNO and ArNHOH are thought to be intermediates but could not be isolated.[660]   Radical reactions yield coupled products   (Figures 95[661] and 96[662]).

Oxidation of o-phenylene diamine, 371, closely resembles the reactions of catechol.   Many of the same catalysts are active and the mechanism, though less well studied, is similar.   The usual products are cis,cis-muconitrile, 373, and/or 374.   Co(ClO$_4$)$_2$ gives 100% 374,[663] whilst CuCl$_2$/MeOH/py gives 90% 373.[664]

$$\underline{359} + O_2 \xrightarrow{\text{VO(acac)}_2} \underline{368} + \underline{360} + \underline{370}$$

41%   27%   15%

655

$$\underline{359} + O_2 \xrightarrow{\text{VO(salen)}} 42\% \quad 24\% \quad 11\%$$

656

$$\underline{348} + O_2 \xrightarrow{\text{Mn(II)}} \underline{349}$$

639

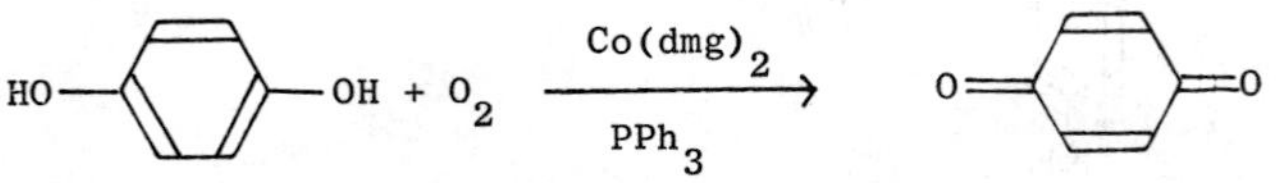

$$\text{HO}-\!\!\!\!\!\!\!\!\!\bigcirc\!\!\!\!\!\!\!\!\!-\text{OH} + O_2 \xrightarrow[\text{PPh}_3]{\text{Co(dmg)}_2} O=\!\!\!\!\!\!\!\!\!\bigcirc\!\!\!\!\!\!\!\!\!=O$$

657

$$\underline{359} + O_2 \xrightarrow[\text{M = Mn, Co, Cu}]{\text{M(acac)}_3} \underline{360}$$

658

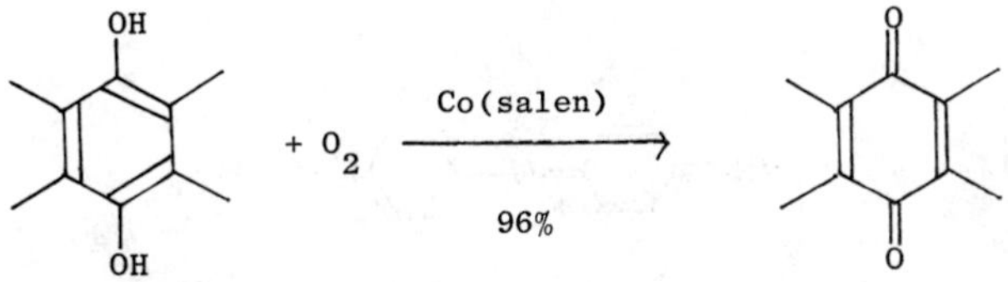

$$+ O_2 \xrightarrow[96\%]{\text{Co(salen)}}$$

659

Figure 94   Catalytic oxidation of catechols.

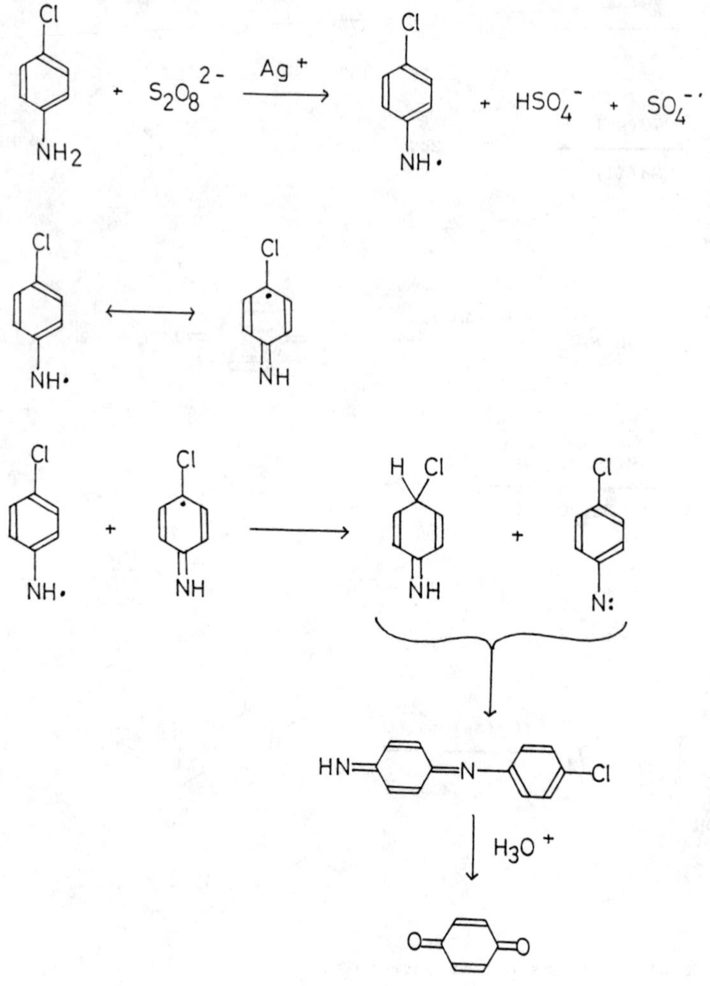

Figure 95 Catalytic oxidation of p-chloroaniline.

$$[Cu(py)_4]^{2+} + PhNH_2 \longrightarrow [Cu(py)_3]^+ + PhNH_2^{+\cdot} + py$$

$$[Cu(py)_3]^+ \cdot [PhNH_2]^{+\cdot} + O_2 \rightleftharpoons [Cu(py)_3]^{2+} + [PhN=O] + H_2O$$

$$2[Ph\overset{\cdot}{N}=O] \longrightarrow PhN=NPh + O_2$$

Figure 96   Catalytic oxidation of aniline.

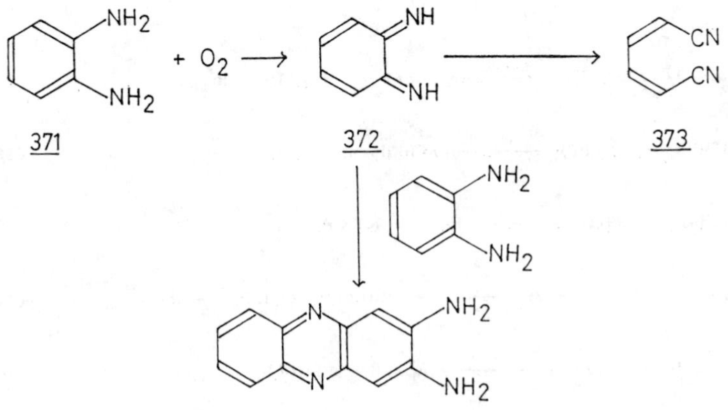

371   372   373

374

6.16 Oxidation of nitrogen containing compounds

6.16.1 Oxidation of amines

6.16.1.1 Primary amines

Hydroperoxides react with primary amines in the absence of a catalyst to give ketinimines or aldimines.   However, with V, Mo or Ti

$$\text{cyclohexylamine} + \underline{t}\text{-BuOOH} \xrightarrow{\text{M(naphthenate)}_n} \underline{375}$$

naphthenate, cyclohexylamine is converted to $\underline{375}$ in good yield.[665]

    Oxidative deamination occurs under radical conditions, the routes given by equations (189) - (194)[666] and (195) - (200)[667] being typical. An alternative reoxidant for the first scheme is $PhI(OAc)_2$,

$$RCH_2NH_2 + Ru(III) \rightleftharpoons RCH\text{-}H\cdots Ru(III) \quad (189)$$

where the intermediate bears an NH_2 substituent.

$$RCH\text{-}H\cdots Ru(III) \xrightarrow{slow} R\overset{\bullet}{C}HNH_2 + Ru(II) + H^+ \quad (190)$$

where the starting species bears an NH_2 substituent.

$$R\overset{\bullet}{C}HNH_2 + Ru(II) \longrightarrow R\overset{+}{C}HNH_2 + Ru(I) \quad (191)$$

$$R\overset{+}{C}HNH_2 + 2H_2O \longrightarrow RCHO + NH_4OH + H^+ \quad (192)$$

$$Ru(I) + BrO_3^- + 2H^+ \longrightarrow Ru(III) + BrO_2^- + H_2O \quad (193)$$

$$R\overset{\bullet}{C}HNH_2 + Ru(III) \longrightarrow R\overset{+}{C}HNH_2 + Ru(II) \quad (194)$$

$$RCH_2NH_2 + Ag^+ \rightleftharpoons [Ag\cdot RCH_2NH_2]^+ \quad (195)$$

$$[Ag\cdot RCH_2NH_2]^+ + Ce(IV) \xrightarrow{slow} [Ag\cdot RCH_2NH_2]^{2+} + Ce(III) \quad (196)$$

$$[Ag\cdot RCH_2NH_2]^{2+} \xrightarrow{fast} [RCH_2NH_2]^{+\bullet} + Ag^+ \quad (197)$$

$$[RCH_2NH_2]^{+\bullet} \longrightarrow R\overset{\bullet}{C}HNH_2 + H^+ \quad (198)$$

$$R\overset{\bullet}{C}HNH_2 + Ce(IV) \longrightarrow R\overset{+}{C}HNH_2 + Ce(III) \quad (199)$$

$$R\overset{+}{C}HNH_2 + 2H_2O \longrightarrow RCHO + NH_4OH + H^+ \quad (200)$$

though in this case a non-radical route is also possible.[666]

With Ru(III)/O$_2$ catalytic dehydrogenation to nitriles is the usual reaction. Some amide is also formed (equation (201)).[489] Using (Ph$_3$P)$_2$Ru(RNH$_2$)$_2$Cl$_2$ as catalyst gives better selectivity to the nitrile.

$$PhCH_2NH_2 + O_2 \xrightarrow{\quad RuCl_3 \quad} PhCN + PhCONH_2 \qquad (201)$$

$$53\% \qquad 30\%$$

Aliphatic amines give lower rates and selectivities.[669] Oxidation of 376 in the presence of Cu$_2$Cl$_2$ gives 377 in good yield,[670] and allylamine gives acrylonitrile,[671] but the reactions have not been generalised.

376   377   90·4 %

6.16.1.2 Secondary amines

The oxidation of secondary amines has not been systematically studied. A variety of products may be obtained but few mechanisms are understood (Figure 97).

6.16.1.3 Tertiary amines

Two groups of reactions of tertiary amines are known, the first being the usual synthesis of amine oxides. Using t-BuOOH as oxidant, vanadium and molybdenum complexes both give excellent selectivity.[676] Tungsten, niobium, tantalum and rhodium complexes are poorer catalysts.[677] t-Butyl, amyl and cumyl hydroperoxides have all been used and the reaction proceeds faster in dipolar aprotic solvents. Some examples are given in Figure 98. Alternatively, in the presence of PhIO/Fe(TPP)Cl, a monooxygenase type cleavage of PhNMe$_2$ to PhNHMe and CH$_2$O occurs.[168]

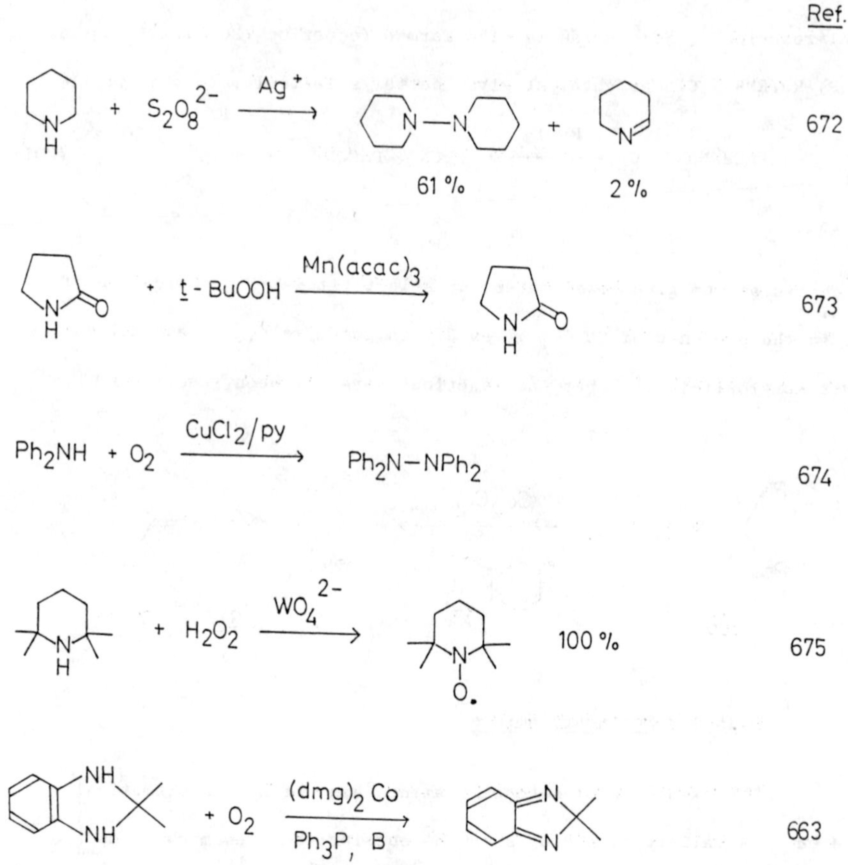
Ref.
Ag+
61 %
2 %
672
Mn(acac)3
t-BuOOH
673
Ph2NH + O2
CuCl2/py
Ph2N–NPh2
674
H2O2
WO4 2-
100 %
675
(dmg)2 Co
Ph3P, B
O2
663

<u>Figure 97</u> Catalytic oxidation of secondary amines.

<u>6.16.2 Oxidation of isocyanides</u>

Oxidation of alkyl isocyanides to their isocyanates is catalysed by various metal complexes. With ($\underline{t}$-BuNC)$_4$Ni, ($\underline{t}$-BuNC)$_2$NiO$_2$ may be crystallised from solution.[679] Other catalysts include (Ph$_3$P)$_3$RhCl, (cod)$_2$Co, (cod)$_2$Ni and (dmg)$_2$Co.[663,679-681]

$$Me_2NC_{12}H_{26} + \underline{t}-BuOOH \xrightarrow[\text{acetone, }60°]{VO(acac)_2} Me_2\overset{\overset{\displaystyle O}{\uparrow}}{N}C_{12}H_{26}$$

676

94%

$$Me_2NC_{12}H_{26} + \underline{t}-BuOOH \xrightarrow[\underline{t}-BuOH, 86\%]{Fe(acac)_2} Me_2\overset{\overset{\displaystyle O}{\uparrow}}{N}C_{12}H_{26}$$

676

10%

$$(HOCH_2CH_2)_3N + \underline{t}-BuOOH \xrightarrow{Ti(O-\underline{i}-C_4H_9)_4} (HOCH_2CH_2)_3N \longrightarrow O$$

678

Figure 98 Oxidation of tertiary amines to yield amine oxides.

6.16.3 Oxidation of nitroso compounds

The reaction of PhNO with $\underline{t}$-BuOOH is initiated by various M(acac)$_n$ complexes.   PhNO$_2$ is formed rapidly and quantitatively (equations (202) - (205)) but the rate equation is complex.[682]

$$M(II) + \underline{t}-BuOOH \longrightarrow M(III)OH + \underline{t}-BuO^{\bullet} \tag{202}$$

$$\underline{t}BuO^{\bullet} + \underline{t}-BuOOH \longrightarrow \underline{t}-BuOH + \underline{t}-BuOO^{\bullet} \tag{203}$$

$$\underline{t}-BuOO^{\bullet} + PhNO \longrightarrow PhNO_2 + \underline{t}-BuO^{\bullet} \tag{204}$$

$$M(III) + \underline{t}-BuOOH \longrightarrow M(II) + \underline{t}-BuOO^{\bullet} + H+ \tag{205}$$

(dmg)$_2$Co/base is also reactive in catalysing oxidation by O$_2$.   Nitrosamines, R^1R^2NNO, are oxidised to nitramines using ROOH and molybdenum catalysts.[683]

6.16.4 Oxidation of azobenzenes

Molybdenum complexes catalyse hydroperoxide oxidation of

azobenzenes to azoxy compounds (equation (206)).[682]   Regioselectivity is

$$Ar^1-N=N-Ar^2 \xrightarrow[\text{MoO}_2(\text{dipivaloylmethane})_2]{\text{t-BuOOH}} \underset{\substack{\downarrow \\ O}}{Ar^1-N=N-Ar^2} \qquad (206)$$

fair with $MO_2(dpm)_2$ with electrophilic attack at the nitrogen on the

unsubstituted benzene ring. $Mo(CO)_6$ is less selective.   The mechanism is

supposed to involve 378.

378

379

6.16.5 Oxidation of hydrazines, hydrazides and hydrazones

Oxidation of simple hydrazines has not been widely studied.

Tosyl hydrazine undergoes oxidative dimerisation to 379 using H_2O_2/Co^{2+}

via a radical pathway.[684]   Benzidine is oxidatively dehydrogenated to

azobenzene in the presence of $(dmg)_2Co/O_2/base$.[663]   Acyl hydrazides are

converted to carboxylic acids by $O_2/Cu(OAc)_2$ (equation (207)).[685]   Vicinal

$$RCONHNH_2 + O_2 \xrightarrow[\text{MeOH 95\%}]{\text{Cu(OAc)}_2} RCOOH + N_2+H_2O \qquad (207)$$

dihydrazones are oxidised to acetylenes by Cu_2Cl_2/O_2,[686] $CuI/O_2/py$ [687]

or $Co(salen)O_2$.[688]   Yields and selectivities are excellent.

6.16.6 Oxidative cleavage.

A range of compounds containing the structural unit $C=C-NR_2$

are oxidatively cleaved in the presence of transition metal complexes.

The reaction with enamines, 380, is particularly successful in the

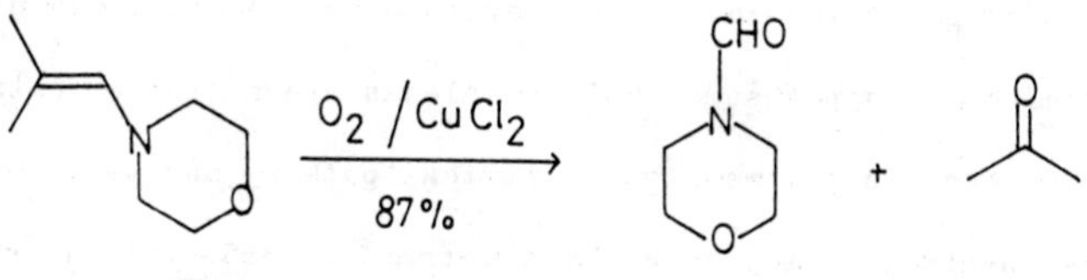

380   381

presence of copper salts.[689] The speed of the reaction permits selective

cleavage of enamines even in the presence of other unsaturation (382 to

383).[690] Mechanistic studies suggest the intermediate is a dioxetane, 384.

3-Membered rings are similarly cleaved (385 to 387), the reaction

proceeding via initial electron transfer to give 386.[691]

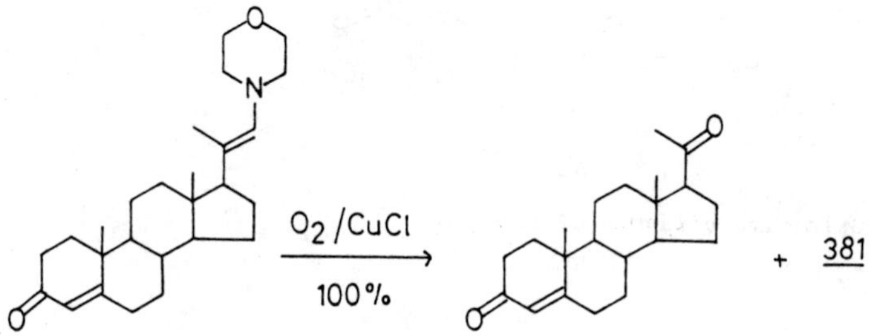

382   383

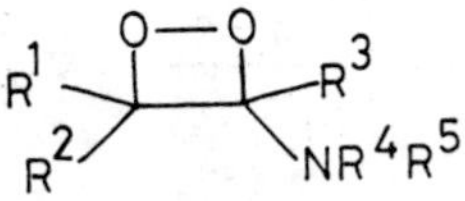

384

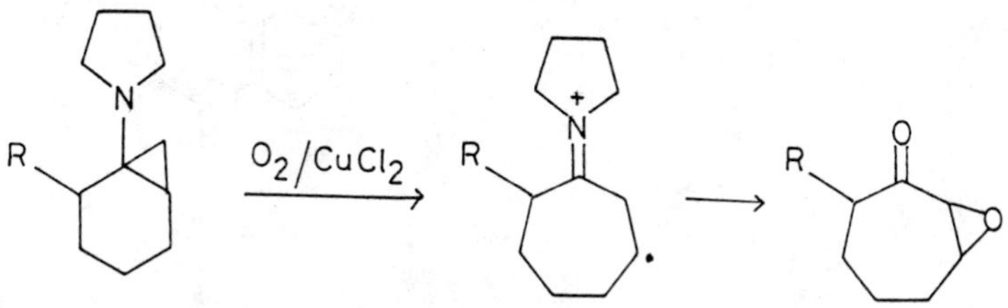

385   386   387

The cleavage of indoles may be regarded as a special example of this reaction; both copper and cobalt complexes are active as catalysts (Figure 99). The reaction proceeds via a radical pathway and is a model for tryptophan-2,3-dioxygenase. When the substrate is 388, the intermediate, 389, cyclises to give 390 in up to 62% yield.[694]

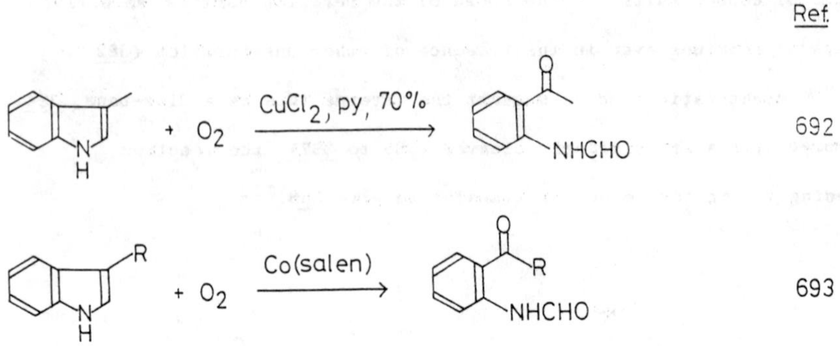

Figure 99 Oxidative cleavage of indoles.

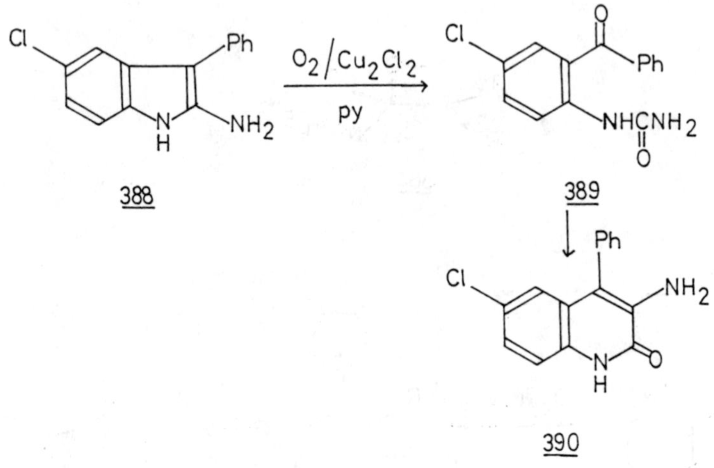

388

389

390

6.17 Oxidation of phosphorus containing compounds

Oxidation of phosphines to phosphine oxides is catalysed by many transition metal complexes.[17] The most thoroughly studied system uses $M(PPh_3)_4$ where M = Ni, Pd or Pt. The nickel complex operates at -35°C and those of Pd and Pt at 90°C.[695] The steps involved are shown in equations (208) - (211).[6] Halpern suggested __391__ as the transition state

$$Pt(PPh_3)_4 \rightleftharpoons Pt(PPh_3)_3 + PPh_3 \qquad (208)$$

$$Pt(PPh_3)_3 + O_2 \longrightarrow (Ph_3P)_2PtO_2 + PPh_3 \qquad (209)$$

$$(Ph_3P)_2PtO_2 + PPh_3 \longrightarrow (Ph_3P)_2PtO_2 \qquad (210)$$

$$(Ph_3P)_2PtO_2 + 2PPh_3 \longrightarrow (Ph_3P)_3Pt + 2Ph_3PO \qquad (211)$$

but other workers invoke a two step pathway via __392__, __393__ and __394__.[696] Other phosphine oxidations catalysed by complexes of these metals are shown in equations (212)[697] and (213).[698]

$$Bu_3P \xrightarrow[\text{Pd(diphos)}_2]{\text{air}} Bu_3PO \qquad (212)$$

$$Ph_3P + O_2 \xrightarrow{Cp_2Ni} Ph_3PO \qquad (213)$$

__391__

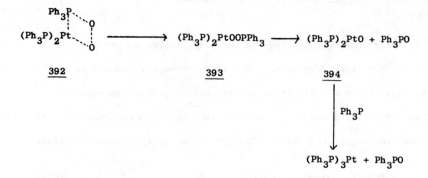

392   393   394

Ph₃P

(Ph₃P)₃Pt + Ph₃PO

Wilkinson's catalyst , $(Ph_3P)_3RhCl$, catalyses complete oxidation of Ph_3P in refluxing toluene (equations (214) - (217)).[699] ESR studies

$$(Ph_3P)_3RhCl \rightleftharpoons (Ph_3P)_2RhCl + Ph_3P \qquad (214)$$

$$(Ph_3P)_2RhCl + O_2 \rightleftharpoons (Ph_3P)_2Rh(II)O_2Cl \qquad (215)$$

$$(Ph_3P)_2Rh(II)O_2Cl + Cl^- \rightleftharpoons [(Ph_3P)_2Rh(II)O_2Cl_2]^- \qquad (216)$$

$$[(Ph_3P)_2Rh(II)O_2Cl_2]^- \rightleftharpoons [(Ph_3P)_2Rh(III)Cl_2(O_2^{2-})] \quad (217)$$

show that both Rh(II) and O_2^{2-} are present.[700] $(Ph_3P)_2Rh(CO)Cl$[701] and $Rh_6(CO)_{16}$[702] may also be used as catalysts. CO_2 can act as an oxidant for R_3P in the presence of $(Ph_3P)_3RhCl$ (equations (218) - (221)).

$$(Ph_3P)_3RhCl + CO_2 \longrightarrow (Ph_3P)_3RhCl(CO_2) \qquad (218)$$

$$(Ph_3P)_3RhCl(CO_2) \dashrightarrow (Ph_3P)_2(Ph_3PO)Rh(CO)Cl \qquad (219)$$

$$(Ph_3P)_2(Ph_3PO)Rh(CO)Cl \longrightarrow (Ph_3P)_2(Ph_3PO)RhCl + CO$$

$$(220)$$

$$(Ph_3P)_2(Ph_3PO)RhCl + Ph_3P \longrightarrow (Ph_3P)_3RhCl + Ph_3PO \quad (221)$$

Iridium complexes, such as $(Ph_3P)_2Ir(CO)X$, catalyse oxidation of Ph_3P with low turnover via 395 and 396,[251] and $[(Ph_2MeP)_3Ir(CO)]ClO_4$ is also a relatively inefficient catalyst.[704]

395   396

Among catalytically active ruthenium complexes we may include $Ru(O_2)(CN)(NO)(PPh_3)_2$ and $Ru(NCS)(CO)(NO)(PPh_3)_2$.[705] A metal dioxygen complex is involved and the mechanism postulated involves simultaneous transfer of oxygen atoms to two coordinated phosphines. $(Ph_3P)_3RuCl_2$[57] and $(Ph_3As)_3RuCl_2$[706] are also good catalysts at 1 atm./25°C. Oxidation of Ph_3P has been studied in the presence of $[Fe(mnt)_2]^-$ and $[Fe(mnt)_3]^{2-}$ where mnt^{2-} is cis-1,2-dicyanoethylene-1,2-dithiolate, but the reaction is slow.[707] Simple inorganic salts such as $FeCl_3$ and $Fe(NCS)_3$ are also effective, but the species isolated after reaction are complexes of Ph_3PO.[708]

Cobalt superoxo complexes are also active in phosphine oxidation. $(Et_3P)_2CoCl_2$ gives stoichiometric oxidation of the coordinated ligands[709] and an unusual trinuclear species formed from $(Ph_3P)_2CoCl_2$, allylamine and benztriazole is also weakly active.[710] On treating $(PhMe_2P)_3Co(CN)_2$ with O_2, 397 is formed[711] and excess $PhMe_2P$ is oxidised catalytically. Finally oxidation may occur by oxygen transfer from a coordinated ligand without a change in oxidation state of the metal. Thus, $(saloph)Co(NO)$ is oxidised by O_2 in the presence of pyridine to $(saloph)Co(py)NO_2$. This transfers oxygen from the NO_2 to Ph_3P, regenerating $(saloph)Co(NO)$. It was established that neither radicals nor MO_2 complexes were involved.[712]

Molybdenum oxo-complexes, such as $MoO_2(S_2CNR_2)_2$, catalyse oxidation of R_3P by a non-radical route (equations (222) - (224)).[713]

$$\underline{397}$$

$$\text{MoO}_2(\text{S}_2\text{CNR}_2)_2 + \text{PPh}_3 \longrightarrow \text{MoO}(\text{S}_2\text{CNR}_2)_2 + \text{Ph}_3\text{PO} \qquad (222)$$

$$\text{MoO}(\text{S}_2\text{CNR}_2)_2 + \tfrac{1}{2}\text{O}_2 \longrightarrow \text{MoO}_2(\text{S}_2\text{CNR}_2)_2 \qquad (223)$$

$$2\text{MoO}_2(\text{S}_2\text{CNR}_2)_2 + \text{Ph}_3\text{P} \longrightarrow \text{Mo}_2\text{O}_3(\text{S}_2\text{CNR}_2)_4 + \text{Ph}_3\text{PO} \qquad (224)$$

Similar reactions occur using cis-MoO_2(ethyl-L-cysteinate)$_2$ and
cis-MoO_2(dipropyldithiocarbamate)$_2$.[528]   Other active oxo complexes involve
MeOMo=O(mesoporphyrin IX dimethyl ester) and $\text{Cp}_2\text{Nb}=\text{O}$.[714]

More limited studies involve oxidation of arsines to their
oxides, though many of the same complexes are active.   For example,
$\text{Ru}(\text{O}_2)(\text{NO})(\text{PPh}_3)_2\text{Cl}$ does not catalyse Ph_3P oxidation but is effective for
Ph_3As.[705] $(\text{Ph}_3\text{P})_3\text{RhCl}$[57] and $[\text{Fe}(\text{mnt})_2]^-$[707] have also been used.

6.18 Oxidation of sulphur containing compounds

The important oxidations of sulphur containing compounds are
given in equations (225) - (227).   Of these thiol oxidation is the

$$2\text{RSH} + \tfrac{1}{2}\text{O}_2 \longrightarrow \text{RSSR} + \text{H}_2\text{O} \qquad (225)$$

$$\text{RSR}^1 + \tfrac{1}{2}\text{O}_2 \longrightarrow \text{RSOR}^1 \qquad (226)$$

$$\text{RSOR}^1 + \tfrac{1}{2}\text{O}_2 \longrightarrow \text{RSO}_2\text{R}^1 \qquad (227)$$

simplest both practically and mechanistically, since the reaction usually
involves radical coupling and overoxidation is uncommon.   Selective
oxidation of sulphides to sulphoxides may also be achieved.

6.18.1 Oxidation of thiols to disulphides

Both one and two electron oxidants are active for oxidative dimerisation of thiols to disulphides. The process is important; removal of thiols from petroleum is desirable because of their offensive smell and the formation of disulphide bridges between cysteine molecules is significant in establishing protein structure. The critical step is electron transfer and the reaction with Mn(III) (equations (228) - (231)) is typical.[715] A second reaction type is typical of metal phthalocyanine

$$RSH + Mn(III) \longrightarrow RS^{\cdot} + Mn(II) + H+ \tag{228}$$

$$RS^{\cdot} + Mn(III) \longrightarrow RS^{+} + Mn(II) \tag{229}$$

$$RS^{+} + RSH \longrightarrow RSSR + H^{+} \tag{230}$$

$$2Mn(II) + 2H^{+} + \tfrac{1}{2}O_2 \longrightarrow 2Mn(III) + H_2O \tag{231}$$

catalysts (equations (232) - (235)); in this case RS^{-} becomes metal coordinated and a base is essential.[16]

$$RSH + HO^{-} \longrightarrow RS^{-} + H_2O \tag{232}$$

$$RS^{-} + M(II)Pc \longrightarrow M(I)Pc + RS^{\cdot} \tag{233}$$

$$2RS^{\cdot} \longrightarrow RSSR \tag{234}$$

$$4M(I)Pc + O_2 + 4H^{+} \longrightarrow 4M(II)Pc + 2H_2O \tag{235}$$

Copper ions have been particularly important catalysts since the electron transfer is easy. Thus PhSH is oxidised by the route of equations (236) - (241); PhSSOPh is a significant side product, but it is formed directly and not via PhSSPh.[716] Cysteine is also dimerised in the presence of Cu^{2+}/L, at a rate dependent on the stability of the

$$Cu^{2+} + PhSH \longrightarrow Cu^{+} + PhS^{\cdot} + H^{+} \tag{236}$$

$$2PhS^{\cdot} \longrightarrow PhSSPh \tag{237}$$

$$Cu^+ + \tfrac{1}{2}O_2 + 2H^+ \longrightarrow Cu^{2+} + H_2O \qquad\qquad (238)$$

$$PhS^{\cdot} + O_2 \longrightarrow PhSO_2^{\cdot} \qquad\qquad (239)$$

$$PhSO_2^{\cdot} + PhS^{\cdot} \longrightarrow 2PhSO^{\cdot} \qquad\qquad (2400$$

$$PhSO^{\cdot} + PhS^{\cdot} \longrightarrow PhSSOPh \qquad\qquad (241)$$

LCu(cysteine) complex.[717]   As well as oxygen, $Fe(CN)_6^{3-}$ may be used as reoxidant.   The reaction is rather involved and two mechanisms have been proposed.   (Figure 100)[718-720]

Mechanism A

$$Cu^{2+} + RS^- \;\rightleftharpoons\; [Cu(SR)]^+$$

$$[Cu(SR)]^+ + RS^- \;\rightleftharpoons\; [Cu(SR)_2]$$

$$[Cu(SR)_2] + [Fe(CN)_6]^{3-} \longrightarrow [Cu(III)(SR)_2]^+ + [Fe(CN)_6]^{4-}$$

$$[Cu(III)(SR)_2]^+ \longrightarrow Cu(I) + RSSR$$

$$Cu(I) + [Fe(CN)_6]^{3-} \longrightarrow Cu(II) + Fe(CN)_6^{4-}$$

Mechanism B

$$2[Cu(SR)_2] + O_2 \longrightarrow [(RS)_2Cu{-}O{-}O{-}Cu(SR)_2]$$

$$[(RS)_2Cu{-}O{-}O{-}Cu(SR)_2] \longrightarrow [(RS^{\cdot})(RS^-)Cu(II)(O_2)^{2-}]^-$$

$$+ [(RS^{\cdot})(RS^-)Cu(II)]^+$$

$$[(RS^{\cdot})(RS^-)Cu(II)(O_2)^{2-}]^- + O_2 \longrightarrow [(RSSR)Cu(II)(O_2)^{-\cdot}]^- + HO_2^-$$

$$[(RSSR)Cu(II)(O_2)^{-\cdot}]^- + 2RS^- \longrightarrow [(RS^-)_2Cu(II)(O_2)^{-\cdot}]^- + RSSR$$

$$[(RS^-)_2Cu(II)(O_2)^{-\cdot}]^- \longrightarrow [(RS^{\cdot})(RS^-)Cu(II)(O_2)^{2-}]^-$$

Figure 100 Mechanism of oxidation of cysteine in the presence of Cu^{2+}.

Cobalt complexes of macrocycles have been popular catalysts. Co(III) tetradehydrocorrinate(CN)$_2$ has been used to oxidise PhSH. The precise mechanism is unknown but the cobalt acts as an electron transport mediator.[721] With **398** thiolate becomes coordinated to cobalt in **399**. **399** binds oxygen with Brønsted acid assistance to give **400**, which dissociates to products.[722] Cobalt phthalocyanine bound on to a polymer bearing basic groups combines the two functions,[723] and the polymer also inhibits the formation of inactive μ-dioxo complexes.

Heteropolyacids, $H_{3+n}PMo_{12-n}V_nO_{40}$ have been used in oxidation of thiols to disulphides. The mechanism is unknown but is not thought to involve radicals.[76]

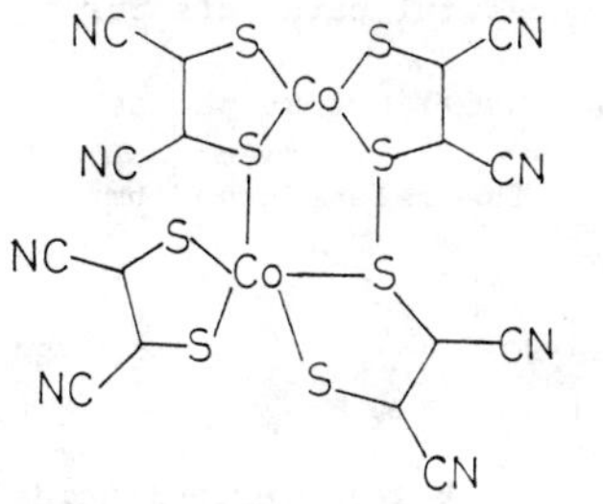

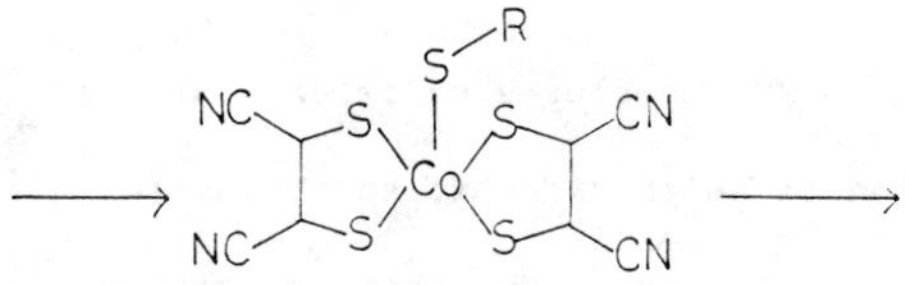

398   399

$$RS^{\cdot} + HO_2^{\cdot} + B$$

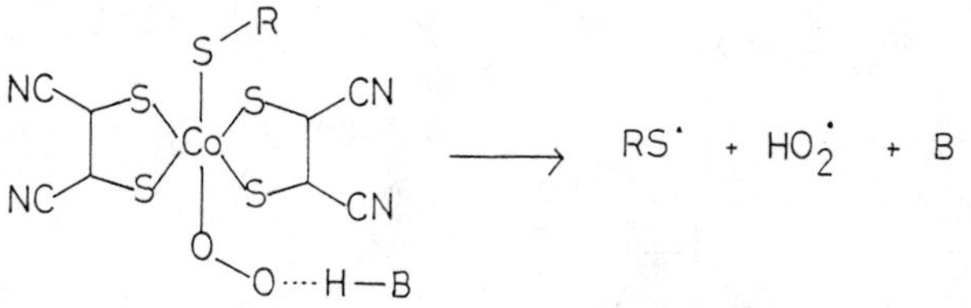

400

6.18.2 Oxidation of sulphides and sulphoxides

Oxidation of sulphides to sulphoxides by O_2 occurs in the presence of many metal complexes. Ruthenium complexes have been popular but early work gave poor selectivity. For example, in the presence of $RuCl_3$, Bu_2S is oxidised to Bu_2SO and Bu_2SO_2, so the method is useful only to prepare sulphones.[724] More recently, $(Me_2SO)_4RuX_2$ has given excellent selectivity for sulphoxides. Kinetic and isotope labelling studies show the mechanism of equations (242) - (244).[725]

$$Ru(II) + O_2 \rightleftharpoons Ru(IV) + [O_2]^{2-} \tag{242}$$

$$R_2S + [O_2]^{2-} + H^+ \longrightarrow R_2SO + HO^- \tag{243}$$

$$Ru(IV) (SR_2) + H_2O \longrightarrow Ru(II) + R_2SO + 2H^+ \tag{244}$$

DMSO complexes of iridium are very successful catalysts for oxidation of sulphoxides to sulphones (equation (245)) but sulphides are not easily oxidised under these conditions.[611] The related rhodium

$$Me_2SO + O_2 \xrightarrow{\text{i-PrOH/HIrCl}_2\text{(DMSO)}_3} Me_2SO_2 \tag{245}$$

complex, $H[Rh(DMSO)_2Cl_4]$, is similarly effective.[18] Both heteropolyacids[76] and $Cu(II)$[726] catalyse unselective oxidation of sulphides.

Certain classes of sulphides are oxidatively cleaved in the presence of O_2/Co(II)bzacen, 401.[727] In all cases the sulphides have acidic α-hydrogens (e.g. 402 to 403).

401

$$\text{PhSCH}_2\text{COPh} + \text{O}_2 \xrightarrow{\quad \underline{401} \quad} \text{PhSSPh} + \text{PhCOCHO} + \text{PhCOCOOH} + \text{PhCOOH}$$

$$\underline{402} \qquad\qquad\qquad\qquad \underline{403}$$

Catalytic oxidation of sulphides to sulphoxides by RO_2H is facile in the presence of V, Ti and Mo complexes (Figure 101). There are numerous species present in the reaction mixture, since many exchange processes and acid-base equilibria occur. $VO(OR)_3$ is the major active species and the reactions proceed via the transition state, $\underline{404}$.[728,731,732] The relative rates of oxidation are $Bu_2S > PhSBu > Bu_2SO > $ cyclohexene[733] and allyl sulphides, $\underline{405}$, are selectively oxidised at sulphur. In the presence of dialkyl tartrates and $Ti(O-\underline{i}-Pr)_4$ sulphides may be oxidised to sulphoxides in up to 93% optical yield[734,735] (e.g. $\underline{407}$ to $\underline{408}$).

Ref.

$$Bu_2S + \underline{t}\text{-BuOOH} \xrightarrow[90\%]{\quad VO(acac)_2 \quad} Bu_2SO \qquad\qquad 728$$

$$Bu_2S + \underline{t}\text{-BuOOH} \xrightarrow[100\%]{\quad MoO_2(acac)_2 \quad} Bu_2SO \qquad\qquad 729$$

$$Bu_2S + \underline{t}\text{-BuOOH} \xrightarrow{\quad TiO(acac)_2 \quad} Bu_2SO \qquad\qquad 730$$

Figure 101 Oxidation of sulphides by hydroperoxides.

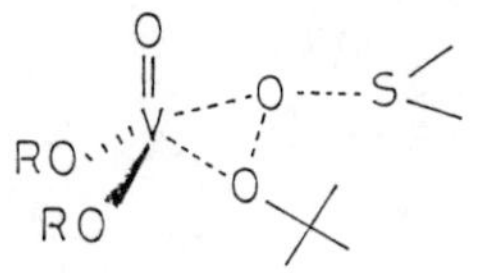

$\underline{404}$

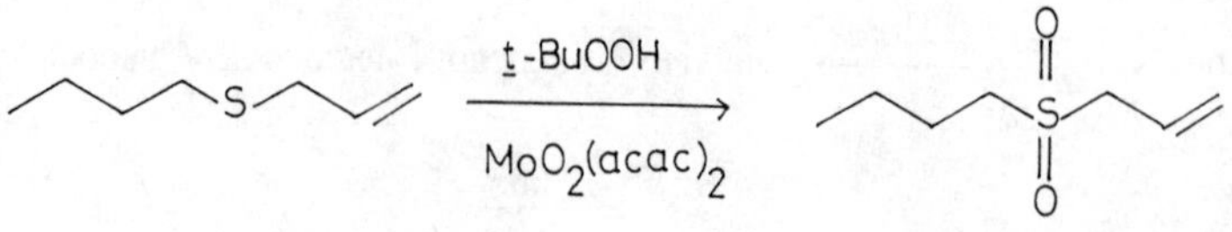

405

t-BuOOH

$\xrightarrow{\quad}$

$MoO_2(acac)_2$

406

407

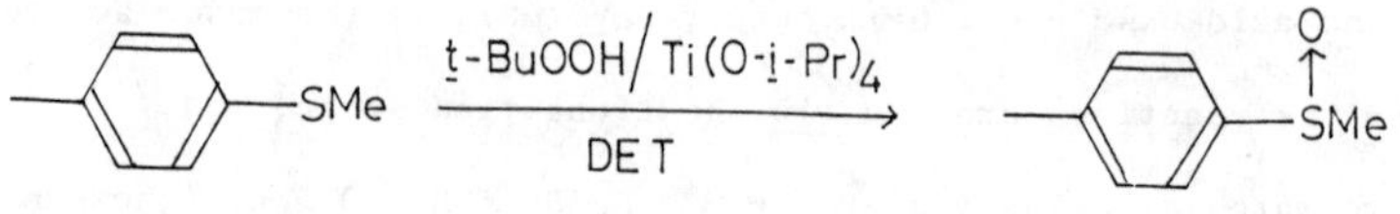

$t\text{-BuOOH}/Ti(O\text{-}i\text{-}Pr)_4$

$\xrightarrow{\qquad\qquad}$

DET

408   88% ee

H_2O_2 may be used as an oxidant instead of RO_2H.   The reaction
is a little less selective, the reagent cheaper, and the range of catalysts
larger.   Some examples are given in Figure 102.   One exception in
respect of mechanism is provided by (TPP)FeCl, which gives an excellent
yield via sulphenium radicals.[737]

Other oxidants have been used, including PhIO in the presence of
TPP complexes.   (TPP)FeCl gives a good initial rate but (TPP)Mn complexes
are longer lived.[738]   Using PhIO/$(Ph_3P)_3RuCl_2$ both sulphoxide and sulphone
are obtained and selenides are also oxidised.[739] BrO_3^-,[740] $MeNO_2$ and
HNO_3[741] have also been used as oxidants.

Ref.

$$RSAr + H_2O_2 \xrightarrow[\text{HMPA}]{WO_5} \overset{\displaystyle O}{\underset{\displaystyle \|}{RSAr}}$$ 736

$$\underline{t}\text{-Bu}_2S + H_2O_2 \xrightarrow{\text{TiO(acac)}_2} \underline{t}\text{-Bu}_2S=O$$ 731

$$\underline{p}\text{-ClC}_6H_4\text{SMe} + H_2O_2 \xrightarrow{\text{VO(acac)}_2} \underline{p}\text{-ClC}_6H_4\overset{\displaystyle O}{\underset{\displaystyle \|}{S}}\text{Me}$$ 45

$$\underline{p}\text{-ClC}_6H_4\text{SMe} + H_2O_2 \xrightarrow{\text{MoO}_2\text{(acac)}_2} \underline{p}\text{-ClC}_6H_4\overset{\displaystyle O}{\underset{\displaystyle \|}{S}}\text{Me}$$ 730

Figure 102   Oxidation of sulphides by hydrogen peroxide.

References

1   R.S.Gall, J.F.Rogers, W.P.Schaefer and G.G.Christoph, J.Amer. Chem. Soc., 98, 5135 (1976).

2   R.Boča, Coord. Chem. Rev., 50, 1 (1983).

3   L.D.Brown and K.N.Raymond, Inorg. Chem., 14, 2595 (1975).

4   A.Nishinaga, H.Tomita, Y.Tarumi and T.Matsura, Tetrahedron Lett., 21, 4849, 4853 (1980).

5   J.P.Collman, R.R.Gagne, C.A.Reed, T.R.Halbert, G.Lang amd W.T.Robinson, J.Amer. Chem. Soc., 97, 1427 (1975).

6   J.Halpern and A.L.Pickard, Inorg. Chem., 9, 2798 (1970).

7   S.J.LaPlaca and J.A.Ibers, J.Amer. Chem. Soc., 87, 2581 (1965).

8   D.Schwarzenbach, Helv. Chim. Acta, 55, 2990 (1972).

9   R.Stomberg and I-B.Ainalem, Acta Chem. Scand., 22, 1439 (1968).

10 R.Stomberg, Acta Chem. Scand., 23, 2755 (1969).

11   J.McGinnety and J.Ibers, J.Chem. Soc., Chem. Commun., 235 (1968).

12   W.P.Schaefer, Inorg. Chem., 7, 725 (1968).

13   J.P.Collman, Acc. Chem. Res., 10, 265 (1977).

14   T.G.Traylor, Acc. Chem. Res., 14, 102 (1981).

15   M.J.Bennett and P.B.Donaldson, J.Amer. Chem. Soc., 93, 3307 (1971).

16   N.S.Enikolopyan, K.A.Bogdanova and K.A.Askarov , Russ. Chem. Rev.,
     52, 13 (1983).

17   J.E.Lyons, Aspects Homogeneous Catalysis, 3, 1 (1977).

18   H.Mimoun, I.S.de Roch and L.Sajus, Tetrahedron, 26, 37 (1970).

19   K.Sharpless, J.M.Townsend and D.R.Williams, J.Amer. Chem. Soc.,
     94, 295 (1972).

20   K.Garbett and R.Gillard, J.Chem. Soc., A, 1725 (1968).

21   C.Walling, Acc. Chem. Res., 8, 125 (1975).

22   N.A.Milas and S.Sussman, J.Amer. Chem. Soc., 58, 1302 (1936), 59,
     2345 (1937).

23   W.Treibs, Angew. Chem., Int. Ed. Engl., 3, 802 (1964).

24   E.J.Eisenbraun, A.R.Bader, J.W.Polachek and E.Reif, J.Org. Chem.,
     28, 2057 (1963).

25   M.Mugdan and D.P.Young, J.Chem. Soc., 2988 (1949).

26   K.A.Saegebarth, J.Org. Chem., 24, 1212 (1959).

27   N.S.Sonoda and S.Tsutsumi, Bull. Chem. Soc. Jpn., 38, 958 (1965).

28   G.B.Payne and C.W.Smith, J.Org. Chem., 22, 1682 (1957).

29   G.B.Payne and P.H.Williams, J.Org. Chem., 24, 54 (1959).

30   Z.Raciszewski, J.Amer. Chem. Soc., 82, 1267 (1960).

31   H.C.Stevens and A.J.Kaman, J.Amer. Chem. Soc., 87, 734 (1965).

32   M.A.Beg and I.Ahmad, J.Org. Chem., 42, 1590 (1977).

33   J.Itakura, H.Tanaka and H.Ito, Bull. Chem. Soc. Jpn., 42, 1604 (1969).

34   J.A.Connor and E.A.V.Ebsworth, Adv. Inorg. Radiochem., 6, 279 (1964).

35   M.Pralus, J.C.Lecoq and J.P.Schirmann, Fund. Res.Homog. Catal.,
     3, 327 (1979).

36 J.P.Schirmann and S.Y.Delavarenne, Ger. Offen., 2,752,626 (1978);

    Chem. Abs., 89, 59832p (1978).

37 T.Yamamoto and M.Kimura, J.Chem. Soc., Chem. Commun., 948 (1977).

38 I.Ahmad and C.M.Ashraf, J.Prakt. Chem., 321, 345 (1979), Ind. J.Chem.,

    17A, 302 (1979).

39 G.Strukul and R.A.Michelin, J.Chem. Soc., Chem. Commun., 1538 (1984).

40 M.A.Beg and I.Ahmad, J.Catal., 39, 260 (1975), Ind. J.Chem., 15B,

    656 (1977).

41 M.Saotome, Y.Itoh and M.Terashi, Japan. Kokai, 73 39,435 (1973);

    Chem. Abs., 79, 78591u (1973).

42 P.Khare and G.L.Agrawal, React. Kinet. Catal. Lett., 23, 207 (1983).

43 M.Igarishi and H.Midorikawa, J.Org. Chem., 29, 2080 (1964).

44 G.Sturtz and A.P.Raphalen, Bull. Soc. Chim. Fr., 125 (1983).

45 O.Bortolini, F.DiFuria, P.Scrimin and G.Modena, J.Mol. Catal.,7,

    59 (1980).

46 B.M.Trost and Y.Masuyama, Is. J.Chem., 24, 134 (1984).

47 P.A.Kilty and W.M.H.Sachtler, Catal. Rev., 10, 1 (1974).

48 H.T.Spath and K.D.Handel, Adv. Chem. Ser., 133, 395 (1974).

49 J.E.Lyons, Tetrahedron Lett., 2737 (1974).

50 I.A.Krylov, M.Yu.Baevskii, I.Yu. Litvintsev, V.N.Sapunov and N.N.Lebedev,

    Kinet. Catal., 23, 717, 721 (1982).

51 A.F.Noels, A.J.Hubert and P.Teyssie, J.Organomet. Chem., 166, 79 (1979).

52 R.A.Sheldon, J.Mol. Catal., 20, 1 (1983).

53 A.P.Filippov, G.A.Konishevskaya and V.M.Belousov, Kinet. Catal., 23,

    287 (1982).

54 L.D.Tyutchenkova and Z.K.Maizus, Bull. Acad. Sci. U.S.S.R., Div. Chem.

    Sci., 28, 2019 (1979).

55 A.A.Blanc, H.Arzoumanian, E.J.Vincent and J.Metzger, Bull. Soc. Chim.

    Fr., 2175 (1974).

56 S.Cenini, A.Fusi and F.Porta, Gazz. Chim. Ital., 108, 109 (1978).

642

57   S.Cenini, A .Fusi and G.Capparella, J.Inorg. Nucl. Chem., 33, 3576
     (1971).

58   M.E.Pudel' and Z.K.Maizus, Bull. Acad. Sci. U.S.S.R., Div. Chem. Sci.,
     24, 36 (1975).

59   R.A.Budnik and J.K.Kochi, J.Org. Chem., 41, 1384 (1976).

60   W.Fuchs and C.Dudeck,Ger. Offen., 2,106,413 (1972); Chem. Abs., 77,
     164435m (1972).

61   N.S.Aprahamian, Can. Pat., 941,835 (1974); Chem. Abs., 82, 73696j (1975).

62   L.N.Khabibullina, V.S.Gumerova, V.P.Yur'ev and S.R.Rafikov, Dokl. Chem.,
     256, 58 (1981).

63   S.Ito, K.Inoue and M.Matsumoto, J.Amer. Chem. Soc., 104, 6450 (1982).

64   R.A.Sheldon and J.A.van Doorn, J.Organomet. Chem., 94, 115 (1975).

65   W.F.Brill, J.Mol. Catal., 19, 69 (1983).

66   B.S.Tovrog, F.Mares and S.E.Diamond, J.Amer. Chem. Soc., 102,
     6616 (1980).

67   M.A.Andrews and C-W.F.Cheng, J.Amer. Chem. Soc., 104, 4268 (1982).

68   O.Leal, M.R.Goldwasser, H.Martinez, M.Garmendia, R.Lopez and
     H.Arzoumanian, J.Mol. Catal., 22, 117 (1983).

69   W.F.Brill, J.Amer. Chem. Soc., 85, 141 (1963).

70   W.F.Brill and N.Indictor, J.Org. Chem.,29, 710 (1964), 30, 2074 (1965).

71   R.A.Sheldon and J.A.vanDoorn, J.Catal., 34, 242 (1974).

72   Mitsui Petrochemical Industries, Ltd., Jpn. Kokai Tokkyo Koho JP,
     81,133,279 (1981); Chem. Abs., 96, 52165j (1982).

73   S.Ozaki, T.Takahashi and I.Sudo, Japan. Kokai, 74 124,003 (1974);
     Chem. Abs., 84, 58590t (1976).

74   J.Sobczak and J.J.Ziołkowski, J.Mol. Catal., 3, 165 (1977/8).

75   H.Arakawa and A.Ozaki, Chem. Lett., 1245 (1975).

76   I.V.Kozhevnikov and K.I.Matveev, Appl. Catal., 5, 135 (1983).

77   P.Forzatti and F.Trifiro, React. Kinet. Catal. Lett., 1, 367 (1974).

78 J.O.Turner and J.E.Lyons, Ger. Offen., 2,231,678 (1973); Chem. Abs.,

    78, 111925k (1973).

79 Halcon International,Inc., Neth. Appl., 75 14,538 (1976); Chem. Abs.,

    86, 189689w (1977).

80 A.L.Stautzenberger, U.S.Patent, 3,931,249 (1976); Chem. Abs., 84,

    105379t (1976).

81 Sumitomo Chemical Co., Ltd., Jpn. Kokai Tokkyo Koho JP, 57,200,375

    [82,200,375] (1982); Chem. Abs., 98, 215469u (1983).

82 A.Tamaki, T.Takahashi, I.Sudo and S.Ozaki, Japan. Kokai, 75 84,504

    (1975); Chem. Abs., 84, 43811s (1976).

83 K.B.Sharpless and T.R.Verhoeven, Aldrichim. Acta, 12, 63 (1979).

84 M.N.Sheng and J.G.Zajacek, Adv. Chem. Ser., 76, 418 (1968).

85 R.A.Sheldon, J.A.van Doorn, C.W.A.Schram and A.J.de Jong, J.Catal.,

    31, 438 (1973).

86 M.I.Farberov, G.A.Stozhkova, A.V.Bondarenko, T.M.Kirik and N.A.

    Ognevskaya, Neftekhimiya, 11, 404 (1971); Chem. Abs., 75, 87877m (1971).

87 R.Nagata, T.Matsuura and I.Saito, Tetrahedron Lett., 25, 2691 (1984).

88 T.Hiyama and M.Obayashi, Tetrahedron Lett., 24, 395 (1983).

89 R.A.Sheldon and J.A.van Doorn, J.Catal., 31, 427 (1973).

90 R.A.Sheldon, Rec., 92, 253, 367 (1973).

91 M.Kimura and T.Muto, Chem. Pharm. Bull., 27, 109 (1979), 28, 1836 (1980).

92 H-F.Boeden, J.Dahlmann, H-J.Hamman and E.Höft, J.Prakt. Chem., 324,

    526 (1982).

93 D.V.Banthorpe and S.E.Barrow, Chem. Ind., 502 (1981).

94 A.M.Trzeciak, J.Sobczak and J.J.Ziolkowski, J.Mol. Catal., 12, 321

    (1981).

95 H.Mimoun, J.Mol. Catal., 7, 1 (1980).

96 N.Takamitsu, S.Hamamoto and I.Nishifuto, Japan. Kokai, 77,113,933

    (1977); Chem. Abs., 89, 146748j (1978).

644

97   M.N.Sheng and J.G.Zajacek, Ger. Offen., 2,148,432 (1972); Chem. Abs., 77, 751113n (1972).

98   T.N.Baker, G.J.Mains, M.N.Sheng and J.G.Zajacek, J.Org. Chem., 38, 1145 (1973).

99   S.M.Kulikov and I.V.Kozhevnikov, Kinet. Catal., 24, 33 (1983).

100   J.Kaloustian, L.Lena and J.Metzger, Bull. Soc. Chim. Fr., 4415 (1971).

101   J.Sobczak and J.J.Ziolkowski, J.Less Common Met., 54, 149 (1977).

102   F.Trifiro, P.Forzatti, S.Preite and I.Pasquon, J.Less Common Met., 36, 319 (1974).

103   A.D.Westland, F.Haque and J-M.Bouchard, Inorg. Chem., 19, 2255 (1980).

104   A.Herbowski, J.M.Sobczak and J.J.Ziolkowski, J.Mol. Catal., 19, 309 (1983).

105   M.Yamazaki, H.Endo, M.Tomoyama and Y.Kurusu, Bull. Chem. Soc. Jpn., 56, 3523 (1983).

106   H.J.Ledon, P.Durbut and F.Varescon, J.Amer. Chem. Soc., 103, 3601 (1981).

107   J.Fleischer, D.Schnurpfeil, K.Seyferth and R.Taube, J.Prakt.Chem., 319, 995 (1977).

108   V.N.Sapunov, I.Yu.Litvintsev, I.Margitfal'vi and N.N.Lebedev, Kinet. Catal., 18, 521 (1977).

109   J.Sobczak and J.J.Ziolkowski, Inorg. Chim. Acta, 19, 15 (1976).

110   I.A.Gailyunas, E.M.Tsyrlina, N.I.Solov'eva, N.G.Komalenkova and V.P.Yur'ev, J.Gen. Chem. U.S.S.R., 47, 2188 (1977).

111   I.A.Gailyunas, E.M.Tsyrlina, L.V.Spirikhin, A.S.Mustafina and V.P.Yur'ev, J.Gen. Chem. U.S.S.R., 49, 1579 (1979).

112   A.O.Kolmakov, V.M.Fomin, T.N.Aizenshtadt and Yu.A.Aleksandrov, J.Gen. Chem. U.S.S.R., 51, 2420 (1981).

113   E.P.Tepenitsyna, N.V.Dormidontova and M.I.Farberov, Uch. Zap., Yaroslav. Tekhnol. Inst., 27, 35 (1971); Chem. Abs.,79, 105011p (1973).

114   S.B.Grinenko, V.M.Belousov, L.A.Oshin, G.A.Shakhovtseva,
      N.I.Kovtyukhova, A.P.Filippov and K.B.Yatsimirskii, Dokl. P.Chem.,
      263, 309 (1982).

115   J.Sobczak and J.J.Ziółkowski, React. Kinet. Catal. Lett., 11, 359
      (1979).

116   G.A.Tolstikov, V.P.Yur'ev and U.M.Dzhemilev, Russ. Chem. Rev., 44,
      319 (1975).

117   M.N.Sheng and J.G.Zajacek, J.Org. Chem., 35, 1839 (1970).

118   G.A.Tolstikov, U.M.Dzhemilev and V.P.Yur'ev, J.Org. Chem. U.S.S.R.,
      8, 1200 (1971).

119   G.A.Tolstikov, V.P.Yur'ev and I.A.Gailyunas, Bull. Acad. Sci. U.S.S.R.,
      Div. Chem. Sci., 22, 1395 (1973).

120   S.Bhaduri, A.Ghosh and H.Khwaja, J.Chem. Soc., Dalton, 447 (1981).

121   G.L.Linden and M.F.Farona, Inorg. Chem., 16, 3170 (1977), J.Catal.,
      48, 284 (1977).

122   C-C.Su, J.W.Reed and E.S.Gould, Inorg. Chem., 12, 337 (1973).

123   J.Sobczak and J.J.Ziółkowski, J.Mol. Catal., 13, 11 (1981).

124   H.Mimoun, Angew. Chem.,Int. Ed. Engl., 21, 734 (1982).

125   K.B.Sharpless and R.C.Michaelson, J.Amer. Chem. Soc., 95, 6136
      (1973).

126   S.Tanaka, H.Yamamoto, H.Nozaki, K.B.Sharpless, R.C.Michaelson and
      J.D.Cutting, J.Amer. Chem. Soc., 96, 5254 (1974).

127   Atlantic Richfield Co., Brit. Pat., 1,214,843 (1970); Chem. Abs.,
      76, 153559h (1972).

128   M.I.Farberov, L.V.Mel'nik, B.N.Bobylev and V.A.Podgornova,
      Kinet. Catal., 12, 1018 (1971).

129   M.Kobayashi, S.Kurozumi, T.Toru, T.Tanaka, T.Miura and S.Ishimoto,
      Japan. Kokai, 77,108,951 (1977); Chem. Abs., 88, 22587r (1978).

130   M.Isobe, M.Kitamura, S.Mio and T.Goto, Tetrahedron Lett., 23, 221
      (1982).

131   K.S.Kirshenbaum, Nouv. J.Chim., 7, 699 (1983).

132   D.Mansuy, P.Battioni and J-P.Renaud, J.Chem. Soc., Chem. Commún.,
       1255 (1984).

133   M.F.Powell, E.F.Pai and T.C.Bruice, J.Amer. Chem. Soc., 106, 3277
       (1984).

134   V.P.Rajan, S.N.Bannore,H.N.Subbarao and S.Dev, Tetrahedron, 40,
       983 (1984).

135   V.P.Yur'ev, I.A.Gailyunas,L.V.Spirikhin and G.A.Tolstikov, J.Gen.
       Chem. U.S.S.R., 45, 2269 (1975).

136   M.B.Groen and F.J.Zeelen, Tetrahedron Lett., 23, 3611 (1982).

137   A.J.Pearson, Y-S.Chen, S-Y.Hsu and T.Ray, Tetrahedron Lett., 25,
       1235 (1984).

138   G.A.Tolstikov, V.P.Yur'ev,I.A.Gailyunas and U.M.Dzhemilev, J.Gen.
       Chem. U.S.S.R., 44, 205 (1974).

139   G.A.Tolstikov,U.M.Dzhemilev and V.P.Yur'ev, J.Org. Chem. U.S.S.R.,
       8, 2253 (1972).

140   H.B.Henbest and R.A.L.Wilson, J.Chem. Soc., 1958 (1957).

141   J.E.Lyons, Adv. Chem. Ser., 132, 64 (1974).

142   A.S.Narula, Tetrahedron Lett., 23, 5579 (1982).

143   R.D.Bach, G.J.Wolber and B.A.Coddens, J.Amer. Chem. Soc., 106,
       6098 (1984).

144   T.Itoh, K.Kaneda and S.Teranishi, J.Chem. Soc., Chem. Commun.,
       421 (1976).

145   T.Itoh, K.Jitsukawa, K.Kaneda and S.Teranishi, Tetrahedron Lett.,
       3157 (1976), J.Amer. Chem. Soc., 101, 159 (1979).

146   B.E.Rossiter, T.R.Verhoeven and K.B.Sharpless, Tetrahedron Lett.,
       4733 (1979).

147   E.D.Mihelich, Tetrahedron Lett., 4729 (1979).

148   H.Tomioka, T.Suzuki, K.Oshima and H.Nozaki, Tetrahedron Lett.,
       23, 3387 (1982).

149 E.D.Mihelich, K.Daniels and D.J.Eickhoff, J.Amer. Chem. Soc., 103, 7690 (1981).

150 F.Sato, M.Kusakabe and Y.Kobayashi, J.Chem. Soc., Chem. Commun., 1130 (1984).

151 S.Danishefsky, M.Hirama, K.Gombatz, T.Harayama, E.Berman and P.Schuda, J.Amer. Chem. Soc., 100, 6536 (1978).

152 T.Kato, M.Suzuki, M.Takahashi and Y.Kitahara, Chem. Lett., 465 (1977).

153 Y.S.Sanghvi and A.S.Rao, J.Het. Chem., 21, 317 (1984).

154 B.Meunier, Bull. Soc. Chim. Fr., 345 (1984).

155 J.T.Groves, T.E.Nemo and R.S.Myers, J.Amer. Chem. Soc., 101, 1032 (1979).

156 J.T.Groves, W.J.Kruper, T.E.Nemo and R.S.Myers, J.Mol. Catal., 7, 169 (1980).

157 J.R.L.Smith and P.R.Sleath, J.Chem. Soc., Perkin II, 1009 (1982).

158 D.Mansuy, J.Leclaire, M.Fontecave and P.Dansette, Tetrahedron, 40, 2847 (1984).

159 M.Fontecave and D.Mansuy, J.Chem. Soc., Chem. Commun., 879 (1984).

160 J.T.Groves and T.E.Nemo, J.Amer. Chem. Soc., 105, 5786, 6243 (1983).

161 C.C.Franklin, R.B.VanAtta, A.Fan Tai and J.S.Valentine, J.Amer. Chem. Soc., 106, 814 (1984).

162 D.Mansuy, J-P.Mahy, A.Dureault, G.Bedi and P.Pattoni, J.Chem. Soc., Chem. Commun., 1161 (1984).

163 I.Willner, D.W.Otvos and M.Calvin, J.Chem. Soc., Chem. Commun., 964 (1980).

164 O.Bortolini and B.Meunier, J.Chem. Soc., Chem. Commun., 1364 (1983).

165 E.Guilmet and B.Meunier, J.Mol. Catal., 23, 115 (1984), Tetrahedron Lett., 25, 789 (1984).

166 J.A.S.J.Razenberg, R.J.M.Nolte and W.Drenth, Tetrahedron Lett., 25, 789 (1984).

648

167   G.Balavoine,C.Eskenazi, F.Meunier and H.Rivière, Tetrahedron Lett.,

       25, 3187 (1984).

168   M.W.Nee and T.L.Bruice, J.Amer. Chem. Soc., 104, 6123 (1982).

169   W.H.Pirkle and P.L.Rinaldi, J.Org. Chem., 42, 2080 (1977).

170   S-i.Yamada, T.Mashiko and S.Terashima, J.Amer. Chem. Soc., 99,1988 (1977).

171   T.Mashiko, S.Terashima and S.Yamada, Yakugaku Zasshi, 100, 319, 328

       (1980); Chem.Abs., 93, 131963j, 192718e (1980).

172   S.Coleman-Kammula and E.T.Duim-Koolstra, J.Organomet. Chem.,

       246, 53 (1983).

173   R.C.Michaelson, R.E.Palermo and K.B.Sharpless, J.Amer. Chem. Soc.,

       99, 1990 (1977).

174   I.D.Williams, S.F.Pederson, K.B.Sharpless and S.J.Lippard, J.Amer.

       Chem. Soc., 106, 6430 (1984).

175   K.B.Sharpless, S.S.Woodard and M.G.Finn, Pure Appl. Chem., 55,

       1823 (1983).

176   B.E.Rossiter, T.Katsuki and K.B.Sharpless, J.Amer. Chem. Soc., 103,

       464 (1981).

177   T.Katsuki and K.B.Sharpless, J.Amer. Chem. Soc., 102,5974 (1980).

178   K.Mori, T.Ebata and S.Takechi, Tetrahedron, 40, 1761 (1984).

179   S.Takano, S.Otaki and K.Ogasawara, J.Chem. Soc., Chem. Commun.,

       1172 (1983).

180   K.Mori and S.Otsuka, Tetrahedron, 39, 3267 (1983).

181   V.S.Martin, S.S.Woodard, T.Katsuki, Y.Yamada, M.Ikeda and K.B.

       Sharpless, J.Amer. Chem. Soc., 103, 6237 (1981).

182   S.Masamune, Heterocycles, 21, 107 (1984).

183   N.Koizumi, M.Ishiguro, M.Yasuda and N.Ikekawa, J.Chem. Soc.,

       Perkin I, 1401 (1983).

184   L.N.Pridgen, S.C.Shilcrat and I.Lantos, Tetrahedron Lett., 25,

       2835 (1984).

185   A.C.Oehschlager and E.Czyzewska, Tetrahedron Lett., 24, 5587

       (1983).

186   B.E.Rossiter and K.B.Sharpless, J.Org. Chem., 49, 3707 (1984).

187   M.J.Farrall, M.Alexio and M.Trecarten, Nouv. J.Chim., 7, 449 (1983).

188   L.D.-L.Lu, R.A.Johnson, M.G.Finn and K.B.Sharpless, J.Org. Chem.,
       49, 728 (1984).

189   K.Tani, M.Hanafusa and S.Otsuka, Tetrahedron Lett., 3017 (1979).

190   J.T.Groves and R.S.Myers, J.Amer. Chem. Soc., 105, 5791 (1983).

191   J.Smidt, W.Hafner, R.Jira, R.Sieber, J.Sedlmeier and A.Sabel,
       Angew. Chem., Int. Ed. Engl., 1, 80 (1962).

192   W.Hafner, R.Jira, J.Sedlmeier and J.Smidt, Chem. Ber., 95, 1575
       (1962).

193   R.Jira, J.Sedlmeier and J.Smidt, Ann., 693, 99 (1966).

194   A.M.Aguiló, Adv. Organomet. Chem., 5, 321 (1967).

195   P.M.Henry, J.Amer. Chem. Soc., 86, 3246 (1964), Adv. Chem. Ser., 70,
       126 (1968).

196   O.G.Levanda and I.I.Moiseev, Kinet. Catal., 12, 501 (1971).

197   R·N·Pandey and P.M.Henry, Canad. J.Chem., 57, 982 (1979).

198   I.I.Moiseev, M.N.Vargaftik and Ya.K.Syrkin, Bull. Acad. Sci. U.S.S.R.,
       Div. Chem. Sci., 1050 (1963).

199   N.Gregor, K.Zaw and P.M.Henry, Organometallics, 3, 1251 (1984).

200   J.E.Bäckvall, B.Åkermark and S.O.Ljunggren, J.Amer. Chem. Soc., 101,
       2411 (1979).

201   A.F.Danilyuk, V.A.Likholobov and Yu.I.Ermakov, Kinet. Catal., 18,
       252 (1977).

202   K.I.Matveev, E.G.Zhizhina, N.B.Shitova and L.I.Kuznetsova, Kinet.
       Catal., 18, 320 (1977).

203   R.Jira and W.Freiesleben, Organometal.React., 3, 1 (1972).

204   J.Tsuji, Synthesis, 369 (1984), Pure Appl. Chem., 53, 2371 (1981),
       51, 1235 (1979).

205   J.Tsuji, I.Shimizu and K.Yamamoto, Tetrahedron Lett., 2975 (1976).

206   G.A.Dzhemileva, V.N.Odinokov, U.M.Dzhemilev and G.A.Tolstikov,
      Bull. Acad. Sci. U.S.S.R., Div. Chem. Sci., 32, 307 (1983).

207   J.Tsuji, T.Yamakawa and T.Mandai, Tetrahedron Lett., 3741 (1979).

208   K.Mori, T.Ebata and S.Takechi, Tetrahedron, 40, 1761 (1984).

209   R.Januskiewicz and H.Alper, Tetrahedron Lett., 24, 5159 (1983).

210   D.R.Fahey and E.A.Zuech, J.Org. Chem., 39, 3276 (1974).

211   J.Tsuji, T.Yamada and I.Shimizu, J.Org. Chem., 45, 5209 (1980).

212   J.Tsuji, I.Shimizu and Y.Kobayashi, Is. J.Chem., 24, 153 (1984).

213   P.D.Magnus and M.S.Nobbs, Synth. Commun., 10, 273 (1980).

214   P.M.Henry, J.Amer. Chem. Soc., 94, 1527, 5200 (1972).

215   J.S.Coe and J.B.J.Unsworth, J.Chem. Soc., Dalton, 645 (1975).

216   E.G.Zhizhina, N.B.Shitova and K.I.Matveev, Kinet. Catal., 22, 1153
      (1981).

217   A.Heumann, F.Chauvet and B.Waegell, Tetrahedron Lett., 23, 2767
      (1982).

218   S.F.Davison, B.E.Mann and P.M.Maitlis, J.Chem. Soc., Dalton, 1223
      (1984).

219   I.V.Kozhevnikov and K.I.Matveev, Russ. Chem. Rev., 51, 1075 (1982).

220   H.Mimoun, Pure Appl. Chem., 53, 2389 (1981).

221   M.Roussell and H.Mimoun, J.Org. Chem., 45, 5387 (1980).

222   F.Igersheim and H.Mimoun, Nouv. J.Chim., 4, 711 (1980).

223   M.Matsumoto and K.Kuroda, Jpn. Kokai Tokkyo Koho, 80 00,317(1980);
      Chem. Abs., 93, 7829f (1980).

224   W.G.Lloyd and B.J.Luberoff, J.Org. Chem., 34, 3949 (1969).

225   T.Hosokawa, T.Ohta and S-I.Murahashi, J.Chem. Soc., Chem. Commun.,
      848 (1983).

226   N.T.Byrom, R.Grigg, B.Kongkathip, G.Reimer and A.R.Wade, J.Chem. Soc.,
      Perkin I, 1643 (1984).

227   B.Pugin and L.M.Venanzi, J.Amer. Chem. Soc., 105, 6877 (1983).

228   H.Ogawa, F.Fujinami, K.Taya and S.Teratani, Bull. Chem. Soc. Jpn.,
       57, 1908 (1984).

229   J.Tsuji, H.Nagashima and K.Hori, Chem. Lett., 257 (1980).

230   H.Nagashima, K.Sakai and J.Tsuji, Chem. Lett., 859 (1982).

231   J.Tsuji, N.Hagashima and K.Hori, Tetrahedron Lett., 23, 2679 (1982).

232   E.Wenkert, B.L.Buckwalter, A.A.Craveiro, E.L.Sanchez and S.S.Sathe,
       J.Amer. Chem. Soc., 100, 1267 (1978).

233   I.Shimizu, Y.Naito and J.Tsuji, Tetrahedron Lett., 21, 487 (1980).

234   B.R.James and M.Kastner, Canad. J.Chem., 50, 1698, 1708 (1972).

235   F.Igersheim and H.Mimoun, J.Chem. Soc., Chem. Commun., 559 (1978),
       Nouv. J.Chim., 4, 161 (1980).

236   R.Tang, F.Mares, N.Neary and D.E.Smith, J.Chem. Soc., Chem. Commun.,
       274 (1979).

237   O.Bortolini, F.DiFuria, G.Modena and R.Seraglia, J.Mol. Catal., 22,
       313 (1984).

238   H.Mimoun, M.M.P.Machirant and I.S.deRoch, J.Amer. Chem. Soc., 100,
       5437 (1978).

239   G.Read and P.J.C.Walker, J.Chem. Soc., Dalton, 883 (1977).

240   J.Farrar, D.Holland and D.J.Milner, J.Chem. Soc., Dalton, 815 (1975).

241   K.Januskiewicz and H.Alper, Tetrahedron Lett., 5163 (1983).

242   H.Bonnemann, W.Nunez and D.M.M.Rohe, Helv. Chim. Acta, 66, 177
       (1983).

243   K.Takao, M.Wayaku, Y.Fujiwara, T.Imanaka and S.Teranishi, Bull. Chem.
       Soc. Jpn., 43, 3898 (1970).

244   L.Carlton and G.Read, J.Mol. Catal., 10, 133 (1981).

245   G.Read, J.Mol. Catal., 4, 83 (1978).

246   F.J.McQuillin and D.G.Parker, J.Chem. Soc., Perkin I, 2092 (1975).

247   M.Bressan, F.Morandini and P.Rigo, J.Organomet. Chem., 247, C8 (1983).

248   D.A.Muccigrosso, F.Mares, S.E.Diamond and J.P.Solar, Inorg. Chem.,
      22, 960 (1983).

249   K.Takao, H.Azuma, Y.Fujiwara, T.Imanaka and S.Teranishi, Bull. Chem.
      Soc. Jpn., 45, 2003 (1972).

250   L.Carlton, G.Read and M.Urgelles, J.Chem. Soc., Chem. Commun., 586
      (1983).

251   K.Takao, Y.Fujiwara, T.Imanaka and S.Teranishi, Bull. Chem. Soc. Jpn.,
      43, 1153 (1970)

252   M.T.Atlay, M.Preece, G.Strukul and B.R.James, J.Chem. Soc., Chem.
      Commun., 406 (1982).

253   H.C.Volger and W.Brackman, Rec., 84, 579, 1017, 1203,1233 (1965), 85,
      817 (1966).

254   D.C.Heckert, U.S.Patent, 3,701,722 (1972); Chem. Abs., 78, 15529p
      (1973).

255   M.Perrée-Fauvet and A.Gaudemer, J.Chem. Soc., Chem. Commun., 874
      (1981).

256   J.R.Gilmore and J.M.Mellor, J.Chem. Soc., C, 2355 (1971).

257   A.B.Solov'eva, E.I.Karakazova, K.A.Bogdanova, V.Mel'nikova, L.V.
      Karmilova, G.A.Nikiforov, K.K.Pivnitskii and N.S.Enikolopyan,
      Dokl. P.Chem., 269, 138 (1983).

258   I.Tabushi and N.Koga, J.Amer. Chem. Soc., 101, 6456 (1979).

259   I.G.Tishchenko, L.S.Novikov and I.I.Korsak, J.Org. Chem. U.S.S.R., 8,
      735 (1972).

260   D.A.Baines and W.Cocker, J.Chem. Soc., Perkin I, 2232 (1975).

261   E.Bordier, Bull. Soc. Chim. Fr., 2621, 3291 (1973).

262   C.S.Sharma, S.C.Sethi and S.Dev, Synthesis, 45 (1974).

263   I.A.Krylov, I.Yu.Litvintsev and V.N.Sapunov, Kinet. Catal., 24, 37
      (1983).

264 W.K.Rybak and J.J.Ziołkowsky, Inorg. Chim. Acta, 24, L69 (1977).

265 I.M.Reibel' and A.F.Sandu, Russ. J.Phys. Chem., 50, 632 (1976), 49,

     946 (1975).

266 J.H.Fuhrhop, M.Baccouche, H.Grabow and H.Arzoumanian, J.Mol. Catal.,

     7, 245 (1980).

267 H.Kropf, S.K.Ivanov and P.Diercks, Ann., 2046 (1974).

268 J.E.Baldwin and J.C.Swallow, Angew. Chem.,Int. Ed. Engl., 8, 601 (1969).

269 D.Holland and D.J.Milner, J.Chem. Soc., Dalton, 2440 (1975).

270 B.R.James, F.T.T.Ng and E.Ochai, Canad. J.Chem., 50, 590 (1972).

271 G.Read and J.Shaw, J.Chem. Soc., Chem. Commun., 1313 (1984).

272 B.H. vanVugt and W.Drenth, Rec., 96, 225 (1977).

273 H.Arzoumanian, A.Blanc, U.Hartig and J.Metzger, Tetrahedron Lett.,

     1011 (1974).

274 J.E.Lyons and J.O.Turner, J.Org. Chem., 37, 2881 (1972).

275 Y.Aoyama, T.Watanabe, H.Onda and H.Ogoshi, Tetrahedron Lett., 24,

     1183 (1983).

276 S.Uemura and S.R.Patil, Tetrahedron Lett., 23, 4353 (1982), Chem.Lett.,

     1743 (1982).

277 C.Y.Chan and B.R.James, Inorg. Nucl. Chem. Lett., 9, 135 (1973).

278 M.T.Atlay, M.Preece, G.Strukul and B.R.James, Canad. J.Chem., 61,

     1332 (1983).

279 A.Fusi, R.Ugo, F.Fox, A.Pasini and S.Cenini, J.Organomet. Chem., 26,

     417 (1971).

280 I.P.Stolyarov, M.N.Vargaftik, O.M.Nefedov and I.I.Moiseeva, Kinet.

     Catal., 23, 313 (1982).

281 A.Heumann and B.Åkermark, Angew. Chem., Int. Ed. Engl., 23, 453 (1984).

282 J.E.McMurray and P.Kočovský, Tetrahedron Lett., 25, 4187 (1984).

283 S.Uemura, S-i.Fukuzawa, A.Toshimitsu and M.Okano, Tetrahedron Lett.,

     23, 87 (1982).

284   J.Muzart, P.Pale and J-P.Pete, Tetrahedron Lett., 23, 3577 (1982).

285   M.Julia and D.Mansuy, Bull. Soc. Chim. Fr., 1678 (1974).

286   A.M.Salimov, A.A.Medzhidov, V.S.Aliev and T.M.Kutovaya, Kinet. Catal.,
       16, 346 (1975).

287   M.Araki and T.Nagase, Ger. Offen., 2,625,030 (1976); Chem.Abs., 86,
       120886r (1977).

288   K.Kaneda, K.Jitsukawa, T.Itoh and S.Teranishi, J.Org. Chem., 45,
       3004 (1980).

289   J.T.Groves and W.J.Kruper, J.Amer. Chem. Soc., 101, 7613 (1979).

290   T.L.Sidall, N.Miyaura, J.C.Huffmann and J.K.Kochi, J.Chem. Soc.,
       Chem. Commun., 1185 (1983).

291   M.Schröder, Chem. Rev., 80, 187 (1980).

292   J.F.Cairns and H.L.Roberts, J.Chem. Soc., C, 640 (1968).

293   A.D.Tait, Steroids, 20, 531 (1972).

294   J.Y.Savoie and P.Brassard, Canad. J.Chem., 49, 3515 (1971).

295   N.A.Milas, J.H.Trepagnier, J.T.Nolan and M.Iliopulos, J.Amer. Chem.
       Soc., 81, 4730 (1959).

296   L.R.Subbaram, J.Subbaramam and E.J.Behrman, Inorg. Chem., 11, 2621 (1972).

297   K.A.Hofmann, O.Ehrhart and O.Schneider, Chem. Ber., 46, 1657 (1913).

298   G.Braun, J.Amer. Chem. Soc., 52, 3188 (1930).

299   I.Ernest, Coll. Czech. Chem. Commun., 29, 266 (1964).

300   H.Muxfeldt and G.Hardtmann, Ann., 669, 113 (1963).

301   G.Büchi and J.A.Carlson, J.Amer. Chem. Soc., 90, 5336 (1968).

302   K.Miescher and J.Schmidlin, Helv. Chim. Acta, 33, 1840 (1950).

303   B.M.Trost, J.M.Timko and J.L.Stanton, J.Chem. Soc., Chem. Commun.,
       436 (1978).

304   FMC Corporation, Neth. Appl., 74,11,150 (1976); Chem. Abs., 86,
       4925d (1976).

305   K.Wiesner and J.Santroch, Tetrahedron Lett., 5939 (1966).

306   T.Hiyama, K.Kobayashi and M.Fujita, Tetrahedron Lett., 25, 4959 (1984).

307   J.E.McMurray, A.Andrus, G.M.Ksander, J.H.Musser and M.A.Johnson,

       J.Amer. Chem. Soc., 101, 1330 (1979).

308   L.A.Mitscher, G.W.Clark and P.B.Hudson, Tetrahedron Lett., 2553 (1978).

309   K.B.Sharpless and K.Akashi, J.Amer. Chem. Soc., 98, 1986 (1976).

310   C.-H.Chang, W.R.Midden, J.S.Deetz and E.J.Behrman, Inorg. Chem., 18,

       1364 (1979).

311   K.Akashi, R.E.Palermo and K.B.Sharpless, J.Org. Chem., 43, 2063 (1978).

312   V. Van Rheenan, R.C.Kelly and D.Y.Cha, Tetrahedron Lett., 1973 (1976).

313   V. Van Rheenan, D.Y.Cha and W.M.Hartley, Org. Synth., 58, 43 (1978).

314   D.P.Curran and Y-G.Suh, Tetrahedron Lett., 25, 4179 (1984).

315   E.J.Corey, R.L.Danheiser, S.Chandrasekaran, P.Siret, G.E.Keck and

       J.L.Gras, J.Amer. Chem. Soc., 100, 8031 (1978).

316   R.Ray and D.S.Matteson, Tetrahedron Lett., 21, 449 (1980), J.Ind. Chem.

       Soc., 59, 119 (1982).

317   M.P.Singh, H.S.Singh, B.S.Arya, A.K.Singh and A.K.Sisodia, Ind. J.Chem.,

       13, 112 (1975).

318   P.S.Radhakrishnamurti and B.K.Panda, Ind. J.Chem., 21A, 128 (1982).

319   K.Behari, J.P.Pachauria, P.Kumar and B.Krishna, Ind. J.Chem., 21A,

       301 (1982).

320   E.I. Du Pont de Nemours, and Co., Brit. Pat., 1138132 (1968);

       Chem. Abs., 70, 46853d (1969).

321   K.B.Sharpless, A.O.Chong and K.Oshima, J.Org. Chem., 41, 177 (1976).

322   P.S.Radhakrishnamurti and S.A.Misra, Int. J.Chem. Kinet., 14, 631

       (1982).

323   S.Wolfe, S.K.Hasan and J.R.Campbell, J.Chem. Soc., Chem. Commun.,

       1420 (1970).

656

324   I.I.Moiseev, M.N.Vargaftik and Ya.K.Syrkin, Dokl. Phys. Chem., 133,
       801 (1960).

325   P.M.Henry, J.Org. Chem., 38, 1631 (1973).

326   A.V.Devekki, D.V.Mushenko and V.S.Fedorof, J.Org. Chem. U.S.S.R.,
       17, 2250 (1981).

327   R.G.Brown and J.M.Davidson, Adv. Chem. Ser., 132, 49 (1974).

328   R.G.Brown, R.V.Chaudhari and J.M.Davidson, J.Chem. Soc., Dalton,
       183 (1977).

329   W.Kitching, Z.Rappoport, S.Winstein and W.G.Young, J.Amer. Chem. Soc.,
       88, 2054 (1966).

330   E.W.Stern, M.L.Spector and H.P.Leftin, J.Catal., 6, 152 (1966).

331   J.K.Stille, R.A.Morgan, D.D.Whitehurst and J.R.Doyle, J.Amer. Chem.
       Soc., 87, 3282 (1965).

332   L.S.Hegedus, Tetrahedron, 40, 2415 (1984).

333   G.Lu, W.Liu, Y.Qi, Y.Zhang, R.Wang and Q.Bai, Huadong Huagong
       Xueyuan Xuebao, 13 (1981); Chem. Abs., 97, 94402k (1982).

334   V.A.Likholobov and Yu.I.Ermakov, Kinet. Catal., 17, 104 (1976),
       Dokl. Chem., 218, 681 (1974).

335   P.M.Henry and R.N.Pandey, Adv. Chem. Ser., 132, 33 (1974).

336   E.G.Lebedeva, A.V.Devekki, Yu.I.Malov, D.V.Mushenko and V.S.Volkova,
       J.Org. Chem. U.S.S.R., 14, 1485 (1978).

337   A.N.Gartsman, M.G.Volkhonskii and V.A.Likholobov, Kinet. Catal.,
       24, 1045 (1983).

338   A.B.Svetlova, S.M.Brailovskii and O.N.Temkin, Kinet. Catal., 19,
       1292 (1978).

339   M.G.Volkhonskii, V.A.Likholobov and Yu.I.Ermakov, Kinet. Catal., 24,
       289, 488 (1983).

340   N.I.Kuznetsova, V.A.Likholobov, M.A.Fedotov and Yu.I.Ermakov, Bull.
       Acad. Sci. U.S.S.R., Div. Chem. Sci., 31, 2475 (1982).

341   T.Urasaki, W.Funakoshi and H.Fujimoto, Japan. Kokai, 76 29,421 (1976);
      Chem. Abs., 85, 93847z (1976).

342   S.M.Brailovskii, L.Elfteriu, O.N.Chernysheva and A.P.Belov,
      Kinet. Catal., 23, 42 (1982).

343   M.S.Shlapak, S.M.Brailovskii and O.N.Temkin, Kinet. Catal., 24,
      1169 (1983).

344   J-E.Bäckvall and R.E.Nordberg, Pure Appl. Chem., 55, 1669 (1983),
      Acc. Chem. Res., 16, 335 (1983).

345   C.E.H.Bawn and J.A.Sharp, J.Chem. Soc., 1854, 1866 (1957).

346   H.Mutoh, T.Okada and Y.Kamiya, Ind. Eng. Chem., Prod. Res. Devel.,
      20, 487 (1981).

347   R.Tang and J.K.Kochi, J.Inorg. Nucl. Chem., 35, 3845 (1973).

348   T.Okamoto and S.Oka, J.Org. Chem., 49, 1589 (1984).

349   E.I.Heiba, R.M.Dessau and P.G.Rodewald, J.Amer. Chem. Soc., 96,
      7977 (1974).

350   J.Kollar, Ger. Offen., 1,931,563 (1970); Chem.Abs., 72, 78448h (1970).

351   C.S.Reddy and E.V.Sundaram, Gazz. Chim. Ital., 113, 177 (1983).

352   W.E.Fristad and J.R.Peterson, Tetrahedron Lett., 24, 4547 (1983).

353   P.H.J.Carlsen, T.Katsuki, V.S.Martin and K.B.Sharpless, J.Org. Chem.,
      46, 3936 (1981).

354   N.S.Ming, U.S.Patent, 3,839,375 (1974); Chem. Abs., 81, 169145j (1974).

355   S.Sondu, B.Sethuram and T.N.Rao, J.Ind. Chem. Soc., 60, 198 (1983).

356   E.V.Gusevskaya, V.A.Likholobov and Yu.I.Ermakov, Kinet. Catal., 21,
      466 (1980), 24, 42, 49 (1983).

357   W.E.Fristad and J.R.Petersen, unpublished data.

358   S.Yoshioka, K.Omae, K.Matsushiro, Y.Kotani, H.Yamamoto and S.
      Nakamura, Japan., 71 38,128 (1971); Chem. Abs., 76, 86371a (1972).

359   W.Gaenzler, K.Kabs and G.Schroeder, Ger. Offen., 2,256,847 (1974);
      Chem. Abs., 81, 49277g (1974).

360   L.Bassignani, A.Brandt, V.Caciagli and L.Re, J.Org. Chem., 43, 4245
(1978).

361   P.Müller and J.Godoy, Helv. Chim.Acta, 64, 2531 (1981), Tetrahedron
Lett., 23, 3661 (1982).

362   H.Gopal and A.J.Gordon, Tetrahedron Lett., 2941 (1971).

363   S.A.Matlin and P.G.Sammes, J.Chem. Soc., Perkin I, 2851 (1973).

364   Y.Ninagawa, T.Nishida, Y.Tamai, Y.Oomura, T.Hosogai, Y.Fujita,
F.Mori and K.Itoi, Japan. Kokai, 75,108,204 (1975); Chem. Abs., 84,
30417t (1976).

365   L.G.Fedenok, V.M.Berdnikov and M.S.Shvartsberg, J.Org. Chem. U.S.S.R.,
12, 1385 (1976), 14, 1328, 1334 (1978).

366   L.Eberson and L.Jönsson, Acta Chem. Scand. B, 28, 771 (1974), 30,
579 (1976), J.Chem. Soc., Chem. Commun., 885 (1974), Ann., 233 (1977).

367   T.Itahara, Chem. Ind., 599 (1982).

368   H.Sugimoto, I.Kawata, H.Taniguchi and Y.Fujiwara, J.Organomet. Chem.,
266, C44 (1984).

369   J.Tsuji and H.Nagashima, Tetrahedron, 40, 2699 (1984).

370   R.Janin and L.Krumenacker, Ger. Offen., 2,422,460 (1974); Chem. Abs.,
83, 61722x (1975).

371   Mitsubishi Chemical Industries Ltd., Jpn. Kokai Tokkyo Koho, JP,
57,139,037 [82,139,037]; Chem. Abs., 98, 53451c (1983).

372   P.Müller and C.Bobillier, Tetrahedron Lett., 22, 5157 (1981),
24, 5499 (1983).

373   P.O'Neill, S.Steenken and D.Schulte-Frohlinde, J.Phys. Chem., 79,
2773 (1975), 81, 26 (1977).

374   C.Giordano, A.Belli, A.Citterio and F.Minisci, J.Org. Chem., 44,
2314 (1979).

375   S.Tamagaki, K.Suzuki, H.Okamoto and W.Tagaki, Tetrahedron Lett.,
24, 4847 (1983).

376   M.A.Brook, L.Castle, J.R.L.Smith, R.Higgins and K.P.Morris, J.Chem.
      Soc., Perkin II, 687 (1982)

377   J.M.Maissant, J.M.Bodroux, C.Bouchoule and M.Blanchard, J.Mol. Catal.,
      9, 237 (1980).

378   A.K.Chakraborti and U.R.Ghatak, Synthesis, 746 (1983).

379   H.Mimoun, L.Saussine, E.Daire, M.Postel, J.Fischer and R.Weiss,
      J.Amer. Chem. Soc., 105, 3101 (1983).

380   E.I.Heiba, R.M.Dessau and W.J.Koehl, J.Amer. Chem. Soc., 91,
      138, 6830 (1969).

381   J.R.Gilmore and J.M.Mellor, J.Chem. Soc., Chem. Commun., 507 (1970),
      Tetrahedron Lett., 3977 (1971).

382   J.M.Anderson and J.K.Kochi, J.Amer. Chem.Soc., 92, 2450 (1970).

383   R.E. van der Ploeg, R.W. de Korte and E.C.C.Kooyman, J.Catal., 10, 52
      (1968).

384   J.Hanotier, M.Hanotier-Bridoux and P. de Radzitzky, J.Chem. Soc.,
      Perkin II, 381, 1035 (1973).

385   E.I.Heiba and R.M.Dessau, J.Amer. Chem. Soc., 93, 995 (1971).

386   A.Onopchenko, J.G.D.Schulz and R.Seekircher, J.Org. Chem., 37,
      1414, 2950 (1972).

387   K.Namie and S.Takeda, Japan. Kokai, 76,108,028 (1976); Chem. Abs.,
      86, 106, 1802 (1977).

388   T.A.Cooper and W.A.Waters, J.Chem. Soc., B, 687 (1967).

389   K.Sakota, Y.Kamiya and N.Ohta, Canad. J.Chem., 47, 387 (1969).

390   Y.Kamiya and M.Kashima, J.Catal., 25, 326 (1972), Bull. Chem. Soc.
      Jpn., 46, 905 (1973).

391   E.J.Y.Scott and A.W.Chester, J.Phys. Chem., 76, 1520 (1972).

392   T.Morimoto and Y.Ogata, J.Chem. Soc., B, 62, 1353 (1967).

393   Y.Kamiya, Tetrahedron, 22, 2029 (1976), Adv. Chem. Ser., 76, 193
      (1968), J.Catal., 33, 480 (1974).

660

394   H.D.Holtz, J.Org. Chem., 37, 2069 (1972), J.Chem. Soc., Chem. Commun.,
      1166 (1971).

395   N.P.Belous, B.N.Tmenov, T.P.Bespalova, O.V.Alimova and F.F.
      Shcherbina, J.Org. Chem. U.S.S.R., 19, 311 (1983).

396   A.W.Chester, E-A.Heiba, R.M.Dessau and W.J.Koehl, Inorg. Nucl.
      Chem. Lett., 5, 277 (1969).

397   J.K.Kochi, R.T.Tang and T.Bernath, J.Amer. Chem. Soc., 95, 7114
      (1973).

398   G.S.Bezhanishvili, N.G.Digurov and N.N.Lebedev, Kinet. Catal.,
      24, 852 (1983).

399   C.H.Bell, U.S.Patent, 3,631,204 (1971); Chem. Abs., 76, 72234w (1972).

400   N.A.Batygina, T.V.Bukharkina and N.G.Digurov, Kinet. Catal.,
      22, 920 (1981).

401   A.Onopchenko, J.G.D.Schulz and R.Seekircher, U.S.Patent, 3,644,512
      (1972); Chem. Abs., 76, 99125y (1972).

402   A.W.Chester, E.J.Y.Scott and P.S.Landis, J.Catal., 46, 308 (1977).

403   M.Shigeyasu and N.Kusano, Japan. Kokai, 73 80,531 (1973); Chem. Abs.,
      80, 59700h (1974).

404   E.Stasek and D.Hlavacova, Fr. Demande, 2,162,007 (1973); Chem. Abs.,
      80, 14746m (1980).

405   E.Katzschmann, Ger. Offen., 2,132,909 (1973); Chem. Abs., 78,
      84032h (1973).

406   Y.Kawai and K.Kida, Japan. Kokai, 75 49,247 (1975); Chem. Abs., 83,
      96762n (1975).

407   Y.Kamiya and T.Ushiba, Bull. Chem. Soc. Jpn., 48, 1563 (1975).

408   T.Suzuki, A.Tateishi, S.Naito and H.Higuchi, Jpn. Kokai Tokkyo Koho,
      78,130,629 (1978); Chem. Abs., 90, 87066f (1979).

409   K.Tsunoi and T.Kato, U.S.Patent, 3,626,000 (1971); Chem. Abs., 76,
      59229d (1972).

410   K.Nakaoka, H.Yoshiwara and S.Wakamatsu, Japan. Kokai, 73 81,829

      (1973); Chem. Abs., 80, 108195h (1974).

411   E.Kwaskowska-Chec and J.J.Ziolkowski, Pol. J.Chem., 52, 593 (1978).

412   M.Hronec, V.Vesely and J.Herain, Coll. Czech. Chem. Commun., 44,

      3362 (1979).

413   T.Fujii and G.Yamashita, Japan., 74 39,663 (1974); Chem. Abs., 83,

      9515t (1975).

414   New Japan Chemical Co., Ltd., Jpn. Kokai Tokkyo Koho, JP, 57,197,243

      (82,197,243)(1982); Chem. Abs., 98, 160426v (1983).

415   T.Matsuda, T.Shirafuji and T.Murata, Jpn. Kokai Tokkyo Koho,

      79 59,243 (1979); Chem. Abs., 91, 192993z (1979).

416   J.D.V.Hanotier and M.G.S.Hanotier-Bridoux, Ger. Offen., 2,242,386

      (1974); Chem. Abs., 80, 120789n (1974).

417   P.A.Konstantinov, T.V.Shchedrinskaya, I.V.Zakharov and M.N.Volkov,

      J.Org. Chem. U.S.S.R., 8, 2639 (1972).

418   V.K.Kondratov and N.D.Rus'yanova, Kinet. Catal., 21, 167 (1980).

419   Y.Ichikawa, Y.Yamanaka, T.Yamaji and H.Tsuruta, Japan., 77 03,377

      (1977); Chem. Abs., 87, 39127t (1977).

420   I.V.Zakharov, Kinet. Catal., 15, 1287 (1974).

421   T.V.Shchedrinskaya, P.A.Konstantinov, V.P.Litvinov, E.G.Ostapenko,

      I.V.Zakharov and M.N.Volkov, J.Gen. Chem. U.S.S.R., 44, 806 (1974).

422   A.Onopchenko, J.G.Schulz and R.Seekircher, Ger. Offen., 2,202,418

      (1973); Chem.Abs., 80, 27638c (1974).

423   K.Ito, H.Kaminaka, H.Suda, H.Sugahara, K.Oie, H.Tomita, M.Ohsu

      and T.Kobayashi, Ger. Offen., 2,138,283 (1973); Chem. Abs., 78,

      97333g (1973).

424   R-K.Bai, H-J.Zong, J-G.He and Y-Y.Jiang, Makromol. Chem., Rapid

      Commun., 5, 501 (1984).

425   E.Baciocchi and R.Ruzziconi, J.Chem. Soc., Chem. Commun., 445 (1984).

426   M.N.Volkov, O.A.Kazakova, É.G.Ostapenko, P.A.Konstantinov and
      R.I.Shupik, J.Org. Chem. U.S.S.R., 15, 1070 (1979).

427   A.J.Chalk. S.A.Magennis and W.E.Newman, Fund.Res. Homog. Catal.,
      3, 445 (1979).

428   N.G.Digurov, V.F.Kashirskii, E.V.Shevyreva and N.N.Lebedev,
      Kinet. Catal., 22, 274 (1981).

429   G.H.Jones, J.Chem. Res.(S), 207 (1982).

430   H.Morimoto, Y.Tohda, H.Torigata and K.Nakaoka, Ger. Offen.,
      2,749,638 (1978); Chem.Abs., 89, 23985b (1978).

431   H.Torikata and K.Nakaoka, Jpn. Kokai Tokkyo Koho, 79 44,630 (1979);
      Chem. Abs., 91, 157450y (1979).

432   E.Kwaskowska-Chec, K.Fried, H.Przywarska-Boniecka and J.J.Ziolkowski,
      React. Kinet. Catal. Lett., 2, 425 (1975).

433   A.Krasuska, E.Treszczanowicz and W.Ormaniec, Przem. Chem., 52,
      91 (1973); Chem. Abs., 78, 147501h (1973).

434   K.Yoshida and T.Horie, Sekiyu Gakkai Shi, 18, 90 (1975); Chem. Abs.,
      87, 201003y (1977).

435   M.S.Brodskii, Yu.A.Yalter, M.Ya.Gervits and A.E.Kruglik, Kinet. Catal.,
      20, 274 (1979).

436   J.Horyna, D.Hlavacova, E.Stasek and V.Kohoutek, Ger. Offen.,
      2,127,844 (1971); Chem. Abs., 76, 72228x (1972).

437   A.M.Kyamkin, G.D.Kharlampovich, N.S.Mulyaeva and Yu.V.Cherkin,
      Neftepererab. Neftekhim. (Moscow), 1973(9) 27; Chem. Abs., 80,
      14694t (1974).

438   Yu.V.Pozdnyakovich, V.V.Borodovitsyn, T.I.Borodovitsyna,V.M.Sysa
      and S.M.Shein, J.Org. Chem. U.S.S.R., 19, 349 (1983).

439   P.Maggioni and F.Minisci, Chim. Ind. (Milan), 61, 101 (1979).

440   A.Citterio and F.Minisci, Chim. Ind. (Milan), 64, 320 (1982).

441   M.V.Bhatt and P.T.Perumal, Tetrahedron Lett., 22, 2605 (1981).

442   A.Belli, C.Giordano and A.Citterio, Synthesis, 477 (1980).

443   C.Walling, C.Zhao and G.M.El-Taliawi, J.Org. Chem., 48, 4910 (1983).

444   G.N.Gvozdovskii, Yu.N.Koshelev, L.L.Rassadina, L.V.Bogdanova,

      I.Ya.Kvitko and L.V.Alam, J.Org. Chem. U.S.S.R., 14, 1021 (1978).

445   G.P.Pavlov, I.D.Sinovich and N.V.Bykova, J.Org. Chem. U.S.S.R.,

      16, 1084 (1980).

446   V.G.Vinogradova and A.N.Zverev, Bull. Acad. Sci. U.S.S.R., Div.

      Chem. Sci., 24, 2103 (1975).

447   D.L.Allara, J.Org. Chem., 37, 2448 (1972).

448   J.E.McIntyre and D.A.S.Ravens, J.Chem. Soc., 4082 (1961).

449   P.S.Radhakrishnamurti and H.P.Panda, Ind. J.Chem., 18A, 419 (1979),

      React. Kinet. Catal. Lett., 14, 193 (1980).

450   J.Blum, H.Rosenman and E.D.Bergmann, Tetrahedron Lett., 3635 (1967).

451   A.J.Birch and G.S.R.Subba Rao, Tetrahedron Lett., 2917 (1968).

452   Yu.V.Geleti and A.E.Shilov, Kinet. Catal., 24, 413 (1983).

453   A.Onopchenko and J.G.D.Schulz, J.Org. Chem., 38, 909, 3729 (1973).

454   D.Mansuy, J-F.Bartoli and M.Momenteau, Tetrahedron Lett., 23,

      2781 (1982).

455   J.A.Smegal and C.L.Hill, J.Amer. Chem. Soc., 105, 3515 (1983).

456   C.L.Hill and J.A.Smegal, Nouv. J.Chim., 6, 287 (1982).

457   J.Hanotier, P.Camerman, M.Hanotier-Bridoux and P. de Radzitzky,

      J.Chem. Soc., Perkin II, 2247 (1972).

458   K.Tanaka, A.C.S. Petrol. Chem. Div., Prepr., 19(1), 103 (1974),

      Chem. Tech., 555 (1974), Hydrocarbon Process., 114 (1974).

459   A.Onopchenko and J.G.D.Schulz, U.S.Patent, 4,032,569 (1977);

      Chem. Abs., 87, 85483d (1977).

460 D.Mansuy, J-F.Bartoli, J.C.Chottard and M.Lange, Angew. Chem., Int. Ed. Engl., 19, 909 (1980).

461 C.L.Hill and B.C.Schardt, J.Amer. Chem. Soc., 102, 6374 (1980).

462 H.J.Callot and F.Metz, Tetrahedron Lett., 23, 4321 (1982).

463 R.Breslow and S.H.Gellman, J.Chem. Soc., Chem. Commun., 1400 (1982).

464 J.T.Groves, G.A.McLusky, R.E.White and M.J.Coon, Biochem. Biophys. Res. Commun., 81, 154 (1978).

465 J.T.Groves, W.J.Kruper and R.C.Haushalter, J.Amer. Chem. Soc., 102, 6379 (1980).

466 P.S.Traylor, D.Dolphin and T.G.Traylor, J.Chem. Soc., Chem. Commun., 279 (1984).

467 S.P.Srivastava, A.Kumar, A.K.Mittal and V.K.Gupta, Oxidn. Commun., 1, 265 (1980).

468 S.P.Srivastava, V.K.Gupta, J.C.Gupta and M.K.Maheshwari, J.Ind. Chem. Soc., 57, 797 (1980).

469 F.Minisci, A.Citterio and C.Giordano, Acc. Chem. Res., 16, 27 (1983).

470 K.Behari, N.N.Pandey, R.S.Shukla and R.K.Dwivedi, Chem. Scripta, 20, 219 (1982).

471 N.Nath, L.P.Singh and R.P.Singh, J.Ind. Chem. Soc., 58, 1204 (1981).

472 M.P.Singh and S.Kothari, Ind. J.Chem., 22A, 697 (1983).

473 R.K.Dwivedi, M.Verma, P.Kumar and K.Behari, Tetrahedron, 39, 815 (1983).

474 M.Schröder and W.P.Griffith, J.Chem. Soc., Chem. Commun., 58 (1979).

475 L.D.Burke and J.F.Healy, J.Chem. Soc., Dalton, 1091 (1982).

476 R.N.Singh, R.K.Singh and H.S.Singh, J.Chem. Res.(S), 249 (1977), 176 (1978).

477 S.Kanemoto, K.Oshima, S.Matsubara, K.Takai and H.Nozaki, Tetrahedron Lett., 24, 2185 (1983).

478   Y.Blum and Y.Shvo, Is. J.Chem., 24, 144 (1984), J.Organomet. Chem., 263, 93 (1984).

479   S.O.Travin, Yu.I.Skurlatov, N.V.Gorbunova and A.P.Purmal', Bull. Acad. Sci. U.S.S.R., Div. Chem. Sci., 28, 1377 (1979).

480   V.P.Tretyakov, G.P.Zimtseva, E.S.Rudakov and A.V.Bogdanov, React. Kinet. Catal. Lett., 19, 263 (1982).

481   P.Camerman and J.Hanotier, Fr. Patent, 2,095,160 (1972); Chem. Abs., 77, 125991v (1972).

482   K.Kaneda, Y.Kawanishi and S.Teranishi, Chem.Lett., 1481 (1984).

483   Y.Masuyama, M.Takahashi and Y.Kurusu, Tetrahedron Lett., 25, 4417 (1984).

484   H.Laurent and R.Wiechert, Ger. Offen., 2,260,303 (1974); Chem. Abs., 81, 91809g (1974).

485   R.N.Singh and H.S.Singh, Ind. J.Chem., 16A, 145 (1978).

486   P.Becker and J.K.Beattie, Aust. J.Chem., 35, 1245 (1982).

487   P.Müller and J.Godoy, Tetrahedron Lett., 22, 2361 (1981).

488   K.Behari, H.Narayan, R.S.Shukla and K.C.Gupta, Int. J.Chem. Kinet., 16, 195 (1984).

489   R.Tang, S.E.Diamond, N.Neary and F.Mares, J.Chem. Soc., Chem. Commun., 562 (1978).

490   M.Tanaka, T.Kobayashi and T.Sakakura, Angew. Chem., Int. Ed. Engl., 23, 518 (1984).

491   J.A.Caputo and R.Fuchs, Tetrahedron Lett., 4729 (1967).

492   V.Uma,B.Sethuram and T.N.Rao, React. Kinet. Catal. Lett., 18, 283 (1981).

493   R.R.Gagné and D.N.Marks, Inorg. Chem., 23, 65 (1984).

494   A.Rosenthal, M.Sprinzl and D.A.Baker, Tetrahedron Lett., 4233 (1970).

495   R.K.Dwivedi, H.Narayan and K.Behari, J.Inorg. Nucl. Chem., 43, 2893 (1981).

496   C.Walling, G.M.El Taliawi and C.Zhao, J.Org. Chem., 48, 4914 (1983).

497   T.F.Blackburn and J.Schwartz, J.Chem. Soc., Chem. Commun., 157 (1977).

498   H.Nagashima and J.Tsuji, Chem. Lett., 1171 (1981).

499   Y.Tamaru, Y.Yamada, K.Inoue, Y.Yamamoto and Z-i. Yoshida, J.Org. Chem., 48, 1286 (1983),Tetrahedron Lett., 22, 1801 (1981).

500   C.Jallabert, C.Lapinte and H.Riviere, J.Mol. Catal., 14, 75 (1982).

501   F.Steinbach and H.Schmidt, J.Catal., 29, 515 (1973), 52, 302 (1978).

502   A.V.Savitskii, J.Gen. Chem. U.S.S.R., 44, 1518 (1974).

503   P.Camerman and J.Hanotier, Fr. Demande, 2,094,808 (1972); Chem. Abs., 77, 125986x (1972).

504   N.G.Digurov, V.F.Kashirskii, M.A.Karyugin and N.N.Lebedev, Kinet. Catal., 20, 1222 (1979).

505   B.S.Tovrog, S.E.Diamond, F.Mares and A.Szalkiewicz, J.Amer. Chem. Soc., 103, 3522 (1981).

506   H.S.Singh, K.K.Singh, S.M.Singh, P.Singh and P.Thakur, Ind. J.Chem., 21A, 816 (1982).

507   J.Martin, C.Martin, M.Faraj and J.M.Brégeault, Nouv. J.Chim., 8, 141 (1984).

508   T.Okamoto, K.Sasaki and S.Oka, Chem. Lett., 1247 (1984).

509   E.S.Huyser and G.W.Hawkins, J.Org. Chem., 48, 1705 (1983).

510   J.T.Groves and M. van der Puy, J.Amer. Chem. Soc., 96, 5274 (1974), 97, 7118 (1975), Tetrahedron Lett., 1949 (1975).

511   H.Tomioka, K.Oshima and H.Nozaki, Tetrahedron Lett., 23, 539 (1982).

512   B.M.Trost and Y.Masuyama, Tetrahedron Lett., 25, 173 (1984), Is. J. Chem., 24, 134 (1984).

513   P.S.Radhakrishnamurti and B.Sahu, Ind. J.Chem., 16A, 259 (1978).

514   M.P.Doyle, W.J.Patrie and S.B.Williams, J.Org. Chem., 44, 2955 (1979).

515   S.E.Jacobson, D.A.Muccigrosso and F.Mares, J.Org. Chem., 44, 921 (1979).

516   S.L.Regen and G.M.Whitesides, J.Org. Chem., 37, 1832 (1972).

517   U.Zeidler, H.Lepper and W.Stein, Fette, Seifen. Anstrichm., 76,
       260 (1974); Chem. Abs., 81, 91007u (1974).

518   G.Schreyer, W.Schwarze and W.Weigert, Ger. Offen., 2,052,815 (1972);
       Chem. Abs., 77, 33963j (1972).

519   K.Kaneda, Y.Kawanishi, K.Jitsukawa and S.Teranishi, Tetrahedron Lett.,
       24, 5009 (1983).

520   S.P.Srivastava and A.Kumar, Ind. J.Chem., 15A, 1061 (1977).

521   R.P.Chaplin, A.S.Walpol, S.Zadro, S.Vorlow and M.S.Wainwright,
       J.Mol.Catal.,22, 269 (1984).

522   S.Tsuruya, T.Masuoka and M.Masai, J.Mol. Catal., 10, 21 (1980).

523   H.Matsumoto and S.Ito, Synth. Commun., 14, 697 (1984).

524   F.Marta, E.Boga and M.Matok, Discuss. Farad. Soc., 46, 173 (1968).

525   M.G.Roelofs, E.Wasserman, J.H.Jensen and A.E.Nader, J.Amer. Chem. Soc.,
       105, 6329 (1983).

526   E.B.Nitteberg and O.A.Rokstad, Acta Chem. Scand., A34, 199 (1980).

527   H.Sakamoto, T.Funabiki and K.Tarama, J.Catal., 48, 427 (1977).

528   G.Speier, Inorg. Chim. Acta, 32, 139 (1979).

529   J-i.Hojo, S.Yuasa, N.Yamazoe, I.Mochida and T.Seiyama, J.Catal., 36,
       93 (1975).

530   P.S.Radhakrishnamurti and P.C.Misra, Ind. J.Chem., 18A, 126 (1979).

531   T.Matsuzaki, J.Imamura and N.Ohta, Kogyo Kagaku Zasshi, 71, 700
       (1968); Chem. Abs., 69, 66848p (1968).

532   C.E.H.Bawn, T.P.Hobin and L.Raphael, Proc. Roy. Soc., A, 257,
       313 (1956).

533   G.C.Allen and A.Aguilo, Adv. Chem. Ser., 76, 363 (1968).

534   N.N.Lebedev, M.N.Manakov and A.P.Litovka, Kinet. Catal., 15, 703,
       791 (1974).

535   Y.Ohkatsu, T.Hara and T.Osa, Bull. Chem. Soc. Jpn., 50, 696 (1977).

536   T.Motoda, T.Koyama and K.Sakai, Japan., 73 01,366 (1973); Chem. Abs.,
       79, 65798s (1973).

537   S.Sakai, N.Kasuga and Y.Ishii, Kogyo Kagaku Zasshi, 72, 1687 (1969);
       Chem. Abs., 72, 30957w (1970).

538   I.Apostol, J.Haber, T.Mlodnicka and J.Poltowicz, J.Mol.Catal., 14
       197 (1982).

539   O-T.Onsager, H.C.A.Swensen and J.E.Johansen, Acta Chem. Scand., B38
       567 (1984).

540   H.Inoue, Y.Kimura and E.Imoto, Bull. Chem. Soc. Jpn., 46, 3303 (1973).

541   W.Brackman, C.J.Gaasbeek and P.J.Smut, Rec., 85, 437 (1966).

542   S.Yamaguchi, H.Nakajima, H.Kimura and H.Nishimaru, Japan. Kokai,
       74 35,317 (1974); Chem. Abs., 81, 119990k (1974), Japan. Kokai,
       73 97,809 (1973); Chem. Abs., 80, 70348v (1974).

543   M.Gay, Ger. Offen., 2,136,481 (1972); Chem. Abs., 76, 112913a (1972).

544   Kuraray Co., Ltd., Jpn. Kokai Tokkyo Koho, JP, 82,112,350 (1982);
       Chem. Abs., 97, 162385j (1982).

545   G.I.Nikishin, M.G.Vinogradov and S.P.Verenchikov, Bull. Acad. Sci.
       U.S.S.R., Div. Chem. Sci., 1698 (1969).

546   K.Eguchi, I.Aso, N.Yamazoe and T.Seiyama, Chem. Lett., 1345 (1979).

547   V.D.Komissarov and E.T.Denisov, Russ. J.Phys. Chem., 43, 426 (1969),
       44, 216 (1970).

548   M.A.Saitova and V.D.Komissaraov, Kinet. Catal., 13, 496 (1972).

549   A.M.Sakharov and I.P.Skibida, Bull. Acad. Sci. U.S.S.R., Div. Chem.
       Sci., 29, 344 (1980).

550   S.Suzuki, T.Tokumaro, E.Ando and Y.Watanabe, Japan. Kokai, 74, 00,235
       (1974); Chem. Abs., 80, 95508k (1974).

551   L.M.Sayre and S-J.Jin, J.Org. Chem., 49, 3498 (1984).

552   S.C.Agarwal, S.P.L.Agarwal, V.B.Agarwal and V.Dwivedi, Rev. Roum.
       Chim., 27, 265 (1982).

553 G.I.Nikishin, E.I.Troyansky and M.I.Lazareva, Tetrahedron Lett.,
25, 4987 (1984).

554 G.P.Panigrahi and P.K.Misro, Ind. J.Chem., 16A, 762 (1978).

555 S.C.Pati and M.Panda, Bull. Soc. Chim. Belg., 91, 271 (1982).

556 M.P.Singh, H.S.Singh and M.K.Verma, J.Phys. Chem., 84, 256 (1980).

557 S.A.Kamath and S.B.Chandalia, J.Appl. Chem. Biotechnol., 23, 469 (1973).

558 S.Senhara and T.Tsuji, Japan. Kokai, 78 05,110 (1978); Chem. Abs.,
88, 121899a (1978).

559 H.B.Charman, Brit. Pat., 1,114,885 (1968); Chem. Abs., 69, 51616n
(1968).

560 Asahi Chemical Industry Co., Ltd., Brit. Pat., 1,103,885 (1968);
Chem. Abs., 69, 10099c (1968).

561 C.Gardner, A.H.Gilbert and W.Morris, Brit. Pat., 1,250,192 (1971);
Chem. Abs., 76, 3404q (1972).

562 I.I.Korsak, V.E.Agabekov and N.I.Mitskevich, Neftekhimiya, 15,
130 (1975); Chem. Abs., 82, 154900y (1975).

563 D.M.Roundhill, M.K.Dickson, N.S.Dixit and B.P.Sudha-Dixit, J.Amer.
Chem. Soc., 102, 5538 (1980).

564 G.D.Mercer, J.S.Shu, T.B.Rauchfuss and D.M.Roundhill, J.Amer. Chem. Soc.,
97, 1967 (1975).

565 J.D.Druliner, J.Org. Chem., 43, 2069 (1978).

566 E.S.Rudakov, V.P.Tret'yakov, V.V.Chudaev, G.P.Zimtseva and M.A.
Simonov, Kinet. Catal., 24, 918 (1983).

567 S.E.Jacobson, R.Tang and F.Mares, J.Chem. Soc., Chem. Commun.,
888 (1978).

568 S.Ito, K.Aihara and M.Matsumoto, Tetrahedron Lett., 25, 3891 (1984).

569 M.G.Vinogradov, V.I.Dolinko and G.I.Nikishin, Bull. Acad. Sci.
U.S.S.R., Div. Chem. Sci., 31, 2036 (1982).

570   S.Aida, H.Ohta and Y.Kamiya, Chem. Lett., 1639 (1981).

571   M.Costantini, A.Dromard, M.Jouffret, B.Brossard and J.Varagnat,

       J.Mol. Catal., 7, 89 (1980).

572   R.A.Sheldon and J.K.Kochi, Org. React., 19, 279 (1972).

573   W.E.Fristad and J.A.Klang, Tetrahedron Lett., 24, 2219 (1983).

574   Yu.N.Ogibin, E.I.Troyanskii, G.I.Nikishin, O.S.Chizhov and V.I.

       Kadentsev, Bull. Acad. Sci. U.S.S.R., Div., Chem. Sci., 24, 884

       (1975), 25, 2345 (1976).

575   P.S.Radhakrishnamurti and B.R.K.Swamy, Ind. J.Chem., 15A, 1115 (1977).

576   C.S.Reddy, V.Laxmi and E.V.Sundaram, Ind. J.Chem., 19A, 544 (1980).

577   H.Singh, J.C.Govil and S.C.Saksena, J.Ind. Chem. Soc., 58, 951

       (1981).

578   S.C.Agarwal and L.K.Saksena, J.Ind. Chem. Soc., 58, 1159 (1981).

579   R.L.Yadav and D.P.Dubey, J.Ind. Chem. Soc., 56, 498 (1979), Egypt.

       J.Chem., 22, 151 (1979).

580   V.Lal, V.N.Singh, H.S.Singh and M.P.Singh, Ind. J.Chem., 10, 392

       (1972).

581   G.C.Soni and G.D.Menghani, Z.Naturforsch.,B27, 908 (1972).

582   S.Jha, P.D.Sharma and Y.K.Gupta, Inorg. Chem., 22, 1393 (1983).

583   S.Kumar and P.C.Mathur, J.Inorg. Nucl. Chem., 40, 581 (1978).

584   S.S.Srivastava and P.C.Mathur, Ind. J.Chem., 15A, 788 (1977).

585   M.A.Rao, B.Sethuram and T.N.Rao, Ind. J.Chem., 17A, 260 (1979).

586   V.Uma, B.Sethuram and T.N.Rao, Ind. J.Chem., 22A, 65 (1983).

587   I.M.Krip, B.I.Chernyak and A.I.Klyuchkivskii, Kinet. Catal., 24,

       65 (1983).

588   N.P.Belous, F.F.Shcherbina, D.N.Tmenov and T.B.Bespalova, J.Org.

       Chem. U.S.S.R., 17, 1656 (1981).

589   Yu.P.Zhukov, R.B.Svitych, N.N.Rzhevskaya, O.P.Yablonskii and

       A.V.Bondarenko, J.Org. Chem.U.S.S.R., 18. 67 (1982).

590 S.K.Dhar, J.Abdelaziz, P.Cozzi, P.Jasien, C.Mason and R.Zalenas, J.Chem. Soc., Chem. Commun., 8 (1984).

591 H.Singh, K.S.Chauhan and U.K.Rathi, J.Ind. Chem. Soc., 57, 809 (1980).

592 S.Yoshifuji, M.Matsumoto, K-i.Tanaka and Y.Nitta, Tetrahedron Lett., 2963 (1980).

593 G.A.Hamilton and A.Revesz, J.Amer. Chem. Soc., 88, 2069 (1966).

594 G.Chandra and S.N.Srivastava, Rev. Roum. Chim., 25, 1149 (1980).

595 R.N.Mehrotra,R.C.Kapoor and S.K.Vajpai, J.Chem. Soc., Dalton, 999 (1984).

596 B.Singh, C.P.S.Gaur, R.B.Singh and B.B.L.Saxena, J.Ind. Chem. Soc., 59, 59 (1982).

597 C.M.Ashraf, I.Ahmad and F.K.N.Lugemwa, Ind. J.Chem., 18A, 373 (1979).

598 R.F.Jameson and N.J.Blackburn, J.Chem. Soc., Dalton, 534, 1596 (1976), 9 (1982).

599 C.Fabre and C.Lapinte, Nouv. J.Chim., 7, 123 (1983).

600 O.S.Fedorova, D.B.Lim and V.M.Berdnikov, React. Kinet. Catal. Lett., 8, 371 (1978).

601 M.M.Taqui Khan and A.E.Martel, J.Amer. Chem. Soc., 89, 4176, 7104 (1967), 90, 6011 (1968).

602 K.Takamura and M.Ito, Chem. Pharm. Bull., 25, 3218 (1977).

603 M.Kimura, A.Kobayashi and K.Boku, Bull. Chem. Soc. Jpn., 55, 2068 (1982).

604 K.Tsukahara, H.Ushio and Y.Yamamoto, Chem. Lett., 1137 (1980).

605 S.Shinkai, Y-i. Ishikawa and O.Manabe, Chem. Lett., 809 (1982).

606 B.Banaś and J.Mroziński, React. Kinet. Catal. Lett., 20, 425 (1982).

607 D.M.Wagnerová, J.Votruba, J.Blanck and J.Veprek-Šiska, Collect. Czech. Chem. Commun., 47, 744, 755 (1982).

608 K.Hirai, A.Nutton and P.M.Maitlis, J.Mol. Catal., 10, 203 (1981).

672

609   A.B.Smith and R.M.Scarborough, Synth. Commun., 10, 205 (1980).

610   H.Niwa, K.Wakamatsu, T.Hida, K.Niiyama, H.Kigoshi, M.Yamada, H.
       Nagase, M.Suzuki and K.Yamada, J.Amer. Chem. Soc., 106, 4547 (1984).

611   H.B.Henbest and J.Trocha-Grimshaw, J.Chem. Soc., Perkin I, 607 (1974).

612   T.Hosokawa, Y.Imada and S-I.Murahashi, J.Chem. Soc., Chem. Commun.,
       1245 (1983).

613   A.Nishinaga and H.Tomita, J.Mol. Catal., 7, 179 (1980).

614   L.H.Vogt, J.G.Wirth and H.J.Finkbeiner, J.Org. Chem., 34, 273 (1969).

615   I.B. Afanas'ev and N.G.Baronova, J.Gen. Chem. U.S.S.R., 52, 846
       (1982).

616   M.Jouffret, Ger. Offen., 2,518,028 (1975); Chem.Abs., 84, 73905t (1976).

617   A.J. De Jong and R. van Helden, Ger. Offen., 2,460,665 (1975);
       Chem. Abs., 83, 192831e (1975), Ger. Offen., 2,517,870 (1975);
       Chem. Abs., 84, 58927h (1976).

618   M.Tada and T.Katsu, Bull. Chem. Soc. Jpn., 45, 2558 (1972).

619   F.Frostin-Rio, D.Pujol, C.Bied-Charreton, M.Perrée-Fauvet and
       A.Gaudemer, J.Chem. Soc., Perkin I, 1971 (1984).

620   M.Costantini and M.Jouffret, Ger. Offen., 2,747,497 (1978); Chem. Abs.,
       89, 42581h (1978).

621   A.Nishinaga, T.Itahara, T.Matsuura, S.Berger, G.Henes and A.Rieker,
       Chem. Ber., 109, 1530 (1976).

622   A.Nishinaga, T.Itahara, T.Shimizu, H.Tomita, K.Nishizawa and T.
       Matsuura, Photochem. Photobiol., 28, 687 (1978).

623   A.Nishinaga, T.Shimizu, Y.Toyoda, T.Matsuura and K.Hirotsu, J.Org.
       Chem., 47, 2278 (1982).

624   M.Matsumoto and K.Kuroda, Tetrahedron Lett., 22, 4437 (1981).

625   S.A.Bedell and A.E.Martell, Inorg. Chem., 22, 364 (1983).

626   A.E.Martell, Pure Appl. Chem., 55, 125 (1983).

627   X-Y.Wang, R.J.Motekaitis and A.E.Martell, Inorg. Chem., 23, 271
      (1984).

628   E.L.Reilly, U.S.Patent, 4,257,968 (1981); Chem. Abs., 94, 174647x
      (1981).

629   M.Tanaka and A.Goh, Japan. Kokai, 75 95,231 (1975); Chem. Abs., 84,
      43638r (1976).

630   J.P.J.Verlaan, P.J.T.Alferink and G.Challa, J.Mol. Catal., 24, 235
      (1984).

631   M.Flintermann, G.Challa, R.Barbucci and P.Ferruti, J.Mol. Catal., 18,
      149 (1983).

632   J.P.J.Verlaan, R.Zweirs and G.Challa, J.Mol. Catal., 19, 223 (1983).

633   D.P.Mobley, J.Polym.Sci., 22, 3203 (1984).

634   P.Capdeviello and M.Maumy, Tetrahedron Lett., 24, 5611 (1983).

635   K.Kushioka, J.Org. Chem., 48, 4948 (1983).

636   S.P.Srivastava and A.K.Shukla, Ind. J.Chem., 14A, 958 (1976), 15A,
      603 (1977).

637   P.S.Radhakrishnamurti and B.V.D.Rao, Ind. J.Chem., 21A, 487 (1982).

638   V.S.Markevich, N.Kh.Shtivel' and Z.S.Smirnova, Kinet. Catal., 15,
      1247 (1974).

639   C.A.Tyson and A.E.Martell, J.Amer. Chem. Soc., 94, 939 (1972).

640   T.R.Demmin, M.D.Swerdloff and M.M.Rogic, J.Amer. Chem. Soc., 103,
      5795 (1981).

641   G.Speier and Z.Tyeklar, J.Mol. Catal., 9, 233 (1980).

642   G.S.Vigee and E.E.Eduok, J.Inorg. Nucl. Chem., 43, 2171 (1981).

643   J.Tsuji, H.Takayanagi and I.Sakai, Tetrahedron Lett., 1245 (1975),
      1365 (1976), J.Amer. Chem. Soc., 96, 7349 (1974).

644   M.M.Rogic, T.R.Demmin and W.B.Hammond, J.Amer. Chem. Soc., 98, 7441
      (1976).

645   N.Oishi, Y.Nishida, K.Ida and S.Kida, Bull. Chem. Soc. Jpn., 53, 2847 (1980).

646   M.Okamoto, N.Oishi, Y.Nishida and S.Kida, Inorg. Chim. Acta, 64, L217 (1982).

647   U.Casellato, S.Tamburini, P.A.Vigato, A. de Stefani, M.Vidali and D.E.Fenton, Inorg. Chim. Acta, 69, 45 (1983).

648   A.B.P.Lever, B.S.Ramaswamy and S.R.Pickens, Inorg. Chim. Acta, 46, L59 (1980).

649   A.J.Pandell, J.Org. Chem., 48, 3908 (1983).

650   K.Ohkubo, K.Miyata and S.Sakaki, J.Mol. Catal., 17, 85 (1982).

651   M.G.Weller and U.Weser, J.Amer. Chem. Soc., 104, 3752 (1982).

652   T.Funabiki, H.Sakamoto, S.Yoshida and K.Tarama, J.Chem. Soc., Chem. Commun., 754 (1979).

653   M.Matsumoto and K.Kuroda, J.Amer. Chem. Soc., 104, 1433 (1982).

654   Y.Tsuji, T.Ohta, T.Ido, H.Minbu and Y.Watanabe, J.Organomet. Chem., 270, 333 (1984).

655   U.Casellato, S.Tamburini, P.A.Vigato, M.Vidali and D.E.Fenton, Inorg. Chim. Acta, 84, 101 (1984).

656   K.Tatsuno, M.Tatsuda, S.Otsuka and K.Tani, Chem. Lett., 1209 (1984).

657   S.Nemeth, A.Fulep-Poszmik and L.I.Simándi, Acta Chim. Acad. Sci. Hung., 110, 461 (1982).

658   H.Sakamoto, T.Funabiki, S.Yoshida and K.Tarama, Bull. Chem. Soc. Jpn., 52, 2760 (1979).

659   A.McKillop and S.J.Ray, Synthesis, 847 (1977).

660   G.R.Howe and R.R.Hiatt, J.Org. Chem., 35, 4007 (1970).

661   S.P.Srivastava and V.K.Gupta, Oxidn. Commun., 2, 19 (1981).

662   E.G.Derouane, J.N.Braham and R.Hubin, J.Catal., 35, 196 (1974).

663   S.Nemeth and L.I.Simándi, J.Mol.Catal., 14, 87,241 (1982), Inorg. Chim. Acta, 64, L21 (1982).

664   T.Kajimoto, H.Takahashi and J.Tsuji, J.Org. Chem., 41, 1389 (1976).

665   J.L.Russell and J.Kollar, U.S.Patent, 3,960,954 (1976); Chem. Abs.,
      85, 108354n (1976).

666   P.S.Radhakrishnamurti and L.D.Sarangi, Ind. J.Chem., 21A, 132 (1982).

667   M.A.Rao, B.Sethuram and T.N.Rao, J.Ind. Chem. Soc., 59, 1040 (1982).

668   S.C.Pati and R.C.Mahaptro, Ind. J.Chem., 19A, 1126 (1980).

669   S.Cenini, F.Porta and M.Pizzotti, J.Mol. Catal., 15, 297 (1982).

670   T.Isshiki, T.Tomita and O.Aoki, Jpn. Kokai Tokkyo Koho, 79,106,455
      (1979); Chem. Abs., 92, 22261n (1980).

671   J.S.McConaghy and D.C.Owsley, U.S.Patent, 3,940,429 (1976); Chem. Abs.,
      84, 151229s (1976).

672   K.Ogawa, Y.Nomura, Y.Takeuchi and S.Tomoda, J.Chem. Soc., Perkin I,
      3031 (1982).

673   J.E.McKeon and D.J.Trecker, U.S.Patent, 3,634,346 (1972); Chem. Abs.,
      76, 72397b (1972).

674   T.Kajimoto, H.Takahashi and J.Tsuji, Bull. Chem. Soc. Jpn., 55,
      3673 (1982).

675   V.D.Sen', V.A.Golubev and N.N.Efremova, Bull. Acad. Sci. U.S.S.R.,
      Div. Chem. Sci., 31, 53 (1982).

676   M.N.Sheng and J.G.Zajacek, J.Org. Chem., 33, 588 (1968).

677   D.P.Riley, J.Chem. Soc., Chem. Commun., 1530 (1983).

678   F.List and L.Kuhnen, Erdoel Kohle, Erdgas, Petrochem., 20, 192 (1967);
      Chem. Abs., 67, 43126w (1967).

679   S.Otsuka, A.Nakamura and Y.Tatsuno, J.Chem. Soc., Chem. Commun.,
      836 (1967), J.Amer. Chem. Soc., 91, 6994 (1969).

680   S.Takahashi, K.Sonogashira and N.Hagihara, Nippon Kagaku Zasshi,
      87, 610 (1966); Chem. Abs., 65, 14885d (1966).

681   R.Ugo, Coord. Chem. Rev., 3, 319 (1968).

682   N.A.Johnson and E.S.Gould, J.Amer. Chem. Soc., 95, 5198 (1973),

       J.Org. Chem., 39, 407 (1974).

683   G.A.Tolstikov, U.M.Jemilev, V.P.Jurjev,F.B.Gershanov and S.R.Rafikov,

       Tetrahedron Lett., 2807 (1971).

684   R.L.Jacobs, J.Org. Chem., 42, 571 (1977).

685   J.Tsuji, S.Hayakawa and H.Takayanagi, Chem. Lett., 437 (1975).

686   J.Tsuji, H.Takahashi and T.Kajimoto, Tetrahedron Lett., 4573 (1973).

687   J.Tsuji, H.Kesuka, Y.Toshida, H.Takayanagi and K.Yamamoto, Tetrahedron,

       39, 3279 (1983).

688   A.Inada, Y.Nakamura and Y.Morita, Chem. Pharm. Bull., 30, 1041 (1982).

689   K.Kaneda, T.Itoh, N.Kii, K.Jitsukawa and S.Teranishi, J.Mol. Catal.,15,349

690   V. Van Rheenen, J.Chem. Soc., Chem. Commun., 314 (1969).

691   T.Itoh, K.Kaneda and S.Teranishi, Tetrahedron Lett., 2801 (1975).

692   J.Tsuji, H.Kezuka, H.Takayanagi and K.Yamamoto, Bull. Chem. Soc. Jpn.,

       54, 2369 (1981).

693   A.Nishinaga, H.Ohara, H.Tomita and T.Matsuura, Tetrahedron Lett.,

       24, 213 (1983).

694   H.Yukimasa, H.Sawai and T.Takizawa, Chem. Pharm.Bull., 27, 551 (1979).

695   G.Wilke, H.Schott and P.Heimbach, Angew. Chem. Int. Ed. Engl., 6,

       92 (1967).

696   E.Stern, Transition Metals in Homogeneous Catalysis, Marcel Dekker

       Inc., New York, 1971, page 138.

697   J.D.Curry, U.S.Patent, 3,760,000 (1973); Chem. Abs., 79, 126609e

       (1973);

698   N.Hagiwara, N.Takahashi and H.Kojima, Japan. Pat., 74 24,900 (1974);

       Chem. Abs., 82, 86404n (1975).

699   R.K.Poddar and U.Agarwala, Inorg. Nucl. Chem. Lett., 9, 785 (1973).

700   H.Arai and J.Halpern, J.Chem. Soc., Chem. Commun., 1571 (1971).

701   W.R.Cullen, B.R.James and G.Strukul, Inorg. Chem., 17, 484 (1978).

702   M.K.Dickson, N.S.Dixit and D.M.Roundhill, Inorg. Chem., 22, 3130 (1983).

703   K.M.Nicholas, J.Organomet. Chem., 188, C10 (1980).

704   V.J.Choy and C.J.O'Connor, J.Chem. Soc., Dalton, 2017 (1972).

705   B.W.Graham, K.R.Laing, C.J.O'Connor and W.R.Roper, J.Chem. Soc.,
       Dalton, 1237 (1972).

706   M.M.Taqui Khan, R.K.Andal and P.T.Manoharan, J.Chem. Soc., Chem.
       Commun., 561 (1971).

707   N.Sutin and J.K.Yandell, J.Amer. Chem. Soc., 95, 4847 (1973).

708   I.Ondrejkovicová, V.Vančová and G.Ondrejovič, Coll. Czech. Chem.
       Commun., 48, 254 (1983).

709   D.D.Schmidt and J.T.Yoke, J.Amer. Chem. Soc., 93, 637 (1971).

710   J.Drapier and A.J.Hubert, J.Organomet. Chem., 64, 385 (1974).

711   J.Halpern, B.L.Goodall, G.P.Khare, H.S.Lim and J.J.Pluth, J.Amer.
       Chem. Soc., 97, 2301 (1975).

712   B.S.Tovrog, S.E.Diamond and F.Mares, J.Amer. Chem. Soc., 101, 270
       (1979).

713   R.Barral, B.Bocard, I.S. de Roch and L.Sajus, Tetrahedron Lett.,
       1693 (1972), Kinet. Catal., 14, 164 (1973).

714   V.P.Fedin and I.P.Romanova, J.Gen. Chem. U.S.S.R., 52, 1105 (1982).

715   T,Nakaya, H.Arabori and M.Imoto, Bull. Chem. Soc. Jpn., 43, 1888 (1970).

716   B.W.Brooks and R.M.Smith, Chem. Ind., 326 (1973).

717   A.Hanaki and H.Kamide, Bull. Chem. Soc. Jpn., 56, 2065 (1983).

718   F.Duke and V.C.Bulgrin, J.Phys. Chem., 79, 2323 (1975).

719   G.L.Bridgart, M.W.Fuller and I.R.Wilson, J.Chem. Soc., Dalton,
       1274 (1973).

720   J.Zwart, J.H.M.C. van Wolput and D.C.Koningsberger, J.Mol. Catal.,
       12, 85 (1981).

721   Y.Murakami, Y.Aoyama and S.Nakanishi, Chem. Lett., 857 (1976).

722   I.G.Dance and R.C.Conrad, Aust. J.Chem., 30, 305 (1977).

723   W.M.Bronwer, P.Piet and A.L.German, J.Mol. Catal., 22, 297 (1984).

724   M.A.Ledlie, K.G.Allum, I.V.Howell and R.C.Pitkethly, J.Chem. Soc.,
      Perkin I, 1734 (1976).

725   D.P.Riley and R.E.Shumate, J.Amer. Chem. Soc., 106, 3179 (1984),
      Inorg. Chem., 22, 1965 (1983).

726   A.V.Mashkina, G.V.Varnakova, L.M.Zagryatskaya, Z.A.Suleimanova,
      R.M.Masagutov, A.Kh.Sharipov, V.N.Yakovleva, L.V.Vlasova and
      N.P.Kirik, Kinet. Catal., 22, 454 (1981).

727   Y.Watanabe, T.Numata and S.Oae, Bull. Chem. Soc. Jpn., 55, 1915
      (1982).

728   F.DiFuria, G.Modena, R.Curci and J.O.Edwards, Gazz. Chim. Ital.,
      109, 571 (1979), 110, 487 (1980), Rec., 98, 181 (1979).

729   R.Curci, F.DiFuria and G.Modena, J.Chem. Soc., Perkin II, 576 (1977).

730   O.Bortolini, C.Campello, F.DiFuria and G.Modena, J.Mol. Catal.,
      14, 53, 63 (1982).

731   O.Bortolini, F.DiFuria and G.Modena, J.Mol. Catal., 16, 61, 69 (1982).

732   S.Cenci, F.DiFuria, G.Modena, R.Curci and J.O.Edwards, J.Chem. Soc.,
      Perkin II, 979 (1978).

733   R.Curci, F.DiFuria, R.Testi and G.Modena, J.Chem. Soc., Perkin II,
      752 (1974).

734   P.Pitchen and H.B.Kagan, Tetrahedron Lett., 25, 1049 (1984).

735   F.DiFuria, G.Modena and R.Seraglia, Synthesis, 325 (1984).

736   A.Arcoria, F.P.Ballistreri, G.A.Tomaselli, F.DiFuria and G.Modena,
      J.Mol. Catal., 18, 177 (1983).

737   S.Oae, Y.Watanabe and K.Fujimori, Tetrahedron Lett., 1189 (1982).

738   W.Ando, R.Tajima and T.Takata, Tetrahedron Lett., 23, 1685 (1982).

739   P.Muller and J.Godoy, <u>Helv. Chim. Acta</u>, <u>66</u>, 1790 (1983).

740   P.S.Radhakrisnamurti and N.C.Sahu, <u>Kinet. Catal.</u>, <u>22</u>, 471 (1981).

741   F.Gasparrini, M.Giovannoli, D.Misiti, G.Natile and G.Palmieri,

      <u>Tetrahedron</u>, <u>39</u>, 3181 (1983), <u>40</u>, 165 (1984).

7 Reactions of the carbonyl group

7.1   Reduction

7.2   Reduction by Hydrosilylation

    7.2.1   Saturated Aldehydes and Ketones

    7.2.2   Acyl Halides

    7.2.3   Carbon Dioxide

    7.2.4   αβ-Unsaturated Carbonyl Compounds

    7.2.5   Hydrosilylation of Carbon Nitrogen Double Bonds

    7.2.6   Asymmetric Hydrosilylation

7.3   Attack of Organometallic Nucleophiles on Carbonyl Groups

    7.3.1   Acyl Halides

    7.3.2   αβ-Unsaturated Carbonyl Compounds

    7.3.3   Other Substrates

7.4   Metal Catalysed Hydrolysis and Interconversion of Carbonyl Derivatives

    7.4.1   Ester Hydrolysis

    7.4.2   Esterification and Transesterification

    7.4.3   Reactions of Thioesters

    7.4.4   Hydrolysis and Formation of Amides and Peptides

    7.4.5   Reactions of Amines with Aldehydes and Ketones

    7.4.6   Conversion of Nitriles to Amides

7.1   Reduction

The reduction of carbonyl compounds to alkanes or alcohols is of major synthetic importance but rather few of the reactions involved are catalytic. Of these the major is hydrogenation, discussed in Chapter 2, involving hetero- and homogeneous processes on a laboratory and an industrial scale. The most common laboratory reduction involves complex metal hydrides such as NaBH$_4$ and LiAlH$_4$. These may be modified by transition metal salts, but these are normally added in stoichiometric quantities, generating a new, if unisolated, reducing agent. For example, amides are normally inert to sodium borohydride, but in the presence of CoCl$_2^1$ or a liquid Lewis acid such as SnCl$_4^2$ a high yield reduction to an amine is achieved.

The reduction of acyl halides to aldehydes has been effected under neutral conditions with (PPh$_3$)$_2$CuBH$_4$, $\underline{1}$, in the presence of PPh$_3$.[3] (PPh$_3$)$_3$CuCl may be recovered quantitatively[4] and converted back to $\underline{1}$ with NaBH$_4$. Under acid conditions aldehydes and ketones are reduced to the corresponding alcohols in 80-100% yield; 4-tert-butylcyclohexanone is reduced with 94% selectivity for the equatorial alcohol.[5] The reaction is not strictly catalytic, but the more expensive material may be recycled, and the process competes favourably with the Rosemund[6] reduction in terms of convenience and selectivity. Complex reducing agents derived from sodium hydride have also been developed. In the presence of catalytic (~10%) amounts of tert-sodium amylate, Ni(OAc)$_2$ and MgBr$_2$, NaH reduces ketones in 90% - 100% yield. 4-tert-butylcyclohexanone is reduced to the alcohol in 95% yield with 75:25 trans-selectivity.[7]

The use of cob(I)alamin as a catalyst for a variety of reductions has been studied by Fischli. Aliphatic nitriles such as $\underline{2}$ may be reduced to secondary amines with good selectivity, the reaction proceeding via an imine.[8] With excess zinc a primary amine is obtained as in the conversion of $\underline{4}$ to $\underline{5}$.[9] Alternatively, in aqueous acid $\underline{2}$ is converted to the corresponding aldehyde in 85% yield. $\alpha\beta$-unsaturated carboxylate esters and amides are reduced to the saturated compounds under the same conditions in about 80% yield.[10] When the substrate is prochiral (as $\underline{6}$), the S-product predominates to the extent of 20%. The analogous E-ester gives a racemic product. The origin of stereoselectivity is not clear, but the reaction is thought to proceed via attack of a nucleophilic Co orbital, perpendicular to the ring system, on the α,β-unsaturated carbonyl, followed by electron transfer from Zn/HOAc.[11]

CN $\xrightarrow[\text{Cob(I)alamin}]{\text{Zn/HOAc}}$ NH$_2$

2

3   81%

CN $\xrightarrow[\text{Cob(I)alamin}]{\text{Zn/HOAc}}$ NH$_2$

4

5   75%

COX $\xrightarrow[\text{Cob(I)alamin}]{\text{Zn/HOAc}}$ H   COX

6a X = OEt

6b X = NH$_2$

7

~20% S

Other reports of catalysed reductions are sparse. Butanal is reduced by formic acid in acetic acid in the presence of IrH$_3$(PPh$_3$)$_3$, complete conversion to butanol and butyl acetate being achieved at 50° in 1 hour.[12] Aldehydes and ketones are reduced by carbon monoxide at 100° in the presence of Et$_3$N, water and catalytic HFe(CO)$_4^{\ominus}$.[13] The reaction is suggested to proceed via attack of iron on the carbonyl group to give 8, protonation and loss of Fe(CO)$_4$. Fe(CO)$_4$ is converted back to HFe(CO)$_4^{\ominus}$ by reaction with CO and OH$^{\ominus}$. The reduction of Schiff's bases to secondary amines by EtMgBr is achieved in excellent yield in the presence of NiBr$_2$; the true catalyst is presumed to be a Ni(0) complex.[14] The presence of FeCl$_3$[15] or bis(cyclopentadienyl)titanium dichloride[16] alters the course of reaction of Grignard reagents with carbonyl compounds from addition to

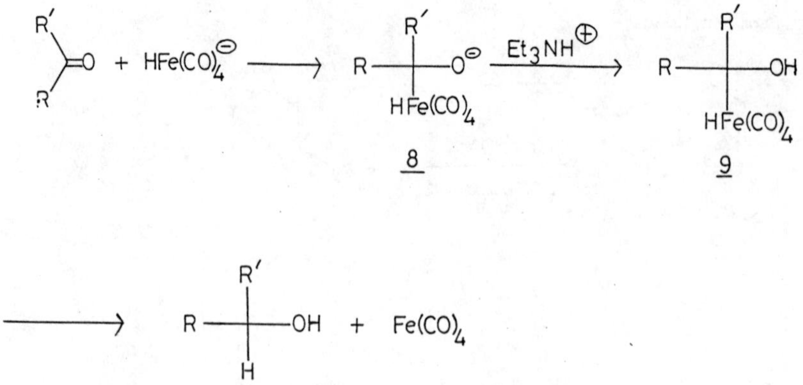

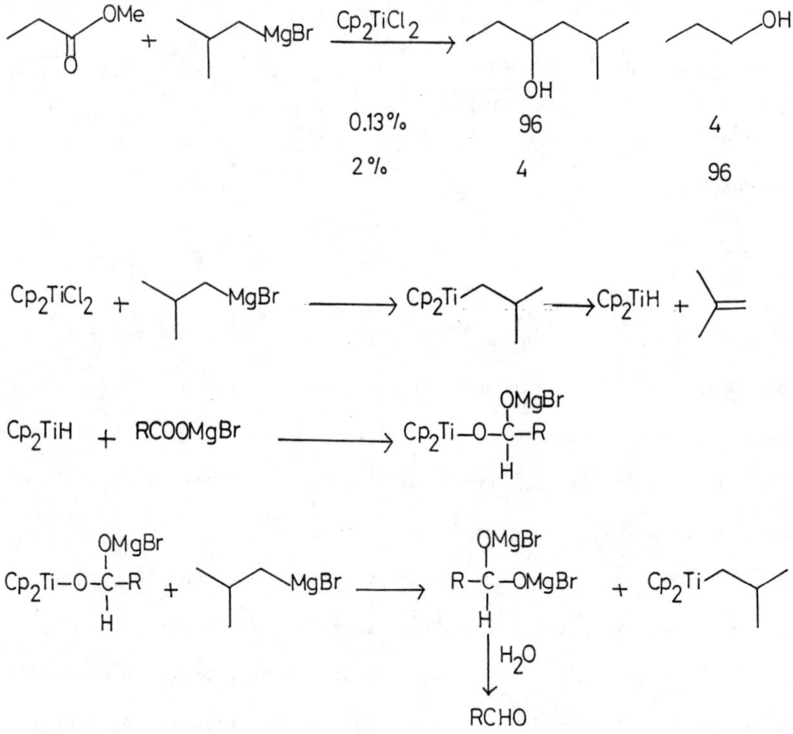

Figure 1 Carbonyl reduction by Grignard reagents catalysed by Cp_2TiCl_2.

reduction. For example, n-propylmagnesium bromide reacts with acetone in 94% total yield in the presence of Cp_2TiCl_2 to give isopropanol and 2-methyl-2-pentanol in the ratio 83:17. In this case the reducing agent is thought to be Cp_2TiH formed by elimination of alkene from Cp_2TiR. When the substrate is an ester reduction to either a primary or a secondary alcohol may be effected, depending on the mole % of catalyst.[17] (Figure 1) Total yields are always in excess of 90%. Carboxylic acids are analogously reduced by 2 moles of Grignard reagent to the corresponding aldehyde which is obtained in 50–70% yield.[18] Acyl halides are converted to a mixture of aldehydes and esters on treatment with Bu_3SnH. In the presence of $(PPh_3)_4Pd$ the reaction is accelerated and only aldehydes are obtained. Benzoyl chloride is reduced in 95% and heptanoyl chloride in 81% yield.[19] The reaction conditions are both particularly mild and neutral and neither double bond migration nor overreduction occur.

7.2   Reduction by Hydrosilylation

7.2.1   Saturated Aldehydes and Ketones

The addition of a hydrosilane to a carbon-oxygen double bond gives an O-silylated alcohol. The silicon oxygen bond is easily hydrolysed, and the overall reaction is equivalent to hydrogenation (Figure 2). It is an attractive alternative, since the conditions are milder and may be adjusted to tolerate a variety of functional groups. The catalytic hydro-silylation of carbonyl compounds was first observed by Petrov[20] and Calas,[21] using $ZnCl_2$, nickel metal and H_2PtCl_6 as catalysts. The conditions were severe, leading to side reactions and isomerisation of both reactants and products. The disproportionation of polyhydrosilanes also occurred under these conditions,[22] restricting the use of these catalysts to addition of monohydrosilanes.

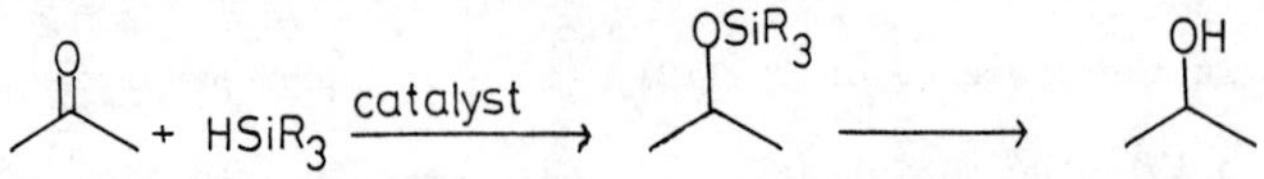

Figure 2 Hydrosilylation of a carbon-oxygen double bond.

A range of catalysts is now available for the hydrosilylation of
ketones in good yield at room temperature. Some examples are given in
Figure 3. A number of general observations may be made. Rh(I) complexes
are the most active catalysts; Ru(II) complexes require higher temperatures
and there has been little work on simple platinum based systems. Aliphatic
aldehydes and ketones are reduced more rapidly than aromatic compounds,[27]
and electron-donating groups on the aromatic ring show little effect.[28]
For unhindered cyclohexanones the attack of Rh(I) is from an axial
direction, as predicted by orbital-distortion arguments, but with hindered
substrates equatorial attack predominates.[23]

There is little evidence as to the nature of the intermediates in
the reaction pathway. The first step is the oxidative addition of the
silane to the Rh(I) complex as in the reaction with alkenes and
$Rh(PPh_3)_2H(SiEt_3)Cl$ has been isolated.[26] Indirect support for the
intermediacy of an α-siloxyalkyl-rhodium complex such as 10 comes from a
study of the reaction between acetophenone and α-naphthylphenylsilane in the
presence of a Rh(I)DIOP catalyst. The process was monitored by e.s.r. with
nitrosodurene as a spin trap. The presence of α-naphthyl SiH(NO·)Ar was
shown to derive from a heteroatom stabilised alkyl rhodium intermediate.[28]

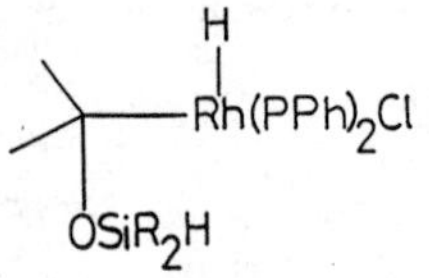

10

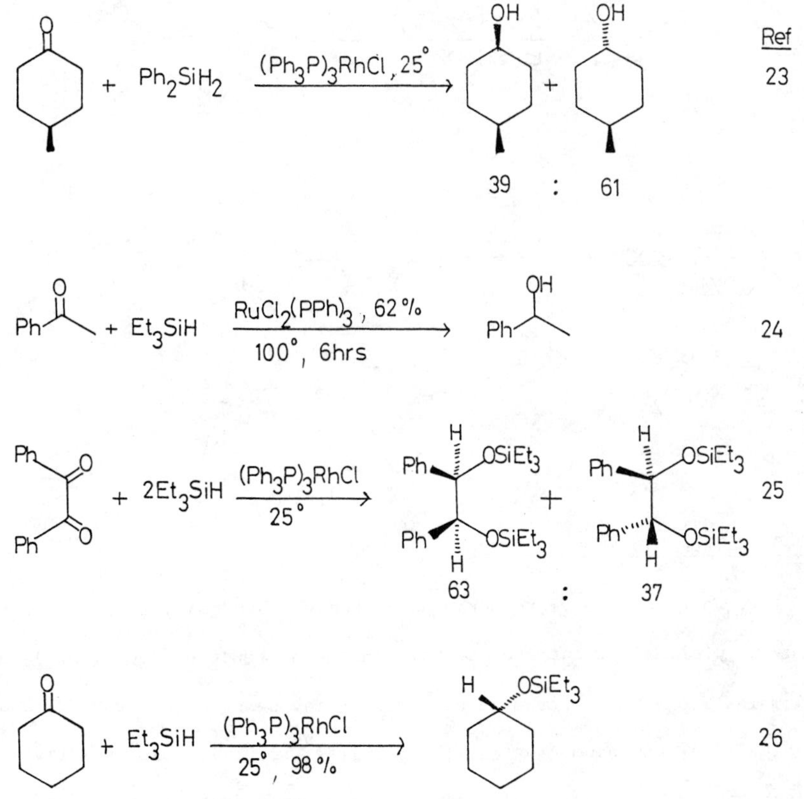

Figure 3 Hydrosilylation of ketones.

7.2.2 Acyl Halides

Hydrosilylation of acyl halides provides a further method for reduction to aldehydes or in some cases the synthesis of ketones. Platinum catalysts are active, but most work has been done with Rh(I) complexes.[29] Aliphatic acyl halides are reduced to aldehydes in good yield in the presence of trans-RhClCO(PEtPh$_2$)$_2$ (e.g. 11 → 12). Aromatic substrates may yield aldehydes or ketones according to substitution. Thus from 13a the ketone 15a is obtained in 63% yield with 2% 14a while from 13b, 14b is the sole product in 66% yield.[30]

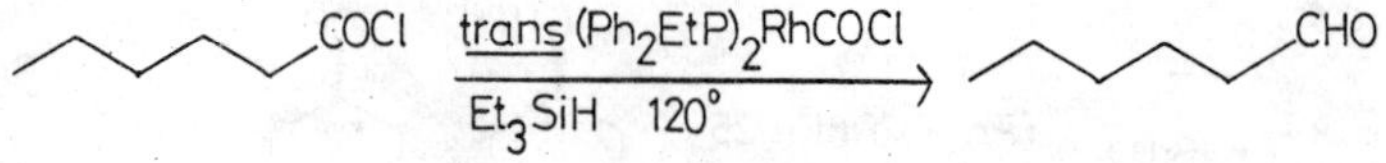

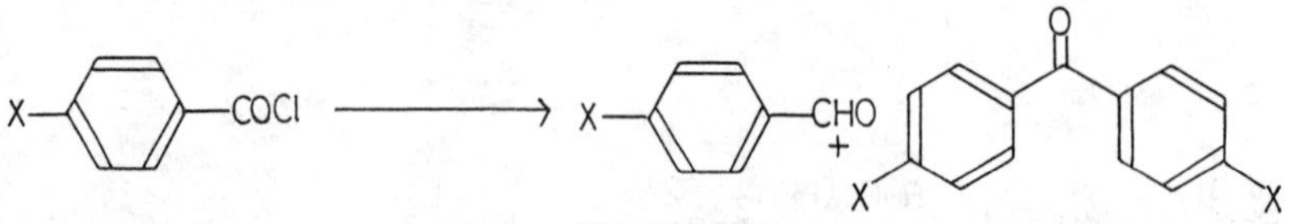

11 ⟶ 12

(reaction: 11 + trans (Ph$_2$EtP)$_2$RhCOCl / Et$_3$SiH, 120° ⟶ 12)

13a X = OMe
13b X = NO$_2$

14

15

7.2.3 Carbon Dioxide

Two reports have recently described the hydrosilyation of carbon dioxide to yield silyl formate esters. In the presence of $RuCl_2(PPh_3)_2$ up to 14 moles $HCO_2SiMeEt_2$ per mole of catalyst may be obtained from CO_2 and Et_2MeSiH.[31] A ruthenium cluster, $[HRu_3(CO)_{11}]^{\ominus}$, 16, is also an active catalyst and evidence for the catalytic cycle comes from isolation of $[H_3Ru_4(CO)_{12}]^{\ominus}$, 17, and IR characterisation of $[HRu_3(CO)_{10}(SiEt_3)_2]^{\ominus}$, 18.[32]

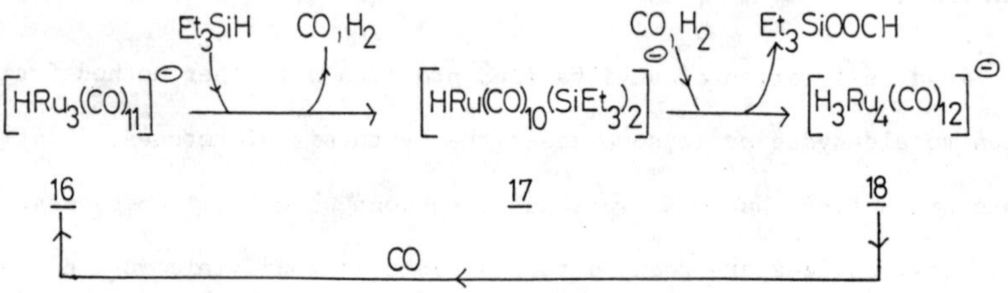

$[HRu_3(CO)_{11}]^{\ominus}$ $\xrightarrow[\text{CO,H}_2]{\text{Et}_3\text{SiH}}$ $[HRu(CO)_{10}(SiEt_3)_2]$ $\xrightarrow[\text{Et}_3\text{SiOOCH}]{\text{CO,H}_2}$ $[H_3Ru_4(CO)_{12}]^{\ominus}$

16   17   18

(⟵ CO)

7.2.4 αβ-Unsaturated Carbonyl Compounds

The hydrosilylation of enones may proceed either by 1,2- or
1,4-addition to give, after hydrolysis, an allyl alcohol or a saturated
ketone (Figure 4). 1,2-addition is comparatively rare; tetracyclone is
reduced specifically in this way, presumably for steric reasons.[33] In the
presence of Ni(cod)$_2$, crotonaldehyde, 19, gives a mixture of cis- and
trans-allyl silyl ethers, 20 and 21, in 40% yield. When PPh$_3$ is added to
the reaction, the yield rises to 85%; Ni(P(OEt)$_3$)$_4$ is also a useful
catalyst.[34]

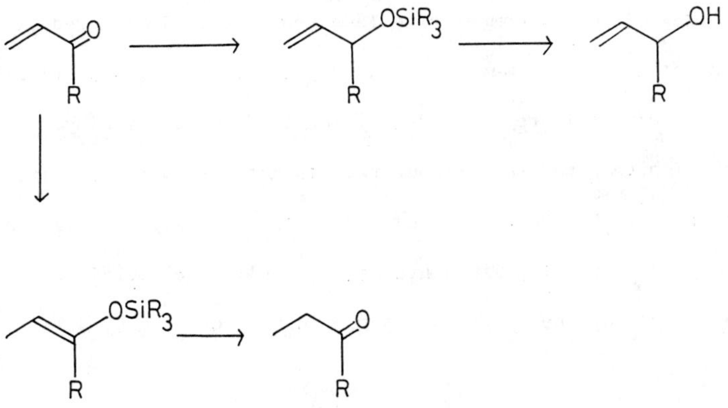

Figure 4 Routes for hydrosilylation of enones.

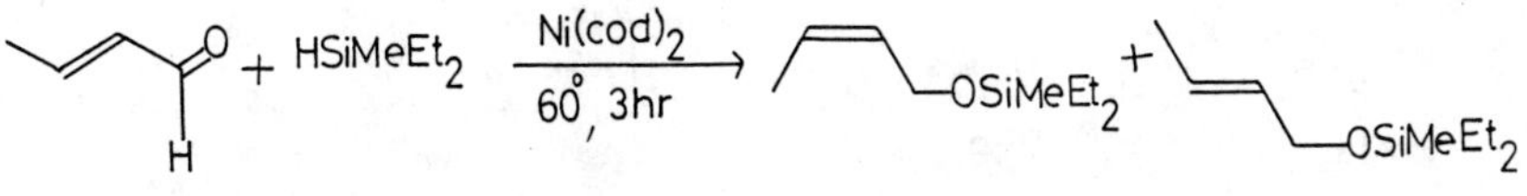

19   20   21

The 1,4-addition of hydrosilanes to αβ-unsaturated carbonyl compounds provides a useful method for the specific reduction of conjugated double bonds. Thus citral, 22, is reduced to citronellal, 23, in 96% yield; hydrogenation gives a product which is extensively decarbonylated.[35]

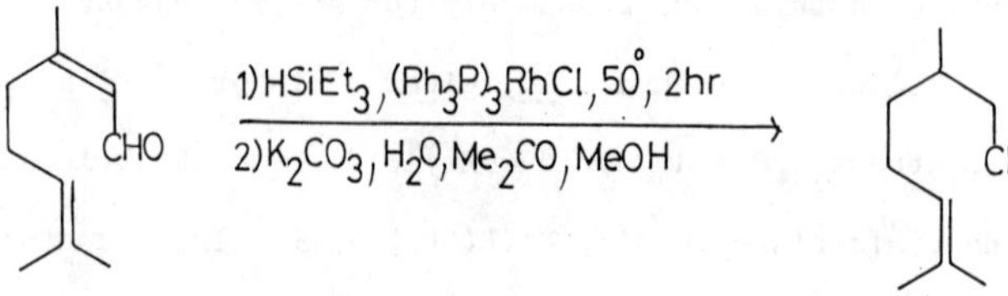

1,2-addition is competitive in a few cases. In the reduction of β-ionone, 24, the reaction may be tuned to give 1,2- or 1,4-addition by choice of silane[35] (Figure 5) 1,2-addition is favoured by dihydrosilanes and 1,4- by high temperatures,[25] but this is not easily rationalised in mechanistic terms. For very hindered enones Rh(I) catalysis fails, and an alternative procedure using TiCl$_4$ has been useful. The steroid 27 is reduced in 81% yield, the 5β product predominating over 5α by 3:1.[36]

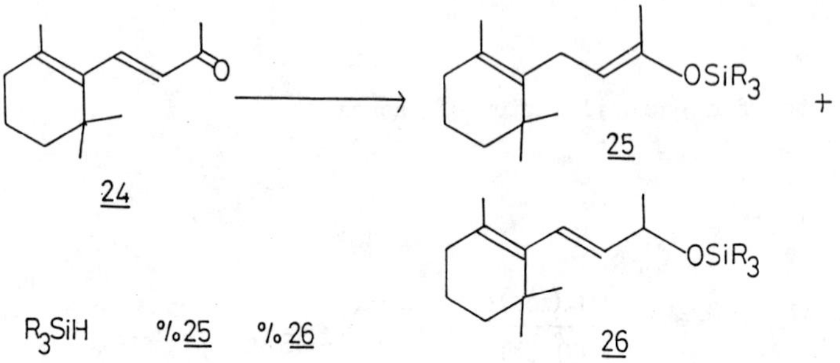

R$_3$SiH	%25	%26
PhMe$_2$SiH	91	9
Et$_3$SiH	44	56
Et$_2$SiH$_2$	0	100

Figure 5   1,2- and 1,4-addition of hydrosilanes to β-ionone.

27 → 28

1) $Me_3SiH/TiCl_4/25°/CH_2Cl_2$
2) $K_2CO_3/H_2O/Me_2CO$

Hydrosilylation of $\alpha\beta$-unsaturated esters gives β-silyl esters by addition to the carbon-carbon double bond, but 1,4-addition gives the more interesting mixed ketene acetals such as 31. These may be converted to the saturated esters in ethanol,[37] dimerised to succinates in the presence of $TiCl_4$,[38] converted to β-lactams,[39] or homologated with dichlorocarbene. Selectivity is variable; with ethyl acrylate only the β-silyl ester is obtained with Me_2PhSiH or Me_2EtSiH. Both α- and β-methyl substitution are found to enhance the 1,4-addition. The details of the mechanism are unknown either for Rh(I)[37] or Pt(II)[41] catalysts, but in the case of Rh(I) a radical mechanism is indicated since the reaction is suppressed in the presence of a radical trap such as the galvinoxyl radical.[*]

$\nearrow$COOEt + Et_3SiH $\xrightarrow{(Ph_3P)_3RhCl}$ $Et_3Si$$\nearrow$COOEt +

29   30

OEt
$\diagup$ OSiEt$_3$

31

[*] The galvinoxyl radical is the 2,6-di-t-butyl-α-(3,5-di-t-butyl-4-oxo-2,5-cyclohexadiene-1-ylidene)-p-tolyloxy radical.

7.2.5 Hydrosilylation of Carbon–Nitrogen Double Bonds

Addition of hydrosilanes to C–N multiple bonds has been studied for imines, isocyanates, and diimides. For imines $ZnCl_2$[42] and H_2PtCl_6 give unselective reactions, but Rh(I) and Pd(II) complexes give more satisfactory results.[43] The product is an N-silylated amine which may be hydrolysed to a secondary amine or converted directly to an amide (Figure 6). Yields are excellent, generally in excess of 90%. The regiochemistry of the hydrosilylation of isocyanates is complex. For an aryl substrate N-silylated formamides 33 are obtained while alkyl isocyanates give acyl silanes 34. Diimides give monosilylated products with $PdCl_2$ or $(PPh_3)_3RhCl$ as catalyst, also in excellent yields.[45] As before, these products may be hydrolysed or acylated.

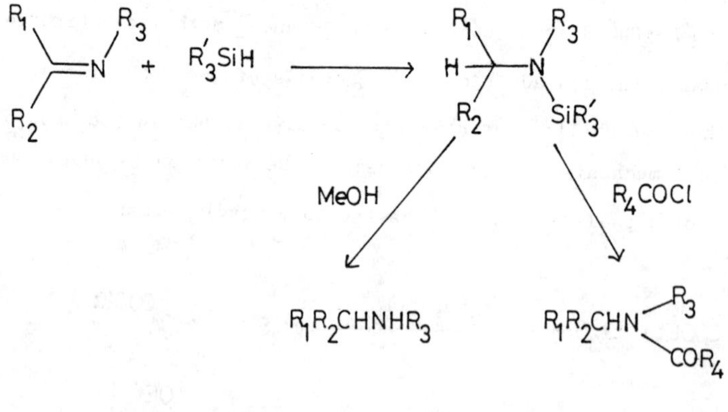

Figure 6 Hydrosilylation of an imide.

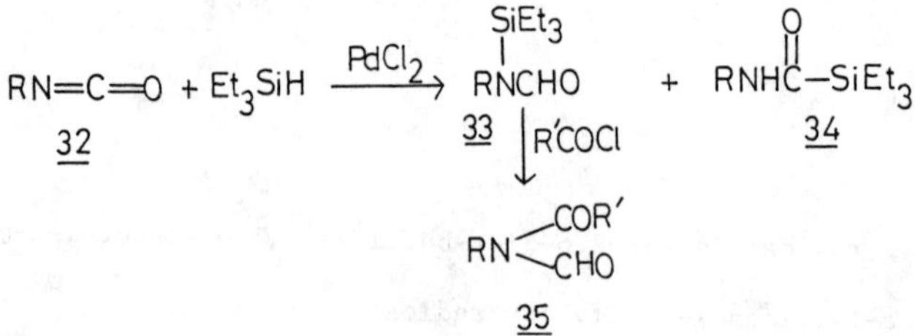

7.2.6 Asymmetric Hydrosilylation

Hydrosilylation is an effective means for the asymmetric reduction of carbon-oxygen and carbon-nitrogen double bonds. Although the high enantiomer excesses obtained routinely with complex chiral hydride reducing agents[46] are lacking, the process is rather milder. Much of the most interesting recent work on ketone hydrosilylation, both synthetic and mechanistic, has been in this area.

The first asymmetric hydrosilylation of prochiral ketones was effected by using a chiral platinum complex of S-BzMePhP.[47] Thus 36 reacted with Cl_2MeSiH to give, after hydrolysis, the alcohol, 37, in 19% optical yield. However, the reaction was limited to Cl_2MeSiH and the chemical yield was poor with bulky ketones. Dialkyl ketones were not suitable substrates, giving extensive side reactions.

36   37

Wilkinson's catalyst and cationic complexes of the type $RhL_2H_2(solvent)_2^+$ are active for the hydrosilylation of ketones and many chiral analogues have been developed. Some examples are given in Figure 7. With cationic complexes as catalysts a marked dependence of enantiomer excess on the structure of the hydrosilane is observed. Bulky silanes are more effective; Kagan found α-naphthylphenylsilane to be the most useful of those examined with a Rh(DIOP) complex.[54] A more wide ranging study by Kagan and his co-workers, where silane, substrate and phosphine were systematically varied suggested that electronic effects involving charge transfer reactions might be important but decisive conclusions were not reached.[55] Chelating biphosphines frequently do not give better results

Ref

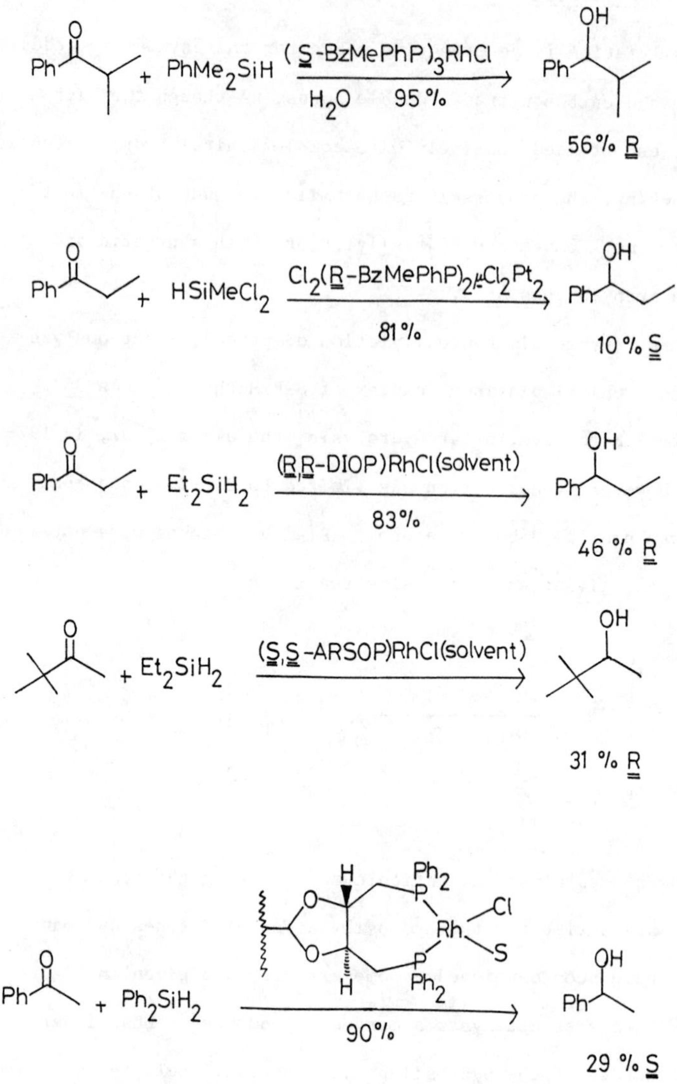

Ph—C(=O)—CH(CH₃)... + PhMe₂SiH $\xrightarrow[\text{H}_2\text{O} \quad 95\%]{(\underline{S}\text{-BzMePhP})_3\text{RhCl}}$ Ph—CH(OH)—CH(CH₃)₂

56% R

48

Ph—C(=O)—CH₂CH₃ + HSiMeCl₂ $\xrightarrow[81\%]{Cl_2(\underline{R}\text{-BzMePhP})_2\mu Cl_2 Pt_2}$ Ph—CH(OH)—CH₂CH₃

10% S

49

Ph—C(=O)—CH₂CH₃ + Et₂SiH₂ $\xrightarrow[83\%]{(\underline{R},\underline{R}\text{-DIOP})\text{RhCl(solvent)}}$ Ph—CH(OH)—CH₂CH₃

46% R

50

(CH₃)₃C—C(=O)—CH₃ + Et₂SiH₂ $\xrightarrow{(\underline{S},\underline{S}\text{-ARSOP})\text{RhCl(solvent)}}$ (CH₃)₃C—CH(OH)—CH₃

31% R

51

Ph—C(=O)—CH₃ + Ph₂SiH₂ $\xrightarrow{90\%}$ Ph—CH(OH)—CH₃

29% S

52

Figure 7 Asymmetric hydrosilylation of ketones.

$$Ph \overset{O}{\underset{}{\parallel}} + \alpha NpPhSiH_2 \xrightarrow[\text{65\%}]{} Ph \overset{OH}{\underset{}{}} \qquad 53$$

65 % enantiomer excess

configuration not specified

Figure 7   cntd.

than monophosphines, and kinetic studies suggest that for DIOP at least the ligand may be acting in a monodentate fashion at critical points in the catalytic cycle.[56]

The best optical yields in the reduction of carbonyl compounds are achieved with substrates bearing an additional polar group which may coordinate to the metal, stabilising one diastereomer in the product determining transition state.   The reduction of α-keto esters, typically alkyl pyruvates and phenyl glyoxylates, has been catalysed by Rh(I) complexes of benzylmethylphenyl phosphine (BMPP), DIOP, and (2$\underline{S}$, 4$\underline{S}$)-N-butoxycarbonyl-4-diphenylphosphino-2-diphenylphosphinomethylpyrrolidine (BPPM) as chiral ligands.   Some results are shown in Table 1.

With β-keto esters optical yields obtained in reduction are similar to those for simple ketones (Table 1).   This suggests that the ester is not exerting any directing effect though it has also been proposed that the low induction may be attributed to a transfer hydrogenation process within an enol form of the substrate coordinated through a C=C double bond.   The asymmetric hydrosilylation of levulinates followed by hydrolysis gives optically active 4-methyl-γ-butyrolactone, 40, a useful

Table 1. Asymmetric Hydrosilylation of Keto-esters

α-Keto ester	Silane	Phosphine	Chemical Yield	ee	Configuration	Ref.
CH_3COCO_2nPr	Ph_2SiH_2	R BMPP[a]	86	60	R	57
	$PhNpSiH_2$	RR DIOP	90	85	R	57
	Ph_2SiH_2	RR BPPM	78	79	R	58
CH_3COCO_2iBu	$\alpha NpPhSiH_2$	SS DIOP	84	72	S	58
$PhCOCO_2Et$	Et_2SiH_2	R BMPP[a]	80	6	S	57
	$\alpha NpPhSiH_2$	SS DIOP	87	39	S	57
$CH_3COCH_2CO_2Me$	$\alpha NpPhSiH_2$	RR DIOP		21	R	57
$CH_3COCH_2CO_2Et$	"	"		26	R	57
$CH_3COCH_2CH_2CO_2Me$	"	"		76	R	57
$CH_3COCH_2CH_2CO_2nPr$	"	"		83	R	57

(a) The enantiomer excesses for reactions in the presence of R-BMPP have
    been "corrected" for the 70% optical purity of the ligand.

synthon for chiral terpenes or alkaloids via optically active amino alcohols
or diols.

For enones asymmetric reduction takes place with similar regioselectivity to that already noted for the achiral reaction. Monohydrosilanes give 1,4-addition in the presence of RhClDIOP(solvent) or
$(BMPP)_2\overset{\oplus}{Rh}H_2(solvent)_2$ but the chiral saturated ketones are obtained in low
optical purity.[59] The reaction with dihydrosilanes yields allyl alcohols

via 1,2-addition in fair to good optical yield.[60]   In the presence of

bis(neomenthyldiphenylphosphinyl(NMDPP))Rh(cod)Cl 41 is reduced by Ph_2SiH_2

to the ketal of the insecticide component allethrolone, 42, in 71%

enantiomer excess.[61]

41   42

        Applying the same procedures to imines leads to a useful synthesis

of chiral secondary amines, the reaction often proceeding more efficiently

than for the related ketones (Figure 8).

7.3.   Attack of Organometallic Nucleophiles on Carbonyl Groups

7.3.1 Acyl Halides

        The reaction of organometals with acid chlorides to give ketones

is well known; the problem associated with the process is that the product

is often sufficiently electrophilic for further, undesired, attack to

occur. Acyl halides undergo ready oxidative addition to suitable metal

complexes to give reactive acyls. If there is an appropriate leaving group,

organometals may substitute at the metal and a reductive elimination then

yields a ketone (Figure 9).

        The stoichiometric reaction has been studied for rhodium;[63] an

organometal is added to $RhClCO(PPh_3)_2$ to give an isolated metal alkyl.

Subsequent addition of an acid chloride yields ketones in good yield, and

the starting complex is recovered. Since the method depends on the

stability of the metal alkyl, it is limited to R groups without a β-hydrogen

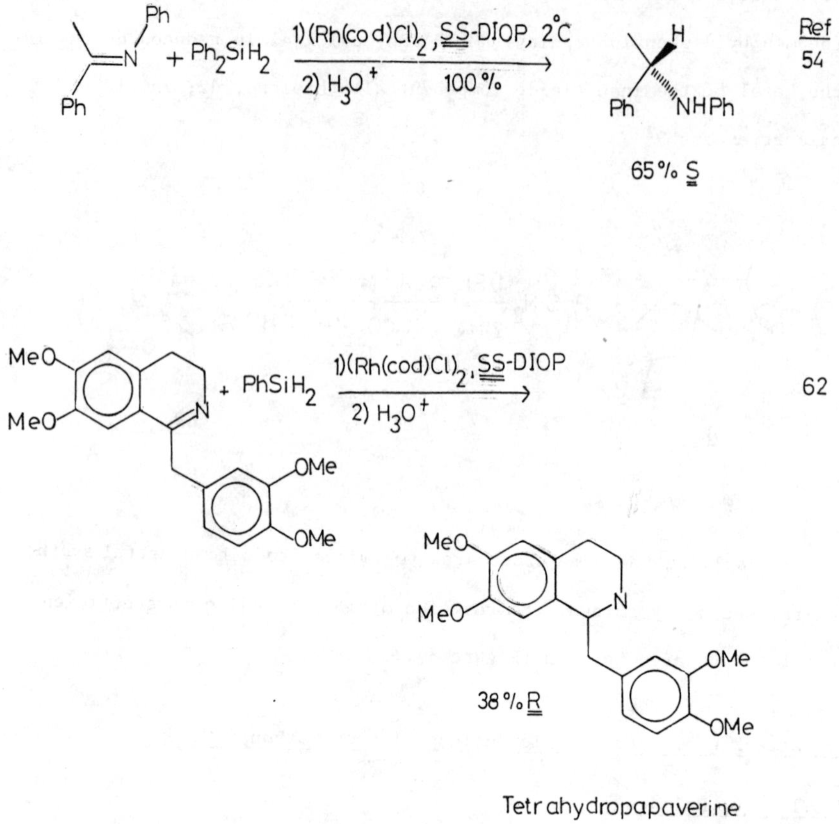

Figure 8 Asymmetric hydrosilylation of imines.

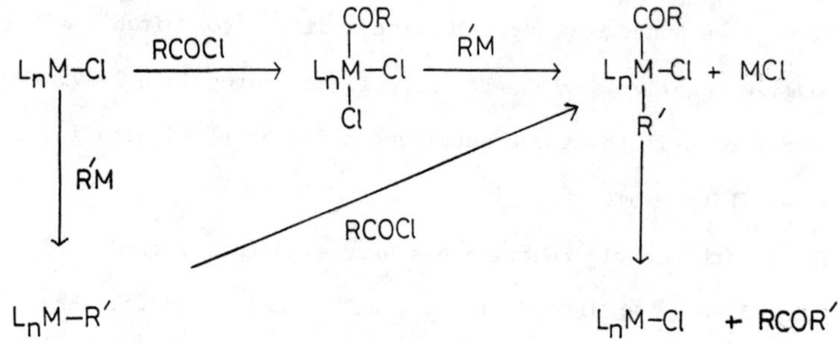

Figure 9 Catalysed reaction of an organometal with an acid chloride.

(e.g. Me, Ph).

The reaction of organomagnesium compounds with acyl halides is modified by the presence of cuprous salts; metal exchange to give the alkyl copper compound occurs, and ketones are produced specifically. The reaction is sensitive to electronic effects, correlating with σ^*, but, unlike the uncatalysed reaction, it is rather insensitive to steric effects.[64] This has led to a useful synthesis of very hindered ketones such as 44.[65]

Allyl tin compounds are not very nucleophilic, but in the presence of transition metal complexes they react with acid chlorides in good yield. Allyl ketones are difficult to isolate because of their acid and base sensitivity but may be obtained efficiently by the action of allyltributyl tin, 46, on acyl halides in the presence of $(PPh_3)_3RhCl$.[66] In a case where allylic rearrangement is possible (e.g. 48 + 49 → 50) this does not occur.[67] This reaction is also catalysed by $(PPh_3)_4Pd$,[68] but the air sensitivity of this catalyst makes it less suitable than $PhCH_2Pd(PPh_3)_2Cl$.[69] With this last a wide range of substrates has been investigated (Figure 10).[70] The method is tolerant of many functional groups (R_2CO, RCHO, CN, COOMe, NO_2), yields are essentially quantitative in 10–15 minutes under neutral conditions, and there is no need for an inert atmosphere. In mixed compounds phenyl, allyl, and benzyl groups are transferred more reaily than alkyls in direct contrast to the strengths of their bonds to tin. Russian workers have utilised this observation to transfer R groups from $RSnMe_3$ to acid chlorides in the presence of π-allyl palladium chloride dimer to give ketones in 65–97% yield.[71]

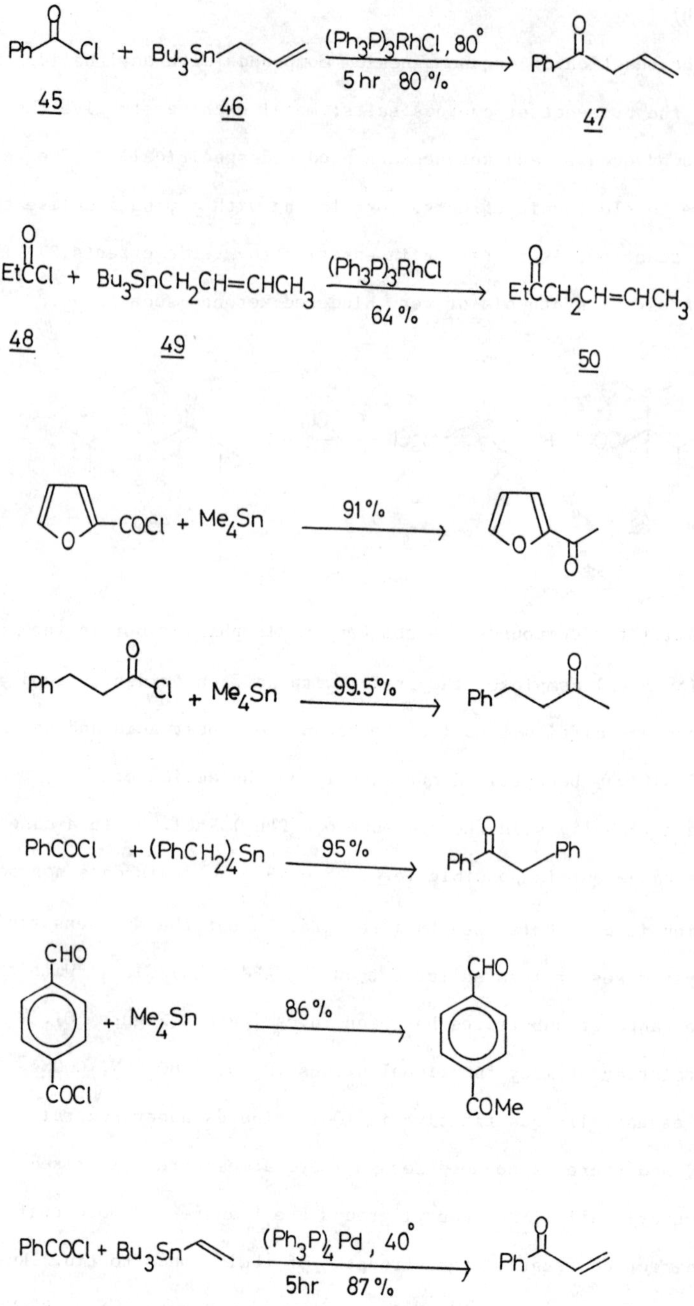

Figure 10 Reaction of an alkyl tin derivative with acyl halides in the presence of $PhCH_2Pd(PPh_3)_2Cl$.

Palladium complexes are also active catalysts for the addition of mercury and copper alkyls to acyl halides. Mercury compounds are easily prepared and relatively insensitive to acid, base, or heat. In the presence of $(PPh_3)_4Pd$ in hexamethylphosphoramide diethyl mercury converts benzoyl bromide to phenyl ethyl ketone in 86% yield.[72] For vinyl

$$\underset{51}{\diagup\diagdown\diagup\diagdown\diagdown HgCl} + \underset{}{\diagup\overset{O}{C}Cl} \xrightarrow[58\%]{(Ph_3P)_4Pd} \underset{52}{\diagup\diagdown\diagup\diagdown\diagdown\diagdown}$$

mercurials the best catalyst is $AlCl_3$, giving a quantitative yield of 52 with only a minor loss of steriochemical integrity; catalysis by $(PPh_3)_4Pd$ gives 58% of the enone but the stereochemistry of the original double bond is completely preserved.[73]

Two other interesting examples of this type of coupling involve the in situ formation of a suitable organometal. The coupling of an alkyne and an acyl halide is achieved in the presence of catalytic amounts of Cu_2I_2 and $(PPh_3)_2PdCl_2$. Thus excellent yields of 55 are obtained with a minimum of side reactions.[74] In the second example a benzyl bromide is converted to the corresponding organozinc compound and in the presence of $(PPh_3)_4Pd$ reacts with acid chlorides to give ketones in 60-90% yield (Figure 11).[75] Competing self coupling to give bibenzyls occurs when Pd(II) or Ni(II) catalysts are used.

$$\underset{53}{R_1-C\equiv C-H} + \underset{54}{R_2COCl} \xrightarrow[Cu_2I_2,\ Et_3N]{(Ph_3P)_2PdCl_2} \underset{55}{R_1-C\equiv C-\overset{O}{\overset{\|}{C}}-R_2}$$

$R_1,R_2 = Ph$ 96%

$R_1=Ph,\ R_2=\underline{t}Bu$ 78%

$R_1=\underline{n}Bu,\ R_2=Ph$ 81%

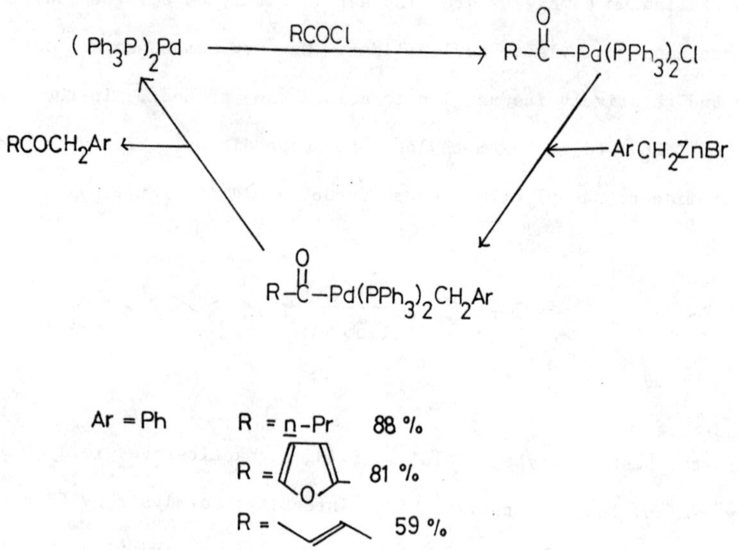

$$(Ph_3P)_2Pd \xrightarrow{\quad RCOCl \quad} R-\overset{\overset{O}{\|}}{C}-Pd(PPh_3)_2Cl$$

RCOCH₂Ar ←

ArCH₂ZnBr →

$$R-\overset{\overset{O}{\|}}{C}-Pd(PPh_3)_2CH_2Ar$$

Ar = Ph   R = n-Pr   88 %

R = (furyl)   81 %

R = (allyl)   59 %

Figure 11 Palladium catalysed reaction of an organozinc compound with an acid chloride.

7.3.2 αβ-Unsaturated Carbonyl Compounds

Attack of nucleophiles on αβ-unsaturated carbonyl compounds may give 1,2- or 1,4-addition (Figure 12). Grignard reagents normally undergo 1,2-addition while organocuprates favour conjugate attack. The presence of copper salts in a reaction with an organomagnesium compound may change its selectivity to give 1,4-addition. Many examples are known (Figure 13) and need little further comment.

The addition of trimethyl aluminium to enones proceeds without catalyst in hydrocarbon solvents to give 100% tertiary alcohol. Ether is not normally suitable as a medium for the reaction, but in the presence of Ni(acac)₂ 1,4-addition occurs exclusively, the uncatalysed reaction being suppressed[80] (Figure 14). The main side reactions are reduction to the allylic alcohol and Michael reactions of the enolate with further enone.

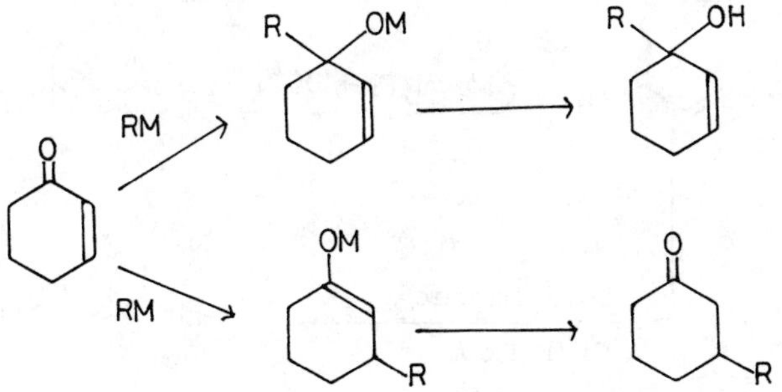

703

Figure 12 Attack of organometallics on αβ-unsaturated carbonyl compounds.

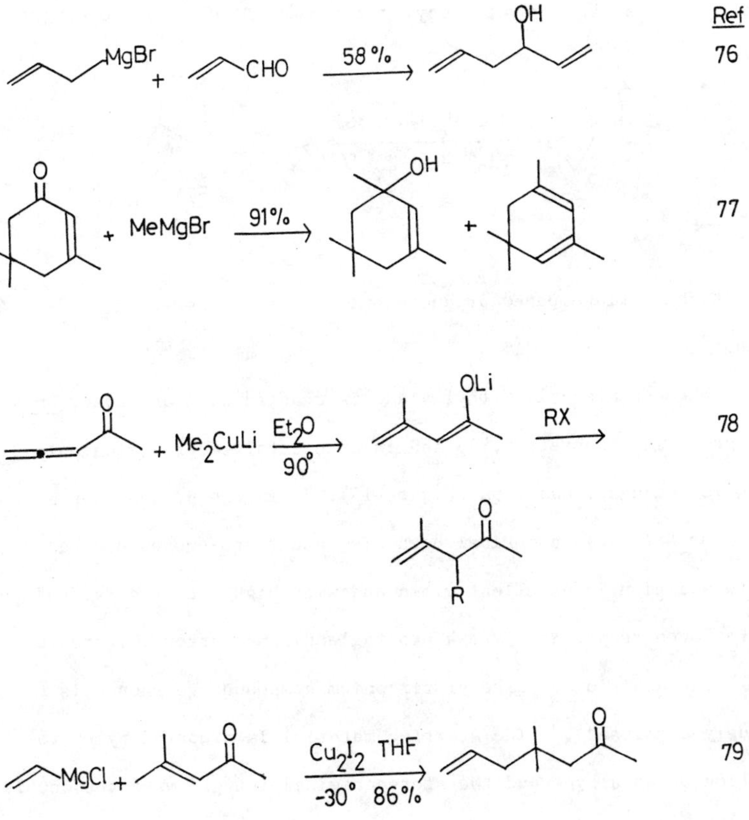

Figure 13 Examples of Grignard and organocuprate addition to enones.

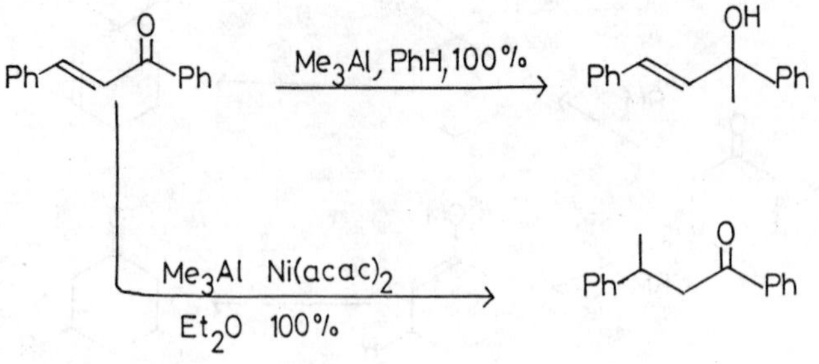

Figure 14 Addition of trimethylaluminium to enones.

Fe(II), Fe(III), Co(III), and Cu(II)[81] acetylacetonates are also catalysts but are less selective. Cyclopropyl ketones (e.g. 56) undergo conjugate

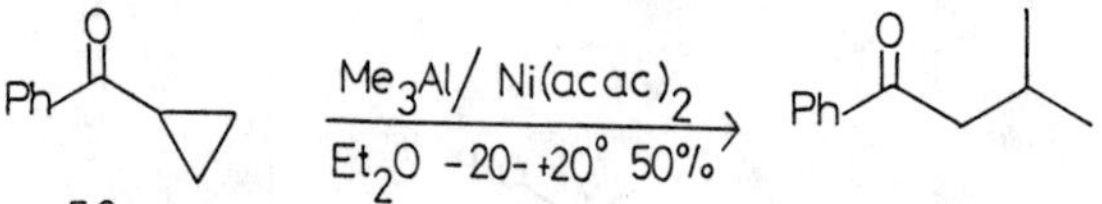

addition to give ring-opened products but yields are lower and 1,2-addition competes.[82]

When an alkynyldimethyl alane is reacted with an enone, only the alkynyl group is transferred,[83] and in an uncatalysed reaction only S-cis-enones give conjugate addition (Figure 15).[84] In the presence of a catalyst derived from Ni(acac)$_2$ and Dibal both cis- and trans-enones undergo conjugate addition in excellent yield and with high specificity, and this method has been used to prepare a prostaglandin precursor (Figure 16).

The addition of alkenyl zirconium compounds to enones is analogously catalysed.[85] The starting material is prepared by hydro-zirconation of an alkyne and the stereochemical integrity of the double bond is maintained in the addition. The reaction is highly stereospecific for substituted enones (Figure 17). The active catalyst is thought to be a Ni(I) species and the mechanism (Figure 18) to be analogous to nickel-mediated aryl coupling.[86,87]

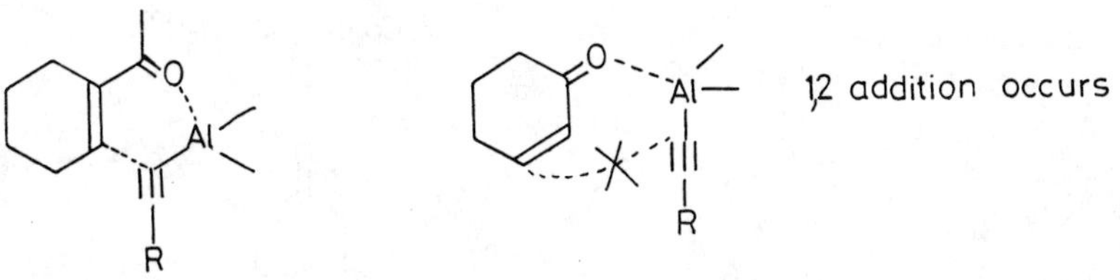

1,2 addition occurs

Figure 15 Addition of alanes to cis- and trans-enones.

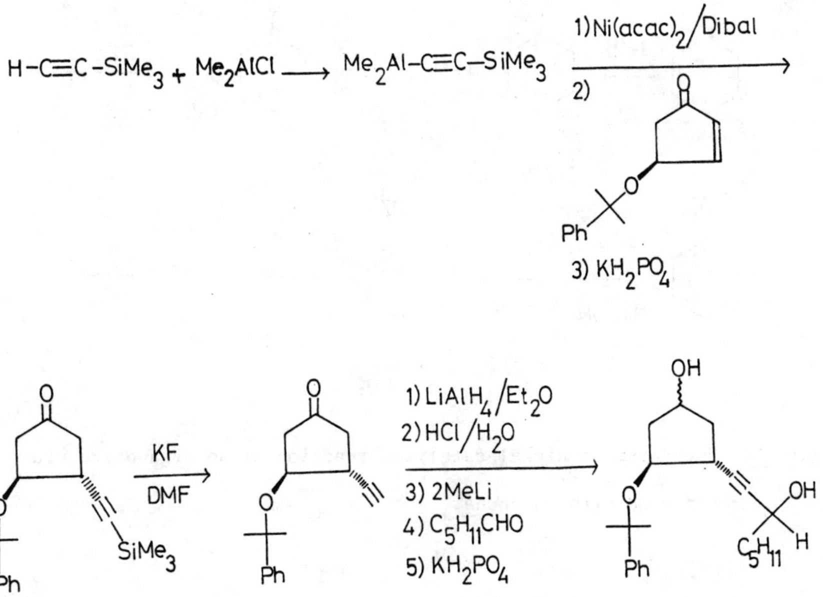

$H-C\equiv C-SiMe_3 + Me_2AlCl \longrightarrow Me_2Al-C\equiv C-SiMe_3$

1) Ni(acac)$_2$/Dibal
2)
3) KH$_2$PO$_4$

KF
DMF

1) LiAlH$_4$/Et$_2$O
2) HCl/H$_2$O
3) 2MeLi
4) C$_5$H$_{11}$CHO
5) KH$_2$PO$_4$

Figure 16 Nickel catalysed conjugate addition to an enone as a route to a prostagalandin precursor.

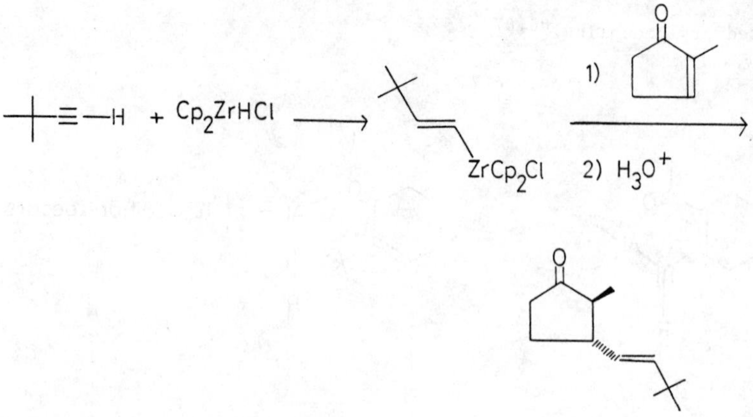

Figure 17 Nickel catalysed addition of a zirconium alkenyl to an enone.

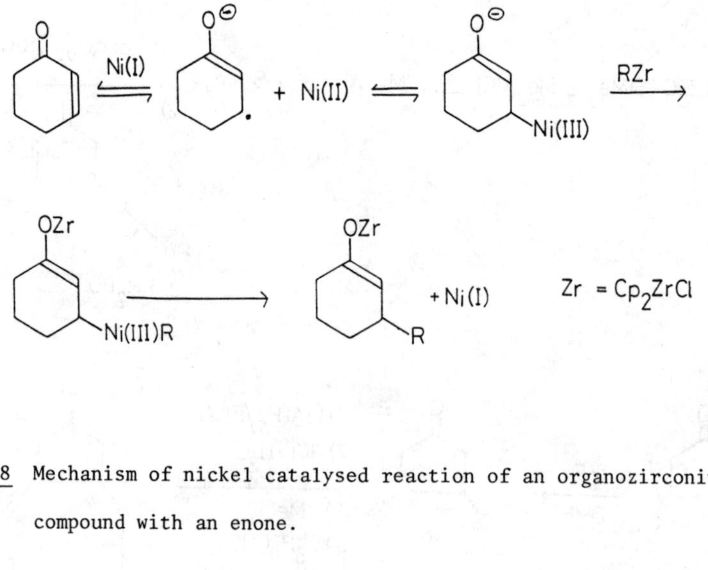

Figure 18 Mechanism of nickel catalysed reaction of an organozirconium compound with an enone.

7.3.3   Other Substrates

The reaction of Me_3Al with saturated carbonyl compounds proceeds only in hydrocarbon solvents to give 1,2-addition together with nonproductive enolisation.   With $Ni(acac)_2$ a greater range of products is found,[88] but mixtures are usually obtained and few predictions of reaction course may be made (Figure 19).

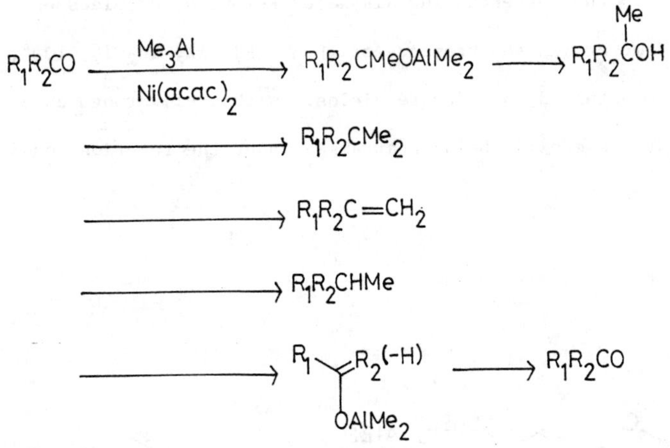

Figure 19   Nickel catalysed reaction of alanes with saturated ketones.

The reaction of alkyl metals with nitriles to give a metallated imine provides a route to ketones.   This is of limited use, however, since removal of the acidic H-atom α to the cyano group competes effectively to yield ketenimines which can undergo extensive side reactions.   In the presence of $Ni(acac)_2$, R_3Al may be added specifically to give, after hydrolysis, ketones in good yield.   Even 58 with its very acidic α-H is converted to 59 in 53% yield.[89]

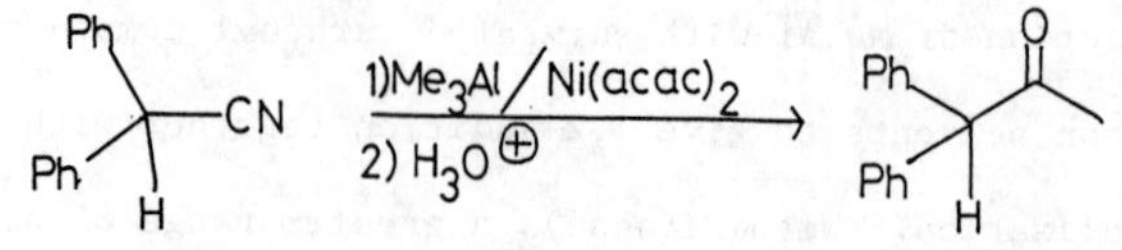

$$ \text{58} \xrightarrow[\text{2) } H_3O^{\oplus}]{\text{1)} Me_3Al/Ni(acac)_2} \text{59} $$

α-haloketones in the presence of AIBN are converted by allyl tributyl tin to the corresponding alkylated ketone by displacement of chloride. However, in the presence of $(PPh_3)_4Pd$ (1%, CH_2Cl_2, 100°, 2 hr) epoxides were obtained in moderate yields. With aryl ketones only aldehydes are isolated, the epoxide being rearranged under the reaction conditions[90] (Figure 20).

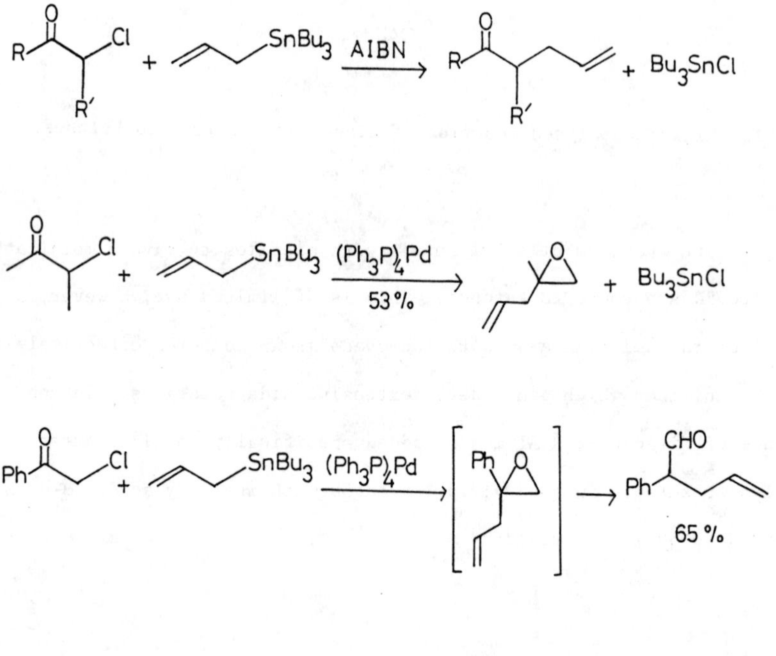

$$ R\text{-}CO\text{-}CHCl\text{-}R' + \text{SnBu}_3 \xrightarrow{AIBN} R\text{-}CO\text{-}CH(R')\text{-}CH_2CH=CH_2 + Bu_3SnCl $$

$$ + \text{SnBu}_3 \xrightarrow[53\%]{(Ph_3P)_4Pd} + Bu_3SnCl $$

$$ Ph\text{-}CO\text{-}CH_2Cl + \text{SnBu}_3 \xrightarrow{(Ph_3P)_4Pd} \left[\quad \right] \longrightarrow Ph\text{-}CH(CHO)\text{-}CH_2CH=CH_2 \quad 65\% $$

Figure 20 Nickel catalysed reaction of organotin compounds with α-haloketones.

7.4. Metal-Catalysed Hydrolysis and Interconversion of Carbonyl Derivatives

7.4.1 Ester Hydrolysis

Solvolysis of esters might be expected to be catalysed by metal ions since coordination increases the electrophilicity of the carbonyl group. In general this is a weak interaction, comparable in magnitude to acid catalysis. However, for certain types of substrate with a second potential coordination site binding and consequently catalysis are greatly enhanced. Many of these reactions have been studied as models for the mechanism of action of certain metalloenzymes.

The hydrolysis of amino acid esters was one of the first observed[91] examples of this type of reaction and is probably the best understood. A variety of mechanisms may operate. The most common involves coordination of the ester to the metal via nitrogen and the ester oxygen atom (60). This causes polarization of the carbonyl group and enhances its susceptibility to nucleophilic attack by external $OH^{\ominus}$ or H_2O. It is not necessary that the species involving the coordinated carbonyl be the major one in solution, but rapid, pre rate-determining step coordination is necessary.[92] This requires that a site must be available for coordination and it is found that for the trifunctional substrate, diethyl-L-aspartate, cobalt complexes such as $Co(trien)X_2$, having only two binding sites, are ineffective.[93] A wide range of metal complexes has been studied as catalysts for the hydrolysis of ethyl glycinate-N,N-diacetic acid; the

60

lanthanides were particularly effective and the mechanism was shown to involve attack of external $OH^{\ominus}$ on a precoordinated carbonyl group.[94] ^{18}O-labelling has shown that acyl-O and alkyl-O fission are both possible;[95,96] for acyl fission the carbonyl is activated toward attack, while for alkyl fission metal coordination makes the carboxyl a better leaving group, thus accelerating carbonium ion formation. Very impressive rate enhancements in the range 10^4 to 10^6 (over uncoordinated substrate) are obtained in this way.

Other groups may act to assist coordination of esters as metal chelates. The methyl ester of γ-aminobutyric acid is hydrolysed in the complex cis-$[CoCl(en)_2H_2N(CH_2)_3COOMe]^{2+}$, though only a modest rate enhancement (7.2) is noted.[97] The solvolysis of 61 is catalysed by acetate ion and the reaction proceeds faster in the presence of Ag^+; it is probable that the metal is coordinated to the C—C double bond and the carbonyl, potentiating $OH^{\ominus}$ attack, though again the effect is not large.[98]

61

Both the hydrolysis and the transesterification of pyridine carboxylic acid esters are catalysed by metal salts.[99] (e.g., 62 → 63. In the presence of the same copper complex the α-ester of glutamic acid diesters is specifically exchanged, providing a useful route to mixed esters. For 64 a third coordination site is available, and a rate acceleration of 10^3-10^6 is obtained in the presence of Cu^{2+}, Co^{2+}, or Ni^{2+}.[99] The hydrolysis of 65 in the presence of Ni^{2+} has been proposed as a model for the action of carboxypeptidase. A rate enhancement of 9300 is observed in the presence of Ni^{2+} salts, but the interaction of the metal and the carboxylate anion is only semi-cooperative.[100]

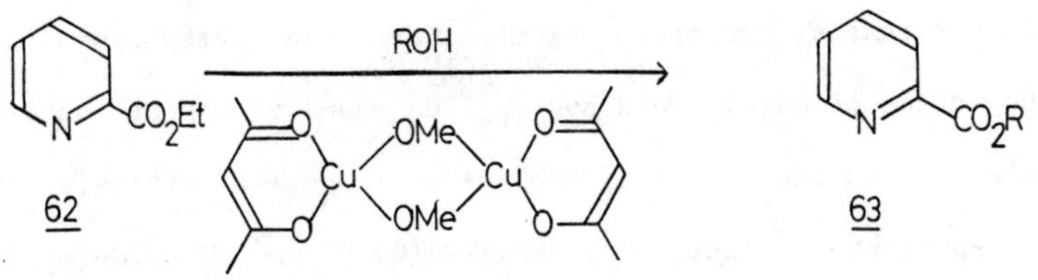

62   63

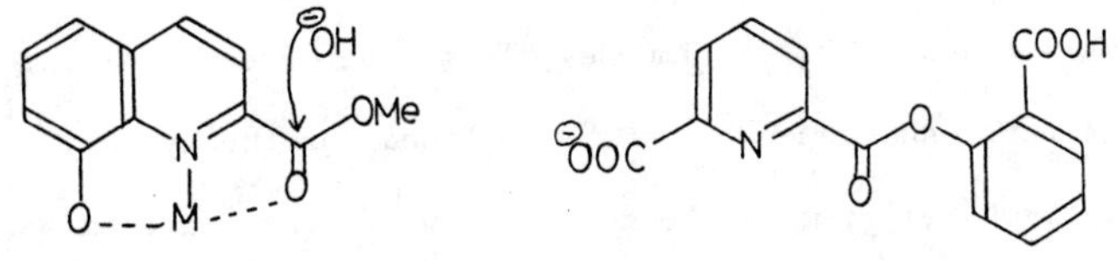

64   65

In the hydrolysis of glycylglycine methyl ester catalysed by Cu(II) between pH 6.5 and 8 the substrate is bound <u>via</u> the terminal anion, the deprotonated amide and, at least transiently, the ester carbonyl. The rate-determining step is attack of $OH^{\ominus}$, which occurs 10^3 times faster than in the absence of Cu(II). Deprotonation and coordination of the amide is essential to catalysis, since no rate enhancement is noted for glycyl-sarcosine methyl ester where the amide is methylated. Zn(II) which does not promote amide dissociation is not a catalyst, although more generally it is capable of promoting ester hydrolysis.[101] The interesting species <u>66</u> promotes ester hydrolysis in the presence of Cu^{2+} by coordination of the carbonyl group to an imidazole bound Cu^{2+} ion in the transition state followed by attack of the unionised oxime. The substrate and catalyst are brought into close proximity by a hydrophobic effect.[102]

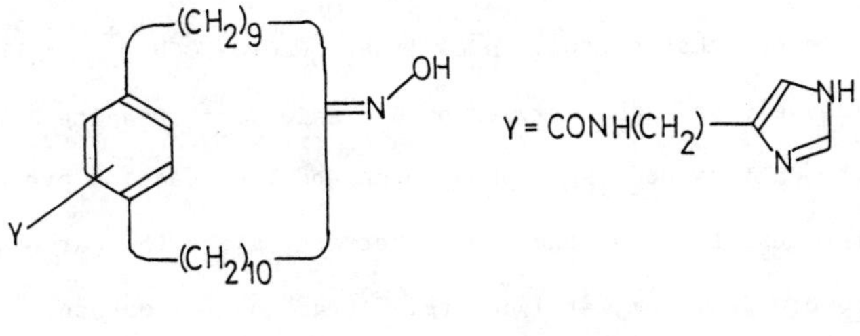

66

An alternative mechanism for metal catalysed ester hydrolysis involves the attack of metal bound hydroxyl on bound or unbound substrate. In the simplest cases the reaction involves nucleophilic catalysis; in the cleavage of p-nitrophenyl acetate by $(NH_3)_5Co(OH)^{2+}$ and $(NH_3)_5Co(Im)^{3+}$ the reaction rates parallel the difference in basicity of the catalysts, and $[(NH_3)_5Co(OCOMe)]^{2+}$ and $[(NH_3)_5Co(ImCOMe)]^{3+}$ may be detected as intermediates.[103] A rate enhancement of 10^{10} is found for the hydrolysis of cis-$[Co(en)_2glyglyOC_3H_7]^{3+}$; metal bound OH or OH_2 is the nucleophile, and the mechanism proposed does not involve carbonyl coordination. When correctly orientated, M—OH and M—OH_2 are potent nucleophiles despite the reduction in basicity caused by metal coordination.[104] An ^{18}O tracer study of the cyclisation of cis-$[Co(en)_2(OH)glyOH]^{3+}$, 67, showed that lactonisation occurred intramolecularly without the loss of coordinated water; a 10^7 – 10^{12}-fold acceleration was noted, comparable to that found for intramolecular lactone formation in purely organic molecules.[105]

The hydrolysis of acetyl phosphate and acetyl phenylphosphate is also catalysed by metal ions[106-111] but until recently the rôle of the metal has been uncertain. It has now been demonstrated (by ^{18}O labelling and product studies) that hydrolysis of $[(NH_3)_5CoOPO_3COCH_3]^+$ in base proceeds exclusively by carbon–oxygen bond fission.[112] Despite metal coordination, no attack occurs at phosphorus and the reaction proceeds only slightly faster than that for the free substrate, since the carbonyl group attacked is remote from the metal centre. Attack at the carbonyl group is also involved in the hydrolysis of acetyl phenylphosphate monoanion by

$[(NH_3)_5CoOH]^{2+}$ since one of the products isolated is $[(NH_3)_5CoOCOCH_3]^{2+}$.
In this latter case hydroxyl ion was more effective than M-OH by a factor
of only 10^2 whereas their basicities differ by 10^9 showing that even
weakly basic metal hydroxides can operate efficiently as nucleophiles at
near neutral pH.

   Two interesting enzyme model systems have been studied in detail.
Zn(II) is present at the active site of Carboxypeptidase A. The metal may
polarise the carbonyl of amide or ester substrates, increasing their
susceptibility to nucleophilic attack; alternatively, attack may involve
zinc coordinated hydroxyl ion. In the hydrolysis of 69 intramolecular base
catalysis was unable to compete with metal ion promoted attack of OH$^{\ominus}$,
suggesting that general base catalysis (by glu-270) is not important in
Carboxypeptidase catalysed ester solvolysis. However, the kinetic model
could not distinguish between inter- and intramolecular OH$^{\ominus}$ attack (69a
or 69b).[113]

69 a   69 b

   Other authors studied the hydrolysis of 70 in the presence of
metal ions, also as a Carboxypeptidase model; with Co^{2+} and Ni^{2+} rate
enhancements are 10^3 - 10^5 in 1:1 complexes. Titrimetric and kinetic
results imply a pre-equilibrium formation of a tetrahedral intermediate

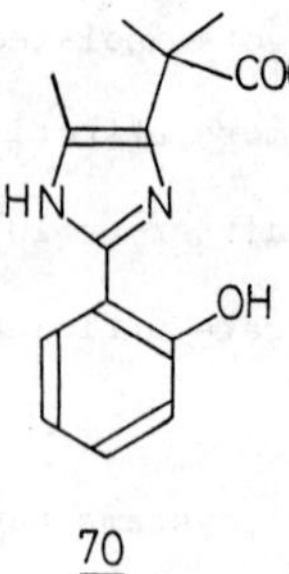

70

accompanying nucleophilic addition of metal bound hydroxyl, followed by
formation of a carboxyl metal anhydride which is subsequently hydrolysed
with acid catalysis (Figure 21).[114]

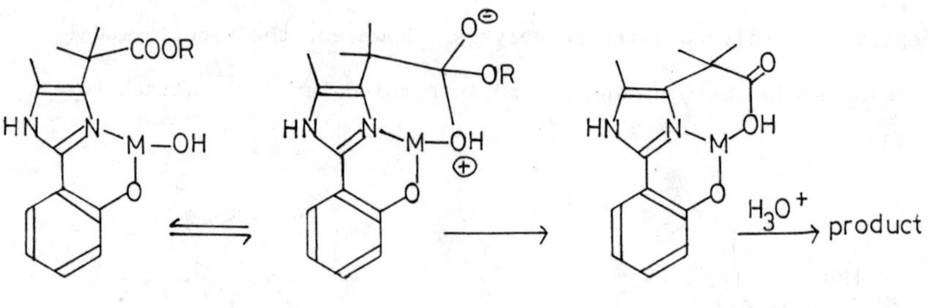

Figure 21 Metal catalysed ester hydrolysis.

7.4.2 Esterification and Transesterification

In cases where the mechanisms of these processes are known, they
appear to be closely analogous to those described for hydrolysis. Zn^{2+}
catalyses transesterification between N-(β-hydroxyethyl)ethylene diamine,71
and p-nitrophenylpicolinate,72. The zinc has a dual rôle, perturbing the
pK_a of the alcohol to give a high concentration of an effective nucleophile and acting as a template to direct intracomplex reaction in a reactive
ternary complex.[115] Another example of a template effect is provided by the
Cu(OAc)$_2$ catalysed lactonisation of the salts 74 and 75. Acetate is

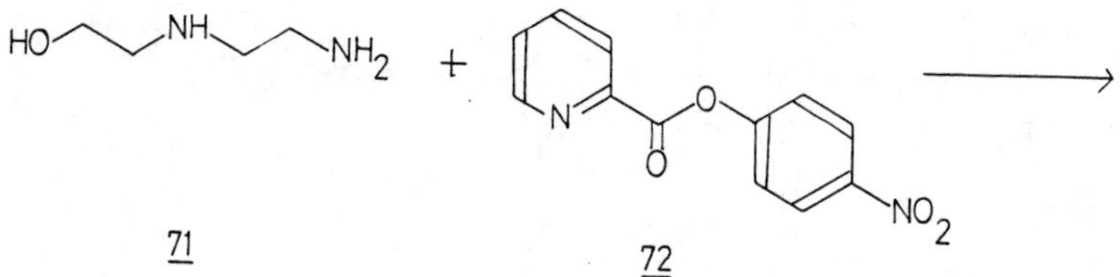

71   72

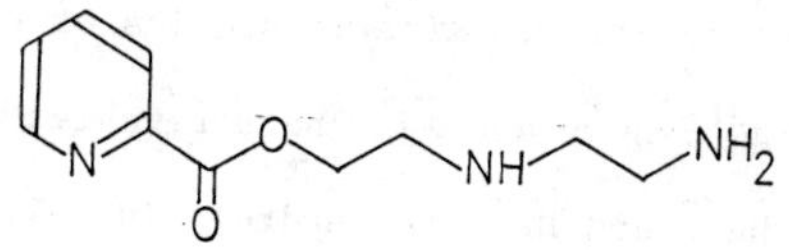

73

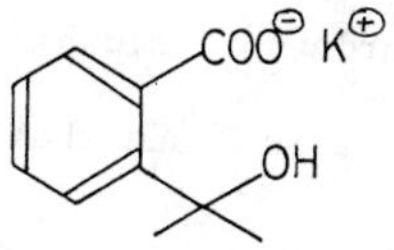

74 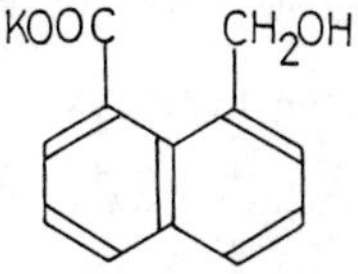   75

displaced from copper by the salt, and the polarised carbonyl is then activated toward intramolecular attack. The rate determining step may be either cyclisation or collapse of the tetrahedral intermediate.[116]

7.4.3 Reactions of Thioesters

A number of metal ions, especially Ag^+ and Hg^{2+}, facilitate the reactions of organosulphur compounds. The hydrolysis and exchange reactions of thiocarbonyls will be dealt with only very briefly here; they are not truly catalytic, since a molar equivalent of the metal ion is required, and the metal is recovered as an insoluble sulphide. They may more strictly be described as metal promoted.[117]

Thioacetals (e.g., 76) are more stable toward protic acid (a "hard" acid) than their oxygen analogues, but they are readily hydrolysed

$$R_1 \underset{R_1}{\overset{SR_2}{\underset{SR_2}{\bigtimes}}} + 2Ag^+ + 3H_2O \longrightarrow \underset{R_1}{\overset{R_1}{\diagdown}}=O + 2R_2SAg + 2H_3O^{\oplus}$$

76

in a metal ion promoted reaction.[118] Similarly, hydrolysis of thioesters
is poorly catalysed by H^+ but rapidly promoted by _soft_ metal ions. Both
thiolo, 77, and thiono, 78, esters are reactive. Kinetic studies of the
hydrolysis of 79 in the presence of Hg^{2+}, Hg^+, and Ag^+ have revealed A_1
and A_2 mechanisms (Figure 22). With Hg^{2+} and Hg^+ the hydrolysis follows
the A_1 pathway for R = OMe, H but A_2 for R = NO_2. For Ag^+ the A_2 pathway
always predominates, and in some cases two Ag^+ ions are coordinated to
sulphur.[119] Hg^{2+} and Hg^+ are 10^6 and 10^3, respectively, times as effective
as H^+ in accelerating hydrolysis. Thus the order of thiophilicity was
established as $Hg^{2+} > Hg^+ > Ag^+$. Occasionally it is found that Hg^{2+} is

$$R_1COSR_2 + H_2O + HgCl_2 \longrightarrow R_1COOH + HCl + R_2SHgCl$$

77

$$R_1CSOR_2 + H_2O + 2AgNO_3 \longrightarrow R_1COOR_2 + Ag_2S + 2HNO_3$$

78

too undiscriminating toward electron rich centres, the cytochalasin
precursor 80 being a case in point. For the conversion of 80 to acetyl
cytochalasin B, 81, Ag(I), Cu(I), or Cu(II) were found to be suitable
thiophiles, catalysing transesterification and ring closure.[120]

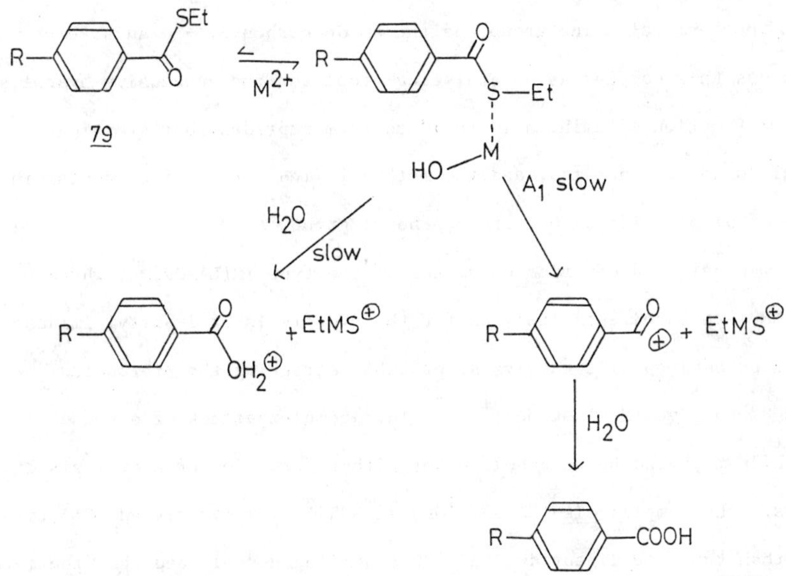

Figure 22 Metal promoted hydrolysis of a thioester.

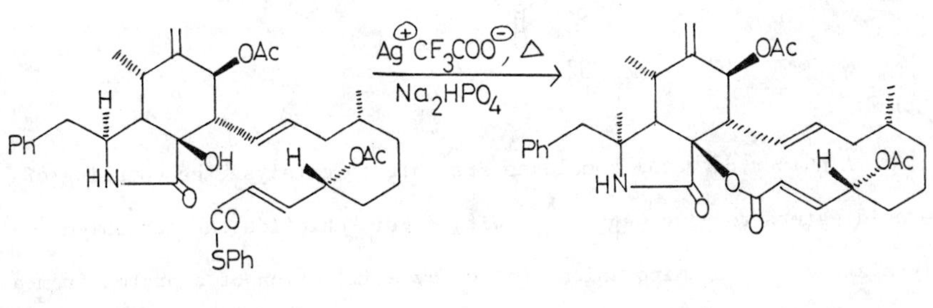

7.4.4 Hydrolysis and Formation of Amides and Peptides

The hydrolysis of metal coordinated peptides is analogous to that of esters. Dimethyl formamide is activated by OH^- attack in $[Co(NH_3)_5DMF]^{3+}$, the reaction being 10^4 times faster than for the free amide.[121] Base catalysed hydrolysis of the peptide bond in the glycyl-glycine complex of β-(trien)Co(II) is enhanced by a factor of 6.5×10^4; a crystal structure of the complex shows that the peptide is bound to metal through the terminal amino group and the amide carbonyl.[122] An attempt was made to use this complex as an analytical tool for the successive hydrolysis and identification of amino acid residues from peptides, but starting material is not regenerated, and the method failed for peptides containing proline or amino acids with a free carboxyl group.[123]

Cobalt and chromium complexes of the type $[M(L)_5OH]^{2+}$, where L = H_2O or NH_3 are excellent catalysts for the hydrolysis of 1-acetyl imidazole. They act as nucleophilic catalysts, probably attacking the protonated species $\underline{82}$ to give $[M(L)_5OCOMe]^{2+}$.[124] Intracomplex attack of a bound nucleophile may also be responsible for either formation or hydrolysis of peptides. The complex $\underline{cis}$-$[Co(en)_2(OH_2)glyNHR]^{3+}$ is hydrolysed 10^{11} times faster than the free dipeptide, and ^{18}O labelling established that the metal bound oxygen is retained in the product $[Co(en)_2glyO]^{2+}$.[125]

$$\underline{82}$$

A number of metal complexes are able to catalyse the coupling of amino acid esters to give peptides. With cupric chloride the mechanism involves attack of an amino anion (formed by elimination of a proton from a copper coordinated amino group) on the unactivated ester carbonyl of the amino acid ester bound to the same copper centre (Figure 23).[126] The aminolysis of amino acid esters coordinated to Co(III) has been used by Buckingham and his co-workers in a synthesis of dipeptides.[127] Attack of the amino group on $\underline{83}$ is accelerated by the binding of the carbonyl to cobalt and the dipeptide is liberated from $\underline{84}$ by electrolytic reduction to Co(II). Yields are typically 80-90% and racemisation is minimal. The preparation of simple amides from amines and esters is markedly accelerated by the presence of $RhCl_3$, though the mechanism of the reaction remains undiscovered.[128] In the presence of $RhCl_3$ amino acid esters may be coupled to give diketopiperazines under conditions mild enough to avoid extensive racemisation.

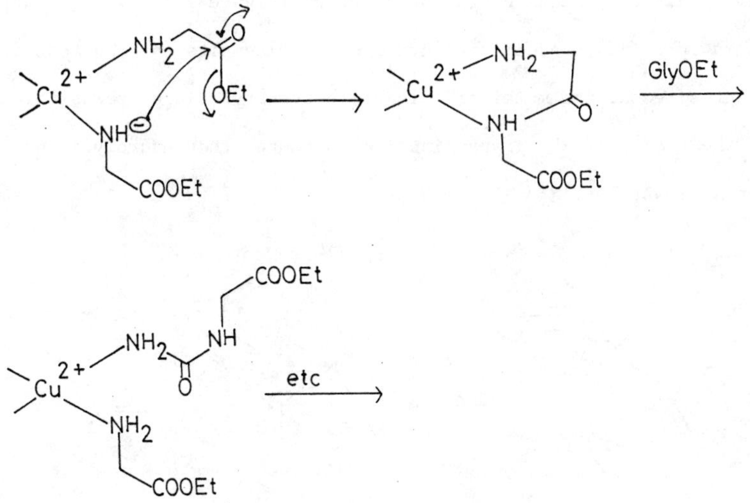

Figure 23 Copper catalysed peptide formation.

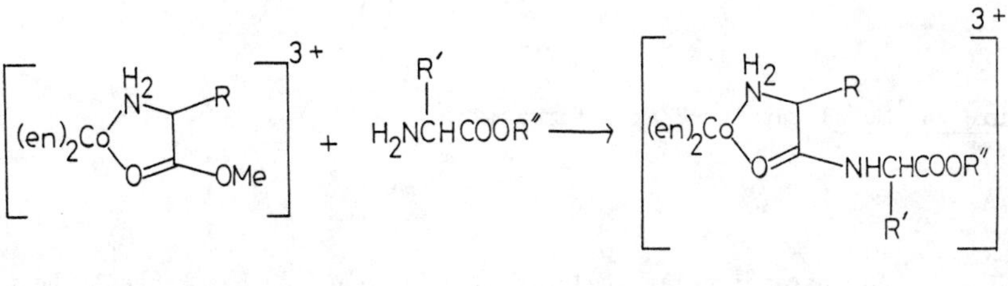

83   84

7.4.5   Reactions of Amines with Aldehydes and Ketones

Anil formation and hydrolysis is catalysed by a variety of metal salts, thiourea complexes particularly effective. The rate-determining step is thought to be attack of amine in the second coordination sphere of the metal on directly metal coordinated ketone (Figure 24). The NH_2 of the thiourea may provide an additional hydrogen bonding site,[129] but one or more vacant sites at the metal are clearly necessary, since excess thiourea inhibits the reaction.[130] Transamination is faster than hydrolysis by a factor of at least 10^2.[131]

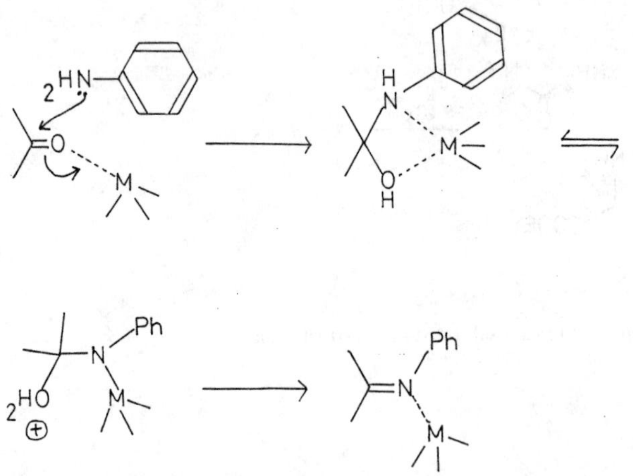

Figure 24   Metal catalysed anil formation.

An intramolecular analogue of this reaction has recently been reported;[132] $[Co(NH_3)_5(H_2NCH_2COCH_3)]^{3+}$ undergoes intramolecular base catalysed cyclisation to a coordinated carbinolamine which undergoes slower dehydration to give a chelated imine, $[Co(NH_3)_4(NH_2-CH_2-C(CH_3)=NH)]^{2+}$.

7.4.6 Conversion of Nitriles to Amides

Like carbonyls, nitriles may be coordinated to metal centres, and the consequent polarisation of the triple bond renders it more susceptible to attack by $OH^{\ominus}$. The most widely studied reactions are those in which the nitrile is coordinated to cobalt in $[(NH_3)_5CoN{\equiv}C{-}R]^{3+}$ and is attacked by external $OH^{\ominus}$. The reaction is instantaneous at room temperature in dilute base, an enhancement of 10^6 to 10^{11} over the uncatalysed reaction, and contrasts sharply with the vigorous conditions usually used for hydrolysis.[133] For hydrolysis catalysed by $[PdCl(OH)(bipy)(H_2O)]$ a full kinetic investigation has been made, giving agreement with the model proposed and accounting for the observed inhibition of the reaction by the product amide.[134] A variety of other metal complexes are active, including trans-Pt(Me)(NHCOMe)(PPh$_3$)$_2$,[135] Na$_2$PdCl$_4$--2,2'-bipyridyl,[136] RuCl$_3$—o—phenantholine,[137] RhCl$_3$/trilauryl trithiosphosphate,[138] and Cu^{2+}.[139] In some cases, unfortunately, the product amide forms a strong adduct with the metal, necessitating the use of a stoichiometric amount of complex, but the starting material may usually be regenerated and free amide obtained (Figure 25). It should be noted, however, that this reaction is industrially important, most of the catalysts in this area being heterogeneous, metals and metal oxides.

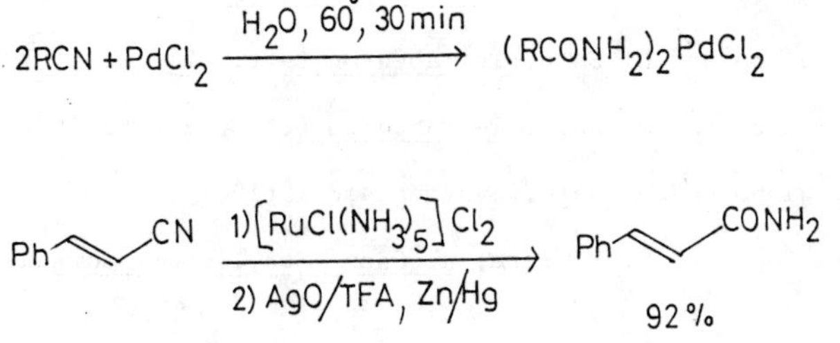

Figure 25 Metal catalysed conversion of nitriles to amides.

References

1. T. Satoh, S. Suzuki, Y. Suzuki, Y. Miyaji and Z. Imai, Tetrahedron Lett., 4555 (1969).

2. T. Kutsuma and I. Nagayama, Jpn. Kokai, 73 92,335 (1973); Chem. Abs., 80, 108162v (1974).

3. G.W.J. Fleet, C.J. Fuller and P.J.C. Harding, Tetrahedron Lett., 1437 (1978).

4. G.W.J. Fleet and P.J.C. Harding, Tetrahedron Lett., 975 (1979).

5. G.W.J. Fleet and P.J.C. Harding, Tetrahedron Lett., 22, 675 (1981).

6. O.H. Wheeler, "The Chemistry of Acyl Halides", edited by S. Patai, p. 231, Interscience, 1972.

7. J.J. Brunet, L.L. Mordenti and P. Caubere, J. Org. Chem., 43, 4804 (1978).

8. A. Fischli, Helv. Chim. Acta, 61, 3028 (1978).

9. A. Fischli, Helv. Chim. Acta, 62, 882 (1979).

10. A. Fischli and D. Süss, Helv. Chim. Acta, 62, 48 (1979).

11. A. Fischli and J.J. Daly, Helv. Chim. Acta, 63, 1628 (1980).

12. R.S. Coffey, Brit. Pat., 1,227,601 (1971); Chem. Abs., 75, 19689b (1971).

13. L. Marko, M.A. Randi and I. Ötvös, J. Organomet. Chem., 218, 369 (1981).

14. J. Thomas, Tetrahedron Lett., 847 (1975).

15. E.C. Ashby and T.L. Weisemann, J. Amer. Chem. Soc., 100, 189 (1978).

16. F. Sato, T. Jinbo and M. Sato, Tetrahedron Lett., 21, 2171 (1980).

17. F. Sato, T. Jinbo and M. Sato, Tetrahedron Lett., 21 , 2175 (1980).

18. F. Sato, J. Jinbo and M. Sato, Synthesis, 871 (1981).

19. F. Guibe, P. Four and H. Riviere, J. Chem. Soc., Chem. Commun., 432 (1980).

    P. Four and F. Guibe, J. Org. Chem., 46, 4439 (1981).

20. S.I. Sadykh-Zade and A.D. Petrov, J. Gen. Chem., U.S.S.R., 29, 3159 (1959).

21. R. Calas, E. Frainnet, and J. Bonastre, Compt. Rend., 251, 2987 (1960).

22. M. Gilman and D.H. Miles, J. Org. Chem., 23, 326 (1958).

23. J-i. Ishiyama, Y. Senda, I. Shinoda and S. Imaizumi, Bull. Chem. Soc. Jpn., 52, 2353 (1979).

24. C. Eaborn, K. Odell and A. Pidcock, J. Organomet. Chem., 63, 93 (1973).

25. I. Ojima, M. Nihonyanagi, T. Kogure, M. Kumagai, S. Horiuchi and K. Nakatsugama, J. Organomet. Chem., 94, 449 (1975).

26. I. Ojima, M. Nihonyanagi and Y. Nagai, J. Chem. Soc., Chem. Commun., 938 (1972).

27. M. Bottrill and M. Green, J. Organomet. Chem., 111, C6 (1976).

28. J.F. Peyronel and H.B. Kagan, Nouv. J. Chim., 2, 211 (1978).

29. S.P. Dent, C. Eaborn and A. Pidcock, J. Chem. Soc., Chem. Commun., 1703 (1970).

30. B. Courtis, S.P. Dent, C. Eaborn and A. Pidcock, J. Chem. Soc., Dalton, 2460 (1975).

31. H. Koinuma, F. Kawakami, H. Kato and H. Hirai, J. Chem. Soc., Chem. Commun., 213 (1981).

32. G. Süss-Fink and J. Reiner, J. Organomet. Chem., 221, C36 (1981).

33. A.P. Barlow, N.M. Boag and F.G.A. Stone, J. Organomet. Chem., 191, 39 (1980).

34. M.F. Lappert and T.A. Nile, J. Organomet. Chem., 102, 543 (1975).

35. I. Ojima, T. Kogure and Y. Nagai, Tetrahedron Lett., 5035 (1972); Y. Nagai, I. Ojima and T. Kogure, Japan Kokai, 74 75,511 (1974); Chem. Abs., 82, 86454d (1975).

36. E. Yoshii, T. Koizumi, I. Hayashi and Y. Hiroi, Chem. Pharm. Bull., 25, 1468 (1977).

37. I. Ojima, M. Kumagai and Y. Nagai, J. Organomet. Chem., 111, 43 (1976).

38. S-i. Inaba and I. Ojima, Tetrahedron Lett., 2009 (1977).

39. I. Ojima, S-i. Inaba and K. Yoshida, Tetrahedron Lett., 3643 (1977).

40. E. Yoshii, T. Koizumi and T. Kawazoe, Chem. Pharm. Bull., 24, 1957 (1976).

41. E. Yoshii, Y. Kobayashi, T. Koizumi and T. Orike, Chem. Pharm. Bull., 22, 2767 (1974).

42. K.A. Adrianov, M. Filimonova and V.I. Sidorov, J. Organomet. Chem., 142, 31 (1977).

43. I. Ojima, T. Kogure and Y. Nagai, Tetrahedron Lett., 2475 (1973).

44. I. Ojima, S. Inaba and Y. Nagai, Tetrahedron Lett., 4363 (1973); I. Ojima and S-i. Inaba, J. Organomet. Chem., 140, 97 (1977).

45. I. Ojima, S-i. Inaba and Y. Nagai, J. Organomet. Chem., 72, C11 (1974).

46. H.B. Kagan and J.C. Fiaud, Topics in Stereochemistry, 10, 175 (1978).

47. K. Yamamoto, T. Hayashi and M. Kumada, J. Organomet. Chem., 46, C65 (1972).

48. I. Ojima, T. Kogure, M. Kumagai, S. Horiuchi and T. Sato, J. Organomet. Chem., 122, 83 (1976).

49. T. Hayashi, K. Yamamoto and M. Kumada, J. Organomet. Chem., 112, 253 (1970).

50. T. Hayashi, K. Yamamoto, K. Kasuga, H. Omizu and M. Kumada, J. Organomet. Chem., 113, 127 (1976).

51. A.D. Calhoun, W.J. Kobos, T.A. Nile and C.A. Smith, J. Organomet. Chem., 170, 175 (1979).

52. J -C. Poulin, W. Dumont, T -P. Dang and H.B. Kagan, C.R. Acad. Sci. Paris, 277C, 41 (1973).

53. T.H. Johnson, K.C. Klein and S. Thomen, J. Mol. Catal., 12, 37 (1981).

54. H.B. Kagan, N. Langlois and T -P. Dang, J. Organomet. Chem., 90, 353 (1975).

55. J -F. Peyronel, J -C. Fiaud and H.B. Kagan, J. Chem. Res., S, 320 (1980).

56. I. Kolb and J. Hetflejš, Coll. Czech. Chem. Commun., 45, 2808, 2224 (1980).

57. I. Ojima, T. Kogure and Y. Nagai, Tetrahedron Lett., 1889 (1974); I. Ojima, T. Kogure and M. Kumagai, J. Org. Chem., 42, 1671 (1977).

58. I. Ojima, K. Yamamoto and M. Kumada, Aspects Homogeneous Catal., 3, 185 (1977).

59. T. Hayashi, K. Yamamoto and M. Kumada, Tetrahedron Lett., 3 (1975).

60. I. Ojima, T. Kogure and Y. Nagai, Chem. Lett., 985 (1975).

61. Y. Mori and T. Akitani, Jpn. Kokai Tokkyo Koho, 79 79,252 (1979); Chem. Abs., 91, 210971x (1979).

62. H.B. Kagan, Pure Appl. Chem., 43, 401 (1975).

63. L.S. Hegedus, S.M. Lo and D.E. Bloss, J. Amer. Chem. Soc., 95, 3040 (1973); L.S. Hegedus, P.M. Kendall, S.M. Lo and J.R. Sheats, J. Amer. Chem. Soc., 97, 5448 (1975).

64. J.A. MacPhee, M. Boussu and J -E. Dubois, J. Chem. Soc., Perkin II, 1525 (1974).

65. C. Lion, J -E. Dubois and Y. Bonzougou, J. Chem. Res. S., 46 (1978).

66. M. Kosugi, Y. Shimizu and T. Migita, J. Organomet. Chem., 129, C36 (1977).

67. M. Kosugi, K. Sasazawa, Y. Shimizu and T. Migita, Chem. Lett., 301 (1977).

68. M. Kosugi, Y. Shimizu and T. Migita, Chem. Lett., 1423 (1977).

69. D. Milstein and J.K. Stille, J. Amer. Chem. Soc., 100, 3636 (1978). J. Org. Chem., 44, 1613 (1979).

70. M. Pereyre and J -P. Quintard, Pure Appl. Chem., 53, 2407 (1981).

71. A.N. Kashin, I.G. Bumagina, N.A. Bumagin, I.P. Beletskaya, Izv. Akad. Nauk S.S.S.R., Ser. Khim., 1433 (1981).

72. K. Takagi, T. Okamoto, Y. Sakakibara, A. Ohno, S. Oka and N. Hayama, Chem. Lett., 951 (1975).

726

73. R.C. Larock and J.C. Bernhardt, J. Org. Chem., 43, 710 (1978).

74. Y. Tohda, K. Sonogashira and N. Hagihara, Synthesis, 777 (1977).

75. T. Sato, K. Naruse, M. Enokiya and T. Fujisawa, Chem. Lett., 1135 (1981).

76. J.C.H. Hwa and H. Sims, Org. Synth., Coll. Vol. V, 608 (1973).

77. M.S. Kharash and P.O. Tawney, J. Amer. Chem. Soc., 63, 2308 (1941).

78. M. Bertrand, G. Gil and J. Viala, Tetrahedron Lett., 1785 (1977).

79. K. Von Fraunberg, Ger. Offen., 2, 432, 232 (1976); Chem. Abs., 84, 164171s (1976).

80. L. Bagnell, E.A. Jeffrey, A. Meisters and T. Mole, Aust. J. Chem., 28, 821 (1975).

81. E.C. Ashby and G. Heinsohn, J. Org. Chem., 39, 3297 (1974).

82. L.A. Bagnell, A. Meisters and T. Mole, Aust. J. Chem., 28, 821 (1975).

83. R.T. Hansen, D.B. Carr and J. Schwartz, J. Amer. Chem. Soc., 100, 2244 (1978).

84. J. Schwartz, D.B. Carr, R.T. Hansen and F.M. Dayrit, J. Org. Chem., 45, 3053 (1980).

85. M.J. Loots and J. Schwartz, J. Amer. Chem. Soc., 99, 8045 (1977).
    J. Schwartz, M.J. Loots and H. Kosugi, J. Amer. Chem. Soc., 102, 1333 (1980).
    F.M. Dayrit and J. Schwartz, J. Amer. Chem. Soc., 103, 4466 (1981).

86. F.M. Dayrit, D.E. Gladkowski and J. Schwartz, J. Amer. Chem. Soc., 102, 3976 (1980).

87. M.J. Loots, F.M. Dayrit and J. Schwartz, Bull. Soc. Chim. Belg., 89, 897 (1980).

88. E.A. Jeffrey, A. Meisters and T. Mole, Aust. J. Chem., 27, 2569 (1975).

89. L. Bagnell, E.A. Jeffrey, A. Meisters and T. Mole, Aust. J. Chem., 27, 2577 (1974).

90. M. Kosugi, H. Arai, A. Yoshino and T. Migita, Chem. Lett., 795 (1978).

91. H. Kroll, J. Amer. Chem. Soc., 74, 2036 (1952); A.E. Martell, Pure
     Appl. Chem., 17, 129 (1968).

92. R.J. Angelici and B.E. Leach, J. Amer. Chem. Soc., 90, 2499 (1968).

93. A.Y. Girgis and J.I. Legg, J. Amer. Chem. Soc., 94, 8420 (1972).

94. B.E. Leach and R.J. Angelici, J. Amer. Chem. Soc., 90, 2504 (1968).

95. D.A. Buckingham, D.M. Foster and A.M. Sargeson, Aust. J. Chem., 22,
     2479 (1969).

96. Y. Wu and D.H. Busch, J. Amer. Chem. Soc., 92, 3326 (1970).

97. R.W. Hay, R. Bennett and D.P. Piplani, J. Chem. Soc., Dalton, 1046
     (1978).

98. M. Takeishi, S. Fujii, S. Niino and S. Hayama, Tetrahedron Lett.,
     3361 (1978).

99. R.W. Hay and C.R. Clark, J. Chem. Soc., Dalton, 1866 (1977).

100. R. Breslow and C. McAllister, J. Amer. Chem. Soc., 93, 7096 (1971).

101. R. Nakon and R.J. Angelici, J. Amer. Chem. Soc., 95, 3170 (1973).

102. Y. Murakami, Y. Aoyama, M. Kida and J-i. Kikuchi, J. Chem. Soc.,
     Chem. Commun. 494 (1978).

103. J. MacB. Harrowfield, V. Norris and A.M. Sargeson, J. Amer. Chem. Soc.,
     98, 7282 (1976).

104. D.A. Buckingham, F.R. Keene and A.M. Sargeson, J. Amer. Chem. Soc., 96,
     4981 (1974).

105. C.J. Boreham, D.A. Buckingham, D.J. Francis, A.M. Sargeson and L.G.
     Warner, J. Amer. Chem. Soc., 103, 1975 (1981).

106. P.J. Briggs, D.P.N. Satchell and G.F. White, J. Chem. Soc. B, 1008
     (1970).

107. C.H. Oestreich and M.M. Jones, Biochemistry, 6, 1515 (1967).

108. C.H. Oestreich and M.M. Jones, Biochemistry, 5, 2926 (1966).

109. D.E. Koshland, J. Amer. Chem. Soc., 74, 2286 (1952).

110. C.H. Oestreich and M.M. Jones, Biochemistry, 5, 3151 (1966).

111. J.P. Klinman and D. Samuel, Biochemistry, 10, 2126 (1971).

112. D.A. Buckingham and C.R. Clark, Aust. J. Chem., 34, 1769 (1981).

113. T.H. Ffe, T.J. Przystas and V.L. Squillacote, J. Amer. Chem. Soc., 101, 3017 (1979).

114. M.A. Wells and T.C. Bruice, J. Amer. Chem. Soc., 99, 5341 (1977).

115. D.S. Sigman and C.T. Jorgensen, J. Amer. Chem. Soc., 94, 1724 (1972).

116. X. Creary and S.D. Fields, J. Org. Chem., 42, 1470 (1977).

117. D.P.N. Satchell, Chem. Soc. Rev., 6, 345 (1977).

118. D. Gravel, V. Vaziri and S. Rahal, J. Chem. Soc., Chem. Commun., 1323 (1972); T -L. Ho and C.M. Wong, Canad. J. Chem., 50, 3740 (1972).

119. D.P.N. Satchell and I.I. Secemski, J. Chem. Soc. B, 1306 (1970), Tetrahedron Lett., 1991 (1969); G. Patel and R.S. Satchell, J. Chem. Soc., Perkin II, 458 (1979).

120. S. Masamune, Y. Hayase, W. Schilling, W.K. Chan and G.S. Bates, J. Amer. Chem. Soc., 99, 6756 (1977).

121. D.A. Buckingham, J. MacB. Harrowfield and A.M. Sargeson, J. Amer. Chem. Soc., 96, 1726 (1974).

122. R.W. Hay and P.J. Morris, J. Chem. Soc., Chem. Commun., 1028 (1969).

123. E. Kimura, Inorg. Chem., 13, 951 (1974).

124. D.A. Buckingham and C.R. Clark, J. Chem. Soc., Dalton, 1757 (1979).

125. C.J. Boreham, D.A. Buckingham and F.R. Keene, J. Amer. Chem. Soc., 101, 1409 (1979).

126. W. Wagatsuma, S. Terashima and S. Yamada, Tetrahedron, 29, 1497 (1973).

127. C.R. Clark, R.F. Tasker, D.A. Buckingham, D.R. Knighton, D.R.K. Harding and W.S. Hancock, J. Amer. Chem. Soc., 103, 7023 (1981).

128. Y. Yamamoto, H. Yatagai and K. Maruyama, J. Chem. Soc., Chem. Commun., 835 (1980).

129. A.R. Boate and D.R. Eaton, Canad. J. Chem., 55, 2432 (1977).

130. A.R. Boate and D.R. Eaton, Canad. J. Chem., 54, 3895 (1976).

131. A.R. Boate and D.R. Eaton, Canad. J. Chem., 55, 2426 (1977).

132. A.R. Gainsford, R.D. Pizer, A.M. Sargeson and P.O. Whimp, J. Amer. Chem. Soc., 103, 792 (1981).

133. D. Pinnel, G.B. Wright and R.B. Jordan, J. Amer. Chem. Soc., 94, 6104 (1972); D.A. Buckingham, F.R. Keene and A.M. Sargeson, J. Amer. Chem. Soc., 95, 5649 (1973); R.J. Balahura, P. Cock and W.L. Purcell, J. Amer. Chem. Soc., 96, 2739 (1974).

134. G. Villain, A. Gaset and P. Kalck, J. Mol. Catal., 12, 103 (1981).

135. M.A. Bennett and T. Yoshida, J. Amer. Chem. Soc., 95, 3030 (1973).

136. R.W. Goetz and I.L. Mador, U.S. 3,670,021 (1972); Chem. Abs., 77, 100857t (1972).

137. D.W. Smith and J. Feldman, U.S. 3,980,61 (1976); Chem. Abs., 85, 158803d (1976).

138. J.L. Greene and M. Godfrey, U.S. Reissue 28,525 (1975); Chem. Abs., 84, 4521c (1976).

139. S. Paraskewas, Synthesis, 574 (1974).

140. S.E. Diamond, B. Grant, G.M. Tomand and H. Taube, Tetrahedron Lett., 4025 (1974).

8 Formation of carbon-carbon bonds

8.1 Polymerisation of Alkenes

  8.1.1. Soluble Titanium Catalysts.

  8.1.2. Vanadium Ziegler Catalysts.

  8.1.3. Complexes of Other Metals as Catalysts.

  8.1.4. Mechanism of Polymerisation.

  8.1.5. Control of Molecular Weight and Chain Branching.

  8.1.6. Stereoselectivity

  8.1.7. Selectivity between Alkenes and Copolymer Formation.

8.2 Oligomerisation and Coupling of Alkenes

  8.2.1. Homooligomerisation of Simple Alkenes.

  8.2.2. Oligomerisation of Functionalised Alkenes.

  8.2.3. Olgiomerisation of Strained Alkenes.

  8.2.4. Codimers and Cooligomers of Alkenes.

  8.2.5. Cyclisation Reactions of Alkenes.

  8.2.6. Coupling of Alkenes and Arenes.

8.3 Polymerisation and Oligomerisation of Dienes

  8.3.1. Polymerisation and Oligomerisation of 1,2-Dienes

  8.3.2. Polymerisation of 1,3-Dienes.

    8.3.2.1. Polymerisation of Substituted 1,3-Dienes.

    8.3.2.2. Copolymers of 1,3-Dienes.

  8.3.3. Oligomerisation of 1,3-Dienes.

    8.3.3.1. Oligomers of Butadiene.

    8.3.3.2. Oligomers of Substituted 1,3-Dienes.

    8.3.3.3. Cooligomers of Dienes.

    8.3.3.4. Cooligomers of Dienes and Alkenes.

    8.3.3.5. Cooligomers of Dienes and Alkynes.

8.4 Telomerisation of Dienes

   8.4.1. Telomerisation with Water and Alcohols.

   8.4.2. Telomerisation with Acids.

   8.4.3. Telomerisation with Ammonia and Amines.

   8.4.4. Telomerisation with Carbon Nucleophiles.

8.5 Polymerisation and Oligomerisation of Alkynes

   8.5.1. Polymerisation of Alkynes.

   8.5.2. Linear Dimers of Alkynes.

   8.5.3. Cyclooligomerisation of Alkynes.

8.6 Coupling Reactions of Arenes and Alkynes

   8.6.1. Coupling Reactions of Aryl Halides.

   8.6.2. Oxidative Coupling of Arenes by Palladium (II).

   8.6.3. Phenol Oxidative Coupling.

   8.6.4. Oxidative Coupling of 1-Alkynes.

   8.6.5. Coupling of Arenes with Alkynes.

8.7 Coupling Reactions of Carbanions

   8.7.1. Coupling of Grignard Reagents.

      8.7.1.1. Homocoupling.

      8.7.1.2. Coupling with Aryl Derivatives.

      8.7.1.3. Coupling with Vinyl Derivatives.

      8.7.1.4. Coupling with Allyl Derivatives.

      8.7.1.5. Coupling with Alkyl and Silyl Derivatives.

   8.7.2. Homocoupling of Other Organometals.

   8.7.3. Coupling of Other Organometals with Aryl Derivatives.

   8.7.4. Coupling of Other Organometals with Vinyl Derivatives.

   8.7.5. Coupling of Other Organometals with Alkyl Derivatives.

   8.7.6. Coupling of Other Organometals with Palladium Allyl Complexes.

      8.7.6.1. The Orgins of the Palladium Allyl Complex.

      8.7.6.2. Nature of the Nucleophile

      8.7.6.3. Reaction Regioselectivity.

      8.7.6.4. Stereoselectivity.

      8.7.6.5. Uses in Synthesis.

   8.7.7. Allyl Complexes of Other Metals.

   8.7.8. Attack of Anions on Metal Coordinated Alkenes.

8.8 Coupling of Halides

The formation of new carbon carbon bonds is the most important process in organic synthesis and consequently its realisation in transition metal catalysed reactions has been well studied. Semmelhack[1] provides a convenient classification of the fundamental reactions involved. (Figure 1). We have already encountered α-insertion in carbonylations, and the attack of nucleophiles on metal coordinated arene rings is rarely seen in catalytic reactions, but the remaining groups give scope for a wide variety of reactivities and selectivities.

8.1 Polymerisation of Alkenes

Whilst this reaction is extremely important economically few of the commercially useful catalysts are homogeneous and the homogeneous/ heterogeneous nature of a number of systems is not established. We shall be concerned with the heterogeneous catalysts only when their relevance to the homogeneous ones is clear; a proper study would require a book in itself. Ziegler catalysts were discovered in 1953[2] and normally consist of a transition metal halide, alkoxide, alkyl or aryl, with a main group element alkyl or alkyl halide. Ethylene and α-alkenes are converted under mild conditions to long chain polymers. With α-alkenes stereo-regular isotactic, 1, or syndiotactic, 2, structures may be obtained under suitable conditions. The reaction mechanism has been subject to intense investigation[3] but definitive experiments have only been performed on systems which are relatively poor catalysts. The best known types of homogeneous catalyst are shown in Table 1, though even a few of these do not give fully homogeneous solutions throughout the reaction.

734

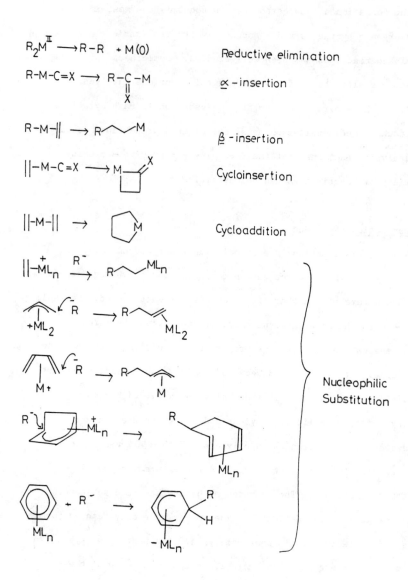

Reductive elimination

α - insertion

β - insertion

Cycloinsertion

Cycloaddition

Nucleophilic
Substitution

<u>Figure 1</u> Classification of carbon carbon bond forming reactions in metal
complexes.

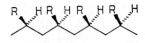

1 isotactic polymer

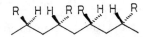

2. syndiotactic polymer

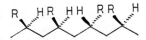

3 atactic or random polymer

Table 1   Common Homogeneous Alkene Polymerisation Catalysts

Catalyst System	References
$Cp_2TiCl_2	Me_2AlCl$	4
$Cp_2TiCl_2	Me_2AlCl	H_2O$	5
$Cp_2TiMe_2	Me_3Al	H_2O$	6
$Ti(OBu)_4	Et_3Al$	7, 8
$VO(acac)_2	Et_2AlCl$	9
$Cp_2VCl_2	Me_3Al_2Cl_3$	10
$VCl_4	Et_3Al$	11
$Cr(allyl)_3$	12, 13
$Ti(benzyl)_4$	14
$Ti(benzyl)_3Cl$	12
$Ti(benzyl)_4	Et_2AlCl$	15
$Zr(benzyl)_4$	16
$Zr(benzyl)_3Cl$	12
$Cp_4Zr	Et_3Al$	17

8.1.1. Soluble Titanium Catalysts

The literature in both homogeneous and heterogeneous Ziegler-Natta chemistry is dominated by reports on titanium complexes. The active species is usually formed by reaction of a titanium (IV) species with an aluminium alkyl. Examples are shown in Table 2. The most complete data as to the active complex in catalysis derive from the Cp_2TiCl_2 system, discovered by Breslow.[27] Measurements of the initial rate of polymerisation[28] suggest that the primary complex is not the active species and an induction period is normally observed. There is an equilibrium between the reactants and the complex 6; 6 may not be isolated but both it and a 1:2 complex[29] have been observed spectroscopically.[30-32] When the catalyst $Cp_2TiEtCl|EtAlCl_2$ is used, a complex 7 is identified by [13]C n.m.r. spectroscopy[33] but kinetic studies suggest that two equilibria are involved in the generation of the active species.[34] Yet another suggestion involves the in situ production of $AlCl_3$, which, in the presence of moisture, protonates $Cp_2TiEtClAlCl_3$ to labilise the Ti-C bond.[35] ESR studies indicate that various Ti(III) species are present (e.g. 8 and 9) but they are unlikely to be involved in polymerisation, the rate of which is independent of their concentration.[36] A detailed discussion of alkene coordination and insertion will be deferred, since the comments apply equally to most coordination polymerisations.

$$2 \; \underset{4}{\overset{Cp}{\underset{Cp}{>}}Ti\overset{Cl}{\underset{Cl}{<}}} \; + \; \underset{5}{Et\overset{Cl}{\underset{Cl}{>}}Al\overset{Cl}{\underset{Cl}{<}}Al\overset{Et}{\underset{Et}{<}}} \; \rightleftharpoons \; 2 \; \underset{6}{\overset{Cp}{\underset{Cp}{>}}Ti\overset{Cl}{\underset{Cl\;Cl}{<}}Al\overset{Et}{\underset{Et}{<}}}$$

$$\underset{7}{Cp-\overset{\overset{Cp}{|}}{\underset{\underset{Et}{|}}{Ti}}-Cl-\overset{\overset{Cl}{|}}{\underset{\underset{Cl}{|}}{Al}}-Et}$$

Table 2   Homogeneous Titanium Polymerisation Catalysts

Catalyst System	Special Features	Ref
$Cp_2TiCl_2 \| Et_2AlCl$	Long lived catalyst for C_2H_4, C_3H_6 and styrene polymerisation. Styrene is polymerised with high isotacticity.	4, 18, 19
$Ti(OBu)_4 \| Et_3Al(Al:Ti \geq 6)$	The truly soluble catalyst gives an iso-tactic polymer	20
$Ti(NEt_2)_4 \| Me_3Al(Al:Ti < 6)$	$MeTi(NEt_2)_3$ is the active species and probably reacts via an anionic route.	8, 21
$TiCl_4 \| Et_2AlCl$	Small amounts of ethers increase reaction rate and molecular weight of polymer.	22
$TiCl_4 \| Ph_2Mg \| i\text{-}Bu_3Al$	Polymerises C_2H_4 at 1 atm. to give very high molecular weight polymer.	23
$TiCl_4 \| (i\text{-}Bu)_2Zn$	$\underline{S}$-4-methylhexene can be copolymerised with excess racemic alkene to give a polymer of defined helicity.	24
$Ti(Benzyl)_4$	Styrene, α-methyl methacrylate and methylene cyclo-butane are all good substrates.	25, 26

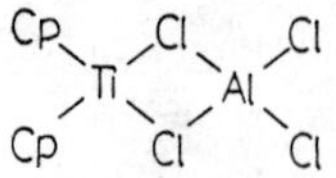

8

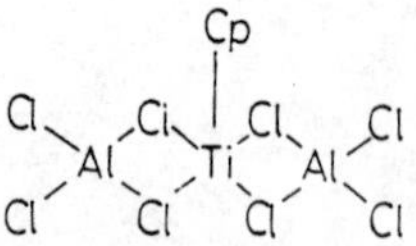

9

Other important titanium containing systems are formed by reaction of TiX_4 (X=Cl, OR) with aluminium alkyls. With $TiCl_4$ the molecular weight of the polymer is increased with increase in catalyst concentration, Al:Ti ratio and temperature, but in the last case efficiency is decreased.[37] Small amounts of ether increase both reaction rate and molecular weight,[22] an effect previously noted with water. With $Ti(OBu)_4$ based systems, suggestions as to the reaction mechanism have been more varied including a cationic route[7] (from electrodialysis measurements in styrene polymerisation) and a radical mechanism.[38] The presence of the Ti(III) intermediate, 10, was confirmed by I.R. and e.s.r. spectroscopy[8,39] and a bipyramidal structure predicted to be its most stable form.[40] Recent work has invalidated many early conclusions, however, since it is clear that both homogeneous and heterogeneous mechanisms are operating in most cases. The homogeneous catalyst (most important at low temperatures) gives an isotatic polymer with a narrow, unimodal molecular weight distribution.[20]

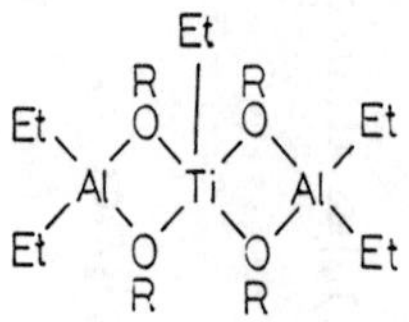

10

8.1.2. Vanadium Ziegler Catalysts

These systems bear a close resemblance to their titanium analogues

except that the polymers produced are usually syndiotactic and have rather

narrow molecular weight distributions[41] On treating Cp_2VCl_2 with Et_2AlCl

up to 3 paramagnetic species are formed[10,42], variously suggested as

structures 11 to 14. Again, however, it is unclear which, if any, of

these is involved in catalysis. Both VCl_3[43] and $VOCl_3$[44] give active

catalysts on treatment with organoaluminium compounds, though once again

both homogeneous and heterogeneous pathways are operating.[44-47] Truly

homogeneous systems are formed on alkylation of $V(acac)_3$; this polymerises

propene to a syndiotactic product with a narrow molecular weight distri-

bution.[48] The active species is suggested to be $[RV(acac)ClAl_2R_4Cl_2]$.[49]

11   X = Cl, Et   12

13   14

8.1.3. Complexes of Other Metals as Catalysts

$(Acac)_3Cr$ yields an alkene polymerisation catalyst in the presence

of R_3Al and the active species is probably an anionic initiator.[50,51]

Allyl chromium complexes are the most effective of a range of metal allyls[12]; the nature of the active complex is unknown[52], but it is thought that there are few active centers and that water is necessary for their production[13]. Cp_2ZrCl_2 shows behaviour in catalysis similar to its titanium analogue[53] but the most commonly studied precursor is $(benzyl)_4Zr$.[25] The mechanism is thought to be a coordination-anionic type in which one of the C-Zr bonds is activated[54] but only a small fraction of the zirconium is involved in catalysis. A long-lived zirconium catalyst is derived from Cp_4Zr and Et_3Al.[55] Catalyst activation was shown to involve the equilibria of Figure 2.

$$Cp_4Zr + Et_3Al \longrightarrow Cp_3ZrEt + CpAlEt_2$$

$$Cp_3Zr\ Et + Et_3Al \longrightarrow C_2H_6 + Cp_3ZrCH_2CH_2AlEt_2$$

$$Cp_3ZrCH_2CH_2AlEt_2 + Et_3Al \longrightarrow C_2H_6 + Cp_3Zr-CH(AlEt_2)_2$$

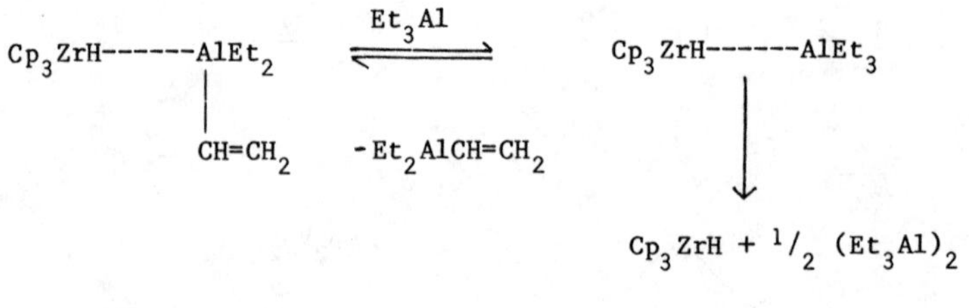

$$
\begin{array}{ccc}
Cp_3ZrH\text{------}AlEt_2 & \underset{-Et_2AlCH=CH_2}{\overset{Et_3Al}{\rightleftharpoons}} & Cp_3ZrH\text{------}AlEt_3 \\
\big| & & \big\downarrow \\
CH=CH_2 & & \\
& & Cp_3ZrH + {}^1/_2\ (Et_3Al)_2
\end{array}
$$

Figure 2   Activation of Cp_4Zr/Et_3Al for alkene polymerisation.

8.1.4. Mechanism of Polymerisation

Two main mechanisms have been proposed to account for alkene polymerisation by Ziegler-Natta catalysts. The first was proposed by Cossée,[56] and it is in terms of this that most subsequent discussion has been conducted. It involves coordination of a molecule of alkene to

a vacant site at a metal alkyl, 15. This is followed by activation of

the alkene and insertion into the metal-alkyl bond to give 16. The theory

may be used to describe both homogeneous and heterogeneous systems[57] and

was extended by Olivé.[36,58] However, it was pointed out by Green[59] that

there existed, in 1978, no unambiguous models for the insertion of alkenes

into metal-carbon bonds, in sharp contrast to the many examples of their

insertion into metal-hydrogen bonds. There are, however, many precedents

for the formation of metal carbene complexes and their highly stereo-

selective conversion to metallocyclobutanes (Figure 3). Titanium

metallocycles are well characterised and the reaction stereospecificity

was convincingly explained. A theoretical study of the system

$MeTiCl_3|C_2H_4$[60] suggested that the use of Cossée mechanism is generally

unjustified for soluble catalysts. $MeTiCl_3$ may produce a vacant site at

titanium by distortion to a trional bipyramid and ethylene is then

coordinated to an axial site. The necessary 4-centre transition state

is then rather disfavoured.

Three recent experiments have shed some light on the matter.

Alkene insertion into a metal-carbon bond has finally been achieved.[61]

$Cp_2MeLu(Et_2O)$, 17, is prepared and treated with propene at -30°. After

a few minutes the insertion product, 19, is observed, in the n.m.r.

spectrum, to account for 30-50% of the total metal. Longer chain metal

alkyls are subsequently noted and 17 is a catalyst for polymerisation of

ethylene. On the other hand 21 is also a catalyst for ethylene poly-

merisation at a relatively slow rate. Treatment of 21 with 1-5 moles of

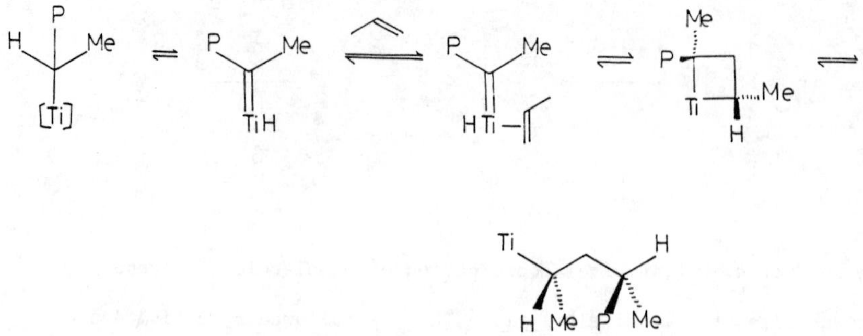

Figure 3 Carbene mechanism for alkene polymerisation.

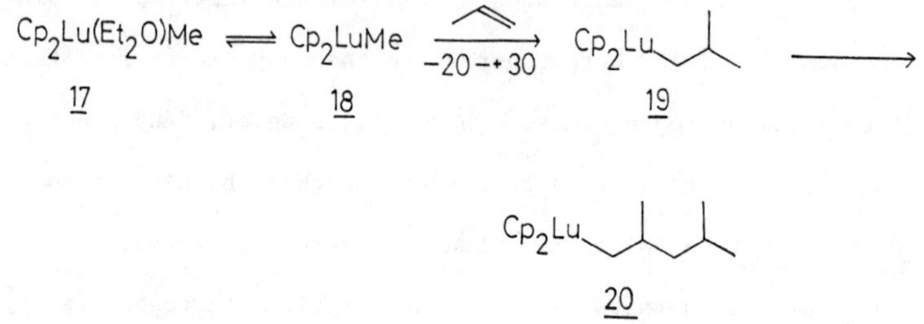

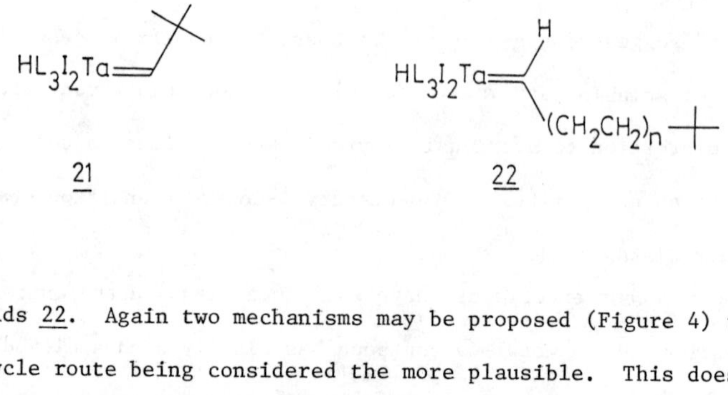

C_2H_4 yields 22. Again two mechanisms may be proposed (Figure 4) the
metallocycle route being considered the more plausible. This does not
require that β-elimination be slow compared with α-elimination, merely
that displacement of alkene by ethylene be relatively unfavourable.[62]
Grubbs[63] considered the two schemes in terms of their potential deuterium
isotope effects which should be present for the carbene mechanism
(A, Figure 5) but absent in the insertion route (B). In the system he

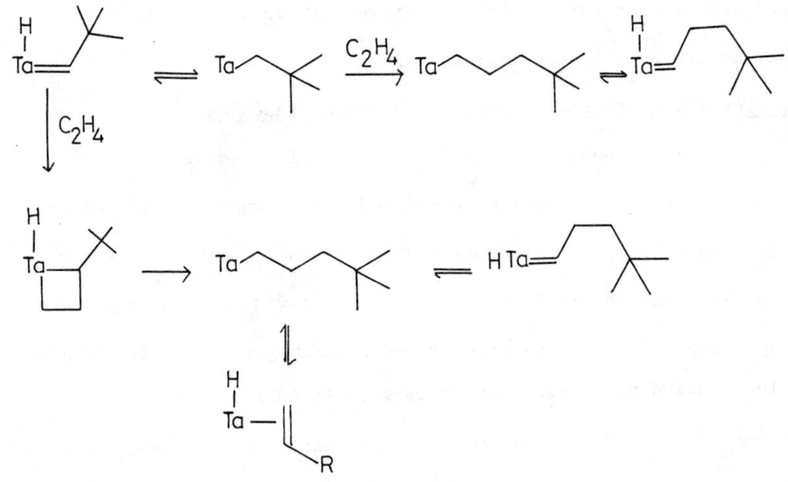

Figure 4 Mechanism of ethylene polymerisation by a metal carbene complex.

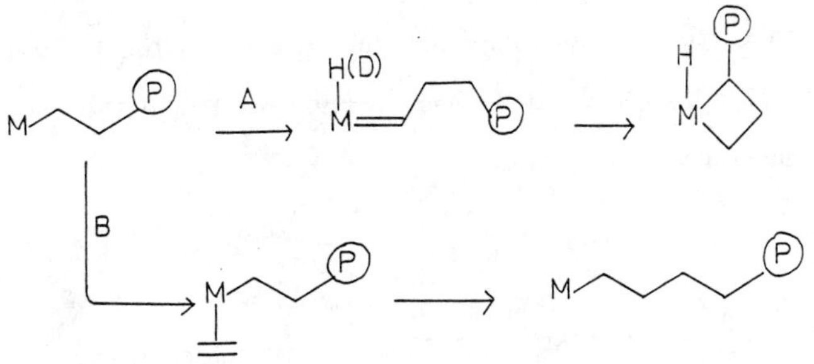

Figure 5 Deuterium isotope effect in polymerisation mechanisms.

chose $(Cp_2TiEtCl|EtAlCl_2)$ the catalyst was formed faster than chain growth, and chain transfer is slow. $k_H|k_D$ was found to be 1.04 ± 0.03 implying that no hydrogen migration occurs in the rate determining step. This does not entirely rule out route A, however, since alkene addition may be rate-determining and it must be noted that the catalyst is atypical in that it does not polymerise propene. At this point the question remains unresolved but it is clear that the experimental sophistication of the approaches to its solution is increasing.

8.1.5. Control of Molecular Weight and Chain Branching

Controlling these parameters is essential to ensure desireable physical properties for the polymer produced. For example, high density polyethylene is a linear polymer, melting point 136°C, whereas low density polyethylene is highly branched and has a wide melting range. Chain length depends on the relative rates of propagation and termination steps. Termination may occur in several ways (Figure 6). Metal hydrides may add alkene to regenerate active centres and in some cases when monomer is used up catalysis may be restarted by adding monomer. This is described as a living polymer. Where evolution of the catalyst system involves a series of reactions, different numbers and types of active centres are formed. Several related catalysts are present, each giving its own characteristic polymer and hence a broad molecular weight distribution.[45] This is generally a less serious problem with homogeneous than analogous heterogeneous systems.

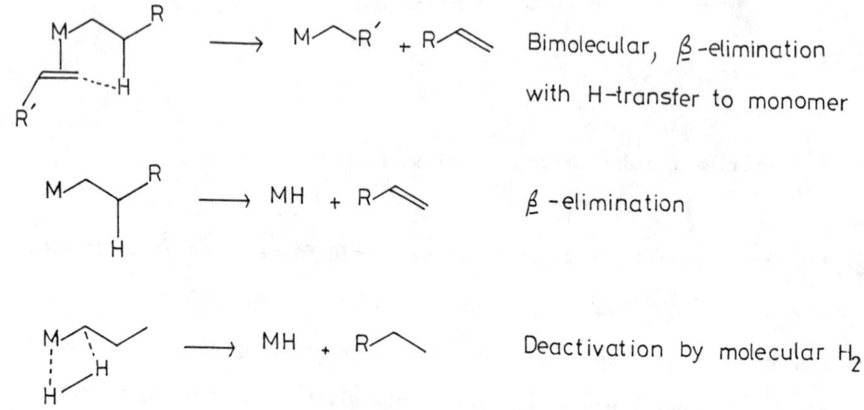

Figure 6 Termination steps in alkene polymerisation.

Methods to tailor catalysts to yield a polymer with particularly desireable physical characteristics are mainly technical in nature and have largely been developed for heterogeneous systems. Narrow molecular weight distributions are best obtained at low temperatures where termination and transfer of polymer chains from one metal atom to another are unimportant.[64] Molecular weight may be reduced by using excess monomer so that bimolecular as well as unimolecular termination routes are available,[65] or by hydrogenation.

8.1.6. Stereoselectivity

It was long believed that stereoselective polymerisation could only be achieved in the defined environment provided by heterogeneous catalysts and that homogeneous systems would always yield atactic products.[56,66-68] Commercial polypropylene, produced by modified heterogeneous $TiCl_3|R_3Al$ catalysts,[69] is highly stereoregular and a model of non-bonded interactions between the growing chain and incoming monomer at a site composed of three titanium atoms suffices to explain the selectivity.[70] Isotactic specificity arises from the favoured complexation of one face of the α-alkene followed by a stereospecific cis-addition. In 23 steric hindrance determines the mode of monomer approach. A few isospecific homogeneous catalysts are known. Styrene gives a largely isotactic polymer with $Cp_2TiCl_2|Et_2AlCl$ as does the homogeneous portion of the catalyst formed from $Ti(OBu)_4|Et_3Al_2Cl_3$.

By contrast soluble vanadium catalysts give syndiotactic polymers where they are at all specific.[64] Syndiospecific propagation takes place only when the proper ligands are bonded to the metal and stereo-control derives from the asymmetric configuration of the last chain unit.[71] The propagation is largely regiospecific with insertion into M-C (second-ary) bonds.[72] (24 to 25 not 26 to 27). The model proposed involves a V-atom coordinated to three chlorines, a secondary carbon and a π-bonded alkene, in a trigonal bipyramidal arrangement. Four modes of alkene coordination are possible. Insertion should occur via a four centre complex, 28, requiring a short C*-C* distance which is achieved by alkene rotation with the growing chain occupying an axial site. A theoretical study then gives good agreement with experimental data, the rate being controlled by non-bonded interactions between monomer and the ligands at vanadium.

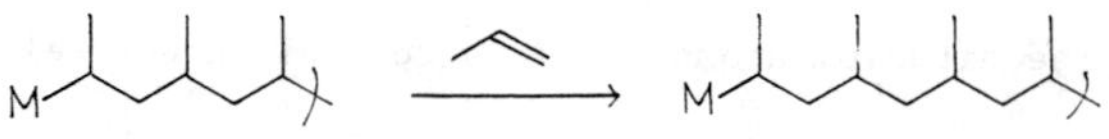

24

25

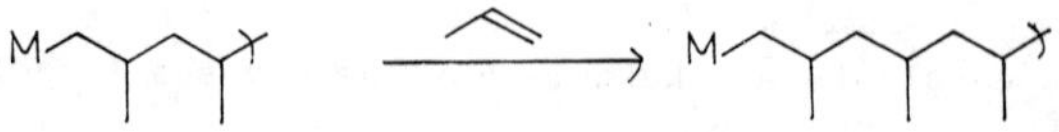

26

27

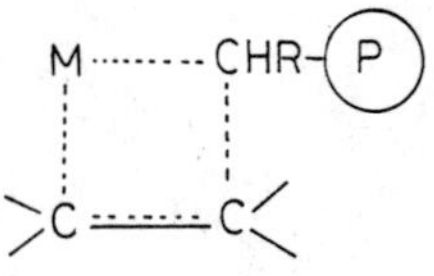

28

8.1.7. Selectivity between Alkenes and Copolymer Formation

Homopolymers of ethylene and propene are usually plastics but their copolymers are elastic and may be used to replace rubber. Some include small amounts of dienes to allow for cross-linking. The relative reactivity of the components is $C_2H_4 > C_3H_6 >$ dienes. It is generally the case that the more substituted an alkene the slower its polymerisation[73], though strained alkenes, such as norbornene, have anomalously high reactivity.[74] Other factors are steric hindrance and stability of the monomer metal complex. Vanadium catalysts are successful for the random incorporation of α-alkenes into polyethylene[75], the strictly soluble ones giving a syndiospecific alternating sequence with little compositional inhomogeneity.[76] Insoluble catalysts yield a broader compositional range.[44] Titanium catalysts are less successful for copolymerisation; two types of copolymer result from the different types of catalytic centre.[77,78]

Some interesting results have been obtained in the copolymerisation of acrylonitrile and methylmethacrylate. With a radical initiator the composition is about 1:1 in the two monomers but in the presence of Ni(bipy) there is an almost linear relationship between log (monomer reactivity) and the stability constant of the Ni(bipy)(monomer) complex.[79,80] Similarly in the copolymerisation of styrene in the presence of $(\underline{\eta}^3 2-CH_3-C_3H_4)_3Cr$, only the methyl methacrylate is incorporated into a predominently syndiotactic polymer.[52]

8.2 Oligomerisation and Coupling of Alkenes

8.2.1. Homooligomerisation of Simple Alkenes

Some oligomerisations are analogous to polymerisations, merely less efficient. A number of types of selectivity must be considered; the first is the production of dimers and trimers versus longer oligomers.

Even with dimers of substituted alkenes regioselection is a problem, and
additionally both starting materials and products may isomerise under
the reaction conditions.[81]

In the analogue of Ziegler polymerisation termination occurs
after most insertion steps by β-hydride abstraction (Figure 7). The
simplest example is ethylene dimerisation. A commerical aluminium
catalyst gave mixtures of C_2 to C_{40} linear alkenes, important in the
synthesis of fatty acids, aldehydes and alcohols.[82,83] This method of
butene production is well studied, though relative costs make it
commercially unattractive. Nickel complexes are most prominent as
catalysts. For example $(\underline{n}^3C_3H_5NiBr)_2|TiCl_4$ gives 88% dimers and 15%
trimers.[84] The dimer fraction consists mainly of butenes but much of

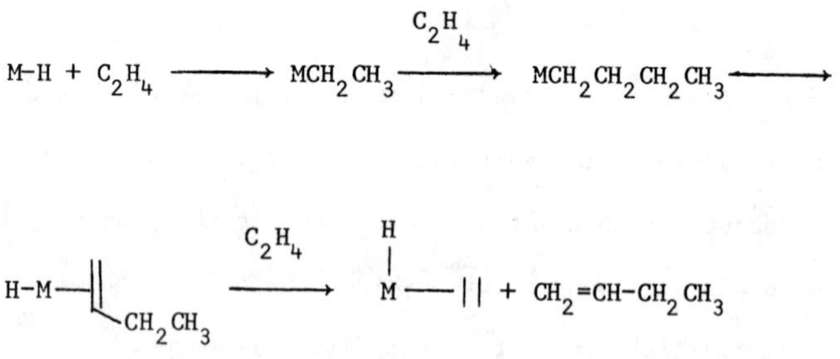

$$M{-}H + C_2H_4 \xrightarrow{} MCH_2CH_3 \xrightarrow{C_2H_4} MCH_2CH_2CH_2CH_3 \longleftarrow$$

$$H{-}M \overset{C_2H_4}{\xrightarrow{}} M\overset{H}{-}|| + CH_2{=}CH{-}CH_2CH_3$$

Figure 7   Mechanism of ethylene dimerisation.

the 1-butene has been isomerised to cis- and trans-2-butenes.
$(\underline{n}^3C_3H_5NiCl)_2|EtAlCl_2$ gives a similar result.[81] Clearly displacement is
favoured over further insertions. Added ligands may alter the reaction's
course, the highest selectivity for dimers (>98%) being achieved with the
basic, but sterically undemanding, Me_3P. More bulky phosphines give
higher oligomers, and in the presence of excess $(\underline{t}{-}Bu)_3P$ polyethylene is
obtained.[85] The decrease in displacement rate is explained by a penta-

coordinate <u>bis</u>(alkene) intermediate, which is disfavoured by bulky phosphines.[86] Other Ni(0) and Ni(II) complexes are active catalysts, including (cod)$_2$Ni[87], Ni(P(OPh)$_3$)$_4$[88,89], (C$_2$H$_4$)Ni(P(<u>i</u>-Pr)$_3$)$_3$ and (C$_2$H$_4$)$_2$Ni(P(<u>i</u>-Pr)$_3$)$_2$[90] in the presence of Bronsted or Lewis acids, Cp$_2$Ni[91,92] and ((Ph$_3$P)$_2$NiCl(C$_6$Cl$_5$))|Ag$^+$.[93] In most cases the active species is thought to be HNiLY since they give, under comparable conditions, products of the same composition, which changes in a predictable way on phosphine addition.[94-97] Further evidence is lack of allyl incorporation into the oligomers and measurement of the highest rates when no metal allyls are observed.[98] Spectroscopically observable cationic nickel hydrides are formed on addition of strong Bronsted acids to Ni(0) complexes.[99,100] Reactions of Co(I) in the presence of Lewis acids[101,102] give apparently analogous results but in these cases the key step is insertion of Co(I) into a C-H bond of ethylene.

An interesting titanium catalysed reaction throws light on Ziegler polymerisation.[103] Ti(OBu)$_4$ is treated with Me$_3$Al yielding methane during catalyst evolution. It is proposed (Figure 8) that a titanium carbene, <u>29a</u>, is formed and dimerises to <u>29b</u>. Treatment with D$_2$ gives CH$_2$DCH$_2$D. Subsequent steps are suggested to be coordination of two moles of ethylene, hydrogen transfer between them, insertion and ligand exchange. The catalytic cycle accounts for the deuterium labelling studies but must clearly join other variants awaiting more definitive verification.

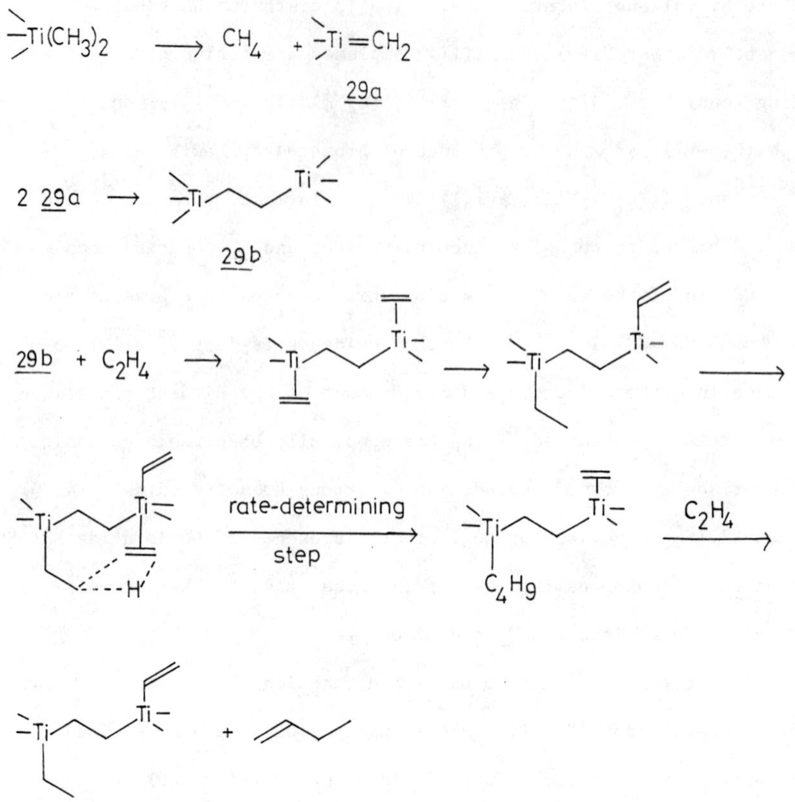

$$2\ \underline{29a} \longrightarrow$$

$$\underline{29b}$$

$$\underline{29b}\ +\ C_2H_4 \longrightarrow$$

rate-determining

step

C_4H_9

C_2H_4

+

Figure 8 Ethylene dimerisation via a titanium carbene.

A different mechanism operates for ethylene dimerisation by
tantalum carbene complexes such as $\underline{30a}$[104]. Study of catalyst evolution
shows that the first formed complex is $\underline{30b}$; on treatment with C_2H_4

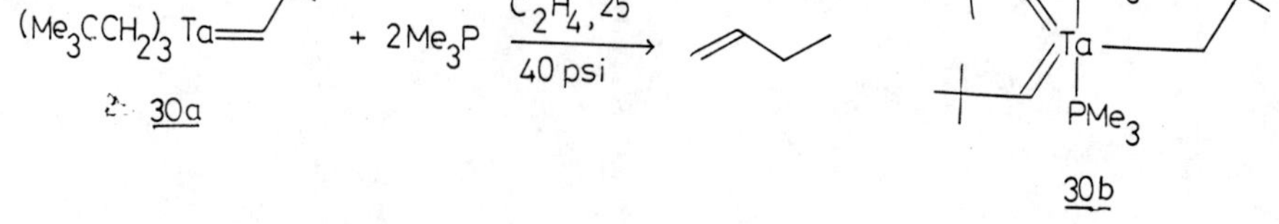

$(Me_3CCH_2)_3$ Ta $+$ 2 Me$_3$P $\xrightarrow[\text{40 psi}]{C_2H_4, 25^\circ}$

$\underline{30a}$

$\underline{30b}$

3 moles 3,3-dimethylbutene are formed, together with an active catalyst.
This consists of two compounds in the ratio 7:3, which may be observed
but not isolated. The major component was prepared independently from
31 and shown to be 33. A niobium analogue of 32 can be formed but is
inactive; this led to a proposal of a metallocyclic intermediate, 34,
which would be less favourable with niobium. Notably these complexes
are not isomerisation catalysts so 1-butene is obtained selectively. A
metallocyclopentane is also invoked as an intermediate in ethylene oligomerisation by zirconium and hafnium complexes; elimination of butadiene
(to yield Cp_2MH_2) and subsequent reinsertion, yields 37, which is
inactive and explains the rapid deterioration of the catalyst.[105]

$$Cp_2M-\| \quad + \quad \diagup\!\!\!\diagup \quad \longrightarrow \quad Cp_2M\diagdown$$

36   37

Dimerisation and oligomerisation of 1-alkenes introduces the further problem of regioselectivity. Consider the reaction of propene with a metal hydride. The products are hexenes, methylpentenes and dimethylbutenes, the proportions depending on the direction of each insertion and the extent of isomerisation. Nickel hydride catalysts abound and phosphine modification can yield predominently any of the desired products. (Table 3). Analysis of the results suggest that changes in product mixture result from changes in the second insertion step, with addition largely to C_2 in the first step. Selectivity is steric in origin[108]; the second step is more sensitive to hindrance and only the very hindered $P\underline{i}\text{-}Pr(\underline{t}\text{-}Bu)_2$ changes the selectivity of the first step.

Other complexes are active in dimerisation but have been less systematically studied. For example $Pd(acac)_2|EtAlCl_2|Bu_3P$ gives linear hexenes in low conversion but 95% selection.[109] Coordination cationic polymerisation occurs in the presence of $\underline{cis}\text{-}(M(NO)_2(MeCN)_4)(BF_4)_2$ (M=Mo,W). α-Methylstyrene is dimerised and trimerised (Figure 10) in a reaction which differs from that under acid catalysis.[110] Tantalum carbene complexes give slower reaction with $RCH=CH_2$ than C_2H_4; 38 and propene at 0° give 3,4,4-trimethylpentene and a metallocycle 39.[111] Longer chain alkenes give more "head to tail" dimerisation at still slower rates. 40 yields only 41 and the metallocycle 42 is not observable. Labelling and isotope effect studies show that the rate determining step is β-hydride elimination from 43.[112]

Table 3   Propene Dimerisation with Modified Nickel Hydride Catalysts

Catalyst Precursor	Phosphine	% hexenes	% methyl pentenes	% dimethyl butenes	Ref.
Ni oleate$\mid$R$_3$Al$_2$Cl$_3$	-	22.4	70.2	7.3	106
(η^{3}C$_3$H$_5$NiX)$_2$$\mid$Et$_3Al_2Cl_3$	-	19.8	76.0	4.2	81
(η^{3}C$_3$H$_5$NiX)$_2$$\mid$Et$_3Al_2Cl_3$	PPh$_3$	21.6	73.9	4.5	81
(η^{3}C$_3$H$_5$NiX)$_2$$\mid$Et$_3Al_2Cl_3$	PMe$_3$	9.9	80.3	9.8	81
(η^{3}C$_3$H$_5$NiX)$_2$$\mid$Et$_3Al_2Cl_3$	PBu$_3$	7.1	69.6	23.3	81
(η^{3}C$_3$H$_5$NiX)$_2$$\mid$Et$_3Al_2Cl_3$	PCy$_3$	3.3	37.9	58.8	81
(η^{3}C$_3$H$_5$)NiBr(PCy$_3$)$\mid$EtAlCl$_2$	PCy$_3$(-55°)	1.0	23.0	76.0	107
(C$_2$H$_4$)$_2$Ni(P$\underline{i}$Pr$_3$)$\mid$BF$_3$	P($\underline{i}$-Pr)$_3$	3.8	34.8	61.4	90
(η^{3}C$_3$H$_5$NiX)$_2$$\mid$Et$_3Al_2Cl_3$	P($\underline{i}$-Pr)$_3$	1.8	30.3	67.9	81
(η^{3}C$_3$H$_5$NiX)$_2$$\mid$Et$_3Al_2Cl_3$	P($\underline{i}$-Pr)$_2$$\underline{t}$-Bu	0.1	19.0	80.9	81
(η^{3}C$_3$H$_5$NiX)$_2$$\mid$Et$_3Al_2Cl_3$	P($\underline{i}$-Pr)($\underline{t}$-Bu)$_2$	0.6	70.1	29.2	81

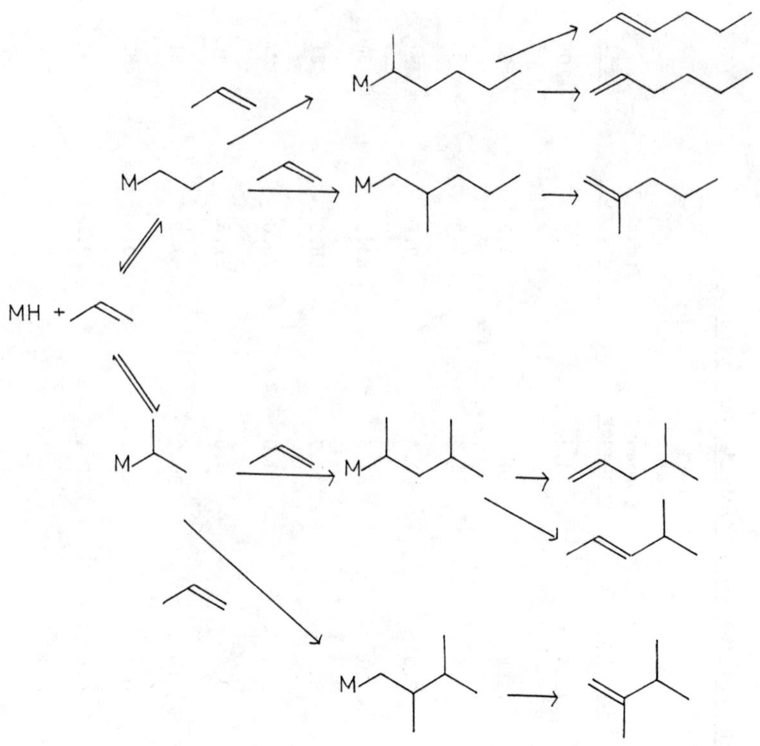

754

Figure 9   Reaction of propene with a metal hydride.

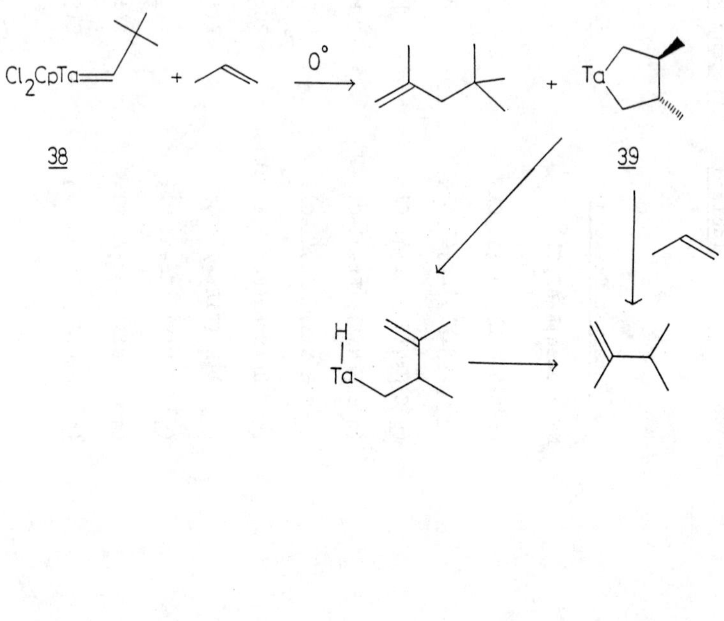

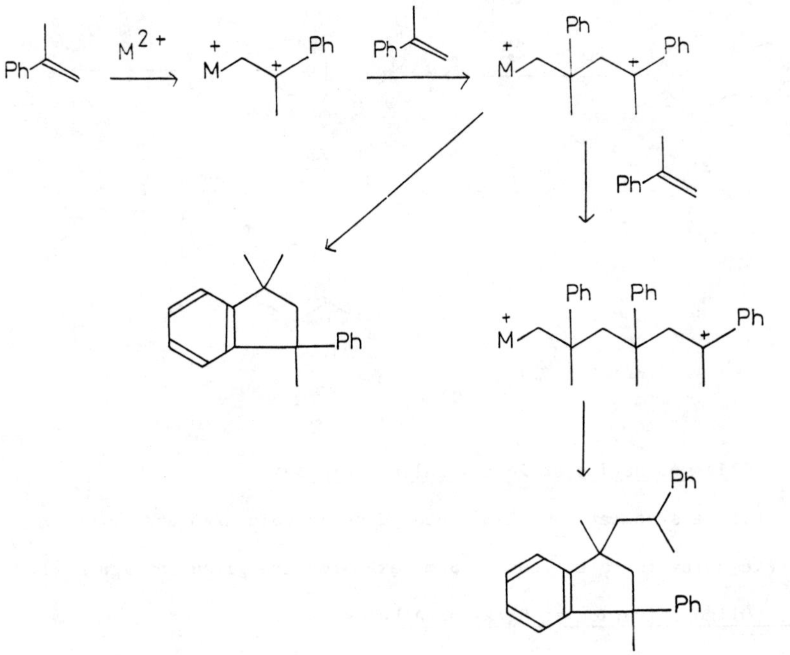

Figure 10 Coordination cationic polymerisation of α-methyl styrene.

40

41

42

$k_H/k_D = 3.3$

43   44

45

8.2.2. Oligomerisation of Functionalised Alkenes

Little systematic work has been done in this area and no mechanistic unity is to be found. Some examples are given in Figure 11.

8.2.3. Oligomerisation of Strained Alkenes

Alkenes containing strained double bonds undergo a richer variety of reactions than simple alkenes. The dimers of norbornadiene, 46, have been particularly intensely studied. The simplest are 47 and 48 ostensibly formed from $2\pi+2\pi$ cycloaddition. With bis(acrylonitrile)Ni as catalyst their proportions are related to the electronic character of the phosphine additive.[119] The intermediate proposed is a tricoordinate bis(alkene) nickel phosphine complex. $Ni(CO)_4$ as catalyst yeilds 47 and 48, together with the endo, endo-dimer[120] whilst with $(Ph_3P)_2Ni(CO)_2$ the exo-trans-exo-trans-exo-trimer, 49, is the main product.[121]

Rhodium (I)-catalysed reactions give a bewildering array of products. For example, $(nbd)_2Rh^+$ catalyses formation of 50 to 53[122] and Wilkinson's catalyst gives 51, 4 other dimers, of which the major is 54, and three trimers including 53.[123] With $Rh_2(OOCR)_4(R=CH_3,CF_3)$ as catalyst, 51 is the major product in ethanol whilst Binor-S, 55, is

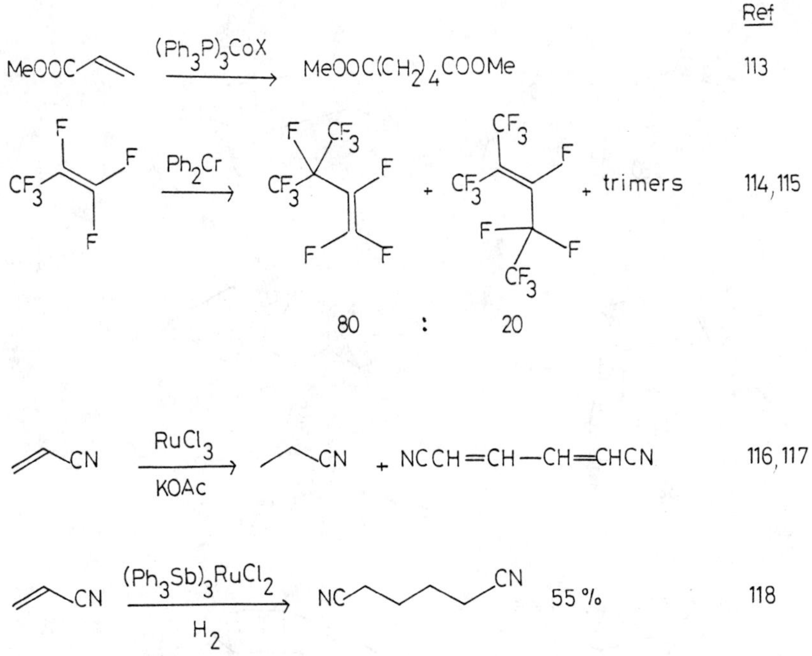

Ref

MeOOC⌒ $\xrightarrow{(Ph_3P)_3CoX}$ MeOOC$(CH_2)_4$COOMe   113

80 : 20   114, 115

⌒CN $\xrightarrow[\text{KOAc}]{\text{RuCl}_3}$ ⌒CN + NCCH=CH—CH=CHCN   116, 117

⌒CN $\xrightarrow[\text{H}_2]{(Ph_3Sb)_3RuCl_2}$ NC⌒⌒CN 55%   118

Figure 11   Oligomerisation of functionalised alkenes.

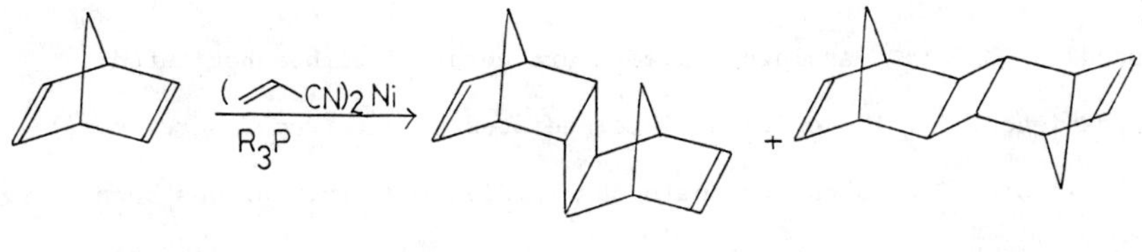

$\underline{46}$ $\underline{47}$ (endo-exo) $\underline{48}$ (exo-exo)

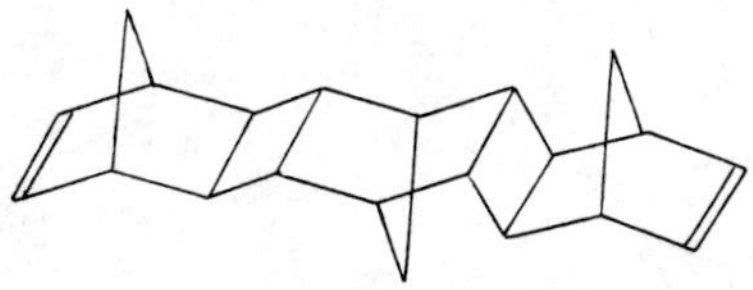

$\underline{49}$

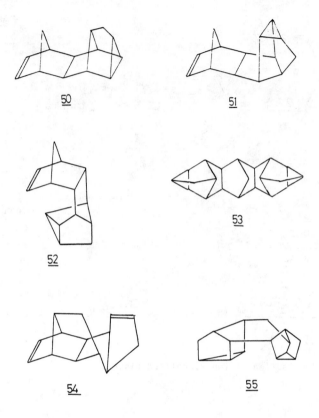

50

51

52

53

54

55

produced in DMSO.[124] Analogous dimers are obtained with substituted norbornadienes.[125] The critical intermediates result from insertion of rhodium into a C-C bond and an analogous iridium complex, 56, has been characterised by x-ray crystallography.[126] Cobalt complexes give Binor-S, 55, as the major product whatever their ligands.[127-131]

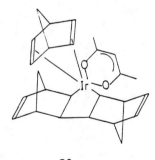

56

Cyclopropenes are readily dimerised and oligomerised in the presence of metal complexes. For 57 the anti-dimer, 58, is the major product in the presence of (cod)Ni(bipy), with a small amount of trimer 59.[132,133] When no nitrogen or phosphorus ligand is present a benzene soluble oligomer, 60, is formed.[134] With Pt(0) as catalyst, however, the phosphine free species gives 80% dimer whereas when $(R_1R_2CH)_3P$:Pt is 1:1 the trimer is the sole product in >90% yield.[135]

Methylenecyclopropane, 61, also gives a variety of products under conditions of metal catalysis, which are quite different from those produced thermally. Thermolysis (200°) gives 62 as the main product (77% at 63% conversion) together with a small amount of 63.[136] At −15° in the presence of $(cod)_2Ni$ the major product is 64 with a small amount of 63. The mechanism proposed involves a bis-alkene complex, 65, which on coupling gives 66. Reductive elimination yields 67 whereas rearrangement followed by elimination gives 64.[137] Some higher oligomers are formed and their yield is enchanced with $TiCl_4|Et_2AlCl$ as catalyst. If the reaction is run in the presence of an activated alkene the intermediates may be intercepted (Figure 12).[138] The differing products from Pd(0) and Ni(0) reflect the difference in stabilities of the trimethylene methane complex.

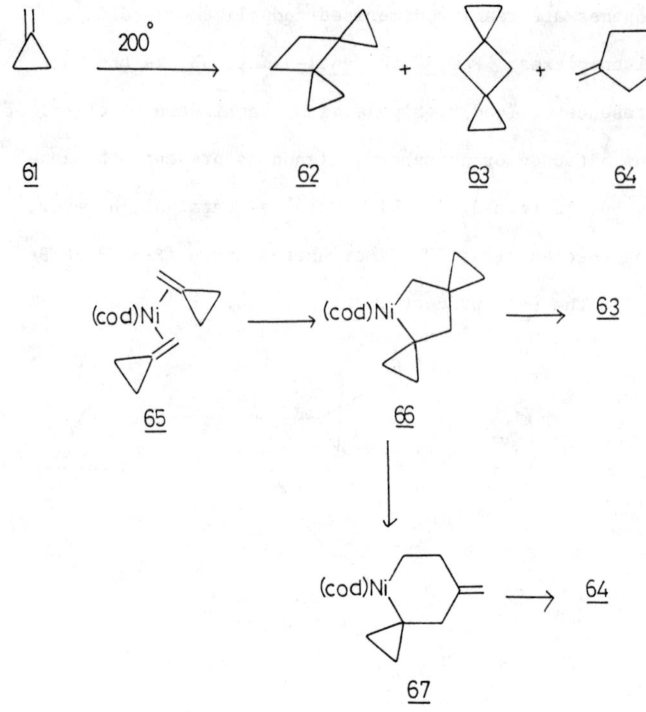

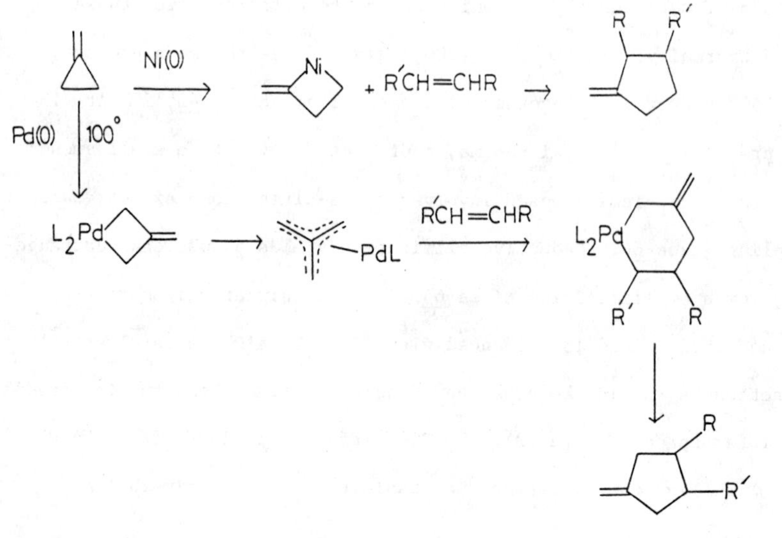

Figure 12 Cycloaddition of methylenecyclopropane catalysed by Ni(0) and Pd(0) complexes.

8.2.4. Codimers and Cooligomers of Alkenes

Successful codimer preparation requires supression of homooligo-merisation and few useful systems are known. $[R_3R'P^+][R_3PNiCl_2]^-$ (R=iPr,|R'=Bz)|Et$_3$Al$_2$Cl$_3$ catalyses formation of ethylene|butene codimers with up to 85% selection. The major product is E-3-methyl-2-pentene in 52% yield and the active species is probably a nickel hydride. Codimerisa-tion of styrene and ethylene occurs with Ni and Pd complexes as catalyst. 68 (Ar=mesityl) gives up to 91% selectivity for 3-aryl-1-butene.[140] It is equally useful for reaction with propene (giving 4-phenyl-1-pentene) and butene (giving 4-phenyl-2-hexene).[141] An analogous palladium complex was less efficient.[142]

68

69

Enantioselective cooligomerisation has been achieved; norbornene and ethylene yield 69 in the presence of a nickel complex of dimenthyl-methylphosphine, the optical yield rising to 80% at -100°.[143] Fortunately the catalyst is sufficiently active to allow the reaction to occur at the low temperatures which give the maximum selectivity advantage. In codimerisation of styrene and ethylene to chiral 3-phenyl-1-butene the optical yield is dependent on the counterion.[144] An alternative approach using aminophosphine ligands, 70 and 71, gives coupling of cyclohexadiene and ethylene to 72 (97% selectivity at 90% conversion) in up to 74% enantiomer excess.[145] This is clearly a reaction with future potential but has not yet been studied systematically.

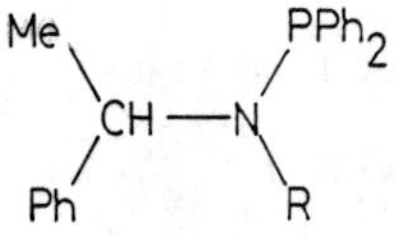

70 R=H,Me

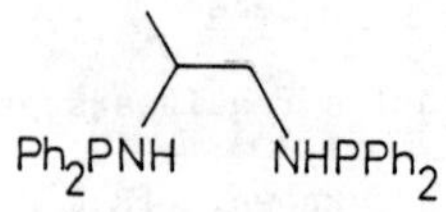

71

72

8.2.5. Cyclisation Reactions of Alkenes

$2\pi+2\pi$ cycloadditions are Woodward Hoffman forbidden but are relatively common in coordination catalysis. Some intramolecular reactions of norbornadienes were considered earlier, as isomerisations. Ethylene is dimerised in the presence of 73 to yield cyclobutane, and propene gives 1,2-dimethylcyclobutane reflecting the relative stabilities of the possible intermediate metallocycles.[146] The cumulene, 74, is dimerised by $(Ph_3P)_2Ni(CO)_2$[147], as is the strained alkene 75.[148]

Other strained alkenes undergo $2\pi+2\pi$ cycloadditions. For example, the norbornene derivative, 76, reacts with butadiene to yield

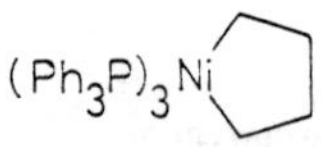

$(Ph_3P)_3Ni$

73

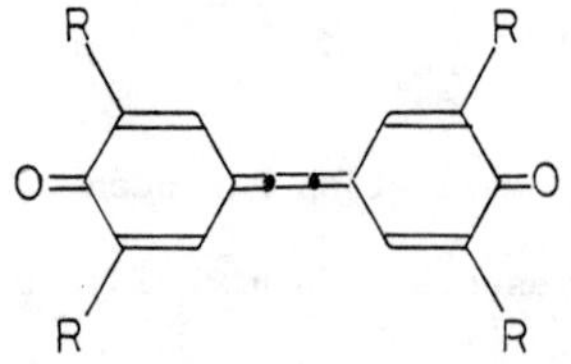

74

75

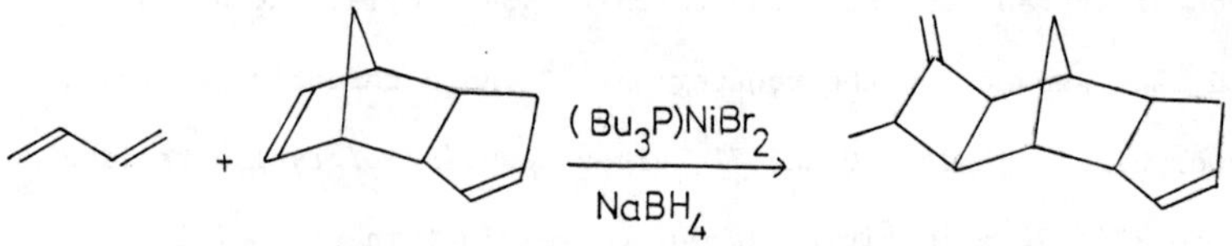

76

$(Bu_3P)NiBr_2$

$NaBH_4$

77

77, _via_ a stepwise process involving a nickel allyl complex.[149,150]
Alkynes also cycloadd to norbornene to give _exo_-cyclobutenes such as 79,
no homodimers being detected.[151,152] Activated alkenes react in the
same way with 80, the process being sufficiently selective that the
three-membered ring is not opened.[153]

78   79   56-87%

80   81   63%

82   37%

$2\pi+2\pi$-addition is also promoted by copper triflate when this is
photoactivated. Norbornene yields the _exo_, _exo_-(2+2)-dimer probably
via photoexcitation of a 2:1 norbornene copper complex.[154] Cyclopentene
gives a (2+2)-dimer in low yield.[155] Intramolecular (2+2)-addition of
1,6 dienes also occurs with this catalyst, particularly when there is
an oxygen functionality to bind copper.[156,157] 83 is converted to 84
in 80-90% yield; the _endo_-product predominated by >20:1 since the complex
85 is favoured over 86.

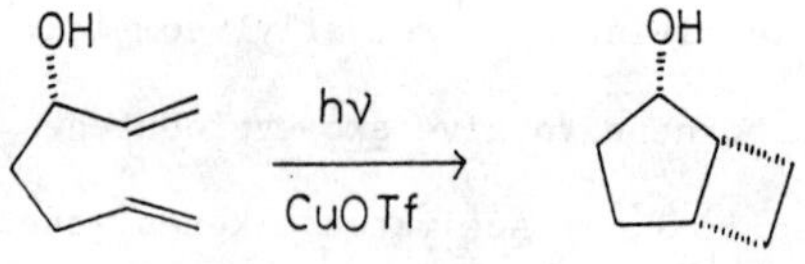

83

84

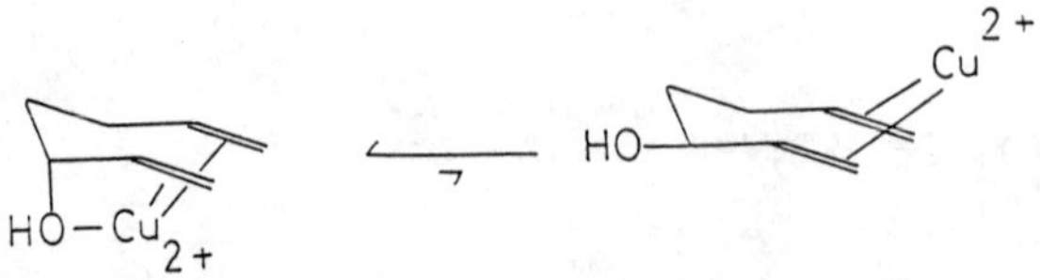

85

86

Methylene cyclopropane undergoes cyclisation with a number of reactive alkenes in the presence of Ni(0)[158,159] or Pd(0)[160] catalysts. (Figure 13) Labelling studies of the bis(acrylonitrile)Ni catalysed reaction with dimethyl fumarate were consistent with the intermediacy of the trimelhylenemethane complex 87.[163] Molecular orbital calculations establish the importance of the zwitterionic form 87b.[164] The analogous palladium complex is more usually generated from an allylic acetate bearing a trimethylsilyl group (88 to 89).[165] It is extremely effective in cycloadditions with alkenes bearing electron-withdrawing groups, and where the complex is non-symmetric, equilibration (90a > 90b) is faster than cycloaddition.[166] The most anionic carbon then attacks the alkene (Figure 14). Bicyclobutanes and bicyclopentanes react with M(0) complexes to give the products of cycloaddition but in these cases deuterium labelling suggests an alkyl metal, 92, as an intermediate.

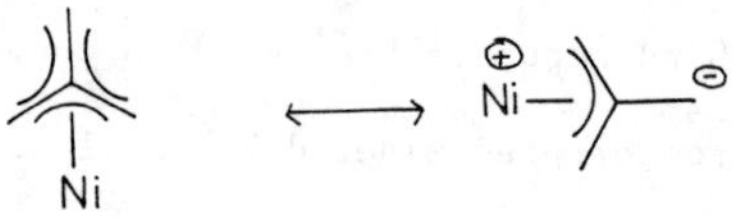

87 a

87 b

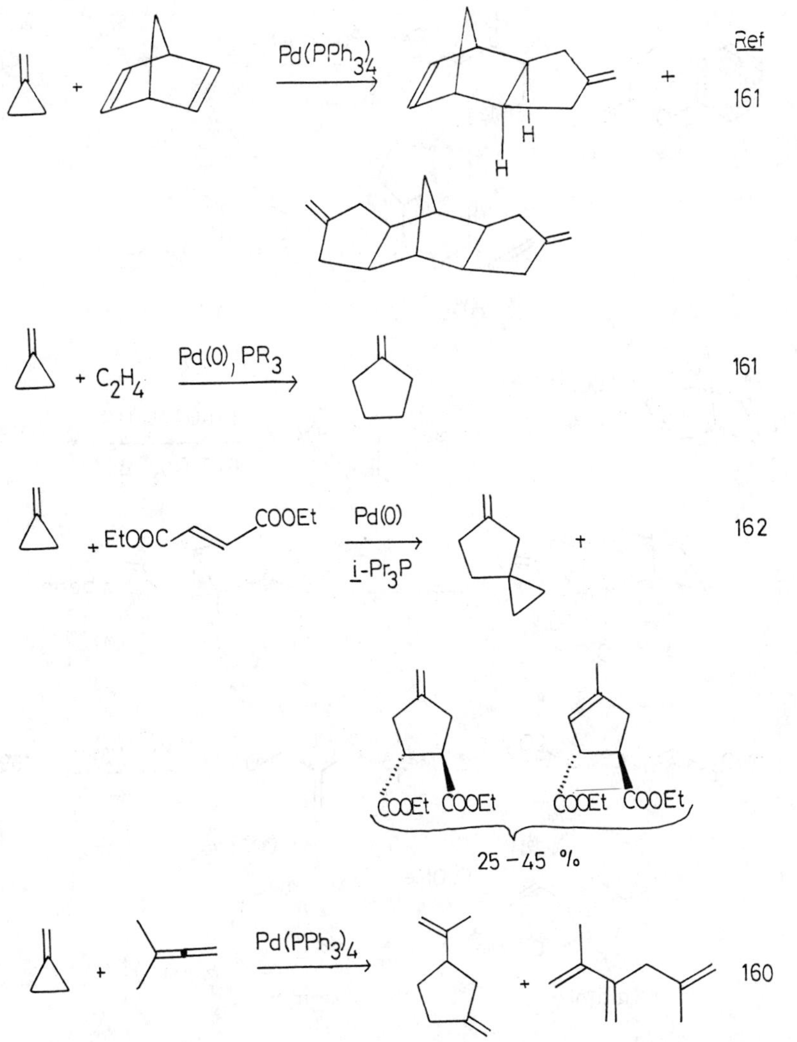

Figure 13 Cycloaddition via trimethylenemethane metal(0) complexes.

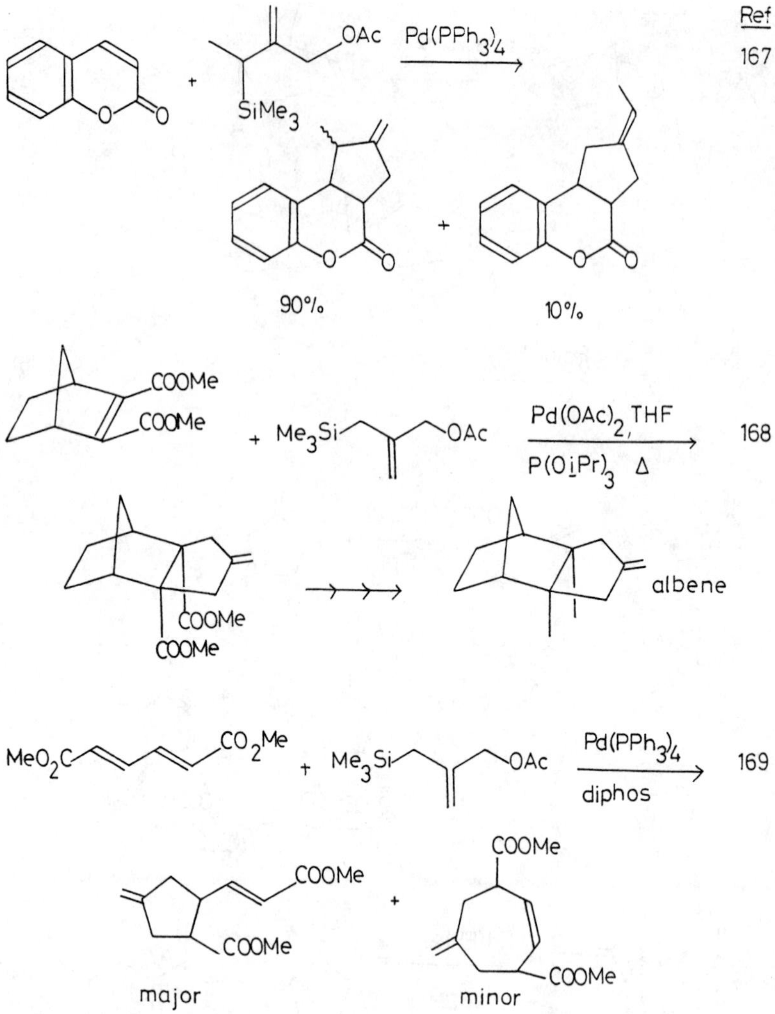

90%   10%

168

albene

169

major   minor

Figure 14   Reactions of palladium trimethylenemethane complexes with
electron-poor alkenes.

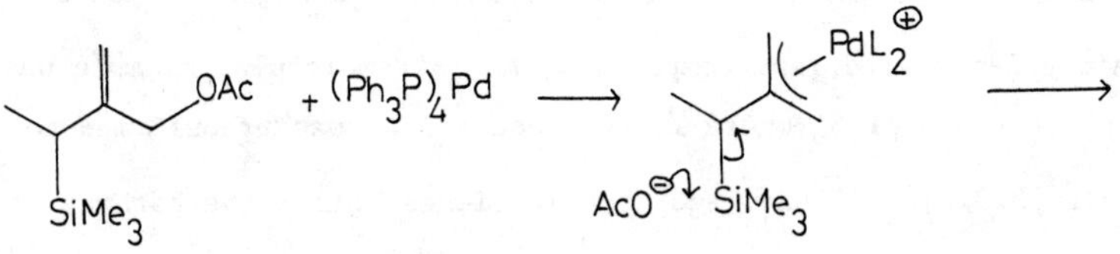

88

89

90a

90b

Ni(0)

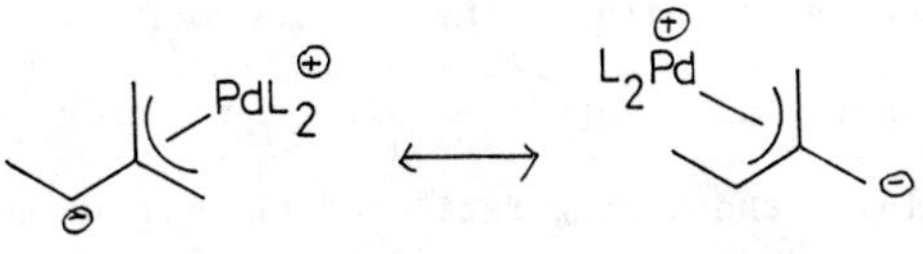

91

92

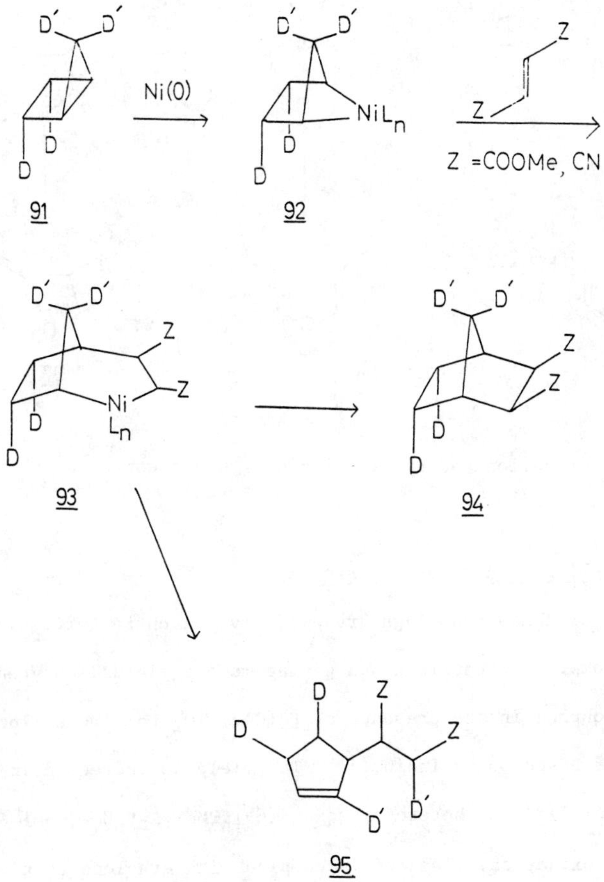

Z = COOMe, CN

93

94

95

1,5-and 1,6-dienes undergo isomerisation and cyclisation to give mainly 5-membered ring compounds. An intramolecular example was seen earlier (Chapter 5) where 1,5-cyclooctadiene was transformed to cis-bicyclo[3.3.0]oct-2-ene, and acyclic dienes follow the pathway of Figure 15. The best yields of 97 are obtained with bulky phosphines such as P($\underline{i}$-Pr)$_3$ and the counterion also exerts an influence, with ClO$_4^-$ being the most satisfactory. If the reaction of 96 or 1,6-heptadiene occurs in the presence of chiral methyl $\underline{t}$-butylmenthylphosphine an optical yield of up to 40% is achieved with the configuration at phosphorus determining the configuration of the product.[81]

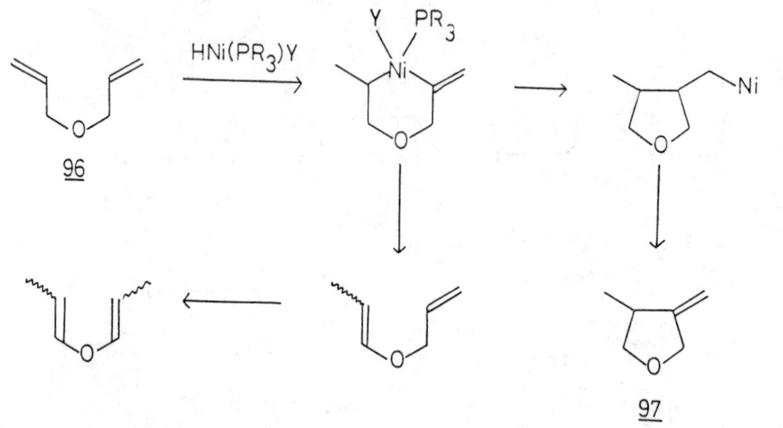

Figure 15   Isomerisation and cyclisation of 1,6-dienes.

8.2.5.   Coupling of Alkenes and Arenes

Most of these couplings are oxidative in character, involving a palladium complex as catalyst and giving modest yields. Styrene and benzene are coupled in the presence of Pd(OAc)$_2$|O$_2$ to give stilbene in 1100% yield based on palladium.[170] The yield is increased in the presence of Cu(II)[171] or heteropolyacids (H$_{3+n}$PMo$_{12-n}$V$_n$O$_{40}$, n=1-6)[172] as relays for oxidation. Benzene is coupled with ethylene to give some styrene but mainly stilbene.[173,174] Mechanistic studies suggest that a trans-(arene)(alkene) Pd(OAc)$_2$ complex is initially formed and isomerised to the cis-species, before losing H$^+$ in a rate determining step to give the σ-aryl, 98.[175]

98

Another useful coupling is exemplified by 99 to 100 (X=Me,OMe, COOEt, Cl, Br, NO$_2$).[176] The diazo compound may be generated in situ using t-butyl nitrite.[177,178] Synthesis of disubstituted alkenes generally gives better results than for 99 to 100, since the products undergo polymerisation less readily. (Figure 16). The reaction is thought to proceed via an aryl Pd(II) complex which adds across the carbon-carbon double bond, followed by cis-elimination of HPdX.[179,180]

99   100

8.3 Polymerisation and Oligomerisation of Dienes

8.3.1. Polymerisation and Oligomerisation of 1,2-Dienes

Allene reacts in the presence of Ni(0) complexes to give mainly cyclooligomers 101, 102 and 103.[181] (Cod)$_2$Ni alone gives mainly the pentamer, 103,[182] addition of phosphine yielding more 102.[183] The mechanism is established; the initial complex is observed with substituted allenes to be 104 and 105 is stable giving rise to 101 by metal extrusion.

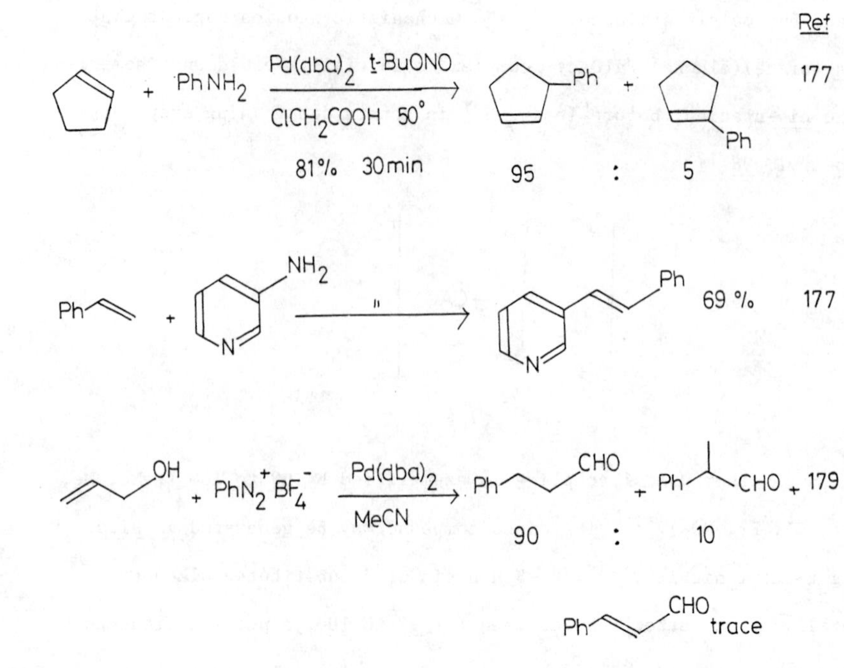

Figure 16 Palladium catalysed coupling of diazo compounds with alkenes.

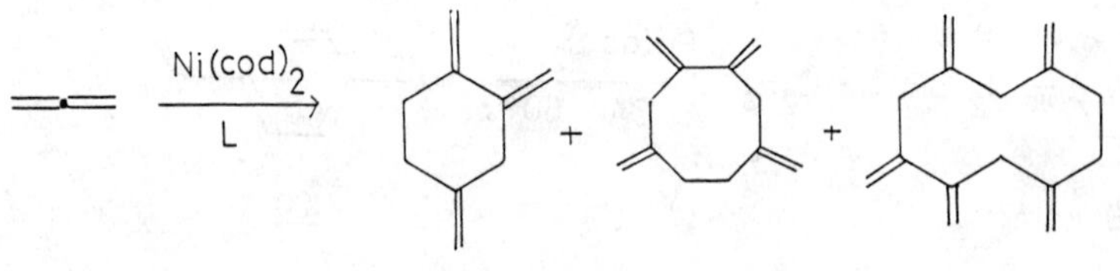

101 102 103

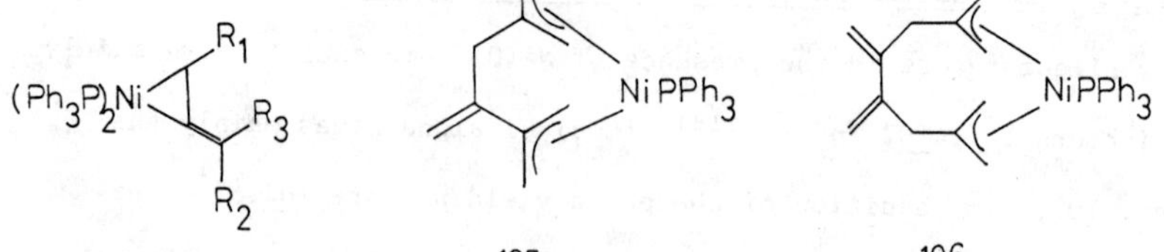

104 105 106

Tetramer and pentamer are formed in the same way after further insertion and 106 has been characterised.[184] If conditions are more vigorous (200°, [$Ph_2PC_6H_4PPh_2Ni(CO)_2$]$_n$) cyclobutane dimers, 107 and 108, are formed in addition to 101.[185] Palladium (0) catalysts have been less exploited, mainly because of their strong preference for reaction with 1,3-dienes.[186] Reactions catalysed by rhodium complexes give contrasting results. In the presence of ($RhCl(C_2H_4)_2$)$_2$ 103 is the major product but when Ph_3P is added tetramer, 109, is formed in 80% yield.[187] There is no convincing mechanistic evidence to explain its origin.

107   108   109

In polymerisation two selectivites are known for allene, viz. 1,2,2,1, 110, and 1,2,1,2, 111. 110 is obtained from catalysts $Pd(NO_3)_2$ | PPh_3 | $TsOH$[188] or $PdCl_2(PhCN)_2$[189] whereas 111 is the main product with $Co(acac)_2$ | i-Bu_3Al[190], $(Ph_3P)_3CoCl$[182,187], $Ni(acac)_2$ | R_nAlX_{3-n}[191], $VOCl_3$ | R_3Al[192] or [$(diphos)Rh(CO)_2$]$^+$.[193] Nickel allyl complexes[188] especially in the presence of pyridine[194] give mixtures. Little is known of the mechanism but ESR and NMR data imply that a Ni(I) intermediate is involved.[195]

110   111

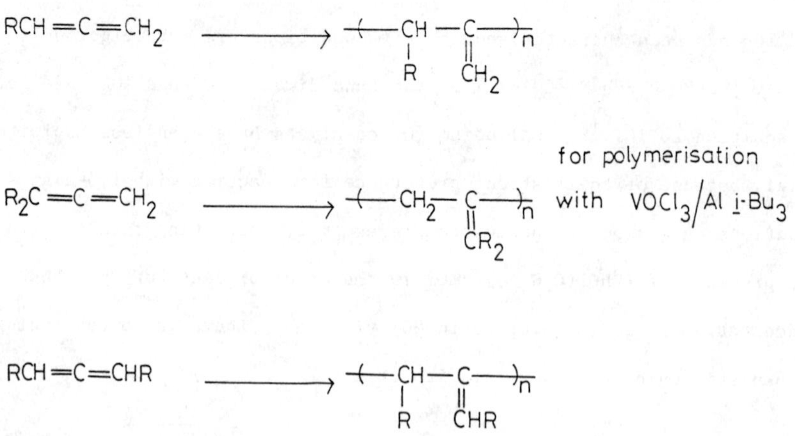

RCH=C=CH$_2$ $\longrightarrow$ $+$CH$-$C$\xrightarrow{}_n$
  $\quad$ R $\quad$ CH$_2$

R$_2$C=C=CH$_2$ $\longrightarrow$ $+$CH$_2$$-C\xrightarrow{}_n$ with VOCl$_3$/Al i-Bu$_3$
  $\quad\quad\quad$ CR$_2$

for polymerisation

RCH=C=CHR $\longrightarrow$ $+$CH$-$C$\xrightarrow{}_n$
  $\quad$ R $\quad$ CHR

Figure 17 Polymerisation of substituted allenes.

Substituted allenes are polymerised more slowly, both reactivity
and selectivity depending on steric considerations (Figure 17).[196]
In the presence of nickel complexes, however, 1,2-butadiene yields the
1,2-polymer, 112, in 80 - 90% selectivity.[197] Chiral allenes give chiral
polymers.[198] Heterosubstituted allenes such as 113 follow similar
selectively rules, 114 being obtained in >90% yield with NiCl$_2$|Et$_3$Al,
(n^{3}C$_3$H$_5$NiX)$_2$ or CoCl$_2$|Et$_3$Al, 115 with TiCl$_4$, VOCl$_3$ or FeCl$_3$|Et$_3$Al and
meso-116 with most palladium catalysts.[199]

$+$CH$_2$$-C\xrightarrow{}_n$
  $\quad\quad$ CH$_3$

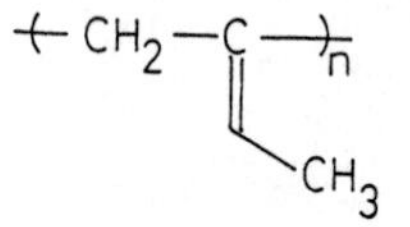

112

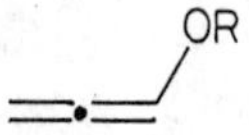

$=$•$=$OR

113

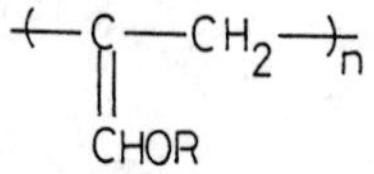

$+$C$-$CH$_2$$\xrightarrow{}_n$
  CHOR

114

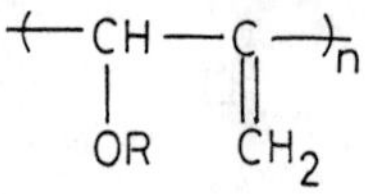

$+$CH$-$C$\xrightarrow{}_n$
  OR $\quad$ CH$_2$

115

$+$CH$-$C$-$C$-$CH$\xrightarrow{}_n$
  OR $\quad$ CH$_2$ CH$_2$ OR

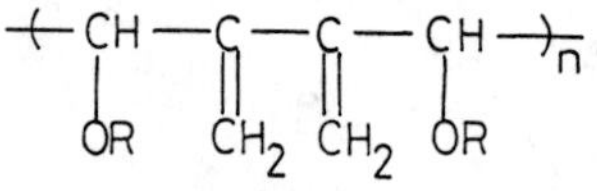

116

8.3.2. Polymerisation of 1,3-dienes

The importance of 1,3-diene polymers lies in their similarity to natural rubber, which arises from a cis-1,4-polymerisation of isoprene to 117. For butadiene polymers three distinct structural types may be produced with appropriate catalysts (118-120). Active complexes fall into several broad groups, the first two of which are related to the Ziegler catalysts used to polymerise alkenes. Titanium Ziegler catalysts are very active and highly selective for 118.[200] Reaction conditions are important and addition of iodide is particularly helpful to selectivity.[201] (Table 4). Most authors consider that the propagating species is a metal allyl[206,207], but e.s.r. studies make it clear that the system is complex.[206-209]

117 118 cis -1,4 119 trans-1,4

120 1,2

Hydrocarbon soluble cobalt salts are truly homogeneous catalysts and are used commercially to give 96-98% selectivity for cis-1,4-poly-butadiene.[210] Industrial catalysts are prepared from Co(II) carboxylates and $R_3Al_2Cl_3$,[211,212] and another useful soluble precursor is $(py)_2CoCl_2 | R_nAlCl_{3-n}$; all give similar selectivities.[213-218] In halogen free systems 1,2-polybutadiene is obtained with high syndiospecificity.

774

Table 4   Butadiene Polymerisation by Titanium Catalysts

Catalyst	Conditions	% cis-1,4	% trans-1,4	% 1,2	Ref
TiCl$_4$|AlCl$_3$|Ni(PCl$_3$)$_4$	toluene, 17 hr 20°	92	-	-	202
Bz$_3$TiI	toluene	73.5	20	6.5	203
Bz$_3$TiI	toluene	94-97%	-	-	204
Bz$_4$Ti	pentane, 6 hr, 23°	20	20	60	205

π-Allyl nickel complexes are of little commercial importance but have been well studied with regard to their specificity of action, which is varied, but predictable.[219] The catalyst is typically a π-allyl nickel halide/additive but analogous species are generated from other nickel salts in the presence of butadiene.[220,221] π-Allyl nickel chlorides give a high percentage of cis-1,4-polymer whereas the iodides give the more stable trans-1,4-product.[222-225] Adding Lewis or protic acids varies this considerably allowing the preparation of "block" polymers with regions of each stereochemistry.[226] (Table 5)

The observed stereoselectivity is explained by the orientation of coordination of butadiene. Initiation occurs by conversion of the π-allyl to a σ-allyl and monomer binding (121 → 122); however, if the catalyst is a π-crotyl derivative only 10-20% of the crotyl groups are incorporated into the polymer.[238-240] The anti-stereochemistry of 123 then derives from the bidentate coordination of butadiene. On further insertion 123 gives cis-1,4-polymer.[241] π-Allyl nickel iodide[224] (or any catalyst with a strong donor) is 5-coordinate and butadiene is bound in a monodentate manner (124) Subsequent insertion gives 125 and 126.[242] That the propagating species is a π-allyl is suggested by NMR studies[243] but other authors[234] interpret the results in terms of σ-allyls and steric control or ligand effects on bond reactivity.[222,233]

121   122   123

Table 5   Polymerisation of Butadiene by $\underline{\pi}$-allyl Nickel Complexes

Catalyst	Additive	Additive/Catalyst	% cis-1,4	% trans-1,4	% 1,2	Comments	Ref
$(\eta^3-C_3H_5NiOCOCF_3)_2$	N-methylphthalimide	5-25	40	50	10	CH_2Cl_2|Solvent	227
$(\eta^3-C_4H_7NiCl)_2$	$CoCl_2$	1	94	5	1		228
$[\underline{\eta}^3-C_4H_7Ni(P(OEt)_3)_2]^+PF_6^-$	-	-	-	>95%	-		229
$(\underline{\eta}^3-C_3H_5NiOCOCF_3)_2$	excess CF_3COOH	>2	50	50	-	PhCl solvent, truly equibinary	230-232
$(\underline{\eta}^3-C_3H_5NiOSO_2p-MeC_6H_4)_2$	-	-	50	50		Truly equibinary	230
$(\underline{\eta}^3-C_3H_5NiOCOCF_3)_2$	pentane	-	>95%	-	-		233, 234
$(\underline{\eta}^3-C_4H_7NiCl)_2$	CCl_3COOH	2	>95%	-	-	10-40°, some radical reaction may occur	235
$(\eta^3-C_4H_7NiI)_2$	CCl_3COCCl_3	2	92	5	3	-	236
Ni-1,2-diyl-2,6,10-dodecatriene	CH_3SO_3H / HCl / HI / CF_3COOH	1 / 1 / 1 / 1	44 / 84 / - / 91	51 / 13 / 100 / 4	5 / 3 / - / 5		237

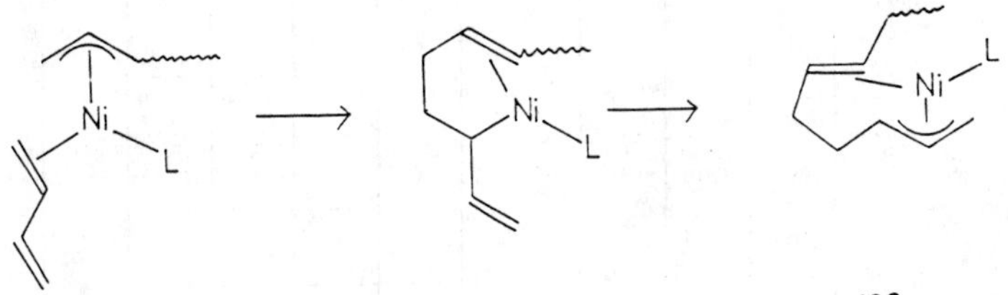

124   125   126

Other metal allyl catalysts are active and some give high
selectivities, but they have not been subjected to systematic scrutiny.
Examples are given in Table 6. A few other useful catalysts are known;
their mechanisms of action are varied and not conveniently classified.
(Table 7)

8.3.2.1. Polymerisation of Substituted 1,3-Dienes

Polymerisation of substituted dienes proceeds more slowly than
for butadiene and lower oligomers are often obtained.[255] Consider the
polymerisation of isoprene, 127. For the 1,4-trans-mode we may have
head to head, head to tail and tail to tail pairings shown respectively
as 128, 129 and 130. Additionally there are two vinylic polymers-1,2
and 3,4, viz. 131 and 132, which may have varying tacticity. The micro-
structure of the polymer has not always been studied in detail but 129
appears to be the most common.[210] Some examples are given in Table 8;
selectivities follow those established for butadiene and the explanations

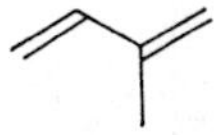

127 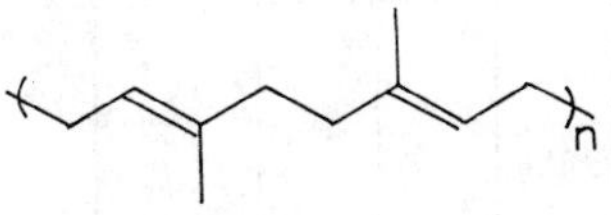   128

Table 6   Polymerisation of Butadiene by Metal Allyl Complexes

Catalyst	Conditions	% cis-1,4	% trans-1,4	% 1,2	Ref.
Cl_2Ru(dodeca-2,6,10-triene-1,12-diyl)	Inactive without added Ph_3P	37-60	20-32	13-34	244
$(\eta^3-C_3H_5)_3Cr$		-	17	83	245
$(\eta-C_3H_5)CrCl_2$	+ 2HCl or $2CCl_3COOH$	90	-	-	245
$(\eta^3C_3H_5)CrCl_2$	O_2	-	93	-	245
$(\eta^3-C_3H_5)_4Mo$	CCl_3CHO or $(CH_3)_2CHCH_2I$ as activator; needs H_2O	-	-	>95	246, 247
$(\eta^3-C_4H_7)_4Co$	+ HCl 22% yield	7.5	10	82.5	248
$(\eta^3-C_3H_5)_4U$	+ HX, $TiCl_4$	>99	-	-	249
$(\eta^3-C_4H_7)_3Nb$	+ HCl, $ZnCl_2$	85	11.5	3.5	250
$(\eta^3-C_4H_7)_3Rh$	+ HCl	-	93	7	250
$(\eta^3-C_3H_5MoC_7H_8)_2$	benzene as solvent	-	-	100	251

Table 7   Miscellaneous Catalysts for Butadiene Polymerization

Catalyst	Conditions	% cis-1,4	% trans-1,4	% 1,2	Ref.
NdCl$_3$|Et$_3$Al	THF solvent	97	-	-	252
Rh(acac)$_3$|Et$_2$AlCl	CH$_3$(CH$_2$)$_3$Cl solvent	-	>98%	-	253
HRuCl(PPh$_3$)$_3$	DMA solvent, 80°	50	50	-	254
[Rh(NO)(NCMe)$_4$](BF$_4$)$_2$	-	-	>95%	-	255
Fe(acac)$_3$|$\underline{i}$-Bu$_3$Al	Phenanthroline	50	-	50	256

Table 8   Polymerisation of Isoprene

Catalyst	Conditions	% cis-1,4	% trans-1,4	% 1,2	% 3,4	Comments	Ref.
Co(naphthenate)|Et$_2$AlOEt		50			50	Equibinary	210
($\underline{\eta}^3$-C$_3$H$_5$)$_3$Cr	toluene, 50°	←—— 25% ——→		34	41		257
R$_3$TiI		87	-	-	13		258
R$_3$TiI|PPh$_3$	P:Ti=1	6	11	7	76		258
R$_4$Ti		-	75	17	18		258

129

130

131

132

again involve relative stabilities and rates of interconversion of allyl isomers.[259]

Penta-1,3-diene is also polymerised[260] but only the _trans_-isomer is reactive and the _cis_-isomer must be converted to it in order to react.[261,262] A predominently _cis_-1,4-polymer is obtained with $(BuO)_4Ti|Et_3A\ell$[263], $Co(acac)_2|Et_2AlCl$[262], $Ni(acac)_2|Et_2AlCl$[264] or $Nd(OCOR)_3|Et_2AlCl$.[264] Titanium catalysts give mainly isotactic structures[263], confirmed to be diisotactic by deuterium labelling[265] whereas cobalt complexes yield mainly syndiotactic polymer.

8.3.2.2. Copolymers of 1,3-Dienes

Diene copolymers have few commercial uses and have not been extensively studied. The proportions of monomers incorporated generally depend on their relative reactivities, but alternation is common, for reasons which are poorly defined. Alternating copolymers are formed from butadiene and 133[227] or propene.[266] In the latter case with $TiCl_4|\underline{i}-Bu_3A\ell$ as catalyst 90-100% of diads are obtained[267] and the butadiene units are mainly 1,4-_trans_.[268,269] Changes in regio- and stereo-selectivity may be rationalised on the basis of allyl stabilities at the growing end of the chain.[270] For example the copolymer of 1,3-pentadiene

133

134

135

and isoprene is enriched in vinyl units relative to the homopolymer[271] and copolymerisation of butadiene and 134 in the presence of $\underline{n}^3$-C_4H_7-Ni catalysts gives up to 30% of the unexpected 1,4-trans structure.[272]

8.3.3. Oligomerisation of 1,3-Dienes

8.3.3.1. Oligomers of Butadiene

Butadiene is dimerised to 1,3,7-octatriene, 135, at 100-120° in aprotic solvents in the presence of a variety of palladium catalysts.[273] Particularly effective as catalysts are 136 and 137[274], although $Pd(PPh_3)_4$ and $Pd(PCy_3)_3$ are also useful.[275] Some protic solvents are unsuitable but the reaction proceeds smoothly in isopropanol. With

136

137

$(CH_3)_2CHOD$ the major deuteration is at C_6 (138). The mechanism involves formation of the bis-π-allyl complex, 139, in equilibrium with the $\underline{\sigma}$-allyl, 140; protonation from the solvent occurs at C_6 while elimination of H occurs at C_4. A slightly different reaction occurs in formic acid[276], which behaves as a reductant, and an octadiene, 141, is the major product.[277] Improved selection for 142 or 143 is obtained by control of

138   139   140

+ HCOOH $\xrightarrow[\text{NaOPr, 80}^{\circ}\text{, 8hr}]{(Ph_3P)_2PdCl_2}$ 141   + CO_2

142   +   143

reaction conditions. $Pd(OAc)_2|Et_3P|Et_3N|DMF$ at 50°-70° gives 95% yield

with 93% selectivity for 142 whereas omitting the phosphine and warming to

90° yields 143 in 99% selectivity[278]. Carbon dioxide pressure raises the

reaction rate and the rate of isomerisation of 135 to octa-2,4,6-tri-

ene.[275,279]

With nickel complexes as catalysts the reaction is more complex

and the main products are the cyclooligomers 144 to 152.[280] Wilke ex-

plained the difference between these and the palladium catalysed reactions

in terms of the atomic volumes of the metals and the tendency of Pd to

abstract H from a coordinated chain to give the intramolecular H-shift

required for the formation of linear oligomers.[95] The pathways proposed

for formation of the cyclodimers are shown in Figure 18. The key discovery

was that $Ni(acac)_2|R_3Al$ yielded a very reactive form of "naked nickel"

with a high affinity for dienes.[281] Two moles of butadiene are coordinated

to give the bis-allyl complex, 153. An analogue, 154, is isolable and

has been studied at low temperatures.[282] In the absence of donors further

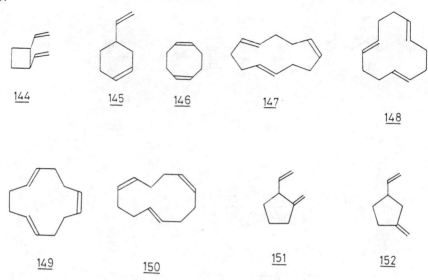

144 145 146 147

148

149 150 151 152

butadiene is coupled to give 155 which cyclises to 156. The outcome of
the reaction is controlled by the addition of coordinating ligands.[282-286]
In a systematic study it was found that at low L:Ni trimers and tetramers
dominated whereas at higher L:Ni further coordination of monomer is
inhibited and dimers predominate.[287] Using cis, cis-1,4-dideuterobutadiene
as substrate the reaction was shown to be highly stereospecific.[288]
However, an attempt to prepare chiral 145 in the presence of a chiral

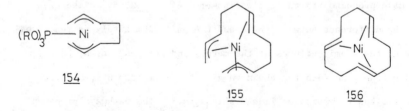

154

155 156

phosphite was only marginally successful (11.6% enantiomer excess).[289]

　　Numerous other metal complexes give analogous results but the
systems are less well characterised. The commercial synthesis of cyclo-
dodecatriene uses $TiCl_4 | Et_3Al_2Cl_3$ to give the cis, trans, trans-isomer,
147, in 70-90% selectivity.[290-292] Some other examples are given in Table 9.

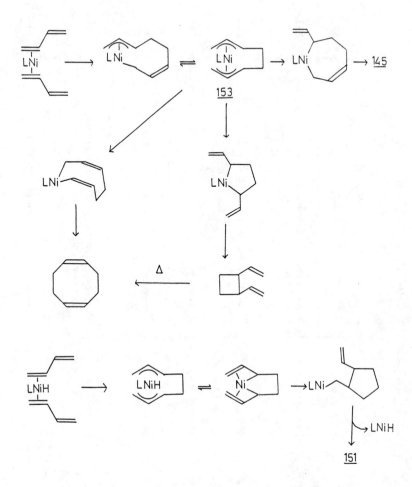

<u>Figure 18</u> Cyclodimerisation of butadiene by nickel catalysts.

Table 9   Catalysts for Oligomerisation of Butadiene

Catalyst	Conditions	Products	Ref.
(Imidazole)Rh(CO)$_2$Cl	MeOH, EtOH or acetone	4-vinylcyclohexene, 1,5-cod, 1,4-trans polymer	293
(Imidazole)Rh(CO)$_2$Cl	DMF	4-vinylcyclohexene	293
Cl(CO)$_2$Rh–L–Rh(CO)$_2$Cl	L=m-phenylenediamine	1,5-cod.	293
Fe(CO)$_2$(NO)	80°	4-vinylcyclohexene, up to 99% selective	294
(cot$_2$)Zr		trans, cis-1,3,6-octatriene; reaction is stereospecific	295
Et$_2$Fe(dipy)$_2$		4-vinylcyclohexene:1,5-cod in ratio 1:3	296

8.3.3.2. Oligomers of Substituted 1,3-Dienes

Linear head to tail dimerisation of isoprene, 127, is a useful model for terpene synthesis. The reaction is slower than for butadiene but some useful selectivities may be achieved (Table 10). Reaction in the presence of formic acid gives partially reduced products, used by Heck in a synthesis of α-and β-citronellals[301] (Figure 19).

127	157	158	159	160
	tail to tail	head to head	head to tail	tail to head

Substitution induces major changes in nickel catalyses cyclo-oligomerisations. Isoprene is less reactive than butadiene and inhibits its dimerisation in the presence of $(Ph_3P)_2NiBr_2|NaBH_4$.[302] However, addition of amines yields a range of linear dimethyloctadienes and trienes as well as dimethylcyclooctadienes and vinylcyclohexenes.[303] Propylamine increases the proportion of dimethyloctatriene, whereas formation of cyclic dimers is promoted by γ-picoline and pyridine, suggesting that the effect is both steric and electronic.[304,305] The best reported selectivity is for 165, 90%, achieved with $Ni(0)|Bu_3P|(CH)_2NH$.[306] Evidence for the reaction mechanism comes from treating (cyclododecatriene nickel (PCy_3)) with isoprene to yield 166, which has been studied crystallographically.[307]

165

Table 10   Dimerisation of Isoprene

Catalyst	157	158	159	160	trimers	Ref.
diphos PdBr$_2$|PhONa| PhOH (trace)	89	-	-	6	12	298
diphos PdBr$_2$|PhONa| PhOH (excess)	9	19	23	34	34	298
136|acetone	98	-	-	-	25	299
(Et$_3$P)$_2$Pd(CO)$_2$	up to 86	-	-	-	-	300

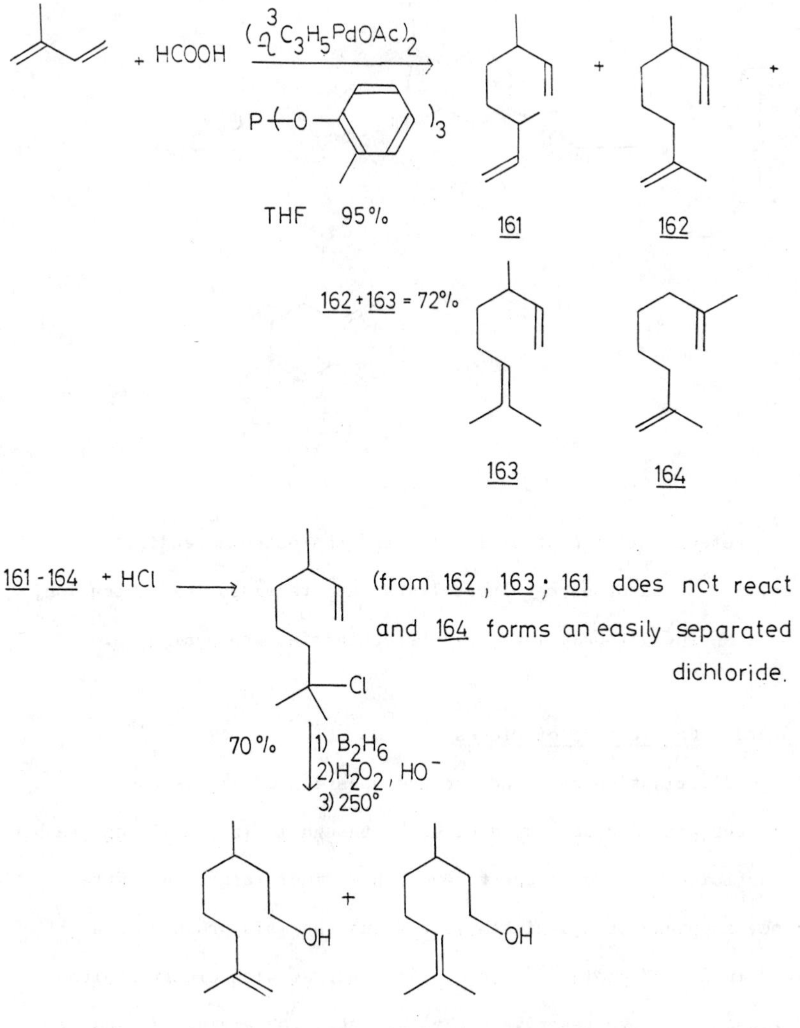

Figure 19 Synthesis of α- and β-citronellals from isoprene dimers.

Other dienes also react with altered selectivities. For example
Ni(0)|P-(O-o-phenyl C_6H_4)$_3$ converts butadiene to cyclooctadiene and 4-
vinylcyclohexene in the ratio 97:3[308-10] but with 2,3-dimethylbutadiene
the ratio is changed to 6:86. Isoprene gives an 8:6 membered ring ratio
of 55:35 whilst piperylene gives 91:5. Cis-piperylene gives four
isomers of dimethylcyclooctadiene; whilst the reaction is stepwise not

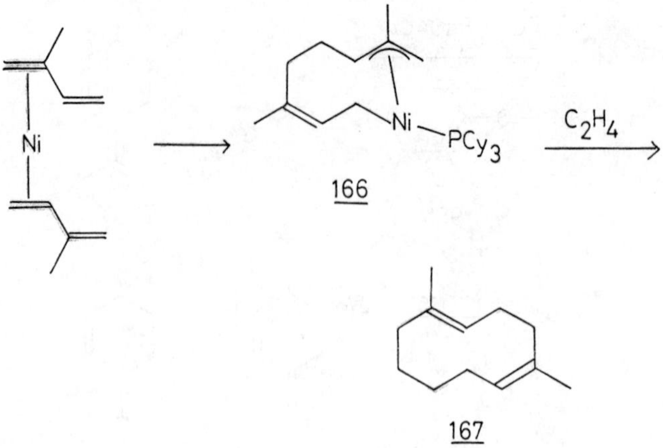

166

167

concerted stereochemical control is high and the outcome readily
predicted.[311] Again other metal catalysts have received less attention,
though some serendipitously impressive slectivities are available.
(Figure 20)

8.3.3.3. Cooligomers of Dienes

Codimerisation of butadiene and substituted dienes occurs
readily in the presence of "naked nickel", though it is usually desireable
to use a deficiency of butadiene to avoid homodimer formation. With
piperylene, isoprene or 2,3-dimethylbutadiene the main product is a 1:1
cyclodimer in 84-92% yield [282] (e.g. 168). Dienes with electron withdrawing groups are also reactive; ethyl sorbate, 169, yields 170 and 171
with the selectivity depending on the conditions.[316,317] Cross-cyclotrimerisation is also facile; butadiene and isoprene yield 172 and isomers
in the presence of Ni(O) and (PhO)$_3$P[318] whilst the heterodiene reaction,
173 to 174 and 175, opens an important synthetic route to macrohetero-
cycles.[319]

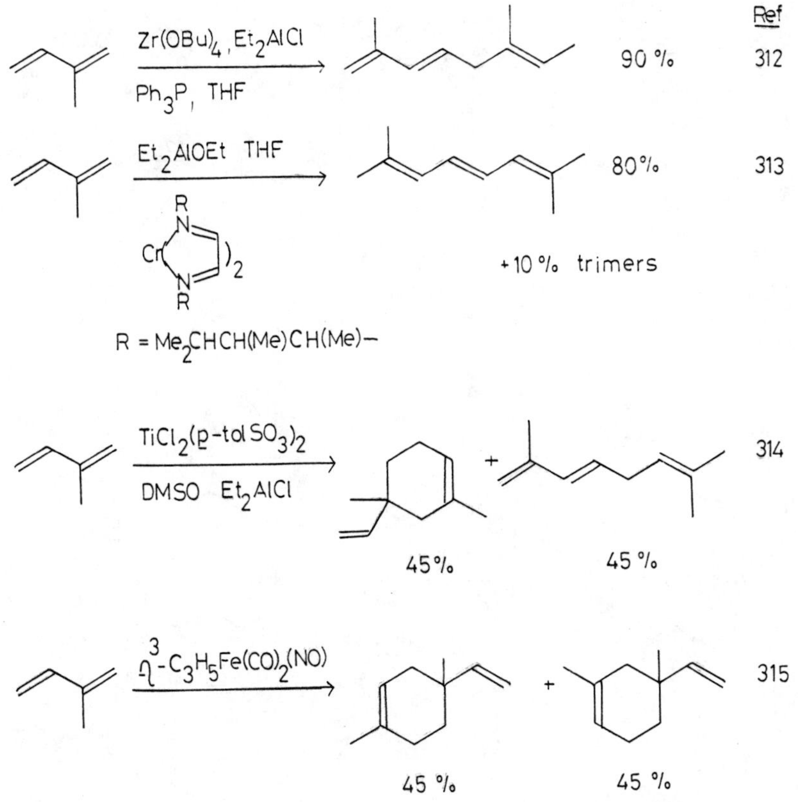

Figure 20 Oligomerisation of substituted dienes.

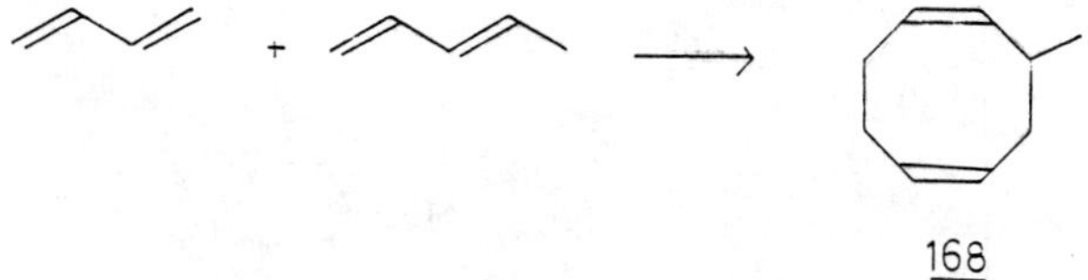

168

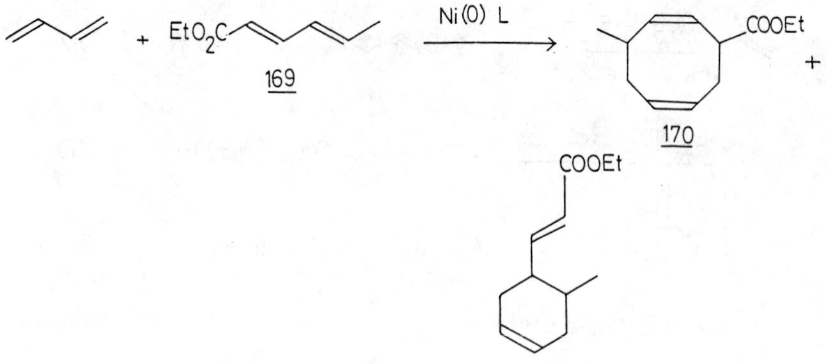

169

170

171

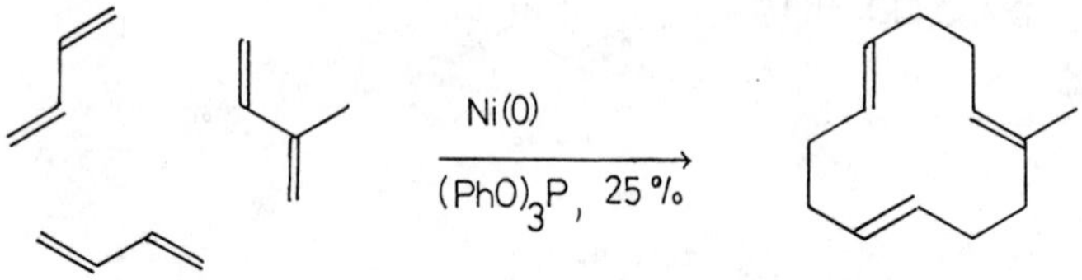

172

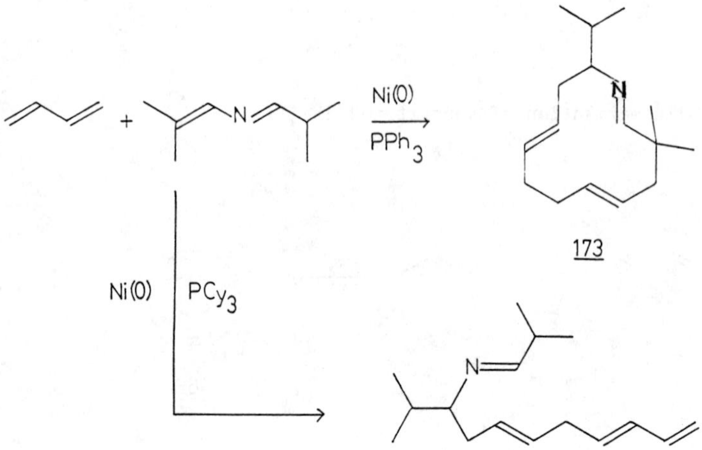

173

174

8.3.3.4. Cooligomers of Dienes and Alkenes

Codimerisation of ehtylene and butadiene to 1,4-hexadiene is important because of the value of the product in cross-linking ethylene-propene copolymers to give synthetic rubber.[320] The earliest catalyst was $RhCl_3$[321] which gave 91% C_6-diene and 80% 1,4-selection. Other rhodium complexes are active[322], and the reaction proceeds via stable π-crotyl rhodium complexes, such as <u>175</u>, which is isolable from the reaction mixture.[323] Product isomerisation is supressed by maintaining a large excess of butadiene. Alcoholic solvents increase the reaction rate and the <u>trans:cis</u> ratio[320] suggesting that the true catalyst is the monomer, <u>176</u>, and that the <u>syn</u>-crotyl (leading to <u>trans</u>-product) predominates when only monodentate coordination of butadiene is possible.

As one might expect, analogous reactions occur in the presence

<u>175</u>

<u>176</u>

of $(Bu_3P)_2NiCl_2 | i-Bu_2A\ell Cl$ to give 1,4-hexadiene in 60-70% yield with a <u>trans:cis</u> ratio of 2.5-2.8[324], together with conjugated diene and the branched product, <u>177</u>. The reaction proceeds via the π-crotyl, <u>178</u>, and <u>179</u> (L=(EtO)$_3$P) was isolated in a model system.[325] Catalyst activity is optimised for P:Ni=1[326] and bidentate phosphines inhibit the reaction.[327] <u>Trans:cis</u> ratio correlates not with phosphine cone angle [328] but with $C\equiv O$ stretching frequencies in nickel phosphine carbonyl complexes[329,330] indicating that the ligand's effect is mainly electronic. Electron acceptors gave increased <u>trans:cis</u> ratios and decreased amounts of unwanted <u>177</u>. It is thought that the electronic nature of phosphine

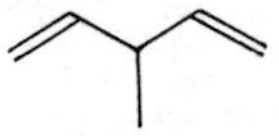

177 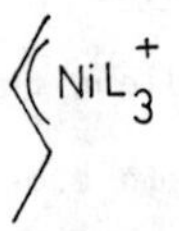   178 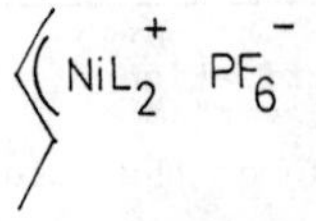   179

and cocatalyst[331] determines the mode of butadiene coordination in much the same manner as for the allyl nickel polymerisation catalysts.

Iron and cobalt complexes (MX_n, R_3P, R_3Al) are also reactive, the cobalt systems being faster and more selective.[332-335] The best phosphines are chelating ones, $R_2P(CH_2)_nPR_2$ (R=aryl, n=2,3). The mechanism proposed involves a cobalt hydride which reacts with coordinated butadiene to give a σ-crotyl, 181, which establishes the highly specific cis-stereochemistry of the final product.[336]

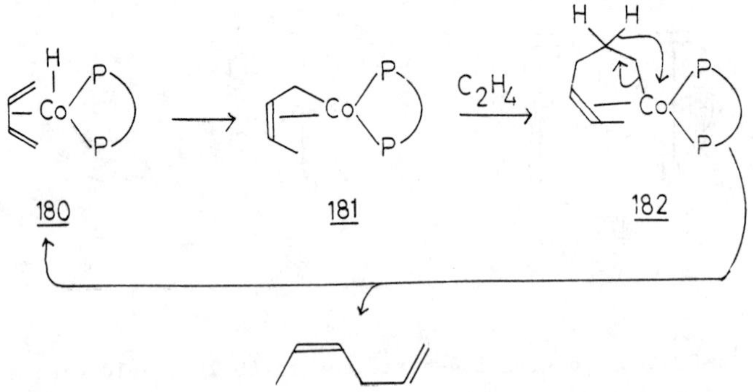

180   181   182

Cooligomerisation of butadiene and propene yields, in the presence of $PdCl_2|Et_2AlCl|R_3P$, 2,5-heptadiene and 2-methyl-1,4-hexadiene, their ratio being dependent on R_3P.[337,338] The codimer with styrene (($n^3C_3H_5PdCl)_2|BF_3|PPh_3$) is 1-phenyl-1,4-hexadiene[338] with largely trans-stereochemistry[339] whilst with dimethylallylamine (($Ph_3P)_2PdCl_2|Et_3Al$) a 2:1 oligomer, 183, is the major product.[148] Again π-crotyl intermediates are likely and insertion of the alkene occurs at the less hindered allyl terminus.

Reactions with nickel catalysts may yield both linear and cyclo-oligomers. Cross-cyclotrimerisation of butadiene and ethylene is six times faster than diene homotrimerisation and so provides a useful synthesis of 10-membered rings.[341] With substituted alkenes isomer mixtures are obtained [282,318,342] the direction of addition depending on the LUMO of the alkene interacting with ψ_2 of the metal allyl; the terminal carbon of the allyl becomes bonded to the more electron deficient carbon of the alkene. Thus styrene gives only 186 on insertion, propene gives 76% 186 and 24% 187, and methylmethacrylate only 188.[343] A number of linear cooligomerisations are also known, but except for cyclic dienes the selectivities are unimpressive (Figure 21).

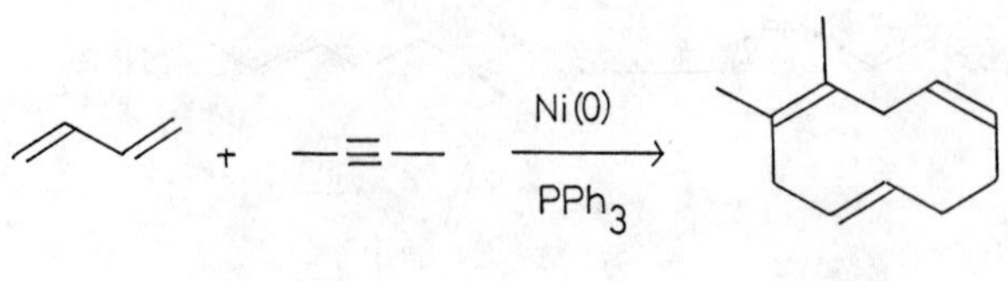

188

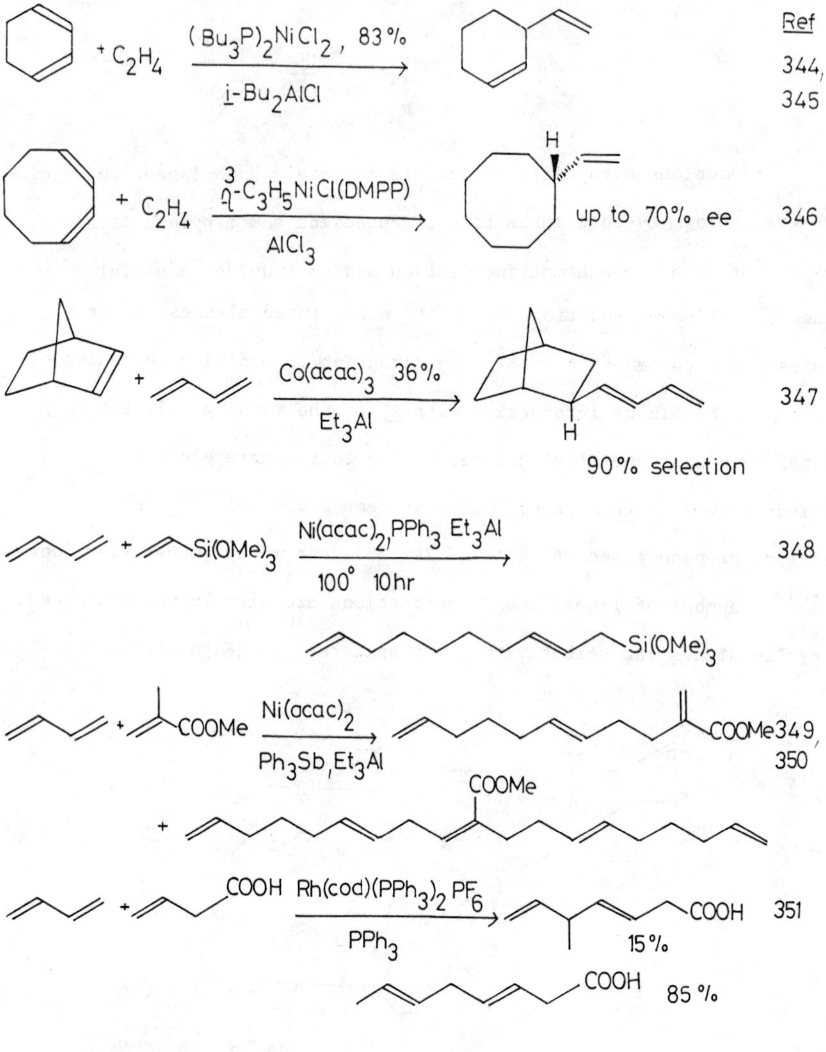

Figure 21 Cooligomerisation of dienes and alkenes.

8.3.3.5. Cooligomers of Dienes and Alkynes

Alkynes are cooligomerised with butadiene to yield 1,2-disubstituted cyclododecatrienes[316,352]; for example 2-butyne yields 188 and the product is contaminated with only traces of butadiene oligomers.[310] With excess alkyne larger rings such as 189 (from 2-butyne)[352] and 190 (from 1,4-dimethoxy-2-butyne) are obtained.[309] The reaction of diethyl acetylene dicarboxylate was studied in detail and 191 isolated in 30% yield.[353]

189   X = H
190   X = OMe

Recently production of a linear codimer was reported for terminal alkynes. Butadiene reacts with 193 to give 194 stereospecifically in 90-100% yield (R=alkyl; R=Ar gives largely homodimer). The mechanism proposed involves insertion of butadiene into a ruthenium alkyne bond.[354]

8.4 Telomerisation of dienes

The most characteristic reaction of dienes in the presence of Pd(0) is dimerisation with incorporation of a nucleophile, YH, called telomerisation[355]. The products are 1- and 3-substituted octadienes, 195a and 195b. Nucleophiles include alcohols, thiols, water, carboxylic acids, amines and carbon acids. Although some nickel complexes do promote the reaction, both activity and selectivity are usually lower.

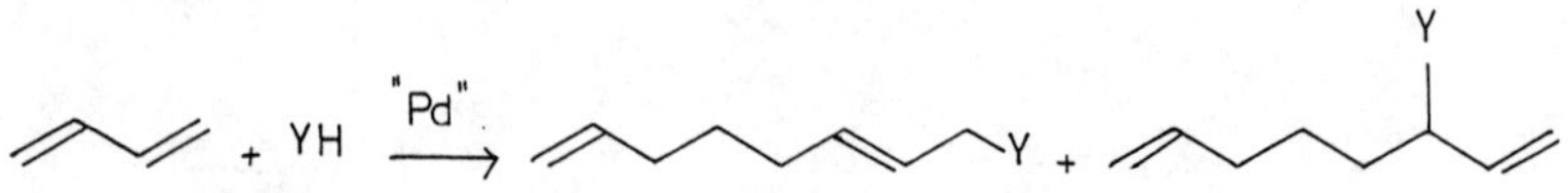

195a   195b

8.4.1. Telomerisation with Water and Alcohols

Telomerisation of butadiene and water gives the alcohols 196 and 197, 1,3,7-octatriene and dioctadienyl ether, 198.[355] Without CO_2 pressure the reaction is slow but the reason for this effect is unclear.[356] 196 is the major product in up to 69% yield[357] the pH of the medium being critical to useful reaction.[358]

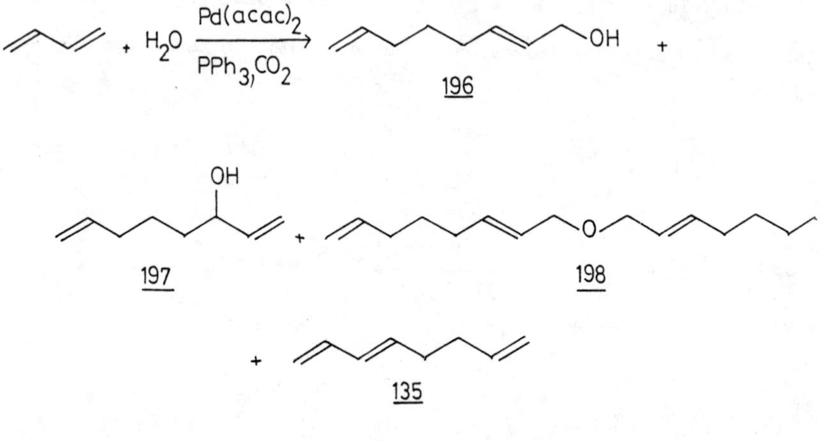

196

197   198

135

Primary alcohols are telomerised to ethers, the range of products including 135 and 199-202.[359] With $(Ph_3P)_2Pd$ (maleic anhydride) at 70° MeOH yielded 201 (85%) 202 (5%) (R=Me) and 135 (3%).[360] With CH_3OD deuterium is incorporated at C_6 of 201, suggesting that the intermediate, 203, is attacked by methanol with concomitant protolysis of the C_6-Pd bond.[361] With $(Ph_3P)_2PdCl_2|PhONa$ as catalyst 201 is obtained in 73% yield with 21% 202; bromides or iodides give lower yields. The base is essential for the predominent formation of 2:1 adducts, which are also favoured by trialkyl rather than triaryl phosphines.[359] Cationic $(\underline{n}^3C_4H_7PdLn)^+$ complexes give further condensation to C_{16}, C_{24} and C_{32} compounds.[362] Higher alcohols give similar results; ethanol is telomerised to 201 and 202 (R=Et)[274], the best selectivity for 201 being obtained in the presence of Bu_3P and the most 202 (49%) with $(Me_2N)_3P$.[363] More hindered alcohols react more slowly requiring forcing conditions and giving lower yields (Figure 22). Phenol reacts smoothly to yield mainly 201 and a little 202 (R=Ph)[367], and the product has been used in the synthesis of the pheno- mones of the bollworm moth[368] and Ceratitis capitula.[369]

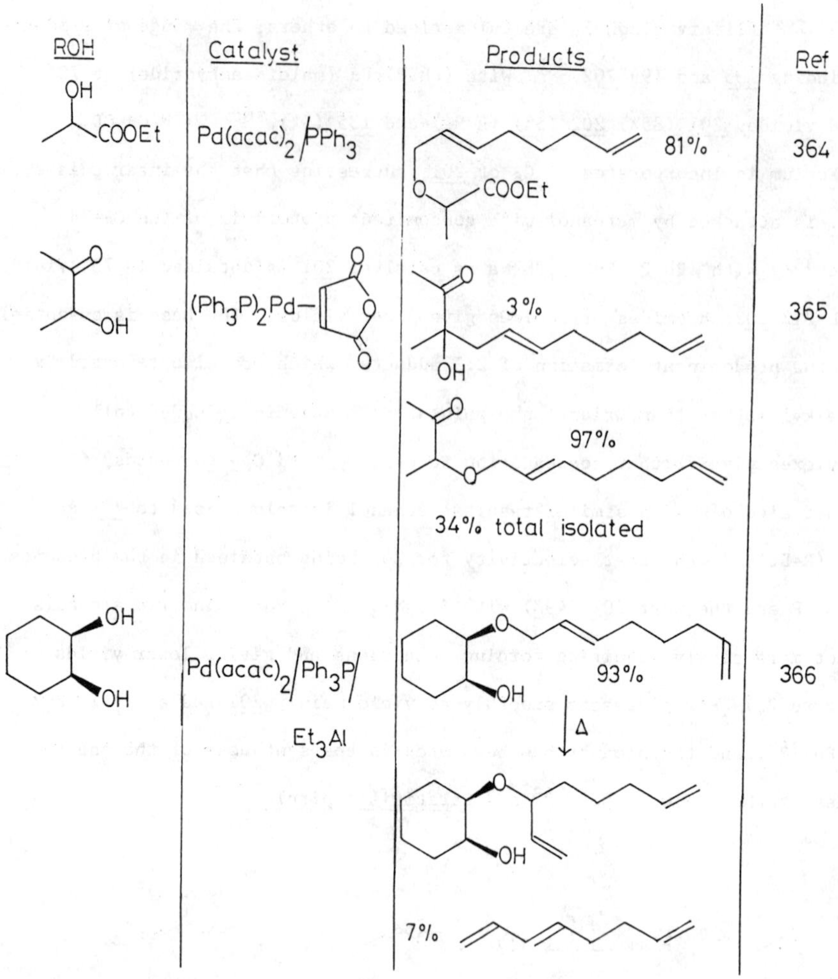

Figure 22 Telomerisation of butadiene and alcohols.

Isoprene is less reactive than butadiene and complex product mixtures are often obtained. The head to tail telomer, 204, may be obtained regioselectively but unfortunately the oxygen function is not in the same position as for most natural products. Other observed selectivities are given in Table 11. Enantioselective telomerisation occurs in the presence of menthyldiphenylphosphine to give 205 in 17.6% optical yield.[374]

Table 11   Telomerisation of Isoprene

Catalyst	Isoprene dimers	(OMe isomer 1)	(OMe isomer 2)	(OMe isomer 3)	Ref
$(Ph_3P)_4Pd\vert O_2$	5	67	23	–	370
$(\eta^3-C_3H_5PdCl)_2\vert PPh_3$	–	90	9	–	371
$(\eta^3-C_3H_5PdCl)_2\vert PBu_3$	2	56	31	–	371
$(\eta^3-C_3H_5PdCl)_2\vert PBu_3\vert$ i-PrOH	12	63	11	–	371
$(Ph_3P)_2PdCl_2\vert$ i-PrOH	trace	80	15	–	371
$Pd(OAc)_2\vert PPh_3\vert 80°$	59	–	21	17	372
$\eta^3-C_3H_5-Pd(PPh_3)OMe$	–	97	–	–	373

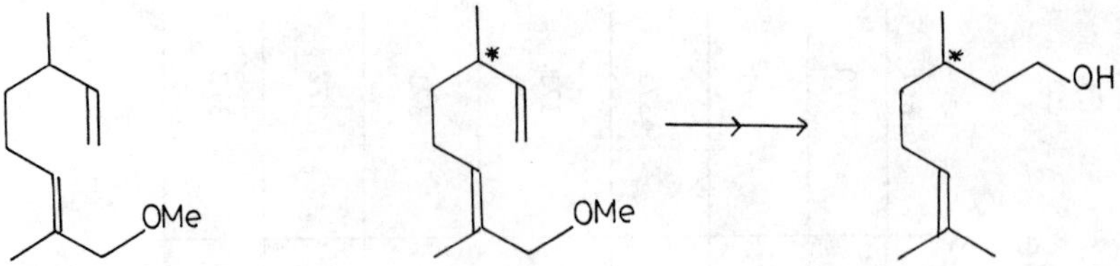

204   205   206

This may be converted into citrononellal, 206 in up to 35% enantiomer

excess.[375]

8.4.2. Telomerisation with Acids

A mixture of acetoxyoctadienes, 207 and 208,is formed on

telomerisation of butadiene with acetic acid.[274],[376] Addition of base,

either trialkylamine[377] or sodium acetate[378],[379], is essential and 207

is generally the major product. Addition of triarylphosphites is

particularly successful in promoting selectivity for 207[377],[380] with 92%

in the best case. Few other acids have been tested but acrylic acid

yields 209 and 210 in 90% overall yield with 98% selectivity for 209

under the most favourable conditions.[381]

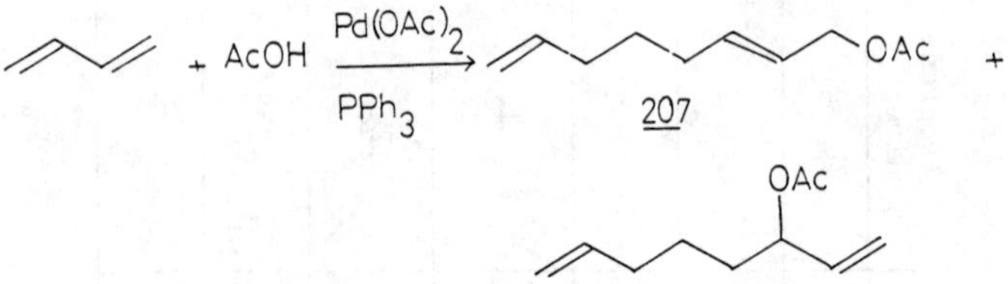

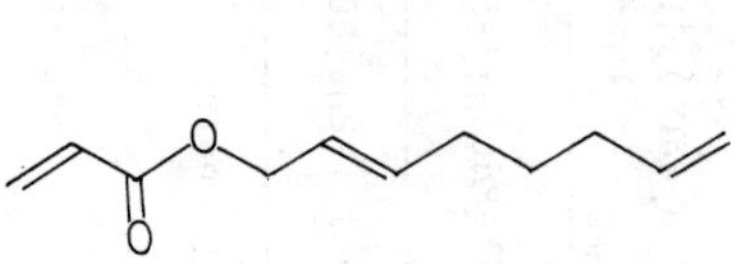

209 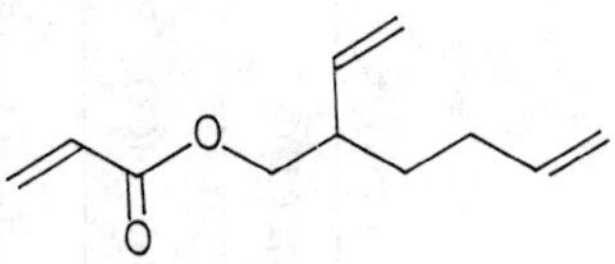   210

Numerous synthetic uses for both 207 and 208 are reported. The synthesis of Matsutako alcohol, 211, derived from a Japanese mushroom, is shown in Figure 23.[382-4] 208 is the starting material for synthesis of lipoic acid[384] and 207 was converted to diplodialide, 216 (Figure 24).[385,386]

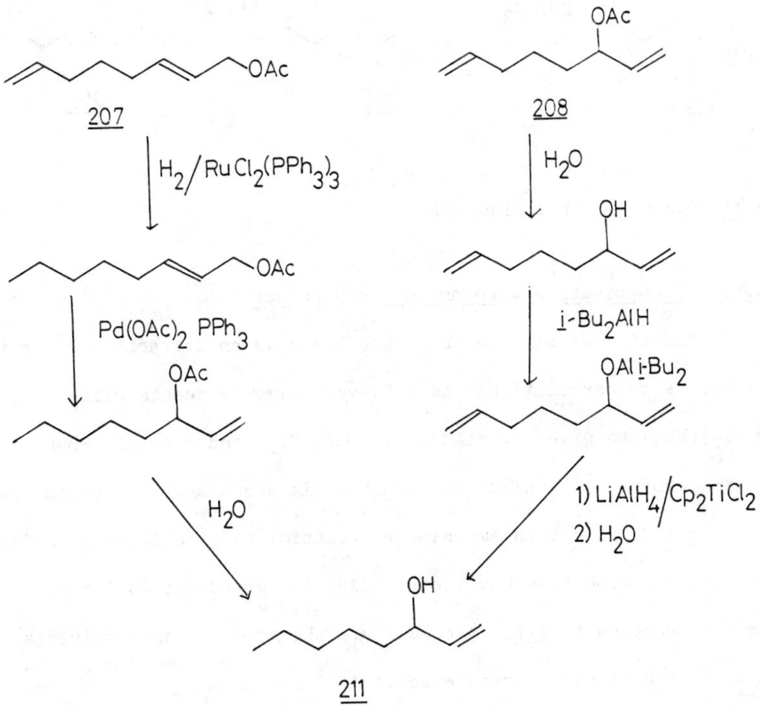

Figure 23 Synthesis of Matsutako alcohol.

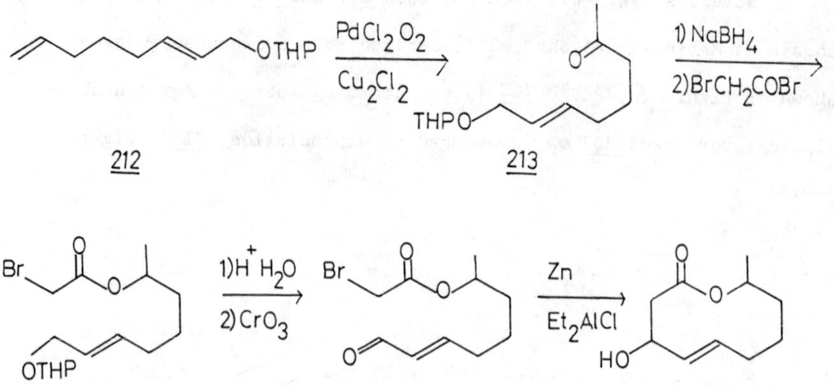

Figure 24 Synthesis of diplodialide.

8.4.4. Telomerisation with Ammonia and Amines

Addition of ammonia to double bonds is an attractive route to amines but is rather difficult to achieve. Ammonia reacts with butadiene $(Pd(OAc)_2 | PPh_3)$ to give trioctadienylamine, 217, and a small amount of dioctadienylamine.[387] Whilst the reaction is stepwise the intermediates are more nucleophilic than ammonia, and control to give 1° or 2°-amines is impossible. Some branched isomer, 218, is formed but is isomerised during the reaction to 217. Alcohols are the most suitable solvents and a small amount of water is essential.[388]

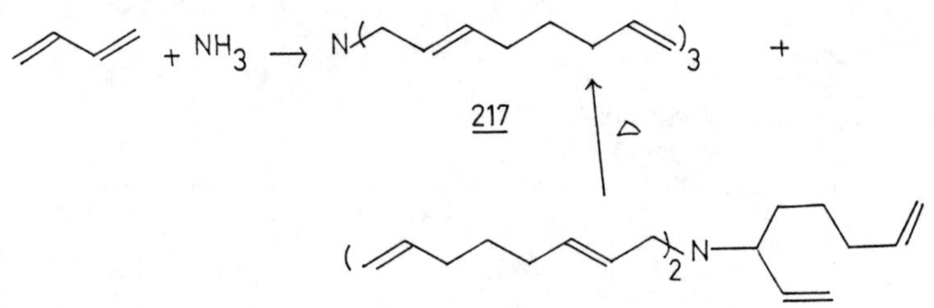

217

218

Primary amines react to yield alkyldioctadienylamines[389] with branched amines kinetically, and linear isomers thermodynamically, favoured.[390] Secondary amines are more useful, however, giving linear and branched tertiary amines (Figure 25). The amino function is a better nucleophile towards the allyl intermediates than alcohols and amino alcohols react only at nitrogen.[397]

Allyl nickel and platinum complexes are less active as catalysts.[391,396,398] Primary amines give 4:1 adducts in yields which depend on conditions but which may be as much as 90%.[393] A study of the reactions of substituted anilines showed that reactivity increased with basicity,[274,399] but that ortho-substitution gave 2° rather than 3°-amines.

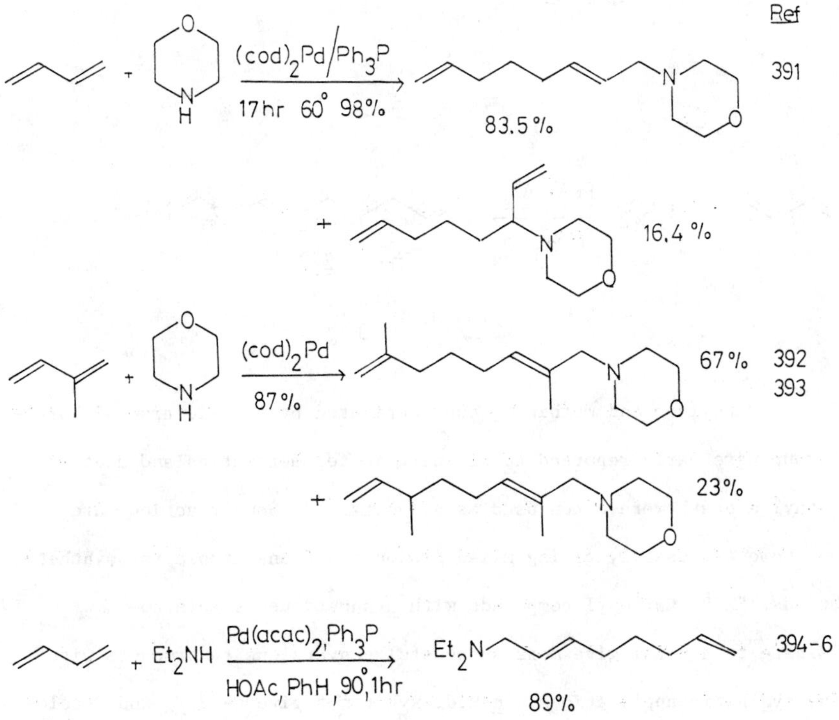

Figure 25 Telomerisation of butadiene with amines.

8.4.4. Telomerisation with Carbon Nucleophiles

The critical step in this reaction is attack of a nucleophile on a metal coordinated allyl, a reaction which we shall consider in more detail later. Enamines are octadienylated and may then be hydrolysed to ketones (e.g. 219).[377] The anions of nitro compounds react smoothly with butadiene to give mono-, di- and trialkylated porducts[387,400], 220, 221 and 222. Careful control of conditions is essential to achieve telomerisation without significant amounts of butadiene dimer. t-BuOH is an excellent solvent and the presence of a base is advantageous. Conversions of nitroethane telomer to reciferolide and nitromethane 2:1 adduct to cis-civetone are shown in Figure 26.[401-3]

219

n = 1   220
    2   221
    3   222

Methylene and methine groups activated by one electron withdrawing group were early reported to be inert in telomerisation and acetone and ethyl acetate were often used as solvents.[273] Some reactions are known, however, usually giving mixed products and unsuitable for synthetic reactions.[404-6] Carbonyl compounds with α-substituents able to coordinate to a metal give more successful conversions to products with useful synthetic applications. α-Hydroxyketones give C-(223) and O-telomerised products (224); for R_1, R_2=Me, 223 is obtained with 89% selectivity. Change of the catalyst to $(Ph_3P)_4Pd$ gives 224 as the major product.[365]

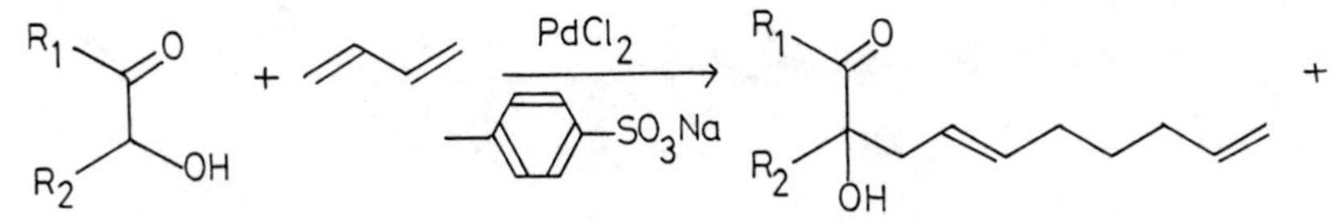

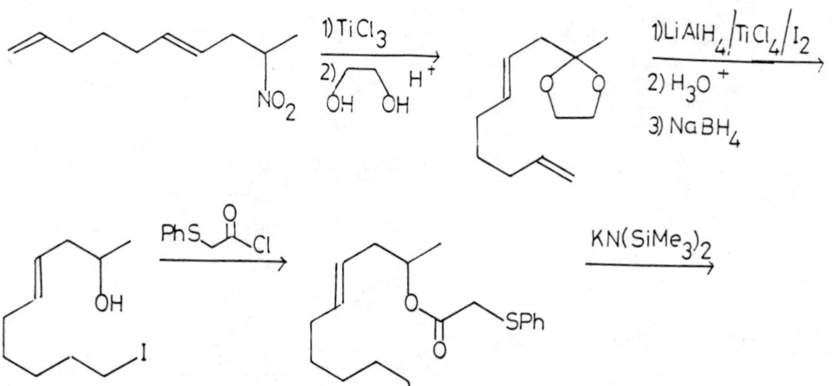

Figure 26 Synthesis of reciferolide.

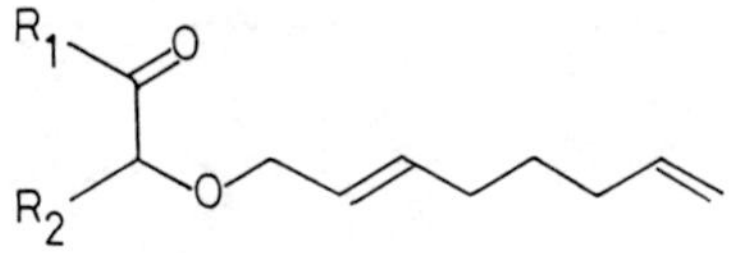

Figure 26 cntd. Synthesis of civetone.

α-Amido ketones react similarly to give only C-alkylated telomers.[407]

Compounds activated by two electron withdrawing groups react with butadiene to give mono- and di-2,7-octadienyl substituted products, together with branched biproducts.[408,409] The reaction occurs with β-keto esters[409], β-diketones, malonates[410], α-formylketones, α-cyano- and α-nitroesters, cyanoacetamide and phenylsulphonyl acetate.[411] $(Ph_3P)_2PdCl_2$ is a good catalyst and a base (PhONa) is essential. $Pd(OAc)_2|PPh_3$ is also useful but addition of chelating biphosphines causes formation of 1:1

adducts, <u>225</u> and <u>226</u>. Some synthetic applications are shown in Figure 27.

<u>225</u>   <u>226</u>

8.5 Polymerisation and Oligomerisation of Alkynes

8.5.1. Polymerisation of Alkynes

Some alkyne polymers are pyrophoric and many are paramagnetic;[413]

their main use is that, when appropriately doped, they give p- or n-

type semiconductors.[414,415] A range of metal complexes act as catalysts

including $VO(OC_5H_7)_2 | 2Et_3Al$[416], $Ti(OBu)_4 | 4Et_3Al$[414], $NiBr_2$[417], $CoCl_2 |$

i-Bu_3Al[418] and naphthenoate complexes of the lanthanides.[419] The pro-

portion of <u>cis</u>- and <u>trans</u>-polyacetylene is widely variable; $NiBr_2$ gives

only the <u>trans</u>-isomer, nickel clusters such as $(Me_4N)_2[Ni_{12}(CO)_{21}H_2]$[413]

give equal proportions and $Dy(naphth)_3 | Et_3Al$ and $Ti(OBu)_4 | Et_3Al$[420] give

<u>cis</u>-rich products. Polymer morphology is also controlled by catalyst

and solvent.[421]

Studies of polymerisation of substituted alkynes have focussed

on phenylacetylene as substrate. Again $TiX_4 | R_3Al$ catalysts are

effective[422], probable intermediates being σ-metal alkynyls and alkenyls.

When $(Ph_3P)_2PtCl_2$ or $(Ph_3P)_2PtHCl$ is treated with phenylacetylene <u>cis</u>-

and <u>trans</u>-$(Ph_3P)_2Pt(-C\equiv C-Ph)_2$ is formed and polymerisation occurs on

warming to 130°.[423] Insertion of an alkyne into a metal carbon bond is

well established in stoichiometric reactions (<u>227</u> to <u>228</u>)[424] and

propagation may take place according to Figure 28 or Figure 29.[429]

The bimetallic catalyst, <u>229</u>, gives <u>230</u> and <u>231</u> with 2-butyne resulting

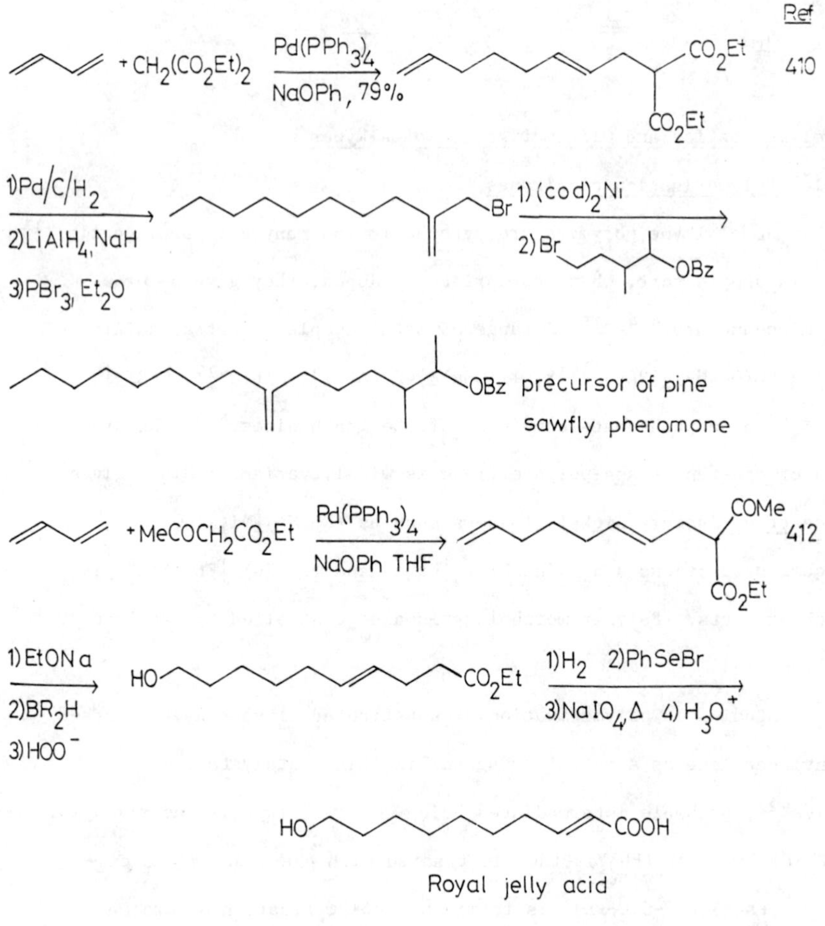

Figure 27 Synthetic applications of butadiene telomers.

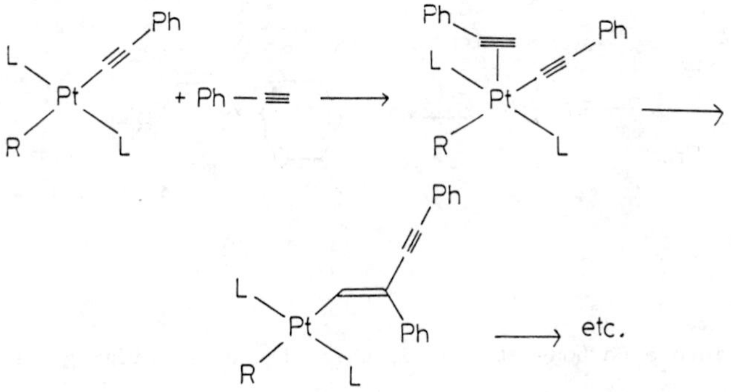

Figure 28   Mechanism of alkyne polymerisation by platinum complexes.

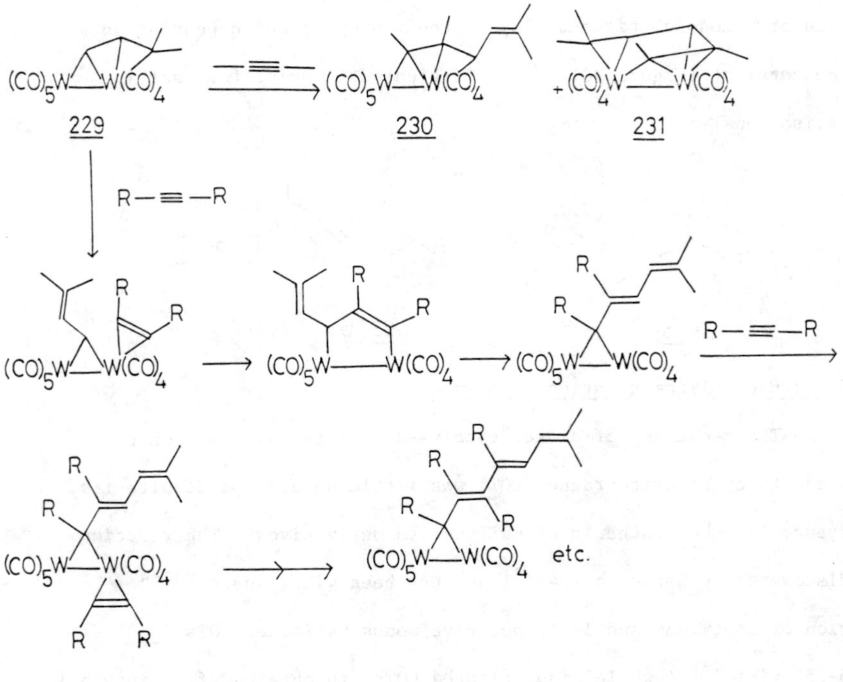

Figure 29   Mechanism of alkyne polymerisation by bimetallic tungsten complexes.

from insertion into a carbon-metal bond, this time of a bridging alkyl-idene. This provides a mechanism for chain propagation in which each step is well established.

8.5.2. Linear Dimers of Alkynes

A number of dimerisations are known and most resemble the polymerisations. **232a** and **232b** represent head to tail and tail to tail reaction of a monosubstituted alkyne, their proportions depending on R and the catalyst (Figure 30). Insufficient data exist for useful generalisations.

8.5.3. Cyclooligomerisation of Alkynes

The mechanism of nickel catalysed cyclotetramerisation of acetylene to cyclooctatetraene (COT) was little understood despite its importance for the synthesis of COT and its derivatives. The reaction was discovered by Reppe in 1948[431] and has been widely used.[432-4] Reaction of acetylene and 1-alkynes give monosubstituted COTs (**233**) in 15-25% yield[435-7] while disubstituted COTs are obtained from internal alkynes, again in modest yields.[438-441] The homo-reaction of 1-alkynes gives a variety of tetrasubstituted COTs with interesting selectivi-ties.[442] (Figure 31)

R	Catalyst	% 232a	% 232b	Ref.
MeOCH$_2$	(Ph$_3$P)$_3$RhCl	97	3	426
t-Bu	(Ph$_3$P)$_3$RhCl	-	100	427
Ph	(Ph$_3$P)$_3$RhCl	-	100	427
C$_6$H$_{11}$	(Ph$_3$P)$_3$RhCl	87	13	428
n-Bu	bis(N-methylsalicylal diamine)Ni/i-Bu$_2$Zn	98	-	429
C$_6$H$_{13}$	Cu$^+$/HOAc	-	100 (Z:E=43:57)	428
Me$_3$Si	Ni(acac)$_2$|Et$_3$Al|Ph$_3$P	-	32(all E)	430

+32%

Me$_3$SiCH=CH-C(SiMe$_3$)=CH-C≡C-SiMe$_3$

+35% higher oligomers

Figure 30 Dimerisation of 1-alkynes.

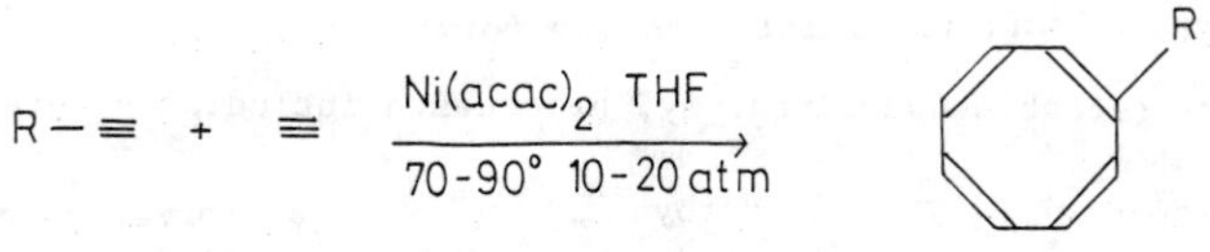

$$R-\equiv \;+\; \equiv \quad \xrightarrow[70-90° \; 10-20\,atm]{Ni(acac)_2 \;\; THF}$$

233

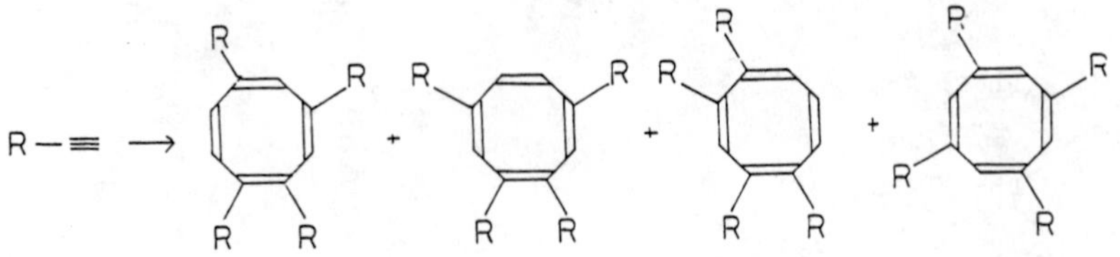

R	Catalyst	Products
CH_2OH	$Ni(acac)_2$	71% 234d
COOMe	$Ni(PCl_3)_4$	83% 234a
COOMe	$PdCl_2/NaOAc$	234c low yield
CH_3	Ni(0)	30% 234a, 15% 234b, 10% 234d

Figure 31 Cyclotetramerisation of 1-alkynes.

Four district topological alternatives have been proposed for the tetramerisation (Figute 32). The reaction was originally proposed to occur via the concerted "zipper" process[443,444] or some stepwise equivalent, particularly one involving two metal centres.[445-449] Stable nickel cyclobutadiene complexes have been isolated and seemed appropriate models for the cyclisation, particularly as some of them may be decomposed to COT under mild conditions.[450,451] The other two mechanisms are topologically inequivalent; the first involves formation of metal carbynes, which are recombined arbitrarily, precedents including several examples of conversion of alkynes to carbynes.[452,453] The formation of benzene derivatives from alkynes is well-established and these might insert a further alkyne unit. The topology of the reaction was elucidated by Vollhardt.[454] Singly (97% ^{13}C) labelled acetylene was converted to COT in the presence of $Ni(acac)_2 | Et_2AlOEt$ and the product degraded to dimethyl phthalate for mass spectroscopic analysis. This showed

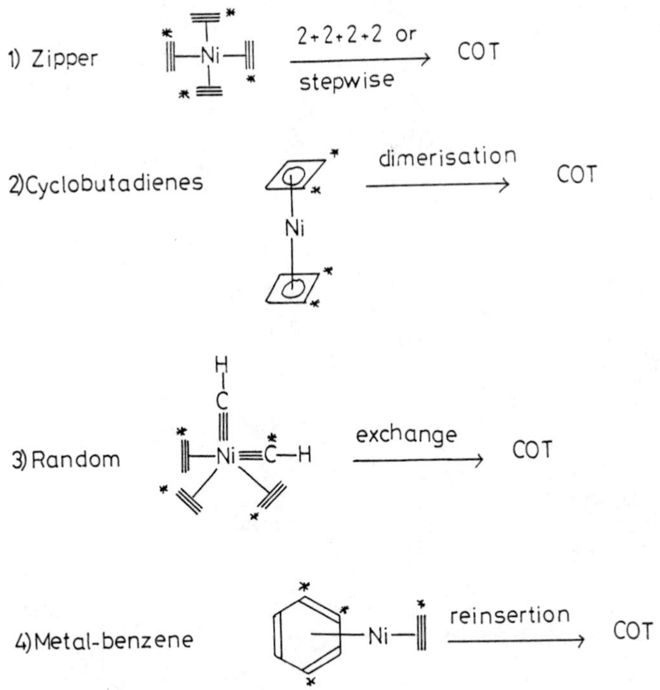

Figure 32 Topology of acetylene cyclotetramerisation.

unequivocally that the topology of the reaction is consistent only with
a "zipper" step-wise mechanism. It should be noted that the results
do not provide a mechanism, only define the topology involved.

Numerous complexes catalyse trimerisation of alkynes to
benzene derivatives (Figure 33) and the reaction is tolerant of many
functional groups, including -OH, -COOMe and NR_2.[460] Although the
mechanism is unproved the metallocyclopropane, 234 is characterised as
the product from $Ta_2Cl_6(SC_4H_8)_3$ and 2-butyne[461] and in the trimerisation
of $MeO_2CC\equiv CCO_2Me$ by iridium complexes the metallocyclopentadiene, 236, was
isolated.[462] Metallocyclopentadienes were independently prepared by
treatment of $(Ph_3P)_2NiCl_2$ with 237 and both stoichiometric and catalytic
insertion of alkynes to give benzenes established.[463] Metallocyclopenta-

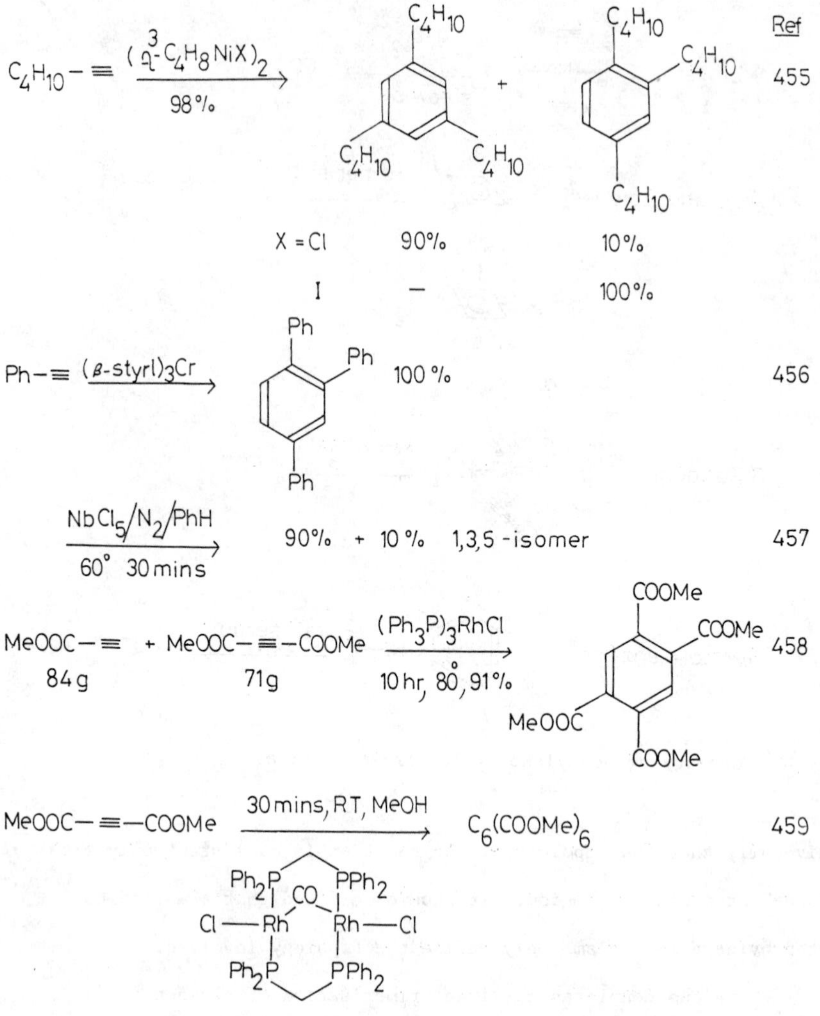

$$C_4H_{10} - \equiv \xrightarrow[98\%]{(\eta^3\text{-}C_4H_8NiX)_2}$$

X = Cl	90%	10%
I	—	100%

455

$$Ph - \equiv \xrightarrow{(\beta\text{-styrl})_3Cr}$$

100%

456

$$\xrightarrow[60° \ 30\,mins]{NbCl_5/N_2/PhH}$$ 90% + 10% 1,3,5 -isomer

457

$$MeOOC - \equiv \ + \ MeOOC - \equiv -COOMe \xrightarrow[10\,hr, \ 80°, 91\%]{(Ph_3P)_3RhCl}$$

84 g   71 g

458

$$MeOOC - \equiv -COOMe \xrightarrow{30\,mins, \ RT, \ MeOH} C_6(COOMe)_6$$

459

Figure 33 Trimerisation of alkynes.

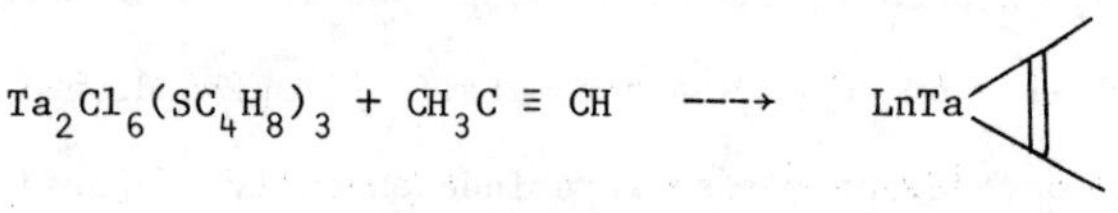

$$Ta_2Cl_6(SC_4H_8)_3 + CH_3C \equiv CH \ \longrightarrow \ LnTa$$

234

235   236   237   238   239

dienes are also implicated in the important intra- and intermolecular

cotrimerisations developed by Vollhardt (Figure 34). Other alkyne

trimerisations and cycloadditions are catalysed by nickel and rhodium

complexes[470]; yields are good but there is little known of the mechanisms

(240 to 241[471], 242 to 243 [472,473]).

Oligomerisation of alkynes by palladium complexes gives a

mixture of products, thoroughly investigated by Maitlis.[474] An overall

scheme has been postulated to account for the formation of 251 and 255

(Figure 35). If the equilibrating system, 249, 250 and 252, is treated

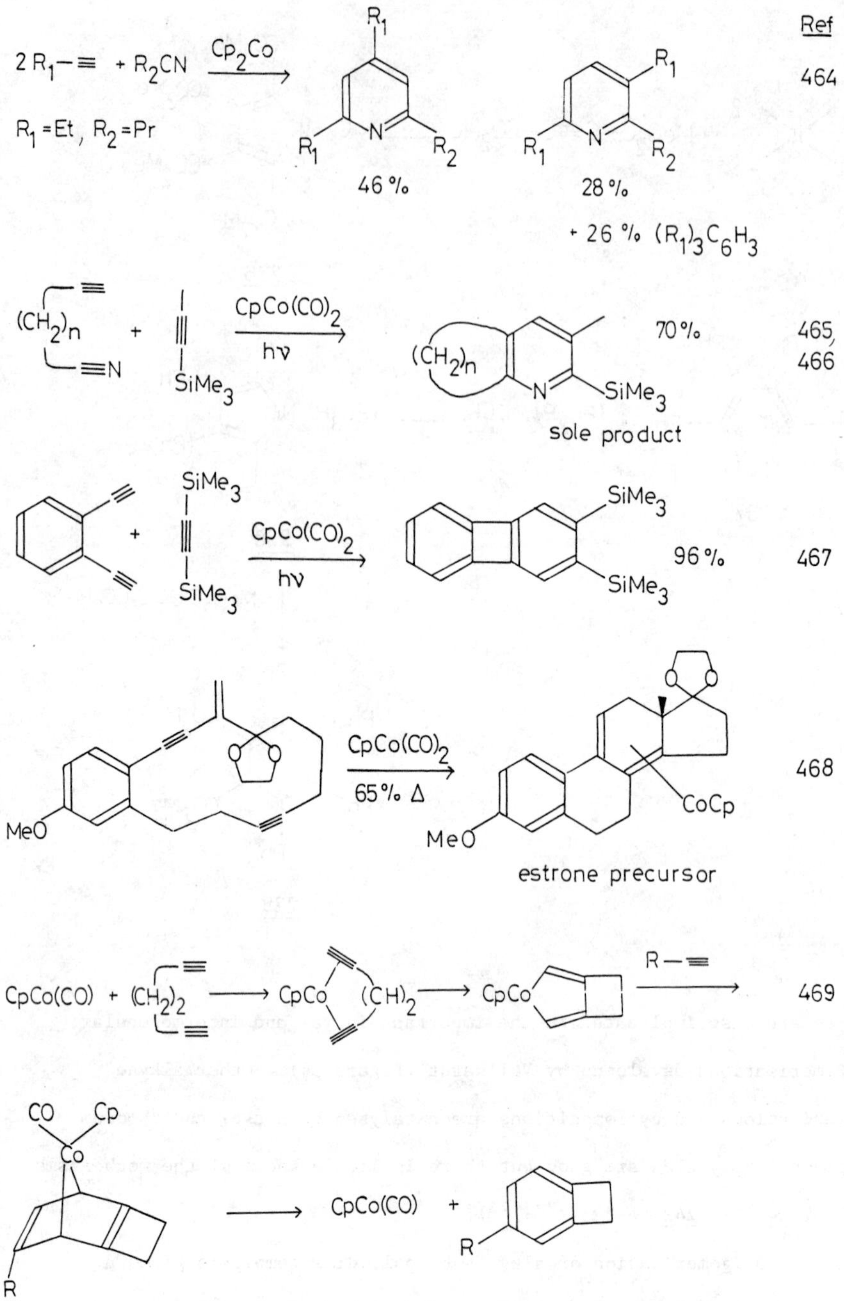

Figure 34 Synthetic applications of cobalt catalysed alkyne trimerisation and cotrimerisation.

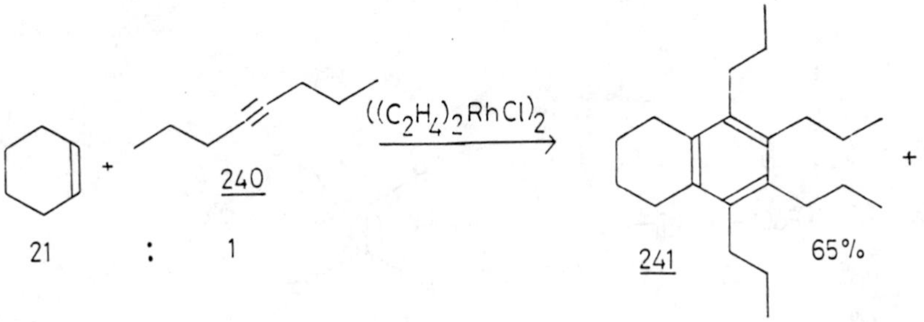

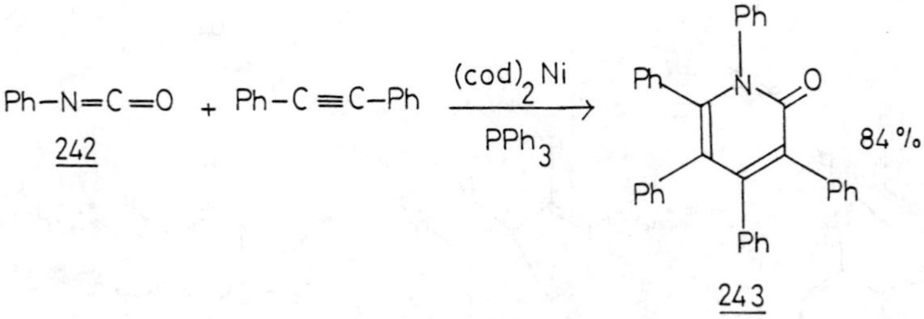

with excess phosphine at 0° a cyclopentadiene 256 (R,R'=Me) is isolated.
Under slightly different conditions tetramer, 257,[475] or fulvene trimers,
258-260[476,477], are formed. Clearly these reactions represent finely
balanced pathways and small changes in conditions or steric effects
cause large changes in product proportions.

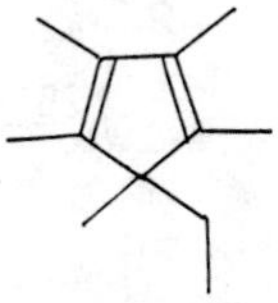

256

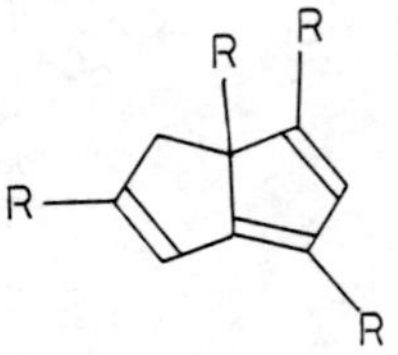

257

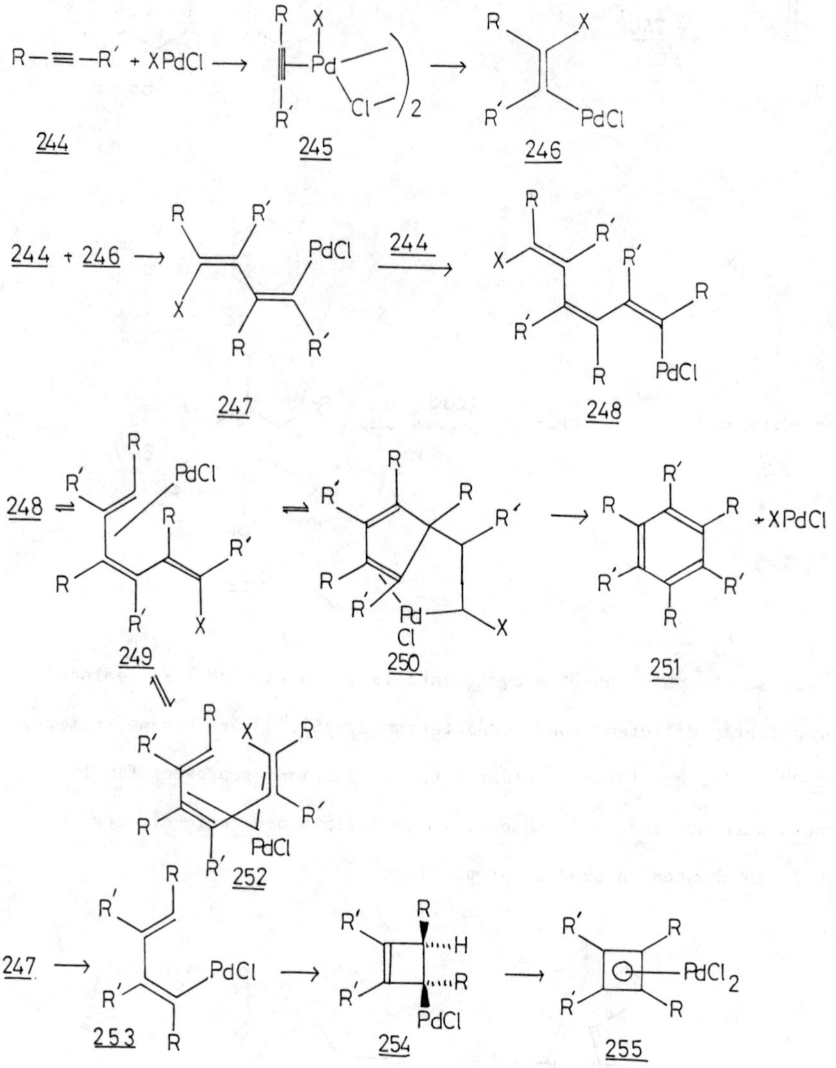

Figure 35 Palladium catalysed oligomerisation of alkynes.

$$Ph-C\equiv C-H \xrightarrow[80-85°]{Pd(OAc)_2}$$

258   259   260

8.6 Coupling Reactions of Arenes and Alkynes

8.6.1. Coupling of Aryl Halides

The Ullmann reaction has been used to generate aryl-aryl bonds

for almost 80 years[478] and continues to be widely applied in synthesis.[479]

Typically 2 moles aryl halide react with one of finely divided copper

to give a biaryl and a copper halide; this is not normally a catalytic

reaction. Generally, electron-withdrawing groups ortho to the halogen

are strongly activating but large substituents in this position may

inhibit the reaction. Amino, hydroxyl and carboxyl groups prevent

satisfactory coupling by providing alternative pathways for the inter-

mediates and protecting groups must be used.[480]

Numerous variants are known.[481] In its simplest form symmetric

biaryls are formed but cross-coupling is also possible. Optimum yields

are achieved when one of the aryl halides is activated and the other is

relatively unreactive. Heterocyclic halides are also useful. Iodoarenes

give the best reactions followed by the bromo compounds, chlorides

reacting only under forcing conditions and fluorides being quite inert.

Intramolecular reactions are particularly successful and have been used

in several natural product syntheses. (Figure 36)

Both reaction rate and yield are dependent on the physical state

of the copper used. Commercial copper powder is activated by washing

with the disodium salt of EDTA giving a reactant at least as good

as freshly precipitated copper.[485] Very active reagent is prepared

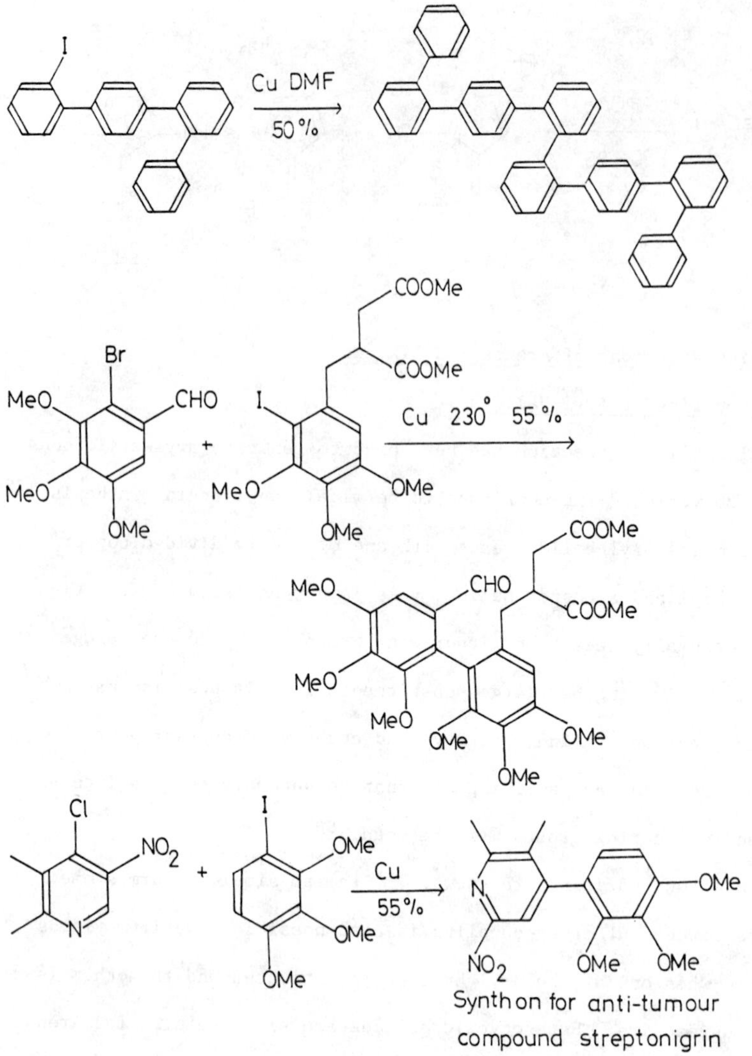

Ref

482

483

484

Figure 36 Synthetic uses of the Ullmann reaction.

electrolytically[486] or by reduction with potassium.[487] In all these cases yields comparable with the standard procedure (DMF, >200°) are obtained at much lower temperatures. There is strong evidence that the reaction involves an aryl copper compound, probably solvent stabilised. This suggests that a reliable synthesis of unsymmetrical biaryls free from self-coupling products could be effected by treating a preformed

aryl copper compound with an aryl halide (Figure 37).[488,489]

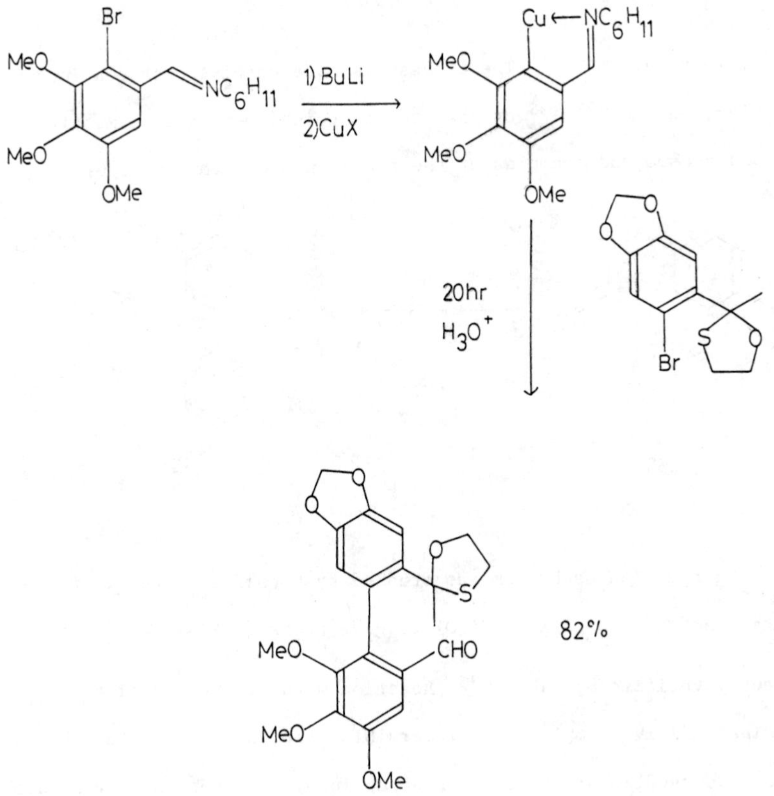

Figure 37 Coupling of an arylcopper compound with an aryl halide to yield
a precursor to (±) steganacin.

More interesting are reactions catalysed by Ni(0) complexes. The

reaction was discovered by Semmelhack using (cod)$_2$Ni in stoichiometric

quantities[490] (261 to 262). High yields are obtained under mild conditions

and the reaction is tolerant of ketone, aldehyde, ester and cyano groups,

but is inhibited by -OH and -COOH (which give reduction) -NO$_2$ and large

ortho groups. (Ph$_3$P)$_4$Ni is also a catalyst and was used in a synthesis

$$(\text{cod})_2\text{Ni} + 2\ \text{ArX} \xrightarrow[\text{25 - 40}°]{\text{DMF}} \text{Ar} - \text{Ar} + \text{NiX}_2 + 2\ \text{cod}$$

261   262

of dimethylalnusone, 264.[491] The process is made catalytic by continuous electrolytic reduction[492,493] or the use of stoichiometric zinc metal.[494] The proposed scheme and examples of the reaction are given in Figure 38.

263   264

The reaction mechanism was studied by Kochi.[497] The first step is oxidative addition of ArX to Ni(0) to give trans-$(R_3P)_2$NiArX, a step independently verified by Fahey.[498] Reaction with Ar'X could give a hexacoordinate Ni(IV) complex in a reversible step, accounting for the scrambling observed in attempted cross-couplings. Although aryl radicals were shown rigorously not to be involved in the reaction, the process is inhibited by radical scavengers, suggesting that an odd electron Ni(I) species may participate (Figure 39). Any process involving elimination of ArAr' from ArAr'Ni$(PR_3)_2$ was excluded as a significant step since its measured rate is much slower than that of the catalytic reaction. However, these observations relate to reaction in hydrocarbon solvents and decomposition of ArAr'Ni$(PR_3)_2$ in DMF is much faster.[499]

$$2ArX + Ni(0) \longrightarrow Ar_2 + NiX_2$$

$$NiX_2 + Zn \longrightarrow Ni(0) + ZnX_2$$

$$PhBr \xrightarrow[\text{Ph}_3\text{P/Zn}]{(\text{Ph}_3\text{P})_2\text{NiCl}_2} Ph-Ph$$

Ref

495

MeO—⟨⟩—Cl $\xrightarrow[\text{Zn} \mid \text{DMF}]{\text{NiCl}_2 \mid \text{PPh}_3 \mid \text{bipy}}$

MeO—⟨⟩—⟨⟩—OMe   496

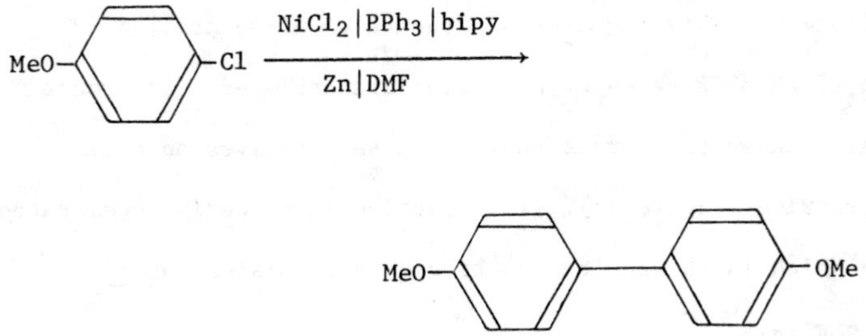

Figure 38 Nickel (0) catalysed coupling of aryl halides.

$$Ni(I)X + ArX \rightleftharpoons ArNi(III)X_2$$

$$ArNi(III)X_2 + Ar'Ni(II)X \rightleftharpoons NiX_2 + ArAr'Ni(III)X$$

$$ArAr'Ni(III)X \longrightarrow Ar-Ar' + Ni(I)X.$$

Figure 39 Mechanism of nickel catalysed coupling of aryl halides.

8.6.2. Oxidative Coupling of Arenes by Palladium (II)

Oxidative coupling of arenes to biphenyls is achieved by Pd(II) under mild conditions[500] (265 to 266).[501] As the reaction proceeds Pd metal is deposited and stoichiometric Pd(II) is needed. The rate-determining step is the formation of a σ-bonded aryl palladium (II) complex. Electron donors accelerate the reaction and C_6D_6 shows a marked kinetic isotope effect. The scheme of Figure 40 is proposed. Breakdown of 267 is initiated by AcO^- but the mechanism of its formation is unknown. If Pd(0) may be reoxidised with O_2 the reaction becomes catalytic in palladium. This requires a pressure of 25-45 atm.[502-504] and yields of up to 20,600% based on palladium are achieved. The route of reoxidation is unknown but a reasonable proposal involves an aryl palladium hydroperoxide (Figure 41). The reaction is markedly accelerated by acetylacetone which is thought to function by stabilising the cis-aryl palladium complex.

$$2 \; \text{C}_6\text{H}_6 \; + \; \text{Pd (II)} \; + \; 2\,\text{Cl}^- \; \xrightarrow[\text{NaOAc}]{\text{HOAc } 70^\circ} \; \text{C}_6\text{H}_5\text{-C}_6\text{H}_5 \; +$$

265   266

$$2 \, \text{HCl} + \text{Pd(0)}$$

Direct oxidation of Pd(0) by O_2 is relatively difficult and the customary cooxidant, Cu(I), is not useful because of its low solubility in organic solvents. The most promising cooxidants are heteropolyacids (HPA) of general formula $H_{3+n}PMo_{2-n}V_nO_{40}$ (n=1 to 8). These act as reversible oxidants under mild conditions (50-90°, 1.5 atm O_2, Figure 42) and although rates and yields are low some useful transformations have been accomplished.[505-507]

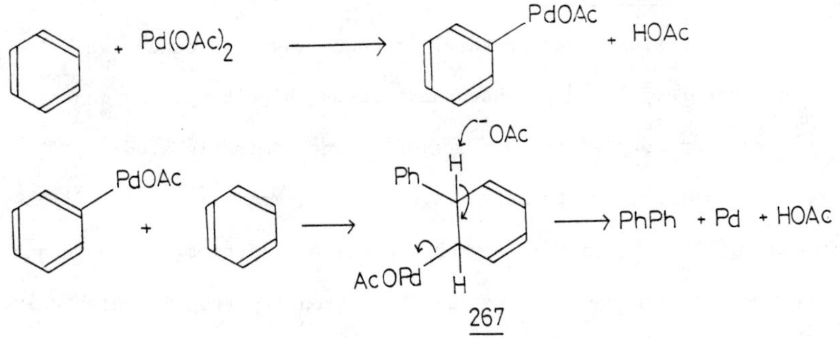

Figure 40   Mechanism of stoichiometric oxidative coupling of arenes by Pd(II).

PdX_2 + 2ArH $\longrightarrow$ Ar_2Pd + 2HX

Ar_2Pd $\longrightarrow$ ArAr + Pd(0)

Pd(0) + O_2 + ArH $\longrightarrow$ ArPdOOH

ArPdOOH + acacH $\longrightarrow$ ArPd acac + HOOH

ArPdacac + ArH $\longrightarrow$ Ar_2Pd + acacH

Figure 41   Rôle of oxygen and acetylacetone in palladium catalysed arene coupling.

2ArH + Pd(II) $\longrightarrow$ ArAr + Pd(0) + $2H^+$

Pd(0) + HPA + $2H^+$ $\longrightarrow$ Pd(II) + H_2(HPA)

H_2(HPA) + 1/2 O_2 $\longrightarrow$ HPA + H_2O

Figure 42   Heteropolyacids as reoxidants in palladium catalysed arene coupling.

8.6.3. Phenol Oxidative Coupling

Biosynthesis of many natural products involves oxidative coupling of phenols.[508,509] In natural systems the ultimate oxidant is O_2, _via_ various coupled redox reactions. In biomimetic studies, however, high oxidation state metals (Fe(III)[510], Mn(IV)[511], V(IV)[512], V(V)[512] and Cu(II)[513]) are more commonly used as stoichiometric oxidants. Recently, however, a number of examples have been reported in which only a catalytic amount of metal is employed and the reaction is carried out in an oxygen atmosphere (Figure 43).

8.6.4. Oxidative Coupling of 1-Alkynes

The Glaser reaction for coupling 1-alkynes with Cu(II) salts has been known for over a century. Cu(II) is simultaneously reduced to Cu(I)[517], pyridine is essential for a good yield and e.p.r. studies suggest that $Cu(Py)_4^{2+}$ is the active species.[518] Many applications are known, the chief being the synthesis of polyacetylenes and intramolecular coupling in annulene synthesis (Figure 44). If Cu(I) could be reoxidised _in situ_ by air the reaction would be catalytic and examples are known where less than stoichiometric amounts of Cu(II) are required. In the reaction of 268 with Cu_2Cl_2 in O_2 only a small amount of the salt is needed and stoichiometric Cu(I) actually causes cleavage of the C-Si bond.[520] By a series of cross-couplings and deprotection steps compounds containing up to 12 C-C triple bonds were synthesised and their U.V. spectra studied.[521] However, since Cu(II) is inexpensive and Cu(I) easily recovered, most synthetic applications continue to use stoichiometric reactions.

$$H - C \equiv C - SiEt_3 \quad \xrightarrow[\text{TMEDA/Acetone}]{Cu_2Cl_2/O_2} \quad Et_3Si - C \equiv C - C \equiv C - SiEt_3$$

268   269

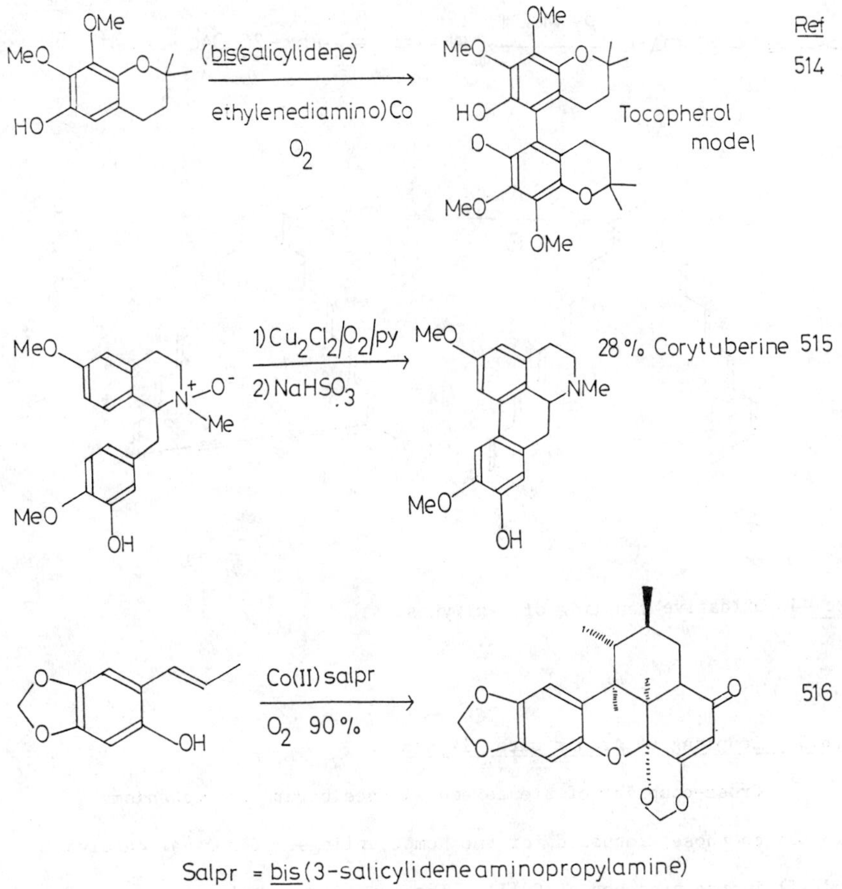

Figure 43   Catalytic phenol oxidative coupling.

$$\text{Ph}\!-\!\!\equiv \;+\; 2\text{Cu(OAc)}_2 \xrightarrow{\text{py, MeOH}} \text{Ph}\!-\!\!\equiv\!-\!\equiv\!-\text{Ph} \;+\; 2\text{CuOAc} \;+2\text{AcOH} \qquad \text{517}$$

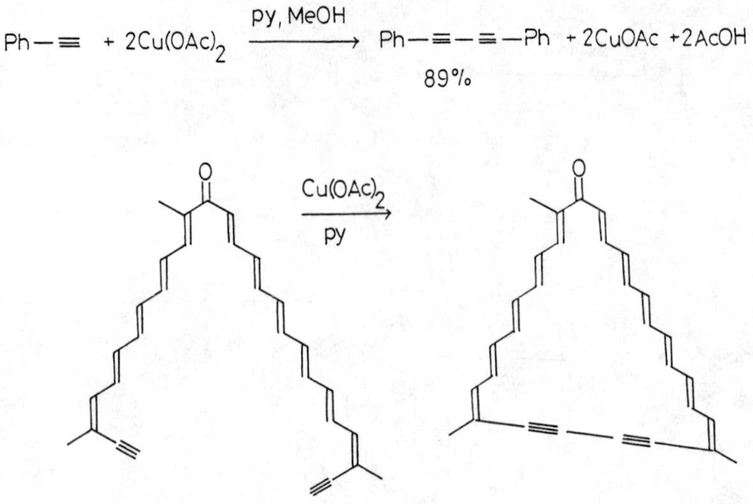

$$89\%$$

Ref

519

Figure 44 Oxidative coupling of 1-alkynes.

8.6.5. Coupling of Arenes with Alkynes

Cross-coupling of arenes and alkynes occurs by mechanisms analogous to those discussed for the homocouplings. The usual catalyst is Pd(II) in the presence of Cu(I). The reaction is tolerant of a wide range of functional groups including -CHO, -COOMe, -COR, -NH$_2$ and -NO$_2$ [522] (Figure 45). The function of the Cu(I) is unknown since it does not act as an oxidant and Pd(0) complexes are also catalytically active.[526] It has been postulated that coupling occurs via oxidative addition to a Pd(0) complex (Figure 46).

523

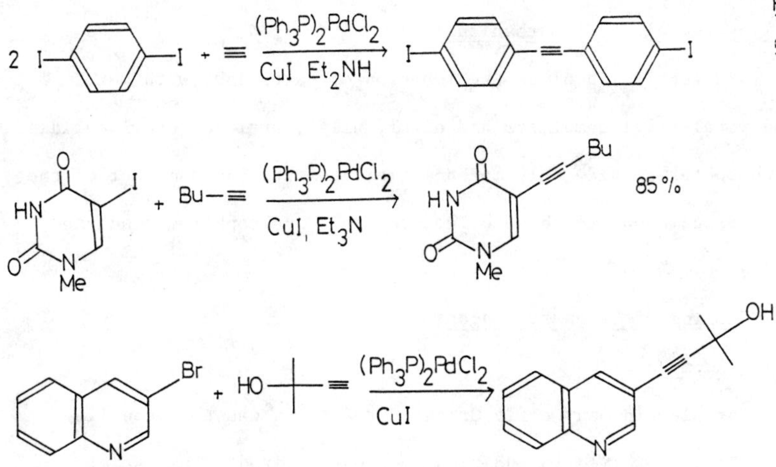

85 %   524

525

Figure 45 Arene alkyne coupling.

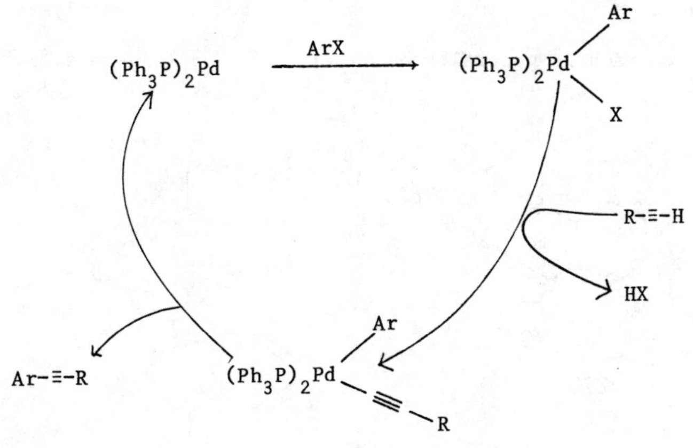

Figure 46 Mechanism of arene alkyne coupling.

8.7   Coupling Reactions of Carbanions

This section involves the reaction of carbanions with metal alkene and metal allyl complexes and alkyl, allyl, aryl and vinyl halides. We shall encounter considerable mechanistic diversity and a number of the reactions are analogous to those of heteroatomic nucleophiles discussed in Chapter 4.

8.7.1. Coupling of Grignard Reagents

8.7.1.1 Homocoupling

A problem in many early Grignard reactions was that coupling products, R-R, were formed in addition to those desired. The use of homocoupling in synthesis has its origin in the 1940's with Kharash, who formed biaryls from cobalt catalysed coupling of RMgBr.[527]   The reaction is still used and is largely insensitive to steric effects (270 to 271).[528] It is also effective with alkyl Grignards; cyclohexylmagnesium bromide is coupled to give bicyclohexyl in 62% yield.[529]   A few other catalysts are known but the reaction is generally an unwanted one[530] (Figure 47).

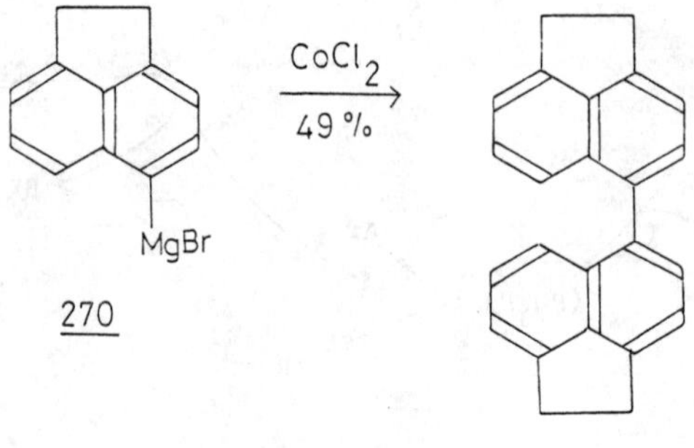

270

271

$$2 \ ArMgBr \quad \xrightarrow{Mn(acac)_3} \quad Ar\text{-}Ar \quad 98\% \qquad\qquad 531$$

$$PhMgBr \ + \ PhCH_2Cl \quad \xrightarrow{Cp(Me_2PCH_2CH_2PMe_2)FeBr} \quad \begin{array}{c} Ph\text{-}Ph \\ 92\% \end{array} \qquad\qquad 532$$

$$+ \ PhCH_2CH_2Ph$$
$$85\%$$

$$\underline{n}\text{-}C_4H_9MgBr \quad \xrightarrow[79\%]{Ag^+} \quad \underline{n}C_8H_{18} \qquad\qquad 533$$

$$PhO(CH_2)_4MgBr \quad \xrightarrow{TiCl_4} \quad PhO(CH_2)_8OPh \qquad\qquad 529$$

Figure 47 Homocoupling of Grignard reagents.

8.7.1.2. Coupling with Aryl Derivatives

Cross-coupling of Grignards with aryl halides is a useful route
to biaryls. The Karasch method is successful for intramolecular
reactions[534] but most modern syntheses employ Ni or Pd complexes as
catalysts.[535] The procedure involves addition of a preformed Grignard
in the presence of $(R_3P)_2MX_2$ (M=Pd,Ni,X=Cl,Br) (Figure 48). Iodides react
faster than bromides, chlorides with difficulty and fluorides not at
all.[538,540] This allows selective reaction of p-chlorobromobenzene with
PhMgBr to give p-chlorobiphenyl in good yield with palladium, though not
nickel, complex catalysts. With branched Grignards (e.g. i-PrMgCl) are
used both branched and linear products are obtained in the presence of
L_2NiX_2 (273 and 274). 273 is the major product with non-basic phosphines
such as diphos but 274 predominates with stronger electron donors
(273:274=4:94 L=Me_2PCH_2CH_2PMe_2).[541,542]

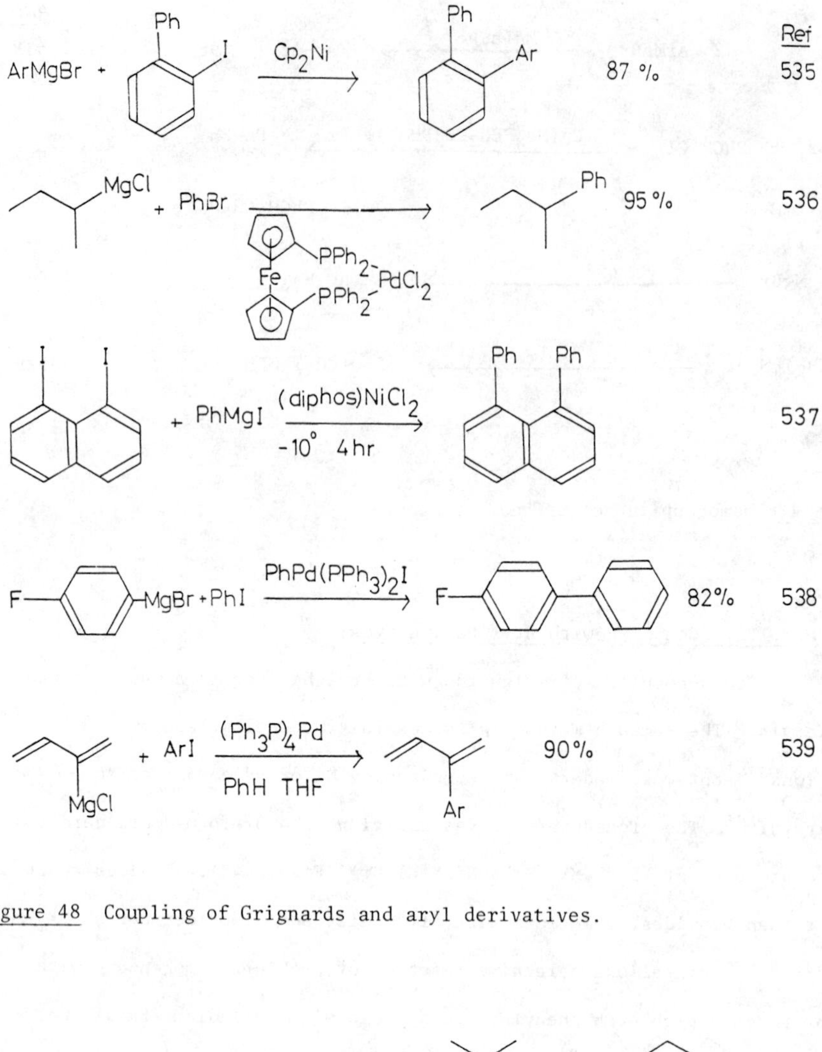

ArMgBr + (2-phenyl-iodobenzene) $\xrightarrow{\text{Cp}_2\text{Ni}}$ (2-phenyl-aryl-benzene) 87 % Ref
535

(sec-butyl)MgCl + PhBr $\xrightarrow{\text{Fe(Cp-PPh}_2)_2\text{PdCl}_2}$ (sec-butyl-Ph) 95 % 536

(1,8-diiodonaphthalene) + PhMgI $\xrightarrow[-10°\ 4\,hr]{\text{(diphos)NiCl}_2}$ (1,8-diphenylnaphthalene) 537

F—C$_6$H$_4$—MgBr + PhI $\xrightarrow{\text{PhPd(PPh}_3)_2\text{I}}$ F—C$_6$H$_4$—C$_6$H$_5$ 82% 538

(2-(isopropenyl with MgCl)-1,3-butadiene) + ArI $\xrightarrow[\text{PhH THF}]{(\text{Ph}_3\text{P})_4\text{Pd}}$ (Ar-substituted diene) 90% 539

Figure 48 Coupling of Grignards and aryl derivatives.

(isopropyl)MgCl + PhCl $\longrightarrow$ (isopropylbenzene) + (n-propylbenzene)
272   273   274

Early mechanistic studies suggested that the important steps in the reaction were oxidative addition of aryl halide to the metal, nucleophilic substitution by the Grignard and reductive elimination from the dialkyl metal.[538,543,544] However, reductive elimination from isolated 275 is too slow to account for the observed catalysis[545], and Kochi proposes an electron transfer route to accomodate both this, and the observed acceleration of the reaction by oxygen (Figure 49).

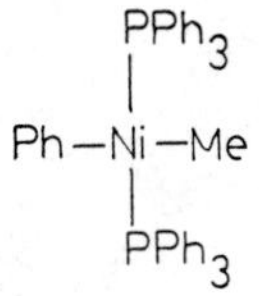

$$
\begin{array}{c}
PPh_3 \\
| \\
Ph-Ni-Me \\
| \\
PPh_3
\end{array}
$$

275

$$O_2 + R-M \longrightarrow O_2^{\cdot-} + R-M^{+\cdot}$$

$$L_2NiArMe + ArBr \longrightarrow L_2\overset{+\cdot}{Ni}ArMe + ArBr^{-\cdot}$$

$$L_2\overset{+\cdot}{Ni}ArMe + ArBr^{-\cdot} \longrightarrow L_2\overset{+}{Ni}\overset{Ar}{\underset{Me}{\diagdown}} + Br^-$$

$$\downarrow$$

$$ArMe + L_2NiArBr$$

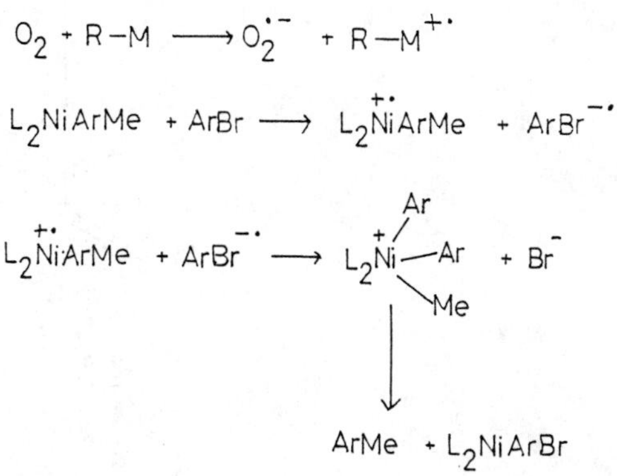

Figure 49   Mechanism for Grignard/aryl halide coupling.

Many synthetic applications are in heterocyclic chemistry, using both heterocyclic Grignards and halides (Figure 50). Substrates other than halides are also reactive, notably aryl thioethers[554,555] and aryl phosphates.[556] Their reactions are slower than those of the halides allowing the selective transformation, 276 to 278. Enantioselective versions of the reaction are known in the presence of nickel complexes of chiral phosphines (279 to 280).[540,554] Enantiomer excesses have generally been modest and there is clearly scope for improvement.[558,559]

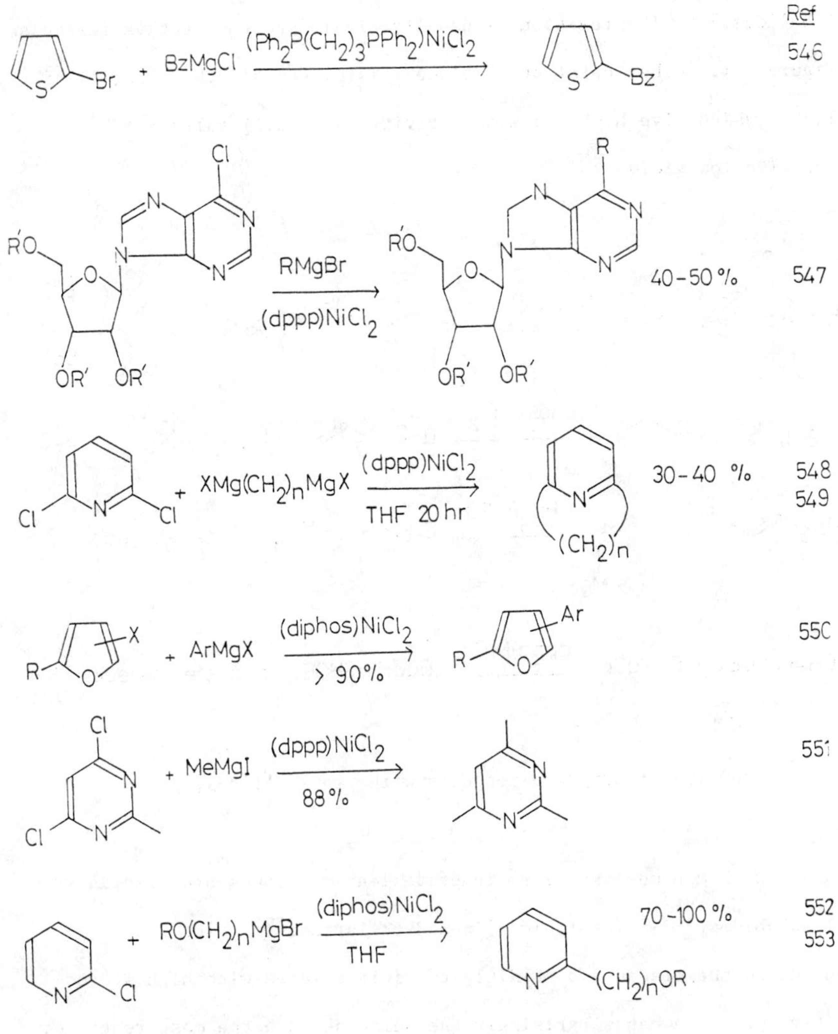

Figure 50 Synthetic uses of Grignard/aryl halide coupling.

8.7.1.3. Coupling with Vinyl Derivatives

Vinyl halides may be coupled with alkyl, vinyl and aryl Grignards in the presence of palladium and nickel complexes. The mechanism has not been meticulously studied but is presumed to be analogous to that with aryl halides.[560] The reaction is usually highly stereoselective (examples in Figure 51). Other catalysts include tris(dibenzoylmethido)Fe(III)[564], FeCl$_3$[565] (which give high stereoselectivity) and Cu(I) salts[566,567] (which give low yields).

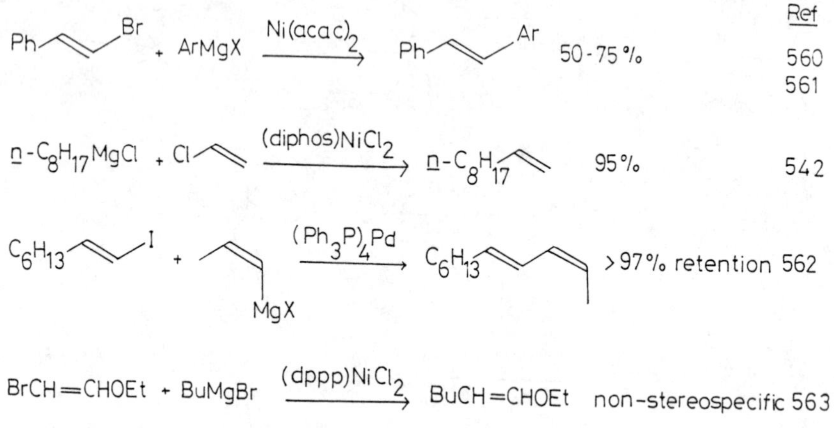

Ref

Ph⌒Br + ArMgX →[Ni(acac)$_2$] Ph⌒Ar 50-75%   560
                                                   561

n-C$_8$H$_{17}$MgCl + Cl⌒ →[(diphos)NiCl$_2$] n-C$_8$H$_{17}$⌒ 95%   542

C$_6$H$_{13}$⌒I + ⌒MgX →[(Ph$_3$P)$_4$Pd] C$_6$H$_{13}$⌒⌒ >97% retention 562

BrCH=CHOEt + BuMgBr →[(dppp)NiCl$_2$] BuCH=CHOEt non-stereospecific 563

Figure 51 Coupling of Grignard reagents with vinyl halides.

Double bonds with less reactive leaving groups are also in vogue. Vinyl sulphides,[568,569] selenides[570] and tellurides[571] react with Grignards in the presence of L$_2$NiCl$_2$ to yield alkenes with high stereoselectivity. Somewhat surprisingly the selenides are the most reactive, surpassing the corresponding chlorides. Only Grignards without β-H are suitable, reduction being the main reaction with i-PrMgBr.[572] The differential reactivity of bromides and thioethers is exploited in the conversion of 281 to 283 in excellent yield and >99% stereoselection.[573]

Br⌒⌒SPh + PhMgBr —diphos NiCl₂→ Ph⌒⌒SPh —MeMgBr→
281   **282** (Ph₃P)₄ Pd

Ph⌒⌒
283

Sulphones are reactive in the presence of Ni(acac)$_2$[574] and Cu(acac)$_2$[575],
again giving substition with high retention of double bond geometry.
Vinylsilylenol ethers[576] and vinyl phosphates have also been studied[577]
and it is clear that the range of reactive functionalities will be extended, and systematically studied for the selectivities they may afford.

Asymmetric coupling of Grignards with vinyl halides is probably
the best studied of their enantioselective reactions.[578] 1-Phenethyl
magnesium bromide is the reagent of choice and selectivity depends on the
ability of the catalyst to effect kinetic resolution of R- and S- **284**.
Typically, successful catalysts contain a chiral aminophosphine ligand
and the function of the amino groups is to complex magnesium at some
stage of the catalytic cycle. Some ligands and enantioner excesses are
given in Figure 52. The presence of an amino function is not entirely
critical, since (NORPHOS)NiCl$_2$ gives **285** in a 65% enantiomer excess.[585]
Palladium complexes are equally useful catalysts as the synthesis of **288**
shows.[586]

Me H   H Me
Ph⌒MgBr   ⇌   Ph⌒MgBr   + ⌒X   —L*₂NiCl₂→   Ph⌒
284a   **284b**   **285**

Ligand	Conditions	Enantiomer excess	Ref

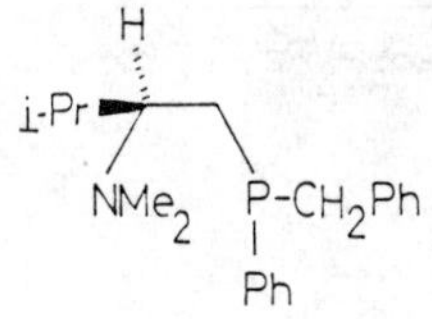

i-Pr, H, NMe₂, P-CH₂Ph, Ph — L₂NiCl₂, Et₂O, 0° — 53% S — 579

tBu, H, NMe₂, PPh₂ — R-t-Leuphos — L₂NiCl₂, Et₂O, 0° — 83% R — 580

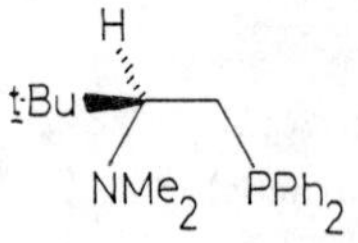

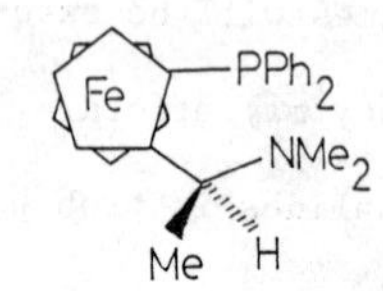

Fe, PPh₂, NMe₂, Me, H — S,R-PPFA — L₂NiCl₂ — 66% R — 581-584

Figure 52 Asymmetric coupling of α-phenethyl magnesium chloride.

Me₃Si, Ph, MgBr + Ph-CH=CH-Br →[R S-PPFA PdCl₂] Ph-CH=CH-C(Ph)(SiMe₃)(H) 95% ee

286   287

287 + F⁻ + t-BuBr ⟶ t-Bu-C(Ph)(H)-CH=CH-Ph   93% ee

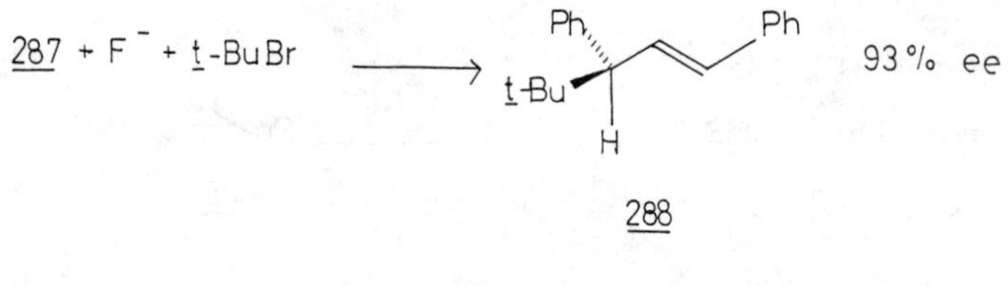

288

8.7.1.4. Coupling with Allyl Derivatives

The mechanism of this reaction potentially differs in certain respects from those in the previous two sections. When allyl chloride reacts with a Ni or Pd complex the result is a π- (289) rather than a σ-allyl (290) complex. Reductive elimination from 291 then gives two products, 292 and 293, readily distinguished in the substituted case. Allyl halides themselves are poor substrates giving mainly reduced[587-9] or homocoupling products.[590,591] This is no especial disadvantage, however, since the range of allyl derivatives which yield metal allyl complexes is extensive. Allyl ethers[592-5] and esters[596,597] give

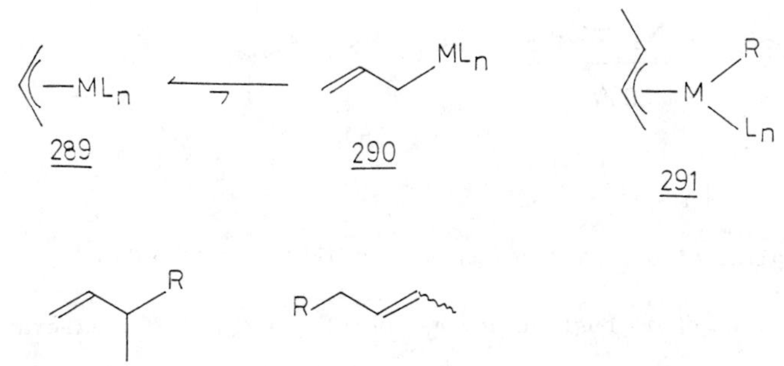

289   290   291

292   293

coupling in excellent yield and varying regioselection (e.g. 294 to 295 and 296). Control of regioselectivity is complex and we shall return to this topic later. Other derivatives are also reactive (Figure 53), though less exploited to date.

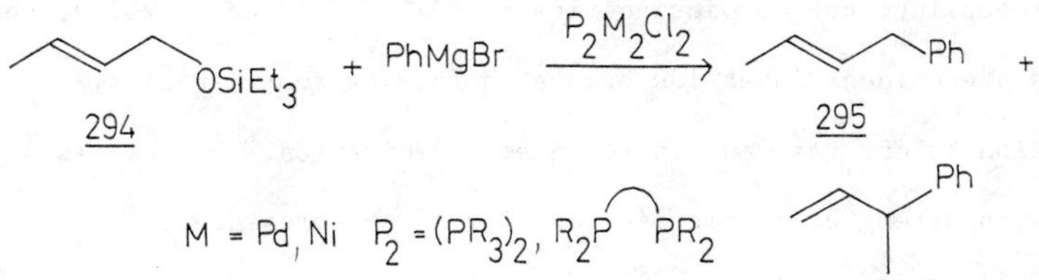

294   295

$M = Pd, Ni \quad P_2 = (PR_3)_2, R_2P \frown PR_2$

296

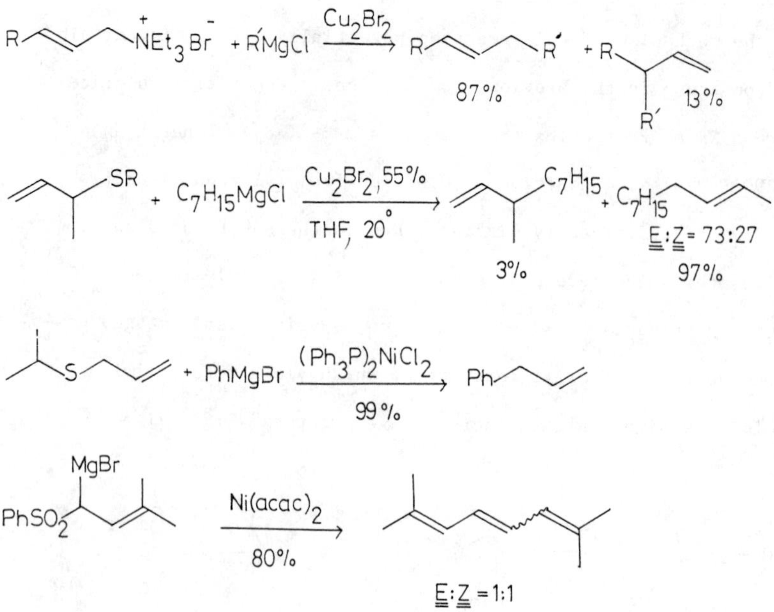

Figure 53 Coupling of Grignard reagents with allyl derivatives.

Allyl alcohols must occupy a special place in our deliberations, if only for the attention lavished on them. When 297 reacts with MeMgBr or PhMgBr in the presence of $(R_3P)_2NiCl_2$ a mixture of alkenes is obtained, deriving from substitution at both ends of the postulated metal allyl intermediate.[602] If the cis-alcohol, 303, is the substrate the regioselectivity is changed to give attack mainly at the more substituted carbon.[603] Careful analysis of these results led the authors to propose a metal allyl mechanism (Figure 54) where syn-anti isomerisation is slower than coupling, but not insignificant.[604,605] If the chiral alcohol, 305, is used predominent inversion occurs in forming 306; σ-π-allyl interconversion is not fast enough to cause racemisation.[606] This is exploited in an attempted asymmetric coupling in the presence of $(DIOP)NiCl_2$[607]; 307, 308 and 309 give 310 and isomers but enantiomer excesses are modest.

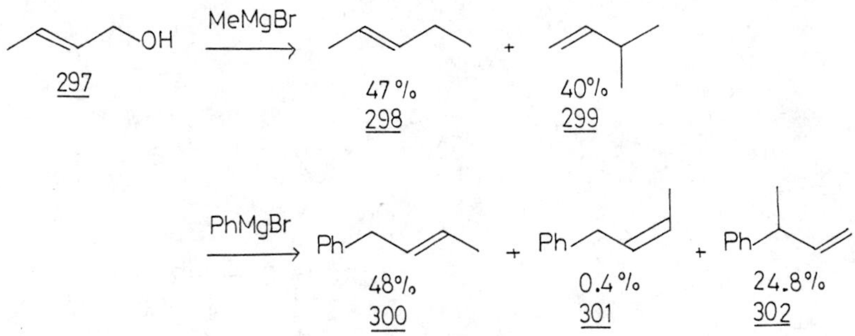

297
MeMgBr

298 47% + 299 40%

PhMgBr

Ph 300 48% + Ph 301 0.4% + Ph 302 24.8%

303
MeMgBr

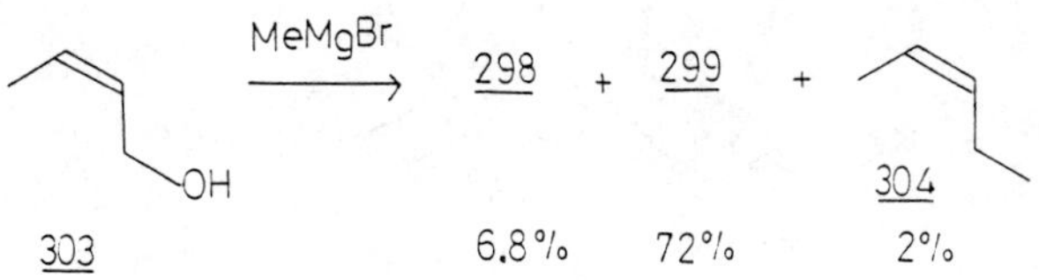

298 6.8% + 299 72% + 304 2%

305 OH → 306 Ph

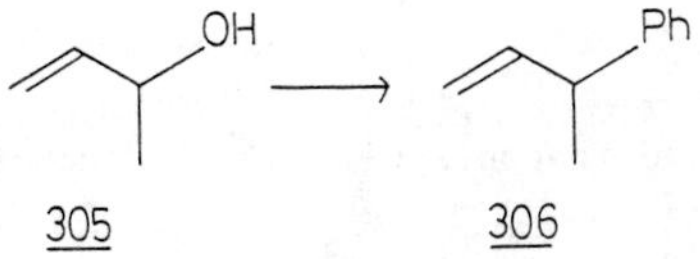

307 OH

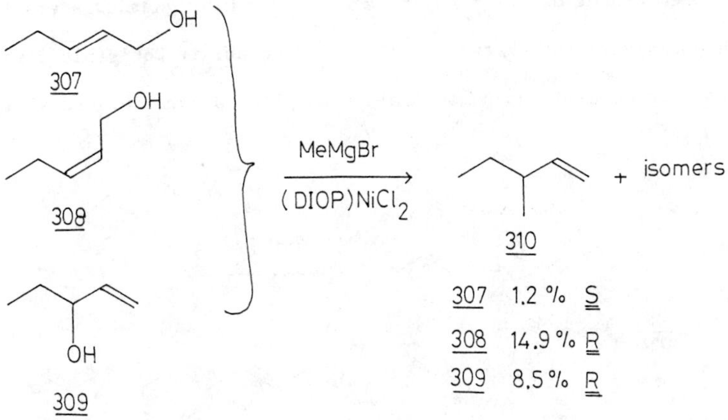

308 OH

309 OH

MeMgBr
(DIOP)NiCl$_2$

310 + isomers

307 1.2% S
308 14.9% R
309 8.5% R

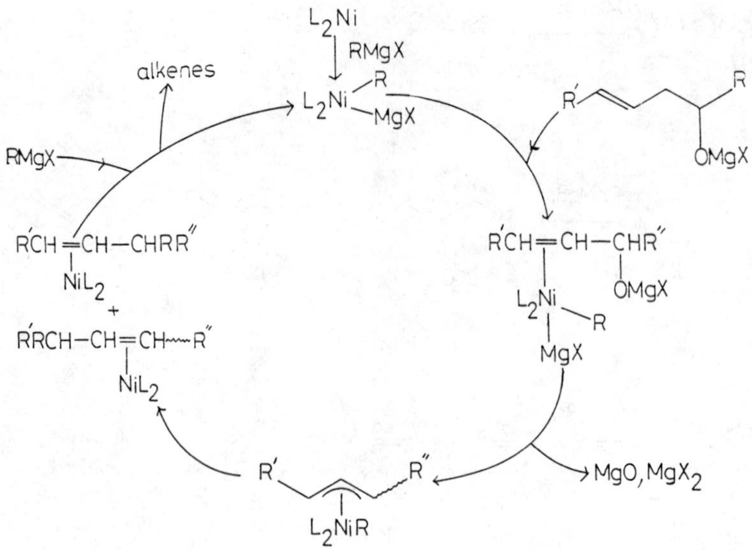

alkenes

L$_2$Ni

RMgX

L$_2$Ni—R / MgX

RMgX

R'CH=CH—CHRR"
 |
 NiL$_2$
 +
R"RCH—CH=CH—R"
 |
 NiL$_2$

R' / R"
 |
L$_2$NiR

MgO, MgX$_2$

R'CH=CH—CHR"
 | OMgX
L$_2$Ni—R
 |
MgX

R'—CH=CH—CH$_2$—CHR
              |
             OMgX

Figure 54 Mechanism of coupling of Grignard reagents with allyl alcohols.

All the reactions discussed so far have involved Grignards with no β-H, which cannot act as reducing agents. Such Grignards generally give hydrogenolysis rather than coupling. However, if PdCl$_2$(bis(diphenylphosphino)ferrocene) is used as catalyst coupling occurs in good yield (311 to 312).[608]

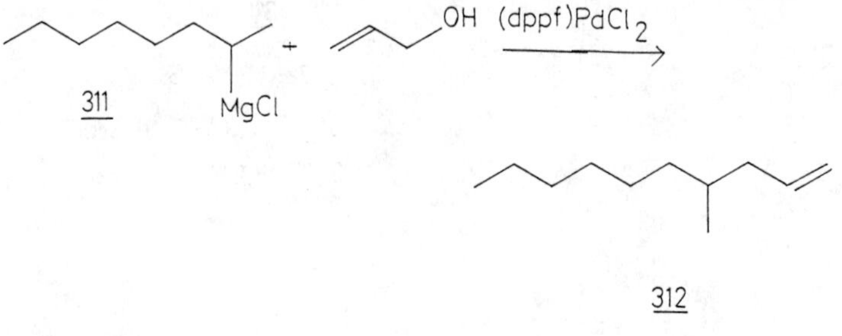

311 MgCl

+ OH (dppf)PdCl$_2$

312

8.7.1.5. Coupling with Alkyl and Silyl Derivatives

Despite their apparent simplicity these reactions are rare, since alkyl halides do not undergo facile oxidative addition to metals. The usual catalyst is a Cu(I) salt and the mechanism is though to involve a multimetallic intermediate.[596] Akyl[609], allyl[610] vinyl[611] and aryl Grignards[612] react and substrates include alkyl iodides[613], bromides[612] and tosylates[614] (Figure 55).

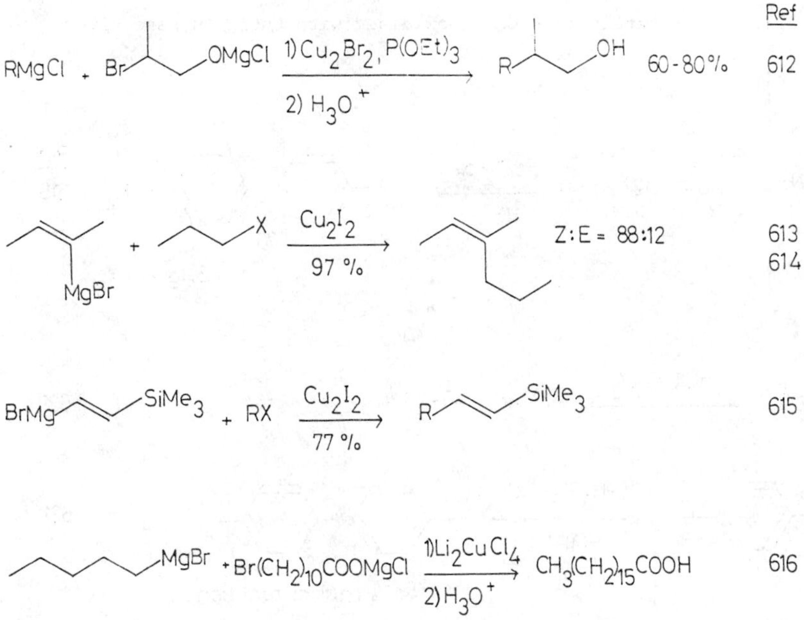

Figure 55 Coupling of Grignard reagents with alkyl halides.

8.7.2. Homocoupling of Other Organometals

Aryl and vinyl mercury (II) derivatives are readily prepared, stable and easily purified. Palladium and rhodium complexes[617] catalyse coupling to dienes and biaryls. Either ArHgX or Ar_2Hg may be used, and unlike Ullmann coupling, the reaction is compatible with amino groups as well as many others[618] (free -OH or -COOH still cause deactivation) (Figure 56). Coupling of vinyl mercurials is generally both regio and stereospecific, the Rh(I) catalysed reaction giving the tail to tail product, 313, in excellent yield.[620,621] Stoichiometric Li_2PdCl_4 also yields 313,[623] but catalytic $PdCl_2$ reoxidised with $CuCl_2$ in base gives mainly 314.[624]

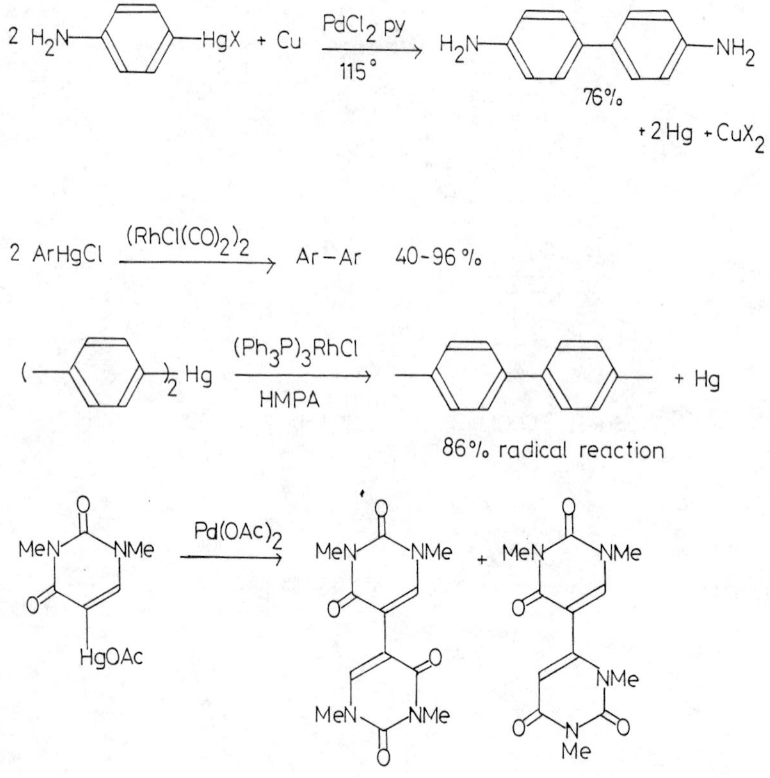

Ref
619
620
621
622

Figure 56 Homocoupling of aryl mercury compounds.

313   314

8.7.3. Coupling of Other Organometals with Aryl Derivatives

Carbanions, both hard and soft, are easily coupled with aryl bromides and iodides in the presence of metal complex catalysts; chlorides are relatively unreactive. The sodium salts of stabilised carbanions such as acac and dimethyl malonate react with ortho-bromosodium benzoates in the presence of Cu(I) to give substitution; the product is then cyclised to 318.[625,626] Sodium acetylides may be coupled with α-amino aryl halides in the presence of $(Ph_3P)_4Pd$;[627] in both these cases it seems likely that the polar functional group provides an additional catalyst binding site.

315   316   317

318

Coupling reactions of vinyl alanes[628] and zirconium[629,630] compounds were mentioned earlier when discussing the carbometallation reactions from which they derive. E-1-Hexenyldiisobutylalane reacts with aryl iodides and bromides in the presence of $(Ph_3P)_4Ni$ to give coupling in good yield with >99% retention of double bond stereochemistry.

The reactions of organozinc compounds closely parallel those of Grignard reagents. Alkyl, vinyl, aryl, alkynyl and allenyl compounds all react in excellent yield with ArX in the presence of Ni or Pd complexes[630] (Figure 57). Of comparable electronegativity are Cu and Hg derivatives[636,637] which couple to ArX under similar conditions. Organotin compounds[638,639] have also found applications, Pd(0) catalysts proving most suitable (Figure 58). Even the weakest of "organometals",

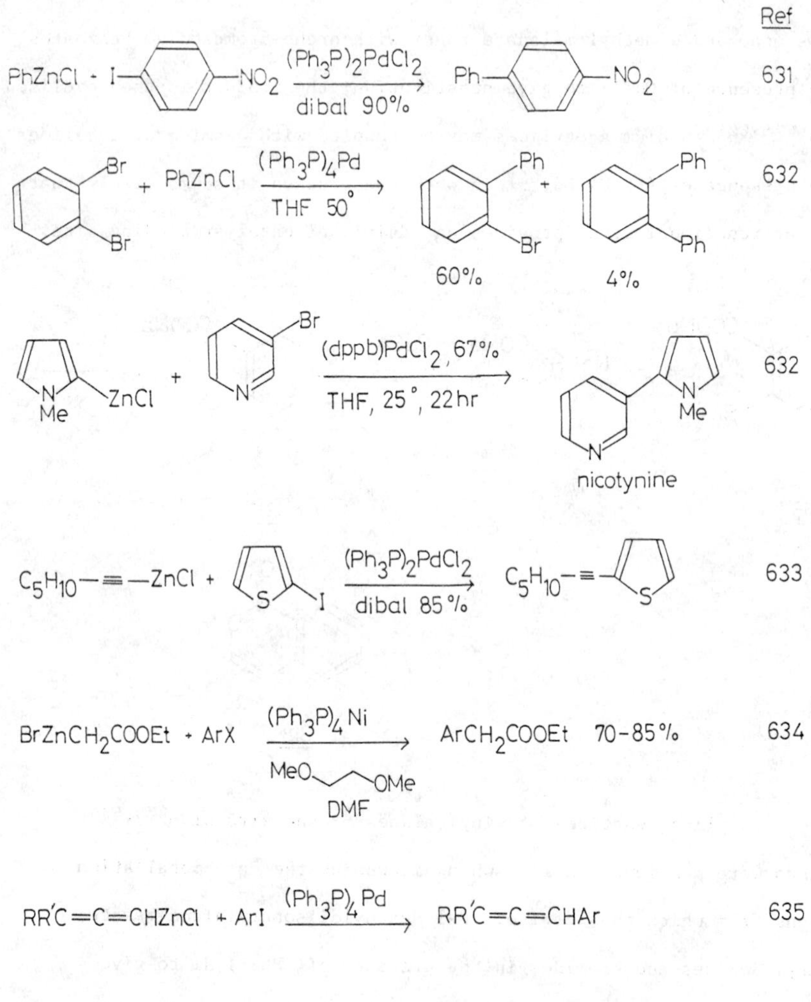
Ref
PhZnCl + I–C6H4–NO2
(Ph3P)2PdCl2
dibal 90%
Ph–C6H4–NO2
631
Br
Br
+ PhZnCl
(Ph3P)4Pd
THF 50°
Ph
Br
60%
+
Ph
Ph
4%
632
N
Me
ZnCl
+
Br
N
(dppb)PdCl2, 67%
THF, 25°, 22hr
N
Me
N
nicotynine
632
C5H10—≡—ZnCl +
S
I
(Ph3P)2PdCl2
dibal 85%
C5H10—≡—
S
633
BrZnCH2COOEt + ArX
(Ph3P)4Ni
MeO OMe
DMF
ArCH2COOEt 70-85%
634
RR'C=C=CHZnCl + ArI
(Ph3P)4Pd
RR'C=C=CHAr
635

Figure 57 Coupling of organozinc compounds with aryl halides.

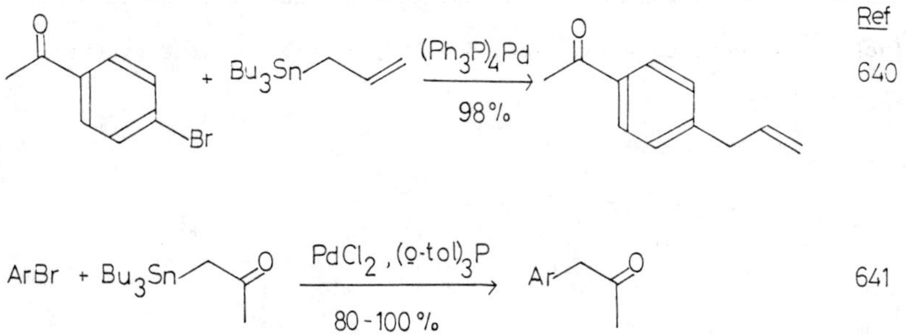

Ref

640

641

Figure 58   Coupling of organotin compounds with aryl halides.

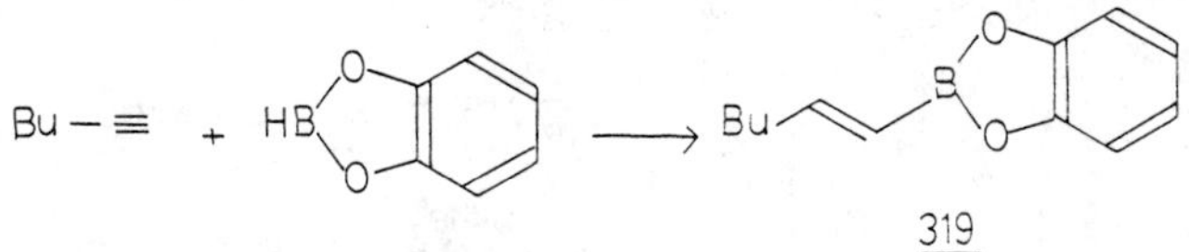

the boron derivative, 319, reacts with ArX to give stereospecifically

substituted alkenes such as 320,[642,643]

8.7.4.   Coupling of Other Organometals with Vinyl Derivatives

These reactions are closely mechanistically related to those

of aryl halides, our additional concern here being for the preservation of

stereochemistry about the double bond.   Derivatives of Li[644], Zr[630], Al[645],

Zn[646] and B[647] are reactive and Pd(0) complexes are the catalysts of

choice   (Figure 59).   Whatever the organometal excellent stereoselection

is achieved with Pd(0) complexes[652], Ni(0)[653] and Fe(II) complexes[654]

giving only slightly poorer results.   The reaction mechanism is as before;

oxidative addition of the halide to palladium is followed by nucleophilic

substitution by the organometal and reductive elimination. Applications
have included a route to mokupalide, 324,[655,656] and an allene synthesis
(325 to 327).[657-9]

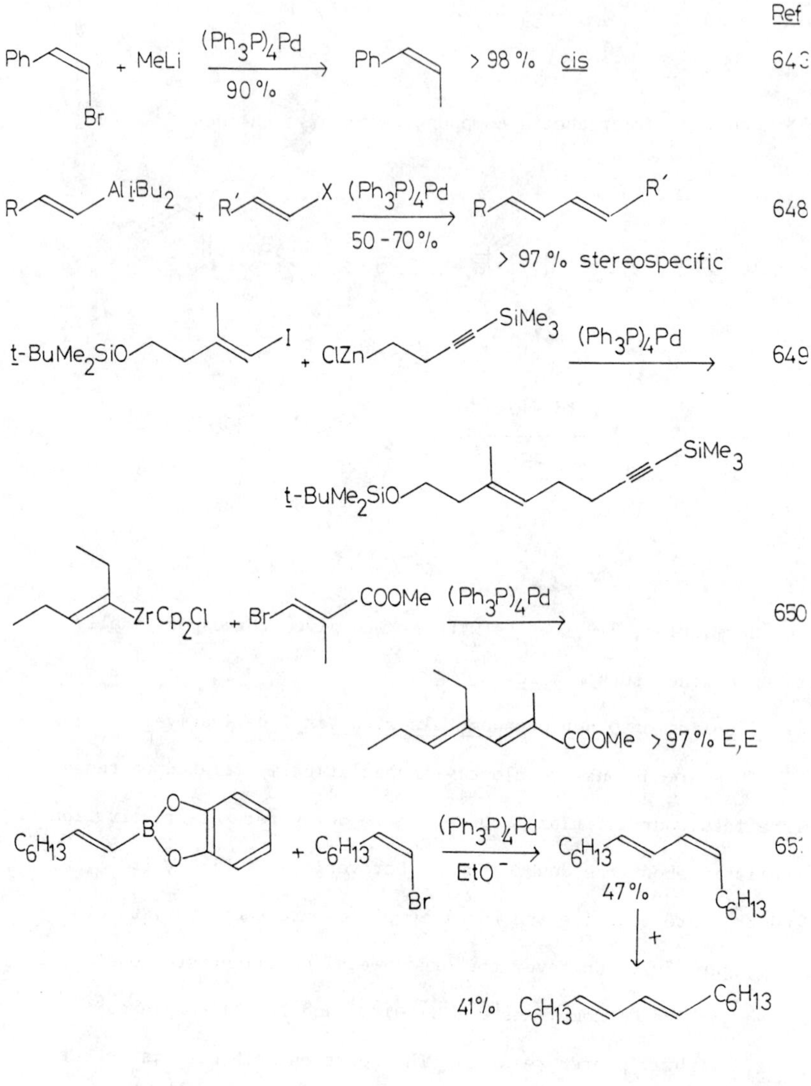

Figure 59 Coupling of carbanions with vinyl halides.

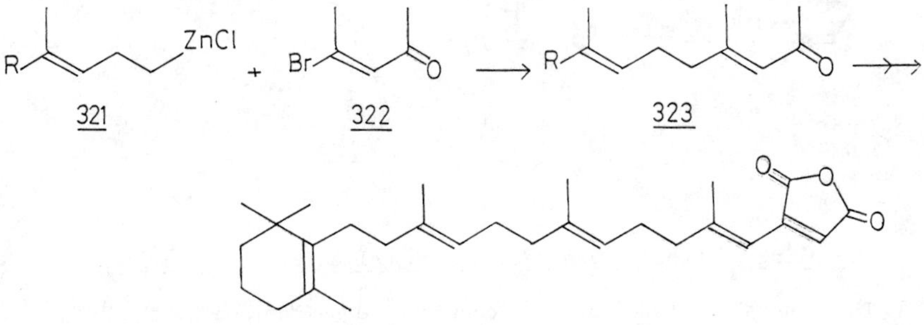

321 + 322 → 323 →→

324

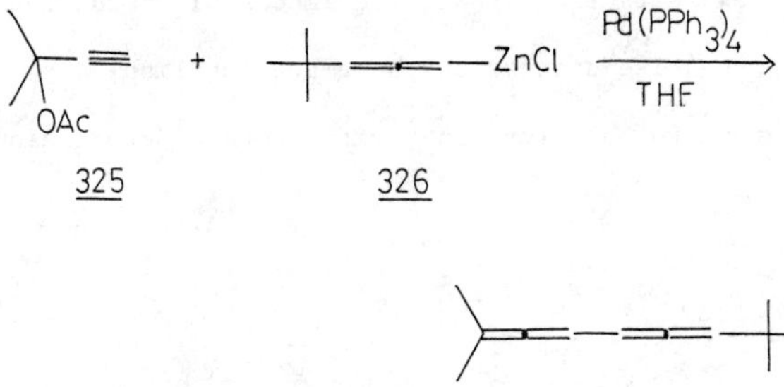

325 + 326

327

8.7.5. Coupling of Other Organometals with Alkyl Derivatives

As for the analogous Grignard reactions these are relatively
rare. Organolithium compounds react with benzyl halides in the presence
of Pd(0) complexes to give coupling, with mainly inversion of configuration
at the benzylic position.[660] The reaction is not a clean one, however,
and organotin derivatives give better yields, with retention. Vinyl
alanes also react to give good yields of 329 with retention of stereochemistry about the double bond.[653] Organomercurials are methylated in
the presence of $(Ph_3P)_3RhCl$[661] and vinylboranes react with cyclohexyliodide[662] but these are curiosities at present rather than general
strategies.

$$\underline{328} \quad + \text{ Ar CH}_2\text{X} \xrightarrow{\text{Pd L}_4} \quad \underline{329}$$

8.7.6. Coupling of Other Organometals with Palladium Allyl Complexes

The general features of the reaction are summarised by the
conversion 330 to 332. This apparently simple pathway has generated so
many studies over the last few years that it will be impossible to do them
justice here. Several comprehensive discussions have been published[165,663,664] and we shall concern ourselves only with factors determining
reaction selectivity.

$$\underline{330} \quad \xrightarrow{\text{Pd(0)}} \quad \underline{331} \quad \xrightarrow{\text{Nu}^-} \quad \underline{332} \quad + \text{ Pd(0)}$$

8.7.6.1. The Origins of the Palladium Allyl Complex

Early syntheses involved generation from alkenes in the presence
of an oxidant and a mild base (333 to 334).[665] Whilst such complexes
are useful[666], stoichiometric palladium is necessary. The ready formation
of allyl complexes from dienes does lead to catalytic reactions; the
palladium allyls we have seen involved in butadiene oligomerisation
react with suitable carbanions (e.g. 335 and 338).[667,668]

$$\underline{333} \quad \xrightarrow[\substack{\text{NaCl NaOAc} \\ \text{HOAc}}]{\text{PdCl}_2, \text{CuCl}_2} \quad \underline{334}$$

COMe

COMe

$\diagdown\!\!\diagdown\!\!\diagdown$ + $\diagup$COOEt $\xrightarrow[\text{NaOPh}]{\text{Pd(OAc)}_2}$ $\diagdown\!\!\diagdown$COOEt +

335 336

COOEt

337 COMe

COOEt

$\diagdown\!\!\diagdown\!\!\diagdown$ + NaC–NHAc $\xrightarrow[\text{PhONa, 91\%}]{\text{Pd(OAc)}_2, \text{PPh}_3}$

COOEt

338

COOEt

AcNH—$\diagdown\!\!\diagdown\!\!\diagdown$

COOEt

339

    The most important method of generating allyl complexes is oxidative addition of an allyl derivative to Pd(0) to give 331. Acetates are the most popular leaving group[669-673], because of their stability and ease of preparation, but are by no means unique. Other leaving groups include $-OCO_2R$[674], $-SO_2Ph$[675], $-NEt_3^{+}$[676], $-SMe_2^{+}$[676], $-OPh$[677], $-Cl$[678], $-OH$[679], $-CN$[680], $-OCR=NR'$[681], $-OPO(OEt)_2$[682] and $-NO_2$.[683-685] There are currently insufficient data to relate the reactivities of all these groups but some sequences are known (Figure 60).[685-7] The difference in reactivity was exploited in the synthesis of the chrysanthemic acid derivative 343 from 340.[688] Some interesting and reactive substrates have an intramolecular leaving group; this may involve epoxide opening, as for 345.[689] An added advantage is that the alkoxide generated acts as a base, so no added base is necessary (Figure 61).

$$\text{halogens} > \text{OAc} > \text{OAlR}_2 > \overset{\overset{\displaystyle O}{\|}}{\text{OP(OR)}_2} > \text{OSiR}_3$$

$$\text{OPh} > \text{OAc} > \text{OBz} > \text{OMe}$$

$$\text{OAc} > \text{OH}$$

$$\text{OAc} > \text{CN}$$

<u>Figure 60</u> Leaving group reactivity in allyl derivatives.

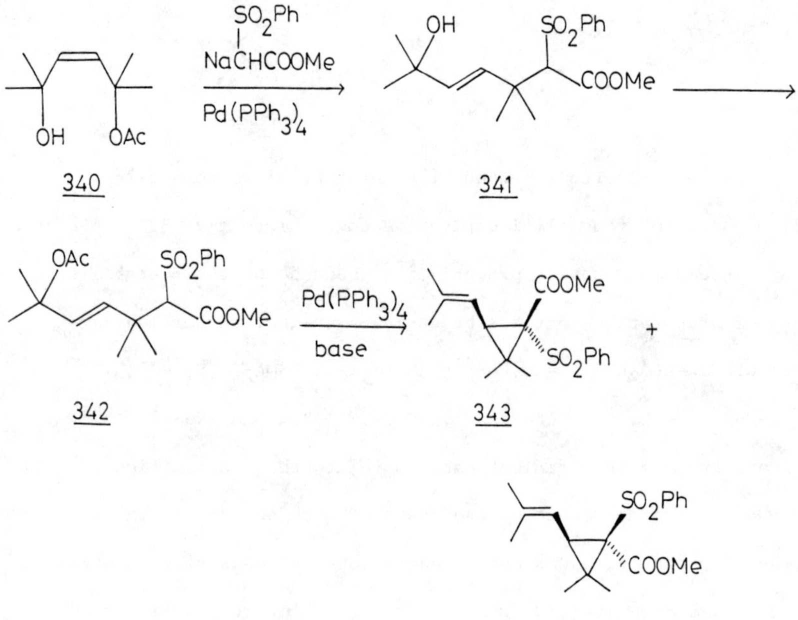

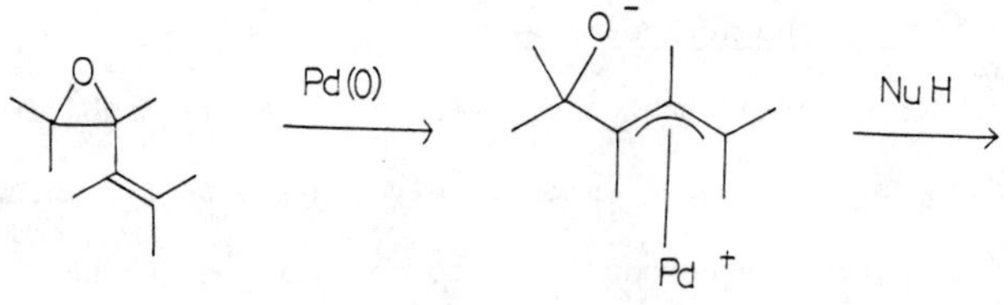

345 346

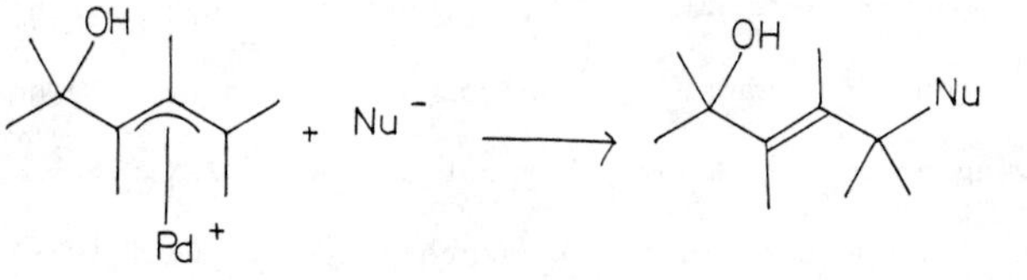

347 348

Ref

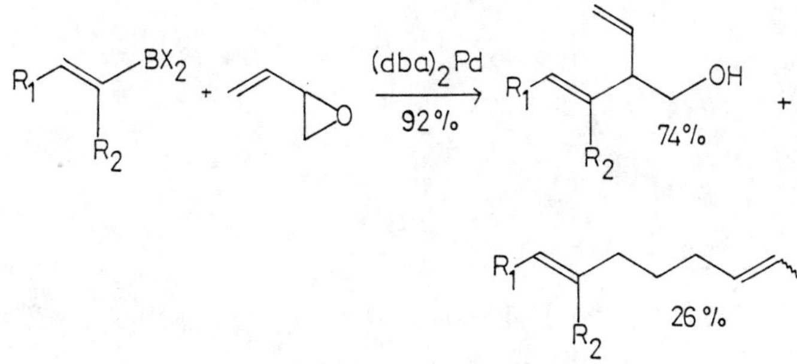

690

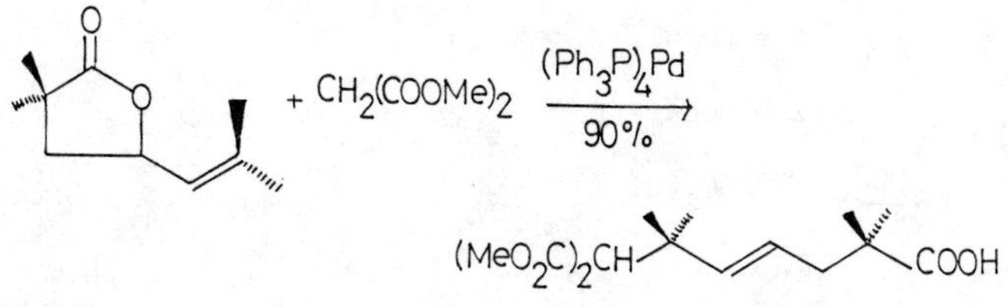

691

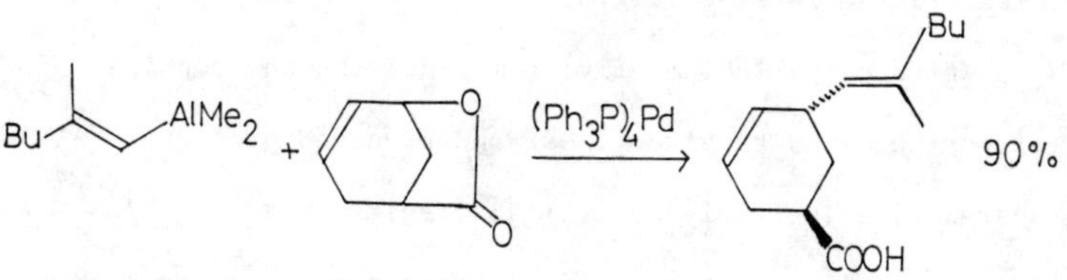

692

Figure 61 Formation of palladium allyls from cyclic oxygen compounds.

8.7.6.2. Nature of the Nucleophile

It has been observed that successful nucleophilic substitutions generally take place only when the anion involved has a $pK_a < 17$; otherwise attack at palladium predominates and leads to reduction.[694] Thus, the most popular nucleophiles have been active methylene compounds 349 to 354.[679,695-698] A number of organic derivatives of weak metals are also useful including compounds of aluminum[693], mercury[699], zinc[669], tin[700], zirconium[701] and boron.[702] Reaction of anions stabilised by only one electron withdrawing group less common but not unknown. 2-Nitropropane anion reacts with allyl acetate in the presence of $(Ph_3P)_2PdCl_2 | Ph_3P$ in moderate yield[686] a better result being achieved with the dianion.[703] Lithium[704] and tributyltin[705] enolates of ketones and esters are also useful[706] and even aryl rings may be allylated under forcing conditions.[672,707]

$$X-CH_2-Y$$

349   X = Y = COOMe

350   X = Y = COMe

351   X = COOMe, Y = COMe

352   X = COOMe, Y = SO_2Ph

353   X = COMe,   Y = SO_2Ph

354   X = SO_2Ph, Y = NO_2

8.7.6.3. Reaction Regioselectivity

With substituted palladium allyl complexes the nucleophile may attack at either terminus to give linear and branched products. Attack at the less hindered site is fairly general and regioselectivity high (Figure 62). Intramolecular reactions give interesting and unexpected results. For example a key step in the synthesis of exaltolide is the

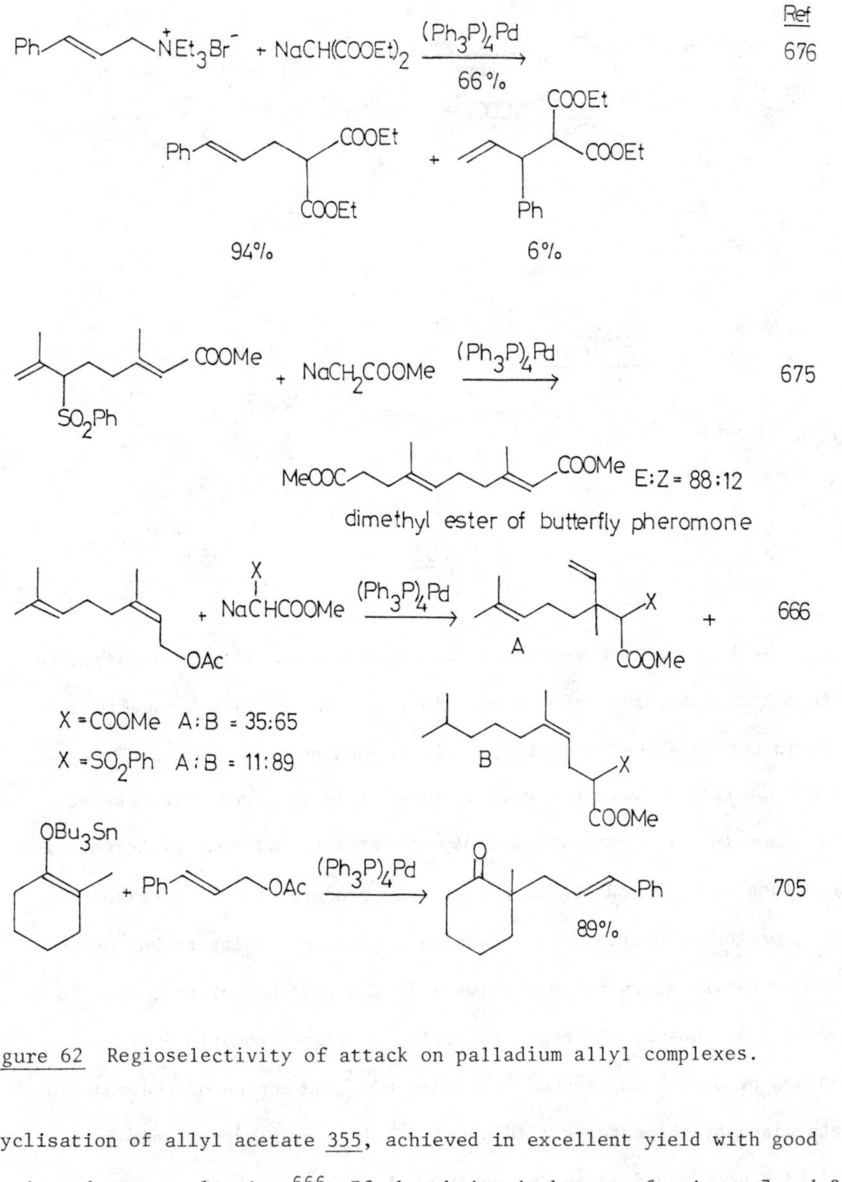

Figure 62   Regioselectivity of attack on palladium allyl complexes.

cyclisation of allyl acetate 355, achieved in excellent yield with good

regio and stereoselection.[666]   If the choice is between forming a 7-and 9-

membered ring the macrocyclisation gives only 358 as opposed to the

thermodynamically favoured 359. The selectivity for an 8- over a 6-membered

ring is 93:7 though this is dependent both on the catalyst used and the

precise substitution pattern of the ring.[165]

Pd(PPh₃)₄
NaH

355

356

Pd(PPh₃)₄
NaH

357

358

+

359

We have not yet mentioned the problem which plagues many simple
enolate alkylations, that of O-alkylation, because it is not usually a
problem in the catalysed reaction, at least in acyclic systems. This is
not to imply that it does not occur, but that it is reversible, whereas
C-alkylation is not. However, for 360 the stereoelectronic preference for
O-alkylation is high and 361 is the kinetic product of the reaction.[708]
However, prolonged heating, with catalyst, gives reopening to 362 and
closure by C-alkylation to 363. Similarly the tetrahydrofuran, 364, is
opened to 365; closure may occur to yield two stereoisomeric cyclo-
pentanones or a cycloheptenone[708] (Figure 63). The course of the reaction
is determined by the catalyst; (diphos)₂Pd gives mainly cycloheptenone,
(Ph₃P)₄Pd gives 5- and 7-ring products whilst the most bulky catalyst,
a polymer bound species, gives only cyclopentenone (369 to 370). The
thermal 3,3-sigmatropic shift gives cycloheptenone. An analogous scheme was
used by Tsuji to convert an acyclic precursor to methyl dihydrojasmonate.
In the reaction of 371 (Pd(OAc)₂|R₃P|base) an unhindered phosphite gives

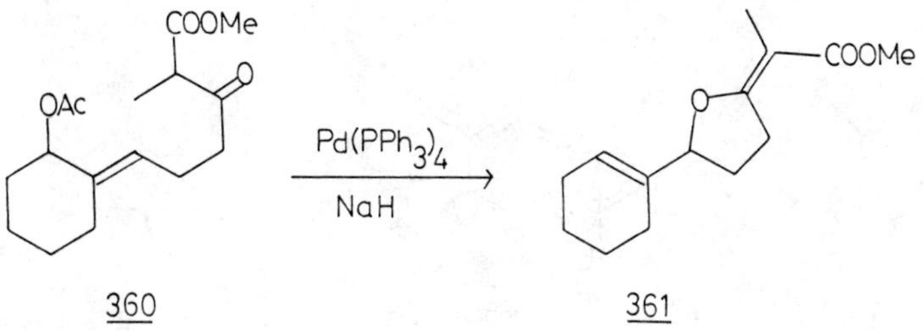

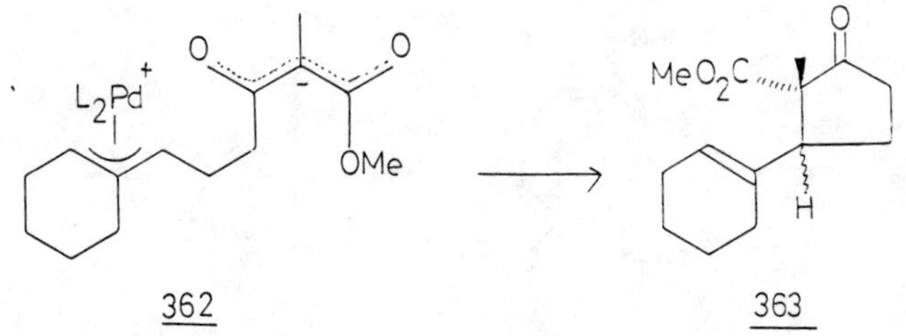

360

361

362

363

the tetrahydrofuran 374 with 100% selectivity but with a bulky phosphine
372 was obtained in 85% yield. With 375 as substrate 87% 376 was
obtained and was smoothly converted to 377.[664,695,709]

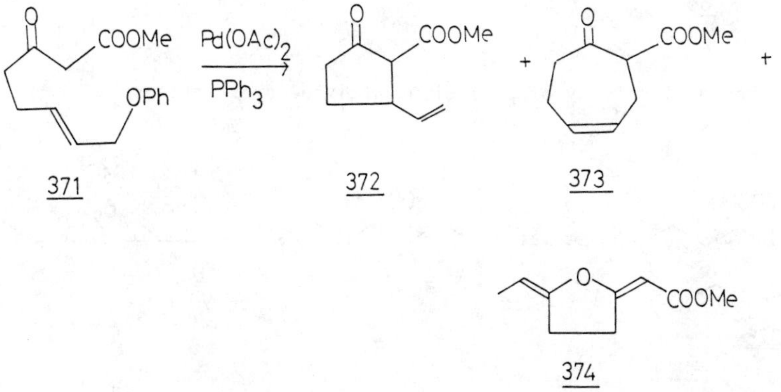

371

372

373

374

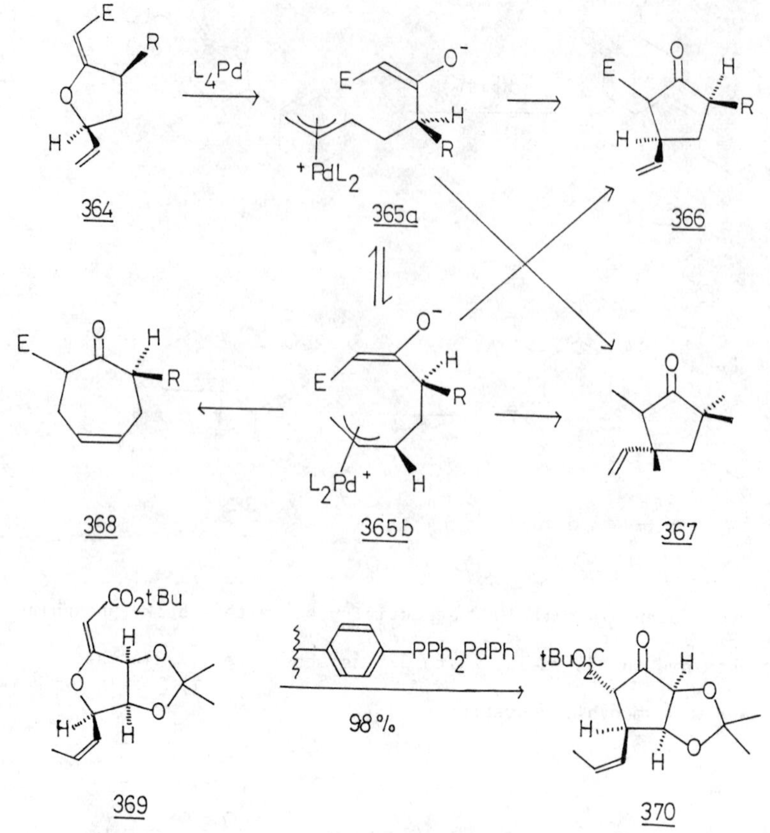

Figure 63 Regioselectivity of palladium catalysed cyclisations.

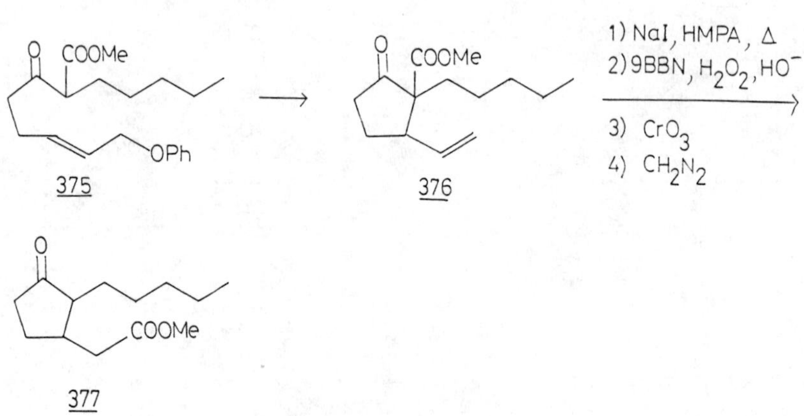

8.7.6.4. Stereoselectivity

Nucleophilic attack on allyl complexes is usually faster than syn ⇌ anti isomerisation so that the stereochemistry of the original double bond is maintained. Thus, geranyl acetate, 378, is converted to 379 and 380 with complete maintenance of stereochemistry[666] and α-farnesene, 383, is synthesised with retention in both components.[678]

The stereochemistry of substitution at the allylic carbon is well studied, to provide information on the reaction mechanism. Substitution of 384 occurs with retention[670,710] of configuration, resulting from a double inversion[666,704] (Figure 64). Some loss of stereochemical control occurs when the nucleophile is $NaCH(SO_2Ph)_2$ or when excess Ph_3P is omitted; the slower attack allows time for allyl isomerisation.[711]

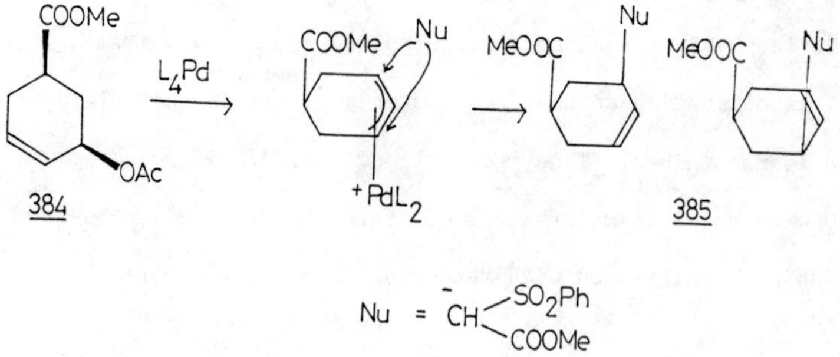

Figure 64 Stereoselectivity of palladium catalysed allyl substitution.

Recently a reaction has been noted in which clean inversion of stereochemistry occurs[693] (386 to 387); the same substrate with an active
methylene compound gives the expected retention. Clearly the nature of
the nucleophile is critical and it may be that delivery via transmetalla-
tion with palladium, and hence retention, is possible.

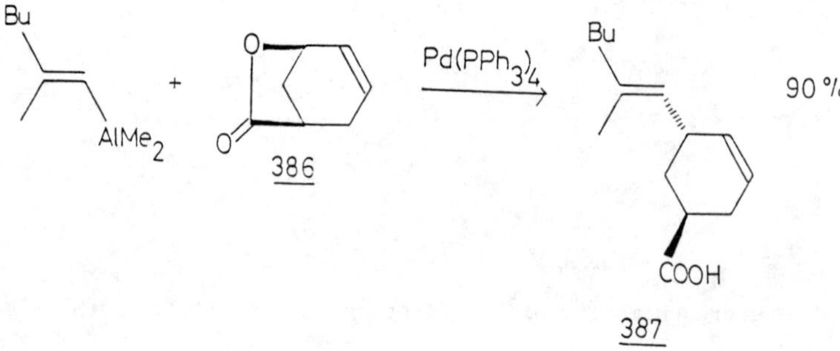

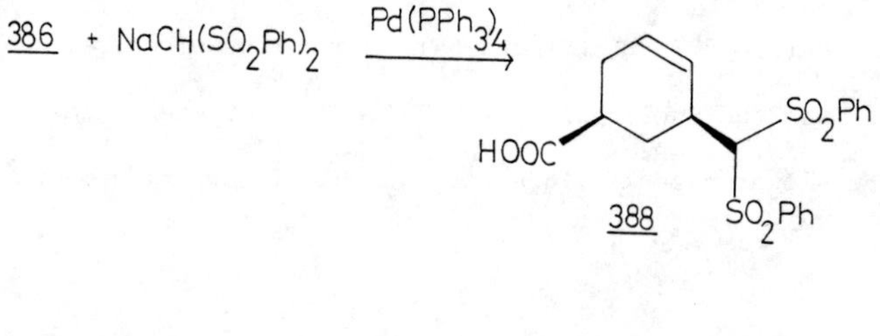

Enantioselective carbon-carbon bond formation is a very attractive prospect, but unfortunately the mechanism of the reaction, with
nucleophile delivery from the face of the allyl remote from palladium (and
its asymmetric ligands), is unhelpful in achieving control. The reaction
of cis($\pm$)-384 with nucleophiles is achieved in up to 46% enantiomer
excess with DIOP as ligand.[670] Some doubts were thrown on the mechanism
proposed[712] since the isolated complex 389 gives a lower optical yield
than the corresponding acetate. Allylation of 390 has been attempted
with a variety of chiral catalysts[713,714] giving up to 52% enantioselectivity with exotic phosphines. Also 391 is cyclised in 48% optical
yield.[714] The most thorough study is due to Bosnich.[715] He found that
equilibration between the diastereotopic complexes 394a and 394b is fast
and that its position is determined by the size of the anti-group(s).
Enantioselection substantially greater than the ratios of the complexes
may be achieved. Thus, the selectivity is a kinetic phenomenon in the
rate determining alkylation.

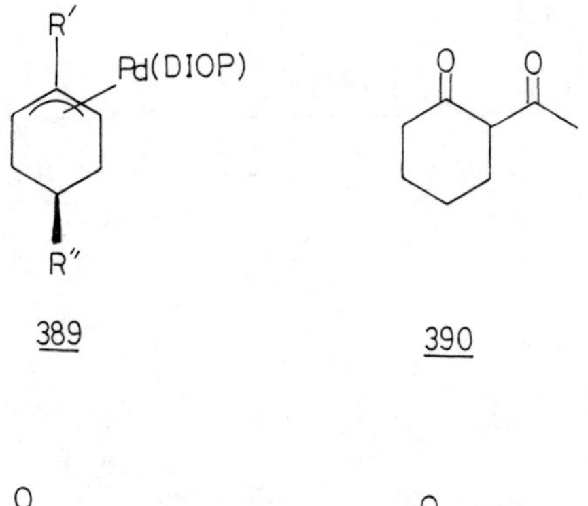

389

390

391

392

393

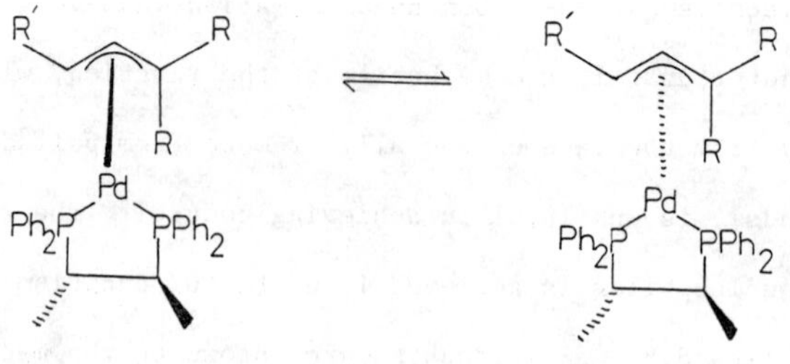

394a   394b

8.7.6.5. Uses in Synthesis

There is time to consider only a few of many of the ways in which this remarkable reaction may be applied to natural product syntheses. Cyclisation to cyclopropanes is achieved in the presence of (diphos)$_2$Pd, allowing the synthesis of chrysanthemic acid derivatives, 396.[673] A critical step in the synthesis of humulene, 399, is another cyclisation (397 to 398)[671] whilst the conversion of 400 to 401 gives a precursor to the anti-tumour compound sarkomycin, 402.[716]

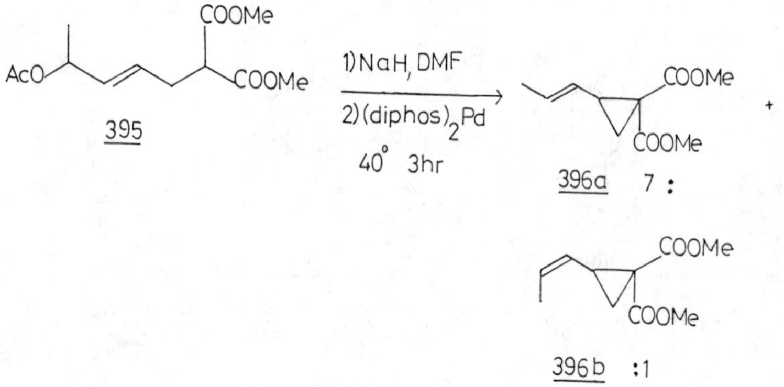

395

1) NaH, DMF
2) (diphos)$_2$Pd
40° 3hr

396a   7 :

396b   :1

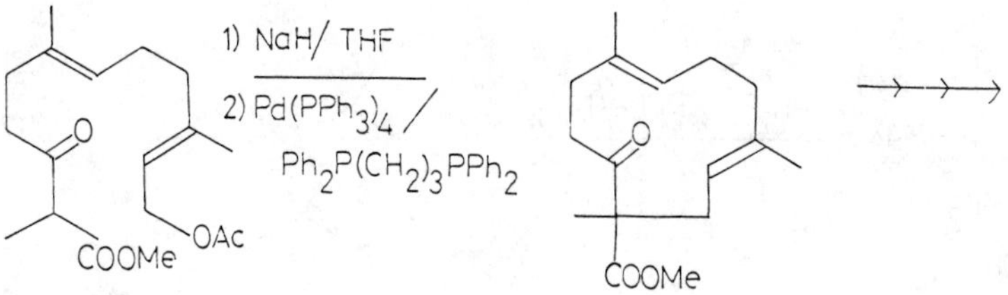

397

398

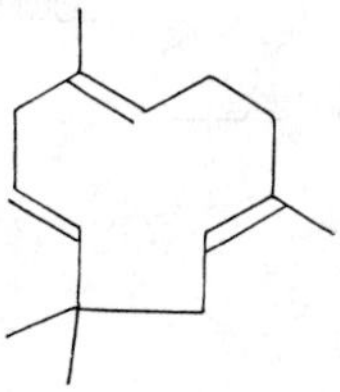

399

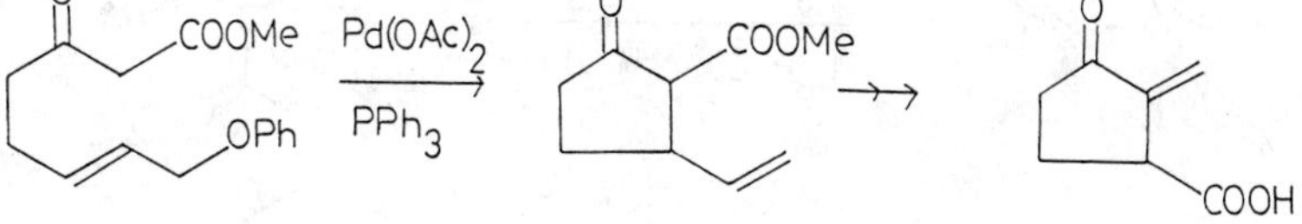

400   401   402

8.7.7. Allyl Complexes of Other Metals

One might expect the other metals of the palladium group,
platinum and nickel, to exhibit similar catalytic behaviour. Whilst this
is true to a certain extent, there are a number of problems. Platinum
allyls, particularly, tend to exist as σ-rather than π-allyls. A few
examples are known using $(Ph_3P)_3Pt$ as catalyst, giving somewhat different
regiochemistry; σ-allyl complexes were isolated. Some examples are given
in Figure 65.

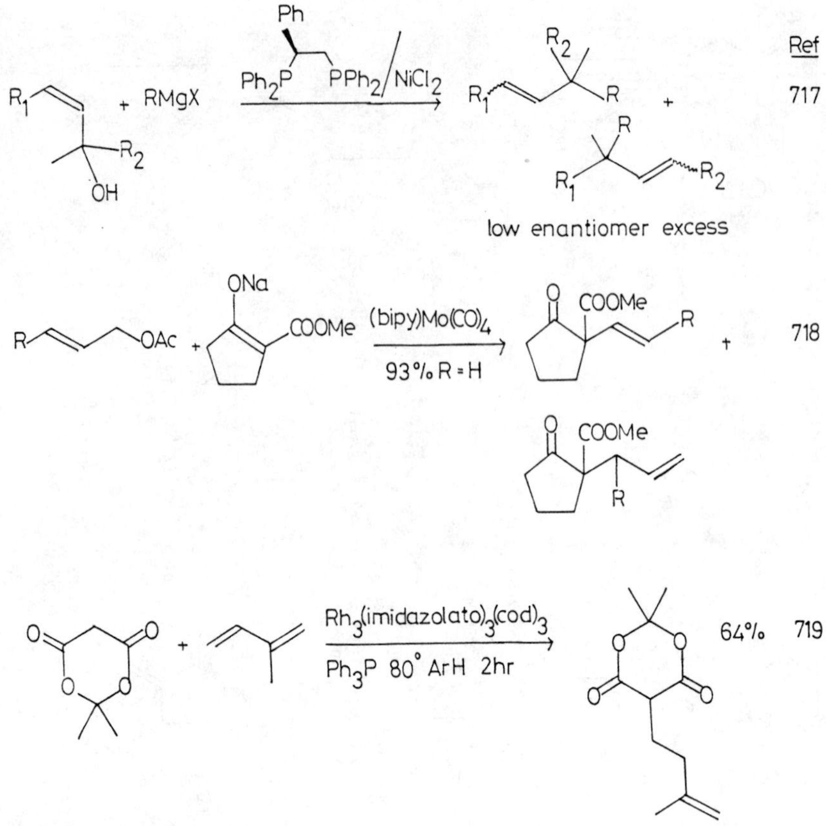

866

Ref

717

low enantiomer excess

718

Figure 65   Coupling of nucleophiles with other metal allyl complexes.

8.7.8.   Attack of Anions on Metal Coordinated Alkenes

This group is less mechanistically simple and less well defined
than its predecessor. The stoichiometric reaction is common and well
studied; similar considerations arise to those we considered when dealing
with heteroatomic nucleophiles (Chapter 4).[720-23] Examples include the
reactions of 403 and 405.[724,725] In order for the reaction of 405 to
proceed catalytically the Pd(0) released in the last step must be reoxidised with Cu_2Cl_2, but few of such reactions have general applications.[726,727]

403

404

1) 2 HMPA, 25°
2) 2 Et$_3$N, -78°
3) 2 BzMgCl, -60°
4) H$_2$

70 %

405

406

8.8 Coupling of Halides

Once again we are encountering a mechanistically and synthetically diverse group of reactions. Many reactions involve an intermediate
organometal and are distinguished from the previous group only by the
fact that this is not observable. Others are oxidative and resemble the
aryl couplings discussed earlier. At some stage a new carbon-metal bond
is formed, usually by oxidative addition of a halide to a metal complex.
The Heck reaction (Figure 66) is typical and one of the best understood.[728]
The organic group of RX adds regioselectively to the less substituted
end of the double bond and the reaction is stereospecific with cis-
addition and cis-elimination. The most stable alkene is formed and the
most hydridic H eliminated. The conditions are mild, the catalyst is
unaffected by H$_2$O or O$_2$, and the reaction is tolerant of most functional
groups.

Aryl[729-31], heterocyclic[732-4], vinyl[735-7] and allyl halides[738]
are all suitable sources of the carbon palladium bond, with the familiar
reactivity series I>Br>Cl. Early work[739] focussed on reactions with

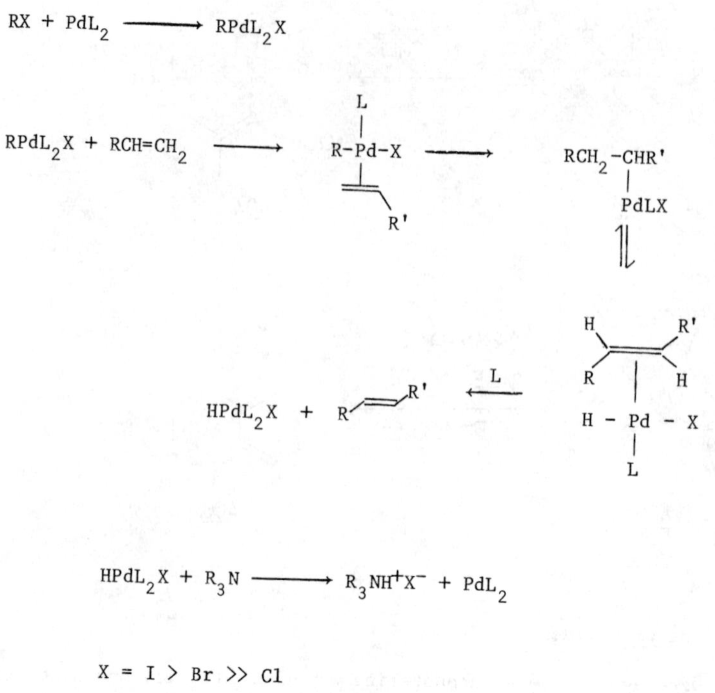

$$RX + PdL_2 \longrightarrow RPdL_2X$$

$$RPdL_2X + RCH=CH_2 \longrightarrow \underset{\overset{|}{R'}}{\overset{\overset{L}{|}}{R-Pd-X}} \longrightarrow \underset{\overset{|}{PdLX}}{RCH_2-CHR'}$$

$$HPdL_2X + R\diagup\!\!\!=\!\!\!\diagdown^{R'} \longleftarrow \overset{L}{} \quad \underset{\overset{|}{L}}{H-Pd-X} \text{ with } \overset{H}{R}\diagup\!\!\!=\!\!\!\diagdown\overset{R'}{H}$$

$$HPdL_2X + R_3N \longrightarrow R_3NH^+X^- + PdL_2$$

$$X = I > Br \gg Cl$$

Figure 66   Mechanism of the Heck reaction.

simple alkenes[740] to give anti-Markovnikov products[741] with later extensions to coupling with dienes[736] and alkynes[738,742] (Figure 67).   Reaction with enones and other double bonds bearing electron withdrawing groups is excellent[732], the only problem being loss of stereochemistry due to allyl formation or isomerisation of substrate or product.   This is also a common difficulty with the vinyl halide reactants we have so far discussed. Regiochemical control is, however, rather better than with simple alkenes[748] (Figure 68).

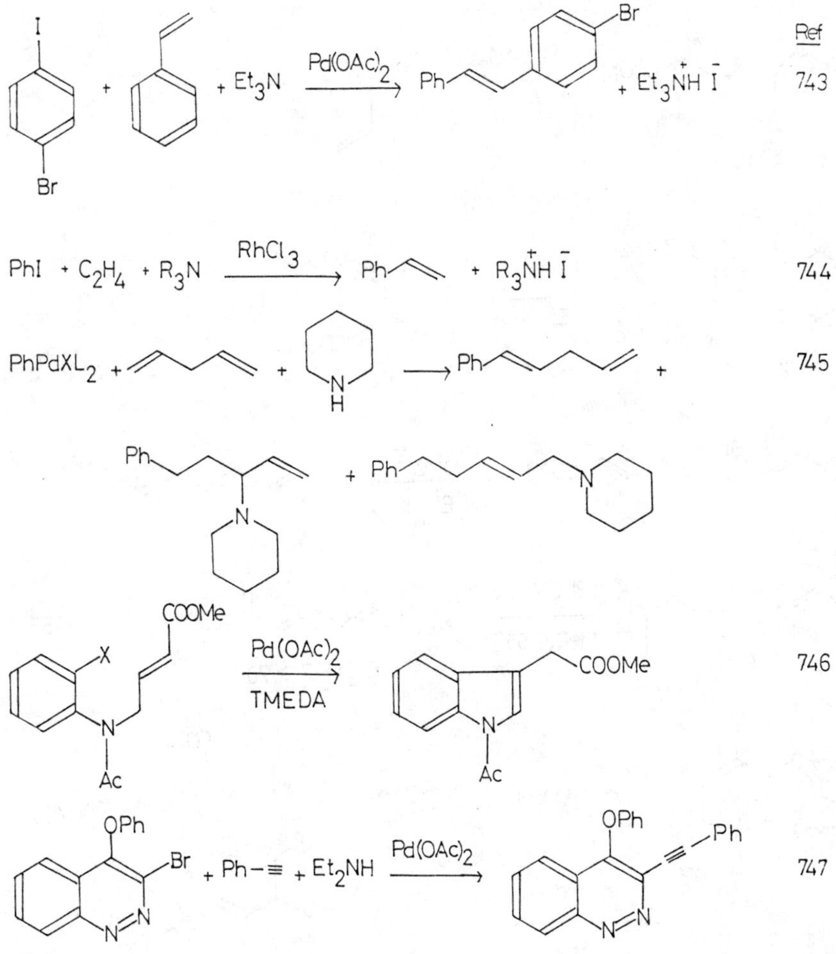

Figure 67   Examples of the Heck reaction.

Electron rich alkenes have attracted less attention and give rather unsatisfactory results.   Methyl[754] and ethyl[755] vinyl ethers give mixtures of regioisomers and vinyl acetate gives further substition (to give, for example, stilbene with iodobenzene).[729]   N-vinylpyrrolidone gives 407 and 408, steric and electronic factors working in opposite directions in this case.[756]   This has been used in the synthesis of phenylcyclidine analogues (e.g. 409).[757]

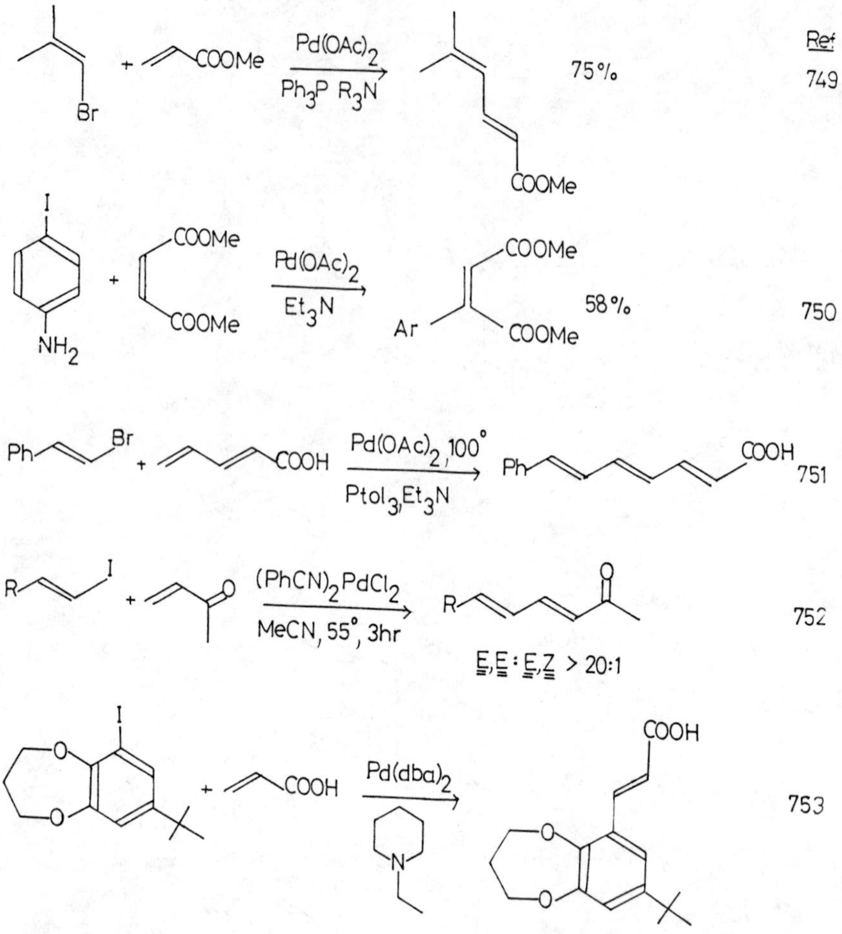

Figure 68 Heck reaction with alkenes bearing electron withdrawing groups.

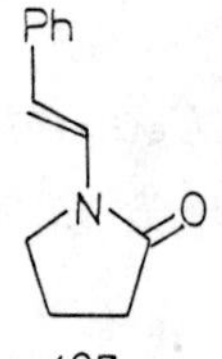

407

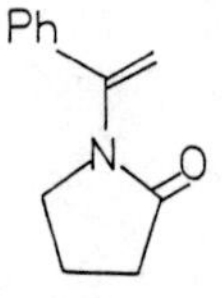

408

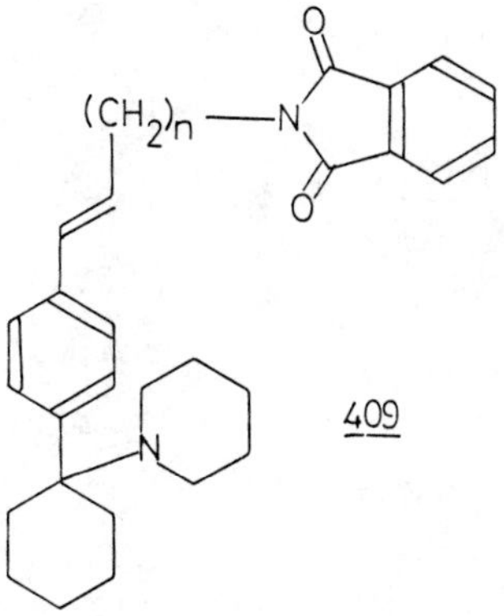

409

Allyl alcohols also react in this way giving, with aryl iodides, β-aryl aldehydes and ketones in excellent yield[758] (Figure 69). Regiochemical control is quite good, depending on the substituents on the allyl alcohol.[733,759] [760] Other authors suggest an allyl mechanism to explain the transfer of chirality when R-410 is converted to 411.[761] (Figure 70) Silyl ethers of allyl alcohols give products with higher regioselectivity but unfortunately lower yields.[762] Vinyl halide derived palladium nucleophiles are less useful the possibilities for isomerisation _via_ allyls being much increased.[737,764]

Coupling of allyl halides in the presence of nickel salts is well known, initially as a stoichiometric reaction (e.g. of 412 and 414).[765] It proceeds _via_ formation and coupling of a _bis_(allyl)Ni complex.[282] The intramolecular reaction has been used in the synthesis of annulenes[766] and humulene[767] and may be made catalytic by changing the ligands at nickel.[768]

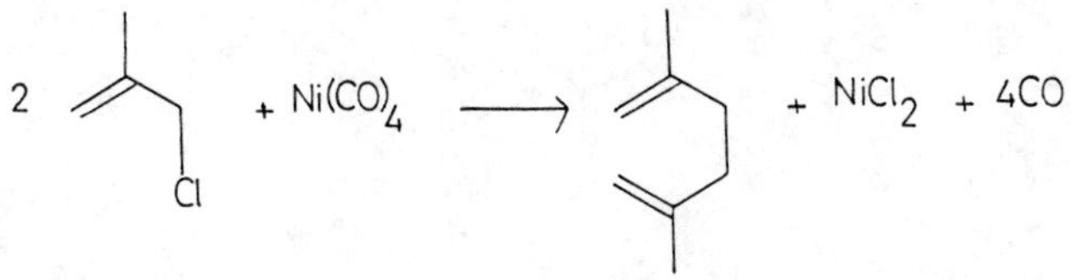

412   413

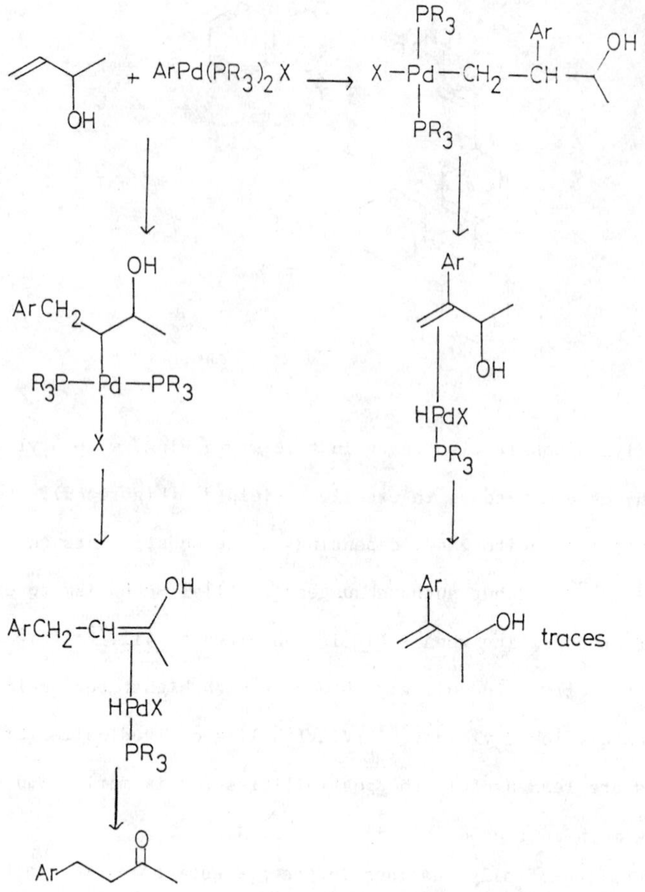

Figure 69 Heck reaction of aryl iodides and allyl alcohols.

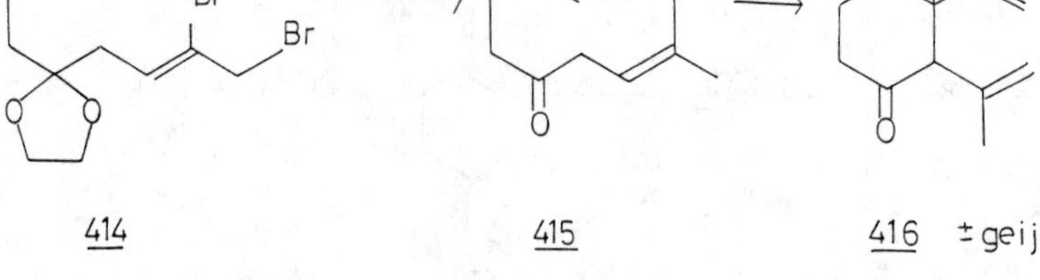

414   415   416 ± geijerone

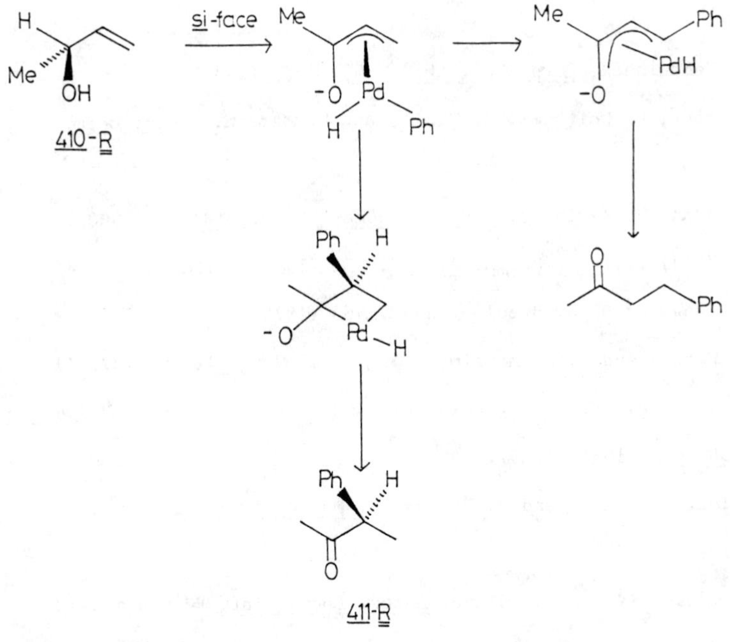

Figure 70 Allyl mechanism for the Heck reaction of allyl alcohols.

References

1. M. F. Semmelhack, Pure Appl. Chem., 53, 2379 (1981).

2. K. Ziegler, E. Holzkamp, H. Breil, and H. Margin, Angew. Chem., 67, 541 (1955).

3. H. Sinn and W. Kaminsky, Adv. Organomet. Chem., 18, 99 (1980).

4. A. E. Yildirim, Eur. Polym. J., 17, 551 (1981).

5. W. P. Long and D. S. Breslow, Ann., 463 (1975).

6. J. A. Waters and G. A. Mortimer, J. Polym. Sci. A1, 10, 1827 (1972).

7. T. S. Dzhabiev, F. S. D'yachkovskii and A. L. Khamrayeva, Polym. Sci. U.S.S.R., 11, 1881 (1969).

8. M. Koide, K. Iimura and M. Takeda, J. Macromol. Sci., Chem., A9, 961 (1975).

9. G. Henrici-Olivé and S. Olivé, Angew. Chem., Int. Ed. Engl., 10, 776 (1971).

10. A. G. Evans, J. C. Evans and J. Mortimer, J. Amer. Chem. Soc., 101, 3204 (1979).

11. Y. Doi, M. Takada and T. Keii, Bull Chem. Soc. Jpn., 52, 1802 (1979).

12. D. G. H. Ballard, E. Jones, T. Medinger and A. J. P. Pioli, Makromol. Chem., 148, 175 (1971).

13. Yu. I. Ermakov, É. A. Demin, A. M. Lazutkin and T. E. Kuzyaeva, Kinet. Catal., 13, 703 (1972).

14. U. Gianinni, U. Zuchini and E. Albizzati, J. Polym. Sci., B8, 405 (1970).

15. J. E. W. Chien and J. T. T. Hsieh, J. Polym. Sci., Polym. Chem. Ed., 14, 1915 (1976).

16. D. G. H. Ballard, E. Jones, R. J. Wyatt, R. T. Murray and P. A. Robinson, Polymer, 15, 169 (1974).

17. H. Sinn and W. Kaminsky, Makromol. Kolloq., Freiburg, (1978).

18. K. H. Reichert and J. Berthold, Makromol. Chem., 124, 103 (1969).

19. G. P. Belov, Z. Ya. Latypov and N. M. Chirkov, Polym. Sci. U.S.S.R., 16, 1541 (1974).

20. Y. Doi, Y. Nishimura and T. Keii, Polymer, 22, 469 (1981).

21. N. Koide, K. Iimura and M. Takeda, J. Polym. Sci., Polym. Chem. Ed., 11, 3161 (1973).

22. Z. V. Arkhipova, V. K. Badaev and B. V. Erofeev, Dokl. Chem., 183, 1097 (1968).

23. Ph. Radenkov, T. Petrova, L. Petkov and D. Jelyazkova, Eur. Polym. J., 11, 313 (1975).

24. C. Carlini, F. Ciardelli and P. Pino, Makromol. Chem., 119, 244 (1968).

25. D. G. H. Ballard and P. W. van Lienden, Makromol. Chem., 154, 177 (1972).

26. R. Rossi, P. Diversi and L. Porri, Macromolecules, 5, 247 (1972).

27. D. S. Breslow, U. S. Patent 2,827,446 (1958); Chem. Abs., 53, 1840i (1959).

28. K. Meyer and K. H. Reichert, Angew. Makromol. Chem., 12, 175 (1970).

29. G. Fink, R. Rottler, D. Schnell and W. Zoller, J. Appl. Polym. Sci., 20, 2779 (1976).

30. A. Andresen, H-G. Cordes, J. Herwig, W. Kaminsky, A. Merck, R. Mottweiler, J. Pein, H. Sinn and H-J. Vollmer, Angew. Chem., Int. Ed. Engl., 15, 630 (1976).

31. W. P. Long, J. Amer. Chem. Soc., 81, 5312 (1959).

32. W. P. Long and D. S. Breslow, J. Amer. Chem. Soc., 82, 1953 (1960).

33. G. Fink, R. Rottler and C. G. Kreiler, Angew. Makromol. Chem., 96, 1 (1981)

34. G. Fink and D. Schnell, Angew. Makromol. Chem., 105, 15 (1982).

35. E. A. Fushman, A. N. Shupik, L. F. Borisova, V. E. L'vovskii and
    F. S. D'yachkovskii, Dokl. P. Chem., 264, 403 (1982).

36. G. Henrici-Olivé and S. Olivé, Angew. Chem., Int. Ed. Engl., 6, 790
    (1967).

37. A. A. Semonov, P. Ye. Matkovskii and F-S. D'yachkovskii, Polym. Sci.
    U.S.S.R., 20, 1973 (1978).

38. G. P. Budanova and V. V. Mazurek, Polym. Sci. U.S.S.R., 9, 2703 (1967).

39. K. Hiraki, S. Ikeda, S. Kaneko and H. Hirai, J. Polym. Sci., Polym.
    Chem. Ed., 17, 2363 (1979).

40. O. Novaro, S. Chow and P. Magnouat, J. Catal., 42, 131 (1976).

41. H. Emde, Angew. Makromol. Chem., 60, 1 (1977)

42. A. E. Yildirim, Eur. Polym. J., 17, 1029 (1981).

43. A. Zambelli, I. Pasquon, R. Signorini and G. Natta, Makromol. Chem.,
    112, 160 (1968).

44. Ye. G. Erenburg, I. A. Livshits, Ye. O. Osipchuk, Ye. R. Gershtein,
    S. P. Yevdokimova and I. Ya.Poddubnyi, Polym. Sci. U.S.S.R., 22,
    1841 (1980).

45. I. L. Dubnikova, I. N. Meshkova, Ye. I. Vizen and N. M. Chirkov,
    Polym. Sci. U.S.S.R., 15, 1835 (1973).

46. I. N. Meshkova, N. G. Kudryakova and N. M. Chirkov, Polym. Sci.
    U.S.S.R., 15, 1220 (1973).

47. Y. Doi, M. Takada and T. Keii, Makromol. Chem., 180, 57 (1979).

48. Y. Doi, S. Ueki and T. Kaii, Polymer, 21, 1352 (1980).

49. Y. Doi, S. Ueki, S. Tamura, S. Nagahara and T. Keii, Polymer, 23,
    258 (1982).

50. A. B. Deshpande, S. M. Kabe and S. L. Kapur, J. Polymer. Sci.,
    Polym. Chem. Ed., 11, 2105 (1973).

51. A. M. Korshun and V. V. Mazurek, Polym. Sci. U.S.S.R., 17, 3055 (1975).

52. D. G. H. Ballard, W. H. Janes and T. Medinger, J. Chem. Soc., B, 1168 (1968).

53. W. Kaminsky, H-J. Vollmer, E. Heins and H. Sinn, Makromol. Chem., 175, 443 (1974).

54. D. G. H. Ballard, J. V. Dawkins, J. M. Key and P. W. van Lienden, Makromol. Chem., 165, 173 (1973).

55. H. Sinn, W. Kaminsky, H-J. Vollmer and R. Woldt, Angew. Chem., Int. Ed. Engl., 19, 390 (1980).

56. P. Cossée, J. Catal., 3, 80 (1964).

57. L. L. Bohm, Polymer, 19, 545 (1978).

58. G. Henrici-Olivé and S. Olivé, J. Organomet. Chem., 16, 339 (1969).

59. K. J. Ivin, J. J. Rooney, C. D. Stewart, M.L.H. Green and R. Mahtab, J. Chem. Soc., Chem. Commun., 604 (1978).

60. P. Cassoux, F. Crasnier and J.-F. Labarre, J. Organomet. Chem., 165, 303 (1979).

61. P. L. Watson, J. Amer. Chem. Soc., 104, 337 (1982).

62. H. W. Turner and R. R. Schrock, J. Amer. Chem. Soc., 104, 2331 (1982).

63. J. Soto, M. L. Steigerwald and R. H. Grubbs, J. Am. Chem. Soc., 104, 4479 (1982).

64. Y. Doi, S. Uekii and T. Keii, Makromol. Chem., 180, 1359 (1979); Macromolecules, 12, 814 (1979).

65. D. G. H. Ballard and T. Medinger, J. Chem. Soc. B, 1176 (1968).

66. P. Cossée, Tetraheadron Lett., (17) 12 (1960).

67. E. J. Arlman and P. Cossée, J. Catal., 3, 99 (1964).

68. G. Allegra, Makromol. Chem., 145, 235 (1971).

69. E. J. Vandenberg, Macromol. Synth., 5, 95 (1974).

70. P. Corradini, V. Barone, R. Fusco and G. Guerra, J. Catal., 77, 32 (1982).

71. A. Zambelli, C. Tosi and C. Sacchi, Macromolecules, 5, 649 (1972).

72. A. Zambelli and G. Allegra, Macromolecules, 13, 42 (1980).

73. G. Heublein, H. Hartung, M. Helbig and D. Stadermann, J. Macromol. Sci. Chem., A17, 821 (1982).

74. D. L. Christman and G. I. Keim, Macromolecules, 1, 358 (1968).

75. M. V. Favorskaya, U. V. Belyaev, D. V. Sokol'skii, N. D. Zavorokhin and T. A. Vlasova, Kinet. Catal., 14, 295 (1973).

76. A. Zambelli, A Lety, C. Tosi and I. Pasquon, Makromol. Chem., 115, 73 (1968).

77. I. D. Rubin, J. Polym. Sci. Al, 5, 1119 (1967).

78. M. G. Marti and K. H. Reichert, Makromol. Chem., 144, 17 (1971).

79. A. Yamamoto, T. Shimizu and S. Ikeda, Makromol. Chem., 136, 297 (1970).

80. A. Yamamoto and S. Ikeda, Prog. Polym. Sci. Japan, 3, 49 (1972); Chem. Abs., 78, 30238m (1973).

81. B. Bogdanović, Adv. Organomet. Chem., 17, 105 (1979).

82. K. Ziegler in Organometallic Chemistry, H Zeiss ed., p 194, 229, Reinhold, 1960.

83. A. J. Lundeen and J. E. Yates, Brit. Patent 1,144,390 (1969); Chem. Abs., 70, 105941f (1969).

84. S. G. Abasova, A. I. Leshcheva, E. A. Mushina, V. Sh. Fel'dblyum and B. A. Krentsel', Bull. Acad. Sci. U.S.S.R., Div. Chem. Sci., 21, 608 (1972).

85. B. Bogdanović, B. Henc, H.-G. Karmann, H.-G. Nüssel, D. Walker and G. Wilke, Ind. Eng. Chem., 62, 34 (1970); Chem. Abs., 74, 52937h (1971).

86. G. Henrici-Olivé and S. Olivé, Coordination and Catalysis p. 134, Verlag-Chemie, Weinheim, 1977.

87. B. Bogdanović, J. Galle, N. Hoffman and G. Wilke, unpublished data.

88. F. K. Shmidt, L. V. Mironova, G. A. Kalabin, A. G. Proidakov and
    A. V. Kalabina, Neftekhimiya, 16, 547 (1976); Chem. Abs., 85,
    176747t (1976).

89. W. C. Drinkard and R. V. Lindsey, Ger. Offen. 1,808,434 (1969);
    Chem. Abs., 71, 70093x (1969).

90. N. V. Petrushanskaya, A. I. Kurapova and V. Sh. Fel'dblyun,
    Dokl. Chem., 211, 593 (1973).

91. J. D. McClure and K. W. Barnett, J. Organomet. Chem., 80, 385 (1974).

92. M. Tsutsui and T. Koyano, J. Polym. Sci. Al, 5, 681 (1967).

93. T. Koike, K. Kawakami, K-i. Maruya, T. Mizoroki and A. Ozaki,
    Chem. Lett., 551 (1977).

94. R. B. A. Pardy and I. Tkatchenko, J. Chem. Soc., Chem. Commun., 49
    (1981).

95. G. Wilke, B. Bogdanović, P. Hardt, P. Heimbach, W. Keim, M. Kröner,
    W Oberkirch, K. Tanaka, E. Steinrücke, D. Walter and H. Zimmermann,
    Angew. Chem., Int. Ed. Engl., 5, 151 (1966).

96. G. G. Eberhardt and W. P. Griffin, J. Catal., 16, 245 (1970).

97. O-T. Onsager, H. Wang and U. Blindheim, Helv. Chim. Acta, 52, 187,
    196, 215, 224, 230 (1969).

98. B. Henc, Ph.D. Thesis, University of Bochum (1971).

99. C. A. Tolman, Inorg. Chem., 11, 3128 (1972).

100. C. A. Tolman, J. Amer. Chem. Soc., 92, 4217 (1970); 94, 2994 (1972).

101. K. Kawakami, T. Mizoroki and A. Ozaki, Chem. Lett., 903 (1975).

102. K. Kawakami, T. Mizoroki and A. Ozaki, Bull. Chem. Soc. Jpn., 51,
     21 (1978).

103. G. P. Belov, T. S. Dzhabiev and I. M. Kolesnikov, J. Mol. Catal.,
     14, 105 (1982).

104. J. D. Fellmann, G. A. Rupprecht and R. R. Schrock, J. Amer. Chem. Soc., 101, 5099 (1979).

105. U. Dorf, K. Engel and G. Erker, Angew. Chem., Int. Ed. Engl., 18, 914 (1982).

106. V. Sh. Fel'dblyum, N. V. Obeshchalova and A. I. Leshcheva. Dokl. Chem., 172, 19 (1907).

107. B. Bogdanović, B. Spliethoff and G. Wilke, Angew. Chem., Int. Ed. Engl., 19, 622 (1980).

108. B. Bogdanović Abstr. Int. Conf. Organomet. Chem., 5th 1 201 (1971).

109. G. Henrici-Olivé and S. Olivé, Angew. Chem., Int. Ed. Engl., 14, 104 (1975).

110. A. Sen and R. R. Thomas, Organometallics, 1, 1251 (1982).

111. S. J. McLain and R. R. Schrock, J. Amer. Chem. Soc., 100, 1315 (1978).

112. S. J. McLain, J. Sancho and R. R. Schrock, J. Amer. Chem. Soc., 102, 5610 (1980).

113. H. Kanai and K. Ishii, Bull. Chem. Soc. Jpn., 54, 1015 (1981).

114. Y. Huang, J. Li, J. Zhou, Z. Zhu and G. Hou, J. Organomet. Chem., 205, 185 (1981).

115. Y. Huang, J. Li, J. Zhou, Q. Wang and M. Gui, J. Organomet. Chem., 218, 169 (1981).

116. D. J. Milner and R. Whelan, J. Organomet. Chem., 152, 193 (1978).

117. Y. Watanabe and M. Takeda, Nippon Kagaku Kaishi, 2023 (1972); Chem. Abs., 78, 57347h (1973).

118. W. Strohmeier and A. Kaiser, J. Organomet. Chem., 114, 273 (1976).

119. S. Yoshikawa, J. Kiji and J. Furukawa, Bull. Chem. Soc. Jpn., 49, 1093 (1976).

120. G. E. Voecks, P. W. Jennings, G. D. Smith and C. N. Caughlan, J. Org. Chem., 37, 1460 (1972).

121. P. W. Jennings, G. E. Voecks and D. G. Pillsbury, J. Org. Chem., 40, 260 (1975).

122. M. Green and T. A. Kuc, J. Chem. Soc., Dalton, 832 (1972).

123. N. Acton, R. J. Roth, T. J. Katz, J. K. Frank, C. A. Maier and I. C. Paul, J. Amer. Chem. Soc., 94, 5446 (1972).

124. N. F. Gol'dshleger, B. I. Azbel', A. A. Grigor'ev, I. G. Sirotina and M. L. Khidekel', Bull. Acad. Sci. U.S.S.R., Div. Chem. Sci., 31, 561 (1982).

125. U. M. Zhemilev, R. I. Khusnutdinov, Z. S. Muslimov, L. V. Spirikhin, G. A. Tolstikov and O. M. Nefedov, Bull. Acad. Sci. U.S.S.R., Div. Chem. Sci., 30, 1889 (1981).

126. A. R. Fraser, P. H. Bird, S. A. Bezman, J. R. Shapley, R. White and J. A. Osborn, J. Amer. Chem. Soc., 95, 597 (1973).

127. G. A. Catton, G. F. C. Jones, M. J. Mays and J. A. S. Howell, Inorg. Chim. Acta, 20, L41 (1976).

128. T. Kamijo, T. Kitamura, N. Sakamoto and T.Joh, J. Organomet. Chem., 54, 265 (1973).

129. J. E. Lyons, K. H. Myers and A. Schneider, J. Chem.Soc., Chem. Commun., 638 (1978).

130. M. Ennis and A. R. Manning, J. Organomet. Chem., 116, C31 (1976).

131. M. Ennis, R. M. Foley and A. R. Manning, J. Organomet. Chem., 166, C18 (1979).

132. M. J. Doyle, J. McMeeking and P. Binger, J. Chem. Soc., Chem. Commun., 376 (1976).

133. P. Binger, G. Schroth and J. McMeeking, Angew. Chem., Int. Ed. Engl., 13, 465 (1974).

134. P. Binger and J. McMeeking, Angew. Chem., Int. Ed. Engl., 13, 466 (1970).

135. P. Binger, J. McMeeking and U. Schuchardt, Chem. Ber., 113, 2372 (1980).

136. P. Binger, Angew. Chem., Int. Ed. Engl., 11, 433 (1972).

137. P. Binger, Angew. Chem., Int. Ed. Engl., 11, 309 (1972).

138. P. Binger and U. Schuchardt, Angew. Chem., Int. Ed. Engl., 16, 249 (1977).

139. G. G. Eberhardt and H. K. Myers, J. Catal., 26, 459 (1972).

140. N. Kawata, K-i. Maruya, T. Mizoroki and A. Ozaki, Bull. Chem. Soc. Jpn., 44, 3217 (1971), 47, 413 (1974).

141. K. Kawakami, N. Kawata, K-i. Maruya, T. Mizoroki and A. Ozaki, J. Catal., 39, 134 (1975).

142. K. Kawamoto, T. Imanaka and S. Teranishi, Bull. Chem. Soc. Jpn., 43, 2512 (1970).

143. B. Bogdanović, Angew. Chem., Int. Ed. Engl., 12, 954 (1973).

144. A. Lösler, Ph.D. Thesis, University of Bochum (1973).

145. G. Buono, G. Pieffer, A.Mortreux and F. Petit, J. Chem. Soc., Chem. Commun., 937 (1980).

146. R. H. Grubbs and A. Miyashita, J. Amer. Chem. Soc., 100, 7416 (1978).

147. L. Hagelee, R. West, J. Calabrese and J. Norman, J. Amer. Chem. Soc., 101, 4888 (1979).

148. L. J. Kricka and A. Ledwith, Synthesis, 539 (1974).

149. T. Kiji, S. Yoshikawa, E. Sasakawa, S. Nishimura and J. Furukawa, J. Organomet. Chem., 80, 267 (1974).

150. S. Yoshikawa, S. Nishimura, J. Kiji and J. Furukawa, Tetrahedron Lett., 3071 (1973).

151. T-a. Mitsudo, K. Kokuryo, T. Shinsugi, Y. Nakagawa, Y. Watanabe and Y. Takegami, J. Org. Chem., 44, 4492 (1979).

152. T. A. Mitsudo, K. Kokuryo and Y. Takegami, J. Chem. Soc., Chem. Commun., 722 (1976).

153. H. Takaya, M. Yamakawa and R. Noyori, Bull. Chem. Soc. Jpn., 55, 852 (1982).

154. R. G. Saloman and J. K. Kochi, J. Amer. Chem. Soc., 96, 1137 (1974).

155. R. G. Saloman, K. Folting, W. E. Streib and J. K. Kochi, J. Amer. Chem. Soc., 96, 1145 (1974).

156. R. G. Salomon, D. J. Coughlin, S. Ghosh and M. G. Zagorski, J. Amer. Chem. Soc., 104, 998 (1982).

157. S. R. Raychaudhuri, S. Ghosh and R. G. Salomon, J. Amer. Chem. Soc., 104, 6841 (1982).

158. P. Binger and U. Schuchardt, Angew. Chem., Int. Ed. Engl., 16, 249 (1977).

159. R. Noyori, T. Ishigami, N. Hayashi and H. Takaya, J. Amer. Chem. Soc., 95, 1674 (1973).

160. P. Binger and U. Schuchardt, Chem. Ber., 113, 1063 (1980).

161. P. Binger and U. Schuchardt, Chem. Ber., 113, 3334 (1980).

162. P. Binger and U. Schuchardt, Chem. Ber., 114, 3313 (1981).

163. R. Noyori, M. Yamakawa and H. Takaya, Tetrahedron Lett., 4823 (1978).

164. T. A. Albright, J. Organomet. Chem., 198, 159 (1980).

165. B. M. Trost, Pure Appl. Chem., 53, 2357 (1981).

166. D. J. Gordon, R. F. Fenske, N. Nanninga and B. M. Trost, J. Amer. Chem. Soc., 103, 5974 (1981).

167. B. M. Trost and D. M. T. Chan, J. Amer. Chem. Soc., 103, 5972 (1981).

168. B. M. Trost and P. Renaut, J. Amer. Chem. Soc., 104, 6668 (1982).

169. B. M. Trost, T. N. Nanninga and D. M. T. Chan, Organometallics, 1, 1543 (1982).

170. R. S. Shue, J. Chem. Soc., Chem. Commun., 1510 (1971).

884

171. Y. Fujiwara, O. Maruyama, M. Yoshidomi and H. Taniguchi, J. Org. Chem., 46, 851 (1981).

172. V. E. Taraban'ko, I. V. Kozhevnikov and K. I. Matveev, Kinet. Catal., 19, 937 (1978).

173. Y. Fujiwara, I. Moritani, M. Matsuda and S. Teranishi, Tetrahedron Lett., 3863 (196).

174. Y. Fujiwara, I. Moritani, S. Danno, R. Asano and S. Teranishi, J. Amer. Chem. Soc., 91, 7166 (1969).

175. I. V. Kozhevnikov, S. Ts. Sharapova and L. I. Kurteeva, Kinet. Catal., 23, 292 (1982).

176. K. Kikukawa, K. Nagira, N. Terao, F. Wada and T. Matsuda, Bull. Chem. Soc. Jpn., 52, 2609 (1979).

177. K. Kikukawa, K. Maemura, K. Nagira, F. Wada and T. Matsuda, Chem. Lett., 551 (1980).

178. K. Kikukawa, K. Maemura, Y. Kiseki, F. Wada, T. Matsuda and C. S. Giam, J. Org. Chem., 46, 4885 (1981).

179. K. Kikukawa, K. Nagira, F. Wada and T. Matsuda, Tetrahedron, 37, 31 (1981).

180. K. Kikukawa and T. Matsuda, Chem. Lett., 159 (1977).

181. S. Otsuka, K. Tani and T. Yamagata, J. Chem. Soc., Dalton, 2491 (1973).

182. S. Otsuka, A. Nakamura, S. Ueda and H. Minamida, Kogyo Kagaku Zasshi, 72, 1809 (1969); Chem. Abs., 72, 22012u (1970).

183. B. A. Krentsel, E. A. Mushina, E. M. Khar'kova and M. V. Shishkina, Eur. Polym. J., 11, 865 (1975).

184. S. Otsuka, A. Nakamura, T. Yamagata and K. Tani, J. Amer. Chem. Soc., 94, 1037 (1972).

185. F. W. Hoover and R. V. Lindsey, J. Org. Chem., 34, 3051 (1969).

186. H. Siegel, H. Hopf, A. Germer and P. Binger, Chem. Ber., 111, 3112 (1978).

187. S. Otsuka, A. Nakamura and H. Minamida, J. Chem. Soc., Chem. Commun., 191 (1969).

188. J. E. van den Enk and H. J. van der Ploeg, J. Polym. Sci. A1, 9, 2395 (1971).

189. G. D. Shier, J. Organomet. Chem., 10, P15 (1967).

190. J. G. van Ommen, J. Stijntjes and P. Mars, J. Mol. Catal., 5, 1 (1979).

191. J. G. van Ommen, P. C. J. M. van Berkel and P. J. Gellings, Eur. Polym. J., 16, 745 (1980).

192. R. Havinga and A. Schors, J. Macromol. Sci., Chem., A2, 1 (1968).

193. P. Albano and M. Aresta, J. Organomet. Chem., 190, 243 (1980).

194. J. G. van Ommen, H. J. van der Ploeg, P. J. C. M. van Berkel and P. Mars, J. Mol. Catal., 2, 409 (1977).

195. J. G. van Ommen, J. G. M. van Rens and P. J. Gellings, J. Mol. Catal., 13, 313 (1981).

196. R. Havinga and A. Schors, J. Macromol. Sci., A2, 31 (1968).

197. S. Otsuka, K. Mori, T. Suminoe and F. Imaizumi, Eur. Polym. J., 3, 73 (1967).

198. L. Porri, R. Rossi and G. Ingrosso, Tetrahedron Lett., 1083 (1971).

199. M. Ghalamkar-Moazzam and T. L. Jacobs, J. Polym. Sc., Polym. Chem. Ed., 16, 615 (1978).

200. W. M. Saltman and E. Schoenberg, Macromol. Synth., 2, 50 (1966).

201. E. W. Duck and J. M. Locke, "Polydienes by Anionic Catalysts" in W. M. Saltman, ed., The Stereorubbers, Wiley 1977.

202. D. K. Jenkins and D. G. Timms, U. S. Patent 3,414,555 (1968); Chem. Abs., 70, 29553p (1969).

203. I. Sh. Guzman, O. K. Sharaev, E. I. Tinyakova and B. A. Dolgoplosk, Bull. Acad. Sci. U.S.S.R., Div. Chem. Sci., 20, 599 (1971).

204. B. A. Dolgoplosk, E. I. Tinyakova, O. K. Sharaev, I. Sh. Guzman and N. N. Chigir, Eur. Polym. J., 11, 829 (1975).

205. I. Sh. Guzman, O. K. Sharaev, E. I. Tinyakova and B. A. Dolgoplosk, Dokl. Chem., 202, 165 (1972).

206. K. Hiraki, T. Inoue and H. Hirai, J. Polym. Sci., A1, 8, 2543 (1970).

207. H. Hirai, K. Hiraki, I. Noguchi, T. Inoue and S. Makishima, J. Polym. Sci., A1, 8, 2393 (1970).

208. Yu. B. Monakov, S. R. Rafikov, A. M. Ivanova, A. A. Panasenko, G. A. Tolstikov, A. A. Pozdeyeva, Ye. Ye. Zayev, R. Z. Lukmanova and G. G. Igoshkina, Polym. Sci. U.S.S.R., 17, 3025 (1975).

209. K. Matsuzaki and T. Yasukawa, J. Polym. Sci., A1, 5, 511, 521 (1967).

210. F. Dawans and P. Teyssie, Eur. Polym. J., 5, 541 (1969).

211. C. F. Gibbs, V. L. Folt, E. J. Carlson, S. E. Horne and H. Tucker, Brit. Patent 916,383 (1963); Chem. Abs., 58, 9311h (1963).

212. M. Gippin, Macromol. Synth., 2, 42 (1966).

213. B. Veruovič and J. Křepelka, Coll. Czech. Chem. Commun., 36, 3387 (1971).

214. L. A. Volkov, I. D. Chernova, I. L. Rachushnova, V. T. Robysheva and V. I. Lysova, Polym. Sci. U.S.S.R., 18, 1701 (1976).

215. G. U. Timofeyeva, N. A. Kokorina and S. S. Medvedev, Polym. Sci. U.S.S.R., 11, 677 (1969).

216. G. V. Timofeeva, N. D. Seregina, N. A. Kokorina and S. S. Medvedev, Dokl. P. Chem., 187, 475 (1969).

217. L. A. Volkov and S. S. Medvedev, Dokl. P. Chem., 184, 6, 186 305 (1969).

218. L. A. Volkov, E. F. Dudko and S. S. Medvedev, Dokl. P. Chem., 185, 241 (1969).

219. B. A. Dolgoplosk, Polym. Sci. U.S.S.R., 13, 367 (1971).

220. F. Dawans and P. Teyssie, C. R. Acad. Sci. Paris, C263, 1512 (1966).

221. B. S. Turov, T. A. Nuzhdina and G. N. Shilova, Polym. Sci. U.S.S.R., 22, 2827 (1980).

222. T. Matsumoto and J. Furukawa, J. Macromol. Sci. Chem., A6, 281 (1972).

223. A. I. Kadantseva, V. S. Byrikhin and S. S. Medvedev, Polym. Sci. U.S.S.R., 14, 3104 (1972).

224. V. A. Kormer, B. B. Babitskiy and M. I. Lobach, ACS, Adv. Chem. Ser., 91, 306 (1968).

225. A. M. Lazutkin, V. A. Vashkevich, S. S. Medvedev and V. N. Vasil'eva, Dokl. P. Chem., 175, 583 (1967).

226. P. Teyssié, Proc. 1st. Int. Symp. Homog. Catal., Corpus Christi, 1978.

227. F. Borg-Visse, F. Dawans and E. Maréchal, J. Polym. Sci., Polym. Chem. Ed., 18, 2491 (1980).

228. E. A. Mushina, T. K. Vydrina, E. V. Sakharova, E. I. Tinyakova and B. A. Dolgoplosk, Dokl. Chem., 170, 881 (1966).

229. J. F. Harrod and A. Navarre, Macromolecules, 10, 579 (1977).

230. V. A. Yakovlev, B. A. Dolgoplosk, K. L. Makovetskii and E. I. Tinyakova, Dokl. Chem., 187, 557 (1969).

231. R. Warin, M. Julémont and P. Teyssie, J. Organomet. Chem., 185, 413 (1980).

232. J. C. Marechal, F. Dawans and P. Teyssie, J. Polym. Sci., A1, 8, 1993 (1970).

233. M. Julémont, R. Warin and P. Teyssié, J. Mol. Catal., 7, 523 (1980).

234. J. Furukawa, Acc. Chem. Res., 13, 1 (1980).

235. V. M. Gorelik, E. S. Novikova, O. P. Parenago, V. M. Frolov and B. A. Dolgoplosk, Kinet. Catal., 13, 1274 (1972).

236. O. K. Sharaev, A. V. Alferov, E. I. Tinyakova and B. A. Dolgoplosk, Bull. Acad. Sci. U.S.S.R., Div. Chem. Sci., 2469 (1967).

237. J. P. Durand, F. Dawans and P. Teyssie, J. Polym. Sci., A1, 8 979 (1970).

238. V. M. Gorelik, O. P. Parenago, V. M. Frolov and B. A. Dolgoplosk, Kinet. Catal., 14, 1281 (1973).

239. R. V. Rabovskaya, Ye. I. Tinyakova and G. A. Parfenova, Polym. Sci. U.S.S.R., 19, 2701 (1977).

240. E. S. Novikova, A. P. Klimov, O. P. Parenago, V. M. Frolov, G. V. Isagulyants and B. A. Dolgoplosk, Dokl. P. Chem., 220, 140 (1975).

241. J. M. Thomassin, E. Walckiers, R. Warin and P. Teyssié, J. Polym. Sci., Polym. Chem. Ed., 13, 1147 (1975).

242. J. F. Harrod and L. R. Wallace, Macromolecules, 5, 685 (1972).

243. V. A. Kormer and M. I. Lobach, Macromolecules, 10, 572 (1977).

244. K. Hiraki and H. Hirai, Macromolecules, 3, 382 (1970).

245. B. A. Dolgoplosk and E. I. Tinyakova, Bull. Acad. Sci. U.S.S.R., Div. Chem. Sci., 291 (1970).

246. E. S. Novikova and V. S. Stroganov, Kinet. Catal., 22, 1178 (1981).

247. R. I. Ter-Minasyan, O. P. Parenago, V. M. Frolov and B. A. Dolgoplosk, Dokl. P. Chem., 194, 821 (1970).

248. R. I. Ter-Minasyan, O. P. Parenago, V. M. Frolov and B. A. Dolgoplosk, Dokl. Chem., 214, 106 (1974).

249. G. Lugli, A. Mazzei and S. Poggio, Makromol. Chem., 175, 2021 (1974).

250. I. A. Oreshkin, I. Ya. Ostrovskaya, V. A. Yakovlev, E. I. Tinyakova and B. A. Dolgoplosk, Dokl. Chem., 173, 402 (1967).

251. M. L. H. Green, J. Knight, L. C. Mitchard, G. G. Roberts and W. E. Silverthorn, J. Chem. Soc., Chem. Commun., 987 (1972).

252. J-H. Yang, M. Tsutsui, Z. Chen and D. E. Bergbreiter, Macromolecules, 15, 230 (1982).

253. B. Veruovič, J. Zachoval and S. Bittner, Coll. Czech. Chem. Commun., 33, 3026 (1968).

254. B. R. James and L. D. Markham, J. Catal., 27, 442 (1972).

255. N. G. Connelly, P. T. Draggett and M. Green, J. Organomet. Chem., 140, C10 (1979).

256. Z. Y. Zhang, H. J. Zhang, H. M. Ma and Y. Wu, J. Mol. Catal., 17, 65 (1982).

257. I. F. Gavrilenko, A. L. Lazutskii, N. N. Stefanovskaya, E. I. Tinyakova, V. F. Chirkova and B. A. Dolgoplosk, Dokl. Chem., 249, 544 (1979).

258. T. K. Vydrina, I. Sh. Guzman, B. A. Dolgoplosk, E. I. Tinyakova, O. K. Sharaev and O. N. Yakovleva, Dokl. Chem., 230, 582 (1976).

259. A. A. Entezami, A. Deluzarche, B. Kaempf and F. Schué, Eur. Polym. J., 13, 203 (1977).

260. K. Bujadoux, J. Josefonvicz and J. Néel, Eur. Polym. J., 6, 1233 (1970).

261. K. Bujadoux R. Clement, J. Josefonvicz and J. Néel, Eur. Polym. J., 9, 189 (1973).

262. L. Porri, A. di Corato and G. Natta, Eur. Polym. J., 5, 1 (1969).

263. R. Clement, Eur. Polym. J., 17, 1293 (1981).

264. A. Bolognesi, S. Destri, L. Porri and F. Wang, Makromol. Chem., Rapid Commun., 3, 187 (1982).

265. S. Destri, G. Gatti and L. Porri, Makromol. Chem., Rapid Commun., 2, 605 (1981).

266. W. Wieder and J. Witte, J. Appl. Polym. Sci., 26, 2503 (1981).

267. L. A. Kazaryan and E. N. Kropacheva, Kinet. Catal., 23, 411 (1982).

268. L. A. Kazaryan, Ye. N. Kropacheva and Yu. G. Kamevev, Polym. Sci. U.S.S.R., 23, 2426 (1981).

269. L. K. Kurnosova, Ye. N. Kropacheva, I. G. Zhuchikhina, N. A. Konovalenko, V. I. Anosov and I. A. Ikonitskii, Polym. Sci. U.S.S.R., 23, 2418 (1981).

270. B. A. Dolgoplosk, S. I. Beilin, Y. U. Korshak, G. M. Chernenko, L. M. Vardanyun and M. P. Teterina, Eur. Polym. J., 9, 895 (1973).

271. K. Bujadoux, M. Galin, J. Josefonvicz and A. Szubarga, Eur. Polym. J., 10, 1 (1974).

272. T. I. Bevza, N. A. Pokatilo, M. P. Teterina and B. A. Dolgoplosk, Polym. Sci. U.S.S.R., 10, 248 (1968).

273. J. Tsuji, Adv. Organomet. Chem., 17, 141 (1979).

274. S. Takahashi, T. Shibano and N. Hagihara, Tetrahedron Lett., 2451 (1967); Bull. Chem. Soc. Jpn., 41, 454 (1968).

275. A. Musco and A. Silvani, J. Organomet. Chem., 88, C41 (1975).

276. S. Gardner and D. Wright, Tetrahedron Lett., 163 (1972).

277. P. Roffia, G. Gregorio, F. Conti, G. F. Pregaglia and R. Ugo, J. Organomet. Chem., 55, 405 (1973).

278. C. U. Pittman, R. M. Hanes and J. J. Yang, J. Mol. Catal., 15, 377 (1982).

279. J. F. Kohnle, L. H. Slaugh and K. L. Nakamaye, J. Amer. Chem. Soc., 91, 5904 (1969).

280. P. Heimbach, Angew. Chem., Int. Ed. Engl., 12, 975 (1973).

281. G. Schomburg, D. Henneberg, P. Heimbach, E. Janssen, H Lehmkuhl and G. Wilke, Ann., 1667 (1975).

282. M. F. Semmelhack, Org. React., 19, 115 (1972) and references therein.

283. G. A. Tolstikov, U. M. Dzhemilev and S. S. Shavanov, Bull. Acad. Sci. U.S.S.R., Div. Chem. Sci., 24, 2518 (1975).

284. J. Thivolle-Cazat and I. Tkatchenko, J. Chem. Soc., Chem. Commun., 377 (1979).

285. M. L. H. Green and H. Munakata, J. Chem. Soc., Dalton, 269 (1974).

286. A. Miyake, H. Kondo and M. Nishino, Angew. Chem., Int. Ed. Engl., 10, 802 (1971).

287. F. Brille, P. Heimbach, J. Kluth and H. Schenkluhn, Angew. Chem., Int. Ed. Engl., 18, 400 (1979).

288. C. R. Graham and L. M. Stephenson, J. Amer. Chem. Soc., 99, 7098 (1977).

289. W. J. Richter, J. Mol. Catal., 13, 201 (1981).

290. W. Ring and J. Gaube, Chem. Ing. Tech., 36, 1041 (1966).

291. H. Morikawa, Ger. Offen. 1,942,729 (1970); Chem. Abs., 73, 14304w (1970).

292. A. P. Bakhonin, A. N. Pudovik and O. O. Sidorova, Dokl. Chem., 176, 866 (1967).

293. P. S. Chekrii, M. L. Khidekel', I. V. Kalechits, O. N. Eremenko, G. I. Karyakina and A. S. Todozhokova, Bull. Acad. Sci. U.S.S.R., Div. Chem. Sci., 21, 1521 (1972).

294. D. Huchette, B. Thery and F. Petit, J. Mol. Catal., 4, 433 (1978).

295. I. Tkatchenko, J. Organomet. Chem., 124, C39 (1977).

296. H-J. Kablitz and G. Wilke, J. Organomet. Chem., 51, 241 (1973).

297. A. Yamamoto, K. Morifuji, S. Ikeda, T. Saito, Y. Uchida and A. Misono, J. Amer. Chem. Soc., 90, 1878 (1968).

298. K. Takahashi, G. Hata and A. Miyake, Bull. Chem. Soc. Jpn., 46, 600 (1973).

299. A. D. Josey, J. Org. Chem., 39, 139 (1974).

300. A. Musco, J. Mol. Catal., 1, 443 (1975/6).

301. J. P. Neilan, R. M. Laine, N. Cortese and R. F. Heck, J. Org. Chem., 41, 3455 (1976).

302. C. U. Pittman and L. R. Smith. J. Amer. Chem. Soc., 97, 341 (1975).

303. I. Mochida, S. Yuasa and T. Seiyama, J. Catal., 41, 101 (1976).

304. I. Mochida, K. Kitagawa, H. Fujitsu and K. Takeshita, Chem. Lett., 417 (1977).

305. I. Mochida, K. Kitagawa, H. Fujitsu and K. Takeshitia, J. Catal., 54, 175 (1978).

306. U. M. Dzhemilev, A. Z. Yakupova and G. A. Tolstikov, J. Org. Chem. U.S.S.R., 15, 1037 (1979).

307. B. Barnett, B. Büssemeier, P. Heimbach, P. W. Jolly, C. Krüger, I. Tkatchenko and G. Wilke, Tetrahedron Lett., 1457 (1972).

308. P. Heimbach and W. Brenner, Angew. Chem., Int. Ed. Engl., 5, 961 (1966).

309. P. Heimbach and R. Schimpf, Angew. Chem., Int. Ed. Engl., 7, 727 (1968).

310. W. Brenner, P. Heimbach and G. Wilke, Ann., 727, 194 (1969).

311. H. Buchholz, P. Heimbach, H.-J. Hey, H. Selbeck and W. Wiese, Coord. Chem. Rev., 8, 129 (1972).

312. A. Misono, Y. Uchida, K-i. Furuhata and S. Yoshida, Bull. Chem. Soc., Jpn., 42, 1383, 2303 (1969).

313. H. tom Dieck and A. Kinzel, Angew. Chem., Int. Ed. Engl., 18, 324 (1979).

314. J. Itakura and H. Tanaka, Makromol. Chem., 123, 274 (1969).

315. J. P. Candlin and W. H. Janes, Brit. Patent 1,148,177 (1969); Chem. Abs., 71, 49409v (1969).

316. P. Heimbach, P. W. Jolly and G. Wilke, Adv. Organomet. Chem., 8, 29 (1969).

317. P. J. Garratt and M. Wyatt, J. Chem. Soc., Chem. Commun., 251 (1974).

318. P. Heimbach, Aspects Homog. Catal., 2, 79 (1974).

319. P. Brun, A. Tenaglia and B. Waegell, J. Mol. Catal., 9, 453 (1980), 17, 105 (1982).

320. A. C. L. Su, Adv. Organomet. Chem., 17, 269 (1979).

321. T. Alderson, U. S. Patent 3,013,066 (1961); Chem. Abs., 57, 11016h (1962). T. Alderson, E. L. Jenner and R. V. Lindsey, J. Amer. Chem. Soc., 87, 5638 (1965).

322. R. Cramer, Inorg. Chem., 1, 722 (1962), J. Amer. Chem. Soc., 86, 217 (1964).

323. R. Cramer, J. Amer. Chem. Soc., 89, 1663 (1967); Acc. Chem. Res., 1, 186 (1968).

324. R. G. Miller, T. J. Kealy and A. L. Barney, J. Amer. Chem. Soc., 89, 3756 (1967).

325. C. A. Tolman, J. Amer. Chem. Soc., 92, 6777 (1970).

326. A. C. L. Su and J. W. Collette, J. Organomet. Chem., 36, 177 (1972).

327. Y. Inoue, T. Kagawa and H. Hashimoto, Tetrahedron Lett., 1099 (1970).

328. C. A. Tolman, J. Amer. Chem. Soc., 92, 2956 (1970).

329. L. S. Meriwether and M. L. Fiene, J. Amer. Chem. Soc., 81, 4200 (1959).

330. C. A. Tolman, *J. Amer. Chem. Soc.*, <u>92</u>, 2953 (1970).

331. N. Kawata, K-i. Maruya, T. Mizoroki and A. Ozaki, <u>Bull. Chem. Soc. Jpn.</u>, <u>47</u>, 2003 (1974).

332. R. J. Harder, <u>Fr. Patent</u> 1,536,670 (1968); <u>Chem. Abs.</u>, <u>71</u>, 101272m (1969).

333. C. Sarafidis, <u>Fr. Patent</u> 1,561,485 (1969); <u>Chem. Abs.</u>, <u>71</u>, 101273n (1969).

334. G. Henrici-Olive and M. Olive, <u>U. S. Patent</u> 3,647,902 (1972); <u>Chem. Abs.</u>, <u>76</u>, 112641k (1972).

335. J. S. Yoo, <u>U. S. Patent</u> 3,669,949 (1972); <u>Chem. Abs.</u>, <u>77</u>, 100717x (1972).

336. G. Hata, <u>J. Amer. Chem. Soc.</u>, <u>86</u>, 3903 (1964).

337. T. Ito, T. Kawai and Y. Takami, <u>Tetrahedron Lett.</u>, 4775 (1972)

338. T. Ito, K. Takahashi and Y. Takami, <u>Tetrahedron Lett.</u>, 5049 (1973).

339. T. Ito and Y. Takami, <u>Bull. Chem. Soc. Jpn.</u>, <u>51</u>, 1220 (1978).

340. C. Moberg, <u>Tetrahedron Lett.</u>, <u>22</u>, 4827 (1981).

341. P. Heimbach and G. Wilke, <u>Ann.</u>, <u>727</u>, 182 (1969).

342. U. M. Dzhemilev, L. Yu. Gubaidullum and G. A. Tolstikov, <u>Bull. Acad. Sci. U.S.S.R.</u>, <u>Div. Chem. Sci.</u>, <u>27</u>, 1287 (1978).

343. P. Heimbach, A. Roloff and H. Schenkluhn, <u>Angew. Chem., Int. Ed. Engl.</u>, <u>16</u>, 252 (1977).

344. R. G. Miller, T. J. Kealy and A. L. Barney, <u>J. Amer. Chem. Soc.</u>, <u>89</u>, 3756 (1967).

345. G. Peiffer, X. Cochet and F. Petit, <u>Bull. Soc. Chim, Fr. II</u>, 415 (1979).

346. B. Bogdanović, B. Henc, B. Meister, H. Pauling and G. Wilke, <u>Angew. Chem., Int. Ed. Engl.</u>, <u>11</u>, 1023 (1972).

347. H. Sato and T. Inukai, <u>Bull. Chem. Soc. Jpn.</u>, <u>45</u>, 944 (1972).

348. V. P. Yur'ev, F. G. Yusupova, G. A. Gailyunas and V. D. Sheludyakov, Bull. Acad. Sci. U.S.S.R., Div. Chem. Sci., 26, 1563 (1977).

349. H. Singer, Synthesis, 189 (1974).

350. H. Singer, W. Umbach and M. Dohr, Synthesis, 42 (1972).

351. G. P. Chiusoli, L. Pallini and G. Salerno, J. Organomet. Chem., 238, C85 (1982).

352. W. Brenner, P. Heimbach, K.-J. Ploner and F. Thömel, Angew. Chem., Int. Ed. Engl., 8, 753 (1969).

353. B. Büssemeier, P. W. Jolly and G. Wilke, J. Amer. Chem. Soc., 96, 4726 (1974).

354. T-A. Mitsudo, Y. Nakagawa, H. Watanabe, K. Watanabe, H. Misawa and Y. Watanabe, J. Chem. Soc., Chem. Commun., 496 (1981).

355. T. Hosokawa, K. Maeda, K. Koga and I. Moritani, Tetrahedron Lett., 739 (1973).

356. K. E. Atkins, W. E. Walker and R. M. Manyik, J. Chem. Soc., Chem. Commun., 330 (1971).

357. R. Jira, Tetrahedron Lett., 1225 (1971).

358. J. Tsuji and T. Mitsuyasu, Japan Patent 74 35,603; Chem. Abs., 82, 124743g (1975).

359. D. Commereuc and Y. Chauvin, Bull. Soc. Chim. Fr., 652 (1974).

360. S. Takahashi, H. Yamazaki and N. Hagihara, Bull. Chem. Soc. Jpn., 41, 254 (1968).

361. I. I. Moiseev and M. N. Vargaftik, Bull. Acad. Sci. U.S.S.R., Div. Chem. Sci., 744 (1965).

362. P. Grenouillet, D. Neibecker, J. Poirier and I. Tkatchenko, Angew. Chem., Int. Ed. Engl., 21, 767 (1982).

363. J. Beger and H. Reichel, J. Prakt. Chem., 315, 1067 (1973).

364. R. Klüter, M. Bernd and H. Singer, J. Organomet. Chem., 137, 309 (1977).

365. Y. Tamaru, R. Suzuki, M. Kagontani and Z. Yoshida, Tetrahedron Lett., 21, 3787 (1980).

366. U. M. Dzhemilev, R. V. Kunakova, N. Z. Baibulatova, G. A. Tolstikov and L. M. Zelanova, Bull. Acad. Sci. U.S.S.R., Div. Chem. Sci., 30, 1506 (1981).

367. E. J. Smutny, J. Amer. Chem. Soc., 89, 6793 (1967).

368. T. Mandai, H. Yasuda, M. Kaito, J. Tsuji, R. Yamaoka and H. Fukami, Tetrahedron, 35, 309 (1979).

369. L. I. Zakharkin and E. A. Petrushkina, Bull. Acad. Sci. U.S.S.R., Div. Chem. Sci., 31, 1054 (1982).

370. H. Yamazaki, Japan. Kokai 74 48,613 (1974); Chem. Abs., 81, 119953a (1974).

371. H. Yagi, E. Tanaka, H. Ishiwatari, M. Hidai and Y. Uchida, Synthesis, 334 (1977).

372. L. I. Zakharkin and S. A. Babich, Bull. Acad. Sci. U.S.S.R., Div. Chem. Sci., 25, 1967 (1976).

373. K. Sawatari and E. Tanaka, Japan. Kokai 74,125,313 (1974); Chem. Abs., 82, 155323z (1975).

374. M. Hidai, H. Ishiwatari, H. Yagi, E. Tanaka, K. Onozawa and Y. Uchida, J. Chem. Soc., Chem. Commun., 170 (1975)

375. M. Hidai, H. Mizuta, H. Yagi, Y. Nagai, K. Hata and Y. Uchida, J. Organomet. Chem., 232, 89 (1982).

376. J. Tsuji, Pure. Appl. Chem., 51, 1235 (1979).

377. W. E. Walker, R. M. Manyik, K. E. Atkins and M. L. Farmer, Tetrahedron Lett., 3817 (1970).

378. T. Arakawa and H. Miyake, Kogyo Kagaku Zasshi, 74, 1143 (1971); Chem. Abs., 75, 75669g (1971).

379. T. Mitsuyasu and J. Tsuji, Ger. Offen. 2,040,708 (1971);

     Chem. Abs., 76, 3411q (1972).

380. D. Rose and H. Lepper, J. Organomet. Chem., 49, 473 (1973).

381. N. Z. Baibulatova, R. V. Kunakova, S. I. Lomakina and U. M.

     Dzhemilev, J. Org. Chem., U.S.S.R., 18, 39 (1982).

382. J. Tsuji, K. Tsuroaka and K. Yamamoto, Bull. Chem. Soc. Jpn., 49,

     1701 (1976).

383. H. Munekata and N. Imaki, Japan. Kokai 75 96,510, 75 96,511 (1975);

     Chem, Abs., 84, 4460g, 4461h (1976).

     J. Tsuji and T. Mandai, Chem. Lett., 975 (1977).

384. J. Tsuji, H. Yasuda and T. Mandai, J. Org. Chem., 43, 3606 (1978).

385. T. Ishida and K. Wada, J. Chem. Soc., Chem. Commun., 209 (1975).

386. J. Tsuji and T. Mandai, Tetrahedron Lett., 1817 (1978).

387. T. Mitsuyasu, M. Hara and J. Tsuji, J. Chem. Soc., Chem. Commun.,

     345 (1971).

388. J. Tsuji and M. Takahashi, J. Mol. Catal., 10, 107 (1981).

389. U. M. Dzhemilev, F. A. Selimov, A. Z. Yakupova and G. A. Tolstikov,

     Bull. Acad. Sci. U.S.S.R., Div. Chem. Sci., 27, 1230 (1978).

390. C. Moberg and B. Åkermark, J. Organomet. Chem., 209, 101 (1981).

391. F. G. A. Stone, M. Green, G. Scholes and J. L. Spencer, U. S. Patent

     4,104,471 (1978); Chem. Abs., 90, 87479t (1979).

392. M. Green, G. Scholes and F. G. A. Stone, J. Chem. Soc., Dalton,

     309 (1978).

393. K. Kaneda, H. Kurosaki, M. Terasewa, T. Imanaka and S. Teranishi,

     J. Org. Chem., 46, 2356 (1981).

394. S. Miyamoto, Japan. 76 04, 968 (1976); Chem. Abs., 85, 77631u

     (1976).

395. A. M. Lazutkin, V. M. Mastikhin and A. I. Lazutkina, Kinet. Catal., 19, 857 (1978).

396. A. M. Lazutkin, A. I. Lazutkina and Yu. I. Ermakov, Kinet. Catal., 14 , 1411 (1973).

397. J. D. Umpleby, Helv. Chim. Acta, 61, 2243 (1978).

398. B. Åkermark, G. Åkermark, C. Moberg, C. Björklund and K. Siirala-Hansen, J. Organomet. Chem., 164, 97 (1979).

399. U. M. Dzhemilev, R. N. Fakhretdinov, A. G. Telin, M. Yu. Dolomatov, E. G. Galkin and G. A. Tolstikov, Bull. Acad. Sci. U.S.S.R., Div. Chem. Sci., 29, 148 (1980).

400. T. Mitsuyasu and J. Tsuji, Tetrahedron, 30, 831 (1974).

401. J. Tsuji, T. Mitsuyasu and K. Ohno, Japan. 72 11,204 (1972); Chem. Abs., 77, 19164t (1972).

402. J. Tsuji, T. Yamakawa and T. Mandai, Tetrahedron Lett., 565 (1978).

403. J. Tsuji and T. Mandai, Tetrahedron Lett., 3285 (1977).

404. A. Musco, Inorg. Chim. Acta, 11, L11 (1974).

405. K. Ohno, T. Mitsuyasu and J. Tsuji, Tetrahedron, 28, 3705 (1972).

406. R. Bortolini, G. Gatti and A. Musco, J. Mol. Catal., 14, 95 (1982).

407. Y. Tamaru, R. Suzuki, M. Kagotani and Z. Yoshida, Tetrahedron Lett., 21, 3791 (1980).

408. K. Takahashi, A. Miyake and G. Hata, Chem. Ind., 488 (1971).

409. G. Hata, K. Takahashi and A. Miyake, J. Org. Chem., 36, 2116 (1971).

410. R. Baker, P. M. Winton and R. W. Turner, Tetrahedron Lett., 21, 1175 (1980).

411. R. V. Kunakova, G. A. Tolstikov, U. M. Dzhemilev, F. V. Sharipova and D. L. Sazikova, Bull. Acad. Sci. U.S.S.R., Div. Chem. Sci., 27, 806 (1978).

412. J. Tsuji, M. Masaoka, T. Takahashi, A. Suzuki and N. Miyaura, Bull. Chem. Soc. Jpn., 50, 2507 (1977).

413. S. Ceriotti, G. Longoni and P. Chini, J. Organomet. Chem., 174, C27 (1979).

414. M. Aldissi, F. Schué, L. Giral and M. Rolland, Polymer, 23, 246 (1982).

415. C. K. Chiang, C. R. Fincher, Y. W. Park, A. J. Heeger, H. Shirakawa, E. J. Louis, S. C. Gau and A. G. MacDiarmid Phys. Rev. Lett., 39, 1098 (1977).

416. I. V. Nicolescu and E. M. Angelescu, J. Polym. Sci. Al, 4, 2963 (1966).

417. W. A. Kornicker, U. S. Patent 3,578,626 (1971); Chem. Abs., 75, 49875b (1971).

418. G. A. Chukhadzhyan, Zh. I. Abramyan and M. Sh. Grigoryan, Polym. Sci. U.S.S.R., 10, 2337 (1968).

419. S. Zhiquan, Y. Mujie, S. Mingxiao and C. Yiping, J. Polym. Sci., Polym. Lett. Ed., 20, 411 (1982).

420. Y. Cao, R. Qian, F. Wang and X. Zhao, Makromol. Chem., Rapid Commun., 3, 687 (1982).

421. F. Wang, X. Zhao, Z. Gong, Y. Cao, Q. Yang and R. Qian, Makromol. Chem., Rapid Commun., 3, 929 (1982).

422. A. C. Chiang, P. F. Waters and M. H. Aldridge, J. Polym. Sci., Polym. Chem. Ed., 20, 1807 (1982).

423. A. Furlani, I. Collamati and G. Sartori, J. Organomet. Chem., 17, 463 (1969).

424. J. M. Huggins and R. G. Bergman, J. Amer. Chem. Soc., 103, 3002 (1981).

425. J. Levisalles, F. Rose-Munch, H. Rudler, J-C. Daran, Y. Dromzee, Y. Jeannin, D. Ades and M. Fontanille, J. Chem. Soc., Chem. Commun., 1055 (1981).

426. S. Yoshikawa, J. Kiji and J. Furukawa, Makromol. Chem., 178, 1077 (1977).

427. H. J. Schmitt and H. Singer, J. Organomet. Chem., 153, 165 (1978).

428. L. Carlton and G. Read, J. Chem. Soc., Perkin I, 1631 (1978).

429. G. Giacomelli, F. Marcacci, A. M. Caporusso and L. Lardicci, Tetrahedron Lett., 3217 (1979)

430. F. G. Yusupova, G. A. Gailyunas, G. V. Nurtdinova and V. P. Yur'ev, Bull. Acad. Sci. U.S.S.R., Div. Chem. Sci., 27, 2160 (1978).

431. W. Reppe, O. Schlichting, K. Klager and T. Toepel, Ann., 560, 1 (1948).

432. W. Reppe, Experientia, 5, 93 (1949).

433. G. N. Schrauzer, P. Glockner and S. Eichler, Angew. Chem., Int. Ed. Engl., 3, 185 (1964).

434. L. P. Yur'eva, Russ. Chem. Rev., 43, 48 (1974).

435. N. Hagihara, J. Chem. Soc. Japan, Pure Chem. Sect., 73, 323 (1952); Chem. Abs., 47 10490i (1953).

436. A. C. Cope and R. M. Pike, J. Amer. Chem. Soc., 75, 3220 (1953).

437. A. C. Cope and D. F. Rugen, J. Amer. Chem. Soc., 75, 3215 (1953).

438. A. C. Cope and J. E. Meili, J. Amer. Chem. Soc., 89, 1883 (1967).

439. A. C. Cope and W. R. Moore, J. Amer. Chem. Soc., 77, 4939 (1955).

440. R. S. H. Liu and C. G. Krespan, J. Org. Chem., 34, 1271 (1969).

441. A. C. Cope and H. C. Campbell, J. Amer. Chem. Soc., 73, 3536 (1951), 74, 179 (1952).

442. L. H. Simons and J. J. Lagowski, J. Org. Chem., 43, 3247 (1978).

443. G. N. Schrauzer and S. Eichler, Chem. Ber., 95, 550 (1962).

444. G. N. Schrauzer, Adv. Organomet. Chem., 2, 1 (1964).

445. D. J. Brauer and C. Krüger, J. Organomet. Chem., 122, 265 (1976).

446. G. Wilke, Pure Appl. Chem., 50, 677 (1978).

447. S. P. Kolesnikov, J. E. Dobson and P. S. Skell, J. Amer. Chem. Soc., 100, 999 (1978).

448. S. A. R. Knox, R. F. D. Stansfield, F. G. A. Stone, M. J. Winter and P. Woodward, J. Chem. Soc., Chem. Commun., 221 (1978).

449. M. Green, N. C. Norman and A. G. Orpen. J. Amer. Chem. Soc., 103, 1269 (1981).

450. C. Fröhlich and H. Hoberg, J. Organomet. Chem., 204, 131 (1981), Angew. Chem., Int. Ed. Engl., 19, 145 (1980).

451. H. Hoberg and W. Richter, J. Organomet. Chem., 195, 347, 355 (1980).

452. J. R. Fritch, K. P. C. Vollhardt, M. R. Thompson and V. W. Day, J. Amer. Chem. Soc., 101, 2768 (1979).

453. J. R. Fritch and K. P. C. Vollhardt, Angew. Chem., Int. Ed. Engl., 19, 559 (1980).

454. R. E. Colborn and K. P. C. Vollhardt, J. Amer. Chem. Soc., 103, 6259 (1981).

455. V. O. Reikhsfel'd, B. E. Lein and L. K. Makovetskii Dokl. Chem., 190, 31 (1970).

456. M. Sato and Y. Ishida, Bull. Chem. Soc. Jpn., 41, 730 (1968).

457. T. Higashimura, Jpn. Kokai Tokkyo Koho, 80 94,232; Chem. Abs., 94, 174596e (1981).

458. Mitsui Toatsu Chemicals, Inc., Jpn. Kokai Tokkyo Koho, 80 160,727 (1980); Chem. Abs., 95, 24575d (1981).

459. M. Cowie and T. G. Southern, Inorg. Chem., 21, 246 (1982).

460. W. Reppe, N. v. Kupetov and A. Magin, Angew. Chem., Int. Ed. Engl., 8, 727 (1969).

461. F. A. Cotton, W. T. Hall, K. J. Cann and F. J. Karol, Macromolecules, 14, 233 (1981).

462. J. P. Collman, J. W. Kang, W. F. Little and M. F. Sullivan, Inorg. Chem., 7, 1298 (1968).

463. J. J. Eisch and J. E. Galle, J. Organomet. Chem., 96, C23 (1975).

464. Y. Wakatsuki and H. Yamazaki, Synthesis, 26 (1976).

465. C-A. Chang, J. A. King and K. P. C. Vollhardt, J. Chem. Soc., Chem. Commun., 53 (1981).

466. D. J. Brien, A. Naiman and K. P. C. Vollhardt, J. Chem. Soc., Chem. Commun., 133 (1982).

467. B. C. Berris, Y-H. Lai and K. P. C. Vollhardt, J. Chem. Soc., Chem. Commun., 953 (1982).

468. E. D. Sternberg and K. P. C. Vollhardt, J. Org. Chem., 47, 3447 (1982).

469. K. P. C. Vollhardt and R. G. Bergman, J. Amer. Chem. Soc., 96, 4996 (1974).

470. R. Grigg, R. Scott and P. Stevenson, Tetrahedron Lett., 23, 2691 (1982).

471. D. M. Singleton, Tetrahedron Lett., 1245 (1973).

472. H. Hoberg and G. Burkhart, Synthesis, 525 (1979).

473. H. Hoberg and B. W. Oster, Synthesis, 324 (1982).

474. P. M. Maitlis, Pure Appl. Chem., 33, 489 (1973), Acc. Chem. Res., 9, 93 (1976).

475. P. M. Bailey, B. E. Mann, I. D. Brown and P. M. Maitlis, J. Chem. Soc., Chem. Commun., 238 (1976).

476. G. A. Chukhadzhyan, Zh. I. Abramyan, G. M. Tonyan and V. A. Matosyan, J. Org. Chem. U.S.S.R., 10, 2007 (1974).

477. G. A. Chukhadzhyan, Zh. I. Abramyan, G. M. Tonyan L. I. Sagradyan and T. S. Élbakyan, J. Org. Chem. U.S.S.R., 17, 1636 (1981).

478. F. Ullmann and J. Bielecki, Chem. Ber., 34, 2174 (1901) F. Ullmann, Ann., 332, 38 (1904), 366, 79 (1909).

479. P. E. Fanta, Chem. Rev., 38, 139 (1946), 64, 613 (1964),
     Synthesis, 9 (1974).

480. F. D. King and D. R. M. Walton, Synthesis, 40 (1976).

481. M. Sainsbury, Tetrahedron, 36, 3327 (1980).

482. S. Ozasa, Y. Fujioka, M. Tsukada and E. Ikubi, Chem. Pharm. Bull.,
     29, 344 (1981).

483. E. Brown and J-P. Robin, Tetrahedron Lett., 2015 (1977).

484. P. J. Wittek, T. K. Liao and C. C. Cheng, J. Org. Chem., 44,
     870 (1979).

485. A. H. Lewin, M. J. Zovko, W. H. Rosewater and T. Cohen, J. Chem.
     Soc., Chem. Commun., 80 (1967).

486. Z. Kulicki and W. Karminski, Zeszyty Nauk. Politech. Slask., Chem.,
     16, 11 (1963); Chem. Abs., 62, 4001b (1965).

487. R. D. Rieke and L. D. Rhyne, J. Org. Chem., 44, 3445 (1979).

488. F. E. Ziegler, K. W. Fowler and S. Kaufer, J. Amer. Chem. Soc., 98,
     8282 (1976).

489. F. E. Ziegler, I. Chliwner, K. W. Fowler, S. J. Kanfer, S. J. Kuo
     and N. D.Sintra, J. Amer. Chem. Soc., 102, 790 (1980).

490. M. F. Semmelhack, P. M. Helquist and L. D. Jones, J. Amer. Chem.
     Soc., 93, 5908 (1971).

491. M. F. Semmelhack and L. S. Ryono, J. Amer. Chem. Soc., 97, 3873
     (1975).

492. M. Mori, Y.Hashimoto and Y. Ban, Tetrahedron Lett., 21, 631 (1980).

493. G. Schiavon, G. Bontempelli and B. Corain, J. Chem. Soc., Dalton,
     1074 (1981).

494. K. Takagi, N. Hayama and S. Inokawa, Bull. Chem. Soc. Jpn., 53,
     3691 (1980).

495. M. Zembayashi, K. Tamao, J-i. Yoshida and M. Kumada, Tetrahedron Lett., 4089 (1977).

496. I. Colon, L. M. Maresca and G. T. Kwiatkowski, Eur. Pat. Appl., 12,201 (1980); Chem. Abs., 94, 65293h (1981).

497. T. T. Tsou and J. K. Kochi, J. Amer. Chem. Soc., 100, 1634 (1978), 101, 7547 (1979).

498. D. R. Fahey and J. E. Mahan, J. Amer. Chem. Soc., 99, 2501 (1977).

499. M. F. Semmelhack, P. Helquist, L. D. Jones, L. Keller, L. Mendelson, L. S. Ryono, J. G. Smith and R. D. Stauffer, J. Amer. Chem. Soc., 103, 6460 (1981).

500. I. V. Kozhevnikov and K. I. Matveev, Russ. Chem. Rev., 47, 649 (1978).

501. R. van Helden and G. Verberg, Rec., 84, 1263 (1965).

502. M. Kashima, H. Yoshimoto and H. Itatani, J. Catal., 29, 92 (1973).

503. H. Iataaki and H. Yoshimoto, J. Org. Chem., 38, 76 (1973).

504. H. Yoshimoto and H. Itatani, Bull Chem. Soc. Jpn., 46, 2490 (1973), Chem. Ind., 674 (1971).

505. V. E. Tarabanko, I. V. Kozhevnikov and K. I. Matveev, React. Kinet. Catal. Lett., 8, 77 (1978).

506. A. I. Rudenkov, G. U. Mennenga, L. N. Rachkovskaya, K. I. Matveev and I. V. Kozhevnikov, Kinet Catal., 18, 758 (1977).

507. G. U. Mennenga, A. I. Rudenkov, K. I. Matveev and I. V. Kozhevnikov, React. Kinet. Catal. Lett., 5, 401 (1976).

508. W. I. Taylor and A. R. Battersby, Oxidative Coupling of Phenols, Marcel Dekker, New York, 1967.

509. T. Kametami, Lect. Heterocycl. Chem., 2, 57 (1974).

510. E. McDonald and A. Suksamrarn J. Chem. Soc., Perkin I, 440 (1978).

511. B. Franck, G. Dunkelmann and H. J. Lubs, Angew. Chem., Int. Ed. Engl., 6, 1075 (1967).

512. W. L. Carick, G. L. Karapinka and G. T. Kwiatkowski, J. Org. Chem., 34, 2388 (1969).

M. A. Schwartz, R. A. Holton and S. W. Scott, J. Amer. Chem. Soc., 91, 2800 (1969).

513. W. W. Kaeding, J. Org. Chem., 28, 1063 (1963).

A. Tkac, R. Prikryl and L. Malik, J. Elastoplast., 5 (Jan), 20 (1973); Chem. Abs., 79, 32338p (1973).

514. N. Minami and S. Kijima, Yakagaku Zasshi, 98, 433 (1978); Chem. Abs., 89, 75337 m (1978).

515. T. Kametani and M. Ihara, J. Chem. Soc., Perkin I, 629 (1980).

516. M. Matsumoto and K. Kuroda, Tetrahedron Lett., 22, 4437 (1981).

517. F. D. Campbell and G. Eglinton, Org. Synth., 45, 39 (1965).

518. E. G. Derouane, J. N. Brahan and R. Hubin, Chem. Phys. Lett., 25, 243 (1974).

519. J. Ojima, K. Wada and M. Terasaki, Tetrahedron Lett., 22, 457 (1981).

520. R. Eastmond, T. R. Johnson and D. R. M. Walton, Tetrahedron, 28, 4601 (1972).

521. T. R. Johnson and D. R. M. Walton, Tetrahedron, 28, 5221 (1972).

522. W. B. Austin, N. Bilow, W. J. Kelleghan and K. S. Y. Lau, J. Org. Chem., 46, 2280 (1981).

523. A. N. Novikov and M. G. Grigor'ev, J. Org. Chem. U.S.S.R., 14, 1662 (1978).

524. M. J. Robins and P. J. Barr, Tetrahedron Lett., 22, 421 (1981).

525. D. E. Ames, D. Bull and C. Takundwa, Synthesis, 364 (1981).

526. L. Cassar, J. Organomet. Chem., 93, 253 (1975).

527. J-P. Morizur, Bull. Soc. Chim. Fr., 1331 (1964).

528. L. A. Jones and R. Watson, Canad. J. Chem., 51, 1833 (1973).

529. R. Pallaud and J-M. Pleau, C. R. Acad. Sci. Paris, C265, 316 (1967).

530. L. M. Zubritskii, T. N. Fomina and Kh. V. Bal'yan, J. Org. Chem. U.S.S.R., 17, 63 (1981).

531. T. Nakaya, H. Arabori and M. Imoto, Bull. Chem. Soc. Jpn., 44, 1422 (1971).

532. H. Felkin and B. Meunier, J. Organomet. Chem., 146, 169 (1978).

533. M. Tamura and J. Kochi, J. Amer. Chem. Soc., 93, 1483, 1485, 1487 (1971); Synthesis, 303 (1971).

534. B. F. Bonini and M. Tiecco, Gazz. Chim. Ital., 96, 1792 (1966)

535. E. Ibuki, S. Ozasa, Y. Fujioka, M. Okada and K. Terada, Bull. Chem. Soc. Jpn., 53,, 821 (1980).

536. T. Hayashi, M. Konishi and M. Kumada, Tetrahedron Lett., 1871 (1979).

537. R. L. Clough, P. Mison and J. D. Roberts, J. Org. Chem., 41, 2252 (1976).

538. A. Sekiya and N. Ishikawa, J. Organomet. Chem., 118, 349 (1976), 125, 281 (1977).

539. S. Nunomoto, Y. Kawakami and Y. Yamashita, Bull. Chem. Soc. Jpn., 54, 2831 (1981).

540. K. Tamao, A. Minato, N. Miyake, T. Matsuda, Y. Kiso and M. Kumada, Chem. Lett., 133 (1975).

541. Y. Kiso, K. Tamao and M. Kumada, J. Organomet. Chem., 50, C12 (1973).

542. K. Tamao, Y. Kiso, K. Sumitani and M. Kumada, J. Amer. Chem. Soc., 94, 9268 (1972).

543. K. Tamao, K. Sumitani, Y. Kiso, M. Zembayashi, A. Fujioka, S-i. Kodama, I. Nakajima, A. Minato and M. Kumada, Bull. Chem. Soc. Jpn., 49, 1958 (1976).

544. L. Farády, L. Bencze and L. Markó, J. Organomet. Chem., 17, 107 (1969).

545. D. G. Morrell and J. K. Kochi, J. Amer. Chem. Soc., 97, 7262 (1975).

546. K. Tamao, S. Kodama, I. Nakajima, M. Kumada, A. Minato and K. Suzuki, Tetrahedron., 38, 3347 (1982).

547. D. E. Bergstrom and P. A. Reday, Tetrahedron Lett., 23, 4191 (1982).

548. K. Tamao, S-i. Kodama, T. Nakatsuka, Y. Kiso and M. Kumada, J. Amer. Chem. Soc., 97, 4405 (1975).

549. M. Kumada, K. Tamao and K. Sumitani, Org. Synth., 58, 127 (1978).

550. L. N. Pridgen and S. S. Jones, J. Org. Chem., 47, 1590 (1982).

551. H. Yamanaka, K. Edo, F. Shoji, S. Konno, T. Sakamoto and M. Mizugaki, Chem. Pharm. Bull., 26, 2160 (1978).

552. O. Piccolo and T. Martinengo, Synth. Commun., 11, 497 (1981).

553. L. N. Pridgen, J. Het. Chem., 12, 443 (1975).

554. L. N. Pridgen and L. B. Killmer, J. Org. Chem., 46, 5402 (1981).

555. M. Tiecco, L. Testaferri, M. Tingoli, D. Chianelli and E. Wenkert, Tetrahedron Lett., 23, 4629 (1982).

556. T. Hayashi, Y. Katsuro, Y. Okamoto and M. Kumada, Tetrahedron Lett., 22, 4449 (1981).

557. G. Consiglio and C. Botteghi, Helv. Chim. Acta, 56, 460 (1973).

558. Y. Kiso, K. Tamao, N. Miyake, K. Yamamoto and M. Kumada, Tetrahedron Lett., 3 (1974).

559. G. Consiglio, O. Piccolo and F. Morandini, J. Organomet. Chem., 177, C13 (1979).

560. K. Tamao, M. Zembayashi, Y. Kiso and M. Kumada, J. Organomet. Chem., 55, C91 (1973).

561. R. J. P. Corriu and J. P. Masse, J. Chem. Soc., Chem. Commun., 144 (1972).

562. H. P. Dang and G. Linstrumelle, Tetrahedron Lett., 191 (1978).

563. K. Tamao, M. Zembayashi and M. Kumada, Chem. Lett., 1237, 1239 (1976).

564. R. S. Smith and J. K. Kochi, J. Org. Chem., 41, 502 (1976).

565. S. M. Neumann and J. K. Kochi, J. Org. Chem., 40, 599 (1975).

566. R. H. Mitchell, B. N. Ghose and M. E. Williams, Canad. J. Chem., 55, 210 (1977).

567. A. Commercon, J. F. Normant and J. Villieras, J. Organomet. Chem., 128, 1 (1977).

568. H. Takei, H. Sugimura, M. Miura and H. Okamura, Chem. Lett., 1209 (1980).

569. H. Okamura, M. Miura and H. Takei, Tetrahedron Lett., 43 (1979).

570. H. Okamura, M. Miura, K. Kosugi and H. Takei, Tetrahedron Lett., 21, 87 (1980).

571. S. Uemura and S-i. Fukuzawa, Tetrahedron Lett., 23, 1181 (1982).

572. E. Wenkert and T. W. Ferreira, J. Chem. Soc., Chem. Commun., 840 (1982).

573. V. Fiandanese, G. Marchese, F. Naso and L. Ronzini, J. Chem. Soc., Chem. Soc., Chem. Commun., 647 (1982).

574. J-L. Fabre, M. Julia and J-N. Verpeaux, Tetrahedron Lett., 23, 2469 (1982).

575. M. Julia, A. Righini and J-N. Verpeaux, Tetrahedron Lett., 2393 (1979).

576. T. Hayashi, Y. Katsuro and M. Kumada, Tetrahedron Lett., 21, 3915 (1980).

577. T. Hayashi, T. Fujiwa, Y. Okamoto, Y.Katsuro and M. Kumada, Synthesis, 1001 (1981).

578. T. Hayashi and M. Kumada, Acc. Chem. Res., 15, 395 (1982).

579. T. Hayashi, N. Nagashima and M. Kumada, Tetrahedron Lett., 21, 4623 (1980).

580. T. Hayashi, M. Fukushima, M. Konishi and M. Kumada, Tetrahedron Lett., 21, 79 (1980).

581. K. Tamao, T. Hayashi, H. Matsumoto, K. Yamamoto and M. Kumada, Tetrahedron Lett., 2155 (1979).

582. T. Hayashi, M. Konishi, T. Hioki, M. Kumada, A. Ratajczak and H. Niedbała, Bull. Chem. Soc. Jpn., 54, 3615 (1981).

583. T. Hayashi, M. Konishi, M. Fukushima, T. Mise, M. Kagotani, M Tajika and M. Kumada, J. Amer. Chem. Soc., 104, 180 (1982).

584. T. Hayashi, M. Tajika, K. Tamao and M. Kumada, J. Amer. Chem. Soc., 98, 3718 (1976).

585. H. Brunner and M. Pröbster, J. Organomet. Chem., 209, C1 (1981).

586. T. Hayashi, M. Konishi, H. Ito and M. Kumada, J. Amer. Chem. Soc., 104, 4962 (1982).

587. Y. Ohbe, M. Takagi and T. Matsuda, Tetrahedron, 30, 2669 (1974).

588. Y. Ohbe and T. Matsuda, Tetrahedron, 29, 2989 (1973).

589. Y. Ohbe, K. Doi and T. Matsuda, Nippon Kagaku Kaishi, 193 (1974); Chem. Abs., 80, 83209a (1974).

590. H. Yasuda, M. Yamauchi, A. Nakamura, T. Sei, Y. Kai, N. Yasuoka and N. Kasai, Bull. Chem. Soc. Jpn., 53, 1089 (1980).

591. Y. Kajihara, K. Ishikawa, H. Yasuda and A. Nakamura, Bull. Chem. Soc. Jpn., 53, 3035 (1980).

592. A. Commercon, M. Bourgain, M. Delaumeny, J. F. Normant and J. Villieras, Tetrahedron Lett., 3837 (1975).

593. J. F. Normant, A. Commercon, Y. Gendreau, M. Bourgain and J. Villieras, Bull. Soc. Chim. Fr., II, 309 (1979).

594. T. Hayashi, M. Konishi, K-i. Yokota and M. Kumada, _J. Chem. Soc._,
Chem. Commun., 313 (1981).

595. E. Wenkert and T. W. Ferreira, _Organometallics_, _1_, 1670 (1982).

596. M. Schlosser, _Angew. Chem., Int. Ed. Engl._, _13_, 701 (1974).

597. A. Claesson and L - I. Olsson, _J. Chem. Soc., Chem. Commun._, 621
(1978).

598. M. Commercon-Bourgain, J. F. Normant and J. Villieras, _C. R. Acad._
Sci. Paris, _C285_, 211 (1977).

599. Y. Gendreau, J. F. Normant and J. Villieras, _J. Organomet. Chem._,
_142_, 1 (1977).

600. H. Okamura and H. Takei, _Tetrahedron Lett._, 3425 (1979).

601. M. Julia and J-N. Verpeaux, _Tetrahedron Lett._, _23_, 2457 (1982).

602. C. Chuit, H. Felkin, C. Frajerman, G. Roussi and G. Swierczewski,
J. Chem. Soc., Chem. Commun., 1604 (1968).

603. H. Felkin and G. Swierczewski, _Tetrahedron Lett._, 1433 (1972).

604. H. Felkin and G. Swierczewski, _Tetrahedron_, _31_, 2735 (1975).

605. C. Chuit, H. Felkin, C. Frajerman, G. Roussi and G. Swierczewski,
J. Organomet. Chem., _127_, 371 (1977).

606. H. Felkin, M. Joly-Goudket and S. G. Davies, _Tetrahedron Lett._,
_22_ , 1157 (1981).

607. M. Chérest, H. Felkin, J. D. Umpelby and S. G. Davies, _J. Chem._
Soc., Chem. Commun., 681 (1981).

608. T. Hayashi, M. Konishi and M. Kumada, _J Organomet. Chem._, _186_,
Cl (1980).

609. J. F. Normant, J. Villieras and F. Scott, _Tetrahedron Lett._, 3263
(1977).

610. T. Ishihara, A. Yamamoto and K. Taguchi, _Eur. Pat. Appl._ EP 44,558
(1982); _Chem. Abs._, _96_, 199062g (1982).

611. S. Nunomoto, Y. Kawakami and Y. Yamashita, <u>Bull. Chem. Soc. Jpn.</u>, <u>54</u>, 2831 (1981).

612. J. F. Normant, T. Mulamba, F. Scott, A. Alexakis and G. Cahiez, <u>Tetrahedron Lett.</u>, 3711 (1978).

613. F. Derguini-Boumechal and G. Linstrumelle, <u>Tetrahedron Lett.</u>, 3225 (1976).

614. G. Fouquet and M. Schlosser, <u>Angew. Chem., Int, Ed. Engl.</u>, <u>13</u>, 82 (1974).

615. C. Huynh and G. Linstrumelle, <u>Tetrahedron Lett.</u>, 1073 (1979).

616. T. A. Baer and R. L. Carney, <u>Tetrahedron Lett.</u>, 4697 (1976).

617. N. A. Bumagin, I. O. Kalinovskii and I. P. Beletskaya, <u>J. Org. Chem. U.S.S.R.</u>, <u>18</u>, 1151 (1982).

618. R. C. Larock, <u>Tetrahedron</u>, <u>38</u>, 1713 (1982).

619. R. A. Kretchner and R. Glowinski, <u>J. Org. Chem.</u>, <u>41</u>, 2661 (1976).

620. R. C. Larock and J. C. Bernhardt, <u>J. Org. Chem.</u>, <u>42</u>, 1680 (1977).

621. K. Takagi, N. Hayama, T. Okamoto, Y. Sakakibara and S. Oka, <u>Bull. Chem. Soc. Jpn.</u>, <u>50</u>, 2741 (1977).

622. I. Arai, R. Hanna and G. D. Daves, <u>J. Amer. Chem. Soc.</u>, <u>103</u>, 7684 (1981).

623. R. C. Larock, <u>J. Org. Chem.</u>, <u>41</u>, 2241 (1976).

624. R. C. Larock and B. Riefling, <u>J. Org. Chem.</u>, <u>43</u>, 1468 (1978).

625. A. Bruggink, S. J. Ray and A. McKillop, <u>Org. Synth.</u>, <u>58</u>, 52 (1978).

626. R. G. R. Bacon and J. C. F. Murray, <u>J. Chem. Soc., Perkin I</u>, 1267 (1975).

627. R. W. M. Ten Hoedt, G. Van Koten and J. G. Noltes, <u>J. Organomet. Chem.</u>, <u>170</u>, 131 (1979).

628. E-i. Negishi and S. Baba, <u>J. Chem. Soc., Chem. Commun.</u>, 596 (1976).

629. P. Vincent, J-P. Beaucourt and L. Pichat, <u>Tetrahedron Lett.</u>, <u>23</u>, 63 (1982).

630. E-i. Negishi, Acc. Chem. Res., 15, 340 (1982).

631. E-i. Negishi, A. O. King and N. Okukado, J. Org. Chem., 42, 1821 (1977).

632. A. Minato, K. Tamao, T. Hayashi, K. Suzuki and M. Kumada, Tetrahedron Lett., 21, 845 (1980), 22, 5319 (1981).

633. A. O. King, E-i. Negishi, F. J. Villani and A. Silveira, J. Org. Chem., 43, 358 (1978).

634. J. F. Fauvargue and A. Jutand, J.Organomet. Chem., 177, 273 (1979).

635. K. Rivtenberg, H.Kleijn, J. Meijer, E. A. Oostveen and P. Vermeer, J. Organomet. Chem., 224, 399 (1982).

636. N. Jabri, A. Alexakis and J. F. Normant, Tetrahedron Lett., 22, 3851 (1981).

637. N. A. Bumagin, I. O. Kalinovskii and I. P. Beletskaya, Bull. Acad. Sci. U.S.S.R., Div. Chem. Sci., 30, 1993, 2366 (1981).

638. M. Pereyre and J-P. Quintard, Pure Appl. Chem., 53, 2401 (1981).

639. M. Kosugi, Y. Kato, K. Kiuchi and T. Migita, Chem. Lett., 69 (1981).

640. M. Kosugi, K.Sasazawa, Y.Shimizu and T. Migita, Chem. Lett., 301 (1977).

641. M. Kosugi, M. Suzuki, I. Hagiwara, K. Goto, K. Saitoh and T. Migita, Chem. Lett., 939 (1982).

642. N. Miyaura and A. Suzuki, J. Chem. Soc., Chem. Commun., 866 (1979).

643. N. Miyaura, K. Maeda, H.Suginome and A. Suzuki, J. Org. Chem., 47, 2117 (1982).

644. S-I. Murahashi, M. Yamamura, K-i. Yanagisawa, N. Mita and K. Kondo, J. Org. Chem., 44, 2408 (1979).

645. F. Sato, H. Kodama and M. Sato, Chem. Lett., 789 (1978).

646. E-i. Negishi, H. Matsushita and N. Okudado, Tetrahedron Lett., 22, 2715 (1981).

647. N. Miyaura and A. Suzuki, J. Organomet. Chem., 213, C53 (1981).

648. S. Baba and E-i. Negishi, J. Amer. Chem. Soc., 98, 6729 (1976).

649. C. L. Rand, D. E. Van Horn, M. W. Moore and E-i. Negishi, J. Org. Chem., 46, 4093 (1981).

650. E-i. Negishi, N. Okudado, A. O. King, D. E. Van Horn and B. I. Spiegel, J. Amer. Chem. Soc., 100, 2254 (1978).

651. N. Miyaura, K. Yamada and A. Suzuki, Tetrahedron Lett., 3437 (1979).

652. J. F. Fauvarque and A. Jutand, J. Organomet. Chem., 209, 109 (1981).

653. E-i. Negishi, Pure Appl. Chem., 53, 2333 (1981).

654. H. M. Walborsky and R. B. Banks, J. Org. Chem., 46, 5074 (1981).

655. M. Kobayashi and E-i. Negishi, J. Org. Chem., 45, 5223 (1980).

656. E-i. Negishi, L. F. Valente and M. Kobayashi, J. Amer. Chem. Soc., 102, 3298 (1980).

657. K. Ruitenberg, H. Kleijn, C. J. Elsevier, J. Meijer and P. Vermeer, Tetrahedron Lett., 22, 1451 (1981).

658. K. Ruitenberg, H. Kleijn, H. Westmijze, J. Meijer and P. Vermeer, Rec., 101, 405 (1982)

659. T. Jeffrey-Luong and G. Linstrumelle, Synthesis, 738 (1982).

660. D. Milstein and J. K. Stille, J. Amer. Chem. Soc., 101, 4981, 4992 (1979).

661. R. C. Larock and S. S. Hershberger, J. Organomet. Chem., 225, 31 (1982).

662. H. Yatagi, Bull. Chem. Soc. Jpn., 53, 1670 (1980).

663. B. M. Trost, Pure Appl. Chem., 51, 787 (1979), Acc. Chem. Res., 13 385 (1980).

664. J. Tsuji, Pure Appl. Chem., 53, 2371 (1981).

665. B. M. Trost, L. Weber, P. E. Strege, T. J. Fullerton and T. J. Dietsche, J. Amer. Chem. Soc., 100, 3426, 3416, 3407 (1978).

666. B. M. Trost and T. R. Verhoeven, J. Amer. Chem. Soc., 98, 630 (1976), 99, 3867 (1977), 100, 3435 (1978) 102, 4730 (1980).

667. K. Takahashi, A. Miyake and G. Hata, Bull. Chem. Soc. Jpn., 45, 1183 (1972); Japan. 74 28,172 (1974); Chem. Abs., 82, 155389a (1975); Japan. 72 40, 775 (1972); Chem. Abs., 78, 3731k (1973).

668. J-P. Haudegond, Y. Chauvin and D. Commereuc, J. Org. Chem., 44, 3063 (1979).

669. H. Matsushita and E-i. Negishi, J. Org. Chem., 47, 4161 (1982).

670. B. M. Trost and P. E. Strege, J. Amer. Chem. Soc., 99, 1649 (1977).

671. Y. Kitagawa, A. Itoh, S. Hashimoto, H. Yamamoto and H.Nozaki, J. Amer. Chem. Soc., 99, 3864 (1977).

672. W. E. Billups, R. S. Erkes and L. E. Reed, Synth. Commun., 10, 147 (1980).

673. J.-P. Genet, M. Balabane and F. Charbonnier, Tetrahedron Lett., 23, 5027 (1982).

674. J. Tsuji, I. Shimizu, I. Minami and Y. Ohashi, Tetrahedron Lett., 23, 4809 (1982).

675. B. M. Trost, N. R. Schmuff and M. J. Miller, J. Amer. Chem. Soc., 102, 5979 (1980).

676. T. Hirao, N. Yamada, Y.Oshiro and T. Agawa, J. Organomet. Chem., 236, 409 (1982).

677. H.Onoue, I. Moritani and S-I.Murahashi, Tetrahedron Lett., 121 (1973).

678. H. Matsushita and E-i. Negishi, J. Amer. Chem. Soc., 103, 2882 (1981).

679. M. Moreno-Mañas and A.Trius, Tetrahedron, 37, 3009 (1981).

680. F. Guibe, D. S. Grierson and H-P. Husson Tetrahedron Lett., 23,
      5055 (1982).

681. T. Ikariya, Y. Ishikawa, H. Hirai and S. Yoshikawa, Chem. Lett.,
      1815 (1982).

682. Y. Tanigawa, K. Nishimura, A. Kawasaki and S-i. Murahashi,
      Tetrahedron Lett., 23, 5549 (1982).

683. N. Ono, I. Hamamoto and A. Kaji, J. Chem. Soc., Chem. Commun.,
      821 (1982).

684. R. Tamura and L. S. Hegedus, J. Amer. Chem. Soc., 104, 3727 (1982).

685. E-i. Negishi, S. Chatterjee and H. Matsushita, Tetrahedron Lett.,
      22, 3737 (1981).

686. P. Aleksandrowicz, H. Piotrowska and W. Sas, Tetrahedron, 38,
      1321 (1982), Monatsh. Chem., 113, 1221 (1982).

687. J. C. Fiaud, A. Hibon de Gournay, M. Larcheveque and H. B. Kagan,
      J. Organomet. Chem., 154, 175 (1978).

688. J. P. Genêt, F. Piau and J. Ficini, Tetrahedron Lett., 21, 3183
      (1980).

689. B. M. Trost and G. A. Molander, J. Amer. Chem. Soc., 103, 5969
      (1981).

690. N. Miyaura, Y. Tanabe, H. Suginome and A. Suzuki, J. Organomet.
      Chem., 233, C13 (1982).

691. B. M. Trost and T. P. Klun, J. Amer. Chem. Soc., 101,
      6756 (1979).

692. B. M. Trost and S. J. Brickner, J. Amer. Chem. Soc., 105, 568 (1983).

693. H. Matsushita and E-i. Negishi, J. Chem. Soc., Chem. Commun.,
      160 (1982).

694. L. S. Hegedus, W. H. Darlington and C. E. Russell, J. Org. Chem.,
      45, 5193 (1980).

695. J. Tsuji, Pure Appl. Chem., 54, 197 (1982).

696. P. A. Wade, S. D. Morrow, S. A. Hardinger, M. S. Saft and H. R. Hinney, J. Chem. Soc. Chem. Commun., 287 (1980).

697. J. Tsuji, H. Ueno, Y. Kobayashi and H. Okumoto, Tetrahedron Lett., 22, 2573 (1981).

698. J. P. Genet, M. Balabane and Y. Legras, Tetrahedron Lett., 23, 331 (1982).

699. R. C. Larock, J. C. Bernhardt and R. J. Driggs, J. Organomet. Chem., 156, 45 (1978).

700. J. Godschalx and J. K. Stille, Tetrahedron Lett., 21, 2599 (1980).

701. Y. Hayasi, M. Riediker, J. S. Temple and J. Schwartz, Tetrahedron Lett., 22, 2629 (1981).

702. N. Miyaura, T. Yano and A. Suzuki, Tetrahedron Lett., 21, 2865 (1980).

703. P. A. Wade, S. D. Morrow and S. A. Hardinger, J. Org. Chem., 47, 365 (1982).

704. J-C. Fiaud and J-L. Malleron, J. Chem. Soc., Chem. Commun., 1159 (1981).

705. B. M. Trost and E. Keinan, Tetrahedron Lett., 21, 2591 (1980).

706. J. Tsuji, Y. Kobayashi, H. Katoaka and T. Takahashi, Tetrahedron Lett., 21, 3393 (1980).

707. Y. Fujiwara, M. Yoshidomi, H. Kuromaru and H. Taniguchi, J. Organomet. Chem., 226, C36 (1982).

708. B. M. Trost and T. A. Runge, J. Amer. Chem. Soc., 103, 7550, 7559 (1981).

709. J. Tsuji, Y. Kobayashi, H. Katoaka and T. Takahashi, Tetrahedron Lett., 21, 1475 (1980).

710. B. M. Trost and T. R. Verhoeven, J. Org. Chem., 41, 3215 (1976).

711. B. M. Trost, T. R. Verhoeven and J. M. Fortunak, Tetrahedron Lett., 2301 (1979).

712. J. C. Fiaud and J. L. Malleron, Tetrahedron Lett., 22 1399 (1981).

713. T. Hayashi, K. Kanehira, H. Tsuchiya and M. Kumada, J. Chem. Soc., Chem. Commun., 1162 (1982).

714. K. Yamamoto and J. Tsuji, Tetrahedron Lett., 23, 3089 (1982).

715. B. Bosnich and P. B. Mackenzie, Pure Appl. Chem., 54, 189 (1982).

716. H. Kurosawa, J. Chem. Soc., Dalton, 939 (1979).

717. G. Consiglio, F. Morandini and O. Piccolo, Helv. Chim. Acta, 63, 987 (1980).

718. B. M. Trost and M. Lautens, J. Amer. Chem. Soc., 104, 5543 (1982).

719. Mitsui Petrochemical Industries, Ltd., Jpn. Kokai Tokkyo Koho 80 59,181 (1980); Chem. Abs., 94, 15745m (1981).

720. A. Solladié-Cavallo, J. L. Haesslein and J. E. Bäckvall, Tetrahedron Lett., 23, 939 (1982).

721. L. L. Wright, R. M. Wing and M. F. Rettig, J. Amer. Chem. Soc., 104, 610 (1982).

722. L. S. Hegedus, R. E. Williams, M. A. McGuire and T. Hayashi, J. Amer. Chem. Soc., 102, 4973 (1980).

723. R. F. Heck, J. Amer. Chem. Soc., 90, 5518 (1968).

724. L. S. Hegedus and M. A. McGuire, Organometallics 1, 1175 (1982).

725. D. L. Reger, P. J. McElligott, N. G. Charles, E. A. H. Griffith and E. L. Amma, Organometallics, 1, 443 (1982).

726. B. M. Trost and Y. Tanigawa, J. Amer. Chem. Soc., 101, 4743 (1979).

727. N-T. Luong-Thi and H. Riviere, Tetrahedron Lett., 587 (1971).

728. R. F. Heck, Pure Appl. Chem., 50, 691 (1978).

729. A. Kasahara, T. Izumi and N. Fukuda, Bull. Chem. Soc. Jpn., 50, 551 (1977).

730. R. F. Heck and J. P. Nolley, J. Org. Chem., 37, 2320 (1972).

731. N. J. Malek and A. E. Moormann, J. Org. Chem., 47, 5395 (1982).

732. R. F. Heck, Pure Appl. Chem., 53, 2323 (1981).

733. Y. Tamaru, Y. Yamada and Z-i. Yoshida, Tetrahedron Lett., 3365 (1977).

734. K. Edo, T. Sakamoto and H. Yamanaka, Heterocycles, 12, 383 (1979).

735. H. Horino, N. Inoue and T. Asao, Tetrahedron Lett., 22, 741 (1981).

736. B. A. Patel, L-C. Kao, N. A. Cortese, J. V. Minkiewicz and R. F. Heck, J. Org. Chem., 44, 918 (1979).

737. B. A. Patel and R. F. Heck, J. Org. Chem., 43, 3898 (1978).

738. K. Kaneda, F. Kawamoto, Y. Fujiwara, T. Imanaka and S. Teranishi, Tetrahedron Lett., 1067 (1974).

739. R. F. Heck, J. Amer. Chem. Soc., 91, 6707 (1969).

740. T. Mizoroki, K. Mori and A. Ozaki, Bull. Chem. Soc. Jpn., 44, 581 (1971).

741. W. C. Frank, Y. C. Kim and R. F. Heck, J. Org. Chem., 43, 2947 (1978).

742. R. Rossi, A. Carpita, M. G. Qurici and M. L. Gaudenzi, Tetrahedron, 38, 631 (1982).

743. J. E. Plevyak, J. E. Dickerson and R. F. Heck, J. Org. Chem., 44, 4078 (1979).

744. K. Mori, T. Mizoroki and A. Ozaki, Bull. Chem. Soc. Jpn., 49, 758 (1976).

745. D. D. Bender, F. G. Stakem and R. F. Heck, J. Org. Chem., 47, 1278 (1982).

746. M. Mori, K. Chiba and Y. Ban, Tetrahedron Lett., 1037 (1977).

747.   D. E. Ames and D. Bull, Tetrahedron, 38, 383 (1982).

748.   R. F. Heck, Acc. Chem. Res., 12, 146 (1979), Adv. Chem. Ser., 196, 213 (1982).

749.   H. A. Dieck and R. F. Heck, J. Amer. Chem. Soc., 96, 1133 (1974).

750.   N. A. Cortese, C. B. Ziegler, B. J. Hrnjez and R. F. Heck, J. Org. Chem., 43, 2952 (1978).

751.   B. A. Patel, J. E. Dickerson and R. F. Heck, J. Org. Chem., 43 5018 (1978).

752.   F. E. Ziegler, U. R. Chakraborty and R. B. Weisenfeld, Tetrahedron 37, 4035 (1981).

753.   K. Kikukawa, S. Takamura, H. Hirayama, H. Namiki, F. Wada and T. Matsuda, Chem. Lett., 511 (1980).

754.   A. Hallberg, L. Westfelt and B. Holm, J. Org. Chem., 46, 5414 (1981).

755.   I. Arai and G. D. Daves, J. Org. Chem., 44, 21 (1979).

756.   C. B. Ziegler and R. F. Heck, J. Org. Chem., 43, 2949 (1978).

757.   P. Y. Johnson and J. Q. Wen, J. Org. Chem., 46, 2767 (1981).

758.   J. B. Melpolder and R. F. Heck, J. Org. Chem., 41, 265 (1976).

759.   Y. Tamaru, Y. Yamada, T. Arimoto and Z-i. Yoshida, Chem. Lett., 975 (1978).

760.   Z-i. Yoshida, Y. Yamada and Y. Tamaru, Chem. Lett., 423 (1977).

761.   W. Smadja, S. Czernecki, G. Ville and C. Georgoulis, Tetrahedron Lett., 22, 2479 (1981).

762.   T. Hirao, J. Enda, Y. Ohshiro and T. Agawa, Chem. Lett., 403 (1981).

763.   L-C. Kao, F. G. Stakem, B. A. Patel and R. F. Heck, J. Org. Chem., 47, 1267 (1982).

764.   I. D. Webb and G. T. Borcherdt, J. Amer. Chem. Soc., 73, 2654 (1951).

765. O. P. Vig, J. Ind. Chem. Soc., 59, 609 (1982).

766. I. T. Storie and F. Sondheimer, Tetrahedron Lett., 4567 (1978).

767. E. J. Corey and E. Hamanaka, J. Amer. Chem. Soc., 86, 1641 (1964).

768. P. M. Reijnders, J. F. Blankert and H. M. Buck, Rec., 97, 30 (1978).

9 Alkene metathesis

9.1 Introduction

9.2 The Catalysts

9.3 The Mechanism of Metathesis

9.4 Initiation of Metathesis; the Formation of a Metal Carbene

9.5 Propagation of Metathesis - Selectivity between Alkenes

9.6 Stereochemical Implications of Metathesis

9.7 Cyclopropanes

9.8 Alkyne Metathesis

9.9 Synthetic Applications of Metathesis

9.1 <u>Introduction</u>

This reaction was discovered by Banks and Bailey in 1964[1] and has been well studied and well used. It involves the interchange of groups on double bonds in the presence of a catalyst (<u>1</u> to <u>2</u> and <u>3</u>). Most simple alkenes are substrates, together with some bearing polar functional groups. Catalysts are usually complexes of Mo, W or Re. The reaction is very fast with turnover numbers up to 1000 sec^{-1} in the most favourable cases. Little selectivity is found if the reaction goes to completion with proportions of alkenes reflecting their thermodynamic stabilities.

$$2RCH=CHR' \rightleftharpoons RCH=CHR \quad + \quad R'CH=CHR'$$

$$\underline{1} \qquad\qquad \underline{2} \qquad\qquad \underline{3}$$

9.2 <u>The Catalysts</u>

Methathesis occurs with homo- and heterogeneous catalysts. The heterogeneous catalysts are mainly modified metal oxides[2], for example, $Mo(CO)_6$ or Re_2O_7 with $\underline{\gamma}-Al_2O_3$[3] and are outside the scope of this work. The term homogeneous will be used to refer to catalyst systems formed from interaction of soluble components, but it should be noted that in some *cases solid is* deposited and the true catalyst may not be soluble.[4] Also many of the *solutions are highly* coloured, making determination of their solubility difficult. A small class of catalysts are those supported on modified polystyrene, but which are structurally related to the homogeneous systems.[5-7] The elements whose complexes display the highest activity are Mo, W, Re and Ti and the procedures involved in catalyst generation are diverse. Some examples are given in Table 1. Precise methods of preparation differ and accurately reproducible results are difficult to obtain. In particular, the activity of many systems depends on the presence of a trace of oxygen; the normal precautions (septa, syringes and gas streams) were inadequate to exclude this.[8]

9.3 The Mechanism of Metathesis

It was first necessary to establish which bonds were made and broken. Reaction might take place by cleavage of C—R (in 1) or C=C bonds. In the degenerate metathesis of a mixture of 2-butene, 4, and perdeutero-2-butene, 5, only products with zero, four and eight deuterium atoms were obtained, this being consistent with double bond cleavage.[40] The two cleavages needed might proceed simultaneously or consecutively. Early speculations favoured simultaneous cleavage with intermediates variously proposed as cyclobutane metal complexes,[41] metallocyclopentanes[42] and tetracarbenes.[43] Calculations and orbital symmetry considerations sug-gested that concerted exchange via a cyclobutane complex might be favour-able.[44] However, cyclobutanes are not normally observed in metathesis reactions, a sole exception being provided by the conversion 7 to 8.[45] The cyclobutanes, 9 and 11, revert to dienes with the same catalyst but other cyclobutanes are unreactive under metathesis conditions.[46]

Table I Alkene Metathesis Catalysts

Catalyst	Comments	Ref.
WCl_6/R_4Sn	R=Me, Et, Bu, Ph	9-12
$WCl_6/EtAlCl_2$	CO,ROH beneficial	13,14
WCl_6/Me_2Zn		15
$WCl_6/LiAlH_4$	Other hydride reductants also effective	16,17
$WCl_6/\underline{n}\text{-BuLi}$	May be heterogeneous	4,18
WCl_6/C_5H_5Na	PhC_2Na, RMgBr, R_3Al,R_3B also effective	19-21
$W(CO)_6/RAlCl_2/O_2$	Oxygen free catalyst is inactive	8
$W(CO)_6 \xrightarrow[CCl_4]{h\nu}$		22
$(CO)_5W=CPh_2$		23
$W(OPh)_6/EtAlCl_2$	$WO(OPh)_4$ also active	8,24
$W(CO)_5L/EtAlCl_2/O_2$	$L=PPh_3$, CO,MeCN	25
$Mo(CO)_5py/RAlCl_2/Bu_4N^+Cl$		26
$MoCl_2(NO)_2L_2/EtAlCl_2$	L=py, PPh_3	27,28,29
$MoN_2(PPh_3)_2PhMe$		30
$MoCl(NO)(CO)_2(PPh_3)_2/RAlCl_2$	R=Et, Me	31
$ReCl_5/Et_3Al/O_2$	rather inactive	32
$ReCl_4(PPh_3)/EtAlCl_2$	rather inactive	33
$Ir_2\mu Cl_2(cyclooctene)_4$	only for strained alkenes	34
$Rh(PPh_3)_2LCl$	$L=PPh_3$, CO; only for electron rich alkenes	35
$M(CHCMe_3)(O\underline{t}Bu)_2XL$	M=Nb, Ta; L= R_3P X=Cl, Br	36
$Cp_2Ti\langle\rangle AlMe_2$ (Cl bridge)		37-39

$$CH_3CH=CHCH_3 \quad + \quad CD_3CD=CDCD_3 \rightleftharpoons CH_3CH=CDCD_3$$

$\underline{4}$ $\quad D_0$	$\underline{5}$ $\quad D_8$	$\underline{6}$ $\quad D_4$

Metallocyclopentanes and metal alkene complexes are known to be interconvertable, providing a model for the first step of metathesis[47] (Figure 1). Tungstacyctopentanes could not, however, be generated under metathesis conditions[42] and tantalocyclopentanes are dimerisation rather than metathesis catalysts.[48] The preparation of a tungstacyclobutane was unsuccessful but metathesis did occur, providing the first clue that these might be important intermediates in the reaction.[49]

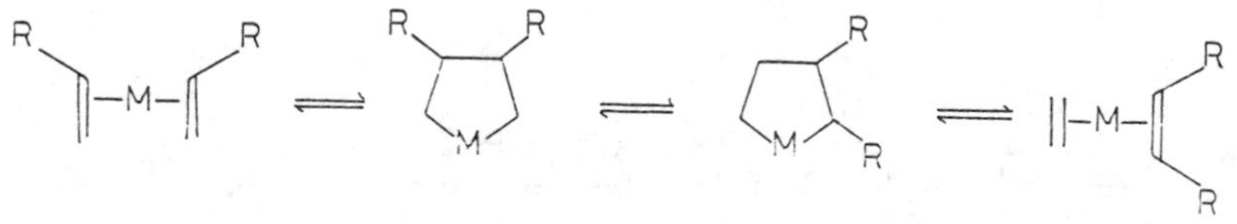

Figure 1 Proposed mechanism of metathesis via a metallocyclopentane.

The mechanism now accepted involves a metal carbene and a metallocyclobutane[50] (Figure 2). A considerable body of evidence supports this scheme. Early work showed that a number of stable metal carbene complexes are metathesis catalysts. The rhodium complex, 16, is isolated from the mixture obtained from metathesis of 13 and 14.[35] Casey showed

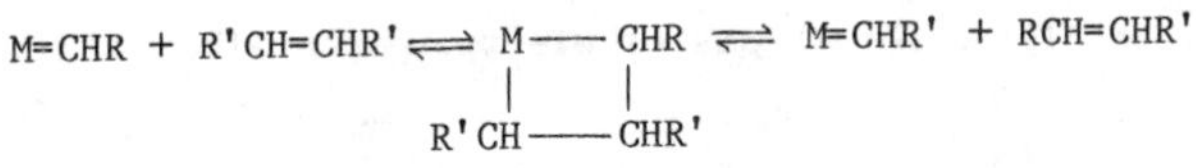

$$M{=}CHR + R'CH{=}CHR' \rightleftharpoons M{-}CHR \rightleftharpoons M{=}CHR' + RCH{=}CHR'$$
$$R'CH{-}CHR'$$

Figure 2 Mechanism of alkene metathesis.

that $(CO)_5W{=}CPh_2$ could undergo alkylidene exchange with alkenes[51] and it was subsequently found to be a sluggish metathesis catalyst.[23,52] The problem seems to be loss of CO to provide a site for alkene coordination. This is supported by a study of the reactivity of $(CO)_5W{=}C(OEt)C_4H_9$, which is inert until $TiCl_4$ is added when CO is evolved and metathesis begins.[53]

The most conclusive evidence for non-pairwise exchange of alkylidene units comes from a careful study of the initial products of the metathesis of 1,7-octadiene, 17, and its tetradeutero analogue, 18. This reaction is particularly favourable since the products undergo no reverse reaction. Cyclohexene is inert to most metathesis catalysts and undergoes

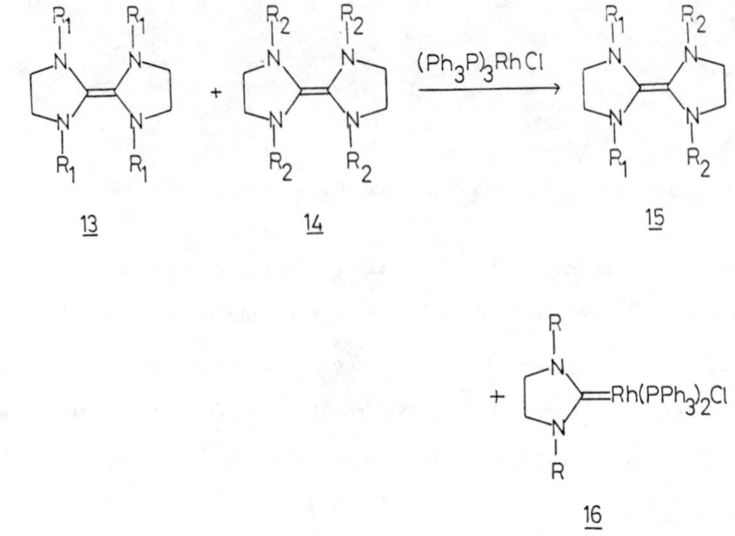

Figure 3   Rhodium catalysed metathesis of electron-rich alkenes.

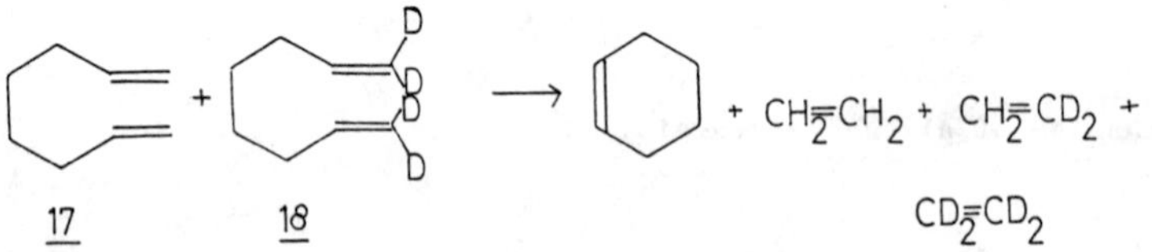

ring opening polymerisation only under special conditions.[54,55] The ratio

of ethylene D_o:D_2:D_4 can be calculated both for pairwise (1:1.6:1) and

carbene mechanisms (1:2:1). The ethylene was analysed by mass spectro-

metry and the ratios found to be in exact accord with those predicted for

the carbene mechanism. Control experiments were rigorous; there was no

scrambling of label in either starting material or products and isotope

effects were shown to be insignificant.[56-58] The mechanism seems to be

general for "homogeneous" catalysts; three separate types were used, all

giving similar results. These represented the main groups known at the time viz. (a) a soluble catalyst not requiring an alkyl metal cocatalyst

(PhWCl$_3$/AlCl$_3$) (b) a soluble catalyst requiring an alkylaluminium cocatalyst

cocatalyst (MoCl$_2$(PR$_3$)$_2$(NO)$_2$/Me$_3$Al$_2$Cl$_3$) and (c) a partially insoluble

catalyst (WCl$_6$/n-BuLi).

More recent work on metathesis mechanisms has used metal carbene

complexes, μ-carbene complexes and metallocycles as catalysts. For

example, a range of stable niobium and tantalum alkylidene complexes were

prepared by Schrock and coworkers.[59] These react stoichiometrically with

alkenes to give alkylidene exchange via metallocycles (e.g. 19 to 22

23 and 24). However, their metathesis reactions are slow by comparison

with "ordinary" catalysts.

Greater progress in understanding this reaction has been made by

a study of the catalyst, 25, formed on reaction of Cp$_2$TiCl$_2$ and 2Me$_3$Al.[60]

It is an extremely active catalyst and reacts by generation of Cp$_2$Ti=CH$_2$

in the presence of a Lewis base. Subsequent addition of alkenes yields

stable metallocyclobutanes, such as 26,[61] which are also good catalysts.

Exchange occurs with other alkenes, the position of the equilibrium

(27 + 28 to 29 + 30) depending on the steric bulk of the R substituents.[62]

25   26

27   28   29   30

The rate determining step in all cases seems to be breakdown of the metallocycle to yield an alkene alkylidene complex,[39] the resting state of the catalyst being the metallocycle.

A number of metathesis catalysts are inactive in a rigorously anaerobic atmosphere, whilst another group contain a stable metal oxygen double bond, which is essential, but does not appear to participate in the reaction. The function of external oxygen would seem to be to generate such a "spectator" oxygen. The role of the oxygen has been convincingly explained by the theoretical studies of Goddard[63] (Figure 4). Addition of an alkene to $Cl_4M=CH_2$ is unfavourable but the reaction with $Cl_2M(=O)=CH_2$ is favourable; a π orbital becomes available to give a partial triple bond to oxygen in the product. This explains why $WOCl_4$ or $((CH_3)_3CH_2)_3WOCl$ are inactive separately but on mixing undergo exchange to yield $Cl_2W(=O)=CH(CH_3)_3$ which is an excellent catalyst. Crystal structural studies have been carried out on several analogues ($\underline{31}$[64] and $\underline{32}$[65]). The reaction of the oxo-complexes is improved by the addition of Lewis acids which complex oxygen and enhance the advantage gained on formation of the third bond.[66,67] Imino spectator ligands are adequate to ensure metathesis but both theory[63] and experiment show them to be less successful than the oxo-species.[67]

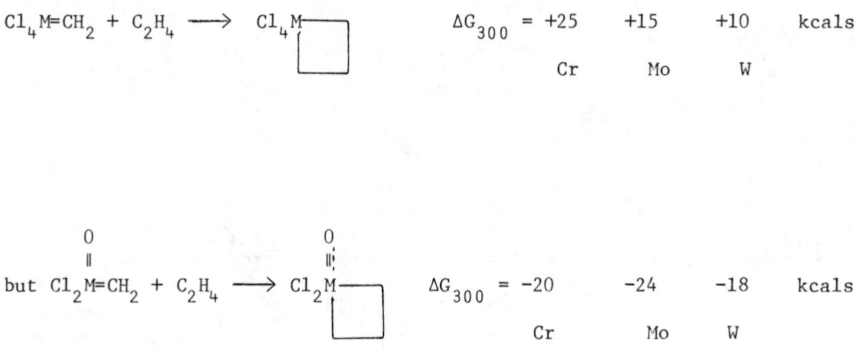

$$Cl_4M{=}CH_2 + C_2H_4 \longrightarrow Cl_4M\rule[-3mm]{0pt}{3mm}$$

ΔG_{300} = +25   +15   +10   kcals

Cr   Mo   W

$$but\ \ Cl_2M{=}CH_2 + C_2H_4 \longrightarrow Cl_2M\rule[-3mm]{0pt}{3mm}$$

ΔG_{300} = -20   -24   -18   kcals

Cr   Mo   W

Figure 4 Thermochemistry of alkene metathesis.

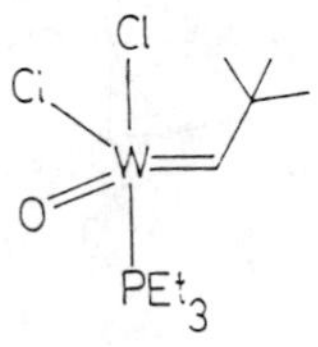

31 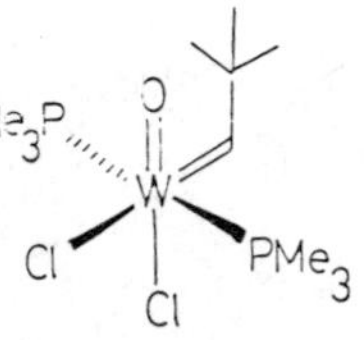   32

9.4 Initiation of Metathesis; the Formation of a Metal Carbene

Whilst the origin of carbene or metallocycle in the previous
groups of catalysts is rather clear, this is not the case for most
metathesis catalysts. Another group of catalysts require organometals as
cocatalysts. Among these are WCl_6/R_2AlCl, WCl_6/R_4Sn, $(Ph_3P)_2Mo(NO)_2Cl_2/$
$Me_3Al_2Cl_3$[68] and $ReCl_4(PPh_3)/EtAlCl_2$[33]. Organometals react with metal
halides to give $\underline{\sigma}$-alkyl complexes which may give carbene hydrides and
alkanes (Figure 5).[15] This $\underline{\alpha}$-hydrogen abstraction has precedents in
tantalum chemistry where it is thought to proceed $\underline{via}$ a radical
route.[69,70] No deuterium is incorporated from deuterated solvent,
precluding this as an origin for the metal hydride. Similarly, on
treatment of $(Ph_3P)_2MoCl_2(NO)_2$ with $Me_3Al_2Cl_3$, methane and ethylene are
evolved and a catalytically active species generated. The ethylene is

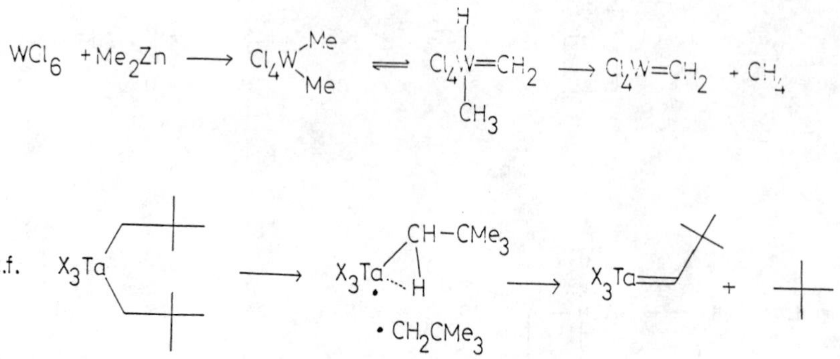

Figure 5 Postulated mechanism for initiation of metathesis.

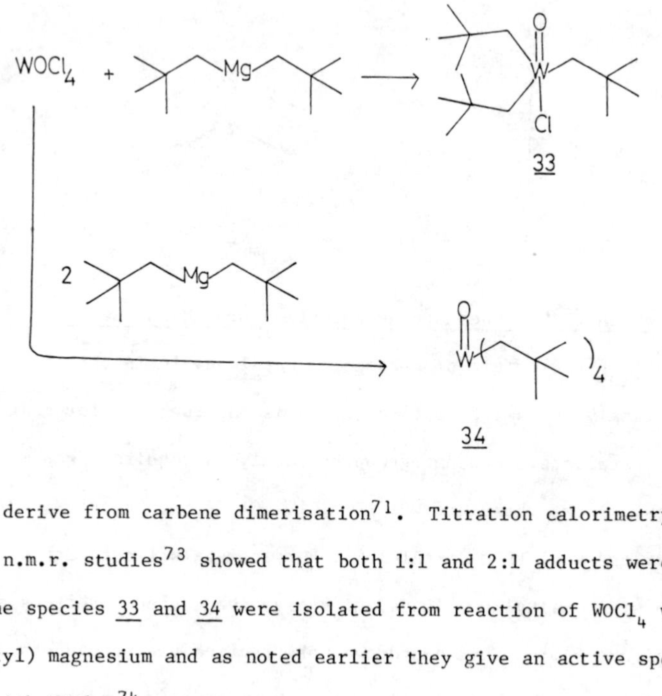

thought to derive from carbene dimerisation[71]. Titration calorimetry,[72] IR and [31]P n.m.r. studies[73] showed that both 1:1 and 2:1 adducts were formed. The species 33 and 34 were isolated from reaction of $WOCl_4$ with bis(neopentyl) magnesium and as noted earlier they give an active species on mixing with $WOCl_4$.[74]

The reaction of catalyst precursors bearing NO or CO ligands with organoaluminium compounds may follow a different path. These are generally metals in a low oxidation state, typically W(0). A Lewis acid role may be proposed for the organoaluminium compound; in the reaction of

$(CO)_5WPPh_3$ with $EtAlCl_2$, CO dissociation is assumed to result leaving a vacant site for alkene coordination.[25,75] An analogy is provided by photolytic dissociation of $W(CO)_6$ to $W(CO)_5$, which is also active. The optimum ratio of Al to W, Mo or Re varies with the system used and two distinct schemes are postulated.

In the reaction of $Mo(CO)_5py$ with $RAlCl_2$, RH and CO are evolved in equal amounts. Reaction of the product (R=Et) with octa-1,7-diene yields 1 mole propene per mole of catalyst. For the catalyst, R=Me, no propene is obtained; the ethylene expected is indistingushable from the metathesis product (Figure 6).[26]

In an alternative scheme migration of the alkyl group from aluminium is to the carbonyl rather than the metal (Figure 7). Its purpose is to account for the formation of oxygenated products and 1-butene in the early stages of octa-1,7-diene metathesis (requiring that the initiating carbene be propylidene rather than ethylidene).[76] Certain molybdenum catalysts follow an analogous route when the solvent is heptane rather than chlorobenzene. The role of the solvent is to stablise the charged species.

The mechanism of carbene formation in reactions not employing alkylaluminium cocatalysts is less easy to discern. Osborn showed that $Mo(C_2H_4)_2(diphos)_2$, 35, undergoes protonation and rapid reversible insertion to give an alkylmetal. With 38, however, the π-allyl hydride, 39, is the stable product.[78] Green[78] showed that π-allyl complexes of W and Mo are converted to metallocyclobutanes by hydride attack at C_2 (Figure 8). The thermal stability of the products was dependent on the substituents; those bearing Ph or Me decomposed faster than the unsubstituted ones, all giving cyclopropanes and alkenes derived from hydrogen migration. On photolysis, however, the products seem to derive from metal carbenes (Figure 9).[79] We may thus propose a route for the direct generation of metallocycles, 42, from alkenes. This implies that

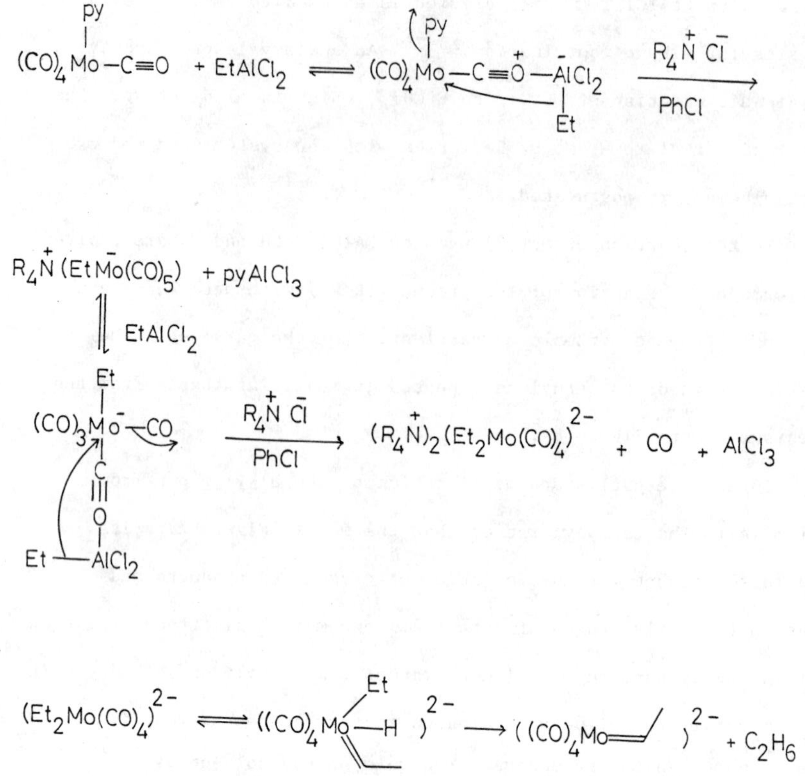

$$(CO)_4 Mo \overset{\overset{\displaystyle py}{|}}{-} C \equiv O \;+\; EtAlCl_2 \;\rightleftharpoons\; (CO)_4 Mo \overset{\overset{\displaystyle py}{|}}{-} C \equiv \overset{+}{O} \overset{-}{-} \underset{\underset{\displaystyle Et}{|}}{\overset{-}{A}lCl_2} \;\xrightarrow[PhCl]{R_4\overset{+}{N}\,\overset{-}{Cl}}$$

$$R_4\overset{+}{N}\,(Et\,\overset{-}{Mo}(CO)_5) \;+\; py\,AlCl_3$$

$$\big\Updownarrow \; EtAlCl_2$$

$$(CO)_3 Mo \underset{\underset{\displaystyle Et-AlCl_2}{|}}{\overset{\overset{\displaystyle Et}{|}}{\overset{-}{}}} \!\!\!\overset{\displaystyle C \atop {\parallel\parallel} \atop O}{\longleftarrow} CO \;\xrightarrow[PhCl]{R_4\overset{+}{N}\,\overset{-}{Cl}}\; (R_4\overset{+}{N})_2 (Et_2 Mo(CO)_4)^{2-} \;+\; CO \;+\; AlCl_3$$

$$(Et_2 Mo(CO)_4)^{2-} \;\rightleftharpoons\; ((CO)_4 Mo\overset{\overset{\displaystyle Et}{|}}{\underset{\diagdown}{-}}H)^{2-} \;\longrightarrow\; ((CO)_4 Mo{=}\!\!\diagup)^{2-} \;+\; C_2H_6$$

Figure 6 Reaction of EtAlCl$_2$ with a molybdenum carbonyl complex.

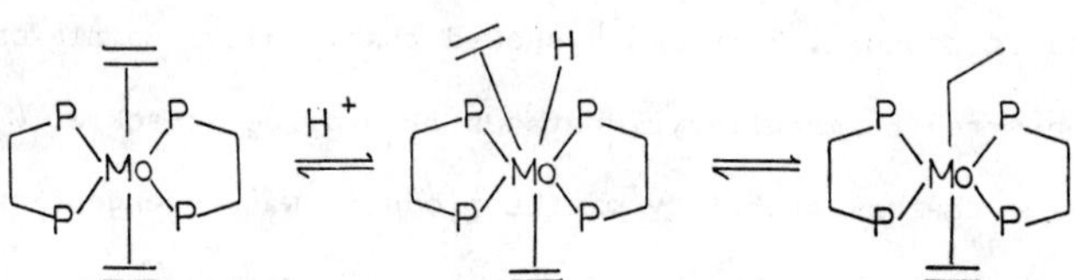

35   36   37

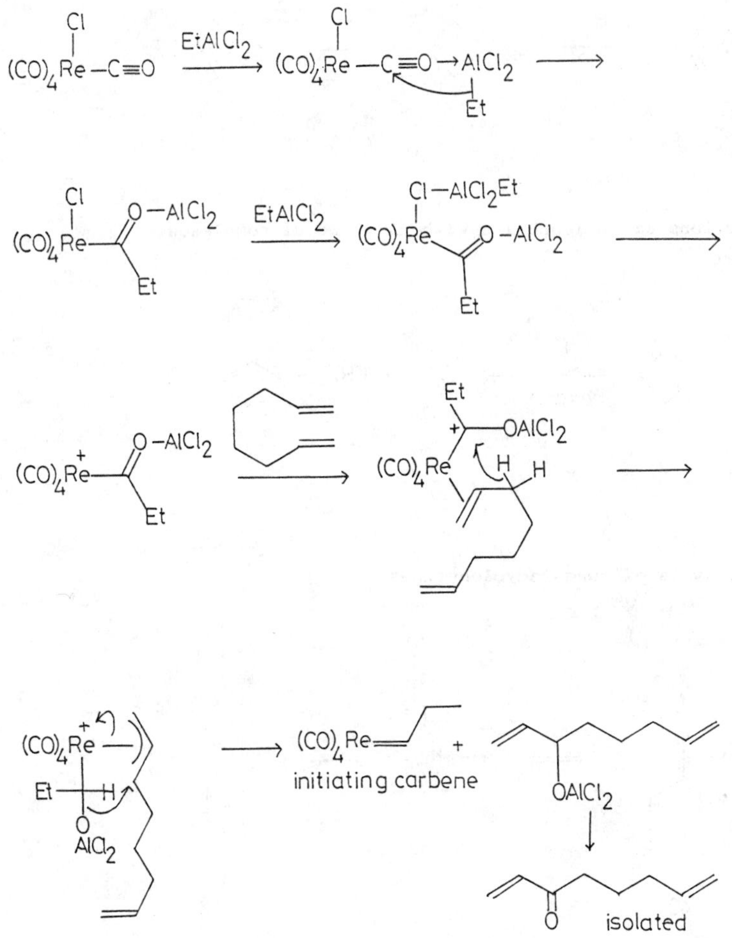

Figure 7 Reaction of EtAlCl$_2$ with a rhenium carbonyl complex.

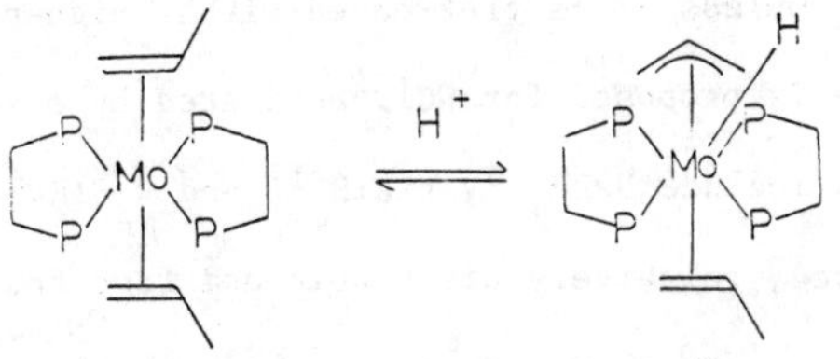

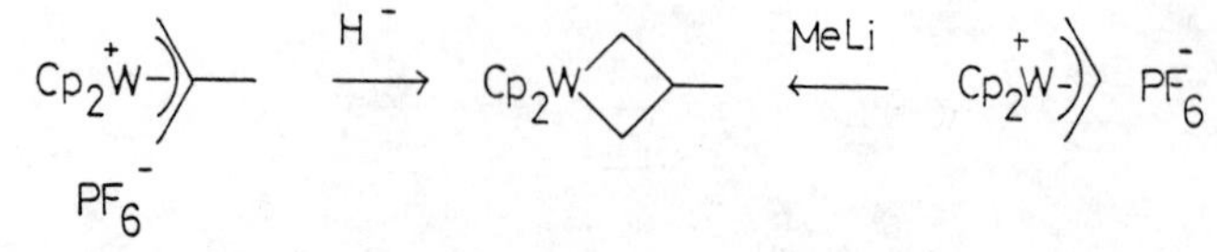

Figure 8 Reactions of nucleophiles with allyl metal complexes.

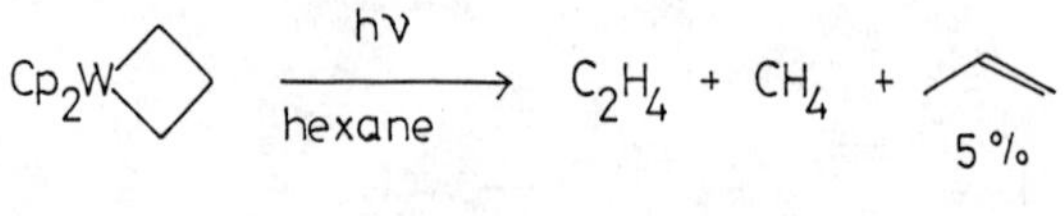

Figure 9 Photolysis of tungstacyclobutanes.

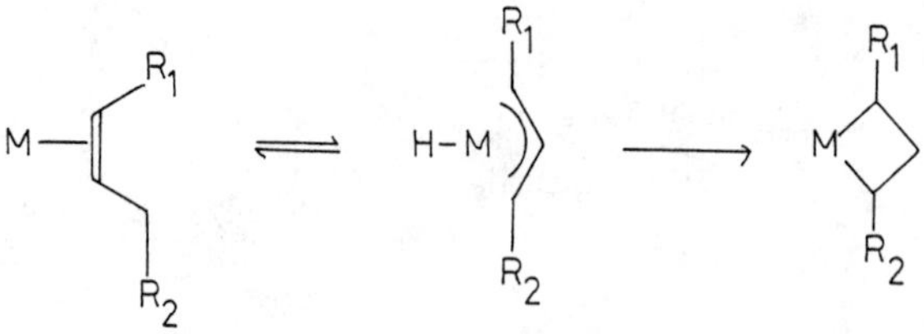

ethylene, which cannot generate the metallocycle should be inert to self metathesis; examples are few but no scrambling of C_2D_4 and C_2H_4 takes place over Re_2O_7/Al_2O_3 unless it is pretreated with a higher alkene.[80]

Another route is proposed for WCl_6 activated by a source of H^-. Successful cocatalysts include $NaBH_4$[16], $LiAlH_4$[17] and R_3SiH.[81] The catalysts are long lived, relatively air stable and free from side reactions.[16] However, there is an induction period which increases with the alkene/W ratio and some alkene is consumed in catalyst activation. It is suggested that W(II) hydrides are the active species (Figure 10).[82]

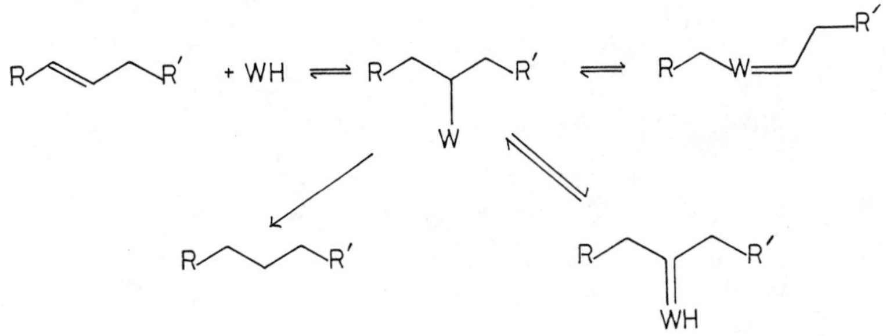

Figure 10 depicts a metal hydride initiation of metathesis scheme.

Figure 10 Metal hydride initiation of metathesis.

$$CH_3C \equiv CCH_3 + WCl_6 \longrightarrow \underset{\underset{WCl_5}{|}}{CH_3-C=CClCH_3} \longrightarrow \underset{\underset{WCl_4}{\|}}{CH_3-C-CCl_2CH_3}$$

<u>43</u>   <u>44</u>

WCl_6 may also be activated by addition of alkynes,[83] the reaction proceeding <u>via</u> species such as <u>43</u> and <u>44</u>.[84]

The route to carbene or metallocycle formation with catalysts formed from transition metal salts and Lewis acids is unclear, though it is likely that reduction of the metal takes place[18] and a good catalyst is obtained by electrolytic reduction of WCl_6.[85] Creation of a vacant coordination site is established in some cases.[86] Metathesis is also catalysed by the species formed on photolysis of $W(CO)_6$ in CCl_4. $W(CO)_5$ is produced initially and reacts fast with CCl_4 to give $W(CO)_5Cl$.[87] Further photolysis yields $W(CO)_4Cl_2$.[88] In the presence of 2-butene small amounts of 1,1-dichloropropene are isolated suggesting that the active species is a tungsten dichlorocarbene complex[89] (Figure 11).

$$W(CO)_6 \xrightarrow{h\nu} W(CO)_5 + CO$$

$$W(CO)_5 + CCl_4 \longrightarrow W(CO)_5Cl + \cdot CCl_3$$

$$W(CO)_5Cl + \cdot CCl_3 \longrightarrow W(CO)_4Cl_2 + :CCl_2 + CO$$

$$W(CO)_4Cl_2 \rightleftharpoons W(CO)_3Cl_2 + CO$$

$$W(CO)_3Cl_2 + :CCl_2 \longrightarrow (CO)_3Cl_2W = CCl_2$$

Figure 11 Formation of a metathesis catalyst by photolysis.

9.5 Propagation of Metathesis-Selectivity Between Alkenes

The chemistry of stable metal carbene complexes is frequently reviewed[90],[91] and only features relevant to metathesis will be discussed here. Three resonance forms 45a–c may be considered. The contribution of each form depends on the oxidation state of the metal, the steric and electronic nature of the ligands and the substituents R_1 and R_2. The first carbene complexes, prepared by Fischer[92] are stabilised by an electron donor R_1 (usually OMe) and are best described in terms of 45c. However, stable tantalum and niobium carbenes such as 46 undergo Wittig type reactions and more closely resemble 45a.[93]

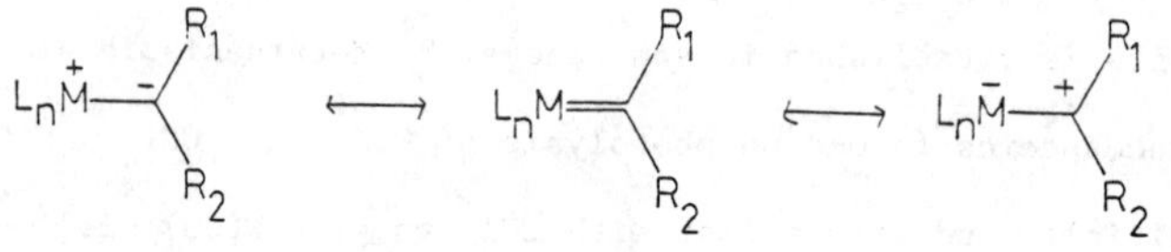

45a   45b   45c

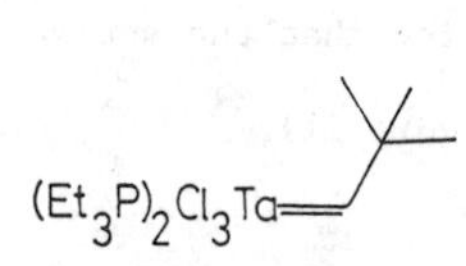

46

Consider the decomposition of metallocyclobutane, 47, formed from a methylene complex and a terminal alkene (Figure 12). If the polarization of the carbene complex resembles that of 45c, 49 would be the favoured carbene, since substitution would stabilise the positive centre. If 45a is the more important contributer, however, 48 will be the more stable carbene. The nature of the resting state of the catalyst is not invariably clear so that studies of the relative stabilities of metallocycles are also pertinent.[61,62] The stability of the metallocycle increases with the steric bulk of the substituents at the β-position; cleavage involves a transition state where the relief of steric strain due to the rocking of the β-methylene is lost.

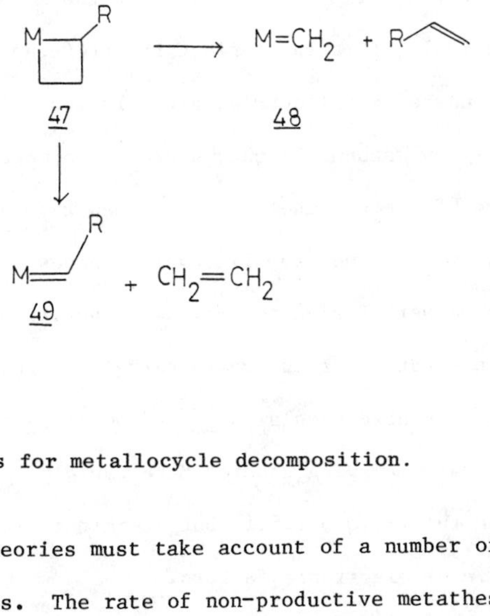

Figure 12   Pathways for metallocycle decomposition.

These theories must take account of a number of well established experimental facts.   The rate of non-productive metathesis of terminal alkenes through exchange of methylene groups is up to 1000 times faster than productive metathesis to ethylene and internal alkenes.[52,60,94] This in turn is usually faster than cross metathesis of a terminal alkene with an internal alkene[95] and reactions of internal alkenes are slower still.[96]   Casey studied the reaction of $(CO)_5W=C(\underline{p}-tolyl)_2$, 50, with alkenes to see which alkylidene fragment is transferred to the carbene.

Isolation of 51 as the major product on reaction with isobutene was interpreted in terms of a preference for the formation of the more substituted metal carbene. Alternatively, this may reflect the preference for the formation of the less hindered metallocycle, 52. Platinocyclobutanes are known to rearrange easily by mechanisms unconnected with metathesis.[97,98] The electrophilic nature of the carbenes was confirmed by studying the reaction of $(CO)_5W=CHPh$ with alkenes to give cyclopropanes. The reaction rate increases with the number of electron releasing substituents on the double bond.[99] Direct observation of the chain carrying carbenes was recently achieved for the related complexes 53 and 54[66]; 54 is the more active catalyst giving chain carrying species more rapidly. The ratios for R=Me, Et, iPr are in the ratio 6:1.5:1 but it is not clear if the effects are due to steric or electronic factors.

These data all indicate that M=CHR rather than $M=CH_2$ is the chain carrier. Further support comes from a study of the cross metathesis of 1,1-dideutero-1-octene and Z-1-deutero-1-decene. If the chain carrier were $M=CH_2$, Z-and E-1-deutero-1-octene would be produced in equal amounts (Figure 13, Sequence A). In Sequence B, however, the diequatorially substituted cyclobutane, 55, is thought to be preferred and yields

Sequence A

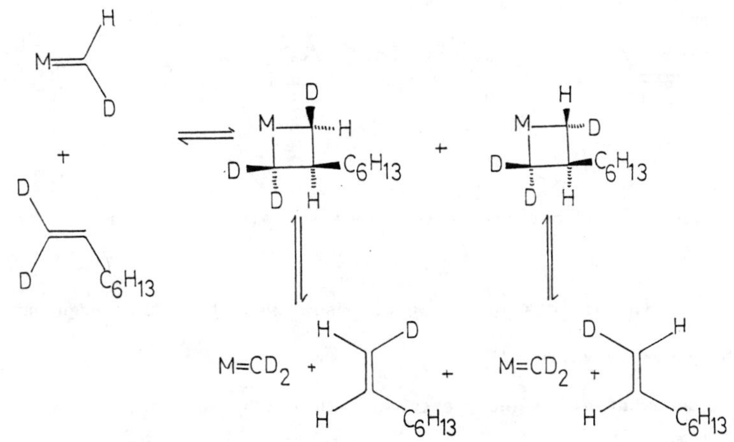

Sequence B

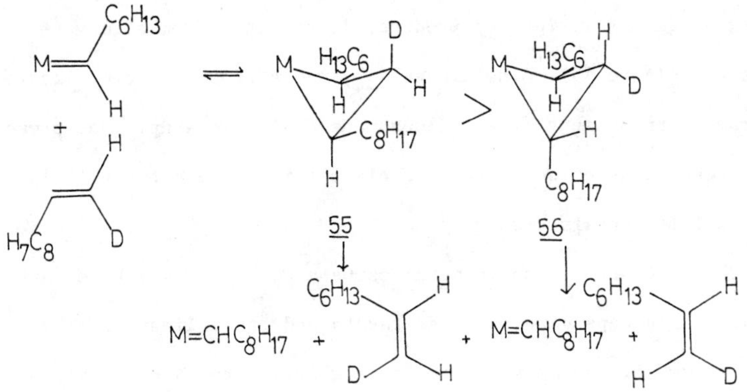

Figure 13 Schemes for the metathesis of 1,1-dideutero-1-octene and Z-1-deutero-1-decene.

Z-1-deutero-1-octene. With $MoCl_2(NO)_2(PPh_3)_2/Me_3Al_2Cl_3$ as catalyst, Quenching the reaction after 0.7% productive metathesis gave a Z:E ratio of 2 for 1-octene-d_1. This is interpreted in favour of substituted

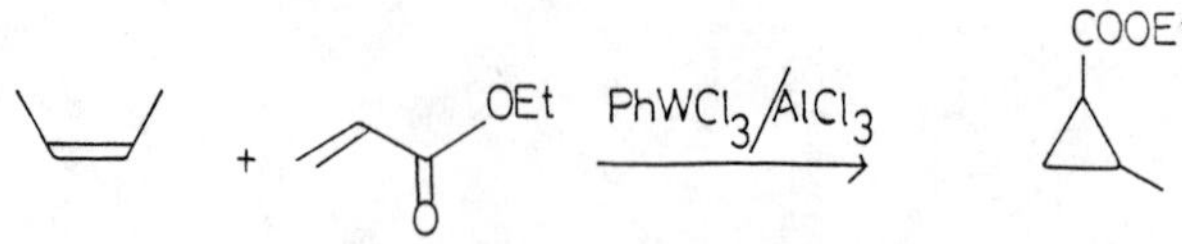

<u>Figure 14</u> Reaction of metathesis intermediates with Michael acceptors.

carbenes as chain carriers but as we will see later that this arguement

may not be entirely valid.[10]

A somewhat different system was studied by Gassman. It was shown

that Michael acceptors could intercept one-carbon fragments in meta-

thesis[101] (Figure 14). However, yields were low. The reaction of enol

ethers with $(CO)_5W=CPh_2$ gives rise to $(CO)_5W=CH(OEt)$ and $Ph_2C=CH_2$ by

carbene exchange. A variety of products is obtained from ethoxy

cyclopentene (Figure 15). The alkene, <u>57</u>, comes from a Wittig reaction

and <u>59</u> from rearrangement of a cyclopropane during work-up. The presence

of <u>57</u> is regarded as evidence for a nucleophilic carbene but this is far

from a normal metathesis system.[102]

Charge distribution in metal carbene complexes may have varying

bond polarisation depending on substituents and other ligands. The

stability of metallocycles seems to be largely goverened by steric

factors. We may expect that the balances controlling selectivity are

rather fine ones.

9.6 Stereochemical Implications of Metathesis

If metathesis proceeds to completion the ratio of alkenes closely

reflects their thermodynamic stabilities. <u>Cis</u>- and <u>trans</u>-alkenes are

produced, the <u>trans</u> normally predominating at equilibrium and the reaction

providing a route for <u>cis-trans</u> isomerisation.[104] However, at low

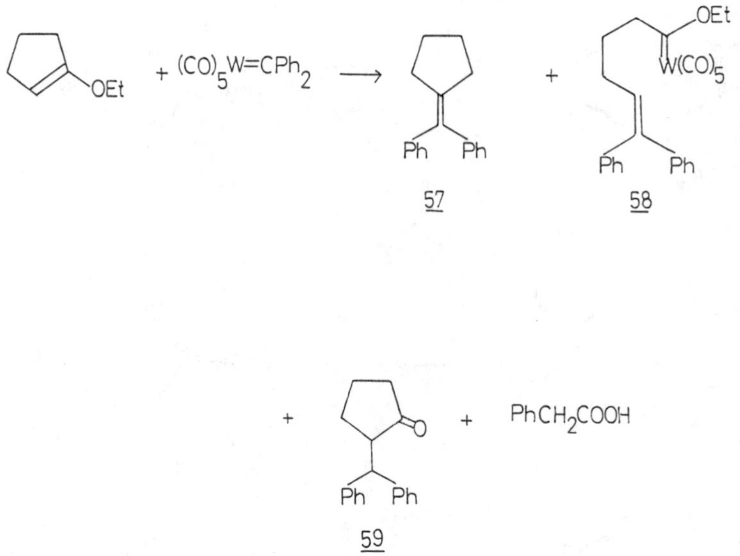

Figure 15 Reaction of $(OC)_5W=CPh_2$ with ethoxycyclopentene.

conversion considerable stereoselectivity is noted. For example, in the metathesis of 2-pentene by $Mo(NO)_2pyCl_2/EtAlCl_2$[103] the cis-isomer gives a ratio of cis:trans-2-butene of 4.4 and 3-hexene 2.3 after 2 minutes. The corresponding trans-isomer gives cis:trans-2-butene = 0.1 and 3-hexene < 0.05 at the same time. Similar, if less striking, results are obtained with tungsten catalysts.[12,23,28,105] Basset developed a very detailed model to account for these observations.[106] In this addition of a disubstituted unsymmetrical alkene to a carbene leads to a metallocyclobutane of predetermined structure. The interaction of cis-2-pentene with an ethylidene tungsten complex is represented by reactions 1 to 4 (Figure 16). They suggested that for $W(CO)_5PPh_3/EtAlCl_2/O_2$ as catalyst certain pathways were marginally preferred for the cis-alkene but steric interactions were greater in the trans-case, in accord with the observed results.

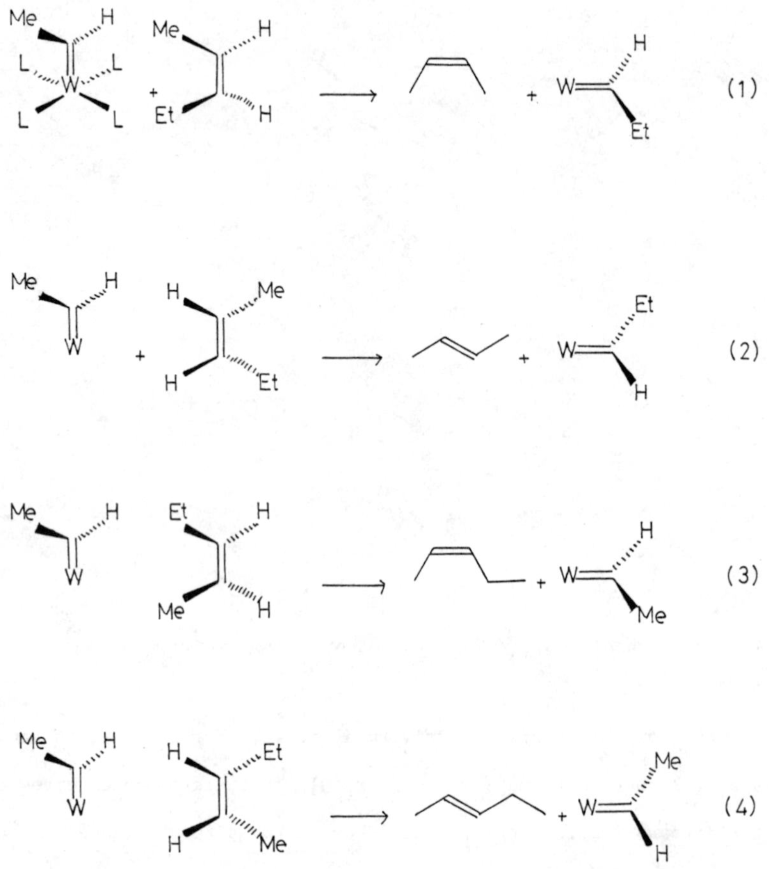

942

Figure 16 Approach of cis-2-pentene to a tungsten ethylidene complex.

Many authors adopted an alternative model considering steric interactions in the intermediate metallocyclobutane which they treated as very puckered. Thus, 60a is slightly preferred over 60b but 61a is strongly preferred over 61b. The scheme rationalises the observed results for pentene but more recent data show that the argument is specious. Stable titanocyclobutanes, 62, were studied crystallographically; they are clearly flat[108] and are not distorted towards an alkene-carbene complex as

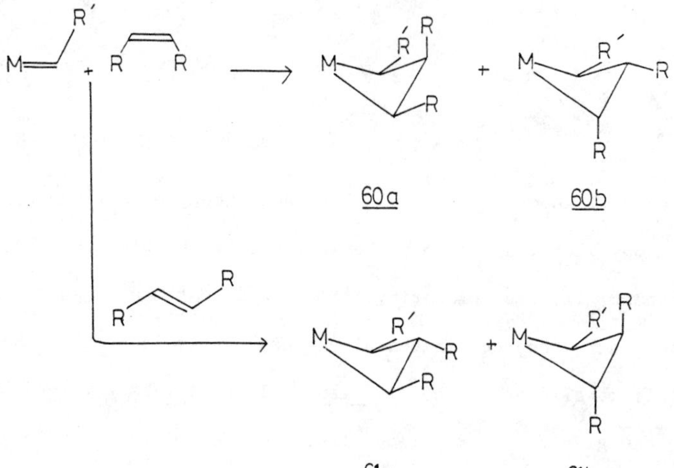

60a   60b

61a   61b

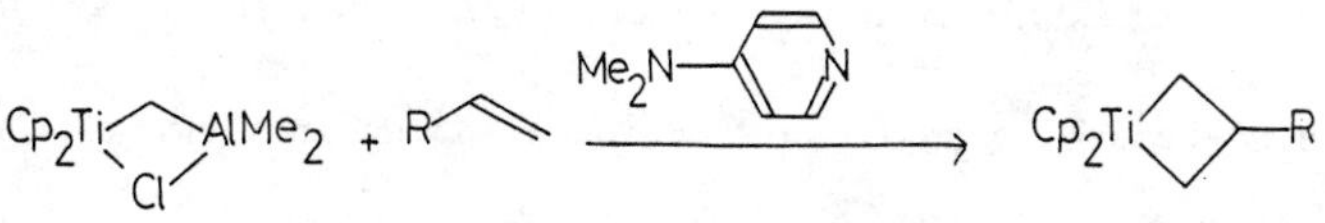

62

calculations had suggested.[109] In the presence of Me_2AlCl rapid cis-trans-isomerisation of α-substutents occurs (63 to 64).[39] The system is a metathesis catalyst and shows impressive selectivity in the reaction of cis, cis-2,8-decadiene, 65. In the 2-butene produced the ratio of cis:trans-isomers falls from 35 after 35 minutes to 13 after 285 minutes. Extrapolation to more active systems may be premature but in this case the ring is known to be flat and the origin of the stereoselection is unclear.

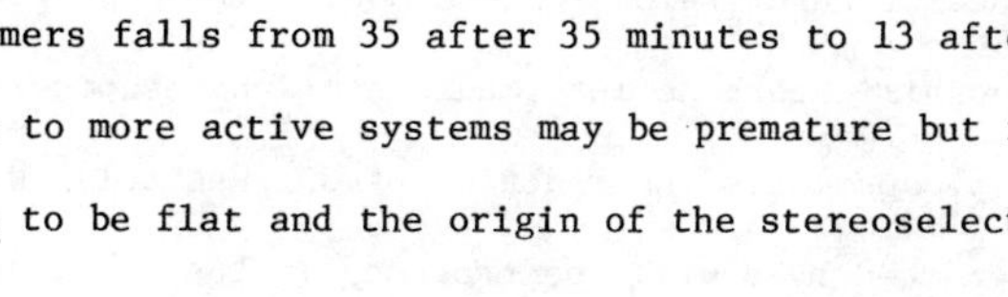

63   64

Another titanocycle, 66, has also been shown to be flat[110] but relatively floppy with respect to puckering, whilst the platinocycle, 67, deviates substantially from planarity.[111] The latest theory suggests that for cases where the coordinated alkene has energy lower than that of either possible metallocycle, the cis/trans selectivity is governed by the relative energies of the metallocycles. However, if the alkene complex is higher in energy, stereoselection is lost.[29] Yet another model favours bimetallic intermediates with the carbene on one metal and the alkene on the other.[22,112] This is very clearly an area of current confusion with no generally applicable rationalisation in view.

9.7 Cyclopropanes

Formation of cyclopropanes is rare in metathesis despite the fact that some zerovalent carbene complexes react readily with alkenes. Cyclopropane is formed in trace amounts in ethylene metathesis[113] and traces of 68 were isolated (after hydrolysis) from metathesis of cis-2-butene in the presence of $PhWCl_3/AlCl_3/CH_2$=CHCOOEt. This was cited as evidence for a nucleophilic carbene intermediate[101] and as previously noted 49 is a minor product from the reaction of$(CO)_5W$=C(tol)$_2$ with isobutene. 70 is detected by n.m.r. spectroscopy in the reaction of $(CO)_5W$=C(OMe)p-tolyl with CH_2=CH(CH_2)$_2$OH in the presence of base. At 35° it collapses to 71.[114,115] A comparable complex formed from allylamine, 72, is isolable and has been studied by X-ray crystallography.[116] Two reasons may be advanced for the lack of a stable metallocycle in this series. Firstly, in contrast to the titanium complexes, this would be a

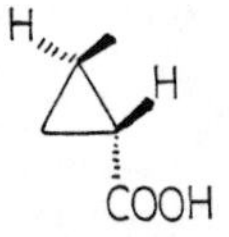

68

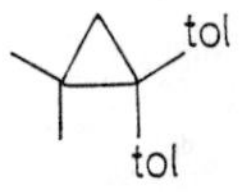

69

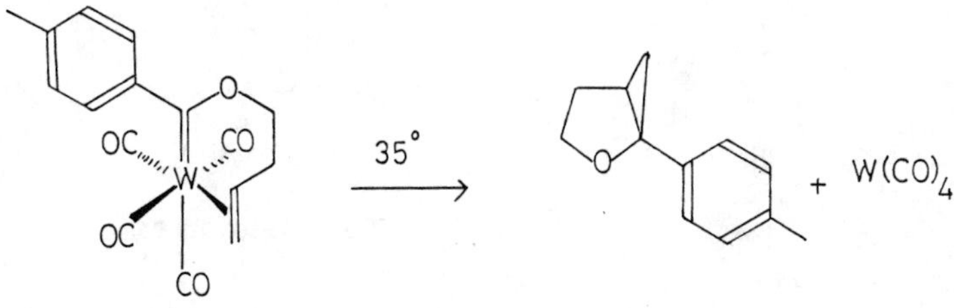

35°

70

71

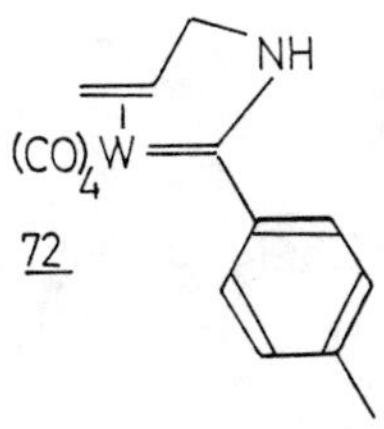

72

20e rather than an 18e complex (Figure 17) and secondly the X-ray of 72

shows that the carbene and alkene are at right angles, a most unsuitable

arrangement for cycloaddition.

Interconversion of cyclopropanes and metallocyclobutanes is well

documented for platinum complexes as in the conversion of 73 to 75 and 76,

but extrapolation to earlier transition metals is questionable.[97,115]

Methathesis of an alkylcyclopropane with $PhWCl_3/AlCl_3$ gives alkenes and

ethylene but the system is unusual, since although bicyclo [2.1.0] pen-

tane, 77, is opened to cyclobutene in a formal retrocarbene addition, the

cyclobutene does not polymerise.[107] Other alkylcyclopropanes yield

alkenes or rearranged cyclopropanes with $PhWCl_3/EtAlCl_2$ (e.g. 79 to 80) but the mechanism is thought to involve metal hydrides rather than

carbenes.[117]

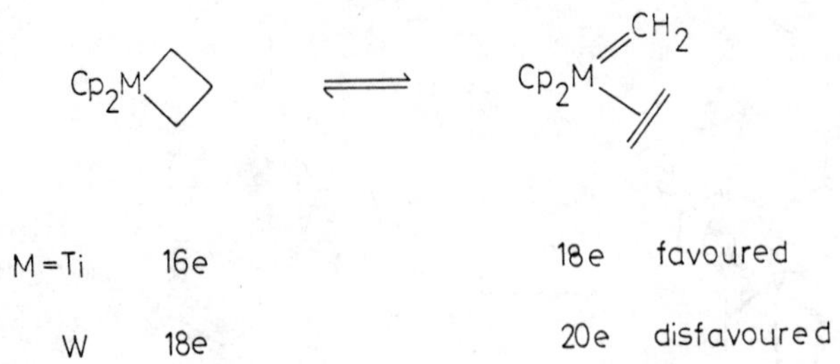

M = Ti	16e	18e	favoured
W	18e	20e	disfavoured

Figure 17 Equilibria between carbene-alkene and metallocyclobutane complexes.

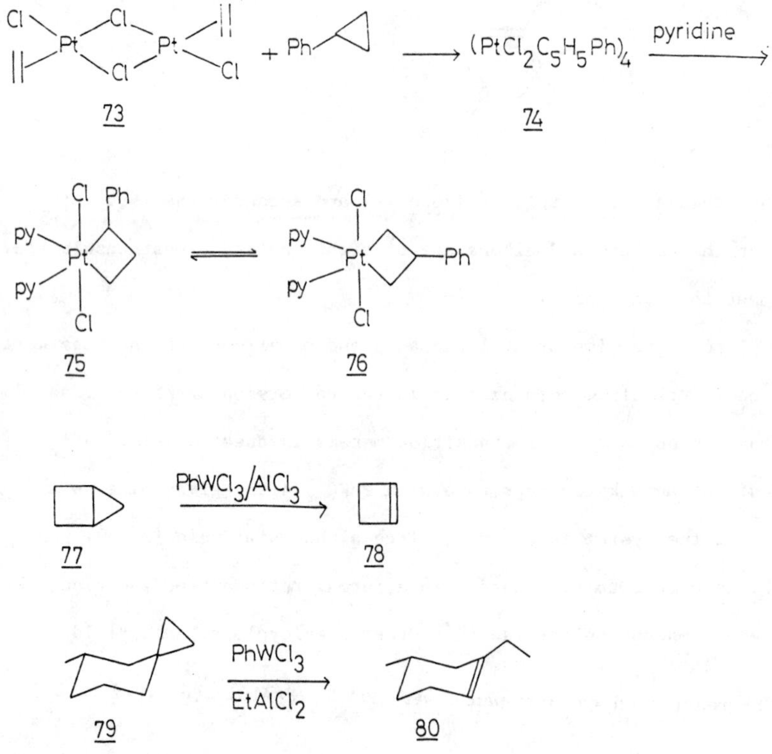

73   74

75   76

77   78

79   80

Mango[118] considered the relative thermodynamic stabilities of the components involved in the interconversions of cyclopropanes and alkenes. He suggested that the results reported (little cyclopropane generated in metathesis and conversion of cyclopropane to ethylene) were contrary to thermodynamic expectations for the carbene-metallocyclobutane scheme. However, Grubbs[119] refuted the arguement by the scheme of Figure 18. In order for catalytic interconversion of cyclopropanes and alkenes to occur, the final conversion of X' to X is essential. The position of this equilibrium is not determined by the relative thermodynamic stabilities of alkenes and cyclopropanes. Thus, while their interconversion may occur under metathesis conditions, it is not on the main reaction pathway.

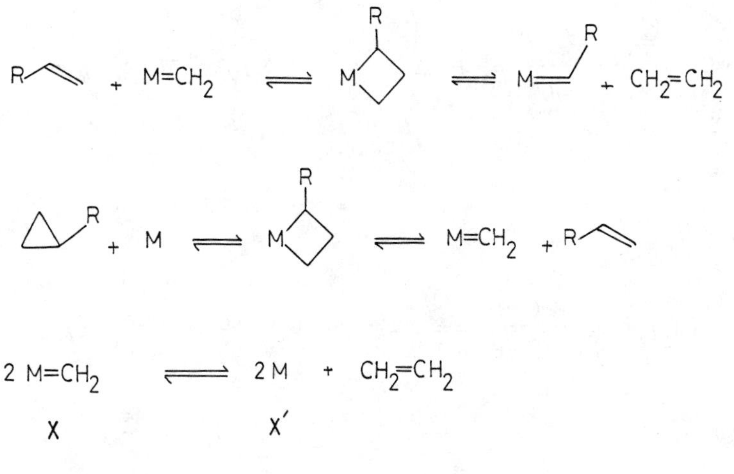

Figure 18   Scheme for the interconversion of cyclopropanes and alkenes under metathesis conditions.

9.8 Alkyne Metathesis

The metathesis of alkynes is less well studied and it only comparatively recently that the non-pairwise nature of the exchange reaction has been proven. Phenyl tolyl acetylene, 81, with a ^{13}C label, is treated with a carbyne complex to yield 81, 82 and 83 with a labelling pattern consistent only with a triple bond splitting mechanism.[120] Photolysis of a mixture of $Mo(CO)_6$, p-chlorophenol and 4-nonyne gives low conversion but rather selective reaction.[121]   Other molybdenum carbonyl complexes including $(nonbornadiene)Mo(CO)_4$, $(cycloheptatriene)Mo(CO)_3$ and $(cyclohexadiene)_2Mo(CO)_2$ are even more active[122], and a similar system was used for the first reported metathesis of a functionalised alkyne[123] (84 to 85 and 86). This has a number of advantages in synthetic schemes

$Y = OH, OAc, Br, COOH, COOMe, CN$

since the cis/trans mixtures of double bonds are avoided. The only report available suggests that alkynes are more reactive towards metathesis than alkenes; 87 reacts selectively at the triple bond in the presence of $Mo(O_2)(acac)_2/AlEt_3/PhOH$.[124] Carbyne complexes such as 88 and 89[126] are also important initiators of alkyne metathesis.

$$CH_2=CHCH_2C\equiv C(CH_2)_3CH_3$$

<u>87</u>

$(\underline{t}\text{-BuO})_3W\equiv C\text{-}\underline{t}\text{-Bu}$ $((\underline{i}\text{-Pr})_2N)_3W\equiv C\text{-}\underline{t}\text{-Bu}$

<u>88</u>   <u>89</u>

Three mechanisms have been proposed for the reaction. The first invokes a metal cyclobutadiene complex such as <u>90</u>. It is based on the fact that flash vacuum pyrolysis of <u>90</u> gives alkynes in good yields.[127] An alternative proposal suggests that the reaction proceeds _via_ similar intermediates to those invoked for alkyne trimerisation [128] (<u>91</u>, <u>92</u> and <u>93</u>). Many alkynes polymerise readily under metathesis conditions.

$$R-C\equiv C-R \quad (89\% \text{ at } 47\% \text{ conversion}$$
$$R=Ph)$$

0.005 sec, 10^{-5} torr 726°

<u>90</u>

The most likely pathway, however, invovles interconversion of a metal carbyne complex and a metallocyclobutadiene in a manner entirely analogous to alkene metathesis[128] (Figure 19). In support of this it has been found that $(\underline{t}BuO)_3W\equiv C\text{-}CMe_3$ catalyses metathesis of 3-heptyne and a ^{13}C n.m.r. of the reaction mixture shows the carbyne complexes $W\equiv C\text{-}Pr$ and $W\equiv C\text{-}Et$.[129] The carbene equivalent, <u>25</u>, reacts irreversibly with diphenyl acetylene to give the metallocyclobutene, <u>94</u>.[108] The metallocyclobutadiene, <u>95</u>,[125,130] has been isolated and its crystal structures determined; it is relatively stable and has a particularly short metal to C-β distance.

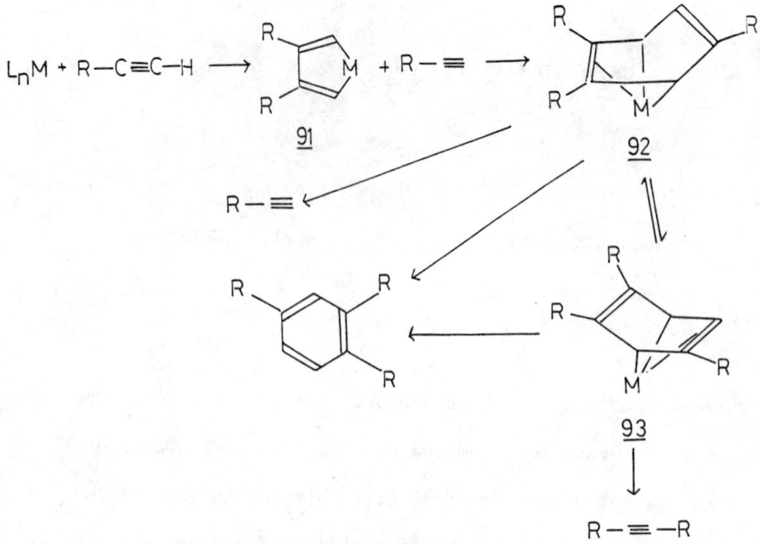

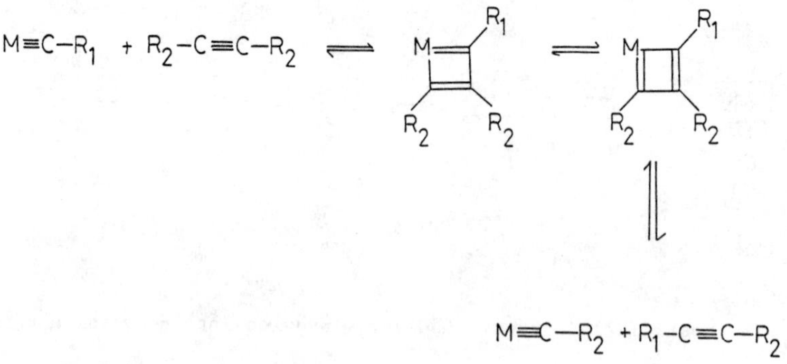

$L_nM + R-C\equiv C-H \longrightarrow$ **91** $+ R-\equiv \longrightarrow$ **92**

$R-\equiv$

93

$R-\equiv-R$

$M\equiv C-R_1 + R_2-C\equiv C-R_2 \rightleftharpoons M \overset{R_1}{\underset{R_2 \quad R_2}{\rightleftharpoons}} M \overset{R_1}{\underset{R_2 \quad R_2}{}}$

$M\equiv C-R_2 + R_1-C\equiv C-R_2$

Figure 19 Postulated mechanism of alkyne metathesis.

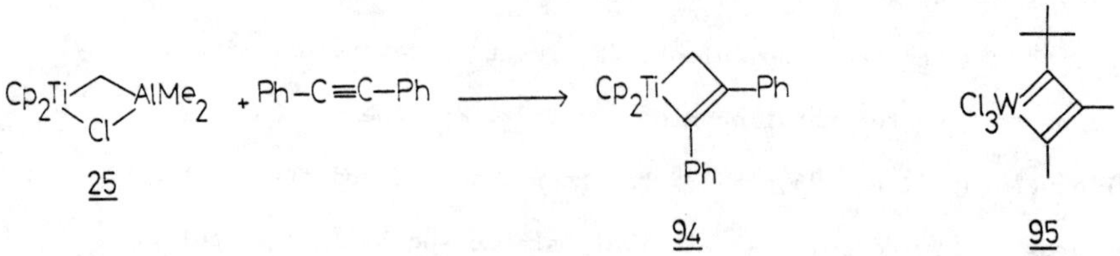

$Cp_2Ti\overset{}{\underset{Cl}{}}AlMe_2 + Ph-C\equiv C-Ph \longrightarrow Cp_2Ti\overset{}{\underset{Ph}{}}Ph$

25

94

Cl_3W **95**

The applications of alkyne metathesis are almost entirely in the field of polymerisation and it is not clear that the mechanism followed is always that of metathesis. Some examples of polymerisation under metathesis conditions are shown in Table 2.

Table 2 Polymerisation of Alkynes by Metathesis Catalysts

Catalyst	Substrate	Polymer	Ref.
$W(CO)_6/h\nu/CCl_4$	$Ph-C\equiv C-H$	10^5 molecular weight methanol insoluble	84
$Cl_4Mo=O$	C_2H_2 or $R-C\equiv C-H$	Conversion ~98% <1 hr, 20°C .01-.001 mol% catalyst	131
WCl_6/Ph_4Sn	$Ph-C\equiv C-H$	10^5 molecular weight in dioxan, 10^4 in ethers. Trans-rich structure	132
WCl_6	$Ph-C\equiv C-H$	Addition H_2O increases rate and mean molecular weight	133,134
$NbCl_5$	$Ph-C\equiv CCH_3$	Molecular weight ~10^6	135

9.9 Synthetic Applications of Metathesis

Metathesis of simple alkenes is potentially useful industrially to balance the alkene production from naphtha cracking. Thus, the Triolefin process, developed by Phillips Petroleum Company[136] is used to convert propylene to ethylene and 2-butenes, the ethylene being of a high purity suitable for polymerisation. In the neohexene process diisobut- ylene, 96, gives neohexene, 97, which is used in the manufacture of a synthetic musk.[2] Other alkene interconversions lead to the stepwise building up of C_6 to C_8- alkenes for plasticiser alcohols and C_{12} to C_{16}- alkenes for surfactants in the "Shell Higher olefins Process".[136,137]

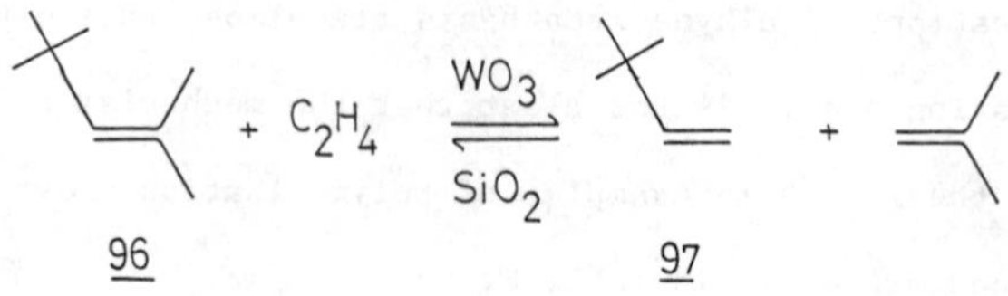

$$96 \qquad\qquad 97$$

Additionally, cometathesis of iso-butene and 2-butene yields iso-amylene, important for the synthesis of polyisoprene.[138] However, the catalysts involved in these reactions are heterogeneous and so fall outside the scope of this work.

Metathesis of cycloalkenes leads to the production of oligomers or high molecular weight polymers. Early evidence for the carbene mechanism was adduced from a study of cyclopentene polymerisation. The polymer produced early in the reaction is of high molecular weight and most of it is linear. This is consistent with a chain growing scheme involving a metal carbene intermediate[14] (Figure 20). Further examples are given in Figure 21.

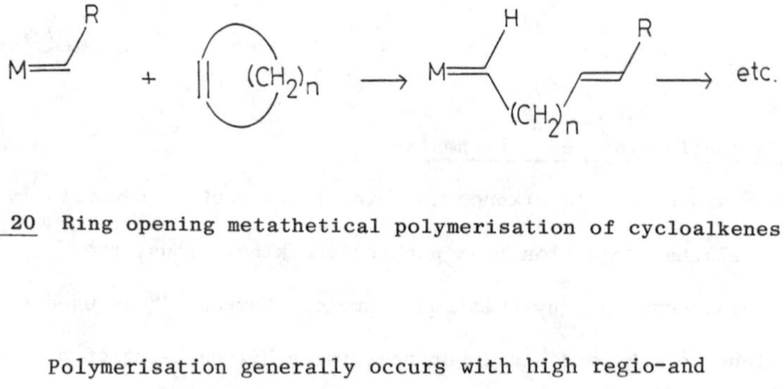

Figure 20 Ring opening metathetical polymerisation of cycloalkenes.

Polymerisation generally occurs with high regio-and stereoselectivity. Polymersation of 1-methylcyclobutene in the presence of $(CO)_5W=CPh_2$ gives a polymer which is largely translationally invariant and has a high selectivity for the formation of cis-double bonds. With 1-trimethylsilylcyclobutene as substrate the polymer is fully translationally and geometrically invariant. Similar cis-selectivity is

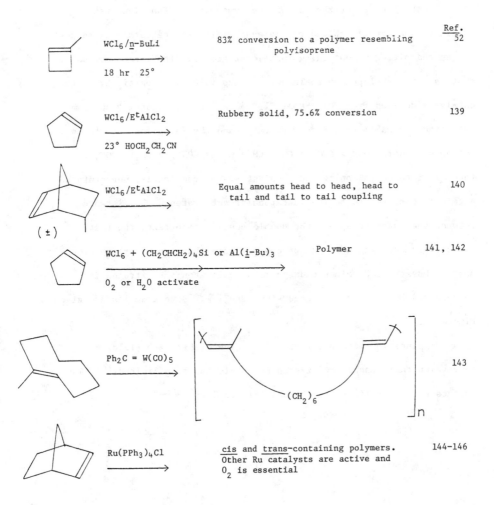

			Ref.
	WCl$_6$/n-BuLi → 18 hr 25°	83% conversion to a polymer resembling polyisoprene	52
	WCl$_6$/E^{t}AlCl$_2$ → 23° HOCH$_2$CH$_2$CN	Rubbery solid, 75.6% conversion	139
(±)	WCl$_6$/E^{t}AlCl$_2$ →	Equal amounts head to head, head to tail and tail to tail coupling	140
	WCl$_6$ + (CH$_2$CHCH$_2$)$_4$Si or Al(i-Bu)$_3$ → O$_2$ or H$_2$O activate	Polymer	141, 142
	Ph$_2$C = W(CO)$_5$ →		143
	Ru(PPh$_3$)$_4$Cl →	cis and trans-containing polymers. Other Ru catalysts are active and O$_2$ is essential	144-146

Figure 21   Examples of cycloalkene metathesis.

found for cyclopentene, cycloheptene, cyclooctene and norbornene with this catalyst.[148] With other catalysts including ReCl$_5$[149] WCl$_6$/Me$_4$Sn[150], MoCl$_5$/Bu$_4$Sn[151] and LW(CO)$_5$[9], norbornene and its substituted analogues give a range of structures.

The use of metathesis for the synthesis of functionalised molecules was long inhibited by the sensitivity of the catalysts to most oxygen and nitrogen containing functional groups. The first successful systems were developed by Boelhouwer.[9] The relative sensitivity of the catalysts derived from $WCl_6/Me_3Al_2Cl_3$ to various functional groups was determined by Nakamura and co-workers[152] and the rates found to be in the following order; COOR > OCOR > OR > CN > COR > $CONR_2$ > NH_2, COOH, OH, these last three poisoning the catalyst system completely. Unsaturated esters have proved to be the most popular substrates. The distance between the ester group and the double bond is important, the best conversions occuring when the groups are more than two methylene groups apart. Unsaturated amines undergo metathesis when the nitrogen is quaternised but yields and conversions are low. Some examples are given Figure 22.

Since few metathesis reactions give either high yields or conversion their use in multi-step synthesis has been limited.[161] Two syntheses of macrolides are shown in Figure 23.[154,162]

$$CH_2=CH(CH_2)_8COOMe \underset{80°/18\ hr/PhCl}{\overset{WCl_6/Me_4Sn}{\rightleftharpoons}} C_2H_4 + MeOOC(CH_2)_8CH=CH(CH_2)_8COOMe \qquad 153,\ 154$$

$$CH_2=CH(CH_2)_8CH_2OAc \overset{WCl_6/Me_4Sn}{\rightleftharpoons} C_2H_4 + AcOCH_2(CH_2)_8CH=CH(CH_2)_8CH_2OAc \qquad 155$$

$$\underset{\underline{cis}}{CH_3(CH_2)_7CH=CH(CH_2)_7COOMe} \underset{Me_4Sn}{\overset{WCl_6}{\rightleftharpoons}} \underset{\underline{cis}}{CH_3(CH_2)_7C=CH(CH_2)_7CH_3} \qquad 156,\ 157$$

$$+\ MeOOC(CH_2)_7CH=CH(CH_2)_7COOMe$$

$$\underline{cis}$$

$$acid,\ \Delta$$

$$\underset{\underline{trans}}{CH_3CH=CH(CH_2)_3\overset{+}{N}Me_3\ I^-} \underset{EtAlCl_2/O_2}{\overset{(Mesitylene)W(CO)_3}{\longrightarrow}} \underset{\underline{trans}}{[Me_3\overset{+}{N}(CH_2)_3CH=CH(CH_2)_3\overset{+}{N}Me_3]2I^-} \qquad 158$$

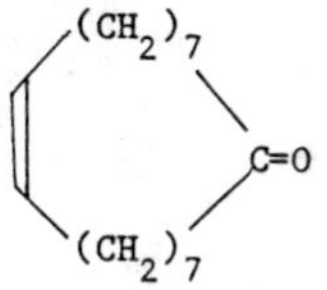

$$CH_2=CH(CH_2)_9OTs \underset{PhCl}{\overset{Me_3SnCl/WCl_6}{\longrightarrow}} 98\%\ \underset{cis:trans\ \sim\ 1:1}{TsO(CH_2)_9CH=CH(CH_2)_9OTs} \qquad 159$$

$$CH_2=CH(CH_2)_nCN \underset{373K}{\overset{WCl_6/Me_4Sn}{\longrightarrow}} C_2H_4 + NC(CH_2)_nCH=CH(CH_2)_nCN \qquad 160$$

$$1 \leqslant n \leqslant 4$$

Figure 22 Metathesis of functionalised alkenes.

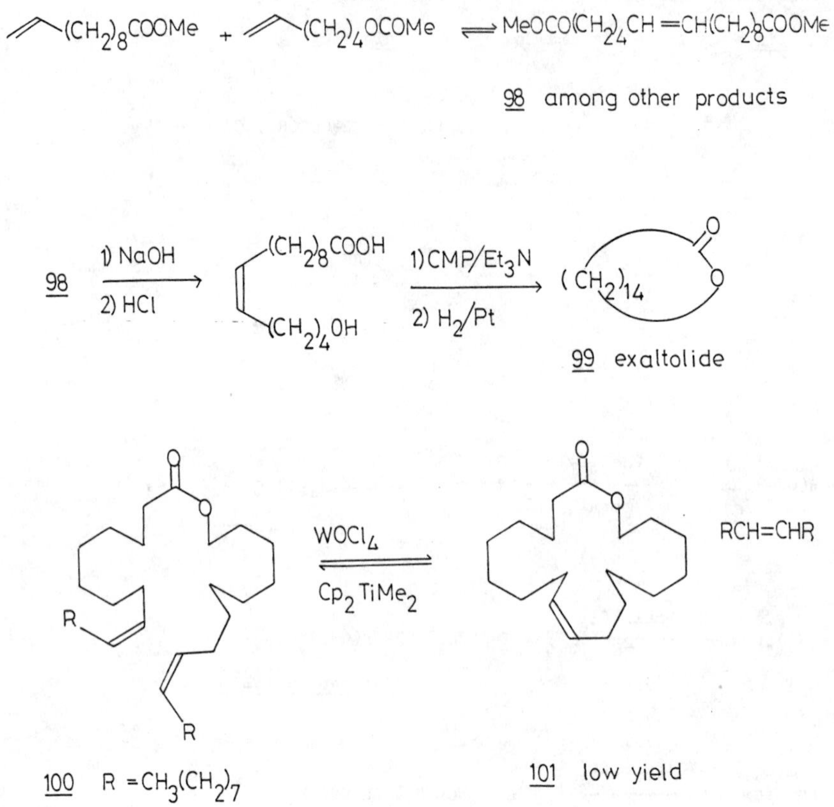

$$\diagup\!\!\diagdown (CH_2)_8COOMe \ + \ \diagup\!\!\diagdown (CH_2)_4OCOMe \ \rightleftharpoons \ MeOCO(CH_2)_4CH=CH(CH_2)_8COOMe$$

$\underline{98}$ among other products

$$\underline{98} \quad \xrightarrow[\text{2) HCl}]{\text{1) NaOH}} \quad \begin{array}{c} (CH_2)_8COOH \\ \\ (CH_2)_4OH \end{array} \quad \xrightarrow[\text{2) } H_2/Pt]{\text{1) CMP/Et}_3N} \quad (CH_2)_{14} \!\!\!\!\! \begin{array}{c} O \\ \diagdown \\ \diagup \end{array}$$

$\underline{99}$ exaltolide

$\underline{100} \quad R = CH_3(CH_2)_7$

$\underline{101}$ low yield

Figure 23   Macrolide synthesis by metathesis.

References

1. R.L. Banks and G.C. Bailey, Ind. Eng. Chem., Prod. Res. Develop., 3, 170 (1964).

2. R.L. Banks, D.S. Banasiak, P.S. Hudson and J.R. Norell, J. Mol. Catal., 15, 21 (1982).

3. R.L. Burwell and R. Brenner, J. Mol. Catal., 1, 77 (1975/76).

4. E.L. Meutterties and M.A. Busch, J. Chem. Soc., Chem. Commun., 754 (1974).

5. S. Warwel and P. Buschmeyer, Angew. Chem., Int. Ed. Engl., 17, 131 (1978).

6. R.H. Grubbs, S. Swetnick and S. C-H. Ha, J. Mol. Catal., 3, 11 (1977/78).

7. S. Tamagaki, R.J. Card and D.C. Neckers, J. Amer. Chem. Soc., 100, 6635 (1978).

8. M.T. Mocella, R. Rovner and E.L. Muetterties, J. Amer. Chem. Soc., 98, 4689 (1976).

9. P.B. van Dam, M.C. Mittelmeijer and C. Boelhouwer, J. Chem. Soc., Chem. Commun., 1221 (1972).

10. R. Rossi, Gazz. Chim. Ital., 106, 1103 (1976).

11. K. Ichikawa, T. Takagi and K. Fukuzumi, Yukagaku, 25, 136 (1976); Chem. Abs., 85, 4789s (1976).

12. R.H.A. Bosma, X.D. Xu and J.C. Mol., J. Mol. Catal., 15, 187 (1982).

13. L. Bencze and L. Markó, J. Organomet. Chem., 28, 271 (1971).

14. N. Calderon, H.Y. Chen and K.W. Scott, Tetrahedron Lett., 3327 (1967).

15. E.L. Muetterties, Inorg. Chem., 14, 951 (1975).

958

16. S.A. Matlin and P.G. Sammes, J. Chem. Soc., Perkin I, 624 (1978).

17. J. Chatt, R.J. Haines and G.J. Leigh, J. Chem. Soc., Chem. Commun., 1202 (1972).

18. P.B. van Dam, M.C. Mittelmeijer and C. Boelhouwer, React. Kinet. Catal. Lett., 1, 486 (1974).

19. K. Ichikawa, T. Takagi and K. Fukuzumi, Bull. Chem. Soc. Jpn., 49, 750 (1976).

20. L. Bencze and L. Marko, Hung. Teljes, 8274 (1974); Chem. Abs., 82, 35379a (1975).

21. R. Nakamura, S. Fukuhara, S. Matsumoto and K. Komatsu, Chem. Lett., 253 (1976).

22. F. Garnier, P. Krausz and J-E. Dubois, J. Organomet. Chem., 170, 195 (1979).

23. T.J. Katz and W.H. Hersch, Tetrahedron Lett., 585 (1977).

24. A. Uchida, M. Hinenoya and T. Yamamoto, J. Chem. Soc., Dalton, 1089 (1981).

25. J.L. Bihou, J.M. Basset and R. Mutin, J. Organomet. Chem., 87, C4 (1975).

26. V.W. Motz and M.F. Farona, Inorg. Chem., 16, 2545 (1977).

27. R. Taube and K. Seyferth, Z. Chem., 14, 284 (1974).

28. J.M. Basset and M. Leconte, Fund. Res. Hom. Catal., 3, 285 (1979).

29. C. Larroche, J.P. Laval, A. Lattes, M. Leconte, F. Quignard and J.M. Basset, J. Chem. Soc., Chem. Commun., 220 (1983).

30. M. Hidai, T. Tatsumi and Y. Uchida, Bull. Chem. Soc. Jpn., 47, 3177 (1974).

31. K. Seyferth and R. Taube, J. Organomet. Chem., 229, 275 (1982).

32. Y. Uchida, M. Hidai and T. Tatsumi, Bull. Chem. Soc. Jpn., 45, 1158 (1972).

33. E.T. Kittleman and E.A. Zeuch, Fr. Patent 1,561,025 (1969); Chem. Abs., 72, 31193f (1970).

34. R. Rossi, P. Diversi, A. Lucherini and L. Porri, Tetrahedron Lett., 879 (1974).

35. D.J. Cardin, M.J. Doyle and M.F. Lappert, J. Chem. Soc., Chem. Commun., 927 (1972).

36. R. Schrock, S. Rocklage, J. Wengrovius, G. Rupprecht and J. Fellmann, J. Mol. Catal., 8, 73 (1980).

37. U. Klabunde, F.N. Tebbe, G.W. Parshall and R.L. Harlow, J. Mol. Catal., 8, 37 (1980).

38. P.J. Krusic and F.N. Tebbe, Inorg. Chem., 21, 2900 (1982).

39. K.C. Ott, J.B. Lee and R.H. Grubbs, J. Amer. Chem. Soc., 104, 2942 (1982).

40. N. Calderon, E.A. Ofstead, J.P. Ward, W.A. Judy and K.W. Scott, J. Amer. Chem. Soc., 90, 4133 (1968).

41. C.P.C. Bradshaw, E.J. Howman and L. Turner, J. Catal., 7, 269 (1967).

42. R.H. Grubbs and T.K. Brunck, J. Amer. Chem. Soc., 94, 2538 (1972).

43. G.S. Lewandos and R. Pettit, Tetrahedron Lett., 789 (1971).

44. F.D. Mango and J. Schachtschneider, J. Amer. Chem. Soc., 93, 1123 (1971).

45. P.G. Gassman and T.H. Johnson, J. Amer. Chem. Soc., 98, 861 (1976).

46. G.S. Lewandos and R. Pettit, J. Amer. Chem. Soc., 93, 7087 (1971).

47. J.X. McDermott, J.F. White and G.M. Whitesides, J. Amer. Chem. Soc., 95, 4451 (1973).

48. J.D. Fellman, G.A. Rupprecht and R.R. Schrock, J. Amer. Chem. Soc., 101, 5099 (1979).

49. J. Levisalles, H. Rudler and D. Villemin, J. Organomet. Chem., 193, 69 (1980).

960

50. J-L. Hérisson and Y. Chauvin, <u>Makromol. Chem.</u>, <u>141</u>, 161 (1971).

51. C.P. Casey and T.J. Burkhardt, <u>J. Amer. Chem. Soc.</u>, <u>96</u>, 7808 (1974).

52. T.J. Katz, J. McGinnis and C. Altus, <u>J. Amer. Chem. Soc.</u>, <u>98</u>, 605, 606 (1976).

53. J. Chauvin and D. Commereuc, Int. Symp. Metathesis, Mainz, (1976).

54. J. Levisalles, F. Rose-Munch, H. Rudler, J-C. Daran, Y. Dromzée and Y. Jeannin, <u>J. Chem. Soc., Chem. Commun.</u>, 152 (1981).

55. R. Giezyński and A. Korda, <u>J. Mol. Catal.</u>, <u>7</u>, 349 (1980).

56. T.J. Katz and R. Rothchild, <u>J. Amer. Chem. Soc.</u>, <u>98</u>, 2519 (1976).

57. R.H. Grubbs, P.L. Burk and D.D. Carr, <u>J. Amer. Chem. Soc.</u>, <u>97</u>, 3265 (1975).

58. R.H. Grubbs, <u>Prog. Inorg. Chem.</u>, <u>24</u>, 1 (1977).

59. S.M. Rocklage, J.D. Fellman, G.A. Rupprecht, L.W. Messerle and R.R. Schrock, <u>J. Amer. Chem. Soc.</u>, <u>103</u>, 1440 (1981).

60. F.N. Tebbe, G.W. Parshall and D.W. Ovenall, <u>J. Amer. Chem. Soc.</u>, <u>101</u>, 5074 (1979).

61. J.B. Lee, K.C. Ott and R.H. Grubbs, <u>J. Amer. Chem. Soc.</u>, <u>104</u>, 7491 (1982).

62. D.A. Straus and R.H. Grubbs, <u>Organometallics</u>, <u>1</u>, 1658 (1982).

63. A.K. Rappé and W.A. Goddard, <u>J. Amer. Chem. Soc.</u>, <u>104</u>, 448 (1982).

64. M.R. Churchill, J.R. Missert and W.J. Youngs, <u>Inorg. Chem.</u>, <u>20</u>, 3388 (1981).

65. M.R. Churchill and A.L. Rheingold, <u>Inorg. Chem.</u>, <u>21</u>, 1357 (1982).

66. J. Kress, M. Wesolek and J.A. Osborn, <u>J. Chem. Soc., Chem. Commun.</u>, 514 (1982).

67. J. Kress, M. Wesolek, J-P. Le Ny and J.A. Osborn, <u>J. Chem. Soc., Chem. Commun.</u>, 1039 (1981).

68. E.A. Zeuch, Fr. Patent 1,575,778 (1969); Chem. Abs., 72, 99988z (1970).

69. R.R. Schrock, unpublished work.

70. J.C. Hayes, G.D.N. Pearson and N.J. Cooper, J. Amer. Chem. Soc., 103, 4648 (1981).

71. R.H. Grubbs and C.R. Hoppin, J. Chem. Soc., Chem. Commun., 634 (1977).

72. D.M. Singleton and D.J. Eatough, J. Mol. Catal., 8, 175 (1980).

73. M. Leconte, Y. Ben Taarit, J.L. Bilhou and J.M. Basset, J. Mol. Catal., 8, 263 (1980).

74. J.R.M. Kress, M.J.M. Russel, M.G. Wesolek and J.A. Osborn, J. Chem. Soc., Chem. Commun., 431 (1980).

75. J.M. Basset, G. Coudurier, R. Mutin, H. Praliaud and Y. Trambouze, J. Catal., 34, 196 (1974).

76. W.S. Greenlee and M.F. Farona, Inorg. Chem., 15, 2129 (1976).

77. J.W. Byrne, H.U. Blaser and J.A. Osborn, J. Amer. Chem. Soc., 97, 3871 (1975).

78. G.J.A. Adam, S.G. Davies, K.A. Ford, M. Ephritikhine, P.F. Todd and M.L.H. Green, J. Mol. Catal., 8, 15 (1980).

79. M. Ephritikhine and M.L.H. Green, J. Chem. Soc., Chem. Commun., 926 (1976).

80. A.A. Olsthoorn and C. Boelhouwer, J. Catal., 44, 207 (1976).

81. N.S. Nametkin, V.M. Vdovin, É.D. Babich, V.N. Karel'skii and B.V. Kacharmin, Dokl. Chem., 213, 872 (1973).

82. J. Levisalles, H. Rudler and D. Villemin, J. Organomet. Chem., 192, 195 (1980), 193, 235 (1980).

83. T. Masuda, K. Yamamoto and T. Higashimura, Polymer, 23, 1663 (1982).

962

84. T.J. Katz and C-C. Han, Organometallics, 1, 1093 (1982).

85. M. Gilet, A. Mortreux, J. Nicole and F. Pettit, J. Chem. Soc., Chem. Commun., 522 (1979).

86. Y. Ben Taarit, J.L. Bilhou, M. Lecomte and J.M. Basset, J. Chem. Soc., Chem. Commun., 38 (1978).

87. P. Krausz, F. Garnier and J-E. Dubois, J. Organomet. Chem., 108, 197 (1976).

88. A. Agapiou and E. McNelis, J. Organomet. Chem., 99, C47 (1975).

89. F. Garnier, P. Krausz and H. Rudler, J. Organomet. Chem., 186, 77 (1980).

90. E.O. Fischer, Adv. Organomet. Chem., 14, 1 (1976).

91. F.J. Brown, Prog. Inorg. Chem., 27, 1 (1980).

92. E.O. Fischer, Rev. Pure Appl. Chem., 24, 404 (1976).

93. R.R. Schrock, J. Amer. Chem. Soc., 96, 6796 (1974), 98, 5399 (1976).

94. M.T. Mocella, M.A. Busch and E.L. Muetterties, J. Amer. Chem. Soc., 98, 1283 (1976).

95. H.T. Dodd and K.J. Rutt, J. Mol. Catal., 15, 103 (1982).

96. C.P. Casey, H.E. Tuinstra and M.C. Saeman, J. Amer. Chem. Soc., 98, 608 (1976).

97. R.J. Puddephatt, M.A. Quyser and C.F.H. Tipper, J. Chem. Soc., Chem. Commun., 626 (1976).

98. S.S.M. Ling and R.J. Puddephatt, J. Chem. Soc., Chem. Commun., 412 (1982).

99. C.P. Casey, S.W. Polichnowski, A.J. Shusterman and C.R. Jones, J. Amer. Chem. Soc., 101, 7282 (1979).

100. C.P. Casey and H.E. Tuinstra, J. Amer. Chem. Soc., 100, 2270 (1978).

101. P.G. Gassman and T.H. Johnson, J. Amer. Chem. Soc., 98, 6055, 6057 (1976).

102. H. Rudler, J. Mol. Catal., 8, 53 (1980).

103. W.B. Hughes, J. Chem. Soc., Chem. Commun., 431 (1969).

104. P. Krausz, F. Garnier and J-E. Dubois, J. Organomet. Chem., 146, 125 (1978).

105. M. Leconte and J.M. Basset, Nouv. J. Chim., 3, 429 (1979).

106. J.L. Bilhou, J.M. Basset, R. Mutin and W.F. Graydon, J. Amer. Chem. Soc., 99, 4083 (1977).

107. C.P. Casey, L.D. Albin and T.J. Burkhardt, J. Amer. Chem. Soc., 99, 2533 (1977).

108. R.H. Grubbs, unpublished work.

109. O. Eisenstein, R. Hoffmann and A.R. Rossi, J. Amer. Chem. Soc., 103, 5582 (1981).

110. A.K. Rappé and W.A. Goddard, J. Amer. Chem. Soc., 104, 297 (1982).

111. J.A. Ibers, R. DiCosimo and G.M. Whitesides, Organometallics, 1, 13 (1982).

112. F. Garnier and P. Krausz, J. Mol. Catal., 8, 91 (1980).

113. R.H. Grubbs, D.D. Carr and P.L. Burk, Organotransition Metal Chemistry, V. Ishii and M. Tsutsui, Eds., Plenum Press, New York, 1975, p. 135.

114. C.P. Casey and A.J. Schusterman, J. Mol. Catal., 8, 1 (1980).

115. C.P. Casey, D.M. Scheck and A.J. Schusterman, Fund. Res. Hom. Catal., 3, 141 (1979).

116. C.P. Casey, A.J. Schusterman, N.W. Vollendorf and K.J. Haller, J. Amer. Chem. Soc., 104, 2417 (1982).

117. A. Uchida and K. Hata, J. Mol. Catal. 15, 111 (1982).

118. F.D. Mango, J. Amer. Chem. Soc., 99, 6117 (1977).

119. R.H. Grubbs, Inorg. Chem., 18, 2623 (1979).

120. G.J. Leigh, M.T. Rahman and D.R.M. Walton, J. Chem. Soc., Chem. Commun., 541 (1982).

121. A. Mortreux, J.C. Delgrange, M. Blanchard and B. Lubochinsky, J. Mol. Catal., 2, 73 (1977).

122. A. Bencheick, M. Petit, A. Mortreux and F. Petit, J. Mol. Catal., 15, 93 (1982).

123. D. Villemin and P. Cadiot, Tetrahedron Lett., 23, 5139 (1982).

124. M. Petit, A. Mortreux and F. Petit, J. Chem. Soc., Chem. Commun., 1385 (1982).

125. S.F. Pederson, R.R. Schrock, M.R. Churchill and H.J. Wasserman, J. Amer. Chem. Soc., 104, 6808 (1982).

126. J. Sancho and R.R. Schrock, J. Mol.Catal., 15, 75 (1982).

127. J.R. Fritch and K.P.C. Vollhardt, Augrew. Chem., Int. Ed. Engl., 18, 409 (1979).

128. T.J. Katz and J. McGinnis, J. Amer. Chem. Soc., 97, 1592 (1975).

129. J.H. Wengrovius, J. Sancho and R.R. Schrock, J. Amer. Chem. Soc., 103, 3932 (1981).

130. B.E. Bursten, J. Amer. Chem. Soc., 105, 121 (1983).

131. M.G. Voronkov, V.B. Puknarevich, S.P. Sushchinskaya, V.Z. Annenkova, V.M. Annenkova and N.J. Andreeva, J. Polm. Sci., Polym. Chem. Ed., 18, 53 (1980).

132. T. Masuda, T. Takahashi, K. Yamamoto and T. Higashimura, J. Polym. Sci., Polym. Chem. Ed. 20, 2603 (1982).

133. T. Masuda, K-i. Hasegawa and T. Higashimura, Macromolecules, 7, 728 (1974).

134. K-i. Hasegawa, T. Masuda and T. Higashimura, Macromolecules, 8, 255 (1975).

135. T. Masuda, T. Takahashi and T. Higashimura, J. Chem. Soc., Chem. Commun., 1297 (1982).

136. R.L. Banks, Hydrocarbon Process., 46, 232 (1967), Chem. Tech., 9, 494 (1979).

137. R.L. Banks, J. Mol. Catal., 8, 269 (1980).

138. S. Miyata, Japan. Kokai, 73 92,304 (1973); Chem. Abs., 80, 120213v (1974).

139. E. Ofstead, U.S. Patent, 3, 997,471 (1976); Chem. Abs., 86, 56559g (1977).

140. K.J. Ivin, D.T. Lavertey, J.J. Rooney and P. Watt, Rec, 96, M54 (1977).

141. N.I. Pakuro, K.L. Makovertskii, A.R. Gantmakher and B.A. Dolgoplosk, Bull. Acad. Sci. U.S.S.R., Div. Chem. Sci., 31, 456 (1982).

142. A.J. Amass and J.A. Zurimendi, Eur. Polym. J., 17, 1 (1981).

143. S.J. Lee, J. McGinnis and T.J. Katz, J. Amer. Chem. Soc., 98, 7818 (1976).

144. K. Hiraki, A. Kuroiwa and H. Hirai, J. Polym. Sci., A1, 9, 2323 (1971).

145. K.J. Ivin, B.S.R. Reddy and J.J. Rooney, J. Chem. Soc., Chem. Commun., 1062 (1981).

146. A.J. Amass, Canad. J. Chem., 60, 6 (1982).

147. T.J. Katz, S.J. Lee and M.A. Shippey, J. Mol. Catal., 8, 219 (1980).

148. T.J. Katz, S.J. Lee and N. Acton, Tetrahedron Lett., 4247 (1976).

149. G.I. Devine, H.T. Ho, K.J. Ivin, M.A. Mohamed and J.J. Rooney, J. Chem. Soc., Chem. Commun., 1229 (1982).

150. H.T. Ho, K.J. Ivin and J.J. Rooney, Makromol. Chem., 183, 1629 (1982).

966

151. C. Larroche, J.P. Laval, A. Lattes, M. Leconte, F. Quignard and J.M. Basset, J. Org. Chem., 47, 2019 (1982).

152. R. Nakamura, S. Matsumoto and E. Echigoya, Chem. Lett., 1019 (1976).

153. D. Villemin, Tetrahedron Lett., 21, 1715 (1980).

154. E. Verkuijlen and C. Boelhouwer, Chem. Phys. Lipids, 24, 305 (1979).

155. R. Baker and M.J. Crimmin, Tetrahedron Lett., 441 (1977).

156. P.B. van Dam, M.C. Mittelmeijer and C. Boelhouwer, Fette, Seifen Ansmichmittel, 76, 264 (1974).

157. J. Tsuji and S. Hashiguchi, J. Organomet. Chem., 218, 69 (1981).

158. J-P. Laval, A. Lattes, R. Mutin and J.M. Basset, J. Chem. Soc., Chem. Commun., 502 (1977).

159. D.G. Daly and M.A. McKervey, Tetrahedron Lett., 23, 2997 (1982).

160. R.H.A. Bosma, A.P. Kouwenhoven and J.C. Mol, J. Chem. Soc. Chem. Commun., 1081 (1987).

161. J.C. Mol, J. Mol. Catal., 15, 35 (1982).

162. J. Tsuji and S. Hashiguchi, Tetrahedron Lett., 21, 2955 (1980).

Glossary of terms and abbreviations

A   dehydroascorbate

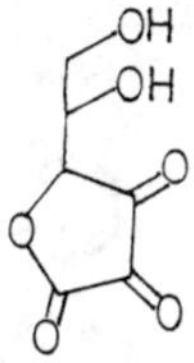

acacH   acetylacetone, pentane-2,4-dione

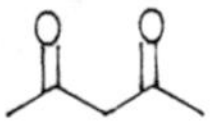

AcOH   acetic acid, CH_3COOH

AIBN   azoisobutyronitrile

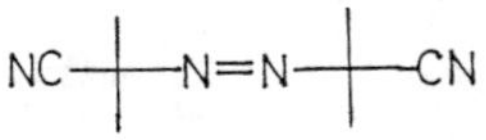

AlaH   alanine, $CH_3CH(NH_2)COOH$

Aliquat 336   trioctylmethylammonium chloride, $((C_8H_{17})_3NCH_3)^+Cl$

                     (surfactant)

Ar   aryl

R,R-ARSOP   (-)-R,R-trans-4,5-bis(diphenylarsinomethyl)-2,2

                     -dimethyl- 1,3-dioxolan

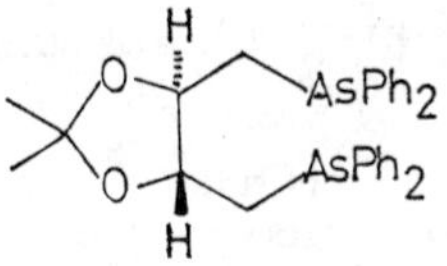

atm   atmosphere

B   base

9-BBN   9-borabicyclo[3.3.1]nonane

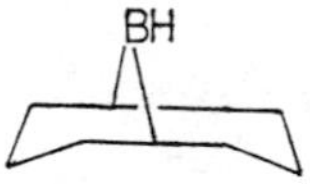

967

BINAP   R-2,2'-bis(diphenylphosphino)-1,1'-binaphthyl

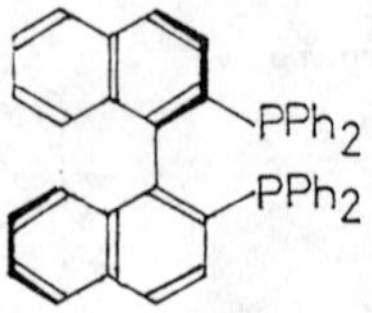

bipy   2,2'-bipyridine

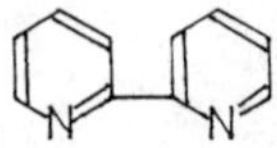

BMPP   benzylmethylphenylphosphine, $PhCH_2P(Me)Ph$

S,R-BPPFA   (S)-N,N-dimethyl-1-1-[(R)-1',2-bis(diphenylphosphino)

ferrocenyl]ethylamine

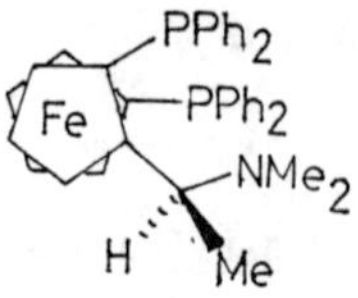

S,R-BPPFOH   (S)-1-[(R)-1',2-bis(diphenylphosphino)ferrocenyl]ethanol

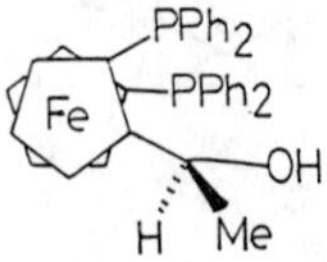

BPPM   2S,4S-N-t-butoxycarbonyl-4-diphenylphosphino-2-

diphenylphosphinomethylprrrolidine

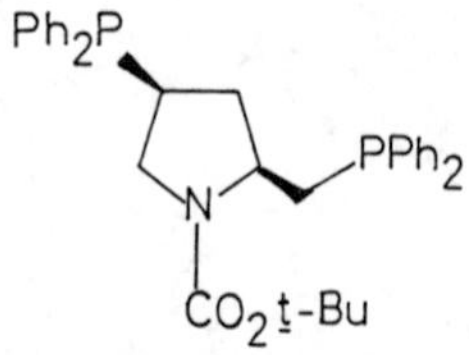

Brij-35   polyoxyethylated-n-dodecyl alcohol (surfactant)

Bu   butyl, $-CH_2CH_2CH_2CH_3$

i-Bu   iso-butyl, $-CH_2CH(CH_3)_2$

t-Bu   tert-butyl, $-C(CH_3)_3$

Bz   benzyl, $-CH_2Ph$

Bzacen   N,N'-ethylene bis(benzoylacetoneiminato) dianion

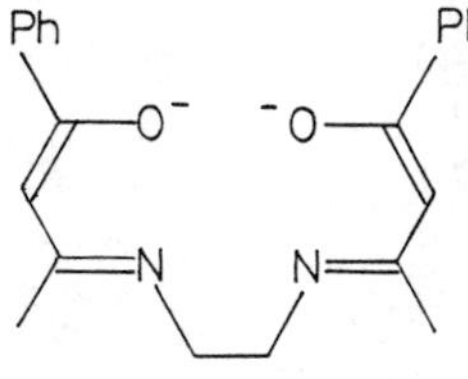

CAMP   cyclohexyl-ortho-methoxyphenylmethyl phosphine

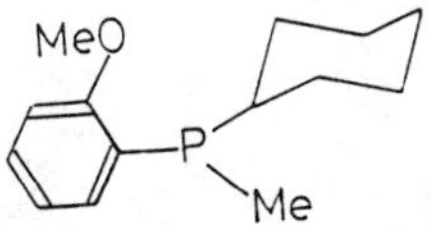

carb   1,2-dicarba-closo-decaborane

c.f.   confer, compare

R-CHAIRPHOS   R-1,3-bis(diphenylphosphino)butane

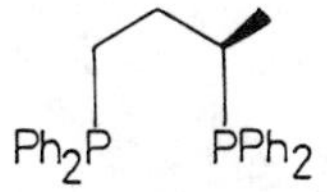

S,S-CHIRAPHOS   S,S-2,3-bis(diphenylphosphino)butane

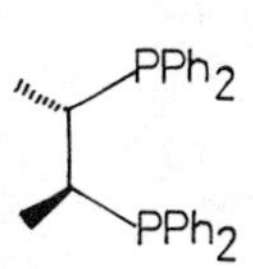

Chloramine-T   N-chloro-4-toluenesulphonamide, sodium salt

$$\text{CH}_3\text{-}\text{C}_6\text{H}_4\text{-}SO_2NClNa$$

CMP   Cytidine-3'-monophosphoric acid

Cob(I)alamin

cod   1,5-cyclooctadiene

cot   1,3,5,7-cycloöctatetraene

Cp $\underline{n}^5$-cyclopentadienyl

$\underline{\alpha}$-cqDH α-camphorquinone dioxime

CTAB   hexadecyltrimethylammonium bromide, $C_{16}H_{33}N(CH_3)_3{}^+Br^-$

Cy   cyclohexyl

$\underline{\underline{R}}$-CYCPHOS $\underline{\underline{R}}$-1-cyclohexyl-1,2-$\underline{bis}$(diphenylphosphino)ethane

dba   dibenzylidene acetone, 1,5-diphenyl-1,4-pentadiene-3-one

DBP   5-phenyl-5-$\underline{H}$-dibenzophosphole

DBPDIOP $\quad$ R,R-trans-4,5-bis(dibenzophospholomethyl-2,2-dimethyl-1,3-dioxolan

DCC $\quad$ N,N'-dicyclohexylcarbodiimide

DDIOS $\quad$ 2R,3R-2,3-dihydroxy-1,4-bis(methylsulphinyl)butane

DET $\quad$ R,R- or S,S-diethyl-2,3-dihydroxybutane dioate, diethyl tartrate

dibal $\quad$ diisobutylaluminium hydride, $((CH_3)_2CHCH_2)_2AlH$

R,R-DIOP $\quad$ R,R-trans-4,5-bis(diphenylphosphinomethyl)-2,2-dimethyl-1,3-dioxolan

R,R-DIOS   R,R-trans-4,5-bis(methylsulphinyl)-2,2-dimethyl-1,3-

dioxolan

DIOXOP   2R,4R-bis(diphenylphosphinomethyl)-1,3-dioxolan

R,R-DIPAMP   R,R-1,2-ethanediylbis(o-methoxyphenylphenylphosphine)

diphos   1,2-bis(diphenylphosphino)ethane, $Ph_2PCH_2CH_2PPh_2$

dipy   2,2'-bipyridine

DMA   N,N-dimethylacetamide, $CH_3CON(CH_3)_2$

DMF   N,N-dimethylformamide, $HCON(CH_3)_2$

dmg   dimethylglyoxime,

DMPP   dimenthylphenylphosphine

DMSO   dimethylsulphoxide, $(CH_3)_2SO$

dpm   dipivaloylmethane, 2,2,6,6-tetramethyl-3,5-heptanedione

dppb   1,4-<u>bis</u>(diphenylphosphino)butane, $Ph_2P(CH_2)_4PPh_2$

dppf   1,1'-<u>bis</u>(diphenylphosphino)ferrocene

dppp   1,3-<u>bis</u>(diphenylphosphino)propane, $Ph_2P(CH_2)_3PPh_2$

EDTA   ethylene diamine tetraacetic acid,

                 $(HO_2CCH_2)_2NCH_2CH_2N(CH_2CO_2H)$

ee   enantiomer excess, $|\%\underline{R} - \%\underline{S}|$

en   1,2-diaminoethane, $H_2NCH_2CH_2NH_2$

ESCA   X-ray photoelectron spectroscopy

esr   electron spin resonance

Et   ethyl, $-CH_2CH_3$

GLUCPHOS 3-O-benzyl-1,2-O-isopropylidene-5,6-dideoxy-5,6-<u>bis</u>

(diphenylphosphino)-<u>α</u>-<u>D</u>-glucofuranose

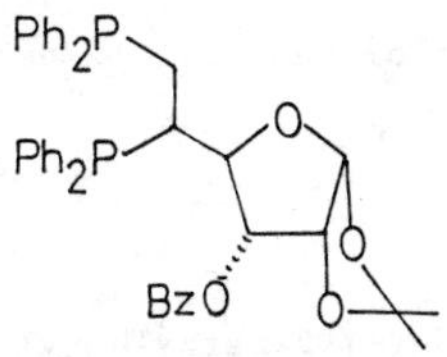

gluH glutamic acid, $HO_2CCH_2CH_2CH(NH_2)COOH$

glyH glycine, $H_2NCH_2CO_2H$

H_2A ascorbic acid

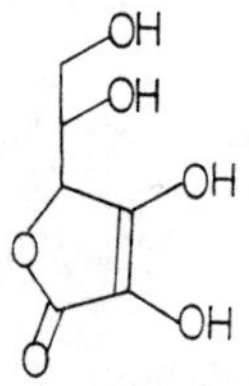

HMPA hexamethylphosphoric triamide, $((CH_3)_2N)_3PO$

HOBT 1-hydroxybenzotriazole

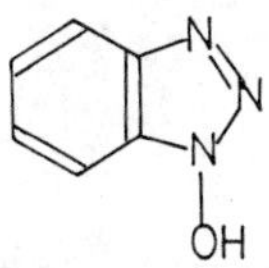

HPA-n heterepolyacid, $H_{3+n}PMo_{12-n}V_nO_{40}$, n = 2 - 8

hν photolysis

Im imidazole

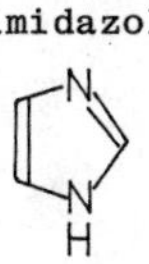

IR infra-red

k rate constant

L	a two electron ligand
L*	a chiral two electron ligand
leuH	leucine, $(CH_3)_2CHCH_2CH(NH_2)CO_2H$
M	an atom of interest, usually a metal
M	molar
m	meta
MBMSO	S,R:S,S-(+)-2-methylbutylmethylsulphoxide

MCPBA	meta-chloroperbenzoic acid

MDPP	menthyldiphenylphosphine

Me	methyl, $-CH_3$
Mem	2-methoxyethoxymethyl, $-CH_2OCH_2CH_2OCH_3$
4,7-Me$_2$phen	4,7-dimethyl-1,10-phenanthroline

3,4,7,8-Me$_4$phen   3,4,7,8-tetramethyl-1,10-phenanthroline

MePNNPMe $\underline{S},\underline{S}$-N,N'-dimethyl-N,N'-bis(diphenylphosphino)-1,2-

diphenyl-1,2-diaminoethane

mnt^{2-} cis-1,2-dicyanoethane-1,2-dithiolate

NAD(H)   nicotinamide adenine dinucleotide

nbd   norbornadiene, bicyclo[2.2.1]-hepta-2,5-diene

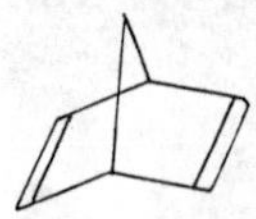

NMDPP   (+)-neomenthyldiphenylphosphine

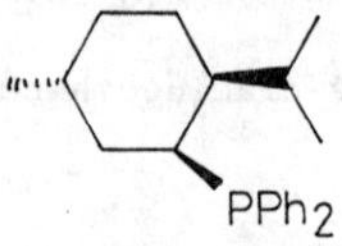

NMO   N-methylmorpholine-N-oxide

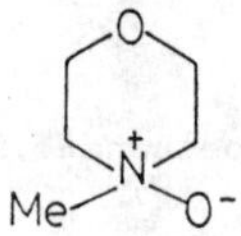

nmr   nuclear magnetic resonance

NORPHOS   R,R-5-endo-6-exo-bis(diphenylphosphino)bicyclo[2.2.1]
hept-2-ene

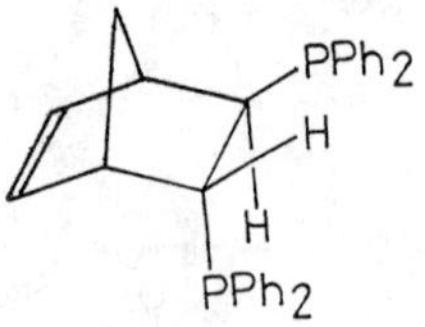

α-Np   α-naphthyl

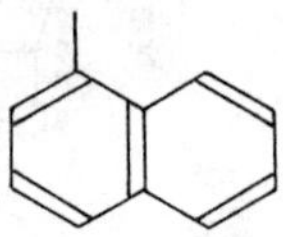

Nu   nucleophile

o   ortho

oxine   8-hydroxyquinoline

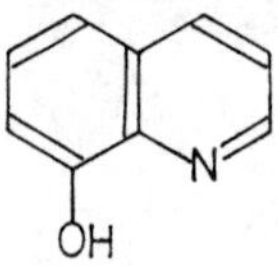

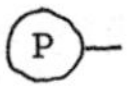

(P)—   polymer backbone

p   *para*

PAMP   phenyl-*ortho*-methoxyphenylmethylphosphine

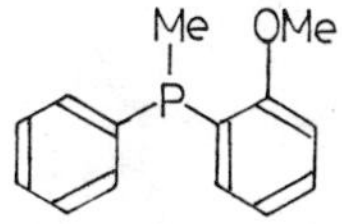

PAMPOP $\underline{\underline{R}},\underline{\underline{R}}$-*trans*-4,5-*bis*(*bis*(ortho-methoxyphenyl)phosphino-

            methyl)-2,2-dimethyl-1,3-dioxolan

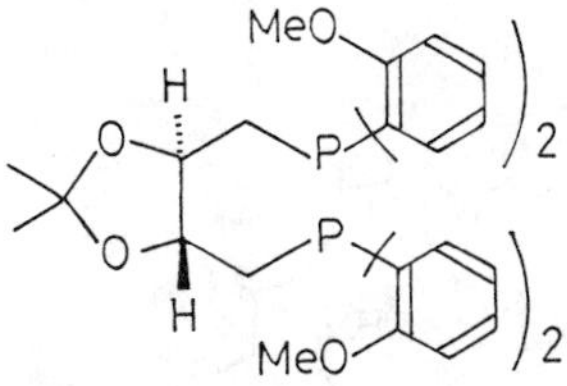

pic   picolinate

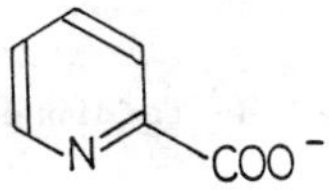

piv   pivalate, 2,2-dimethylpropionate, $(CH_3)_3CCOO^-$

PNNP $\underline{\underline{S}},\underline{\underline{S}}$-N,N'-*bis*(diphenylphosphino)-1,2-diphenyl-1,2-

            diaminoethane

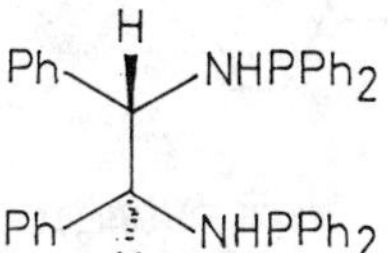

S,R-PPFA   (S)-N,N-dimethyl-1-[(R)-2-diphenylphosphinoferrocenyl]
ethylamine

ppm   parts per million

n-Pr   n-propyl, $-CH_2CH_2CH_3$

i-Pr   iso-propyl, $-CH(CH_3)_2$

PcH_2 phthalocyanine

PCCP   S,S-2,3-diphenyl-1,4-bis(diphenylphosphino)butane

Ph   phenyl

PheH   phenylalanine, $PhCH_2CH(NH_2)COOH$

PHELLANPHOS   5R,7S,8S-2-methyl-5-endo-iso-propyl-7,8-bis

(diphenylphosphino)bicyclo[2.2.2]oct-2-ene

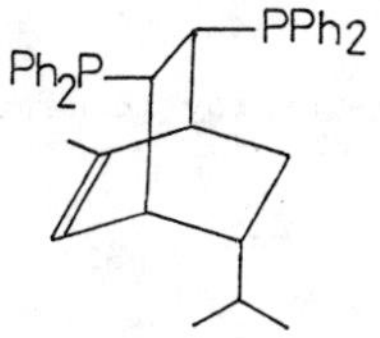

phen   1,10-phenanthroline

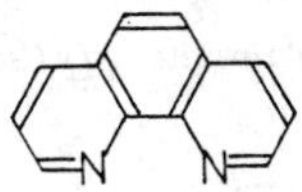

PHEPHOS   R-1-phenyl-1,2-bis(diphenylphosphino)ethane

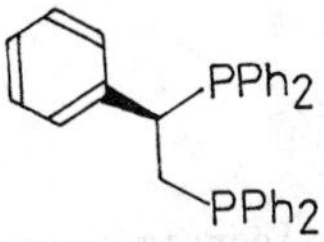

PROPHOS   R-1,2-bis(diphenylphosphino)propane

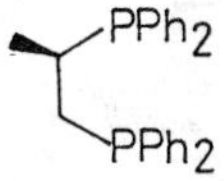

psi   pounds per square inch

py   pyridine

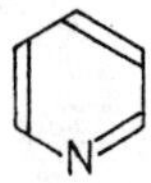

Q   quinine

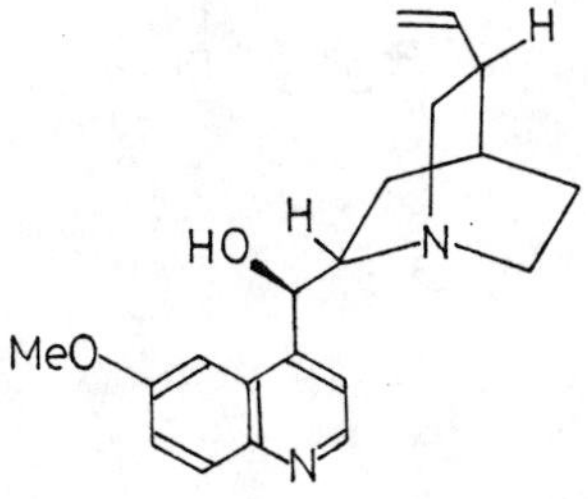

R	alkyl
rt	room temperature

SalenH$_2$ N,N'-ethylene*bis*(salicylideneimine)

SalophH$_2$ *ortho*-phenylene diamine *bis*(salicylideneimine)

SalpnH$_2$ 1,3-*bis*(3-salicylideneiminopropyl)amine

Sia   secondary *iso*-amyl, 2-pentyl, -CH(CH$_3$)CH$_2$CH$_2$CH$_3$

SKEWPHOS   S,S-2,4-*bis*(diphenylphosphino)pentane

T	temperature
$t_{\frac{1}{2}}$	half life
TFA	trifluoroacetic acid, CF_3CO_2H
TfOH	trifluoromethane sulphonic acid, CF_3SO_3H
THF	tetrahydrofuran

THP	tetrahydropyranyl

TMEDA	N,N,N',N',-tetramethyl-1,2-diaminoethane, $(CH_3)_2NCH_2CH_2N(CH_3)_2$
tol	tolyl, 4-methylphenyl
tpp	<u>bis</u>(3-diphenylphosphinopropyl)phenylphosphine

TPPH$_2$	tetraphenylporphyrin

984

trien	triethylenetetramine, 1,8-diamino-3,6-diazaoctane,

$H_2NCH_2CH_2NHCH_2CH_2NHCH_2CH_2NH_2$

TsOH   p-toluene sulphonic acid

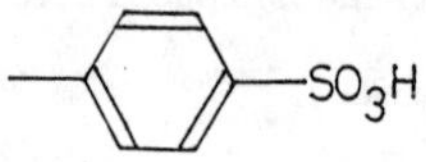

$CH_3-C_6H_4-SO_3H$

UV   ultra violet

VALPHOS   R-2-methyl-3,4-bis(diphenylphosphino)butane

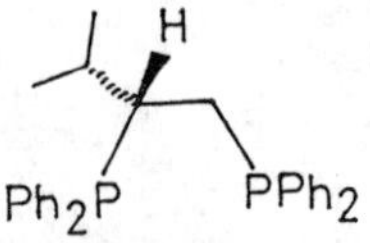

Vaska's compound   trans-bis(triphenylphosphine) iridium chlorocarbonyl

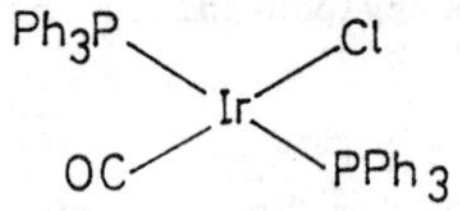

Wilkinson's catalyst   tris(triphenylphosphine)rhodium chloride, $(Ph_3P)_3RhCl$

X   one electron ligand, usually halide

δ chemical shift

Index

Types of metal complex catalysing particular reactions are indexed under that reaction by the name of the metal only.

Acetamido cinnamic acid 131, 135, 142, 164
Acetic acid
  addition to alkenes 366-7
  synthesis by methanol carbonylation 254-61
Acyl halides
  coupling with alkynes 701
  decarbonylation 281-4
  hydrogenolysis 117, 121
  hydrosilylation 687-8
  reaction with Grignards 699
  reaction with organomercury compounds 701
  reaction with organometallics 697-702
  reaction with organotin compounds 699-700
  reaction with organozinc compounds 702
  reduction 682, 685, 687-8
  synthesis by carbonylation 269, 275
Addition
  to alkenes 365-9
  to alkynes 373
  to allenes 369-70
  to dienes 370-1
Adiponitrile 348
A-frame complexes 84, 816
Alcohols
  addition to alkenes 368
  allylic - see allyl alcohols
  by-products in hydroformylation 223-4, 230
  carbonylation 254-61
  dehydrogenation 167-171
  hydrogenolysis 116-8
  oxidation 568-80
  synthesis by aldehyde reduction 682
  synthesis by ketone hydro-silylation 685-7, 693-5
  unsaturated, hydroesterification 246-7
  unsaturated, hydroformylation 225
  unsaturated, hydrogenation 63-7
  unsaturated, isomerisation 225
  unsaturated, oxidation 463-4, 534, 568-80
Aldehydes
  amination 720
  decarbonylation 95, 280, 284-8

Aldehydes, cntd.
  hydrogenation 95-6, 99, 101, 112 222-3, 230, 237
  hydrosilylation 685-7
  oxidation 358, 581-5
  reduction 682-3
  synthesis by acyl halide reduction 682, 685, 687-8
  synthesis by alcohol oxidation 568-71
  synthesis by carboxylic acid reduction 685
  synthesis by hydroformylation 219-43
  unsaturated, hydrogenation 49-51
Alkenes
  addition 365-9
  amination 365-74
  carbohalogenation 353
  carbonylation 248, 272
  codimerisation 761
  cooligomerisation 761-2
  coupling 866-7
  coupling to arenes 768-9
  cycloaddition 762-7
  cyclopropanation 334-45
  dehydrogenation 169
  epoxidation 461-99
  hydration 365
  hydrocarboxylation 245
  hydrocyanation 345-8, 351-3
  hydroesterification 244
  hydroformylation 219-43
  hydrogenation 23-48, 112, 114, 124-5, 129, 222, 803
  hydrohalogenation 352
  hydrometallation 355-7, 803
  hydrosilylation 3, 308-15, 326-32
  isomerisation 24-5, 32, 39, 78, 85, 112-5, 223-4, 230, 244, 311, 322, 331, 405-17
  metal complexes 403-4
  metathesis 921-47, 951-6
  oligomerisation 747-60
  oxidation 460-543, 804
  polymerisation 733-47
  synthesis by acyl halide decarbonyl-ation 283-4
  Wacker oxidation 366, 459, 500-11, 532-3
Alkyl arenes, oxidation 548-64
Alkyl derivatives
  coupling to Grignards 845
  coupling to organometals 851

Alkyl hydroperoxides 460-1, 465, 468-90, 493-500, 505-7, 517-20, 522, 524, 528-30, 542, 545, 564-7, 570-1, 577, 623-6, 637-8
Alkynes
  addition 373
  amination 374
  carbometallation 358-64
  cooligomerisation 796-7
  coupling 543, 828-31
  coupling to acyl halides 701
  coupling to arenes 830
  cycloaddition 763
  cyclopropanation 340
  cyclotetramerisation 812-5
  cyclotrimerisation 815-20
  dimerisation 812-3
  hydration 373
  hydrocarboxylation 249
  hydrocyanation 350-1
  hydroesterification 249-50
  hydrogenation 81-7, 115, 127
  hydrohalogenation 352-3
  hydrometallation 357-8
  hydrosilylation 322-7
  isomerisation 443-4
  metathesis 948-51
  oligomerisation 85, 812-21
  oxidation 542-4
  polymerisation 809-12, 951
  reduction 1-2
Allenes
  addition 369-70
  hydrogenation 67-8
  oligomerisation 369-70, 769-71
  polymerisation 369-70, 769-72
  synthesis 850-1
  synthesis by carbometallation 360
  synthesis by rearrangement 443-4
Allyl acetates
  coupling 841, 853-66
  isomerisation 803
  oxidation 504-5
  synthesis by alkene oxidation 504-5
Allyl alcohols
  carbohalogenation 353
  carbonylation 261
  coupling 842-4, 871
  cyclopropanation 336-7
  decarbonylation 288
  isomerisation 225, 417-20
  oxidation 464-5, 478-97, 504, 528
  synthesis by enone hydrosilylation 2, 689-91, 696-7
  synthesis by enone reduction 2, 101
  tungsten complex 944-5
Allyl amides, isomerisation 421-3

Allyl amines
  isomerisation 6, 421-3
  tungsten complex 944-5
Allyl complexes 404, 407, 418
  amination 369-72
  cyclisation 856-60
  palladium, coupling 852-65
Allyl derivatives
  coupling 853-66
  coupling with Grignards 841-4
Allyl ethers
  coupling 841
  isomerisation 417, 420-4, 442-4
Allyl halides
  carbonylation 268-9
  coupling 841-4, 853-71
  oxidation 463, 478
Allylic oxidation 473, 492, 514, 517-25
Allylic substitution 369-74, 852-66
  molybdenum 866
  nickel 370
  palladium 4,5, 369-74, 852-65
  platinum 370, 865
  rhodium 370, 866
  ruthenium 370
Allylic transposition 441-4, 803
Allyl sulphides, coupling 842
Allyl sulphoxides, coupling 842
Aluminium alkyls (see also vinyl alanes) 355-7, 702-5
  reaction with enones 702-5
  reaction with ketones 707
Aluminium compounds for alkene oligomerisation 748
Amides
  hydrolysis 717-8
  oxidation 596
  reduction 682
  synthesis by alkene carbonylation 248, 256, 262-4, 267, 272
  synthesis by nitrile hydrolysis 721
  unsaturated, reduction 683
Amination
  aldehydes 720
  alkenes 365-74
  alkynes 374
  allyl complexes 369-72
  cobalt 720
  copper 369
  ketones 103, 720
  palladium 367-9
  reductive 102-3
  rhodium 669
  ruthenium 369
Amine oxides
  as oxidising agents 493, 530-1, 544
  synthesis by amine oxidation 623-5

Amines
  carbonylation 272,275
  dehydrogenation 623
  oxiadtion 621-5
  reaction with carbonyl groups 720
  synthesis by alkene amination
  365-74
  synthesis by amide reduction 682
  synthesis by imine hydrosilylation
  692, 697-8
  synthesis by imine reduction 683
  synthesis by telomerisation of
  NH$_3$ and dienes 333-4
  unsaturated, hydrogenation 63-7
Amino acids
  oxidation 596-7
  synthesis 102, 130-4
Anhydrides
decarbonylation 289
  hydrogenation 156-7
  hydrogenolysis 117-8
  synthesis by aldehyde oxidation
  583
  synthesis by alkene carbonylation
  248, 272
Anilines, oxidation 618-21
Anthracene, oxidation 545-7
Arenes
  coupling 544, 826-7
  coupling to alkenes 768-9
  coupling to alkynes 830-1
  hydrogenation 87-93, 127-8
  oxidation 544-9
  synthesis by alkyne cyclotrimerisation 815-20
Arsines, oxidation 632
Aryl halides
  carbonylation 262-5
  coupling 821-7, 847-9, 867-91
  coupling to Grignards 833-7
  hydrogenolysis 90-1
  substitution 375-7
Aryl mercury compounds (see also
organomercurials, vinyl mercurials)
  coupling 846, 851
Ascorbic acid, oxidation 598-600
Asymmetric synthesis (see also
enantioselectivity)
  codimerisation 761-2
  cooligomerisation 796
  coupling 6, 836, 839-40, 842-3,
  863
  cyclisation 768, 788
  cyclopropanation 342-5
  dehydrogenation 170
  epoxidation 493-9
  hydrocarboxylation 252
  hydrocyanation 347, 351
  hydroesterification 252-4

Asymmetric synthesis cntd,
  hydroformylation 238-44
  hydrogenation 5-6, 129-62
  hydrosilylation 326-9, 693-8
  isomerisation 6, 419-22
  oxidation 524
  reduction 682-3
  telomerisation 800
  transfer hydrogenation 162-4
Atactic polymers 735, 745
Ate complexes 364
Atropic acid 131, 136, 147, 152
Autoxidation 459-60, 524, 539,
585-8
Aziridines
  carbonylation 273-4
  isomerisation 440-2
  synthesis 491-2
Azirines 440-2
Azobenzene
  hydrogenation 104
  oxiadtion 625-6
  synthesis 106
Azo compounds 104-5
Azoxybenzene 106, 626

Basketene 428-9
Benzene (see also arenes)
  hydrogenation 87-8, 93-4
  reaction with diazo compounds 340
Benzoin 580
Benzyl halides
  carbonylation 266-8
  coupling 851
Biaryls, synthesis
  by arene coupling 826-7
  by aryl halide coupling 846
  by aryl mercury compound coupling
  846
  by Grignard coupling 832-7
  by oxidative coupling 544
Bicyclobutanes, isomerisation 436-9
BINAP 141-2
β-Blockers 156-7
Boranes 271, 849-51, 855
BPPFA 150, 156, 158, 863
BPPFOH 150, 155-8
BPPM 141-2, 144-5, 156, 158, 165-6,
695-6
Butadiene
  carbonylation 251-2
  cooligomerisation 790-7
  dimerisation 251-2
  hydrocyanation 348-9
  hydrogenation 69-70
  hydrosilylation 317,320
  oligomerisation 782-6
  polymerisation 774-9, 781

Butadiene cntd.
  telomerisation 320-1, 333-4,
  798-800, 802-10, 852-4

CAMP 130, 159
Carbanions, coupling 832-67
Carbene complexes 277-8, 436, 438,
742-4, 750, 927-47, 952
Carboalumination 362-4
Carbohalogenation 353-4
Carbometallation 358-64
Carbon dioxide, hydrosilylation
688
Carbon monoxide
  hydrogenation 276-80
  insertion 218-9
Carbonylation
  alcohols 254-61
  alkenes 248, 272
  allyl alcohols 261
  allyl halides 268-9
  amines 272, 275
  aryl halides 262-5
  aziridines 273-4
  benzyl halides 266-8
  butadiene 251-2
  cobalt 248, 254-6, 268, 270, 275
  cyclopropanes 275
  diazo compounds 273-4
  dienes 251-2
  epoxides 275
  halogenated compounds 261-8
  iridium 258-9
  iron 272
  methanol 254-61
  nickel 258-60, 263, 265-9, 275
  nitro groups 272-3
  organometals 268-9
  oxetanes 275
  oxidative 260, 545
  palladium 248, 260-75
  platinum 248
  polymer supported catalysts 258
  rhodium 255-8, 270-5
  ruthenium 261, 272
  vinyl halides 266-7
Carbonyl compounds
  amination 720
  hydrogenation 90, 94-101, 111,
  128, 155-60
  reduction 681-90
  synthesis by alkene oxidation
  500-11
  unsaturated, hydrogenation 101,
  128, 153, 163-4
  unsaturated, hydrosilylation
  689-91
  unsaturated, oxidation 507-8
  unsaturated, reduction 2

Carboxylation 545
  epoxides 275
  Grignard reagents 268
  nickel 275
Carboxylic acids
  hydrogenation 100
  oxidation 591-2
  reduction 685
  synthesis by alcohol oxidation 571,
  579
  synthesis by aldehyde oxidation 458,
  581-5
  synthesis by alkyl arene oxidation
  548-64
  synthesis by carbonylation 258,
  261-3, 269
  synthesis by catechol oxidation
  614-6
  synthesis by hydrazide oxidation 626
  synthesis by hydrocarboxylation
  245-7
  synthesis by oxidative cleavage
  alkenes 541
  unsaturated, hydrogenation 56-61,
  131-2, 147-50, 153-4, 159, 162-3
  unsaturated, oxidation 463-4
Carboxypeptidase 713-4
Carbyne complexes 948-51
Catechols, oxidation 613-9
CHAIRPHOS 139-40
Chemoselectivity 1-2, 107-8, 219, 230
CHIRAPHOS 134-5, 137, 241
Chloramine T 530
Chloranilic acid 107
Chlorates 527-8, 544
Chromium complexes
  hydrogenation 19, 47-8, 70-1, 73-5,
  77-8, 81, 127
  hydrosilylation 319
  metathesis 929
  oligomerisation 791, 816
  oxidation 490, 525, 530
  polymerisation 735, 739, 747, 778,
  780
Chrysanthemic acid 343-4, 864
Citronellal 789, 802
Cleavage, diols 532, 534-5, 582-9
Cobalt complexes
  amination 720
  carbonylation 248, 254-6, 268, 270,
  275
  $[Co(CN)_5]^{3-}$, coupling 704
  $[Co(CN)_5]^{3-}$, hydrogenation 22, 48,
  50-3, 56-9, 61-3, 71-2, 74-5, 77, 81,
  85, 102-4, 106, 108, 116
  $[Co(CN)_5]^{3-}$, hydrosilylation 350-2

Cobalt complexes, cntd.

$[Co(CN)_5]^{3-}$, oxidation 609

$[Co(CN)_5]^{3-}$, reduction 682-3

$Co_2(CO)_8$, hydrocyanation 346-7, 350

$Co_2(CO)_8$, hydrogenation 22, 38, 48, 51-2, 59, 87, 98, 104, 106, 114, 117, 220-6, 235, 248, 275

$Co(dmgH)_2$, hydrogenation 106-8, 160

$Co(dmgH)_2$, oxidation 619, 624-6

cooligomerisation 794, 796

coupling 704, 829, 832

cyclopropanation 337-8, 344-5

dehydrogenation 169

Fischer Tropsch 280

hydrocyanation 346-7, 349-51

hydroesterification 244, 246, 251-3

hydroformylation 220-6, 235, 238-9

hydrogenation 22, 38-9, 48, 50-63, 68, 71-5, 77, 79-81, 83, 85-90, 92, 94, 98-9, 102-8, 110, 112, 114, 152, 158, 160-1

hydrogenolysis 114, 116

hydrolysis 709-10, 712-3, 717-8, 721

isomerisation 749, 757-8, 818

oxidation 458, 467-70, 473, 516-8, 539, 545, 551-66, 571, 574-5, 579-85, 588, 590-1, 595, 600, 602-9, 618-9, 624, 626, 631-2, 635-6

polymerisation 771-3, 780-1, 809

SALEN 605-6, 619

transesterification 710

Codimerisation 761-2

Cooligomerisation 790-7

Copolymers 747, 781-2

Copper complexes

amination 369

carbometallation 359-61

coupling 543, 701, 704, 821-3, 828-30, 838-9, 842, 845, 847

cycloaddition 763-4

cyclopropanation 337-8, 344-5

dimerisation 813

hydration 366, 373

hydrocarboxylation 245-7

hydrocyanation 349-50

hydrogenation 54, 105

hydrohalogenation 352-3

hydrolysis 711

hydrometallation 356

Copper complexes, cntd.

isomerisation 443

oxidation 469, 491, 500-10, 512, 514, 516, 523, 533-8, 541-7, 560-3, 568-9, 571, 574, 580, 582-6, 590, 592-6, 598-9, 610-5, 618-9, 621-7, 633-6

polymerisation 610-1

reduction 682

transesterification 710-1

Ullmann coupling 375-6

Coupling 832-73

acyl halides/alkynes 701

alkenes 866-7

alkenes/arenes 768-9

alkyl derivatives 851

alkyl derivatives/Grignards 845

alkynes 543, 828-31

alkynes/acyl halides 701

alkynes/arenes 830-1

allyl alcohols 842-4, 871

allyl derivatives 853-66

allyl derivatives/Grignards 841-4

allyl esters 841, 853-66

allyl ethers 841

allyl halides 841-4, 853-71

allyl sulphides 842

allyl sulphoxides 842

arenes 544, 826-7

arenes/alkenes 768-9

arenes/alkynes 830-1

aryl halides 821-5, 847-9, 867-71

aryl halides/Grignards 833-7

aryl mercurials 846, 851

asymmetric 836, 839-40, 842-3, 863

benzyl halides 851

boranes 849-51, 855

carbanions 832-67

cobalt 704, 829, 832

copper 543, 701, 704, 821-3, 828-30, 838-9, 842, 845, 847

diazo compounds 769-70

epoxides 853-5

Grignard reagents 832-45

Grignards/alkyl derivatives 845

Grignards/allyl derivatives 841-4

Grignards/aryl halides 833-7

Grignards/vinyl halides 838-40

iron 704, 828, 833, 838, 849, 867

manganese 828, 833

molybdenum 866

nickel 702-8, 823-5, 833-44, 847-50, 866, 871-2, 878-81

organotin compounds 849

organozinc compounds 848, 850-1

oxidative 768, 826-30

palladium 544, 701, 768-70, 826-7, 830, 833-4, 838-41, 844-72

palladium allyl complexes 852-65

phenols 828-9

Coupling, cntd.
  platinum 865
  regioselectivity 856-60
  rhodium 846, 851, 866, 869
  silver 833
  stereoselectivity 861-4
  styrene 768
  titanium 833
  Ullmann 375-6, 821-3
  vanadium 828
  vinyl alanes 847, 850, 852, 855, 862
  vinyl ethers 869
  vinyl halides 849-51, 867-70
  vinyl halides/Grignards 838-40
  vinyl mercurials 847
  vinyl phosphates 839
  vinyl sulphides 838
  vinyl zirconium compounds 847, 850
Cubane 425-8
Cuneane 425-8
Cuprates 702-3
Cyclisation
  allyl complexes 856-60, 864
  asymmetric 768, 788
  dienes 768
  nickel 768
  palladium 856-60, 864
Cycloaddition 734, 762-7
Cycloalkenes
  metathesis 952
  polymerisation 952-4
Cyclobutanes
  from metathesis 924
Cyclobutenes, isomerisation 424-5
Cyclododecatriene
  hydrogenation 78-80, 129
  synthesis 784-5
Cycloheptatriene, hydrogenation 78
Cyclohexadiene
  hydrogenation 76
  disproportionation 109
Cyclohexane, oxidation 566-7
Cyclohexanol, oxidation 576
Cyclohexanone
  hydrogenation 97, 110
  hydrosilylation 687
  oxidation 588-90
Cyclohexene
  carbohalogenation 354
  oxidation 456-7, 465-6, 470, 472-5
  477, 479, 491-2, 517-8, 520-3, 530-1
Cyclooctadiene
  hydrogenation 78-9
  isomerisation 78-9
Cyclooctatetraene, synthesis 812-5
Cyclooligomerisation 783-6, 812-21
Cyclopentadiene, hydrogenation 77

Cyclopropanation 334-45
  alkenes 334-45
  alkynes 340
  allyl alcohols 336-7
  asymmetric 342-5
  benzene 340
  cobalt 342-3
  copper 337-8, 344-5
  dienes 343-5
  esters, unsaturated 339-43
  palladium 338-9, 341
  rhodium 338-41
  silver 335
Cyclopropanes
  by-product in metathesis 940, 944-7
  carbonylation 275
  isomerisation 431-5
Cyclopropenes
  isomerisation 432-5
  oligomerisation 759
Cyclotetramerisation alkynes 812-5
Cyclotrimerisation alkynes 815-8
CYCPHOS 134-5, 137

Decarbonylation 280-9
  acyl halides 281-4
  aldehydes 95, 280, 284-8
  allyl alcohols 288
  anhydrides 289
  dicarbonyl compounds 289
  iridium 282, 284-5
  ketones 285, 289
  nickel 289
  palladium 283
  rhodium 95, 281-9
  ruthenium 95
  stereochemistry 287
  Wilkinson's catalyst 24-5, 282-6, 288-9
Decarboxylation 591-3, 597
Dehydroamino acids, hydrogenation 130-52, 159
Dehydrogenation 166-71, 623
Dehydropeptides, hydrogenation 146-8
Deuteration 24-6, 60, 70, 81, 83-4, 89-90, 96, 133, 250, 356-60, 407-12, 749-50, 781-3, 799
Dewar benzene 424-5
Diastereoselectivity 3-5, 25-7, 64-5, 98, 110-1, 336-8, 372, 682
Diazo compounds 337-45
  carbonylation 273-4
  coupling 769-70
Diazomethane 276
Dicarbonyl compounds
  decarbonylation 289
  hydrogenation 99-101, 156, 160-1
  oxidation 589-90

Dicarboxylic acids
  oxidation 592-3
Dicobalt octacarbonyl - see Co$_2$(CO)$_8$
Dienes
  addition 370-1
  carbonylation 251-2
  coololigomerisation 790-7
  cyclisation 768
  cyclopropanation 343-5
  dimerisation 251-2
  hydrocyanation 348-50
  hydroesterification 250-2
  hydroformylation 233-4
  hydrogenation 68-87, 127
  hydrosilylation 315-21, 332-3
  isomerisation 768
  oligomerisation 370-1, 769-71,
  782-91
  oxidation 537-41
  polymerisation 769-82
  telomerisation 320-1, 333-4,
  798-810
Diimides, hydrosilylation 692
Diketones
  hydrosilylation 687
  oxidation 589-90
  synthesis by alkyne oxidation
  542-4
Dimerisation
  alkynes 812-3
  butadiene 251-2
  copper 813
  ethylene 748-51
  nickel 813
  norbornadiene 756-8
  rhodium 813
Diols
  cleavage 528-9, 532, 534-5
  oxidation 578
  synthesis by alkene oxidation
  524-32, 539-42
DIOP 141-7, 149, 153-8, 163-5,
170-1, 241-3, 254, 347, 351, 354,
417, 419, 422, 693-8, 842-3, 863
DIOXOP 146, 151
DIPAMP 133-7, 147, 158, 241
Disproportionation 109, 169
Disulphides
  synthesis by thiol oxidation
  632-5

Electron transfer 105-7, 539, 548,
553, 559-60, 610
Elimination
  β-hydride 752
  reductive 734
Enamide rhodium complexes 136-8,
144-6

Enamines
  synthesis by allyl amine isomerisation
  421-3
Enantioface discrimination 164-4,
496-7
Enantiomer discrimination 162-4, 496-7
Enantioselectivity (see also asymmetric
synthesis) 5-6, 129-62, 220, 238-43,
326-9
Eneynes, carbometallation 359-61
Enol acetates, oxidation 479
Enol ethers, oxidation 478
Enones (see also aldehydes,unsaturated
and ketones, unsaturated)
  hydroformylation 222, 225, 234
  hydrogenation 112, 115, 128, 153,
  163-4, 222, 225, 234
  hydrosilylation 2, 689-91, 696-7
  isomerisation 423-4
  oxidation 508, 515
  reaction with alanes 702-7
  reaction with cuprates 702-3
  reaction with Grignards 702-3
  reaction with organozirconium
  compounds 704-6
  reduction 2, 101
Epoxidation 461-99
  asymmetric 493-9
  iodosobenzene 490-2
  sodium hypochlorite 492-4, 528, 530,
  544
Epoxides 266
  carbonylation 275
  carboxylation 275
  coupling 853-5
  hydrogenolysis 117-8
  isomerisation 439-41
  synthesis from alkenes 461-99
Esterification 714-5
Esters
  hydrogenation 100-1
  hydrogenolysis 116, 120-1
  hydrolysis 709-14
  oxidation 595
  reduction 683-4
  synthesis by carbonylation 256-63,
  267, 269, 271
  synthesis by ether oxidation 600-2
  synthesis by hydroesterification
  244-54
  transesterification 710-1, 714-5
  unsaturated, cyclopropanation 339-43
  unsaturated, hydrocyanation 348
  unsaturated, hydroformylation 225,
  234-6
  unsaturated, hydrogenation 56-61,
  162
  unsaturated, hydrosilylation 315,
  691

Esters, cntd.
  unsaturated, oxidation 463-4,
  478, 508
  unsaturated, reduction 682-3
Ethers
  hydrogenolysis 117, 121
  oxidation 600-2
  unsaturated, hydrogenation 63-7
  vinyl, hydration 366
Ethylene
  codimerisation 761-2
  cooligomerisation 793-5
  cycloaddition 762
  dimerisation 748-51

Farmer's Rule 308
Fats, unsaturated, hydrogenation
73-6
Ferricyanide, as oxidising agent
593-4, 634
Fischer Tropsch synthesis 276-80

GLUCPHOS 134
Grignard coupling 6, 832-45
  alkyl derivatives 845
  allyl derivatives 841-4
  aryl halides 833-6
  asymmetric 6, 838-40
  iron 833, 838
  manganese 833
  nickel 6, 833-44
  palladium 833-4, 838-41, 844
  silver 833
  titanium 833
  vinyl halides 838-40
Grignard reagents
  carbometallation 359-62
  carboxylation 268
  hydrogenolysis 122
  hydrometallation 357-8
  reaction with acyl halides 699
  reaction with enones 702-3
  reducing agent 683-4

Hafnium
  carbometallation 363
  oligomerisation 751
Halogenated compounds
  carbonylation 261-8
  hydrogenolysis 116-7, 119-20
α-Haloketones
  reaction with organotin compounds
  708
Heck reaction 867-73
Heterocycles
  hydrogenation 88-9, 91, 93
  synthesis by cyclotrimerisation
  818
Heterolysis, hydrogen 19-21

Heteropolyacids 366, 502, 506, 563,
573, 577, 584, 588, 635-6, 768, 826-7
Hexachloroplatinic acid - see H_2PtCl_6
1,4-Hexadiene 793-4

$[HFe(CO)_4]^-$ 102, 683-4

Homoallyl alcohols, epoxidation
488-9, 498
Homocubane, isomerisation 428
Homolysis, hydrogen 22
H_2PtCl_6
  hydrogenation 45-6, 48-9
  hydrosilylation 308-10, 313-6, 318-9
  322-32, 692
$HRh(CO)(PPh_3)_3$ 32, 47-8, 62, 67,
226-36
$HRh(PPh_3)_4$ 30-1, 114-5, 311, 323, 422
Hydration 365-6, 373-4
Hydrazines
  hydrogenolysis 104, 119
  oxidation 626
Hydride
  β-hydride elimination 752
  hydrogenolysis 119-20
Hydroalumination 355-8
Hydroaromatics, dehydrogenation 169
Hydrocarbons
  dehydrogenation 166-8, 170-1
  oxidation 563-7
  synthesis by aldehyde decarbonylation
  280
Hydrocarboxylation 243-5
  alkenes 245
  alkynes 249
  asymmetric 252
  copper 245-7
  palladium 245
  platinum 246
  silver 245-7
Hydrocyanation 345-53
Hydroesterification 244-54
  alcohols, unsaturated 246-7
  alkenes 244
  alkynes 249-50
  asymmetric 252-4
  cobalt 244, 246, 251-3
  dienes 250-2
  intramolecular 246-7
  nickel 244, 249
  palladium 244-6, 249-53
  platinum 245-6
  rhodium 246-7
  styrenes 253-4
Hydroformylation 219-43
  alcohols, unsaturated 225
  alkenes 219-43
  asymmetric 238-44
  cobalt 220-6, 235, 238-9

Hydroformylation, cntd.
  dienes 233-4
  enones 222, 225, 234
  esters, unsaturated 225, 234-6
  platinum 236-8, 242
  polymer supported catalysts 238-9
  rhodium 3, 225-36, 239-42
  ruthenium 236-7
  styrene 3, 229, 231-2, 238-43
Hydrogen
  activation 12-22
  complex 13
  heterolysis 19-21
  homolysis 22
  o,p-interconversion 12-17,
  138-9
Hydrogenation
  alcohols, unsaturated 63-7
  aldehydes 95-6, 99, 101, 112,
  222-2, 230, 237
  aldehydes, unsaturated 49-51
  alkenes 23-48, 112-29, 222, 803
  alkynes 81-7, 115, 127
  allenes 67-8
  amines, unsaturated 63-7
  anhydrides 156-7
  arenes 87-93, 127-8
  asymmetric 5-6, 129-62
  azobenzene 104
  azo compounds 104-5
  benzene 87-8, 93-4
  butadiene 69-70
  carbon monoxide 276-80
  carbonyl groups 90, 94-101, 111,
  128, 155-60
  carbonyl compounds, unsaturated
  101, 128, 153, 163-4
  carboxylic acids 100
  carboxylic acids, unsaturated
  56-61, 131-2, 147-50, 153-4, 159,
  162-3
  chromium 19, 47-8, 70-1, 73-5,
  77-8, 81, 127
  cobalt 22, 38-9, 48, 50-63, 68,
  71-5, 77, 79, 80-1, 83, 85-90, 92,
  94, 98-9, 102-8, 110, 112, 114,
  152, 158, 160-1
  copper 54, 105
  cyclododecatriene 79-80
  cycloheptatriene 78
  cyclohexadiene 76
  cyclohexanone 97, 110
  cyclooctadiene 78-9
  cyclopentadiene 77
  dehydroamino acids 130-52, 159
  dehydropeptides 146-8
  dicarbonyl compounds 99-101, 156,
  160-1
  dienes 67-87

Hydrogenation, cntd.
  enones 112, 115, 128, 153, 163-4,
  222, 225, 234
  esters 100-1
  esters, unsaturated 56-61, 162
  ethers, unsaturated 63-7
  fats, unsaturated 73-6
  heterocycles 88-9, 91, 93
  H_2PtCl_6 45-6, 48-9
  imines 101-2, 115, 157-8, 164
  iridium 3-5, 15-19, 35-8, 47-8, 50-1,
  54, 57-8, 60, 64-6, 68-70, 77-9, 81-2,
  84-5, 95-6, 107, 110-3, 115, 126-7,
  154, 162
  iron 43, 48, 57, 75, 83, 86, 93-4,
  102, 104
  ketones 90, 95-9, 110, 112, 154-6,
  160
  ketones, unsaturated 51-2, 90, 112,
  115, 128, 153
  manganese 19, 47, 93
  molybdenum 19, 47, 50, 57, 75, 77-8,
  114
  naphthalene 87
  nickel 43-4, 48, 50-1, 57, 73-80,
  82, 92, 94, 104, 106
  nitriles 104, 129
  nitriles, unsaturated 62
  nitro compounds, unsaturated 62-3
  nitro groups 105-8, 115, 129
  nitroso groups 108
  norbornadiene 79, 81
  osmium 43, 50, 86, 115
  oximes 102-3
  palladium 20, 44-5, 48, 50, 52, 54,
  59, 63, 74, 77, 79, 81, 83, 85-6,
  102, 106-7, 127, 129, 154, 159, 164
  platinum 21, 44-5, 48-9, 52, 74, 76,
  80, 82-3, 85-6, 106, 127, 237
  polymer supported catalysts 121-9,
  164-6
  pyruvates 156
  quinones 54-5
  rhenium 47
  rhodium 5-6, 12-16, 21-35, 47-70,
  75-84, 87, 90-1, 95-9, 101-6, 110-5,
  125-62
  ruthenium 19-21, 39-43, 48, 50, 54,
  57, 59-61, 63, 68-9, 73, 75, 78-82,
  85, 91-101, 103-4, 106, 108-10, 112,
  115, 126-8, 153, 156-7, 160, 162-4
  sorbate 73-6, 127
  steroids 54-5, 110, 125
  styrene 47-8, 153-4, 158, 161, 164
  thiophene 89
  titanium 46, 48-9, 67-8, 73, 79, 82,
  85, 161
  transfer 2-3, 109-16, 126, 162-4
  tungsten 47

Hydrogenation, cntd.
  vanadium 73
  Ziegler 46-8, 52, 57, 77-9,
  82-3, 85, 92, 94, 97, 106, 161
  zirconium 48, 82
Hydrogenolysis 88-91, 116-121
Hydrogen peroxide 460-4, 525-6,
530, 549, 577-8, 597, 601, 615,
624, 638-9
Hydrohalogenation 352-3
Hydrolysis
  amides 717-8
  cobalt 709-10, 712-3, 717-8,
  721
  copper 711
  esters 709-14
  lanthanides 710
  mercury 715-6
  nickel 710, 713
  nitriles 1, 721
  palladium 721
  peptides 717-8
  platinum 721
  rhodium 721
  ruthenium 721
  silver 710, 715-7
  thioacetals 715-6
  thioesters 715-7
  zinc 711, 713
Hydrometallation 355-8, 704-6,
803
Hydroperoxides, synthesis by
alkyl arene oxidation 562
Hydrosilylation 308-34, 685-98
  acyl halides 687-8
  aldehydes 685-7
  alkenes 3, 308-15, 326-32
  alkynes 322-7
  asymmetric 326-9, 693-8
  butadiene 317, 320
  carbon dioxide 688
  carbonyl compounds, unsaturated
  689-91, 696-7
  chromium 319
  cobalt 3, 350-2
  cyclohexanone 687
  dienes 315-21, 332-3
  diimides 692
  diketones 687
  enones 2, 689-91, 696-7
  esters, unsaturated 315, 691
  H$_2$PtCl$_6$ 308-10, 313-6, 318-9,
        322-32, 692
  imines 692, 697-8
  isocyanates 692
  keto esters 695-6
  ketones 685-7, 693-5
  nickel 310-2, 314-7, 319-21,
  327, 332-3, 689

Hydrosilylation, cntd.
  palladium 315-20, 328, 330, 692
  platinum 308-20, 322-32, 685, 691-6
  polymer supported catalysts 329-31,
  694
  rhodium 2, 308, 310-5, 317, 320,
  325-7, 329-30, 332, 686-7, 690-7
  ruthenium 687-8
  styrene 328-9, 331
  titanium 690-1
  vanadium 313, 315
  Ziegler 320
Hydroxy acids, oxidation 593-4
Hydroxylation 576-7
Hydrozirconation 704-5

Imines
  hydrogenation 101-2, 115, 157-8, 164
  hydrosilylation 692, 697-8
  reduction 683
  synthesis 721
Indoles, oxidation 628
Insertion 404-5, 734, 741-2
  carbon monoxide 218-9
Iodosobenzene
  epoxidation 490-2
  oxidising agent 525, 544, 564-7,
  570-2, 623, 638
Iodosobenzene diacetate, oxidising
agent 622
Iridium complexes
  carbonyl 37-8
  carbonylation 258-9
  cationic 36, 65
  decarbonylation 282, 284-5
  dehydrogenation 167, 169
  Fischer Tropsch 276
  hydrogenation 3-5, 15-19, 35-8, 47-8,
  50-1, 54, 57-8, 60-1, 64-6, 68-70,
  77-9, 81-2, 84-5, 95-6, 107, 110-3,
  115, 126-7, 154, 162
  hydrogenolysis 118
  isomerisation 415, 418, 420-1,
  423-5, 436
  metathesis 923
  oligomerisation 758, 817
  oxidation 515-6, 522, 577, 582,
  630-1, 636
  reduction 683
Iron complexes
  carbonylation 272
  cooligomerisation 794
  coupling 704, 828, 833, 838, 848, 867
  Fischer Tropsch 276
  hydrogenation 43, 48, 57, 75, 83,
  93-4, 102, 104
  isomerisation 408-9, 415-6
  oligomerisation 786, 791

Iron complexes, cntd.
  oxidation 463, 465-6, 469, 473,
  480, 490-2, 499, 516, 543, 548-9,
  566-7, 576, 581-2, 584, 586-7,
  589-90, 596-7, 599, 611, 615-7,
  623, 625, 631-2, 638
  polymerisation 772, 779
  reduction 683-4
Isocyanates
  hydrosilylation 692
  synthesis by carbonylation 272-3
  synthesis by isocyanide oxidation
  624
Isocyanides, oxidation 624
Isomerisation 326, 403-50, 768,
803
  alcohols, unsaturated 225
  alkenes 24-5, 32, 39, 76, 78,
  85, 112-5, 223-4, 230, 244, 311,
  322, 331, 405-17
  alkynes 443-4
  allyl acetates 803
  allyl alcohols 225, 417-20
  allyl amides 421-3
  allyl amines 6, 421-3
  allyl ethers 417, 420-4, 442-4
  asymmetric 6, 419-22
  aziridines 440-2
  azirines 440-2
  bicyclobutanes 436-9
  cobalt 407-8, 410, 412, 414,
  416, 422, 431
  copper 443
  cubane 425-8
  cyclobutenes 424-5
  cyclooctadiene 78-9
  cyclopropanes 431-3, 435
  cyclopropenes 432-5
  dienes 768
  enones 423-4
  epoxides 439-41
  homocubane 428
  iridium 415, 418, 420-1, 423-5,
  436
  iron 408-9, 415-6
  kinetics 409
  mechanism 406-14
  nickel 407-8, 410-1, 413-9, 423,
  429, 431, 768
  osmium 418-20
  palladium 414, 425, 428, 441-2,
  803
  quadricyclane 429-31
  rhodium 6, 408, 415-34, 438-41
  ruthenium 85, 414, 420, 422,
  443-4
  skeletal 411-4
  silver 426-8, 430, 432-40,
  443-4
  titanium 415, 417

Isoprene
  hydrosilylation 318-9
  oligomerisation 787-9
  polymerisation 773, 777, 780-1
  telomerisation 800-1
Isotactic polymer 735, 738, 745, 781
Isotope effect 283, 501, 742-4, 756,
826
Itaconic acid 131, 136, 145, 147, 149,
151

Keto esters, hydrosilylation 695-6
Ketones
  amination 103, 720
  decarbonylation 285, 289
  dehydrogenation 169
  hydrogenation 90, 95-9, 110-2, 154-6,
  160
  hydrosilylation 685-7, 693-5
  oxidation 585-9
  reaction with aluminium alkyls 707
  reduction 682-4
  synthesis by acyl halide alkylation
  699
  synthesis by acyl halide hydrosilylation 687-8
  synthesis by alcohol carbonylation
  265-6
  synthesis by alcohol oxidation
  570-8
  synthesis by carbonylation of diazo
  compounds 273-4
  synthesis by carbonylation of
  organomercurial 269-70
  synthesis by enone hydrosilylation
  696
  unsaturated, hydrogenation 51-2, 90,
  112, 115, 128, 153
Kinetic resolution 170

Lactams 248
Lactones
  synthesis by carbonylation 265, 267,
  269-70
  synthesis by hydroesterification
  246-7
Lanthanide complexes
  hydrolysis 710
  oxidation 480
  polymerisation 809
LEUPHOS 840
Lindlaar catalyst, for alkyne reduction
83
Linoleate 73-6
Linolenate 73-6
Lutetium complexes
  polymerisation 741-2

Manganese complexes
  coupling 828, 833

Manganese complexes, cntd.
  hydrogenation 19, 47, 93
  oxidation 480, 490-2, 516-7,
  540, 548-51, 559-61, 567, 574,
  583-6, 590-1, 593, 596, 619,
  624, 633, 638
Mercury complexes (see also organomercurials, aryl mercury compounds
and vinyl mercury compounds)
  hydration 373
  hydrohalogenation 352-3
  hydrolysis 715-6
Metal alkyls 404-5
Metallocyclobutadienes 949-51
Metallocyclobutanes 741-2, 924-7,
931, 937-9, 942-7
Metalloenzymes 713-4
Metathesis 921-56
  alkenes 921-47, 951-6
  alkynes 948-51
  chromium 929
  cycloalkenes 952
  iridium 923
  mechanism 922-36
  molybdenum 921, 923, 927, 929,
  931-3, 941, 948
  niobium 923, 927, 936
  norbornadiene 924
  platinum 945-6
  polymer supported catalysts 922
  rhenium 921, 923, 929, 931,
  933-5, 953
  rhodium 923, 925-6
  ruthenium 953
  stereoselectivity 939-44
  tantalum 923, 927, 930, 936
  titanium 928, 943, 946, 950,
  956
  tungsten 921, 923-31, 934-56
Methanol, carbonylation 254-61
Methylene cyclopropane
  cycloaddition 764-5
  oligomerisation 759-60
α-Methylene lactones 249
Methyl ketones, synthesis 502-9,
512-6, 804
Molybdenum complexes
  allylic substitution 866
  coupling 866
  hydration 365
  hydrogenation 19, 47, 50, 57,
  75, 77-8, 114
  metathesis 921, 923, 927, 929,
  931-3, 941, 948
  oligomerisation 752, 755
  oxidation 457, 462-79, 481-8,
  493, 499, 542, 571, 577, 584-5,
  588, 613, 618, 623, 625-6,
  631-2, 637-9

Molybdenum complexes, cntd.
  polymerisation 778, 951, 953
Monohydride complexes 19-21

Naphthalene, hydrogenation 87
Neodymium complexes 779, 781
Nickel complexes
  allylic substitution 370
  carbometallation 359-60, 362
  carbonylation 258-60, 263, 265-9,
  275
  carboxylation 275
  codimerisation 761
  cooligomerisation 790, 792-7
  coupling 6, 702-8, 823-5, 833-44,
  847-50, 866, 871-2, 878-81
  cyclisation 768
  cycloaddition 762-4, 767
  decarbonylation 289
  dimerisation 813
  hydrocyanation 346-53
  hydroesterification 244, 249
  hydrogenation 43-4, 48, 50-1, 57,
  73-80, 82, 92, 94, 104, 106
  hydrogenolysis 119, 122
  hydrolysis 710, 713
  hydrometallation 357
  hydrosilylation 310-2, 314-7, 319-21,
  327, 332-3, 689
  isomerisation 407-8, 410-1, 413-9,
  423, 429, 431, 768
  oligomerisation 748-9, 752-7,
  759-60, 769-71, 783-90, 812-5, 817,
  819
  oxidation 458, 473, 578, 580-1, 624,
  629
  polymerisation 747, 771-2, 775-7,
  781-2, 809, 812
  reduction 682-3
  telomerisation 320-1, 333, 805
Niobium complexes
  metathesis 923, 927, 936
  oligomerisation 751, 816
  oxidation 471, 473, 623, 632
  polymerisation 778, 951
Nitriles
  hydrogenation 104, 129
  hydrolysis 1, 721
  reaction with organometallics 707-8
  reduction 682-3
  synthesis by amine dehydrogenation
  623
  synthesis by hydrocyanation 345-52
  unsaturated, hydrogenation 62
  unsaturated, reduction 682
Nitro compounds
  carbonylation 272-3
  hydrogenation 105-8, 115, 129
  unsaturated, hydrogenation 62-3

Nitroso groups
  hydrogenation 108
  oxidation 625
Nitrosyl complexes 468-9, 477,
506, 514, 535-6, 575
NMDPP 130-2, 152, 162, 697
Norbornadiene
  dimerisation 756-8
  hydrogenation 79, 81
  metathesis 924
  oligomerisation 756-8
NORPHOS 134-7, 839

Oligomerisation 747-60, 769-71,
782-91, 812-21
  alkenes 747-60
  alkynes 85, 812-21
  allenes 369-70, 769-71
  aluminium 748
  butadiene 782-6
  chromium 791, 816
  cobalt 749, 757-8, 818
  cyclooligomerisation 783-6,
  812-21
  cyclopropenes 759
  dienes 370-1, 769-71, 782-91
  hafnium 751
  iridium 758, 817
  iron 786, 791
  isoprene 787-9
  methylenecyclopropane 759-60
  molybdenum 752, 755
  nickel 748-9, 752-60, 769-71,
  783-5, 787, 789-90, 812-5, 817,
  819
  niobium 751, 816
  norbornadiene 756-8
  palladium 369-70, 752, 759-60,
  771, 782-3, 788-9, 814, 817-21
  platinum 759
  propene 753
  regioselectivity 752
  rhodium 756, 771, 786, 816-7, 819
  ruthenium 757
  tantalum 750-2, 754-5, 815-6
  titanium 748-9, 759, 784, 791
  tungsten 752, 755
  zirconium 751, 786, 791
Organoboranes - see boranes
Organomercurials (see also aryl and
vinyl mercury compounds) 270
  coupling 846
  reaction with acyl halides 701
Organometals
  carbonylation 268-9
  reaction with acyl halides
  69 7-70 2
  reaction with nitriles 707-8

Organotin compounds
  coupling 849
  reaction with acyl halides 699-700
  reaction with α-haloketones 708
Organozinc compounds
  coupling 848, 850-1
  reaction with acyl halides 702
Organozirconium compounds
  coupling 847, 850
  reaction with enones 704-6
Orthometallation 40, 50, 70, 77, 83,
93, 96
Osmium complexes
  hydrogenation 43, 50, 86, 115
  isomerisation 418-20
  oxidation 524-31, 542, 544, 569-70,
  578, 587, 593-4, 597
Oxetanes, carbonylation 275
Oxidation 451-680, 804
  alcohols 568-80
  alcohols, allylic 464-5, 478-97,
  504, 528
  alcohols, homoallylic 488-9, 498
  alcohols, unsaturated 463-4, 534,
  568-80
  aldehydes 458, 581-5
  alkenes 460-543, 804
  alkyl arenes 548-60
  alkynes 542-4
  allyl acetates 504-5
  allyl halides 473, 492, 514, 517-27
  amides 596
  amines 621-5
  amino acids 596-8
  anilines 618-21
  arenes 544-9
  arsines 632
  ascorbic acid 598-600
  asymmetric 493-9, 524
  azobenzene 625-6
  carbonyl compounds, unsaturated 507-8
  carboxylic acids 591-2
  carboxylic acids, unsaturated 463-4
  catechols 613-9
  chromium 490, 525, 530
  cobalt 458, 467-70, 473, 516-8, 539,
  545, 555-61, 563-6, 571, 574-5,
  579-85, 588, 590-1, 595, 600, 602-9,
  618-9, 624, 626, 631-2, 635-6
  copper 469, 491, 500-12, 514, 516,
  523, 533-8, 541-7, 560-3, 568-9, 571,
  574, 580, 582-6, 590, 592-6, 598-9,
  610-5, 618-9, 621-7, 633-6
  cyclohexane 566-7
  cyclohexanol 576
  cyclohexanone 588-90
  cyclohexene 456-7, 465-6, 470, 472-5,
  477, 479, 491-2, 517-8, 520-3, 530-1

Oxidation, cntd.
  dicarboxylic acids 592-3
  dienes 537-41
  diketones 589-90
  diols 578
  enol acetates 479
  enol ethers 478
  enones 508, 515
  esters 595
  esters, unsaturated 463-4, 478,
  508
  ethers 600-2
  hydrazines 626
  hydrocarbons 563-7
  hydroxy acids 593-4
  indoles 628
  iridium 515-6, 522, 577, 582,
  630-1, 636
  iron 463, 465-6, 469, 473, 480,
  490-2, 499, 516, 543, 548-9,
  566-7, 576, 581-2, 584-90, 596-7,
  599, 611, 615-7, 623, 625, 631-2,
  638
  isocyanides 624
  ketones 585-90
  lanthanides 480
  lead 591
  manganese 480, 490-2, 516-7, 540,
  548-51, 559-61, 567, 574, 583-6,
  590-1, 593, 596, 619, 624, 633,
  638
  mechanism 454-60, 475
  molybdenum 457, 462-79, 481-8,
  493, 499, 542, 571, 577, 584-5,
  588, 613, 618, 623, 625-6,
  631-2, 637-9
  nickel 458, 473, 578, 580-1,
  624, 629
  niobium 471, 473, 623, 632
  nitroso compounds 625
  osmium 524-31, 542, 544, 569-70,
  578, 587, 593-4, 597
  palladium 458-9, 468-9, 474,
  500-10, 514, 522-3, 532-8, 544-5,
  573, 582, 584-5, 588, 602, 629,
  804
  peracids 482-5, 541, 576
  phenols 602-13
  phosphines 458, 509, 512-3, 520,
  629-32
  platinum 458, 463, 466, 468-9, 563,
  582, 590, 629-30
  polymer supported catalysts
  468-70, 479, 499
  quinones, as oxidising agents 505,
  522, 537-8
  rhenium 590

Oxidation, cntd.
  rhodium 467, 509-15, 519-21, 530-2,
  543, 545, 564, 574-6, 582, 585,
  590, 600-2, 623-4, 630
  ruthenium 467, 473, 493, 525, 530,
  541-2, 544, 549, 568-70, 572, 578,
  580-2, 587-8, 595-6, 600, 612-3,
  617, 622, 631-2, 636, 638
  silver 562, 568, 571, 579, 592-7,
  612, 622, 624
  stereoselectivity 481-90, 537-8
  styrene 473-6, 491, 512
  sulphides 636
  sulphoxides 636
  tantalum 543, 623
  thiols 632-5
  titanium 471, 480, 489, 494-9, 625,
  637-9
  tungsten 463-4, 470-1, 578, 623-4,
  639
  vanadium 463, 465-6, 469-74, 479-90,
  525, 530, 549, 563-4, 579, 590,
  618-9, 623, 625, 637, 639,
  Wacker 366, 459, 500-11, 532-3
  zirconium 571, 600
Oxidative addition 12-19
Oxidative carbonylation 260, 545
Oxidative cleavage 626-9, 636
Oxidative coupling 768, 826-30
Oximes
  hydrogenation 102-3
  hydrogenolysis 102-3
  synthesis by amine oxidation 622
  synthesis by nitro group hydrogenation
  105
Oxo complexes 490-2
Oxygen activation 454-7

Palladium complexes
  allyl complexes, coupling 852-65
  allylic substitution 4-5, 369-74,
  852-65
  amination 367-9
  carbohalogenation 353-4
  carbonylation 248, 260-75
  clusters 45
  codimerisation 761
  cooligomerisation 794
  coupling 544, 701, 768-70, 826-7,
  830, 833-4, 838-41, 844-7, 866-72
  cycloaddition 764-7
  cyclopropanation 338-9, 341
  decarbonylation 283
  dehydrogenation 169
  hydration 366, 373-4
  hydrocarboxylation 245
  hydrocyanation 346-9
  hydroesterification 244-6, 249-53

Palladium complexes, cntd.
 hydrogenation 20, 44-5, 48, 50,
 52, 54, 63, 74, 77, 79, 81, 83,
 85-6, 102, 106-7, 127, 129, 154,
 159, 164
 hydrogenolysis 117, 119-20, 122
 hydrolysis 721
 hydrosilylation 315-20, 328, 330,
 333, 692
 isomerisation 414, 425, 428,
 441-2, 803
 oligomerisation 369-70, 752,
 759-60, 771, 782-3, 788-9, 814,
 817-21
 oxidation 458-9, 468-9, 474,
 500-10, 514, 522-3, 532-8, 544-5,
 573, 582, 584-5, 588, 602, 629,
 804
 Pd(PPh$_3$)$_4$ 4-5, 120-1, 244, 263,
 346, 369-74, 376-7, 441, 443,
 699-701, 708, 852-5
 Pd(SALEN) 20, 44
 polymerisation 771
 reduction 685
 telomerisation 321-33, 789-810
PAMP 130
PAMPOP 140-1
Pantolactone 156
PCCP 141-2, 149
Peptides
 hydrolysis 717-8
 synthesis 718-9
Peracids
 as oxidising agents 482-5, 541,
 576
 synthesis by aldehyde oxidation
 583-5
Peresters 472-4, 545
Perethers 472-4
Periodate 528-30, 532, 541, 549,
570, 572, 582
Peroxo complexes 454-8, 505, 510,
513, 576
Peroxodisulphate 523, 540, 545,
548, 560-2, 568-71, 579, 586-7,
592-3, 596, 599-600, 612, 620, 624
PHELLANPHOS 134-6
Phenols
 oxidation 602-13
 oxidative coupling 828-9
 polymerisation 610-1
PHEPHOS 134-5, 156, 866
Phosphines, oxidation 458, 509,
512-3, 520, 629-32
Platinum complexes
 allylic substitution 370, 865
 carbonylation 248
 coupling 865
 hydration 366

Platinum complexes, cntd.
 hydrocarboxylation 246
 hydrocyanation 351
 hydroesterification 245-6
 hydroformylation 236-8, 242
 hydrogenation 21, 44-6, 48-9, 74,
 76, 80, 82-3, 85-6, 106, 127, 237
 hydrolysis 721
 hydrosilylation 308-32, 685, 691-6
 metathesis 945-6
 oligomerisation 759
 oxidation 458, 463, 466, 468-9, 563,
 582, 590, 629-30
 polymerisation 809-11
 telomerisation 805
PNNP 141-2
Polyacrylic acid 123
Polymerisation 733-47, 769-82, 809-12,
951-3
 alkenes 733-47
 alkynes 809-12, 951
 allenes 369-70, 769-72
 butadiene 774-9, 781
 chain branching 744-5
 chromium 735, 739, 747, 778, 780
 cobalt 771-3, 780-1, 809
 copolymers 747, 781-2
 copper 610-1
 cycloalkenes 952-3
 dienes 769-82
 iron 772, 779
 isoprene 773, 777, 780-1
 lanthanides 809
 lutetium 741-2
 mechanism 740-4
 molecular weight control 744-5
 molybdenum 778, 951, 953
 neodymium 779, 781
 nickel 747, 771-2, 775-7, 781-2,
 809, 812
 niobium 778, 951
 palladium 771
 phenol 610-1
 platinum 809, 811
 propene 781
 rhenium 953
 rhodium 771, 778-9
 ruthenium 778-9
 stereochemistry 775-9
 stereoselection 745-6
 styrene 737-8, 745
 tantalum 742-3
 termination 744-5
 titanium 735-8, 742-3, 745, 747,
 772-4, 780-1, 809
 tungsten 811, 951, 953
 uranium 778
 vanadium 735, 739, 746-7, 771-2, 809
 zirconium 735, 740

Polymer supported catalysts
  carbonylation 258
  hydroformylation 238-9
  hydrogenation 121-9, 164-6
  hydrosilylation 329-31, 694
  metathesis 922
  oxidation 468-70, 479, 499
Polystyrene 122-3, 125, 164
Polyvinylpyridine 123
PPFA 132, 328, 840
PROLOPHOS 141-2, 149
Propene
  cooligomerisation 794
  cycloaddition 762
  oligomerisation 753
  polymerisation 781
PROPHOS 134-6, 836
Pyrethrins 342-6
Pyrrole, synthesis 374-5
Pyruvic acid 99, 156

Quadricyclane 924
  isomerisation 429-31
Quinine 160-1
Quinones
  hydrogenation 54-5
  oxidising agent 505, 522, 537-8
  synthesis by catechol oxidation
  613-9
  synthesis by phenol oxidation
  602-13

Radical
  addition 3
  rearrangement 565-6
Reduction (see also hydrogenation)
  acyl halides 682, 685, 687-8
  aldehydes 682-3
  alkynes 1-2
  amides 682
  amides, unsaturated 683
  asymmetric 682-3
  carbonyl compounds, unsaturated
  2
  carbonyl groups 681-90
  carboxylic acids 685
  cobalt 682-3
  copper 682
  enones 2, 101
  esters 683-4
  esters, unsaturated 682-3
  imines 683
  iridium 683
  iron 683-4
  ketones 682-4
  nickel 682-3
  nitriles 682-3
  palladium 685
  titanium 683-5

Reductive amination 102-3
Reductive elimination 734
Regioselectivity 3, 69-76, 219-20,
  231-2, 244, 311-6, 323-5
Rhenium complexes
  carbene complexes 277-9
  dehydrogenation 116, 166-7
  Fischer Tropsch 279
  hydrogenation 47
  metathesis 921, 923, 929, 931,
  933-5, 953
  oxidation 590
  polymerisation 953
Rhodium complexes
  allylic substitution 370, 866
  amination 669
  carbohalogenation 354
  carbonyl 32-3, 49, 52
  carbonylation 255-8, 270-5
  carboxylates 160
  cationic 30-2, 47, 49, 65, 69, 76,
  95, 111, 155, 287-8
  cooligomerisation 793
  coupling 846, 851, 866, 869
  cyclopropanation 338-41
  decarbonylation 95, 281-9
  dehydrogenation 166-70
  dimerisation 813
  enamides 136-8, 144-6
  Fischer Tropsch 276
  hydration 366
  hydrocyanation 349-52
  hydroesterification 246-7
  hydroformylation 225-36, 239-42
  hydrogenation 12-16, 21-35, 47-70,
  75-84, 87, 90-1, 95-9, 101-6, 110-5,
  125-62
  hydrogenolysis 117-20
  hydrolysis 721
  hydrosilylation 2, 308, 310-20,
  325-32, 686-7, 690-7
  isomerisation 6, 408, 415-34, 438-41
  metathesis 923, 925-6
  oligomerisation 756, 771, 786, 816-7,
  819
  oxidation 467, 509-15, 519-21, 530-2,
  543, 545, 564, 574-6, 582, 585, 590,
  600-2, 623-4, 630
  polymerisation 771, 778-9
  sulphoxides 160
Ruthenium complexes
  allylic substitution 370
  amination 369
  carbohalogenation 353-4
  carbonyl 42
  carbonylation 261, 272
  cooligomerisation 797
  cycloaddition 763
  decarbonylation 95

Ruthenium complexes, cntd.
  dehydrogenation 168, 170-1,
  623
  hydrocyanation 349
  hydroformylation 236-7
  hydrogenation 19-21, 39-43, 48,
  50, 54, 57, 59-61, 63, 68-9, 73,
  75, 78-82, 85, 91-101, 103-4, 106,
  108-10, 112, 115, 126-8, 153,
  156-7, 160, 162-4
  hydrogenolysis 118-9
  hydrolysis 721
  hydrosilylation 687-8
  isomerisation 85, 414, 420, 422,
  443-4
  metathesis 953
  oligomerisation 757
  oxidation 467, 473, 493, 525,
  530, 541-2, 544, 549, 568-70, 572,
  578, 580-2, 587-8, 595-6, 600,
  612-3, 617, 622, 631-2, 636, 638,
  polymerisation 778-9

SALEN
  cobalt 605-6, 619
  palladium 20, 44
Schiff's bases 158-9, 162
Silanes
  chiral 310
  synthesis by hydrosilylation
  308-24
Silver
  coupling 833
  cyclopropanation 335
  hydrocarboxylation 245-7
  hydrolysis 710, 715-7
  isomerisation 426-8, 430-40, 443-4
  oxidation 562, 568, 571, 579,
  592-3, 595-7, 612, 622, 624
Simmons Smith reaction 334-7
SKEWPHOS 139-40
Snoutene 428-9
Sodium hypochlorite, epoxidation
492-3, 528, 530, 544
Sorbate, hydrogenation 73-6, 127
Steroids, hydrogenation 54-5, 110,
125
Styrene
  carbohalogenation 354
  codimerisation 761
  cooligomerisation 695
  coupling 768
  hydrocyanation 346
  hydroesterification 253-4
  hydroformylation 3, 229, 231-2,
  238-43
  hydrogenation 47-8, 153-4, 158,
  161, 164
  hydrosilylation 328-9, 331

Styrene, cntd.
  oxidation 473-6, 491, 512
  polymerisation 737-8, 745
Substitution
  allylic - see allylic substitution
  aryl halides 375-7
  vinyl halides 375-7
Sugar derived phosphines 144, 151, 154
Sulphides, oxidation 636
Sulphoxides, oxidation 636
Superoxo complexes 454-7
Syndiotactic polymer 735, 739, 746-7,
751
Synthesis gas 219-44

Tantalum complexes
  hydrohalogenation 352
  metathesis 923, 927, 930, 936
  oligomerisation 750-2, 754-5, 815-6
  oxidation 543, 623
  polymerisation 742-3
Telomerisation 798-810
  asymmetric 800
  butadiene 320-1, 333-4, 798-800,
  802-10, 852-3
  dienes 320-1, 333-4, 798-810
  isoprene 800-1
  nickel 320-1, 333, 805
  palladium 321-33, 798-810
  platinum 805
Tetrakis(triphenylphosphine)palladium -
see $Pd(PPh_3)_4$
Thioacetals, hydrolysis 715-6
Thioesters, hydrolysis 715-7
Thiols
  hydrogenolysis 118
  oxidation 632-5
Thiophene, hydrogenation 89
Titanium complexes
  carbometallation 363
  coupling 833
  Fischer Tropsch 278
  hydration 365
  hydrogenation 46-9, 67-8, 73, 79,
  82, 85, 161
  hydrogenolysis 120
  hydrometallation 355-8, 863
  hydrosilylation 690-1
  isomerisation 415, 417
  metathesis 928, 943, 946, 950, 956
  oligomerisation 748-9, 759, 784,
  791
  oxidation 471, 480, 489, 494-9, 625,
  637-9
  polymerisation 735-8, 742-3, 745,
  747, 772-4, 780-1, 809
Transesterification 710-1, 714-5
Transfer dehydrogenation 169-70

Transfer hydrogenation 2-3, 109-16, 126, 162-4
Transfer hydrogenolysis 118-20
Tributyl tin hydride 2, 121, 685
Trimethylene methane complexes 764-7
Tungsten complexes
  hydrocyanation 349
  hydrogenation 47
  metathesis 921, 923-31, 934-56
  oligomerisation 752, 755
  oxidation 463-4, 470-1, 578, 623-4, 639
  polymerisation 811, 951, 953

Ullmann coupling 375-6, 821-3
Uranium complexes, polymerisation 778
Ureas, synthesis 272-3
Unsaturated compounds - indexed under compound type

VALPHOS 157-8
Vanadium complexes
  coupling 828
  hydrogenation 73
  hydrosilylation 313, 315
  oxidation 463, 465-6, 469-74, 479-90, 525, 530, 549, 563-4, 579, 590, 618-9, 623, 625, 637, 639
  polymerisation 735, 739, 746-7, 771-2, 809
Vaska's compound $((Ph_3P)_2Ir(CO)Cl)$
  hydrocyanation 351-2
  hydrogenation 36-7, 48, 61, 77, 126
Vinyl alanes (see also alkyl aluminium compounds)
  coupling 847, 850, 852, 855, 862
  synthesis 362-4
Vinyl ethers
  coupling 869
  hydration 366
Vinyl halides
  carbonylation 266-7
  coupling 849-51, 867-70
  coupling to Grignards 838-40
  substitution 375-7
Vinyl mercury compounds (see also aryl mercury compounds, organo-mercurials)
  coupling 846
Vinyl phosphates, coupling 839
Vinyl sulphides, coupling 838

Wacker oxidation 366, 459, 500-11, 532-3
Water gas shift 93, 106

Wilkinson's catalyst $((Ph_3P)_3RhCl)$
  coupling 851
  cyclotrimerisation alkynes 816
  decarbonylation 24-5, 282-9
  hydrocyanation 349-52
  hydrogenation 12-15, 23-30, 47-9, 51, 53-69, 76, 78, 99, 110, 115, 118-20, 125-6, 158-9
  hydrosilylation 308, 312-3, 315, 325, 330, 332
  isomerisation 421-2
  metathesis 926
  oxidation 509, 512, 515, 520-6, 564, 574, 576, 630, 632
  substitution of acyl halides 699-700

Ziegler catalysts
  hydrogenation 46-8, 52, 57, 73, 77-9, 82-3, 85, 92, 94, 96, 106, 161
  hydrosilylation 320
  polymerisation 733-47, 773
Zinc complexes (see also organozinc compounds)
  hydrolysis 711, 713
  transesterification 714
Zirconium complexes (see also organo-zirconium compounds)
  carbometallation 362-3
  Fischer Tropsch 278
  hydrogenation 48,82
  hydrometallation 355, 357
  oligomerisation 751, 786, 791
  oxidation 571, 600
  polymerisation 735, 740